AF809959

Veterinary Psychiatry of the Cat

Stéphane Bleuer-Elsner • Sylvia Masson
Marie Lacoste • Gérard Muller

Veterinary Psychiatry of the Cat

Diagnosis and Treatment of Behavioral Disorders

Springer

Stéphane Bleuer-Elsner
Private practice
Tel Aviv, Israel

Sylvia Masson
Clinique de la Tivollière
Voreppe, France

Marie Lacoste
Clinique de la Tivollière
Voreppe, France

Gérard Muller
Biarritz, France

ISBN 978-3-032-06259-8 ISBN 978-3-032-06260-4 (eBook)
https://doi.org/10.1007/978-3-032-06260-4

© The Editor(s) (if applicable) and The Author(s), under exclusive license to Springer Nature Switzerland AG 2026

This work is subject to copyright. All rights are solely and exclusively licensed by the Publisher, whether the whole or part of the material is concerned, specifically the rights of reprinting, reuse of illustrations, recitation, broadcasting, reproduction on microfilms or in any other physical way, and transmission or information storage and retrieval, electronic adaptation, computer software, or by similar or dissimilar methodology now known or hereafter developed.
The use of general descriptive names, registered names, trademarks, service marks, etc. in this publication does not imply, even in the absence of a specific statement, that such names are exempt from the relevant protective laws and regulations and therefore free for general use.
The publisher, the authors and the editors are safe to assume that the advice and information in this book are believed to be true and accurate at the date of publication. Neither the publisher nor the authors or the editors give a warranty, expressed or implied, with respect to the material contained herein or for any errors or omissions that may have been made. The publisher remains neutral with regard to jurisdictional claims in published maps and institutional affiliations.

Editorial Contact: Annette Klaus

Cover Image: Cover picture taken by Markus Photo, purchased on Shutterstock (uploaded to the site on April 1, 2022)

This Springer imprint is published by the registered company Springer Nature Switzerland AG
The registered company address is: Gewerbestrasse 11, 6330 Cham, Switzerland

If disposing of this product, please recycle the paper.

Foreword 1

Thinking, Naming, Treating, Evolving: Foundations of Modern Veterinary Psychiatry

This book builds on the pioneering work that laid the foundations of veterinary psychiatry—a young yet essential discipline that has steadily gained clinical and scientific recognition in recent decades. By paying tribute to the trailblazers of the field, it reminds us that true innovation often begins with the courage to think differently.

Veterinary psychiatry has not developed in isolation. It has evolved through dialogue with other theories, models, and disciplines. This ambitious and intellectually honest volume breaks down disciplinary boundaries to offer a cohesive, science-driven perspective. Its goal is not to impose dogma but to provide a robust methodological framework—richly supported by scientific literature—that helps prevent the drift into oversimplification and dogmatism that can threaten emerging sciences.

In this sense, the book is both a practical guide and a scientific statement. It proposes a clear and thoughtful nosography—essential for building shared knowledge—while acknowledging that our understanding will continue to evolve. It is enriched by numerous clinical case examples drawn from the authors' daily practice. These authors, who are both seasoned clinicians and active researchers, continuously test their methods against the rigor of scientific inquiry. This dual lens lends the work both academic depth and immediate clinical relevance.

Whether you are a student exploring a lesser-known field, a general practitioner seeking to deepen your understanding, or a seasoned veterinarian curious about the latest developments, this book offers a path forward—at your own pace, grounded in science.

We hope this book becomes a trusted reference—not as a final word but as a foundation on which to build, to question, and to help the discipline grow. Like all living sciences, veterinary psychiatry must evolve. It thrives on the exchange of ideas, the precision of language, and the imperative to care.

Wishing you a stimulating and insightful read,
DVM, PhD, Dip.ACVIM, Dip.ECVIM
Professeur de médecine interne – Oniris

Nantes, France — Juan Hernandez

All veterinary professionals will find invaluable resources here: practical solutions to enhance feline welfare in clinics; guidance on managing geriatric patients; the intersections of nutrition, endocrinology, and behavior; the latest findings in feline psychopharmacology; the scientific foundations of behavioral therapies; and even the historical evolution of those therapies—including, for the curious, a look at hypnotherapy.

In my view, this book deserves to become a foundational reference in feline medicine.

As Colette once wrote, "There are no ordinary cats." Every cat is exceptional in its own way—just like this remarkable and meticulously structured book, which will with no doubt contribute to elevate the standard of care we provide to our feline patients.

DVM, MSc, DACZM, DECZM (ZHM), DACVB

Associate Professor, Faculté de médecine vétérinaire, Université de Montréal

Montreal, Canada **Marion Desmarchelier**

Acknowledgments

The authors would like to thank their loved ones for their support: Babeth, Guy, Pierre, Yvonne, Laurent, Loïc, Simon, and Adrien.

Special thanks go to Pierre Médam, who handled the layout of the French version of this book, for his availability and insightful advice and to Dr Ariane Maillard Verhagen, whose careful review and constructive comments contributed to improving this book.

Competing Interests The authors have no competing interests to declare that are relevant to the content of this manuscript.

Introduction

Veterinary psychiatry—also referred to as animal psychiatry or zoopsychiatry—is a relatively recent discipline within veterinary medicine. Its necessity was first publicly advocated in France in 1925, through Fernand Méry's thesis *Animal Psychology and Veterinary Psychiatry* (1925). In 1964, the publication of *Animal Psychiatry* by veterinarian Abel Brion and neurologist Henri Ey (1964) further asserted that, if animal psychology is acknowledged, the existence of animal psychopathology must logically be recognized as well.

The field began to take shape only in the 1980s, when "behavioral medicine" or "veterinary behavioral medicine" gained traction. Two major research areas emerged: the first clinical trials of psychotropic medications in animals, and the first attempts to classify behavioral disorders. Pioneering studies, such as Nicholas Dodman's demonstration in 1987 of naloxone halting cribbing in horses, were soon followed by the foundational works of Pageat (1998), Landsberg (1997), and Overall (1997), who established reference manuals on behavioral disorders in companion animals.

The 2000s marked a period of rapid expansion, propelled by advances in internal medicine, imaging, genetics, and neurology. Alongside, the concept of animal welfare gained societal importance, now encompassing psychological and behavioral health. Institutional recognition strengthened through the establishment of the American and European Colleges of Veterinary Behavioral Medicine (ACVBM and ECAWBM), formally integrating behavioral medicine into veterinary curricula worldwide.

Today, most countries possess authoritative texts cataloging animal behavioral disorders and therapeutic interventions. However, variations in conceptual frameworks and terminology remain common. Few references openly adopt the terms "veterinary psychiatry" or "psychiatric disease," favoring the more cautious "behavioral problems." Yet this distinction is significant. Current standards for evaluating animal welfare, including the Five Freedoms defined in 1965 by the UK Farm Animal Welfare Council, are essential but insufficient to fully capture psychological suffering.

Within this evolving context, French veterinary psychiatry offers a distinct perspective. It recognizes that psychological suffering in animals may arise independently of environmental or human factors, as the consequence of intrinsic cerebral dysfunction. A cat living in a secure, enriched environment may nevertheless exhibit signs of emotional distress, underscoring the need for an individualized, multidisciplinary approach that integrates neuroscience, psychology, ethology, neurophysiology, psychopharmacology, and internal medicine.

French veterinary psychiatry laid its foundations with Patrick Pageat's *Behavioral Pathology of the Dog* (1995), which proposed a clinical classification inspired by the DSM-5 framework. This approach, refined through clinical practice, was recently updated in our previous work *Veterinary Psychiatry of the Dog* (Masson et al. 2023).

The present volume seeks to adapt this model to the domestic cat (*Felis catus*), now the most widely kept companion animal, surpassing the dog. Unlike the actively domesticated dog, whose evolution was shaped by cooperative tasks with humans, the cat underwent passive domestication, driven by proximity to human food resources. This evolutionary path accounts for the cat's persistent solitary tendencies and its need for territorial stability to maintain psychological balance.

A recent attempt at structuring feline veterinary psychiatry was proposed by Jacky Ley (2023). In this work, we present our own approach, grounded in the French model, which differs markedly from Anglo-American paradigms. Anglo-American approaches often focus on symptoms and underlying emotional states—such as fear, frustration, or anxiety—without necessarily categorizing psychiatric diseases. Consequently, interventions often target symptoms in isolation. In contrast, the French model adopts a holistic, syndromic framework, aiming for comprehensive diagnosis and integrated treatment of psychiatric disorders.

This book aims to provide an updated and scientifically grounded view of feline psychiatric nosography, synthesizing recent advances and bridging French and Anglo-American perspectives. We hope to contribute toward the emergence of a unified classification of feline psychiatric diseases.

The volume is organized into four main sections. The first explores psychophysiology, detailing neurobiological and behavioral foundations. The second addresses feline psychopathology, analyzing major psychiatric syndromes. The third focuses on psychopharmacology, presenting therapeutic options and clinical guidelines. The final section is devoted to psychotherapy and prevention, emphasizing behavioral interventions and preventive strategies.

By offering a scientifically rigorous yet clinically practical framework, *Veterinary Psychiatry of the Cat* is intended for a wide audience: veterinary students, clinical practitioners, animal behavior specialists, researchers, and cat owners seeking deeper insights into the complex behaviors of their feline companions.

References

Dodman NH, Shuster L, Court MH, Dixon R (1987) Investigation into the use of narcotic antagonists in the treatment of a stereotypic behavior pattern (crib-biting) in the horse. Am J Vet Res 48:311–319

Ey H, Brion A (1964) Psychiatrie animale. Desclée de Brouwer

Landsberg GM, Hunthausen WL, Ackerman LJ (1997) Handbook of behaviour problems of the dog and cat, Veterinary handbook series. Butterworth Heinenann, Oxford, Boston

Ley J (2023) The veterinary psychiatry of cats. Academic Press, S.l.

Masson S, Bleuer-Elsner S, Muller G, Medam T, Chevallier J, Gaultier E (2024) Veterinary psychiatry of the dog: diagnosis and treatment of behavioral disorders. Springer Nature Switzerland, Cham. ► https://doi.org/10.1007/978-3-031-53012-8

Méry F (1925) Psychologie animale et psychiatrie vétérinaire. Rev de Pathol comparée

Overall KL (1997) Clinical behavioral medicine for small animals. Mosby, St. Louis

Pageat P (1998) Pathologie du comportement du chien. Editions du Point Veterinaire, Maisons-Alfort

About the Book Conclusion

The cat's behavior is shaped more by its environment than by social relationships (Turner, D.C. 2000)

Cats are not antisocial; they are selectively social (Bradshaw, J.W.S. 2013)

Often hailed as the companion animal of the twenty-first century, the cat holds a unique place at the crossroads between prey and predator. While generally less social than other species, cats are remarkably capable of forming strong bonds both within their own species and across species lines. Their emotional equilibrium, however, depends critically on the structure and stability of their territory, not only spatially but also through time.

In this book, we introduce an original framework for understanding feline psychiatric disorders, rooted in major functional dimensions. Fear plays a dominant role, easily overwhelming the cat's behavior given its evolutionary position as prey. The development of self-control, another major axis, is notably slower in cats than in dogs. Relationships, both within and across species, form another essential domain, while territoriality remains a central pillar of feline well-being. These aspects do not operate independently but are continually influenced by the overall quality of the cat's environment and communication with its surroundings.

Our approach builds on a modernized interpretation of what is often called the "French model" of feline psychiatry. Rather than attributing emotional and cognitive disorders to the simplistic imbalance of a single neurotransmitter, we now understand these conditions as stemming from dysfunctions within broader neural circuits. This perspective allows for a deeper, more precise understanding of feline brain dysfunctions and opens the door to more targeted and nuanced pharmacological strategies. By combining different medications thoughtfully, we can more effectively modulate the various affected circuits and address a wider range of clinical signs.

The therapeutic relationship between veterinarians and cat owners is pivotal to successful outcomes. When families are confronted with a cat in emotional distress, they need a coherent and understandable narrative that explains the symptoms they observe and clarifies the treatment decisions being recommended. Constructing this narrative is not simply a matter of providing information; it re-establishes a sense of control, offers emotional anchors, and lays the groundwork for a strong, enduring therapeutic alliance.

This alliance is essential because behavioral therapy often requires owners to engage deeply, both by modifying the physical environment their cat inhabits and by adjusting the nature of their daily interactions. Supporting owners throughout this process demands not only clinical knowledge but also a high level of communication skills from the veterinary psychiatrist, who must help families adapt and persevere over time as the cat's clinical condition evolves.

The approach we present is intended to be both practical and faithful to the unique behavioral, physiological, and emotional nature of the feline species. Given the cat's status today as the most popular companion animal in France, a deep and nuanced understanding of feline behavior is more important than ever. Accordingly, the tools and strategies we propose are firmly grounded in contemporary veterinary science and neuroscience.

Nevertheless, this model remains a work in progress. Our diagnostic and therapeutic hypotheses, although structured, must remain open and adaptable to advances in scientific knowledge. As research continues to deepen our understanding of the feline brain, it is inevitable that our classifications and treatment approaches will evolve accordingly.

Ultimately, our goal is to promote a more targeted, ethical, and welfare-centered approach to feline psychiatry. By empowering veterinarians to better understand and treat psychiatric disorders in cats, we hope to enhance the quality of life for both the cats themselves and the families who care for them.

Contents

I Psychophysiology

1	Foundations of Veterinary Psychiatry	3
1.1	**Introduction to the Study of Behavior**	4
1.1.1	What Is a Behavior?	4
1.1.2	The Behavioral Approach in Veterinary Psychiatry	6
1.1.3	Factors Influencing Behavior	10
1.2	**Brain Factors Influencing Behavior**	13
1.2.1	Emotions	13
1.2.2	Mood	15
1.2.3	Cognition	15
1.2.4	Temperament and Personality	26
1.3	**Particularities of Feline Behavior**	29
1.3.1	The Domestic Cat	29
1.3.2	Feline Communication	32
1.3.3	Cognition in Cats	46
	References	49
2	Neuroanatomy and Neurophysiology of the Cat	57
2.1	**The Brain**	58
2.1.1	Major Brain Regions	58
2.1.2	The Brain's Unique Vascular Structures	61
2.2	**Neurons, Synapses and Neurotransmitters**	64
2.2.1	Neurons and Glial Cells	64
2.2.2	The Electoral Activity of Neurons	65
2.2.3	The Synapse and Its Functioning	68
2.2.4	Neurotransmitters	71
2.3	**Neurophysiology of Behaviors**	82
2.3.1	Emotions	83
2.3.2	Mood: From the Neurotransmitter Theory to the Receptor Theory	90
2.3.3	Cognition	91
	References	106
3	Major Biological Functions and Endocrinological Mechanisms of Behavior	111
3.1	**Endocrinology**	112
3.1.1	Overview of the Endocrine System	112
3.1.2	Endocrine System and Veterinary Psychiatry	114
3.2	**Biological Functions of the Body**	120
3.2.1	Biological Rhythms	120
3.2.2	Drinking and Eating Behaviors	123
3.2.3	Sexual, Reproductive, and Parental Behaviors	127
3.2.4	Affiliative Behaviors	129
3.3	**Pain**	131
3.3.1	Types of Pain	132
3.3.2	Pain and Veterinary Psychiatry	135
	References	137

4 **Brain Development and Behavioral Ontogenesis** 143
4.1 **Neurogenesis and Brain Connections** 144
4.2 **Chronological Description of Behavioral Development** 148
4.2.1 The Prenatal Period 148
4.2.2 The Neonatal Period (0–9 Days) 148
4.2.3 The Transitional Period (9–15 Days) 149
4.2.4 The Socialization Period (18 Days—2 Months) 150
4.2.5 The Juvenile Period (2–6 Months) 155
4.2.6 The Pubertal Period (6 Months to 1 Year) 158
4.2.7 The Adult Period (1 Year—8/10 Years) 161
4.2.8 Senescent Period (from 8 to 10 Years) 161
4.3 **Axial Description of Behavioral Development** 167
4.3.1 Fear and Alarm System 167
4.3.2 Acquisition of Self-Controls 170
4.3.3 Relationships and Attachment 174
4.3.4 Territoriality 177
References 182

5 **Concepts of Welfare and Quality of Life** 187
5.1 **Definitions and History** 188
5.1.1 Welfare 188
5.1.2 Welfare, Lifestyle, and Normal Species Behaviors 191
5.1.3 Welfare and Selection 193
5.2 **Individual Welfare Assessment** 195
5.2.1 Welfare and Veterinary Care 195
5.2.2 Preliminary Considerations on Quantitative Welfare Assessment 196
5.2.3 The Welfare Consultation 197
References 203

II Psychopathology

6 **The Architecture of Our Veterinary Psychiatry Model** 207
6.1 **Presentation of the Veterinary Psychiatry Model** 208
6.2 **The Initial Consultation** 209
6.2.1 The Collection of Semiological Data: From Sign to Symptom 211
6.2.2 Behavioral Process and Pathological Pathway 212
6.2.3 Direct Observation 213
6.2.4 Clinical Examination 214
6.2.5 Symptom Assessment and Differential Diagnosis 214
6.2.6 Complementary Examinations 214
6.2.7 Diagnosis 215
6.2.8 Treatments 218
6.3 **Follow-Up and Evaluation of Care** 222
6.4 **Clinical Vignettes** 223
6.4.1 Suki 223
6.4.2 Octavius 225
References 226

7 **Psychological States** 227
7.1 **Reactive State** 229
7.1.1 Definition and Description 229
7.1.2 Evolution 230
7.1.3 Pathogenesis 230
7.1.4 Symptoms 230
7.1.5 Diagnosis 232
7.1.6 Treatments 233
7.1.7 Prevention 234
7.2 **Emotional Pathological States** 235
7.2.1 Phobic State 235
7.2.2 Anxious State 241
7.3 **Mood Pathological States** 251
7.3.1 Depressive State 251
7.3.2 Manic State 256
7.3.3 Bipolar State 260
7.4 **Cognitive Pathological States** 260
7.4.1 Self-Controls Pathological States 260
7.4.2 Derealization Pathological States 277
References 287

8 **Disorders Related to Dysregulation of the Fear Axis** 289
8.1 **Categorical Phobias** 290
8.1.1 Description 290
8.1.2 Diagnosis 298
8.1.3 Prognosis 303
8.1.4 Treatments 303
8.1.5 Therapeutic Follow-Up 307
8.1.6 Prevention 308
8.1.7 Clinical Vignettes 308
8.2 **Sensory Deprivation Syndrome** 311
8.2.1 Description 311
8.2.2 Diagnosis 314
8.2.3 Prognosis 316
8.2.4 Treatments 317
8.2.5 Therapeutic Follow-Up 317
8.2.6 Prevention 317
8.2.7 Clinical Vignettes 318
8.3 **Post-traumatic Phobias or PTSD** 320
8.3.1 Description 320
8.3.2 Diagnosis 321
8.3.3 Prognosis 323
8.3.4 Treatments 323
8.3.5 Therapeutic Follow-Up 323
8.3.6 Prevention 323
8.3.7 Clinical Vignettes 323
References 326

9 **Disorders Related to Dysregulation of the Self-Control Axis** 329
9.1 **Hypersensitivity-Hyperactivity Syndrome (HSHA) or Feline ADHD** 332
9.1.1 Description 332
9.1.2 Diagnosis 339
9.1.3 Prognosis 345
9.1.4 Treatments 346
9.1.5 Therapeutic Follow-Up 350
9.1.6 Prevention 351
9.1.7 Clinical Vignettes 351
9.2 **Addictions** 353
9.2.1 Description 353
9.2.2 Diagnosis 356
9.2.3 Prognosis 358
9.2.4 Treatments 358
9.2.5 Therapeutic Follow-Up 359
9.2.6 Prevention 359
9.2.7 Clinical Vignettes 359
References 361

10 **Disorders Related to Dysregulation of the Territoriality Axis** 363
10.1 **Disorders Resulting from Territorial Poverty** 365
10.1.1 Description 365
10.1.2 Diagnosis 367
10.1.3 Prognosis 369
10.1.4 Treatments 369
10.1.5 Therapeutic Follow-Up 370
10.1.6 Prevention 371
10.1.7 Clinical Vignette 371
10.2 **Disorders Resulting from Territorial Disrupting** 372
10.2.1 Description 372
10.2.2 Diagnosis 375
10.2.3 Prognosis 377
10.2.4 Treatments 377
10.2.5 Therapeutic Follow-Up 378
10.2.6 Prevention 378
10.2.7 Clinical Vignette 379
10.3 **Territoriality Disorders Secondary to Another Behavioral Disorder** 381
10.3.1 Description 381
10.3.2 Diagnosis 382
10.3.3 Prognosis 382
10.3.4 Treatments 384
10.3.5 Therapeutic Follow-Up 384
10.3.6 Prevention 384
10.3.7 Clinical Vignette 384
References 388

11 **Disorders Related to Dysregulation of the Relationships Axis** 391
11.1 **Interspecific Relational Disorders** 392
11.1.1 Description 393
11.1.2 Diagnosis 395

11.1.3 Prognosis ... 399
11.1.4 Treatments ... 399
11.1.5 Therapeutic Follow-Up ... 400
11.1.6 Prevention ... 400
11.1.7 Clinical Vignettes ... 400
11.2 **Intraspecific Relational Disorders** ... 402
11.2.1 Description ... 402
11.2.2 Diagnosis ... 406
11.2.3 Prognosis ... 406
11.2.4 Treatments ... 409
11.2.5 Therapeutic Follow-Up ... 410
11.2.6 Prevention ... 410
11.2.7 Clinical Vignette ... 411
References ... 414

12 **Essential Psychiatric Disorders** ... 417
12.1 **Dissociative Syndrome** ... 418
12.1.1 Description ... 418
12.1.2 Diagnosis ... 420
12.1.3 Prognosis ... 422
12.1.4 Treatments ... 422
12.1.5 Therapeutic Follow-Up ... 423
12.1.6 Prevention ... 423
12.1.7 Clinical Vignette ... 423
12.2 **Dysthymia** ... 424
12.2.1 Description ... 424
12.2.2 Diagnosis ... 426
12.2.3 Prognosis ... 428
12.2.4 Treatments ... 429
12.2.5 Therapeutic Follow-Up ... 429
12.2.6 Prevention ... 430
12.2.7 Clinical Vignette ... 430
References ... 432

13 **Aging Disorders** ... 435
13.1 **Common Points to All Aging Disorders** ... 438
13.1.1 Description ... 438
13.1.2 Diagnosis ... 440
13.1.3 Prognosis ... 441
13.1.4 Treatments ... 441
13.1.5 Therapeutic Follow-Up ... 441
13.2 **Cognitive Dysregulation in Elderly Cats** ... 443
13.2.1 Description ... 443
13.2.2 Diagnosis ... 447
13.2.3 Prognosis ... 449
13.2.4 Treatments ... 449
13.2.5 Therapeutic Follow-Up ... 450
13.2.6 Prevention ... 450
13.2.7 Clinical Vignette ... 450
13.3 **Emotional Dysregulation in Elderly Cats** ... 451

13.3.1 Description ... 451
13.3.2 Diagnosis ... 453
13.3.3 Prognosis ... 453
13.3.4 Treatments ... 456
13.3.5 Therapeutic Follow-Up ... 457
13.3.6 Prevention ... 457
13.3.7 Clinical Vignette ... 457
13.4 **Mood Dysregulation in Elderly Cats** ... 458
13.4.1 Description ... 458
13.4.2 Diagnosis ... 459
13.4.3 Prognosis ... 462
13.4.4 Treatments ... 462
13.4.5 Therapeutic Follow-Up ... 463
13.4.6 Prevention ... 463
13.4.7 Clinical Vignette ... 463
13.5 **Self-Controls Dysregulation in Elderly Cats** ... 464
13.5.1 Description ... 464
13.5.2 Diagnosis ... 466
13.5.3 Prognosis ... 466
13.5.4 Treatments ... 466
13.5.5 Therapeutic Follow-Up ... 468
13.5.6 Prevention ... 468
13.5.7 Clinical Vignette ... 468
References ... 471

14 The Link Between Psychiatric Disorders and Somatic Disorders ... 473
14.1 **Pain** ... 476
14.1.1 Psychiatric Disorders Secondary to a Painful Condition ... 476
14.1.2 The Consequences of Psychiatric Disorders on Pain ... 478
14.2 **Dermatology** ... 479
14.3 **Neurology** ... 481
14.4 **Nutrition** ... 486
14.5 **Internal Medicine** ... 489
14.5.1 Lower Urinary Tract Diseases ... 489
14.5.2 Digestive Diseases ... 490
14.5.3 Feline Hypertension ... 491
14.5.4 Metabolic Diseases ... 491
14.6 **Reproduction** ... 492
14.7 **Endocrinology** ... 493
14.7.1 Hyperthyroidism ... 493
14.7.2 Diabetes ... 494
14.7.3 Hypothyroidism ... 494
14.7.4 Cushing's Syndrome or Hyperadrenocorticism ... 495
14.7.5 Addison's Disease ... 496
14.7.6 Sexual Dysendocrinopathies ... 496
14.7.7 Acromegaly ... 497
14.8 **Surgery** ... 497
14.9 **Aging** ... 498
References ... 498

15 **Semiology in Feline Psychiatry** 503
15.1 **Centripetal Behaviors** 506
15.1.1 Semiology of Feeding Behavior 506
15.1.2 Semiology of Drinking Behavior 510
15.1.3 Semiology of Sleeping Behavior 510
15.1.4 Semiology of Elimination Behavior 511
15.1.5 Semiology of Repetitive Behaviors 517
15.2 **Centrifugal Behaviors** 523
15.2.1 Semiology of Injurious Sequences 523
15.2.2 Semiology of Sexual and Maternal Behavior 528
15.2.3 Semiology of Play and Mutual Grooming 529
15.2.4 Semiology of Exploratory Behavior 531
15.3 **Other Elements of Semiology** 535
15.3.1 Semiology of Behavioral Development 535
15.3.2 Semiology of Direct Observation 537
15.3.3 Semiology of Behavior in Aging Cats 539
15.4 **Semiology by Behavioral Axis** 542
15.4.1 Disturbances Related to the Fear Axis 542
15.4.2 Disturbances Related to the Self-Control Axis 542
15.4.3 Disturbances Related to the Territoriality Axis 543
15.4.4 Disturbances Related to the Relational Axis 544
References 544

III Psychopharmacology

16 **Introduction to Psychopharmacology** 549
16.1 **The Mechanism of Action of Psychotropic Drugs** 550
16.1.1 Actions on Ionotropic Receptors 550
16.1.2 Actions on Metabotropic Receptors 553
16.1.3 Actions on Reuptake Transporters 553
16.1.4 Actions on Voltage-Dependent Ion Channels 556
16.1.5 Actions on Enzymes 556
16.2 **Prescription of Psychotropic Drugs** 559
16.2.1 Basics of Prescription 559
16.2.2 Monotherapy and Polytherapy 560
16.2.3 Duration of Treatment and Withdrawal 561
References 562

17 **Selective Serotonin Reuptake Inhibitors (SSRIs)** 565
17.1 **General Characteristics of All SSRIs** 567
17.1.1 General Pharmacology of SSRIs 567
17.1.2 Common Side Effects and Usage Precautions for All SSRIs 568
17.1.3 Clinical Use and Dosage 569
17.2 **Fluoxetine** 569
17.2.1 Description, Chemical Formula, and Brand Names 569
17.2.2 Pharmacology 569
17.2.3 Common Side Effects and Usage Precautions 570
17.2.4 Clinical Use and Dosage 572

17.3 **Fluvoxamine** 573
17.3.1 Description, Chemical Formula, and Brand Names 573
17.3.2 Pharmacology 573
17.3.3 Common Side Effects and Usage Precautions 574
17.3.4 Clinical Use and Dosage 575
17.4 **Paroxetine** 576
17.4.1 Description, Chemical Formula, and Brand Names 576
17.4.2 Pharmacology 576
17.4.3 Common Side Effects and Usage Precautions 577
17.4.4 Clinical Use and Dosage 578
17.5 **Sertraline** 578
17.5.1 Description, Chemical Formula, and Brand Names 578
17.5.2 Pharmacology 578
17.5.3 Common Side Effects and Usage Precautions 579
17.5.4 Clinical Use and Dosage 580
References 582

18 Tricyclic Antidepressants 585
18.1 **General Characteristics of All Tricyclics** 586
18.1.1 General Pharmacology of TCAs 586
18.1.2 Common Side Effects and Usage Precautions for All TCAs 587
18.1.3 Clinical Use and Dosage 588
18.2 **Clomipramine** 588
18.2.1 Description, Chemical Formula, and Brand Names 588
18.2.2 Pharmacology 588
18.2.3 Side Effects and Usage Precautions 589
18.2.4 Clinical Use and Dosage 590
18.3 **Amitriptyline** 591
18.3.1 Description, Chemical Formula, and Brand Names 591
18.3.2 Pharmacology 591
18.3.3 Side Effects and Usage Precautions 592
18.3.4 Clinical Use and Dosage 593
References 594

19 Monoamine Oxidase Inhibitors (MAOIs) 597
19.1 **General Characteristics of All IMAO** 598
19.2 **Selegiline** 599
19.2.1 Description, Chemical Formula, and Brand Names 599
19.2.2 Pharmacology 599
19.2.3 Side Effects and Usage Precautions 600
19.2.4 Clinical Use and Dosage 600
References 601

20 Benzodiazepines 603
20.1 **General Characteristics of All Benzodiazepines** 604
20.1.1 Pharmacology 604
20.1.2 Side Effects and Usage Precautions 605
20.1.3 Clinical Use and Dosage 606
20.2 **Alprazolam** 607

20.2.1 Description, Chemical Formula, and Brand Names 607
20.2.2 Pharmacology 607
20.2.3 Side Effects and Usage Precautions 607
20.2.4 Clinical Use and Dosage 607
20.3 **Other Benzodiazepines** 608
20.3.1 Pharmacology 608
20.3.2 Side Effects and Usage Precautions 608
20.3.3 Clinical Use and Dosage 608
References 609

21 Sympatholytics 611
21.1 **Propranolol** 612
21.1.1 Description, Chemical Formula, and Brand Names 612
21.1.2 Pharmacology 612
21.1.3 Side Effects and Usage Precautions 613
21.1.4 Clinical Use and Dosage 614
21.2 **Clonidine** 614
21.2.1 Description, Chemical Formula, and Brand Names 614
21.2.2 Pharmacology 614
21.2.3 Side Effects and Usage Precautions 615
21.2.4 Clinical Use and Dosage 616
References 617

22 Antiepileptic Drugs 619
22.1 **General Characteristics of All Antiepileptic Drugs** 621
22.2 **Gabapentin** 621
22.2.1 Description, Chemical Formula, and Brand Names 621
22.2.2 Pharmacology 621
22.2.3 Side Effects and Usage Precautions 622
22.2.4 Clinical Use and Dosage 623
22.3 **Pregabalin** 624
22.3.1 Description, Chemical Formula, and Brand Names 624
22.3.2 Pharmacology 624
22.3.3 Side Effects and Usage Precautions 625
22.3.4 Clinical Use and Dosage 625
22.4 **Imepitoin** 626
22.4.1 Description, Chemical Formula, and Brand Names 626
22.4.2 Pharmacology 626
22.4.3 Side Effects and Usage Precautions 627
22.4.4 Clinical Use and Dosage 627
22.5 **Phenobarbital** 628
22.5.1 Description, Chemical Formula, and Brand Names 628
22.5.2 Pharmacology 628
22.5.3 Side Effects and Usage Precautions 629
22.5.4 Clinical Use and Dosage 630
22.6 **Levetiracetam** 631
22.6.1 Description, Chemical Formula, and Brand Names 631
22.6.2 Pharmacology 631

22.6.3 Side Effects and Usage Precautions ... 631
22.6.4 Clinical Use and Dosage ... 632
References ... **634**

23 Tetracyclic Antidepressants or NaSSAs ... 637
23.1 **Mianserin** ... 638
23.1.1 Description, Chemical Formula, and Brand Names ... 638
23.1.2 Pharmacology ... 638
23.1.3 Side Effects and Usage Precautions ... 638
23.1.4 Clinical Use and Dosage ... 639
23.2 **Mirtazapine** ... 639
23.2.1 Description, Chemical Formula, and Brand Names ... 639
23.2.2 Pharmacology ... 639
23.2.3 Side Effects and Usage Precautions ... 640
23.2.4 Clinical Use and Dosage ... 642
References ... **643**

24 Neuroleptics or Antipsychotics ... 645
24.1 **General Characteristics of All Neuroleptics** ... 646
24.1.1 Description ... 646
24.1.2 General Pharmacology of Neuroleptics ... 646
24.1.3 Side Effects and Usage Precautions ... 652
24.1.4 Clinical Use and Dosage ... 654
References ... **656**

25 Other Psychotropic Drugs ... 657
25.1 **Venlafaxine** ... 658
25.1.1 Description, Chemical Formula, and Brand Names ... 658
25.1.2 Pharmacology ... 658
25.1.3 Side Effects and Usage Precautions ... 659
25.1.4 Clinical Use and Dosage ... 660
25.2 **Trazodone** ... 661
25.2.1 Description, Chemical Formula, and Brand Names ... 661
25.2.2 Pharmacology ... 661
25.2.3 Side Effects and Usage Precautions ... 662
25.2.4 Clinical Use and Dosage ... 663
25.3 **Buspirone** ... 663
25.3.1 Description, Chemical Formula, and Brand Names ... 663
25.3.2 Pharmacology ... 664
25.3.3 Side Effects and Usage Precautions ... 664
25.3.4 Clinical Use and Dosage ... 665
25.4 **Cannabinoids** ... 666
25.4.1 Cannabidiol ... 669
25.4.2 Dronabinol ... 672
References ... **675**

26 Hormonal Therapies ... 679
26.1 **Thyroid Hormones** ... 680
26.1.1 Description, Chemical Formula, and Brand Names ... 680
26.1.2 Pharmacology ... 680
26.1.3 Side Effects and Usage Precautions ... 681

26.1.4	Clinical Use and Dosage	681
26.2	**Sexual Hormones**	681
26.2.1	Megestrol Acetate	681
26.2.2	Oxytocin	681
26.2.3	Deslorelin	682
	References	684
27	Over-the-Counter Treatments	687
27.1	**Pheromone-Based Treatments**	688
27.1.1	Maternal Pheromone	688
27.1.2	Facial Pheromones	688
27.2	**Nutraceuticals and Dietary Supplements**	689
27.2.1	Alpha-Casozepine	689
27.2.2	α-Lactalbumine and Magnesium Combination	690
27.2.3	L-Tryptophan	691
27.2.4	Fish Hydrolysate and Superoxide Dismutase Extracted from Melon Combination	691
27.2.5	Other Dietary Supplements	691
27.3	**Phytotherapy**	692
27.3.1	L-Theanine	692
27.3.2	Crataegus Monogyna—Passiflora Incarnata—Griffonia Simplicifolia Combination	692
27.3.3	Aromatherapy	692
27.3.4	Bach Flowers and Homeopathy	693
	References	694

IV Psychotherapy

28	Introduction to Veterinary Psychotherapies	699
28.1	**Types of Therapies in Human Medicine**	700
28.1.1	Historical Context	700
28.1.2	Cognitive-Behavioral Therapies	701
28.1.3	Systemic Therapies	701
28.2	**Practical Tools for the Veterinary Practitioner: Consultation Techniques**	709
28.2.1	The System in Veterinary Medicine	710
28.2.2	Alliance and Resistances	710
28.2.3	The Therapist Positioning	711
28.2.4	Consultation Techniques	713
28.2.5	Constructing the Therapy	718
	References	721
29	Practical Application of Veterinary Psychotherapies	723
29.1	**General Information**	724
29.1.1	General Principles of Therapies	724
29.1.2	General Principles of Conditioning	731
29.1.3	How to Modify the Probability of a Behavior Occurring?	733
29.2	**Organizing the Cat's Environment and Respectful Communication: Prevention Tools for All Cats**	740
29.2.1	Organizing the Environment	740

29.2.2 Respectful Communication ... 757
29.2.3 Learning and Useful Exercises for All Cats ... 759
29.3 **Behavioral Axis Therapies** ... 768
29.3.1 Disorders Related to a Dysregulation of the Fear Axis ... 768
29.3.2 Disorders Related to a Dysregulation of the Self-Controls Axis ... 774
29.3.3 Disorders Related to a Dysregulation of the Territoriality Axis ... 782
29.3.4 Disorders Related to a Dysregulation of the Relationships Axis ... 786
29.3.5 Other Psychiatric Disorders ... 789
References ... 792

30 **The Causes and Consequences of Veterinary Care** ... 797
30.1 **Veterinary Care** ... 798
30.1.1 The Paradox of Veterinary Care ... 798
30.1.2 Healthcare Phobia ... 803
30.1.3 Consequences of Healthcare Phobia ... 810
30.2 **Practical Solutions in Veterinary Clinics** ... 813
30.2.1 Available Tools ... 814
30.2.2 Medical Training in Veterinary Practices ... 825
30.3 **The Care Journey in Practice** ... 829
30.3.1 Reception and Waiting Area ... 830
30.3.2 The Consultation Room ... 831
30.3.3 Hospitalization ... 833
References ... 837

31 **Veterinary Psychiatry in the Daily Life of the General Practitioner** ... 841
31.1 **Preventive Care** ... 842
31.1.1 Vaccination Visits for Kittens ... 842
31.1.2 Puberty Consultations ... 849
31.1.3 Adult Cat Visits ... 851
31.1.4 Senior Cat Visits ... 853
31.1.5 Behavioral Balance Assessment Questionnaire ... 853
31.2 **General Medicine** ... 857
31.2.1 Painful Conditions ... 857
31.2.2 Chronic Conditions ... 859
31.2.3 End of Life, Palliative Care, and Euthanasia ... 861
31.3 **French Legislation (for informational purposes)** ... 865
References ... 867

Supplementary Information
Appendix 1 Diagnosis Criteria for Phobic and Anxious States ... 871
Appendix 2 Diagnostic Criteria for Impulsive and Compulsive States ... 873
Appendix 3 Diagnostic Criteria for Depressive, Manic, and Bipolar States ... 875
Appendix 4 Diagnostic Criteria for Dissociative and Senile States ... 877
Appendix 5 Disorders Associated with Dysregulation of the Fear Axis ... 879

Appendix 6 Disorders Associated with Dysregulation of the Self-Controls Axis 880
Appendix 7 Disorders Associated with Dysregulation of the Territoriality Axis 882
Appendix 8 Disorders Associated with Dysregulation of the Relationship Axis 884
Appendix 9 Essential Psychiatric Disorders 885
Appendix 10 Aging Disorders 886
Appendix 11 Elements of Differential Diagnosis Between Neurology and Psychiatry 888
Appendix 12 Body Condition Score (BCS) Chart for Cats 889
Appendix 13 Reported Signs in the Semiology of Feeding Behavior and Possible Interpretations 890
Appendix 14 Reported Signs in the Semiology of Sleeping Behavior and Possible Interpretations 891
Appendix 15 Differential Diagnosis of Urinary and Fecal Inappropriate Elimination and Their Associated Characteristics 892
Appendix 16 Reported Signs in the Semiology of Repetitive Behaviors and Possible Interpretations 893
Appendix 17 Classification of Injurious Sequences in Cats Based on the Modified Moyer Classification 894
Appendix 18 Semiology of Play and Mutual Grooming 895
Appendix 19 Semiology of Exploratory Behavior and Marking 896
Appendix 20 Summary Table of SSRIs 897
Appendix 21 Summary Table of Tricyclic Antidepressants 898
Appendix 22 Summary Table of Monoamine Oxidase Inhibitors (MAOIs) 899
Appendix 23 Summary Table of Benzodiazepines 899
Appendix 24 Summary Table of Sympatholytics 900
Appendix 25 Summary Table for Antiepileptics 900
Appendix 26 Summary Table of Tetracyclic NaSSAs 902
Appendix 27 Summary Table of Neuroleptics 903
Appendix 28 Summary Table of Other Psychotropic Medications 905
Appendix 29 Summary Table of Complementary Hormonal Treatments 906
Appendix 30 Psychotropic Medications for Managing Cats with Healthcare Phobia 907
Appendix 31 Fear Behavioral Management General Principles 907
Appendix 32 Self-Control Behavioral Management General Principles 908
Appendix 33 Territoriality Behavioral Management General Principles 909
Appendix 34 Relationships Behavioral Management General Principles 909
Appendix 35 Behavioral Condition Awareness for Owners 910

Index 913

Authors and Contributors

About the Authors

Stéphane Bleuer-Elsner, DVM

Doctor of Veterinary Medicine, holder of the Inter-School Diploma in Veterinary Behavioral Medicine (DIE). Board-certified specialist in Companion Animal Behavioral Medicine (ECAWBM-BM). Member of the board of GECAF (Companion Animal Behavior Study Group / AFVAC) and member of the Examination Committee of the European College of Animal Welfare and Behavioral Medicine (ECAWBM-BM).

After several years practicing general veterinary medicine in France, Dr. Stéphane Bleuer-Elsner specialized in veterinary psychiatry. Since 2011, he has been based in Israel, where he exclusively practices veterinary psychiatry in Tel Aviv, while also offering specialized consultations in France and Switzerland. A frequent speaker in France, Israel, and at international conferences, he has authored and co-authored several scientific publications and contributed to the book *Veterinary Psychiatry of the Dog*.

Contact: vetbehavior.il@gmail.com

Sylvia Masson, DVM

Doctor of Veterinary Medicine, holder of the University Diploma in Veterinary Psychiatry (DUPV). Board-certified specialist in Companion Animal Behavioral Medicine (ECAWBM-BM). President of the Examination Committee of the ECAWBM-BM. Member of the board of GECAF and Secretary of the European Society of Veterinary Clinical Ethology (ESVCE).

After ten years practicing general veterinary medicine, Dr. Sylvia Masson specialized in veterinary psychiatry. She currently practices referral-based behavioral medicine in Voreppe, France. An active speaker in France and internationally, she is the author of several popular science books (*G.E.R.E.R. l'éducation et le comportement de votre chien* and *Comment cohabiter avec votre chat?*) and contributed to the book *Veterinary Psychiatry of the Dog*.

Marie Lacoste, DVM

Doctor of Veterinary Medicine, holder of a Master's degree in Cognitive Neurosciences.

Dr. Marie Lacoste practices both general veterinary medicine and veterinary psychiatry within the Behavioral Medicine Department at Dr. Masson's clinic in Voreppe, France. She has been actively involved in research projects since the beginning of her career.

Gérard Muller, DVM

Doctor of Veterinary Medicine, holder of the Inter-School Diploma in Veterinary Behavioral Medicine (DIE). Board-certified specialist in Companion Animal Behavioral Medicine (ECAWBM-BM).

Founding Diplomate of the European College of Animal Welfare and Behavioral Medicine (ECAWBM-BM) and former President of Zoopsy, the French Veterinary Zoopsychiatry Association, Dr. Gérard Muller practiced for over 30 years in Lille, France, combining general practice with behavioral medicine specialization.

He has trained numerous veterinarians, participated extensively in research, and co-authored the book *Veterinary Psychiatry of the Dog*.

About the Illustrator

Menelia Vasilopoulou Kampitsi

PhD in Biology and Scientific Illustrator.

Holder of a PhD in Biology from the Laboratory of Functional Morphology at the University of Antwerp.

After completing studies in organismal biology and ecology in Greece and France, Dr. Menelia Vasilopoulou Kampitsi pursued a PhD in functional ecology in Belgium, which led her toward science communication and dissemination.

She subsequently trained in scientific illustration in Barcelona and has since specialized professionally in the field.

As a freelance illustrator, she collaborates with PhD students, veterinarians, and researchers on various outreach and dissemination projects. Her illustrations have been published in leading scientific journals and books. She has notably collaborated with institutions such as the CNRS, the University of Antwerp, and the National Museum of Natural History in Paris. She also illustrated the book *Veterinary Psychiatry of the Dog*.

Contact: Menelia VK Illustration
▶ https://meneliavk.wixsite.com/website

Psychophysiology

Contents

Chapter 1 Foundations of veterinary psychiatry – 3

Chapter 2 Neuroanatomy and neurophysiology of the cat – 57

Chapter 3 Major biological functions and endocrinological me – 111

Chapter 4 Brain development and behavioral ontogenesis – 143

Chapter 5 Concepts of welfare and quality of life – 187

Foundations of Veterinary Psychiatry

Contents

1.1 Introduction to the Study of Behavior – 4
1.1.1 What Is a Behavior? – 4
1.1.2 The Behavioral Approach in Veterinary Psychiatry – 6
1.1.3 Factors Influencing Behavior – 10

1.2 Brain Factors Influencing Behavior – 13
1.2.1 Emotions – 13
1.2.2 Mood – 15
1.2.3 Cognition – 15
1.2.4 Temperament and Personality – 26

1.3 Particularities of Feline Behavior – 29
1.3.1 The Domestic Cat – 29
1.3.2 Feline Communication – 32
1.3.3 Cognition in Cats – 46

References – 49

© The Author(s), under exclusive license to Springer Nature Switzerland AG 2026
S. Bleuer-Elsner et al., *Veterinary Psychiatry of the Cat*,
https://doi.org/10.1007/978-3-032-06260-4_1

This chapter introduces and defines the key concepts forming the foundation of this book. The first section explores how behavior can be defined, how it is specifically defined in veterinary psychiatry, and the factors that may influence it.

The second section delves into the neurophysiological aspects of behavior by defining the general concepts of emotion, mood, and cognition, along with the notions of temperament and personality. The final section highlights the unique aspects of communication and cognition in the feline species.

1.1 Introduction to the Study of Behavior

1.1.1 What Is a Behavior?

1.1.1.1 Various Definitions of Behavior

The definition of behavior varies across disciplines. The one adopted for this book is derived from behavioral biology (Levitis et al. 2009) and defines behavior as the set of actions and inactions of an individual that enable it to respond to changes in its internal or external environment. However, depending on the context, the term "behavior" may take on a different meaning.

In a broad sense, as in the phrase "the behavior of the cat...," the term "behavior" refers to the complete behavioral repertoire or *ethogram* of the species in question **(see box below)**. Behaviors categorized in this way can be classified into two groups (Pageat 1995): *centripetal behaviors* and *centrifugal behaviors*. Centripetal behaviors, meaning those "turned inward," include feeding, drinking, sleeping, and elimination behaviors. Centrifugal behaviors, meaning those "turned outward," include exploratory, agonistic, play, sexual, and maternal behaviors. The latter two categories of behaviors can also be described as "mixed."

Important Point of View

The ethogram of an animal species should, in principle, be established through observations of that species in its "natural" environment. However, domestication significantly influences the entire range of behaviors in a species.

Cats were domesticated approximately 10,000 years ago through a process of passive domestication: they remained in proximity to humans because both species benefited from the arrangement. Human food stores were protected from rodents, while cats were provided with an abundant source of prey.

Today's domestic cat is therefore a distinct species from its wild ancestor, *Felis silvestris lybica*, but its ethogram remains close to that of wild felines such as the European wildcat (*Felis silvestris*), the African wildcat (*Felis lybica*), the sand cat (*Felis margarita*), or Biet's cat (*Felis bieti*) (Stanton et al. 2015).

In a strict sense, the term "behavior" can be interpreted in two different ways.

On the one hand, behavior may refer to a *behavioral unit*, such as "stalking behavior," which represents a posture, a movement, or even a signal.

On the other hand, behavior may describe a *behavioral sequence*, such as "predatory behavior." In this case, behavior refers to the functional succession of several behavioral units. For example, in "predatory behavior," the behavioral sequence comprises successive behavioral units, where each preceding unit triggers the next: "waiting," "pouncing," "capturing," "killing," and "ingesting."

1.1.1.2 Different Levels of Behavioral Analysis

The ethogram, or broad definition of an animal species' behavior, is based on both formal and functional descriptions of that species' behaviors (Darmaillacq and Lévy 2019).

Formal analysis of behavior describes what the animal does, where it does it, and when. This involves a physical description of its postures and movements, as well as the spatial and temporal characteristics of the behavior being studied.

Functional analysis, on the other hand, explains how and why the animal performs the behavior. It corresponds to the physiological and causal description of the behavior.

These two approaches to behavior are complementary. The formal description helps understand the shape or form of a behavior, while the functional description aids in predicting its purpose.

Functional analysis itself can be divided, following Tinbergen's framework (Tinbergen 1963), into four additional levels of analysis, grouped into two categories (◘ Fig. 1.1): *distal analysis*, which concerns the species, and *proximal analysis*, which concerns the individual.

Distal analysis explains *why* a behavior exists by determining its function and phylogeny in the species.

The *functional analysis* identifies the role of the behavior within the studied species. The *phylogenetic analysis* determines when and how the behavior appeared during the evolution of species. For instance, predatory behavior in carnivorous species ensures a regular food supply when carrion is scarce (Pereira et al. 2014). This behavior likely dates back to the origins of life itself (Bengtson 2002).

Proximal analysis seeks to explain *how* a behavior exists by describing its mechanisms and ontogeny in the individual.

The *mechanistic analysis* identifies the physiological processes behind the individual's behavior. The *ontogenetic analysis* determines when and how the behavior emerged during the individual's life. For example, predatory behavior in cats is influenced by the internal stimulus of hunger and the sight of

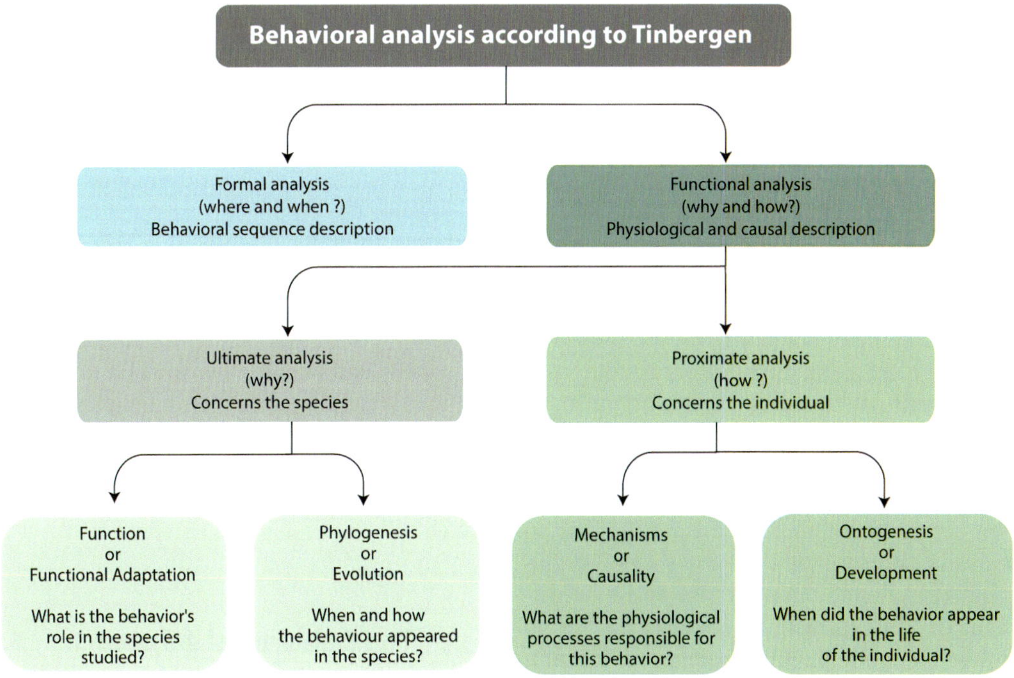

◘ **Fig. 1.1** The levels of functional analysis of behavior according to Tinbergen. (a) Distal analysis concerns the species and focuses on the functional adaptation and phylogeny of behaviors. (b) Proximal analysis concerns the individual and focuses on the mechanisms and ontogeny of behaviors

potential prey (Biben 1979). It emerges during the socialization period.

1.1.1.3 Behavioral Regulation

The function of any behavior is to address an imbalance, whether internal or external, actual or anticipated. It is through the execution of the behavior that internal balance is restored.

An organism's equilibrium relies on two endogenous regulatory processes, known as *homeostasis* (Cannon 1929) and *allostasis* (Sterling 2012), which enable the organism to adaptively respond to stressful stimuli, meaning changes occurring in its internal or external environment. These two concepts, initially developed to describe only the physiological responses of an organism to environmental variations, are now also used to describe its behavioral responses (McEwen and Wingfield 2010).

Homeostasis is the ability of an organism to rebalance or regulate its functioning to preserve a stable internal state despite the stressful stimuli it encounters. It is the process that allows an organism to return to its initial internal state. Homeostasis, or "constancy in stability," is therefore the organism's normal resistance to change, aiming to counter an actual imbalance. For example, *homeostasis* concerning energy inputs involves the mechanisms that enable the storage and release of reserves in the form of glycogen and fat. Homeostatic responses are, of course, not detrimental to the organism. However, their dysfunction can induce a pathological state.

Allostasis is the ability of an organism to destabilize its functioning to achieve a new stable internal state in response to the stressful stimuli it encounters. It is, therefore, a process that prevents an organism from returning to its initial internal state. Allostasis, or "stability after change," represents the active adjustment of the organism to changes, aiming to prevent an anticipated imbalance. For example, when oxygen pressure drops (as occurs at altitude with the reduction of ambient oxygen), the organism initially adapts by increasing blood pressure and subsequently, by raising hemoglobin levels to ensure long-term adaptation to the new oxygen levels. In this example, there are two successive allostatic variations that promote adaptation. Using the example of energy reserve variations, we can understand that this *allostatic* response may occur under the influence of internal factors such as emotional states, or external factors such as preparing for migration. Male seals, for instance, accumulate substantial fat reserves in preparation for extended fasting on the ice pack.

Allostatic responses are always detrimental to the organism to varying degrees. *Allostatic loads* refer to states of relative or temporary imbalance. They are costly but also beneficial to the organism, as they ultimately facilitate adaptation. *Allostatic overloads* refer to states of high or chronic *allostatic* imbalance. These are necessarily harmful to the organism and lead to its collapse (◘ Fig. 1.2). Hypertension associated with renal failure is a typical example.

A behavior is therefore always a sign of a disruption in the equilibrium of the organism that produces it. However, the ability of this behavior to restore the individual's equilibrium distinguishes a normal, adaptive behavior from a pathological one. Similarly, imbalance, being inherent to the functioning of an organism, is normal but becomes pathological if it has lasting detrimental consequences for the organism.

These attempts at adaptation that fail and become pathological are the focus of veterinary psychiatry. This discipline is therefore primarily concerned with proximal analysis and, more specifically, the causal analysis of behavior. It studies the development and functioning of individuals who can no longer adapt to their environment and helps them regain these capacities to return to an adaptive equilibrium.

1.1.2 The Behavioral Approach in Veterinary Psychiatry

1.1.2.1 The Behavioral Sequence

Veterinary psychiatry relies on the ethogram of the species to which the individual belongs to distinguish between adaptive and maladap-

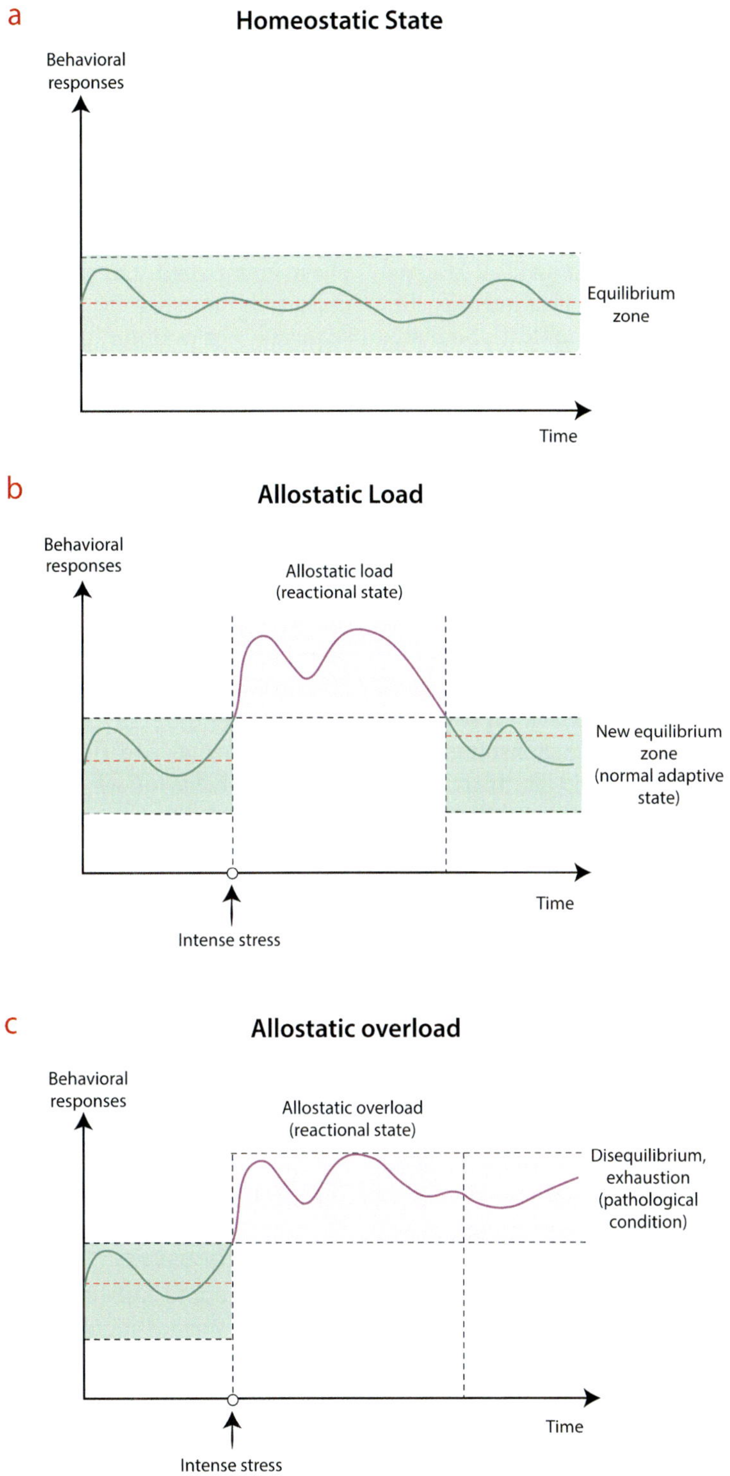

Fig. 1.2 Homeostasis, allostasis, and allostatic load. **a** Homeostatic situation. **b** Allostatic load situation. **c** Allostatic overload situation

tive behaviors. For cats, this means considering the full range of usual behaviors observed in a domestic setting, or at least within the context of a companion animal.

However, abnormal or rather maladaptive behaviors are not limited to those absent from the ethogram. Abnormal behaviors also include behaviors that are normal by their nature but maladaptive for a particular individual in a particular environment due to their intensity, frequency, or irrelevance in context. Similarly, normal or rather adaptive behaviors cannot be reduced to a list of habitual behaviors.

To assess whether behaviors are normal or pathological, veterinary psychiatry must therefore study these behaviors for a given individual and determine whether they are adaptive or not, both in their outcomes and their form. An adaptive behavior must restore the balance of the internal environment as well as provide comfort in the given context.

To study these behaviors, veterinary psychiatry breaks down the behavioral sequence into three phases: appetitive, consummatory, and appeasement (Craig 1917; Sherrington 2023; Tinbergen 1951). The *appetitive phase* encompasses all actions that prepare for the consummatory phase, including the perception of imbalance. The *consummatory phase* corresponds to the specific action that corrects this imbalance. Finally, the *appeasement phase* corresponds to the organism's return to equilibrium.

For example, in the behavioral sequence of predation in cats, the appetitive phase includes internal hunger signals or the sight of prey, triggering stalking behavior with characteristic postures and movements, followed by pursuit and pouncing. The consummatory phase involves the behavioral units of capture, killing, and ingestion. Lastly, the appeasement phase corresponds to internal signals of satiety, often observable as frequent lip licking.

The appetitive phase is not always easy to distinguish from the consummatory phase. It is important to consider that all behaviors triggered by a given imbalance but not compensating for it belong to the appetitive phase. For instance, in the case of urinary elimination behavior, the perception of bladder fullness, initiation, and search for an elimination site are part of the appetitive phase. Only the emptying of the bladder constitutes the consummatory phase. Exploring the urine and scratching the elimination site represent the appeasement phase, marking the return to equilibrium.

Thus, a behavioral sequence is not considered abnormal solely when the consummatory phase is impaired, but also when one of the phases is missing or altered within the sequence. For example, the feeding behavioral sequence in cats is abnormal when ingestion behavior is excessively rapid or frequent, but also when it is not preceded by hunger (appetitive phase) or followed by satiety (appeasement phase) in the individual. There may also be abnormalities in triggers, such as when any movement triggers predation. In this case, the appetitive phase no longer serves the purpose of leading to the consummatory phase of food intake to eliminate hunger but is instead diverted toward another function.

Such alterations in the behavioral sequence are both formal and functional. The causal analysis of behavior, which is the focus of veterinary psychiatry, thus allows maladaptive behaviors in an animal to be described in a comprehensive manner. The behavioral sequences we observe are modified and modulated by what we refer to as elementary processes. The combination of these processes, with varying intensities, constitutes a set of temporarily stable mechanisms that characterize the cerebral functioning of the individual and, consequently, their behaviors. This modeling defines the psychological state of the individual and forms the foundation of our psychiatric approach.

1.1.2.2 Processes

The balance and imbalance of an organism depend on various regulatory processes, that is, different sets of innate adaptive mechanisms that guide the individual's behavioral responses to environmental changes.

Among these processes, we distinguish between *elementary processes*, which correspond to simple neurophysiological mechanisms, and *complex processes*, which involve combinations of several processes that have not yet been fully identified (◘ Table 1.1).

Table 1.1 Elementary processes and complex processes

Elementary processes	Complex processes
Sensitization/Habituation	Acquisition of sensory homeostasis
Anticipation	Acquisition of self-controls
Generalization/Discrimination	Relationships and attachment
Activation/Inhibition	Organization of territory
Instrumentalization	Aging

Elementary processes are revisited and illustrated in ▶ Sect. 1.2.3.4 of this chapter, while complex processes are described in ▶ Sect. 4.3 of ▶ Chap. 4. However, we will briefly define each of the elementary processes in this section.

Sensitization is the mechanism by which a behavioral response to a stimulus intensifies after repeated exposure to that stimulus.

Habituation is the opposite of sensitization; it refers to the progressive disappearance of the alert behavioral response to a stimulus after repeated exposure to that stimulus (Maples-Keller and Rauch 2020).

Extinction is a mechanism similar to habituation but applies to a known stimulus to which the animal is already sensitized, rather than a new stimulus (Bentosela et al. 2008).

Anticipation is the mechanism by which a behavioral response is triggered in the presence of a stimulus that predicts the initial stimulus.

Discrimination is the process through which an individual assigns a different response to stimuli that are similar but not identical (Porritt et al. 2015).

Generalization is the process that allows an animal to group stimuli identified as similar into a single category and produce the same behavioral response to these stimuli (Mclaren and Mackintosh 2002).

Instrumentalization is the mechanism by which a behavioral response becomes automated. This automation is accompanied by a simplification and rigidification of the behavior, which is reduced to its most efficient phase, the consummatory phase, always performed at its maximum intensity.

Activation is the mechanism by which an individual initiates or continues an action because it brings them closer to their goal.

Inhibition is the mechanism by which an individual stops actions or refrains from initiating them because they move them further from their goal or interfere with achieving it.

The processes of activation and inhibition work jointly. It is more a ratio of activation/inhibition tipping toward action or inaction rather than two independent processes.

These elementary processes provide the individual with optimal adaptive capacities in response to environmental changes. They play a role in learning and modulate responses according to the situations encountered by the individual, making them increasingly effective in navigating their environment.

For example, these processes allow an individual to feel fear in a dangerous situation (sensitization) and then react to it earlier by using contextual cues that precede its occurrence (anticipation). Continuing this example, the individual may categorize dangerous elements and generalize their fear response to a group of stimuli resembling the initial danger encountered (generalization). Conversely, they will distinguish harmless stimuli and not respond to them with fear (discrimination).

These processes enable and trigger a behavioral response (activation), while also preventing or stopping it once equilibrium is restored (inhibition). Finally, for repetitive actions, the brain can replace action loops with automated acts, thereby minimizing the energy expended to produce the appropriate response (instrumentalization).

1

However, in some cases, these processes become imbalanced, leading the individual to produce maladaptive responses, making them progressively less effective in dealing with their environment.

Their psychological state reflects this variability in adaptability. The next section focuses on this more global level of analysis.

1.1.2.3 Psychological States

The *psychological state* of an individual encompasses the temporarily stable characteristics of their cerebral functioning and, consequently, their behaviors.

This *psychological state* results from the combination and intensity of the various processes discussed above. It is influenced by the individual's genetic makeup, but also by their life experiences, which can affect cerebral functioning to varying degrees and durations.

An individual's *psychological state* is dynamic and evolves over time. Context and lived experiences modulate the intensity of processes, thereby shaping learning and making behavioral responses increasingly adapted to the environment in which the individual operates.

This *psychological state* is considered normal when it facilitates and promotes adaptation to the environment. Conversely, it is deemed pathological when it prevents the emergence of behavioral responses suited to the context. A *pathological psychological state* is characterized by the production of maladaptive behavioral responses, causing suffering for the individual. Such a state is described as irreversible because spontaneous recovery does not occur. Pathological characteristics can thus be defined by three descriptors: *maladaptive*, *suffering-inducing*, and *spontaneously irreversible*.

Returning to the concepts of *homeostasis* and *allostasis*, a link can be established with the mechanisms leading to a *pathological state*. When an internal or external imbalance is particularly intense, the brain attempts to adapt its reaction. In response to the severity of the imbalance, *elementary processes* function excessively in an effort to address the *allostatic load*. The individual enters a *normal psychological state* referred to as *reactive*. In some cases, they succeed and return to equilibrium. In others, repeated attempts fail to produce an effective solution. The *allostatic load* transforms into an *allostatic overload* as the organism experiences chronic stress that it cannot manage. The brain then generates systematically maladaptive responses, which become detrimental to the individual. At this stage, the *psychological state* has become *pathological*.

This definition of the normal-pathological boundary is a classical concept in medicine and can be applied to all diseases **(see box below)**.

Important Point of View

Normal-pathological limit

Zubin and Spring (1977) proposed an epigenetic model for schizophrenia, which represents the combined influence of genetic vulnerabilities and environmental conditions. This model (◘ Fig. 1.3) is relevant for describing the normal-pathological limit of any disease. Thus, in the green zone, responses are homeostatic, and a return to equilibrium occurs (adaptive normal state). In the intermediate yellow zone, the animal faces environmental stress (*allostatic load*) and attempts to adapt. This is the reactive state. If successful, a return to an adaptive normal state happens. If it cannot adapt, a shift to the orange or red zone occurs, and the animal then finds itself in a pathological state.

In our psychiatric model, we distinguish between various pathological states, which result from the dysregulation of *elementary processes* and correspond to the functional characteristics of different brain circuits (see ▸ Chap. 2, ▸ Sect. 2.3). These pathological states (◘ Fig. 1.4) are detailed in the second part of the book, in ▸ Chap. 7.

1.1.3 Factors Influencing Behavior

Veterinary psychiatry draws on knowledge from ethology, neurophysiology, endocrinology, and genetics to precisely identify not only

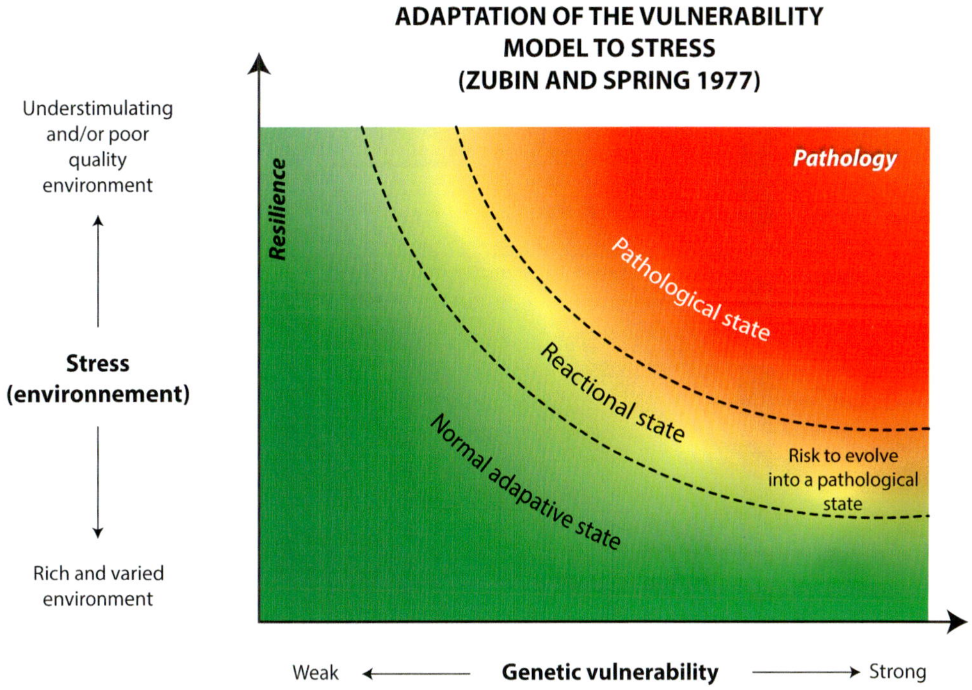

Fig. 1.3 The epigenetic model of Zubin and Spring (1977). (a) Green zone. Adaptive normal state. (b) Orange and red zones. Pathological state. (c) Yellow zone. Environmental stress = *Allostatic load*. This is the Reactive State. Adaptation leads back to the green zone. Failure to adapt leads to a shift to the orange or red zone

external factors but also *internal* factors capable of influencing the structure of a behavior. This modification may affect the form, intensity, or frequency of a behavioral sequence, or even one of its phases.

Internal factors capable of influencing behaviors can be divided into *endocrinological* factors, which include the influence of somatic diseases and pain, and *neurophysiological* factors, whose mechanisms originate in the brain. External factors pertain to the environment and indirectly include all of the individual's experiences, learning processes, and communication.

1.1.3.1 The Influence of Endocrinology on Behavior

Interactions between hormones and behavior are numerous and bidirectional (Roney 2016). Hormones can alter the likelihood of a particular behavior occurring in response to a given stimulus (Nelson 2005). For example, behavior can influence hormone concentrations, as seen in mating in female cats, which triggers ovulation (Brown 2006). Conversely, the concentrations of melatonin, estradiol, and progesterone drive behaviors observed during estrus (Gimenez et al. 2009; Malandain et al. 2011).

Endocrinology is therefore closely linked to the behaviors exhibited, and these connections are explored in detail in ► Chap. 3, ► Sect. 3.1.

The ► Chap. 3 is not limited to endocrinology. An entire section dedicated to pain examines its influence on behavior (see ► Sect. 3.3).

Somatic diseases and their relationship to behavior are detailed in ► Chap. 14.

1.1.3.2 The Influence of Neurophysiology on Behavior

Neurophysiology provides an understanding of the brain mechanisms underlying behavior. Serving as the main foundation of veterinary psychiatry, it is the subject of an entire chapter (see ► Chap. 2).

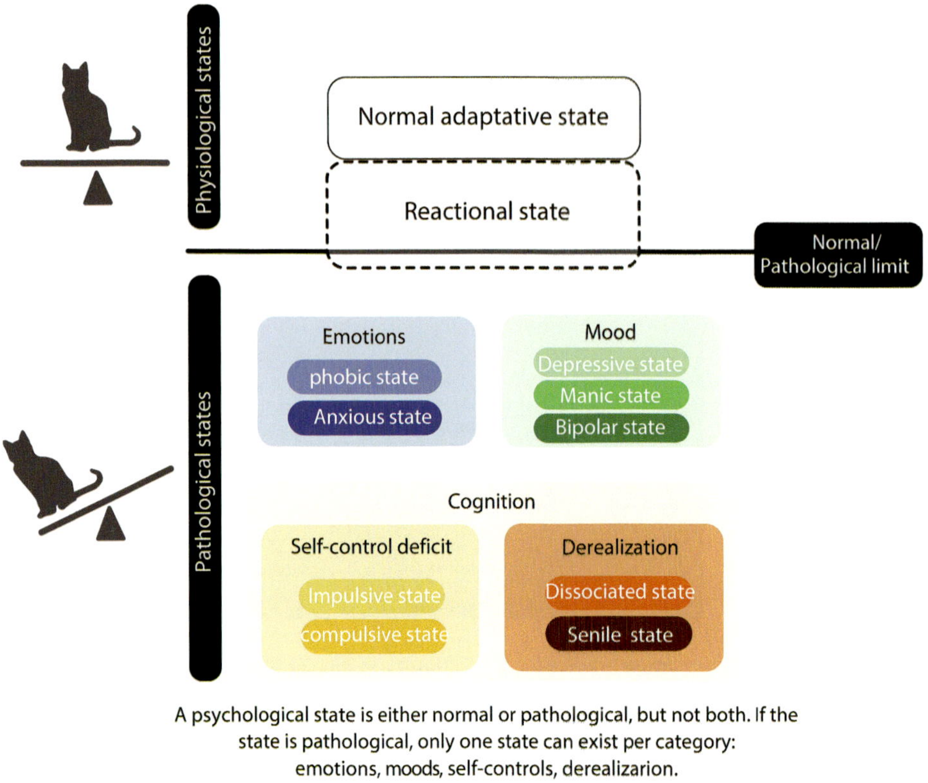

Fig. 1.4 Normal and pathological psychological states. Adaptive and reactive normal states enable the cat to maintain a durable equilibrium. All other states are pathological and spontaneously irreversible, leading to maladaptation and suffering for the individual

However, without delving into the details of brain mechanisms, it is clinically useful to describe the different entities involved in these mechanisms, which, when they malfunction, lead to the pathological states previously mentioned. These include the circuits of emotions, which result in phobic and anxious pathological states; mood circuits, which lead to depressive, manic, and bipolar states; and cognitive circuits, which lead to impulsive, compulsive, dissociative, and senile states.

▶ Section 1.2.1 of this chapter defines the concepts of emotions and mood without addressing their neurophysiological aspects, which are detailed in ▶ Chap. 2 (see ▶ Sect. 2.3, p. 68). Cognition is defined in ▶ Sect. 1.2.3 of this chapter, while its neurophysiological mechanisms are explained in ▶ Chap. 2 (see ▶ Sect. 2.3).

1.1.3.3 The Influence of Environment on Behavior

Beyond endocrinology and neurophysiology, which are internal factors within the individual, there are numerous external factors that can influence behavioral responses.

Among these environmental factors are traumatic factors. For instance, a cat that narrowly escapes being bitten by a dog while crossing a garden is highly unlikely to repeat the behavior of crossing that garden. It may even refuse to leave its home entirely. The trauma has thus altered its behaviors.

There are also non-traumatic factors, which form part of life experiences and influence future behaviors through behavioral development. For example, a kitten that has no contact with humans until the age of 4 months will exhibit intense fear reactions when it sees a human, even though it has never experienced trauma. In contrast, a kitten that grows up in contact with humans is far less likely to display such reactions, even if household children occasionally pick it up or pull its ears, actions it may not enjoy.

No section of this book directly addresses the environment, as it permeates everything the animal does, and its influence is not easily isolated. However, ▶ Sect. 1.2.3.4 of this chapter discusses learning and partially explains this environmental influence on behaviors. We will see that some forms of learning are permanent (notably certain early learning experiences), cannot be undone, and will shape the animal's future responses throughout its life.

In ▶ Sect. 1.3.2 of this chapter, which focuses on feline communication and responses produced by other individuals, we will examine how the forms of this communication significantly influence the responses generated. Finally, ▶ Chap. 4 is entirely devoted to the behavioral development of cats and will detail all its mechanisms.

Summary of ▶ Sect. 1.1

Behavior, defined as the set of actions performed by an individual to respond to environmental changes, results from numerous factors and can be analyzed in various ways. Tinbergen's distal analytical approach situates feline behaviors as a species within their phylogenies and functions. The proximal analytical approach describes both behavioral mechanisms and their ontogenesis, which is of greater interest to clinicians.

Veterinary psychiatry describes behavioral sequences that can be divided into three phases: the appetitive phase, the consummatory phase, and the appeasement phase.

These behavioral sequences are modified and regulated by a set of innate adaptive mechanisms, namely, elementary processes and complex processes. These processes can be homeostatic, helping the animal maintain a normal adaptive state. However, when these processes intensify, they may lead to an *allostatic load* and a reversible reactive state. If the dysregulation of these processes further escalates, it results in spontaneously irreversible pathological states, characterized by maladaptive behaviors and suffering.

Each individual is subjected to various environmental pressures that, when interacting with their genetic predispositions and experiences, define the full range of behavioral variations and the scope of veterinary psychiatry.

1.2 Brain Factors Influencing Behavior

1.2.1 Emotions

Emotion is a psychological and physical reaction that reflects the state of mind of an individual in response to a specific context or the interpretation the individual makes of it. This reaction acts as an internal stimulus, guiding the individual's behavior based on their prior experiences with the context (Thompson 2019). Thus, emotion enables a motor reaction leading to approach (positive emotions), indifference, or avoidance (negative emotions) of the encountered stimulus (Correia-Caeiro et al. 2020; Mitsui et al. 2011; TenHouten 2017).

Emotions are predominantly generated and regulated in the limbic system, or *emotional brain* (Rajagopalan et al. 2017). Any emotional response includes three components integrated within the limbic system: behavioral, neurovegetative, and hormonal dimensions (Jacob-Dazarola et al. 2016).

The *behavioral component* corresponds to the observed postures and movements. For example, a cat afraid of a stranger may hiss and swipe a paw at the stranger, growl, or choose to flee quickly and hide.

The *neurovegetative component* facilitates the behavioral response by rapidly providing energy for vigorous movements. In our example, the cat's sympathetic system is activated while the parasympathetic system is inhibited, causing an acceleration of the heart rate and redirecting blood flow away from the digestive system, which is put on standby, toward the muscles.

The *hormonal component* reinforces the neurovegetative component. Hormones released by the adrenal medulla, specifically adrenaline and noradrenaline, further increase blood flow and stimulate the conversion of nutrients into glucose, thereby enhancing muscle function. The adrenal cortex secretes glucocorticoids, which also make glucose more readily available to muscle cells.

Among emotions, psychologists distinguish between *primary emotions* and *secondary emotions*. *Primary emotions* include fear, surprise, anger, joy, sadness, and disgust (Heath 2018; Rodríguez-Torres et al. 2005). They are considered basic and are shared by many animal species, including cats. *Secondary emotions* are thought to result from a combination of primary emotions, such as shame, which is believed to combine fear and anger turned inward. Cats are likely capable of experiencing a certain number of secondary emotions, although there is no consensus on the full extent of this emotional range (Morris et al. 2008; Pickersgill et al. 2023).

Experiencing emotional responses is adaptive: emotion acts as an internal alarm signal that allows anticipation. It is protective and promotes an appropriate reaction to the context, even before the stimulus triggering the emotion appears. Emotions are associated with specific stimuli or behavioral responses. The animal learns and memorizes these contexts, becoming more effective when faced with them again. In the case of fear, it can react faster and therefore more efficiently if a dangerous stimulus is presented again. In the case of joy and pleasure, it is more likely to perform a behavior to relive the previously experienced pleasant sensation.

However, emotions are not easily controlled by the individual. Their function is to select or produce behaviors that either enable survival in dangerous or perceived dangerous situations or promote pleasure and well-being. When emotions are dysregulated, when fear responses are exaggerated and do not correspond to a truly dangerous situation, they lead to pathological *phobic* and *anxious* states.

Mammals, particularly social species, can read the emotions of an individual (Correia-Caeiro et al. 2020; Thompkins et al. 2018). For example, dogs can accurately interpret the emotions of their conspecifics (Cook 2017) as well as those of humans (Andics and Miklósi 2018; Siniscalchi et al. 2018). Even in cats, which are not considered social species, this ability is present: domestic cats can spontaneously combine vocal and visual signals to recognize the emotion of the emitter, whether it is another cat or a human (Quaranta et al. 2020).

The brain's neurophysiological mecha nisms underlying emotional reactions are detailed in ▶ Chap. 2 (see ▶ Sect. 2.3.1).

Important Point of View

The James-Lange theory

William James and Carl Lange were two psychologists from the late nineteenth century who independently, yet synchronously (one in 1884, the other in 1887), proposed the same theory, now known as the *James-Lange Theory* (James 1994). This theory suggests that the physiological responses associated with emotions, such as trembling, sweating, accelerated heartbeat, and behavioral responses like clenching fists, precede and are responsible for the internal experience of emotion. According to James and Lange, it is the sensory feedback sent to the brain by muscle contractions that generates emotional feelings, rather than the other way around.

This theory is highly counterintuitive: most people believe that they scream or fight because they feel angry, not that the physical actions cause the emotional experience.

The theory poses a significant challenge to demonstrate, as emotional experience is internal to the individual. However, since its inception, no one has been able to disprove it, and many clinical cases have instead supported it. For example, in 1966, Hohman studied patients with spinal cord injuries and questioned them about their emotional experiences. If the *James-Lange Theory* were correct, these patients' emotional experiences would be diminished, especially in cases of higher spinal cord injuries. This is precisely what Hohman found (Leventhal and Scherer 1987).

In 2009, Lewis and Bowler injected botulinum toxin into the foreheads of participants, thereby preventing the muscular expressions of emotional facial movements, particularly those related to negative emotions. Their study showed that these participants experienced these emotions in a much attenuated manner compared to those who had not received the injections (Lewis and Bowler 2009).

1.2.2 Mood

Mood is a persistent psychological state, much longer lasting than *emotion*. Unlike *emotion*, it is not an internal stimulus but a modulator of all internal stimuli. It alters the impact of *emotions* and influences the interpretation of perceptions, and thus behavioral responses. *Mood* is not dependent on recent events or external stimuli, yet it affects all behaviors. It is not governed by the same brain regions and circuits as *emotions* and instead "colors" the animal's overall reactions in a positive, negative, or neutral manner. Thus, we speak of "good mood" and "bad mood." This baseline *mood* level is variable (◘ Fig. 1.5) but with far greater inertia than *emotions* (Broome et al. 2015; House 1988; Marwaha et al. 2014; Mendl et al. 2010).

Mood also manifests over a longer term than emotion and, when dysregulated, leads to pathological mood states (Price and Drevets 2010), namely depressive, manic, and bipolar states (see ► Chap. 7, ► Sect. 7.3).

This distinction between emotion and mood is crucial. These two entities do not correspond to the same level of brain regulation (Beedie et al. 2005). Returning to the example of the emotional fear reaction in a cat mentioned earlier, this reaction could vary depending on its mood. For instance, on a day when the cat is in a "good mood," it might even approach the stranger, whereas on a day when it is in a "bad mood," its fear reaction might be modulated differently, possibly leading to severe biting or scratching.

The brain's neurophysiological mechanisms underlying mood are detailed in ► Chap. 2 (see ► Sect. 2.3.2).

1.2.3 Cognition

Cognition refers to the set of mental processes involved in brain functions related to knowledge. It includes attention, learning, intelligence, language, memory, perception, decision-making, problem-solving, reasoning, and the control of action and emotions.

Let us examine the various elements that make up cognition.

1.2.3.1 Attention

Attention represents the set of mechanisms responsible for selecting and processing information. It involves three systems that are functionally and anatomically distinct but work together to guide learning in a specific direction (Petersen and Posner 2012; Posner and Petersen 1990).

The first attentional system, or alerting network, determines *when* to pay attention. It corresponds to the modulation of vigilance, providing preparatory attention.

When an individual needs to sustain attention over a longer period, *sustained attention* is engaged. This type of attention requires the subject to actively and continuously process information, which is not necessary in simple vigilance.

The second attentional system, or *orienting network*, determines *what* to pay attention to. It corresponds to attention in the strict sense, or *selective attention*. This system enables the selection of specific information among others by amplifying the relevant signal and reducing irrelevant signals (to the point of attentional blindness: Simons 2000). Orienting can be *endogenous* when driven by voluntary focus on a stimulus or *exogenous* when involuntarily attracted to a stimulus. Exogenous orienting can be modulated by endogenous orienting (Raz and Buhle 2006).

Joint attention, the ability of two individuals to attract and maintain shared focus on a

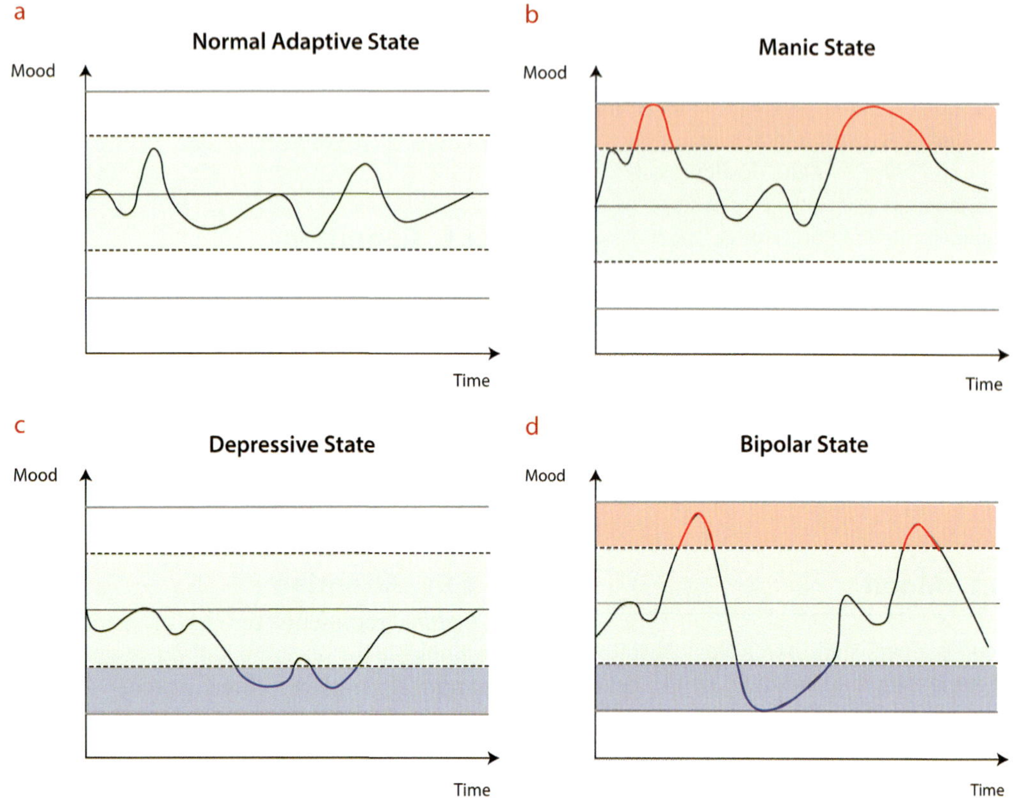

Fig. 1.5 Mood and its variations. **a** Mood fluctuates slowly around a central zone of equilibrium. Dysregulation can lead to either an excessive elevation of mood, resulting in the pathological manic state **b**, a significant decrease in mood, resulting in the pathological depressive state **c**, or an alternation between extremes or between extremes and the normal range, resulting in the pathological bipolar state **d**

stimulus to exchange information or experience related to it, significantly influences selective attention. For instance, this joint attention plays an important role in a cat's ability to follow a human's pointing gesture, though its performance is less effective than that of a dog (Salamon et al. 2023; Mäses et al. 2023).

Finally, the third attentional system, or *executive control*, determines *how* to process the selected information. It corresponds to *executive attention*, encompassing high-level cognitive processes that allow for the selection, planning, and control of the appropriate sequence of operations for a given task. Executive control supervises the functioning of working memory (see "short-term memory or working memory").

It incorporates various complex cognitive processes such as planning, selection, initiation, execution, and monitoring of goal-directed behaviors, as well as the selection of representations, operations, and actions appropriate to achieving the goal.

All of these processes are modulated by *inhibition*, the ability to suppress representations, operations, and actions that divert from the goal, as well as to resist interference from irrelevant information. Consequently, executive control underpins *cognitive flexibility*, the capacity of an individual to adapt both their information processing and voluntary behavior to a given situation.

1.2.3.2 Motivation

Motivation is a very popular concept but difficult to define, with no consensus emerging to date.

Fabien Fenouillet, in his book *La Motivation* (Fenouillet 2017), lists more than a hundred approaches to the concept. One of the most famous is Maslow's, based on the response to needs (Maslow 1943). However, motivation is also intimately linked to *emotions*, as emotions create an imbalance that generates motivation aimed at restoring equilibrium (Berridge 2018). Thus, motivation can be defined as the concept used to describe the forces driving the initiation, direction, intensity, and persistence of behavior. It can also be seen as a vector possessing both a direction or *drive*, which guides the response, and an intensity or *arousal*, which dictates the strength of the response. In practice, motivation precedes and supports the behavioral response.

A *reward* is defined as a stimulus that appears or disappears following the execution of a behavior, increasing the likelihood of that behavior recurring. Rewards have this effect because they produce pleasure through the reward system, which will be detailed in ▸ Chap. 2 (see ▸ Sect. 2.3.3.3). A reward concludes the behavioral response and increases the motivation for future behaviors.

This sequence allows the *motivation-action-reward* system to be conceptualized as follows: once a stimulus is perceived, filtered, and processed, it causes an imbalance in the limbic system and generates an *emotion*. From this imbalance arises the *motivation* to act, aiming to restore equilibrium. This motivation has a given direction (*drive*) and intensity (*arousal*), generating a behavioral response reflecting this drive-arousal pair. The reward then terminates the action, allowing the return to equilibrium. Finally, once equilibrium is restored, memory processes take effect and influence future responses.

Take hunger as an example. It is triggered by internal stimuli or sensations, such as a drop in blood glucose levels, creating an imbalance. The sensation of hunger directs the individual toward seeking food (drive). Once food is located, the amount ingested reflects the intensity of the motivation (arousal), leading to satiety (*reward*), which represents the return to equilibrium.

Based on the modeling of motivation, it is possible to propose methods for altering it. These methods target the indirect consequences of motivation that we can observe, namely the stimulus and the reward. To increase an individual's motivation, rewards must be increased in quality, quantity, or by adding frustration (e.g., quickly removing the reward as soon as the cat shows interest in it). Conversely, to decrease motivation, rewards must be reduced in some way, whether in quantity (*extinction*), quality, or by introducing punishments.

An indirect effect on motivation can also be observed in *phobic* and *anxious states* due to motivational competition with reactions that take precedence in responding to danger. The individual's attention is captured by the *phobogenic stimulus*, diminishing the motivation to perform other actions. In cats, as a prey species, this attentional capture is particularly pronounced.

1.2.3.3 Action Control

Cats possess a less developed prefrontal cortex compared to dogs or humans, yet it still allows them to exercise control over their actions and emotions (Bari and Robbins 2013). This capacity for self-control enables them to refrain from reacting despite feelings of anger or frustration in certain situations (Bunford et al. 2019).

Every behavioral sequence has a beginning and an end. The imbalance created initiates the action, driven by the *elementary process* of *activation*, while the *process of inhibition* ensures that the action does not continue indefinitely but instead stops appropriately.

Activation is the mechanism by which an individual undertakes or continues an action because it brings them closer to their goal, whereas *inhibition* is the mechanism by which they stop actions or refrain from initiating them because they move them further from their goal or interfere with achieving it.

In practice, these two processes are closely interconnected, creating an activation vs. inhibition balance that tilts either toward activation or inhibition.

In humans, there are tests to measure inhibitory capacities, such as "go/no-go" tests, which involve presenting different stimuli to which the individual must respond with an action or inaction. Similar tests have been evaluated in dogs (Bunford et al. 2019), but to our knowledge, these tests have not been rep-

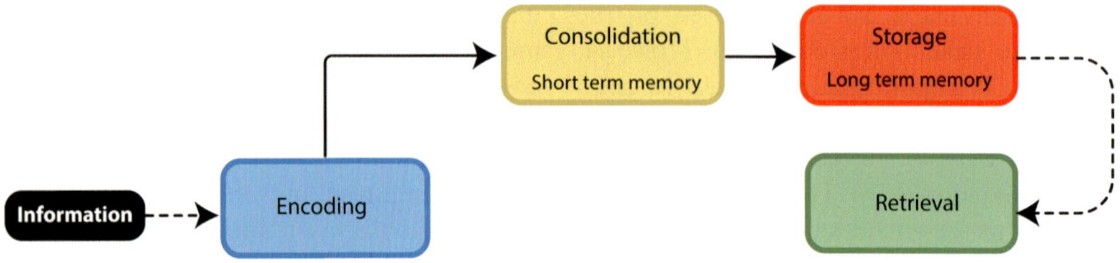

Fig. 1.6 Stages of learning and memorization

licated in cats. However, it is highly likely that such tests would correlate with levels of impulsivity.

1.2.3.4 Learning

Learning encompasses the mechanisms that lead to the creation of a representation of the environment, enabling the orientation of behaviors to make them more adaptive.

It corresponds to a modification of the nervous system, reflected in changes in behavioral responses. *Memorization* is the result of long-term changes induced by learning. These biological changes constitute memories. Learning and memorization occur through various stages: encoding, consolidation, storage, and retrieval (Fig. 1.6), which will be detailed in the next section.

Learning, memorization, and their effects on behavior are only possible due to *brain plasticity*.

There are three main types of learning: perceptual learning, associative learning, and relational learning.

Once learned, information is first consolidated in short-term memory, then stored in long-term memory, from where it can be recalled as needed.

Perceptual Learning

Perceptual learning refers to the ability to learn to recognize stimuli that have been perceived before (Goldstone 1998). The primary function of this type of learning is to identify and categorize objects or situations.

This type of learning is essential for discriminating between different stimuli and for recognizing complex shapes, sounds, or odors. It is a necessary prerequisite for performing many cognitive tasks (Kellman and Massey 2013). It also plays an important role throughout the animal's life, particularly during the socialization period. During this time, it allows the animal to explore its environment and categorize it. Any impairment of the sensory perception systems logically leads to significant learning difficulties, as these systems are involved from the very beginning of the behavioral sequence.

Perceptual learning plays a major role in the processes of *discrimination* and *generalization*.

Discrimination is the process by which an individual produces a different response to stimuli that are similar but not identical. To discriminate, the individual must be able to perceive differences between the two stimuli. For example, a cat's ability to discriminate very effectively between different odors allows it to easily detect the presence of medication in its food.

> **Clinical Illustration**
>
> **Discrimination**
>
> Zora is a cat who is very afraid of dogs. Whenever she hears the neighbor's dog barking, she retreats into the house to hide. However, she has lived with a Leonberger since her adoption and does not react to him at all. She even sleeps in his bed, curled up against him, or lifts his ears and licks them. She discriminates between the dog she lives with and other dogs.

In the process of *generalization*, it is because the animal categorizes two stimuli as being similar that it produces an identical behavioral response, even if the stimuli are not the same (Mclaren and Mackintosh 2002). This is how cats learn to stop fearing all humans, even if they have only encountered a limited variety of humans as kittens.

Clinical Illustration

Generalization

Zebulon is a cat who used to have no fear of dogs. He lives with two Cavalier King Charles Spaniels. His owners often have visitors accompanied by their dogs, and most of the time Zebulon stays in the living room during these visits, sometimes even playing with the dogs as long as they are not too rough with him.

However, two days ago, Zebulon was chased by the neighbor's Border Collie while crossing the garden. He narrowly escaped being bitten by climbing the fence just in time. Since then, he runs upstairs whenever a dog visits the house and tends to stay in high places when the two resident dogs are moving around. He has generalized his fear and categorized all dogs as dangerous, even though only one dog posed a threat.

Associative Learning

Stimulus-response learning is the type of learning we intuitively think of when discussing *associative learning*. It refers to the ability to learn to perform a specific behavior when a particular stimulus is present in the environment. This form of learning involves both the circuits responsible for the perception of stimuli (*perceptual learning*) and those responsible for movement (*conditioned motor learning*).

The behaviors produced during this type of learning can be reflexive, as in classical or *Pavlovian conditioning*, or voluntary, as in *operant conditioning*.

The term *conditioning* is used because repeated exposure to the stimulus is often required to achieve a lasting change in behavior.

Classical Conditioning

Also known as *Pavlovian conditioning*, named after the Russian physiologist Ivan Pavlov, who first demonstrated it, classical conditioning enables the formation of a lasting association between an *unconditioned stimulus* and another stimulus (Bitterman 2006). Pavlov, while observing his laboratory dogs, noticed that when food was presented to a dog, it would immediately begin to salivate. He repeated the experiment by ringing a bell just before presenting the food. After several repetitions, the bell alone caused the dog to salivate, even before the food was presented. The dog had associated the bell with the arrival of food and salivated as if the food were already there. In other words, the *neutral stimulus* (the bell) became a *conditioned stimulus* when it produced in the individual an internal representation of the *unconditioned stimulus* (the food) (Rescorla 1967).

In classical conditioning, an association is formed between a *neutral stimulus*, which initially elicits no response from the individual, and an *unconditioned stimulus*, which already triggers a reflexive, often emotional, reaction (◘ Fig. 1.7). The reactions involved engage glands and smooth muscles and are involuntary, whether physiological (salivation, tachycardia, piloerection) or reflexive (withdrawal reflex, eyelid closure). These reactions underpin the emotions felt by the individual and are referred to as *conditioned emotional responses*.

Sensitization is the mechanism by which a behavioral response to a stimulus increases in intensity and rapidity following repeated exposure to that stimulus. This mechanism operates through classical conditioning and produces *conditioned emotional responses* (Kamin and Brimer 1963). It is through this process of sensitization that we become more effective when encountering a dangerous stimulus repeatedly.

Clinical Illustration

Sensitization

During his first vaccination, Peanuts did not move while the veterinarian administered the injection. His owner was holding him. Peanuts appeared frightened but remained still on the examination table. He clearly did not enjoy the experience but did not attempt to flee and stayed motionless.

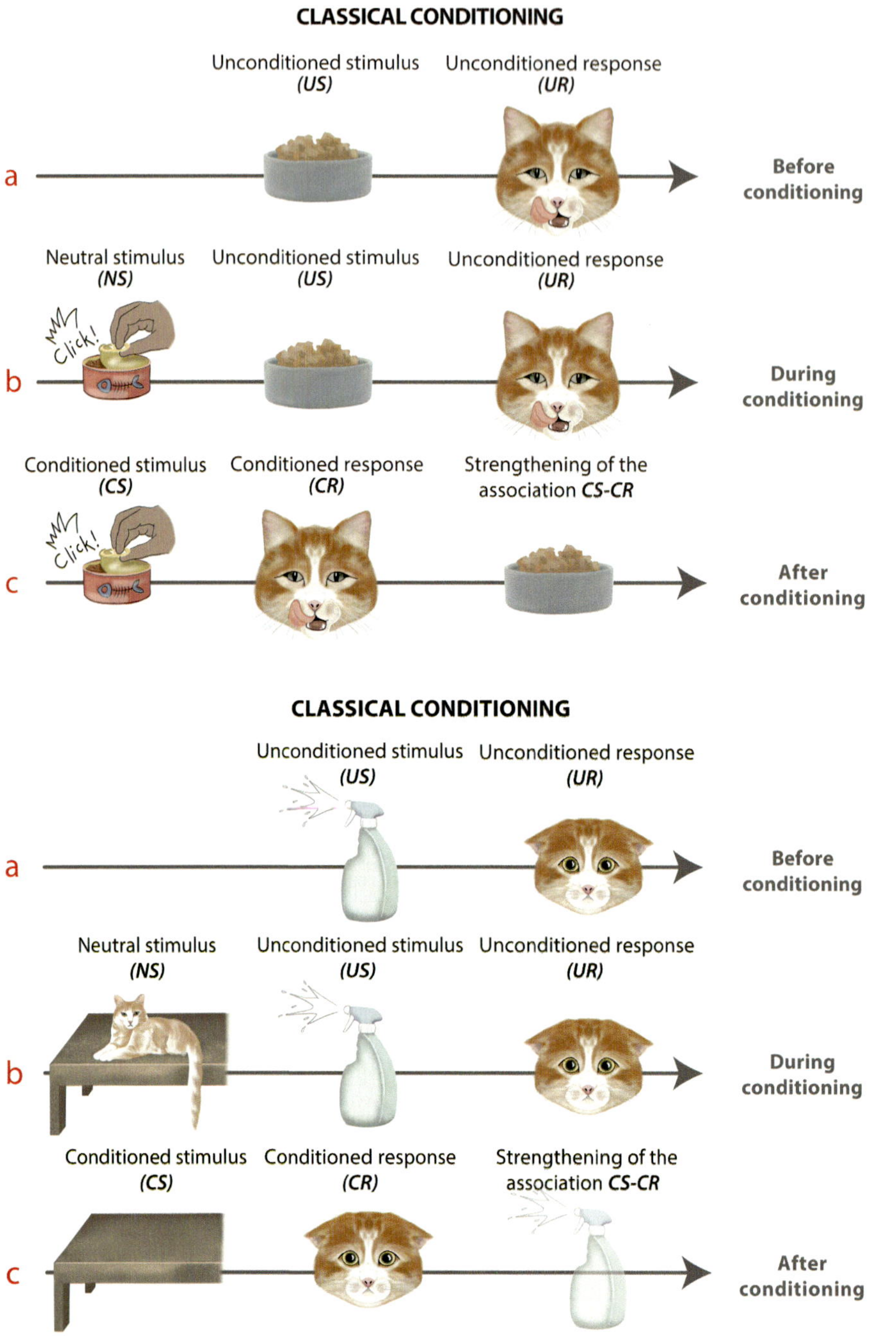

Fig. 1.7 Classical conditioning (**a** associated with pleasure, **b** associated with fear and pain). (a) An unconditioned stimulus triggers an unconditioned response. (b) A neutral stimulus is paired with the unconditioned stimulus. (c) The *neutral stimulus* becomes a *conditioned stimulus*. It alone triggers the response, which becomes a *conditioned response*. The association between the *conditioned stimulus* and the *unconditioned stimulus* is reinforced if the *unconditioned stimulus* reappears after the *conditioned response*

At the next visit, Peanuts' fear reactions were significantly more noticeable. He refused to come out of his carrier voluntarily, and during the clinical examination, the veterinarian observed that Peanuts was sweating from his paw pads. When the injection was administered, the cat tried to escape and sought refuge back in his carrier. Peanuts had become *sensitized* to the injection.

Habituation is the opposite of *sensitization*, meaning the disappearance of the behavioral alert response to a stimulus after repeated exposure to that stimulus (Maples-Keller and Rauch 2020).

Extinction is a mechanism similar to habituation, but it applies to a known stimulus to which the animal is already sensitized, rather than a new stimulus (Quirk and Mueller 2008).

This balance, which determines whether a behavioral response leans toward habituation or sensitization, may seem random, but certain conditions favor one process over the other (◘ Table 1.2).

Clinical Illustration

Habituation

During her first outing in the garden at 5 weeks old, Mya heard the sound of a car arriving in the driveway. She quickly ran to take refuge near her mother. The car stopped, and the noise ceased. She cautiously approached the stationary car, sniffed it, and nothing happened.

The visitor got out, interacted with the kittens, and then returned to the car. The car engine started again. Mya was startled once more and quickly moved away. This time, her fear response was less intense. She turned back and watched the car leave.

Mya repeated this experience several times over the following week. After two or three visits, she no longer reacted when she heard the sound of car engines arriving in the yard: she had become *habituated*.

Pavlovian learning can also be understood as a way to anticipate the arrival of a conditioned stimulus, its positive or negative characteristics, and to prepare the organism for this event. *Anticipation* is the mechanism by which a behavioral response is produced in the presence of a stimulus that predicts the initial stimulus (Bos et al. 2003).

It is important to note that classical conditioning expands the range of stimuli capable of eliciting emotional reactions.

Clinical Illustration

Anticipation

Louna is frightened by the robotic vacuum cleaner. Her first experience with the vacuum was very unpleasant, which sensitized her to the device. When the vacuum starts its cleaning cycle, it emits a distinctive beep. Now, as soon as Louna hears this sound, she runs to her owner's bedroom and hides under the bed, even if the vacuum does not start. She is anticipating.

Classical conditioning is, therefore, a form of learning that allows animals to protect themselves from certain dangers. However, if the conditioned stimulus is presented to the animal without being associated with a dangerous situation (unconditioned stimulus), the animal stops reacting to the conditioned stimulus. This phenomenon is called the extinction of the conditioned emotional response.

Numerous studies have shown that extinction is not synonymous with forgetting. The animal merely learns that the conditioned stimulus is no longer associated with an aversive response and, as a result, actively inhibits the conditioned response (Amano et al. 2010; Goodman and Packard 2018).

▪ Operant Conditioning

Sometimes called *instrumental conditioning*, this type of learning was first demonstrated in 1898 by Edward Thorndike during his doctoral experiments with cats

1

Table 1.2 Criteria favoring habituation or sensitization

Criteria favoring habituation	Criteria favoring sensitization
Open situation Exposure to a low-intensity stimulus Repeated exposure	Closed situation Exposure to a high-intensity stimulus immediately Isolated exposure

(Thorndike 1898a). The cats were placed in cages from which they could see food. Thorndike observed that once the cats managed to open the cage through trial and error, they were able to do so much faster in subsequent attempts. He hypothesized that the pleasure associated with accessing the food outside the cage facilitated the learning of the most effective behavioral response (Thorndike 1898b). This observation led him to formulate the *law of effect*: behavior is determined by the effect it produces.

Building on these foundations, B.F. Skinner defined the different types of reinforcements and punishments and articulated the laws of learning (Jones and Skinner 1939). He developed the "Skinner box," which exposed animals to various instruments (buttons, switches, levers) and stimuli (colored lights, sounds) and allowed the application of different outcomes (treats, electric shocks). This setup enabled him to identify the laws underlying learning, which remain widely used today (Skinner 1953, 1984).

This trial-and-error learning, or more precisely, learning through pleasure and displeasure of behavioral responses, involves the striated musculature, meaning that the behaviors learned are voluntary motor responses. Unlike classical conditioning, which relies on a *stimulus-stimulus* association, operant conditioning is based on a *stimulus-response* association.

Because of its voluntary nature, operant conditioning serves as the foundation for all forms of feline training. The animal actively evaluates the consequences of its experiences, leading it to adjust its behaviors accordingly.

Clinical Illustration

Operant Conditioning

Banjo is a rather food-motivated cat. He enjoys finishing his owner's yogurt pots when allowed. One day, while his owner was eating yogurt, Banjo approached and reached out with his paw toward the pot. Finding this behavior endearing, his owner allowed him to lick an entire spoonful of yogurt.

Now, Banjo routinely taps his owner's hand with his paw whenever yogurt is being eaten. Most of the time, this earns him a spoonful before the pot is finished. Banjo has learned that tapping his owner's hand is associated with getting a spoonful of yogurt.

Clinical Illustration

Importance of the Reinforcer

Neige is a Persian cat who does not enjoy being brushed. When her owner manages to catch her for a grooming session, Neige tolerates it but clearly does not take much pleasure in the experience and leaves as soon as she gets the chance. Her owner previously tried placing kibble on the seat, but Neige showed no interest.

Recently, following a friend's advice, Neige's owner invested in a licking mat and began spreading wet food on it during grooming sessions. Neige loves this wet food, and when the mat is presented, she willingly allows herself to be brushed and no longer tries to leave. After several sessions, brushing has become increasingly easy, and even when the wet food is gone, Neige no longer flees and seems to enjoy the grooming.

The wet food reinforcer was much more appealing to Neige than kibble and successfully taught her to tolerate and even enjoy the brushing sessions.

At the beginning of learning, the brain expends significant energy to determine the optimal way to produce a behavior, both to maximize pleasure over displeasure and to adapt effectively to the environment.

When this type of learning is repeated, it leads to the *instrumentalization* of behaviors. The process of instrumentalization is the mechanism by which a behavioral response becomes automated through operant conditioning. This results in the production of a behavioral sequence reduced to its most effective phase, the *consummatory phase*.

The behavior is then produced in an automated, conditioned manner. It is faster and more efficient because the appetitive and appeasement phases of the behavioral response are shortened or even eliminated. Unlike the earlier stages of learning through conditioning, the phases of analysis and the emotions linked to the stimulus disappear. With instrumentalization, learning does not associate stimuli or contexts with emotions capable of triggering a response. Instead, the stimulus is directly associated with the response, which is triggered automatically.

The energetic cost of producing the behavior becomes minimal. The process of instrumentalization also allows the cortex to be available for analyzing other elements of the environment. As will be detailed in ▶ Chap. 2, the dorsal striatum gradually transitions from a passive observer to the automatic command center for complex behaviors. This shift frees cortical areas for other functions.

Instrumentalized behaviors are more difficult to extinguish due to their connection to the dorsal striatum (Goodman and Packard 2018).

Instrumentalization is the process involved when we learn to ride a bicycle or drive a car. In the context of aggression, the instrumentalization of behavioral sequences leads to rapid and automated triggers, which worsen the prognosis. Instrumentalization is resistant to change, and the danger becomes maximal because the cat no longer gives a warning before escalating to biting or scratching.

As we will see in ▶ Chap. 2 (see ▶ Sect. 2.3.3.3), an unexpected, novel reward with high appetitive value can accelerate this process. Additionally, there are dopaminergic mutants for whom rewards inherently have a high value, making them more vulnerable to addiction. In other words, these individuals are more prone to falling into a compulsive pathological state and developing behavioral sequences that can instrumentalize more quickly.

Motor learning is a specific type of operant conditioning that involves acquiring purely motor behavioral sequences through repetition. It engages different areas of the motor cortex and the basal ganglia, which enable these behaviors to become automated (Ioffe 2004).

Rapid eye movement (REM) sleep plays a central role in motor learning by promoting long-term potentiation and, consequently, the consolidation of information (Li et al. 2017).

This type of learning encompasses all complex movements that cats can master, including the sequences they use for hunting. Through motor learning, their movements become more precise and faster, enhancing their efficiency.

▪ Relational Learning

The previously described types of learning (*perceptual* and *associative*) combine in more complex learning situations. For example, when a cat learns to give a paw in response to a hand placed in front of it or to a verbal command, it uses *perceptual learning* to identify the visual or auditory stimulus, and *stimulus-response learning* to extend its paw and touch the hand.

However, learning is often more complex than this. The individual also learns the relationships between stimuli. For instance, when we hear the sound of a dog barking, we imagine a dog of a certain size, know what it feels like to touch its fur, and, based on the direction of the sound, can deduce that it is a neighbor's dog we have seen before.

Relational learning is the process of linking these various stimuli to memory elements, creating new neural connections that allow the association and coordination of information acquired during the first two types of learning (◘ Fig. 1.8).

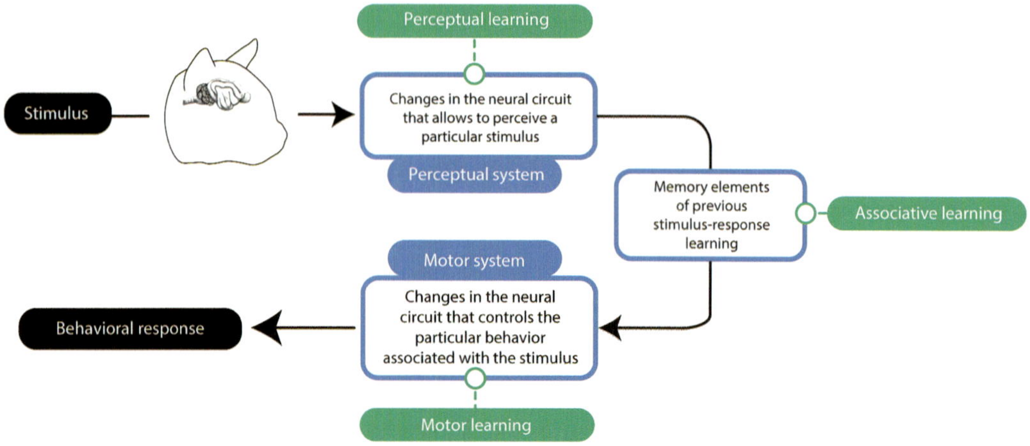

Fig. 1.8 Relational learning based on perceptual and associative learning. A stimulus is recognized through perceptual learning and triggers a motor response reinforced by associative learning

1.2.3.5 Memory

As mentioned in the previous section, the long-term changes induced by learning form memories, which result from the process of memorization. Memories are not stored in a specific area of the brain but rather produce a lasting change in neural pathways. Learning and memorization occur through various stages: encoding, consolidation, storage, and retrieval (Fig. 1.6).

Learning, memorization, and their effects on behavior are only possible thanks to *brain plasticity*.

There are different types of memory, corresponding to varying durations of information retention. These include *sensory memory*, *short-term memory* (or *working memory*), and *long-term memory* (Fig. 1.9). The neurophysiological mechanisms underlying these types of memory are detailed in ▶ Chap. 2 (see ▶ Sect. 2.3.3.1).

Sensory Memory

Information is first processed by *sensory memory*. It is stored for a very brief period, lasting less than a few seconds, allowing the individual to retain the sensation of the stimulus slightly longer than the stimulus itself, giving them time to respond. This memory does not last much beyond the duration of the original stimulus.

For example, sometimes we feel as though we haven't heard a sentence spoken by someone. With a moment of focus, even though we initially thought we didn't hear it, we can recall the sentence resonating in our mind. This is an example of *auditory sensory memory*.

Short-Term Memory or Working Memory

Only a small fraction of sensory memory progresses to the second stage of memorization, known as *short-term memory* or *working memory*. If the information is relevant to the individual—meaning if their preparatory attention is engaged—it will move into *short-term memory*, which has a limited storage capacity. This capacity is restricted to a few distinct elements, such as a sequence of words, a particular intonation, or the presentation of an object.

In humans, this *short-term memory* enables us to remember a phone number or the letters of a name for a short time. This *working memory*, lasting only a few seconds or tens of seconds, can be extended through training and repetition.

In dogs, *working memory*, as tested with a hidden object, functions for several minutes after the object is concealed (Fiset et al. 2003). In cats, however, it is less than 30 s, and the visual elements around the areas where objects are hidden do not appear to help cats retrieve the object (Fiset and Doré 2006).

Attention plays a fundamental role in short-term memorization. To retain a stimu-

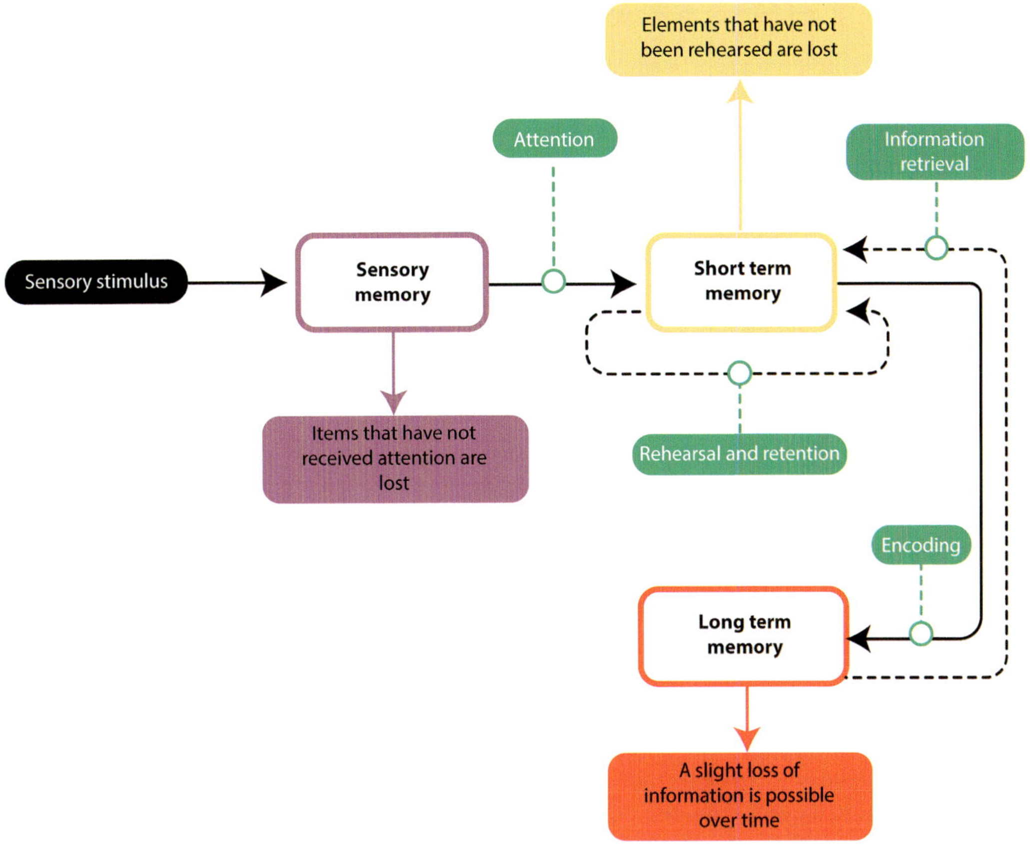

Fig. 1.9 Types of memory. The detection of a stimulus is processed by sensory memory. Attention facilitates its transfer to short-term memory (or working memory), and sufficient repetition enables its recording into long-term memory. Information can then be retrieved as needed

lus, one must focus attention on it. This explains why, in cases of intense emotion, attention is directed toward the stimuli that generated it, leading to highly efficient memorization of the context associated with the emotion (conditioned emotional reaction). Conversely, the individual will have difficulty learning anything else, as their attention is already occupied.

When short-term memorization is repeated, the information can be encoded and stored in *long-term memory*.

Long-Term Memory

The third and final stage of memorization is *long-term memory*. This stage allows information to be stored for extended periods, ranging from minutes to decades. Information from *working memory* undergoes consolidation to transition into long-term storage.

At this stage, there is a distinction between two types of long-term memory: *procedural memory* and *declarative memory* (Fig. 1.10).

Procedural memory includes memories we are not necessarily aware of. It seems to be stored in an automated manner and is the type of memory that enables us to adjust our behaviors, such as when we drive or ride a bicycle. We learn to fine-tune our movements without being able to explain exactly how or even being conscious of each step we execute. Procedural memory is primarily involved in the acquisition of motor behaviors. In cats, training for hunting or agility engages this type of memory.

Declarative memory, on the other hand, encompasses all the facts we remember and can verbalize. This is not limited to verbal memory but includes the ability to recall a specific event, akin to a mental video we can replay at varying speeds.

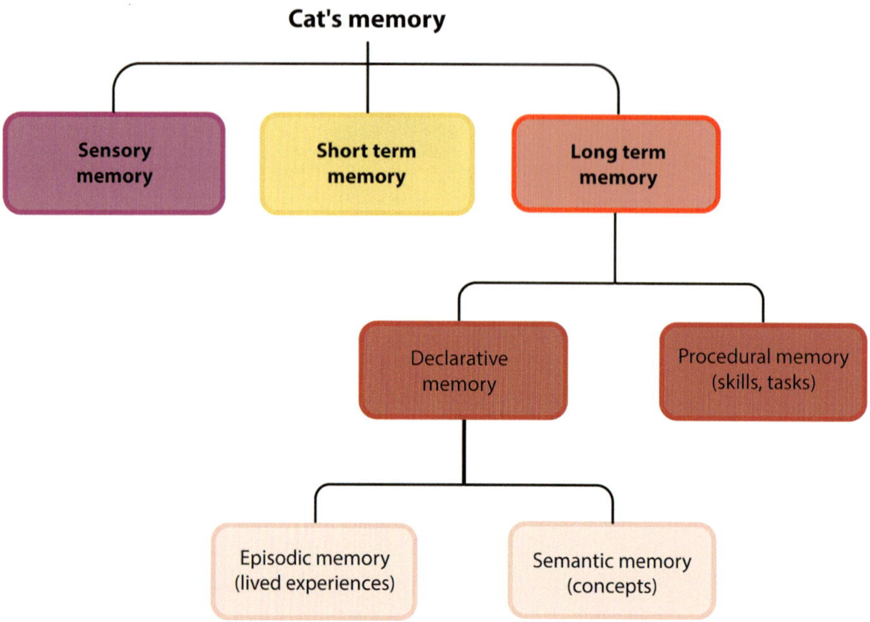

Fig. 1.10 Types of long-term memory. There are two types of long-term memory: procedural memory and declarative memory, the latter further divided into episodic memory (events and past experiences) and semantic memory (general concepts)

Declarative memory can be further divided into episodic memory and semantic memory.

Episodic memory allows individuals to recall specific events, their context, and the order in which they occurred. For example, one might remember an evening at a friend's house, visualize the color of the couch they sat on, and recall the time they left that night.

Semantic memory, which is less specific, pertains to general knowledge without the context of how it was learned. For instance, knowing that a strawberry is a fruit and that the Earth is round is possible through semantic memory. Semantic memory is acquired gradually, whereas episodic memory must often be learned in a single instance.

As research progresses, scientists continue to uncover cognitive abilities in animals. Recent studies have shown that cats recognize their mother's scent at the age of 1 year, even if separated at 8 weeks (Szenczi et al. 2022). Other tests suggest that a cat's memory capacity does not seem to decline over time (McCune et al. 2008).

It also appears that cats, like dogs, are capable of recalling complex events (Fugazza and Miklósi 2015), such as human actions, and can even imitate them (Fugazza et al. 2021). However, whether their *episodic memory* is as effective as that of dogs remains to be determined, as this has not yet been demonstrated (Fugazza et al. 2016).

1.2.4 Temperament and Personality

Temperament is a widely used concept, both in everyday language and in scientific literature (Siegford et al. 2003; Travnik et al. 2020). While its definition varies depending on the source, temperament is generally understood as a set of hereditary traits, stable over time, that reflect the core nature of individuals. Most authors agree that temperament is strongly influenced by an individual's genetics. For this reason, temperament is often associated with breed. For example, it is commonly stated that Ragdolls and Persians are very calm and can be handled without resistance.

Personality, on the other hand, is considered the expression of temperament traits shaped by the individual's environment. Thus, personality applies more to an individual than to a breed or family.

In practice, it is important to note that most authors do not clearly distinguish between these two concepts. In scientific publications, both terms may be used without clarification of their differences (Gartner and Weiss 2013).

In an effort to clarify these often-misused terms, we define temperament here as suggested by Adam Miklosi (2016).

Temperament refers to the set of genetic predispositions and vulnerabilities of an individual that give rise to particular behavioral tendencies, which are stable over time. However, temperament is never directly observable, as from the very first days of development, the individual undergoes profound changes influenced by their environment and learning experiences.

Personality, on the other hand, encompasses the observable characteristics of an individual that cause them to react similarly in repeated contexts. Personality results from the impact of the environment and learning on the individual's initial vulnerabilities. Therefore, personality pertains to a specific individual and defines their behavioral tendencies.

For the remainder of this chapter, we will refer solely to personality traits when discussing observable characteristics, and we will use the term vulnerabilities to describe the inheritable aspects of an individual that predispose them to illness **(see box below)**.

Important Point of View

Zubin's model and vulnerabilities

In ► Sect. 1.1.2.3, we described Zubin's model (Zubin and Spring 1977), which is relevant for understanding any illness (◘ Fig. 1.11). Depending on their genetic vulnerabilities and environmental pressures, an individual may be in a state of behavioral balance (green zone) or experience mild (yellow zone), moderate (orange zone), or severe (red zone) imbalance.

This model also highlights the two extremes frequently observed in consultations. On one end, there are cats raised in poor environmental conditions that show very positive outcomes—resilient cats. On the other end, there are cats raised in favorable environmental conditions, with well-balanced parents, that still develop severe illnesses.

For all these reasons, we choose to use the terms *character* or *personality* only to discuss individual tendencies when illness is not a factor. However, as soon as an individual enters the *orange* or *red zones* of Zubin's model (◘ Fig. 1.11), they are considered ill and must be treated, whether the condition is caused by fragile genetics, a harmful environment, or a combination of both.

The approach to care may differ depending on the circumstances, but as veterinarians, our duty is to provide treatment as soon as we identify the presence of illness.

Clinical Illustration

Joe is an 8-month-old cat who is rather timid: when visitors come over, he hides behind the sofa, but once the guests are seated, he approaches them and, after about 30 min, allows himself to be petted. His owners live in a very quiet apartment and have no children. They believe it is normal for Joe to be cautious and never try to force interactions.

Joe is a cat with vulnerabilities along the fear axis, predisposing him to develop social phobia. However, the conditions of his exposure to unfamiliar people (infrequent, non-forced interactions) are favorable, allowing Joe to avoid tipping into pathological behaviors. A more stimulating environment, such as one with children who might block Joe to pet him, could have pushed him into illness. Yet, in both scenarios, the timid term could still be used to describe Joe's personality.

1

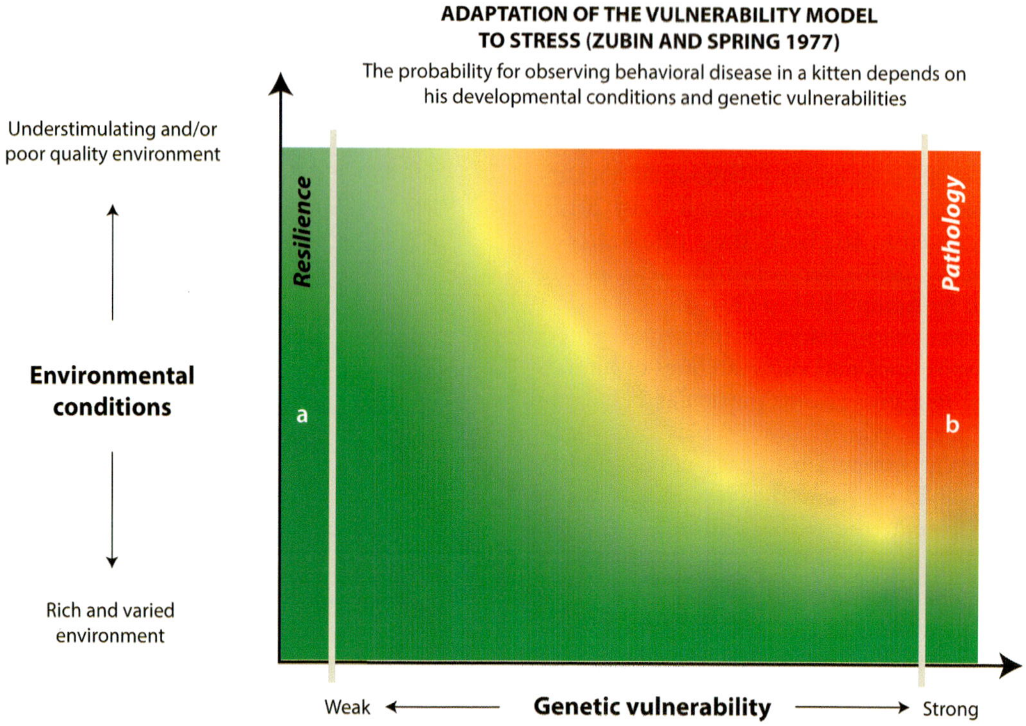

Fig. 1.11 Zubin's model applied to the behavioral development of kittens. At the two extremes: (a) very low genetic vulnerability protects the individual despite poor environmental conditions; (b) high genetic vulnerability promotes pathology despite excellent environmental conditions

Summary of ▶ Sect. 1.2

The behavior observed by the clinician in a cat is the result of three major components: *emotional*, *mood-related*, and *cognitive*.

The behavior observed by the clinician in cats results from three main components: emotional, mood-related, and cognitive.

Emotions are adaptive, enabling cats to quickly identify and respond to changes in their environment. They are difficult to control, and their dysregulation can lead to pathological states such as phobia or anxiety.

Mood is a persistent mental state that modulates internal stimuli. It "colors" behaviors positively—referred to as good mood, or negatively, referred to as bad mood. Mood dysregulation leads to pathological states such as mania, depression, and bipolar disorder.

Cognition encompasses numerous functions, including perception, attention, motivation, action and emotion control, decision-making, learning, memory, problem-solving, and reasoning. Cognitive dysregulation results in pathological states such as impulsivity, compulsivity, dissociation, and senility.

The terms *temperament* and *personality* are often misused, leading to confusion. *Temperament* refers to the set of an individual's genetic vulnerabilities, while *personality* reflects the result of these vulnerabilities interacting with environmental pressures. Personality does not position the individual on the normal-pathological spectrum and has limited value in establishing a diagnosis in veterinary psychiatry.

1.3 Particularities of Feline Behavior

This section does not address the ethogram of the domestic cat, which will be discussed alongside the anomalies that can occur in each behavior in the second part ▶ Chap. 7 to 15. The specific behavioral development of the feline species will be covered in ▶ Chap. 4.

The current section describes communication and cognitive abilities unique to the feline species. This description offers insight into how a cat perceives its world, known as its *umwelt* (von Uexküll 2001). To achieve this, it is essential to understand how cats were domesticated and the behavioral consequences of this process.

What communication channels do cats prefer? How should we behave to ensure cats understand us? Lastly, we describe the cognitive abilities of cats. It is important to note that the cat's advanced cognitive capabilities open the door to sophisticated therapies for treating cerebral dysfunctions, which are explored in the second part, "Psychopathology."

1.3.1 The Domestic Cat

1.3.1.1 Cat Domestication

Evidence of cat domestication has been found in various parts of the world. The earliest traces come from China and date back approximately 10,000 years (Driscoll et al. 2007; Hu et al. 2014), whereas evidence of dog domestication dates back 40,000 years (vonHoldt et al. 2010). Cat domestication is therefore more recent and significantly different from that of dogs.

Cat domestication occurred through *cooperation*, whereas dog domestication followed a *mutualistic* model.

In *cooperation*, there is an exchange of services between the two species. Cats gain access to shelter and small prey within human food stores, while humans benefit from protection against pests. However, this relationship is not essential for the survival of either species.

In *mutualism*, one species actively constructs an environment in which it controls the survival and reproduction of another species to obtain resources or services from it (Purugganan 2022). This results in one species becoming dependent on the other. Such is the case with dogs, which are heavily reliant on humans for survival.

As a result, cat domestication occurred passively, with humans exerting minimal selective pressure. Feline breed clubs emerged in the last century, initially focusing on aesthetic selection criteria. Most cat breeds are less than 100 years old. Today, there seems to be a growing trend toward selecting cats well-suited for family life.

In contrast, dog domestication was active, with strong selective pressures due to their utilitarian roles (e.g., herding, hunting). Canine breed clubs also developed earlier, during the nineteenth century.

The question of whether cats are fully domesticated remains open, as many individuals can spontaneously return to a feral state. These cats are capable of surviving without human intervention by hunting for food (Gosling et al. 2013).

The human-cat relationship also depends heavily on the cat's *socialization* to humans. Early and frequent handling of kittens by multiple people is crucial (Casey and Bradshaw 2008). Such handling reduces fear of humans in adult cats and improves their relationship with their owners (Lowe and Bradshaw 2002). Furthermore, a study showed that handling kittens for 40 min a day, rather than 15 min, increased affiliative behavior later in life (Casey and Bradshaw 2008).

1.3.1.2 A Non-social Species with Solitary Tendencies

The cat is a non-social species, like all felids except the lion (Bradshaw 2016). Unlike social species, cat groups lack a clear hierarchy, do not follow consistent group rules over time, and cats do not need to live in groups.

This solitary nature aligns with that of most wild felines, which prefer to hunt and live alone, except during the breeding season or while raising their young. Examples include species such as the tiger (*Panthera tigris*), leopard (*Panthera pardus*), jaguar (*Panthera onca*), and cheetah (*Acinonyx jubatus*).

For example, hunting is most often carried out solitarily in adult cats, with no evidence of cooperation among cats in this context. Similarly, play directed toward objects or simulated prey, which can mimic predation sequences, is typically performed alone. Play between two partners, mimicking fights, is possible but is predominantly observed during development (Caro 1981) (▶ Chap. 4).

However, the cat's solitary tendency is not absolute: cats can choose to live in proximity to other individuals, particularly when resources are scarce. Cats organize their environment into territorial (see ▶ Sect. 1.3.1.4) and relational fields. This organization involves forming individual relationships with members of their own species as well as with other species.

In open environments, the presence of limited feeding points offering large quantities of food promotes the formation of colonies (MacDonald et al. 2000; Turner and Bateson 2013). It is not uncommon to see groups of stray cats forming long-term colonies around such feeding points. Research indicates that these groups form based on preferential associations rather than randomly (Curtis et al. 2003).

Cats living in colonies are capable of organizing themselves, even though they do not exhibit the hierarchical social structures found in dog groups. Their relationships include recognizing other group members, sharing resources, mutual grooming, allomarking, and participating in the care of young.

There is a wide range of affiliative and agonistic signals in cats (Curtis et al. 2003). These signals increase with group size and help maintain group stability. From an evolutionary perspective, this organization enhances the survival and reproduction rates of cats.

Thus, the cat is a non-social species with solitary tendencies, but it can tolerate group living, whether within a confined environment imposed by humans or in open environments during opportunistic gatherings.

In confined settings, the level of socialization to humans or dogs is crucial, as these species must coexist on a daily basis. When multiple cats share the same living space, socialization with conspecifics is somewhat important, but after puberty, the cat's territorial and solitary nature makes cohabitation with another cat more challenging than with other species. This is particularly true because they share their living environment through pheromonal marking. It is important to note that, in such cases, agonistic interactions cannot be resolved through the flight of one individual.

1.3.1.3 A Prey and Predator Species

The cat is a predator, capable of hunting small mammals, birds, invertebrates, and reptiles for sustenance (Castañeda et al. 2023). Its hunting technique aligns with the "sit-and-wait predator" strategy, where the cat remains still—often crouched on the ground—until prey ventures close enough to be captured (Janetos 1982). This technique is energy-efficient but has a low success rate, making it suitable when prey is small and abundant.

The cat is also prey to predators such as raptors, as well as wild felids and canids (Hull et al. 2022; Nájera et al. 2019).

This dual status as both prey and predator is essential to understanding feline behavior and the behavioral pathologies associated with it. Fear-related behaviors and the various forms of aggression observed in cats are influenced by this duality.

As prey, cats possess a highly developed fear and alarm system. They become sensitized quickly and tend to prefer flight or immobility over attack. This explains why cats are more vulnerable along the *fear axis*, making them more prone to developing phobic or anxious states.

As predators, cats exhibit hunting behaviors and predatory sequences **(see box below)** that are often mistaken for aggression. However, these are frequently manifestations of poorly controlled impulsive states rather than true aggressive intent.

Important Point of View

Effects of cats on wildlife

Numerous studies have demonstrated the significant impact of cats on the decline of wildlife populations (Loss et al. 2013), to the extent that cats have played a key role in the extinction of certain species (Short and Smith 1994). This raises the question of whether confining cats could help mitigate their negative impact on biodiversity.

It is entirely possible to keep a cat happy without granting it outdoor access. This largely depends on the cat's personality (shaped by genetics and experiences) and the environmental enrichments provided.

Alternatively, cats can be allowed outdoors with measures to limit their hunting success, such as wearing a collar with a bell or devices like the *CatBib*® (marketed in the United States). Another option is to take cats outdoors on a harness; however, the cat must be accustomed to the harness and find it enjoyable, as harnesses can prevent escape and may cause anxiety.

The significant impact of the feline species on wildlife is one of the reasons many countries have implemented sterilization policies for feral cats.

1.3.1.4 A Species with a Territorial Tendency

A territory is an area that an animal consistently defends against other individuals of the same or different species. Cats cannot be considered strictly territorial, as they allow other individuals to enter their living space.

However, communication signals enable cats to structure their living space in a spatiotemporal manner and convey their organization to other cats. They designate specific areas or "fields" for particular activities, such as hunting, elimination, feeding, isolation, observation, interaction, or play. These fields are marked using specific pheromonal signals, including facial marking, urine marking, and scratching.

Each type of marking has a distinct purpose. For example, scratch marks are often found around resting and observation areas (McKeown et al. 1988). These are locations where the cat may not tolerate disturbance, and the visual signals alert conspecifics or other individuals to its presence. Urine markings are typically found on the periphery of the territory and indicate to other cats that the area is occupied. Facial markings are deposited on objects within the cat's habitat and along the pathways connecting the different fields of its territory. These markings help the cat navigate and move between its fields (◘ Fig. 1.12). Modern GPS trackers make it amusingly easy to observe how consistently cats follow the same paths when traveling from one point to another.

This spatial organization is complemented by the temporal occupation of fields when two individuals share the same living space. It allows cats to share the area by depositing their calming marks while tolerating those of others. Additionally, the unique markings placed on individuals, known as *allomarking*, create a relational field, explaining how cats can coexist with other individuals.

Many behavioral disorders present with symptoms involving changes in territorial marking, particularly conditions affecting the territoriality axis.

The cat needs to structure its living environment through marking, which gives it a territorial tendency. However, much like its solitary nature, the cat can share its living space with conspecifics by organizing its marks spatially and temporally.

The sense of calm that the cat derives from the organization of its territory is analogous to the sense of calm that a dog achieves through social integration into its group. Put differently, *territory* is to the cat what *hierarchy* is to the dog.

The size of a cat's territory varies depending on sex and environmental constraints. Female territories can extend up to 200 hectares, while male territories can reach up to 1000 hectares (Liberg et al. 2000). The way this territory is occupied differs slightly between feral and non-feral cats. On average, a cat spends 10 h per day sleeping, 5 h resting, and observing, 5 h hunting and playing (though this is highly variable), 3 h grooming, and the remaining hour eating, eliminating, and moving around (Panaman 2010).

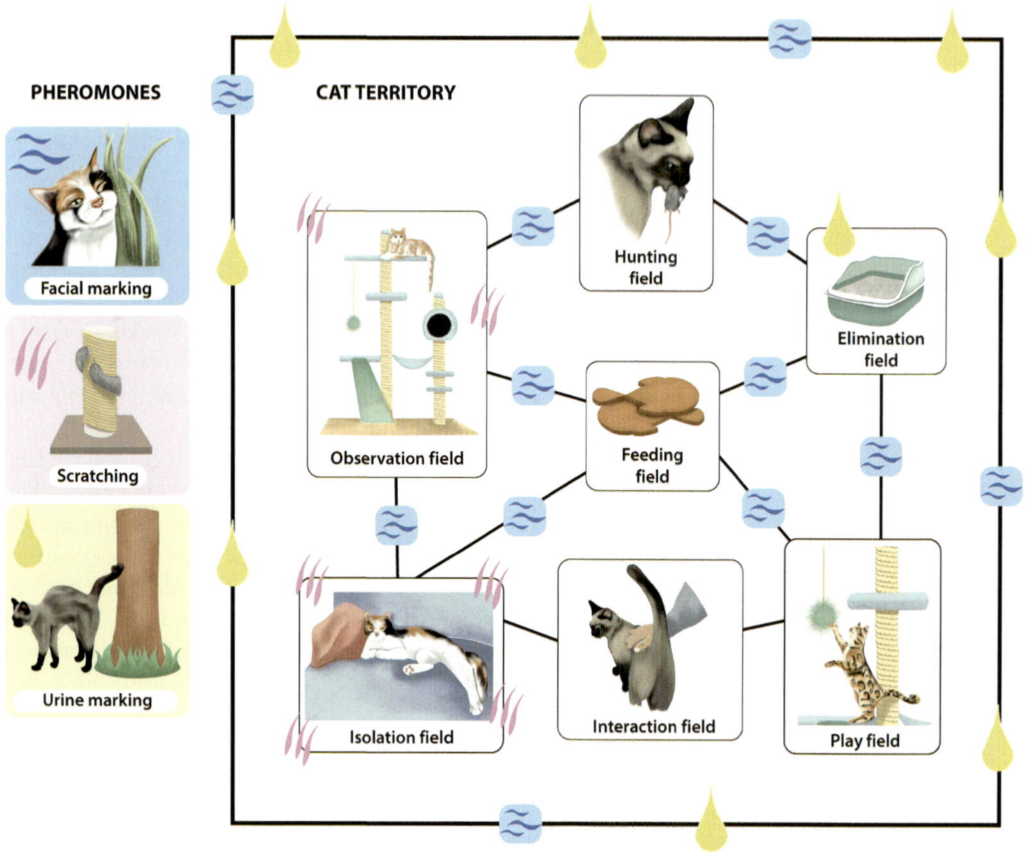

Fig. 1.12 Organization of cat's living environment through the marking of its fields and pathways

1.3.2 Feline Communication

1.3.2.1 Intraspecific Communication in Cats

Communication involves the emission of a signal by a sender animal that elicits a response from a receiver animal, resulting in an advantage for one or both parties. Its functions include species recognition, group cohesion maintenance, and individual exchanges.

To emit and receive communication signals, cats use various sensory channels: tactile, visual, gustatory, auditory, and olfactory. Communication is always multimodal, meaning it involves multiple sensory channels simultaneously. Although each channel is described separately below for clarity, it is essential to remember that, in practice, several dimensions interweave to create a complex message: the feline language.

▪ Sense of Taste

Taste buds develop during gestation, and the sense of taste is already present at birth, although it is not fully mature (Stedman et al. 1983).

The sense of taste is not the cat's most developed sense and is less acute compared to that of dogs (Pekel et al. 2020). Cats are unable to detect sweetness (Li et al. 2007), but they are highly sensitive to bitterness, making it challenging to administer medication (Lei et al. 2015). Additionally, cats may exhibit a "habituation effect," showing a preference for new food over familiar options (Bradshaw 2006).

▪ Tactile Communication

The sense of touch is present at birth and helps kittens orient themselves toward their mother. It likely contributes to the pleasure felt by kittens during the innate kneading behavior observed while nursing, reinforcing this behavior.

The abundance of vibrissae in cats enhances their tactile sense compared to dogs. Vibrissae are specialized mechanoreceptors with a hair-like structure, but their follicles are surrounded by an innervated ring and a venous sinus, connected to a striated skeletal muscle. The nerve fibers join the trigeminal ganglia and then the brainstem. These vibrissae can contract and detect a displacement as small as 5 angstroms, a pressure of 500 milligrams, and a temperature variation of 0.2 degrees Celsius (Kenshalo et al. 1967).

In cats, vibrissae are located above the eyes, around the muzzle, along the cheeks, under the chin, and even on the back of the carpi. They assist cats during jumps, hunting behaviors, and intraspecific communication.

Touch also plays a role when the mother lifts her kittens to move them before they reach 5–6 weeks of age. This particular carrying method triggers a "carrying reflex," characterized by a curled tail, flexed limbs, and a state resembling near-catatonia when held by the scruff of the neck. However, as the cat matures, the tension exerted by this restraint becomes excessive, making the experience painful and aversive, particularly for adults (Moody et al. 2020).

This principle can be utilized in veterinary practice through "clipnosis," where clips are applied to the nape of the cat's neck. In some cases and for certain cats, this method allows procedures to be carried out in a less stressful manner than traditional restraint methods (Moody et al. 2020; Nuti et al. 2016; Pozza et al. 2008). It is crucial to note, however, that lifting a cat by the scruff is painful and should be avoided.

▪ Auditory Communication

Cats are born deaf. They begin to hear around the fifth day of life, but their auditory canals do not fully open until 4 weeks of age.

Kittens emit ultrasonic vocalizations during their first days. Over time, the frequency and amplitude of these sounds decrease. When kittens are deaf, their vocalizations have a higher amplitude, indicating that their production does not depend on learning through imitation (Houpt 2018).

The auditory range of cats is broader than that of humans or even dogs. Cats perceive frequencies from 45 Hz to 65,000 Hz (Strain 2017), whereas humans hear between 20 Hz and 16,000 Hz, and dogs between 70 Hz and 45,000 Hz (Heffner 1983). This range allows them to locate rodents that emit ultrasonic sounds (Caruso et al. 2022).

Auditory communication can be used at short distances but is primarily intended to alert conspecifics to other visual or olfactory signals when visibility is poor, such as in darkness or over long distances.

Auditory communication includes vocal and non-vocal signals. *Non-vocal signals* include teeth chattering and trilling. Teeth chattering is often observed during hunting sequences when the prey is inaccessible and seems to indicate frustration. *Vocal signals* in cats number at least 20 (◘ Table 1.3) (Tavernier et al. 2020).

According to the context, cats will use different vocalizations, as summarized in ◘ Table 1.4.

The *meow* is the most commonly used vocalization and occurs in a variety of circumstances: territorial defense, seeking attention, greetings, play, defense, threats, alerts, or distress. Other vocalizations, such as the mating call emitted by a female in heat to attract males, serve a much more specific function.

The function of *purring* remains poorly understood. It can be used in interspecific or intraspecific communication to soothe individuals around the cat, but this social function is not systematic. Indeed, purring is sometimes emitted in cases of pain and, in such instances, it appears to be used to calm the cat itself.

In general, vocalizations are emitted far more frequently toward humans than toward other cats. Similarly, the audiogram differs between feral cats and cats that have been socialized to humans. This suggests that the social environment has a significant impact on the vocalizations emitted by cats and that they are predominantly used for interspecific communication (Tavernier et al. 2020).

1

Table 1.3 Different types of cat vocalizations and their meanings according to Tavernier et al. (2020)

Vocalization	Context	Definition
Whining	Reproductive (male), threat (female), contact seeking (kittens)	Prolonged sound, variable pitch and amplitude, pronounced slowly with "u" or "o" sounds
Copulatory cry	Reproductive (female)	Low-pitched sound, growl-like, low amplitude
Teeth chattering	Predation with inaccessible prey	Low-amplitude sound, jaw clattering, voiceless
Call	Contact seeking/affiliative, pain	High-pitched, short-duration sound
Trill	Contact seeking/affiliative (female)	Low-intensity sound resembling a gargle, voiceless
Chirping	Contact seeking/affiliative	Resembles trilling with numerous modulations
Gurgling	Contact seeking/affiliative	Resembles trilling but higher pitched and more variable, "wi"-like sound
Murmur	Contact seeking/affiliative	Short sound resembling a squeak or trill
Grunt	Contact seeking/affiliative	Low-pitched, short sound of low amplitude, shorter than purring
Squeak	Contact seeking/affiliative, play	Resembles a meow but more nasal, high-pitched, and hoarse
Growling	Threat, pain	Low-pitched sound, gradually increasing amplitude, guttural
Rumbling	Threat/agonistic, pain	Resembles growling but louder, higher pitched, and shorter
Howling	Threat/agonistic, pain	High-amplitude sound, shorter than a meow
Hissing	Threat/agonistic, pain	Sudden and forceful exhalation, exposing teeth
Spitting	Threat/agonistic, pain	Sudden and forceful exhalation, exposing teeth, more intense than hissing
Pain cry	Pain	Short, high-amplitude, high-pitched sound with a strong exhalation
Meow	Various (contact seeking, reproductive...)	Consistent pronunciation
Purring	Various (contact seeking, reproductive, pain, threat...)	Low-pitched, long, rhythmic sound of low amplitude, produced during breathing
Meowing	Various (reproduction, threat, contact seeking, pain...)	Sound with variable intensity, pitch, and amplitude, prolonged

Olfactory Communication

Olfaction is to animals what writing is to humans: a message that can be transmitted without the presence of the sender. In dogs, the nasal epithelium covers an area of 200 cm^2, in cats 20 cm^2, and in humans up to 4 cm^2, which explains why this sense is likely the most developed among domestic carnivores and plays a more significant role for them than for humans.

This system is functional from birth and allows kittens to locate the nipples even before their eyes open (Raihani et al. 2009).

In addition to this highly efficient olfactory system, cats have an accessory olfactory system located behind the incisors, between the hard palate and the nasal cavity: the vomeronasal organ, or Jacobson's organ. The vomeronasal organ is connected to the acces-

Table 1.4 Types of vocalizations emitted by cats depending on the context

Context of vocalization emission	Types of vocalizations emitted
Reproductive	Whining (male), copulatory cry (female), miaow, purring, meowing
Threat/agonistic	Whining (female), growling, snarling, grinding, hissing, howling, spitting, purring, meowing
Predation	Teeth chattering
Seeking contact/affiliative	Calling, cooing (female), chirping, gurgling, whining (kittens), trill, miaow, murmur, purring, grinding (particularly during play), meowing
Pain	Calling, growling, snarling, grinding, hissing, howling, spitting, pain cry, purring, meowing

sory olfactory bulb via the terminal nerve. The accessory olfactory bulb is, in turn, linked to the amygdala, which explains the effect of pheromones on emotions.

The vomeronasal organ enables many mammals to detect pheromones and decode their messages. In practice, the animal can gather information about the emotional and physiological state of the emitting individual, allowing it to adapt its own behavior accordingly. Interactions with pheromones are facilitated by a specific behavior, *flehmen,* which draws pheromones toward the vomeronasal organ. In cats, flehmen consists of a rapid swallowing motion with the mouth slightly open (Hart 1987), much more discreet than the flehmen observed in horses, which involves completely curling the upper lip while producing loud inhalations (Marinier et al. 1988). Flehmen is observed in cats starting from the fifth week of life.

Cats primarily communicate through pheromones, which are semiochemical substances excreted outside the body that influence the behavior or physiology of other individuals, usually of the same species. Pheromones may be emitted during voluntary marking behaviors, such as urination, or involuntarily, as in the case of anal gland secretions or sweat from paw pads.

Pheromonal marking allows cats to organize their environment by delineating areas and pathways between different zones or *fields*. For example, the isolation field is frequently marked by scratching and the pheromones released from plantar pads. Pheromones also enable cats to recognize other individuals through *allomarking*, during which the cat deposits specific pheromones.

In cats, pheromones are secreted by various organs, as detailed below (Fig. 1.13).

The *facial glands* are located under the chin, lips, cheeks, and ear pinnae. They produce pheromones related to social structure. Cats emit five fractions of facial pheromones, numbered F1 to F5. Depending on the fraction emitted, the signal varies. The roles of fractions F2, F3, and F4 are, respectively, sexual marking, object marking (also referred to as *facial marking*), and marking within interspecific or intraspecific social interactions, or *allomarking*. The roles of fractions F1 and F5 remain unknown (Pageat and Gaultier 2003).

Salivary glands also appear to convey pheromones (Gröschl 2009). Grooming leads to saliva being deposited on the coat, which subsequently enables marking through flank rubbing against environmental elements or during social interactions.

The *glands in the perianal area* include perianal and anal glands. The latter are partially emptied during defecation, giving feces a specific odor (Doty and Dunbar 1974). They may indicate an appropriate elimination area for defecation (Kasbaoui et al. 2022). In cases of intense fear, cats fully empty these glands, producing a foul odor that reinforces the pheromonal message. These olfactory and pheromonal signals constitute a warning sig-

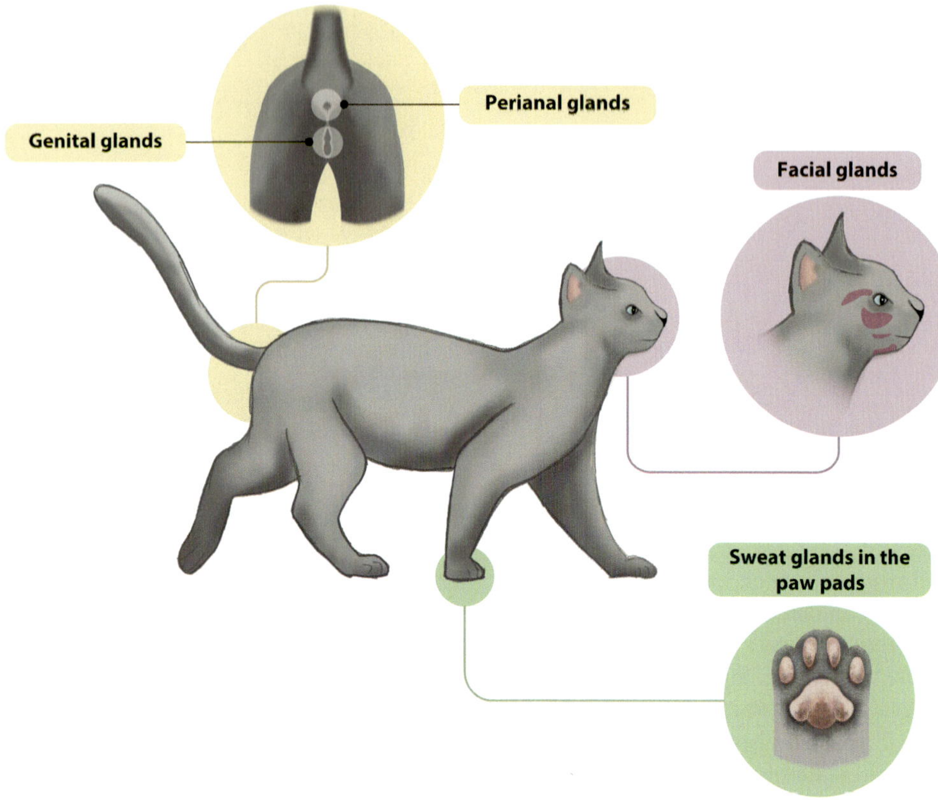

Fig. 1.13 Zones of pheromone secretion in cats. Cats secrete pheromones from specific areas: facial glands, perianal glands, sweat glands in the paw pads, genital glands, feces, and urine

nal to conspecifics, prompting them to avoid the area (Overall 2015).

Sweat glands in the paw pads are activated notably during scratching, territorial marking, and in conjunction with the deposition of feces and urine (Pageat and Gaultier 2003). They leave odors that conspecifics explore extensively. These glands are also activated during intense stress (Mengoli et al. 2013).

Genital glands are located on the prepuce, vulva, genital, and urethral mucosa and play a role in sexual attraction via estrogenic metabolites (Mills 2005).

Feces can also serve as a potential communication signal, though their role is not clearly identified. Some cats appear to intentionally deposit feces along paths or at the edges of territories without covering them. This signal is likely both visual and olfactory (Feldman 1994).

Finally, *urine* is a major communication vector and also carries pheromones. Urine marking differs from urinary elimination. During urine marking, the cat explores the area, searching for olfactory marks, may perform a flehmen response, stands upright with its tail erect and quivering, and does not attempt to cover the urine. The surface is more often vertical, and the volume emitted is limited. During urinary elimination, the cat scratches the surface, squats, and may attempt to cover the urine. The surface is horizontal, and the bladder is fully emptied.

Urine marking begins at puberty and is more frequently observed in intact males, though all cats may engage in urine marking as a physiological behavior. Some studies indicate a reduction in marking behavior in sterilized cats, while others suggest the opposite (Cafazzo et al. 2019; Hart 1980; Horwitz 2019).

Urine marking and scratching are common reasons for consultations and represent significant inconveniences for cat owners

(DePorter and Elzerman 2019; Horwitz 2019). Their pathological consideration, along with other forms of marking, is detailed in Part II, *Psychopathology*. Associated treatments are covered in Part III « Psychopharmacology » and Part IV « Psychotherapies ».

Cats possess an exceptionally strong sense of smell, enabling them to analyze the structure of a territory and the social interactions occurring within it. Their vomeronasal organ grants access to the pheromones emitted by other cats as well as those emitted by other species (Papes et al. 2010). For humans, whose sense of smell is limited and who possess only a vestigial vomeronasal organ, it is challenging to conceptualize the cat's sensory world (Trotier 2011). These perceptual differences are well illustrated in cases of interspecific social phobia. Cats with incomplete sociability may react exclusively to certain types of people, whom we attempt to identify based on characteristics such as age, gender, or other distinguishing features. However, cats likely discriminate these individuals, perceived as hostile or non-hostile, primarily based on their scent or the pheromones they emit. It is often more effective to observe and accept the behavior of these cats without necessarily seeking to understand the exact reasons for their discrimination, as these may be beyond our perceptual reach.

▪ Visual Communication

Kittens are born blind and begin to distinguish shapes around the 7th day of life.

A cat's visual field spans approximately 220°, with a binocular vision of 130°, compared to 83° in dogs. This means that cats can see in front of them, to the sides, and to some extent behind them. Their ability to perceive depth is better than that of herbivores, humans, or dogs, giving them a significant advantage in assessing distances and detecting movement in their environment.

Cats also have far superior night vision compared to humans. This is due to the presence of a reflective layer behind the retina called the *tapetum lucidum*, which reflects unabsorbed light back toward the retina, enhancing vision in low-light conditions. Additionally, the slit shape of their pupils allows cats to better perceive depth and regulate the amount of light entering the eye more effectively: admitting less light during *miosis* and more during *mydriasis* compared to a round pupil. These adaptations are particularly useful for crepuscular vision, optimal during hunting times.

The cat's lens is positioned farther back in the eye than that of humans, allowing it to focus light rays more effectively. Furthermore, cats have a lower cone-to-rod ratio compared to humans, enabling them to detect movement more efficiently but with less ability to perceive colors, particularly red. Their vision is nearly dichromatic (Steinberg et al. 1973).

These visual traits are highly specialized adaptations, enhancing the cat's ability to hunt and locate moving prey under varying light conditions.

Visual communication is often the one we pay the most attention to, as it is easily observable compared to olfactory communication. However, feline visual communication is complex, rapid, precise, and subtle. Even for professionals trained in "reading" feline signals, it is never straightforward to interpret.

Involuntary movements, such as *piloerection*, pupil dilation, emotional urination, and trembling, are the result of neurovegetative manifestations and provide insight into the animal's emotional state.

Voluntary movements include body postures, facial expressions, and other movements. Nearly all body regions can contribute to forming the final visual message: the head, gaze, ear position, vibrissae, and tail. Many models describe the positions of these body parts depending on the cat's emotion or the cat's overall body posture (◘ Fig. 1.14). For instance, a raised tail is often interpreted as a friendly signal. However, depending on the presence of *piloerection* on the tail and the rest of the body posture, it could also indicate surprise or threat.

Body postures broadly refer to sequences of positions produced in a specific order and context, enabling cats to communicate with one another. These postures are innately coded in all cats but require learning through social contact with adults for proper execution. This communication is learned by kittens through interactions with their mother and during encounters with other cats in the first months of life. Below, we describe the primary sequences of body postures, though these may follow one another rapidly and are not always simple to observe (◘ Fig. 1.14).

The body postures of cats are less thoroughly described compared to those of dogs. However, one can observe a cat's gaze, facial expressions, posture, and the kinetics of its movements.

A fixed *gaze* combined with mydriasis indicates fear, predation, or irritation. In contrast, *miosis* signifies attention. The frequency of blinking plays a role in communication, with frequent blinking suggesting a positive interaction (Humphrey et al. 2020).

A raised *tail* signifies interest, frustration, or a friendly approach. A lowered tail is associated with fear, irritation, or predation, particularly if it begins to move. Tail movement varies from gentle swaying to quivering with increasing irritation. When in motion, the tail tends to curve, especially when the cat moves quickly (Ellis 2018).

As for *facial expressions*, forward-facing ears and vibrissae indicate attention or excitement, while backward-facing ears and vibrissae suggest fear, predation, or irritation (Turner and Bateson 2013).

Clinical Illustration

Oreo's owner seeks to understand why his cat, a 4-year-old European cat, exhibits aggressive behavior. Each evening, when the owner returns from work, Oreo follows the same sequence of actions. As the owner enters the door, the cat rushes to the sofa, emits a specific meow, and waits to be petted. Once petted, Oreo runs to his food bowl and waits for food, which he usually receives. If the owner is busy and delays feeding Oreo, the cat becomes very persistent and may even resort to biting, likely out of frustration when the ritual is not completed as expected.

◘ **Fig. 1.14** The various postures of a cat. The combination of gaze, facial expressions, and posture, including the position and movements of the tail, provides information on the cat's emotional and motivational state

A high *posture* with an upright tail suggests excitement or exploration, whereas a low posture with a tucked tail indicates fear or predation (Houpt 2018).

Finally, *movement kinetics* must be considered. A straight and rapid trajectory suggests excitement or readiness to attack. A curved trajectory with pauses indicates an intent to create appeasement. A fragmented kinetic pattern signifies play behavior.

A cat preparing to attack will display miosis, a fixed gaze, flattened ears, a lowered head, retracted vibrissae, a low tail and posture, and may advance on the tips of its toes. It might also show *piloerection*, with claws or teeth visibly exposed.

There are other more subtle indicators of a cat's intentions, such as the shape of its mouth and lips, which contribute to its overall facial expression. More than 200 finely distinct facial signals have been observed in cats during social interactions (Scott and Florkiewicz 2023).

There appear to be *appeasement signals* in cats that can soothe the individual observing them, but there are no submission postures similar to those seen in dogs.

These communication signals are used by cats either when they feel uneasy, to indicate discomfort with a situation, or to send a "peaceful" message to their interlocutor. Such signals are increasingly studied in dogs by the scientific community, and experimental results are gradually confirming their meaning (Mariti et al. 2017). However, there are relatively few studies on appeasement signals in cats (Heath 2018).

Examples of appeasement signals in cats include turning the head and averting the gaze, squinting and blinking, or turning the body away (Bennett et al. 2017).

It is important to practice recognizing these signals because they provide valuable insights into a cat's discomfort. Identifying these signs helps prevent many incidents of aggression during veterinary consultations, as the point at which the animal's adaptive capacities are exceeded can be anticipated. Ideally, one should step back from the stressful situation as soon as appeasement signals are observed. It is equally crucial to make cat owners aware of these signals so they can better understand their pet and adjust their behavior according to the cat's emotions.

Some cats exhibit *communication anomalies.* These may involve an *incongruence in the message being delivered,* where the signals given by the cat do not consistently express the same emotion. For example, a cat may perform a facial marking followed by a retreating movement. Such behavior is ambiguous, and the cat's communication reflects this ambiguity. This may also stem from a lack of opportunities to practice feline communication or from inadequate socialization. In all cases, these message incongruities make interpretation more challenging.

Some cats exhibit dysfunctional communication related to pathological states, such as impulsive states. These cats tend to be rough in their interactions and often escalate a play sequence into aggression. Thus, feline communication can become impaired either because the cat has lacked the opportunity to learn (due to early separation from the mother or social isolation) or because its brain lacks the capacity to correctly acquire the feline language.

Decoding feline communication, therefore, requires considering all these factors, along with the context. This remains a significant challenge since the cat belongs to a species that is not only different from ours but is also non-social. Many cat owners seek explanations about their pet's behavior and frequently ask practitioners what specific reactions mean. It is important to exercise caution when interpreting isolated reactions, especially when the broader context and sequence are unknown.

The widespread use of smartphones facilitates access to numerous videos, simplifying this task. However, it is essential to remain humble in our interpretations. Communication inherently involves multiple individuals, and the response to the cat's signal will significantly influence its subsequent reactions.

Only regular observation and experience enable progress in interpreting feline communication. In many situations, it is not possible to have absolute certainty. Time and repeated observations help refine interpretations, but practitioners should not feel guilty for being unable to draw immediate conclusions from a single isolated behavior. Cats use multimodal communication, often combining multiple communication channels for a recipient who may be near or far and who may receive the message either immediately or with a delay. For example, to drive away a partner, they may hiss (auditory communication) while raising their fur (visual communication) in an immediate exchange with a nearby partner. Alternatively, they may scratch (visual communication) combined with urinary marking (olfactory communication) in a delayed exchange directed at an unseen partner.

Due to their multimodal communication, cats respond better to humans interacting through both vocal and visual communication than to vocal communication alone. They also respond better to visual communication than to vocal communication. Thus, for cats, non-verbal communication appears more important than verbal communication (De Mouzon and Leboucher 2023).

1.3.2.2 Interspecific Communication in Cats

▪ From Cats to Humans

As described in the previous section, cats have their own means of communication. When interacting with individuals of other species, they attempt to communicate as they would with another cat. For instance, when a cat approaches a favored human, it does so as it would approach a familiar cat, with its tail held upright (Turner 2021).

In interspecific communication, each individual must exert greater interpretative effort than they would when interacting with members of their own species. In this context, it is essential for us as humans, possessing more advanced cognition, to make an additional effort to understand.

Certain behaviors between humans and cats may become ritualized. A *ritual* is a behavior that has acquired a communicative function through learning. It is only exhibited in the presence of the intended recipient of the communication and remains highly stable over time, as it has been reinforced repeatedly by the recipient's responses. Rituals between cats and their owners are very common, particularly around meal times, morning routines, or the owner's return home. These behaviors are frequently described during consultations and are often perceived as signs of affection or bonding between the cat and its owner (Turner and Bateson 2013).

While cats have developed skills to communicate with humans both phylogenetically through domestication and ontogenetically through socialization, the reverse is not true. Recent studies have shown that humans struggle to classify the vocalizations emitted by cats according to the emotions the animal might be experiencing, particularly if they are unfamiliar with the cat emitting them (Ellis et al. 2015). However, when vocal communication is supported by visual cues, cats better understand human intentions (De Mouzon and Leboucher 2023).

▪ From Human to Cat

In terms of general principles, certain criteria can increase the likelihood that a message delivered to a cat will be understood.

First, the *message* must be clear, meaning it should use only a few words, spoken in a tone consistent with the request.

It must also be congruent: *body postures* used should align with other communication channels. For example, if expressing anger, a lower pitched voice and an upright posture are more coherent. Using a high-pitched voice while crouching is unlikely to convey anger effectively to the cat.

Studies have shown that both men and women tend to use higher pitched voices and articulate words more clearly when speaking to their cats compared to other humans. This behavior may reflect an effort to enhance communication (Mouzon et al. 2022). Moreover, cats appear to be more receptive to this type of speech and are more responsive to their owner's voice than to a stranger's voice (De Mouzon et al. 2023; Saito and Shinozuka 2013).

Finally, the request embedded in the message must be coherent, meaning the cat must be able to respond within the limits of its ethogram and the given context. In other words, the request should respect the cat's possible reactions within the framework of the exchange. If a cat appears unresponsive to a signal from a human, the primary assumption should be that it does not understand the request. Adjusting the signal to make it more coherent for the cat is recommended.

Additionally, appeasement signals (see ▶ Sect. 1.3.2.1) can be used to reassure the cat. Examples of such signals include averting one's gaze, approaching in a curve, and slow blinking (Heath 2018). Employing these signals, for instance, when the cat is on an examination table, can be an effective way to reduce tension.

Reprimanding a cat differs significantly from disciplining a dog, where negative punishment, such as exclusion from the social group, is recommended. For the cat, a non-social species, such punishment lacks meaning or effect. The primary strategy for correcting behavior should be to interrupt the unwanted behavior, often through disruption, and then redirect the cat to an alternative behavior to calm it down.

Under no circumstances should punishment involve pain or fear, as these constitute positive punishment, which is anxiety-inducing and likely to provoke aggressive reactions.

▪ Non-verbal Communication and Proxemics

Proxemics is a term coined in 1963 by American anthropologist Edward T. Hall to designate the study of the use and meaningful organization of space in relationships between living beings.

For Edward T. Hall, proxemics has a significant impact on our interpersonal communication. He thus defines several distances based on how we position ourselves depending on the person in front of us. It is as if each individual possesses around them spaces of varying size, within which interactions with others are limited in number and duration, depending on the context. The size of these spaces varies according to culture, context, and individuals.

The intimate space is at the physical limit of the individual. This is a zone of actual contact, as in sexual activity or physical combat. Personal space is the sphere that allows the individual to isolate themselves from others. It is the distance we maintain with friends when we are sitting side by side. Social space is the zone where social exchanges occur. It is also the distance we maintain with colleagues during a verbal exchange at work or when talking to an acquaintance in the street. Public space is the zone beyond which the individual no longer feels directly concerned by the interaction.

Hall mainly describes the influence of physical distance between individuals on the relationships established between them (Hall 1963), but notes that this dimension of non-verbal communication is not the only one influencing interactions (Hall et al. 1968): the role of expressions, gestures, postures, and movements in communication has since been widely described in humans (Scheflen 1964) and other animal species, including dogs (Hasegawa et al. 2014).

Moreover, these cues can also be used in interspecies communication: several species of great apes (Kaminski et al. 2004) and domestic animals, such as goats (Nawroth and McElligott 2017) and horses (Proops and McComb 2010), utilize the orientation of human gaze, head, and/or body to assess their attention and even intentions. Proxemics should thus be included in the rules of non-verbal communication governing interactions between humans and other animals, including cats. Miklósi and his team (Miklósi et al. 2005) have also demonstrated that, much like dogs, cats are capable of locating a hidden object or food when pointed at by a human. However, compared to dogs, cats' attention diminishes faster, and their performance decreases over time (Salamon et al. 2023).

Additionally, in cats, interindividual distance plays a significant role in regulating interactions. This is particularly evident during veterinary visits.

For cats, the concept of proxemics defines an *aggression field,* which corresponds to the area of the body or around the body that, if crossed, triggers aggression (Čabraja 2021). This field varies between individuals and, for the same individual, changes over time depending on experiences, particularly socialization, and their emotional and physiological state (◘ Fig. 1.15). For example, for most cats, the head does not fall within the aggression field, and touching this area is tolerated, even during a veterinary consultation. However, touching the belly may never be tolerated as it could be part of the cat's permanent aggression field. Alternatively, it may be tolerated in general but not during veterinary visits or in stressful situations.

Thus, in cats, acceptable contact does not depend on social codes specific to their species intended to initiate polite interaction; instead, it is individual and variable.

Moreover, other feline communication rules that might seem intuitive to humans are often overlooked. For instance, approaching a cat with a curved trajectory rather than a straight one is a way to signal peaceful intentions. The meanings of our gestures, postures, and movements for cats must be considered to communicate with them in a coherent and reassuring manner across various situations.

To prevent aggression, it is recommended to make an olfactory contact before a tactile one, avoid frontal approaches, use calming signals like blinking or averting gaze, allow the cat time to sniff, and pet it in its preferred contact areas (typically the head and the base of the tail). Additionally, a firm, continuous touch is often preferable to multiple intermittent ones.

Over time, an increasing number of veterinarians, veterinary assistants, and feline educators are being trained in this respectful form of communication. These techniques are therefore being applied more frequently, contributing to the well-being of our companion cats.

1.3.2.3 Aggression in Cats

The scientific community has not reached a consensus regarding the sequences of aggression, and different classifications exist depending on the discipline. In cats, a distinction can be made between predation and aggression, as these involve different neurophysiological circuits and neurotransmitters. In predation, the lateral hypothalamus is activated, whereas in aggression, the medial hypothalamus and periaqueductal gray matter are involved. Their activity is modulated by the amygdala, the bed nucleus of the stria terminalis, the cingulate gyrus, and the prefrontal cortex (Gregg and Siegel 2001; Siegel et al. 1999).

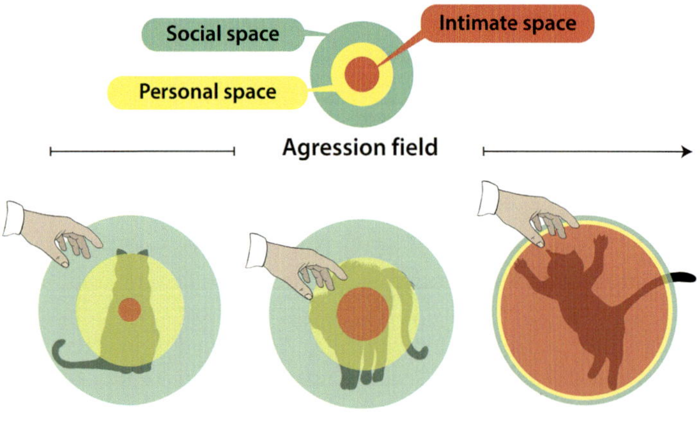

◘ **Fig. 1.15** The cat's aggression field. (a) The spaces defined by E. Hall in the concept of proxemics. In cats, the intimate space can overlap with the aggression field. This field expands or contracts depending on the context and the animal's emotional state. (b) The aggression field is very limited, indicating an emotionally calm cat. Contact with the animal can be made without risk of aggression. (c) The aggression field expands to encompass the cat's body boundaries, revealing a cat with increasing emotional tension. Contact is possible, but tension is noticeable, and aggression may occur if the contact lasts too long. (d) The aggression field reaches its maximum and extends beyond the cat's body boundaries: even approaching the cat triggers aggression. The expansion of the aggression field can be extremely rapid, following the cat's emotional fluctuations

The classification presented below is based on Moyer's modified classification. It considers neurophysiological aspects while maintaining a precise description of behavioral sequences, which is essential for clinical practice (Moyer 1971). Injuries sustained during play are grouped with predation under the category of "injurious sequences that do not fall under aggression."

■ Non-aggressive Injurious Sequences

The intent behind these sequences is not to resolve a conflict, and, therefore, they are not considered agonistic behaviors.

Hunting and predation occupy a significant portion of feral cats' daily activities, up to 50% (Liberg 1984). In predation, the objective is to kill the prey with the intent of consuming it. This behavior is silent. As previously mentioned, cats employ a "sit-and-wait" hunting technique, which involves waiting for prey to come within reach (stalking and ambushing) before pouncing on it (capture). The stalking phase can last several minutes and is characterized by near-complete immobility until the prey is within range. During the ambush phase, the cat crouches, tucks its hind legs under itself, and prepares to pounce. In the capture phase, the cat leaps onto its prey and may, depending on the prey's size, clasp it by the throat to suffocate it, shake it to break its neck, or fight it using its hind legs. After the kill, the cat consumes its prey. The return to equilibrium manifests through behaviors such as grooming, resting, or marking objects with facial pheromones.

Target species are determined primarily during development through maternal teaching. The abundance of certain prey within the cat's hunting range also leads to specialization (Dickman and Newsome 2015).

This behavior can be inhibited with certain familiar individuals belonging to prey species, such as rabbits. However, this inhibition does not generalize to all individuals of the species. It represents familiarization rather than socialization.

Socialization is a learning process that is typically generalized to an entire species and is almost impossible to achieve after the end of the socialization period if the cat has not been exposed to the species in question (see ► Chap. 4). The quality of socialization depends on the living conditions of the kitten during its first 2 months of life: repeated exposures to the target species under positive conditions. Familiarization, on the other hand, pertains to one or more specific individuals and can occur even after this period. However, poor socialization to a species makes familiarization significantly more challenging.

For prey species that the cat has not been socialized with, it will not be possible to prevent an adult cat from preying on an individual of that species, even if it is supposed to be a companion.

Biben (1979) demonstrated that predatory behavior is not directly dependent on hunger, although the likelihood of killing increases with appetite. A cat fed at fixed times is more likely to perform instrumentalized predatory behaviors before feeding, as the brain areas involved in predation become more activated as appetite increases. In cases of low appetite, the cat tends to play with the prey before killing it. Light levels also play a role, with hunting activity peaking at dusk (Horn et al. 2011).

During play, a cat may mimic predatory sequences. For this reason, it is recommended during the kitten's learning period to redirect such play toward objects, avoid playing with hands, and limit the availability of ambush sites where the cat could stalk humans.

Injuries during play are bites and scratches caused by cats with poor control during playful interactions. Initially, there is no intention of aggression in such cases. The sequence lacks any threatening signals, as the cat does not intend to harm. Despite this lack of intent, such incidents are a frequent source of injuries, particularly involving cats exhibiting impulsive or compulsive pathological states.

It is important to note that predation, play, and all aggressive behaviors are expressed in relation to the other party involved, whose response can influence the sequence's progression. For instance, injuries during play may elicit fear responses (such as the human reacting with screams, sudden movements, or attempts to retaliate), potentially escalating the interaction into aggression driven by irritation or causing the cat to flee.

Different Types of Aggression

Aggressive sequences are displayed to resolve a conflict or alleviate discomfort. They can be classified into offensive aggression, where the cat is the aggressor and approaches the target, and defensive aggression, where the cat is responding to an approaching threat. However, this classification does not adequately describe the emotional state of the animal.

We prefer a more detailed classification that describes the contexts in which aggression arises. This approach identifies several types of aggression, linked to varying levels of emotional intensity, resulting in more or less controlled bites and scratches. Additionally, it is important to remember that, as both predator and prey, cats are more prone to sensitization than dogs. Aggression due to irritation, particularly in fearful contexts, is therefore more frequent in cats than in dogs.

Some types of aggression are aimed at enforcing a rule and are associated with a cat experiencing low emotional intensity. These include territorial and maternal aggression. In irritation-based aggression, the cat struggles with growing emotional intensity—whether frustration, fear, or pain—until it loses control. Finally, in some cases, the animal displays aggression with minimal control due to overwhelming emotional states. These instances tend to be more injurious and include uncontrolled emotional aggression and idiopathic aggression associated with derealization states (◘ Table 1.5).

The goal of *territorial aggression* is to establish the boundaries of a territory and prevent intrusions. Cats do not strictly defend their territory; otherwise, they would never allow any individual to enter. However, they may protect it from unknown individuals, most often conspecifics. This type of aggression depends on the defined boundaries of the territory, marked by urinary or facial deposits, and, most importantly, on the perceived threat to the cat.

Territorial aggression occurs when the cat perceives a threat to the integrity of its living space. The threat phase begins before an intrusion and involves vocalizations such as growling and hissing, as well as urinary marking. If the intruder persists, the consummatory phase follows, involving charges, scratches, and bites intended to repel the individual beyond the territory's boundaries. The resolution phase involves exploration and territorial marking after the intruder has left.

This form of aggression targets only unknown or unauthorized individuals. A cat can live in a group with familiar individuals that it has identified and frequently marked with facial deposits. These familiar companions do not trigger territorial aggression.

The purpose of *maternal aggression* is to protect kittens. It can occur in the presence of a litter but may also manifest during a false pregnancy. The threat phase consists of hissing but is often brief. The consummatory phase involves a rapid attack, typically a charge aimed at driving the intruder away, similar to territorial aggression. Resolution occurs only after the intruder has retreated, characterized by the mother returning to the nest and licking her kittens.

In both maternal and territorial aggression, the sequence may be reduced to a few hisses and an aborted chase if the intruder quickly moves away. Although belonging to different social groups, the individuals involved can establish symbolic communication, limited to the early signs of the aggressive sequence. It is akin to the cat putting up a "private" sign to safeguard its territory.

Irritable aggression is very common. It arises when a cat feels threatened, and its tolerance threshold is exceeded. The control exerted by the prefrontal cortex is lower compared to previous types of aggression, and emotions partially take over. Irritable aggression occurs in contexts of pain, fear, frustration, and excitement. The threat phase is often characterized by hissing, attempts to flee, or distancing through paw swipes. If the perceived threat does not dissipate, the cat may escalate to scratching or biting. In confined settings, this type of aggression can quickly evolve into uncontrolled emotional aggression.

Resolution is achieved when the cat feels secure or if the trigger moves away. This type of aggression is well-known to veterinarians

Table 1.5 Classification of aggressions in cats

Name of the sequence	Sub-types	Offen-sive/ Defensive	Target	Trigger	Intention	Level of control
Non-aggressive injurious sequences						
Predation		Offensive	Prey	Prey	Kill to consume	
Vulnerant sequence during play		Offensive	Play partner	Excitement	Play	
Aggressions						
Maternal aggression		Defensive	Outsider to the group	Approach of the kittens	Defend her kittens + repel the threat	
Territorial aggression		Defensive	Outsider to the group	Territorial intrusion	Defend against intrusions into the territory + repel the threat	
Irritation aggression	Frus-tration Excite-ment Fear Pain	Defensive	Indifferent but greater tolerance toward familiars	Approach of the individual	Repel the threat	
Uncontrolled emotional aggression		Defensive	More frequent with strangers but possible with anyone	Approach of the individual	Repel the threat	
Idiopathic aggression		Defensive or offensive	Indifferent	No specific rule	Unknown	

who handle cats experiencing fear or pain during clinical examinations on the consultation table.

Such aggression is also frequently observed during play sessions that escalate due to the cat's heightened frustration.

Uncontrolled emotional aggression, referred to as fear aggression in Moyer's classification, can be misleading when compared to irritation aggression in a context of fear. This type of aggression is marked by a complete loss of control over the prefrontal cortex, triggered by highly intense emotions. It occurs when a cat perceives itself in a confined situation where escape, from its perspective, is impossible. It is crucial to emphasize that the cat's perception is what matters, not the actual circumstances.

The threat phase is very brief, characterized by mydriasis and sometimes hissing and piloerection. The attack is extremely fast, uncontrolled, and involves multiple severe bites and scratches. This aggression may also be accompanied by autonomic manifestations such as urination, defecation, and anal gland expression.

Finally, we must describe certain types of *idiopathic aggression*. These occur in various situations, rare but severe, and are almost always offensive. The first case involves aggression where the trigger is completely unknown. This form of aggression is particularly observed in states of derealization. In such cases, even if the sequence resembles a known one, a return to equilibrium is not always present. For this uncompleted sequence, it is difficult to define an objective that aligns it with any established classification.

In other cases, the sequence is entirely incoherent. Such incoherence often reflects a broader dysfunction in brain function. This may involve aggression that seems to begin with a very high level of intensity, with minimal threat signals.

Moreover, some instrumentalized aggressions cannot be linked to any identifiable sequence from which they might originate. The absence of emotions and the rapidity of execution make any analysis impossible.

1.3.3 Cognition in Cats

This section aims to describe the primary cognitive abilities of cats without claiming to be exhaustive. It highlights abilities relevant for therapeutic approaches and demonstrates that cats cannot simply be reduced to individuals learning solely through conditioning. Our understanding of feline cognition continues to evolve, with research uncovering increasingly complex and sophisticated feline cognitive capabilities. Feline cognition remains less studied than canine cognition, and many studies use canine cognition as a basis for comparison.

1.3.3.1 Non-social Cognition in Cats

Object permanence refers to the ability of an animal to mentally represent an object even when it is no longer visible. This skill is useful in hunting, for instance, when prey disappears behind an obstacle. In human children, this ability develops between 18 and 24 months. Cats succeed in tests involving visible displacement, meaning they search for an object behind the last obstacle where it was seen (Fiset and Doré 2006). This capacity emerges during the socialization period (Dumas 1992). Thus, the cat's representation of the object is stored in memory. However, results are more variable in tests of invisible displacement.

In these tests, an object is placed inside a container, and the container is then moved behind an obstacle. A second, empty container that the cat cannot distinguish is already present behind the obstacle. When the empty container is shown to the cat, it passes the test if it searches for the object behind the obstacle. In most studies, cats fail this test (Goulet et al. 1996), though they succeed when the object is a simulated prey item (Dumas 1992). These results confirm that the feline representation of the world is difficult for us to access and that the tests used are not always appropriate for this species.

Working memory plays a significant role in object permanence tests. Studies involving delayed response tasks have shown that cats' working memory can last up to 1 min, though it typically averages around 30 s in most studies. Performance declines sharply within the first 10–15 s (Fiset and Doré 2006; Goulet et al. 1996).

Regarding **long-term memory**, cats have successfully completed delayed visual matching tests with delays of up to 10 min (Okujava et al. 2005). However, they are significantly less proficient than dogs in both working memory and long-term memory tasks (Doré et al. 1996).

Understanding physical causality refers to the comprehension of the principle: if A implies B and B implies C, then A implies C. This understanding provides numerous adaptive advantages, including the ability to reuse acquired knowledge in new learning situations. Tests have been conducted with cats using strings either obviously or non-obviously connected to food. The way cats pulled the strings did not differ significantly from random behavior. However, the test may not be suited to feline behavior. The authors of the study suggested that pulling a string might be rewarding in itself, thereby biasing the results (Whitt et al. 2009).

No studies have definitively determined whether cats can quantify time. However, they appear capable of discriminating between quantities (Pisa and Agrillo 2009).

1.3.3.2 Social Cognition in Cats

Social cognition focuses on cognitive abilities involving communication with other individuals and learning through observation of others.

Interspecific Social Cognition in Cats

It has been proven that dogs exhibit remarkable sensory subtleties in interpreting human gestures, attention levels, and emotions (Call et al. 2003; Gácsi et al. 2004). Cats, too, are capable of understanding human gestures and will look toward humans for assistance when searching for hidden food (Miklósi et al. 2005; Pongrácz et al. 2019). However, research is inconsistent regarding cats seeking human help by looking at them. In some studies, cats persist in tasks without turning to humans, which has partly been explained by the notion that cats may not use gaze for solving physical problems but instead use it in cases of fear or when seeking affiliative contact (Merola et al. 2015).

Additionally, cats can recognize their owner's voice and respond primarily with facial expressions, head movements, and ear movements rather than vocalizations (Saito and Shinozuka 2013). This behavior might contribute to the communication challenges humans face with cats, as the latter respond with subtle signals that humans often overlook. Cats are also capable of adapting their vocalizations for humans. Studies reveal that cats modify their purring to alter the meaning of their message. For instance, when vocalizing to request food, cats emit a high-frequency meow. Some researchers compare this frequency to the distress cries or wails produced by human infants, suggesting it might be a trait selected through domestication (McComb et al. 2009).

Ainsworth's Strange Situation Test (Ainsworth 1979), traditionally used to evaluate attachment to another individual, has been adapted for cats. Studies have shown that cats behave differently in the presence of a stranger, their owner, or when left alone. They display attachment behaviors toward their owner, such as allomarking, following, and playing, which they do not exhibit with strangers (Edwards et al. 2007). However, when the Strange Situation Test was revised to counterbalance the order of separation and reunion sequences between the cat, its owner, and the stranger (Potter and Mills 2015), the findings were less definitive. While cats vocalized more when their owner, rather than a stranger, left them with the other individual, no additional evidence supported the interpretation that the bond between a cat and its owner fulfills the criteria for secure attachment. Thus, adult cats are generally quite autonomous in their social relationships and do not rely on an attachment figure for a sense of security.

Regarding human emotions, cats turn to their owner in unfamiliar situations and adjust their behavior based on the owner's reactions (Merola et al. 2015). Similarly, when a familiar human is in a depressive mood, cats increase facial and flank marking behaviors directed at them (Rieger and Turner 1999).

Imitative learning refers to the phenomenon in which the behavior of a model individual induces identical behavior in an observing individual. The observer replicates the model's response. This type of learning is instrumental in transmitting behaviors from one generation to the next. For example, a kitten learns early behaviors by mimicking its mother. This can be beneficial, such as learning cleanliness, but can also perpetuate negative behaviors, such as fear responses to strangers if the mother exhibits social phobia.

Observational learning, on the other hand, is a process in which the behavior of the observing individual is modified by watching a model perform the same behavior.

In dogs, the "do as I do" paradigm has been employed to demonstrate their significant ability to imitate actions (Fugazza and Miklósi 2015). In this paradigm, a human performs an action, then gives the command "do it." If the dog understands the pre-taught meaning of the command, it reproduces the action performed by the human. Dogs have demonstrated the ability to replicate actions even after delays of up to 24 h. For instance, when a human picks up a ball with their hands, the dog retrieves it with its mouth. These studies also suggest the dog's ability to form mental representations of human actions (Fugazza et al. 2016; Fugazza and Miklósi 2015).

A similar study was conducted with a cat described as particularly food-motivated. This cat successfully reproduced actions such as turning, standing on its hind legs, touching an object, opening a drawer, biting a string, and lying down after learning the "do as I do" paradigm. Some of these actions were novel to the cat, demonstrating its ability to generalize the command and understand the need to imitate its owner. The cat displayed the capacity to mentally represent the human body parts in relation to its own and flexibly adapt its behavior to mirror that of the human (Fugazza et al. 2021).

These studies suggest that cats possess higher socio-cognitive abilities than might be expected from a non-social species. However, it is important to note that these studies focus exclusively on cats socialized with humans.

▪ Intraspecific Social Cognition in Cats

Just as cats can learn through observing humans, they can also learn from their feline counterparts. Kittens that watch their mother press a lever to obtain food replicate this behavior, whereas they do not imitate the behavior if they observe an unrelated adult performing the same task. This capacity for observational learning or imitation among conspecifics is also observed in familiar adult cats (Herbert and Harsh 1944).

Furthermore, studies examining relationships among cats demonstrate their ability to engage in cooperation and joint attention. Social groups of cats are notably characterized by cooperative behaviors among related adult females in raising kittens from birth (Bradshaw 2016). For example, adult females, referred to as "allomothers," assist the mother in caring for the kittens, particularly when relocating them to a new nesting site.

Summary of ▶ Sect. 1.3

The domestication of cats is significantly more recent than that of dogs and occurred passively through cooperative interactions between the two species.

The domestic cat is a non-social species with solitary and territorial tendencies, yet it is more capable of living in groups within a defined space compared to most other feline species (*Felidae* sp.). Its status as prey makes it particularly vulnerable to disorders of the fear axis, as it is highly prone to sensitization. Conversely, its predatory nature predisposes it to disorders related to self-control.

Predation is not considered aggression. Similarly, injurious sequences during play are not classified as aggressive behaviors, despite their potential to cause significant harm.

Aggression, on the other hand, can be described based on the emotions that trigger it, using the modified Moyer classification. Each sequence reflects specific contextual and emotional triggers. Aggressions can be categorized as territorial, maternal, irritation-induced, emotionally uncontrolled, or idiopathic. The degree of cortical control varies, with minimal control in cases of emotionally uncontrolled aggression, where inflicted injuries can be severe.

Despite being less studied than dogs, feline cognition reveals complex capabilities, including certain forms of social learning.

Summary of This Chapter

More than the domestic cat, it is the companion cat that is the focus of this work—an animal subject to very recent selection pressures, a species that is non-social by phylogeny and behavior, yet whose individuals are not asocial.

The entire repertoire of this cat's behaviors is influenced by hormones, learning, context, and genetics, all of which regulate brain function. These influences can lead to pathological psychological states resulting from dysregulated elementary processes.

These pathological states, which are the focus of veterinary psychiatry, result in a loss of adaptability in behavioral responses and cause suffering. They are spontaneously irreversible.

The domestic cat organizes its territory by using markings to delineate its various activity zones. Research into feline cognition highlights advanced capabilities, demonstrating that a cat's learning cannot be reduced to a mere sequence of conditioned responses.

References

Ainsworth MS (1979) Infant–mother attachment. Am Psychol 34(10):932–937. https://doi.org/10.1037/0003-066X.34.10.932

Amano T, Unal CT, Paré D (2010) Synaptic correlates of fear extinction in the amygdala. Nat Neurosci 13(4):Article 4. https://doi.org/10.1038/nn.2499

Andics A, Miklósi Á (2018) Neural processes of vocal social perception: dog-human comparative fMRI studies. Neurosci Biobehav Rev 85:54–64. https://doi.org/10.1016/j.neubiorev.2017.11.017

Bari A, Robbins TW (2013) Inhibition and impulsivity: Behavioral and neural basis of response control. Prog Neurobiol 108:44–79. https://doi.org/10.1016/j.pneurobio.2013.06.005

Beedie C, Terry P, Lane A (2005) Distinctions between emotion and mood. Cognit Emot 19(6):847–878. https://doi.org/10.1080/02699930541000057

Bengtson S (2002) Origins and early evolution of predation. Paleontol Soc Papers 8:289–318. https://doi.org/10.1017/S1089332600001133

Bennett V, Gourkow N, Mills DS (2017) Facial correlates of emotional behaviour in the domestic cat (Felis catus). Behav Process 141:342–350. https://doi.org/10.1016/j.beproc.2017.03.011

Bentosela M, Barrera G, Jakovcevic A, Elgier AM, Mustaca AE (2008) Effect of reinforcement, reinforcer omission and extinction on a communicative response in domestic dogs (Canis familiaris). Behav Process 78(3):464–469. https://doi.org/10.1016/j.beproc.2008.03.004

Berridge KC (2018) Evolving concepts of emotion and motivation. Front Psychol 9:1647. https://doi.org/10.3389/fpsyg.2018.01647

Biben M (1979) Predation and predatory play behaviour of domestic cats. Anim Behav 27:81–94. https://doi.org/10.1016/0003-3472(79)90129-5

Bitterman ME (2006) Classical conditioning since Pavlov. Rev Gen Psychol 10(4):365–376. https://doi.org/10.1037/1089-2680.10.4.365

Bos RVD, Meijer MK, Van Renselaar JP, Van Der Harst JE, Spruijt BM (2003) Anticipation is differently expressed in rats (Rattus norvegicus) and domestic cats (Felis silvestris catus) in the same Pavlovian conditioning paradigm. Behav Brain Res 141(1):83–89. https://doi.org/10.1016/S0166-4328(02)00318-2

Bradshaw JWS (2006) The evolutionary basis for the feeding behavior of domestic dogs (Canis familiaris) and cats (Felis catus). J Nutr 136(7):1927S–1931S. https://doi.org/10.1093/jn/136.7.1927S

Bradshaw JWS (2016) Sociality in cats: a comparative review. J Vet Behav 11:113–124. https://doi.org/10.1016/j.jveb.2015.09.004

Broome MR, Saunders KEA, Harrison PJ, Marwaha S (2015) Mood instability: significance, definition and measurement. Br J Psychiatry 207(4):283–285. https://doi.org/10.1192/bjp.bp.114.158543

Brown JL (2006) Comparative endocrinology of domestic and nondomestic felids. Theriogenology 66(1):25–36. https://doi.org/10.1016/j.theriogenology.2006.03.011

Bunford N, Csibra B, Peták C, Ferdinandy B, Miklósi Á, Gácsi M (2019) Associations among behavioral inhibition and owner-rated attention, hyperactivity/impulsivity, and personality in the domestic dog (Canis familiaris). J Comp Psychol 133(2):233–243. https://doi.org/10.1037/com0000151

Čabraja J (2021) Nonverbal communication: animals and humans. Sveučilište Josipa Jurja Strossmayera u Osijeku, Filozofski fakultet. https://urn.nsk.hr/urn:nbn:hr:142:147736

Cafazzo S, Bonanni R, Natoli E (2019) Neutering effects on social behaviour of urban unowned free-roaming domestic cats. Animals 9(12):1105. https://doi.org/10.3390/ani9121105

Call J, Bräuer J, Kaminski J, Tomasello M (2003) Domestic dogs (Canis familiaris) are sensitive to the attentional state of humans. J Comp Psychol 117(3):257–263. https://doi.org/10.1037/0735-7036.117.3.257

Cannon WB (1929) Organization for physical homeostasis. Physiol Rev 9(3):399–431. https://doi.org/10.1152/physrev.1929.9.3.399

Caro TM (1981) Predatory behaviour and social play in kittens. Behaviour 76(1–2):1–24. https://doi.org/10.1163/156853981X00013

Caruso A, Marconi MA, Scattoni ML, Ricceri L (2022) Ultrasonic vocalizations in laboratory mice: Strain, age, and sex differences. Genes Brain Behav 21(5):e12815. https://doi.org/10.1111/gbb.12815

Casey RA, Bradshaw JWS (2008) The effects of additional socialisation for kittens in a rescue centre on their behaviour and suitability as a pet. Appl Anim Behav Sci 114(1–2):196–205. https://doi.org/10.1016/j.applanim.2008.01.003

Castañeda I, Forin-Wiart M-A, Pisanu B, Bouillane D, De Lacoste N (2023) Spatiotemporal and individual patterns of domestic cat (Felis catus) hunting behaviour in France. Animals 13(22):3507. https://doi.org/10.3390/ani13223507

Cook P (2017) Studying dog emotion beyond expression and without concern for feeling. Anim Sentien 2(14). https://doi.org/10.51291/2377-7478.1282

Correia-Caeiro C, Guo K, Mills DS (2020) Perception of dynamic facial expressions of emotion between dogs and humans. Anim Cogn 23(3):465–476. https://doi.org/10.1007/s10071-020-01348-5

Craig W (1917) Appetites and Aversions as Constituents of Instincts. Proc. Natl. Acad Sci U.S.A. 3:685–688. https://doi.org/10.1073/pnas.3.12.685

Curtis TM, Knowles RJ, Crowell-Davis SL (2003) Influence of familiarity and relatedness on proximity and allogrooming in domestic cats (Felis catus). Am J Vet Res 64(9):1151–1154. https://doi.org/10.2460/ajvr.2003.64.1151

Darmaillacq A-S, Lévy F (2019) Éthologie animale: Une approche biologique du comportement. De Boeck Superieur

De Mouzon C, Leboucher G (2023) Multimodal communication in the human–cat relationship: a Pilot study. Animals 13(9):1528. https://doi.org/10.3390/ani13091528

De Mouzon C, Gonthier M, Leboucher G (2023) Discrimination of cat-directed speech from human-directed speech in a population of indoor companion cats (Felis catus). Anim Cogn 26(2):611–619. https://doi.org/10.1007/s10071-022-01674-w

DePorter TL, Elzerman AL (2019) Common feline problem behaviors: destructive scratching. J Feline Med Surg 21(3):235–243. https://doi.org/10.1177/1098612X19831205

Dickman CR, Newsome TM (2015) Individual hunting behaviour and prey specialisation in the house cat Felis catus: implications for conservation and management. Appl Anim Behav Sci 173:76–87. https://doi.org/10.1016/j.applanim.2014.09.021

Doré FY, Fiset S, Goulet S, Dumas M-C, Gagnon S (1996) Search behavior in cats and dogs: interspecific differences in working memory and spatial cognition. Anim Learn Behav 24(2):142–149. https://doi.org/10.3758/BF03198962

Doty RL, Dunbar I (1974) Attraction of beagles to conspecific urine, vaginal and anal sac secretion odors. Physiol Behav 12(5):825–833. https://doi.org/10.1016/0031-9384(74)90020-1

Driscoll CA, Menotti-Raymond M, Roca AL, Hupe K, Johnson WE, Geffen E, Harley EH, Delibes M, Pontier D, Kitchener AC, Yamaguchi N, O'Brien SJ, Macdonald DW (2007) The near eastern origin of cat domestication. Science 317(5837):519–523. https://doi.org/10.1126/science.1139518

Dumas C (1992) Object permanence in cats (Felis catus): an ecological approach to the study of invisible displacements. J Compar Psychol (Washington, DC: 1983) 106(4):404–410. https://doi.org/10.1037/0735-7036.106.4.404

Edwards C, Heiblum M, Tejeda A, Galindo F (2007) Experimental evaluation of attachment behaviors in owned cats. J Vet Behav 2(4):119–125. https://doi.org/10.1016/j.jveb.2007.06.004

Ellis SL (2018) Recognising and assessing feline emotions during the consultation: history, body language and behaviour. J Feline Med Surg 20(5):445–456. https://doi.org/10.1177/1098612X18771206

Ellis SLH, Swindell V, Burman OHP (2015) Human classification of context-related vocalizations emitted by familiar and unfamiliar domestic cats: an exploratory study. Anthrozoös 28(4):625–634. https://doi.org/10.1080/08927936.2015.1070005

Feldman HN (1994) Methods of scent marking in the domestic cat. Can J Zool 72(6):1093–1099. https://doi.org/10.1139/z94-147

Fenouillet F (2017) La motivation—3e éd. Dunod

Fiset S, Doré FY (2006) Duration of cats' (Felis catus) working memory for disappearing objects. Anim Cogn 9(1):62–70. https://doi.org/10.1007/s10071-005-0005-4

Fiset S, Beaulieu C, Landry F (2003) Duration of dogs' (Canis familiaris) working memory in search for disappearing objects. Anim Cogn 6(1):1–10. https://doi.org/10.1007/s10071-002-0157-4

Fugazza C, Miklósi Á (2015) Social learning in dog training: the effectiveness of the Do as I do method compared to shaping/clicker training. Appl Anim Behav Sci 171. https://doi.org/10.1016/j.applanim.2015.08.033

Fugazza C, Pogány Á, Miklósi Á (2016) Do as I … Did ! Long-term memory of imitative actions in dogs (Canis familiaris). Anim Cogn 19(2):263–269. https://doi.org/10.1007/s10071-015-0931-8

Fugazza C, Sommese A, Pogány Á, Miklósi Á (2021) Did we find a copycat? Do as I Do in a domestic cat (Felis catus). Anim Cogn 24(1):121–131. https://doi.org/10.1007/s10071-020-01428-6

Gácsi M, Miklósi Á, Varga O, Topál J, Csányi V (2004) Are readers of our face readers of our minds? Dogs (Canis familiaris) show situation-dependent recognition of human's attention. Anim Cogn

7(3):144–153. https://doi.org/10.1007/s10071-003-0205-8

Gartner MC, Weiss A (2013) Personality in felids: a review. Appl Anim Behav Sci 144(1–2):1–13. https://doi.org/10.1016/j.applanim.2012.11.010

Gimenez F, Stornelli MC, Tittarelli CM, Savignone CA, Dorna IV, De La Sota RL, Stornelli MA (2009) Suppression of estrus in cats with melatonin implants. Theriogenology 72(4):493–499. https://doi.org/10.1016/j.theriogenology.2009.04.004

Goldstone R (1998) Perceptual learning. Annu Rev Psychol 49:585–612. https://doi.org/10.1146/annurev.psych.49.1.585

Goodman J, Packard MG (2018) The role of the dorsal striatum in extinction: a memory systems perspective. Neurobiol Learn Mem 150:48–55. https://doi.org/10.1016/j.nlm.2018.02.028

Gosling L, Stavisky J, Dean R (2013) What is a feral cat?: variation in definitions may be associated with different management strategies. J Feline Med Surg 15(9):759–764. https://doi.org/10.1177/1098612X13481034

Goulet S, Doré FY, Lehotkay R (1996) Activation of locations in working memory in cats. Quart J Exp Psychol Sect B 49(1b):81–92. https://doi.org/10.1080/713932612

Gregg TR, Siegel A (2001) Brain structures and neurotansmitters regulating aggression in cats: implications for human aggression. Prog Neuro-Psychopharmacol Biol Psychiatry 25(1):91–140. https://doi.org/10.1016/S0278-5846(00)00150-0

Gröschl M (2009) The physiological role of hormones in saliva. BioEssays 31(8):843–852. https://doi.org/10.1002/bies.200900013

Hall ET (1963) A system for the notation of Proxemic behavior [1]. Am Anthropol 65(5):1003–1026. https://doi.org/10.1525/aa.1963.65.5.02a00020

Hall ET, Birdwhistell RL, Bock B, Bohannan P, Diebold AR, Durbin M, Edmonson MS, Fischer JL, Hymes D, Kimball ST, La Barre W, McClellan JE, Marshall DS, Milner GB, Sarles HB, Trager GL, Vayda AP (1968) Proxemics [and comments and replies]. Curr Anthropol 9(2/3):83–108. https://doi.org/10.1086/200975

Hart BL (1980) Objectionable urine spraying and urine marking in cats: evaluation of progestin treatment in gonadectomized males and females. J Am Vet Med Assoc 177(6):529–533

Hart B (1987) Stimulus and hormonal determinants of flehmen behavior in cats. Horm Behav 21(1):44–52. https://doi.org/10.1016/0018-506X(87)90029-8

Hasegawa M, Ohtani N, Ohta M (2014) Dogs' body language relevant to learning achievement. Animals 4(1):45–58. https://doi.org/10.3390/ani4010045

Heath S (2018) Understanding feline emotions: … and their role in problem behaviours. J Feline Med Surg 20(5):437–444. https://doi.org/10.1177/1098612X18771205

Heffner HE (1983) Hearing in large and small dogs: absolute thresholds and size of the tympanic membrane. Behav Neurosci 97(2):310–318. https://doi.org/10.1037/0735-7044.97.2.310

Herbert MJ, Harsh CM (1944) Observational learning by cats. J Comp Psychol 37(2):81–95. https://doi.org/10.1037/h0062414

Horn JA, Mateus-Pinilla N, Warner RE, Heske EJ (2011) Home range, habitat use, and activity patterns of free-roaming domestic cats. J Wildl Manag 75(5):1177–1185. https://doi.org/10.1002/jwmg.145

Horwitz DF (2019) Common feline problem behaviors: urine spraying. J Feline Med Surg 21(3):209–219. https://doi.org/10.1177/1098612X19831203

Houpt KA (2018) Domestic animal behavior for veterinarians and animal scientists, 6th edn. John Wiley & Sons, Inc., Hoboken, NJ

House A (1988) Mood disorders in the physically ill—problems of definition and measurement. J Psychosom Res 32(4–5):345–353. https://doi.org/10.1016/0022-3999(88)90017-7

Hu Y, Hu S, Wang W, Wu X, Marshall FB, Chen X, Hou L, Wang C (2014) Earliest evidence for commensal processes of cat domestication. Proc Natl Acad Sci 111(1):116–120. https://doi.org/10.1073/pnas.1311439110

Hull KD, Jeckel S, Williams JM, Ciavaglia SA, Webster LMI, Fitzgerald E, Chang Y-M, Martineau HM (2022) Fox (*Vulpes vulpes*) involvement identified in a series of cat carcass mutilations. Vet Pathol 59(2):299–309. https://doi.org/10.1177/03009858211052661

Humphrey T, Proops L, Forman J, Spooner R, McComb K (2020) The role of cat eye narrowing movements in cat–human communication. Sci Rep 10(1):16503. https://doi.org/10.1038/s41598-020-73426-0

Ioffe ME (2004) Brain mechanisms for the formation of new movements during learning: the evolution of classical concepts. Neurosci Behav Physiol 34(1):5–18. https://doi.org/10.1023/B:NEAB.0000003241.12053.47

Jacob-Dazarola R, Ortíz Nicolás JC, Bayona LC (2016) Behavioral measures of emotion. In: Emotion Measurement. Elsevier, pp 197–223. https://doi.org/10.1016/B978-0-12-821124-3.00007-7

James W (1994) The physical basis of emotion. Psychol Rev 101(2):205–210. https://doi.org/10.1037/0033-295X.101.2.205

Janetos AC (1982) Active foragers vs. sit-and-wait predators: a simple model. J Theor Biol 95(2):381–385. https://doi.org/10.1016/0022-5193(82)90252-1

Jones FN, Skinner BF (1939) The behavior of organisms: an experimental analysis. Am J Psychol. https://doi.org/10.2307/1416495

Kamin LJ, Brimer CJ (1963) The effects of intensity of conditioned and unconditioned stimuli on a conditioned emotional response. Canad J Psychol/Revue canadienne de psychologie 17(2):194–200. https://doi.org/10.1037/h0101671

Kaminski J, Call J, Tomasello M (2004) Body orientation and face orientation: two factors controlling apes? begging behavior from humans. Anim Cogn

7(4):216–223. https://doi.org/10.1007/s10071-004-0214-2

Kasbaoui N, Bienboire-Frosini C, Monneret P, Leclercq J, Descout E, Cozzi A, Pageat P (2022) Influencing elimination location in the domestic cat: a Semiochemical approach. Animals 12(7):896. https://doi.org/10.3390/ani12070896

Kellman PJ, Massey CM (2013) Chapter four—perceptual learning, cognition, and expertise. In: Ross BH (ed) Psychology of learning and motivation, vol 58. Academic Press, pp 117–165. https://doi.org/10.1016/B978-0-12-407237-4.00004-9

Kenshalo DR, Duncan DG, Weymark C (1967) Thresholds for thermal stimulation of the inner thigh, footpad, and face of cats. J Comp Physiol Psychol 63(1):133–138. https://doi.org/10.1037/h0024177

Lei W, Ravoninjohary A, Li X, Margolskee RF, Reed DR, Beauchamp GK, Jiang P (2015) Functional analyses of bitter taste receptors in domestic cats (Felis catus). PLoS One 10(10):e0139670. https://doi.org/10.1371/journal.pone.0139670

Leventhal H, Scherer K (1987) The relationship of emotion to cognition: a functional approach to a semantic controversy. Cognit Emot 1(1):3–28. https://doi.org/10.1080/02699938708408361

Levitis DA, Lidicker WZ, Freund G (2009) Behavioural biologists do not agree on what constitutes behaviour. Anim Behav 78(1):103–110. https://doi.org/10.1016/j.anbehav.2009.03.018

Lewis MB, Bowler PJ (2009) Botulinum toxin cosmetic therapy correlates with a more positive mood. J Cosmet Dermatol 8(1):24–26. https://doi.org/10.1111/j.1473-2165.2009.00419.x

Li X, Li W, Wang H, Bayley DL, Cao J, Reed DR, Bachmanov A, Huang L, Legrand-Defretin V, Beauchamp GK, Brand JG (2007) Cats Lack a Sweet Taste Receptor1,2,3

Li W, Ma L, Yang G, Gan W-B (2017) REM sleep selectively prunes and maintains new synapses in development and learning. Nat Neurosci 20(3):Article 3. https://doi.org/10.1038/nn.4479

Liberg O (1984) Food habits and prey impact by Feral and House-based domestic cats in a rural area in Southern Sweden. J Mammal 65(3):424–432. https://doi.org/10.2307/1381089

Liberg O, Sandell M, Pontier D, Natoli E (2000) Density spatial organisation and reproductive tactics in the domestic cat and other felids

Loss SR, Will T, Marra PP (2013) The impact of free-ranging domestic cats on wildlife of the United States. Nat Commun 4(1):1396. https://doi.org/10.1038/ncomms2380

Lowe SE, Bradshaw JWS (2002) Responses of pet cats to being held by an unfamiliar person, from weaning to three years of age. Anthrozoös 15(1):69–79. https://doi.org/10.2752/089279302786992702

Macdonald DW, Yamaguchi N, Kerby G (2000) Group-living in the domestic cat: its sociobiology and epidemiology. The domestic cat: the biology of its behaviour 2:95–118

Malandain E, Rault D, Froment E, Baudon S, Desquilbet L, Begon D, Chastant-Maillard S (2011) Follicular growth monitoring in the female cat during estrus. Theriogenology 76(7):1337–1346. https://doi.org/10.1016/j.theriogenology.2011.06.002

Maples-Keller JL, Rauch SAM (2020) Habituation. In: Clinical handbook of fear and anxiety: maintenance processes and treatment mechanisms. American Psychological Association, pp 249–263. https://doi.org/10.1037/0000150-014

Marinier SL, Alexander AJ, Waring GH (1988) Flehmen behaviour in the domestic horse: discrimination of conspecific odours. Appl Anim Behav Sci 19(3–4):227–237. https://doi.org/10.1016/0168-1591(88)90003-2

Mariti C, Falaschi C, Zilocchi M, Fatjó J, Sighieri C, Ogi A, Gazzano A (2017) Analysis of the intraspecific visual communication in the domestic dog (Canis familiaris): a pilot study on the case of calming signals. J Vet Behav 18:49–55. https://doi.org/10.1016/j.jveb.2016.12.009

Marwaha S, He Z, Broome M, Singh SP, Scott J, Eyden J, Wolke D (2014) How is affective instability defined and measured? A systematic review. Psychol Med 44(9):1793–1808. https://doi.org/10.1017/S0033291713002407

Mäses M, Wascher CAF (2023). Assessing cats' (Felis catus) sensitivity to human pointing gestures. J Comp Psychol 137:38–44. https://doi.org/10.1037/com0000338

Maslow AH (1943) A theory of human motivation. Psychol Rev 50(4). https://doi.org/10.1037/h0054346

McComb K, Taylor AM, Wilson C, Charlton BD (2009) The cry embedded within the purr. Curr Biol 19(13):R507–R508. https://doi.org/10.1016/j.cub.2009.05.033

McCune S, Stevenson J, Fretwell L, Thompson A, Mills DS (2008) Ageing does not significantly affect performance in a spatial learning task in the domestic cat (Felis silvestris catus). Appl Anim Behav Sci 112(3–4):345–356. https://doi.org/10.1016/j.applanim.2007.08.013

McEwen BS, Wingfield JC (2010) What is in a name? Integrating homeostasis, allostasis and stress. Horm Behav 57(2):105–111. https://doi.org/10.1016/j.yhbeh.2009.09.011

McKeown D, Luescher A, Machum M (1988) The problem of destructive scratching by cats. Canadian Vet J = La Revue Veterinaire Canadienne 29(12):1017–1018

Mclaren IPL, Mackintosh NJ (2002) Associative learning and elemental representation: II. Generalization and discrimination. Anim Learn Behav 30(3):177–200. https://doi.org/10.3758/BF03192828

Mendl M, Brooks J, Basse C, Burman O, Paul E, Blackwell E, Casey R (2010) Dogs showing separation-related behaviour exhibit a 'pessimistic' cognitive bias. Curr Biol 20(19):R839–R840. https://doi.org/10.1016/j.cub.2010.08.030

Mengoli M, Mariti C, Cozzi A, Cestarollo E, Lafont-Lecuelle C, Pageat P, Gazzano A (2013) Scratching behaviour and its features: a questionnaire-based study in an Italian sample of domestic cats. J Feline Med Surg 15(10):886–892. https://doi.org/10.1177/1098612X13481468

Merola I, Lazzaroni M, Marshall-Pescini S, Prato-Previde E (2015) Social referencing and cat–human communication. Anim Cogn 18(3):639–648. https://doi.org/10.1007/s10071-014-0832-2

Miklosi A (2016) Dog behaviour, evolution, and cognition ($ {number}nd édition). Oxford biology. Oxford university press, Oxford (GB)

Miklósi Á, Pongrácz P, Lakatos G, Topál J, Csányi V (2005) A comparative study of the use of visual communicative signals in interactions between dogs (Canis familiaris) and humans and cats (Felis catus) and humans. J Comp Psychol 119(2):179–186. https://doi.org/10.1037/0735-7036.119.2.179

Mills D (2005) Pheromonatherapy: theory and applications. In Pract 27(7):368–373. https://doi.org/10.1136/inpract.27.7.368

Mitsui S, Yamamoto M, Nagasawa M, Mogi K, Kikusui T, Ohtani N, Ohta M (2011) Urinary oxytocin as a noninvasive biomarker of positive emotion in dogs. Horm Behav 60(3):239–243. https://doi.org/10.1016/j.yhbeh.2011.05.012

Moody CM, Mason GJ, Dewey CE, Niel L (2020) Getting a grip: cats respond negatively to scruffing and clips. Vet Rec 186(12):385–385. https://doi.org/10.1136/vr.105261

Morris PH, Doe C, Godsell E (2008) Secondary emotions in non-primate species? Behavioural reports and subjective claims by animal owners. Cognit Emot 22(1):3–20. https://doi.org/10.1080/02699930701273716

Mouzon CD, Gilbert C, Di-Stasi R, Leboucher G (2022) How's my kitty? Acoustic parameters of cat-directed speech in human-cat interactions. Behav Process 203:104755. https://doi.org/10.1016/j.beproc.2022.104755

Moyer KE (1971) A preliminary physiological model of aggressive behavior. In: Eleftheriou BE, Scott JP (eds) The physiology of aggression and defeat. Springer US, pp 223–263. https://doi.org/10.1007/978-1-4684-1932-0_7

Nájera F, Sánchez-Cuerda S, López G, Del Rey-Wamba T, Rueda C, Vallverdú-Coll N, Panadero J, Palacios MJ, López-Bao JV, Jiménez J (2019) Lynx eats cat: disease risk assessment during an Iberian lynx intraguild predation. Eur J Wildl Res 65(3):39. https://doi.org/10.1007/s10344-019-1275-5

Nawroth C, McElligott AG (2017) Human head orientation and eye visibility as indicators of attention for goats (*Capra hircus*). PeerJ 5:e3073. https://doi.org/10.7717/peerj.3073

Nelson RJ (2005) An introduction to behavioral endocrinology, 3rd edn. Sinauer Associates, p xvii, 822, Sinauer, Sunderland, Mass.

Nuti V, Cantile C, Gazzano A, Sighieri C, Mariti C (2016) Pinch-induced behavioural inhibition (clipthesia) as a restraint method for cats during veterinary examinations: preliminary results on cat susceptibility and welfare. Anim Welf 25(1):115–123. https://doi.org/10.7120/09627286.25.1.115

Okujava V, Natishvili T, Mishkin M, Gurashvili T, Chipashvili S, Bagashvili T, Andronikashvili G, Kvernadze G (2005) One-trial visual recognition in cats. Acta Neurobiol Exp 65(2):205–211. https://doi.org/10.55782/ane-2005-1557

Overall KL, Rodan I, Beaver V, Carney H, Crowell-Davis S, Hird N, Kudrak S, Wexler-Mitchel E (2005) Feline behavior guidelines from the American Association of Feline Practitioners. J Am Vet Med Assoc 227(1):70–84. https://doi.org/10.2460/javma.2005.227.70

Pageat P (1995) Pathologie comportementale du chien. Editions du point vétérinaire Maison-Alfort

Pageat P, Gaultier E (2003) Current research in canine and feline pheromones. Vet Clin N Am Small Anim Pract 33(2):187–211. https://doi.org/10.1016/S0195-5616(02)00128-6

Panaman R (2010) Behaviour and ecology of free-ranging female farm cats (Felis catus L.). Z Tierpsychol 56(1):59–73. https://doi.org/10.1111/j.1439-0310.1981.tb01284.x

Papes F, Logan DW, Stowers L (2010) The Vomeronasal organ mediates interspecies defensive behaviors through detection of protein pheromone homologs. Cell 141(4):692–703. https://doi.org/10.1016/j.cell.2010.03.037

Pekel AY, Mülazımoğlu SB, Acar N (2020) Taste preferences and diet palatability in cats. Journal of Applied Animal Research 48:281–292. https://doi.org/10.1080/09712119.2020.1786391

Pereira LM, Owen-Smith N, Moleón M (2014) Facultative predation and scavenging by mammalian carnivores: seasonal, regional and intra-guild comparisons. Mammal Rev 44(1):44–55. https://doi.org/10.1111/mam.12005

Petersen SE, Posner MI (2012) The attention system of the human brain: 20 years after. Annu Rev Neurosci 35(1):73–89. https://doi.org/10.1146/annurev-neuro-062111-150525

Pickersgill O, Mills DS, Guo K (2023) Owners' beliefs regarding the emotional capabilities of their dogs and cats. Animals 13(5):820. https://doi.org/10.3390/ani13050820

Pisa PE, Agrillo C (2009) Quantity discrimination in felines: a preliminary investigation of the domestic cat (Felis silvestris catus). J Ethol 27(2):289–293. https://doi.org/10.1007/s10164-008-0121-0

Pongrácz P, Szapu JS, Faragó T (2019) Cats (Felis silvestris catus) read human gaze for referential information. Intelligence 74:43–52. https://doi.org/10.1016/j.intell.2018.11.001

Porritt F, Mansson R, Berry A, Cook N, Sibbald N, Nicklin S (2015) Validation of a short odour discrimination test for working dogs. Appl Anim Behav Sci 165:133–142. https://doi.org/10.1016/j.applanim.2014.11.021

Posner MI, Petersen SE (1990) The attention system of the human brain. Annu Rev Neurosci 13(1):25–42. https://doi.org/10.1146/annurev.ne.13.030190.000325

Potter A, Mills DS (2015) Domestic cats (felis silvestris catus) do not show signs of secure attachment to their owners. PLoS One 10(9):1–17. https://doi.org/10.1371/journal.pone.0135109

Pozza ME, Stella JL, Chappuis-Gagnon A-C, Wagner SO, Buffington CAT (2008) Pinch-induced behavioral inhibition ('clipnosis') in domestic cats. J Feline Med Surg 10(1):82–87. https://doi.org/10.1016/j.jfms.2007.10.008

Price JL, Drevets WC (2010) Neurocircuitry of mood disorders. Neuropsychopharmacology 35(1):Article 1. https://doi.org/10.1038/npp.2009.104

Proops L, McComb K (2010) Attributing attention: the use of human-given cues by domestic horses (Equus caballus). Anim Cogn 13(2):197–205. https://doi.org/10.1007/s10071-009-0257-5

Purugganan MD (2022) What is domestication? Trends Ecol Evol 37(8):663–671. https://doi.org/10.1016/j.tree.2022.04.006

Quaranta A, d'Ingeo S, Amoruso R, Siniscalchi M (2020) Emotion recognition in cats. Animals 10(7):1107. https://doi.org/10.3390/ani10071107

Quirk GJ, Mueller D (2008) Neural mechanisms of extinction learning and retrieval. Neuropsychopharmacology 33(1):56–72. https://doi.org/10.1038/sj.npp.1301555

Raihani G, González D, Arteaga L, Hudson R (2009) Olfactory guidance of nipple attachment and suckling in kittens of the domestic cat: inborn and learned responses. Dev Psychobiol 51(8):662–671. https://doi.org/10.1002/dev.20401

Rajagopalan A, Jinu KV, Sailesh KS, Mishra S, Reddy UK, Mukkadan JK (2017) Understanding the links between vestibular and limbic systems regulating emotions. J Nat Sci Biol Med 8(1):11–15. https://doi.org/10.4103/0976-9668.198350

Raz A, Buhle J (2006) Typologies of attentional networks. Nat Rev Neurosci 7(5):367–379. https://doi.org/10.1038/nrn1903

Rescorla RA (1967) Pavlovian conditioning and its proper control procedures. Psychol Rev 74(1):71–80. https://doi.org/10.1037/h0024109

Rieger G, Turner DC (1999) How depressive moods affect the behavior of singly living persons toward their cats. Anthrozoös 12(4):224–233. https://doi.org/10.2752/089279399787000066

Rodríguez-Torres R, Leyens JP, Pérez AR, Rodriguez VB, Quiles Del Castillo MN, Demoulin S, Cortés B (2005) The lay distinction between primary and secondary emotions: a spontaneous categorization? Int J Psychol 40(2):100–107. https://doi.org/10.1080/00207590444000221

Roney JR (2016) Theoretical frameworks for human behavioral endocrinology. Horm Behav 84:97–110. https://doi.org/10.1016/j.yhbeh.2016.06.004

Saito A, Shinozuka K (2013) Vocal recognition of owners by domestic cats (Felis catus). Anim Cogn 16(4):685–690. https://doi.org/10.1007/s10071-013-0620-4

Salamon A, Uccheddu S, Csepregi M, Miklósi Á, Gácsi M (2023) Dogs outperform cats both in their testability and relying on human pointing gestures: a comparative study. Sci Rep 13(1):17837. https://doi.org/10.1038/s41598-023-45008-3

Scheflen AE (1964) The significance of posture in communication systems. Psychiatry 27:316–331

Scott L, Florkiewicz BN (2023) Feline faces: Unraveling the social function of domestic cat facial signals. Behav Process 213:104959. https://doi.org/10.1016/j.beproc.2023.104959

Sherrington CS (2023) The integrative action of the nervous system, in: Scientific and Medical Knowledge Production, 1796–1918. Routledge, pp. 217–253

Short J, Smith A (1994) Mammal decline and recovery in Australia. J Mammal 75(2):288–297. https://doi.org/10.2307/1382547

Siegel A, Roeling TAP, Gregg TR, Kruk MR (1999) Neuropharmacology of brain-stimulation-evoked aggression. Neurosci Biobehav Rev 23(3):359–389. https://doi.org/10.1016/S0149-7634(98)00040-2

Siegford JM, Walshaw SO, Brunner P, Zanella AJ (2003) Validation of a temperament test for domestic cats. Anthrozoös 16(4):332–351. https://doi.org/10.2752/089279303786991982

Simons DJ (2000) Attentional capture and inattentional blindness. Trends Cogn Sci 4(4):147–155. https://doi.org/10.1016/S1364-6613(00)01455-8

Siniscalchi M, d'Ingeo S, Fornelli S, Quaranta A (2018) Lateralized behavior and cardiac activity of dogs in response to human emotional vocalizations. Sci Rep 8(1):Article 1. https://doi.org/10.1038/s41598-017-18417-4

Skinner BF (1953) Operant behavior. Sci Human Behav. https://doi.org/10.3390/ijerph8093528

Skinner BF (1984) The evolution of behavior. J Exp Anal Behav. https://doi.org/10.1901/jeab.1984.41-217

Stanton LA, Sullivan MS, Fazio JM (2015) A standardized ethogram for the felidae: a tool for behavioral researchers. Appl Anim Behav Sci 173:3–16. https://doi.org/10.1016/j.applanim.2015.04.001

Stedman HM, Mistretta CM, Bradley RM (1983) A quantitative study of cat epiglottal taste buds during development. J Anat 136(Pt 4):821–827

Steinberg RH, Reid M, Lacy PL (1973) The distribution of rods and cones in the retina of the cat (*Felis domesticus*). J Comp Neurol 148(2):229–248. https://doi.org/10.1002/cne.901480209

Sterling P (2012) Allostasis: a model of predictive regulation. Physiol Behav 106(1):5–15. https://doi.org/10.1016/j.physbeh.2011.06.004

Strain GM (2017) Hearing disorders in cats: classification, pathology and diagnosis. J Feline Med Surg 19(3):276–287. https://doi.org/10.1177/1098612X17695062

Szenczi P, Urrutia A, Hudson R, Bánszegi O (2022) Are you my mummy? Long-term olfactory memory of mother's body odour by offspring in the domestic cat. Anim Cogn 25(1):21–26. https://doi.org/10.1007/s10071-021-01537-w

Tavernier C, Ahmed S, Houpt KA, Yeon SC (2020) Feline vocal communication. J Vet Sci 21(1):e18. https://doi.org/10.4142/jvs.2020.21.e18

TenHouten WD (2017) From primary emotions to the Spectrum of affect: an evolutionary neurosociology of the emotions. In: Ibáñez A, Sedeño L, García AM (eds) Neuroscience and social science: the missing link. Springer International Publishing, pp 141–167. https://doi.org/10.1007/978-3-319-68421-5_7

Thompkins AM, Ramaiahgari B, Zhao S, Gotoor SSR, Waggoner P, Denney TS, Deshpande G, Katz JS (2018) Separate brain areas for processing human and dog faces as revealed by awake fMRI in dogs (Canis familiaris). Learn Behav 46(4):561–573. https://doi.org/10.3758/s13420-018-0352-z

Thompson RA (2019) Emotion dysregulation: a theme in search of definition. Dev Psychopathol 31(3):805–815. https://doi.org/10.1017/S0954579419000282

Thorndike E (1898a) Some experiments on animal intelligence. Science. https://doi.org/10.1126/science.7.181.818

Thorndike EL (1898b) Animal intelligence: an experimental study of the associative processes in animals. Psychol Rev Monogr Suppl. https://doi.org/10.1037/h0092987

Tinbergen N (1951) The Study of Instinct. Pygmalion Press, an imprint of Plunkett Lake Press.

Tinbergen N (1963) On aims and methods of ethology. Z Tierpsychol 20(4):410–433. https://doi.org/10.1111/j.1439-0310.1963.tb01161.x

Travnik IDC, Machado DDS, Gonçalves LDS, Ceballos MC, Sant'Anna AC (2020) Temperament in domestic cats: a review of proximate mechanisms, methods of assessment, its effects on human—cat relationships, and one welfare. Animals 10(9):1516. https://doi.org/10.3390/ani10091516

Trotier D (2011) Vomeronasal organ and human pheromones. Eur Ann Otorhinolaryngol Head Neck Dis 128(4):184–190. https://doi.org/10.1016/j.anorl.2010.11.008

Turner DC (2021) Unanswered questions and hypotheses about domestic cat behavior, ecology, and the cat–human relationship. Animals 11(10):2823. https://doi.org/10.3390/ani11102823

Turner DC, Bateson P (eds) (2013) The domestic cat: the biology of its behaviour, 3rd edn. Cambridge University Press. https://doi.org/10.1017/CBO9781139177177

von Uexküll J (2001) An introduction to Umwelt. Semiotica 2001(134). https://doi.org/10.1515/semi.2001.017

vonHoldt BM, Pollinger JP, Lohmueller KE, Han E, Parker HG, Quignon P, Degenhardt JD, Boyko AR, Earl DA, Auton A, Reynolds A, Bryc K, Brisbin A, Knowles JC, Mosher DS, Spady TC, Elkahloun A, Geffen E, Pilot M et al (2010) Genome-wide SNP and haplotype analyses reveal a rich history underlying dog domestication. Nature 464(7290):898–902. https://doi.org/10.1038/nature08837

Whitt E, Douglas M, Osthaus B, Hocking I (2009) Domestic cats (Felis catus) do not show causal understanding in a string-pulling task. Anim Cogn 12(5):739–743. https://doi.org/10.1007/s10071-009-0228-x

Zubin J, Spring B (1977) Vulnerability: a new view of schizophrenia. J Abnorm Psychol. https://doi.org/10.1037/0021-843X.86.2.103

Neuroanatomy and Neurophysiology of the Cat

Contents

2.1 The Brain – 58
2.1.1 Major Brain Regions – 58
2.1.2 The Brain's Unique Vascular Structures – 61

2.2 Neurons, Synapses and Neurotransmitters – 64
2.2.1 Neurons and Glial Cells – 64
2.2.2 The Electoral Activity of Neurons – 65
2.2.3 The Synapse and Its Functioning – 68
2.2.4 Neurotransmitters – 71

2.3 Neurophysiology of Behaviors – 82
2.3.1 Emotions – 83
2.3.2 Mood: From the Neurotransmitter Theory to the Receptor Theory – 90
2.3.3 Cognition – 91

References – 106

© The Author(s), under exclusive license to Springer Nature Switzerland AG 2026
S. Bleuer-Elsner et al., *Veterinary Psychiatry of the Cat*,
https://doi.org/10.1007/978-3-032-06260-4_2

2

The brain is the central organ responsible for the genesis and regulation of behaviors. It is composed of numerous specialized areas interconnected by various circuits that enable information exchange. Understanding the major centers of communication and exchange within the brain itself, as well as between the brain and the rest of the body, is an essential foundation for the practice of veterinary psychiatry.

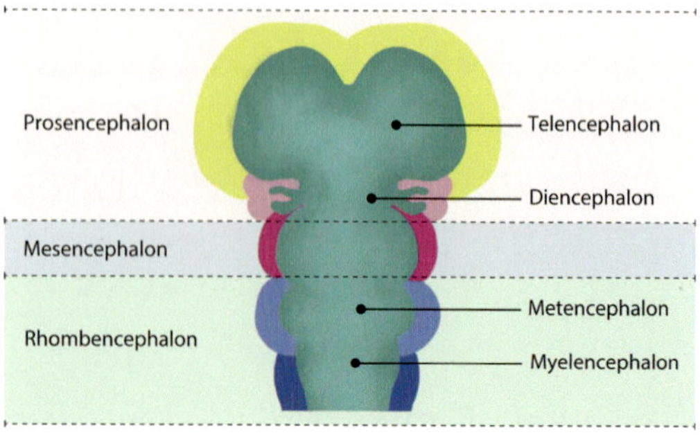

Fig. 2.1 Embryonic vesicles and anatomical regions. From its caudal to its rostral end, the encephalon comprises the rhombencephalon (itself divided into the metencephalon and myelencephalon), the mesencephalon, and the prosencephalon (itself divided into the telencephalon and diencephalon)

2.1 The Brain

2.1.1 Major Brain Regions

2.1.1.1 Principal Anatomical Areas of the Brain

The brain, or *encephalon*, can be divided into three major anatomical sections that develop from embryonic vesicles: the rhombencephalon, mesencephalon, and prosencephalon, which are visible between the 15th and 21st days of gestation (De Morais-Pinto et al. 2021) (Fig. 2.1).

- **The Rhombencephalon**

The *rhombencephalon* is located in the most caudal part of the brain. It is a tubular structure comprising the pons and the medulla oblongata (the most rostral part of the spinal cord), upon which rests the cerebellum (Nepomuceno et al. 2016) (Fig. 2.2).

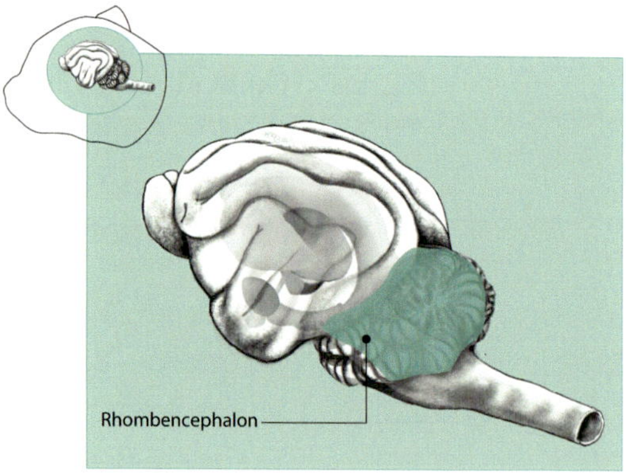

Fig. 2.2 The rhombencephalon in the cat. The rhombencephalon is the most caudal part of the encephalon, located at the level of the cerebellum and rostral to the spinal cord's origin

- **The Mesencephalon**

The *mesencephalon* is located rostro-dorsally to the rhombencephalon (Fig. 2.3). It comprises the tectum and the tegmentum. The tectum is the roof or dorsal part of the mesencephalon, while the tegmentum is its floor or ventral part. The mesencephalon is the origin of noradrenergic, dopaminergic, and serotonergic pathways (Cowie and Holstege 1992) (see ► Sect. 2.2.4.3).

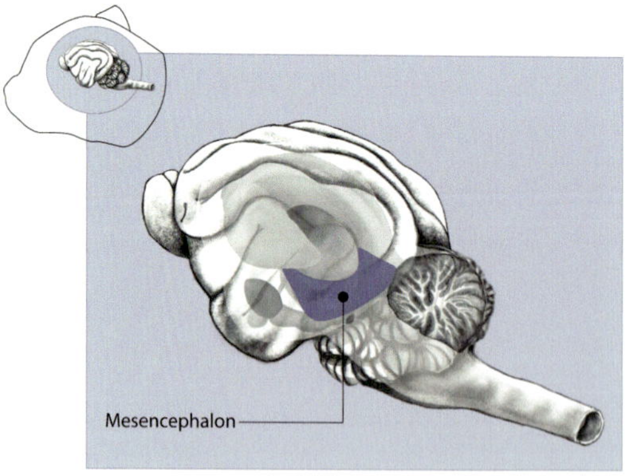

Fig. 2.3 The mesencephalon in the cat. The mesencephalon is located rostro-dorsally to the rhombencephalon and caudo-ventrally to the prosencephalon

- **The Prosencephalon**

The *prosencephalon* encompasses the most developed anatomical structures. It is typically divided into two regions: the telencephalic zone and the diencephalic zone (◘ Fig. 2.4), which will be examined in much greater detail.

2.1.1.2 The Telencephalon in Detail

The telencephalon includes the cortex, the basal ganglia, and the limbic system.

The Cerebral Cortex

The *cerebral cortex* (or cortex) is the outermost layer of the two cerebral hemispheres, the largest and most recognizable of the brain structures due to its folds (sulci) and ridges (gyri). The cerebral cortex is phylogenetically the brain's newest structure. It consists primarily of glial cells and the cell bodies of neurons, which give the cortex surface its gray appearance, known as *gray matter*. Millions of axons connecting neurons run beneath this surface, and their myelin sheaths give the subcortical brain its white appearance, known as *white matter* (◘ Fig. 2.5).

The cortex is the "mastermind" of all brain functions, from perception to communication, decision-making, and consciousness. The most anterior cortical area, the prefrontal cortex (◘ Fig. 2.6), is a regulatory center acting as a sensory filter and focus center (Brutkowski 1965; Murphy et al. 2012). It enables executive functions, allowing for the formulation of strategies based on the memory of past experiences and the prediction of the consequences of future decisions (Uylings et al. 2003).

The Basal Ganglia

The *basal ganglia* consist of two paired subcortical structures (identical and symmetrical) that resemble two spirals (◘ Fig. 2.7). They are composed of six paired sub-elements: the caudate nucleus, the pallidum, the nucleus accumbens, the putamen, the substantia nigra, and the subthalamic nucleus.

The basal ganglia process and integrate advanced cognitive functions of the cortex, sensory information, and mesencephalic pathways. Among other roles, they serve as the center for pleasure and addictions. The striatum, which will be discussed later in this book, is a functional subdivision that

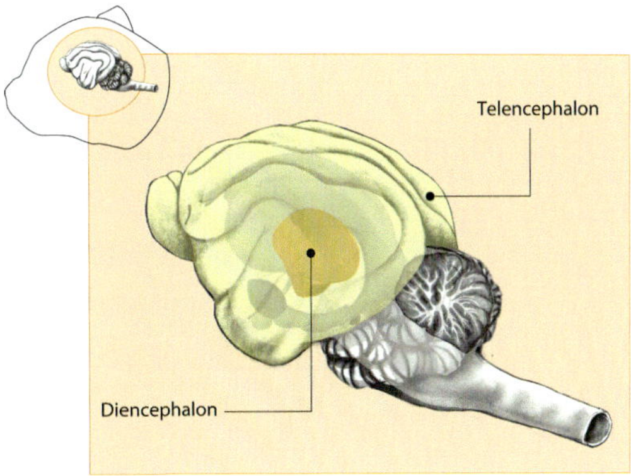

◘ **Fig. 2.4** The prosencephalon in the cat. The prosencephalon is the most developed and rostral part of the encephalon

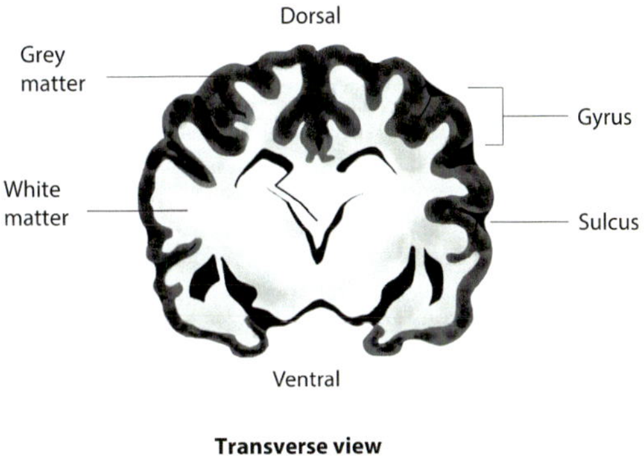

◘ **Fig. 2.5** Gray matter and white matter of the brain (cross section). The cortex is identifiable by its ridged surface composed of gyri. The surface, or gray matter, consists of neuronal cell bodies and astrocytes, while the deeper white matter is composed of axons and their myelin sheaths

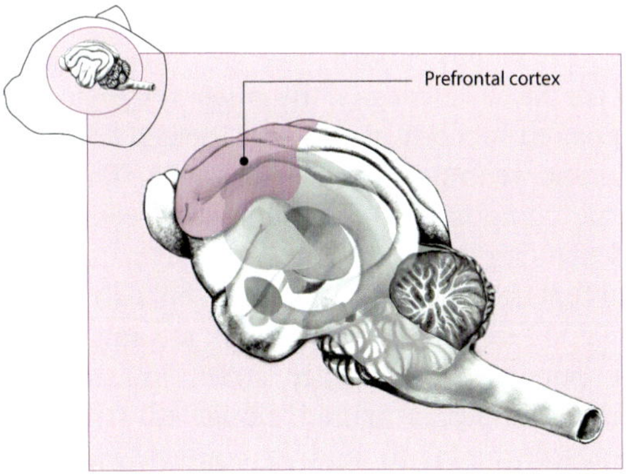

◘ **Fig. 2.6** The prefrontal cortex in the cat. This is the most rostral part of the cortex, responsible for strategy formulation, decision-making, and the regulation of subcortical structures

2

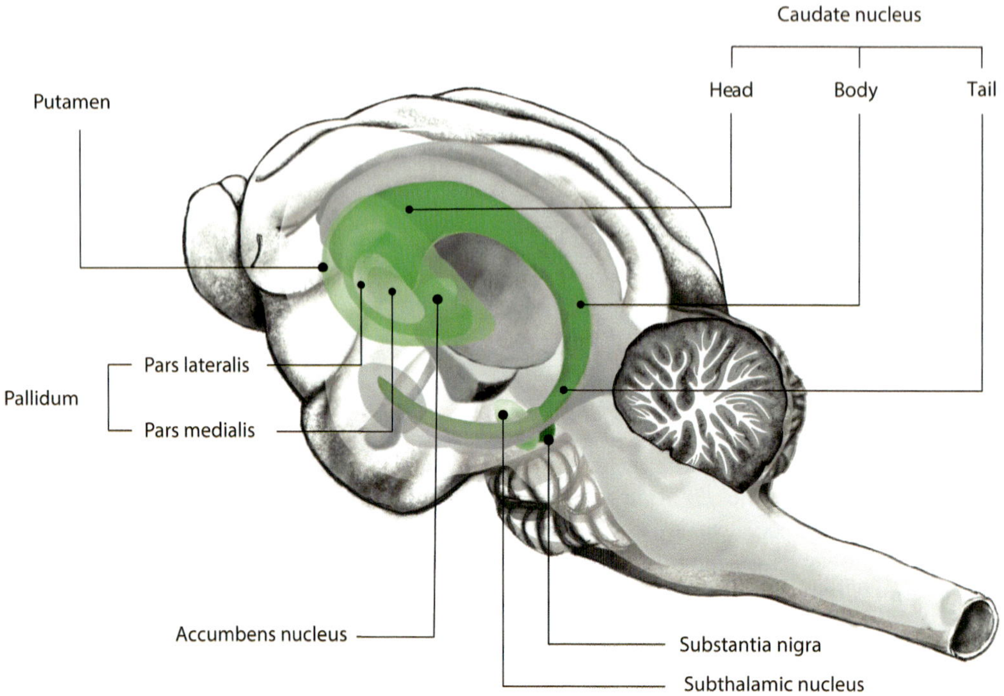

Fig. 2.7 The basal ganglia in the cat. The basal ganglia form two paired spiral-shaped structures. The figure illustrates one of the two pairs, showing from the most rostral to the most caudal: the pallidum with its two components (pars lateralis and pars medialis), the nucleus accumbens, the putamen, and the caudate nucleus, which is divided into three parts (head, body, and tail). Two additional structures, not part of the telencephalon, are also included in the basal ganglia: the substantia nigra, which belongs to the mesencephalon, and the subthalamic nucleus, which is part of the diencephalon

includes the basal ganglia. Anatomically, the dorsal striatum comprises the caudate nucleus and the putamen, while the ventral striatum includes the nucleus accumbens and the olfactory tubercle (Gray-Edwards et al. 2014).

The Limbic System

Like the basal ganglia, the *limbic system* forms a paired subcortical structure located adjacent to and ventral to the basal ganglia. The specific components of the limbic system vary depending on the authors and the anatomical or functional classification used, but two major regions are widely agreed upon: the amygdala, a paired almond-shaped structure, and the hippocampus, a paired elongated structure located caudal to and in contact with the amygdala (Fig. 2.8). The limbic system is involved in motivation and emotions through the amygdala and in learning and memory through the hippocampus (Fonberg 1986).

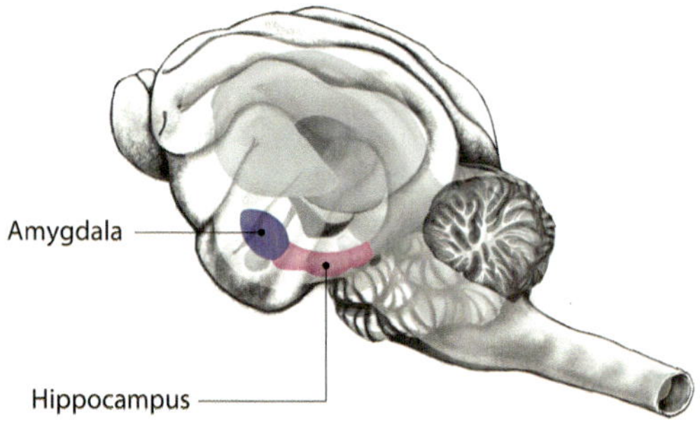

Fig. 2.8 The limbic system in the cat. The limbic system is defined differently by various authors, but it consistently includes two entities: the amygdala and the hippocampus. These paired structures are located rostro-ventrally relative to the basal ganglia

2.1.1.3 The Diencephalon in Detail

The diencephalic structures of the brain include the thalamus and the hypothalamus.

The Thalamus

The *thalamus* is a bilobed structure located at the center of the prosencephalon and sur-

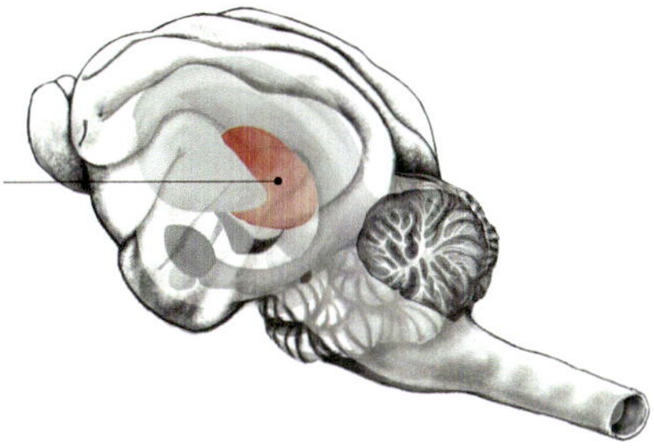

Fig. 2.9 The thalamus in the cat. A central bilobed structure, the thalamus serves as the brain's "hub," receiving all incoming information and transmitting it to the relevant structures

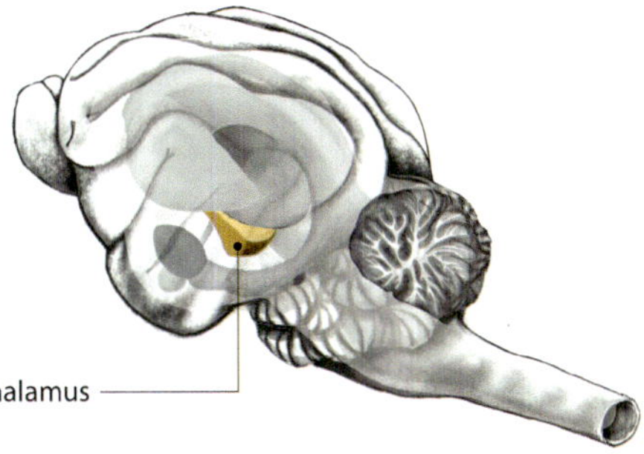

Fig. 2.10 Hypothalamus of the cat. Located just ventral to the thalamus, the hypothalamus is a central bilobed structure and the primary hub for the exchange of neural and hormonal information

rounded by all the telencephalic structures (Fig. 2.9). It functions as the "hub" of the brain, receiving all sensory information before relaying it to the corresponding cortical centers. The thalamus also gathers input from the cortex, enabling it to filter sensory information before passing it on.

Additionally, the thalamus is connected to the limbic system, the center of emotions, and the brainstem, the passageway for motor and sensory pathways. It coordinates signals from cortical areas responsible for regulating attention, particularly the prefrontal cortex (Jayaraman 1985).

The Hypothalamus

As its name suggests, the *hypothalamus* is located ventrally to the thalamus (Fig. 2.10). This structure connects the central nervous system to the endocrine system by controlling the pituitary gland, to which it is linked. The hypothalamus regulates numerous functions, including internal temperature, hunger and thirst, circadian rhythms, stress, motivation, and maternal attachment to offspring.

Figure 2.11 provides a lateral view of all the anatomical parts described in this section. The prosencephalon, mesencephalon, and rhombencephalon extend into the spinal cord and ganglionic systems, which function as both secondary nerve centers (reflex centers) for descending motor pathways and ascending sensory pathways to the brain (DeLahunta and Glass 2009).

2.1.2 The Brain's Unique Vascular Structures

The brain has high nutritional demands. Even a brief disruption in the supply of glucose or oxygen can quickly become harmful or even fatal. Furthermore, as we will see later in this chapter, brain activity relies on a fragile chemical balance. Any disturbance in this equilibrium can lead to varying degrees of dysfunction. Consequently, the brain must be both shielded from the chemical influences of the rest of the body and supplied continuously with large quantities of specific nutrients. Additionally, the brain communicates with the body through electrical and chemical signals, requiring specialized pathways. Three vascular systems unique to the brain meet these requirements: the blood-brain barrier, the choroid plexus, and the pituitary portal system.

2.1.2.1 The Blood-Brain Barrier

The blood-brain barrier exists in all brain capillaries, which, due to their membrane architecture, have much more selective permeability than capillaries in other parts of the body (Fig. 2.12). This creates a relatively impermeable barrier that isolates the brain from the body's overall chemical activity, preventing certain nutrients, toxins, or hormones from disrupting brain function (Fletcher et al. 2006). Therefore, any drug intended to modify brain activity must be capable of crossing the blood-brain barrier to reach its site of action.

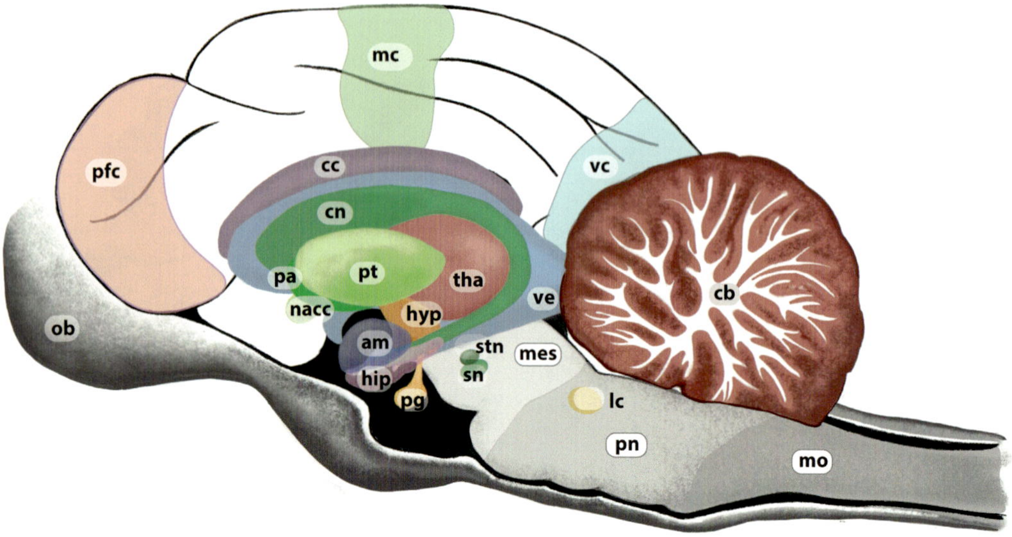

am	amygdala	**mo**	medulla oblongata
mc	motor cortex	**mes**	mesencephalon
vc	visual cortex	**nac**	accumbens nucleus
ob	olfactory bulb	**cn**	caudate nucleus
cc	corpus callossum	**stn**	subthalamic nucleus
pfc	prefrontal cortex	**pa**	pallidum
cb	cerebellum	**pn**	pons
hyp	hypothalamus	**pt**	putamen
hip	hippocampus	**sn**	substantia nigra
pg	pituitary gland	**tha**	thalamus
lc	locus cœruleus	**ve**	ventricles

Fig. 2.11 Lateral view of brain structures. This lateral view situates the various anatomical regions in space and will serve as a foundation for explaining brain pathways and circuits

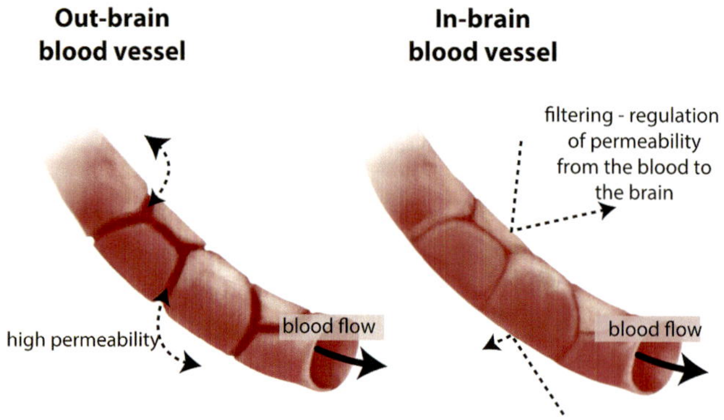

Fig. 2.12 The blood-brain barrier. On the left, a capillary from blood vessels outside the brain with a porous wall allowing extensive exchanges. On the right, a capillary from brain blood vessels with a tightly sealed wall that limits exchanges between the blood and brain cells

Certain restricted areas of the brain, however, have more permeable capillaries. These are "monitoring" zones for the body's homeostasis, enabling appropriate brain responses. For instance, the area postrema, a region in the medulla oblongata, has more permeable vasculature, making it sensitive to various circulating chemical elements related to vomiting, thirst, hunger, and blood pressure control (Rabin et al. 1986).

Brain capillaries are also equipped with transmembrane transport systems, notably for glucose.

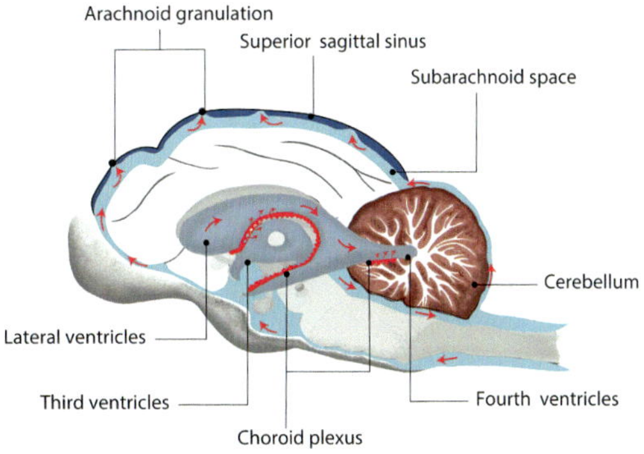

Fig. 2.13 The choroid plexuses and CSF circulation. Cerebrospinal fluid (CSF) is secreted by the choroid plexuses, circulates through the meningeal space, and is subsequently reabsorbed into the bloodstream

2.1.2.2 The Choroid Plexuses

Within each cerebral ventricle, blood capillaries form a network of structures enveloped by epithelial cells. These are the *choroid plexuses*, which secrete cerebrospinal fluid (CSF). The vascularization of the choroid plexuses is part of the blood-brain barrier, allowing only the components necessary for CSF production to pass through.

CSF is a circulating fluid that is continuously renewed and eliminated into the meningeal space before being reabsorbed into the bloodstream (Clementi and Marini 1972) (Fig. 2.13). It protects the brain by distributing its weight and absorbing shocks.

2.1.2.3 The Pituitary Portal System

The hypothalamus serves as the central command for hormonal secretions. It contains specialized neurons known as neurosecretory cells. Hormones secreted by the hypothalamus reach their target cells in the *anterior* pituitary via a communication network between hypothalamic and pituitary capillaries, known as the *hypothalamic-pituitary portal system* (Rijnberk 1996) (Fig. 2.14). For instance, the GnRH-FSH/LH axis for sexual hormones relies on this system.

The *posterior* pituitary, on the other hand, receives the terminal boutons of axons from hypothalamic neurosecretory cells. It directly secretes hormones such as vasopressin and oxytocin into the bloodstream.

Summary of ▶ Sect. 2.1

The brain can be divided into distinct anatomical regions. From caudal to rostral, and in phylogenetic order, it comprises the rhombencephalon, mesencephalon, and prosencephalon. The prosencephalon is further subdivided into the diencephalon and telencephalon.

The rhombencephalon and mesencephalon consist of nuclei that serve as the origins of various pathways associated with specific neurotransmitters. They also connect the brain to the spinal cord. Moving rostrally, the diencephalon acts as the "hub" of brain functions, receiving and redistributing all incoming information. Together with the hypothalamus and pituitary gland, the diencephalon bridges the brain and the endocrine system.

The telencephalon, the most advanced part of the brain, functions as the sensory, motor, emotional, strategic, and memory center.

The brain's vascular system is unique. The blood-brain barrier isolates the brain from the body's chemical fluctuations, which could disrupt its function. However, "monitoring zones" exist for sensory homeostasis. After external or internal stimulation, all brain structures contribute to the processing and integration of information, ultimately generating physiological responses and behaviors via neural and hormonal pathways targeting motor or visceral organs.

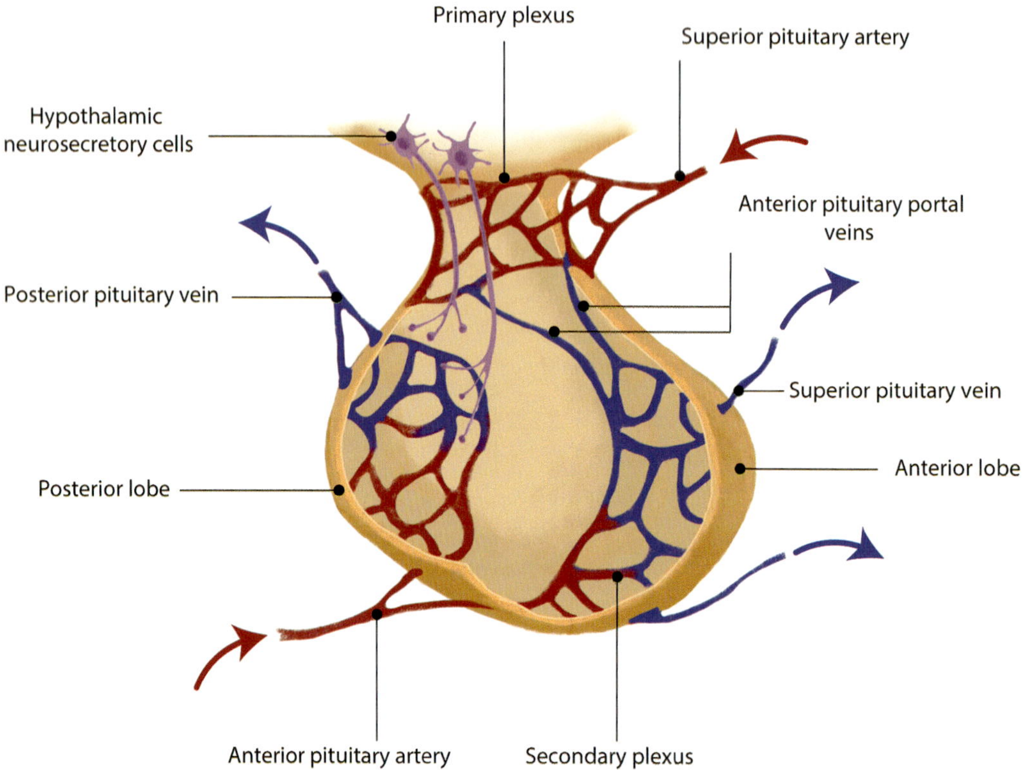

Fig. 2.14 The pituitary portal system in the cat. The pituitary portal system is a network of blood capillaries facilitating hormonal exchange between the hypothalamus and the pituitary gland. et l'hypophyse

2.2 Neurons, Synapses and Neurotransmitters

The major regions of the brain are composed of billions of specialized cells: neurons and glial cells. The electrical activity of neurons, combined with the biochemical activity at interneuronal connections or synapses, allows information to flow from one cell to another and thus from one part of the brain to another along pre-established circuits.

2.2.1 Neurons and Glial Cells

The brain contains two types of cells: nerve cells, or *neurons*, and their supporting cells, the *glial cells*.

Neurons are the primary elements responsible for processing and transmitting information. They vary widely in shape and size, depending on their location and function, but they always consist of the same four components: a *cell body* or *soma*, *dendrites*, *an axon*, and a *terminal bouton* (Carlson and Birkett 2017) (Fig. 2.15).

The soma, or cell body, contains the nucleus and its genetic material as well as all the cytoplasmic structures required for cellular function (mitochondria, ribosomes, etc.). Its shape varies greatly from one neuron to another.

Dendrites form a tree-like network connected to the soma (the term "dendrite" is derived from the Greek word *dendron*, meaning tree). They are the site of connections with the synaptic boutons of other neurons and act like antennas, receiving messages that are subsequently transmitted to the rest of the cell. Each dendritic "branch" is covered with numerous spines of varying sizes and shapes, which enhance connections with the synaptic boutons of neighboring neurons.

The *axon* is a tube-like extension originating from the soma, which can be short or very long, depending on the neuron's position and

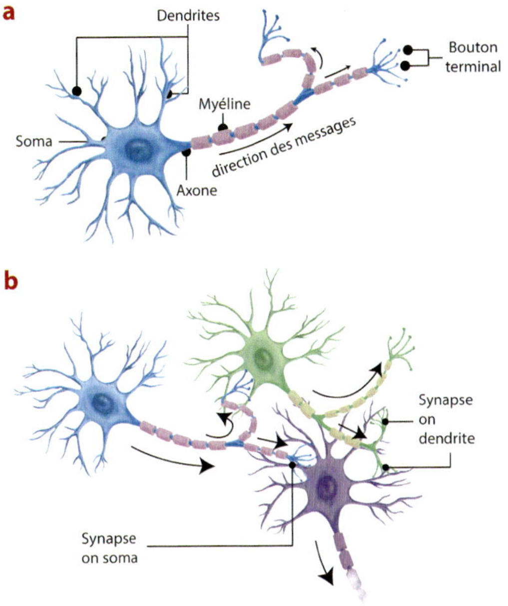

■ **Fig. 2.15** The neuron and its connections. **a** Neuron structures: soma, dendritic trees, axon covered with its myelin sheath, and terminal boutons. **b** Interneuronal connections: terminal boutons can connect to the soma or the dendritic tree of another neuron

function. Along the wall of this tube travels the action potential, a brief electrical signal that transmits information to the terminal bouton. Inside this long cytoplasmic tube are multiple parallel tracks, the microtubules, which transport vesicles containing neurotransmitters.

The *terminal bouton* is the distal end of the axon, preceding its connection with the effector cell, which could be another neuron or a non-neuronal cell (muscle, gland, etc.). This part of the neuron serves as the cell's communicative interface. At the terminal bouton, neurotransmitter vesicles arrive and, in response to an action potential (the neuron's electrical activity), fuse their membranes with the neuron's membrane, releasing their contents into the space between the terminal bouton and the target cell. The terminal bouton may connect with the dendritic tree or directly with the soma of another neuron. Axo-axonic connections also exist.

Glial cells, on the other hand, support neurons. Due to their specialization, neurons are fast-metabolizing cells with limited nutrient storage capacity. They require constant assistance to obtain the nutrients and oxygen essential for survival. This is the role of glial cells. They form a protective sheath around each neuron and are responsible for producing myelin sheaths. Glial cells isolate neurons from one another, preventing interference between neuronal messages. They also act as "undertakers," breaking down and removing the remnants of dead neurons. Additionally, glial cells modulate neurotransmission, though these mechanisms are still not fully understood (Schousboe 2003; Klimczak et al. 2019; Hansson and Rönnbäck 1995).

The network formed by neurons and glial cells is extraordinarily rich and complex, enabling the brain to collect, transport, process, and store information.

2.2.2 The Electoral Activity of Neurons

At rest, the neuron's membrane is not electrically neutral; there is a difference in electrical potential between the extracellular environment, which is positively charged, and the intracellular environment, which is negatively charged. This difference in potential arises from the distribution of ions: Na^+ and Cl^- are predominantly extracellular, while K^+ is primarily intracellular, along with organic anions (negatively charged proteins) that are exclusively intracellular. This distribution opposes osmotic and electrical pressures, which would otherwise equalize ion concentrations across the membrane (■ Fig. 2.16). The membrane potential difference is maintained by the action of a transmembrane ion pump, the sodium-potassium pump. This pump actively removes three Na^+ ions from the intracellular environment to the extracellular environment in exchange for two K^+ ions moving from the extracellular to the intracellular environment. Its action sustains a constant resting membrane potential of −70 mV (■ Fig. 2.17).

The variation in this membrane potential difference (which remains at −70 mV at rest) is the basis for the generation of the *action potential*, or electrical activity, which propa-

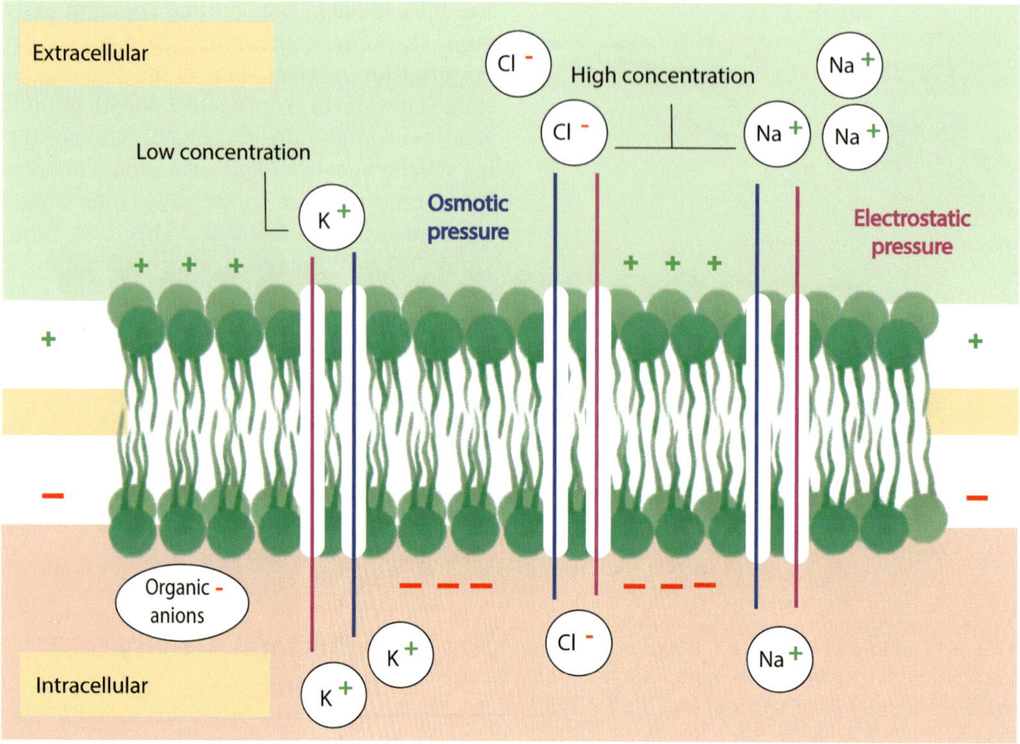

Fig. 2.16 The membrane and its gradients. Osmotic and electrostatic pressures tend to balance ionic distributions. However, there is a high concentration of Na^+ and Cl^- in the extracellular environment and a low concentration in the intracellular environment, while the opposite is true for K^+ ions. This creates a membrane potential of −70 mV. The membrane charge results from the work of a transmembrane ion pump, the sodium-potassium pump (Fig. 2.17)

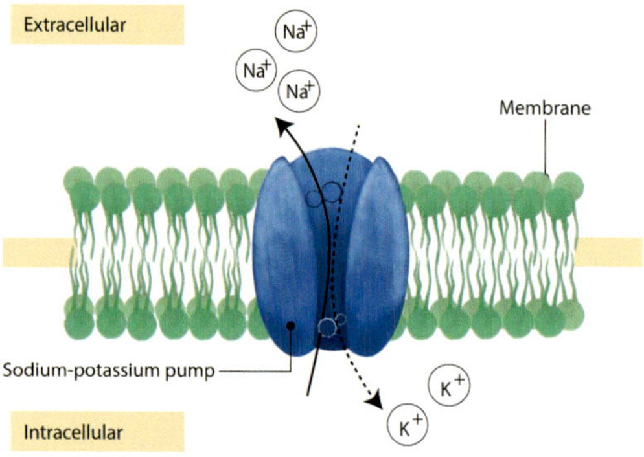

Fig. 2.17 The sodium-potassium pump. By exchanging two K^+ ions into the intracellular environment for three Na^+ ions into the extracellular environment, the membrane potential difference is maintained despite osmotic and electrostatic gradients

gates along the neuron's membrane and constitutes what is known as the nerve impulse.

On the membrane surface, there are channels formed by transmembrane protein assemblies that create millions of pores. When these channels open, an action potential is generated. These channels serve as passive pathways for Na^+ and K^+ ions and open or close depending on changes in membrane potential. Each channel has a specific threshold for opening and closing. Figure 2.18 outlines the various stages of these transmembrane ion fluxes.

An action potential that originates in one part of the neuron (most often at the soma or dendritic level) can result from either a single stimulation strong enough to reach the depolarization threshold or the summation of subthreshold stimulations that, together, trigger the action potential. Thus, a neuron does not respond to stimulation from just one other neuron but rather to the collective input from its connections with neighboring neurons.

The generation of an action potential follows the all-or-nothing law: either the threshold is not reached, and nothing happens, or

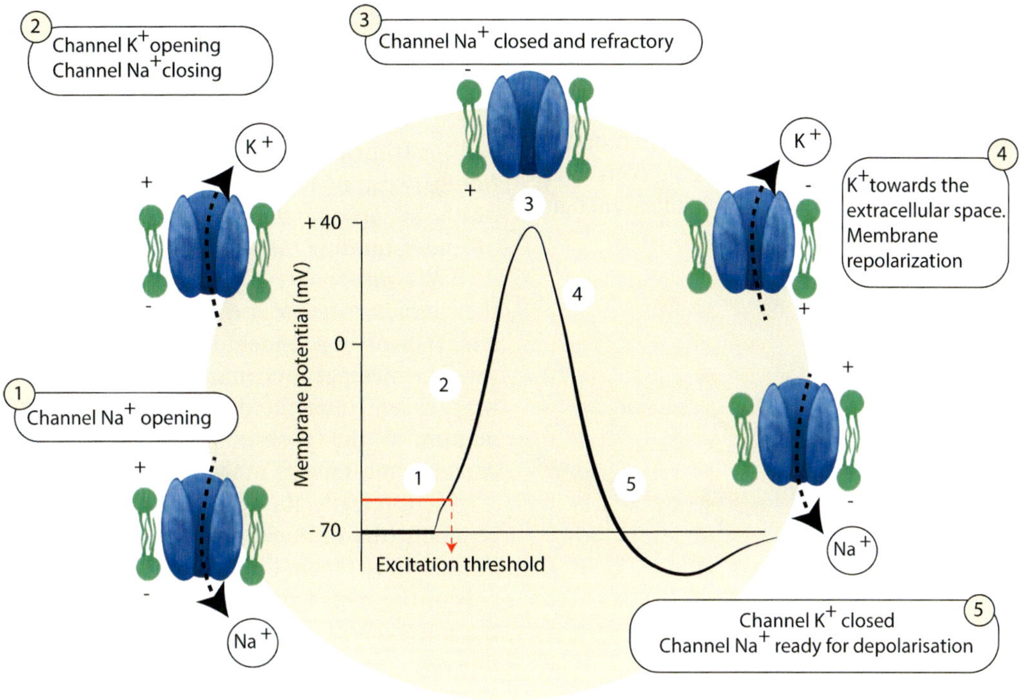

Fig. 2.18 The membrane action potential
1. When the excitation threshold is reached, Na^+ channels are the first to open, allowing a massive influx of Na^+ ions into the cell due to osmotic and electrical gradients. This causes the intracellular environment to become positively charged, reducing the membrane potential difference from −70 mV to +40 mV in approximately 1 millisecond
2. As the membrane action potential rises, Na^+ channels close, and K^+ channels, which have a higher opening threshold, begin to open, leading to the efflux of K^+ ions from the intracellular to the extracellular environment
3. Once the +40 mV peak is reached, sodium channels are completely closed and refractory to further opening
4. K^+ ions continue to exit the intracellular environment, causing a progressive repolarization of the cell membrane, returning it to the resting potential of −70 mV, which marks the threshold for the K^+ channels to close
5. The Na^+ channels then become sensitive again, ready for the next depolarization

the threshold is reached, and an action potential is generated. Once initiated, the action potential propagates with the same magnitude, regardless of the distance traveled. However, a neuron's electrical activity is not continuous. Depending on the degree of stimulation, it emits bursts of action potentials at varying frequencies. The greater the stimulation, the higher the frequency (Fig. 2.19).

Myelin sheaths covering axons isolate them from the extracellular environment, preventing ionic exchanges except at regular interruptions in the sheath called the nodes of Ranvier. The action potential propagates by jumping from one node of Ranvier to the next (Fig. 2.20). This saltatory conduction

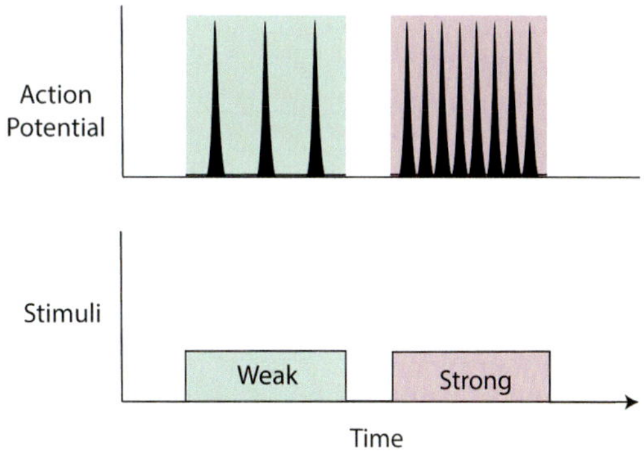

Fig. 2.19 The frequency law. The greater the stimulation, meaning the more the neuron receives inputs that stimulate the opening of sodium channels, the greater the number of action potential bursts

enables the action potential to travel at speeds of up to 120 m/s (432 km/h).

Once the end of the neuron is reached, the action potential, as the neuron's electrical activity, is transmitted to the next neuron through a *chemical message*. This conversion of an electrical signal into a chemical message occurs at the synapse.

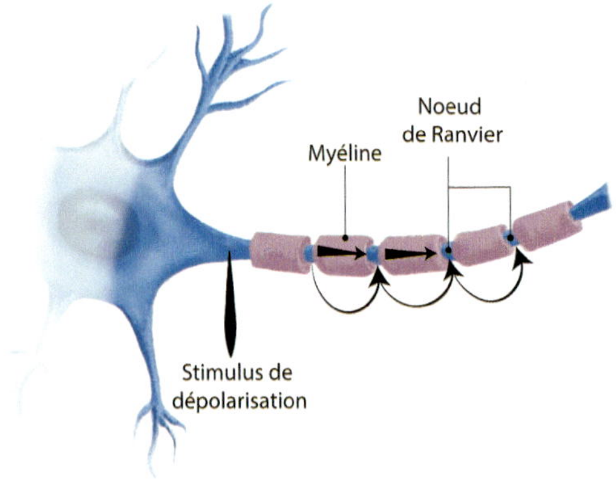

Fig. 2.20 Nodes of Ranvier and saltatory conduction. The presence of myelin sheaths forces the action potential to jump from one node of Ranvier to the next, significantly increasing the speed of information transmission

2.2.3 The Synapse and Its Functioning

The *synapse*, or interneuronal junction, is crucial for transmitting and regulating information between neurons. Psychotropic drugs act on the synapse, underscoring the importance of understanding their mechanisms in detail.

The components of the synapse align with the characteristics of any communication system. It involves a sender and a target, or receptor. The message is chemical in nature and, in most cases, transmitted from the presynaptic neuron to the postsynaptic neuron. These chemical messengers are *neurotransmitters*.

Anatomically, the synapse comprises three elements: the terminal bouton of the presynaptic neuron, including its membrane, which acts as the message sender; the membrane of the postsynaptic neuron, which serves as the message target; and the synaptic cleft, the space between the two, where the message is regulated. The entire structure is enveloped by glial cells (Fig. 2.21).

The sender, or *bouton* of the presynaptic neuron, serves as a reservoir of intracellular

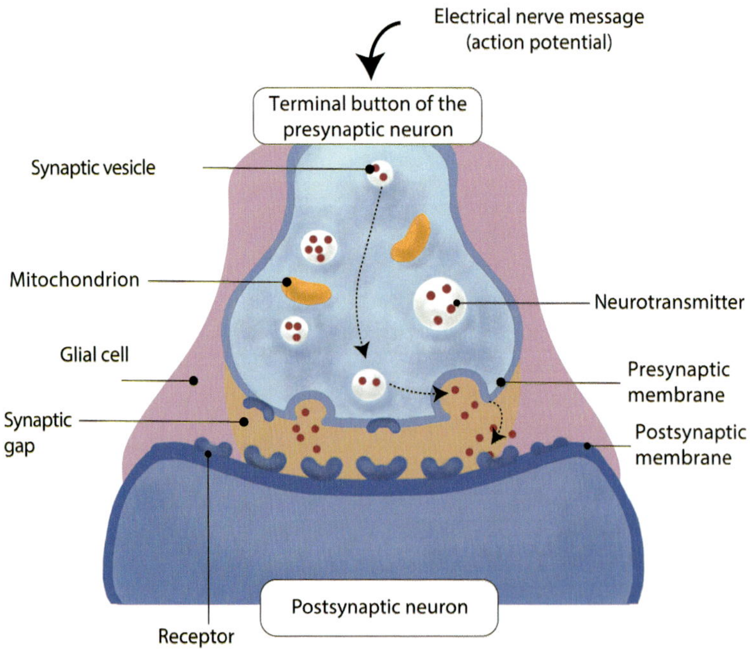

Fig. 2.21 The synapse. The three key components of the synapse are the synaptic bouton, the postsynaptic membrane, and the synaptic cleft. The entire structure is surrounded by glial cells

vesicles containing the neuron's neurotransmitter. A neuron is a specialized cell that produces and secretes only *one type of neurotransmitter*. When the action potential reaches the neuron's terminal, it triggers the migration of vesicles toward the neuron's end and the fusion of their membranes with the presynaptic membrane. The neurotransmitter is then released into the synaptic cleft.

The target of the message is the *postsynaptic neuron's membrane*, which contains numerous receptors, each specific to a particular neurotransmitter. A neuron can have various types of receptors, each specific to a different neurotransmitter. For instance, a dopaminergic neuron may have receptors for dopamine as well as for serotonin or gamma-aminobutyric acid (GABA) on its surface.

The *synaptic cleft* is the site of message regulation. The postsynaptic action potential is a very brief electrical event linked to the activation of receptors by neurotransmitters. To ensure that the action potential remains brief and relevant, the neurotransmitter must quickly disappear from the synaptic cleft. Three systems actively participate in this process: enzyme degradation, reuptake pumps, and autoreceptors. Enzymes neutralize the neurotransmitters released into the synaptic cleft. Reuptake pumps, located on the presynaptic membrane, actively reabsorb the neurotransmitter back into the synaptic bouton. Autoreceptors, also on the presynaptic membrane, are activated by the neurotransmitter and inhibit further neurotransmitter release by the presynaptic neuron. Additionally, some neurotransmitters diffuse toward glial cells outside the synaptic cleft from which they were emitted.

Understanding the different types and mechanisms of membrane receptors is essential, as they are among the biochemical targets used by psychotropic drugs to regulate brain functions. There are two main types of receptors on the postsynaptic membrane: *ionotropic receptors* and *metabotropic receptors*, also known as G-protein-coupled receptors.

Ionotropic receptors open upon binding with a neurotransmitter, allowing a large influx of calcium or sodium ions into the cell (◘ Fig. 2.22). This ionic movement causes depolarization, leading to the generation of a postsynaptic action potential. The action potential enables message transmission. Ionotropic receptors act quickly, with effects lasting from a few seconds to a few minutes.

Metabotropic receptors are complex proteins whose allosteric conformation changes upon binding with a neurotransmitter. This change in receptor shape triggers a cascade of secondary messengers, with the neurotransmitter acting as the primary messenger. The secondary messengers travel within the cell, act on nearby membrane ion channels, and

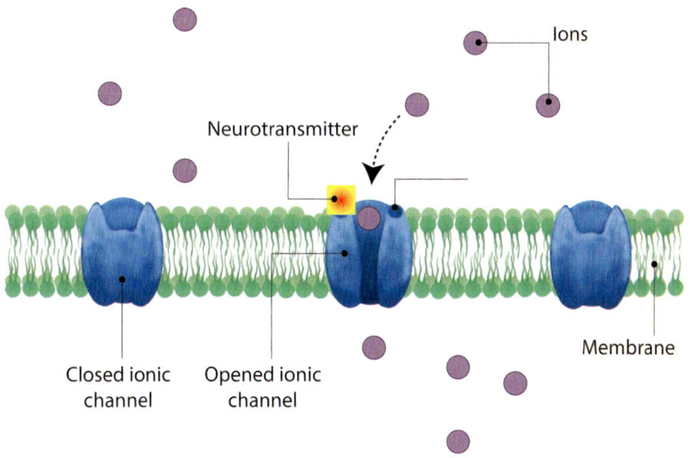

◘ Fig. 2.22 Ionotropic receptors. These are ion channels that open in the presence of a neurotransmitter (a natural ligand or a psychotropic drug). Their opening allows ions to enter or exit the cell, driven by osmotic and electrostatic gradients

trigger an action potential. They can also alter cellular functions by modifying the protein expression of the postsynaptic neuron's DNA (◘ Fig. 2.23).

Since neurotransmitters themselves are a product of the presynaptic neuron's DNA expression, synaptic transmission can be seen as the influence of the presynaptic neuron's genome on the expression of the postsynaptic neuron's genome. Metabotropic receptors, therefore, have a slow action, with effects lasting from a few hours to several weeks.

The electrical consequences of synaptic activation vary from one synapse to another. There are excitatory synapses, whose activity causes depolarization of the postsynaptic membrane, leading to the generation of an action potential. Conversely, there are inhibitory synapses, whose activity either induces repolarization or prevents depolarization.

The bursts of action potentials emitted by a neuron are the result of the integration of its excitatory and inhibitory connections (◘ Fig. 2.24).

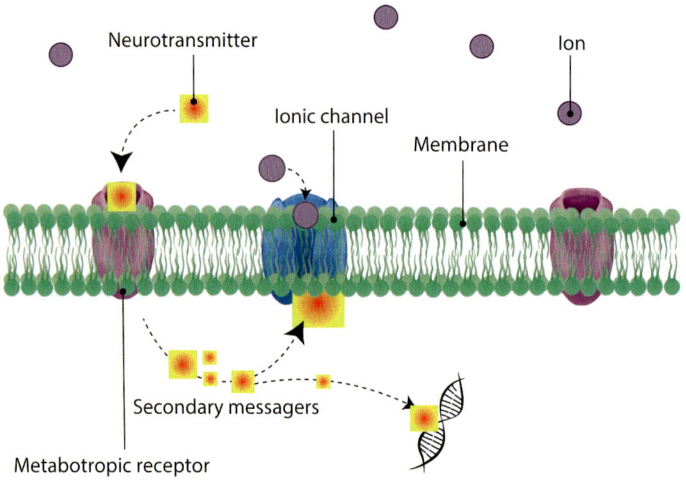

◘ **Fig. 2.23** Metabotropic receptors. The neurotransmitter (a natural ligand or a psychotropic drug) alters the allosteric conformation of the metabotropic receptor, releasing a secondary messenger that opens nearby membrane ion channels and modifies the cell's genetic expression

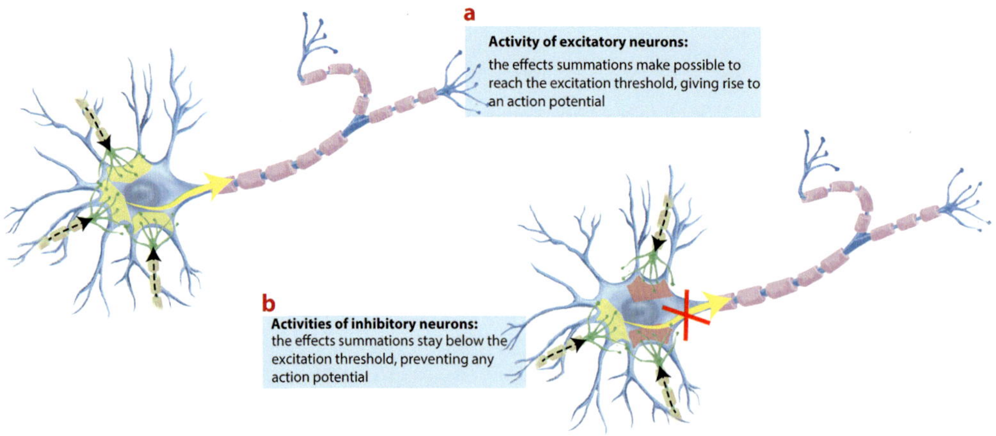

◘ **Fig. 2.24** Neuronal integration. The action potential emitted by a neuron results from the combined inhibitory and excitatory effects of the synapses connected to it

The direction of synaptic transmission is not fixed and can be reversed. There is retrograde communication from the postsynaptic neuron to the presynaptic neuron. This retrograde transmission is particularly relevant for endocannabinoids (endogenous cannabis), gaseous neurotransmitters such as nitric oxide, and neurotrophic factors. The implications of this retrograde transmission are still not fully understood (Iremonger et al. 2011).

Brain information flow is not solely reliant on synaptic transmission. There is also a form of neurotransmission known as volume transmission, or non-synaptic diffusion, which does not depend on a synapse or a specific communication direction. Neurotransmitters released in a synapse can diffuse to neighboring neurons within a certain radius, thereby modulating the activity of neurons beyond their original synapse. The brain is not only an incredibly complex network of neurons but also a highly sophisticated "chemical soup" (Bear et al. 2016; Carlson and Birkett 2017).

2.2.4 Neurotransmitters

2.2.4.1 The Rôle of Neurotransmitters

As discussed earlier, neurotransmitters are the vehicles that enable the transmission and regulation of electrical activity between neurons in the synapse. There are more than a dozen different neurotransmitters, but the majority of brain information is conveyed by two amino acid neurotransmitters: glutamate, which has an excitatory effect, and gamma-aminobutyric acid (GABA), which has an inhibitory effect. Most neuronal circuit activity results from the balance between the excitatory and inhibitory effects of these two neurotransmitters, to the extent that some authors use the term "neuromodulators" to describe other types of neurotransmitters (Carlson and Birkett 2017).

It is likely that no neuron in the brain is devoid of excitatory terminal boutons releasing glutamate and inhibitory terminal boutons releasing GABA. Most sensory information, except for pain, is transmitted from sensory organs through axons with terminal boutons that release glutamate.

Glutamate and GABA are released by neurons throughout the brain, whereas other neurotransmitters are released by neurons belonging to specific circuits. The activation or inactivation of these circuits modulates information related to particular brain functions, behaviors, or mental processes.

Alongside glutamate and GABA, six other neurotransmitters are essential for understanding brain functions and the effects of psychotropic drugs. These neurotransmitters fall into two chemical families: amino acids and monoamines.

2.2.4.2 Amino Acid Neurotransmitters

▪ Glutamate

Synthesis: glutamate is synthesized in a single step from glutamine through the action of an enzyme, glutaminase. It is stored in vesicles and released into the terminal bouton upon the arrival of an action potential.

Receptors: there are four types of glutamate receptors. Three of these are ionotropic, named after the chemical substances that modulate their activity: NMDA, AMPA, and kainate. The fourth receptor is metabotropic and referred to by the acronym mGluRs. Activation of AMPA and kainate receptors results solely in the generation of an action potential in the target cell. Activation of NMDA receptors not only generates an action potential but also alters the structure and biochemistry of the target cell. These changes remodel the synapse, strengthening it and thereby creating a new element in the formation of memory.

Glutamate, through the activation of certain brain cells, is thus involved in learning and various types of memory. Notably, ketamine—a commonly used anesthetic as well as a hallucinogenic substance—and alcohol are antagonists of NMDA receptors, preventing the action of glutamate on them.

Regulation: glutamate is reabsorbed by the presynaptic neuron via a specific amino acid transporter. It is also broken down in the synaptic cleft by an enzyme called glutamine synthetase (Fig. 2.25). This regulation is crucial, as prolonged excitation by glutamate leads to cell death, a phenomenon known as glutamate toxicity or excitotoxicity.

GABA

Synthesis: GABA, or gamma-aminobutyric acid, is synthesized in a single step from glutamic acid through the action of the enzyme glutamic acid decarboxylase. It is stored in vesicles and released into the terminal bouton upon the arrival of an action potential. The inhibitory action of GABAergic neurons is essential. Without GABA, there would be no regulation of excitatory neurons, leading to uncontrolled excitation of neighboring neurons, which would in turn excite the original neuron, resulting in extreme brain instability, such as that observed during epileptic seizures.

Receptors: there are various types of GABA receptors, but most psychotropic drugs used in psychiatry target the $GABA_A$ receptor. This receptor is an ionotropic chloride channel with a primary binding site specific to GABA. It also has additional binding sites which, when occupied by their ligands (molecules capable of interacting with the receptor), enhance GABA's action and induce a greater intracellular influx of chloride ions. These ligands include both endogenous molecules, such as GABA itself, and exogenous molecules, such as barbiturates, benzodiazepines, and alcohol.

Regulation: GABA is reabsorbed by the presynaptic neuron through a GABA-specific transporter. It is also broken down in the syn-

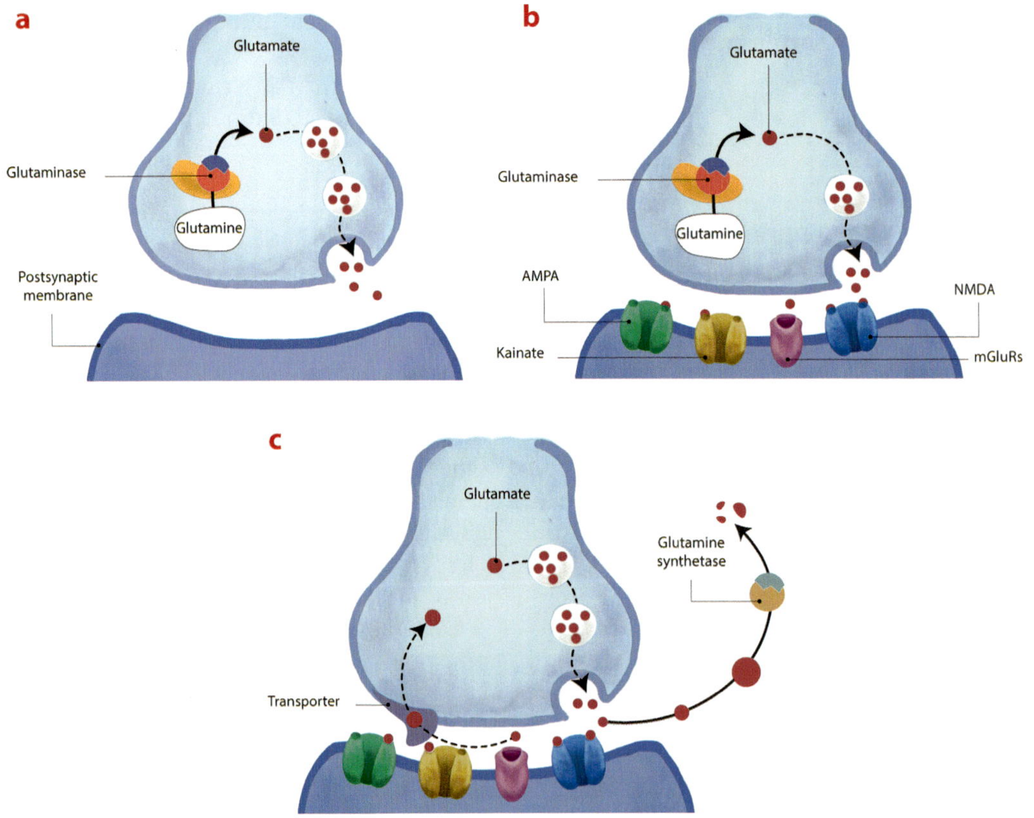

Fig. 2.25 Glutamate. Synthesis, receptors, and regulation. **a** Glutamate is synthesized from glutamine by the enzyme glutaminase. **b** Glutamate has three ionotropic receptors (NMDA, AMPA, and kainate) and one metabotropic receptor (mGluRs). **c** The action of glutamate is regulated through reabsorption by a specific presynaptic transporter and degradation in the synaptic cleft by the enzyme glutamine synthetase

aptic cleft by the enzyme GABA aminotransferase (◘ Fig. 2.26).

▪ Acetylcholine

Synthesis: acetylcholine (ACh) is synthesized from two precursors, choline and acetyl coenzyme A, through the action of the enzyme choline acetyltransferase. It is stored in vesicles and released at the terminal bouton upon the arrival of an action potential.

Receptors: there are two types of ACh receptors. First, ionotropic nicotinic receptors, activated by nicotine, are fast-acting and primarily located at the neuromuscular junction. They are also present in the central nervous system, particularly at axo-axonic junctions, where ACh facilitates transmission. Second, metabotropic muscarinic receptors, activated by muscarine (a toxin from the mushroom *Amanita muscaria*), are the most common cholinergic receptors in the brain. Specific antagonists exist for each receptor type: curare blocks nicotinic ionotropic receptors, while atropine blocks muscarinic metabotropic receptors.

Regulation: ACh is reabsorbed by the presynaptic neuron through a cholinergic transporter and is broken down in the synaptic cleft by the enzyme acetylcholinesterase (◘ Fig. 2.27).

Circuits the effects of acetylcholine in the brain are generally facilitatory. Similar to GABAergic and glutamatergic neurons, cholinergic neurons are diffusely located.

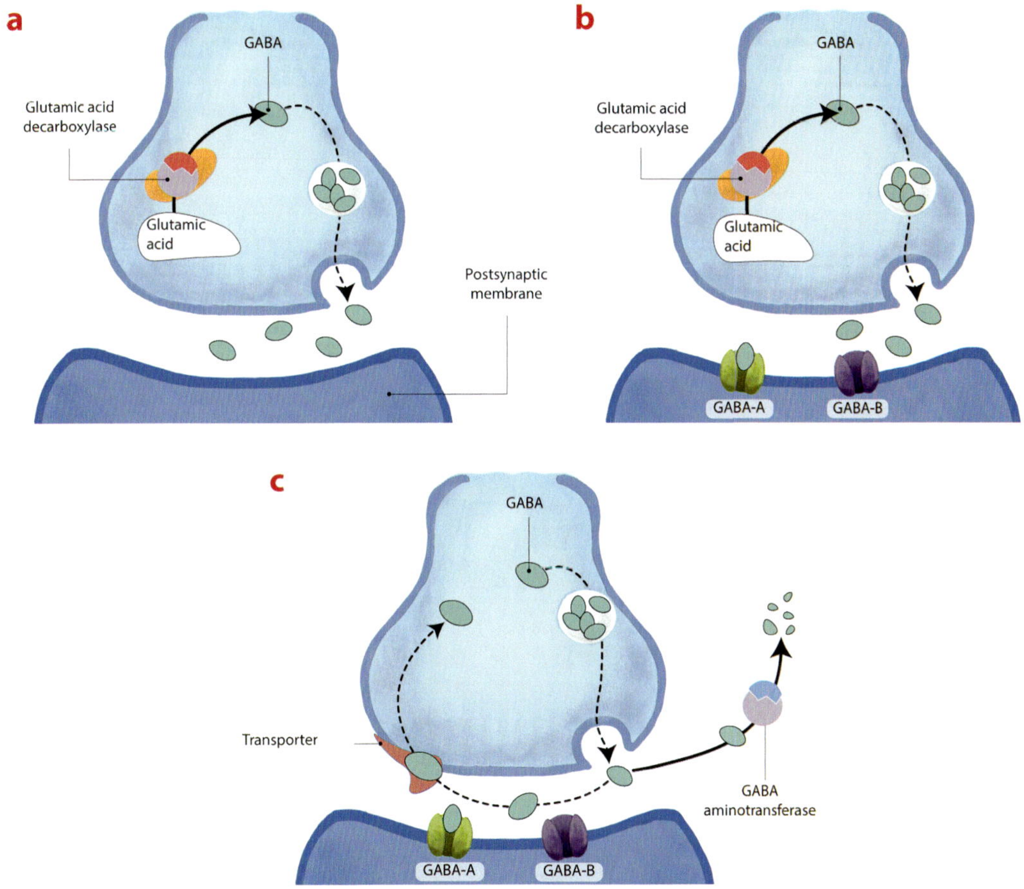

◘ **Fig. 2.26** GABA. Synthesis, receptors, and regulation. **a** GABA is synthesized from glutamic acid by the enzyme glutamic acid decarboxylase. **b** GABA receptors, primarily $GABA_A$, are ionotropic chloride channels that can bind endogenous and exogenous ligands, increasing chloride influx. **c** GABA is regulated by reabsorption via a specific presynaptic transporter and degradation by GABA aminotransferase in the synaptic cleft

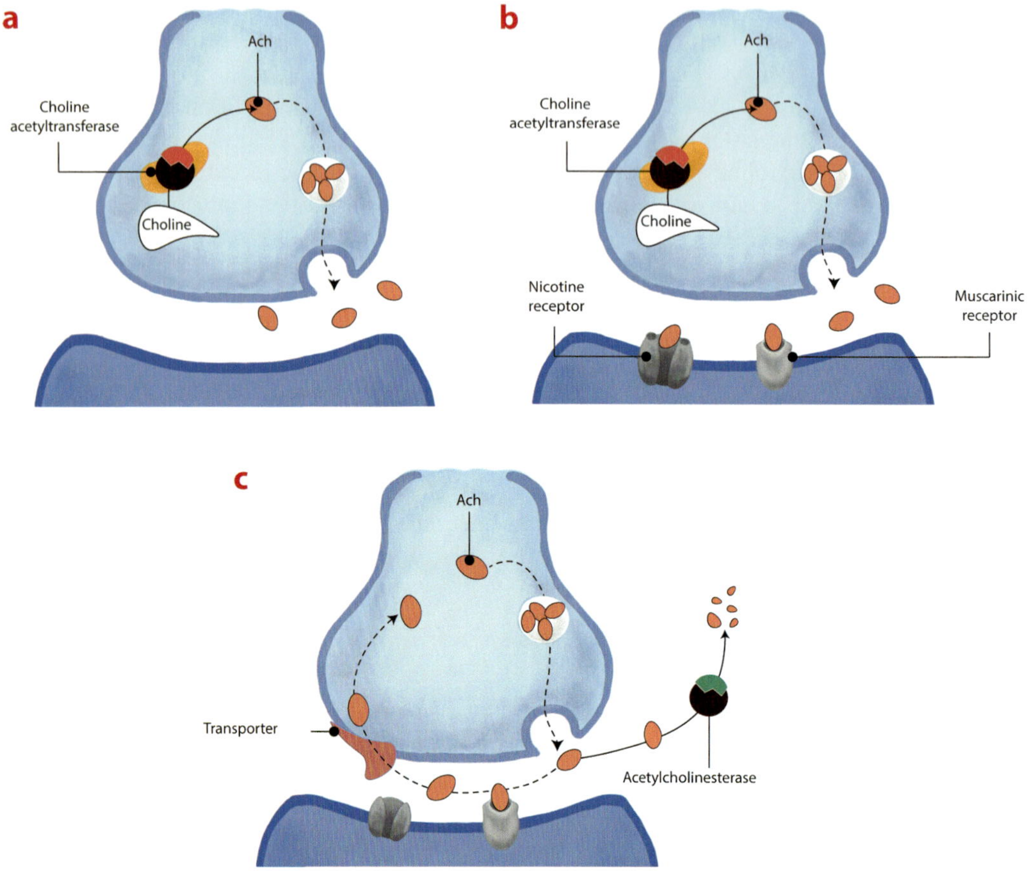

Fig. 2.27 Acetylcholine. Synthesis, receptors, and regulation. **a** ACh is synthesized from choline and acetyl coenzyme A by the enzyme choline acetyltransferase. **b** ACh has two types of receptors: ionotropic receptors called nicotinic and metabotropic receptors called muscarinic. **c** The action of ACh is regulated by reabsorption via a specific presynaptic transporter and degradation in the synaptic cleft by the enzyme acetylcholinesterase

However, three distinct circuits have been identified (Kimura et al. 1981).

The first circuit originates in the pons, a bulge in the brainstem, and projects to the tectum (the dorsal part of the mesencephalon), the thalamus, the cerebellum, and various brainstem regions. This pathway is particularly involved in REM sleep.

The second circuit originates in a nucleus of the ventral forebrain, the basal nucleus of Meynert, and projects to the cortex. This pathway is implicated in learning processes.

The third circuit begins in the lower medial part of the frontal lobe, in the medial septal nuclei, and projects to the limbic system, the hypothalamus, and the cortex (Fig. 2.28). This pathway is particularly involved in memory processes. All monoacid neurotransmitters are summarized in Table 2.1.

2.2.4.3 Monoamine Neurotransmitters

▪ Dopamine

Synthesis: dopamine is synthesized through the conversion of tyrosine into L-DOPA by tyrosine hydroxylase, followed by the transformation of L-DOPA into dopamine by DOPA decarboxylase. It is then stored in vesicles and released at the terminal bouton upon the arrival of an action potential.

Receptors: there are five types of dopamine receptors, all of which are metabotropic. These receptors are classified as D1, D2, D3,

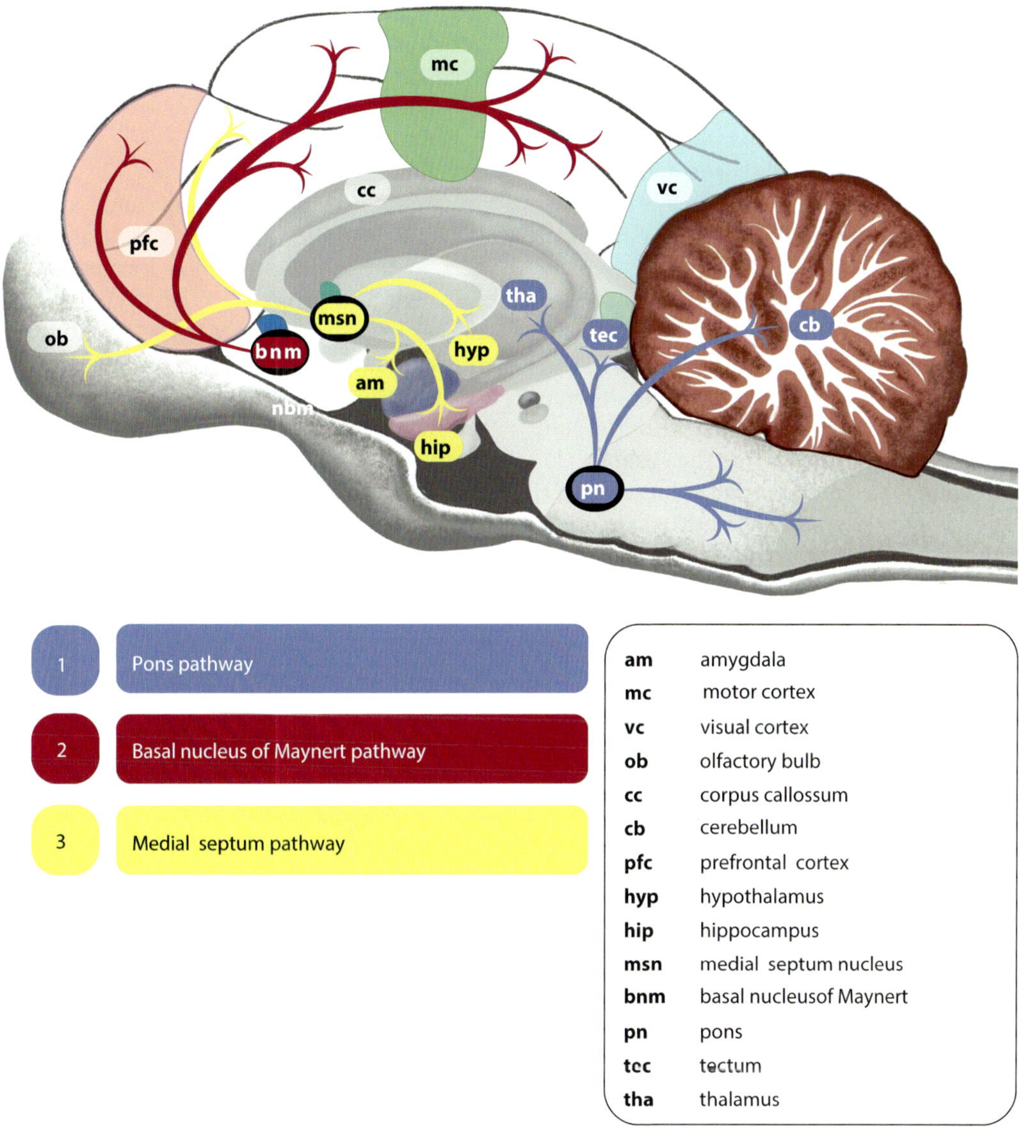

Fig. 2.28 Cholinergic pathways in the cat
Three cholinergic pathways are distinguished:
1. The pathway from the pons, projecting to the thalamus, tectum, and cerebellum, is involved in REM sleep
2. The pathway from the basal nucleus of Meynert, projecting to the cortex, is involved in learning
3. The pathway from the medial septum, projecting to the limbic system, is involved in memory processes

Table 2.1 Summary table of monoacid neurotransmitters

Neurotransmitters	Actions on brain functions
Glutamate	Excitatory; interacts with other neurotransmitter systems.
GABA	Inhibitory; interacts with other neurotransmitter systems.
Acetylcholine (Ach)	Learning, memory, REM sleep.

D4, and D5, with D1 and D2 being the most abundant.

Regulation: the D2 receptor also functions as an autoreceptor located on the dendrites, soma, and axons of dopaminergic neurons. When activated, it inhibits the synthesis and release of dopamine in the presynaptic neuron. Dopamine released into the synaptic cleft is reabsorbed by an active transporter in the presynaptic neuron's membrane. Methylphenidate (Ritalin®), used to treat attention deficit hyperactivity disorder (ADHD), blocks the dopamine reuptake transporter, thereby enhancing dopamine's effects. Additionally, dopamine is broken down in the synaptic cleft by two enzymes: monoamine oxidase (MAO), which is present in the synaptic spaces of all monoamine neurons, and catechol-O-methyltransferase (COMT), which metabolizes all extracellular catecholamines, including dopamine and norepinephrine (◘ Fig. 2.29).

▪▪ Circuits

dopaminergic neurons are distributed throughout the brain in four main circuits (◘ Fig. 2.30). The *nigrostriatal pathway* originates in the substantia nigra located in the brainstem and projects to the striatum (caudate nucleus and putamen). The striatum is a critical component of the basal ganglia involved in motor control. Dysfunctions in this pathway are responsible for the tremors observed in Parkinson's disease.

The *mesolimbic pathway* emerges from the ventral tegmental area and projects to various parts of the limbic system, including the

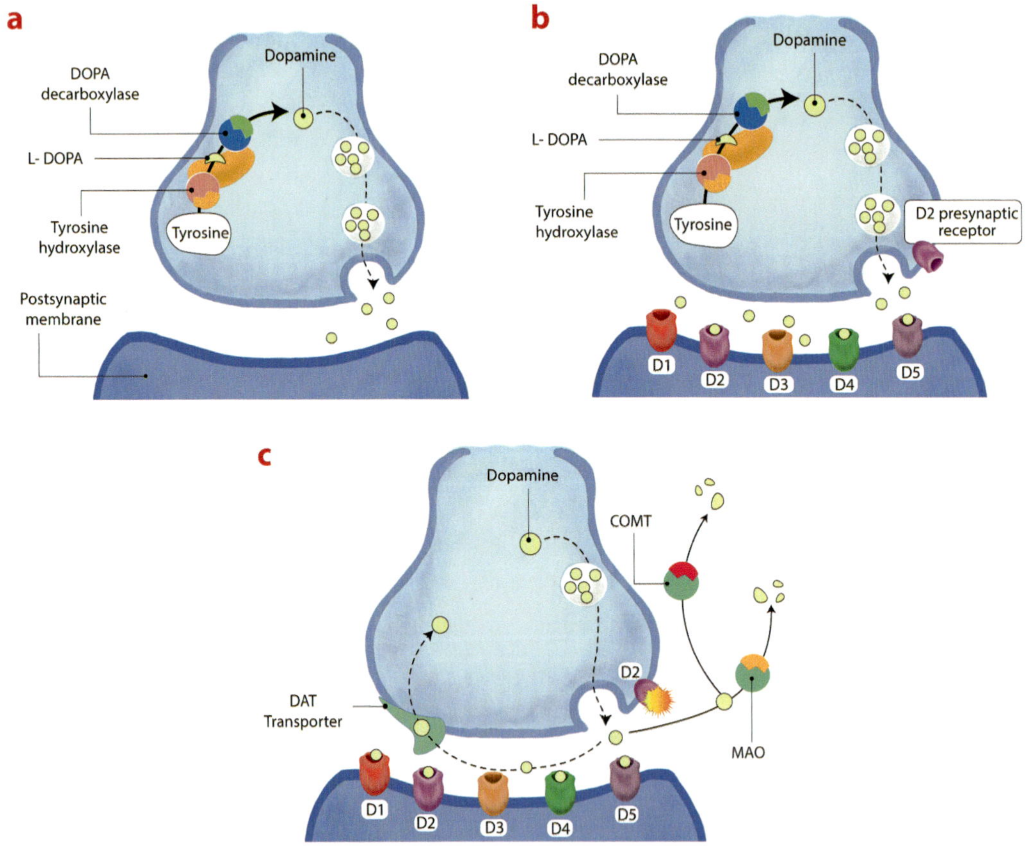

◘ **Fig. 2.29** Dopamine. Synthesis, receptors, and regulation. **a** Dopamine is synthesized from tyrosine by the enzymes tyrosine hydroxylase and DOPA decarboxylase. **b** Dopamine has five types of receptors, all metabotropic. **c** Dopamine's action is regulated through reabsorption by a specific presynaptic transporter (DAT), stimulation of presynaptic D2 receptors, and degradation in the synaptic cleft by the enzymes monoamine oxidase (MAO) and catechol-O-methyltransferase (COMT)

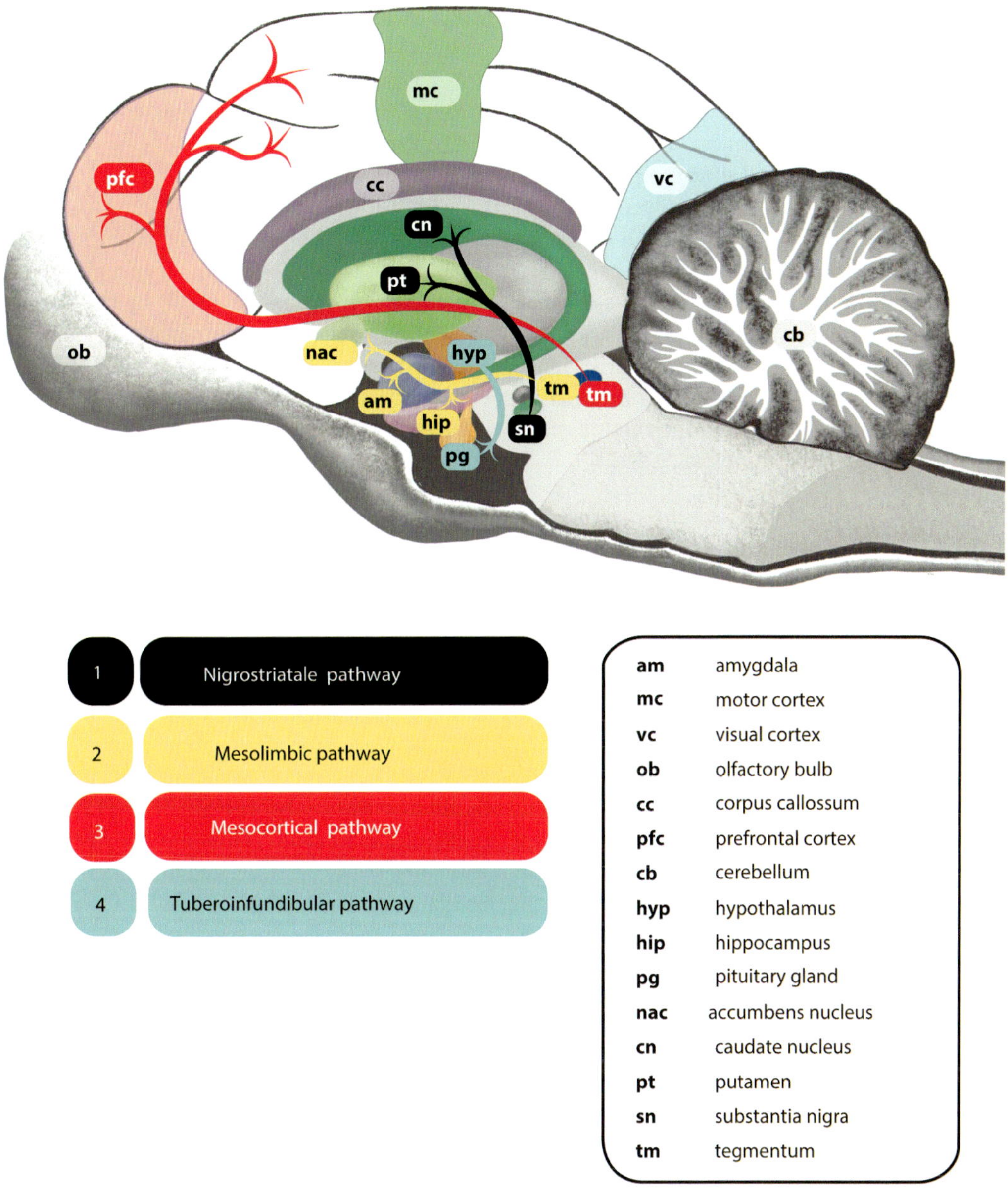

Fig. 2.30 Dopaminergic pathways in the cat
The four dopaminergic pathways are:
1. The nigrostriatal pathway, originating in the substantia nigra and projecting to the dorsal striatum (putamen and caudate nucleus), is involved in motor control
2. The mesolimbic pathway, originating in the ventral tegmental area and projecting to the limbic system (amygdala and hippocampus) and nucleus accumbens, is involved in motivation, learning, pleasure, reward, and addiction
3. The mesocortical pathway, also originating in the ventral tegmental area and projecting to the prefrontal cortex, is involved in short-term memory, organization strategies, and problem-solving
4. The tuberoinfundibular pathway, connecting the hypothalamus to the pituitary gland, is involved in neurohormonal brain functions, particularly prolactin secretion control

amygdala, hippocampus, and nucleus accumbens. This pathway plays a significant role in reinforcing learning, pleasure, reward, and addictive behaviors.

The *mesocortical pathway* also originates from the ventral tegmental area and projects to the prefrontal cortex. It is involved in short-term memory and strategic processes, such as organization and problem-solving.

The *tuberoinfundibular pathway* connects the hypothalamus to the pituitary gland and regulates the secretion of prolactin. This explains the unexpected lactation observed in patients treated with dopamine-related psychotropic drugs, such as neuroleptics.

▪ Norepinephrine

Synthesis: norepinephrine is synthesized from dopamine by the enzyme dopamine-beta-hydroxylase. This final step occurs inside vesicles, where dopamine, the precursor of norepinephrine, is stored before being converted into norepinephrine within the vesicle's cytoplasmic environment.

Receptors: all norepinephrine receptors are metabotropic. There are four types: α1, α2, β1, and β2. These receptors are present in the brain and are also found in many other organs of the body.

Regulation: similar to the D2 dopamine receptor, α2 receptors function as autoreceptors on the dendrites, soma, and axon of noradrenergic neurons. A norepinephrine transporter (NET) enables its active reabsorption by the presynaptic neuron. Additionally, as with dopamine, norepinephrine is metabolized in the synaptic cleft by the enzymes monoamine oxidase (MAO) and catechol-O-methyltransferase (COMT) (◘ Fig. 2.31).

Circuits nearly all brain regions receive noradrenergic inputs (◘ Fig. 2.32). The cell bodies of these neurons are located in the pons, medulla oblongata, or thalamus. The most significant nucleus from which noradrenergic neurons emerge is the *locus coeruleus*, situated in the dorsal part of the pons. The primary effect of activating noradrenergic pathways is an increase in alertness and attention. In addition to ascending pathways projecting to the brain, there is a descending pathway projecting to the spinal cord.

▪ Serotonin

Synthesis: serotonin (5-HT) is synthesized through the conversion of tryptophan into 5-hydroxytryptophan (5-HTP) by the enzyme tryptophan hydroxylase, followed by the transformation of 5-HTP into 5-HT by the enzyme 5-HTP decarboxylase.

Receptors: nine types of serotonergic receptors have been identified, including 5-HT1A, 5-HT1B, 5-HT1D, 5-HT1F, 5-HT2A, 5-HT2C, and 5-HT3. All are metabotropic except for the 5-HT3 receptor, which is ionotropic and associated with a chloride channel. The 5-HT3 receptor inhibits action potentials on the postsynaptic neuron.

Regulation: the 5-HT1A receptor functions as an autoreceptor on the dendrites and soma, and its stimulation halts the release of 5-HT. An active serotonin transporter (SERT) facilitates the reabsorption of 5-HT by the presynaptic neuron. This transporter is targeted by many anti-depressant psychotropic drugs, primarily selective serotonin reuptake inhibitors (SSRIs), such as fluoxetine, which is widely used in veterinary psychiatry. Finally, serotonin is degraded in the synaptic cleft by monoamine oxidase (MAO) enzymes (Stahl and Muntner 2021) (◘ Fig. 2.33).

Circuits serotonergic pathways are similar to noradrenergic pathways, with additional projections to the striatum and nucleus accumbens. The neuron cell bodies are located in a medial zone of the pons, the median raphe nuclei (◘ Fig. 2.34). These pathways regulate mood, appetite, sleep, and alertness. Additionally, a descending pathway to the spinal cord is involved in pain modulation, capable of blocking ascending pain (see ▶ Chap. 3, ▶ Sect. 3.3.1.2). All monoamine neurotransmitters are summarized in ◘ Table 2.2.

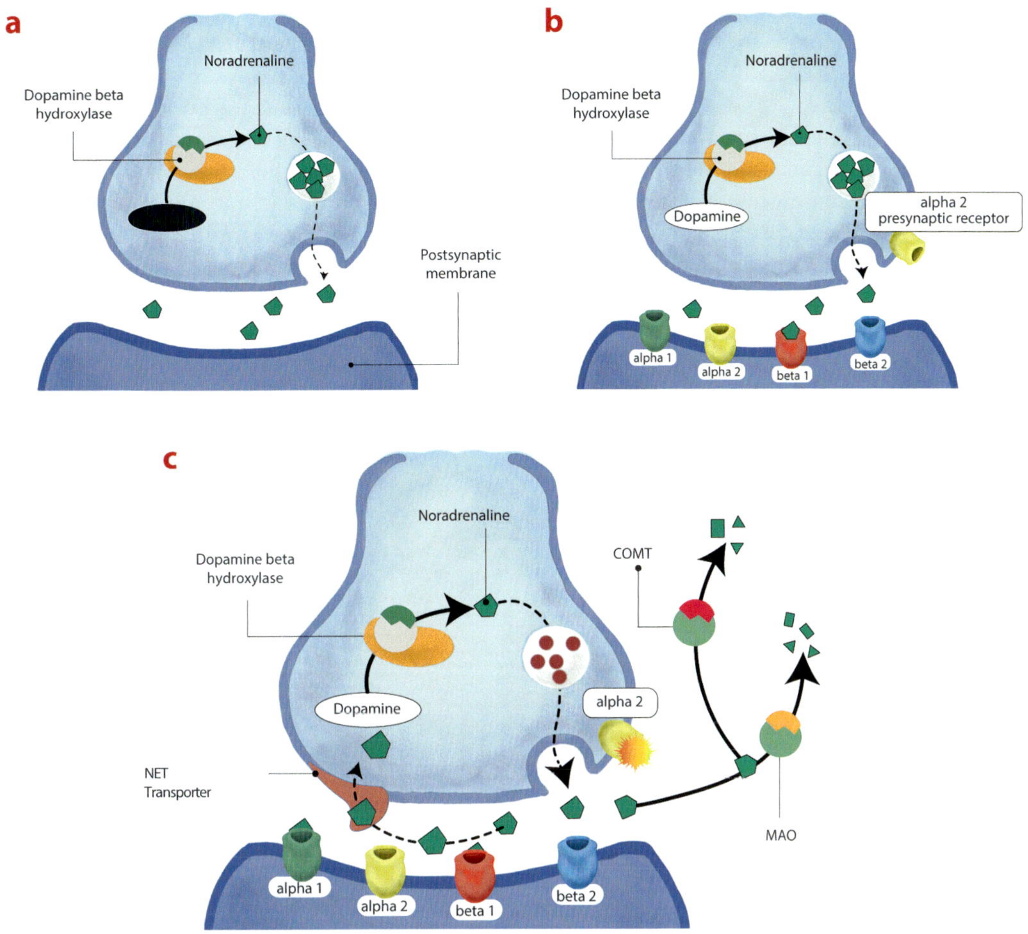

Fig. 2.31 Norepinephrine. Synthesis, receptors, and regulation. **a** Norepinephrine is synthesized from dopamine by the enzyme dopamine-beta-hydroxylase. **b** Norepinephrine has four types of receptors, all metabotropic. **c** Norepinephrine's action is regulated through reabsorption by a specific presynaptic transporter (NET), stimulation of presynaptic α2 receptors, and degradation in the synaptic cleft by the enzymes monoamine oxidase (MAO) and catechol-O-methyltransferase (COMT)

Summary of ▶ Sect. 2.2

Neurons and glial cells are the primary cellular components of the brain. Neurons are the units responsible for transmitting information from one part of the brain to another or from the brain to effector organs. Glial cells are supportive cells closely associated with neurons, providing nutrients and structural support.

Brain activity is electrical, consisting of a series of action potentials traveling along neurons. However, the transfer of electrical information from one neuron to another, or from a neuron to a target organ, is chemical. This process occurs at the synapse—a functional unit comprising the synaptic terminal bouton, the postsynaptic membrane, and the synaptic cleft (the space separating them). Within the synapse, electrical information is converted into a chemical message, which is then reconverted into electrical information in the target neuron.

The chemical messenger between two neurons is called a neurotransmitter. Each neuron produces only one type of neurotransmitter, and each neurotransmitter has a specific affinity for its corresponding receptors. The binding of the neurotransmitter to receptors on the postsynaptic membrane converts the chemical message into an electrical message. On the presynaptic membrane, neurotransmitter-receptor binding regulates the release of the chemical message and its reuptake. Additionally, enzymes in the synaptic cleft neutralize the chemical message, further regulating information transmission.

Each neurotransmitter is associated with specific pathways, originating in distinct regions and projecting to target areas, allowing a better understanding of their respective roles.

In veterinary psychiatry, two major families of neurotransmitters are particularly important, as they represent the primary targets of available medications: amino acids, including glutamate, GABA, and acetylcholine, and monoamines, including dopamine, norepinephrine, and serotonin.

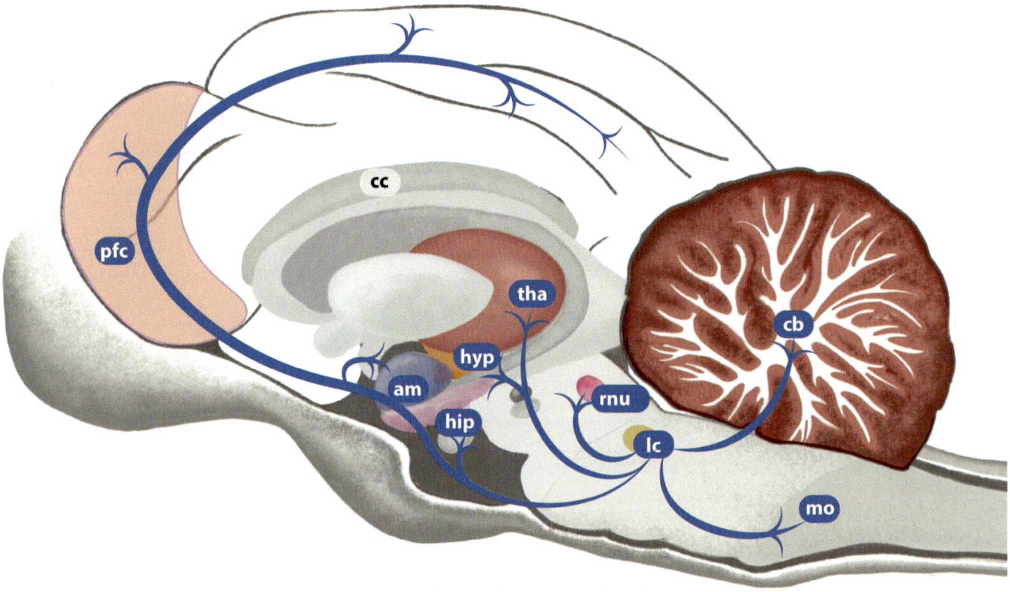

am	amygdala corpus
cc	corpus callossum
pfc	prefrontal cortex
cb	cerebellum
hip	hippocampus
hyp	hypothalamus
lc	locus coeruleus
mo	medulla oblongata
rnu	raphe nuclei
tha	thalamus

Fig. 2.32 Noradrenergic pathways in the cat. The primary pathways originate from the locus coeruleus and project to nearly all parts of the brain, except the basal ganglia, and to the spinal cord. These pathways are all involved in vigilance and attention

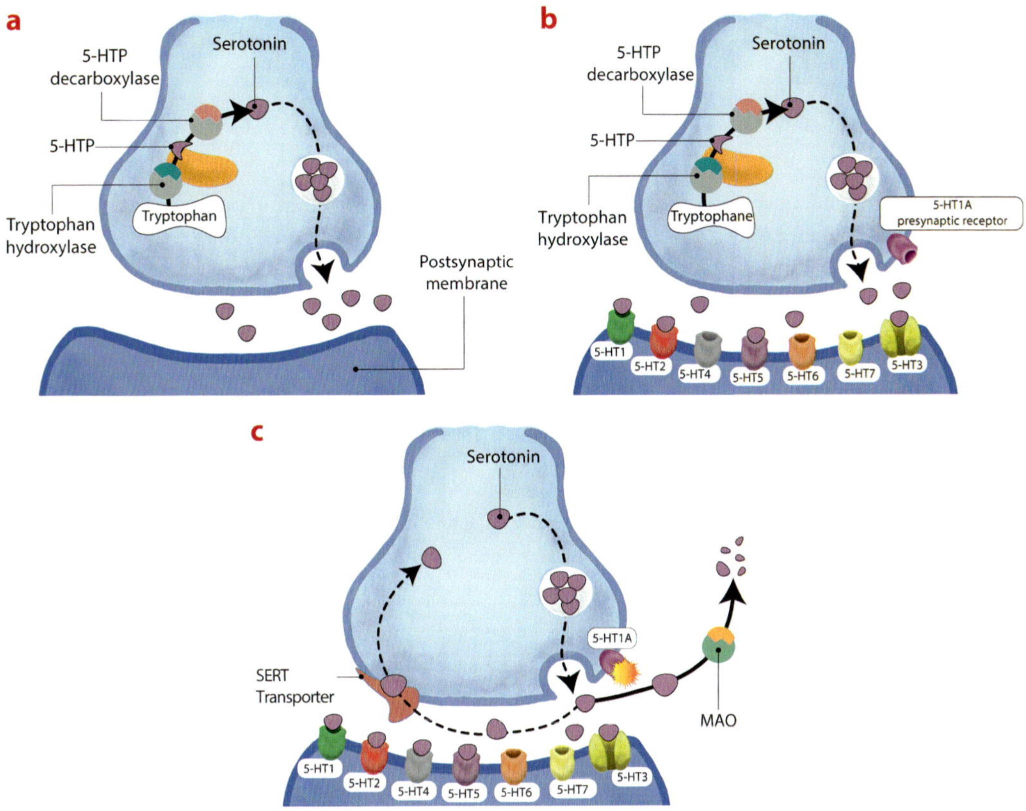

Fig. 2.33 Serotonin. Synthesis, receptors, and regulation. **a** Serotonin is synthesized from tryptophan by the enzymes tryptophan hydroxylase and 5-HTP decarboxylase. **b** Serotonin has nine identified receptor types, all metabotropic except for the ionotropic 5-HT3 receptor. **c** Serotonin action is regulated through reabsorption by a specific presynaptic transporter (SERT), stimulation of presynaptic 5-HT1A autoreceptors, and degradation in the synaptic cleft by monoamine oxidase (MAO) enzymes

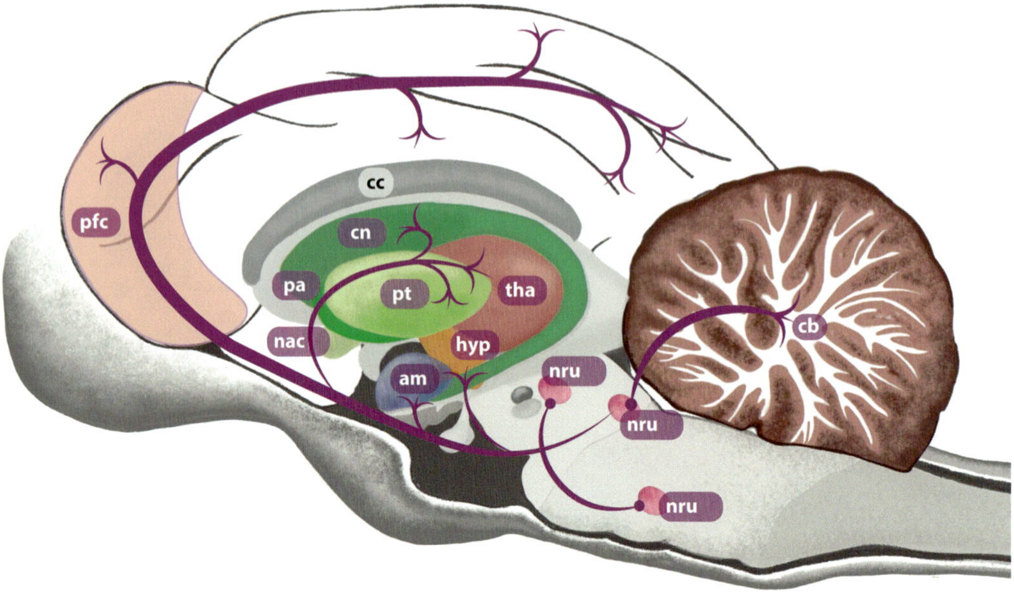

am	amygdala
cc	corpus callossum
pfc	prefrontal cortex
cb	cerebellum
hyp	hypothalamus
nac	accumbens nucleus

cn	caudate nucleus
rnu	raphe nuclei
pa	pallidum
pt	putamen
tha	thalamus

Fig. 2.34 Serotoninergic pathways in the cat. Serotonergic pathways originate from the raphe nuclei and project to all parts of the brain, including the striatum and nucleus accumbens, as well as the spinal cord. These pathways are involved in mood, appetite, sleep, and alertness. The descending medullary pathway plays a role in pain modulation

Table 2.2 Summary table of monoamine transmitters

Neurotransmitters	Actions on brain functions
Dopamine	Voluntary movements, attention, learning, reinforcement, planning, and problem-solving.
Norepinephrine	Alertness.
Serotonin	Regulation of mood, appetite, sleep, dreams, arousal, and impulse control.

2.3 Neurophysiology of Behaviors

This section explores the neurophysiological mechanisms that influence and regulate behaviors. Understanding these mechanisms enables the veterinary psychiatrist to grasp the normal functioning of brain circuits and intervene in cases of dysfunction. To comprehend the neurophysiology of behaviors, it is necessary to make artificial distinctions. We will review the neurophysiological circuits responsible for emotions, mood, and cognition. For cognition, we will delve into the mechanisms underpinning learning and memory, the attentional system, the motivational system, the action control system, and aggression.

2.3.1 Emotions

The neurophysiology of emotions is complex, with much research focusing on fear. The following section details the amygdala circuit, which regulates fear responses and can lead to a phobic state when dysregulated. We will also examine the cortico-striato-thalamo-cortical loop (CSTC), which anticipates danger and can lead to anxiety when dysregulated. Other emotions not covered in this section follow a similar model, involving the limbic system as the source activating the emotion, with numerous afferents, and the prefrontal cortex, which can independently inhibit the emotional response.

2.3.1.1 The Amygdala Circuit

Fear is a protective emotion that helps individuals survive when facing danger. This system is particularly important in feline species due to their dual status as prey and predator. The center that coordinates fear is the amygdala. The amygdala consists of a cluster of almond-shaped brain nuclei and is part of the limbic system. It receives sensory inputs (sound, smell, vision, touch) from the cortex, inputs from the thalamus (a sensory "hub"), and inputs from the hippocampus, which plays a role in memory. After processing these inputs, the amygdala sends information to various parts of the brain, resulting in behavioral reactions (Ulrich-Lai and Herman 2009) (◘ Fig. 2.35).

The connections between the amygdala and the prefrontal cortex enable the regulation of emotions. The amygdala acts as a command center, an alarm system that determines whether a fear response should be triggered. In contrast, the prefrontal cortex analyzes the situation and inhibits the fear response if the perceived danger is not real.

If the fear response is validated—meaning the sum of the activating inputs to the amygdala exceeds the negative regulation from the prefrontal cortex—the amygdala transmits various signals that constitute the fear response:

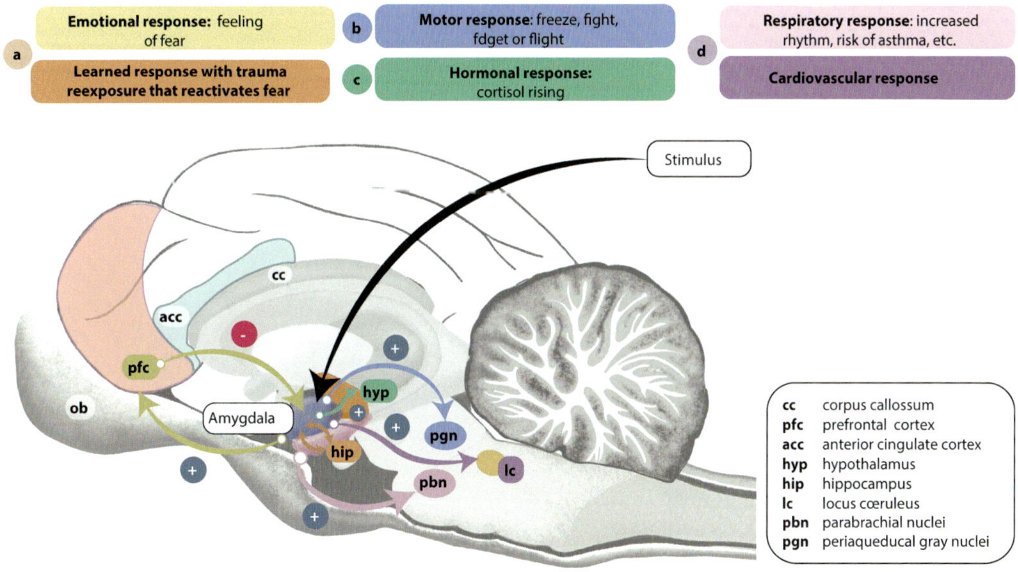

◘ **Fig. 2.35** The amygdala: center of fear regulation. **a** After integrating sensory stimuli (cortical and thalamic inputs), memory elements from the hippocampus (hip), and regulation from the prefrontal cortex (PFC), the amygdala determines whether to trigger fear responses. **b** The motor response to fear (freeze, fight, fidget, flight) results from connections with the periaqueductal gray matter (PAG). **c** The hormonal response to fear (cortisol release) results from connections with the hypothalamus (hyp). **d** The cardiorespiratory response to fear (tachycardia, tachypnea) results from connections with the locus coeruleus (LC) and the parabrachial nucleus (PBN)

- the amygdala's command to the mesencephalon's tegmentum, specifically the periaqueductal gray matter, triggers behavioral fear reactions such as flight, fight, fidget, or freeze.
- the amygdala's command to the locus coeruleus activates descending noradrenergic pathways and the sympathetic system, providing somatic support for behavioral reactions. This includes increased heart and respiratory rates and sweating, which prepare the body to face danger or perceived threats.
- the amygdala's command to the hypothalamus activates the hypothalamic-pituitary-adrenal (HPA) axis, leading to cortisol secretion. This hormonal response initiates a cascade of metabolic and physiological events that support behavioral reactions, such as making glucose available for muscle cells (◘ Fig. 2.35).

The regulation of fear involves numerous inputs to and outputs from the amygdala. These connections are mediated by various neurotransmitters, including GABA, glutamate, serotonin, and norepinephrine. Thus, norepinephrine is not the only neurotransmitter involved in fear. Many psychotropic drugs targeting GABA, glutamate, serotonin, or norepinephrine can be effective in treating phobic pathological states (Stahl and Muntner 2021) (◘ Fig. 2.36).

2.3.1.2 Back to Elementary Processes

In ► Chap. 1, we defined several adaptive processes related to fear: habituation, sensitization, anticipation, and generalization. All these processes correspond to forms of learning, whose general mechanisms were introduced earlier (see ► Chap. 1, ► Sect. 1.1.2.2). At the cellular level, these processes are the result of specific neural circuits, which may dysfunction in pathological states. This section revisits these elementary processes by detailing the neurophysiological mechanisms underlying them and illustrating each with an example.

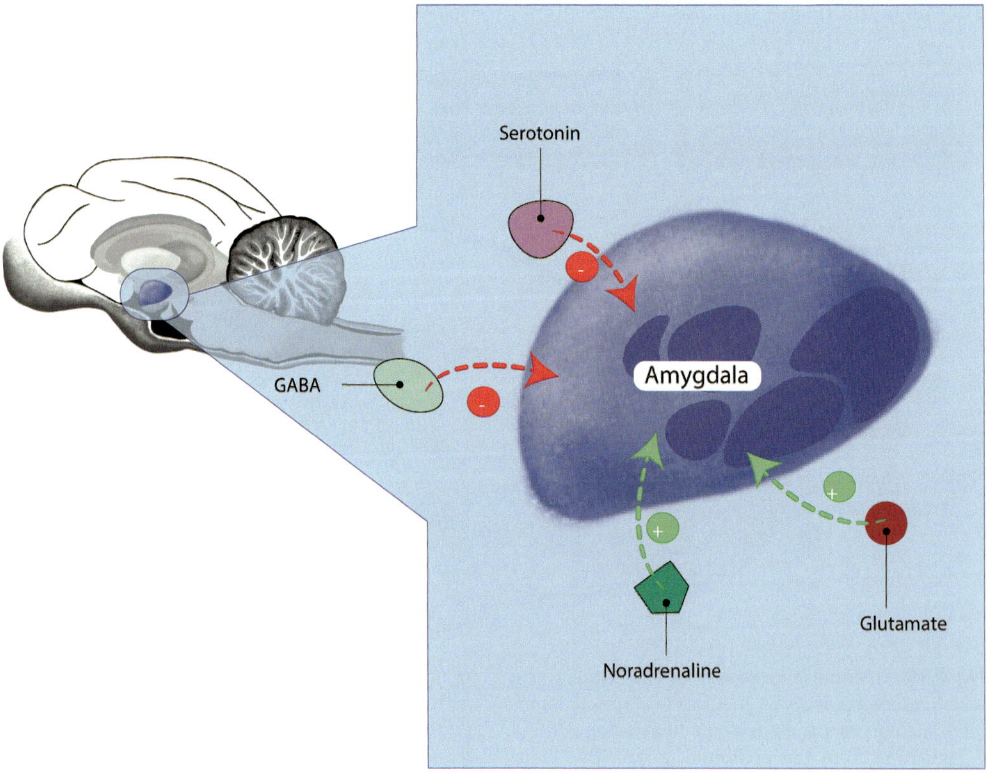

◘ **Fig. 2.36** Neurotransmitters related to fear behaviors. The various neurotransmitters involved in fear responses

Habituation is a relational form of learning resulting from interactions between the amygdala and cortical regions, particularly the prefrontal cortex. Through repeated neutral or positive experiences with a novel stimulus, the amygdala progressively reduces its behavioral, sympathetic, and hormonal fear-related commands (Breiter et al. 1996).

Clinical Illustration

Habituation

When the young couple returned from the maternity ward with their newborn, Nina's initial reaction was to flee and hide under the couch. After an hour or two, as the baby slept in its mother's arms, Nina cautiously approached, sniffed the baby, and lay down against her owner's thigh. Over the following days, the baby's presence and sounds still frightened Nina, but instead of hiding, she distanced herself and observed from afar. Within a week, Nina managed to stay close to the baby and even rubbed against him. She had become habituated to his presence and started engaging in allomarking behavior toward him.

Sensitization is a form of learning in which exposure to a fear-inducing stimulus, if it results in a fear response, strengthens synaptic transmissions in the amygdala, leading to faster and stronger reactions upon subsequent exposure. This process is directly linked to the action of glutamate at the relevant synapse and to long-term potentiation (see ▶ Sect. 2.3.3.1). In other words, if a stimulus is identified by the amygdala as dangerous, the reaction during the next exposure will be quicker, enhancing survival in the event of actual danger. This facilitation, which constitutes the sensitization process, allows for increasingly rapid protective reactions in a dangerous context.

Typically, the amygdala concludes that a stimulus is dangerous because it occurs simultaneously with a stimulus representing an innate fear for cats. This stimulus-stimulus association, characteristic of classical conditioning, is formed through a connection in the lateral nucleus of the amygdala, categorizing the new stimulus as dangerous (◘ Fig. 2.37).

It is worth noting that cats, as a prey species, possess numerous innate fears, which predispose them to sensitization. When sensitization occurs in response to a perceived rather than actual danger, it can lead to the transition from fear, a normal adaptive or reactive state, to a pathological phobic state (Davis et al. 1997).

Clinical Illustration

Sensitization

Fossa and Nimbus had lived together for 2 years in the same apartment and got along cordially. After returning from the veterinarian, where he had been hospitalized for 24 h, Nimbus was threatened by Fossa, who hissed, bristled, and attempted to attack him. Nimbus managed to escape to a high perch. Since then, whenever Fossa enters the room Nimbus occupies, Nimbus runs to hide or climbs as high as possible. Nimbus became sensitized to Fossa's presence after a single frightening encounter.

Anticipation also arises from stimulus-stimulus associative learning through classical conditioning. This process involves the formation of new synapses in the lateral nucleus of the amygdala, linking two inputs: the initial fear-inducing stimulus and a neutral stimulus that precedes the original fear-inducing stimulus in time within the same context. These two inputs merge in the lateral nucleus and produce the same amygdala-driven commands (◘ Fig. 2.37).

As a result, the non-phobic stimulus associated with the original phobic stimulus (now a conditional stimulus) activates the same fear-related amygdala efferents as the original phobic stimulus, even in its absence. This mechanism enables the animal to react more quickly to early warning stimuli that predict a dangerous element (Büchel 2000; Delgado et al. 2006; VanElzakker et al. 2014). Anticipation contributes to the progression of

a.

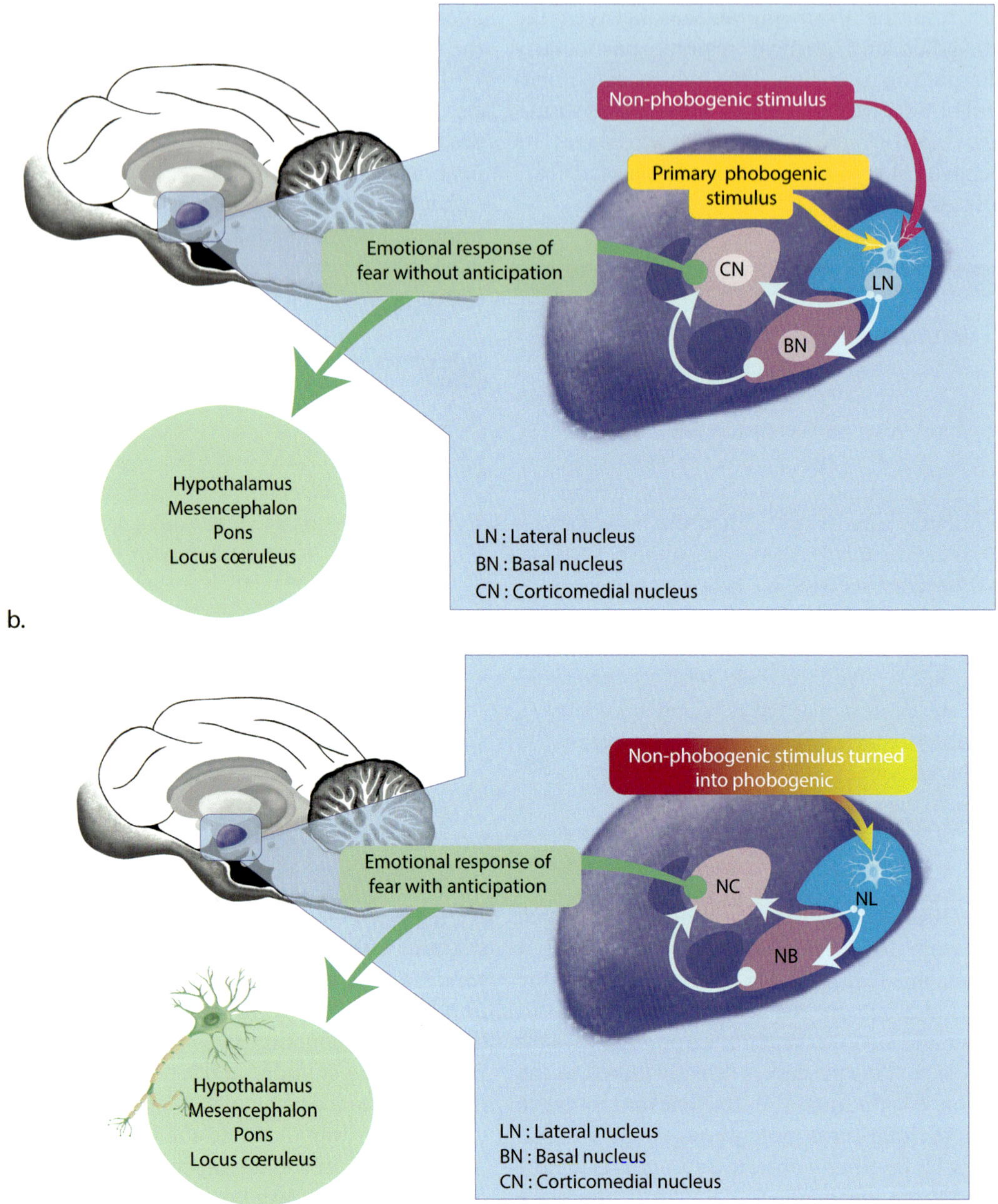

Fig. 2.37 Physiology of sensitization and emotional anticipation in the amygdala. **a** The non-phobic stimulus and the primary phobic stimulus, both occurring in the same context, become associated through new synaptic connections in the lateral nucleus of the amygdala. **b** Once the two stimuli are associated, the initially non-phobic stimulus becomes phobic and triggers the same fear reactions as the primary phobic stimulus

a simple phobic state into a complex phobic state. It is also a process that can escalate a pathological phobic state into an anxious state.

Clinical Illustration

Anticipation

Sony had always been calm at the veterinarian's office. He was quiet and well-behaved in the waiting room, and although slightly tense during the consultation, he allowed himself to be examined without any issues. This behavior was particularly appreciated by his owners and the clinic's staff.

Recently, however, Sony returned home with a painful abscess on his back. Once placed in his carrier, he meowed and even drooled throughout the transport. Upon arrival at the clinic, the veterinarian had to anesthetize him immediately to drain and treat the abscess.

Since this event, as soon as he arrives in the waiting room, Sony meows loudly and hisses at everything that comes near. He anticipates future painful procedures by associating the context of the clinic with the previous treatment.

Generalization is inseparable from discrimination. These two processes rely on the ability to classify similar but distinct stimuli into the same category (generalization) or into different categories (discrimination).

Generalization and discrimination are forms of relational learning involving another part of the limbic system: the hippocampus. The hippocampus receives inputs from the amygdala and various regions of the neocortex via a cortical area close to the limbic system, the limbic cortex. Through its afferent and efferent connections, the hippocampus links emotions, motor behaviors, and memories. In certain situations, it consolidates specific memories. Through this memorization process, negative experiences become associated with particular contexts. Subsequently, contexts resembling those previous threatening contexts may trigger amygdala responses (generalization), even if these new contexts are not inherently dangerous (Dunsmoor et al. 2011).

In discrimination, by contrast, even if stimuli are similar, the fear response is not triggered.

Generalization and anticipation, by broadening the range of stimuli capable of eliciting a fear response, are the processes by which a phobic state (exaggerated fear in response to one or more stimuli) can progress to an anxious state (fear without a specific object). This occurs because they increase the number of stimuli capable of provoking conditioned emotional responses.

Clinical Illustration

Discrimination

José and Brenda live together in the same household and get along very well. They groom each other, sleep together, and play. Recently, new neighbors moved into the adjoining house with a cat that looks almost identical to José. Whenever Brenda sees this new cat, she becomes very aggressive and chases it out of their garden.

However, despite the resemblance, Brenda remains affectionate and playful with José. She perfectly discriminates between the two cats, even though they look very similar.

Clinical Illustration

Generalization

Castor was a generally friendly cat with children but mostly kept his distance. During a birthday party at the household, many children were present. One of them insisted on petting Castor, picking him up, and playing with him, without understanding the signs of distress the cat was displaying. Castor tried to escape or hide, but the child would repeatedly pull him out of his refuge. Eventually, Castor scratched the child.

Since then, Castor has been very wary of all children, including those in the household. He has generalized his fear of children based on one particularly unpleasant experience.

Extinction is a form of learning mediated by the ventromedial prefrontal cortex (vmPFC) (Amano et al. 2010; Sotres-Bayon and Quirk 2010). It actively inhibits previously learned behaviors. The vmPFC integrates environmental information via its afferent connections and issues commands for a variety of behaviors and physiological and emotional responses through its efferent connections (◘ Fig. 2.38).

Repeated exposure to the same conditional stimulus without any painful or fear-inducing consequences leads to changes in vmPFC connections. Over time, the stimulus is recognized as inconsequential and no longer triggers reactions. However, the fear response itself is not erased but actively inhibited by the cortex. As a result, conditioned emotional responses are never completely erased. They may be diminished through repeated neutral experiences, but a single adverse exposure can be enough to reactivate them.

In the process of habituation, the same vmPFC-dependent mechanisms are involved, but the synapses of the amygdala have not been sensitized beforehand, meaning there are no connections to inhibit. Consequently, if a fear response does occur, it is less intense and slower after habituation compared to extinction.

The phenomenon of extinction highlights how it is neurophysiologically impossible to "unlearn" or erase a negative or painful association with an event. Cats, due to their dual status as prey and predator, are particularly sensitive to aversive stimuli. Painful or frightening exposures leave vulnerabilities and indelible traces in their brains.

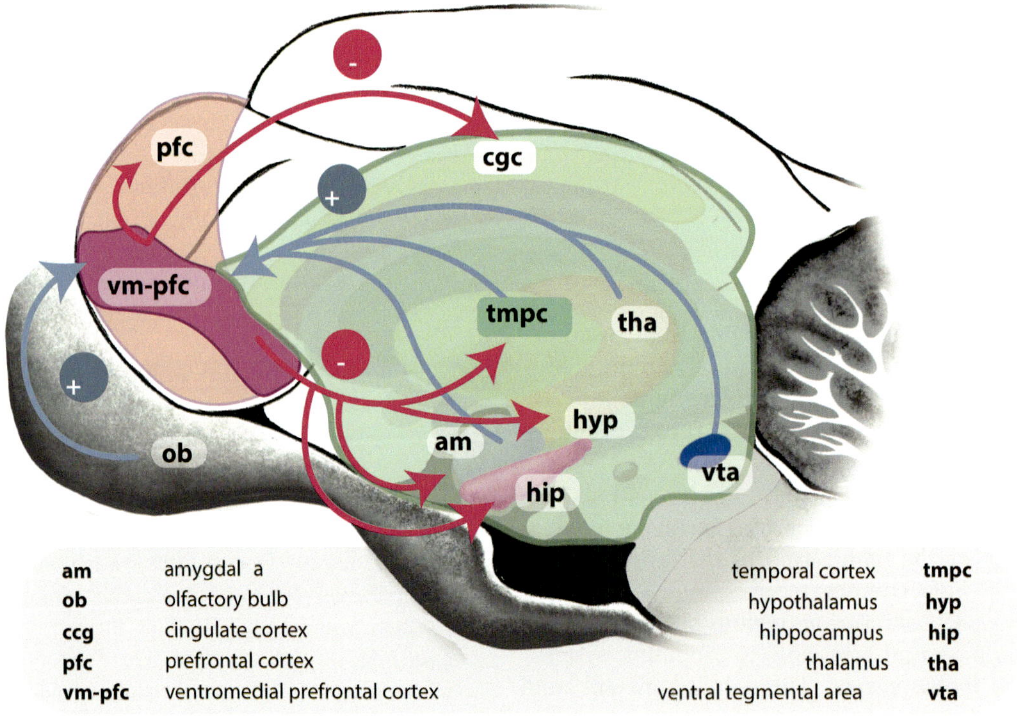

◘ **Fig. 2.38** The prefrontal cortex and its role in habituation and extinction processes. The ventromedial prefrontal cortex (vmPFC) receives inputs from the thalamus, temporal cortex, ventral tegmental area, olfactory system, and amygdala. It projects connections to multiple brain areas, including the cingulate cortex, hippocampus, temporal cortex, hypothalamus, and amygdala, as well as other prefrontal regions. The vmPFC centralizes sensory, emotional, and memory information, with its connections adapting based on the outcomes of exposures

Clinical Illustration

Extinction

Two weeks ago, Flora began meowing and scratching at the bedroom door of her owners during the night. Exhausted, the young couple gave in a few times and opened the door for her. Since then, Flora has returned to meow and scratch at the door every night.

The owners, unwilling to let Flora join them at night, were advised to stop giving in and refrain from opening the door to avoid reinforcing Flora's behavior. This approach, known as behavior extinction, involves not rewarding Flora's scratching at the door. They were warned that at the beginning of the process, Flora would likely escalate her behavior—meowing and scratching more intensely when her efforts prove ineffective. This temporary increase in behavior is referred to as an "extinction burst."

After 10 days of patience and consistency, Flora completely gave up trying to enter the bedroom at night.

The study of the amygdala circuit and the associated elementary processes highlights numerous activating afferent pathways but only one inhibitory pathway: the one involving the ventromedial prefrontal cortex (vmPFC). When an individual has associated a context with danger, the vmPFC is bypassed. The resulting reaction is rapid and uncontrollable, referred to as the *short pathway* of the circuit. To manage an emotional reaction, particularly fear, the individual must engage the so-called *long pathway*, which involves the vmPFC. In this pathway, the cortex analyzes the situation and, if it determines there is no real danger, prevents the emotional reaction from occurring.

The short pathway is favored by processes such as sensitization, anticipation, and generalization. While the preferential use of this short pathway can be advantageous for adapting to actual danger, it becomes a disadvantage when responding to perceived but unreal threats. This is precisely what happens in the transition from fear to a phobic state.

Treating this pathological state requires intervention in the amygdala circuit to reduce its level of activation (as achieved by drugs like gabapentin or clonidine) or to enhance the action of the inhibitory circuit (as achieved by fluoxetine, for instance).

2.3.1.3 The Cortico-Striato-Thalamo-Cortical Loop (CSTC)

In addition to the amygdala circuit, there is another specific loop associated with anxiety. As discussed earlier in this section, the processes of anticipation and generalization facilitate the transition from a phobic state to an anxious state. Indeed, activation of the amygdala circuit also triggers the cortico-striato-thalamo-cortical (CSTC) loop.

However, anxiety is more than an advanced form of a phobic state. It can also emerge independently. Anxiety is not solely linked to the functioning of the limbic system or the amygdala circuit but instead depends on the CSTC loop. This loop originates in the prefrontal cortex, connects to the striatum (part of the basal ganglia), descends to the thalamus, and then returns to the prefrontal cortex (Marchand 2010; Peters et al. 2016) (■ Fig. 2.39). The CSTC loop is also involved in obsessions and hallucinations. It does not respond to actual dangerous situations but anticipates the potential emergence of such situations. It corresponds to a heightened state of vigilance, where the brain is on high alert and actively scanning the environment for potentially threatening stimuli.

As with the amygdala circuit, the CSTC loop is beneficial in enabling an animal to increase its vigilance and reactivity in potentially hostile contexts. However, it becomes maladaptive when overactivated in a non-dangerous environment. This loop allows the individual to anticipate the emergence of a risky situation, and it logically activates after a verified dangerous situation has occurred, that is, once the amygdala circuit has been triggered.

Under intense or repeated stress, the CSTC loop becomes activated, with each activation increasing the likelihood of transitioning into a pathological anxious state. In other words,

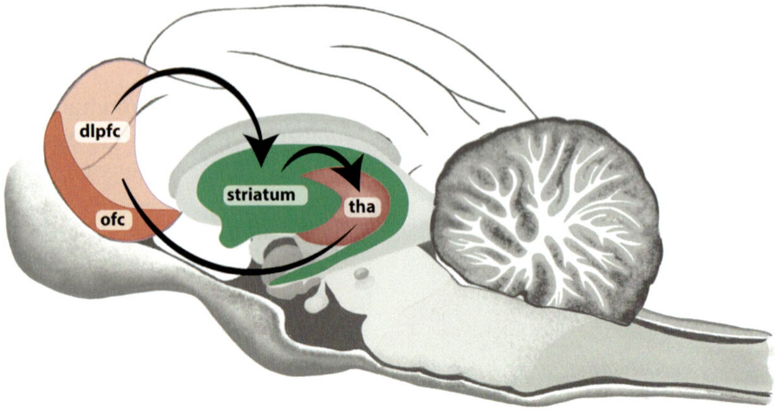

dlpf	dorsolateral prefrontalcortex
ofc	orbitofrontal cortex
striatum	accumbens nucleus caudate nucleus putamen
tha	thalamus

Fig. 2.39 Cortico-striato-thalamo-cortical loop in anxiety states. Anxiety states are linked to a loop that begins in the prefrontal cortex, connects to the striatum, descends to the thalamus, and returns to the prefrontal cortex

when the brain is exposed to chronic stress, the transition to an anxious state is facilitated.

The neurotransmitters regulating the CSTC loop include serotonin, GABA, glutamate, norepinephrine, and dopamine (Fig. 2.40). In humans, studies have shown that the enzyme COMT, which regulates dopamine levels in the prefrontal cortex, exists in different genetic variants. Each variant leads to differences in prefrontal cortex functioning and CSTC loop activity. These differences partly explain individual vulnerabilities to anxiety and other conditions involving the prefrontal cortex, such as ADHD (Stahl and Muntner 2021).

It is likely that similar genetic variants exist in cats. This suggests that some individuals, even when raised in good environmental conditions and without trauma, are more susceptible to transitioning into an anxious state.

2.3.2 Mood: From the Neurotransmitter Theory to the Receptor Theory

Unlike emotions, mood is not easily identifiable as a specific circuit. Instead, it reflects a general balance between neurotransmitters and their receptors throughout the brain. The neurophysiological mechanisms underlying mood are not yet clearly understood, and several hypotheses have been proposed.

The oldest hypothesis suggests that mood results from the action of the three monoamines—norepinephrine, dopamine, and serotonin—and their circuits. Mood and its variations, including pathological changes, are believed to stem from quantitative modifications in monoamine receptors. These generalized changes trigger a series of molecular

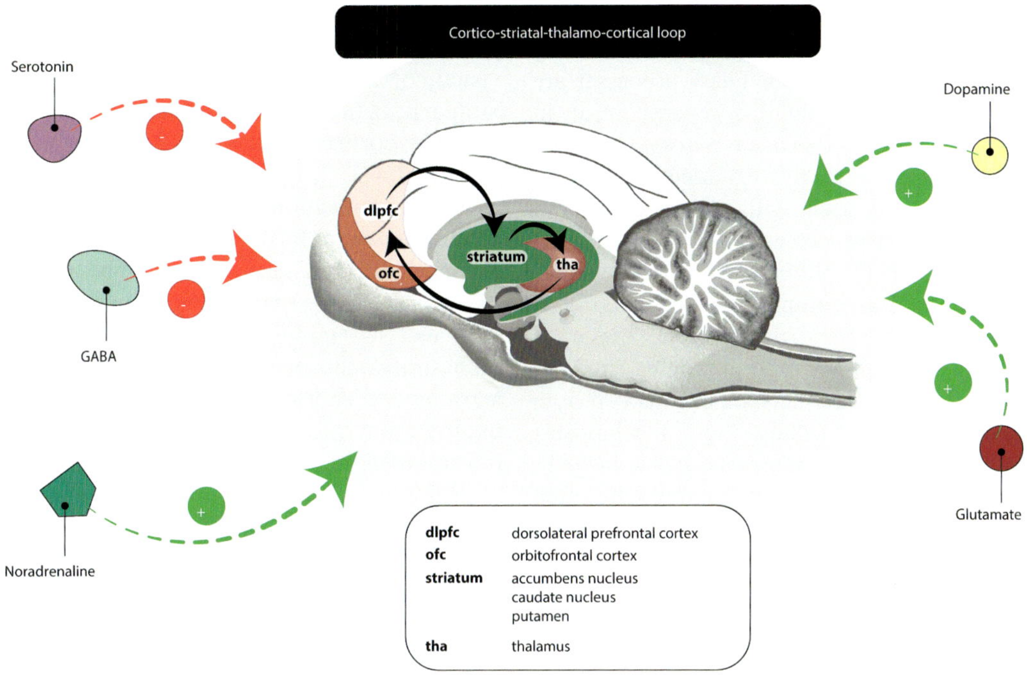

Fig. 2.40 Neurotransmitters involved in an anxious state

events in postsynaptic neurons, including alterations in gene expression within their nuclei. This process also influences growth factors, notably the *Brain-Derived Neurotrophic Factor* (BDNF) (Elhwuegi 2004; Jesulola et al. 2018; Leuchter et al. 2009; Martinowich et al. 2007; Wong and Licinio 2004).

BDNF plays a key role in promoting the growth and survival of monoaminergic neurons and maintaining synaptic connections.

Since the distribution and function of monoamine receptors are governed by the genome, genetic expression and its regulation (epigenetics) significantly influence mood and mood disorders. For example, a study (Caspi et al. 2003) on the genes encoding serotonin receptors (5-HT receptors) demonstrated a genetic vulnerability to depressive symptoms and suicidal thoughts in response to stressful events. The promoter region of the gene encoding 5-HT receptors exists in two allelic forms: short and long. The study found that individuals carrying at least one copy of the short allele were more vulnerable to depression than those with two copies of the long allele.

Other theories highlight the role of glutamate and glial cells in mood disorders (Haroon et al. 2017; Schroeter et al. 2010). This theory is not incompatible with the monoamine receptor theory, as glial cells influence all monoaminergic neurons. It provides a framework for understanding the potential global impact on brain function, as well as the effects on BDNF and neurotransmitters as a whole.

Currently, the neurophysiological mechanisms involved in mood regulation remain debated and controversial. However, their dysfunction is recognized as the basis for the emergence of pathological states such as depressive, manic, and bipolar states (see ► Chap. 7, ► Sect. 7.3).

2.3.3 Cognition

2.3.3.1 Neurophysiology of Learning and Memory

In ► Chap. 1, we explored three types of learning (i.e., perceptual, associative, and relational) (see ► Sect. 1.2.3.4), and three types of memory: sensory, working, and long-term memory (see ▸ Sect. 1.2.3.5). Learning depends on the proper functioning of elementary processes. These processes are typically adaptive, such as generalization in the organization of a cat's territory. However, they can become dysregulated, hindering adaptation, as seen with anticipation and generalization processes in the development of pathological phobic or anxious states when a cat's environment is too unstable or unpredictable.

The following paragraphs delve into the neurophysiological mechanisms involved in learning. Understanding these mechanisms helps practitioners comprehend why, in pathological states, learning and memory are often quickly impaired.

Perceptual learning involves recognizing stimuli and memorizing their characteristics. It relies on synaptic modifications in the associative sensory cortex, which underlie the memorization of the stimulus. Sensory pathways become facilitated, meaning that nerve impulses travel more quickly along these specific routes. For example, if an animal perceives a visual signal, the eye-to-visual-cortex pathway is strengthened. Similarly, if the signal is auditory, the ear-to-auditory-cortex pathway is reinforced (◘ Fig. 2.41).

The modifications occurring during *perceptual learning* form the basis of *sensory memory* and *short-term memory*. These types of memory are fleeting, lasting from a few seconds to a few minutes. The *prefrontal cortex* orchestrates the organization of perceptual learning and short-term memory by discarding irrelevant information and reinforcing significant data.

Stimulus-stimulus *associative learning* through classical conditioning is linked to involuntary reactions, whether associated with positive emotions (pleasure) or negative ones (fear, pain) (Yin 2015) (see ▸ Chap. 1, ▸ Sect. 1.2.3.4).

The neurophysiological basis of this type of emotional conditioned learning lies in the amygdala, as previously described (see ▸ Sect.

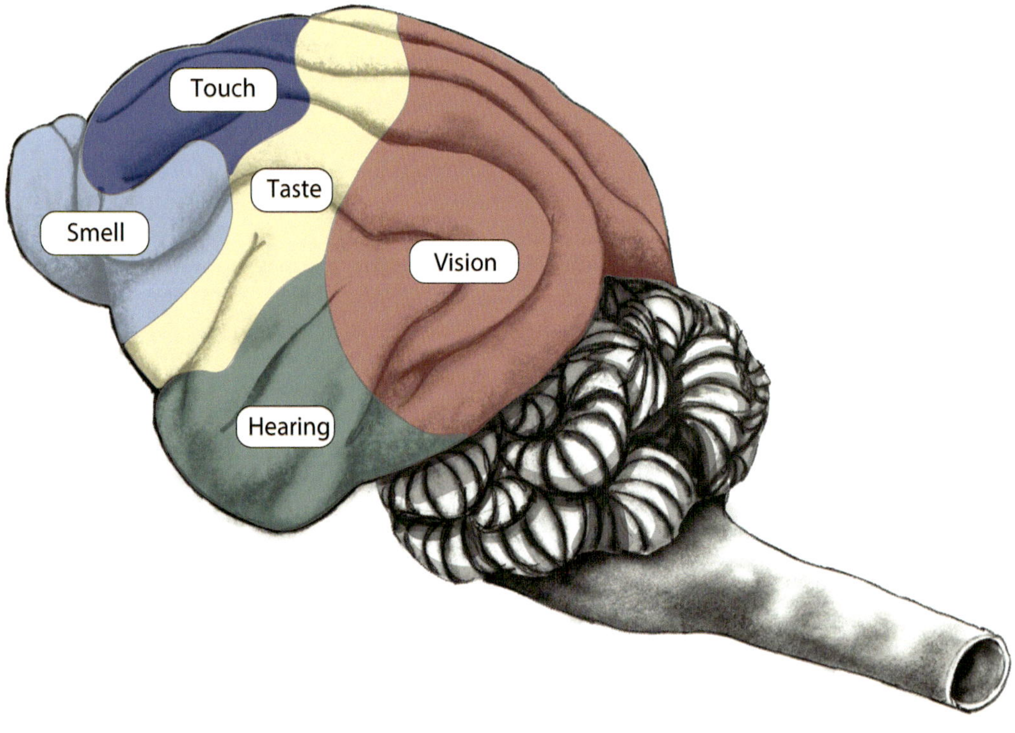

◘ **Fig. 2.41** The different sensory areas in the cat's brain. The more a sensory area is activated and processed by its corresponding associative cortex, the more efficient the corresponding pathway becomes

2.3.1.1). The lateral nucleus of the amygdala receives sensory inputs related to the conditional stimulus and the unconditional stimulus. It then sends axonal projections to the basal and central nuclei of the amygdala. Initially, the sensory inputs corresponding to a neutral stimulus, such as a sound, have little effect on transmission between the lateral nucleus and the effector nuclei. However, associating this neutral stimulus with an unconditional stimulus, like fear or pain, strengthens the axon-dendrite connections between the effector neurons of the unconditional stimulus and the neurons linking the lateral nucleus to the effector nuclei.

This phenomenon is described by Hebb's Law: "*Neurons that fire together, wire together*" (Brown and Milner 2003). The synapse between neurons encoding two different stimuli becomes consolidated, such that stimulating either neuron results in the same postsynaptic activation. Consequently, the conditional stimulus eventually has the same effect on the amygdala as the unconditional stimulus.

As a result, a particular context, even without the presence of the unconditional stimulus, can produce the same effects, inducing fear or pain as long as the learned association exists (◘ Fig. 2.42). For example, in litter box aversion following a painful episode of cystitis in cats, the aversion persists even after the cystitis is treated and the pain subsides. The association between the litter box and the pain results in sensitization to the litter box.

Operant conditioning begins in various parts of the sensory cortex and culminates in the activation of the motor cortex in the frontal lobe. Initially, the *direct transcortical pathway* is used. This pathway involves information processing through the thalamus and the basal ganglia (Balleine et al. 2009) and works in conjunction with the hippocampus, a brain structure involved in memory. This direct pathway is engaged during the learning phase, allowing the acquisition of complex behaviors that require instruction following and deliberation. The speed of information acquisition depends on the emotional valence provided by the amygdala (Prichard et al. 2018; Witte and Kipke 2005).

With repetition, a shorter circuit is formed that bypasses the cortex and relies solely on the basal ganglia (Bayley et al. 2005). The learned behavior becomes automatic and habitual. The dorsal striatum (caudate nucleus + putamen) plays a key role in this transition of control (Devan et al. 2011). It receives inputs from all sensory cortical regions as well as the frontal motor lobe. The behavior is then produced automatically, in a conditioned manner (◘ Fig. 2.43). Operant conditioning can thus be represented as a neuronal pathway that becomes more robust with each use, eventually creating an optimal pathway. This pathway becomes so direct and consolidated that the action can be performed without conscious thought, freeing the cortex from involvement. Thanks to this learning process, we can perform mechanical tasks while thinking about something else, a phenomenon known as the *instrumentalization* of behaviors.

Relational learning is the most complex form of learning and involves all brain regions. It enables the association of sensory elements with memory and emotional components. For example, hearing neighborhood cats fighting may evoke a mental image of the cats, their color, or the texture of their fur if touched.

The hippocampus plays a central role in memory mechanisms. Through its multiple afferent connections, it acts as an observer of the neuronal modifications underlying sensory and short-term memories. These memories are "stored" in the brain regions associated with the event that generated them. For instance, auditory memory is stored in the associative auditory cortex, short-term motor memory in the associative motor cortex, and procedural memory in the basal ganglia.

While the recognition of stimuli is managed by the sensory cortex, their organization and classification as relevant or irrelevant require the involvement of the prefrontal cortex and the basal ganglia (Fukushima et al. 2004). The basal ganglia filter out unnecessary information, transmitting only the useful data to the prefrontal cortex, which then maintains it in memory (Baier et al. 2010).

When an event is repeated, the hippocampus retrieves disparate memory elements, con-

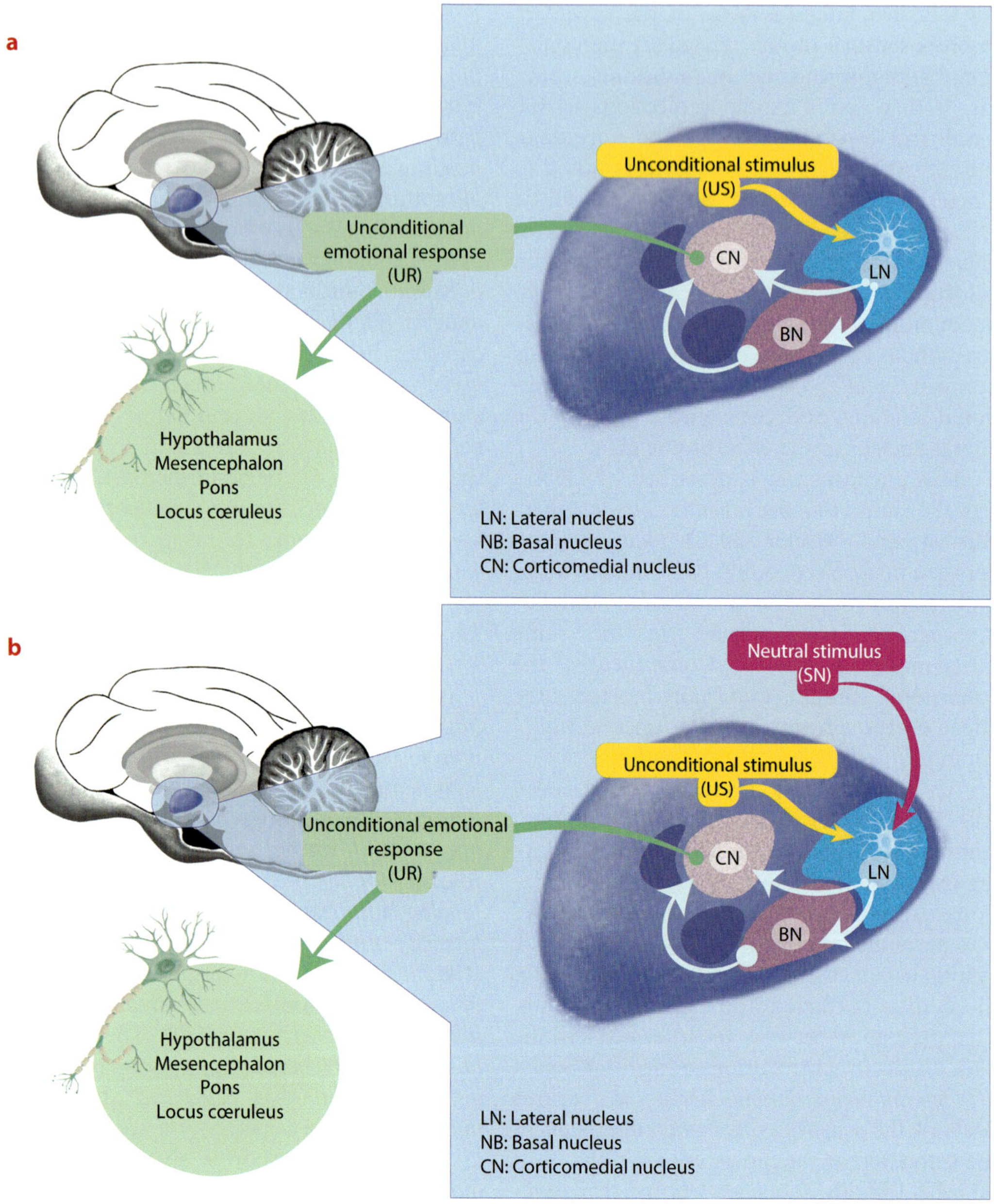

Fig. 2.42 Physiology of associative learning via classical conditioning. **a** The unconditional stimulus triggers an emotional response. **b** The neutral stimulus and the unconditional stimulus, which occur together in the same context, become associated through new synaptic connections in the lateral nucleus of the amygdala. **c** Once associated, the initial neutral stimulus becomes conditional (sensitization) and triggers the same emotional responses as the original unconditional stimulus

solidates them, and establishes new neural connections in the cortex. This process underlies *relational learning* and *long-term memory* formation. The hippocampus, along with the olfactory bulb, is unique in its ability to generate neurons throughout life. Memory information is not "stored" in the hippocampus but depends on its connections to various brain areas.

Memory should not be visualized as a discrete area in the brain, like a hard drive, but rather as a network of grooves that deepen

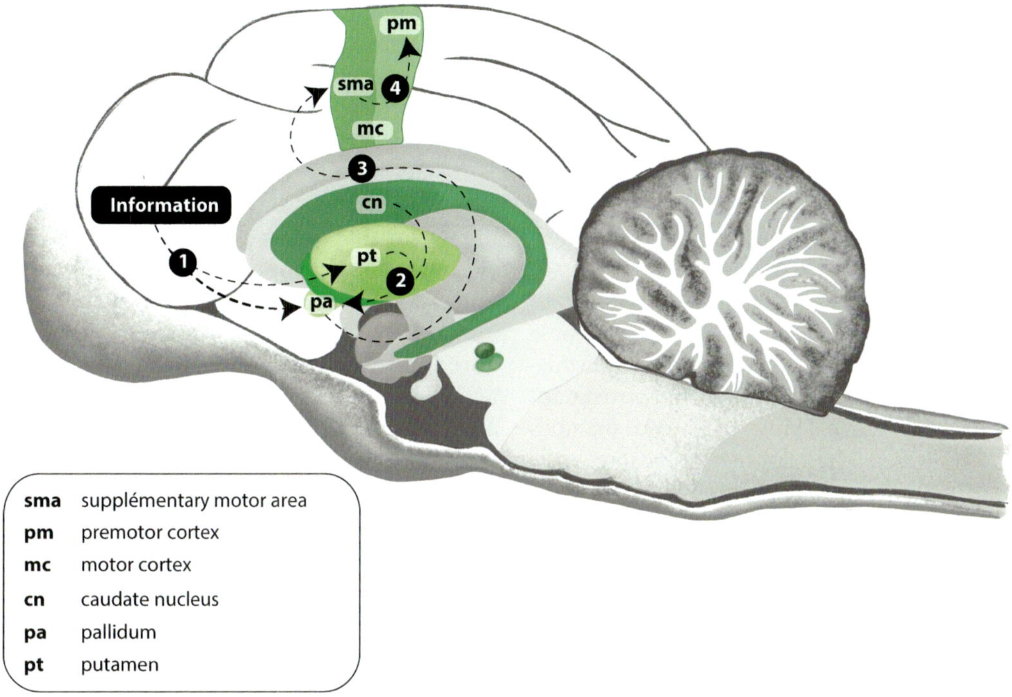

Fig. 2.43 Associative learning and behavior instrumentalization. With repetition, behaviors become automatic. Control gradually shifts from the cortex to the basal ganglia, freeing cortical areas for other functions. This is how we can ride a bicycle while solving complex problems

with repeated information transmission. The deeper the groove, the faster the information can travel through it.

All learning and memory phenomena result from the consolidation of synaptic transmission. At the cellular level, this consolidation is driven by the phenomenon of *long-term potentiation* (LTP). LTP is fundamental to learning and memory because it modifies synaptic connections. These durable changes allow information to be stored (Shors and Matzel 1997).

LTP is directly linked to the neurotransmitter *glutamate* and the unique properties of its receptors. All neurons have glutamate receptors on their dendrites or soma. Glutamate interacts with two major types of ionic receptors: *NMDA receptors* and *AMPA receptors* (see ▶ Sect. 2.2.4.2., « glutamate »).

NMDA receptors are typically blocked by a magnesium ion (Mg^{2+}). Opening the ion channel of these receptors requires two simultaneous conditions. The first is postsynaptic membrane depolarization, which expels the Mg^{2+} ion blocking the channel. The second is the binding of glutamate to the receptor, which becomes accessible after the Mg^{2+} ion is expelled. Once these conditions are fulfilled, the channel opens, allowing calcium ions (Ca^{2+}) to enter the cell.

The influx of Ca^{2+} ions has two critical effects. On one hand, it strengthens the postsynaptic potential, enhancing the excitatory signal. On the other, it triggers the migration of AMPA receptors to the postsynaptic membrane, positioning them near the activated NMDA receptors.

AMPA receptors control sodium channels. When activated by glutamate, they generate excitatory potentials in the postsynaptic membrane, which facilitates the expulsion of Mg^{2+} ions and thereby opens additional calcium channels.

As the postsynaptic membrane becomes increasingly depolarized, NMDA receptors recruit more AMPA receptors and open more readily. In other words, the more a neuron is activated, the more it becomes activatable. This self-reinforcing process of synaptic transmission consolidation is the foundation of long-term potentiation (◘ Fig. 2.44). This mechanism strengthens synapses, deepening the metaphorical grooves in the neural network.

Long-term potentiation (LTP) also involves presynaptic changes in the glutamatergic neuron, resulting in an increased release of glutamate in the terminal bouton. These changes are mediated by retrograde transmission of nitric oxide (NO) from the postsynaptic neuron. Establishing LTP also includes morphological changes in the postsynaptic dendritic spines, which become thicker and, in some cases, even grow new spines. As a result, NMDA receptors become more numerous, further amplifying the phenomenon.

The hippocampus is particularly rich in NMDA receptors, which explains its pivotal role in memory formation. The LTP described here represents an early stage of potentiation. Long-lasting LTP, which endures for more than a few hours, requires the synthesis of a protein called PKM-zeta. This protein facilitates the continuous movement of AMPA receptors via a positive feedback loop (Sacktor 2011). While LTP was initially discovered in the hippocampus, it also occurs in other brain regions.

Additionally, there is a related phenomenon called long-term depression (LTD), which weakens synaptic connections by reducing the number of AMPA receptors in the membrane. The interplay of LTP and LTD underpins synaptic plasticity by strengthening certain synapses while weakening others, thereby enabling the encoding of information in memory.

2.3.3.2 Neurophysiology of the Attentional System: Wake-Activation Circuit

Maintaining an arousal system is essential for the activation of behaviors in cats. This maintenance is both homeostatic and allostatic, as it is crucial for an organism to stay awake in the presence of danger or during activities such as seeking food or water, which are vital for survival.

Glutamate plays a central role in activation, functioning as a diffuse system that engages the entire brain. It is likely that no brain cell is devoid of connections to the glutamatergic activation system.

In addition to the glutamate system, five neuronal circuits are important for maintaining the arousal system. Three of these circuits were described in ► Sect. 2.2.4 on neurotransmitters (see ► Sect. 2.2.4). These include cholinergic circuits involved in cortical activation, noradrenergic circuits responsible for vigilance levels, and serotonergic circuits associated with automatic behaviors such as locomotion and grooming.

Two additional circuits contribute to the arousal system: the histaminergic and orexinergic circuits, both originating in the hypothalamus (◘ Fig. 2.45). The involvement of the histaminergic circuit in wakefulness explains why all antihistaminic drugs, including certain psychotropics, can suppress the animal's arousal state.

The activator system is constantly counterbalanced by an inhibitory system. All brain regions, and therefore all behaviors, are under the constant regulation of this balance between activation and inhibition. For every activation initiated by an excitatory neuron, there is an immediate activation of inhibitory neurons. This inhibition prevents neural overexcitement, also known as runaway excitation.

In 2014, the team led by Professor Scanziani demonstrated that the activation-to-inhibition ratio remains nearly constant over time (Xue et al. 2014). In other words, when the brain presses the accelerator, it simultaneously applies the brake. Inhibitory neurons act as regulators, ensuring that inhibitory synapses are adjusted to the level of excitation received by these cells for every stimulus.

When the activation-to-inhibition ratio is altered, either by excessive activation or insufficient inhibition, the brain's adaptive capacity is overwhelmed, leading to pathology. This imbalance has been demonstrated in human medicine in several disorders, including autism (Nelson and Valakh 2015), epilepsy (Cossart et al. 2001), and schizophrenia (Ford et al. 2019).

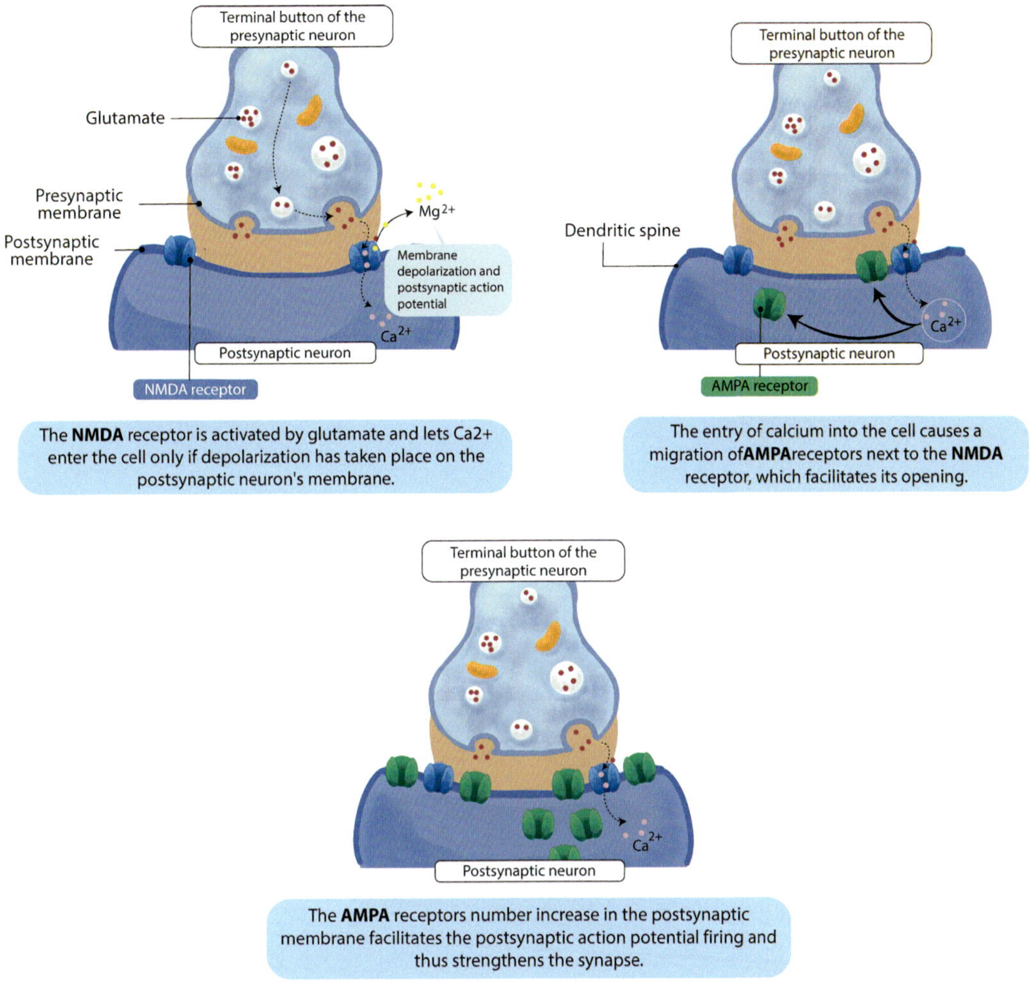

Fig. 2.44 Long-term potentiation (LTP). **a** Depolarization of the postsynaptic membrane releases the Mg^{2+} ion that blocks the NMDA ion channel. Glutamate binding to the NMDA receptor causes the channel to open, allowing an intracellular influx of Ca^{2+}. **b** This influx induces stronger depolarization of the postsynaptic neuron and the recruitment of AMPA membrane receptors, which, in turn, enhance the intracellular influx of Ca^{2+} and the action potential. **c** The synaptic connection becomes progressively stronger and more efficient

The primary inhibitory system in the brain is the GABA system. Like glutamate, it is a diffuse system, and it is unlikely that any brain cell lacks a connection to the inhibitory GABAergic system. Serotonin, by stimulating the GABA system, is also an important factor in inhibition (Lee et al. 2010).

The balance between activation and inhibition is maintained by GABAergic interneurons. These interneurons function as switches for the activator system. When a glutamate neuron activates another neuron, it also activates GABAergic interneurons, which immediately regulate the level of activation (Chiu et al. 2019) (Fig. 2.46).

2.3.3.3 Neurophysiology of the Motivational System: The Reward Circuit

Positive reinforcement learning is a subset of associative learning (see ► Chap. 1, ► Sect. 1.2.3.4, « associative learning »). It forms the basis of many behavioral therapy techniques, making an understanding of its neurophysiological mechanisms important for veterinary psychiatrists.

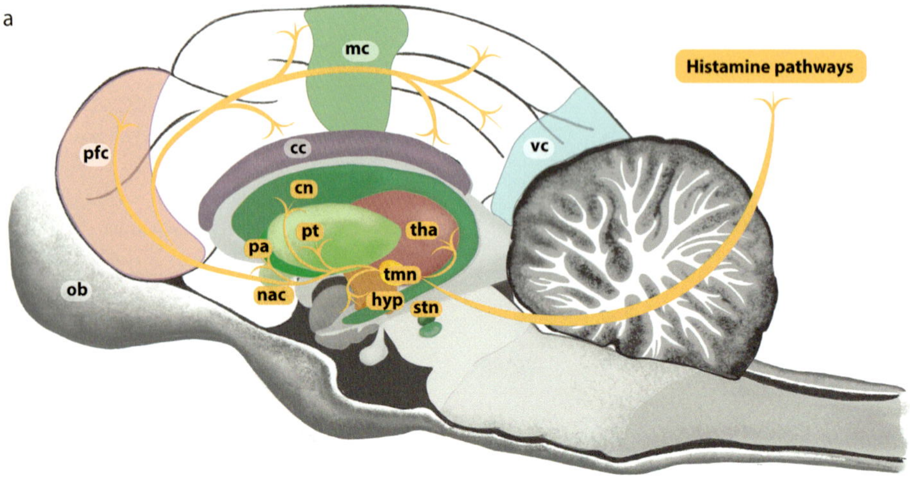

mc	motor cortex	**stn**	subthalamic nucleus
vc	visual cortex	**ntb**	tuberomamilar nucleus
ob	olfactory bulb	**nac**	accumbens nucleus
cc	corpus callossum	**pa**	pallidum
pfc	prefrontal cortex	**pt**	putamen
hyp	hypothalamus	**tha**	thalamus
cn	caudate nucleus		

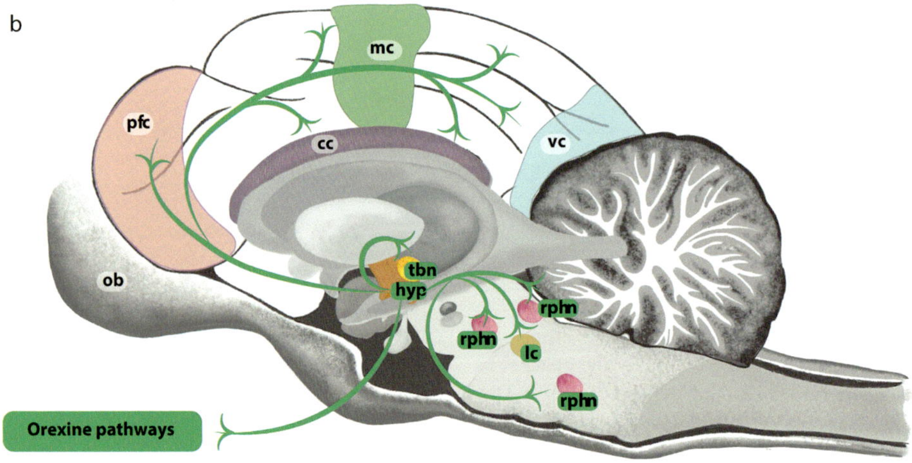

mc	motor cortex	**hyp**	hypothalamus
vc	visual cortex	**lc**	locus cœruleus
ob	olfactory bulb	**rphn**	raphe nuclei
cc	corpus callossum	**tbn**	tuberomamilar nucleus
pfc	prefrontal cortex		

Fig. 2.45 Histaminic and orexigenic pathways. These two pathways are integral to the arousal system. **a** The histaminergic circuits originate from the tuberomammillary nucleus in the hypothalamus and project to the cortex, thalamus, basal ganglia, and other hypothalamic regions. They also interact with cholinergic circuits, enhancing their activity. **b** The orexinergic pathways have their cell bodies in the lateral hypothalamus, with axons projecting to nearly all brain regions, particularly the cortex and areas involved in wakefulness maintenance: the locus coeruleus (origin of noradrenergic pathways), the raphe nuclei (origin of serotonergic pathways), the tuberomammillary nucleus (origin of histaminergic pathways), and cholinergic pathways

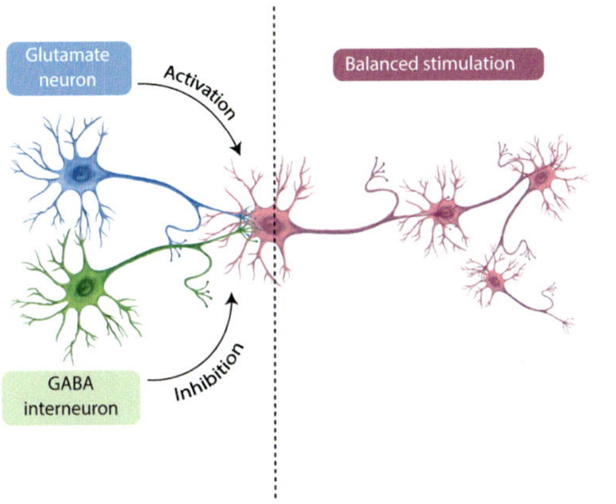

Fig. 2.46 The dual activation-inhibition system. Glutamate neurons stimulate both the target neuron and inhibitory GABAergic interneurons, thereby regulating the level of stimulation

All natural reinforcers, such as eating in response to hunger, drinking in response to thirst, or orgasm in sexual behaviors, share a common physiological effect: they elevate dopamine levels in the nucleus accumbens (a basal ganglia nucleus). This dopamine increase is essential for the establishment of positive reinforcement of a behavior and is responsible for the sensation of pleasure.

Dopaminergic circuits are therefore crucial in operant conditioning, pleasure, and compulsive behaviors that may follow (Di Chiara and Bassareo 2007; Wise 1999). To understand the reward circuit, it is necessary to revisit dopaminergic pathways, particularly the mesolimbic and mesocortical pathways. These pathways originate in the ventral tegmental area.

A potent reinforcer, such as an unexpected and highly valued reward like a new appetizing treat, activates neurons in the ventral tegmental area. These neurons, in turn, activate the nucleus accumbens through the mesolimbic pathway, generating pleasure. Simultaneously, these neurons activate various cortical regions related to memory, including the prefrontal cortex, via the mesocortical pathway (Irle and Markowitsch 1984; Markowitsch and Pritzel 1977) (Fig. 2.47). As a result, the cat experiences significant pleasure from receiving the new treat while associating the reward with memory-related data, such as the circumstances or actions that led to obtaining it (e.g., meowing and rubbing against its owner's legs).

Dopamine acts as a trigger for long-term potentiation (see ▶ Sect. 2.3.3.1), contributing to the formation and maintenance of new synaptic connections within the reward circuits (Dallérac et al. 2011; Wittmann et al. 2005).

Additionally, the mesocortical pathway, which connects the ventral tegmental area to the prefrontal cortex, is reciprocally activated by the prefrontal cortex. This region projects excitatory glutamatergic neurons to the ventral tegmental area, stimulating the dopaminergic neurons located there.

The prefrontal cortex, involved in strategies and evaluations of progress toward goal-directed behavior, activates the reward circuit whenever behaviors bring the animal closer to its goal and the strategy proves effective (Carr and Sesack 2000; Tzschentke 2000; Gariano and Groves 1988; Seong 1996).

Clinical Illustration

Princesse greatly enjoys playing with running water. As soon as the kitchen faucet is turned on, she plays with the stream and begins to drink. When her owner is in the kitchen, Princesse starts meowing and rubbing against her legs. The owner then turns on the faucet, and Princesse jumps onto the counter to play with the water and drink.

This interaction has developed into a ritual of communication between Princesse and her owner, reinforced by the positive feedback of meowing followed by the faucet being turned on. It has now become nearly impossible to cook in the kitchen with Princesse present without turning on the faucet, as her persistent meowing and allomarking behaviors in this context are highly insistent.

All drugs and psychotropics that act on dopamine can stimulate learning and pleasure. For instance, individuals who received L-Dopa, a precursor of dopamine, demonstrated better

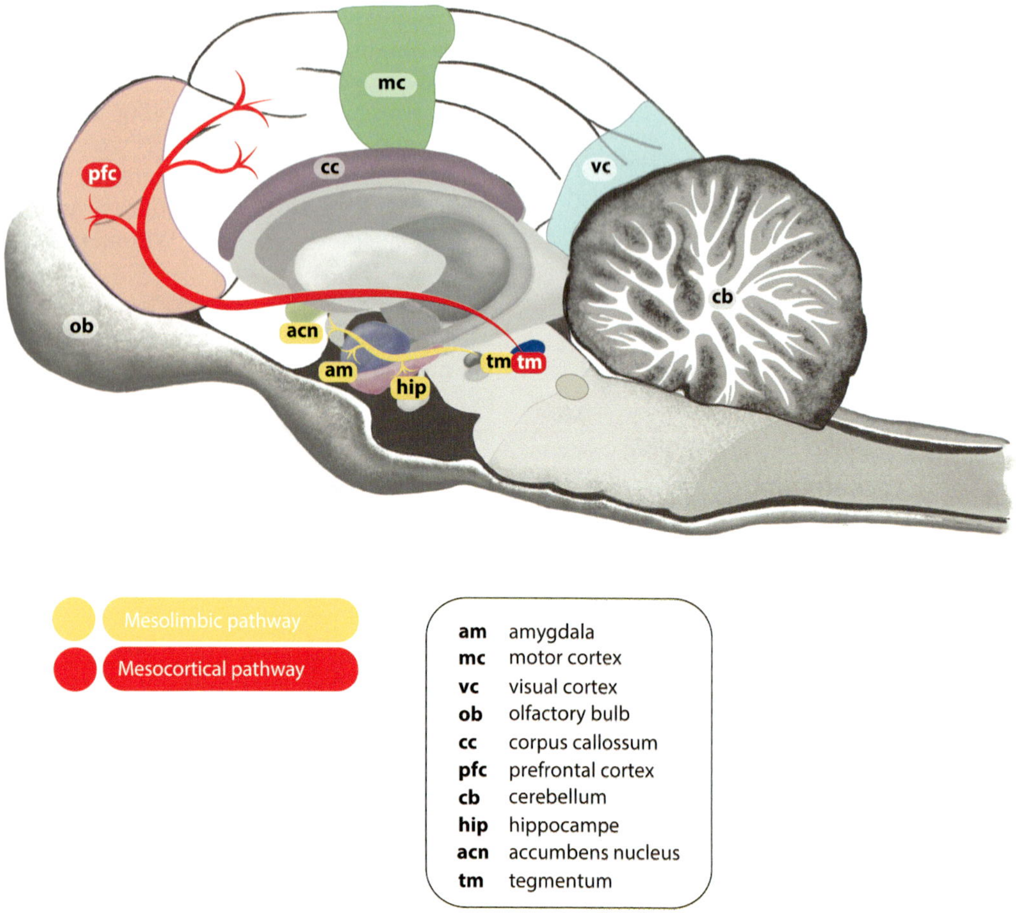

Fig. 2.47 Mesolimbic and mesocortical dopaminergic pathways in cats. These two dopaminergic pathways are involved in positive reinforcement learning by simultaneously activating pleasure sensations and memory formation

performance in learning specific tasks compared to a control group that did not receive it (Knecht et al. 2004).

All addictive substances stimulate the mesolimbic pathway and dopamine secretion in the nucleus accumbens, creating a sensation of pleasure. However, they also positively reinforce the behaviors that produced this pleasure, leading to compulsive repetition of these behaviors and addiction (Schultz 2011; Mameli and Lüscher 2011; Kauer and Malenka 2007; Saal et al. 2003).

In humans, certain mutations in the gene encoding the dopamine D2 receptor have been shown to cause hypersensitivity to reward and almost immediate addictive behaviors (Blum et al. 1996; Noble 2000). This vulnerability to addiction explains the automation of certain behaviors (see the instrumentalization process and compulsive states). An unexpected, novel, and highly appetitive reward can accelerate this process (Gerencsér et al. 2018).

2.3.3.4 Neurophysiology of Action Control: Circuits of Impulsivity and Compulsion

The GABAergic inhibitory system (see ▶ Sect. 2.2.4.2, « GABA ») encompasses the neurophysiological mechanisms that enable the brain to say "no." When these mechanisms fail, certain neurophysiological circuits become overactive. Among these circuits, those governing impulsivity and compulsion warrant specific attention.

Impulsivity is the inability to refrain from initiating a behavior, whereas compulsion is the inability to stop a behavior once it has begun.

The ventromedial prefrontal cortex (vmPFC) plays a central role in regulating impulsivity. For instance, individuals who have survived accidents, such as the famous case of Phineas Gage **(see box below)**, or brain diseases affecting the vmPFC, have been shown to lose the ability to restrain themselves from engaging in harmful or inappropriate actions, whether to themselves or others (Eslinger and Damasio 1985).

Important Point of View

The Phineas Gage Accident

In 1848, Phineas Gage, a railroad foreman, suffered a severe injury due to a mishap during an explosion. A tamping iron pierced his skull, completely passing through it, damaging the left part of his ventromedial prefrontal cortex (vmPFC). This clinical case has become a cornerstone in neurology, as it highlighted the role of this brain region.

Following the accident, Phineas Gage's personality underwent a dramatic transformation. Previously known as serious, attentive, reliable, and sociable, he became unstable, antisocial, moody, impulsive, capricious, and prone to fits of anger (Ratiu et al. 2004).

The ventromedial prefrontal cortex (vmPFC) functions as an interface, connecting mechanisms involved in automatic responses, particularly emotions, with those governing more complex behavior control. Based on the information it receives, it facilitates the development of strategies. Both impulsivity and compulsion are linked to neurobiological loops with *excitatory ascending pathways* or *bottom-up* and *inhibitory descending pathways* or *top-down*.

Ascending pathways originate, in the case of impulsivity, from the ventral striatum (nucleus accumbens and olfactory tubercle), and in the case of compulsion, from the dorsal striatum (caudate nucleus and putamen). Descending pathways originate from different parts of the prefrontal cortex (Ding et al. 2021).

The behavior displayed by the cat is ultimately the result of the interplay between these two opposing pathways: the excitatory ascending pathway and the inhibitory descending pathway. Impulsivity or compulsion arises either from excessive excitation of the ascending pathway or from insufficient inhibition by the descending pathway (◘ Fig. 2.48).

The impulsivity loop originating in the ventral striatum is closely tied to the systems of reward, pleasure, and conditioned learning. Over time and with repetition, through mechanisms of neuroadaptation and neuroplasticity, the loop transitions from the ventral striatum to the dorsal striatum. The dorsal striatum assumes a critical role in selecting appropriate movements to achieve specific goals while discarding superfluous actions. This transition simplifies, automates, and rigidifies behaviors (Stahl and Muntner 2021; Tedford et al. 2015; Jupp et al. 2013; DeCoteau et al. 2007). Effectively, the dorsal striatum evolves from a passive observer to an automatic command center for complex behaviors, freeing cortical regions for other functions. Compulsive behaviors driven by this circuit are particularly resistant to extinction due to their association with the dorsal striatum (Goodman and Packard 2018).

The ascending excitatory pathway is connected to the mesolimbic pathway, making dopamine crucial for promoting both impulsivity and compulsion. Addictive drugs activate the mesolimbic pathway, rapidly transforming impulsivity (the inability to resist taking a drug) into compulsion (the inability to stop using and the development of strategies to obtain drugs, even at personal risk) (Antonini et al. 2011; Dalley et al. 2011; Ondo and Lai 2008; Blum et al. 1995).

The prefrontal cortex, the starting point of the descending inhibitory pathway, receives significant input from serotonergic pathways and is particularly rich in serotonin receptors.

2

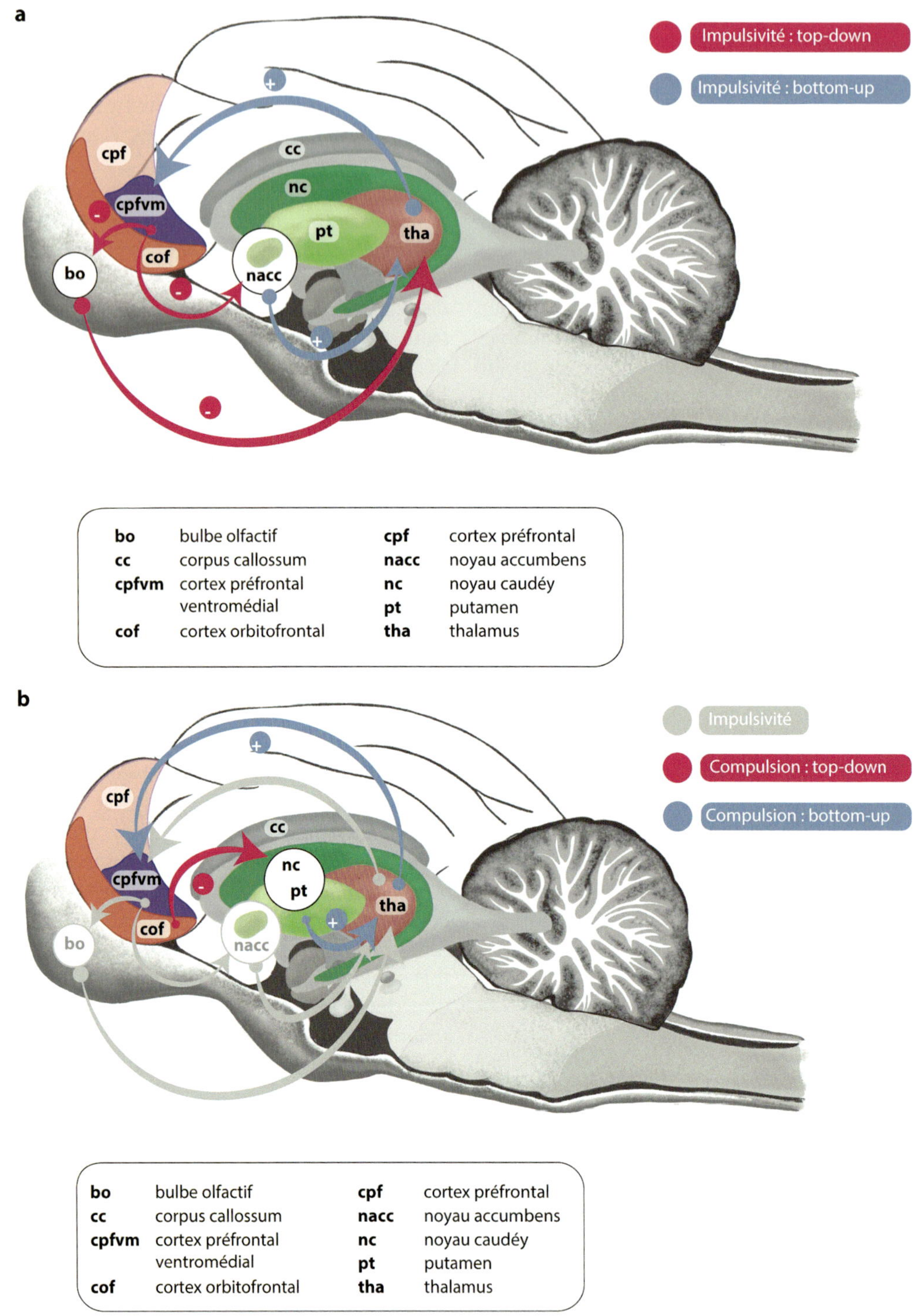

Fig. 2.48 Impulsivity and compulsion loop. **a** The impulsivity loop originates in the ventral striatum (nucleus accumbens and olfactory tubercle), which relays information to the thalamus, subsequently informing the prefrontal cortex. The ventromedial prefrontal cortex inhibits this loop. **b** The compulsion loop originates in the dorsal striatum, which relays information to the thalamus, subsequently informing the prefrontal cortex. The orbitofrontal cortex provides inhibitory control over this loop

Thus, serotonin, by activating the prefrontal cortex and its inhibitory descending pathway, helps control impulsivity and compulsion (Puig and Gulledge 2011; Cools 1973).

As seen in fear circuits, the motivational system and reward circuit can activate through numerous inputs but rely on a single inhibitory regulator—the ventromedial prefrontal cortex (vmPFC).

In cats, the inhibition exerted by the vmPFC may be less pronounced than in species such as dogs. This difference likely stems from the cat's dual role as both predator and prey, necessitating rapid adaptability to both scenarios. However, the precise neurobiological determinants underlying this distinction remain unknown.

2.3.3.5 Neurophysiology of Aggression

The sequences of aggressive behavior in cats have been detailed earlier (see ▶ Chap. 1, ▶ Sect. 1.3.1.3). This section focuses on the neurophysiological mechanisms that trigger and regulate these behaviors. A distinction is made between aggression related to *social interaction* and *predatory behavior*, each of which involves distinct triggers and neural circuits.

Social aggression includes threatening and attacking behaviors aimed at distancing or deterring the source of a perceived threat or nuisance. The neural circuits underlying these behaviors involve the limbic system, which governs emotions, the hypothalamus, and the sympathetic division of the autonomic nervous system (Bear et al. 2016; Haller 2013). A central role in aggressive behaviors is played by the *periaqueductal gray matter* (PAG), a structure within the midbrain. The PAG acts as a hub connecting cerebral centers with the effector organs responsible for aggressive responses (Chou 2020).

The *cerebral aqueduct,* also known as the *aqueduct of Sylvius,* is a narrow channel running through the midbrain, connecting the third ventricle in the diencephalon to the fourth ventricle in the rhombencephalon. The PAG involved in aggression encircles this aqueduct within the tegmentum. It receives inputs from the amygdala and the hypothalamus and projects to descending sympathetic pathways.

The amygdala, hypothalamus, and PAG are further anatomically divided into multiple nuclei. These divisions allow for more specific differentiation between two distinct circuits: one associated with social aggression and the other with predatory behavior.

Unlike social aggression, *predatory behavior* is not aimed at deterring or repelling a nuisance but at acquiring food. It does not constitute aggression in the same sense. Predation is characterized by highly focused, controlled actions that are triggered by specific sensory cues (e.g., movement or sound) and rely on a separate set of neural pathways. These pathways involve the amygdala and hypothalamus, but the PAG activation differs significantly from that in social aggression, reflecting the distinct purpose and emotional context of the behavior.

By distinguishing between these circuits, we gain a better understanding of the triggers and neural dynamics underlying aggression and predation, which are essential for tailoring behavioral interventions and treatments.

Aggressions involve the central nucleus and the basal nucleus of the amygdala, the medial hypothalamus, and the dorsal periaqueductal gray matter, which has orthosympathetic projections. Predatory behavior is part of feeding behavior and is regulated by hunger centers located in the lateral hypothalamus. Predation involves the medial nucleus of the amygdala, the lateral hypothalamus, and the ventral periaqueductal gray matter, whose orthosympathetic projections are significantly less numerous.

These neurophysiological differences (◘ Fig. 2.49) support the distinction between aggression, a social interaction where emotions play an important role, and predation, a behavior devoid of emotional involvement. For instance, a cat is not angry at the mouse it hunts (Gregg and Siegel 2001).

The ventromedial prefrontal cortex (CPFvm) is capable of inhibiting aggressive

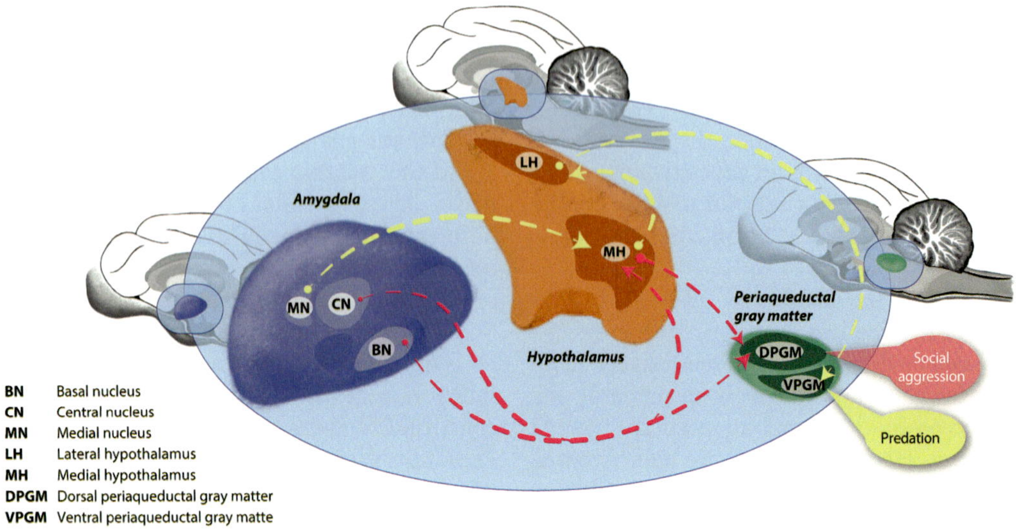

Fig. 2.49 Aggression pathways. Aggression involves the central and basal amygdala nuclei, the medial hypothalamus, and the dorsal PAG. Predation utilizes the medial amygdala, the lateral hypothalamus, and the ventral PAG, highlighting the neurophysiological distinctions between these behaviors

responses through the same pathways that regulate impulsivity in general. In fact, some authors describe aggression as a specific form of impulsivity.

The primary neurotransmitter involved in aggression is serotonin (Amat et al. 2013; Haller 2013). Numerous studies in humans and animals have shown a negative correlation between cerebral serotonergic activity levels and aggressive behavior (Sheard 2021; Denenberg and Dubé 2018; Frank and Dehasse 2003; Hassanain et al. 2003). Lower serotonin levels are associated with higher tendencies toward risk-taking and aggression.

As with previous circuits, the aggression circuit is inhibited by the same key player: the CPFvm, which itself is modulated by serotonergic activity.

Summary of ▸ Sect. 2.3

The identification of specific neurophysiological circuits is essential for understanding behaviors.

Fear is governed by the amygdala, which, based on the information it receives, activates hormonal, cardiorespiratory, and motor responses essential for individual protection. The fear response can be inhibited by the ventromedial prefrontal cortex (CPFvm) when cortical analysis determines no danger (long pathway). Conversely, the fear response is validated and activated by elementary processes such as sensitization, anticipation, and generalization. These processes allow for faster reactions when the dangerous stimulus is encountered again, bypassing the cortex because the danger has already been confirmed (short pathway). Dysregulation of the amygdala circuit can lead to a *phobic state.* Another loop involved in fear reactions is the *cortico-striato-thalamo-cortical loop* (CSTC). This loop alerts the individual to respond to potential dangers and anticipates risky situations. When dysregulated, it results in an *anxious state.*

Mood is not associated with a specific circuit, and research into its neurophysiological basis is ongoing. It appears to be linked to diffuse modifications in monoamines and their receptors, as well as the BDNF (*Brain-Derived Neurotrophic Factor*). Some authors suggest broader regulation by glial cells. Dysregulation of mood can lead to pathological states such as depressive, manic, or bipolar states.

Learning and memory allow animals to adapt to their environment. The *basal ganglia* play a key role in learning by automating behaviors, thereby freeing higher cortical functions. The *hippocampus* is central to memory, creating new connections or reinforcing existing ones. At the cellular level, memory corresponds to the strengthening of two synaptic connections through *long-term potentiation* (LTP), primarily mediated by glutamate.

The *wake-activation system* relies on the balance between glutamate (activating) and GABA (inhibiting). This system underpins attention and prepares the individual for action.

The *motivational system* is regulated by the *reward circuit*, crucial for all forms of positive reinforcement learning. It involves the basal ganglia and dopaminergic pathways. Overactivation of dopaminergic pathways or failure of the GABAergic inhibitory system can activate the impulsivity circuit, which, if persistent, can evolve into compulsive behavior. Dysregulation of these systems can lead to *impulsive and compulsive states*.

Aggression is regulated by circuits involving the amygdala, hypothalamus, and periaqueductal gray (PAG). Social aggression involves the central and basal nuclei of the amygdala, the medial hypothalamus, and the dorsal PAG. Predatory behavior, on the other hand, involves the medial amygdala, the lateral hypothalamus, and the ventral PAG .

Table 2.3 Major functions, dysregulations, and the involved neurotransmitters

	Glutamate	GABA	Acetylcholine	Dopamine	Norepinephrine	Serotonin
Fear/Phobia	•	•			• • •	• •
Anxiety	•	•		• • •	• •	• •
Mood	• •			• •	• •	• •
Learning	•		• •	• • •		
Memory	• • •		• •			
Wakefulness	•		• • •		• • •	• •
Activation	• • •		•		• •	
Inhibition		• • •		•		• • •
Impulsivity	•	•		• •	•	• • •
Compulsion	•	•		• •	•	• • •
Aggression	•	•			•	• • •

2

Summary of This Chapter

Throughout this chapter, the practitioner has become acquainted with the various brain centers, the physiology of neuronal information transmission, and the circuits underlying the neurophysiology of behaviors. In clinical practice, during the collection of semiological data, the practitioner identifies clinical signs to pinpoint the dysfunctional circuit and, consequently, the neurotransmitters involved (◘ Table 2.3).

Understanding the specifics of each circuit and neurotransmitter allows the practitioner to locate the dysfunction and select an appropriate therapeutic approach.

References

Amano T, Unal CT, Paré D (2010) Synaptic correlates of fear extinction in the amygdala. Nat Neurosci 13:489–494. https://doi.org/10.1038/nn.2499

Amat M, Le Brech S, Camps T, Torrente C, Mariotti VM, Ruiz JL, Manteca X (2013) Differences in serotonin serum concentration between aggressive English cocker spaniels and aggressive dogs of other breeds. J Vet Behav Clin Appl Res 8:19–25. https://doi.org/10.1016/j.jveb.2012.04.003

Antonini A, Siri C, Santangelo G, Cilia R, Poletti M, Canesi M, Caporali A, Mancini F, Pezzoli G, Ceravolo R, Bonuccelli U, Barone P (2011) Impulsivity and compulsivity in drug-naïve patients with Parkinson's disease: impulsivity and compulsivity in Drug-Naïve PD. Mov Disord 26:464–468. https://doi.org/10.1002/mds.23501

Baier B, Karnath H-O, Dieterich M, Birklein F, Heinze C, Müller NG (2010) Keeping memory clear and stable—the contribution of human basal Ganglia and Prefrontal Cortex to working memory. J Neurosci 30:9788–9792. https://doi.org/10.1523/JNEUROSCI.1513-10.2010

Balleine BW, Liljeholm M, Ostlund SB (2009) The integrative function of the basal ganglia in instrumental conditioning. Behav Brain Res,Special issue on the role of the basal ganglia in learning and memory 199:43–52. https://doi.org/10.1016/j.bbr.2008.10.034

Bayley PJ, Frascino JC, Squire LR (2005) Robust habit learning in the absence of awareness and independent of the medial temporal lobe. Nature 436:550–553. https://doi.org/10.1038/nature03857

Bear MF, Connors BW, Paradiso MA (2016) Neuroscience: exploring the brain, 4th edn. Wolters Kluwer, Philadelphia

Blum K, Sheridan PJ, Wood RC, Braverman ER, Chen TJH, Comings DE (1995) Dopamine D2 receptor gene variants: association and linkage studies in impulsive-addictive-compulsive behaviour. Pharmacogenetics 5:121–141. https://doi.org/10.1097/00008571-199506000-00001

Blum K, Sheridan PJ, Wood RC, Braverman ER, Chen TJH, Comings DE (1996) The D2 Dopamine Receptor gene as a determinant of reward deficiency syndrome. J R Soc Med 89:396–400. https://doi.org/10.1177/014107689608900711

Breiter HC, Etcoff NL, Whalen PJ, Kennedy WA, Rauch SL, Buckner RL, Strauss MM, Hyman SE, Rosen BR (1996) Response and habituation of the human Amygdala during visual processing of facial expression. Neuron 17:875–887. https://doi.org/10.1016/S0896-6273(00)80219-6

Brown RE, Milner PM (2003) The legacy of Donald O. Hebb: more than the Hebb Synapse. Nat Rev Neurosci 4:1013–1019. https://doi.org/10.1038/nrn1257

Brutkowski S (1965) Functions of prefrontal cortex in animals. Physiol Rev 45:721–746. https://doi.org/10.1152/physrev.1965.45.4.721

Büchel C (2000) Classical fear conditioning in functional neuroimaging. Curr Opin Neurobiol 10:219–223. https://doi.org/10.1016/S0959-4388(00)00078-7

Carlson NR, Birkett MA (2017) Physiology of behavior, 12th edn, global ed, Always learning. Pearson, Boston/Columbus/Indianapolis/New York City/San Francisco/Amsterdam/Cape Town/Dubai/London

Carr DB, Sesack SR (2000) Projections from the rat prefrontal cortex to the ventral tegmental area: target specificity in the synaptic associations with Mesoaccumbens and Mesocortical neurons. J Neurosci 20:3864–3873. https://doi.org/10.1523/JNEUROSCI.20-10-03864.2000

Caspi A, Sugden K, Moffitt TE, Taylor A, Craig IW, Harrington H, McClay J, Mill J, Martin J, Braithwaite A, Poulton R (2003) Influence of life stress on depression: moderation by a polymorphism in the 5-HTT gene. Science 301:386–389. https://doi.org/10.1126/science.1083968

Chiu CQ, Barberis A, Higley MJ (2019) Preserving the balance: diverse forms of long-term GABAergic synaptic plasticity. Nat Rev Neurosci 20:272–281. https://doi.org/10.1038/s41583-019-0141-5

Chou D (2020) Topiramate inhibits offensive aggression through targeting ventrolateral periaqueductal gray. Neuropharmacology 181:108361. https://doi.org/10.1016/j.neuropharm.2020.108361

Clementi F, Marini D (1972) The surface fine structure of the walls of cerebral ventricles and of choroid plexus in cat. Z. Fr Zellforsch. Mikrosk Anat 123:82–95. https://doi.org/10.1007/BF00337675

Cools AR (1973) The caudate nucleus and neurochemical control of behaviour: the function of dopamine and serotonin in the caput nuclei caudati of cats. Nijmegen:[sn]

Cossart R, Dinocourt C, Hirsch JC, Merchan-Perez A, De Felipe J, Ben-Ari Y, Esclapez M, Bernard C (2001) Dendritic but not somatic GABAergic inhibition is decreased in experimental epilepsy. Nat Neurosci 4:52–62. https://doi.org/10.1038/82900

Cowie RJ, Holstege G (1992) Dorsal mesencephalic projections to pons, medulla, and spinal cord in the cat: limbic and non-limbic components. J Comp Neurol 319:536–559. https://doi.org/10.1002/cne.903190406

Dallérac GM, Vatsavayai SC, Cummings DM, Milnerwood AJ, Peddie CJ, Evans KA, Walters SW, Rezaie P, Hirst MC, Murphy KPSJ (2011) Impaired long-term potentiation in the Prefrontal Cortex of Huntington's disease Mouse Models: rescue by D_1 Dopamine Receptor activation. Neurodegener Dis 8:230–239. https://doi.org/10.1159/000322540

Dalley JW, Everitt BJ, Robbins TW (2011) Impulsivity, compulsivity, and top-down cognitive control. Neuron 69:680–694. https://doi.org/10.1016/j.neuron.2011.01.020

Davis M, Walker DL, Lee Y (1997) Amygdala and bed nucleus of the stria terminalis: differential roles in fear and anxiety measured with the acoustic startle reflex. Philos Trans R Soc Lond Ser B Biol Sci 352:1675–1687. https://doi.org/10.1098/rstb.1997.0149

De Morais-Pinto L, Da Veiga ML, Almeida Da Anunciação AR (2021) Central nervous system development of cats (Felis catus L. 1758). Res Vet Sci 141:81–94. https://doi.org/10.1016/j.rvsc.2021.10.015

DeCoteau WE, Thorn C, Gibson DJ, Courtemanche R, Mitra P, Kubota Y, Graybiel AM (2007) Oscillations of local field potentials in the rat dorsal striatum during spontaneous and instructed behaviors. J Neurophysiol 97:3800–3805. https://doi.org/10.1152/jn.00108.2007

DeLahunta A, Glass E (2009) Veterinary neuroanatomy and clinical neurology, 3rd. edn. Saunders Elsevier, St. Louis

Delgado MR, Olsson A, Phelps EA (2006) Extending animal models of fear conditioning to humans. Biol Psychol 73:39–48. https://doi.org/10.1016/j.biopsycho.2006.01.006

Denenberg S, Dubé MB (2018) Tools for managing feline problem behaviours: psychoactive medications. J Feline Med Surg 20:1034–1045

Devan BD, Hong NS, McDonald RJ (2011) Parallel associative processing in the dorsal striatum: segregation of stimulus–response and cognitive control subregions. Neurobiol Learn Mem 96:95–120. https://doi.org/10.1016/j.nlm.2011.06.002

Di Chiara G, Bassareo V (2007) Reward system and addiction: what dopamine does and doesn't do. Curr Opin Pharmacol 7:69–76

Ding J, Hu X, Xu F, Yu H, Ye Z, Zhang S, Pan H, Pan D, Tu Y, Zhang Q (2021) Suppression of top-down influence decreases neuronal excitability and contrast sensitivity in the V1 cortex of cat. Sci Rep 11:16034

Dunsmoor JE, Prince SE, Murty VP, Kragel PA, LaBar KS (2011) Neurobehavioral mechanisms of human fear generalization. NeuroImage 55:1878–1888. https://doi.org/10.1016/j.neuroimage.2011.01.041

Elhwuegi AS (2004) Central monoamines and their role in major depression. Prog Neuro-Psychopharmacol Biol Psychiatry 28:435–451. https://doi.org/10.1016/j.pnpbp.2003.11.018

Eslinger PJ, Damasio AR (1985) Severe disturbance of higher cognition after bilateral frontal lobe ablation: patient EVR. Neurology 35:1731–1731. https://doi.org/10.1212/WNL.35.12.1731

Fletcher NF, Brayden DJ, Brankin B, Worrall S, Callanan JJ (2006) Growth and characterisation of a cell culture model of the feline blood–brain barrier. Vet Immunol Immunopathol 109:233–244. https://doi.org/10.1016/j.vetimm.2005.08.025

Fonberg E (1986) Amygdala, emotions, motivation, and depressive states. In: Biological foundations of emotion. Elsevier, pp 301–331. https://doi.org/10.1016/B978-0-12-558703-7.50019-X

Ford TC, Abu-Akel A, Crewther DP (2019) The association of excitation and inhibition signaling with the relative symptom expression of autism and psychosis-proneness: implications for psychopharmacology. Prog Neuro-Psychopharmacol Biol Psychiatry 88:235–242. https://doi.org/10.1016/j.pnpbp.2018.07.024

Frank D, Dehasse J (2003) Differential diagnosis and management of human-directed aggression in cats. Vet Clin Small Anim Pract 33:269–286

Fukushima T, Hasegawa I, Miyashita Y (2004) Prefrontal neuronal activity encodes spatial target representations sequentially updated after nonspatial target-shift cues. J Neurophysiol 91:1367–1380. https://doi.org/10.1152/jn.00306.2003

Gariano RF, Groves PM (1988) Burst firing induced in midbrain dopamine neurons by stimulation of the medial prefrontal and anterior cingulate cortices. Brain Res 462:194–198. https://doi.org/10.1016/0006-8993(88)90606-3

Gerencsér L, Bunford N, Moesta A, Miklósi Á (2018) Development and validation of the Canine Reward Responsiveness Scale –examining individual differences in reward responsiveness of the domestic dog. Sci Rep 8:4421. https://doi.org/10.1038/s41598-018-22605-1

Goodman J, Packard MG (2018) The role of the dorsal striatum in extinction: a memory systems perspective. Neurobiol Learn Mem 150:48–55. https://doi.org/10.1016/j.nlm.2018.02.028

Gray-Edwards HL, Salibi N, Josephson EM, Hudson JA, Cox NR, Randle AN, McCurdy VJ, Bradbury AM, Wilson DU, Beyers RJ, Denney TS, Martin DR (2014) High resolution MRI anatomy of the cat brain at 3Tesla. J Neurosci Methods 227:10–17. https://doi.org/10.1016/j.jneumeth.2014.01.035

Gregg TR, Siegel A (2001) Brain structures and neurotansmitters regulating aggression in cats: implications for human aggression. Prog Neuro-Psychopharmacol Biol Psychiatry 25:91–140. https://doi.org/10.1016/S0278-5846(00)00150-0

Haller J (2013) The neurobiology of abnormal manifestations of aggression—a review of hypothalamic mechanisms in cats, rodents, and humans. Brain Res

Bull Extrasynaptic ionotropic receptors 93:97–109. https://doi.org/10.1016/j.brainresbull.2012.10.003
Hansson E, Rönnbäck L (1995) Astrocytes in glutamate neurotransmission. FASEB J 9:343–350
Haroon E, Miller AH, Sanacora G (2017) Inflammation, Glutamate, and Glia: a trio of trouble in mood disorders. Neuropsychopharmacology 42:193–215. https://doi.org/10.1038/npp.2016.199
Hassanain M, Bhatt S, Siegel A (2003) Differential modulation of feline defensive rage behavior in the medial hypothalamus by 5-HT1A and 5-HT2 receptors. Brain Res 981:201–209
Iremonger KJ, Kuzmiski JB, Baimoukhametova DV, Bains JS (2011) Dual regulation of anterograde and retrograde transmission by Endocannabinoids. J Neurosci 31:12011–12020. https://doi.org/10.1523/JNEUROSCI.2925-11.2011
Irle E, Markowitsch HJ (1984) Differential effects of prefrontal lesions and combined prefrontal and limbic lesions on subsequent learning performance in the cat. Behav Neurosci 98:884
Jayaraman A (1985) Organization of thalamic projections in the nucleus accumbens and the caudate nucleus in cats and its relation with hippocampal and other subcortical afferents. J Comp Neurol 231:396–420. https://doi.org/10.1002/cne.902310309
Jesulola E, Micalos P, Baguley IJ (2018) Understanding the pathophysiology of depression: from monoamines to the neurogenesis hypothesis model - are we there yet? Behav Brain Res 341:79–90. https://doi.org/10.1016/j.bbr.2017.12.025
Jupp B, Caprioli D, Saigal N, Reverte I, Shrestha S, Cumming P, Everitt BJ, Robbins TW, Dalley JW (2013) Dopaminergic and GABA-ergic markers of impulsivity in rats: evidence for anatomical localisation in ventral striatum and prefrontal cortex. Eur J Neurosci 37:1519–1528. https://doi.org/10.1111/ejn.12146
Kauer JA, Malenka RC (2007) Synaptic plasticity and addiction. Nat Rev Neurosci 8:844–858. https://doi.org/10.1038/nrn2234
Kimura H, McGeer PL, Peng JH, McGeer EG (1981) The central cholinergic system studied by choline acetyltransferase immunohistochemistry in the cat. J Comp Neurol 200:151–201. https://doi.org/10.1002/cne.902000202
Klimczak A, Kozłowska U, Sanford J, Walczak P, Małysz-Cymborska I, Kurpisz M (2019) Immunological characteristics and properties of Glial restricted Progenitors of Mice, Canine primary culture suspensions, and human QSV40 immortalized cell lines for prospective therapies of neurodegenerative disorders. Cell Transplant 28:1140–1154. https://doi.org/10.1177/0963689719848355
Knecht S, Breitenstein C, Bushuven S, Wailke S, Kamping S, Flöel A, Zwitserlood P, Ringelstein EB (2004) Levodopa: faster and better word learning in normal humans: faster and better word learning. Ann Neurol 56:20–26. https://doi.org/10.1002/ana.20125
Lee S, Hjerling-Leffler J, Zagha E, Fishell G, Rudy B (2010) The largest group of Superficial Neocortical GABAergic interneurons expresses ionotropic Serotonin Receptors. J Neurosci 30:16796–16808. https://doi.org/10.1523/JNEUROSCI.1869-10.2010
Leuchter AF, McCracken JT, Hunter AM, Cook IA, Alpert JE (2009) Monoamine oxidase A and Catechol-O-Methyltransferase functional polymorphisms and the Placebo Response in major depressive disorder. J Clin Psychopharmacol 29:372–377. https://doi.org/10.1097/JCP.0b013e3181ac4aaf
Mameli M, Lüscher C (2011) Synaptic plasticity and addiction: learning mechanisms gone awry. Neuropharmacology 61:1052–1059. https://doi.org/10.1016/j.neuropharm.2011.01.036
Marchand WR (2010) Cortico-basal ganglia circuitry: a review of key research and implications for functional connectivity studies of mood and anxiety disorders. Brain Struct Funct 215:73–96. https://doi.org/10.1007/s00429-010-0280-y
Markowitsch HJ, Pritzel M (1977) Comparative analysis of prefrontal learning functions in rats, cats, and monkeys. Psychol Bull 84:817
Martinowich K, Manji H, Lu B (2007) New insights into BDNF function in depression and anxiety. Nat Neurosci 10:1089–1093. https://doi.org/10.1038/nn1971
Murphy ER, Fernando ABP, Urcelay GP, Robinson ESJ, Mar AC, Theobald DEH, Dalley JW, Robbins TW (2012) Impulsive behaviour induced by both NMDA receptor antagonism and GABAA receptor activation in rat ventromedial prefrontal cortex. Psychopharmacology 219:401–410. https://doi.org/10.1007/s00213-011-2572-1
Nelson SB, Valakh V (2015) Excitatory/inhibitory balance and circuit homeostasis in autism spectrum disorders. Neuron 87:684–698. https://doi.org/10.1016/j.neuron.2015.07.033
Nepomuceno AC, Zanatta R, Chung DG, Costa PF, Feliciano MAR, Avante ML, Lopes LS, Canola JC (2016) Brain neuroimaging of domestic cats: correlation between computed tomography and cross-sectional anatomy. Arq Bras Med Veterinária E Zootec 68:1105–1111. https://doi.org/10.1590/1678-4162-8928
Noble EP (2000) Addiction and its reward process through polymorphisms of the D_2 dopamine receptor gene: a review. Eur Psychiatry 15:79–89. https://doi.org/10.1016/S0924-9338(00)00208-X
Ondo WG, Lai D (2008) Predictors of impulsivity and reward seeking behavior with dopamine agonists. Parkinsonism Relat Disord 14:28–32. https://doi.org/10.1016/j.parkreldis.2007.05.006
Peters SK, Dunlop K, Downar J (2016) Cortico-Striatal-Thalamic loop circuits of the Salience Network: a central pathway in psychiatric disease and treatment. Front Syst Neurosci 10. https://doi.org/10.3389/fnsys.2016.00104
Prichard A, Chhibber R, Athanassiades K, Spivak M, Berns GS (2018) Fast neural learning in dogs: a mul-

timodal sensory fMRI study. Sci Rep 8:14614. https://doi.org/10.1038/s41598-018-32990-2
Puig MV, Gulledge AT (2011) Serotonin and Prefrontal Cortex function: neurons, networks, and circuits. Mol Neurobiol 44:449–464. https://doi.org/10.1007/s12035-011-8214-0
Rabin BM, Hunt WA, Chedester AL, Lee J (1986) Role of the area postrema in radiation-induced taste aversion learning and emesis in cats. Physiol Behav 37:815–818. https://doi.org/10.1016/0031-9384(86)90189-7
Ratiu P, Talos I.-F (2004) The Tale of Phineas Gage, Digitally Remastered. N Engl J Med 351, e21. https://doi.org/10.1056/NEJMicm031024
Rijnberk A (1996) Clinical endocrinology of dogs and cats: an illustrated text. Springer, Dordrecht
Saal D, Dong Y, Bonci A, Malenka RC (2003) Drugs of abuse and stress trigger a common synaptic adaptation in Dopamine neurons. Neuron 37:577–582. https://doi.org/10.1016/S0896-6273(03)00021-7
Sacktor TC (2011) How does PKMζ maintain long-term memory? Nat Rev Neurosci 12:9–15. https://doi.org/10.1038/nrn2949
Schousboe A (2003) Role of astrocytes in the maintenance and modulation of glutamatergic and GABAergic neurotransmission. Neurochem Res 28:347–352
Schroeter ML, Abdul-Khaliq H, Sacher J, Steiner J, Blasig IE, Mueller K (2010) Mood disorders are Glial disorders: evidence from *In Vivo* Studies. Cardiovasc Psychiatry Neurol 2010:1–7. https://doi.org/10.1155/2010/780645
Schultz W (2011) Potential vulnerabilities of neuronal reward, risk, and decision mechanisms to addictive drugs. Neuron 69:603–617. https://doi.org/10.1016/j.neuron.2011.02.014
Seong S (1996) Effects of lesions in the medial prefrontal cortex on the activity of midbrain dopamine neurons. Neuropsychopharmacology 15:437–441. https://doi.org/10.1016/S0893-133X(96)00052-8
Sheard MH (2021) Aggressive behavior: effects of neural modulation by serotonin. In: Aggressive Behavior. Routledge, pp 167–181
Shors T, Matzel L (1997) Long-term potentiation: what's learning got to do with it? Behav Brain Sci 20:597–614; discussion 614. https://doi.org/10.1017/s0140525x97001593
Sotres-Bayon F, Quirk GJ (2010) Prefrontal control of fear: more than just extinction. Curr Opin Neurobiol 20:231–235. https://doi.org/10.1016/j.conb.2010.02.005
Stahl SM, Muntner N (2021) Stahl's essential psychopharmacology: neuroscientific basis and practical applications, 5th edn. Medicine. Cambridge University Press, Cambridge, UK/New York/Melbourne/New Delhi/Singapore
Tedford SE, Persons AL, Napier TC (2015) Dopaminergic lesions of the Dorsolateral Striatum in rats increase delay discounting in an impulsive choice task. PLoS One 10:e0122063. https://doi.org/10.1371/journal.pone.0122063
Tzschentke TM (2000) The medial prefrontal cortex as a part of the brain reward system. Amino Acids 19:211–219. https://doi.org/10.1007/s007260070051
Ulrich-Lai YM, Herman JP (2009) Neural regulation of endocrine and autonomic stress responses. Nat Rev Neurosci 10:397–409. https://doi.org/10.1038/nrn2647
Uylings HBM, Groenewegen HJ, Kolb B (2003) Do rats have a prefrontal cortex? Behav Brain Res 146:3–17. https://doi.org/10.1016/j.bbr.2003.09.028
VanElzakker MB, Kathryn Dahlgren M, Caroline Davis F, Dubois S, Shin LM (2014) From Pavlov to PTSD: the extinction of conditioned fear in rodents, humans, and anxiety disorders. Neurobiol Learn Mem 113:3–18. https://doi.org/10.1016/j.nlm.2013.11.014
Wise RA (1999) Cognitive factors in addiction and nucleus accumbens function: some hints from rodent models. Psychobiology 27:300–310. https://doi.org/10.3758/BF03332124
Witte RS, Kipke DR (2005) Enhanced contrast sensitivity in auditory cortex as cats learn to discriminate sound frequencies. Cogn Brain Res 23:171–184
Wittmann BC, Schott BH, Guderian S, Frey JU, Heinze H-J, Düzel E (2005) Reward-related fMRI activation of dopaminergic midbrain is associated with enhanced hippocampus- dependent long-term memory formation. Neuron 45:459–467. https://doi.org/10.1016/j.neuron.2005.01.010
Wong M-L, Licinio J (2004) From monoamines to genomic targets: a paradigm shift for drug discovery in depression. Nat Rev Drug Discov 3:136–151. https://doi.org/10.1038/nrd1303
Xue M, Atallah BV, Scanziani M (2014) Equalizing excitation–inhibition ratios across visual cortical neurons. Nature 511:596–600. https://doi.org/10.1038/nature13321
Yin S (2015) Feline learning. Feline Behav. Health Welf. 41

Major Biological Functions and Endocrinological Mechanisms of Behavior

Contents

3.1 Endocrinology – 112
3.1.1 Overview of the Endocrine System – 112
3.1.2 Endocrine System and Veterinary Psychiatry – 114

3.2 Biological Functions of the Body – 120
3.2.1 Biological Rhythms – 120
3.2.2 Drinking and Eating Behaviors – 123
3.2.3 Sexual, Reproductive, and Parental Behaviors – 127
3.2.4 Affiliative Behaviors – 129

3.3 Pain – 131
3.3.1 Types of Pain – 132
3.3.2 Pain and Veterinary Psychiatry – 135

References – 137

© The Author(s), under exclusive license to Springer Nature Switzerland AG 2026
S. Bleuer-Elsner et al., *Veterinary Psychiatry of the Cat*,
https://doi.org/10.1007/978-3-032-06260-4_3

3

This chapter explores the influence of hormones, pain, and other major biological functions on behavior. In particular, numerous connections between endocrine systems and neurophysiology will be highlighted through the description of key physiological functions.

► Section 3.1 examines endocrinological functions and their behavioral influences. ► Section 3.2 delves into the major biological functions common to all mammals, with a focus on their expression in the feline species. ► Section 3.3 addresses pain and its consequences.

While clinical veterinary psychiatry primarily focuses on brain functions for diagnostic purposes, understanding these major biological functions remains essential to maintaining a comprehensive approach to feline behavior.

3.1 Endocrinology

3.1.1 Overview of the Endocrine System

Hormones are produced by endocrine glands, such as the hypothalamus, thyroid, or adrenal glands, which secrete hormones into the bloodstream, or *exocrine glands*, such as salivary, pancreatic, or mammary glands, which release their secretions into specific ducts. Hormones from *endocrine glands* travel through the bloodstream to target tissues containing specific hormone receptors. Activation of these receptors triggers a series of biochemical events that directly or indirectly activate genes, inducing specific biological responses (Stahn and Buttgereit 2008). In some cases, these interactions occur very rapidly, leading to so-called "non-genomic" effects that are immediately visible (Falkenstein et al. 2000).

3.1.1.1 Key Hormones in Cats

There are four chemical classes of hormones: peptide hormones, steroid hormones, lipid-derived hormones, and monoamines.

- **Proteins and Peptides**

Peptide hormones in cats are all secreted by the hypothalamus and are listed in Table 3.1.

- **Steroids**

Steroids, or steroid hormones, encompass a group that includes androgens, estrogens, progestogens, mineralocorticoids, and glucocorticoids. These hormones are all derived from cholesterol (Fig. 3.1), are liposoluble, and can easily cross cellular membranes. Steroid receptors are located within the cells.

Although androgens and estrogens are often referred to as sex hormones, they are not exclusive to one sex. However, circulating levels of androgens and estrogens differ between sexes and vary cyclically (Blottner and Jewgenow 2007). These specific concentrations are determined by the relative proportion of steroid-specific enzymes produced in the gonads.

Steroids are primarily synthesized in the gonads but can also be produced in the adrenal glands, which are responsible for mineralocorticoids and glucocorticoids. A single defective enzyme in the adrenal glands can result in the secretion of large amounts of androgens, leading to numerous behavioral and physiological changes (Erger et al. 2023).

Table 3.1 Main peptide hormones in cats

Production site	Hormones
Hypothalamic hormones	Thyrotropin-releasing hormone (TRH) Growth hormone-releasing hormone (GHRH) Somatostatin (GHIH) Gonadotropin-releasing hormone (GnRH) Melanotropin-releasing hormone (MRH) Corticotropin-releasing hormone (CRH)

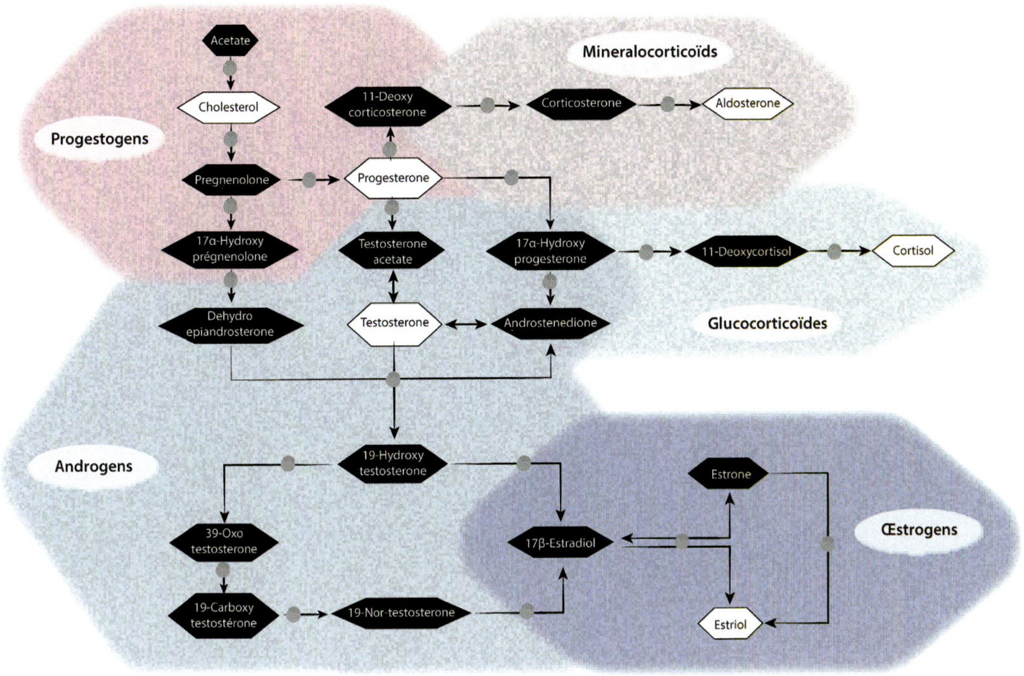

Fig. 3.1 Synthesis pathway of steroids in cats. Steroids include androgens, estrogens, progestogens, mineralocorticoids, and glucocorticoids. All of these hormones derive from cholesterol

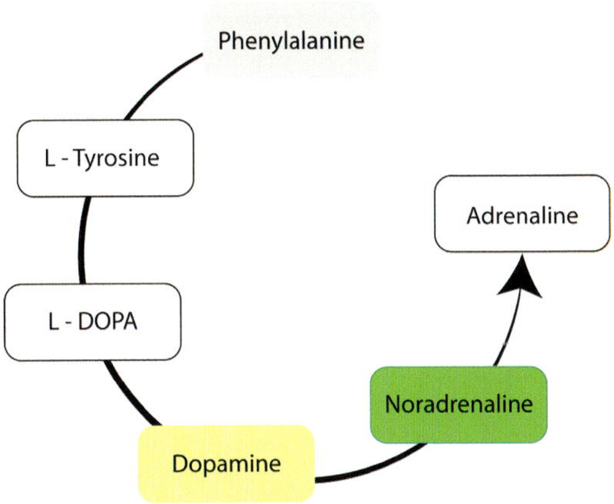

Fig. 3.2 Synthesis pathway of dopamine and norepinephrine. Norepinephrine and dopamine are catecholamines synthesized from phenylalanine

Monoamines

Norepinephrine and *dopamine* are two catecholamines that serve dual roles as both hormones and neurotransmitters. They share a common synthesis pathway (Fig. 3.2). Melatonin and serotonin are two other monoamines secreted by the epiphysis.

Lipid Hormones

Prostaglandins are cytokine metabolites derived from membrane phospholipids . They play critical roles in various processes, including inflammation, vasodilation, and pain (Fig. 3.3). Numerous subtypes of prostaglandins exist (Jones and Budsberg 2000).

3

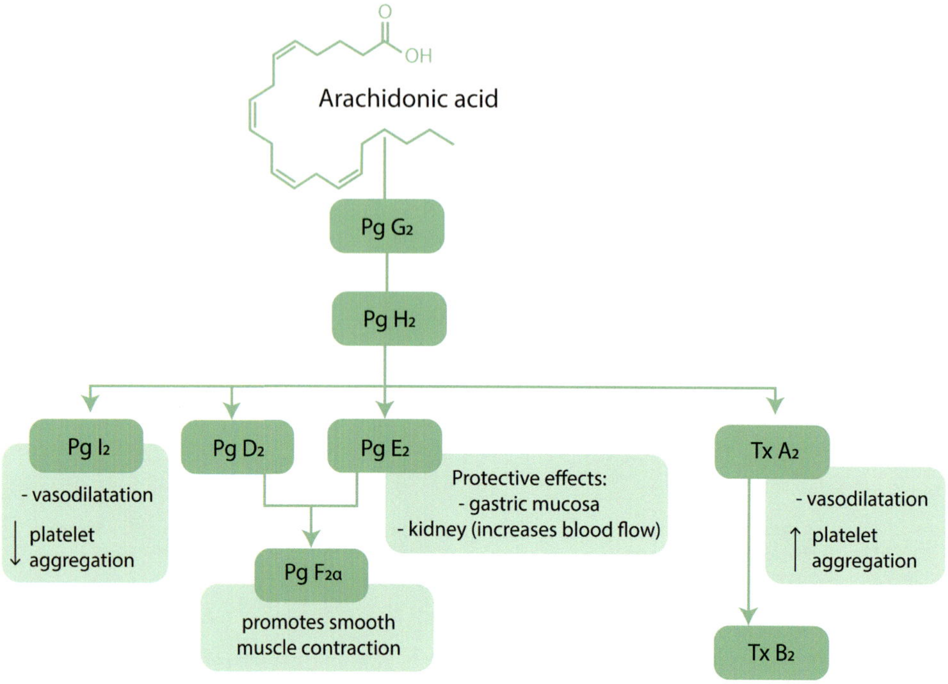

Fig. 3.3 Synthesis of different prostaglandin subtypes. Prostaglandins are cytokines derived from membrane phospholipids, involved in inflammation, vasodilation, and pain

3.1.1.2 Hormonal Regulation Mechanisms

Hormone levels in circulation are predominantly regulated by *negative feedback* mechanisms, either directly through their own concentration or indirectly via other hormones whose secretion is induced by the hormone to be regulated (Fig. 3.4). In negative feedback, the secretion of a hormone decreases the production of its precursors, leading to the regulation of the final hormone secretion. This mechanism ensures that hormone levels remain within a specific range. Numerous environmental stimuli also influence hormone production.

A *positive feedback* mechanism exists for certain hormones that require a rapid biological response. In these cases, hormone secretion increases the production of its precursor. For instance, such a system regulates hormones involved in stress response or ovulation.

In addition to feedback loops, hormone levels often influence the production of their own receptors. For example, membrane receptors for prolactin may undergo "*up-regulation*" (an increase in available receptor numbers) in response to estrogens or insulin, or "*down-regulation*" in response to progesterone or an acute rise in prolactin levels.

There are also interactions between individuals. For example, the secretion of sex hormones in young animals during puberty promotes active detachment from the mother. This detachment aids in the maturation of the young animal's gonads. Similarly, glucocorticoid production facilitates pulmonary maturation to prepare for breathing at birth. It is also the release of glucocorticoids that triggers the onset of parturition.

3.1.2 Endocrine System and Veterinary Psychiatry

3.1.2.1 Neurovegetative and Hormonal Stress Systems

- **The Stress Response**

Any modification of an organism's equilibrium in response to a stimulus perceived as

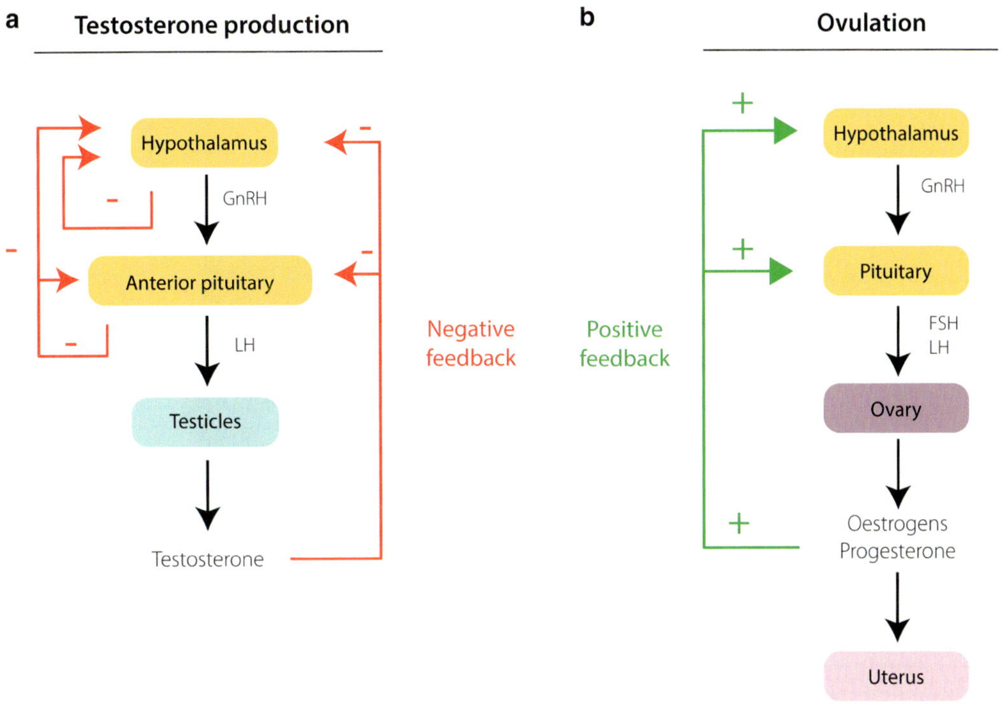

Fig. 3.4 Classical mechanisms of negative and positive hormonal feedback regulation. **a** In negative feedback, hormone secretion decreases the production of its precursors, regulating the secretion of the final hormone. **b** In positive feedback, hormone secretion increases the production of its precursor

potentially threatening can be classified as stress (Selye 1955). Stress factors can be environmental, such as the appearance of a predator; physiological, such as a drop in blood glucose levels; or psychological, such as the presence of an anxiety-related pathological state. To address and restore balance, the organism engages a series of physiological and behavioral reactions (see ▶ Chap. 1, ▶ Sect. 1.1.1.3).

Hans Selye (1950) studied the biological processes underlying the management of stressors, which he termed the General Adaptation Syndrome. This syndrome encompasses three stages: the alarm reaction, resistance, and exhaustion.

Short-Term Effects and Alarm Reaction

As discussed in ▶ Chap. 1 (see ▶ Sect. 1.2.1), emotions comprise three components: behavioral, autonomic (neurovegetative), and hormonal.

Stress activates the autonomic nervous system (ANS, neurovegetative component) and triggers the release of adrenaline from the adrenal medulla and noradrenaline from the sympathetic nervous system. This immediate stress-response pathway, dependent on the sympathetic nervous system, is referred to as the sympathetic nervous system (SNS) axis. Activation of this axis results in increased immediate energy availability, heightened oxygen consumption, reduced blood flow to non-essential areas for movement, and the inhibition of digestion, growth, immune function, reproduction, and pain. Initially, environmental perception and memory are enhanced. These effects facilitate rapid adaptation to the situation (Fig. 3.5). Sympathetic activation is regulated by the amygdala, the locus coeruleus, and the nuclei of noradrenergic neurons.

On the other hand, the hypothalamic-pituitary-adrenal (HPA) axis is responsible for releasing several hormones in response to stress (hormonal component).

First, the hypothalamus secretes corticotropin-releasing hormone (CRH) and

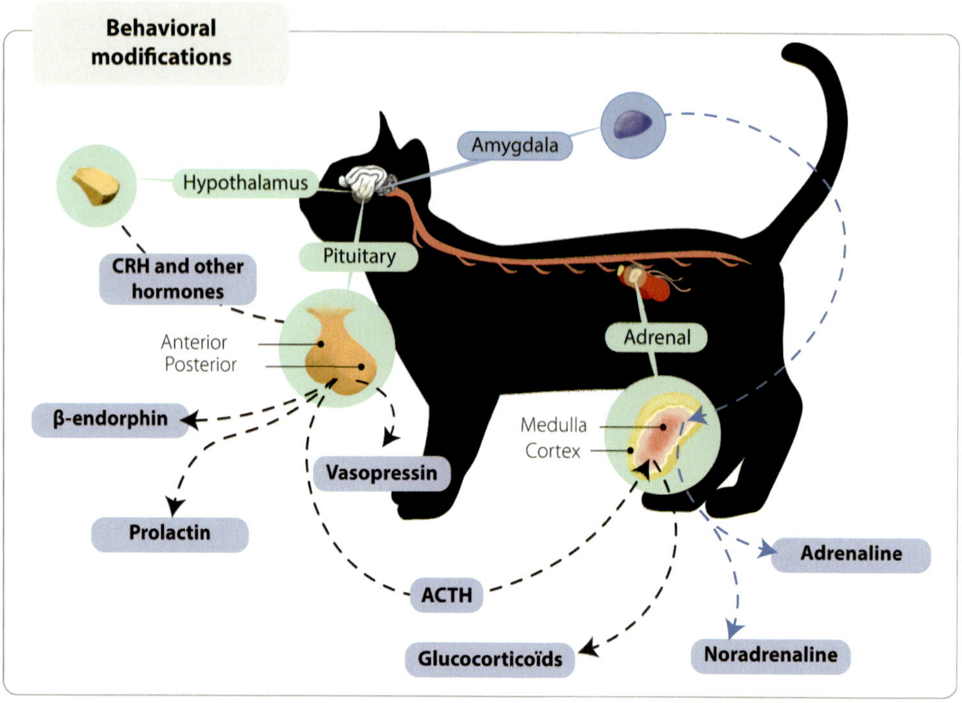

Fig. 3.5 Rapid stress response through the HPA and SNS axes. In acute stress, the hypothalamic-pituitary-adrenal (HPA) axis and the sympathetic nervous system (SNS) are activated. This leads to the production of glucocorticoids, adrenaline, and noradrenaline, enabling the organism to respond swiftly by mobilizing its reserves

other hormones, which stimulate the anterior pituitary to release adrenocorticotropic hormone (ACTH) and β-endorphin. ACTH then promotes the secretion of corticosterone by the adrenal cortex. Additionally, prolactin is released from the anterior pituitary, while vasopressin is secreted by the posterior pituitary (Beerda et al. 1999).

Glucocorticoids, besides aiding the animal in reacting to stressful situations, also facilitate memory formation. In rats, adrenal removal leads to massive destruction of cells in the dentate gyrus of the hippocampus, a key structure for memory, unless glucocorticoids are administered (Sloviter et al. 1989).

HPA axis activation is regulated by neurons in the paraventricular nucleus of the hypothalamus (Herman et al. 2008). CRH also functions as a neurotransmitter in regions of the limbic system involved in emotional responses, such as the periaqueductal gray, the locus coeruleus, and the central nucleus of the amygdala (Kolber et al. 2008).

Long-Term Effects: Resistance and Exhaus tion

Over time, the excessive secretion of glucocorticoids leads to harmful effects. These include hypertension, muscle tissue damage, diabetes, infertility, growth inhibition, suppression of inflammatory responses, and immune system suppression (Baid and Nieman 2004).

Although acute stress factors enhance immediate learning and memory of danger-associated contexts, glucocorticoids, when elevated over a long period due to chronic stress, impair learning and memory. The initial synthesis of glucose, which supports immediate responses to danger, cannot be sustained indefinitely. This temporary mobilization of glucose to motor cells allows a rapid reaction but decreases glucose availability in the hippocampus. This glucose deficit results in massive neuronal death in the hippocampus, a critical region for long-term memory formation **(see box below)**.

In ► Sect. 3.3 of this chapter, dedicated to pain, we will explore how the mechanisms underlying acute and chronic pain mirror those of stress. Acute pain is protective and adaptive, while chronic pain becomes harmful. The deleterious effects of stress and pain thus contribute to the development of pathological states.

Important Point of View

Glucocorticoids and memory

When stress persists, glucocorticoids reduce glucose uptake and glutamate reabsorption in neurons of the CA1 region of the hippocampus, a key area involved in long-term memory (Krugers et al. 2006). Excess extracellular glutamate triggers calcium influx into cells via NMDA receptors. The resulting intracellular calcium overload leads to neuronal death.

It has been demonstrated that this neurodegeneration in the hippocampus plays a significant role in the development of depressive states. An experiment conducted on monkeys exposed to chronic stress (harassment by other monkeys) revealed, upon autopsy, gastric ulcers, adrenal gland enlargement, and complete destruction of the CA1 region of the hippocampus (Conrad 2008).

Important Point of View

Effects of chronic stress in humans

During prenatal stress, moderate to severe stress can lead to a reduction in the volume of the hippocampus and amygdala due to corticosteroid secretion by the mother (Weinstock 2008). Additionally, glucocorticoids interfere with the chemical messages transmitted by cytokines, thereby reducing the long-term efficiency of the immune system (Almawi et al. 1996).

Psychological dwarfism is a rare syndrome found in children raised in highly stressful environments. Growth hormone (GH) secretion is suppressed in these individuals (Silver and Finkelstein 1967). When removed from their stressful environment, they can recover and experience catch-up growth. While chronic stress is a factor, it appears that the absence of an attachment figure is particularly responsible for the inhibition of GH secretion (Hofer 2005).

Conversely, researchers have shown that certain hormonal factors, such as testosterone, neuropeptide Y, and the hormone DHEA, can protect against excessive cortisol levels, promoting resilience (Russo et al. 2012).

▪ Hormones and Psychiatric Disorders

In humans, the use of steroid hormones can lead to immune, hepatic, renal, and cardiovascular dysfunctions, as well as psychiatric symptoms such as manic states (Hall and Hall 2005). During withdrawal, athletes who had taken these hormones exhibited depression and suicidal thoughts, despite having no prior history of mental illness (Piacentino et al. 2015).

In women, premenstrual syndrome (PMS) is characterized by physiological, behavioral, and mood symptoms correlated with changes in steroid hormone concentrations during the menstrual cycle (Hofmeister 2016). Similarly, female cats may exhibit manic or depressive states during estrus or diestrus.

The HPA axis is also implicated in the development of depression. Half of all depressive patients show elevated cortisol levels (Burke et al. 2005). Estrogen deficiencies have also been associated with depression (Slopien et al. 2018).

In cats, the harmful effects of endogenous glucocorticoids on behavior have been documented as well (Finkler and Terkel 2010; Stella et al. 2013).

3.1.2.2 Metabolism and Hypothalamic-Pituitary-Thyroid Axis

The thyroid gland, located along the trachea, secretes the thyroid hormones thyroxine (T4) and triiodothyronine (T3). T4 is converted into T3 in the liver. While T4 exists predominantly in a protein-bound form, only free T4 and T3 are biologically active (◘ Fig. 3.6).

The central role of the thyroid is to regulate the intensity of all organ functions, essentially the basal metabolic rate. Thyroid hormones stimulate nearly all tissues in the body and increase cellular oxygen consumption. Consequently, they affect many vital functions, including heart rate, calorie utilization rate, skin integrity, growth, heat production, fertility, and digestion (Khare et al. 2018; Peterson 1984).

The production of these hormones is regulated by the hypothalamic-pituitary-thyroid (HPT) axis. The hypothalamus activates the production of TRH upon receiving noradrenergic signals. TRH, in turn, stimulates the production of TSH by the anterior pituitary gland. T4, T3, somatostatin, and dopamine have inhibitory effects on TSH production, whereas estrogens act as activators. TSH stimulates the thyroid gland to synthesize T3 and T4. Catecholamines (adrenaline and noradrenaline) and prostaglandins enhance this synthesis, while somatostatin (a growth hormone-inhibiting hormone secreted by the hypothalamus) and acetylcholine inhibit it. Fasting promotes the conversion of T4 to T3, thereby activating the general metabolism.

The regulation of this axis is maintained through various negative feedback mechanisms. Overproduction of T3 and T4 caused by an adenoma can lead to primary hyperthyroidism in cats (Peterson 1984).

Additionally, numerous substances, including most psychotropics, barbiturates, progestagens, and glucocorticoids, alter thyroid hormone concentrations. Hypothyroidism is very rare in cats and is often the result of hyperthyroidism treatments, such as bilateral thyroidectomy, radiotherapy, or excessive use of antithyroid medications (Galgano et al. 2014) (◘ Fig. 3.7).

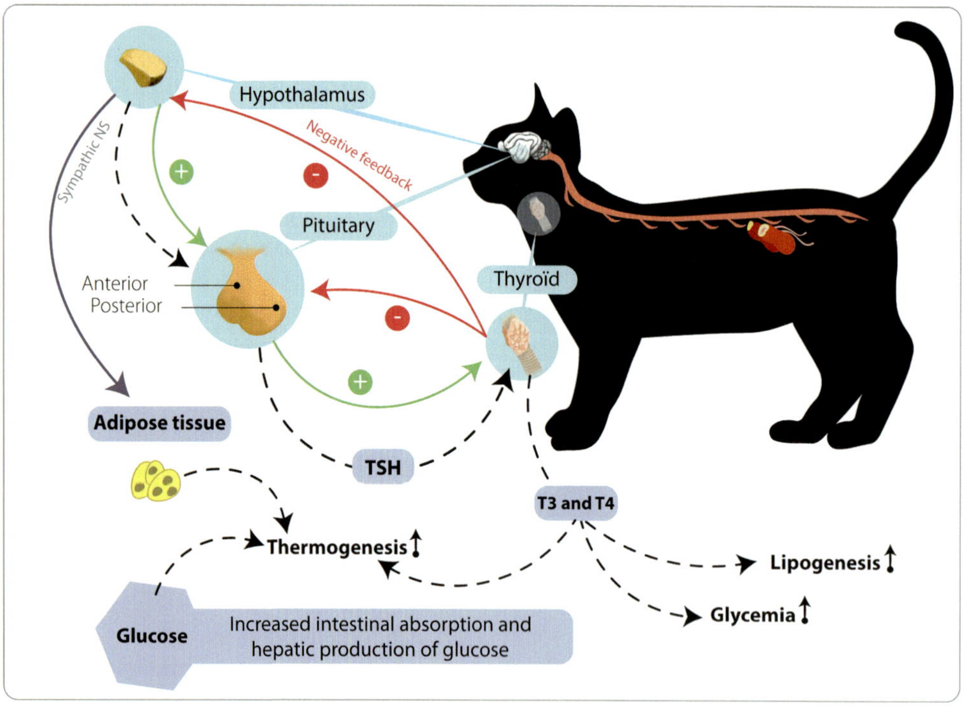

◘ **Fig. 3.6** Hypothalamic-pituitary-thyroid axis and thyroid hormone synthesis. The thyroid gland, located along the trachea, secretes the thyroid hormones thyroxine (T4) and triiodothyronine (T3). T4 is primarily converted into T3 in the liver. Most T4 exists in a protein-bound form, but only free T4 and T3 are biologically active. In cats, T3 is predominantly derived from the conversion of T4 to T3 in the liver or target cells, whereas in humans, part of T3 is already synthesized directly in the thyroid

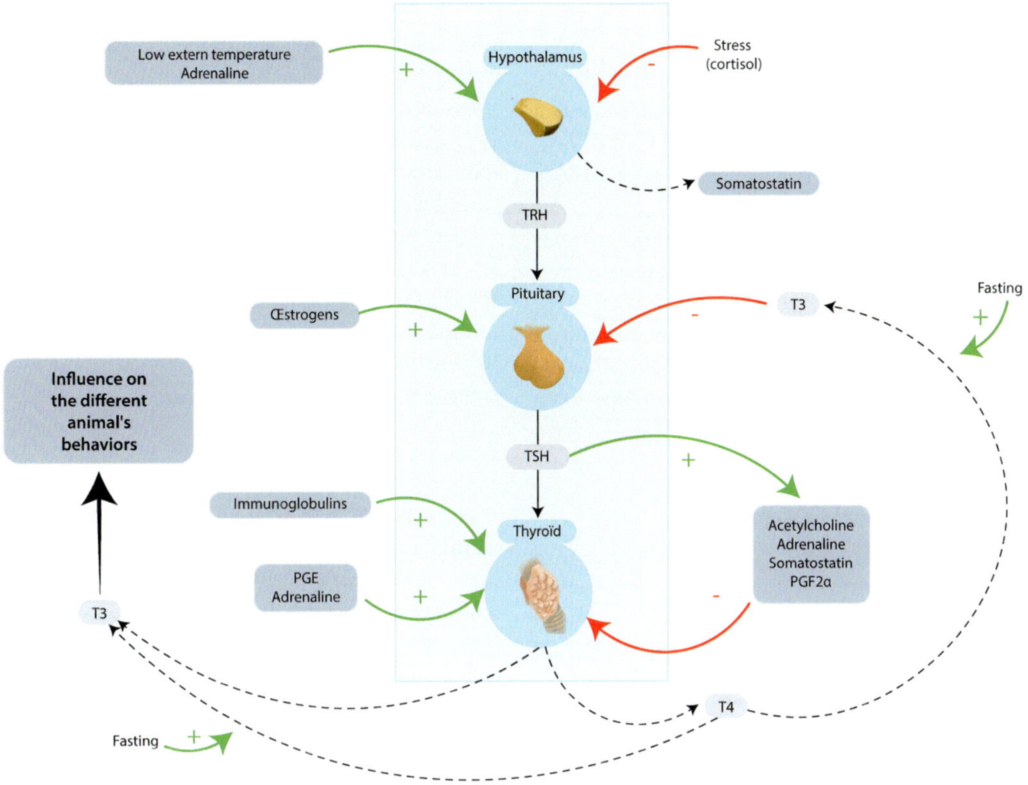

Fig. 3.7 Factors influencing the hypothalamic-pituitary-thyroid axis. The regulation of the hypothalamic-pituitary-thyroid axis occurs through various negative feedback loops. Additionally, many substances, including most psychotropics, barbiturates, progestagens, glucocorticoids, and chronic stress, induce a decrease in thyroid hormone levels

Thyroid hormone levels vary significantly throughout the day, and reference values can differ greatly between individuals.

Primary hyperthyroidism is a common condition among older cats. Its average prevalence is 10% in cats over 10 years old, with significant variation across countries (Peterson 2012). Hyperthyroidism is mainly caused by adenomas of the thyroid gland, leading to excessive production of thyroid hormones. The initial symptoms usually appear in older cats, and due to the thyroid's extensive metabolic role, clinical signs may involve multiple systems, reflecting a general acceleration of metabolism. Observable changes include weight loss despite increased appetite, increased water intake, agitation, diarrhea, vomiting, elevated heart rate, breathing difficulties, changes in vocalizations, and skin abnormalities linked to excessive grooming (dull fur, poor regrowth, thinning, scaling, hyperpigmentation, fragile claws) (Khare et al. 2018).

Behavioral manifestations are also numerous and characteristic. These include fear or aggression reactions that were previously absent, frequent solicitation of owners with vocalizations even at night, stress intolerance up to panic attacks during restraint, and hyperactivity (Khare et al. 2018; Peterson 1984).

Hypothyroidism is very rare in cats and often results from hyperthyroidism treatments such as bilateral thyroidectomy, radiotherapy, or excessive use of antithyroid medications (Galgano et al. 2014).

In humans, certain psychiatric conditions such as anxiety and impulsivity may increase the risk of developing hyperthyroidism (Xia et al. 2012).

3

Summary of ▶ Sect. 3.1

Hormones are produced by endocrine glands. They travel through the bloodstream to reach a target organ, where they exert their effects. Hormone levels are most often regulated by negative feedback mechanisms.

Stress activates the autonomic nervous system and the hypothalamic-pituitary-adrenal (HPA) axis. Acute stress triggers activation and the release of norepinephrine and glucocorticoids, promoting a rapid and adaptive behavioral response corresponding to the allostatic load of the reactive state. However, if this response fails to restore equilibrium, stress becomes chronic, and glucocorticoids have detrimental effects on the body, corresponding to the allostatic overload of the pathological state.

The hypothalamic-pituitary-thyroid (HP T) axis influences most organs, including behaviors. Hyperthyroidism is one of the most common endocrine disorders in cats over 10 years old. It is often caused by adenomas and significantly impacts behavior.

3.2 Biological Functions of the Body

3.2.1 Biological Rhythms

3.2.1.1 Biological Clocks

Endocrine functions vary across a broad spectrum of temporal scales, most of which are governed by *biological rhythms* generated by internal clocks. Many homeostatic processes exhibit temporal fluctuations over hours, days, months, or years (Stevenson 2018). These rhythms are synchronized with geophysical cycles, such as the day/night (circadian) or annual (circannual) cycles. Some biological rhythms, termed ultradian, are shorter than circadian cycles and do not appear to align with any identified geophysical cycle, such as the frequency of daily meals (Goh et al. 2019).

Biological rhythms persist even in the absence of geophysical cycles, maintaining periodicity close to that of the corresponding geophysical cycle. These rhythms have evolved to synchronize individuals' activities with environmental changes and their internal physiological processes, enhancing functional efficiency.

Circadian clocks exist at the cellular level in all plants and animals. In mammals, these clocks operate hierarchically, coordinated by a master clock located in the suprachiasmatic nuclei of the hypothalamus (Carmona-Alcocer et al. 2018; Hastings et al. 2018).

Hormones can influence these cycles. For instance, estrogens accelerate the clock's function, shortening daily locomotor activity, while progesterone lengthens activity cycles (Hatcher et al. 2020). Light plays a critical role in entraining biological rhythms across species, notably by regulating the secretion of melatonin by the pineal gland. This hormone is only produced in the absence of light, leading to high melatonin levels during the long nights of winter and low levels during the short nights of summer. Melatonin aids in adapting to work schedule changes or jet lag. Its relaxing effects have made it a component of the "Chill protocol" for managing phobic and aggressive dogs (Costa and Renata 2019). Recent research has shown that melatonin facilitates restraint during cardiac examinations in cats without sedating them, unlike gabapentin (Ruviaro Tuleski et al. 2022).

Seasonal changes observed in cats are primarily linked to the rhythm of reproductive functions. Seasonal rhythms influence mating behavior, parental behavior, territoriality, aggression, and learning in various species, all regulated by gonadal steroid hormones (Blottner and Jewgenow 2007; Leyva et al. 1984; Spindler and Wildt 1999).

3.2.1.2 The Sleep-Wake Cycle

The sleep-wake cycle is regulated not by melatonin, which influences slower rhythms, but by adenosine (Huang et al. 2011). Adenosine is produced by astrocytes, a type of glial cell that stores glycogen to supply neurons with glucose. As glycogen levels in astrocytes decrease, adenosine production increases. Adenosine is therefore secreted whenever neurons consume glucose, which occurs throughout the day. This mechanism promotes sleep and creates the need for rest. Adenosine also activates neurons in the ventrolateral preoptic area, which inhibit the wakefulness system.

During *light sleep*, triggered by high adenosine levels, glucose reserves are restored, and adenosine levels decrease. This reduction allows for a new cycle, enabling wakefulness. Glucose is consumed again, adenosine is produced, and a new phase of light sleep follows. Caffeine acts as a competitive inhibitor of adenosine by blocking its receptors, thereby reducing the sensation of sleepiness (Portas et al. 1997).

Slow-wave sleep depends on a complex dual-circuit system in the hypothalamus, as well as the actions of adenosine and other neurotransmitters (see accompanying box) (Saper et al. 2005). Slow-wave sleep, characterized by significantly reduced brain activity, minimizes cerebral metabolism and allows the brain to rest while astrocytic glycogen reserves are replenished.

These phases of slow-wave sleep are interspersed with periods of *paradoxical sleep*, during which brain activity resembles that of wakefulness. REM sleep is marked by muscle paralysis except for occasional twitches and rapid eye movements, hence the term paradoxical sleep or rapid eye movement (REM) sleep. The function of REM sleep, associated with dreaming, remains partially understood (Peever and Fuller 2017). One hypothesis suggests that it facilitates synaptic changes that consolidate the day's learning, mediated by the hippocampus (Todorova and Zugaro 2019).

Sleep serves two primary functions. Firstly, it acts as a protective mechanism for cells. By regularly resting cerebral metabolism, sleep limits free radical production and reduces cellular damage. Total sleep deprivation leads to irreversible brain damage and death (Montagna et al. 2003). Secondly, sleep plays a crucial role in learning, enabling the consolidation of daily experiences into long-term memory.

Important Point of View

The dual circuit system governing wakefulness and sleep phases

Among the neurotransmitters involved in maintaining wakefulness, norepinephrine from the locus coeruleus ensures vigilance, acetylcholine induces cortical activation, and serotonin promotes automatic behaviors such as locomotion and grooming (see ▸ Chap. 2, ▸ Sect. 2.2.4.2, "Acetylcholine,"). Histamine and orexin support sustained wakefulness (Kayama and Koyama 2003). A primary circuit dependent on these neurotransmitters facilitates the initiation of slow-wave sleep.

REM sleep, on the other hand, is promoted by circuits located in the hypothalamus and periaqueductal gray matter (Chen et al. 2018; Lu et al. 2006). These circuits activate only during slow-wave sleep. Thus, the activation of the primary circuit enables the secondary circuit to initiate the transition from slow-wave sleep to REM sleep (◘ Fig. 3.8). Consequently, it is impossible to transition directly from wakefulness to REM sleep.

In addition to these two primary types of sleep, sleep can be further subdivided into distinct phases characterized by different electrical activities in the brain (◘ Fig. 3.9). These brain activities are recorded using electroencephalograms (EEGs). The sleep-wake cycles of cats are comparable to those observed in other mammals (Ursin 1968; Tobler and Scherschlicht 1990).

The different phases of sleep correspond to distinct patterns of brain activity. Beta waves are associated with an active waking state. Alpha waves are observed during light relaxation. Theta waves appear during light slow-wave sleep, delta waves are characteristic of deep slow-wave sleep, and gamma waves correspond to REM sleep, marked by intense activity similar to wakefulness.

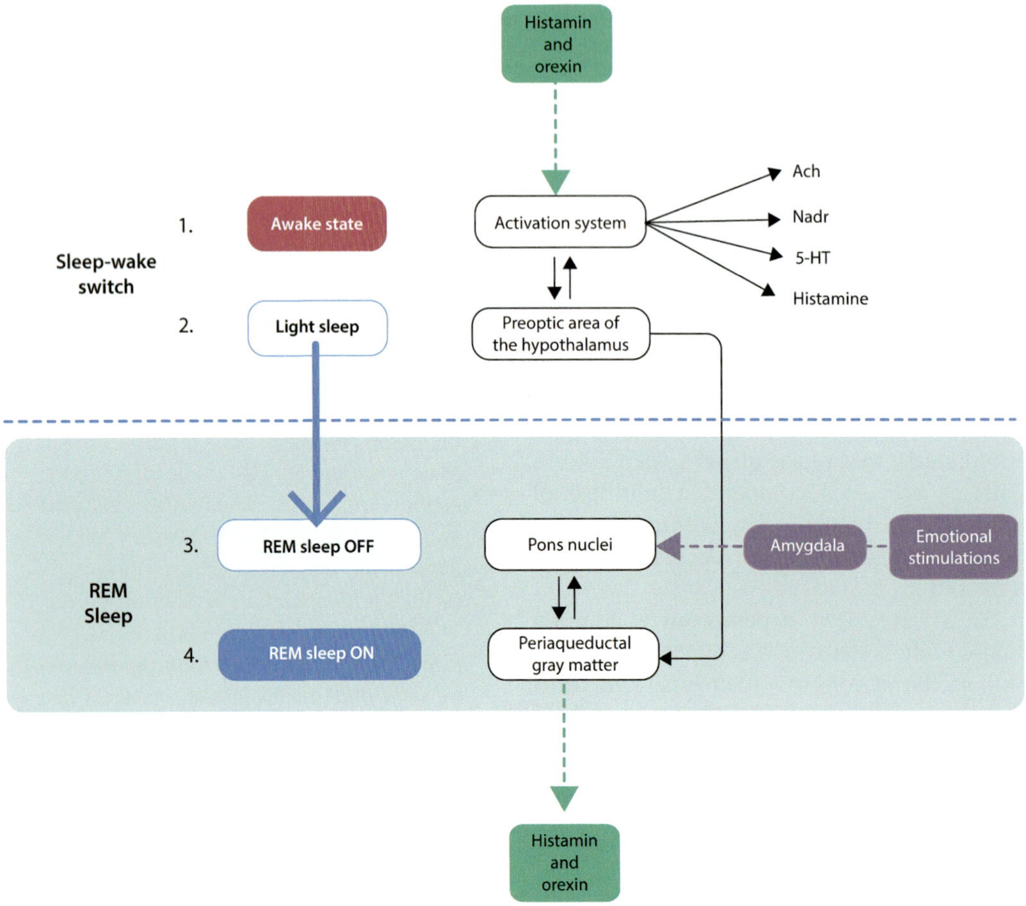

Fig. 3.8 The bidirectional circuit system allowing different phases of wakefulness and sleep

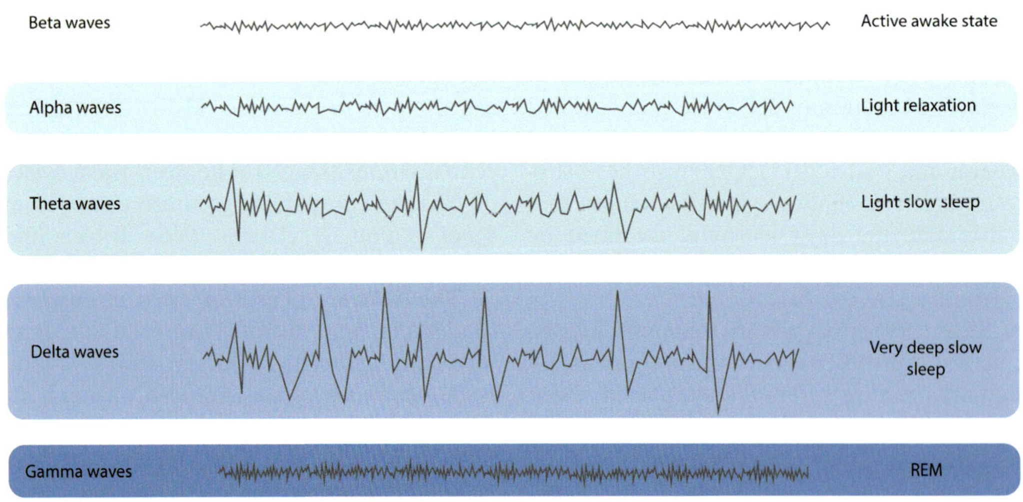

Fig. 3.9 Brain waves observed during different phases of wakefulness and sleep in cats (based on Tobler and Scherschlicht 1990). This figure illustrates the characteristic brainwave patterns during the various stages of wakefulness and sleep, as recorded in cats

Bálint et al. (2024) compared the sleep of cats and dogs. The study found no macrostructural differences between the sleep phases of the two species. However, minor differences in brainwave patterns were observed during the different phases, attributed to varying stress levels, circadian rhythms, and potentially reflecting neuronal and behavioral differences between the species (Bálint et al. 2024).

Indeed, the circadian rhythm and locomotor activity of dogs depend on their owner's daily schedule and presence, unlike cats (Piccione et al. 2014). Dogs exhibit a peak of sleep during the night, similar to humans, whereas cats show a daytime peak (Lucas 1978).

Differences in theta waves suggest lighter sleep in cats. Additionally, the reappearance of alpha waves during deep slow-wave sleep implies that cats have more superficial sleep than dogs and may be more receptive to external stimuli during sleep. Furthermore, cats display high-frequency beta bands, generally interpreted as markers of cortical hyperarousal, often observed in individuals suffering from insomnia (Feige et al. 2013). Feige et al. (2013) concluded that cats experience lighter sleep than dogs. This aligns logically with the cat's status as a prey species and, more significantly, its nature as a solitary animal, relying only on itself rather than a social group to respond to danger.

3.2.2 Drinking and Eating Behaviors

Most biological mechanisms that contribute to maintaining the internal stability of the organism function like thermostats (Damasio and Damasio 2016), ensuring the homeostasis of the body. A detection system identifies any deviation from the norm (appetitive phase) and activates mechanisms that mobilize the organism to make changes that bring the deviated value back to normal (consummatory phase). Another system halts the process once the normal state is restored (termination phase). These mechanisms are not always observable from a behavioral perspective, but they function in the same way as the behavioral phases described in ► Chap. 1 (see ► Sect. 1.1.2.1).

3.2.2.1 Drinking Behavior

Thirst or *drinking behavior* is regulated by the sodium concentrations in bodily fluids. Most of the water in multicellular organisms is located inside the cells. The water in the extracellular environment is distributed between blood vessels and interstitial fluid. Sodium plays a major role in the movement of water between the intracellular and extracellular environments.

There are two types of thirst: osmotic and hypovolemic (Burton and Hopper 2019) (◘ Fig. 3.10).

Osmotic thirst corresponds to intracellular dehydration, whether direct, due to a lack of water intake and evaporation, or indirect, through the ingestion of salty or sugary foods. This dehydration affects osmoreceptors in the brain, leading to the secretion of antidiuretic hormone (ADH) or vasopressin by the posterior pituitary, which promotes water conservation by the kidneys (Kozłowski and Szczepańska-Sadowska 1975). Drinking water stops osmotic thirst.

Hypovolemic thirst is caused by a decrease in blood volume (Szczepanska-Sadowska et al. 1974). This hypovolemia stimulates the production of a powerful vasoconstrictor, angiotensin II. Angiotensin II stimulates aldosterone production by the adrenal glands, which promotes sodium and water retention by the kidneys. Restoring hypovolemic thirst requires the ingestion of water as well as sodium and other solutes (Davis et al. 2013; Zoran et al. 1992).

3.2.2.2 Eating Behavior

Cats expend energy continuously throughout the day, while their food intake is episodic. This pattern causes the body to be in a constant state of imbalance, either in surplus immediately after a meal or in deficit between meals. To cope with fasting periods, the body has developed the ability to store energy either short-term in the liver as glycogen (Verbrugghe and Hesta 2017) or long term in adipose tissue as triglycerides (Zoran 2002).

During *fasting periods* between meals, the sympathetic nervous system is activated. Insulin levels decrease while glucagon and adrenal catecholamine levels increase. These changes lead to the conversion of hepatic gly-

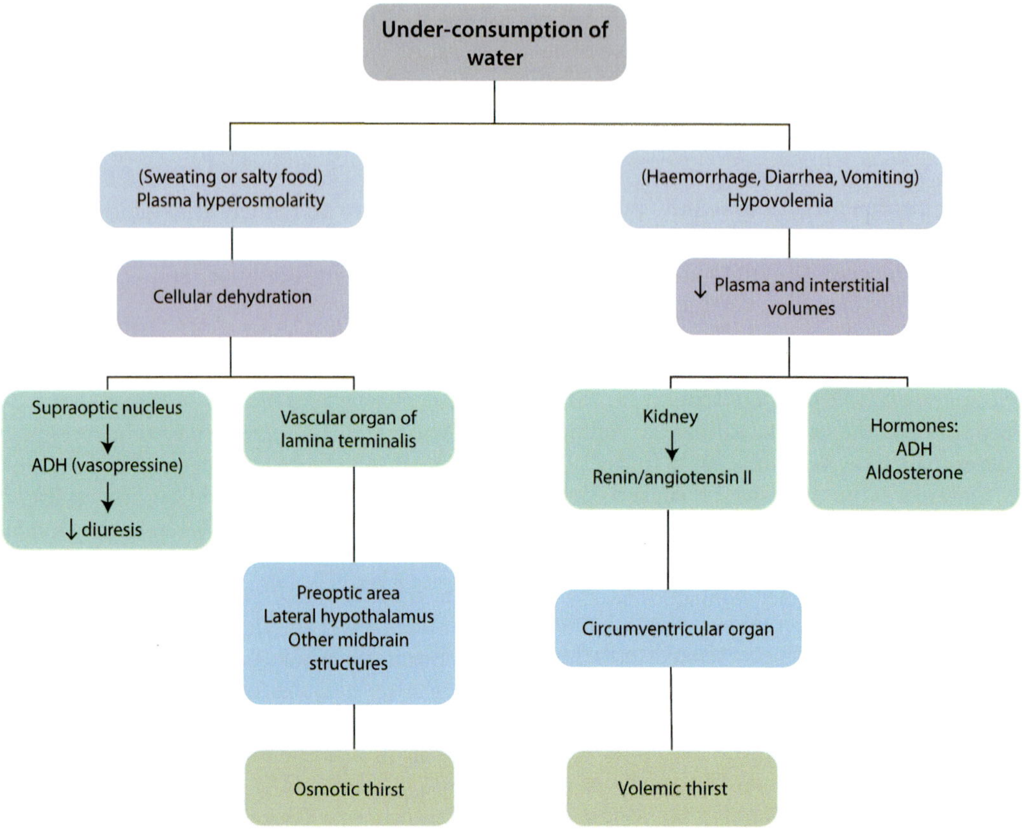

Fig. 3.10 Osmotic and hypovolemic thirst. Osmotic thirst corresponds to intracellular dehydration, whether direct, due to a lack of water intake and evaporation, or indirect, via the consumption of salty or sugary foods. Hypovolemic thirst is caused by a decrease in blood volume

cogen into glucose and triglycerides into fatty acids. When insulin levels are low, glucose is primarily consumed by the brain (Hirvonen et al. 2011), while the rest of the body relies on fatty acids.

In cases of prolonged fasting, a hormone-sensitive enzyme called lipase is stimulated by various hormones, including catecholamines. This enzyme is inhibited by insulin (Kraemer and Shen 2002). When activated, it induces significant lipolysis, releasing fatty acids that are partially absorbed by the liver. In the presence of L-carnitine, these fatty acids are oxidized to provide energy. In its absence, they are esterified into triglycerides and accumulate. This esterification process is responsible for hepatic triglyceride accumulation or hepatic lipidosis, a well-known consequence of prolonged fasting in cats.

Several theories have been proposed to explain this phenomenon, including L-carnitine deficiency, apolipoprotein deficiency, hypoinsulinemia caused by anorexia affecting hepatic enzyme activity, insulin resistance, and pre-existing hepatic abnormalities caused by obesity (Armstrong and Blanchard 2009).

Conversely, during the *feeding phase*, the parasympathetic nervous system is activated. The digestive tract fills, triggering storage processes such as the conversion of glucose into glycogen and fatty acids into triglycerides. Insulin levels are elevated, allowing all cells to metabolize glucose. Both short-term and long-term energy stores are replenished.

Food intake is triggered by hunger (the appetitive phase of eating behavior), which itself is induced by an increase in ghrelin, a hormone secreted by the stomach when the small intestine is empty (Rindi et al. 2002). Additionally, during a fasting period covering several meals, receptors in the liver detect glucose and lipid deprivation and transmit this information to the brain via the vagus nerve. The brain also has an independent system for detecting glucose deprivation located in the hypothalamus (Heisler and Lam 2017).

Thus, the stimuli capable of inducing eating behavior are numerous and include physiological signals from fat reserves, glycogen stores, the nature and quality of food, as well as neural and hormonal signals from the digestive tract (Agnew and Korman 2014).

Short-term *satiety* (the stopping phase of eating behavior) is generated by various *sensory factors* such as the appearance, texture, smell, and nutritional value of the food, along with *digestive factors* such as stomach receptors, a reduction in ghrelin secretion in the stomach, or the secretion of cholecystokinin and peptide YY in the intestine, which help terminate the ongoing meal (Bosch et al. 2022).

There are also longer term signals that modulate hunger and food intake across all meals. This is the role of adipose tissue when it is present in excess. Adipose tissue secretes leptin, a hormone that increases the individual's metabolism and decreases hunger. The development of leptin resistance is one of the causes of obesity in cats (Appleton et al. 2000; Zoran 2010).

Gonadal steroid hormones also affect food intake and body mass. Estrogens reduce food intake and body mass while increasing voluntary exercise. Progesterone counteracts these effects of estrogens. Androgens generally increase hunger and consequently body mass (Zoran 2010). These factors explain why spaying female cats promotes weight gain.

All peripheral hormonal information is centrally transmitted to the hypothalamus, which is the regulatory center for eating behavior. Each of the aforementioned hormones has specific receptors located in the lateral hypothalamus to trigger eating behavior or in the medial hypothalamus to induce satiety **(see box below)**.

Important Point of View

Biological determinants of eating behavior

The *lateral hypothalamus* secretes the peptides orexin (Sakurai 2006) and MCH (melanin-concentrating hormone) (Pissios et al. 2006), which, in addition to promoting eating behavior, help keep the animal awake (Ohno and Sakurai 2008). It distributes axons to cortical regions involved in motivation (DiLeone et al. 2003), movement, and metabolism. The release of neuropeptide Y (NPY) in the lateral hypothalamus triggers frenzied eating by stimulating orexin and MCH neurons (Edwards et al. 1999). Ghrelin is the main activator of NPY synthesis in the lateral hypothalamus. It also acts in the ventral tegmental area, where it activates the reward system via the dopaminergic neurons of the mesolimbic pathway, causing dopamine release in the nucleus accumbens, which leads to the pleasure experienced during meals (Abizaid et al. 2006). Blocking NPY receptors induces anorexia.

Endogenous opioids also alter the hedonic value of food (Caref and Nicola 2018; Nogueiras et al. 2012). Treatment with opioid antagonists thus reduces food intake (Kirkham and Williams 2001). Furthermore, endocannabinoids stimulate appetite by increasing the release of orexin and MCH (Morello et al. 2016). Glucocorticoids are also orexigenic agents as they decrease insulin sensitivity in cells (Lowe et al. 2009).

Conversely, insulin, leptin, and peptide YY block NPY receptors in the arcuate nucleus of the ventromedial hypothalamus, reducing food intake and energy expenditure while stimulating the expression of pro-opiomelanocortin (POMC). This neurotransmitter inhibits neurons possessing cocaine- and amphetamine-related transcript (CART) receptors, which promote catabolism (Wakshlag and Loftus 2014). The appetite-suppressing effects of cocaine and amphetamines are

3

due to their blockade of this mechanism (Germano et al. 2007). Finally, serotonin (Heisler et al. 2006) and β-catecholamines also inhibit food intake (Gálosi et al. 2015) (Fig. 3.11).

The involvement of neurotransmitters such as dopamine and serotonin in the regulation of hunger explains the link between chronic stress or pathological anxiety and dysregulation of the feeding system (Volkow et al. 2011), including compulsive states such as addictive behaviors (Rothman et al. 2008). Some cats exhibit eating behavior disorders, which are a significant cause of distress (Delgado and Dantas 2020; McMillan 2013). Ethically, it is essential not to rely solely on mechanical measures to limit or prevent food intake, as such measures fail to improve the animal's well-being. Moreover, prolonged fasting in cats carries a substantial risk of hepatic lipidosis. When a lack of food-related satiety is evident, anorexigenic agents should be prescribed to help the animal achieve an acceptable level of well-being, rather than recommending severe rationing. Since serotonin reduces appetite, selective serotonin reuptake inhibitors (SSRIs) are effective agents for regulating food intake (Halford et al. 2005).

It is important to note that in cats, as in most mammals, the systems that activate feeding behavior are much more effective than those that induce satiety. This can be seen as a biological adaptation to environments where food scarcity is a risk. However, in modern contexts where cats often receive sufficient nutritional intake, these activation systems remain active and are sometimes exacerbated by the palatability of offered meals, while the systems responsible for food satiety are insufficient or even deficient (Jerjen et al. 2023).

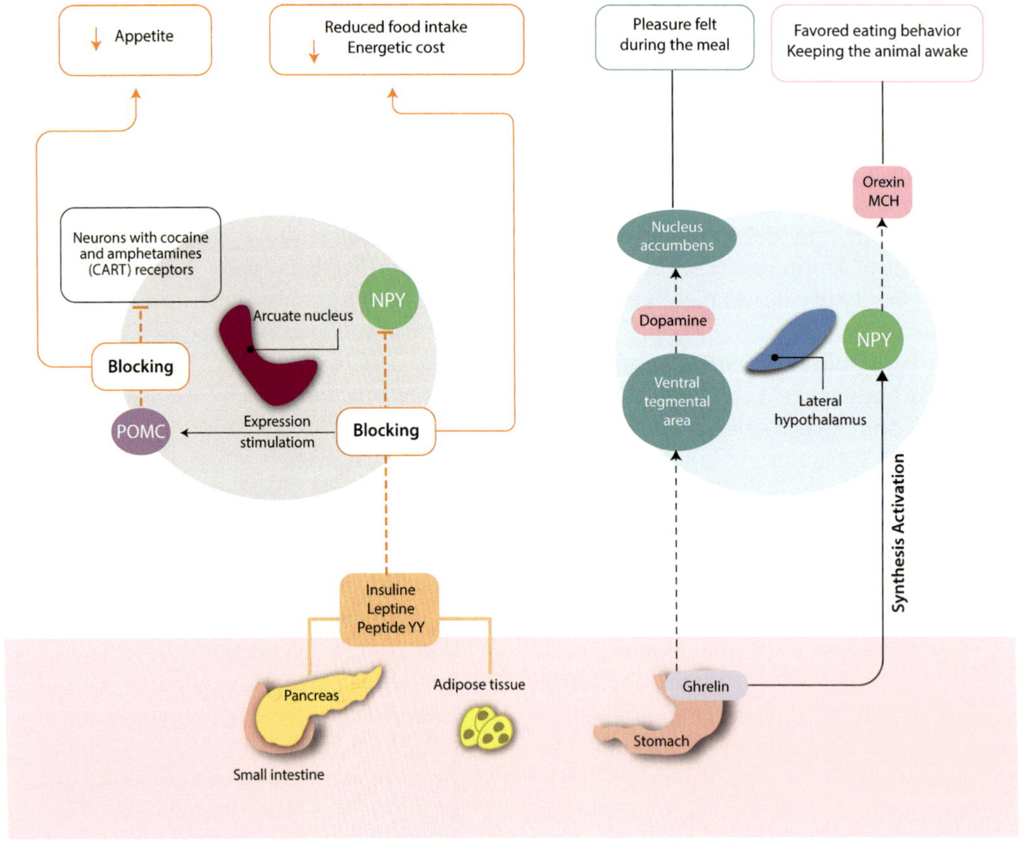

Fig. 3.11 Biological determinants of eating behavior

Consequently, in both cats and humans, there is a heightened food motivation combined with a less effective inhibitory system for food intake compared to the activation system, leading to a significantly higher prevalence of obesity than underweight cases (Agnew and Korman 2014).

3.2.3 Sexual, Reproductive, and Parental Behaviors

In contrast to the previous section, which emphasized the distal purpose of certain behaviors, sexual, reproductive, and parental behaviors are not related to the survival of the individual but rather to the survival of the species.

3.2.3.1 Sexual Differentiation

In cats, as in all mammals, chromosomal sex is XX in females and XY in males. This chromosomal sex determines the gonadal sex.

Certain processes of sexual differentiation depend solely on chromosomes and not on hormones. The expression of certain genes present on the sex chromosomes occurs early in development and independently of the concentration of gonadal hormones such as testosterone, estrogen, and progesterone (Arnold 2022; Arnold et al. 2004).

The development of secondary sexual organs in females requires feminization through the development of Müllerian ducts and defeminization through the regression of Wolffian ducts. Conversely, in males, defeminization occurs through the regression of Müllerian ducts, and masculinization occurs via the development of Wolffian ducts (Lyle 2007).

These masculinizing and feminizing effects are referred to as the organizational effects of hormones because they influence the sex of sexual organs and are present from birth.

Sexual differentiation is completed under the action of sex steroid hormones at puberty, leading to sexual dimorphism. These are referred to as activational effects because they require the action of hormones produced by the gonads during puberty. Morphological sexual differences ultimately result in behavioral sexual differences.

This process of sexual differentiation is complex, and errors can occur at any stage. For instance, postnatal exposure to androgens is a cause of infertility in females (Demaldé et al. 2016).

3.2.3.2 Sexual Behavior

Male cats are capable of mating year-round. However, testosterone levels and sperm quality vary with photoperiod, leading to fluctuations in fertility (Nuñez-Favre et al. 2012).

In contrast, female cats will only accept mating during their heat period. Their cycle is regulated by the cyclic production of GnRH from the anterior hypothalamus (Dissen et al. 2017).

In females, sexual behavior can be divided into three components: attractivity, proceptivity, and receptivity (Beach 1976). Attractivity refers to the degree of sexual motivation a female elicits in a given male. Proceptivity encompasses the behaviors through which the female initiates copulation, reflecting an underlying motivational state. Receptivity represents her ability to trigger ejaculation in the male, i.e., her responsiveness to the sexual behaviors of her feline partner. Estrogens enhance attractivity, proceptivity, and receptivity, whereas progestogens reduce them (Beach 1976). When receptive to the male, female cats deviate their tail and adopt a mating posture known as lordosis (Turner and Bateson 2013).

▪ Sexual Behavior of Male Cats

Puberty begins between 5 and 8 months in male cats (Tsutsui et al. 2004). After puberty, most males develop a high motivation to seek a sexual partner as the testes begin secreting testosterone.

Sexual reflexes, including the mating posture, erection, and ejaculation, are organized in the spinal cord (Allard et al. 2005). Sexual motivation and copulatory behavior are controlled by the anterior preoptic area of the hypothalamus (Paredes 2003). Destruction of this region permanently abolishes copulatory behavior. During ejaculation, the reward circuit involving dopamine is activated (see ▶ Chap. 2, ▶ Sect. 2.3.3.3).

■ Sexual Behavior of Female Cats

Female cats reach puberty between 6 and 10 months on average, though it may be delayed up to 18 months in wild-born females born in the spring (Kutzler 2022).

The feline sexual cycle is seasonal and regulated by the retino-hypothalamic pathway. Melatonin, secreted during periods of darkness by the pineal gland, inhibits the secretion of kisspeptin and GnRH from the hypothalamus, thereby suppressing the central regulation of sexual hormones (Revel et al. 2007; Roy et al. 2001). Consequently, during winter and periods of low light, most female cats enter anestrus.

The estrous cycle in cats consists of four phases: proestrus, estrus, metestrus (if ovulation and fertilization occur), or diestrus (if ovulation does not occur), and anestrus. The cycle may be limited to proestrus followed by diestrus in cases of anovulation. This cycle is governed by the sexual hormones of the hypothalamus-pituitary-ovarian axis. During proestrus, an increase in estradiol levels produced by the ovaries and a peak in the neuropeptide kisspeptin produced in the hypothalamus induce GnRH production in the hypothalamus (Oakley et al. 2009; Santos et al. 2022; Santos and Silva 2023; Tanyapanyachon et al. 2018).

GnRH, in turn, induces a surge of LH and FSH in the pituitary gland, which triggers a peak in estrogen levels in the ovaries, leading to estrus. Ovulation, and therefore estrus, is induced by copulation, which stimulates sensory receptors, causing GnRH secretion, which in turn induces a surge of pituitary LH, resulting in ovulatory discharge (Goericke-Pesch 2010). The LH surge also stimulates progesterone production. This is referred to as a neuro-hormonal reflex. Spontaneous ovulation has also been observed in female cats (Gudermuth et al. 1997).

Multiple copulations may be required to induce the LH surge, meaning that sperm from several males can be present in the genital tract. Consequently, not all oocytes are fertilized by the same male, and kittens in a litter may have different fathers (Wildt 1991).

In the absence of ovulation, ovarian follicles degenerate, and estrous cycles repeat, driven by the retino-hypothalamic pathway. For this reason, the female cat is a seasonally polyestrous species. Estrus is recognizable through specific behaviors: the female purrs, solicits attention from owners, crouches or rolls onto her back, raises her perineal region (lordosis posture), and scratches the ground. These behaviors are induced by increased estradiol levels (Santos and Silva 2023). If a male is present, the female accepts mating during estrus but rejects him during proestrus.

If ovulation occurs without fertilization, progesterone levels gradually decrease during diestrus and into anestrus if it occurs, whereas progesterone levels remain elevated during pregnancy (Santos and Silva 2023).

Parturition typically occurs around 63 to 66 days after fertilization and is associated with progesterone withdrawal. Fetal cortisol surge, a well-known mechanism in some other species, has not yet been clearly demonstrated in cats.

Important Point of View

Neurophysiology of sexual behaviors

The most critical brain region for sexual behavior in females is the ventromedial nucleus of the hypothalamus. In rats, destruction of this nucleus abolishes copulatory behavior, while its stimulation facilitates it (Pfaff and Sakuma 1979).

Orgasm in females is accompanied by increased activity in regions similar to those activated during ejaculation in males, including the nucleus accumbens of the reward circuit and nuclei in the periaqueductal gray matter (Lodé 2020).

In voles, oxytocin has been shown to play a role in affiliative behaviors, including sexual behavior, parental behavior, and caregiving for young (Ahern 2009). Field voles are monogamous, whereas mountain voles are polygamous. Researchers have demonstrated that the distribution of oxytocin receptors differs between these species: in monogamous voles, receptors are more abundant in the prelimbic cortex, nucleus accumbens, medial nucleus of the hypothalamus, and lateral regions of the amygdala. In contrast, in their polygamous relatives, these receptors are more concentrated in the lateral septum, ventromedial nucleus of the hypothalamus, and cortical nucleus of the amygdala (Insel and Shapiro 1992).

Even more remarkable, these receptors redistribute differently in the postpartum period in mountain voles, the only time maternal behavior is observed in this species.

3.2.3.3 Le Comportement Parental

As in most mammals, paternal behavior is absent in male cats during the first few weeks postpartum. However, male cats may later become playmates for kittens once they are capable of moving independently.

The onset of maternal behavior is hormonally mediated. The abrupt drop in progesterone levels at parturition, combined with high concentrations of estradiol, prolactin, and oxytocin, triggers maternal behavior (Stern 1989). This complex hormonal profile suppresses the fear response the queen might otherwise feel toward her kittens (see ▶ Chap. 4, ▶ Sect. 4.3.3). During the first 2–3 days postpartum, the queen spends nearly all her time with her kittens. This time diminishes quickly thereafter, and nursing sessions become shorter (Turner and Bateson 2013).

This change is tied to prolactin production, which is stimulated whenever the kittens suckle. Prolactin, in turn, stimulates corticosteroid secretion, which increases the metabolic rate—i.e., the amount of energy consumed at rest—and indirectly raises the mother's body temperature. As the kittens grow, their close proximity during nursing hinders the mother's ability to dissipate heat, making the situation increasingly uncomfortable. When the brain's thermoreceptors detect excessive temperature elevation, the mother leaves the nest to prevent hyperthermia.

Maternal behavior also includes defending the kittens. This is evident in maternal aggression directed toward individuals approaching the nest (see ▶ Chap. 1, ▶ Sect. 1.3.2.3). The queen will hiss, claw, and drive away intruders until she perceives her kittens as safe. Maternal aggression is most intense and frequent in the days immediately following parturition. It gradually decreases as the kittens grow but may intensify if the mother is frequently disturbed, underscoring the importance of minimizing interventions during the first 2–3 weeks after birth.

Important Point of View

Neurophysiology of Parental Behaviors

Parental behaviors are governed by the medial preoptic area of the hypothalamus, which contains receptors for estradiol, prolactin, and oxytocin (Tsuneoka 2019). Lesions in this region abolish maternal and paternal behaviors in rats (Brown et al. 2017). The neuronal connection between the hypothalamus and the amygdala is responsible for suppressing fear responses toward kittens during parturition.

Additionally, a circuit originating in the medial preoptic area passes through the ventral tegmental area and connects to the nucleus accumbens, activating the reward circuit. This activation induces pleasure in the mother while caring for her kittens (Rincón-Cortés and Grace 2020).

3.2.4 Affiliative Behaviors

In cats, we refer to affiliative behaviors rather than social behaviors, as the domestic cat (*Felis catus*) is not a social species.

A species is considered social when reciprocal attraction between individuals, such as group gathering, is not driven solely by environmental pressures, such as limited resources, and results in cooperative behaviors. Group cohesion in social species is maintained by establishing hierarchies, which minimize agonistic interactions among group members. The term *social* pertains to a species, while *sociality* refers to the tendency of a species to exhibit social behavior.

Some authors suggest that temporary association during reproduction and rearing of young is sufficient to classify a species as social. In this book, we argue that this is insufficient; evidence of adult cooperation is required to define a species as social. By this definition, the domestic cat is not a social species (Bradshaw 2016).

Sociability, on the other hand, refers to an individual's ability to interact with other individuals (Vitale and Udell 2019). This trait can be observed even in non-social species like cats and may also include interspecific relationships, such as between a cat and a human.

Regardless of its history, a cat will never become social because it does not belong to a social species. However, some cats can be friendly toward humans or other cats (Vitale and Udell 2019). During development, socialization occurs very early (Bradshaw 2016). For adult cats exhibiting relational difficulties, the aim is to *socialize* them, or in more severe cases, to *familiarize* them—facilitating interaction with a select few familiar individuals.

3.2.4.1 Attachment and Affiliative Behaviors

Affiliative behaviors encompass all interactions between individuals, regardless of whether they belong to a social species or even the same species.

The earliest affiliative behaviors are those observed between a mother and her kittens, established through the process of attachment (see ▶ Chap. 4, ▶ Sect. 4.3.3). These behaviors are influenced by oxytocin and vasopressin (Carter 2017). Parental behaviors also rely on these hormones, as well as glucocorticoids, which affect the secretion of dopamine and serotonin. Pheromones contribute to regulating attachment, parental care, and social affiliations.

Among members of the same group, affiliative and agonistic behaviors are common. Grooming and mutual rubbing, in particular, facilitate the identification of group members. Cats sniff each other and may verify identity by touching noses face to face (Crowell-Davis et al. 2004). These behaviors may be influenced by opioids and dopamine, as observed in social species (Machin and Dunbar 2011), though no conclusive evidence exists for cats.

It has been demonstrated, however, that oxytocin plays a role in forming affiliative bonds between cats and humans (Rault et al. 2017) as well as between cats and other cats (Nagasawa et al. 2022).

3.2.4.2 Behavior and Pheromones

The behavior of individuals is regulated by the brain and hormones, which together govern an individual's behavioral choices based on the context. However, an external system is also necessary to mediate interactions among individuals, enabling reproduction and social interactions. This is the pheromonal system, which likely exists throughout the animal kingdom. In cats, this system is crucial for organizing their territory. Cats possess numerous regions capable of secreting pheromones (see ▶ Chap. 1, ▶ sect. 1.3.2). These secretions are often accompanied by odors, and associative learning frequently leads individuals to respond intentionally to scents associated with pheromones (see ▶ Chap. 1 ▶ sect. 1.3.2.1). Pheromones emitted by one individual are, by definition, capable of altering another individual's behavior independently of their will.

Reproductive behaviors are heavily influenced by pheromones. Some pheromones play a role in sexual attraction, while others signal ovulation, gestation, or anestrus. These include genital pheromones emitted by the genital glands (Mills 2005). The F2 fraction, emitted during facial marking, is also associated with male sexual behavior (Pageat and Gaultier 2003).

Pheromones also influence agonistic interactions. The F4 fraction emitted by facial glands is essential for allomarking and individual recognition (Pageat and Gaultier 2003). These pheromones facilitate the identification of group members and may trigger agonistic behaviors when recognition fails.

A maternal pheromone, the Cat Appeasing Pheromone (CAP), analogous to the canine appeasing pheromone, has been synthesized in Feliway Friends® (CEVA). It helps improve relationships between cats (DePorter and Elzerman 2019) and even between cats and dogs in shared households (Prior and Mills 2020).

Pheromones also structure a cat's territory through urine marking, scratching, and facial marking (emitting the F3 fraction). The F3 pheromonal fraction has been synthesized and marketed, demonstrating effectiveness in reducing urine marking (Mills et al. 2011). Commercial products such as Feliway® (CEVA) and Zenifel® (Virbac), the latter enhanced with nepetalactone (a catnip extract), are examples. Territorial defense, however, may also lead to aggressive behaviors (see the next section).

Finally, pheromones indicate the emotions experienced by a cat. In cases of intense fear, the discharge of anal glands and sweat glands releases pheromones that signal an alarm to other cats (Overall 2015).

3.2.4.3 Aggression

Aggression is always associated with a strong activation of the sympathetic nervous system (Hine et al. 2004). Its purpose is to drive away a perceived threat, whether direct or indirect. Aggressive behaviors include threats, such as hissing, and attacks, such as scratching or biting. Predation, however, is an attack that is not considered aggression, as it involves no activation of the sympathetic system; a predator is not angry at its prey.

The different types of aggression and their neurophysiological mechanisms are detailed in ▶ Chap. 1 (see ▶ Sect. 1.3.2.3) and ▶ 2 (see ▶ Sect. 2.3.3.5).

In cats, studies suggest that serotonergic synapse activity inhibits aggression (Hine et al. 2004; Romaniuk et al. 1989). Androgens are also long recognized as major modulators of aggressive behavior (Bouissou 1983). Testosterone, secreted beginning at puberty, has a potentiating effect on aggression (Giammanco et al. 2005). By increasing the motivation for reproductive behavior, it heightens aggression linked to sexual competition.

Therefore, for castration to have a positive effect on aggressive behavior between individuals, the aggression must have begun at puberty and be associated with some form of sexually motivated competition.

Summary of ▶ Sect. 3.2

Adenosine induces the transition from wakefulness to slow-wave sleep, while a second circuit facilitates the transition to REM sleep.

Thirst behavior is regulated by sodium concentration in the body's compartments and blood volume.

Feeding behavior includes numerous mechanisms to activate hunger but fewer regulatory mechanisms to achieve satiety, which explains why obesity is a more significant issue than underweight conditions in cats.

Sexual, maternal, and parental behaviors are influenced by sex hormones. Oxytocin, vasopressin, and pheromones play a crucial role in establishing maternal behaviors and facilitating social interactions.

Pheromones are essential for cats to structure their territory and create a calming environment. They also influence relationships with individuals the cat encounters, integrating these beings into a coherent territorial framework for the cat.

3.3 Pain

Pain is a common feature in many organic diseases, and its psychological consequences have been widely studied. Conversely, pathological psychological states can lead to somatization.

For these reasons, this section details the mechanisms underlying pain, whether it is acute or chronic, nociceptive or neuropathic. ▶ Chapter 15 will address the clinical aspects of pain management in feline psychiatric consultations.

Pain is defined as an unpleasant sensory and emotional experience associated with, or resembling that associated with, actual or potential tissue damage (IASP definition,

Raja et al. 2020). It serves as a stop signal for the body, particularly when acute and intense, warning the animal of potential danger and prompting the cessation of the ongoing behavioral sequence. Failure to heed this signal could result in severe, even fatal, damage to tissues or organs.

Pain achieves this by generating intense fear, which enhances contextual memory. This mechanism often sensitizes cats to veterinary care; pain experienced during visits for somatic conditions frequently conditions the animal to associate intense fear with the veterinary context, leading to emotional responses that complicate future interactions.

3.3.1 Types of Pain

Pain can be classified in several ways, including by its duration (acute or chronic) or by the mechanisms involved in its production.

3.3.1.1 Acute or Chronic Pain

Acute pain is adaptive, biologically beneficial, and triggers behavioral changes aimed at preserving the integrity of the organism. For the animal, it represents a form of stress and corresponds to an allostatic charge. When the behavioral response to this pain fails to eliminate it, the pain transitions to *chronic pain,* which becomes detrimental to the organism, representing an allostatic overload.

Acute pain is easy to recognize because it arises suddenly and manifests with highly visible symptoms, such as limping or a sudden refusal to move. It is often accompanied by vocalizations. The abrupt behavioral changes associated with acute pain make it readily noticeable to pet owners.

Chronic pain, on the other hand, is maladaptive and offers no long-term benefit to the animal. Although it does not immediately compromise the integrity of the organism, it maintains the animal in a state of long-term harm. Chronic pain is more challenging to detect. The animal adjusts its behavior gradually, making these pains less apparent to owners. However, clinicians should investigate the possibility of chronic pain when an animal exhibits disproportionate irritability relative to the context or compared to its usual behavior.

3.3.1.2 Nociceptive Pain

Nociceptive pain is the type of pain we instinctively think of when we hear the term "pain." It is caused by the activation of nociceptive nerve fibers (◘ Fig. 3.12).

This type of pain is generated by an *ascending pathway,* originating from sensory receptors and traveling to the cortex and limbic system. It encompasses four primary components: *sensory (discriminative), emotional, cognitive, and motor.*

An external stimulus activates nociceptive receptors. The signal is transmitted via neurons in the dorsal horn of the spinal cord to the thalamus and then to the somatosensory cortex. This *sensory pathway,* referred to as the discriminative pathway, carries information about the location and intensity of the painful stimulus. A subset of these dorsal horn neurons also projects to the limbic system, contributing to the *emotional component* of the pain experience, essentially, the sensation of suffering.

Only after these two streams of information reach the cortex is the pain perceived on a *cognitive level.* This triggers an *appropriate motor reaction* based on the context (Koltzenburg 2000).

Important Point of View

Cellular neurophysiology of acute pain

The degree of nociceptive neuronal activity within pain pathways determines whether acute pain is experienced. An action potential in a presynaptic neuron triggers an influx of sodium ions, which subsequently leads to an influx of calcium ions and ultimately the release of neurotransmitters.

In some instances, the action potential generated in the presynaptic neuron results in minimal neurotransmitter release. Consequently, the postsynaptic neuron is not significantly stimulated, and the nociceptive input does not reach the brain. Put simply, *pain is not perceived.*

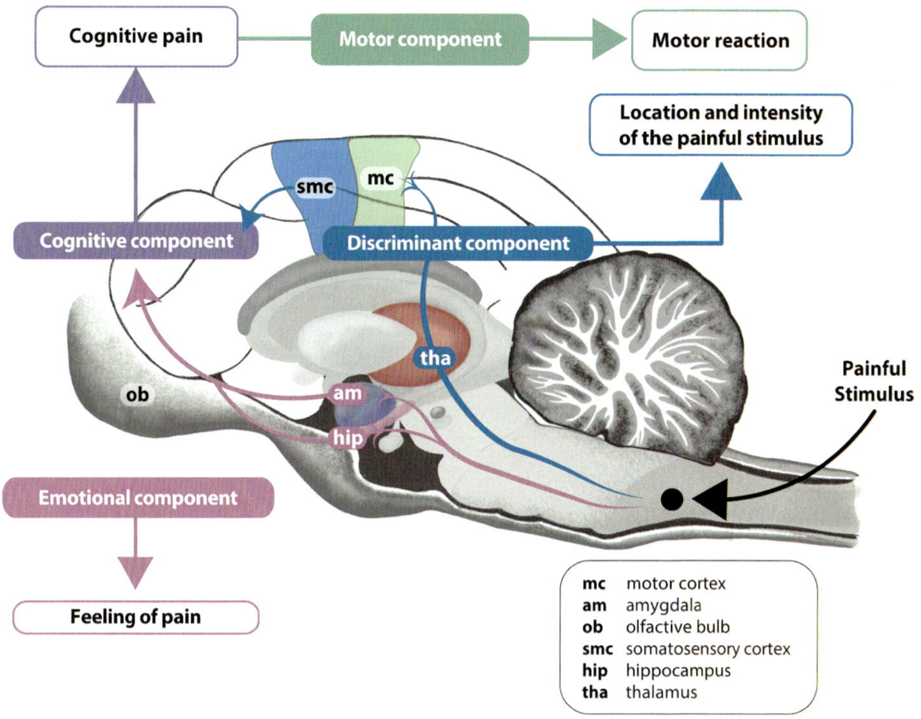

Fig. 3.12 Ascending pathways of pain. The ascending pathways of pain travel through the dorsal horn of the spinal cord to reach the somatosensory cortex (sensory pathway) and the limbic system (emotional pathway). The information is subsequently processed in the cortex (cognitive component) and leads to a motor response (motor component)

In other cases, a stronger action potential in the presynaptic neuron causes voltage-dependent calcium channels to remain open for a longer duration. As a result, a greater amount of neurotransmitters is released, providing increased stimulation to the postsynaptic neuron. The nociceptive input is transmitted to the brain, and **acute** *pain is perceived.*

The *descending pain pathway* originates in the cortex and can inhibit the transmission of ascending pain signals, effectively blocking the sensation of pain before it is consciously perceived.

Descending opioid fibers function in this way during severe injuries or stimuli perceived as highly dangerous. They inhibit nociceptive neurotransmission in the dorsal horn, allowing the individual to act quickly without fully experiencing the intensity of the pain. In addition to opioids, noradrenergic and serotonergic neurons also inhibit dorsal horn activity through postsynaptic α2-adrenergic receptors and 5-HT1B and 5-HT1D receptors. This inhibition suppresses all signals from muscles or organs unrelated to the painful stimulus, permitting only those related to the stimulus to pass through.

A deficiency in the inhibition of monoaminergic neurons can result in neuropathic pain. Selective serotonin and norepinephrine reuptake inhibitors (SNRIs) enhance α2-adrenergic and serotonergic activity, thereby reducing the individual's perception of pain. This occurs because the brain no longer interprets certain signals as painful.

Pain modulation is primarily governed by descending pathways, but supraspinal pathways influenced by the prefrontal cortex also play a role (Fig. 3.13). Cats, however, lack the capacity to understand the purpose of pain, which often exacerbates their fear

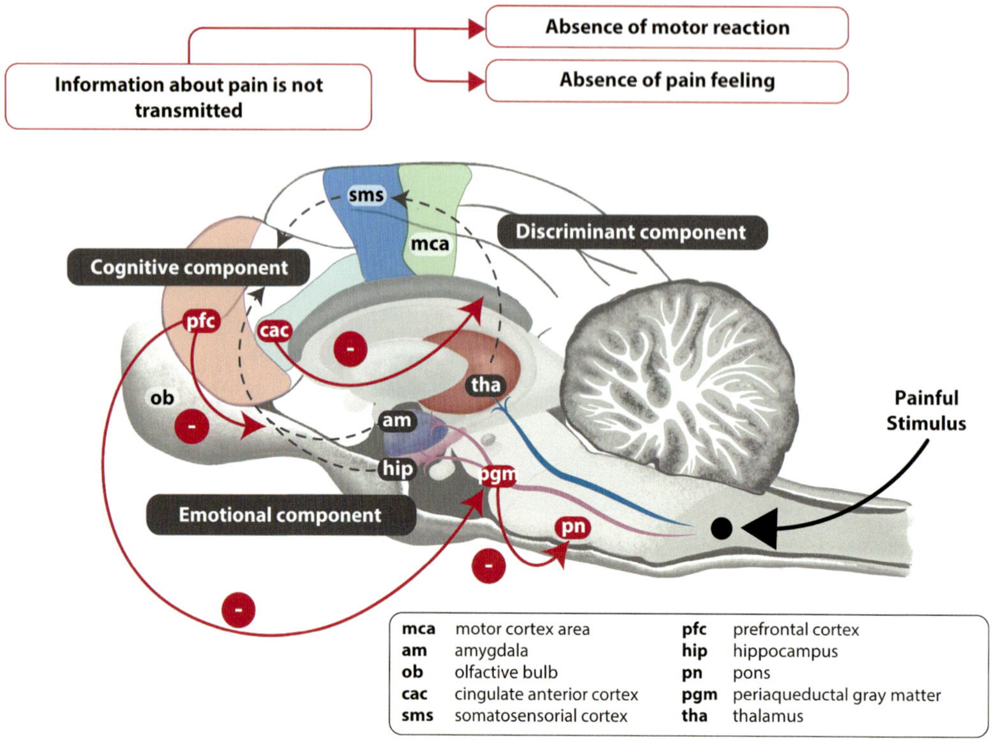

Fig. 3.13 Ascending and descending pain pathways and their modulation by emotions. Pain is transmitted to the cortex via ascending pathways and is primarily modulated by descending pathways. It can also be influenced by supraspinal pathways, which are themselves affected by the prefrontal cortex

responses to it. For this reason, analgesia is more frequently required for cats than humans, even for procedures that involve minimal pain.

Additionally, MRI studies have shown that chronic pain activates the same areas of the limbic system as chronic stress (Fomberstein et al. 2013).

3.3.1.3 Neuropathic Pain

Neuropathic pain arises from damage to or dysfunction of any part of the peripheral or central nervous system.

In some cases, changes within certain brain regions along the nociceptive pathway can lead to sensitization at the level of the thalamus or sensory cortex. This phenomenon is known as *suprasegmental central sensitization.*

At the synaptic level, acute pain is primarily mediated by AMPA glutamate receptors. Similar to conditioned emotional responses, these receptors facilitate the opening of NMDA glutamate receptors. Consequently, subsequent activation of the receptors becomes easier, enhancing synaptic transmission. This heightened synaptic activity amplifies the perceived pain. Such amplification may occur following resolved pain or even in the absence of identifiable triggers. This mechanism underpins conditions like fibromyalgia, generalized chronic pain, and pain symptoms sometimes observed in depressive or anxious states.

Chronic neuropathic pain has implications far beyond the sensation of pain itself (Apkarian et al. 2009). It can cause brain dysfunctions that lead to anxiety, depression, sleep disturbances, and difficulties in decision-making (Baliki et al. 2008). Some studies suggest that chronic pain may even reduce brain volume, particularly in the dorsolateral prefrontal cortex, contributing to the cognitive dysfunction seen in painful conditions such as fibromyalgia (Seminowicz and Moayedi 2017).

Important Point of View

Neuropathic pain and gabapentin

Repeated action potentials can lead to prolonged opening of calcium channels, resulting in excessive neurotransmitter release into the synaptic cleft and overstimulation of postsynaptic neurons. Ultimately, this overstimulation may cause molecular, synaptic, and structural changes that underlie neuropathic pain.

α2δ ligands, such as gabapentin, bind to the α2δ subunit of voltage-gated calcium channels. By altering their conformation, they reduce calcium influx, thereby decreasing neurotransmitter release and limiting excessive stimulation of postsynaptic receptors. This mechanism underlies their analgesic properties.

Important Point of View

Cannabinoids and pain

The endocannabinoid system serves as a modulator for various functions within the body, including the regulation of pain and its associated emotional perception (Seminowicz and Moayedi 2017). In cases of chronic pain, the endocannabinoid system may become dysregulated, losing its ability to modulate pain effectively, which can lead to chronic pathological states.

Research has shown that CBD does not act on type 1 cannabinoid receptors (CB1) like THC does. Instead, it modulates serotonin transmission by desensitizing 5HT1A receptors, which are involved in anxiety and pain (De Gregorio et al. 2019). The resulting increase in serotonin levels contributes to pain modulation.

3.3.2 Pain and Veterinary Psychiatry

Acute stress affects the HPA axis and the sympathetic nervous system (◘ Fig. 3.5). Through their activation, pain perception is diminished to enable rapid motor responses, as immediate action against danger is critical. Glucocorticoids released during this process induce the secretion of opioids, which inhibit pain signal transmission. Consequently, acute stress has an analgesic effect.

Conversely, during chronic stress, the prolonged secretion of glucocorticoids exacerbates pain by increasing hyperalgesia. In this context, the threat is perceived as low, but vigilance, enhanced by glucocorticoids, is heightened to detect potential problems.

Chronic stress worsens chronic pain, and chronic pain exacerbates chronic stress. Both conditions induce similar changes in the brain (Murison 2016).

Evaluating pain in cats is a standard part of routine clinical examinations, but meticulous investigation during veterinary psychiatry consultations is crucial (see ► Chap. 15), particularly when symptoms appear inconsistent with expected pathological patterns.

Summary of ► Sect. 3.3

Pain is an unpleasant sensory and emotional experience associated with actual or potential tissue damage. It acts as a mechanism to signal the body to stop harmful actions or behaviors.

Pain can be acute and adaptive, promoting behavioral changes that safeguard the body's integrity. Conversely, chronic pain is maladaptive and offers no long-term benefit to the animal.

Nociceptive pain arises from the activation of nociceptive fibers in the dorsal horn of the spinal cord, which project to the thalamus and limbic system via ascending pathways. These pathways convey the intensity and localization of the painful stimulus as well as the associated fear and suffering. Descending pathways, originating in the cortex, can inhibit the activation of ascending pathways. These are modulated by opioids, serotonin, and the endocannabinoid system.

In some cases, sensitization mediated by AMPA and NMDA glutamate receptors amplifies the perceived pain, even when the primary lesions are no longer present. This phenomenon explains certain neuropathic pain conditions.

Pain is also influenced by chronic stress, which leads to pain amplification or hyperalgesia. Chronic stress and chronic pain are closely linked at the cerebral level, resulting in mutual exacerbation and similar detrimental outcomes.

Summary of This Chapter

This chapter highlighted the significance of endocrinology and its profound influence on behavior. The hypothalamic-pituitary-adrenal (HPA) axis plays a crucial role in responding to stress; when adaptation fails, it contributes to pathological states, as glucocorticoids have long-term detrimental effects on the body. These glucocorticoids, whether endogenous or exogenous, also regulate thyroid hormone production, thereby influencing numerous behaviors.

The study of biological cycles provides insights into the specific determinants of essential functions such as feeding, sleeping, thirst, sexual behaviors, affiliative behaviors, and agonistic interactions.

Wakefulness and sleep behaviors, along with dipsic and feeding behaviors, are governed by complex neuroendocrine systems. Understanding these systems helps veterinarians address dysregulation and apply appropriate pharmacological solutions.

Sexual and affiliative behaviors depend on both neuroendocrine circuits and the cat's pheromonal system. Territoriality, including its defense and structuring, is a distinctive feature of feline species. While cats are not inherently social, they can develop varying degrees of intra- and interspecific sociability and individual affiliations.

Finally, examining the mechanisms underlying pain reveals similarities with stress. Acute pain is adaptive, helping the organism respond to its environment, whereas chronic pain, like chronic stress, induces damage mediated by glutamate and its NMDA receptors, which are harmful to the individual.

Understanding the major biological functions and endocrinological mechanisms of behavior equips veterinary behaviorists with a holistic view of the animal, enabling them to identify the sources of imbalance and provide tailored solutions.

References

Abizaid A, Liu Z-W, Andrews ZB, Shanabrough M, Borok E, Elsworth JD, Roth RH, Sleeman MW, Picciotto MR, Tschöp MH, Gao X-B, Horvath TL (2006) Ghrelin modulates the activity and synaptic input organization of midbrain dopamine neurons while promoting appetite. J Clin Invest 116(12):3229–3239. https://doi.org/10.1172/JCI29867

Agnew W, Korman R (2014) Pharmacological appetite stimulation: rational choices in the inappetent cat. J Feline Med Surg 16(9):749–756. https://doi.org/10.1177/1098612X14545273

Ahern TH (2009) The impact of early life family structure on adult social attachment, alloparental behavior, and the neuropeptide systems regulating affiliative behaviors in the monogamous prairie vole (Microtus ochrogaster). Front Behav Neurosci 3. https://doi.org/10.3389/neuro.08.017.2009

Allard J, Truitt WA, McKenna KE, Coolen LM (2005) Spinal cord control of ejaculation. World J Urol 23(2):119–126. https://doi.org/10.1007/s00345-004-0494-9

Almawi WY, Beyhum HN, Rahme AA, Rieder MJ (1996) Regulation of cytokine and cytokine receptor expression by glucocorticoids. J Leukoc Biol 60(5):563–572. https://doi.org/10.1002/jlb.60.5.563

Apkarian AV, Baliki MN, Geha PY (2009) Towards a theory of chronic pain. Prog Neurobiol 87(2):81–97. https://doi.org/10.1016/j.pneurobio.2008.09.018

Appleton DJ, Rand JS, Sunvold GD (2000) Plasma leptin concentrations in cats: reference range, effect of weight gain and relationship with adiposity as measured by dual energy X-ray absorptiometry. J Feline Med Surg 2(4):191–199. https://doi.org/10.1053/jfms.2000.0103

Armstrong PJ, Blanchard G (2009) Hepatic Lipidosis in cats. Vet Clin N Am Small Anim Pract 39(3):599–616. https://doi.org/10.1016/j.cvsm.2009.03.003

Arnold AP (2022) X chromosome agents of sexual differentiation. Nat Rev Endocrinol 18(9):574–583. https://doi.org/10.1038/s41574-022-00697-0

Arnold AP, Xu J, Grisham W, Chen X, Kim Y-H, Itoh Y (2004) Minireview: sex chromosomes and brain sexual differentiation. Endocrinology 145(3):1057–1062. https://doi.org/10.1210/en.2003-1491

Baid S, Nieman LK (2004) Glucocorticoid excess and hypertension. Curr Hypertens Rep 6(6):493–499. https://doi.org/10.1007/s11906-004-0046-0

Baliki MN, Geha PY, Apkarian AV, Chialvo DR (2008) Beyond feeling: chronic pain hurts the brain, disrupting the default-mode network dynamics. J Neurosci 28(6):1398–1403. https://doi.org/10.1523/JNEUROSCI.4123-07.2008

Bálint A, Reicher V, Csibra B, Gácsi M (2024) Noninvasive EEG measurement of sleep in the family cat and comparison with the dog. J Mammal gyad122. https://doi.org/10.1093/jmammal/gyad122

Beach FA (1976) Sexual attractivity, proceptivity, and receptivity in female mammals. Horm Behav 7(1):105–138. https://doi.org/10.1016/0018-506X(76)90008-8

Beerda B, Schilder MB, van Hooff JA, de Vries HW, Mol JA (1999) Chronic stress in dogs subjected to social and spatial restriction. I. Behavioral responses. Physiol Behav 66:233–242. https://doi.org/10.1016/s0031-9384(98)00289-3

Blottner S, Jewgenow K (2007) Moderate seasonality in testis function of domestic cat. Reprod Domest Anim 42(5):536–540. https://doi.org/10.1111/j.1439-0531.2006.00817.x

Bosch G, Gilbert M, Beerda B (2022) Properties of foods that impact appetite regulation in cats. Front Anim Sci 3:873924. https://doi.org/10.3389/fanim.2022.873924

Bouissou M-F (1983) Androgens, aggressive behaviour and social relationships in higher mammals. Horm Res 18(1–3):43–61. https://doi.org/10.1159/000179778

Bradshaw JWS (2016) Sociality in cats: a comparative review. J Vet Behav 11:113–124. https://doi.org/10.1016/j.jveb.2015.09.004

Brown RSE, Aoki M, Ladyman SR, Phillipps HR, Wyatt A, Boehm U, Grattan DR (2017) Prolactin action in the medial preoptic area is necessary for postpartum maternal nursing behavior. Proc Natl Acad Sci 114(40):10779–10784. https://doi.org/10.1073/pnas.1708025114

Burke HM, Davis MC, Otte C, Mohr DC (2005) Depression and cortisol responses to psychological stress: a meta-analysis. Psychoneuroendocrinology 30(9):846–856. https://doi.org/10.1016/j.psyneuen.2005.02.010

Burton AG, Hopper K (2019) Hyponatremia in dogs and cats. J Vet Emerg Crit Care 29(5):461–471. https://doi.org/10.1111/vec.12881

Caref K, Nicola SM (2018) Endogenous opioids in the nucleus accumbens promote approach to high-fat food in the absence of caloric need. elife 7:e34955. https://doi.org/10.7554/eLife.34955

Carmona-Alcocer V, Abel JH, Sun TC, Petzold LR, Doyle FJ, Simms CL, Herzog ED (2018) Ontogeny of circadian rhythms and synchrony in the Suprachiasmatic nucleus. J Neurosci 38(6):1326–1334. https://doi.org/10.1523/JNEUROSCI.2006-17.2017

Carter CS (2017) The role of oxytocin and vasopressin in attachment. Psychodynam Psychiatr 45(4):499–517. https://doi.org/10.1521/pdps.2017.45.4.499

Chen K-S, Xu M, Zhang Z, Chang W-C, Gaj T, Schaffer DV, Dan Y (2018) A hypothalamic switch for REM and non-REM sleep. Neuron 97(5):1168–1176.e4. https://doi.org/10.1016/j.neuron.2018.02.005

Conrad CD (2008) Chronic stress-induced hippocampal vulnerability: the glucocorticoid vulnerability hypothesis. Rev Neurosci 19(6). https://doi.org/10.1515/REVNEURO.2008.19.6.395

Costa S, Renata D (2019) Chill protocol to manage aggressive & fearful dogs. Cliniciansbrief.Com
Crowell-Davis SL, Curtis TM, Knowles RJ (2004) Social organization in the cat: a modern understanding. J Feline Med Surg 6(1):19–28. https://doi.org/10.1016/j.jfms.2003.09.013
Damasio A, Damasio H (2016) Exploring the concept of homeostasis and considering its implications for economics. J Econ Behav Organ 126:125–129. https://doi.org/10.1016/j.jebo.2015.12.003
Davis H, Jensen T, Johnson A, Knowles P, Meyer R, Rucinsky R, Shafford H (2013) 2013 AAHA/AAFP fluid therapy guidelines for dogs and cats*. J Am Anim Hosp Assoc 49(3):149–159. https://doi.org/10.5326/JAAHA-MS-5868
De Gregorio D, McLaughlin RJ, Posa L, Ochoa-Sanchez R, Enns J, Lopez-Canul M, Aboud M, Maione S, Comai S, Gobbi G (2019) Cannabidiol modulates serotonergic transmission and reverses both allodynia and anxiety-like behavior in a model of neuropathic pain. Pain 160(1):136–150. https://doi.org/10.1097/j.pain.0000000000001386
Delgado M, Dantas LMS (2020) Feeding cats for optimal mental and behavioral well-being. Vet Clin N Am Small Anim Pract 50(5):939–953. https://doi.org/10.1016/j.cvsm.2020.05.003
Demaldé L, Lopez Merlo M, Vercellini R, Barbeito CG, Fernandez P, Gobello C (2016) Disrupting effect of androgens in postnatal female domestic cats. Anim Reprod Sci 171:65–71. https://doi.org/10.1016/j.anireprosci.2016.05.015
DePorter TL, Elzerman AL (2019) Common feline problem behaviors: destructive scratching. J Feline Med Surg 21(3):235–243. https://doi.org/10.1177/1098612X19831205
DiLeone RJ, Georgescu D, Nestler EJ (2003) Lateral hypothalamic neuropeptides in reward and drug addiction. Life Sci 73(6):759–768. https://doi.org/10.1016/S0024-3205(03)00408-9
Dissen G, Adachi K, Lomniczi A, Chatkupt T, Davidson B, Nakai H, Ojeda S (2017) Engineering a gene silencing viral construct that targets the cat hypothalamus to induce permanent sterility: an update. Reprod Domest Anim 52(S2):354–358. https://doi.org/10.1111/rda.12834
Edwards C, Abusnana S, Sunter D, Murphy K, Ghatei M, Bloom S (1999) The effect of the orexins on food intake: comparison with neuropeptide Y, melanin-concentrating hormone and galanin. J Endocrinol 160(3):R7–R12. https://doi.org/10.1677/joe.0.160r007
Erger CG, Gerras AL, Conley AJ, Gilor C, Burns DeMarle K, Refsal KR, Fleming JM, Sledge DG, Langlois DK (2023) Case report: androgen-secreting adrenocortical tumors in eight cats. Front Vet Sci 10:1158142. https://doi.org/10.3389/fvets.2023.1158142
Falkenstein E, Norman AW, Wehling M (2000) Mannheim classification of nongenomically initiated (rapid) steroid action(s). J Clin Endocrinol Metabol 85(5):2072–2075. https://doi.org/10.1210/jcem.85.5.6516
Feige B, Baglioni C, Spiegelhalder K, Hirscher V, Nissen C, Riemann D (2013) The microstructure of sleep in primary insomnia: an overview and extension. Int J Psychophysiol 89(2):171–180. https://doi.org/10.1016/j.ijpsycho.2013.04.002
Finkler H, Terkel J (2010) Cortisol levels and aggression in neutered and intact free-roaming female cats living in urban social groups. Physiol Behav 99(3):343–347. https://doi.org/10.1016/j.physbeh.2009.11.014
Fomberstein K, Qadri S, Ramani R (2013) Functional MRI and pain. Curr Opin Anaesthesiol 26(5):588–593. https://doi.org/10.1097/01.aco.0000433060.59939.fe
Galgano M, Spalla I, Callegari C, Patruno M, Auriemma E, Zanna G, Ferro S, Zini E (2014) Primary hypothyroidism and thyroid Goiter in an adult cat. J Vet Intern Med 28(2):682–686. https://doi.org/10.1111/jvim.12283
Gálosi R, Hajnal A, Petykó Z, Hartmann G, Karádi Z, Lénárd L (2015) The role of catecholamine innervation in the medial prefrontal cortex on the regulation of body weight and food intake. Behav Brain Res 286:318–327. https://doi.org/10.1016/j.bbr.2015.03.017
Germano CMR, De Castro M, Rorato R, Laguna MTC, Antunes-Rodrigues J, Elias CF, Elias LLK (2007) Time course effects of adrenalectomy and food intake on cocaine- and amphetamine-regulated transcript expression in the hypothalamus. Brain Res 1166:55–64. https://doi.org/10.1016/j.brainres.2007.05.077
Giammanco M, Tabacchi G, Giammanco S, Di Majo D, La Guardia M (2005) Testosterone and aggressiveness. Med Sci Monit: Int Med J Exp Clin Res 11(4):RA136-145
Goericke-Pesch S (2010) Reproduction control in cats: new developments in non-surgical methods. J Feline Med Surg 12(7):539–546. https://doi.org/10.1016/j.jfms.2010.05.005
Goh G, Maloney S, Mark P, Blache D (2019) Episodic Ultradian events—Ultradian rhythms. Biology 8(1):15. https://doi.org/10.3390/biology8010015
Gudermuth DF, Newton L, Daels P, Concannon P (1997) Incidence of spontaneous ovulation in young, group-housed cats based on serum and faecal concentrations of progesterone. J Reprod Fertil Suppl 51:177–184
Halford J, Harrold J, Lawton C, Blundell J (2005) Serotonin (5-HT) drugs: effects on appetite expression and use for the treatment of obesity. Curr Drug Targets 6(2):201–213. https://doi.org/10.2174/1389450053174550
Hall RCW, Hall RCW (2005) Abuse of Supraphysiologic doses of anabolic steroids. South Med J 98(5):550–555. https://doi.org/10.1097/01.SMJ.0000157531.04472.B2

Hastings MH, Maywood ES, Brancaccio M (2018) Generation of circadian rhythms in the suprachiasmatic nucleus. Nat Rev Neurosci 19(8):453–469. https://doi.org/10.1038/s41583-018-0026-z
Hatcher KM, Royston SE, Mahoney MM (2020) Modulation of circadian rhythms through estrogen receptor signaling. Eur J Neurosci 51(1):217–228. https://doi.org/10.1111/ejn.14184
Heisler LK, Lam DD (2017) An appetite for life: brain regulation of hunger and satiety. Curr Opin Pharmacol 37:100–106. https://doi.org/10.1016/j.coph.2017.09.002
Heisler LK, Jobst EE, Sutton GM, Zhou L, Borok E, Thornton-Jones Z, Liu HY, Zigman JM, Balthasar N, Kishi T, Lee CE, Aschkenasi CJ, Zhang C-Y, Yu J, Boss O, Mountjoy KG, Clifton PG, Lowell BB, Friedman JM et al (2006) Serotonin reciprocally regulates Melanocortin neurons to modulate food intake. Neuron 51(2):239–249. https://doi.org/10.1016/j.neuron.2006.06.004
Herman J, Flak J, Jankord R (2008) Chronic stress plasticity in the hypothalamic paraventricular nucleus, in: Progress in Brain Research. Elsevier, pp. 353–364. https://doi.org/10.1016/S0079-6123(08)00429-9
Hine FR, Maddox GL, Williams RB, Carson RC, Williams RB (2004) The neurobiology of agression. In: Hine FR, Maddox GL, Williams RB, Carson RC, Thompson RJ (eds) Introduction to behavioral science in medicine. Springer, New York, pp 281–288. https://doi.org/10.1007/978-1-4612-5452-2_20
Hirvonen J, Virtanen KA, Nummenmaa L, Hannukainen JC, Honka M-J, Bucci M, Nesterov SV, Parkkola R, Rinne J, Iozzo P, Nuutila P (2011) Effects of insulin on brain glucose metabolism in impaired glucose tolerance. Diabetes 60(2):443–447. https://doi.org/10.2337/db10-0940
Hofer MA (2005) The psychobiology of early attachment. Clin Neurosci Res 4(5–6):291–300. https://doi.org/10.1016/j.cnr.2005.03.007
Hofmeister S (2016) Premenstrual syndrome and premenstrual dysphoric disorder. Am Fam Phtsician 94(3):236
Huang Z-L, Urade Y, Hayaishi O (2011) The role of Adenosine in the regulation of sleep. Curr Top Med Chem 11(8):1047–1057. https://doi.org/10.2174/156802611795347654
Insel TR, Shapiro LE (1992) Oxytocin receptor distribution reflects social organization in monogamous and polygamous voles. Proc Natl Acad Sci 89(13):5981–5985. https://doi.org/10.1073/pnas.89.13.5981
Jerjen CP, Kumaran SJ, Liesegang A, Hall E, Wichert B, Haase B (2023) *Melanocortin-4 receptor* and *proopiomelanocortin*: candidate genes for obesity in domestic shorthair cats. Anim Genet 54(5):637–642. https://doi.org/10.1111/age.13335
Jones CJ, Budsberg SC (2000) Physiologic characteristics and clinical importance of the cyclooxygenase isoforms in dogs and cats. J Am Vet Med Assoc 217(5):721–729. https://doi.org/10.2460/javma.2000.217.721
Kayama Y, Koyama Y (2003) Control of sleep and wakefulness by brainstem monoaminergic and cholinergic neurons. In: Katayama Y (ed) Neurosurgical re-engineering of the damaged brain and spinal cord. Springer, Vienna, pp 3–6. https://doi.org/10.1007/978-3-7091-6081-7_1
Khare D, Gupta D, Shukla P, Meena N, Khare R (2018) Feline hyperthyroidism: an overview. J Entomol Zool Stud 6:418
Kirkham TC, Williams CM (2001) Synergistic efects of opioid and cannabinoid antagonists on food intake. Psychopharmacology 153(2):267–270. https://doi.org/10.1007/s002130000596
Kolber BJ, Roberts MS, Howell MP, Wozniak DF, Sands MS, Muglia LJ (2008) Central amygdala glucocorticoid receptor action promotes fear-associated CRH activation and conditioning. Proc Natl Acad Sci USA. 105:12004–12009. https://doi.org/10.1073/pnas.0803216105
Koltzenburg M (2000) Neural mechanisms of cutaneous nociceptive pain. Clin J Pain 16(Supplement):S131–S138. https://doi.org/10.1097/00002508-200009001-00004
Kozłowski S, Szczepańska-Sadowska E (1975) Antagonistic effects of vasopressin and hypervolemia on osmotic reactivity of the thirst mechanism in dogs. Pflügers Arch Eur J Physiol 353(1):59–65. https://doi.org/10.1007/BF00584511
Kraemer FB, Shen W-J (2002) Hormone-sensitive lipase. J Lipid Res 43(10):1585–1594. https://doi.org/10.1194/jlr.R200009-JLR200
Krugers HJ, Goltstein PM, Van Der Linden S, Joëls M (2006) Blockade of glucocorticoid receptors rapidly restores hippocampal CA1 synaptic plasticity after exposure to chronic stress. Eur J Neurosci 23(11):3051–3055. https://doi.org/10.1111/j.1460-9568.2006.04842.x
Kutzler MA (2022) Reproductive anatomy and puberty in the queen. In: Johnson A, Kutzler M (eds) Feline reproduction. CABI, pp 1–10. https://doi.org/10.1079/9781789247107.0001
Leyva H, Addiego L, Stabenfeldt G (1984) The effect of different photoperiods on plasma concentrations of Melatonin, Prolactin, and Cortisol in the domestic cat*. Endocrinology 115(5):1729–1736. https://doi.org/10.1210/endo-115-5-1729
Lodé T (2020) A brief natural history of the orgasm. All Life 13(1):34–44. https://doi.org/10.1080/21553769.2019.1664642
Lowe AD, Graves TK, Campbell KL, Schaeffer DJ (2009) A pilot study comparing the Diabetogenic effects of dexamethasone and Prednisolone in cats. J Am Anim Hosp Assoc 45(5):215–224. https://doi.org/10.5326/0450215
Lu J, Sherman D, Devor M, Saper CB (2006) A putative flip–flop switch for control of REM sleep. Nature 441(7093):589–594. https://doi.org/10.1038/nature04767

3

Lucas EA (1978) Effects of a Short Light-Dark cycle on the sleep-wake patterns of the cat. Sleep 1(3):219–317. https://doi.org/10.1093/sleep/1.3.299
Lyle SK (2007) Disorders of sexual development in the dog and cat. Theriogenology 68(3):338–343. https://doi.org/10.1016/j.theriogenology.2007.04.015
Machin AJ, Dunbar RIM (2011) The brain opioid theory of social attachment: a review of the evidence. Behaviour 148(9/10):985–1025
McMillan FD (2013) Stress-induced and emotional eating in animals: a review of the experimental evidence and implications for companion animal obesity. J Vet Behav 8(5):376–385. https://doi.org/10.1016/j.jveb.2012.11.001
Mills D (2005) Pheromonatherapy: theory and applications. In Pract 27(7):368–373. https://doi.org/10.1136/inpract.27.7.368
Mills DS, Redgate SE, Landsberg GM (2011) A meta-analysis of studies of treatments for feline urine spraying. PLoS One 6(4):e18448. https://doi.org/10.1371/journal.pone.0018448
Montagna P, Gambetti P, Cortelli P, Lugaresi E (2003) Familial and sporadic fatal insomnia. Lancet Neurol 2(3):167–176. https://doi.org/10.1016/S1474-4422(03)00323-5
Morello G, Imperatore R, Palomba L, Finelli C, Labruna G, Pasanisi F, Sacchetti L, Buono L, Piscitelli F, Orlando P, Di Marzo V, Cristino L (2016) Orexin-A represses satiety-inducing POMC neurons and contributes to obesity via stimulation of endocannabinoid signaling. Proc Natl Acad Sci 113(17):4759–4764. https://doi.org/10.1073/pnas.1521304113
Murison R (2016) The neurobiology of stress. In: Neuroscience of pain, stress, and emotion. Elsevier, pp 29–49. https://doi.org/10.1016/B978-0-12-800538-5.00002-9
Nagasawa T, Kimura Y, Masuda K, Uchiyama H (2022) Physiological assessment of the health and welfare of domestic cats—an exploration of factors affecting urinary Cortisol and Oxytocin. Animals 12(23):3330. https://doi.org/10.3390/ani12233330
Nogueiras R, Romero-Picó A, Vazquez MJ, Novelle MG, López M, Diéguez C (2012) The opioid system and food intake: homeostatic and Hedonic mechanisms. Obes Facts 5(2):196 207. https://doi.org/10.1159/000338163
Nuñez-Favre R, Bonaura M, Tittarelli C, Mansilla-Hermann D, De La Sota R, Stornelli M (2012) Effect of natural photoperiod on Epididymal sperm quality and testosterone serum concentration in domestic cat (F *elis silvestris catus*). Reprod Domest Anim 47(s6):232–234. https://doi.org/10.1111/rda.12050
Oakley AE, Clifton DK, Steiner RA (2009) Kisspeptin signaling in the brain. Endocr Rev 30(6):713–743. https://doi.org/10.1210/er.2009-0005
Ohno K, Sakurai T (2008) Orexin neuronal circuitry: role in the regulation of sleep and wakefulness. Front Neuroendocrinol 29(1):70–87. https://doi.org/10.1016/j.yfrne.2007.08.001
Overall KL (2015) Manual of clinical behavioral medicine for dogs and cats ([Enhanced Credo edition]). Elsevier, Credo Reference
Pageat P, Gaultier E (2003) Current research in canine and feline pheromones. Vet Clin N Am Small Anim Pract 33(2):187–211. https://doi.org/10.1016/S0195--5616(02)00128-6
Paredes RG (2003) Medial preoptic area/anterior hypothalamus and sexual motivation. Scand J Psychol 44(3):203–212. https://doi.org/10.1111/1467-9450.00337
Peever J, Fuller PM (2017) The biology of REM sleep. Curr Biol 27(22):R1237–R1248. https://doi.org/10.1016/j.cub.2017.10.026
Peterson ME (1984) Feline hyperthyroidism. Vet Clin N Am Small Anim Pract 14(4):809–826. https://doi.org/10.1016/S0195-5616(84)50082-5
Peterson M (2012) Hyperthyroidism in cats: what's causing this epidemic of thyroid disease and can we prevent it? J Feline Med Surg 14(11):804–818. https://doi.org/10.1177/1098612X12464462
Pfaff DW, Sakuma Y (1979) Deficit in the lordosis reflex of female rats caused by lesions in the ventromedial nucleus of the hypothalamus. J Physiol 288(1):203–210. https://doi.org/10.1113/jphysiol.1979.sp012691
Piacentino D, Kotzalidis GD, Pomara C, Girardi P, Sani G (2015) Steroid use and psychopathology in athletes. Mood Disorders
Piccione G, Marafioti S, Giannetto C, Di Pietro S, Quartuccio M, Fazio F (2014) Comparison of daily distribution of rest/activity in companion cats and dogs. Biol Rhythm Res 45(4):615–623. https://doi.org/10.1080/09291016.2014.884303
Pissios P, Bradley RL, Maratos-Flier E (2006) Expanding the scales: the multiple roles of MCH in regulating energy balance and other biological functions. Endocr Rev 27(6):606–620. https://doi.org/10.1210/er.2006-0021
Portas CM, Thakkar M, Rainnie DG, Greene RW, McCarley RW (1997) Role of adenosine in behavioral state modulation: a microdialysis study in the freely moving cat. Neuroscience 79(1):225–235. https://doi.org/10.1016/S0306-4522(96)00640-9
Prior MR, Mills DS (2020) Cats vs. dogs: the efficacy of Feliway FriendsTM and AdaptilTM products in multispecies homes. Front Vet Sci 7:399. https://doi.org/10.3389/fvets.2020.00399
Raja SN, Carr DB, Cohen M, Finnerup NB, Flor H, Gibson S, Keefe FJ, Mogil JS, Ringkamp M, Sluka KA, Song X-J, Stevens B, Sullivan MD, Tutelman PR, Ushida T, Vader K (2020) The revised International Association for the Study of Pain definition of pain: concepts, challenges, and compromises. Pain 161(9):1976–1982. https://doi.org/10.1097/j.pain.0000000000001939
Rault J-L, Van Den Munkhof M, Buisman-Pijlman FTA (2017) Oxytocin as an indicator of psychologi-

cal and social well-being in domesticated animals: a critical review. Front Psychol 8:1521. https://doi.org/10.3389/fpsyg.2017.01521
Revel FG, Ansel L, Klosen P, Saboureau M, Pévet P, Mikkelsen JD, Simonneaux V (2007) Kisspeptin: a key link to seasonal breeding. Rev Endocrine Metabol Disord 8(1):57–65. https://doi.org/10.1007/s11154-007-9031-7
Rincón-Cortés M, Grace AA (2020) Adaptations in reward-related behaviors and mesolimbic dopamine function during motherhood and the postpartum period. Front Neuroendocrinol 57:100839. https://doi.org/10.1016/j.yfrne.2020.100839
Rindi G, Necchi V, Savio A, Torsello A, Zoli M, Locatelli V, Raimondo F, Cocchi D, Solcia E (2002) Characterisation of gastric ghrelin cells in man and other mammals: studies in adult and fetal tissues. Histochem Cell Biol 117(6):511–519. https://doi.org/10.1007/s00418-002-0415-1
Romaniuk A, Strzelczuk M, Wieczorek M (1989) Serotonin Depletion with P-chlorophenylalanine in the cat: effects on carbachol-induced defensive behavior and regional brain amine content
Rothman RB, Blough BE, Baumann MH (2008) Dual dopamine/serotonin releasers: potential treatment agents for stimulant addiction. Exp Clin Psychopharmacol 16(6):458–474. https://doi.org/10.1037/a0014103
Roy D, Angelini NL, Fujieda H, Brown GM, Belsham DD (2001) Cyclical regulation of GnRH gene expression in GT1–7 GnRH-secreting neurons by melatonin. Endocrinology 142(11):4711–4720. https://doi.org/10.1210/endo.142.11.8464
Russo SJ, Murrough JW, Han M-H, Charney DS, Nestler EJ (2012) Neurobiology of resilience. Nat Neurosci 15(11):1475–1484. https://doi.org/10.1038/nn.3234
Ruviaro Tuleski GL, Silveira MF, Bastos RF, Pscheidt MJGR, Prieto WDS, Sousa MG (2022) Behavioral and cardiovascular effects of a single dose of gabapentin or melatonin in cats: a randomized, double-blind, placebo-controlled trial. J Feline Med Surg 24(12):e524–e534. https://doi.org/10.1177/1098612X221124359
Sakurai T (2006) Roles of Orexins and Orexin receptors in central regulation of feeding behavior and energy homeostasis. CNS Neurol Disord Drug Targets 5(3):313–325. https://doi.org/10.2174/187152706777452218
Santos LC, Silva JF (2023) Molecular factors involved in the reproductive morphophysiology of female domestic cat (Felis catus). Animals 13(19):3153. https://doi.org/10.3390/ani13193153
Santos LC, Dos Anjos Cordeiro JM, Santana LDS, Barbosa EM, Santos BR, Da Silva TQM, De Souza SS, Corrêa JMX, Lavor MSL, Da Silva EB, Silva JF (2022) Expression profile of the Kisspeptin/Kiss1r system and angiogenic and immunological mediators in the ovary of cyclic and pregnant cats. Domest Anim Endocrinol 78:106650. https://doi.org/10.1016/j.domaniend.2021.106650
Saper CB, Scammell TE, Lu J (2005) Hypothalamic regulation of sleep and circadian rhythms. Nature 437(7063):1257–1263. https://doi.org/10.1038/nature04284
Selye H (1950) Stress and the general adaptation syndrome. BMJ 1(4667):1383–1392. https://doi.org/10.1136/bmj.1.4667.1383
Selye H (1955) Stress and disease. Science 122(3171):625–631. https://doi.org/10.1126/science.122.3171.625
Seminowicz DA, Moayedi M (2017) The Dorsolateral Prefrontal cortex in acute and chronic pain. J Pain 18(9):1027–1035. https://doi.org/10.1016/j.jpain.2017.03.008
Silver HK, Finkelstein M (1967) Deprivation dwarfism. J Pediatr 70(3):317–324. https://doi.org/10.1016/S0022-3476(67)80127-6
Slopien R, Pluchino N, Warenik-Szymankiewicz A, Sajdak S, Luisi M, Drakopoulos P, Genazzani AR (2018) Correlation between allopregnanolone levels and depressive symptoms during late menopausal transition and early postmenopause. Gynecol Endocrinol 34(2):144–147. https://doi.org/10.1080/09513590.2017.1371129
Sloviter RS, Valiquette G, Abrams GM, Ronk EC, Sollas AL, Paul LA, Neubort S (1989) Selective Loss of Hippocampal Granule Cells in the Mature Rat Brain After Adrenalectomy. Science 243:535–538. https://doi.org/10.1126/science.2911756
Spindler RE, Wildt DE (1999) Circannual variations in Intraovarian Oocyte but not epididymal sperm quality in the domestic Cat1. Biol Reprod 61(1):188–194. https://doi.org/10.1095/biolreprod61.1.188
Stahn C, Buttgereit F (2008) Genomic and nongenomic effects of glucocorticoids. Nat Clin Pract Rheumatol 4(10):525–533. https://doi.org/10.1038/ncprheum0898
Stella J, Croney C, Buffington T (2013) Effects of stressors on the behavior and physiology of domestic cats. Appl Anim Behav Sci 143(2–4):157–163. https://doi.org/10.1016/j.applanim.2012.10.014
Stern JM (1989) Maternal behavior: sensory, hormonal, and neural determinants. In: Psychoendocrinology. Elsevier, pp 105–226. https://doi.org/10.1016/B978--0-12-137952-0.50008-2
Stevenson TJ (2018) Epigenetic regulation of biological rhythms: an evolutionary ancient molecular timer. Trends Genet 34(2):90–100. https://doi.org/10.1016/j.tig.2017.11.003
Szczepanska-Sadowska E, Kozlowski S, Sobocinska J (1974) Blood antidiuretic hormone level and osmotic reactivity of thirst mechanism in dogs. Am J Physiol Legacy Content 227(4):766–770. https://doi.org/10.1152/ajplegacy.1974.227.4.766
Tanyapanyachon P, Amelkina O, Chatdarong K (2018) Localization of Kisspeptin immunoreactivity in the

cat ovary on different reproductive stages. Reprod Fertil Dev 30(1):193. https://doi.org/10.1071/RDv30n1Ab107

Tobler I, Scherschlicht R (1990) Sleep and EEG slow-wave activity in the domestic cat: effect of sleep deprivation. Behav Brain Res 37(2):109–118. https://doi.org/10.1016/0166-4328(90)90086-T

Todorova R, Zugaro M (2019) Isolated cortical computations during delta waves support memory consolidation. Science 366(6463):377–381. https://doi.org/10.1126/science.aay0616

Tsuneoka Y (2019) Molecular neuroanatomy of the mouse medial preoptic area with reference to parental behavior. Anat Sci Int 94(1):39–52. https://doi.org/10.1007/s12565-018-0468-4

Tsutsui T, Kuwabara S, Kuwabara K, Kugota Y, Kinjo T, Hori T (2004) Development of spermatogenic function in the sex maturation process in male cats. J Vet Med Sci 66(9):1125–1127. https://doi.org/10.1292/jvms.66.1125

Turner DC, Bateson P (eds) (2013) The domestic cat: the biology of its behaviour, 3rd edn. Cambridge University Press. https://doi.org/10.1017/CBO9781139177177

Ursin R (1968) The two stages of slow wave sleep in the cat and their relation to REM sleep. Brain Res 11(2):347–356

Verbrugghe A, Hesta M (2017) Cats and carbohydrates: the carnivore fantasy? Vet Sci 4(4):55. https://doi.org/10.3390/vetsci4040055

Vitale KR, Udell MAR (2019) The quality of being sociable: the influence of human attentional state, population, and human familiarity on domestic cat sociability. Behav Process 158:11–17. https://doi.org/10.1016/j.beproc.2018.10.026

Volkow ND, Wang G-J, Baler RD (2011) Reward, dopamine and the control of food intake: implications for obesity. Trends Cogn Sci 15(1):37–46. https://doi.org/10.1016/j.tics.2010.11.001

Wakshlag J, Loftus J (2014) Canine and feline obesity: a review of pathophysiology, epidemiology, and clinical management. Vet Med Res Rep 49. https://doi.org/10.2147/VMRR.S40868

Weinstock M (2008) The long-term behavioural consequences of prenatal stress. Neurosci Biobehav Rev 32(6):1073–1086. https://doi.org/10.1016/j.neubiorev.2008.03.002

Wildt DE (1991) Fertilization in cats. In: Dunbar BS, O'Rand MG (eds) A comparative overview of mammalian fertilization. Springer US, pp 299–328. https://doi.org/10.1007/978-1-4757-8982-9_16

Xia C, Zhai H, Tong Y (2012) Correlation between mental health and impulsive personality in hyperthyroidism patients. Chin J Public Health 28(11):1491–1493. https://doi.org/10.11847/zgggws2012-28-11-34

Zoran DL (2002) The carnivore connection to nutrition in cats. J Am Vet Med Assoc 221(11):1559–1567. https://doi.org/10.2460/javma.2002.221.1559

Zoran DL (2010) Obesity in dogs and cats: a metabolic and endocrine disorder. Vet Clin N Am Small Anim Pract 40(2):221–239. https://doi.org/10.1016/j.cvsm.2009.10.009

Zoran DL, Jergens AE, Riedesel DH, Johnson GS, Bailey TB, Martin SD (1992) Evaluation of hemostatic analytes after use of hypertonic saline solution combined with colloids for resuscitation of dogs with hypovolemia. Am J Vet Res 53(10):1791–1796

Brain Development and Behavioral Ontogenesis

Contents

4.1 Neurogenesis and Brain Connections – 144

4.2 Chronological Description of Behavioral Development – 148

4.2.1 The Prenatal Period – 148
4.2.2 The Neonatal Period (0–9 Days) – 148
4.2.3 The Transitional Period (9–15 Days) – 149
4.2.4 The Socialization Period (18 Days—2 Months) – 150
4.2.5 The Juvenile Period (2–6 Months) – 155
4.2.6 The Pubertal Period (6 Months to 1 Year) – 158
4.2.7 The Adult Period (1 Year—8/10 Years) – 161
4.2.8 Senescent Period (from 8 to 10 Years) – 161

4.3 Axial Description of Behavioral Development – 167

4.3.1 Fear and Alarm System – 167
4.3.2 Acquisition of Self-Controls – 170
4.3.3 Relationships and Attachment – 174
4.3.4 Territoriality – 177

References – 182

© The Author(s), under exclusive license to Springer Nature Switzerland AG 2026
S. Bleuer-Elsner et al., *Veterinary Psychiatry of the Cat*,
https://doi.org/10.1007/978-3-032-06260-4_4

This chapter addresses neurophysiological development and the ontogenesis of behaviors. It explores key periods of behavioral development and aging, with a special focus on the ontogenesis of fear responses, the acquisition of self-control, territorial organization, and relationships, including attachment.

These four dimensions, namely fear, self-control, territory, and relationships, form the foundational pillars of development. Throughout this work, diseases and behavioral abnormalities will be categorized within these four axes.

4

4.1 Neurogenesis and Brain Connections

Neurons are specialized cells that originate, differentiate, and migrate from the walls of the brain vesicles. The axons of these cells grow, guided by chemical signals that are either attractive or repulsive. Once the target cell is reached, a connection, called a synapse, forms between the two cells.

During the embryonic phase, neurogenesis is excessive and leads to a very dense network of connections, so dense that it becomes dysfunctional and chaotic. After birth, based on sensory stimulation and through a selection process, the entanglement of neurons and synapses simplifies, enabling the establishment of organized and functional brain pathways and circuits. This selection is based on the reduction of neurons and synapses through cell death and a decrease in the number of synapses attached to each neuron. A functional brain, therefore, results from a fine balance between the formation and elimination of neurons and synapses.

Neuronal growth and cell survival are closely dependent on the presence of growth factors in sufficient quantities within the cellular environment. These growth factors or *neurotrophins* are secreted by target cells and stimulate neuronal growth until a synapse is established. Once the synapse is formed, it appears that all other neurons competing to establish this connection vanish due to a lack of neurotrophins. This cell death, also known as apoptosis, is a genetically programmed self-destruction process (Changeux and Mikoshiba 1978; Yuan and Horvitz 2004). The presence of neurotrophins acts as a switch preventing the "self-destruction" gene from being expressed, thus keeping the neuron alive. Through competition, only certain neurons benefit from this genetic switch. These form the organized networks.

This selection is not random but is based on the individual's experiences. Each new experience strengthens preexisting synaptic connections, while unstimulated synapses disappear (Dichio and De Vico Fallani 2024). Therefore, a brain exposed to a rich environment will have a denser network of neurons and more numerous connections than one that develops in a less varied environment (◘ Fig. 4.1).

This massive synaptic pruning early in development occurs during a period known as the critical period (see sidebar below). During adolescence, a new phase of selective neuronal reduction also takes place, enabling new behavioral changes characteristic of adolescence.

Each neuron has a finite capacity for synaptic connections on its soma and dendrites. This fine selection is based on postsynaptic activity. As previously mentioned, the most active synapses will persist, while the less active ones will disappear. This type of selection is particularly active during adolescence. Studies in macaques have shown that the number of synapses in the visual cortex remains stable during infancy but decreases by 50% over a 2-year period during adolescence. A quick calculation estimates a staggering destruction of 5000 synapses per second during this time (Rakic et al. 1986).

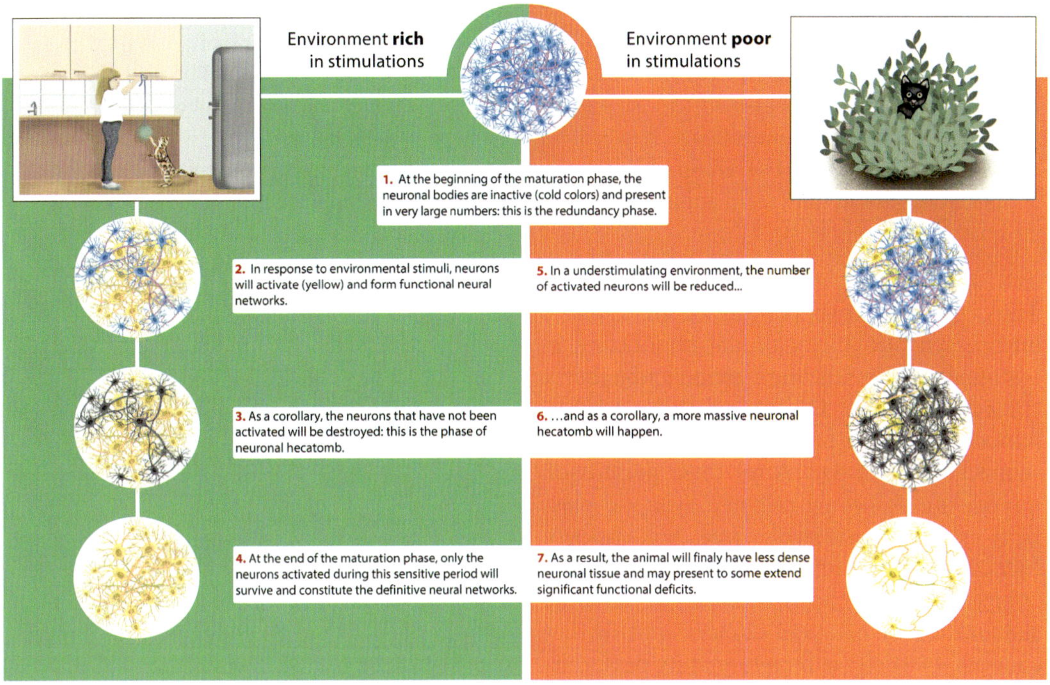

Fig. 4.1 Synaptic pruning. Environmental stimulation activates neurons from birth. A richly stimulating environment (left column) leads to the establishment of numerous effective and adaptive networks, while unstimulated neurons are eliminated. Conversely, a poorly stimulating environment (right column) results in a higher mortality rate of unstimulated neurons, ultimately leading to significant synaptic deficits

Important Point of View

The Concept of the "Critical Period"

The studies by Wiesel and Hubel, conducted on cats and later replicated on monkeys, demonstrated the notion of a critical period (Blakemore et al. 1978; Hubel and Wiesel 1964). They focused on the circuits of binocular vision. By suturing the eyelids of one eye, they observed that the neuronal layers connected to the sutured eye thinned, while those connected to the unsutured eye thickened. Thus, neuronal development is indeed dependent on neurosensory stimulation. Upon releasing the sutures of the first eye and suturing the second, they noticed a return to balance, with thinning of the overly thick layers and thickening of the overly thin ones. However, this reverse effect was only observed during a specific period, beyond which neuronal plasticity disappeared, resulting in permanent monocular blindness in the animal. This period is therefore referred to as the critical period.

As in Wiesel and Hubel's study, the concept of the *critical period* is fundamental to understanding the ontogenesis of behaviors. The phenotypic expression of feline behaviors is linked to early sensory experiences. Certain processes, such as species imprinting, are irreversible (or nearly so). An orphaned kitten raised away from other cats develops an irreversible imprint to another species (e.g., to the human caregiver). Such a cat will later be unable to exhibit appropriate reproductive behaviors with other cats, as it will not recognize them as members of its species. The fate of neurons and cerebral circuits is therefore heavily influenced by these early experiences, which shape the future behaviors of the cat.

The brain's plasticity during the critical period is so pronounced that recent studies have shown that in cases of congenital deafness in cats, connections between auditory areas and those of other senses can compensate for the handicap very quickly (Sacco et al. 2024).

Thus, there are two periods of particularly intense synaptic and neuronal remodeling in mammals: the socialization period and puberty (Blakemore 2008). Outside of these periods, remodeling is less pronounced and decreases significantly with age. After these periods of remodeling, following puberty, synaptic plasticity slows considerably (see box below). Indeed, the cessation of axonal growth, the completion of synaptic maturation, and the stabilization of neuronal circuits limit further changes. Plasticity is therefore at its peak during the first weeks of life through adolescence. Beyond these periods of behavioral development, certain types of learning become nearly impossible or, at the very least, very slow and non-generalizable (Juraska 2024).

Important Point of View

Brain Plasticity and Long-Term Potentiation

The cellular and biochemical processes involved in synaptic plasticity are well understood. The persistence of a synapse depends on the simultaneous activation of both the presynaptic and postsynaptic neurons. A single neuron receives synapses from numerous other neurons, and the persistence of a synapse is determined by the synchronization of its activity with others. Synapses that are synchronized are strengthened, while those that are not become weakened and eventually disappear.

At the molecular level, this phenomenon is based on the action of glutamate via its N-methyl-D-aspartate (NMDA) (ionotropic) and α-amino-3-hydroxy-5-methyl-4-isoxazolepropionic acid (AMPA) (metabotropic) receptors. The more glutamate that is released simultaneously, the more NMDA channels will open. One consequence of this activation is the production of numerous additional AMPA receptors on the surface of the postsynaptic membrane. The presence of AMPA receptors further facilitates the opening of NMDA receptors, making the synapse more reactive and sensitive to glutamate.

This phenomenon, known as long-term potentiation (LTP) (see ▶ Chap. 2, ▶ Sect. 2.3.3.1), plays a central role in all processes of learning and memory formation.

The notions of synaptic selection or pruning, cellular plasticity, and critical periods emphasize how strongly the environment shapes the brain and how crucial a rich sensory environment is during intense periods of brain maturation. An inadequate sensory environment—whether too impoverished or overly frightening—during the critical developmental periods of a kitten can lead to irreversible disorders that become lifelong handicaps for the animal.

Some individuals, even those raised in severely impoverished conditions, are capable of rapid recovery when placed in a richer environment. This overcompensation demonstrates that not all subjects possess equal abilities to strengthen synapses. Consequently, there are genetic predispositions that facilitate synapse formation—and thus more effective learning—and vulnerabilities that render learning fragile, even in high-quality environments.

Genetic vulnerabilities and the environmental influence on development were discussed in ▶ Chap. 1 ("Zubin's model," see ▶ Sect. 1.2.4). Let us recall the Zubin and Spring model (◘ Fig. 4.2), which illustrates the unequal responses of individuals to favorable or unfavorable environmental conditions, and their varying tendencies to develop pathological states depending on their genetic vulnerabilities.

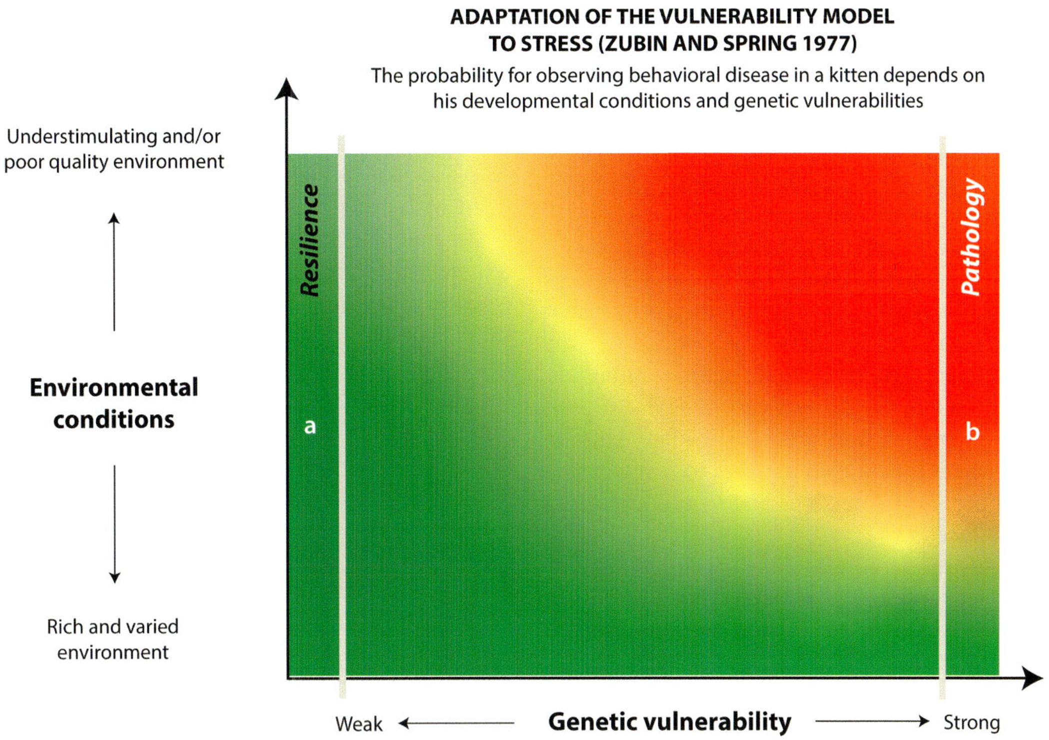

Fig. 4.2 Zubin's model applied to behavioral development in kittens. At the two extremes: (a) Very low genetic vulnerability protects the individual despite poor environmental conditions. (b) High genetic vulnerability predisposes the individual to pathology despite excellent environmental conditions

Summary of ▸ Sect. 4.1

From birth to adulthood, the cat's brain evolves and develops under the influence of neurobiological factors. Changes in neurons occur during specific periods that mark the stages of development. At birth, neurons are abundant but undergo remodeling during two critical or sensitive periods: the socialization period and puberty.

These remodelings involve two processes: strengthening synaptic connections that have been utilized by the individual, reflecting their experiences in their specific environment, and eliminating neurons and synapses that have not been used. This process, known as synaptic pruning, is influenced by neurotrophins and enables brain plasticity. Through these cellular eliminations, the individual becomes specialized and better adapted to the particular environment in which they live.

4.2 Chronological Description of Behavioral Development

We have outlined the neurophysiological basis of neurogenesis that occurs during behavioral development. Let us now examine its chronology. The behavioral development of cats can indeed be divided into several periods based on neurological development. This development affects sensory organs, which guide environmental exploration, as well as the individual's motor and cognitive abilities. These interactions between sensory organs and brain function are also evident later in life, where the process appears to reverse: sensory perception declines and synaptic plasticity reduces.

Before detailing these periods one by one, let us recall that cats are a nidicolous species, with offspring that are underdeveloped at birth.

The work of Scott and Fuller in the 1960s (Scott 1962; Scott and Marston 1950) and Bateson in the late 1970s (Bateson 1979) identified several so-called *sensitive* or *critical periods*. During these phases, neurosensory influences are crucial to shaping the cat's future behavior. During these periods, the animal can learn very quickly despite limited exposure to a situation. Conversely, deprivation of exposure during these times is difficult to overcome, as the individual's brain no longer retains the exceptional plasticity characteristic of critical periods.

The periods studied by Scott and Fuller represent extensive work conducted over more than a decade. Their research provided valuable insights into the behavioral development of dogs and mammals in general. However, species-specific differences are evident. Replicating this research on cats would be valuable, but such studies are not easily feasible today due to ethical concerns. Indeed, Scott and Fuller employed electric shocks and sensory deprivation during their experiments, practices that would no longer be permitted today.

It is also essential to consider the individual nature of development. For instance, late adoption is not necessarily a disadvantage as long as the kitten grows up in a high-quality environment rich in diverse stimuli. Conversely, if the environment is understimulating, the brain cannot develop optimally, and removing the kitten before 8 weeks to place it in a richer, more varied environment with other adult cats may be beneficial.

In practice, late-developing kittens should remain with their mother for a longer period, while earlier-developing kittens may be separated sooner. The mother cat's behavior toward her kittens is also an important indicator. Indeed, mothers need to teach their kittens certain skills; however, if they become pregnant again shortly after giving birth, their educational role may be shortened.

4.2.1 The Prenatal Period

The *prenatal period* corresponds to the gestation period, which lasts approximately 63 days in domestic cats. Births are seasonally distributed due to the estrous cycle of female cats, which primarily occurs during periods of increased daylight. As a result, most litters are born in summer, with births being less common in winter (Jennett et al. 2016).

During this period, the embryo's brain undergoes active neurogenesis, which is minimally influenced by the environment (Knospe 2002). However, in the second half of gestation, the fetus begins to develop gustatory abilities. Its tactile sensory pathways and emotions also start to develop, although they are not yet mature. Interactions with the mother or exposure to stressful events can logically be perceived by the fetus and may contribute to its development. Nonetheless, the level of evidence supporting these claims remains very low (Braastad 1998).

4.2.2 The Neonatal Period (0–9 Days)

The *neonatal period*, spanning from birth to eye opening, marks the progressive development of sensory channels. At birth, the brain weighs approximately 20% of its adult weight, which is reached around 3 months of age (Smith and Jansen 1977).

Kittens are born deaf and blind, unable to regulate their internal temperature, with motor skills limited to crawling. They are guided to their mother's nipple by thermotropism and olfactory signals. Remarkably, inexperienced kittens find the nipple within minutes (Raihani et al. 2009). Removing the olfactory bulb prevents a kitten from locating the nipple, highlighting the critical role of olfaction in this process (Kovach and Kling 1967). Kittens also quickly learn the specific olfactory cues of their mother and their designated nipple (Raihani et al. 2009) (see the box "Suckling Order").

The ability to suckle is independent of milk production, as kittens fed artificially will still attempt to suckle both lactating mothers and non-lactating adults, even after repeated experiences (Koepke and Pribram 1971). However, if kittens are exclusively fed artificially via a nasogastric tube from birth, they lose the suckling reflex by approximately 3 weeks of age.

Important Point of View

Suckling Order

Within 12 h of birth, a suckling order is established among kittens (Hudson et al. 2009). While this order is generally adhered to, it is not absolute. However, there is no correlation between a kitten's specific position on a nipple and its weight gain, the amount of milk ingested per feeding, or its behavior during suckling. On average, kittens consume 1.4 g of maternal milk per feeding during the first week and 4 g by 4 weeks of age.

Kittens' eyes open between 6 and 10 days after birth, but visual acuity develops gradually between 2 and 10 weeks. Interestingly, kittens raised in darkness open their eyes earlier (Braastad and Heggelund 1984).

During the neonatal period, several reflexes are present, disappearing as the brain matures (Houpt 2018). For example, the *suckling reflex* is active from birth, with kittens suckling on various surfaces, including artificial ones, until about 3 weeks of age.

Kittens predominantly exhibit flexor muscle tone during the first 2 weeks of life, which forces them to crawl by pulling themselves forward with their forelimbs.

Urination and defecation are not spontaneous during this period. Instead, they rely on the *urogenital reflex*, usually triggered by the mother's licking of the genital region.

Behavior is limited to seeking the nipple for feeding and sleeping. Sleep during this stage is characterized by a high proportion of REM sleep, and kittens' sleep cycles are significantly shorter than those of adult cats (Hoppenbrouwers and Sterman 1975).

4.2.3 The Transitional Period (9–15 Days)

The *transitional period*, spanning from eye opening to the maturation of the auditory system, corresponds to the refinement of sensory systems. Sensory stimulation intensifies as the sensory organs develop. In cats, the startle reflex appears very early, 1–2 days before the eyes open, but the auditory canals do not fully open until 4 weeks of age (Villablanca and Olmstead 1979).

The kittens' visual acuity becomes sufficient for them to orient toward their nest using visual rather than olfactory cues as soon as their eyes open. By the end of the transitional period, kittens can visually locate their mother to approach her (Freeman and Rosenblatt 1978).

At the conclusion of this period, urination and defecation become spontaneous, although the mother continues to clean the perineal area.

From birth, kittens knead their mother's nipples with pawing movements during nursing. This kneading behavior on soft surfaces can persist into adulthood when associated with pleasurable experiences, such as being petted by their owners. In many cases, the behavior continues spontaneously, likely because it remains a source of comfort for the kitten and evolves into a learned behavior reinforced by positive experiences.

Teeth begin to emerge just before 2 weeks of age and continue to develop until about 5 weeks.

4.2.4 The Socialization Period (18 Days—2 Months)

4

In ▶ Chap. 1, we explained that cats do not belong to a social species. (See ▶ Chap. 1, ▶ Sect. 1.3.1.2). The term "socialization period" is odd because strictly speaking, particularly for a non-social species like cats, it would be more accurate to call it the "sociabilization period." However, we have chosen to retain the term "socialization period" while discussing potential misunderstandings (see the box "Socialization—Sociabilization—Familiarization").

Important Point of View

Socialization—Sociabilization—Familiarization

These three concepts can be confusing and require clear definitions. Socialization refers to a social species, like dogs, that prefer living in the company of conspecifics. These species develop interindividual communication to establish organization and group living rules. They are "social obligates."

Sociabilization refers to an individual and the relationships they form with other individuals. This concept applies to non-social species like cats.

Familiarization describes sociabilization limited to one or a few individuals, where relationships with the rest of the species are not possible.

During the so-called socialization period, it is essential to accept that cats are, in fact, sociabilizing.

The *socialization period* is a crucial phase for behavioral development. Many critical learning processes take place during this early stage of life. Synaptic pruning, which occurs around 2 months of age, leaves a definitive and almost irreversible imprint on the brain's organization.

As we have seen (see ▶ Chap. 2, ▶ Sect. 1.3.3.1), cats categorize their environment. The term "environment" should be understood broadly to include interactions: interactions with the mother, interactions with littermates, and interactions with other species, including humans, who play a central role for domestic cats.

Attachment to the mother is highly significant during this period (Martínez-Byer et al. 2023). Through this attachment, the mother serves as a calming base, a model, and a guide. She is the primary reference figure, and without her, kittens remain in the nest, unable to explore and discover.

The feline nursing period can be divided into several stages. The first stage, from birth to 14 days, is characterized by the mother initiating nursing. During the second stage, from 2 to 3 weeks, both the mother and the kittens initiate nursing. From 3 to 5 weeks, nursing is predominantly initiated by the kittens (Landsberg and Ley 2012). Beyond 5 weeks, the mother gradually reduces her acceptance of nursing, though she does not exhibit aggressive behavior toward her kittens. It is also at this age, around 4–5 weeks, that kittens begin to eat solid food, marking the start of weaning (Martin 1986). The mother increasingly moves away and distances herself when her kittens attempt to nurse. By 7–8 weeks, nursing frequency drops significantly, and the majority of the kittens' diet becomes solid. At this point, weaning is considered complete (Martin 1986).

In groups where it is possible, kittens are often allowed to nurse from females other than their mother (Bradshaw 2016; Feldman 1993).

Alongside this gradual weaning, the mother may begin bringing live prey to her kittens if she has access to the outdoors. The most skilled kittens can start killing mice as early as 5 weeks old (Baerends-Van Roon and Baerends 1979). Kittens' food preferences tend to mirror those of their mother, whether for solid food or prey types, though these preferences can evolve in adulthood.

Weaning begins earlier in larger litters (Martin 1986) and is also influenced by the mother's pregnancy status. When the mother is pregnant again, weaning is completed more rapidly, but without any particular aggression toward the kittens (Bateson 1994).

In urban areas where many cats, some more or less sociabilized to humans, are left to reproduce freely, it is common to observe females having 2–4 litters per year. These repeated pregnancies require a high density of cats combined with easy access to food. During such successive litters, the mother often cannot adequately manage the development of her kittens. Early separations between the mother and her kittens are common in these cases, with the kittens being forced to fend for themselves for food as early as 6 weeks of age.

Among these kittens, many will be able to approach humans to secure shelter and food, especially if they are less than 8 weeks old. It is often the mothers who lead their kittens toward resource-rich areas, frequently located near humans. Some of these humans may decide to take responsibility for the kittens, providing them with care and a chance of survival.

Kittens' *sleep patterns* also evolve during the socialization period. At the beginning of this phase, their sleep is predominantly REM sleep, but by the end of the period, around 8 weeks of age, their sleep begins to follow the cycles of adult cats (McGinty et al. 1977).

The vocalizations of kittens and their mothers are highly varied (see ▶ Chap. 1, ▶ Sect. 1.3.2.1). Purring first appears during the socialization period, often during nursing, but rarely before 6 weeks of age. While purring is commonly associated with moments of pleasure, it is not unusual for very ill cats to purr. Purring serves as a calming signal both for the cat emitting it and for the receiver perceiving it. Its low-frequency sound allows communication that is less likely to be detected by predators (Tavernier et al. 2020). Recently, it was discovered that cats can produce purring without any active muscular or neural contractions, thanks to specific cells in their larynx. This finding contradicts previously held beliefs (Herbst et al. 2023).

Urination and defecation become voluntary around 5–6 weeks of age, at which point the mother no longer needs to lick the perianal region (Fox 1970). From this age, most kittens already dig a hole before squatting to urinate when provided with a litter box filled with soil. Most kittens also scratch after elimination (Sung and Crowell-Davis 2006).

The mother instills caution regarding dangerous areas, both by example and through direct interventions. Most commonly, she does this by carrying overly adventurous kittens in her mouth. During transport, the kitten adopts a reflexive posture, curling its tail against its abdomen and remaining completely immobile. This posture, known as the carrying reflex, facilitates its transport.

During the socialization period, the mother frequently relocates the nest, carrying her kittens if she perceives the area as unsafe. This behavior is even more common in natural conditions where multiple mothers manage communal litters (Bradshaw 2016; Feldman 1993). In domestic cats, this behavior may occur if the mother is disturbed too often or if the kittens are handled carelessly, causing them to vocalize.

Carrying by the mother begins as early as 2 weeks of age and decreases rapidly as the kittens gain weight. It is very rare to observe this behavior beyond 6–8 weeks. The increasing weight of the kittens likely explains why mothers carry them less frequently. In addition, as the kittens grow, physical guidance becomes unnecessary; they readily follow their mother in response to a simple vocalization, thanks to the maternal bond developed during this period.

Interestingly, in feral cat groups, kittens quickly show a strong preference for their mother and are more accepting of care from other adults within their group. This dynamic facilitates the kittens' relational learning, as they observe and imitate their mother's behavior (Zentall 2013).

Important Point of View

The Cat that Lands on Its Feet

Kittens begin developing extensor muscle tone starting in the third week of life, while flexor tone is already established during the first 2 weeks. This progression is reflected in their movement: initially, kittens pull themselves forward using their forelimbs, gradually incorporating their hind limbs over time (Villablanca and Olmstead 1979). This explains their very unsteady gait at the beginning of the socialization period. By the fifth week, they are capable of brief bursts of running, and by the sixth or seventh week, they can utilize all the forms of movement typical of adult cat locomotion (Peters 1983).

The ability of cats to "land on their feet," known as the aerial righting reflex, is acquired between the fourth and sixth weeks of life (Martin and Bateson 1985).

During the socialization period, kittens establish their sensory homeostasis threshold. Their reactivity to environmental stimuli depends on their experiences and maternal influence. The more diverse their experiences and the more appropriately the kitten responds, the broader the boundaries of their sensory homeostasis will be (see the box "Sensory Homeostasis, Development, and End Threshold Concepts").

Once the socialization period ends, the ability to habituate to unfamiliar stimuli, while still present, diminishes significantly, particularly the capacity to generalize (see box "The Generalization Process and Its Importance").

Important Point of View

Sensory Homeostasis, Development, and End Threshold Concepts

As an extension of the concept of homeostasis discussed in ▸ Chap. 1 (see ▸ Sect. 1.1.3), sensory homeostasis refers to the set of processes, whether conscious or not, that enable an individual to remain below an emotional threshold compatible with cognitive functions and adaptive reactions. Responses are triggered when stimulation reaches the lower homeostatic threshold. These normal responses remain effective up to a certain level of stimulation (intensity and quantity), known as the upper homeostatic threshold. Beyond this threshold, cognitive responses are bypassed, and the individual reacts solely with avoidance or survival responses.

Optimal brain function, or at least a comfortable state of being, occurs between two stimulation boundaries: the lower limit, which corresponds to the absence of perception of environmental and internal changes, and the upper limit, beyond which reactions to stimuli become automatic, intense, and aimed at removing the individual from a perceived dangerous environment.

The thresholds defining sensory homeostasis are determined by the influence of early life conditions on brain development. Only high-quality early exposure allows the maximum expression of the genetic potential established at birth.

The first wave of synaptic and neuronal pruning, described in ▸ Sect. 4.1 of this chapter, occurs at the end of the socialization period. During this process, *neurons that have not formed numerous connections are eliminated.*

Numerous studies on sensory or affective deprivation highlight the profound impact of the early environment on kitten development. For instance, kittens artificially separated from their mother at 2 weeks of age subsequently develop severe behavioral abnormalities. They typically become fearful and aggressive toward both cats and humans, often exhibit poor locomotor control, and their learning abilities are impaired (Seitz 1959).

In other words, a kitten raised in an understimulating environment cannot compensate for its developmental delay simply by later exposure to novel stimuli.

While synaptic pruning may appear maladaptive, as the kitten loses a significant number of neurons, it should instead be viewed as an adaptive process. It enables the kitten to specialize in response to the environment it is expected to inhabit. This process refines the skills necessary for the specific conditions of its life.

In practice, however, the place of birth does not always correspond to the kitten's eventual living environment. If the disparity between these environments is too great, the risk of developing a behavioral disorder significantly increases.

A kitten's friendliness toward humans is heavily influenced by genetic factors. Experiments conducted with groups of kittens sired by friendly and unfriendly fathers demonstrated that the father's temperament significantly affected the kittens' friendliness at 1 year of age, even though the kittens had never met their father. For example, kittens from unfriendly fathers were more likely to exhibit unfriendly behaviors, even if they were handled by humans during their development (McCune 1995; Reisner et al. 1994). This suggests that kittens born to feral mothers, who are often less friendly toward humans, may also be at a higher risk of being less friendly compared to those born to known and friendly parents.

However, early handling has a notable positive effect on development and is always beneficial. Kittens handled for just 5 min a day during the socialization period develop faster, are more amicable toward humans, and learn more quickly (Meier 1961; Wilson et al. 1965). In both cases, the observed effects are attributed to a reduction in fear responses.

Friendly relationships between cats are also significantly facilitated and maintained into adulthood if they begin before the end of the socialization period. Generally, cats respond to a silhouette of their own species as they would to a real cat, but this ability develops between 6 and 8 weeks of age.

Similarly, the fear response of an adult cat—characterized by piloerection and a puffed-up tail—when confronted with a predator or its silhouette is not present at the youngest ages. For example, kittens aged 5 weeks do not react to a predator's silhouette, those at 6 weeks react minimally, but by 8 weeks, kittens exhibit the typical fear response identical to that of an adult (Kolb and Nonneman 1975).

It is also around this age that kittens begin to perform the flehmen response, characterized in cats by a slightly open mouth, upturned nose, and flared base of the vibrissae. This posture, more easily observed in horses, allows pheromones to be drawn into the vomeronasal organ (see ▸ Chap. 1, ▸ Sect. 1.3.2.1).

Important Point of View

The Generalization Process and Its Importance

The process of generalization, previously described in ▸ Chaps. 1 (see ▸ Sect. 1.2.3.4, "Perceptual Learning,") and ▸ Chap 2 (see ▸ Sect. 2.3.3.1), becomes particularly significant during the socialization period.

Generalization allows an animal to create categories of stimuli or individuals. Elements within a category will subsequently trigger the same type of behavioral response (Mclaren and Mackintosh 2002). For example, this is how cats learn not to fear all humans, even if they have only encountered a few individuals as kittens.

The ability to generalize experiences and form groups diminishes after the socialization period. Consequently, the chances of rehabilitation for cats suffering from sensory deprivation syndrome (see ▸ Chap. 8, ▸ Sect. 8.2) remain limited.

Play provides kittens with numerous opportunities for relational experiences. The mother often remains present to monitor interactions, minimize risks, and enforce controls on locomotion, as well as the intensity of scratching and biting. Without her intervention, self-control mechanisms fail to develop (Seitz 1959). Bottle-fed kittens, for example, frequently exhibit significant deficits in self-control (Bateson and Young 1981). These maternal controls are often exerted through licking or physical restraint between her paws, aimed at halting the kitten's movements to teach them to regulate the intensity of their emotions.

Play among littermates begins to emerge around 3–4 weeks of age. Initially, it consists of gentle pawing, but as kittens improve their motor coordination, biting, scratching, and chasing replace the gentle swats. Combat games are observed among siblings and increase in both frequency and intensity up to 3 months of age. The most intense and rough play occurs around 3 months, during the juvenile period (Kolb and Nonneman 1975). In single-kitten litters, the mother compensates for the lack of sibling play by engaging more actively with her sole offspring (Caro 1981; Mendl 1988).

Object play does not appear until the end of the socialization period, coinciding with weaning. This suggests a transition from social play to object play as the kitten becomes independent and begins to distance itself from its mother and the nest (Barrett and Bateson 1978).

Important Point of View

Nipping and Scratching

The common belief that kittens nip to "relieve teething pain" is incorrect. This misconception is often reported by owners of kittens over 2 months old. While the eruption of milk teeth can indeed cause discomfort when breaking through the gums, this process is completed by 5 weeks of age, well before most kittens are adopted. Adult teeth emerge between 3.5 and 6 months, but the level of nipping observed and its timing do not align with this process.

Biting and scratching are often signs of excessive excitement during play, reflecting the kitten's lack of emotional control. These behaviors indicate immaturity in the ventromedial prefrontal cortex (vmPFC), which has not yet developed the capacity to regulate the kitten's motor, emotional, and cognitive impulses. By the end of the juvenile period, around 6 months of age, kittens should no longer bite or scratch, even during excited play.

This lack of control is much more common in bottle-fed kittens, who miss out on the social play interactions with siblings and the maternal guidance necessary to learn self-regulation.

Predatory behavior develops alongside object play and begins to emerge during the socialization period, displaying motor patterns like those observed in play. Play is often described as a way for kittens to practice adult life skills (Bateson and Martin 2013; Moelk 1979). However, experiments show that kittens deprived of play are still capable of hunting by 11 weeks of age. In general, most cats become proficient hunters as adults, almost regardless of the type of experiences they had during their early development (Caro 1980).

At around 5 weeks of age, the mother starts bringing dead prey to her kittens. Gradually, she introduces live prey, which she releases near the kittens (Biben 1979).

A species with which a kitten has been socialized before the age of 2 months will not typically be hunted later. For example, in 1930, Kuo demonstrated that kittens isolated from their littermates but raised with rats developed an attachment to the rats and did not hunt them. If littermates were also present, the kittens bonded primarily with their siblings but still exhibited friendly behaviors toward the rats and showed little inclination to hunt them later (Kuo 1930).

The socialization period is considered critical because learning during this time will not, or only minimally, occur later. It establishes a dialogue between the kitten's environment and its genetic predispositions. Behavioral disorders or, conversely, resilience, result from the interaction between these two factors.

As illustrated in ◘ Fig. 4.3, optimal developmental conditions significantly reduce the likelihood of pathology, except for disorders

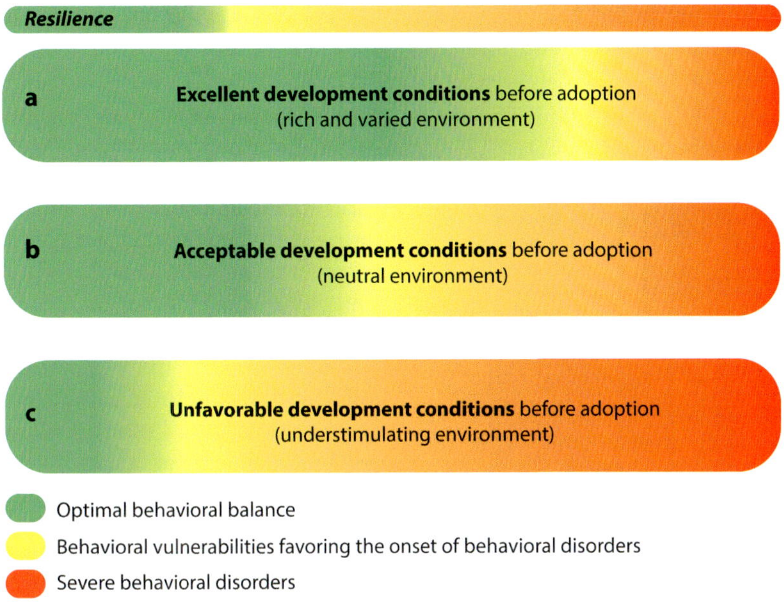

Fig. 4.3 Probability of normal behavior development based on the kitten's environmental conditions before adoption. (a) A rich and varied environment minimizes the emergence of pathology (extensive green zone, limited red zone). (b) Intermediate situation. (c) An under-stimulating environment predisposes the kitten to severe behavioral disorders, with exceptions for a few resilient individuals (limited green zone, extensive red zone)

with a strong genetic component. Conversely, poor developmental conditions increase the risk of pathological states, with only a small number of kittens escaping due to their individual resilience capacities.

4.2.5 The Juvenile Period (2–6 Months)

During the juvenile period, kittens consolidate and structure their learning. They refine their motor skills, such as the ability to walk and turn on a narrow plank by around 11 weeks of age (Villablanca and Olmstead 1979). Neurological improvements, like the development of eye-paw coordination, enable them to engage in object play (Barrett and Bateson 1978) and chasing games, which begin to take precedence over the social play more prominently during the socialization period (Martin and Bateson 1985). Play often begins with a pawing movement at an object, followed by a leap when the object moves. This sequence mirrors the behavior of adult cats manipulating their prey after capture.

The juvenile period is also when kittens' playtime reaches its peak, around 12 weeks, with more than 2 h of daily activity. Interactions among littermates are most intense and rough at this age (Kolb and Nonneman 1975). Young male cats play with objects more frequently than females, while females with at least one brother play more than those without (Bateson and Young 1979). Playtime between kittens gradually decreases by 4 months, and solitary play, although more prominent, also significantly diminishes by the end of the juvenile period (Martin 1984).

The juvenile period often takes place in the adoptive family (see the box "The Ideal Age for Kitten Adoption"). During this time, fear responses dominate the kitten's reactions to unfamiliar stimuli. Positive exposures during this stage are crucial to prevent the formation of negative associations. This is also the age when vaccinations are administered, making it essential to use minimal restraint and provide food rewards to prevent the development of conditioned emotional responses and a fear of veterinary care.

Important Point of View

The Ideal Age for Kitten Adoption

In natural conditions, the mother completes weaning by 8 weeks, but kittens remain in contact with her until about 4 months of age (Bradshaw et al. 2012).

In France, the legal adoption age is 8 weeks. However, many kittens are found and adopted before this age, often before the start of the juvenile period. For purebred cats, it is more common for breeders to release kittens only after 3 or even 4 months of age, often already spayed or neutered.

A study investigated the effects of adoption age on adult cat behavior. The study analyzed data from 5726 cats of 40 different breeds and found that adoption before 8 weeks significantly increases the risk of aggressive behavior but does not exacerbate fearfulness. In addition, cats adopted at 14 weeks have a lower risk of displaying aggressive or compulsive behaviors in adulthood compared to those adopted at 12 weeks (Ahola et al. 2017).

These findings must be contextualized with the environment the kitten experiences before adoption. An understimulating environment (absence of a mother, lack of exposure to stimuli) negatively impacts a kitten's behavioral future. However, in the case of a family-raised litter, recommending adoption between 12 and 14 weeks allows the mother sufficient time to complete the kittens' education and support their behavioral development during the early juvenile period. This approach appears favorable for the kitten's future behavioral balance.

Reactions during the juvenile period are heavily influenced by the learning experiences established during the socialization period. A kitten lacking *social interactions* during socialization is already at a disadvantage during the juvenile stage. The transition to a new environment requires significant adaptation, particularly if the new setting differs greatly from the previous one. For instance, moving from a rural area to an urban environment or encountering unfamiliar species, such as a dog, places considerable demands on the kitten's adaptive capacities.

Similarly, *self-control* mechanisms are not fully developed at the beginning of the juvenile period. In natural conditions, the mother's presence until 4 months of age promotes the development of self-control (Bradshaw et al. 2012). In France, where adoption typically occurs before 4 months, owners must compensate by exposing the kitten to diverse stimuli to enhance its sensory homeostasis and teach it better control during play and interactions to mitigate the absence of maternal guidance.

When it comes to interactions with humans, it is well established that allowing the kitten to initiate contact fosters higher-quality interactions and increases their frequency. Generally, respecting the cat's body language, including its desire to avoid certain types of contact, improves the overall quality of human-cat interactions (Turner 1991).

Oral exploration is significantly less pronounced in cats than in dogs, and it is rare for kittens to ingest non-food objects. Cats cause minimal damage to furniture with their mouths, and their lack of self-control is more evident in play interactions or their tolerance for contact. During the juvenile period, kittens improve their control over biting and scratching. By the end of this phase, around 6 months, they no longer injure others during interactions. They also manage frustration more effectively, as the maturation of their prefrontal cortex allows them to progressively regulate both their motor activity and their emotions, such as anger, joy, and frustration.

Weaning is completed before the juvenile period begins, and food preferences tend to mirror those of the mother (see ► Chap. 3, ► Sect. 3.2.2.2). During the juvenile stage, a kitten's diet gradually transitions to resemble that of adult cats. They eat numerous small meals throughout the day, maintaining a regular rhythm, with higher feeding frequencies at dawn and dusk—times corresponding to their natural predatory activity (Bradshaw 2006; Kane et al. 1981).

If food is not available ad libitum, cats adapt their behavior by eating fewer but more calorie-dense meals. External factors such as food availability and the cat's psychological state play a significant role in feeding behavior (Thorne 1982).

Cleanliness is established at the beginning of the juvenile period, as elimination becomes voluntary and locomotion is controlled. Litter training is greatly facilitated by the presence of an adult role model, often the mother, and substrate preferences develop based on the type of litter presented (Casey et al. 2022). However, even in the absence of an adult, kittens naturally prefer sandy and absorbent materials for elimination (Neilson 2004).

During the urinary elimination sequence, the cat sniffs the area, typically digs a hole with its forepaws, squats over the hole with its tail raised, and completely empties its bladder. Afterward, the cat turns around, may sniff its urine, and then covers it with its forepaws. For the fecal elimination sequence, the cat also digs a hole, squats less deeply, and does not always cover its feces (Sung and Crowell-Davis 2006).

Important Point of View

Maternal and Territorial Attachment

Natural detachment, or the age at which kittens separate from their mother, typically occurs during the juvenile period, around 4 months of age. As the kitten grows, its dependence on the mother is replaced by the comfort it derives from structuring its territory using territorial markers, including pheromones. Toward the end of the juvenile period, kittens begin marking their territory with pheromones.

When young (up to 5 or 6 weeks), kittens primarily use allomarking, which they initially apply to their mother and, gradually, to their siblings. Territorial facial marking starts only after weaning. The distress associated with maternal rejection likely motivates the kitten to establish its first isolation zone and mark it with its pheromones. Pheromones help structure the environment by delineating different areas, as described in ► Chap. 1, ► Sect. 1.3.1.4. This behavior manifests through facial marking, visible as cheek rubbing against objects.

Allomarking of other individuals becomes increasingly evident, performed on individuals the kitten appreciates. Scratching on objects also becomes more organized, typically occurring around resting or observation areas.

Some cats naturally engage in less marking than others without this being abnormal. However, practitioners should be concerned about the complete absence of marking at puberty or signs of disorganized marking, such as inappropriate locations, excessively high intensity or frequency, or sudden changes in marking patterns (e.g., a significant increase or decrease). These changes can indicate underlying behavioral or medical issues that warrant attention.

The juvenile phase is also a time of heightened vulnerability, particularly if the preceding socialization phase was of poor quality or if the kitten has genetic predispositions to behavioral challenges. It is therefore crucial that this period is dominated by the most positive interactions possible. Identifying behavioral weaknesses early is essential to compensate for developmental delays.

Early detection of vulnerabilities offers the best chance for the kitten to achieve normal behavior through simple behavioral therapy. Unfortunately, mild vulnerabilities often go unnoticed, as owners naturally assume these minor issues will resolve with time. This highlights the critical role of professionals during this period. They can guide interventions and prevent the progression to pathological states (Gazzano et al. 2015).

There is also a significant risk of decompensation during adolescence. A "shy" kitten may become aggressive as an adult when frightened. A hyperactive kitten, which injures during interactions and is subsequently punished, may see its relationship with the owner deteriorate. Similarly, a kitten that poorly structures its territory may develop marking abnormalities, resulting in reduced comfort and anxiety in adulthood.

The proactive involvement of professionals during this phase can make a pivotal difference in preventing these potential outcomes and ensuring the kitten's successful behavioral development.

4.2.6 The Pubertal Period (6 Months to 1 Year)

The *pubertal period* marks the maturation of the reproductive system. Sexual hormones begin to be secreted, influencing behavior and triggering the development of the sexual pheromonal system. This phase represents a transition as the young cat transforms into an adolescent. However, it is also a period of instability and vulnerability.

During this time, a new wave of synaptic pruning occurs, which consolidates and rigidifies behavioral tendencies established during the socialization and juvenile periods. Kittens that exhibited fragility in earlier stages, such as shyness, difficulties in acquiring self-control, strained relationships with their owners, or poorly structured environments, are at risk of experiencing worsening deficiencies during the pubertal period.

The onset of the pubertal period is marked by urine marking in intact males and the first estrus in females. As cats are a seasonally polyestrous species, the age of sexual maturity varies, ranging from 4–5 months for late-born kittens to 10 months for those born earlier in the year (Jemmett and Evans 1977). In addition, the breeding season varies significantly, as cats near the equator, where daylight is longest, can reproduce year-round. The further from the equator, the shorter the breeding season becomes, with reproduction limited to 6 months near the poles (Ng et al. 2023).

Kittens neutered at a very young age still go through a pubertal period, even if the associated behavioral signs appear invisible. Current studies show that the majority of cats are spayed or neutered by 6 months (60%), with 85% sterilized by 1 year in countries like the United States and Australia (Mazeau et al. 2021).

Neutering or spaying typically occurs during the pubertal period. Cats vulnerable along the fear axis may develop a phobic state associated with veterinary care during this time (Mariti et al. 2016). Adolescence is a period of fragility, and cats that exhibited shy tendencies or little enjoyment of interactions may shift toward more aggressive strategies. Owners who attempt to force contact or resort to physical punishment are often the targets of heightened aggression.

Unlike dogs, where *detachment* occurs during the pubertal period in natural conditions, adolescent cats are already detached and independent. In cat colonies, this is often the period when male cats leave the group, while females tend to stay and raise their litters together, frequently remaining within the group. Food resources play a significant role in these group dynamics, and as food availability decreases, cats tend to adopt more solitary behaviors (Crowell-Davis et al. 2004).

Adolescent cats complete their *acquisition of self-control* during this period: biting and scratching are fully controlled, and they no longer injure during play interactions. Overall, they play less frequently (Henning et al. 2022). They also gain better emotional regulation, becoming capable of tolerating handling, waiting, and managing frustration.

Adolescent cats learn to structure their activity zones through pheromonal and olfactory marking.

Urine marking begins with puberty, and intact male cats can mark more than ten times per hour. This behavior is the primary reason for neutering male domestic cats at puberty. Females in estrus also mark with urine, although less frequently and only during estrus. For owners, the continuous vocalizations and disturbances associated with estrus are the main motivations for spaying.

Urine marking is both olfactory and visual. Cats tend to inspect these marks olfactorily, especially if they were deposited less than 4 h earlier. The flehmen response allows them to gather information about the individual who left the mark (De Boer 1977). Urine marks are typically placed on the periphery of the cat's territory, far from feeding areas. Fecal marking, characterized by conspicuous, uncovered feces visible to other cats, is rarer but occurs on the periphery to delineate territory boundaries.

Facial marking also becomes more structured during puberty. It involves the cat rubbing the outer corners of its mouth against objects or individuals. Depending on the action, different pheromones are deposited. Five fractions have been identified on the cheeks, namely F1, F2, F3, F4, and F5 (see ◘ Fig. 4.4).

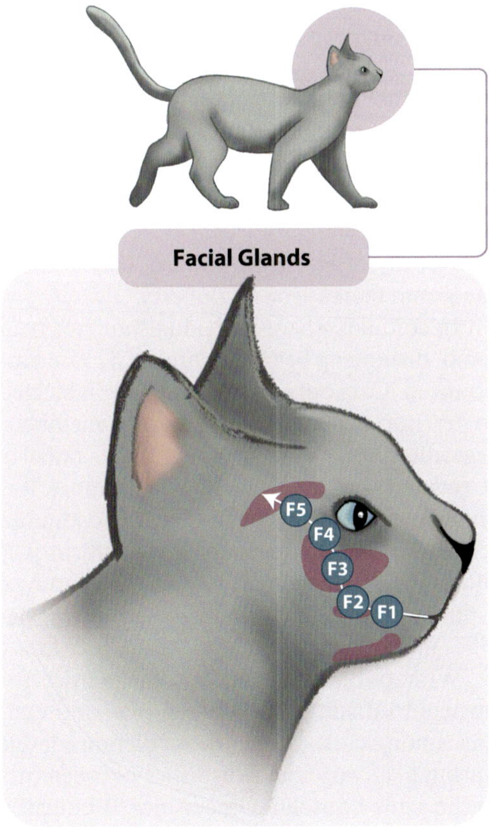

◘ **Fig. 4.4** The facial glands of the cat

The F3 fraction is the most studied and is used in certain pheromonal diffusers, such as Feliway Classic®. F3 allows cats to leave territorial marks on objects. The F4 fraction, on the other hand, is used to mark social partners with whom the cat is at least familiar. This fraction is found in other synthetic pheromonal products, such as Felifriend®. It is common to see cats rubbing their faces against human faces or legs to deposit these pheromones.

During adolescence, grooming and mutual allomarking behaviors become almost exclusively reserved for cats and humans within the same group (Curtis et al. 2003).

The F2 fraction plays a role in reproductive behavior and is deposited during sexual excitement. The roles of F1 and F5 fractions remain unknown.

Scratching marks serve a territorial marking function when made on vertical surfaces. They act as a visual signal, indicating the cat's isolation, resting, or observation zones, and also involve the deposition of pheromones. These pheromones are secreted by the exocrine glands in the cat's paw pads during the scratching sequence. When maintaining their claws, cats may also scratch on horizontal surfaces.

Scratching marks are intended for individuals outside the group, whereas facial markings are primarily for the cat itself and members of the group with whom it has familiarized or socialized.

The frequency and intensity of these markings vary greatly between individual cats. For cats in psychological balance at puberty, the organization of facial marking sites delineates the boundaries of their activity zones and helps them structure their territory (◘ Fig. 4.5). Later, changes in this marking organization often serve as clinical indicators of psychological disturbances.

During the pubertal period, cats in impulsive or compulsive states struggle to structure their territory, and their markings tend to be intense but disorganized due to their lack of control. It is also likely that the pheromones produced by a given cat are influenced by its emotional state. An anxious animal is likely less capable of producing calming marking

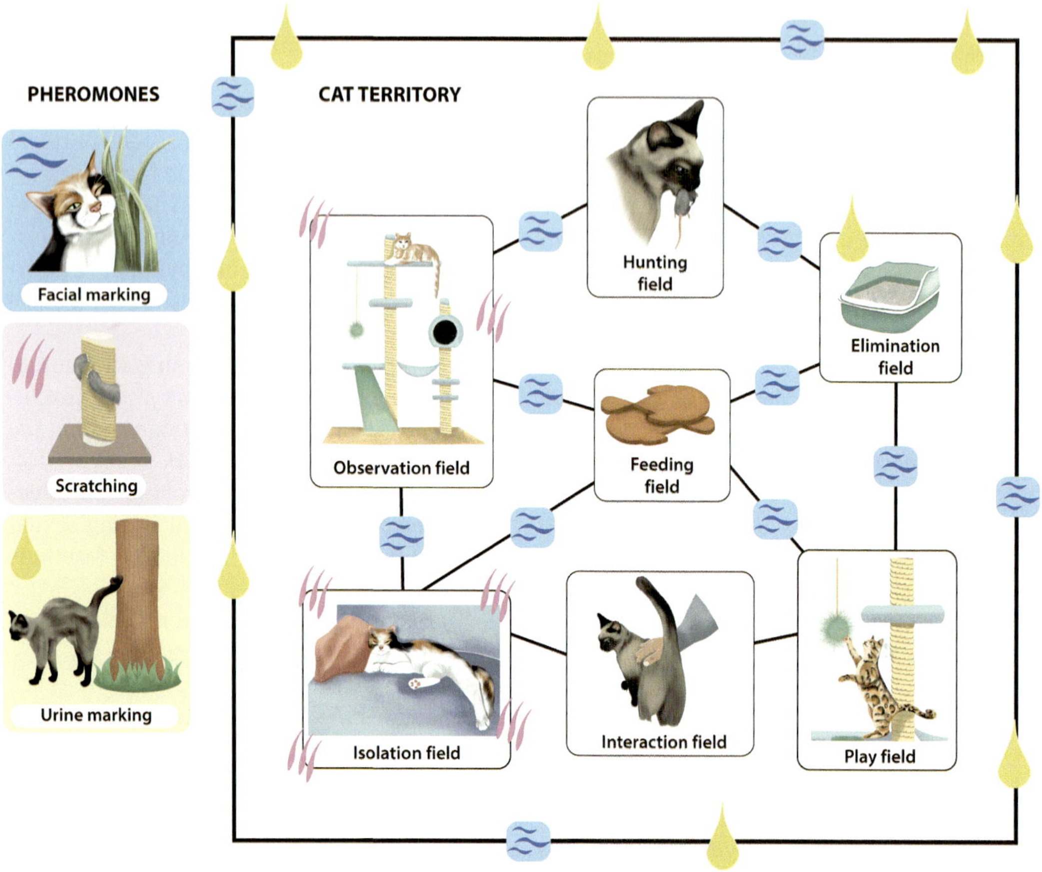

Fig. 4.5 Organization of pathways and territorial zones in cats

pheromones and, as a result, struggles to organize a soothing territory. Just as an anxious dog is less likely to foster calming social interactions, an anxious cat builds an anxiety-inducing territory.

Relationships with other cats also become more mature: play is no longer the primary means of communication, and increased aggression toward conspecifics is common, particularly in the presence of females in heat and among individuals of the same sex. The cat's sense of smell plays a major role in establishing these interactions (Vitale Shreve and Udell 2017). Two cats approaching each other amicably will hold their tails upright and greet one another by touching noses, a behavior frequently observed between cats in the same household (Gajdoš-Kmecová et al. 2023). Conversely, aggression toward cats outside the group increases with puberty.

In colonies where individuals are not neutered, distancing behaviors are even more pronounced. Cats outside the group are subjected to territorial aggression by group members, regardless of sex, starting from puberty (Crowell-Davis et al. 2004). Furthermore, it is challenging for an outside cat to join an already established group. This difficulty is often observed when attempting to introduce a pubescent cat to one or more adult conspecifics in the same household.

With puberty, it can become harder for owners to distinguish between play and conflict among cats, especially as tolerance levels during play vary and skirmishes between cats in the same household occur more frequently than during earlier developmental stages (Gajdoš-Kmecová et al. 2023).

4.2.7 The Adult Period (1 Year—8/10 Years)

This is a period of behavioral stability, although learning continues throughout the cat's life. The cat reaches its social, sexual, and behavioral maturity. It exercises self-control in interactions with members of its group, whether they are cats, dogs, or humans, ensuring it does not cause injury even during moments of high excitement.

However, personal experiences can disrupt this balance, especially if the foundations established in early life are fragile. This is particularly true for phobias, which may develop following traumatic exposures. Due to its status as a prey species and its tendency to structure its territory independently rather than relying on group living, as dogs do, the cat is significantly predisposed to sensitization. This makes it vulnerable to the development of phobic disorders.

Preventive measures are therefore essential before veterinary visits. These include habituation to the transport carrier, gentle handling during visits, and other supportive measures (see ► Chap. 31). Without such precautions, sensitization and progression to a fear of handling and veterinary care are almost inevitable (Mariti et al. 2016).

Adult cats properly structure their territory through the organization of pheromonal markings that delineate pathways between their various activity zones (see ◘ Fig. 4.5). They are particularly sensitive to environmental changes, far more so than dogs.

It is important to note that an adult cat's intraspecific relationships differ significantly from those of a kitten. When multiple cats share the same household, they often organize themselves to avoid using the same areas at the same times of day. A spatio-temporal organization emerges, functioning like a perfectly managed shared living arrangement. This allows each cat to structure its own living space despite the presence of conspecifics. However, as discussed in ► Chap. 1, domestic cats retain a territorial tendency, which makes introducing another adult cat far more challenging. For this reason, when owners wish to introduce a new cat into a household with an existing adult cat, adopting a kitten is recommended, as it is far more likely to be accepted than an adult.

The introduction should be gradual, beginning with olfactory and visual contact. This allows each cat to interact with the pheromonal and olfactory markings of the other before progressing to physical contact.

Interspecific relationships are similarly stable over time, maintained by the establishment of rituals that foster harmonious interactions. These rituals are facilitated by the adult cat's ability to engage in social learning, both with other cats (Barnett 1968) and with humans (Fugazza et al. 2021). For example, cats may learn to come drink from a faucet when their owner brushes their teeth, incorporating this behavior into their structured territory alongside group members and their routines. Observing such rituals makes it easier for an observer cat to learn and adopt the same behavior.

Negative experiences, such as punishments causing fear or pain, or somatic illnesses, can naturally interfere with relationships and the cat's territory, disrupting the functioning of the adult cat's brain. However, if the cat's development has been harmonious and enriched, these life challenges are much easier to overcome.

4.2.8 Senescent Period (from 8 to 10 Years)

Aging is a progressive, normal, and inevitable process. Even in the absence of disease, organs gradually weaken due to senescence. An internal biological clock defines each stage of life: embryogenesis, puberty, maturation, and senescence. This theory of a programmed process involving hormones and certain neurotransmitters acting on their target organs suggests that each species has a theoretical maximum lifespan. Naturally, any genetic or acquired alterations reduce this theoretical maximum. This programmed obsolescence is a mechanism that offers little advantage to the individual but is essential for the survival and evolution of the species. It remains an indispensable corollary of evolution.

The *senescent period* is characterized by cognitive decline that accompanies aging. The senses become less efficient. Cats may socially isolate themselves. Pain is common and contributes to the progressive withdrawal of the cat from its usual activities, including rituals established with its owners. Isolation leads to a loss of motivation and cognitive decline. The mood becomes more fragile. Overall, with aging, the cat's territorial range decreases.

This period is also characterized by the phenomenon of *involution*. In its primary sense, involution refers to a specific mode of aging in which an organ regresses to an earlier developmental state. This is the case for the thymus or ovaries. However, a complete return to a prior state never occurs; there is always partial or total loss of function.

This process of involution can also become dysregulated, targeting organs that should not undergo involution. In such cases, involution and aging combine in a harmful spiral. In the brain, this manifests as a dysregulation of the relationships between cognition, mood, and emotions. Regardless of the initial trigger for dysregulation, one or more vicious cycles emerge, leading to a runaway system (◘ Fig. 4.6).

The aging process renders all organs more vulnerable, leading to a decline in both physical and cognitive capacities that are considered "normal." This results in performance issues, demotivation, and avoidance of effort, which in turn accelerate the loss of abilities. Cognitive impairments also lead to mood disorders, which exacerbate performance issues and contribute to relational dif-

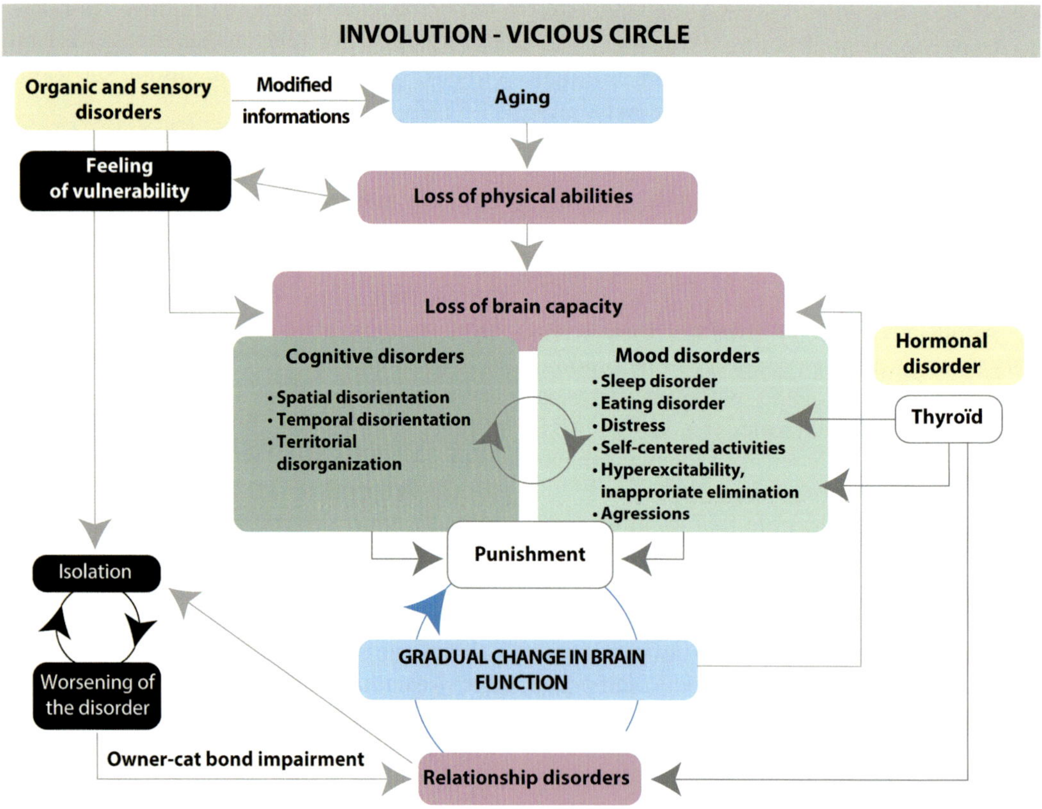

◘ **Fig. 4.6** Processes of aging. The figure illustrates the complex interplay of factors involved in aging and their escalation over time

ficulties. These relational difficulties further weaken the bond with others, worsening mood and cognitive impairments. The animal becomes increasingly isolated, and the problems intensify. Added to this are hormonal disruptions, which also impact mood and cognitive disorders.

In practice, all behavioral axes can be affected, as aging causes a rigidity in behavior. Vulnerabilities observed throughout the animal's development, similar to those seen during puberty, can predispose the emergence of behavioral disorders. For instance, some cats with a stable phobic state may deteriorate into a more generalized anxious state. Sensory losses can disrupt territorial marking and lead to intra- or interspecific relational difficulties, as aging acts as an additional aggravating factor, pushing the animal into pathological states.

When examining an older cat, practitioners must consider the impact of aging on behavior, which includes a general slowing down. However, they must also remember that normal aging does not explain sudden learning deficits, demotivation, cognitive impairments, or loss of self-control (see the boxes "Neurophysiology of Aging" and "Telomerase and Aging"). In addition, as cats are not highly expressive animals, organic illnesses often present initially as subtle behavioral changes or discomfort. Thus, organic issues must be meticulously investigated during consultations for behavioral changes in senior cats.

Important Point of View

Neurophysiology of Aging

In the brain, cortical atrophy is observed with aging. However, it is much more pronounced in certain pathological forms of aging, such as those associated with conditions like Alzheimer's disease, where cortical sulci and the hippocampus are particularly affected by the involution process.

In "normal" aging, there is no destruction of neurons, only fatigue: mitochondria become fewer in number and abnormal in structure.

At the cellular level, basal metabolism and all cellular deaths produce free radicals via mitochondria, which are, in turn, destroyed by these same free radicals. The loss of membrane fluidity is linked to the production of these free radicals. Through a chain reaction, they cause the oxidation of mitochondrial membrane lipids, further amplifying free radical production.

Free radicals can be neutralized by *superoxide dismutase* (SOD), which requires vitamin E and selenium to function effectively. Therefore, vitamin E and selenium are excellent agents for combating cellular aging.

During the involution process, SOD levels decrease and lose efficiency. Interestingly, β-amyloid deposits appear to be SOD precipitates. Cats can serve as models for Alzheimer's disease research (Zadik-Weiss et al. 2020), as β-amyloid and tau deposits (Fiock et al. 2020) observed in cats are very similar to those seen in human medicine. While their presence is not inherently pathological, these deposits are significantly increased in Alzheimer-like conditions.

Inflammation also plays a major role in aging, and non-steroidal anti-inflammatory drugs (NSAIDs) have proven significantly effective in slowing the process.

At the molecular level, *involution* is characterized by a *decrease in the synthesis of all neurotransmitters*. These disruptions are partly due to changes in lipid membranes, which become more rigid, thereby interfering with the release, reuptake, and mobility of postsynaptic receptors.

4

Important Point of View

Telomerase and Aging

Telomerase is an enzyme that maintains chromosome length during DNA replication by adding a specific structure to each end. Telomeres function much like the plastic tips at the ends of shoelaces, protecting the DNA during cell division. However, with each cell division, the telomeres become shorter. Once they are completely worn down, the DNA is exposed, and replication errors lead to cell apoptosis.

As highlighted by Elizabeth Blackburn and her team, who were awarded the Nobel Prize in Medicine in 2009 for their discovery, telomeres have become a key area of research in regenerative medicine. Their length appears to be a highly precise indicator of a cell's age. When telomeres are reduced to zero, even minor stress can trigger cell apoptosis. In theory, if telomeres could remain intact, cellular aging could be halted.

In November 2010, researchers caused a sensation with a publication in *Nature*, showing that they had reversed mammalian senescence for the first time. The study suggested that activating telomerase could rejuvenate cells in the brain, spleen, and reproductive organs, effectively reducing the biological age of older mice to that of young adults. However, their survival was not optimal, and practical applications remain out of reach (Jaskelioff et al. 2011).

In addition to this chronological description (see ◘ Table 4.1), the behavior and development of cats can also be analyzed through the four axes we have previously discussed: *fear, acquisition of self-control, territory,* and *relationships*.

Summary of ▶ Sect. 4.2

The behavioral development of cats is divided into chronological periods corresponding to neurobiological changes. The *neonatal period*, from birth to the opening of the eyes at around 7 days, is followed by the *transitional period*, which spans from eye opening to the opening of the auditory canals at around 2 weeks. During this time, kittens are very inactive and remain in constant contact with their mother, who should not be disturbed. Maternal attachment is established as soon as the eyes open.

The *socialization period* lasts from 2 weeks to 2 months and represents a crucial phase for behavioral development. During this period, kittens establish their sensory homeostasis thresholds; by exploring their environment, they learn to recognize and categorize elements as non-threatening. In particular, socialization with humans must occur during this time for the cat to be friendly as an adult. Kittens learn to control their movements, primarily through play with their littermates and mother. The mother progressively refuses nursing and contact, preparing for detachment. Food weaning is completed around 6–8 weeks. At the end of the socialization period, a massive *synaptic pruning* occurs.

The *juvenile period* extends the socialization phase, with the kitten continuing to explore the world and refine its self-control. In natural conditions, detachment is completed by the mother at around 14 weeks. The kitten becomes accustomed to new stimuli as long as they are not unpleasant. The pheromonal system enabling the creation of a calming territory completes its maturation.

The *pubertal period* is another critical phase, during which vulnerabilities acquired earlier may push the cat into a pathological state. The adolescent cat structures its territory using facial markings, scratching, and urine marking at the territory's periphery. *Adulthood* marks behavioral stability and is followed by the senescent period, during which cognitive decline begins.

Finally, the *senescent period* is characterized by sensory and cognitive decline, an increase in mood disorders, and a risk of territorial disorganization, with markings becoming less effective.

Table 4.1 Key formal and functional milestones in feline behavioral development

Period	Age	Formal analysis	Functional analysis
Prenatal		Development of gustatory and tactile competencies	
Neonatal	0–7 days (eyes opening)	Behavior dominated by nutrition and sleep; auditory canals begin to open (not mature) around 5–6 days	Mother-kitten attachment (*care-giving*)
Transitional	7–15 days	Eye opening, maturation of neurosensory systems; emergence of first teeth	Establishment of maternal attachment
Socialization	2 weeks–2 months	Social interactions (intra- and interspecific) and exposure to non-social stimuli (sounds, objects, diverse environments); play primarily among littermates, shifting to object play toward the end of the period; onset of predatory behavior	Maternal attachment complete; detachment begins with refusal to nurse and full weaning before 8 weeks
			First fear reactions, alarm system maturation, early sensory homeostasis establishment
			Autocontrols begin to develop through play with siblings and the mother
			Territory not yet structured
Juvenile	2–6 months (variable by light)	Reinforcement of social and non-social interactions and resulting learning	Detachment occurs naturally around 14 weeks in optimal conditions (common in purebred kittens) but may occur at 8 weeks in early adoptions (e.g., found or gifted kittens)
			Alarm system matures; non-positive experiences lead to rapid sensitization
			Autocontrols progress through object play and maternal interventions
			Territorial organization begins with facial marking

(continued)

4

Table 4.1 (continued)

Period	Age	Formal analysis	Functional analysis
Pubertal	6 months–1 year	First heat in females and initial sexual expressions/ markings in males; hormonal changes increase vulnerability across all axes	The cat becomes independent, no longer needing humans or conspecifics. If interspecific socialization occurred, relations with humans are friendly. Sexual pheromonal system matures
			New fears may emerge if the alarm system is fragile, with intense sensitization following negative experiences
			Adolescent cats control scratching, biting, and locomotion
			Territorial organization matures with allomarking, facial marking, scratching, and (for intact animals) urine marking
Adult	1–8/10 years	Stability, with possible disruptions due to vulnerabilities or exposures	The adult cat is autonomous
			Reliable alarm system and reactions only to real dangers
			Acquired and stable self-controls
			Soothing through territorial marking (facial marking, scratching, absence of urine marking except at the territory's periphery)
Senescent	After 8–10 years (varies)	Sensory and cognitive declines; predisposition to mood, emotional, or cognitive disorders	Possible development of dependency
			Alarm system may alter, leading to new fears
			Some individuals may lose autocontrols
			Territorial disorganization can occur, with markings becoming less effective

4.3 Axial Description of Behavioral Development

We have chosen to revisit behavioral development from an axial perspective, rather than a chronological one, as we believe this approach is critical for understanding the genesis of behavioral disorders, which we will detail in the second part of this work.

The axes we have selected logically correspond to the major dysfunctions that we will elaborate upon in the nosography. Each of these axes represents a biologically more complex process than the elementary processes but provides a framework during consultations for explaining pathological pathways in addition to the elementary processes. For this reason, we refer to them as complex processes.

These processes can be represented as shown in ◘ Fig. 4.7, with a notion of severity in deviation from the norm increasing as one moves toward the red zone (often indicative of aggressive behaviors or situations endangering the animal). This highlights the urgency of addressing the behavior in question.

4.3.1 Fear and Alarm System

4.3.1.1 Definition: Fear, Sensorial Homeostasis, and Comfort Zone

Fear and its neurophysiological mechanisms were detailed in ► Chap. 2 (► Sect. 2.3.1.1). Fear is a protective mechanism that emerges in cats during the socialization period. Its purpose is to shield the individual from environmental dangers. The threshold for triggering fear depends on the *sensory homeostasis* the kitten develops during its growth. When this threshold is at a normal level, the individual fears genuinely dangerous elements in its environment but does not fear familiar aspects of its surroundings. Such a cat is *well adapted to its environment.*

As discussed in ► Chap. 1 (► Sect. 1.1.3), sensory homeostasis refers to a set of processes that keep the organism within a range of comfort and balance determined by its behavioral development. This homeostasis establishes the acceptable intensity range of environmental changes. Beyond the upper threshold, change becomes unacceptable and

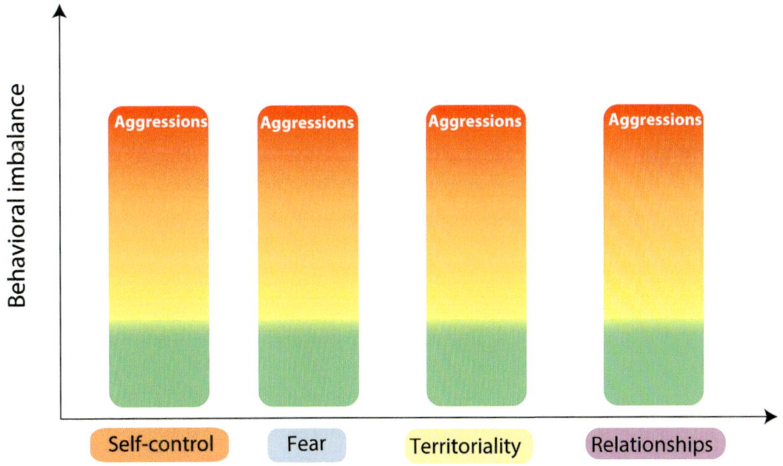

◘ **Fig. 4.7** The four axes to evaluate during each consultation. Every consultation should include an assessment of the following four axes: acquisition of self-control, fears, Territoriality, and relationships

triggers fear. Below the lower threshold, change is imperceptible and elicits no reaction. Sensory homeostasis reflects the cat's ability to adapt calmly to changes in its environment.

Sensory homeostasis is predominantly acquired during the socialization and juvenile periods of a kitten's life, depending on the richness of its sensory experiences (◘ Fig. 4.3).

The concept of sensory homeostasis allows us to envision an optimal level of stimulation corresponding to a *comfort zone* for the cat. The lower threshold is necessary to trigger a behavioral response, whereas exceeding the upper threshold elicits a strong emotional fear response. This intense fear reaction prevents the cat from adapting to its environment.

As the young cat grows, it can adapt if slightly pushed outside its comfort zone (yellow zone). However, it will be unable to adapt if it is taken too far beyond this zone (orange zone). Adaptation becomes increasingly difficult as the cat ages and the further it is pushed outside its comfort zone. A narrow sensory homeostasis amplitude results in a very limited comfort zone, often placing the individual outside this zone. This creates a vicious cycle that promotes sensitization rather than habituation, thereby hindering adaptation (◘ Fig. 4.8).

The more experiences a kitten has, the better it will be able to face varied environments without overreacting with fear or excitement. Notably, a mother's reaction to danger (real or perceived) helps kittens categorize situations to avoid. This explains why fearful

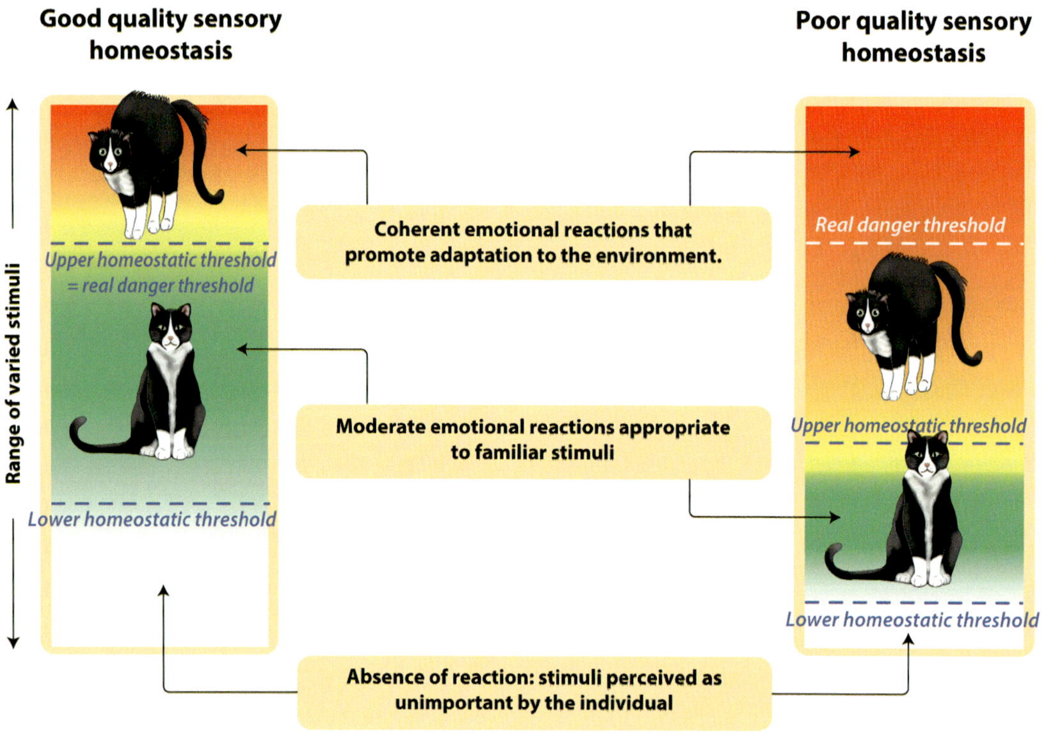

◘ **Fig. 4.8** Sensory homeostasis. (a) High-quality sensory homeostasis with a broad comfort zone. The low homeostatic threshold allows for a significant absence of perception of non-threatening stimuli. The high homeostatic threshold, the trigger for emotional reactions, is elevated and corresponds to actual danger. (b) Low-quality sensory homeostasis with a restricted comfort zone. Fear triggers at lower thresholds. The low homeostatic threshold permits minimal absence of perception, causing too many stimuli to be perceived and triggering reactions. The low and high homeostatic thresholds fail to correspond to the real danger level of stimuli mothers, who perceive danger in too many situations, pass on these fears to their kittens.

The socialization period should provide the kitten with an environment sufficiently similar to its future living conditions, allowing it to build a rich database that enables adaptation in any circumstance. Through generalization and categorization, these elements, along with those closely related, become normal parts of the environment for the kitten. As a result, its brain no longer reacts to these stimuli, allowing it to focus on genuinely relevant information in its surroundings. Stimuli categorized as non-threatening are automatically ignored and filtered out by the brain.

As mentioned earlier, synaptic pruning occurs at the end of the socialization period, eliminating neurons that have not formed connections. While this does not mean that neurogenesis or learning is impossible later, it becomes much more challenging. Contact with humans is particularly crucial before 8 weeks of age, as the lack of such interaction can lead to sensory deprivation syndrome or, at the very least, severe interspecific social phobia (Ahola et al. 2017; Lowe and Bradshaw 2002; Martin 1986).

At the neurophysiological level, critical periods such as the socialization phase and puberty facilitate rapid learning. During the socialization phase, experiences may be less pleasant, whereas in puberty, it becomes increasingly important to remain within the intermediate zone. Exceeding this zone risks *sensitizing* the cat. The socialization period can be described as a truly facilitating phase, while puberty is a period of vulnerability.

The first encounter with a stimulus is particularly significant, as it will shape future reactions and the level of sensitization. During the development period, veterinarians must ensure that encounters do not provoke excessively intense fear reactions (see the box "Desensitization and Sociabilization to Humans").

A young kitten exhibiting frequent, intense fear responses and failing to habituate to stimuli within a few weeks should be addressed promptly, before puberty. Behavioral therapy based on desensitization and counterconditioning can potentially mitigate this issue. If preventive measures are not swiftly effective or if the juvenile dysfunction is left unaddressed, the condition risks worsening, and aggression may emerge during puberty.

Important Point of View

Desensitization and Sociabilization to Humans

The idea that a fearful animal should be habituated or desensitized by repeated exposure to the richest and most varied environments possible must be nuanced. A common recommendation to owners of phobic cats is to be gentle and patient. However, as we have seen, a cat's sociability toward humans becomes nearly impossible to develop after 14 weeks of age. For a fearful kitten, stimuli must be introduced gradually, but medical intervention should be immediate and assertive if the cat is already older than 10–12 weeks, as recovery becomes increasingly difficult with time.

Owners should be trained to observe fear signals and postures to keep the kitten in its mild discomfort zone (yellow zone in ◘ Fig. 4.8). If the kitten is placed in the orange zone, it will instead become sensitized and exhibit increasingly intense fear reactions. When the development of the alarm system (amygdala circuit) is poor, the comfort zone is too narrow, and the animal exhibits inappropriate reactions (phobic or anxious pathological states). The role of medication is precisely to expand this comfort zone, enabling the animal to analyze situations and habituate to stimuli.

In other words, if habituation attempts do not show results after a few days of training, owners should be referred promptly to a behavioral consultation to maximize the chances of recovery for the kitten, as its cerebral plasticity diminishes over time.

4

4.3.1.2 Evolution of the Alarm System

The juvenile and pubertal periods establish a "fear repertoire," making learning increasingly challenging. Fear reactions observed in kittens indicate a behavioral vulnerability that can predispose them to future disorders. Typically, these are the cats that are easily traumatized by seemingly minor events, such as a neutering procedure. Even when staff have been entirely considerate and pain has been managed appropriately, a cat with high fear-axis vulnerability may develop an apparently inexplicable phobia of the veterinarian.

When the fear system, particularly the amygdala, activates too quickly, a trivial event is perceived as a major trauma. What ultimately matters is not the event itself but how the cat perceives it. More so than in dogs, this system is often fragile in cats, which are prey animals and rely on their territory for reassurance rather than a group dynamic. This makes them particularly vulnerable to sensitization in the context of care.

When the fear axis develops normally, cats reach adulthood without exhibiting fear reactions in the vast majority of situations they encounter. Of course, an alert reaction to an unfamiliar situation is entirely appropriate, but a healthy cat will adapt within minutes.

In adulthood, behavioral patterns become stable, and unless a trauma occurs, a cat has no reason to spontaneously develop phobic or anxious states. However, environmental changes—such as moving, new neighboring cats, or the arrival of a baby—are harder for cats to cope with than for dogs, and these changes can push a cat into a pathological state.

In aging cats, the rigidity of behavioral responses can lead to phobic reactions. In our experience, these cases almost always involve cats with a history of significant vulnerability. Sometimes, lifelong stable phobias may generalize or evolve into anxious or depressive states with age. Such developments indicate pathological aging.

4.3.2 Acquisition of Self-Controls

4.3.2.1 Definition: Self-Controls, Motivation, and Inhibition

Self-controls encompass the set of processes involved in the ability to regulate one's behaviors and emotions. It results from the balance between activation and inhibition processes.

Motivation is a system that drives engagement in action, closely linked to the pleasure and reward circuits (Robbins and Everitt 1996) (see ▶ Chap. 2, ▶ Sect. 2.3.3.3). Engagement in action also depends on sensory homeostasis, as discussed in the previous section. The activation threshold applies to not only fear but also other reactions. A stimulus of insufficient intensity fails to trigger any behavior. It is not perceived by the attentional system, and the motivational system does not activate. Conversely, an excessively intense stimulus relative to the upper level of homeostasis results in a reaction of very high intensity that bypasses control by the prefrontal cortex. This short-circuited pathway skips the ventromedial prefrontal cortex (vmPFC), leading to an automatic reaction in which the individual loses the ability to analyze the situation. The individual becomes engaged in the action and cannot easily stop it.

Behavioral motivation, which triggers an appropriate response, operates within the limits set by early development. Positive experiences help shape the direction of actions through rewards, which enhance the intensity and guide the focus of motivation. This is exemplified when a cat catches a mouse. Even without prior training from its mother, the reward sensation motivates the cat to repeat the behavior (Biben 1979).

Behavioral inhibition is an essential process for regulating behaviors and emotions. It actively enables the stopping phase that concludes any behavioral sequence; without this mechanism, behavior would continue in a loop. Indeed, every behavioral output is counterbalanced by behavioral inhibition. The brain maintains a constant activation/inhibition ratio within its circuits (Xue et al. 2014).

Self-control results from this activation/inhibition balance and reflects an individual's ability to regulate themselves spontaneously when left to their own devices. Self-control is acquired through interactions with adults, particularly the mother, who corrects uncontrolled biting or scratching, overly rough play, or excessive excitement. This process involves learning to regulate emotions, motor control, and impulsivity.

Sensory homeostasis also plays a crucial role in acquiring self-control. As illustrated in ◘ Fig. 4.9, when sensory homeostasis is of poor quality, the cat tends to exhibit overly intense reactions, indicative of hyperreactivity. In addition, the cat may respond to a stimuli that it should ignore, a sign of hypersensitivity.

From the socialization period onward, inhibition is learned through maternal behaviors aimed at immobilizing the kitten until it calms down. The mother restrains the kitten with her forelegs and may "knead" its abdomen with her hind legs. This is always done without causing injury and without extending her claws. The mother stops as soon as the kitten relaxes. If the kitten immediately gets back up, it is restrained again until it manages to stay calm for several seconds. This method of external control allows the kitten to experience a state of calm and the associated soothing effect. After a few attempts, the kitten learns to self-regulate, with decreasing intervention from its mother. It then develops internal control, which it can spontaneously reproduce in similar contexts without external input.

These interactions with the mother teach kittens to gradually inhibit biting, meaning they learn to control the strength of their jaws (see the box "Controlled Biting and Scratching"); control scratching, avoiding injury during interactions; and regulate locomotion, ensuring their movements do not harm others during interactions. Play serves as a primary means of learning during this period, provided it is supervised by an adult, most often the mother, but also other adults in the group.

Important Point of View

Controlled Biting and Scratching

As kittens grow, they gradually develop increasing inhibitory capacities until the end of puberty, facilitated by the maturation of their ventromedial prefrontal cortex (vmPFC). This enables them to control their movements, avoid abrupt gestures, and manage their emotions, allow them to tolerate frustration, and restrain excessive joy and excitement.

For biting control, kittens first learn to release their grip in response to the high-pitched cries of their siblings and the active intervention of their mother, who reacts to these cries. This ability should be acquired by around 8 weeks, even though, at this stage, the kitten may resume nibbling shortly after a brief pause. The critical point is that the kitten stops as soon as a disruptive stimulus is presented. This control is further refined during the juvenile period, by the end of which bites should be completely painless, even in situations of high excitement.

The same applies to claw control. To achieve this, owners should progressively interrupt biting and clawing earlier, even as their intensity decreases. The development of controlled biting and scratching is highly dependent on the mother. The longer she stays with her kittens, the better their self-control abilities. Kittens separated from their mother before 12–14 weeks show significantly reduced self-control capacities (Bateson and Martin 2013).

Thus, kittens adopted well before weaning are particularly at risk of developing poor self-control due to the lack of competence of humans or even adult cats other than the mother.

Weaning, which occurs around 5–8 weeks of age, marks the first step in detachment and independence. However, maternal presence remains essential for regulation, at least until 8 weeks, which is the legal adoption age in France, but ideally until 12 or even 14 weeks in cats (Lowe and Bradshaw 2002).

4

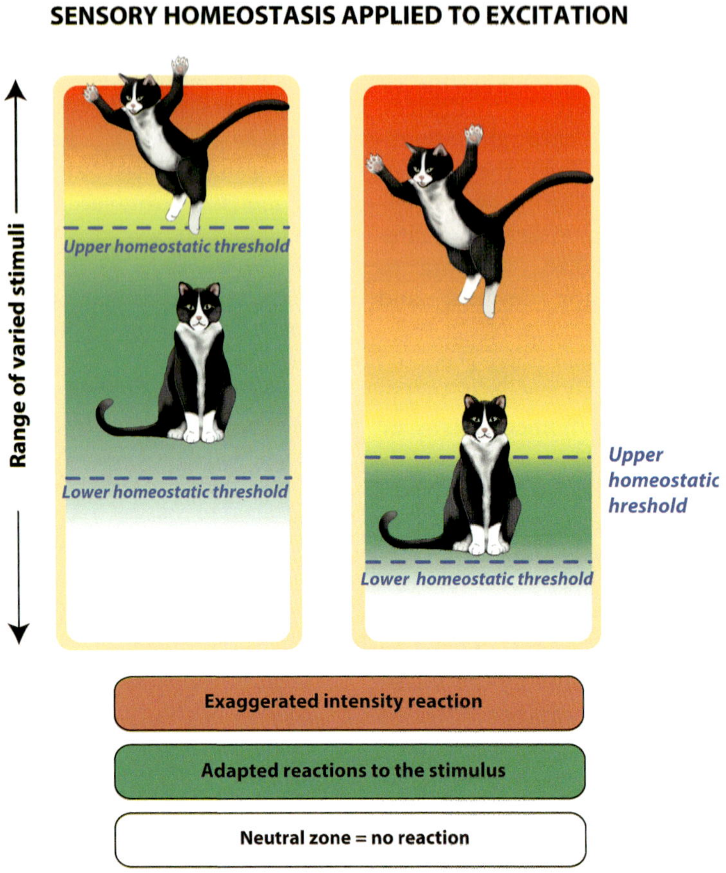

Fig. 4.9 Sensory homeostasis applied to self-control. (a) Extended sensory homeostasis allows a wide range of stimuli to elicit no reaction. Only stimuli of very high intensity trigger exaggerated responses. (b) Restricted sensory homeostasis: almost all stimuli, even those of very low intensity, provoke reactions

The separation age of 14 weeks observed in free-ranging cats is particularly significant when comparing purebred cats adopted much later (3–4 months) to family or farm kittens adopted earlier, around 8–10 weeks. In theory, later adoption could predispose kittens to poorer adaptation to their future living conditions, as they are well beyond the socialization period. However, in practice, these purebred kittens do not appear more fearful or poorly controlled. This may be attributed to the greater care breeders take in fostering optimal development conditions, but it seems more likely that the later adoption age (12–14 weeks) supports better development of self-control (Ahola et al. 2017).

Regarding fear, later adoption may also help prevent excessive sensitization to novelty at the start of the juvenile period, a phase during which sensitization to mildly unpleasant experiences occurs rapidly. In addition, breeder-raised kittens are generally well socialized to humans and often experience a lifestyle similar to their future family environment, making "late" adoption less problematic for the development of their alarm system (see ▶ Sect. 4.3.1 in this chapter).

Regardless of the kitten's adoption age, biting and scratching behaviors should diminish in response to the first signs of distress from a play partner. Redirecting these behaviors toward toys and using negative punishment by stopping the interaction can typically achieve good control by 4 months, the middle of the juvenile period. By puberty, around 6 months of age, the cat should no longer cause any harm during excitement, neither with its mouth nor its claws.

4.3.2.2 The Evolution of Self-Control Acquisition

At the end of the socialization period, kittens may still be restless, playful, and prone to nibbling, but they can stop if interrupted with a firm "no." They sleep extensively, well over 12 h a day, and their playtime is significant, although it begins to decrease at around 12 weeks and continues to decline until about 5 months, after which play remains present but much shorter. By the end of the juvenile period, around 6 months of age, the cat should be fully capable of controlling its claws and bites to avoid causing injury during interactions with its owners, even when highly excited.

At sunrise and sunset, cats often exhibit what is colloquially referred to as "crazy episodes," characterized by bursts of running without any apparent goal or trigger, regardless of their living environment. However, they should be able to stop these episodes after a few minutes and calm themselves spontaneously. A kitten that reacts to every movement in the house, still causes injuries during excitement, and sleeps very little at 6 months old requires special attention, as this behavior is not normal.

As previously emphasized, motor and emotional controls are heavily influenced by the presence of the mother, even into later stages of development. Kittens adopted well before weaning are particularly at risk of having deficits in self-control as adults, even if raised with other adult cats or if owners attempt to compensate for the absence of the mother.

During the pubertal phase, behavioral problems can worsen in cats with incomplete self-control acquisition. As with the fear axis, this is the age when aggression may emerge, often driven by frustration or pain. In cats, which are inherently predatory, frustration can lead to aggression initiated through sequences related to either play or predatory behavior. These sequences are often too quickly or intensely triggered and may be directed at their owners. Before aggressive episodes caused by irritation, cats often display signs of heightened excitement, such as a lashing tail, a low crouching posture, and waiting for a triggering movement.

Some cats may also begin attacking during interactions, often labeled as "petting-induced aggression." These cats struggle with prolonged contact, particularly in sensitive areas like the abdomen. The cat's increasing size can exacerbate these behaviors, making them more painful and prompting owners to seek help or consultation as they grow concerned about their inability to manage the behavior despite their patience.

This situation often leads to a deterioration in the owner-cat relationship. Owners may find themselves being attacked and respond defensively by pushing the cat away, yelling, or moving, which in turn triggers more aggression. This symmetrical escalation can damage the relationship further: the cat becomes unable to stop jumping, biting, or scratching, while the owners, increasingly defensive and frustrated, start to fear their cat's actions, perpetuating the cycle.

If an adult cat has acquired strong self-control, these abilities remain stable until the senescent phase. Such a cat will almost always choose flight over defensive action and will generally be easy to handle in most living contexts. Conversely, incomplete self-control affects all aspects of the cat's behavior, leading it to react too quickly or too strongly in specific situations. For example, these cats may rapidly resort to aggression in scenarios of frustration, such as restraint during medical care or confinement in a small space.

Although the senescent phase is not typically associated with the self-control axis, pathological aging in some cats can lead to deregulation, resulting in impulsive behaviors and a loss of the stopping signal in certain actions. These disruptions can be challenging to distinguish from cognitive disorders, which may manifest as a loss of learned behaviors, such as inappropriate elimination, excessive vocalization, or episodes of motor agitation (Bellows et al. 2016).

4.3.3 Relationships and Attachment

Maternal attachment is a vital process present in all mammals that influences the quality of an individual's future relationships. The term *attachment* refers to the active behaviors of a kitten aimed at seeking proximity to a preferred and protective figure, typically its mother. The attachment figure provides comfort and pleasure, acting as a refuge for the kitten to return to when it feels fear. Attachment thus serves as a security base that facilitates exploration. Maternal reactions serve as a model for the kitten, which mirrors its behavior on that of its mother.

In French, the term *attachment* also encompasses the caregiving behaviors of the mother, such as seeking and attending to her kittens. While the kitten's attachment to its mother and the mother's attachment to her kittens (*caregiving*) share hormonal bases, such as oxytocin, vasopressin, and cortisol, they are distinct processes (Mota-Rojas et al. 2023; Rybicka et al. 2021).

Beyond providing safety and courage for exploration, the mother's presence is a key source of learning for the kittens' adult life, especially in terms of hunting. Through play and interaction with the mother, kittens learn foundational postures and behaviors that they will later use in predation.

4.3.3.1 Types of Attachment

The theory of attachment—developed by H. F. Harlow (Harlow and Zimmermann 1959), J. Bowlby (Bowlby 1958), and later expanded by M. Ainsworth (Ainsworth and Bell 1970; Ainsworth et al. 1978)—rests on the principle that normal social and emotional development requires a specific relationship with at least one other individual, most often the mother, who provides consistent and continuous care. This attachment figure plays a dual role for the offspring: acting as a refuge to provide reassurance in fearful situations and as a secure base for exploring the world.

The Strange Situation Test (Ainsworth and Bell 1970; Ainsworth et al. 1978) identified different types of attachment by observing the reactions of young children in various scenarios: in the presence of their mother, during separation, in the presence of a stranger, and upon reunion. Mary Ainsworth identified several types of attachment: secure attachment, insecure-avoidant attachment, and insecure-ambivalent attachment. Later, Mary Main and Judith Solomon added disorganized attachment to this classification (Main and Solomon 1990).

While four types of attachment, similar to those observed in children, have been demonstrated in human-dog relationships (Solomon et al. 2019), no comparable studies have confirmed such attachment types in cats.

Understanding the different types of attachment is crucial beyond mere classification. Children (and likely kittens) who experience a particular type of attachment will have adult relationships significantly influenced by the relationship developed with their mother.

Important Point of View

The Strange Situation Test in Cats

While attachment between kittens and their mothers is natural and easily observable, can we genuinely discuss attachment in an interspecific human-cat relationship?

Several studies have addressed this question by adapting the Strange Situation Test to evaluate cats' attachment to their owners. Findings indicate that cats exhibit attachment behaviors toward their owners (Edwards et al. 2007; Vitale et al. 2019). For instance, cats vocalize more when left with a stranger than with their owner. However, other studies emphasize that cats show few signs of distress when left alone, despite these indications of attachment (Potter and Mills 2015).

These findings align with the fact that cats are not a social species and, from puberty onward, tend to find comfort in their environment rather than relying on humans, making them more independent than dogs, for example.

The cited research was conducted under experimental conditions, not in the cats' home environment. This raises questions about the calming role of the environment, which is particularly significant for a species like the cat. Conducting the same test with cats in their usual environment and in an unfamiliar one could provide valuable insights into the role of environmental factors in their attachment behaviors.

Attachment is a complex process that is observed and reinforced in both distressing situations and the comfort and pleasure provided by proximity to the attachment figure. In social species like dogs, maternal attachment is naturally transferred to the social group during puberty. However, in cats, natural detachment occurs around 4 months of age, coinciding with the development of territorial marking pheromones. At this stage, a kitten that can no longer seek comfort by cuddling with its mother organizes its first isolation zone, which it marks with these pheromones.

From then on, the young cat finds comfort in its environment, which it maps out with facial markings and protects from intrusions with urine marking and scratching. Accepted individuals bear traces of allomarking and, in turn, become sources of comfort.

When humans are considered part of the cat's environment, the concept of territory can be expanded to include owners as an additional dimension. Like other accepted and marked living beings, they play a comforting role for the cat through the mediation of allomarking. Within this framework, the cat is not comforted by its group as a whole but by the individual members of the group, each as a familiar component of its environment.

This hypothesis also helps explain why the vast majority of cats show little distress when their owners are absent, unlike the visible signs of separation distress seen in dogs with autonomopathies. This does not mean that the absence of owners cannot create an imbalance, but the biological basis and its expression are fundamentally different.

4

Important Point of View

Neurophysiological Foundations of Attachment

Attachment is fundamentally sensory-based, with smell, taste, and touch playing central roles in its activation during the first weeks of life. *Cat appeasing pheromones*, secreted by the skin along the mammary sulcus of the mother, contribute to the establishment of attachment. For the mother, *oxytocin* secretion during parturition facilitates rapid attachment to her kittens from birth. *Prolactin* release during nursing further strengthens this bond. For kittens, due to the slow maturation of their sensory organs, attachment develops more gradually, taking several days after their eyes open.

On a neuroanatomical level, attachment is supported by the *inactivation of negative emotions* and the *activation of the reward circuit* in the presence of the attachment figure (Bartels and Zeki 2004). Attachment is a *complex process* that reduces fear and induces pleasure. The regions involved are rich in *oxytocin* and *vasopressin*. Oxytocin acts on the amygdala (Viviani et al. 2011), inactivating cortisol receptors and thereby suppressing the fear response (Moriceau and Sullivan 2004).

The reward circuit is also activated, leading to the release of *opioids* that strengthen the attachment bond, providing pleasure, comfort, and dependence (Moles 2004). Adding to this complex system are neuromodulator like norepinephrine, which influences fear and the memory of specific details about the attachment figure. *Dopamine* is also released, playing a role in the pleasure and reward circuits and regulating prolactin secretion.

4.3.3.2 Detachment and Evolution of Relationships

The maternal attachment period concludes with the process of detachment, which is complete by 3–4 months in feral cats (Bradshaw et al. 2012). This process begins with the mother's gradual refusal of contact during the socialization period. The initial refusals to allow nursing mark the start of detachment. While weaning is completed before the end of the socialization period, detachment progresses gradually during the juvenile period. Interactions among cats evolve, with olfactory contact (nose-to-nose greetings) serving as the initial step in their interactions, teaching them a form of "greeting" behavior, even if repeated several times daily.

In cases of late adoption (beyond 3 months), the kitten is already naturally detached from its mother. This is significantly different from dogs, where adoption occurs before natural detachment (during the pubertal period). Early detachment can impact the acquisition of self-control but has minimal practical implications: owners do not need to actively manage detachment. For kittens adopted well before weaning, however, this responsibility falls to the owners and should occur around 3–4 months by refusing constant interactions with the kitten. This often happens naturally, as owners typically decline behaviors like kneading or licking when seeking quiet moments.

During the pubertal period, cats become increasingly independent, structuring their territories, and their interactions shift. With other cats, they adopt more "territorial" behaviors and exhibit distancing reactions—hissing, avoidance, or even aggression—toward unknown conspecifics. These behaviors are normal. For this reason, introducing a new pubescent cat often presents more challenges than introducing a kitten into the household (Crowell-Davis et al. 2004). Some cats, particularly neutered males, show greater tolerance toward other males (Cafazzo et al. 2019; Barry and Crowell-Davis 1999). However, occasional hissing or non-injurious fights among cohabiting cats are normal due to the species' territorial nature.

Regarding interspecific relationships, particularly with humans, a cat's friendliness largely depends on proper socialization before the critical age of 8 weeks (Seitz 1959). Despite being an animal with solitary tendencies, cats often enjoy the presence of humans, though they do not typically seek physical contact as consistently as dogs do. This difference can lead to frustration among owners who expect their cats to be more affectionate, even when the cat is exhibiting normal behavior.

Well-socialized cats enjoy being around humans, and while some may crave direct contact, others prefer being "next to" rather than engaging in direct physical interaction. However, balanced cats do not avoid humans, even unfamiliar ones. On the contrary, they are curious and interested in exploring the scents of new individuals in their environment. Interestingly, for a well-socialized adult cat, interspecific relationships with unfamiliar humans are often more easily friendly than intraspecific relationships with unknown cats.

In cases where humans feel frustrated by a perceived lack of affection, forced interactions can quickly deteriorate the relationship and lead to interspecific relational disorders (see ▶ Chap. 11, "Disorders of the Relational Axis").

As cats age, emotional and cognitive disorders may emerge, potentially causing relational issues and a loss of autonomy. The extent of these effects depends on the cat's history and individual vulnerabilities.

4.3.4 Territoriality

Cats are a territorial species that mark their environment to create a sense of security (De Boer 1977). This behavior is quite different from the social dynamics of humans, and our limited olfactory abilities, especially our inability to perceive cat pheromones, make it difficult to fully understand the territorial tendencies of this species.

From birth, the sense of smell plays a significant role in a cat's behavior. Newborn kittens orient themselves to nurse using this well-developed sense (Raihani et al. 2009). They quickly learn to recognize their mother and locate their nest using their sense of smell (Freeman and Rosenblatt 1978).

4.3.4.1 Markings and Territory Organization

Olfaction, particularly the perception of pheromones, plays a central role for cats. Facial markings, which deposit the F3 fraction of their pheromonal complex (◘ Fig. 4.4) on objects, begin at the end of the juvenile period and intensify and organize during puberty. Similarly, allomarking, involving the F4 fraction, and scratching, which includes pheromones secreted by the paw pads, follow the same developmental timeline. As maternal detachment occurs during the juvenile period, kittens compensate by structuring their environment to restore a sense of calm (Feldman 1994).

Studies observing feline activity have identified specific paths that cats frequently traverse, likely guided by previously laid pheromonal trails (Horn et al. 2011). Cats exhibit a predominantly crepuscular activity pattern, optimized for hunting (Delmar Cerutti et al. 2019).

Domestic cats with outdoor access spend approximately 80% of their time within a 50-m radius of their home (Bischof et al. 2022). These zones, though relatively small, are extensively marked to establish boundaries and provide reassurance. Different types of markings become noticeable by the end of the juvenile period, and the organization of distinct activity zones matures fully during puberty. This maturation coincides with the development of aggression toward unfamiliar conspecifics, highlighting the link between feline territoriality and non-sociality.

Some cats focus more on marking objects within their environment. These cats are particularly sensitive to cleaning by their owners and to the introduction of new objects. They often exhibit heightened interest in exploring shopping bags or their owner's clothing upon their return. These cats may also be more affected by changes in their living situation, such as a move or a temporary relocation during holidays.

Other cats are more oriented toward living elements in their environment, such as family members. These cats structure their environment by establishing numerous rituals with their owners and are more likely to position themselves close to them.

In both cases, a cat's environment, composed of living or inanimate elements, functions as a network marked by pheromones. This marking provides daily comfort and reassurance, underscoring the importance of a stable and predictable environment for feline well-being.

By observing an adult cat, it is possible to determine whether they are more sensitive to changes in objects or people. This insight can help guide decisions about whether to take the cat on vacation or leave them in their usual environment.

The cat's territory can be broken down into distinct activity zones, as illustrated in ◘ Fig. 4.10. These zones include the resting or isolation zone, observation zone, play zone, hunting zone, feeding zone, elimination zone, and interaction zone.

The paths traveled by a cat between each zone are almost always consistent, marking reassuring routes with pheromones produced by the exocrine glands located in the corners of the lips and paw pads.

It is important to note that facial pheromones and those involved in allomarking are primarily intended for the cat itself.

In contrast, pheromones associated with urinary marking are directed at cats unfamiliar with the territory. This type of marking may include a sexual message, in which case the urinary spot is often reinforced by facial marking with sexual significance.

Finally, pheromones deposited through scratching serve a mixed purpose, as the cat uses them to protect its isolation zone.

4.3.4.2 Evolution of the Territoriality

The territorial organization established during the pubertal period is not expected to change in adulthood. However, in practice, events such as moving to a new home, the arrival of new individuals in the household, or the presence of new cats in the neighborhood can disrupt the cat's initial territorial structure.

When the Territoriality axis is balanced, an adult cat demonstrates flexibility and can adapt to such changes relatively easily, even if it initially experiences a reactive state. Conversely, a cat with emotional vulnerabilities and poor territorial soothing may become highly disturbed, even by minor changes. This distress is often expressed through alterations in marking behavior, typically by increasing pheromone deposits. Furthermore, if the cat is unsettled, the composition of its pheromones may not be normal, and these deposits might exacerbate adversity rather than alleviate it.

In response, territorial markings may shift toward urinary marking, typically performed against a vertical surface, with the tail raised and twitching slightly. This urinary marking reflects the cat's discomfort within its territory but does not provide relief. Owners often react strongly to this behavior, as urine marking outside litter boxes quickly becomes intolerable. This issue is one of the most common reasons for consultations in feline behavioral medicine, often central to Territoriality axis disorders.

Finally, during aging, territorial disorganization can occur for various reasons: pain, sensory decline, or organic diseases may hinder the cat's ability to maintain proper marking behavior, leading to territorial affections.

Important Point of View

The Zone of Mild Fear

It is common to observe the existence of an activity zone near the cat's territory that the cat visits occasionally but which is not fully integrated into its territory. In this zone, the cat remains on alert and is prepared to quickly retreat to its protected territory.

For an indoor cat, this might typically be the stairwell. The cat enjoys going there, possibly leaving urinary spots, but at the slightest noise, it rushes back to the safety of its apartment.

This zone adds richness to the cat's territory and is often valuable to describe during therapeutic interventions.

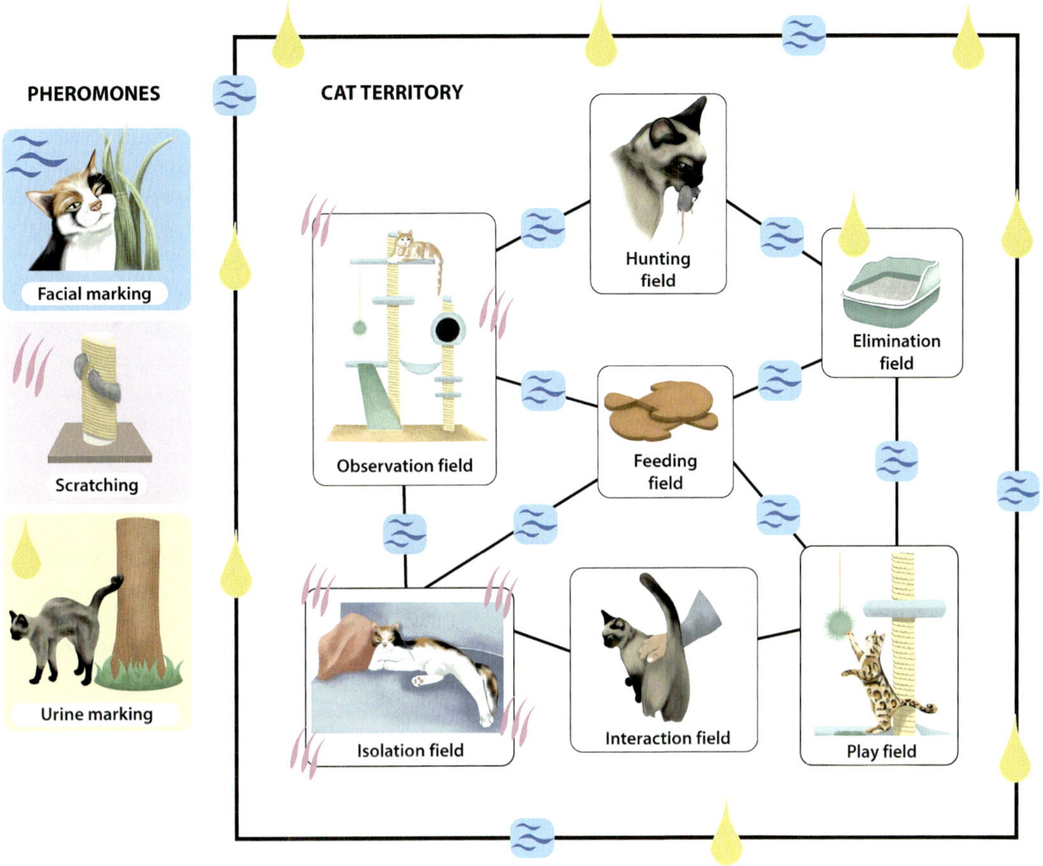

Fig. 4.10 Organization of a cat's territory through various activity zones and markings, including facial, urinary, and scratching markings

4

Summary of ▶ Sect. 4.3

The behavioral development of cats can also be analyzed along specific behavioral axes.

Fear begins to develop around 5–6 weeks of age. During the socialization period, the kitten learns to accept its environment without fear, creating a database of stimuli, enriched by exposure to diverse surroundings. Early interactions with other species are critical for fostering interspecies sociability in adulthood. During juvenile and pubertal stages, positive exposure to novel stimuli promotes habituation and prevents sensitization. In adulthood, behavior stabilizes, with trauma being the primary cause of phobias in well-developed individuals.

Self-controls develop progressively from birth to adulthood as the ventromedial prefrontal cortex matures. Maternal influence is critical for acquiring self-control up to natural weaning, which occurs at 3–4 months. Kittens adopted before weaning are at greater risk of deficits in self-control. By puberty, self-control matures, resulting in fully regulated scratching, biting, and locomotion.

Attachment begins with eye opening, forming a unique bond between the kitten and its mother, who acts as a secure base, allowing the kitten to explore external stimuli. Detachment follows in the early juvenile period, and by 4 months, kittens are largely autonomous. Interspecies relationships, facilitated by early socialization, become significant during the juvenile and pubertal periods. High-quality socialization ensures easier interactions with family members and strangers. Conversely, *intra-species relationships* mature differently, as adolescence reduces tolerance for unfamiliar conspecifics. This intolerance underscores the need for gradual introductions when integrating adult cats.

Territorial organization begins in the late juvenile period, marked by facial marking of objects, allomarking on individuals, scratching near resting areas, and urinary marking at the territory's periphery. These markings provide the cat with a sense of predictability and comfort in its living space. Territorial changes can lead to unwanted urinary or scratching marks, often causing frustration among owners and straining their relationship with the cat.

An additional dimension is the *communication axis*, which influences all other behavioral axes. Effective communication between owners and their cats reduces behavioral vulnerabilities, while poor communication or coercive methods exacerbate fragilities and increase the likelihood of pathological behaviors in the affected axis (◘ Fig. 4.11).

Summary of This Chapter

In this chapter, we have described how behavioral development is divided into critical periods during which specific types of learning are facilitated. These periods can be viewed chronologically, each characterized by particular neural reorganizations that promote the selection of behaviors best suited to the environment in which the animal is raised. Synapses that are most stimulated are strengthened, while neurons that fail to establish connections are eliminated through synaptic pruning. This process allows the individual to specialize and become optimally adapted to their developmental environment.

Behavioral development can also be analyzed through the evolution of four major behavioral axes, which will serve as the framework for describing behavioral disorders in the second part of this book. These axes include fear and the alarm system, the acquisition of self-control, attachment and relationships, and territorial organization. Each of these axes follows a maturation process with specific milestones, which must be understood to detect deviations from the norm as early as possible. This early detection enables the implementation of appropriate therapeutic measures to correct the deviations. When vulnerabilities are too severe or the environment is inadequate, the cat may develop pathological conditions, potentially leading to debilitating psychiatric disorders.

Lastly, the communication axis of the owner interacts with these four behavioral axes, either improving or exacerbating initial deviations.

A comprehensive understanding of the normal course of behavioral development across these axes and over time enables practitioners to identify at-risk individuals during their growth period and throughout their lives (◘ Fig. 4.11).

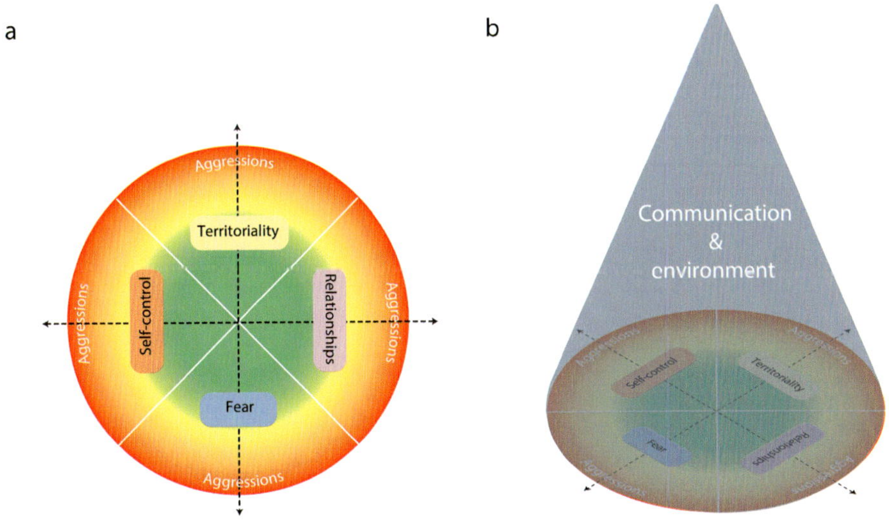

◘ **Fig. 4.11** The four behavioral dimensions and the communication and environment axis. **a** The four axes: fear, self-control, attachment/relationships, and territory/marking feature a severity zone at their extremes that signals the need for prompt intervention. This red zone is often associated with aggression directed toward others. **b** A fifth dimension, the quality of communication and environment, modulates the trajectories of the first four axes

4

References

Ahola MK, Vapalahti K, Lohi H (2017) Early weaning increases aggression and stereotypic behaviour in cats. Sci Rep 7(1):10412. https://doi.org/10.1038/s41598-017-11173-5

Ainsworth MD, Bell SM (1970) Attachment, exploration, and separation: illustrated by the behavior of one-year-olds in a strange situation. Child Dev 41(1):49–67. https://doi.org/10.2307/1127388

Ainsworth MDS, Blehar MC, Waters E, Wall S (1978) In: Bowlby J (ed) Patterns of attachment: a psychological study of the strange situation. Erlbaum, Hillsdale NJ

Baerends-Van Roon JM, Baerends GP (1979) The morphogenesis of the behaviour of the domestic cat, with a special emphasis on the development of prey-catching. North-Holland Pub. Co, Amsterdam

Barnett SA (1968) The "instinct to teach". Nature 220(5169):747–749. https://doi.org/10.1038/220747a0

Barrett P, Bateson P (1978) The development of play in cats. Behaviour 66(1–2):106–120. https://doi.org/10.1163/156853978X00422

Barry KJ, Crowell-Davis SL (1999) Gender differences in the social behavior of the neutered indoor-only domestic cat. Appl Anim Behav Sci 64(3):193–211. https://doi.org/10.1016/S0168-1591(99)00030-1

Bartels A, Zeki S (2004) The neural correlates of maternal and romantic love. Neuroimage 21(3):1155–1166. https://doi.org/10.1016/j.neuroimage.2003.11.003

Bateson P (1979) How do sensitive periods arise and what are they for? Anim Behav 27:470–486. https://doi.org/10.1016/0003-3472(79)90184-2

Bateson P (1994) The dynamics of parent-offspring relationships in mammals. Trends Ecol Evol 9(10):399–403. https://doi.org/10.1016/0169-5347(94)90066-3

Bateson P, Martin P (2013) Play, playfulness, creativity and innovation, 1st edn. Cambridge University Press, New York. https://doi.org/10.1017/CBO9781139057691

Bateson P, Young M (1979) The influence of male kittens on the object play of their female siblings. Behav Neural Biol 27(3):374–378. https://doi.org/10.1016/S0163-1047(79)92468-3

Bateson P, Young M (1981) Separation from the mother and the development of play in cats. Anim Behav 29(1):173–180. https://doi.org/10.1016/S0003-3472(81)80163-7

Bellows J, Center S, Daristotle L, Estrada AH, Flickinger EA, Horwitz DF, Lascelles BDX, Lepine A, Perea S, Scherk M (2016) Evaluating aging in cats: how to determine what is healthy and what is disease. J Feline Med Surg 18(7):551–570

Biben M (1979) Predation and predatory play behaviour of domestic cats. Anim Behav 27:81–94. https://doi.org/10.1016/0003-3472(79)90129-5

Bischof R, Hansen NR, Nyheim ØS, Kisen A, Prestmoen L, Haugaasen T (2022) Mapping the "catscape" formed by a population of pet cats with outdoor access. Sci Rep 12(1):5964. https://doi.org/10.1038/s41598-022-09694-9

Blakemore C, Garey LJ, Vital-Durand F (1978) The physiological effects of monocular deprivation and their reversal in the monkey's visual cortex. J Physiol 283(1):223–262. https://doi.org/10.1113/jphysiol.1978.sp012498

Blakemore S-J (2008) The social brain in adolescence. Nat Rev Neurosci 9(4):267–277. https://doi.org/10.1038/nrn2353

Bowlby J (1958) The nature of the child's tie to his mother. Int J Psychoanal 33(6):282–293. https://doi.org/10.1016/j.appdev.2012.08.002

Braastad BO (1998) Effects of prenatal stress on behaviour of offspring of laboratory and farmed mammals. Appl Anim Behav Sci 61(2):159–180. https://doi.org/10.1016/S0168-1591(98)00188-9

Braastad BO, Heggelund P (1984) Eye-opening in kittens: effects of light and some biological factors. Dev Psychobiol 17(6):675–681. https://doi.org/10.1002/dev.420170609

Bradshaw JW (2016) Sociality in cats: a comparative review. J Vet Behav 11:113–124

Bradshaw JWS (2006) The evolutionary basis for the feeding behavior of domestic dogs (*Canis familiaris*) and cats (*Felis catus*). J Nutr 136(7):1927S–1931S. https://doi.org/10.1093/jn/136.7.1927S

Cafazzo S, Bonanni R, Natoli E (2019) Neutering effects on social behaviour of urban unowned free-roaming domestic cats. Animals 9(12):1105. https://doi.org/10.3390/ani9121105

Caro TM (1980) Effects of the mother, object play, and adult experience on predation in cats. Behav Neural Biol 29(1):29–51. https://doi.org/10.1016/S0163-1047(80)92456-5

Caro TM (1981) Sex differences in the termination of social play in cats. Anim Behav 29(1):271–279. https://doi.org/10.1016/S0003-3472(81)80174-1

Casey RA, Heath S, Zulch H (2022) Companion animal behaviour problems: prevention and management of behaviour problems in veterinary practice. CAB International, Wallingford

Changeux J-P, Mikoshiba K (1978) Genetic and 'epigenetic' factors regulating synapse formation in vertebrate cerebellum and neuromuscular junction. In: Progress in brain research, vol 48. Elsevier, pp 43–66. https://doi.org/10.1016/S0079-6123(08)61015-8

Crowell-Davis SL, Curtis TM, Knowles RJ (2004) Social organization in the cat : a modern understanding. J Feline Med Surg 6(1):19–28. https://doi.org/10.1016/j.jfms.2003.09.013

Curtis TM, Knowles RJ, Crowell-Davis SL (2003) Influence of familiarity and relatedness on proximity and allogrooming in domestic cats (*Felis catus*). Am J Vet Res 64(9):1151–1154. https://doi.org/10.2460/ajvr.2003.64.1151

De Boer JN (1977) The age of olfactory cues functioning in chemocommunication among male domestic cats. Behav Process 2(3):209–225. https://doi.org/10.1016/0376-6357(77)90026-2

Delmar Cerutti R, Rizzo M, Alberghina D, Cristina Scaglione M, Piccione G (2019) Locomotor activity patterns of domestic cat (*Felis silvestris catus*) modulated by different light/dark cycles. Biol Rhythm Res 50(6):838–844. https://doi.org/10.1080/09291016.2018.1499372

Dichio V, De Vico Fallani F (2024) Exploration-exploitation paradigm for networked biological systems. Phys Rev Lett 132(9):098402. https://doi.org/10.1103/PhysRevLett.132.098402

Edwards C, Heiblum M, Tejeda A, Galindo F (2007) Experimental evaluation of attachment behaviors in owned cats. J Vet Behav 2(4):119–125. https://doi.org/10.1016/j.jveb.2007.06.004

Feldman HN (1993) Maternal care and differences in the use of nests in the domestic cat. Anim Behav 45(1):13–23. https://doi.org/10.1006/anbe.1993.1002

Feldman HN (1994) Methods of scent marking in the domestic cat. Can J Zool 72(6):1093–1099. https://doi.org/10.1139/z94-147

Fiock KL, Smith JD, Crary JF, Hefti MM (2020) B-amyloid and tau pathology in the aging feline brain. J Comp Neurol 528(1):112–117. https://doi.org/10.1002/cne.24741

Fox MW (1970) Neurobehavioral development and the genotype-environment interaction. Q Rev Biol 45(2):131–147. https://doi.org/10.1086/406491

Freeman NCG, Rosenblatt JS (1978) The interrelationship between thermal and olfactory stimulation in the development of home orientation in newborn kittens. Dev Psychobiol 11(5):437–457. https://doi.org/10.1002/dev.420110508

Fugazza C, Sommese A, Pogány Á, Miklósi Á (2021) Did we find a copycat? Do as I Do in a domestic cat (*Felis catus*). Anim Cogn 24(1):121–131. https://doi.org/10.1007/s10071-020-01428-6

Gajdoš-Kmecová N, Peťková B, Kottferová J, Halls V, Haddon C, De Assis LS, Mills DS (2023) An ethological analysis of close-contact inter-cat interactions determining if cats are playing, fighting, or something in between. Sci Rep 13(1):92. https://doi.org/10.1038/s41598-022-26121-1

Gazzano A, Bianchi L, Campa S, Mariti C (2015) The prevention of undesirable behaviors in cats: effectiveness of veterinary behaviorists' advice given to kitten owners. J Vet Behav 10(6):535–542. https://doi.org/10.1016/j.jveb.2015.07.042

Harlow HF, Zimmermann RR (1959) Affectional responses in the infant monkey. Science 130(3373):421–432

Henning J, Fernandez E, Nielsen T, Hazel S (2022) Play and welfare in domestic cats: current knowledge and future directions. Anim Welf 31(4):407–421. https://doi.org/10.7120/09627286.31.4.005

Herbst CT, Prigge T, Garcia M, Hampala V, Hofer R, Weissengruber GE, Svec JG, Fitch WT (2023) Domestic cat larynges can produce purring frequencies without neural input. Curr Biol 33(21):4727–4732.e4. https://doi.org/10.1016/j.cub.2023.09.014

Hoppenbrouwers T, Sterman MB (1975) Development of sleep state patterns in the kitten. Exp Neurol 49(3):822–838. https://doi.org/10.1016/0014-4886(75)90062-X

Horn JA, Mateus-Pinilla N, Warner RE, Heske EJ (2011) Home range, habitat use, and activity patterns of free-roaming domestic cats. J Wildl Manag 75(5):1177–1185. https://doi.org/10.1002/jwmg.145

Houpt KA (2018) Domestic animal behavior for veterinarians and animal scientists, 6th edn. Wiley, Hoboken

Hubel DH, Wiesel TN (1964) Effects of monocular deprivation in kittens. Naunyn Schmiedebergs Arch Exp Pathol Pharmakol 248(6):492–497

Hudson R, Raihani G, González D, Bautista A, Distel H (2009) Nipple preference and contests in suckling kittens of the domestic cat are unrelated to presumed nipple quality. Dev Psychobiol 51(4):322–332. https://doi.org/10.1002/dev.20371

Jaskelioff M, Muller FL, Paik J-H, Thomas E, Jiang S, Adams AC, Sahin E, Kost-Alimova M, Protopopov A, Cadiñanos J, Horner JW, Maratos-Flier E, DePinho RA (2011) Telomerase reactivation reverses tissue degeneration in aged telomerase-deficient mice. Nature 469(7328):102–106. https://doi.org/10.1038/nature09603

Jemmett JE, Evans JM (1977) A survey of sexual behaviour and reproduction of female cats. J Small Anim Pract 18(1):31–37. https://doi.org/10.1111/j.1748-5827.1977.tb05821.x

Jennett AL, Jennett NM, Hopping J, Yates D (2016) Evidence for seasonal reproduction in UK domestic cats. J Feline Med Surg 18(10):804–808. https://doi.org/10.1177/1098612X15595665

Juraska JM (2024) The last stage of development: the restructuring and plasticity of the cortex during adolescence especially at puberty. Dev Psychobiol 66(2):e22468. https://doi.org/10.1002/dev.22468

Bradshaw JWS, Casey R, Brown S (2012) The behaviour of the domestic cat (CABI). CABI, Wallingford

Kane E, Rogers QR, Morris JG (1981) Feeding behavior of the cat fed laboratory and commercial diets. Nutr Res 1(5):499–507. https://doi.org/10.1016/S0271-5317(81)80053-X

Knospe C (2002) Periods and stages of the prenatal development of the domestic cat. Anat Histol Embryol 31(1):37–51. https://doi.org/10.1046/j.1439-0264.2002.00360.x

Koepke JE, Pribram KH (1971) Effect of milk on the maintenance of sucking behavior in kittens from birth to six months. J Comp Physiol Psychol 75(3):363–377. https://doi.org/10.1037/h0030941

Kolb B, Nonneman AJ (1975) The development of social responsiveness in kittens. Anim Behav 23:368–374. https://doi.org/10.1016/0003-3472(75)90085-8

Kovach JK, Kling A (1967) Mechanisms of neonate sucking behaviour in the kitten. Anim Behav 15(1):91–101. https://doi.org/10.1016/S0003-3472(67)80017-4

Kuo ZY (1930) The genesis of the cat's responses to the rat. J Comp Psychol 11(1):1–36. https://doi.org/10.1037/h0075723

Landsberg G, Ley JM (2012) Kitten development. In: The cat. Elsevier, pp 182–190. https://doi.org/10.1016/B978-1-4377-0660-4.00009-0

Lowe SE, Bradshaw JWS (2002) Responses of pet cats to being held by an unfamiliar person, from weaning to three years of age. Anthrozoös 15(1):69–79. https://doi.org/10.2752/089279302786992702

Main M, Solomon J (1990) Procedures for identifying infants as disorganized/disoriented during the Ainsworth strange situation. In: Attachment in the preschool years: theory, research, and intervention. University of Chicago Press, Chicago

Mariti C, Bowen JE, Campa S, Grebe G, Sighieri C, Gazzano A (2016) Guardians' perceptions of cats' welfare and behavior regarding visiting veterinary clinics. J Appl Anim Welf Sci 19(4):375–384. https://doi.org/10.1080/10888705.2016.1173548

Martin P (1984) The time and energy costs of play behaviour in the cat. Z Tierpsychol 64(3–4):298–312. https://doi.org/10.1111/j.1439-0310.1984.tb00365.x

Martin P (1986) An experimental study of weaning in the domestic cat. Behaviour 3:221–249

Martin P, Bateson P (1985) The ontogeny of locomotor play behaviour in the domestic cat. Anim Behav 33(2):502–510. https://doi.org/10.1016/S0003-3472(85)80073-7

Martínez-Byer S, Hudson R, Bánszegi O, Szenczi P (2023) Effects of early social separation on the behaviour of kittens of the domestic cat. Appl Anim Behav Sci 259:105849. https://doi.org/10.1016/j.applanim.2023.105849

Mazeau L, Wylie C, Boland L, Beatty JA (2021) A shift towards early-age desexing of cats under veterinary care in Australia. Sci Rep 11(1):811. https://doi.org/10.1038/s41598-020-79513-6

McCune S (1995) The impact of paternity and early socialisation on the development of cats' behaviour to people and novel objects. Appl Anim Behav Sci 45(1–2):109–124. https://doi.org/10.1016/0168-1591(95)00603-P

McGinty DJ, Stevenson M, Hoppenbrouwers T, Harper RM, Sterman MB, Hodgman J (1977) Polygraphic studies of kitten development: sleep state patterns. Dev Psychobiol 10(5):455–469. https://doi.org/10.1002/dev.420100506

Mclaren IPL, Mackintosh NJ (2002) Associative learning and elemental representation: II. Generalization and discrimination. Anim Learn Behav 30(3):177–200. https://doi.org/10.3758/BF03192828

Meier GW (1961) Infantile handling and development in Siamese kittens. J Comp Physiol Psychol 54(3):284–286. https://doi.org/10.1037/h0045008

Mendl M (1988) The effects of litter-size variation on the development of play behaviour in the domestic cat: litters of one and two. Anim Behav 36(1):20–34. https://doi.org/10.1016/S0003-3472(88)80246-X

Moelk M (1979) The development of friendly approach behavior in the cat: a study of kitten-mother relations and the cognitive development of the kitten from birth to eight weeks. In: Advances in the study of behavior, vol 10. Elsevier, pp 163–224. https://doi.org/10.1016/S0065-3454(08)60095-9

Moles A (2004) Deficit in attachment behavior in mice lacking the -opioid receptor gene. Science 304(5679):1983–1986. https://doi.org/10.1126/science.1095943

Moriceau S, Sullivan RM (2004) Corticosterone influences on mammalian neonatal sensitive-period learning. Behav Neurosci 118(2):274–281. https://doi.org/10.1037/0735-7044.118.2.274

Mota-Rojas D, Marcet-Rius M, Domínguez-Oliva A, Martínez-Burnes J, Lezama-García K, Hernández-Ávalos I, Rodríguez-González D, Bienboire-Frosini C (2023) The role of oxytocin in domestic animal's maternal care : parturition, bonding, and lactation. Animals 13(7):1207. https://doi.org/10.3390/ani13071207

Neilson J (2004) Thinking outside the box: feline elimination. J Feline Med Surg 6(1):5–11. https://doi.org/10.1016/j.jfms.2003.09.008

Ng TT, Fascetti AJ, Larsen JA (2023) Reproduction of domestic cats in laboratories, catteries, and feral colonies: a review. Top Companion Anim Med 55:100780. https://doi.org/10.1016/j.tcam.2023.100780

Peters SE (1983) Postnatal development of gait behaviour and functional allometry in the domestic cat (*Felis catus*). J Zool 199(4):461–486. https://doi.org/10.1111/j.1469-7998.1983.tb05100.x

Potter A, Mills DS (2015) Domestic cats (*Felis silvestris catus*) do not show signs of secure attachment to their owners. PLoS One 10(9):e0135109. https://doi.org/10.1371/journal.pone.0135109

Raihani G, González D, Arteaga L, Hudson R (2009) Olfactory guidance of nipple attachment and suckling in kittens of the domestic cat: inborn and learned responses. Dev Psychobiol 51(8):662–671. https://doi.org/10.1002/dev.20401

Rakic P, Bourgeois J, Eckenhoff M, Zecevic N, Goldman-Rakic P (1986) Concurrent overproduction of synapses in diverse regions of the primate cerebral cortex. Science 232(4747):232–235. https://doi.org/10.1126/science.3952506

Reisner IR, Houpt KA, Erb HN, Quimby FW (1994) Friendliness to humans and defensive aggression in cats: the influence of handling and paternity. Physiol Behav 55(6):1119–1124. https://doi.org/10.1016/0031-9384(94)90396-4

Robbins TW, Everitt BJ (1996) Neurobehavioural mechanisms of reward and motivation. Curr Opin Neurobiol 6(2):228–236. https://doi.org/10.1016/S0959-4388(96)80077-8

Rybicka M, Kaźmierczak M, Pawlicka P, Łada-Maśko AB, Anikiej-Wiczenbach P, Bielawski KP (2021) (Re-)activity in the caregiving situation: genetic diversity within Oxytocin–Vasopressin Pathway is associated with salivary oxytocin and vasopressin concentrations in response to contact with a crying infant-simulator. Psychoneuroendocrinology

4

131:105294. https://doi.org/10.1016/j.psyneuen.2021.105294

Sacco A, Gordon SG, Lomber SG (2024) Connectome alterations following perinatal deafness in the cat. NeuroImage 290:120554. https://doi.org/10.1016/j.neuroimage.2024.120554

Scott JP (1962) Critical periods in behavioral development: critical periods determine the direction of social, intellectual, and emotional development. Science 138(3544):949–958. https://doi.org/10.1126/science.138.3544.949

Scott JP, Marston MV (1950) Critical periods affecting the development of normal and mal-adjustive social behavior of puppies. J Genet Psychol 77(1):25–60. https://doi.org/10.1080/08856559.1950.10533536

Seitz P (1959) Infantile experience and adult behavior in animal subjects. Psychosom Med 5(21):353–378

Smith BA, Jansen GR (1977) Brain development in the feline. Nutr Rep Int 4:487–495

Solomon J, Beetz A, Schöberl I, Gee N, Kotrschal K (2019) Attachment security in companion dogs: adaptation of Ainsworth's strange situation and classification procedures to dogs and their human caregivers. Attach Hum Dev 21(4):389–417. https://doi.org/10.1080/14616734.2018.1517812

Sung W, Crowell-Davis SL (2006) Elimination behavior patterns of domestic cats (*Felis catus*) with and without elimination behavior problems. Am J Vet Res 67(9):1500–1504. https://doi.org/10.2460/ajvr.67.9.1500

Tavernier C, Ahmed S, Houpt KA, Yeon SC (2020) Feline vocal communication. J Vet Sci 21(1):e18. https://doi.org/10.4142/jvs.2020.21.e18

Thorne CJ (1982) Feeding behaviour in the cat—recent advances. J Small Anim Pract 23(9):555–562. https://doi.org/10.1111/j.1748-5827.1982.tb02516.x

Turner DC (1991) The ethology of the human-cat relationship. Schweiz Arch Tierheilkd 133(2):63–70

Villablanca JR, Olmstead CE (1979) Neurological development of kittens. Dev Psychobiol 12(2):101–127. https://doi.org/10.1002/dev.420120204

Vitale KR, Behnke AC, Udell MAR (2019) Attachment bonds between domestic cats and humans. Curr Biol 29(18):R864–R865. https://doi.org/10.1016/j.cub.2019.08.036

Vitale Shreve KR, Udell MAR (2017) Stress, security, and scent: the influence of chemical signals on the social lives of domestic cats and implications for applied settings. Appl Anim Behav Sci 187:69–76. https://doi.org/10.1016/j.applanim.2016.11.011

Viviani D, Charlet A, Van Den Burg E, Robinet C, Hurni N, Abatis M, Magara F, Stoop R (2011) Oxytocin selectively gates fear responses through distinct outputs from the central amygdala. Science 333(6038):104–107. https://doi.org/10.1126/science.1201043

Wilson M, Warren JM, Abbott L (1965) Infantile stimulation, activity, and learning by cats. Child Dev 36(4):843. https://doi.org/10.2307/1126928

Xue M, Atallah BV, Scanziani M (2014) Equalizing excitation–inhibition ratios across visual cortical neurons. Nature 511(7511):596–600. https://doi.org/10.1038/nature13321

Yuan J, Horvitz HR (2004) A first insight into the molecular mechanisms of apoptosis. Cell 116:S53–S56

Zadik-Weiss L, Ritter S, Hermush V, Asher N, Avital A, Or R (2020) Feline cognitive dysfunction as a model for Alzheimer's disease in the research of CBD as a potential treatment—a narrative review. J Cannabis Res 2(1):43. https://doi.org/10.1186/s42238-020-00054-w

Zentall TR (2013) Social learning psychological and biological perspectives. Psychology Press, Hove

Concepts of Welfare and Quality of Life

Contents

5.1 Definitions and History – 188
5.1.1 Welfare – 188
5.1.2 Welfare, Lifestyle, and Normal Species Behaviors – 191
5.1.3 Welfare and Selection – 193

5.2 Individual Welfare Assessment – 195
5.2.1 Welfare and Veterinary Care – 195
5.2.2 Preliminary Considerations on Quantitative Welfare Assessment – 196
5.2.3 The Welfare Consultation – 197

References – 203

© The Author(s), under exclusive license to Springer Nature Switzerland AG 2026
S. Bleuer-Elsner et al., *Veterinary Psychiatry of the Cat*,
https://doi.org/10.1007/978-3-032-06260-4_5

Animal welfare is a complex topic, far exceeding the scope of this book. In our role as caregivers, our primary objective is to improve the quality of life of both the animal and its owner, fostering a harmonious and high-quality relationship.

By defining animal welfare through Maslow's hierarchy of needs, we are led to differentiate between livestock and companion animals.

5

The concept of a captive animal does not seem relevant for cats, domesticated for over ten millennia, especially since many of them have free outdoor access yet choose to spend the majority of their time with their owners. However, achieving psychological balance in individuals must be based on the alignment between their developmental environment and their final living environment. Breeder selection has a significant impact on the welfare of companion cats, as it can determine how well a cat is adapted to life as a companion animal, enhancing the well-being of both the cat and its owner.

The practitioner's role should extend to an assessment of this welfare, and it is therefore necessary to develop recommendations for a consultation dedicated to this evaluation.

5.1 Definitions and History

5.1.1 Welfare

In 1943, Maslow constructed his theory of motivation based on a hierarchy of needs, illustrated by a pyramid of needs (Maslow 1943) (◘ Fig. 5.1). According to Maslow, to achieve an optimal level of welfare, humans must fulfill a succession of needs organized hierarchically in their order of appearance: physiological needs, safety needs, needs for belonging, esteem needs, and finally, self-actualization. Fulfilling a lower level of the pyramid transforms the needs of the next level into essential requirements.

Although the ordered and linear aspect of this hierarchy has been debated, this theory

◘ **Fig. 5.1** Maslow's hierarchy of needs. The five levels of the pyramid are illustrated in the figure. Fulfilling one level makes the next one essential

serves as a basis for most evaluations of animal welfare. These evaluations are recent and are founded on the five freedoms articulated by the UK Farm Animal Welfare Council in 1979 (Council, F.A.W. 1992), which are also embedded in the World Organisation for Animal Health (OIE) definition of animal welfare. The five fundamental freedoms are: *freedom from hunger and thirst*, *freedom from discomfort*, *freedom from pain, injury, or disease*, *freedom from fear and distress*, and finally, the *freedom to express normal species-specific behaviors*. These freedoms clearly echo Maslow's pyramid and aim to determine the essential needs—variable by species—that an animal must fulfill to achieve an optimal state of welfare.

The choices made to adapt Maslow's pyramid to the five freedoms are debatable. Indeed, *welfare*, which refers to the quality of life of the *individual*, encompasses both *physical health* and welfare as well as *psychological health* and welfare. Some of the elements proposed by Maslow have been overlooked, yet they significantly contribute to individual welfare, such as safety. Let us revisit this adaptation of the pyramid.

Hunger and thirst are constraints linked to environmental conditions. We can extend this "first freedom" to include all physiological needs—for instance, ensuring that a cat can eliminate in a dedicated space distinct from other activity or resting areas, as well as providing the opportunity to sleep undisturbed.

Freedom from discomfort due to physical constraints pertains to the animal's living comfort and was originally meant to ensure sufficiently large and comfortable resting spaces for each individual. To the comfort encompassed by this category, safety could also be added. Safety is not purely environmental: for example, when we prevent our cat from going out at night, we are ensuring its safety, though this involves removing an *immediate form of freedom*. Similarly, the decision to neuter or spay a cat can be seen as a protective measure since it significantly reduces the risk of abscesses from fights with other cats and the likelihood of contracting diseases such as FIV or FeLV. Restricting physical exercise when a cat is injured also decreases immediate comfort but contributes to improving the animal's quality of life in the medium and long term.

The *diseases* mentioned alongside pain and injuries refer to somatic illnesses. These are to be investigated during our routine clinical examinations. However, the question of *welfare* requires a different approach, one that focuses our attention on aspects beyond the scope of our usual technical tools. The key inquiry becomes: what is the impact of a given disease on an individual's welfare? To what extent does the animal experience pain? Can it still find moments of pleasure despite its chronic condition? Can this physical suffering be minimized, whether in the short, medium, or long term?

The *protection from fear and distress* should more broadly correspond to the absence of psychological suffering, though these definitions predate the emergence of veterinary psychiatry. They refer to environmental conditions that are not stressful for the animal, without necessarily aiming to achieve or guarantee an *optimal psychological state*. For example, a cat may be in an optimal environment but still exhibit a developmental disorder, such as sensory deprivation syndrome (see ▶ Chap. 8, ▶ Sect. 8.2), and experience depressive states despite the absence of identifiable phobogenic stimuli or traumatic events.

Finally, the fifth freedom refers to "normal species-specific behaviors." Unfortunately, these behaviors are often defined ethologically, meaning they are considered normal if they are typically displayed by the species in the absence of constraints. Ethologists frequently regard environmental evolution as a constraint, leading to a common confusion between *ancestral behaviors* and *normal behaviors*. This issue is particularly relevant when dealing with domestic species that live alongside humans (see ▶ Chap. 1, ▶ Sect. 1.1.1.1). For example, can we consider hunting with a bow a "normal human behavior" simply because it occupied much of our time in a certain era? Similarly, can we claim that playing video games does not bring pleasure or is not normal because it was not part of the human ethogram five centuries ago?

This question is critical because it touches on the discovery of repetitive behaviors induced by severely impoverished environmental conditions (Shyne 2006). For instance, researchers have shown that pigs bite the bars of their enclosures compulsively to alleviate the discomfort caused by confinement without any engaging activity (Rushen et al. 1990). Similarly, a horse kept in a stall without varied and sufficient forage develops repetitive behaviors, termed "stereotypies," associated with the release of endorphins, which provide temporary relief from discomfort (Dodman et al. 1987). However, this adaptation is short-lived and eventually leads to an anxious state (see ▶ Chap. 7, ▶ Sect. 7.2.2).

The shortcut that has been made, a major reasoning bias, is as follows: *since stereotypical behaviors are markers of poor welfare and can be caused by a low-quality environment, a high-quality environment necessarily provides welfare*. This error has led to the promotion of environmental enrichment techniques, which involve adding elements to the animal's environment to attempt to compensate for behaviors the environment prevents them from expressing. This approach has become common practice for farm animals (Casal-Plana et al. 2017) and also for companion animals (Herron and Buffington 2012). However, while advocating for a rich environment is *necessary*, it is not *sufficient*.

This proposal, which is almost entirely environmental, is limited for two reasons: the existence of counterexamples and the reliance on a species-specific rather than an individual-specific standard. On one hand, there are animals in optimal environments that still experience significant psychological suffering. On the other hand, imposing the idea of "normal species-specific behavior" to evaluate individual welfare suggests, wrongly, that there is no variability in the sources of pleasure within the species. The animal's pleasure is regarded, incorrectly, as achieved solely through the fulfillment of "normal behaviors."

Yet there is no universal individual norm for pleasure: what kind of welfare would a human who enjoys running and reading experience if they were required to play an hour of tennis and an hour of violin daily? Can we impose on a cat that dislikes going outside the obligation to remain outdoors on the assumption that cats, as a predatory species, inherently enjoy hunting? The more pertinent question for animals would therefore be: can they fulfill their *own individual preferences and desires*?

In 2018, the French Agency for Food, Environmental and Occupational Health & Safety (Anses) proposed a new definition of animal welfare that aligns with a more individual-centered perception of the animal. This definition is grounded in a philosophical, societal, and legal perspective of the concept: "The welfare of an animal is the positive mental and physical state related to the satisfaction of its physiological and behavioral needs, as well as its expectations" (Anses 2018). This ambitious definition takes the animal's expectations into account and encourages owners to fulfill them.

To address the animal's expectations, we are thus faced with the need to look not only for markers of distress (to minimize them as much as possible) but also for positive markers of welfare (to aim for their fulfillment) (Boissy et al. 2007). Moreover, this effort to meet the animal's expectations must respect its environment and not come at the expense of other individuals in the group. This is where the notion of quality of life becomes crucial. Quality of life encompasses all elements that contribute to creating conditions conducive to the individual's well-being. While absolute welfare may not be realistic or achievable in a cat's life, striving to provide optimal quality of life is an attainable goal.

Our proposed approach to evaluating feline welfare involves reformulating the five freedoms into five key domains: respect for physiological needs, environmental comfort and security, control of physical suffering, control of psychological suffering, and fulfillment of the animal's expectations (◘ Fig. 5.2).

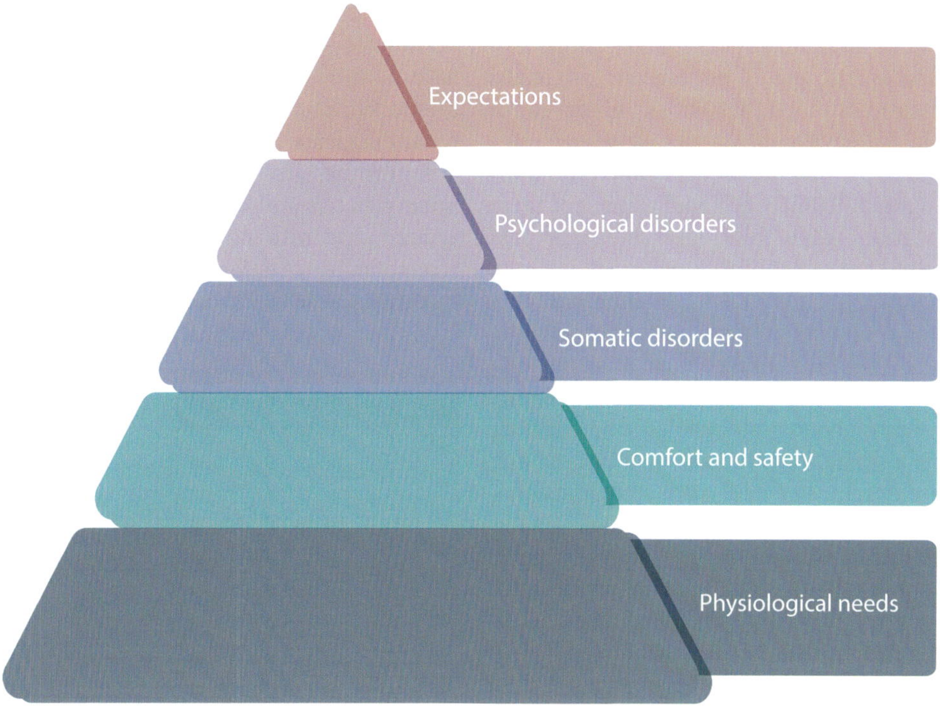

Fig. 5.2 Maslow's pyramid adapted to cat

Important Point of View

Cat's welfare and access to outdoor

Some ethologists argue that domestic animals cannot express the normal behaviors of their species within a "human environment" and suggest modifying that environment to allow such expression. However, a perfectly free-ranging cat often chooses to return home or even to stay indoors. This aspect of feline behavior highlights its attachment to its territory, which includes the humans it lives with.

Thus, the notion that a cat living in an apartment cannot have its needs met is as mistaken as assuming that a cat with outdoor access will automatically thrive. In fact, it is not uncommon to see psychologically balanced indoor cats whose well-being deteriorates after a move that provides outdoor access, as they may struggle to adapt to the overwhelming stimuli in their new environment.

In understanding an individual cat's behaviors, it is essential to consider the balance between desire and freedom, as protection can sometimes equate to a reduction in liberty.

We must contribute to the selection and breeding of cats to ensure their capacities and needs align with the lifestyles we offer them.

5.1.2 Welfare, Lifestyle, and Normal Species Behaviors

Our highly urbanized Western lifestyle challenges the idea that studying cats in their ancestral natural environment provides a reliable reference for defining optimal species behaviors. Since our world is no longer "natural," the expression of these ancestral behaviors is often neither feasible nor suitable. We must therefore address the significant mismatch between the species being studied (evolved to live in a world that no longer exists) and its current environment (an urbanized planet inhabited by eight billion humans).

Moreover, there is no evidence that the expression of such behaviors, observed in their initial environment, guarantees an optimal level of welfare. For instance, humans lived very differently millennia ago, yet there is no proof that their level of happiness exceeded that of modern humans. If we were to propose returning to the living conditions and environments of thousands of years ago, most people would likely refuse or struggle to survive. On the contrary, many indicators (life expectancy, access to technological advances, quality of life) suggest an overall increase in welfare despite this departure from the original environment.

Extending this reasoning further, a given environment holds specific value for an individual. For example, a human raised in an urban setting might become depressed if suddenly relocated to a rural environment, even though such a setting may appear closer to their so-called natural habitat.

Welfare must therefore align with the Anses definition and rely on individual criteria. The notion of a "natural environment" should be abandoned. Instead, it would be more logical to speak of an *optimal environment* tailored to each individual's adaptive capacities. We are predisposed to thrive in conditions similar to those in which we developed (see ▶ Chap. 4, ▶ Sect. 4.1). For instance, some cats that dislike social contact or are phobic or anxious will experience higher welfare if they are not subjected to daily interactions with humans or other cats. Rabbits born near airfields tolerate planes passing overhead without fear. English blue tits have learned to pierce the foil caps of milk bottles to access the cream. These behaviors are neither ancestral nor natural.

Furthermore, the modern world poses significantly greater dangers to animals, primarily due to transportation infrastructures built by human societies. What might appear to be a restriction of freedom—preventing an animal from going outside if the home is near a busy road, or sterilizing the animal to reduce its motivation to roam—can often be an act of protection, preventing accidents. This logic extends to inhibiting some natural behaviors, such as reproduction. In certain cases, allowing a natural need to go unchecked can be a form of neglect. Failing to sterilize an urban cat, for instance, can compromise its welfare. Allowing sexual behaviors to manifest increases frustration in indoor cats with no access to mates, while free-roaming cats face risks of accidents or infectious diseases from fights with conspecifics over access to females.

Clinical Illustration

To a client who was bringing in kittens for euthanasia for the second time, I suggested sterilizing the mother. He gave me a troubling response: *"Don't you think about the joy of motherhood?"*

Finally, in our daily practice, we frequently witness *harmonious* and *highly positive relationships* between cats and their owners, even in conditions that may not align with the optimal ethological standards for stray cats. The relationship cats maintain with humans often involves numerous rituals adapted to urban lifestyles, yet these rituals are not inherently synonymous with poor welfare. In many cases, owners genuinely strive to care for their cats well-being. Most instances of diminished welfare caused by owners are *unintentional* and stem from a *lack of knowledge* rather than an *intent to harm*. Similar patterns can be observed in areas such as nutrition or dermatology.

Our role as caregivers places us in a position where we must inform owners and provide them with tools to improve their cat's quality of life, while considering the constraints they face, be they financial, organizational, or time-related. This dual responsibility entails, on the one hand, a technical role as educators who assess the cat's quality of life and, on the other, a supportive role in decision-making, working collaboratively with owners to forge a path toward better welfare for their companion. Within this framework, the decision to abandon or rehome a cat represents an extreme measure, signaling a failure in our ability to restore balance to the situation presented to us *"here and now."*

One of the first technical points we can share with owners is the importance of choosing developmental conditions for their future kitten that closely align with the animal's eventual living environment. Unfortunately, preventive consultations conducted before acquiring a kitten are even rarer than those for dogs. Most cat adoptions are still driven by the notion that a cat is "free" and that purchasing one would be unnecessary. While the growth of pedigree cat breeding has slightly shifted this perspective, the vast majority of kitten adoptions remain gratuitous or limited to covering vaccination and identification costs. This often leads to disappointment and discouragement among owners, and in some cases, abandonment, as the adopted kitten differs significantly from the one they had envisioned.

For example, it is common for a family to adopt a kitten for one of their children, only to find that the kitten was not properly socialized, was weaned too early, and came from an environment vastly different from their own. Such a kitten, frightened by the children and unable to control its biting or scratching, is far removed from the cuddly ball of fluff they had imagined. What was intended to be a companion for the children becomes a source of fear and injuries. Educating the public on the importance of ensuring compatibility between the development environment and the living environment after adoption, as well as on the ideal adoption age (see ▶ Chap. 4, box « The ideal age for kitten adoption »), is therefore essential to help owners select kittens more likely to meet their expectations.

However, our guidance extends beyond breeding conditions and also considers the presence of genetic predispositions to certain behaviors in specific breeds or lines (Malik 2001). The following section logically explores the impact of selection on the behavior of our domestic cats.

5.1.3 Welfare and Selection

We define pedigree cats as those registered in a feline registry, such as the *Livre Officiel des Origines Félines* (LOOF n.d.) in France. These cats contrast with non-pedigree cats, commonly referred to as European Shorthairs or domestic cats, whose lineage lacks official documentation. As of now, the International Cat Association (TICA n.d.) recognizes 73 breeds.

In France, statistics from 2022 indicate that of the 15 million companion cats, only 180,000 hold pedigrees (source: LOOF – *Livre Officiel des Origines Félines*). Additionally, studies estimate that eight to ten million cats are stray and do not belong to any recognized breed (source: SFEPM—*Société Française pour l'Étude et la Protection des Mammifères* n.d.). These figures highlight that pedigree cats represent only a small fraction of the overall cat population in France.

The selection of domestic cats, particularly pedigree cats, has historically focused on both morphological traits, such as coat color or facial structure, and behavioral traits, such as ease of handling and tolerance of care. Several studies have explored the genetics of pedigree cats, confirming this dual selection approach (Montague et al. 2014; Salonen et al. 2019; Turner and Bateson 2013).

Additionally, in the 2000s, selective breeding aimed to produce hypoallergenic cat breeds, most notably the Siberian cat (Butt et al. 2012). While this hypoallergenic status remains controversial, these breeds exhibit a genetic mutation that reduces the production of *Fel d 1*, an allergen found in saliva, sebaceous glands, and anal glands (Ligabue-Braun 2020; Satorina et al. 2014).

However, the proportion of pedigree cats remains low, and free-ranging cats reproduce without the influence of artificial selection pressure. Surveys conducted among cat owners reveal behavioral differences between pedigree and European cats. Pedigree cats exhibit lower levels of activity and predation while displaying higher frequencies of vocalization

(Duffy et al. 2017). Similarly, aggression toward family members is less common in pedigree cats compared to European cats (Hart and Hart 2013). Nevertheless, according to a study based on insurance company statistics, the proportion of veterinary visits for behavioral reasons and the age at which these visits occur do not differ significantly between pedigree and European cats (Hadar et al. 2023).

Our field has unique complexities that make investigations into breed-specific behaviors challenging. Firstly, environmental factors and learning experiences significantly influence behavior, making it difficult to establish a direct correlation between observed symptoms and *genotype*. Additionally, most behavioral disorders are polygenic, with intricate epigenetic factors complicating genetic identification. Finally, the measurement of behavior is predominantly clinical, and the tools available for objectively quantifying behaviors are limited and lack clear thresholds for distinguishing normal from pathological.

To enable meaningful welfare assessments, it is essential to determine the point at which deviations from the norm become *impairing for the animal*, to the extent that medical intervention is warranted.

The selection of behavioral and morphological traits has also inadvertently selected for genetic vulnerabilities to somatic diseases. Inherited factors have been linked to the development of osteochondrodysplasia in Scottish Folds, polycystic kidney disease in Persians, myopathy and muscular dystrophy in Devon Rex cats, and an increased risk of feline infectious peritonitis in Abyssinians, Bengals, Birmans, Himalayans, Ragdolls, and Rex breeds (Pesteanu-Somogyi et al. 2006).

These vulnerabilities are thought to stem from inbreeding and the limited number of founder individuals within breeds. Additionally, stressful breeding environments, frequent introduction of new cats, regular reproduction cycles, and high population density, common in breeding operations, contribute to the emergence of somatic diseases (Pedersen et al. 1981).

Although this issue is less discussed in the context of behavioral disorders than somatic ones, it is highly likely that similar patterns exist. However, studies and data in this area are even scarcer for cats than for dogs.

In practice, we increasingly have scientific tools at our disposal to enable high-quality selection, such as genetic tests that identify mutations and allow breeders to choose individuals free from certain diseases. Selecting solely based on aesthetic criteria results in the same pitfalls as any overly specialized selection: an increased risk of vulnerabilities in other areas. The future will undoubtedly bring even more technical resources to assist us, but the priority must remain a selection process oriented toward cats adapted to our modern world, suited to their future lifestyle, and in good physical and psychological health. This selection can complement breeding conditions to ensure optimal behavioral development.

These considerations about welfare inevitably lead to the need for methods to measure and evaluate it. Several approaches have already been proposed to assess feline welfare in various contexts: veterinary visits (Mariti et al. 2016), shelters (Vojtkovská et al. 2020), and broader assessments (Delgado and Reevy 2018).

In the following section, we outline the essential foundations for any practitioner aiming to conduct an *individualized welfare evaluation*.

Summary of ▶ Sect. 5.1

Welfare is a contemporary concept rooted in Maslow's hierarchy of fundamental needs and was first applied to farm animals by the UK Farm Animal Welfare Council in 1979. These group-focused proposals must be adapted for the individual evaluation of companion animals. Thus, we propose five essential domains for the welfare of domestic cats: respect for physiological needs, environmental comfort and security, management of physical suffering, management of psychological suffering, and the fulfillment of the animal's expectations. The concept of species-typical behaviors cannot use the ancestors of domestic cats as a model for normality. Consideration of domestic cat welfare should focus on meeting the individual's expectations within its living environment—namely, in close contact with humans. Therefore, ideal living conditions should prioritize alignment between the developmental environment and the cat's final living environment.

Cats have been selected based on morphological and behavioral traits. While behavioral traits aim to adapt cats to family life, selection merely steers general tendencies, and it is crucial to account for the cat's developmental conditions.

The veterinarian's role is to inform owners about their cat's welfare level without judgment and to support them in achieving an optimal quality of life within the reasonable and feasible limits of their current situation.

5.2 Individual Welfare Assessment

5.2.1 Welfare and Veterinary Care

In our veterinary profession, we are often required to cause pain or physical and psychological discomfort to provide treatment. It is crucial to ensure that, despite the potentially distressing or painful manipulations involved, the relationship with the animal remains intact.

During treatment, closely observing the animal's *posture* and *reactions* is vital to detect early signs of discomfort. This enables us either to propose medical training to avoid *sensitization* or to choose anesthesia instead of continuing a *particularly distressing procedure* that might provoke aggression or result in post-traumatic phobia (see ▶ Chap. 30, ▶ Sect. 30.2). Many resources discuss restraint and handling techniques. Without delving deeply into this topic, we emphasize our agreement with the principle that these methods must be gentle and respectful, even within the highly demanding veterinary context, which sometimes necessitates causing pain or fear to ultimately improve the animal's quality of life.

On a broader timeline, more and more owners seek veterinary advice to determine whether their cat is happy or suffering and, if so, to learn how they can improve its quality of life. In medical follow-ups, whether addressing somatic or psychological suffering, welfare assessment is also a key consideration. If treatments fail to improve or show promise of improving the animal's welfare in the medium or long term, a decision for euthanasia may be reached in agreement with the owner.

The quality of life experienced by owners with their pets is another crucial point to evaluate. Studies suggest that a cat's welfare is more significantly influenced by the owner's characteristics (e.g., family size, age) than the cat's own traits (e.g., sterilization status, relationships with other cats) (Adamelli et al. 2005). A decline in the owner's quality of life often results in a degraded human-cat relationship, substantially increasing the likelihood of rehoming, abandonment, or even euthanasia (Salman et al. 2000; Scarlett et al. 2002; Siracusa et al. 2017).

Given the growing public interest in animal welfare, veterinarians are increasingly drawn to the subject. Scheduling an appointment for a "welfare consultation" might seem innovative today, but it could soon become a standard request from owners.

In the following section, we will outline the critical considerations that we believe are essential for conducting a consultation specifically aimed at *assessing and improving a cat's quality of life.*

5

5.2.2 Preliminary Considerations on Quantitative Welfare Assessment

To assess welfare, practitioners must gather information on the five domains we defined: physiological needs, environmental comfort and security, physical suffering, psychological suffering, and the animal's expectations. While these aspects will be explored in detail in the following section, it is essential first to consider the broader elements that must be integrated into this assessment.

The importance of accounting for different perspectives in this evaluation must be emphasized, just as we do when making decisions regarding euthanasia or managing chronic illnesses.

The *animal's perspective* is inferred from the signs observed during the clinical examination, the presence or absence of diseases, its living conditions, and the fulfillment of its expectations, as observed through its postures and moments of pleasure.

The *veterinarian's perspective* is the expert opinion provided, explaining what can be done to improve the various aspects identified as influencing the animal's quality of life. This perspective must also consider technical capabilities and material constraints.

The *owner's perspective* consists of defining the limits within which they can and are willing to follow the proposed recommendations.

Taking these three perspectives into account therefore requires discussion and compromise aimed at improving the quality of life of the cat presented in consultation.

For this reason, even if the use of welfare assessment grids may seem appealing, they cannot be used in isolation. There is a significant risk that these grids could turn into tools for judgment and reprimand. As with most clinical tools, they should be considered complementary instruments. They allow for the illustration and objectification of observations but should not replace clinical reasoning or hinder discussions with the owner.

A welfare assessment should take the *form of a consultation*, include a *clinical examination*, and result in *recommendations*. A collaborative dialogue and negotiations with the owner are necessary to constructively improve the animal's welfare.

Regarding the quantification of welfare, we have proposed five domains for evaluation. Reflecting on these five domains leads to the following observation: welfare is measured on a *logarithmic scale* (◘ Fig. 5.3). Although rarely emphasized, regardless of the domain in question, it is evident that addressing the most basic elements is relatively straightforward. However, achieving perfect balance is an endless pursuit. Absolute welfare can only be approached, never fully attained.

Take nutrition as an example: most cats have enough food to avoid starvation, but achieving the perfect balance in their diet remains challenging. For this reason, it seems reasonable to establish a threshold for each domain, below which discussions and compromises with the owner must continue before moving on to the next domain. For instance, if nutritional needs are not being met and the cat is emaciated, we should not be discussing its psychological welfare or playtime in the apartment but should instead first work on providing a balanced diet.

The domains are constructed in reference to the pyramid of feline needs, and this order is no coincidence. We can only focus on the higher levels of the pyramid once the lower levels are sufficiently solidified.

However, despite the importance of the order, it is equally evident that welfare follows the "*zero contagion law,*" which states that if one domain falls below a certain threshold, overall welfare becomes impossible. For example, when a cat suffers from an incurable and painful chronic disease, its welfare is so severely affected that some owners may consider euthanasia, almost regardless of the state of the other domains. The implication of

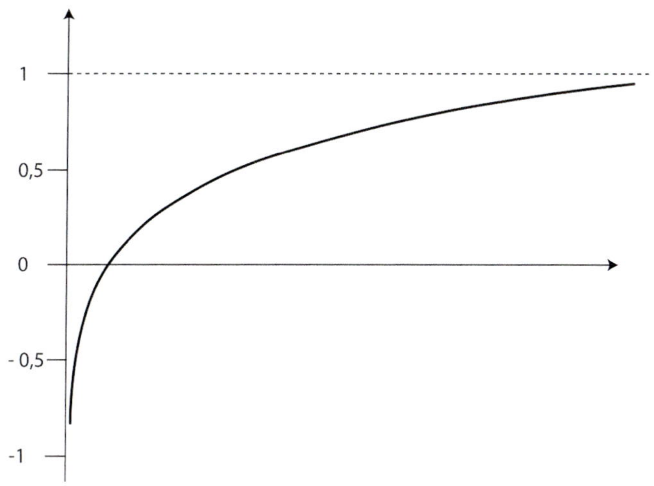

Fig. 5.3 Welfare progresses from its lowest to its highest level following a logarithmic scale

this rule is that any grid designed to measure welfare should be a product of the sub-scores for each domain, and each sub-score should follow a logarithmic scale. While we have not yet developed, finalized, or validated such a grid, these considerations should inform the creation of a usable tool.

Setting aside numerical evaluation, let us now explore how discussions with the owner can be conducted in alignment with these preliminary reflections.

5.2.3 The Welfare Consultation

During a welfare consultation, the veterinarian reviews the five main domains. For each domain, their approach should involve identifying and quantifying the factors affecting the welfare of the cat presented to them, positioning these factors on a logarithmic scale (i.e., assessing whether they fall below or above the acceptable welfare threshold), presenting this information to the owner along with existing means to improve welfare, and discussing these until an acceptable consensus is reached (bringing the domain above the acceptable threshold). The veterinarian then proceeds to the next domain, and so on (Fig. 5.4).

At the end of the consultation, providing a report summarizing the agreements discussed is an effective way to validate the veterinarian-owner alliance that should have been built during the consultation.

Let us now review each domain to better define them and understand how they affect the cat.

First, the veterinarian must ensure that the cat's *physiological needs* are met. Food and water must be available, but it is also important to assess the quality of the diet and how it is provided. Advances in feline nutrition often allow for the animal's energy requirements to be met adequately, even though a perfectly balanced diet is not common. The veterinarian can recommend dietary adjustments to optimize this domain. Obesity is a frequent condition in cats and significantly impacts their quality of life. Its management falls under a subsequent domain (treatment of somatic illnesses). If there are gaps in meeting the cat's energy and protein needs, a discussion should be initiated with the owner to restore balance.

The method of food delivery is a critical consideration for cats and will be addressed alongside psychological needs. However, a minimum standard involves ensuring free-access feeding (ad libitum) that aligns with a cat's natural feeding behavior.

Additionally, the ability to eliminate within reasonable intervals must be verified. A litter box should always be accessible, clean enough, and placed in a location that

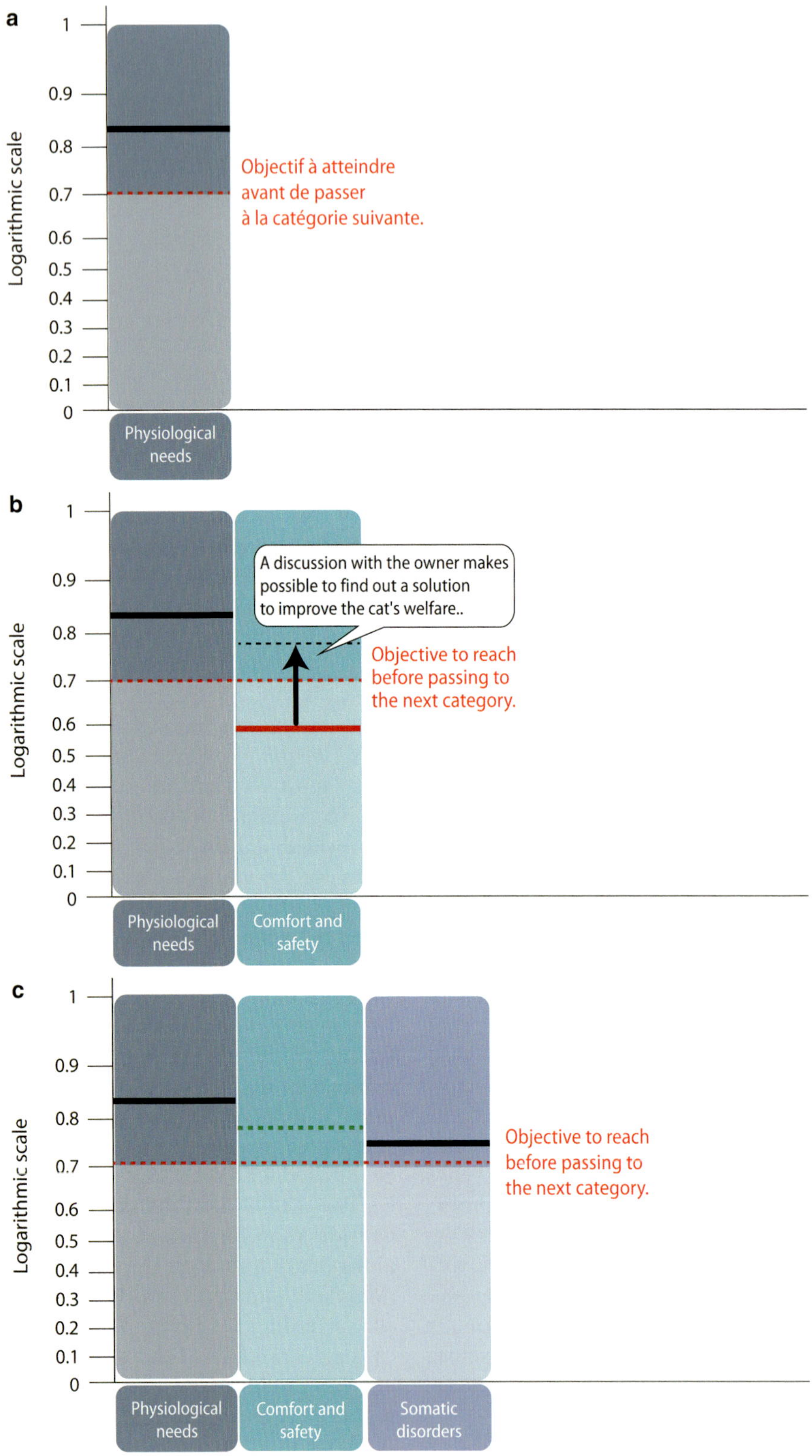

Fig. 5.4 The overall welfare score. This score follows the "zero contagion law," meaning it must be the product rather than the sum of the five sub-scores, each of which follows a logarithmic progression

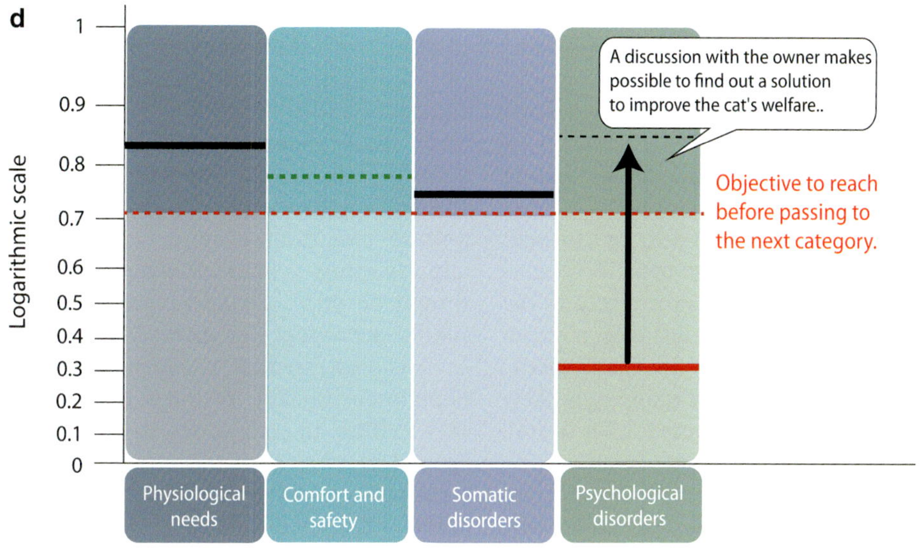

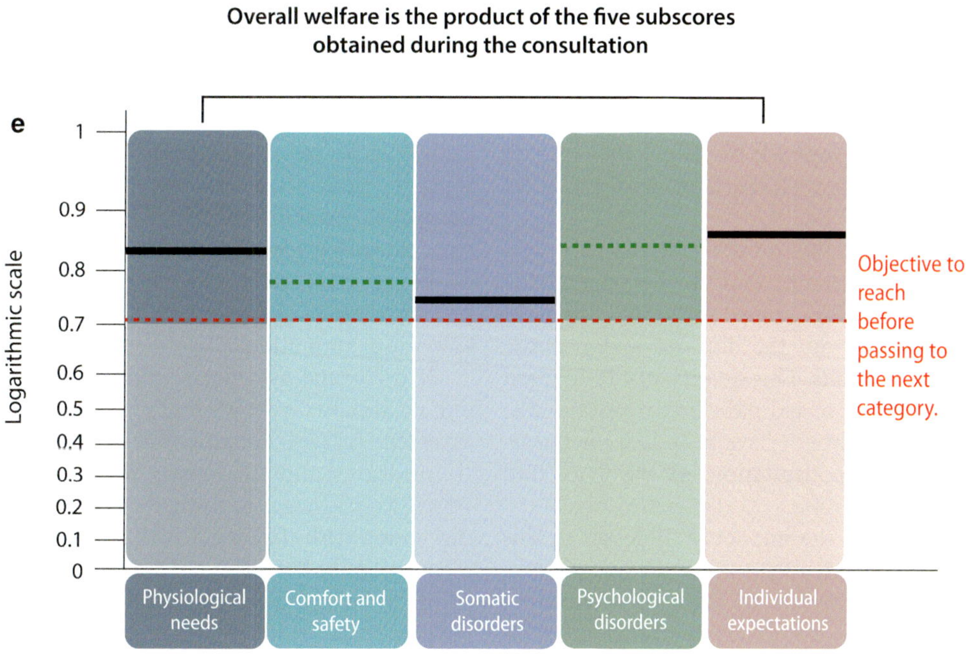

Fig. 5.4 (continued)

suits the cat's preferences. Finally, the quality of sleep must be checked, with the cat having access to a quiet area and a sufficient amount of sleep, which should be at least 10 h per day.

Comfort related to the environment and safety is the next domain to explore. This involves ensuring that the cat has an appropriate resting area and observation spot and can seek shelter from adverse weather conditions if desired. Observing the cat's posture in its resting place and its ability to relax are key indicators of comfort. Ambient temperature is another factor to assess. When significant discomfort is identified, discussions with the owner should lead to practical measures to improve the cat's comfort level before proceeding to the next domain.

It is important to emphasize that this assessment should not be anthropomorphic. For example, some cats can tolerate conditions that humans might deem uncomfortable, such as lying near heaters in winter. This evaluation requires a thorough understanding of feline needs.

The animal's safety must also be assessed. This includes considering the benefits of sterilizing a cat that reproduces excessively, to the detriment of its health. For instance, a queen that has two litters per year for five consecutive years due to unrestricted roaming is likely to suffer negative consequences for her overall well-being.

The domain of *physical suffering*, encompassing somatic diseases, is quintessentially within the veterinarian's expertise. The practitioner must catalog all the animal's ailments and evaluate their impact on quality of life. Much of this exploration is subjective, as diseases affect individuals differently. Pain, which is present in many conditions, should be prioritized, with emphasis on its negative impact on the cat's life (Robertson and Lascelles 2010).

Sometimes, this process will lead the veterinarian to recommend further diagnostic tests or specialized consultations to gain more information about the illnesses and propose tailored solutions. This quality-of-life-focused approach may reveal pain or early-stage diseases that might not have been identified during a routine consultation, as the focus and search criteria differ.

Evaluating the impact of illnesses in this context is not merely a clinical examination; it aligns more closely with the reasoning applied when discussing euthanasia decisions, emphasizing the profound influence of disease and pain on an individual cat's daily existence.

This evaluation of discomfort must be complemented by an investigation of *positive signs of welfare*. This means assessing whether, despite its illness or illnesses, the cat exhibits moments of relaxation, has not altered its marking behaviors, and has maintained many rituals with its owners. Naturally, when diseases are severe, chronic, or multiple, the impact on welfare is significant.

The practitioner should engage in discussions with the owner to propose technical solutions, outline their benefits, risks, and costs, and collaboratively design a treatment plan aimed at improving the cat's quality of life while respecting the family's constraints. Perfection is not mandatory: welfare can exist despite certain somatic imperfections. The veterinarian's focus should be on *assessing welfare* rather than simply the absence of disease.

It is also essential to consider the impact of an intensive daily treatment regimen on the cat's welfare. In some cases, the decision may be made *not to treat* if the burden of the treatment outweighs the potential benefits for the animal's quality of life.

This leads us to the concept of managing *psychological suffering*, and we emphasize this term rather than "absence of fear and distress." In our psychiatric model, it is straightforward to reason that an animal should be in a normal adaptive or reactive state, and any pathological state that hinders the animal must be addressed. If the practitioner is unable to establish a behavioral diagnosis, they can often identify psychological distress or, if necessary, refer the case to a colleague, just as they might for certain aspects of a somatic examination.

If a chronic psychiatric condition is present, its impact on the animal's welfare must be assessed. As with somatic illnesses, therapeutic measures and treatments should be proposed to the owner to mitigate this impact. A discussion with the owner is essential to reach a compromise that allows the animal's quality of life to be optimized despite the existence of psychological suffering. For example, a cat that is phobic of humans could be isolated when the owners host many guests, rather than left to hide under furniture with visitors, if significant signs of distress are observed. A cat that has always feared veterinary visits could benefit from gabapentin or pregabalin administration prior to the appointment.

These interventions require the active participation of owners and involve discussions aimed at raising awareness about this concept of welfare. It is also worth emphasizing that, as with somatic conditions, failing to offer available medical treatments when necessary—when the animal is in a pathological state—constitutes a breach of our duty of care and is detrimental to the animal's welfare. Although psychiatry is still regarded differently from other disciplines by many, there is no justification for not treating a malfunctioning brain if we find it logical to treat the kidneys, skin, or any other organ.

Finally, at the highest level, we address the *animal's expectations*. This involves discussing what the specific cat in question enjoys. Owners are often able to provide highly nuanced and precise answers. Based on the cat's preferences, it is essential to determine the extent to which the animal can engage in its favorite activities (see sidebar below). This category is more ambitious than simply verifying physiological needs. It aims to identify markers of pleasure, increase positive moments, and enhance the quality of the human-cat relationship.

Important Point of View

Predictability, boredom, and comfort zone

Numerous studies highlight the importance of predictability in reducing stress in animals, particularly captive animals (Bassett and Buchanan-Smith 2007), as well as in cats (Stella et al. 2014). This concept is also prominent in behavioral therapies for dogs (Amat et al. 2014). However, predictability should not be absolute. A perfectly predictable environment risks leading to boredom. The pursuit of pleasure also involves variations around the comfort zone (Fig. 5.5). Achievements that require effort bring us far more pride than those that are trivial or guaranteed in advance.

It is, therefore, essential to emphasize that taking an animal out of its comfort zone is important, as is providing it with

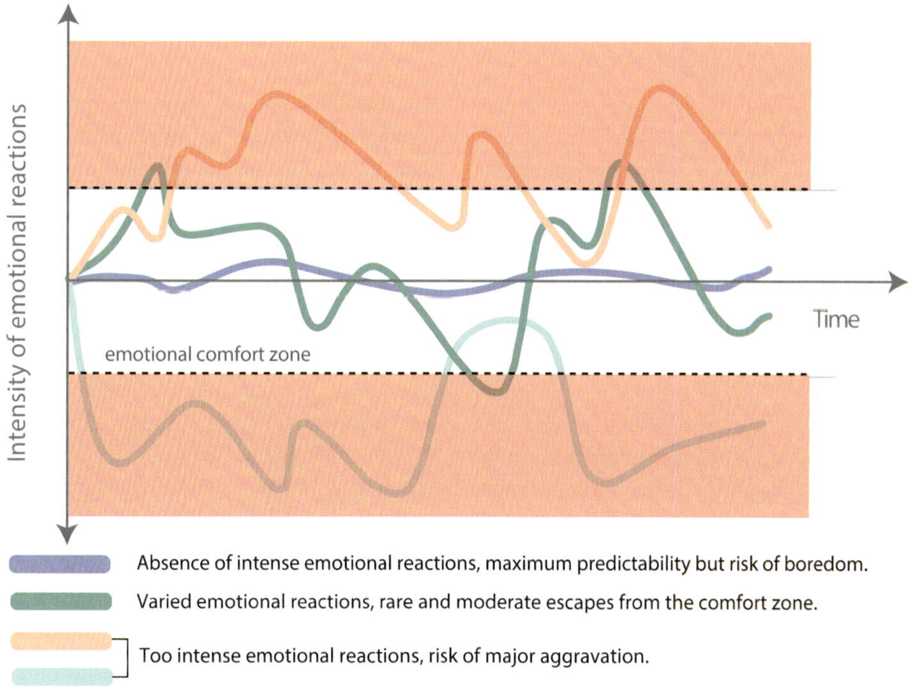

Fig. 5.5 Predictability, boredom, and comfort zone

the opportunity to make choices. The key is to limit variations around the comfort zone without reaching extremes. This goal will again strongly depend on the individual, leading to the same conclusion: our welfare prescriptions must respect the individual's adaptive capacities. The more the animal is in a normal state, the more adaptable, optimistic, and willing it is to accept and even enjoy variations.

Even if 1 day we have analytical tools providing numerical data on pain assessment or pleasure markers to establish welfare scores, it is essential that the co-construction of a cat's welfare be achieved through a compassionate negotiation process that maintains the alliance with the owner. This *therapeutic alliance* (see ▶ Chap. 28, ▶ Sect. 28.2.2) involves a triple agreement on the therapeutic relationship, the therapy goals, and the means to achieve them.

In conclusion, we must emphasize the risk that assessing an animal's welfare might lead us to pass judgment on the owners, who are partly responsible for their pet's well-being. Our role is to provide support while respecting individual accountability for their choices. It is also crucial to remember that welfare is a subjective and relative concept, making it inherently challenging to evaluate. We can only interpret what the animal feels, and this interpretation is shaped by our culture and personal experiences.

Positive welfare markers, which we may 1 day quantify, can be helpful. Regardless, the proposed approach for a welfare evaluation consultation allows us to provide owners with essential insights and support them in improving their pet's quality of life *without being judgmental.*

Summary of ▶ Sect. 5.2

The *evaluation of individual welfare* involves the five domains mentioned in ▶ Sect. 5.1 of this chapter. Each domain follows a logarithmic scale, and overall welfare is calculated as the product, not the sum, of these five domains because it adheres to the "zero contagion rule": if one domain is significantly compromised, it prevents the achievement of an acceptable overall welfare score.

The evaluation of individual welfare cannot be reduced to a mere checklist. A discussion with the owner is essential, and this evaluation should be conceptualized as a *welfare consultation* rather than a simple measurement.

This consultation is tripartite, involving the cat, its owner, and the veterinarian. The veterinarian, as a specialist, combines their observations with the owner's account to assess the cat's experience. Based on this, the veterinarian provides technical advice and suggests possible improvements. The owner, in turn, adjusts these proposed changes according to their practical capabilities and their unique knowledge of their pet.

This consultation must be a *compassionate negotiation*, aligned with our ethics of care and concern, free from judgment, and aimed at maintaining the alliance with the owner. The goal is to assist the owner in enhancing their animal's welfare effectively.

Summary of This Chapter

The evaluation of farm animal welfare, initially conducted at a *collective* level, has progressively shifted toward an *individual* approach. This evolution is reflected in the definition by Anses, which describes animal welfare as a positive mental and physical state linked to the satisfaction of physiological and behavioral needs, as well as expectations. The phrase "normal species-specific behaviors" seems problematic to us, as it fails to account for individual variability. Environmental norms cannot be ideal for every individual. Therefore, we emphasize two crucial points for improving the welfare of companion cats: raising awareness about the importance of matching developmental and living environments, and focusing on the selection of cats by limiting the reproduction of free-roaming or feral cats and promoting behavioral criteria.

Our approach focuses on five modified domains: physiological needs, environmental comfort and safety, physical suffering control, psychological suffering control, and the animal's expectations, all while considering the constraints imposed by other individuals sharing its living environment. In this framework, the practitioner must identify *positive welfare markers* and raise awareness among owners regarding their observations to achieve an improvement in individual welfare.

This evaluation should be conducted through dialogue; while scores or assessment grids may serve as tools to support the practitioner, they should *never replace* the *co-construction of improved welfare*. This process involves three stakeholders: cat, owner, and veterinarian, each with their unique perspectives. The ultimate goal is a compromise reached through a compassionate and collaborative negotiation.

References

Adamelli S, Marinelli L, Normando S, Bono G (2005) Owner and cat features influence the quality of life of the cat. Appl Anim Behav Sci 94(1–2):89–98. https://doi.org/10.1016/j.applanim.2005.02.003

Amat M, Camps T, Le Brech S, Manteca X (2014) Separation anxiety in dogs: the implications of predictability and contextual fear for behavioural treatment. Anim Welf 23(3):263–266. https://doi.org/10.7120/09627286.23.3.263

Anses (2018) Opinion of the French Agency for Food, Environmental and Occupational Health & Safety (Anses) on the definition of animal welfare. *Agence nationale de sécurité sanitaire de l'alimentation, de l'environnement et du travail (Anses)*, Report No. 2016-SA-0283, November 28 2018

Bassett L, Buchanan-Smith HM (2007) Effects of predictability on the welfare of captive animals. Appl Anim Behav Sci 102(3–4):223–245. https://doi.org/10.1016/j.applanim.2006.05.029

Boissy A, Arnould C, Chaillou E, Désiré L, Duvaux-Ponter C, Greiveldinger L, Leterrier C, Richard S, Roussel S, Saint-Dizier H, Meunier-Salaün M, Valance D, Veissier I (2007) Emotions and cognition: a new approach to animal welfare. Anim Welf 16(S1):37–43. https://doi.org/10.1017/S0962728600031717

Butt A, Rashid D, Lockey RF (2012) Do hypoallergenic cats and dogs exist? Ann Allergy Asthma Immunol 108(2):74–76. https://doi.org/10.1016/j.anai.2011.12.005

Casal-Plana N, Manteca X, Dalmau A, Fàbrega E (2017) Influence of enrichment material and herbal compounds in the behaviour and performance of growing pigs. Appl Anim Behav Sci 195:38–43. https://doi.org/10.1016/j.applanim.2017.06.002

Council, F.A.W (1992) FAWC updates the five freedom. Vet Rec 131:357

Delgado MM, Reevy GM (2018) Development and psychometric evaluation of the cat Care and needs scale (CCANS). Anthrozoös 31(1):89–100. https://doi.org/10.1080/08927936.2018.1406203

Dodman NH, Shuster L, Court MH, Dixon R (1987) Investigation into the use of narcotic antagonists in the treatment of a stereotypic behavior pattern (crib-biting) in the horse. Am J Vet Res 48(2):311–319

Duffy DL, De Moura RTD, Serpell JA (2017) Development and evaluation of the Fe-BARQ: a new survey instrument for measuring behavior in domestic cats (Felis s. catus). Behav Process 141:329–341. https://doi.org/10.1016/j.beproc.2017.02.010

5

Hadar BN, Bonnett BN, Poljak Z, Bernardo TM (2023) Morbidity of insured Swedish cats between 2011 and 2016: comparing disease risk in domestic crosses and purebreds. Vet Rec 192(12):e2778. https://doi.org/10.1002/vetr.2778

Hart BL, Hart LA (2013) Your ideal cat: insights into breed and gender differences in cat behavior. Purdue University Press. https://doi.org/10.2307/j.ctt6wq4zv

Herron ME, Buffington CAT (2012) Environmental enrichment for indoor cats: implementing enrichment. Compendium (Yardley, PA) 34(1):E3

Ligabue-Braun R (2020) Hello, kitty: could cat allergy be a form of intoxication? J Venomous Anim Toxins Incl Trop Dis 26:e20200051. https://doi.org/10.1590/1678-9199-jvatitd-2020-0051

LOOF – Livre Officiel des Origines Félines (n.d.) LOOF – Races/annee. Consulté 8 juin 2024, à l'adresse https://bis.loof.asso.fr/stats/recap.php

Malik R (2001) Genetic diseases of cats. J Feline Med Surg 3(2):109–113. https://doi.org/10.1053/jfms.2001.0121

Mariti C, Bowen JE, Campa S, Grebe G, Sighieri C, Gazzano A (2016) Guardians' perceptions of cats' welfare and behavior regarding visiting veterinary clinics. J Appl Anim Welf Sci 19(4):375–384. https://doi.org/10.1080/10888705.2016.1173548

Maslow AH (1943) A theory of human motivation. Psychol Rev 50(4):370–396. https://doi.org/10.1037/h0054346

Montague MJ, Li G, Gandolfi B, Khan R, Aken BL, Searle SMJ, Minx P, Hillier LW, Koboldt DC, Davis BW, Driscoll CA, Barr CS, Blackistone K, Quilez J, Lorente-Galdos B, Marques-Bonet T, Alkan C, Thomas GWC, Hahn MW et al (2014) Comparative analysis of the domestic cat genome reveals genetic signatures underlying feline biology and domestication. Proc Natl Acad Sci USA 111(48):17230–17235. https://doi.org/10.1073/pnas.1410083111

Pedersen NC, Boyle JF, Floyd K, Fudge A, Barker J (1981) An enteric coronavirus infection of cats and its relationship to feline infectious peritonitis. Am J Vet Res 42(3):368–377

Pesteanu-Somogyi LD, Radzai C, Pressler BM (2006) Prevalence of feline infectious peritonitis in specific cat breeds. J Feline Med Surg 8(1):1–5. https://doi.org/10.1016/j.jfms.2005.04.003

Robertson SA, Lascelles BDX (2010) Long-term pain in cats: how much do we know about this important welfare issue? J Feline Med Surg 12(3):188–199. https://doi.org/10.1016/j.jfms.2010.01.002

Rushen J, De Passillé AMB, Schouten W (1990) Stereotypic behavior, endogenous opioids, and post-feeding hypoalgesia in pigs. Physiol Behav 48(1):91–96. https://doi.org/10.1016/0031-9384(90)90267-8

Salman MD, Hutchison J, Ruch-Gallie R, Kogan L, New JC, Kass PH, Scarlett JM (2000) Behavioral reasons for relinquishment of dogs and cats to 12 shelters. J Appl Anim Welf Sci 3(2):93–106. https://doi.org/10.1207/S15327604JAWS0302_2

Salonen M, Vapalahti K, Tiira K, Mäki-Tanila A, Lohi H (2019) Breed differences of heritable behaviour traits in cats. Sci Rep 9(1):7949. https://doi.org/10.1038/s41598-019-44324-x

Satorina J, Szalai K, Willensdorfer A, Mothes-Luksch N, Lukschal A, Jensen-Jarolim E (2014) Do hypoallergenic cats exist? – Determination of major cat allergen Fel d 1 production in normal and hypoallergenic cat breeds. Clin Transl Allergy 4(S2):P11. https://doi.org/10.1186/2045-7022-4-S2-P11

Scarlett JM, Salman MD, New JG, Kass PH (2002) The role of veterinary practitioners in reducing dog and cat relinquishments and euthanasias. J Am Vet Med Assoc 220(3):306–311. https://doi.org/10.2460/javma.2002.220.306

SFEPM – Société Française pour l'Étude et la Protection des Mammifères. (n.d.) Le nombre de chats en France. Consulté 10 juin 2024, à l'adresse https://www.chat-biodiversite.fr/combien-de-chats-en-france.html

Shyne A (2006) Meta-analytic review of the effects of enrichment on stereotypic behavior in zoo mammals. Zoo Biol 25(4):317–337. https://doi.org/10.1002/zoo.20091

Siracusa C, Provoost L, Reisner IR (2017) Dog- and owner-related risk factors for consideration of euthanasia or rehoming before a referral behavioral consultation and for euthanizing or rehoming the dog after the consultation. J Vet Behav 22:46–56. https://doi.org/10.1016/j.jveb.2017.09.007

Stella J, Croney C, Buffington T (2014) Environmental factors that affect the behavior and welfare of domestic cats (Felis silvestris catus) housed in cages. Appl Anim Behav Sci 160:94–105. https://doi.org/10.1016/j.applanim.2014.08.006

TICA – The International Cat Association (n.d.) Browse all breeds. Consulté 8 juin 2024, à l'adresse https://www.tica.org/breeds/browse-all-breeds

Turner, D. C., & Bateson, P. (Éds.). (2013). The domestic cat: the biology of its behaviour (3rd). Cambridge University Press. https://doi.org/10.1017/CBO9781139177177

Vojtkovská V, Voslářová E, Večerek V (2020) Methods of assessment of the welfare of shelter cats: a review. Animals 10(9):1527. https://doi.org/10.3390/ani10091527

Psychopathology

Contents

Chapter 6 The architecture of our veterinary psychiatry mode – 207

Chapter 7 Psychological states – 227

Chapter 8 Disorders related to dysregulation of the fear axi– 289

Chapter 9 Disorders related to dysregulation of the self- co – 329

Chapter 10 Disorders related to dysregulation of the territor – 363

Chapter 11 Disorders related to dysregulation of the relation – 391

Chapter 12 Essential psychiatric disorders – 417

Chapter 13 Aging disorders – 435

Chapter 14 The link between psychiatric disorders and somatic – 473

Chapter 15 Semiology in feline psychiatry – 503

The Architecture of Our Veterinary Psychiatry Model

Contents

6.1 Presentation of the Veterinary Psychiatry Model – 208

6.2 The Initial Consultation – 209
6.2.1 The Collection of Semiological Data: From Sign to Symptom – 211
6.2.2 Behavioral Process and Pathological Pathway – 212
6.2.3 Direct Observation – 213
6.2.4 Clinical Examination – 214
6.2.5 Symptom Assessment and Differential Diagnosis – 214
6.2.6 Complementary Examinations – 214
6.2.7 Diagnosis – 215
6.2.8 Treatments – 218

6.3 Follow-Up and Evaluation of Care – 222

6.4 Clinical Vignettes – 223
6.4.1 Suki – 223
6.4.2 Octavius – 225

References – 226

© The Author(s), under exclusive license to Springer Nature Switzerland AG 2026
S. Bleuer-Elsner et al., *Veterinary Psychiatry of the Cat*,
https://doi.org/10.1007/978-3-032-06260-4_6

Our veterinary psychiatry model is a psychopathological model. It is based on data collected from neurophysiology, ethology, psychopharmacology, and comparative psychiatry. In this chapter, we present our reasoning through the progression of a consultation. We describe the collection of semiological elements, diagnosis, prognosis, treatments, and their main principles. Follow-up elements and the conclusion of care are also discussed.

6.1 Presentation of the Veterinary Psychiatry Model

The model presented in this book has been adopted by many French practitioners. Initiated by Patrick Pageat in 1995 for dogs, it has proven its clinical effectiveness and has been used in the writing of other French works on cats (Beata et al. 2015) and dogs (Masson et al. 2023). Moreover, recent findings in the fields of psychopharmacology (Stahl 2013), behavioral physiology (Carlson and Birkett 2022), and endocrinology applied to behavior (Nelson 2011) have made it possible to link brain function to neural circuits. These advances complement the initial model, which was solely based on identifying one or more neurotransmitter dysregulations, a limitation in many cases.

The brain mechanisms underlying psychopathology are still far from being fully understood, and this book will undoubtedly become outdated quickly. Our goal is to synthesize the scientific elements available to our knowledge on feline veterinary psychiatry, which form the basis of our working hypotheses. Moreover, while the categories proposed in this book are open to discussion and should be tested in clinical practice and research across many cases, they offer the advantage of categorizing cats based on comprehensive clinical profiles of diseases that appear to evolve similarly from case to case. The primary objective of this book is to better distinguish the boundary between normal and pathological states, thereby enabling earlier intervention for at-risk cats.

The choices made in this psychopathological model are based on extensive evidence from psychological and psychiatric research on animals (rodents, primates, dogs) and humans, combined with field knowledge in feline clinical ethology. Paradoxically, while the neurophysiological and psychopharmacological foundations of this model are well-documented, the disease nomenclature and classification choices remain grounded in our clinical experience. Ideally, we would have precise biological tests to quantify and optimize medical treatments, but this is not yet within our reach. Therefore, we must continue to base our proposals and practices on the clinical experience of specialists. Studies in the coming decades will confirm or refute these propositions but naming what we currently observe already seems essential to make clinical research relevant and effective.

Important Point of View

Differences between models and the uniqueness of the french model

French, English, Canadian, and American consultations can differ significantly in terms of duration (ranging from one to 4 h depending on the case), the nature of the professionals involved (whether only the veterinarian is present or accompanied by a trainer), or even the content (with a different expression of the diagnosis). However, they also share many commonalities. For instance, most treatment plans include a behavioral prescription in the form of exercises, often combined with a medical prescription. While the approach to selecting these prescriptions may vary depending on the underlying framework, the overall treatment is largely similar.

When it comes to diagnosis, a major difference makes the French model unique. In the Anglo-Saxon model, the diagnosis is presented as a list that nearly mirrors the symptom assessment in the French model. For example, a cat might receive a diagnosis such as fear, genetics, learning, redirected aggression due to frustration, or generalized anxiety disorder. In this approach, symptoms are assessed, and the emotions, motivations, and contexts involved in the disorder are highlighted. The symptoms are then linked to neurotransmitters to select the appropriate molecule, but no specific disease is ultimately identified.

This choice reflects a form of precautionary principle: without certainty that the observed phenotypes correspond to a single nosographic entity, it is deemed better to remain at the stage of listing observed symptoms rather than risk miscategorization. As a result, therapy is organized to address the treatment of each symptom or undesirable behavior observed.

In the French model, we establish a state diagnosis that corresponds to the animal's brain dysfunctions. This diagnosis is equivalent to the emotional state identified by Anglo-Saxon practitioners and similarly allows us to select a molecule linked to the dysfunction of neural circuits and neurotransmitter systems. However, in our model, we go one step further by naming the disease. For example, we might say that a cat is in a generalized anxious state and exhibits a deprivation syndrome.

The obvious risk is misclassification within our list of diseases. For instance, it is possible that deprivation syndrome encompasses three distinct diseases that have been awkwardly grouped together. Nonetheless, this categorization offers the advantage of helping practitioners understand the genesis and progression of the disorder. This approach enables us to refer to diseases whose progression is known and predictable.

To transpose the Anglo-Saxon approach into another discipline, it would be as if some gastroenterologists chose to stop at a symptom assessment such as "acute vomiting, anterior abdominal pain, moderate diarrhea" without diagnosing a specific disease such as acute gastritis or chronic intestinal disease until certainty could be achieved regarding the diseases potentially associated with the observed symptoms. We believe it is more prudent to classify diseases rather than stopping at a symptom inventory.

This distinction appears to us as a significant difference. Based on our combined experiences, it results in a notable difference in treatment efficacy, and it is for this reason that we decided to produce this book. Our nomenclature has never been published in English or presented internationally, and we wish to share it and invite critique from other approaches to enrich everyone's perspectives.

6.2 The Initial Consultation

Describing a consultation seems to us the most appropriate way to depict veterinary psychiatric care. We will therefore outline the initial consultation and the follow-up, illustrating them with a clinical vignette.

Feline veterinary psychiatry consultations are lengthy, typically lasting 1–2 h. In this chapter, we focus on explaining the theoretical framework of a veterinary psychiatry consultation to demonstrate its functionality. In ▶ Chap. 15, we will describe the technical aspects of semiological data collection, behavior by behavior.

The purpose of a veterinary psychiatry consultation is to determine whether the cat's brain is functioning optimally or if a dysfunction can be identified. To achieve this objective, the practitioner reviews all behaviors reported by the owner, compares them with their own direct observations, and conducts a clinical examination of the animal (◘ Fig. 6.1).

By doing so, the veterinarian identifies a range of symptoms, which may include abnormal behaviors (e.g., tail chasing) or normal behaviors exhibited with abnormal frequency or intensity (e.g., hyposomnia, binge eating,

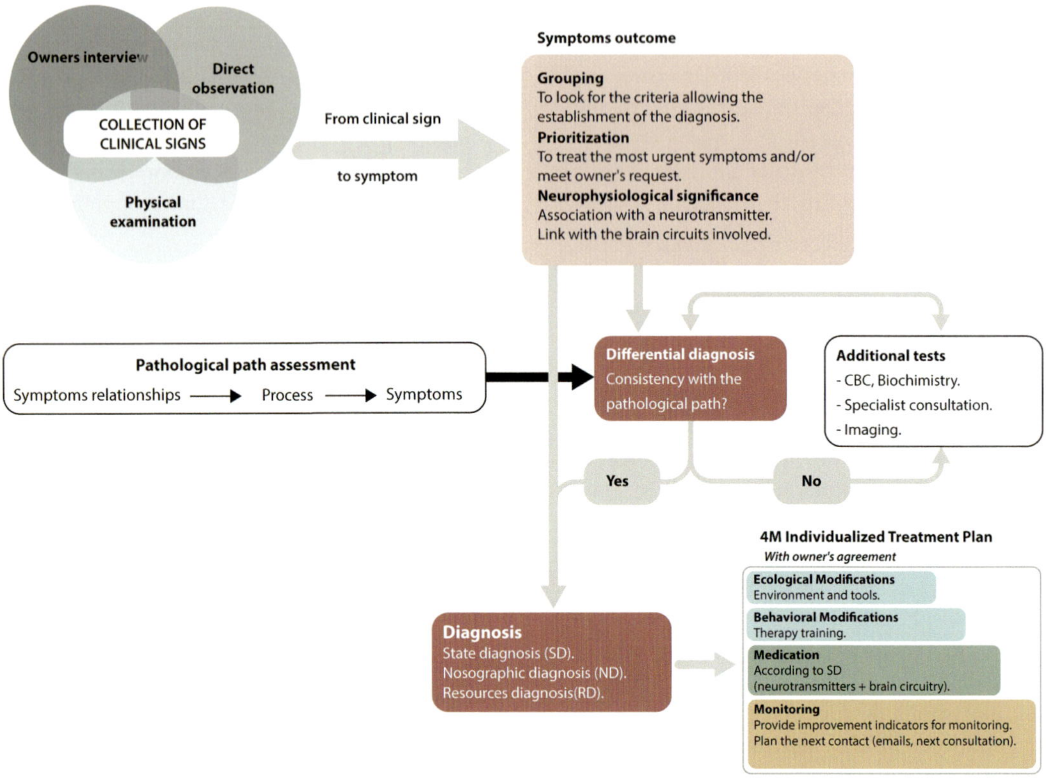

Fig. 6.1 Process of the initial consultation. A veterinary psychiatry consultation consists of three components: the interview with the owners, the clinical examination, and direct observation. These three dimensions of the consultation enable the practitioner to establish a symptom assessment, followed by a differential diagnosis. The latter is narrowed down through additional tests and validation of diagnostic criteria. The veterinarian then evaluates the pathological pathway to identify the most likely condition, which becomes the psychiatric diagnosis. The individualized treatment plan is directly derived from the established diagnosis

compulsive licking of the abdomen). Each symptom, as highlighted in ▶ Chap. 2, can be linked to brain circuits involving multiple neurotransmitters. Together, these symptoms form the symptom assessment.

The practitioner organizes this symptom assessment (through grouping, prioritization, and neurophysiological interpretation) to generate a list of possible diseases in which these symptoms are observed, forming the differential diagnosis. In some cases, the veterinarian may identify a non-behavioral, partial, or complete origin of the observed disorders. These organic conditions are naturally included in the list of diseases outlined in the differential diagnosis.

This list is then analyzed, reorganized, and prioritized to narrow it down to a smaller set of potential conditions relevant to the clinical case at hand. Some diseases may be quickly ruled out due to the absence of specific diagnostic criteria, while others can only be confirmed or excluded after conducting additional tests.

The selected diagnoses are those that account for the entire clinical presentation, including the pathological pathway. The practitioner narrows the diagnostic hypotheses to a limited number of conditions, ranking them from most to least probable. Special attention is given to potential conditions that involve *significant danger* or *high urgency*, as these factors can influence the order of prescriptions.

This diagnostic approach is the standard method used across all veterinary disciplines.

The practitioner must also determine the *pathological pathway* of the presented case. This pathway links symptoms to both elementary and *complex processes*, providing an explanation for the progression of observed symptoms over time. When the pathological pathway accounts for the entire clinical presentation, the practitioner can further refine the diagnosis by excluding additional hypotheses and proceed to develop an individualized treatment plan.

The *psychiatric diagnosis* consists of three levels: the state diagnosis, the nosographic diagnosis, and the diagnosis of the resources within the owner-cat-environment system. It provides both a *prognosis* and an *individualized treatment plan*. This plan always includes behavioral modifications and, when necessary, medical intervention. It is accompanied by a detailed follow-up to maintain the *therapeutic alliance* essential for effective care.

The duration of treatment is often long, ranging from a few weeks for mild cases to several months or even years for more severe cases. In certain diseases, fortunately rare, such as dysthymia or dissociative syndrome, lifelong management may be considered.

These extended durations are explained by the therapeutic necessity of incorporating learning exercises into the treatment. As discussed in ► Chap. 2 (see ► Sect. 2.3.3.1), learning requires the synthesis of specific proteins to transition into long-term memory, a process that takes several months to ensure lasting effects.

6.2.1 The Collection of Semiological Data: From Sign to Symptom

The practitioner conducts the classic work of anamnesis, with the primary difference from other medical disciplines being the large number of items to explore.

A *sign* is the behavior as described by the owner. For example, the owner might report that their cat "*gets stressed around guests,*" "*hisses at certain people, especially children,*" or "*becomes aggressive and scratches us when we try to play.*" The practitioner's task is to transform these signs into symptoms, objectifiable elements of semiology that can be linked to neurophysiological circuits and neurotransmitters (see ► Chap. 2, ► Sect. 2.2.4) during the symptom assessment.

In these examples, the symptoms are intense fear reactions with autonomic manifestations triggered by humans, irritation-based aggression in a fearful context, and a delay in acquiring control over scratching behavior.

This *transformation of a sign into a symptom* is a critical aspect of behavioral consultations, as it allows the practitioner to identify key elements that will contribute to the symptom assessment, the differential diagnosis, and ultimately the final diagnosis. This step inevitably involves a degree of interpretation by the practitioner.

For instance, a play sequence that ends with the cat scratching the owner's arm can be interpreted in various ways (see ► Chap. 15). It could indicate an injurious sequence during play in a cat lacking motor control, or it could be irritation-based aggression in a context of fear or frustration. What guides the practitioner in selecting one hypothesis over another is the coherence of the observed signs.

In this example, if the cat displays visible locomotion control deficits during the consultation, has disorganized exploration, never sleeps while there are external stimuli, can play for hours alone with its toys, and did not hiss before biting in the described sequence, the hypothesis of an injurious play sequence would be favored. Conversely, if the cat exhibits frequent fear reactions toward strangers, and the play sequence occurred with a visitor during which the cat occasionally fled, returned, and hissed when the person insisted on physical contact, the hypothesis of irritation-based aggression in a fearful context would be prioritized.

Note that while semiology can, as in other disciplines, be collected sequentially, behavior by behavior, experienced practitioners will often adopt a more targeted semiological approach. They navigate from one behavior to another based on the elements they wish to confirm or rule out. Typically, the practitioner formulates a hypothesis regarding the potential dysfunction of one of the four axes (self-control, fear, relationships, and territoriality) and seeks additional elements within that axis to assess its coherence. In every case, all axes are systematically reviewed to test the validated hypothesis. In addition to these four axes, communication, organic diseases, and the cat's environmental conditions are also systematically explored.

6

The investigative method leading to diagnosis can be modeled using decision trees, with the practitioner formulating hypotheses and verifying them. This approach often makes the interview more pleasant for the owner, as they transition from one behavior to another following their logic and the animal's history, rather than adhering to a strictly linear, behavior-by-behavior path. This interview method, combined with *open-ended questioning*, enables the collection of richer information. It also allows the practitioner to begin grouping and prioritizing symptoms during the questioning phase.

Finally, *direct observation* and the *clinical examination* of the animal, detailed in ▶ Sects. 6.2.3 and 6.2.4, complement the list of symptoms, forming the animal's *symptom assessment.*

6.2.2 Behavioral Process and Pathological Pathway

A behavioral process is a set of innate learning mechanisms that enables the orientation and regulation of behavioral responses.

We differentiate between elementary processes and complex processes. *Elementary processes* can be explained at the neurophysiological level through simple cellular mechanisms, whereas *complex processes* involve a combination of numerous processes, the details of which are not always fully understood (◻ Table 6.1).

◻ **Table 6.1** Overview of elementary and complex processes

Elementary processes	Complex processes
Sensitization/ habituation	Acquisition of sensory homeostasis
Anticipation	Acquisition of self-control
Generalization/ discrimination	Relationships and attachment
Activation/inhibition	Territorial organization
Instrumentalization	Aging

Elementary and complex processes allow the symptoms to be structured chronologically, forming the pathological pathway

The clinical value of naming behavioral processes lies in linking symptoms together and providing reasoning to explain how the initial symptom evolved logically. This approach allows the practitioner to verify the coherence of the case in relation to the history reported by the owners. The pathological pathway helps identify a non-behavioral origin of the illness and situates the symptoms within a nosography with a known progression. Consequently, the practitioner is better equipped to provide a prognosis.

In practice, the symptoms observed and their evolution throughout the cat's life result from the combination of various processes that explain the disease's observed progression **(see box below).** With experience, the veterinary psychiatrist becomes adept at identifying inconsistencies in history. Any point of inconsistency in the pathological pathway should raise an alert and prompt the investigation of an underlying somatic disease or comorbidity. Ultimately, the proposed diagnosis must be able to account for all the presented symptoms.

Providing a description of the normal evolution of symptomatology for each specific condition is one of the key strengths of this *psychopathological model.* Conversely, any deviation from this logical progression serves as an excellent tool for detecting a non-behavioral origin.

Clinical Illustration

Ms. S. seeks consultation because her 10-month-old female cat, Naya, a Bengal-European crossbreed, was "traumatized during sterilization" at the age of 7 months. The clinic that performed the procedure is known to the veterinarian and follows a very compassionate protocol for handling hospitalized animals. This suggested link between sterilization and a sudden change in behavior therefore raises the veterinarian's concern.

Semiological data reveal that Naya has always been a "shy" cat, tending to flee when people reach out to her. However, during visits at home, she would eventually approach after about 10 min and, once she accepted the visitor, would become friendly. This behavior was never a concern for Ms. S., as she considered it part of Naya's "personality." The history further reveals that Naya was adopted at 4 months from a shelter after being found wandering in the countryside around 3 months old, with no information about her origins.

The clinical examination is difficult because Naya exhibits numerous signs of fear toward the veterinarian. She tries to flee, avoids contact, shows paw pad sweating, and trembles. A thorough examination is not possible without restraining her in a towel, so it is postponed to avoid worsening her condition. A comprehensive blood panel was conducted during her sterilization anesthesia.

The pathological pathway highlights vulnerabilities in the fear axis since Naya's adoption. Her developmental conditions were likely suboptimal, with an understimulating environment and possibly very little human contact before her arrival at the shelter during the socialization period. Her interspecific social phobia was moderate but present. Sterilization occurred at puberty, a sensitive period during which subtle symptoms often worsen. Despite the care team's kindness, Naya's fear axis vulnerability led to sensitization to handling during the procedure.

At subsequent visits, anticipation of handling becomes evident, as Naya displays signs of fear when her owners try to place her in her carrier. What initially appeared to be a post-traumatic care-related phobia is more accurately explained as an exacerbation of her pre-existing social phobia triggered by puberty. Differentiating between these conditions is crucial in clinical practice, as the prognosis for the latter is less favorable.

6.2.3 Direct Observation

During the consultation and the collection of semiological data, the practitioner must observe the cat's reactions. However, observing a cat in an unfamiliar environment is markedly different from observing it in its own territory. Fear reactions may be pronounced, and many behaviors may be absent. Once the cat reaches puberty, it is unlikely to exhibit its typical behaviors in a new environment, as it needs to make the unfamiliar space more reassuring, for example, by performing facial marking. Additionally, the consultation room is often already associated with unpleasant memories, such as vaccinations or sterilization during adolescence, a particularly vulnerable period. As a result, few adult cats are genuinely comfortable and relaxed in a consultation room.

Follow-up consultations can be conducted remotely via video or at the owner's home. These approaches facilitate observation in the cat's usual environment and eliminate the stress of transportation and clinic visits.

Observing the cat allows the practitioner to correct potential misinterpretations by the owner in preparation for therapy. For instance, if the owner has not noticed signs of fear in their cat, the practitioner can highlight the animal's postures during fearful reactions, teaching the owner to recognize and address these signs earlier in the behavioral sequence.

Finally, direct observation provides an opportunity to assess the dynamics of the owner-cat relationship and the owner's communication with the animal.

6.2.4 Clinical Examination

The *clinical examination* is conducted as in any veterinary consultation. However, practitioners specializing in behavioral medicine focus particularly on identifying targeted pain and physical or sensory impairments (e.g., motor function, sensory organs) that might explain the observed behaviors. Additionally, the practitioner's in-depth knowledge of the possible connections between internal medicine, dermatology, neurology, and endocrinology serves as a valuable tool for ruling out a behavioral origin of the disorder.

Many texts portray behavior as a discipline to consider only after all others have failed. Our intention is to position it at the same level as other disciplines, complementing them, as there is no logical reason to prioritize the brain and its dysfunctions after addressing all other organs. It is worth noting that behavior remains the leading cause of euthanasia and abandonment (Casey et al. 2009; Scarlett et al. 1999). It is therefore difficult to be a veterinarian committed to the health and welfare of animals while relegating behavioral concerns to a secondary status within veterinary medicine.

Regardless of the order of examinations, we believe that a behavioral consultation should take place as soon as the general practitioner suspects emotional, cognitive, or mood-related involvement. Sometimes the origin is behavioral, sometimes it is not, but in all cases, the psychological aspect of the condition must be addressed. For example, a chronic disease causing pruritus or pain may lead to an anxious state that does not necessarily resolve, even after the primary condition has been cured.

6.2.5 Symptom Assessment and Differential Diagnosis

The *symptom assessment* is the cornerstone that enables the practitioner to establish a differential diagnosis. Following the collection of semiological data, direct observation, and clinical examination, the practitioner compiles and organizes the list of identified symptoms.

First, the symptoms are categorized to group the elements characteristic of a given state diagnosis. As with the pathological pathway, any inconsistencies should act as alerts. The collected elements must always enable the practitioner to determine the *state diagnosis* by the end of the consultation, whether the diagnosis is normal or pathological.

Once the state diagnosis is established, the practitioner prioritizes the symptoms by categorizing them within *behavioral axes* to validate or refute nosographic criteria. The practitioner then creates a list of possible conditions, ranked from most to least probable. This list constitutes the *differential diagnosis* and includes suspected somatic diseases. In cases where multiple conditions could explain the observed state, the practitioner must prioritize them (◘ Fig. 6.2).

This medical aspect of the consultation is a key factor that makes the process unachievable by a non-veterinary professional.

6.2.6 Complementary Examinations

Complementary examinations are essential for ruling out or confirming the hypotheses included in the differential diagnosis. In ► Chap. 14, which focuses on the link between psychiatry and other disciplines, we will revisit the complementary examinations most commonly used in feline veterinary psychiatry.

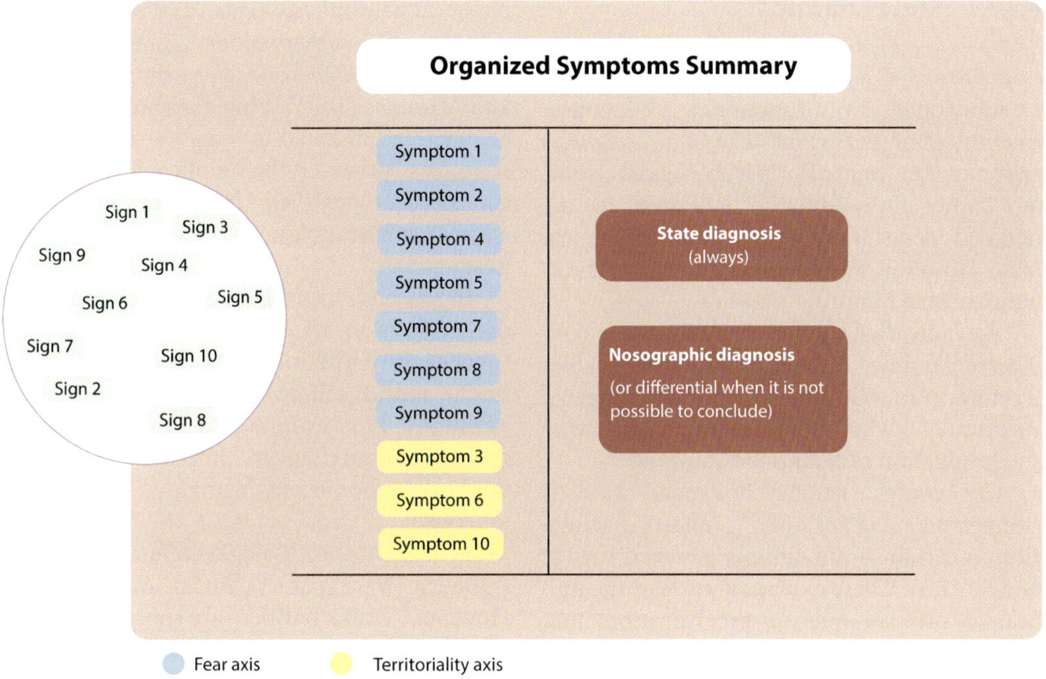

Fig. 6.2 Symptom assessment and prioritization. During the consultation, the practitioner collects a series of symptoms indicative of potential brain dysfunctions and behavioral disorders. These symptoms are organized according to the practitioner's diagnostic hypotheses to highlight the coherence of the clinical picture. This process leads to the differential diagnosis and, ultimately, to the final state diagnosis (always) and nosographic diagnosis

Some of the most frequently prescribed tests include urinalysis, blood pressure measurement, and standard or more specific blood panels that assess thyroid function. We also often rely on specialized consultations in other disciplines (neurology, dermatology, orthopedics) when their areas of expertise appear relevant to the observed behavioral abnormalities (e.g., anxiety-induced licking triggered by atopic dermatitis) or when such abnormalities result from the behavioral disorder (e.g., anxiety-induced licking dermatitis).

Medical imaging, including MRI, is also frequently prescribed to rule out certain organic diseases, such as compressive tumors or hydrocephalus. Soon, additional genetic and biological tests are expected to become available and will likely become standard complementary examinations.

6.2.7 Diagnosis

In veterinary psychiatry, the *diagnosis* consists of three levels. The *state diagnosis* reflects the functioning of the brain. The *nosographic diagnosis*, or the name of the disease, provides insight into its pathogenesis, progression, common comorbidities, differential diagnosis, possible treatments, and prognosis. Finally, the third level is the *assessment of the system's resources*. While this aspect can complement the pathological pathway and provide insight into the genesis of the disorders, it also serves to develop tailored treatment plans that account for the owners' circumstances. The therapist's role is not to reshape the world to suit the cat but to help the cat integrate into the world around it.

6.2.7.1 State Diagnosis

The *psychological state* encompasses the temporarily stable mechanisms that characterize an individual's brain functioning and, consequently, their behaviors. The *state diagnosis* reflects the animal's psychological state. ► Chapter 7 is entirely dedicated to the detailed description of each psychological state. However, we can already define and categorize them here.

An individual's psychological state is considered *physiological* or *normal* when it enables a return to equilibrium in response to internal or external environmental changes—allowing the production of adaptive behaviors.

Conversely, the psychological state is deemed *pathological* when it reflects dysfunction in certain brain circuits (see ► Chap. 2, ► Sect. 2.3). These dysfunctions lead the animal to produce *maladaptive* behaviors that cause *suffering* and are *not spontaneously reversible*.

Our working hypothesis is as follows: by understanding the effect of each circuit on behavior, it is possible to link observed symptoms to dysfunctions in these circuits. The state diagnosis, therefore, corresponds to a description of the presumed alteration in certain neurophysiological circuits. This model is based on neurophysiological and psychopharmacological observations presented in *Essential Psychopharmacology* by Stahl (Stahl and Muntner 2021). While these observations are well-documented in humans, they are less so in cats. However, the significant number of similarities identified in other mammals allows for extrapolation to model state diagnoses in cats.

In practice, the rule for establishing a state diagnosis is to select at most one diagnosis from each of the following categories: emotional, mood, self-control deficits, and derealization. For example, it is possible to diagnose an anxious and impulsive state, but not both a phobic and an anxious state simultaneously.

In many cases, a single state diagnosis is sufficient to explain clinical observations. However, comorbidities are possible when multiple circuits are affected. The diagnosis listed first is the one most prominent for the animal in question (◘ Table 6.2).

6.2.7.2 Nosographic Diagnosis

The nosographic diagnosis is established by naming the disease. ► Chapters 8, 9, 10, 11, 12 and 13 provide a systematic description of each disease. However, the different categories

◘ **Table 6.2** Pathological state diagnosis. In our model, there are nine pathological states, categorized into three types of dysregulation: emotional, mood, and cognitive. Cognitive dysregulations include self-control deficits and derealizations. The state diagnosis can be plural but must include only one diagnosis per type of dysregulation (one choice per color in the table). For instance, an anxious state can coexist with a depressive state and a senile state, but a choice must be made between a phobic state and an anxious state, or between an impulsive state and a compulsive state

EMOTIONAL PATHOLOGICAL STATES		MOOD PATHOLOGICAL STATES	
Phobic	Simple /complex	Depressive	Acute/chronic
Anxious	Productive/inhibited	Manic	Contextual/invasive
	Intermittent/generalised	Bipolar	
PATHOLOGICAL STATES OF COGNITION			
Pathological states of lack of self-controls		Pathological states of lack of derealization	
Impulsive	Contextual/invasive	Dissociated	
Compulsive	Contextual/invasive	Senile	

and their exhaustive list are presented in Table 6.3. These diseases are categorized according to the four axes: fear, self-control, relationships, and territory, along with additional categories for aging-related diseases and essential psychiatric disorders.

Table 6.3 Calssification of behavioral disorders in feline veterinary psychiatry

	Fear axis disorders
1	Categorical phobia
2	Deprivation syndrome
3	Post-traumatic phobia
	Self-control axis disorders
4	HSHA syndrome or feline ADHD
5	Addictive behaviors
	Territoriality axis disorders
6	Disorders resulting from territorial poverty
7	Disorders resulting from territorial disruption
8	Territorial disorders secondary to another behavioral disorder
	Relationship axis disorders
9	Interspecific relational disorders Linked to dysfunctional communication Secondary to another behavioral disorder
10	Intraspecific relational disorders Linked to dysfunctional communication Secondary to another behavioral disorder
	Aging disorders
11	Cognitive dysregulation
12	Emotional dysregulation
13	Mood dysregulation
14	Self-controls dysregulation
	Essential psychiatric disorders
15	Dissociative syndrome
16	Dysthymia
	Psychiatric disorders and organic diseases
17	Pain
	Internal medicine
	Dermatology
	Neurology
	Nutrition
	Reproduction
	Endocrinology
	Orthopedics
	Aging

The different behavioral disorders are organized into several categories: the four behavioral axes, aging-related disorders, essential psychiatric disorders, and organic diseases linked to behavioral disorders

The relationship between psychiatric disorders—whether as a cause or a consequence—and somatic diseases is explored in detail in ► Chap. 14.

We do not yet have all the necessary elements to classify psychiatric disorders based on objective biological markers, such as genetic tests, blood analyses, or medical imaging. However, the collection of symptoms and their similar progression observed in clinical practice allows us to propose a categorization.

We acknowledge the inevitable inaccuracy of this model, but having a suspected etiopathogenesis, frequent comorbidities, defined treatments, and a prognosis for each disorder provides a valuable medical foundation that allows us to optimize clinical care. This nosography, which goes beyond merely listing symptoms, enables us to better understand the potential progression of the cats we see in consultations.

Additionally, it facilitates the use of a shared terminology for large-scale studies in cats or dogs, similar to those already conducted in human psychiatry.

Finally, as veterinarians, this approach aligns with the clinical reasoning we apply to all medical disciplines. There is no clear reason to treat brain functioning differently from that of other organs.

6.2.7.3 Resource Diagnosis

This level of diagnosis is likely emphasized more in our discipline than in others due to the long durations of treatment, but it is important in all areas of veterinary medicine. The resource diagnosis identifies the constraints specific to the clinical case and plays a crucial role in providing an accurate prognosis to the owner. It evaluates the *living conditions* and *contextual factors* that may have a positive or negative influence on the disease's progression.

6

At the conclusion of the consultation, the clinician compiles a list of the owner's strengths and weaknesses in terms of resources to assess the potential impacts, positive or negative, of implementing the treatment plan. These resources, listed in ◘ Fig. 6.3, include the owner's financial capacity, availability of time, communication quality, and limitations related to their context (e.g., disabilities, environmental constraints, presence of children, presence of other animals).

The diagnosis also considers the cat's characteristics, such as its physical abilities, coexisting illnesses, or other treatments that might impede behavioral care (e.g., chronic pain, treatment incompatibilities, inability to administer pills).

This level of diagnosis is critical for treatment success, as even a mild condition cannot be effectively addressed if the owner is unable to adhere to the individualized treatment plan proposed by the clinician.

For this reason, the clinician dedicates part of the consultation to verifying the feasibility of the proposed plan. The very existence of this diagnostic level underscores the *responsibility of the care team*. The proposed therapy must account for these contextual characteristics, and if it is not followed, it indicates that this aspect of the diagnosis was poorly established.

6.2.8 Treatments

Treatments in veterinary psychiatry include a traditional medical component, as commonly practiced, but also a behavioral therapy component, which is far less typical in veterinary medicine.

6.2.8.1 Pharmacological Treatments

The role of pharmacological treatment is not to cure the condition in the strict sense but to *facilitate behavioral therapy*, which is the primary driver of desired improvements.

As discussed earlier, when a practitioner establishes a pathological state diagnosis, they hypothesize dysfunction in one or more brain circuits. Practically, this means the cat is repeatedly unable to produce adaptive behav-

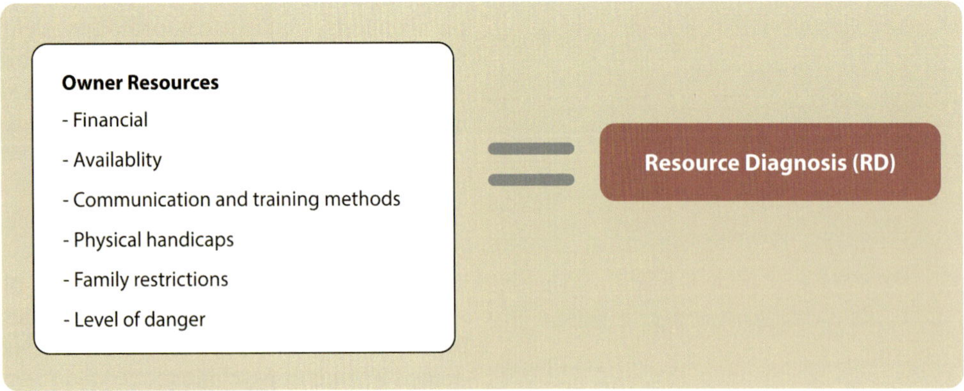

◘ **Fig. 6.3** Resource diagnosis. The resource diagnosis encompasses all factors not directly related to the animal's condition that influence the treatment process

ioral responses to its environment and is suffering as a result. The medication's role is to address brain dysfunction, at least partially, by improving the brain's plasticity. Several studies have demonstrated that individuals learn poorly when their sleep is disrupted (Hobson and Pace-Schott 2002; Walker and Stickgold 2004) and that psychotropic drugs promote neurogenesis (Khodanovich et al. 2018; Micheli et al. 2018). Recognizing a pathological state should, therefore, lead the practitioner to prescribe behavioral therapy alongside medication.

Behavioral therapy requires the animal to acquire new learning. For this to succeed, the animal must have strong adaptability and mental flexibility. This is the primary role of medication and explains why it typically precedes behavioral therapy. In practice, it is possible to implement a purely behavioral approach if the psychological state allows and if behavioral rigidity remains relatively mild, provided time and goal limits are established. This approach parallels cases of mild lameness or isolated diarrhea, which can resolve spontaneously when rest or a restricted diet is introduced.

Pharmacological treatment often raises concerns among owners, yet in our experience, it is a significant facilitator of therapy. It often helps prevent owner burnout caused by the cat's problematic behavior, which can lead to demotivation and a vicious cycle that damages the human-animal bond. Medication also helps avoid behavioral worsening, a risk factor for abandonment or euthanasia. Without medication, learning often fails to occur or progresses too slowly, and the cat's exaggerated emotional reactions may prevent it from exhibiting the appropriate behaviors that could be reinforced.

The connection between the state diagnosis and the dysfunction of a specific neurophysiological circuit, based on current psychopharmacological knowledge (Stahl 2013), identifies the neurotransmitters involved in the implicated circuit. By comparing this theoretical framework with the predominant neurotransmitters involved in the symptom assessment conducted during the consultation, the practitioner selects the appropriate medication (◘ Fig. 6.4). These principles are detailed in the third part of this book.

Pharmacological treatment is theoretically guided by the implicated circuits but is personalized and refined according to the symptoms the practitioner prioritizes for treatment.

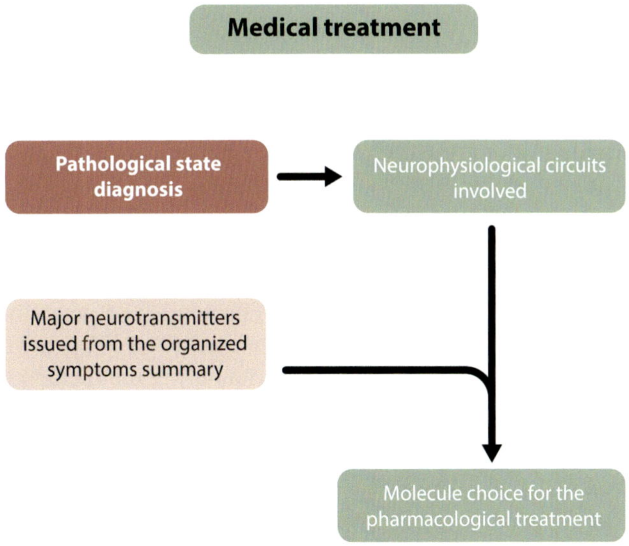

◘ **Fig. 6.4** Selection of pharmacological treatment. The state diagnosis identifies neurophysiological dysfunctions and the neurotransmitters involved. This enables a reasoned choice of psychotropic medication to prescribe in cases of pathological states

It is important to note that, in some cases, it is impossible to address all symptoms with a single medication. The veterinarian must then choose between successive monotherapies or an immediate combination of several drugs. As a general rule, monotherapy is preferred, particularly during the first consultation. However, with experience, combining multiple medications is not uncommon, as discussed in ▶ Chap. 16 (▶ Sect. 16.2.2). The golden rule remains the evaluation of the benefit-risk balance.

Treatment durations in veterinary psychiatry are often longer than in other disciplines, as they always include a behavioral therapy component requiring repetition and learning. As noted in ▶ Chap. 2 (see ▶ Sect. 2.3.3.1), long-term memory formation necessitates the synthesis of specific proteins, making it unrealistic to achieve durable behavioral changes within just a few weeks.

These treatment plans, which can span several months or even years depending on the disorder, often require discussion before prescribing psychotropic medications. This reassures owners about the expected effects and potential risks, which are minor compared to many other commonly used medications, such as glucocorticoids (Nerhagen et al. 2021) or antiepileptics (Charalambous et al. 2018).

6.2.8.2 Behavioral Therapy

Behavioral therapy encompasses two components. The first is the *environmental aspect*, involving changes decided during the consultation that do not require training. For example, the practitioner might recommend altering the diet, introducing a new litter box, or using specific toys or feeding bowls.

In addition to these ecological or *environmental modifications*, *behavioral modifications*, i.e., exercises, are introduced to achieve long-term changes in the cat's behaviors through new learning. Owners must regularly repeat these prescribed exercises following the consultation until these behaviors become habits, almost automatic, and less susceptible to further changes (◘ Fig. 6.5).

As previously discussed, the choice of pharmacological treatment is guided by the state diagnosis. In contrast, therapy depends directly on the nosographic diagnosis. The final part of this book is entirely dedicated to therapies, detailing the most common ones.

Our nosography is built around the four primary behavioral axes: the acquisition of self-control, the management of fear, relationships, and territorial organization. This classification directly guides the category of therapies the practitioner will employ, as these are also organized around these four pillars. Depending on the condition, the practitioner may use different types of therapy: systemic, cognitive, or behavioral. Generally, therapies for disorders related to relationships and territorial organization are more likely to involve *systemic approaches*. In contrast, disorders of the self-control and fear axes are typically addressed through *cognitive-behavioral therapies*.

The communication axis is also addressed, as improving communication consistently benefits the overall clinical picture, regardless of whether the cat is in a physiological or pathological state.

Based on the symptoms identified in the symptom assessment, the clinician selects exercises within the relevant behavioral axis. Simpler exercises are prescribed initially, and during follow-up consultations, more complex exercises are added as the cat progresses, aiming for the eventual elimination of symptoms. As emphasized earlier, it is crucial to ensure the feasibility of the proposed therapy.

The number of prescribed exercises should be limited to avoid overwhelming the owner and rendering the therapy impractical.

Therapeutic prescription is also an ideal time to select specific signs for the owner to monitor and document regularly throughout

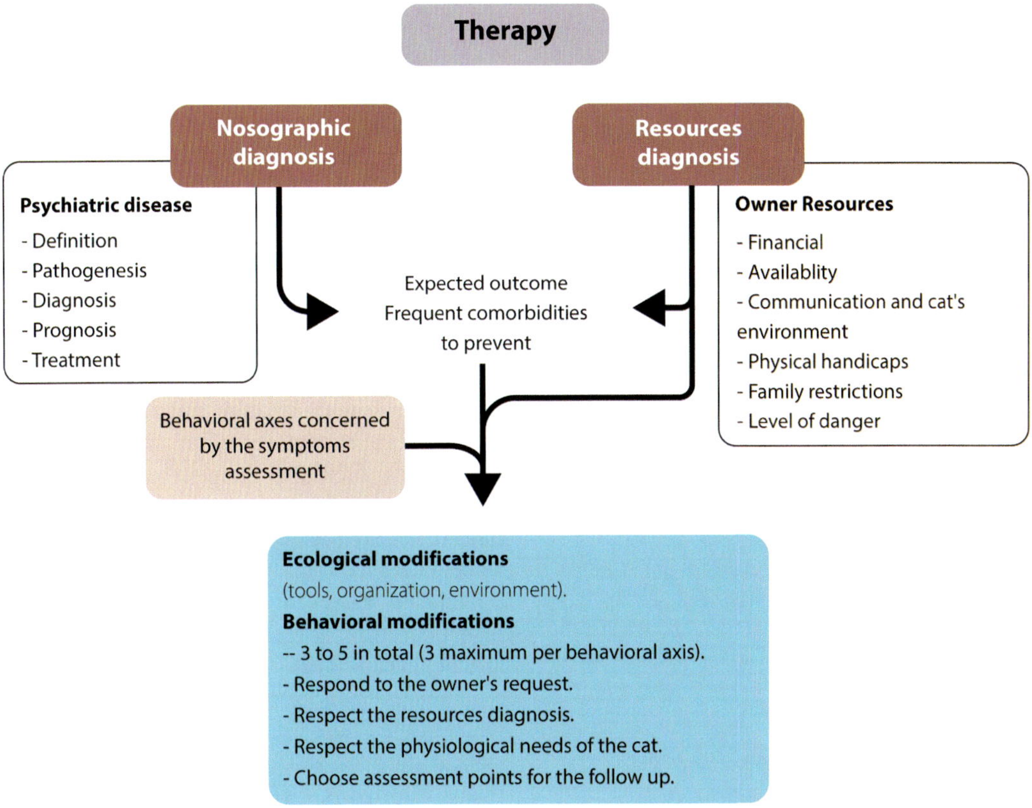

Fig. 6.5 Principles of constructing behavioral therapy. Therapy is primarily derived from the nosographic diagnosis and is built around the resources of the cat-owner system, the behavioral axes involved, and the desired progression, while preventing potential comorbidities. Therapy is divided into ecological and behavioral measures, limiting the latter to three per behavioral axis and a total of five measures. The therapy must address the owner's concerns, respect the cat's physiological needs, and include measurable elements for follow-up

the treatment. For instance, the practitioner may ask the owner to track tolerated playtime or the time it takes for the cat to approach visitors. This not only evaluates the treatment's effectiveness but also strengthens the *therapeutic alliance* by creating a mutual understanding of the *objectives* and the *steps* to achieve them.

The *individual treatment plan* is the personalized outcome of the care proposed to the client. It includes four components, which are easily remembered as they all begin with the letter "M": **m**edicalization, ecological **m**odification, behavioral **m**odification, and **m**onitoring (Fig. 6.6).

The risk of *relinquishment* or *euthanasia* must be addressed by the practitioner to gauge the urgency of the consultation. *Risk assessment* is a critical factor, and when a severe diagnosis is made, it is essential to set limits for the treatment plan. Practitioners should also remember their social responsibility, which extends beyond meeting the owner's requests and protecting animal welfare.

Fig. 6.6 Establishing the individual treatment plan. The symptom assessment and prioritization lead to the establishment of state and nosographic diagnoses. The state diagnosis identifies the implicated brain circuits and the neurotransmitters regulating them, guiding the selection of pharmacological treatment. The nosographic diagnosis provides the basis for constructing the therapy while taking the resource diagnosis into account. All these elements are presented to the owner to assess the therapeutic alliance and to determine in advance the treatment duration, evaluation milestones, as well as the frequency and format of follow-up sessions

6.3 Follow-Up and Evaluation of Care

The initial consultation marks the beginning of the care process. During this consultation, the practitioner defines the follow-up modalities (e.g., email, phone) and the frequency of contact leading up to the follow-up consultation.

Between the initial and follow-up consultations, medication dosages are often adjusted to optimize effects. A cat undergoing psychotropic treatment should not experience a change in personality. The medication is intended to facilitate therapy but should not make the cat lethargic. Conversely, if no positive effects are observed, the dosage may need to be increased.

During the initial consultation, the practitioner carefully notes the baseline of problematic behaviors. This allows them to determine whether the symptom is improving, stable, or worsening. This observation is critical, as the stabilization of a symptom that was previously improving may indicate a regression compared to the start of treatment.

At the follow-up consultation, the practitioner focuses on assessing the progression of various symptoms and the owner's adherence to the treatment plan. This enables adjustments to the individualized treatment plan before the next consultation (Fig. 6.7).

Owners are often asked to provide an *improvement rating*, which is an effective tool for determining when to begin tapering off medication, where possible. For example, the practitioner might decide to discontinue treatment after 3 months of ratings equal to or greater than 7/10. Treatment concludes with the completion of the tapering process, provided the owner agrees to recontact the veterinarian if symptoms reappear after the complete discontinuation of medication.

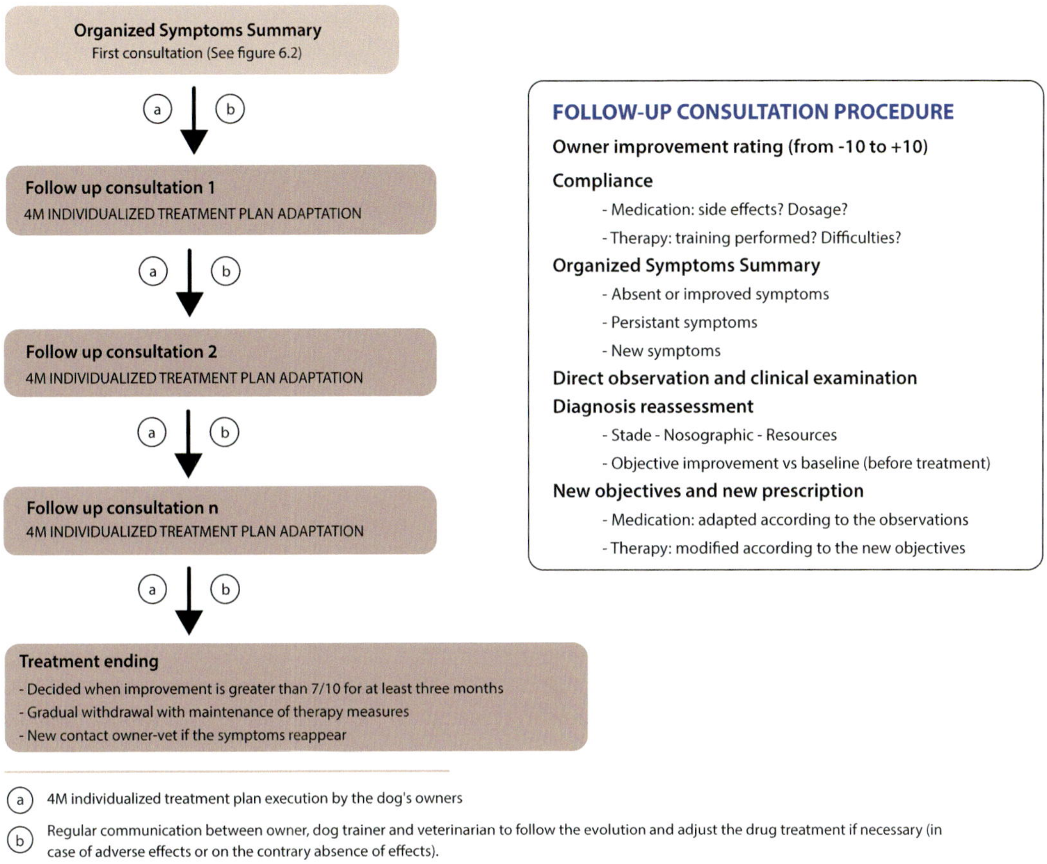

Fig. 6.7 Follow-up and evaluation of care. The follow-up consultation primarily involves evaluating the cat's behavior as reported by the owners (which may show signs of worsening) and assessing the quality of their adherence to the treatment plan. The symptom assessment is then updated based on the cat's progress, leading to a reassessment of the diagnosis and treatment, along with an agreement on any new objectives. A series of follow-up consultations may be required before concluding the care process. Between consultations, regular communication with the owner is essential to ensure the treatment plan is being implemented, to monitor pharmacological effects, and to track the overall progress of the case

For some cats, medical treatment is maintained for life. In such cases, annual blood tests are conducted to assess the long-term impact of medication, particularly on renal and hepatic functions.

6.4 Clinical Vignettes

We have chosen to present two complete clinical cases, without detailing the technical aspects of the consultations, but highlighting the evolution of these cases over several years to emphasize the parallels with treatment in other veterinary disciplines.

For this comparison, we selected dermatology and the management of atopy, though many other comparisons could be drawn.

6.4.1 Suki

Suki, a European shorthair cat, grew up in an apartment with ample stimulation, including the presence of children, numerous toys, cat trees with window views, and a balcony. Adopted at 2 months old, she was seen by her veterinarian for regular checkups starting at 3 months.

The veterinarian noted Suki's high impulsivity early on. The owner reported injurious play sequences and irritation-induced aggression in frustrating contexts. The owner appeared to use punitive methods, such as coercion, when bitten or scratched. Over time, Suki was increasingly isolated or ignored to minimize these behaviors, leading to a poor owner-pet relationship.

The adopter began considering rehoming Suki, realizing they had underestimated the time and attention she required. The family, which included children aged 5 and 8, became apprehensive about interacting with Suki due to her violent play, her habit of ambushing them outside rooms, and her tendency to pounce on them. Attempts to redirect her play were unsuccessful, and scaring her often triggered hissing or aggression unless she chose to flee.

The veterinarian was contacted when Suki was 1 year old. The owner decided to rehome her due to the worsening situation. A rescue organization took her in, and she was placed in foster care.

When a volunteer collected her, Suki hissed at strangers who tried to approach her directly. This behavior was more pronounced with taller males and children. In the foster home, her intense fear reactions overshadowed her impulsivity, which was not reported by the foster family.

During the first psychiatric consultation, Suki had been in foster care for a month. The veterinary diagnosis was: *pervasive impulsive state* and *generalized inhibited anxious state*. The nosographic diagnosis was *HSHA syndrome or feline ADHD*, with a *communication disorder leading to post-traumatic phobia* (resulting from coercive methods).

The foster family was hesitant to initiate medical treatment, opting instead to evaluate Suki's response to behavioral therapy first. The resource diagnosis was good due to the family's high motivation, competence, and financial stability, though not excellent due to their provisional refusal of medication, which could help Suki manage her impulsivity and exaggerated fear, facilitating her learning progress.

The family implemented a rigorous behavioral therapy plan. Suki's anxiety resolved relatively quickly through positive reinforcement methods and strict adherence to respecting her personal space (proxemics). During a veterinary visit for pruritic alopecia, the owner noted that Suki had become sensitized to handling, hissing at the veterinarian during the clinical examination. This prompted the initiation of a desensitization program for handling and care.

Despite these efforts, Suki's impulsivity remained significant. She still struggled with claw control and could injure during play when highly excited. Additionally, she exhibited hyposomnia (7 h of sleep per 24 h), difficulty remaining calm even in the absence of stimuli, and hypersensitivity, reacting inappropriately to normal environmental stimuli. These challenges led to repeated failures during exercises conducted in stimulating environments, causing progress to stagnate.

A follow-up consultation at 18 months resulted in the initiation of pharmacological treatment with *venlafaxine* at a dosage of 1 mg/kg/day.

Suki showed significant improvement following the implementation of medical treatment and behavioral therapy. She became calmer and successfully performed exercises even in stimulating environments. She learned to play without unsheathing her claws or biting and stopped chasing members of the foster family as they moved around the apartment. The owner continued to diligently follow the prescribed exercises, such as inducing excitement during play and then pausing to encourage a return to calmness.

The medical treatment was maintained for 12 months, during which time Suki's progress was sustained. The dose of venlafaxine was gradually reduced to 0.5 mg/kg. A tapering process was attempted after 12 months, gradually discontinuing the medication. During the first month of tapering, Suki's impulsivity slightly increased but remained manageable, and she continued to exhibit normal behavior, successfully controlling her excitement. The learning appeared robust.

However, 1 month after the tapering began, symptoms that had been absent for over 6 months began to re-emerge. These included ambushing family members outside rooms, injurious scratching during play, and extended periods of solitary play exceeding 1 h per day.

The rapid deterioration of Suki's relationship with the foster family and her noticeable distress, marked by a decline in positive interactions, prompted a decision to resume medical treatment in consultation with the veterinarian.

After 6 more months on medication, another tapering attempt was made, this time successfully. Suki maintained her progress without medication, demonstrating long-term stability in her behavior and a solid adaptation to her environment.

Today, 4 years after the cessation of treatment, Suki continues to exhibit self-control. Her relationship with her foster family, who decided to adopt her permanently, remains strong, as Suki can now manage her excitement effectively.

Suki has shown impulsivity throughout her life, and her genetic predisposition undoubtedly contributed to the emergence of her HSHA syndrome. Suki's case highlights how vulnerabilities manifest depending on the favorability of the environmental conditions she encountered. Exposed to inappropriate methods and interactions with children that heightened her excitement, Suki's psychological state deteriorated rapidly.

This case illustrates the respective roles of medical and behavioral treatments. Just as in other disciplines, when a pathology is identified, it is our responsibility to treat it using all available tools, including pharmacological treatments. These medications improve prognosis and enable behavioral therapy to be both possible and effective. Failing to utilize them represents a missed opportunity for the animal and unnecessarily prolongs its state of suffering.

6.4.2 Octavius

Octavius, a Russian Blue cat, was adopted as a neutered kitten at 3 months old from a professional breeder. According to the adopters, there is no known history of allergies in his lineage.

At 8 months of age, Octavius was brought to a consultation for diffuse pruritic alopecia. He had been allowed outside since the age of 6 months and had not received any antiparasitic treatment since his last vaccination at 4 months. Flea droppings were found in the dorso-lumbar region during the clinical examination. Octavius also displayed deep pyoderma lesions on his neck caused by scratching.

A 10-day antipruritic treatment (prednisolone at 1 mg/kg for 5 days, then 0.5 mg/kg for 5 days) and a 20-day antibiotic course (amoxicillin-clavulanic acid at 10 mg/kg amoxicillin and 2.5 mg/kg clavulanic acid) were prescribed. Octavius recovered quickly from this episode, and external antiparasitic treatment was strictly applied monthly thereafter.

A year later, despite continued antiparasitic measures, Octavius experienced another episode of pruritic alopecia. Flea allergy dermatitis could not be ruled out as the cause, as a visiting dog, which was not treated for fleas, had stayed in the household for a few days.

A symptomatic treatment with *prednisolone* resolved the episode once again, but a few days after discontinuation, pruritus reappeared. The owner treated the environment for fleas, including areas with old wooden floors and numerous bedding spots for Octavius that were not frequently cleaned.

Although the pruritus decreased, it did not fully resolve. A blood test and allergological testing were performed, revealing sensitivities to fleas, grasses, and dust mites. A continuous treatment with *cyclosporine* (7 mg/kg, followed by a tapered dose) was initiated, which nearly resolved the pruritus.

The following spring, with continuous treatment in place, the owners expressed interest in trying desensitization therapy. This approach reduced the required doses of cyclosporine, but each attempt to discontinue the medication resulted in recurrent pruritus.

Today, Octavius, now 6 years old, is managed with consistent ecological measures, including meticulous external parasite control, and a continuous immunosuppressive treatment with cyclosporine. These measures have proven effective in maintaining a high level of well-being for Octavius. It is likely that this treatment will need to be maintained for life.

Summary of Chapter

Veterinary psychiatry has developed from data drawn from ethology, psychology, neuroscience, and psychopharmacology, alongside expertise in interview techniques to foster a strong therapeutic alliance.

The architecture of our model is primarily clinical and functional, built on the accumulated experience of specialized clinicians. A veterinary psychiatry consultation relies on observing the animal, conducting a clinical examination, and direct observation. This process leads to a diagnosis composed of three levels: state, nosographic, and resource diagnoses.

The state diagnosis refers to the animal's psychological state. When pathological, this state is underpinned by dysfunctions in certain brain circuits. The symptom assessment, combined with knowledge of brain circuits, identifies the dysfunctional circuits and the neurotransmitters involved, guiding the practitioner's pharmacological choices.

The nosographic diagnosis provides the name of the psychiatric disorder. This level of diagnosis allows the practitioner to explain the causes, expected progression, and prognosis.

The resource diagnosis assesses the context and the capacities of the owner and the cat to follow the proposed treatment steps. It also influences the prognosis.

The treatment plan always includes behavioral therapy, supplemented by biological treatment when the pathological state necessitates it.

References

Beata C, Muller G, Arpaillange C, Graff E, Lachapèle D, Marion M, Marlois N, Massal N, Mège C (2015) Pathologie comportementale du chat (AFVAC). AFVAC

Carlson NR, Birkett MA (2022) Physiology of behavior, 13 edition, global edn. Pearson Education Limited

Casey RA, Vandenbussche S, Bradshaw JWS, Roberts MA (2009) Reasons for relinquishment and return of domestic cats (*Felis Silvestris Catus*) to rescue shelters in the UK. Anthrozoös 22(4):347–358. https://doi.org/10.2752/089279309X12538695316185

Charalambous M, Pakozdy A, Bhatti SFM, Volk HA (2018) Systematic review of antiepileptic drugs' safety and effectiveness in feline epilepsy. BMC Vet Res 14(1):64. https://doi.org/10.1186/s12917-018-1386-3

Hobson JA, Pace-Schott EF (2002) The cognitive neuroscience of sleep: neuronal systems, consciousness and learning. Nat Rev Neurosci 3(9):679–693. https://doi.org/10.1038/nrn915

Khodanovich M, Kisel A, Kudabaeva M, Chernysheva G, Smolyakova V, Krutenkova E, Wasserlauf I, Plotnikov M, Yarnykh V (2018) Effects of fluoxetine on hippocampal neurogenesis and neuroprotection in the model of global cerebral ischemia in rats. Int J Mol Sci 19(1):162. https://doi.org/10.3390/ijms19010162

Masson S, Bleuer-Elsner S, Muller G, Médam T (2023) Psychiatrie vétérinaire du chien. No ledge éditions

Micheli L, Ceccarelli M, D'Andrea G, Tirone F (2018) Depression and adult neurogenesis: positive effects of the antidepressant fluoxetine and of physical exercise. Brain Res Bull 143:181–193. https://doi.org/10.1016/j.brainresbull.2018.09.002

Nelson RJ (2011) An introduction to behavioral endocrinology, 4th edn. Sinauer Associates

Nerhagen S, Moberg HL, Boge GS, Glanemann B (2021) Prednisolone-induced diabetes mellitus in the cat: a historical cohort. J Feline Med Surg 23(2):175–180. https://doi.org/10.1177/1098612X20943522

Scarlett JM, Salman MD, New JG Jr, Kass PH (1999) Reasons for relinquishment of companion animals in U.S. Animal shelters: selected health and personal issues. J Appl Anim Welf Sci 2(1):41–57. https://doi.org/10.1207/s15327604jaws0201_4

Stahl SM (2013) *Mood disorders and antidepressants*, 4th edn. Cambridge University Press

Stahl SM, Muntner N (2021) In: Grady MM (ed) Stahl's essential psychopharmacology: neuroscientific basis and practical applications, 5th edn. Cambridge University Press

Walker MP, Stickgold R (2004) Sleep-dependent learning and memory consolidation. Neuron 44(1):121–133. https://doi.org/10.1016/j.neuron.2004.08.031

Psychological States

Contents

7.1 Reactive State – 229
7.1.1 Definition and Description – 229
7.1.2 Evolution – 230
7.1.3 Pathogenesis – 230
7.1.4 Symptoms – 230
7.1.5 Diagnosis – 232
7.1.6 Treatments – 233
7.1.7 Prevention – 234

7.2 Emotional Pathological States – 235
7.2.1 Phobic State – 235
7.2.2 Anxious State – 241

7.3 Mood Pathological States – 251
7.3.1 Depressive State – 251
7.3.2 Manic State – 256
7.3.3 Bipolar State – 260

7.4 Cognitive Pathological States – 260
7.4.1 Self-Controls Pathological States – 260
7.4.2 Derealization Pathological States – 277

References – 287

© The Author(s), under exclusive license to Springer Nature Switzerland AG 2026
S. Bleuer-Elsner et al., *Veterinary Psychiatry of the Cat*,
https://doi.org/10.1007/978-3-032-06260-4_7

In this chapter, we describe normal and pathological psychological states. The latter can be categorized into four groups: emotional states, which include phobic and anxious states; mood pathological states, comprising manic, depressive, and bipolar states; cognitive pathological states, concerning executive functions, which include impulsive and compulsive self-control deficits; and derealization states, which encompass senile and dissociative states.

The veterinary psychiatry consultation aims to establish a diagnosis that will lead to a prognosis and treatment. The first level of diagnosis is called state diagnosis, the objective of which is to assess the animal's behavioral dysfunction, identify as much as possible the dysfunctional brain circuits and neurotransmitters involved, and guide therapeutic choices, particularly the choice of medication. Indeed, identifying a pathological state must lead the practitioner to prescribe medication. If the brain is dysfunctional, in addition to the distress and suffering it causes, the cat's representations, learning abilities, and decision-making and action capacities are permanently altered, making any behavioral therapy difficult or even impossible.

We will start by describing the reactive state, which must be distinguished from the normal adaptive state. It can be seen as a temporary imbalance, without actual brain dysfunction, which may or may not cause the animal to shift toward one or more pathological states. It is important to note that a pathological state can appear without going through the reactive stage.

We will then describe the pathological states presented in ▶ Chap. 6 by groups (◘ Table 7.1): emotional pathological states, mood pathological states, and cognitive pathological states, which are subdivided into self-control deficit states and derealization states.

◘ **Table 7.1** Pathological states

Emotionally pathological states		**Mood pathological states**	
Phobic	Simple /complex	Depressive	Acute/chronic
Anxious	Productive/inhibited, intermittent/generalized	Manic	Contextual/pervasive
		Bipolar	
Cognitive pathological states			
Self-controls pathological states		Derealization pathological states	
Impulsive	Contextual/pervasive	Dissociative	
Compulsive	Contextual/pervasive	Senile	

7.1 Reactive State

7.1.1 Definition and Description

This refers to a non-pathological psychological state in which the animal attempts to respond to an environment perceived as hostile. The animal is confronted with stress, and its body attempts to adapt to it. This state differs from the *normal adaptive state* in that the animal is engaged in a struggle with its environment, which it perceives as requiring *heightened vigilance due to potential danger*. In other words, in the reactive state, the body endures an allostatic load that it must address (see ▶ Chap. 1, ▶ Sect. 1.1.1.3).

We have defined pathological states by three characteristics: the maladaptation of behavioral responses, the inability to spontaneously return to a stable equilibrium, and the suffering of the organism. In contrast, the *reactive state* does not exhibit the hallmark components of irreversibility, suffering, or behavioral maladaptation. While the responses may be temporarily maladaptive, they can still be modified by the animal, which retains a high degree of mental flexibility and, in fact, uses all its resources to develop appropriate responses.

If the subject does not have any existing vulnerabilities and the environmental hostility remains manageable, this state favorably progresses back to normal, i.e., a normal adaptive state. Stress in this context is only temporary and contributes to motivation and coping efforts. At most, there may be some transient suffering, usually limited to discomfort. The *allostatic load* is thus absorbed, and a return to equilibrium occurs quickly. However, when the organism fails to cope, the allostatic load persists and becomes an *allostatic overload*, which exhausts the organism. It is at this point that the shift to a *pathological state* occurs.

Clinical Illustration

Nono is an adult cat living in a house with a young couple. A major change has occurred with the arrival of a baby, and Nono is faced with many disruptions. First, he must get used to this new being who cries and screams frequently. The peaceful nighttime hours have vanished, and he now encounters his owners up in the middle of the night attending to the baby.

In the first few days, Nono expressed fear. He then tried to reclaim the room that is now off-limits. Whenever possible, he sneaks in with his owners and sniffs this space, which now has a new function. He attempts to leave his scent and deposit pheromonal marks. He even made a urinary marking spot in the baby's room, which reassures him. The situation remains tense, and Nono stays on high alert, often retreating to the garden.

To his owners, he seems a bit jealous. He demands cuddles and tries to capture their attention during feedings or other interactions with the baby.

Over time, however, Nono gradually adjusts. He takes advantage of the baby's nap times to reclaim his place. He manages to avoid the baby, and everything starts to settle down.

This change was experienced as a stressor for Nono, and he became hypervigilant. During this period, his reactive state enabled him to respond to all the variations in his environment. Two outcomes were possible: if the changes had exceeded Nono's adaptability, he might have transitioned into a pathological state, such as a phobic, anxious, or depressive state. In this case, however, Nono found new comfort zones, adapted to the novelty, and returned to a normal adaptive psychological state.

7.1.2 Evolution

The reactive state is, by definition, a *transient state* that is not intended to persist beyond a finite and relatively brief period (a few weeks). It allows the animal to develop responses to the stressful changes in its environment. Once these responses are established and the sources of stress are eliminated, the cat returns to a normal life that does not push it outside its comfort zone too frequently or too intensely.

If the animal is unable to achieve this response, it will repeatedly exhibit maladaptive behaviors that reinforce the hostility of the environment and prevent it from finding balance. Evolution will then progress toward a pathological state.

7.1.3 Pathogenesis

As soon as the living environment becomes hostile or even simply changes, a reaction occurs to facilitate adaptation. The mechanisms that come into play are of the same nature as those that drive any reaction. The perception of change constitutes an *acute stress* that stimulates adrenergic pathways and cortisol secretion, activating the organism. Unlike what occurs in the normal adaptive state, the animal must find new, effective responses.

Referring to the concept of allostasis, the reactive state is triggered by an allostatic load on the organism, which can follow two *evolutionary* paths (◘ Fig. 7.1): either it adapts and returns to a *normal adaptive state* (homeostasis), or it fails to do so and shifts into a *pathological state* (allostatic overload).

7.1.4 Symptoms

It is not possible to refer to *symptoms* in this non-pathological state. The signs described here belong to the animal's physiological behavior.

The most notable characteristic is *vigilance*. In this state, the organism feels exposed to dangers because its living environment has changed. This heightened vigilance is accompanied by exacerbated reactivity, driven by intense emotions. It is almost an exaggerated, yet temporary, *emotional reactivity*. The popular expression "to have raw nerves" aptly illustrates this reactivity. This hypervigilance is accompanied by an intensification of facial marking and, in the case of intrusions, urinary marking. This increased marking helps the cat structure its territory and create multiple points of reassurance within it.

This state represents a unique cerebral and physical condition. In the reactive state, the individual is capable of learning particular behaviors that it would not establish under other circumstances. However, maintaining such a high level of vigilance exhausts the organism, and this state cannot persist for more than a few weeks. In other words, the animal must either quickly return to a balanced state or transition into a pathological state.

Clinical Illustration

Willy is a country cat accustomed to freedom. Recently, his owners moved to the Paris region. Now, to go outside, he has to meow at the door. Once outside, he finds himself in a tiny courtyard with very high walls. By climbing the tree, he has learned how to access the top of the wall. The neighbor's garden is easy to reach because there's a small shed he can use to climb onto the roof. He even knows a path that leads directly to the office window on the first floor, which is often left open by his owner.

Willy goes out less often now, but in return, his owner is frequently home, and he has discovered the joy of sharing the office. In his previous life, he spent little time indoors, and the bedrooms were off-limits.

At first, Willy couldn't stay still; he wanted to go out all the time, only to demand to come back in right away. But now he enjoys curling up on the sofa, venturing outside only to watch the neighbor's little bichon. To his owners, it seems Willy has aged, but he has quickly adapted. His owner understands him well, as he, too, has experienced the change.

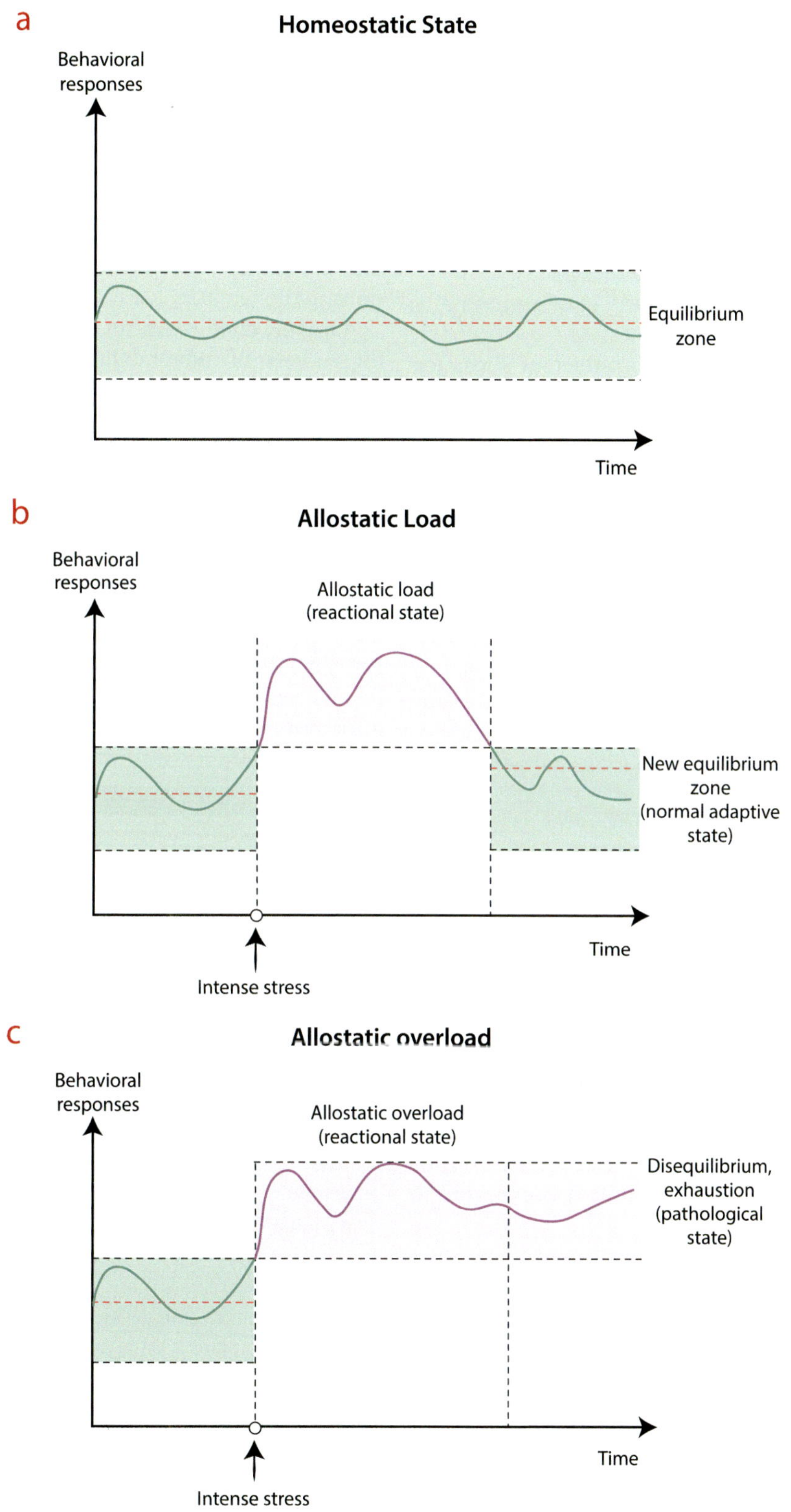

Fig. 7.1 Psychological states. (**a**) Normal adaptive state, homeostatic situation. (**b**) Reactive state, allostatic load situation. (**c**) Pathological state, allostatic overload situation

7.1.5 Diagnosis

For the practitioner, establishing the diagnosis of a reactive state is not straightforward. By considering the animal as "normal," the practitioner implies the possibility of a return to perfect equilibrium. This indirectly signifies that the cat's brain is functioning properly and that a rapid recovery is achievable.

The criteria for establishing the diagnosis of a reactive state are as follows:

- Identification of environmental stress for the cat.
- Heightened vigilance.
- Maladaptive behavioral responses to a new situation.
- Intensification of pheromonal marking.
- Rapid evolution toward improvement or worsening.

The key element of this diagnosis is the presence of maladaptive behavioral responses to a new situation. Often, identifying the contextual change presents the greatest challenge. This may involve an obvious change, such as adoption or a move. In more subtle cases, the change may simply occur too frequently, preventing the animal from finding a coherent response based on prior learning.

Let us focus on this *inconsistency in the environment*. For an animal, the living environment must be predictable, which does not mean devoid of stimulation. Changes in territorial reference points are challenging for cats. The territory is both spatial and emotional. The individuals who frequent it are familiar and predictable. Daily routines should remain consistent, and the behavior of those in the cat's environment is expected to be stable.

In some cases, the environment becomes too variable, whether due to frequent changes in the physical surroundings or the absence or inconsistency of key individuals. Under such circumstances, animals may remain in a reactive state for an extended period (up to 1–2 months) before transitioning into an anxious state.

This reactive state can, of course, arise following a territorial change. However, it is also important to consider that the same living environment can become stressful due to seemingly minor modifications, such as redecorating, restructuring the use of rooms, or changes in the individuals who usually share the space. Humans, in particular, can alter the environment through life events such as moving in with a partner, separating, or the birth of a child.

The practitioner must be able to recognize that life changes can stem from subtle evolutions in the territory across all its dimensions.

Organic conditions can also be responsible for a loss of adaptability in the animal. Sensory disturbances or physical disabilities can place the animal in a reactive state, requiring readjustment. Among the maladaptive behavioral responses, some may be particularly disruptive for owners and constitute presenting complaints.

Typically, characteristic fear reactions are observed, sometimes even aggressive behaviors, in response to situations the animal has not yet fully processed. This fear may manifest as seeking protection, inhibition, avoidance movements, or threatening behaviors.

Clinical Illustration

Pilou is an elderly cat who doesn't enjoy going outside much. He spends many hours at the first-floor window, watching the street and the park. Over the past few weeks, he has started sitting under the window, meowing. Occasionally, he asks to go out but then stays by the door and hides at the slightest movement… His owner is exhausted by this behavior!

Sometimes, the owner gets frustrated and "puts him back in his basket by the window," and everything calms down. Gradually, this became a new habit: whenever Pilou meows, the owner lifts him onto the windowsill.

To make things easier, the owner decided to place a chair under the window, and everything returned to normal. The arthritis that prevented Pilou from climbing to his favorite spot almost destabilized his psychological state. Fortunately, his attentive owner managed to find an appropriate solution.

7.1.6 Treatments

Once again, it is important to emphasize that the term "*treatment*" is somewhat misleading, as the boundaries of pathology, as we have defined them, have not yet been reached. The practitioner's objective is to promote evolution toward a return to equilibrium rather than a shift into a pathological state. This constitutes secondary prevention aimed at avoiding such a shift.

The intervention focuses, when possible, on the *context*, the *conditions of exposure* to change, and the animal's *health status*.

The practitioner's ability to intervene in the *context* is often limited. However, they are frequently required to reassess "territorial" coherence and interpret it through a "feline" perspective. Advising on the *conditions of exposure to change* and helping owners better understand their cat's needs is a typical part of their role. These recommendations are even more effective when owners consult preventively, such as when adopting an animal, preparing for the arrival of a baby, or moving to a new home.

Owners often try to eliminate adversity to increase their pet's immediate well-being, but *avoiding exposure* is not a viable solution for long-term adaptation. *Partial avoidance*, on the other hand, can be protective, giving the animal the opportunity to adapt to less intense changes and then gradually adjust to the complete transformation.

Respecting pheromonal markings and using diffusers helps facilitate a return to normalcy (Hargrave 2014; Zhang et al. 2022).

Clinical Illustration

Mr. and Mrs. Z have reluctantly taken in their daughter's cat. The cat is very afraid of their dog, despite the fact that the dog is extremely calm and placid. The reactive state might have allowed the cat to adapt, but the new owners are worried about their dog and prefer to avoid any trouble. For safety, they establish a private living area for the cat upstairs. A cat flap is installed on the first-floor window, ensuring that the cat does not have to encounter the dog.

Two years later, the situation remains unchanged: the cat never attempts to come downstairs, and since the dog never goes upstairs, the cat's reactive state has resolved, returning to a normal adaptive state. However, the cat has still not adjusted to the presence of the dog. Any future interaction would likely trigger a new reactive state, which could either resolve through habituation or progress to a phobic state.

Finally, although it may seem obvious, the reactive state requires particular attention to optimizing the animal's capacities. Even the slightest *organic impairment* can jeopardize the successful return to equilibrium. *Pain*, once again, must be carefully considered.

7.1.6.1 Pharmacological Treatment

The use of medication is not warranted unless no other therapy is feasible and the evolution is likely to take an unfavorable course.

As previously mentioned, addressing the animal's capacities is essential. Naturally, any organic issues must be treated. However, understanding the *conditions of the animal's development* can help predict its evolution.

Indeed, during a change in context, the adaptive effort required to navigate the reactive state can exacerbate a pre-existing pathological state. Cats that are already pathological have very limited adaptive capacities. For example, they are at risk of transitioning from an initial phobic state to an anxious state.

Clinical Illustration

Pantone is a cat who can be described as socially phobic. Whenever guests visit, he practices discreet avoidance. He managed, with great difficulty, to adapt to his owner's partner and the children in the household, but he does not tolerate anyone else. Financial setbacks forced the family to move to a new region, and now Pantone has a new home.

Although the situation is very similar to the previous one, Pantone is struggling to cope with the change. From the very first days, he displays heightened vigilance and reactivity. He hisses at his owners when startled and hides at the slightest trigger.

Before long, he becomes unapproachable by anyone except his owner. He is irritable and no longer allows himself to be petted. He has transitioned into an anxious state.

7.1.6.2 Behavioral Therapy

7

Behavioral modification is the key factor in guiding the cat back to a normal adaptive state.

On one hand, environmental measures such as organizing the cat's territorial zones and providing opportunities for three-dimensional exploration can address part of the presented problem. On the other hand, re-establishing familiar landmarks or continuing existing rituals can also contribute to improvement. Furthermore, clarifying communication, stabilizing territorial elements, and ensuring the absence of physical punishment are essential. Combined with therapies targeted at the dysfunctional axis, these steps can lead to a favorable evolution.

In cases of a reactive state, the general practitioner can, through simple therapeutic measures, help the animal return to equilibrium without prescribing medication. It is important to remember that because this state is transient, inappropriate measures will not result in resolution. In such cases, it becomes crucial to refer the case to a more specialized veterinarian, whose role will be to refine the diagnosis.

7.1.7 Prevention

The quality of an individual's adaptability largely depends on their developmental period. An animal that has experienced favorable developmental conditions will be better equipped to adapt to numerous changes. The more physiologically healthy the brain is, the more capable it is of responding to environmental changes and the traumas encountered by the individual.

The core of our discipline lies in ensuring that the animal possesses a brain capable of adapting to life's uncertainties. This gives the animal the best chance of successfully adjusting to its environment. It is important to note that adaptability decreases significantly with aging, and age must be considered when planning any major changes to the living context.

Summary of ▶ Sect. 7.1

The reactive state is a physiological, psychological state that differs from the normal adaptive or homeostatic state. In the reactive state, acute stress, referred to as allostatic load, appears. This stress may stem from environmental factors, such as moving to a new home or the arrival of a baby, or internal factors, such as intense fear or pain.

In this state, the cat faces an imbalance, and its body attempts to adapt. To do so, it employs behavioral strategies that may or may not be suited to the situation. The reactive state is therefore a transient state during which the cat strives to return to a normal adaptive state of equilibrium.

In some cases, behavioral responses are effective and enable the animal to return quickly to normal. The reactive state then resolves without medication. However, behavioral therapy is essential to increase the likelihood of a return to a normal adaptive state and to prevent the risk of a shift into a pathological state.

In other cases, these attempts are unsuccessful, and the organism becomes exhausted in its efforts to adapt. The allostatic load then becomes an allostatic overload, and the cat shifts into a pathological state, characterized by a spontaneous, irreversible inability to adapt and the presence of suffering.

The outcome, either transitioning to a pathological state or returning to a normal adaptive state, depends on the individual's vulnerabilities and the intensity of the allostatic load.

7.2 Emotional Pathological States

We have chosen to group under the term *emotional pathological states* the dysregulation of negative emotions caused by an aversive, phobogenic, or painful stimulus (fear, surprise, anger, disgust). The amygdala and the limbic cortex are central to the regulation of these emotions, and dysregulation leads to various forms of phobias and anxieties discussed in this section.

Emotions linked to an appetitive stimulus (joy, pleasure), while partially associated with the same circuits as negative emotions, also involve other structures, such as the basal ganglia and mesolimbic pathways. Their dysregulation primarily impacts executive functions and self-control, which will be addressed in the section on cognitive pathological states (see ► Sect. 7.4).

It should be noted, however, that the nervous system of the cat is less developed than that of the dog. In this species, fear circuits are much more prominent, while inhibitory processes are less pronounced.

7.2.1 Phobic State

7.2.1.1 Definition and Description

The *phobic state* is a pathological psychological condition characterized by the production of *disproportionate* fear responses upon repeated exposure to a specific stimulus or group of stimuli.

A phobic cat exhibits *intense* fear reactions to the phobogenic stimulus, which manifest, as shown in ◘ Fig. 7.2, in one of four ways: an attempt to flee (*flight*), complete immobility due to stupor (*freeze*), aggressive behavior (*fight*), or, as observed in phobic cats, a fourth category involving significant motor agitation (*fidget*).

Clinical Illustration

Loustic is a 5-year-old neutered male Chartreux cat. He cannot tolerate the vacuum cleaner when it is in use. As soon as the vacuum is turned on, he freezes on top of the refrigerator and starts drooling excessively. The salivation is so severe that the owner must mop up every time.

The owners attempted forced exposure, but they were scratched in the process. Once the vacuum cleaner is turned off, Loustic returns to normal. He is not afraid of the vacuum when it is off but avoids it, and his fear is strictly limited to this specific vacuum cleaner. This is a clear case of a simple, non-progressive phobic state.

7.2.1.2 State Variations

In the *simple phobic state*, the animal reacts to one or more isolated stimuli. The cat displays intense signs of fear during exposure but returns to equilibrium as soon as the phobogenic stimulus disappears. A cat may suffer from several simple phobias, such as storm phobia or interspecific social phobia, which is an exaggerated fear of individuals of another species, most often humans.

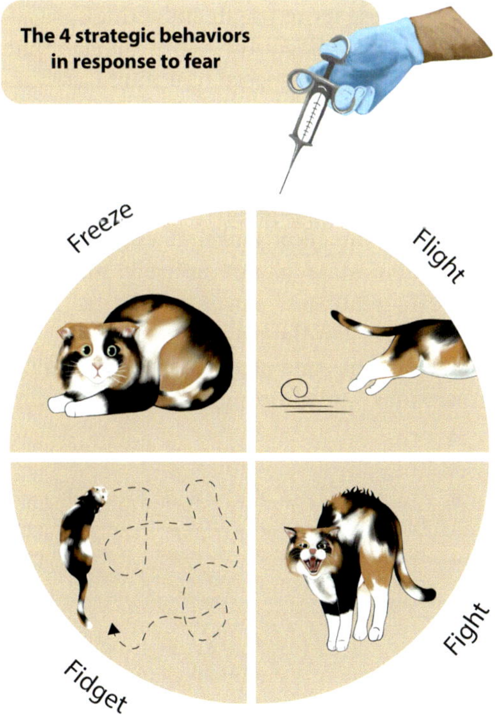

◘ **Fig. 7.2** The four responses to fear

However, it is relatively rare for this state to remain stable. *Anticipation* and *generalization* processes often extend the phobia to events or contexts associated with the *primary phobogenic stimulus*. From a simple phobia, the condition evolves into what is known as a complex phobic state. In some cases, it can even become difficult to trace the source of the phobia and identify the primary phobogenic stimulus.

7.2.1.3 Evolution

The phobic state results from the evolution of a reactive fear state through the process of *sensitization*. It is essential to remember that fear, along with its behavioral expressions—such as flight, immobilization, or aggressive responses—is part of normal physiological reactions. These normal reactions are protective mechanisms.

Environmental factors can promote sensitization, including exposure in *confined conditions*, exposure to a *highly intense phobogenic stimulus*, or *frequent* exposure. Internal factors, such as the animal's individual genetic vulnerabilities, also increase the likelihood of transitioning to a phobic state. In these more fragile individuals, *sensitization* is significantly more likely than *habituation* when exposed to unfamiliar stimuli (see Table 7.2).

The phobic state can remain stable but does not improve spontaneously. Desensitization is possible, but the associated cerebral dysfunction makes it challenging if the phobic state is not actively managed. Medical treatment can facilitate a much quicker return to the reactive state.

The phobic state often worsens, leading to an anxious state through anticipation and generalization. The animal's fear reactions become increasingly pervasive due to heightened vigilance and the intensification of the reaction itself. While minor fear reactions may occur, the phobic reaction is always intense. The cat scans its environment for cues that signal the phobogenic stimulus, becoming *hypervigilant*. At this stage, the source of its fears becomes progressively harder to identify, ultimately causing the cat to transition into an anxious state.

Table 7.2 Criteria favoring the transition to a phobic state

Criteria favoring habituation	Criteria favoring sensitization
• Open situation • Exposure to a low-intensity stimulus • Repeated exposure	• Confined situation • Exposure to a high-intensity stimulus • Isolated exposure

7.2.1.4 Pathogenesis

Fear is a reaction to the environment aimed at preserving the individual's integrity. It always involves three components:

- a neurovegetative component: Activation of the sympathetic nervous system leading to increased heart rate, vasoconstriction of blood vessels supplying the digestive system, and vasodilation of blood vessels supplying the muscles.
- a hormonal component: Secretion of adrenaline and cortisol, which amplify the neurovegetative reactions.
- a motor behavioral component, known as the "4 Fs": *fight*, *freeze*, *flight*, or *fidget*.

The phobic state involves the fear circuit (see ► Chap. 2, ► Sect. 2.3.1.1), which is centered on the amygdala (Fig. 7.3). This circuit, initially adaptive and designed to help the cat ensure its safety when a potential danger arises, can malfunction at various levels. This explains the complexity and wide range of symptoms observed by clinicians.

When the amygdala circuit malfunctions, fear reactions are triggered in non-dangerous contexts, much like a home fire alarm going off in situations far removed from actual fires. This *excessive activation* corresponds to the overactivation of glutamatergic and noradrenergic neurons or the inhibition of GABAergic neurons in areas involved in this circuit (Fig. 7.4).

Additionally, the phobic state can result from a failure of *inhibition in the ventromedial prefrontal cortex* (vmPFC) (Fig. 7.3). In such cases, the fear response is initially appropriate, but cortical inhibition, which prevents a fear response when the context is not dan-

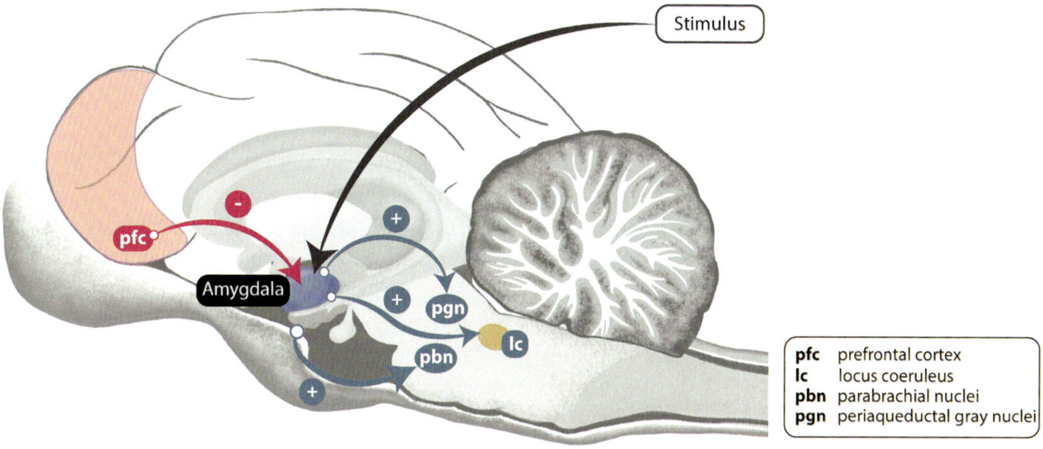

Fig. 7.3 Pathogenesis of phobic state. The fear circuit, centered on the amygdala, leads to the phobic state when there is excessive activation of the locus coeruleus, the periaqueductal gray matter, and the parabrachial nucleus, or when there is a failure of inhibition from the ventromedial prefrontal cortex (vmPFC)

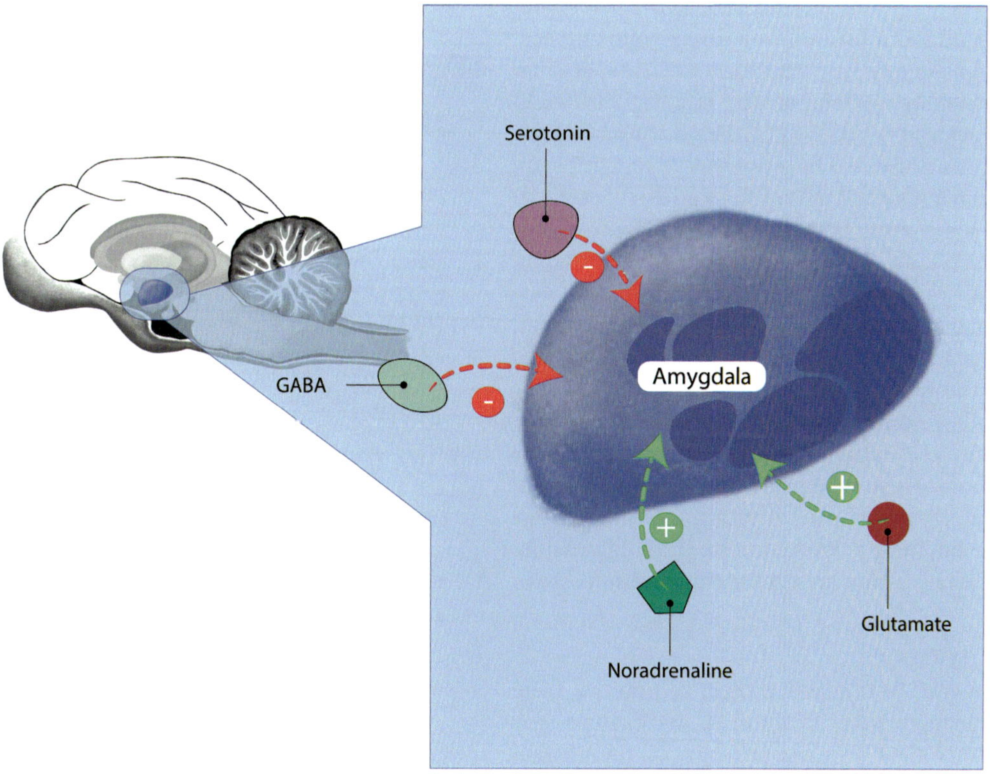

Fig. 7.4 The main neurotransmitters involved in phobia. Phobia results from hyperactivation of the amygdala, involving increased glutamatergic and noradrenergic activity at the expense of GABAergic activity. The reduced activity of the vmPFC is primarily due to decreased serotonergic activity

gerous, does not occur. The primary neurotransmitter involved in this feedback mechanism is serotonin (◘ Fig. 7.4).

7.2.1.5 Symptomatology

▪ Presenting Complaints

The signs reported by owners correspond to the reactions described by the "4 Fs." When the cat *flees*, owners often worry it might put itself in danger, which becomes the reason for consultation. In some cases, the cat may become restless without actually fleeing. Its distress is palpable, and owners consult because they perceive the animal's *suffering* but cannot alleviate it. In these situations, the cat is often irritable, and its pheromonal marking may become disorganized, with *urinary marking* sometimes occurring.

In cases of social phobias, the cat avoids contact and may become *aggressive* if it is prevented from escaping or forced into interaction. Aggressive behaviors, such as hissing, scratching, attempts to bite, or outright attacks, may occur when the phobogenic stimuli are present. These aggressive reactions are often a result of irritation in a fearful context. In some cases, fear reactions escalate to panic, which can include anal gland expression, emotional urination and defecation, and violent aggression. These uncontrolled emotional responses often prompt owners to consult due to the *perceived danger*.

For some cats, fear is expressed through complete immobility. This paralysis may be accompanied by emotional defecation, urination, or excessive salivation. Although this presentation is less common than aggression, it can also be a reason for consultation.

▪ Symptoms

We can distinguish motor symptoms from neurovegetative symptoms. Symptoms are observed only in the presence of the phobogenic stimulus. Once the cat is no longer exposed, it returns to a state of equilibrium.

Motor behavioral symptoms:

- Global motor inhibition
- Avoidance and anticipation characterized by hypervigilance
- Hyperactivity related to flight
- Aggression due to irritation in a context of fear or pain
- Uncontrolled emotional aggression
- Instrumentalized aggression

Organo-vegetative manifestations (OVM):

- Anal gland expression
- Tachycardia, tachypnea, trembling
- Emotional urination and defecation
- Salivation
- Sweating of the paw pads
- Appeasement signals
- Fear postures

7.2.1.6 Affected Conditions

The phobic state is found in several nosographic entities, primarily in *sensory deprivation syndrome*, but also in *categorical phobic disorders* and *post-traumatic phobias*.

> **Important Point of View**
>
> **Phobic state and phobic disorder**
>
> It is essential to distinguish between the *phobic state*, which is a pathological state, and *phobic disorder*, which is a nosographic entity. Renaming this pathological state would have been beneficial, but the term *phobic* is so widely used that we decided against introducing a new term.

7.2.1.7 Diagnosis

The criteria for diagnosing a phobic state are as follows (see ◘ Table 7.4):

- symptoms of exaggerated fear in response to one or more identifiable phobogenic stimuli;
- lack of evolution toward adaptation despite repeated exposures.

The observed symptoms correspond to the four possible fear reactions (4 Fs) and are categorized as:

Motor behavioral symptoms:

- Global motor inhibition
- Avoidance and anticipation characterized by hypervigilance
- Hyperactivity related to flight

- Aggression due to irritation in a context of fear or pain
- Uncontrolled emotional aggression
- Instrumentalized aggression

Organo-vegetative manifestations (OVM):
- Anal gland expression
- Tachycardia, tachypnea, trembling
- Emotional urination and defecation
- Salivation
- Aweating of the paw pads
- Appeasement signals
- Fear postures

7.2.1.8 Differential Diagnosis

The phobic state must be differentiated from the *reactive state*. The reactive state is a *transient* condition that resolves with the disappearance of fear symptoms through habituation or desensitization. The distinction can be made by asking the owner about the evolution of the animal's reactions to the phobogenic stimulus over time. When habituation occurs, progress is noticeable from one exposure to the next, and resolution typically occurs within a maximum of ten sessions.

It is also crucial to distinguish the phobic state from the *anxious state*. In the latter, identifying the phobogenic stimuli is no longer possible. Furthermore, the phobic state manifests only in specific situations, whereas anxiety can be constant. The anxious state results from a different and more severe neurobiological activity than the phobic state. The clinician can distinguish between the two by analyzing the animal's reactions when all phobogenic stimuli are removed: a rapid return to normal indicates a phobic state, whereas a sustained level of alertness over time suggests progression to an anxious state.

In cases where a phobic cat remains immobilized in inhibition, it is necessary to distinguish this state from a *depressive state*, particularly *acute depression*. This distinction is also made by observing the animal's reactions when all phobogenic stimuli are removed: a cat in a phobic state will return to normal behavior quickly, whereas a cat in a depressive state will remain unmotivated.

7.2.1.9 Prognosis

The prognosis depends on several factors: the presence or absence of anticipation and generalization processes, which complicate the phobia; the frequency and intensity of exposures; the possibilities for the animal to avoid exposure; and the age of onset (Table 7.2). Congenital phobias linked to behavioral development have a worse prognosis than phobias that appear in adulthood. Phobias with minimal anticipation and generalization have a better prognosis.

There are also significant genetic vulnerabilities that can negatively impact the prognosis. Additionally, the quality of behavioral development plays a critical role in the development and evolution of phobias (see ▶ Chap. 4, ▶ Sect. 4.3.1).

7.2.1.10 Treatments

Pharmacological Treatment

As we have seen, the phobic state is closely linked to the *amygdala circuit*. The primary neurotransmitters involved are norepinephrine, glutamate, GABA, and serotonin.

In simple phobias, pharmacological agents such as gabapentin, pregabalin, or trazodone can be used exclusively during exposure. This occasional use is referred to as pro re nata (PRN), meaning "as needed."

Glutamate plays a key role in memory formation and, consequently, in the processes of generalization and anticipation. Even occasional use of gabapentin, with its inhibitory effect on glutamate, can significantly accelerate desensitization.

Medication in these cases helps prevent panic episodes, which could *worsen the phobia*. This is often necessary during exposure to fireworks, thunderstorms, or very brief encounters with phobogenic stimuli, such as visiting the veterinarian.

Medication can also help accelerate *desensitization and counter-conditioning therapies*. In cases of phobia related to care or specific objects, such as a brush or a transport carrier, the use of PRN medications reduces the number of required sessions and contributes to a faster improvement in the animal's well-being. These medications lower the emotional response during encounters with the stimulus, restoring the cat's ability to produce physiological reactions. This allows new learning processes, such as extinction, to take place.

When phobogenic stimuli are encountered frequently, medication may be administered *continuously over several weeks or months* until desensitization is achieved. Drugs such as *propranolol* can be used. However, it should be noted that *clonidine* often causes vomiting in cats, making its use impractical.

When anticipation dominates (see ▶ Chap. 2, ▶ Sect. 2.3.1.2), drugs with dopaminergic effects are preferable. Molecules acting on the norepinephrine transporter (NET), which increase dopamine release in the prefrontal cortex, are suitable. Examples include *clomipramine* and venlafaxine.

Finally, when a lack of inhibition is the primary symptom, a selective serotonin reuptake inhibitor (SSRI) such as fluoxetine or fluvoxamine is appropriate.

Pheromones such as Cat Appeasing Pheromone (CAP) appear to have a beneficial effect in early interventions during the reactive state, particularly when there is a risk of transitioning into the phobic state (DePorter et al. 2019).

▪ Behavioral Therapy

The essence of therapy for phobic states is to desensitize the animal to the object of its fear(s). To achieve successful desensitization, the clinician must guide the owner in identifying the cat's early fear signals. This ensures that the work is conducted within a *zone of discomfort* without triggering an *excessive* fear reaction.

The concurrent use of psychotropic medications can significantly accelerate these desensitization therapies.

7.2.1.11 Prevention

Prevention involves exposing the cat to a variety of stimuli during the *socialization period*, allowing exploration at its own pace without being forced or restrained. The presence of a calm and non-fearful mother or adult cat plays an important role in successful habituation. Similarly, habituating the cat to common handling or to a transport carrier are protective measures against transitioning into a phobic state.

Once the cat is adopted, *observing* and responding appropriately to its *postures* is key to avoiding sensitization. Veterinarians play a critical role during initial vaccinations by teaching owners to recognize subtle postures indicative of fear in their animal. They can also provide advice on appropriate procedures, such as utilizing territorial refuges and exposing the cat to phobogenic stimuli in a reasoned manner without forcing or restraining it.

For certain stimuli that are inherently intense and uncontrollable—such as thunderstorms, fireworks, or painful veterinary procedures—it is challenging to create conditions conducive to habituation, and a phobic state may develop quickly.

Genetic vulnerabilities and a mother who is herself phobic are aggravating factors that are difficult to prevent.

Summary of "Phobic States"
A phobic state is a pathological psychological condition characterized by the production of disproportionate fear responses upon repeated exposure to a specific stimulus or group of stimuli.

There are two types of phobic states: simple and complex. In simple phobic states, the cat reacts to identifiable and isolated stimuli through sensitization. In contrast, complex phobic states involve anticipation and generalization, where fear responses extend to a broader group of stimuli.

A phobic state may progress to a more pervasive anxious state, where the triggering stimuli are no longer always identifiable.

In phobic states, the observed symptoms are caused by either excessive activation of the fear circuit or insufficient inhibition from the ventromedial prefrontal cortex (vmPFC) over the amygdala.

Medical treatment is selected based on the cat's behavioral responses (freeze, fight, flight, fidget) and the presence of anticipation or generalization. It may involve episodic treatment aimed at reducing amygdala transmission before and during exposure, such as gabapentinoids, or daily treatment to restore balance by addressing excessive amygdala activation or insufficient inhibition.

Behavioral therapy is specific to the underlying disorder and relies on desensitization and counter-conditioning. Its goal is to develop cortical control over fear responses.

7.2.2 Anxious State

7.2.2.1 Definition and Description

The *anxious state* is a pathological psychological condition in which a cat exhibits inappropriate, recurrent fear reactions without a specific object, triggered by *changes in its internal or external environment.*

Anxiety is often summarized by the phrase *"fear without an object."* In some cases, the cat's reactions can be highly productive, escalating to seizures. More commonly, motor agitation and heightened reactivity—either situational or constant—are observed. For other cats, significant inhibition may occur, accompanied by substitute activities. In all cases, common clinical signs can be identified.

As noted for phobic states, cats, due to their potential prey status, display intense and rapidly escalating fear reactions. Consequently, anxiety is frequently observed in this species.

7.2.2.2 State Variations

Just as fear reactions can manifest in different behavioral expressions, anxiety presents differently from one individual to another.

Anxious reactions can vary in intensity and scope. They may be *intermittent* and confined to specific contexts or extend to the entire environment, leaving the cat unable to find respite. In such cases, the anxiety becomes *generalized.*

Some cats exhibit more *productive* and agitated behaviors, sometimes even aggression, while others tend toward *inhibition* or *avoidance*. In rarer instances, anxiety manifests exclusively in paroxysmal episodes, appearing as intense anxiety attacks or panic episodes. These range from mild trembling to full seizures. This particular form, known as a *paroxysmal crisis,* is merely a specific clinical manifestation of the same underlying pathological anxious state.

We can therefore distinguish different types of anxiety based on their expression: *intermittent* or *generalized,* with *paroxysmal, productive,* or *inhibited* forms (◘ Fig. 7.5).

The clinical expression of anxiety is influenced by genetic factors. It is likely, as observed in dogs, that a genetic component linked to certain breeds promotes specific manifestations of anxiety.

Our clinical experience suggests that breeds such as Persians and Chartreux are more prone to inhibited expressions of anxi-

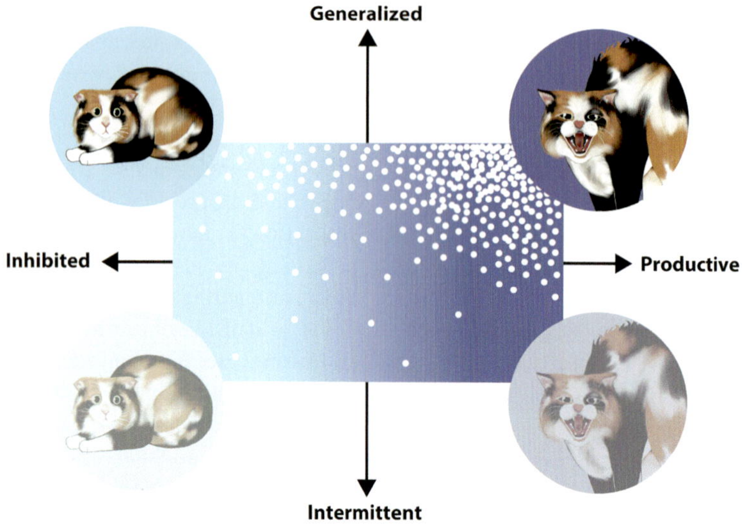

7

Fig. 7.5 Different types of anxiety. Anxieties can be categorized along two axes. The first axis represents their level of generalization, while the second axis represents their expression, ranging from inhibited to productive. Paroxysmal crises are independent of these axes and can occur in any form

ety, whereas breeds such as Bengals and Abyssinians often exhibit more productive forms of this condition.

7.2.2.3 Evolution

The anxious state is relatively stable. It can arise spontaneously or develop as a worsening of a phobic state. In some cases, the anxiety-inducing situation is relatively specific, allowing the cat to experience long periods of calm. This is referred to as intermittent anxiety. Over time, anxiety-inducing situations may multiply, becoming pervasive. Clinically, it is common to observe animals with generalized anxiety. These cats cannot find true rest, are constantly subjected to fear (startling at the slightest noise, scanning their environment, never fully sleeping), and exhibit progressively less effective behavioral responses. The pheromones they produce cease to be calming, and territorial marking becomes disorganized, creating a hostile environment. It can be said that an anxious cat weaves an anxiety-inducing environment around itself.

The *paroxysmal form* of anxiety, when observed, is often associated with subtle signs of anxiety between episodes. However, this form appears very stable over time and does not typically progress to *generalized anxiety.*

Intermittent anxiety generally worsens as the initial symptoms intensify. Increased urinary or digestive issues and self-focused activities are frequently noted. This anxiety can become so pervasive that it evolves into *generalized anxiety,* where the animal remains hypervigilant most of the time. The constant struggle is exhausting, and in some cases, chronic depression may develop, characterized by a significant drop in motivation (refusal of social interactions, lack of play, changes in sleep and appetite, and restriction of territory to an isolated area).

In the anxious state, the animal's defense mechanisms are activated to combat adversity. In the depressive state, the fight ceases, and the animal endures its environment with reduced motivation. The evolution from an anxious to a depressive state is not systematic, just as anxiety is not a mandatory precursor to depression.

Anxiety also increases the risk of senility and accelerates the onset of age-related conditions by weakening the individual.

Important Point of View

Classification of anxious states

We have slightly modified the model of anxious states proposed by Pageat (1995) and followed by Mège (Mège et al. 2003) and Beata et al. (2015). Permanent anxiety has been removed from the classification, as the inhibition it describes corresponds to a lack of motivation, is not influenced by environmental variations, and therefore aligns more closely with a depressive state. Inhibitions observed in anxiety are responses to changes in the internal or external environment.

We have also abandoned the anxiety-depression continuum, as the progression from one state to the other is not a rule. Clinically, complex presentations are sometimes observed, in which symptoms of both anxiety and depression are present simultaneously.

The modeling based on neurophysiological circuits suggests that each circuit can dysfunction independently, supporting the possibility of such comorbidity.

7.2.2.4 Pathogenesis

When faced with an unfavorable environment or context, a cat attempts to adapt. If the threatening stimulus can be identified, it becomes avoidable, and a reactive fear state may develop, allowing the cat to return to equilibrium. However, if the stressful situation is neither tangible nor avoidable (e.g., material or relational territorial disruptions, pain, frustration), flight or fight strategies are no longer applicable. The cat must then find an alternative behavioral response.

This effort to adapt demands the full attention and energy of the individual. The reactive state period (allostatic load) cannot be sustained indefinitely. If the cat fails to adapt, it transitions into a pathological state of phobia or anxiety (allostatic overload).

This explanation helps identify the main triggers of anxiety. All unavoidable forms of stress contribute to its emergence. These include relational contexts that cannot be altered, unresolved pain, and accumulating frustration that cannot be mitigated. The general hostility of the environment resulting from a phobic state, combined with the inability to find a resting zone, further increases the likelihood of transitioning into an anxious state.

The *process of anticipation* sets in, and the cat, unable to find an appropriate response to environmental threats, resorts to avoidance. The *hypervigilance* caused by this anticipation renders the environment increasingly hostile, and discomfort becomes constant. Anxiety transitions into a generalized state. Through emotional economy, fear responses become instrumentalized, leading to heightened irritability and automated reactions to all stimuli perceived as dangerous, and eventually to all stimuli. The fear pathways become increasingly permeable as *cortical control progressively diminishes*. At the ultimate stage, the environment becomes overwhelmingly threatening. Aggressive reactions provide decreasing relief (as the fears are internally driven), and responses shift toward withdrawal accompanied by self-centered compensatory behaviors.

In reaction to the cat's anxious behavior, the family environment also becomes hostile. The organism deteriorates, losing its ability to recover and adapt further.

The neurobiological circuits involved include the fear circuit centered on the amygdala, as in phobias, but especially the *cortico-striato-thalamo-cortical* (CSTC) circuit. This circuit exacerbates fear responses and leads to objectless fears (**◘** Fig. 7.6). Reactions become increasingly invasive and automated.

The neurotransmitters involved in phobias, such as GABA, norepinephrine, and glutamate, are also implicated in anxiety states. Inhibitory serotonergic controls are diminished, and dopamine plays a role in anticipation and reinforcement (**◘** Fig. 7.7).

The treatment of these anxious states involves selective serotonin reuptake inhibitors (SSRIs), selective serotonin and norepinephrine reuptake inhibitors (SSNRIs), or tricyclic anti-depressants (TCAs), sometimes combined with adjunct treatments to manage crises (gabapentin, pregabalin, benzodiazepines). Episodic treatment is insufficient to cure anxious states, whereas it may be adequate for phobic states.

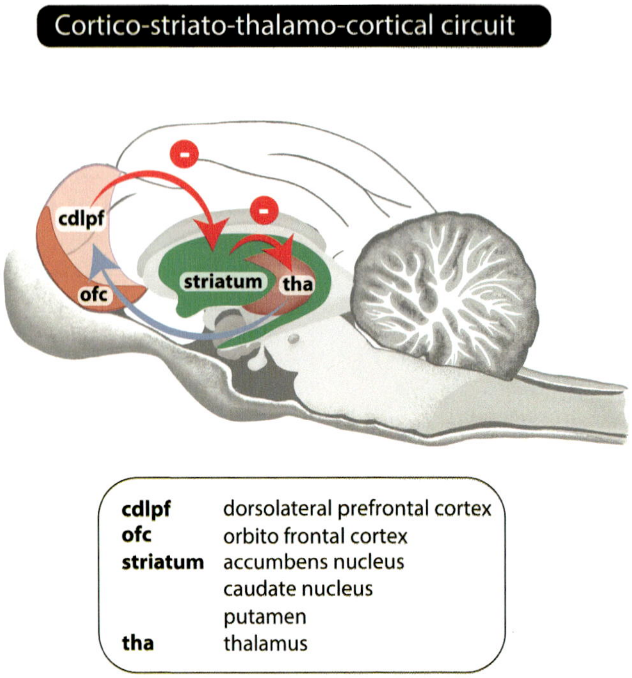

Fig. 7.6 Cortico-striato-thalamo-cortical circuit (CSTC). This CSTC is responsible for the onset of anxiety states when it malfunctions

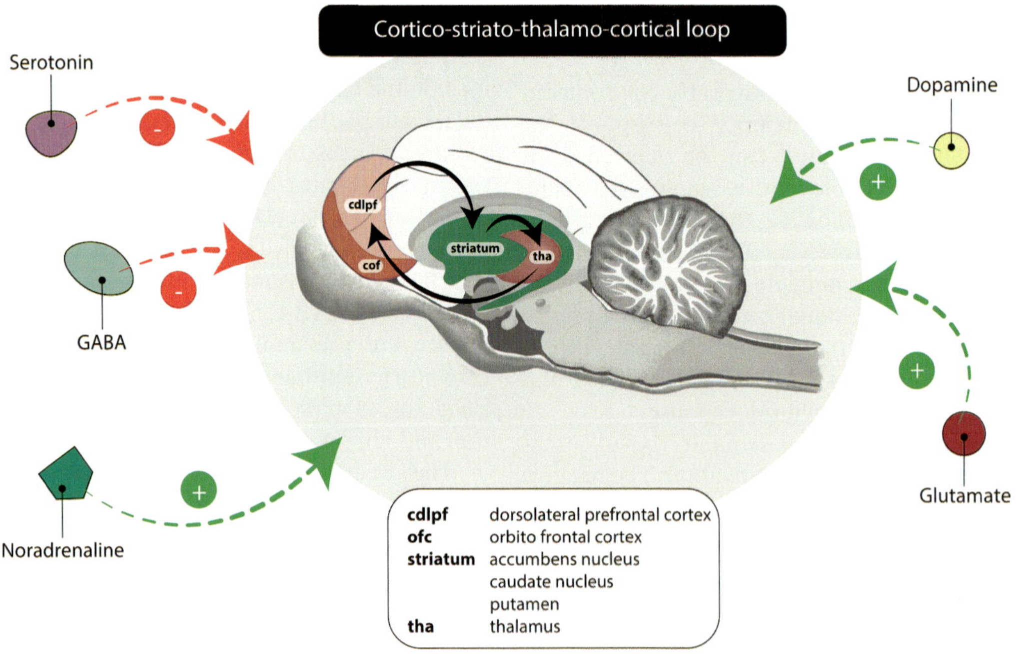

Fig. 7.7 The main neurotransmitters involved in the anxious state

7.2.2.5 Symptomatology

Presenting Complaints

What often troubles owners is one of the symptoms of anxiety. Productive anxious states are far more frequently a reason for consultation than inhibited states. The clinician must take care not to limit themselves to addressing only this specific complaint but rather propose a comprehensive treatment. Symptomatic treatment of this state is not an option.

Clinical Illustration

Norbert is a 7-year-old cat suffering from urine soiling. While he regularly uses his litter box, he also urinates on various surfaces. This is not a case of urine marking but rather urination on horizontal surfaces. The couch is frequently “targeted,” but he has also urinated on armchairs, rugs, and, more recently, on a printer, which is now nonfunctional.

An initial treatment for cystitis, combining NSAIDs and broad-spectrum antibiotics, improved the situation temporarily, but the symptoms quickly returned. Multiple attempts to use pheromone diffusers yielded no results.

A semiological assessment revealed that the cat-owner relationship had significantly deteriorated, making it urgent to find a solution. A treatment combining NSAIDs and venlafaxine produced rapid results. With continued use of venlafaxine, the anxiety disappeared, and cleanliness was restored within 2 weeks. Rare relapses occurred, requiring a short course of NSAIDs for 1 or 2 days. Venlafaxine was maintained for 18 months and then discontinued without any further relapse.

In the anxious state, the pathways of GABA, norepinephrine, and glutamate are enhanced, while the inhibitory pathways of serotonin become less effective. Dopamine plays a major role in anticipation.

Symptoms

The core of the symptomatology consists of fear reactions with anticipation and hypervigilance. Unlike in the phobic state, the object of fear is not identifiable. However, it is worth noting that an anxious cat may also exhibit phobic state reactions, where a specific phobogenic stimulus can be identified.

Hypervigilance is a reactive attitude in cats, during which the animal actively monitors its environment. In this state, the cat searches for particular stimuli and the events that signal them. This behavior enables quicker reactions to danger and more efficient responses to acute stress in general. However, hypervigilance cannot be sustained for extended periods without exhausting the organism, which then transitions into the anxious state.

The symptomatology corresponds primarily to dopaminergic-type disorders. Classically, this manifests as heightened vigilance and anticipation. Avoidant behaviors are observed in the form of withdrawal and restricted use of the territory. Additional disturbances, such as urine marking, can escalate into chronic urine soiling.

In this state, aggression toward owners or other cats is not uncommon. The animal becomes irritable due to its discomfort. Situations that it would typically tolerate add to existing stressors, leading to exaggerated reactions. Pain, as well as frustrations and relational constraints, frequently triggers a reactivity that often surprises owners.

Fear without a specific object leads to *aggression driven by irritation*, due to heightened anticipation and increased cortical control. However, the anxious animal may also exhibit uncontrolled aggressive reactions in fear-inducing situations triggered by specific phobogenic stimuli, particularly when it perceives the situation as closed or inescapable.

Anxiety is often accompanied by organovegetative symptoms. These are frequently digestive or urinary, presenting as chronic disturbances that are resistant to conventional treatments. Urinary disorders, such as idiopathic cystitis, are common and may be associated with increased urine marking.

A specific clinical manifestation in cats is cutaneous hyperesthesia, referred to in Anglo-Saxon literature as "rolling skin syndrome." This condition is characterized by discomfort expressed as shivering along the spine. In severe cases, these dorsolumbar contraction waves can be accompanied by self-mutilation behaviors, such as tail attacks. In milder forms, the symptom may be limited to varying degrees of tail flicking. It can sometimes be challenging to distinguish pain-related causes from purely anxiety-related ones associated with these symptoms, and therapeutic trials often help refine the problem's origin.

In older cats, severe hypertension is frequently observed, independent of thyroid disorders. The potential role of anxiety in the etiology of elevated blood pressure should be carefully considered. Some individuals also exhibit episodes of intense salivation.

The dietary expression of anxiety should also be noted. Some animals develop true compulsive overeating, with food becoming an obsession. These appetite disorders quickly lead to obesity. If owners strictly control food access for such cats, the resulting frustration can itself become a source of anxiety.

Fear reactions can also manifest as inhibited responses, and it is not uncommon to encounter anxious cats that react in this manner. Anxiety in these cases is often accompanied by *substitute activities*, as the cat attempts to adapt and cope with discomfort, though these activities fail to resolve the underlying issue. Substitute behaviors frequently involve self-contact, with variations influenced by the genetic characteristics of the individual. These may include nail-biting (*onychophagia*) or excessive licking of the abdomen. In some cases, this excessive grooming results in complete hair loss except along the dorsal line, which helps differentiate it from alopecia caused by flea infestations.

During paroxysmal anxiety episodes, typical adrenergic symptoms are observed, such as dilated pupils (mydriasis), trembling, seizures, panting, and tachycardia. The animal may also empty its anal glands. These episodes are often accompanied by urinary accidents triggered by episodes of skin shivering (rolling skin syndrome). Eating can also become a substitute activity, leading to severe obesity and its inevitable complications.

In prolonged and invasive anxious states, cognitive performance often declines. It remains unclear whether this is due to a direct reduction in learning abilities or an indirect consequence of the alert state preventing appropriate responses.

Territorial marking is disrupted, as the pheromones produced are of poor quality. This leads to a disorganized territory, which exacerbates anxiety.

Regardless of the clinical form, anxiety tends to become invasive. Sleep disturbances appear rapidly, with sudden awakenings in response to routine environmental stimuli. These symptoms, though not always easy to observe, contribute to reduced cognitive capacity and increased irritability.

7.2.2.6 Affected Conditions

Anxious states are encountered in numerous conditions, including categorical phobias, deprivation syndrome, disorders resulting from territorial or relational disturbances, feline ADHD, and conditions associated with aging. Practitioners should also remember that this state can result from deteriorated somatic conditions, such as pain, sensory disturbances, or hormonal diseases.

Additionally, incoherent communication, an understimulating environment, failure to meet physiological needs, and the use of coercive training methods all promote the development of anxious states (◘ Table 7.3). Therefore, even when anxiety is not the primary condition, it can appear as a secondary state in any illness.

7.2.2.7 Diagnosis

◘ Table 7.4 provides a comprehensive overview of the diagnostic criteria for anxious states, summarizing the key symptoms and their relevance in identifying this condition.

The diagnosis of an anxious state is based on the combination of several symptoms. No single symptom is pathognomonic, but their combination reflects a specific type of functioning in the individual. The progressive

Table 7.3 Factors favoring the onset of anxious states in various conditions

Condition	Factors favoring the onset of anxious states
Phobic disorders	Frequent exposure to the phobogenic stimulus with no possibility of escape
Deprivation syndrome	A fundamental deficiency in mechanisms enabling generalization to different contexts during habituation.
Disorders caused by territorial disruptions	Frequent changes in the environment (e.g., moving, introduction of a new companion)
Relational disorders	Deterioration of the bond with the owner and punishments.
Feline ADHD	Frustration in the cat and the use of aversive training methods in response to learning and concentration difficulties
Aging-related disorders	Loss of familiar reference points, and therefore predictability (e.g., pain, sensory, or cognitive losses). Distinguishing cause from consequence is often challenging
Somatic diseases	All chronic somatic conditions are implicated, particularly those involving chronic pain

escalation of dysfunction is also a precise warning sign.

After identifying the pathological nature of the animal's responses, the practitioner aims to characterize these maladaptive responses: Are they panic attacks, agitation, or more inhibited behaviors? These overall characteristics guide the choice of treatment.

Across all forms of anxiety, general signs are observed, though not all are necessarily present in every case:

- Hypervigilance
- Loss and disorganization of facial marking
- Tachypnea
- Cutaneous hyperesthesia
- Extensive alopecia
- Rolling skin syndrome (RSS)
- Digestive disturbances: dyspepsia, chronic intestinal disease, bulimia
- Bladder dysfunctions: idiopathic cystitis and urine marking
- Irregular learning responses
- Irritability: aggression due to irritation

For some individuals, the anxious reaction manifests as withdrawal and *inhibition*. In these inhibited forms of anxiety, substitute activities are common. General signs of anxiety (digestive issues, loss of autonomy, sensitivity) are also present. *Generalized anxiety* is characterized by significant *territorial degradation,* which is expressed consistently. Marking behaviors and sleep patterns are disrupted. In clinical practice, there are no simple methods for studying sleep; only awakenings can be reported by owners.

The relationship with the owner is also deteriorated due to the consequences of disturbances such as urine soiling and aggression. In some cases, punishments may worsen the disorganization of the living environment. Owners are often aware of the animal's distress. This discomfort is more acutely perceived by owners than in phobic states, particularly in cases of productive anxiety and paroxysmal crises.

Some *owner-based questionnaires* are available and can help raise awareness about behavioral disorders (Bennett et al. 2017; Duffy et al. 2017; McPeake et al. 2023). *Diagnostic questionnaires* can also be used, assisting less experienced practitioners in validating their clinical conclusions. These tools offer the additional benefit of measuring the anxious state and the extent to which anxiety responses dominate the individual's functioning. However, current questionnaires have not been validated to numerically distinguish between reactive, phobic, or anxious states.

7

Table 7.4 Diagnosis criteria for phobic and anxious states

Patho-logical state	Diagnostic criteria	State variations	Differential diagnosis
Phobic	*Phobic state* is a pathological mental state characterized by disproportionate fear responses during repeated exposure to a specific stimulus or group of stimuli No adaptation occurs despite repeated exposures **Behavioral symptoms:** • Global motor inhibition • Avoidance and anticipation with hypervigilance • Hyperactivity related to flight • Irritation-based aggression in the context of fear or pain • Uncontrolled emotional aggression • Instrumental aggression • Appeasement signals • Fear postures **Organo-vegetative symptoms (OVS):** • Anal gland evacuation • Tachycardia • Tachypnea • Tremors • Paw pad sweating • Emotional urination and defecation	*Simple phobia* Limited number of phobogenic stimuli Minimal anticipation or generalization *Complex phobia* Expanded phobogenic stimuli Anticipation and generalization processes are starting to worsen the phobic state	*Reactive state* Transient hypervigilance disappears with habituation Quick return to equilibrium after exposure *Anxious state* Exaggerated fear symptoms without identifiable stimuli Frequent organic symptoms. *Depressive state* Prolonged inhibition without stimuli Reduced motivation rather than fear *Impulsive state* High motivation Motor agitation linked to the desire to react, not exaggerated fear
Anx-ious	*Anxious state* is a pathological mental state where the cat exhibits inappropriate, recurrent, objectless fear reactions in response to changes in internal or external environments **Common symptoms** (in whole or in part): • Objectless fear reactions (4F) • Hypervigilance • Tachypnea • Cutaneous hyperesthesia • Digestive issues: Dyspepsia, recurring gastritis, chronic intestinal disease • Territorial marking anomalies • Urine soiling, emotional cystitis • Alopecia	Paroxysmal crises Highly productive episodes (e.g., syncope, rolling skin syndrome), not attributable to organic causes *Intermittent to generalized forms* Varying levels of anxiety from short episodes (intermittent) to pervasive behaviors affecting relationships and frequent startled awakenings (generalized). *Inhibited type* Withdrawal, self-centered activities *Productive type:* Agitation, irritability	*Reactive state* Transient hypervigilance resolved through habituation. *Phobic state* Fear reactions tied to identifiable stimuli *Epilepsy* Organo-vegetative symptoms (e.g., polypnea, tremors) with or without convulsions; diagnosis relies on therapeutic outcomes. *Depressive state* Context-independent anhedonia and indifference to the environment; grooming may cease entirely. *Impulsive state* High motivation, motor agitation unrelated to hypervigilance. *Compulsive state* Repetitive or stereotyped behavior triggered by identifiable supports (e.g., play, frustration, food)

Important Point of View

Definition of anxiety across different nomenclatures

Feline anxiety is described by all behaviorists, regardless of the model they use. However, no validated scale currently exists to establish a quantified diagnosis of this state. Existing scales often feature a symptom gradient that places the phobic state before anxiety.

In our model, anxiety is considered an inherently pathological state. In the Anglo-Saxon model, however, anxiety is regarded as a stress-related emotion, whether positive or negative. It can therefore correspond either to physiological anticipation or pathological anticipation.

These differences in nomenclature hinder the development of international consensus and the collection of more standardized data. It is, therefore, urgent to unify these fundamental definitions.

7.2.2.8 Differential Diagnosis

Here, we are dealing with a state diagnosis, which must not be confused with a nosographic diagnosis. Establishing a diagnosis of anxiety involves describing the type of reactions exhibited by an animal. In our model, the diagnosis of anxiety is not a nosographic diagnosis and can apply to various conditions.

The primary challenge lies in distinguishing this anxious state from *a reactive state*, which involves determining the pathological nature of the reactions. When a cat is placed in a challenging situation or faces adversity, it will react and attempt to produce an appropriate response. The adversity of the situation inevitably leads to vigilance and a certain level of alertness, enabling the animal to generate adaptive responses. These characteristics define the reactive state, which allows for responses to evolve toward increasing adaptation. Conversely, in the anxious state, behavioral responses fail to return to normal. They remain inappropriate in intensity, duration, or focus. For clinicians, a single instance of fear in a given context is not sufficient to conclude anxiety. However, the repeated occurrence of such situations, despite the absence of objectively identifiable danger, along with a sense that the cat is rarely relaxed, helps make the distinction.

The second difficulty is differentiating this state from a *phobic state*. In the latter, if all phobogenic stimuli are removed, the animal returns to a normal adaptive state. Between episodes of fear, the phobic animal regains balance, which the anxious animal does not achieve.

Chronic depressive state must also be considered in the differential diagnosis. In depressive states, the animal's motivation is reduced, whereas in anxiety, even in inhibited forms, the cat reacts to an environment perceived as hostile.

In *generalized anxiety*, it may appear as though the animal lives in an imaginary world of non-existent dangers. However, this is not a dissociative state.

Finally, *paroxysmal crises* may be considered separately. These episodes are dominated by crises whose amplification may resemble primary epilepsy. Diagnosis will rely on accompanying signs and therapeutic outcomes.

Clinical Illustration

Lesti is an adult cat who lived harmoniously in a large house with free access to the garden through a cat flap. His owners decided to carry out renovations, and during the 3 weeks of construction, Lesti was placed in a boarding facility. Upon his return, his familiar scents had disappeared, and the cat flap had been replaced with a smaller, electronically controlled model that emitted a surprising clicking sound. Lesti was hesitant to use the new flap and required assistance, as the noise caused him to retreat.

For a week, Lesti meowed excessively and wandered through every room, rubbing against various surfaces. At one point, he nearly scratched the couch but ultimately went outside and used an old outdoor scratching post instead. On several occasions, he left small urine spots on the garden wall and the cat flap. He also marked the back of an armchair, prompting his owner to scold him.

Aware of potential behavioral issues, the owners quickly set up pheromone diffusers and left the cat flap open for some time. Despite these efforts, Lesti became increasingly reluctant to go outside and began marking indoors daily. He started showing signs of heightened reactivity, such as startling easily, and his owner observed shivering along his back over the past month. Concerned, she wondered if he might have fleas, as his behavior had changed significantly.

She consulted their veterinarian, who conducted a behavioral evaluation and diagnosed an anxious state stemming from a disorder related to territorial disruption.

7.2.2.9 Prognosis

The prognosis for anxiety varies. It is often favorable in cases of intermittent anxiety but more guarded in generalized anxiety. However, it is important to avoid assuming that a straightforward diagnosis guarantees an equally simple treatment. Some cats are unable to fully recover from the anxious state and may require ongoing treatment with varying levels of calming therapy.

In general, treating anxiety alongside the underlying condition causing the anxious state (essentially addressing its etiology) should help resolve the pathological state. However, clinical experience indicates that the etiology of anxiety is multifactorial. Anxious individuals are often inherently vulnerable, and relapses are common for the same animal. For some, it seems as though there is always a reason to be anxious. Their vulnerability makes them unable to live in a state of comfort without treatment to control their anxiety.

7.2.2.10 Treatments

The treatment of an anxious state naturally involves psychotropic medications, which aim to restore balance within the various brain response pathways. In anxiety, there is an overload of the cortico-striato-thalamo-cortical (CSTC) circuit, accompanied by a progressive loss of cortical control. Both medication and therapy are designed to reverse this progression.

Pharmacological Treatment

Anxiolytic anti-depressants are the cornerstone of treatment. These may be combined with short-term use of anxiolytics in specific situations known to be challenging for the cat, or during the initial phase of treatment before the anti-depressant reaches its therapeutic threshold. In cases of generalized anxiety, the combination of multiple psychotropic medications is often necessary to stabilize the animal in a state of sufficient *well-being*.

Possible combinations include:

- An anxiolytic anti-depressant as the primary treatment
- A gabapentinoid or trazodone on an as-needed basis (PRN)
- Benzodiazepines to manage panic attacks that do not respond to other medications.

The treatment of anxiety requires restoring cortical control and the capacity for learning. For this reason, the use of benzodiazepines should be limited. While they provide immediate comfort, they do not lead to a lasting change in the cat's perception of the world and can interfere with the ability to restore a calming territorial organization.

Physical complications may arise with the onset of anxiety and often require specific additional treatments. Idiopathic cystitis may necessitate anti-inflammatory treatment or even antibiotics if an infectious complication is present. Similarly, some cases of alopecia can progress to skin lesions, sometimes requiring appropriate local treatment. However, the use of an Elizabethan collar to limit licking is often counterproductive, as it can exacerbate the cat's anxiety.

Behavioral Therapy

Behavioral therapy will primarily depend on the underlying condition responsible for the anxiety. The therapy focuses on restoring a stable and predictable environment and improving relationships with the household's occupants, reducing sources of frustration that may contribute to anxiety, and identifying and treating all forms of somatic discomfort to improve the cat's overall well-being.

Summary of ▶ Sect. 7.2.2

Anxious states are common pathological conditions that often accompany various illnesses.

In an anxious state, the cat exhibits fear in numerous situations, without an always identifiable trigger. Anxiety is characterized by both the nature of the reactions, which may be more or less productive, and the degree of its pervasiveness. Thus, the state is described in the diagnostic context as either inhibited or productive, intermittent or generalized.

Exaggerated fear reactions without identifiable triggers differentiate the anxious state from the phobic state. The persistence of hypervigilance reactions without leading to contextual adaptation distinguishes the anxious state from the reactive state.

Pharmacological treatment relies on anxiolytic agents and must be administered continuously for several months or even for life in the most severe cases, which often stem from a high genetic predisposition to vulnerability.

Behavioral therapy depends on the underlying condition responsible for the anxiety. It includes specific exercises targeting the behavioral axis involved in the psychiatric disorder being addressed, such as self-control, fear, territory, or relationships.

7.3 Mood Pathological States

7.3.1 Depressive State

7.3.1.1 Definition and Description

The depressive state is a psychopathological condition in which the animal stops responding to its environment and withdraws into itself. Clinically, this state is characterized by intense inhibition, social withdrawal, sleep and appetite disturbances, and anhedonia. The depressive state is a mood disorder where the individual's *motivation* is significantly diminished, affecting all behaviors.

7.3.1.2 State Variations

There are two distinct forms of the depressive state: the acute form and the chronic form. These forms can occur consecutively, although the chronic form may also arise without being preceded by an acute form.

In the *acute depressive state*, the cat appears to abandon any effort to cope. This form may arise following intense stress perceived as traumatic, such as being confined to an apartment after living a life on the streets without human contact. A cat in an acute depressive state may stop eating and sometimes drinking, putting its life at immediate risk.

The manifestations of the *chronic depressive state* are less intense and evolve in a fluctuating pattern.

7.3.1.3 Evolution

The depressive state is the individual's ultimate reaction to an environment that they cannot adapt to. While the anxious state is marked by a struggle against adversity, the depressive state corresponds to an absence of reaction. A depressed animal primarily exhibits a *significant decrease in motivation*.

Acute depressive states generally appear a few days after a severe trauma. The animal abruptly ceases all activity, withdraws entirely, and stops interacting. Appetite is often nonexistent, leading to complete anorexia. In young animals, this state severely jeopardizes survival, with potential long-term consequences such as brain malformations or cognitive deficits. In adult cats, this state typically progresses to the chronic form.

In *chronic depressive states*, phases of agitation occur, reflecting *intense discomfort*. During these phases, an animal that usually isolates itself may instead exhibit aggressive behaviors.

Interactions with a depressed animal are challenging. The cat tends to stay in its isolated zone and no longer properly structures its territory. It interacts less with its owners and displays irritability.

On the organic level, it is essential to rule out any diseases that could cause apathy and general demotivation, as well as chronic painful conditions.

7.3.1.4 Pathogenesis

Unlike phobic, anxious, impulsive, or compulsive states, the depressive state cannot be easily linked to a specific neural circuit. Instead, it represents a general imbalance between neurotransmitters and their receptors *throughout the brain*. The neurophysiological mechanisms underlying this condition remain unclear to this day.

Depressive states were initially attributed to a dysregulation of all monoamines, followed by a quantitative alteration in *monoamine receptors*. This widespread modification triggers a cascade of molecular events in postsynaptic neurons, including changes to their *nuclei and gene expression*. It also impacts growth factors, particularly *Brain-Derived Neurotrophic Factor* (BDNF) (Elhwuegi 2004; Jesulola et al. 2018; Leuchter et al. 2009; Martinowich et al. 2007; Wong and Licinio 2004).

A study by Caspi et al. (2003) on the genes encoding 5-HT receptors demonstrated a genetic vulnerability to developing depressive states and suicidal thoughts in response to stressful events. The results revealed that individuals with at least one copy of the short allele are more vulnerable to depression than those with two copies of the long allele.

Some theories suggest a significant role of *glutamate* and *glial cells* in mood disorders (Haroon et al. 2017; Schroeter et al. 2010). Such a theory is not incompatible with the *monoamine receptor theory*, as glial cells influence all monoaminergic neurons. This theory provides a potential explanation for the global impact on brain function, including the consequences on BDNF and various neurotransmitters.

It is important to note that, for now, the neurophysiological mechanisms involved in depressive states remain under debate. However, these states are associated with a *broad and diffuse* alteration of most monoaminergic transmissions.

7.3.1.5 Symptomatology

▪ Presenting Complaints

Despite its severity, the depressive state often goes unnoticed by owners. This is because it does not involve disruptive productive behaviors and is instead perceived as an unusually calm temperament.

In acute forms, the condition closely resembles a febrile state, with consultations often prompted by dysorexia and increased sleep.

Certain substitute activities accompanying this state may become the reason for seeking veterinary advice. These include alopecia from excessive licking, polyphagia, and cessation of grooming, which can lead to matting of the fur.

▪ Symptoms

In acute forms, the clinical presentation is characterized by hypersomnia, anorexia, and a lack of motivation for social interactions or environmental stimuli. Chronic forms display more variable symptoms:

- Sleep disturbances, including sudden awakenings and anxiety related to these episodes; for indoor cats, can be particularly distressing for owners, often disrupting their sleep.
- Appetite variations, frequently with polyphagia and food-related obsessions.
- Territorial disorganization and urine soiling.
- General disinterest, particularly in stimuli the cat previously enjoyed.
- Reduced immune defenses.

Owners often identify a stressful trigger for the depressive state, but this information is not decisive for the practitioner. Moreover, as with anxious states, the depressive state develops in vulnerable animals; its etiology is likely more related to the individual's predisposition combined with an unfavorable context than to the triggering event alone.

It is worth noting that the depressive state can coexist with the anxious state. In such cases, it appears as if the cat temporarily resumes the struggle against adversity. Unfortunately, since anxiety is itself a pathological state, the prognosis is poor. The overlap of symptoms from both states complicates the differential diagnosis.

7.3.1.6 Affected Conditions

Several conditions are associated with a depressive state, often referred to as "depression" in cats.

In its acute form, depressive states can occur following trauma, as seen in *post-traumatic phobias*.

In adult cats, the development of chronic depression may result from untreated anxiety and, more rarely, from trauma.

In older animals, mood dysregulation may arise spontaneously, but depressive states can also be a consequence of emotional or cognitive dysregulation, both of which are common conditions associated with aging (see ▶ Chap. 14).

7.3.1.7 Diagnosis

▫ Table 7.5 summarizes the diagnostic criteria for depressive states.

The symptoms that indicate an *acute depressive state* are:

- hyporexia or anorexia.
- hypersomnia.
- generalized decrease in motivation.
- exclusion of organic conditions that could explain these symptoms.

The symptoms that indicate a *chronic depressive state* are:

- sleep disturbances with variations in sleep quality and quantity.
- generalized decrease in motivation associated with withdrawal behavior.
- irritability, which may lead to irritation-based aggression.

7.3.1.8 Differential Diagnosis

The practitioner must first ensure that no somatic condition is overlooked. Particular attention should be given to thyroid and cortisol functions, as abrupt dysfunctions can mimic a depressive state, although in cats, hyperthyroidism tends to provoke productive behaviors rather than depressive ones. Feline viral diseases, such as FeLV and FIV, should also be ruled out.

Once somatic conditions are excluded, the primary challenge is distinguishing depressive states from *anxious states*. In the anxious state, the individual is alert and responds, often excessively, to their environment. In a depressive state, mood is the primary issue, with demotivation affecting all behaviors. A cat that once enjoyed certain activities now avoids them. Spatiotemporal organization may be altered or entirely absent.

Lastly, aging can sometimes be accompanied by depressive states. However, distinguishing between normal and pathological aging is not always straightforward for the practitioner.

Table 7.5 Diagnostic criteria for depressive, manic, and bipolar States

Pathological state	Diagnostic criteria	State variations	Differential diagnosis
Depressive	*The depressive state* is a psychopathological condition where the animal stops responding to its environment and withdraws. Clinically, it is characterized by intense inhibition, social withdrawal, sleep and appetite disturbances, and anhedonia	Acute depressive state • Hyporexia or anorexia • Hypersomnia • Generalized decrease in motivation • Exclusion of organic conditions explaining these symptoms Chronic depressive state • Appetite variations, with phases of hyporexia • Sleep disturbances, including quality and quantity changes • Generalized decrease in motivation associated with withdrawal • Irritability that may lead to irritation-based aggression • Exclusion of organic conditions explaining these symptoms	Organic diseases Conditions presenting symptoms like hyperthermia, hypothermia, intense pain, or acute anemia must be ruled out Anxious state Hypervigilance in response to the environment, associated with productive behaviors or inhibition. Contexts exist where pleasure and motivation remain intact Senile state Older animal with cognitive or emotional changes unrelated to the context
Manic	*The manic state* is a psychopathological condition where the individual experiences episodic crises of invariable obsessive behaviors, either triggered by the context or created by the animal itself Common symptoms: • Onset of obsessive episodes, referred to as manic crises, during puberty or early adulthood, alternating with periods of normalcy; • During manic phases, the animal's perception of reality is altered, leading to disproportionate reactions to external stimuli; • Unpredictable onset and end of episodes, independent of the context	*Contextual to invasive form* The level of pervasiveness depends on the frequency and intensity of symptoms: • Hypervigilance et agitation motrice; • Gaze changes due to near-constant mydriasis. • Hypervigilance and motor agitation. • Obsessive motivation. • Idiopathic aggression, often irritation-based and highly intense without control from the first episodes. This intensity is abnormally disproportionate to the triggering context • Hyperphagia • Hyposomnia	Phobic state Triggering elements are identifiable and generate fear. Anxious state Hypervigilance is constant, making crises predictable Compulsive state The cat is continuously symptomatic; there are no normal phases when the object of compulsion is present Dissociative state Behaviors are incoherent and unrelated to the context Organic diseases *Hormonal diseases* causing increased hunger (e.g., hypercortisolism, diabetes) or heightened food motivation; diagnosed biologically *Neurological tumors.* Other symptoms are usually observable, diagnosed through imaging
Bipolar	Alternation between depressive and manic states		

Clinical Illustration

Diesel, a 6-year-old neutered male European cat, is brought in for a behavioral consultation due to depressive behavior. Although Diesel usually loves going outside, he has been spending most of his time prostrate. His appetite is erratic, and for the past 2 months, he has tended to isolate himself and no longer seeks contact with his owners. The only positive note from the owners is that he occasionally purrs when comfortably settled.

All tests conducted so far, including bloodwork, radiographs, biochemistry, and FIV/FeLV tests, have returned normal results.

However, one detail raises concern for the veterinary behaviorist: Diesel has no apparent reason for being in a depressive state. He is not old, and his living environment has not undergone any changes.

As a result, further clinical investigations are recommended. While hormonal assessments are normal, abdominal ultrasound reveals an enlarged mesenteric lymph node. Cytology confirms a probable intestinal lymphoma, later fully diagnosed.

Diesel's case highlights that the behavior of a sick cat can closely resemble a depressive state. The practitioner must seek clinical coherence and establish a diagnosis that considers the broader context.

7.3.1.9 Prognosis

The prognosis for depressive states is generally poor, particularly in older animals.

For acute depressive states, the prognosis is better if the animal can be quickly removed from the triggering factor. However, it is important to note that in acute forms, the risk of death is significant if recovery is not rapid.

Achieving complete recovery is often challenging, but treatment consistently improves the *animal's well-being*. Even if it does not lead to full resolution, treatment should always be initiated and maintained as long as necessary.

Genetic vulnerabilities are common, and some individuals experience recurrent relapses, particularly during medication withdrawal or when faced with environmental challenges that most other cats tolerate without issue.

7.3.1.10 Treatments

▪ Pharmacological Treatment

The treatment of depressive states relies on anti-depressants. The alleviation of inhibition is often accompanied by significant discomfort, including anxious episodes. For this reason, it is advisable to include anxiolytics at the beginning of the treatment.

For chronic depressive states, mirtazapine can lead to a dramatic resurgence of behaviors at a dose of 1 to 2 mg/kg once daily. In cases of chronic kidney disease, the dose can be reduced by half. Its strong appetite-stimulating effect increases food motivation, improving the efficacy of behavioral therapies. However, caution is necessary to monitor for excessive disinhibition.

Venlafaxine, available in capsules containing granules (Effexor™), is advantageous due to its rapid efficacy and easy administration (Metz et al. 2021; Masson et al. 2024).

SSRIs and TCAs can generally be used at low doses. Clomipramine, in its veterinary formulation (Clomicalm 5 ND), is commonly appreciated due to its veterinary marketing authorization (for dogs) and ease of administration. It is typically used at a dose of 0.252–1 mg/kg once daily.

▪ Behavioral Therapy

The behavioral therapy for depressive states depends on the nosographic diagnosis but primarily relies on techniques to stimulate activity. It must be adapted to the cat's abilities and preferences. Knowledge of the cat's personality before the onset of the pathological state is valuable. The general approach is to introduce very simple exercises and reinforce success to increase motivation.

Depressed cats often seek isolation. Once the initial effects of medication are established, it is often necessary to gently encourage the cat to leave its isolation zone and re-engage in activity. This process can be challenging, as the animal may still need its safe space, but what serves as a refuge should not become a barrier to returning to normal behavior.

Clinical Illustration

Luna is a 6-year-old sterilized adult cat whose living situation has forced her to coexist with an exasperating dog for over a year. She constantly feels watched, and every move she makes is monitored and followed closely. If she happens to run, she is chased like prey. Fortunately, she can retreat to her owner's office, where the dog is less present. However, over time, she chooses to stay upstairs and refuses to confront the "terrorist" dog.

Eventually, Luna confines herself to the office and even finds refuge in the attic, where an old mattress is located. She stops coming downstairs entirely. As a precaution, a litter box is placed near the attic door, and the staff kindly leave kibble on the staircase. Luna no longer ventures outside, spends all her time on a sofa, and her appetite seems inconsistent, although it is not closely monitored. Her coat becomes dirty, and she hardly allows anyone to pet her.

These symptoms suggest depression. Complementary examinations, including biochemistry and complete blood count, return normal results. Clomipramine is prescribed at a dose of 2.5 mg/day, but it shows little effect.

Approximately 2 months later, renovations begin, and the staircase leading to the attic is removed. Luna's bedding is relocated downstairs, forcing her to engage in minimal activity and re-establish contact. Soon after this change, her activity levels return to normal. Remarkably, she is even observed hissing at her canine nemesis.

7

Summary of ▶ Sect 7.3.1

The depressive state is a mood disorder characterized by a loss of motivation and behavioral inhibition.

In the acute form, all behaviors collapse, including feeding and drinking behaviors, requiring urgent and intensive care. In the chronic form, in addition to disturbances in feeding and sleeping behaviors, the cat exhibits anhedonia and varying degrees of irritability.

Pharmacological treatment relies on anti-depressants, which help revive behaviors and regulate sleep and appetite. Mirtazapine, with its strong orexigenic effect, supports behavioral therapy and often helps alleviate inhibition.

Behavioral therapy involves simple changes that encourage the cat to reconnect with its environment and resume activity.

7.3.2 Manic State

7.3.2.1 Definition and Description

The *manic state* is a psychopathological condition classified as a mood disorder. In this state, the individual experiences episodes during which they focus all their energy on a specific objective that does not correspond to an appropriate response to their needs. The onset of these episodes can be triggered by contextual or internal factors. In cats, the trigger is most often of internal origin. To an observer, the reasons behind these mood variations are often unexplained.

7.3.2.2 State Variations

Each manic individual has their own obsession, and as with dogs, obsessive categories can be identified.

In cats, episodes manifest in two primary forms: uncontrolled aggression associated with protecting a refuge area or pronounced sexual behaviors, such as mate-calling. Most commonly, the affected individual becomes aggressive toward anything that might interfere with their goal. Non-aggressive forms likely exist but rarely concern owners.

As with other pathological states, the manic state can vary in its level of pervasiveness. When the condition becomes more invasive, it leads to a degradation of the individual's well-being and adaptability to their environment. Most frequently, the individual expresses their obsession through *episodes* lasting from a few hours to several days. Between episodes, the individual appears perfectly normal.

Clinical Illustration

Marguerite is a 2-year-old spayed Abyssinian mix cat in good health. The consultation is prompted by unexplained episodes of aggression. To date, she is experiencing her third episode (approximately one episode per month), and the severity of the behavioral abnormalities appears to be increasing.

The episodes are preceded by vocalizations accompanied by noticeable mydriasis. At the onset of an episode, Marguerite becomes agitated and roams around the house, strongly refusing contact. After a few hours or even a full day, she typically isolates herself under the couch, becoming extremely aggressive if approached. During this isolation period, which lasted nearly 2 days during the most recent episode, she remains calm but occasionally emits loud howls.

Her owner perceives that she is in pain, but clinical examinations reveal no abnormalities.

7.3.2.3 Evolution

The onset of the manic state typically occurs at the *end of puberty* or *early adulthood.* Initially, episodes may be sporadic and widely spaced over time, but the manic state generally stabilizes. Most reported episodes involve aggressive sequences that may initially appear context-dependent but quickly become independent of environmental triggers. In some cases, manic episodes seem to increase in intensity over time.

The disruption caused to those around the affected individual is significant, and the intense reactions from others often complicate the investigation of the underlying cause.

Occasionally, treatment or the reactions of those around the animal can lead to an extended period without episodes, giving the impression of recovery. However, the deviant behavior typically reappears once the triggering context resurfaces or treatment is discontinued. The progression of the condition almost always trends toward worsening. Interrupting obsessive behaviors during episodes, through control or prevention, may slow the progression of the disorder but does not eliminate the obsession.

Some animals transition into a depressive state between episodes. It appears as though their sole motivation is their obsession, and life holds little interest beyond it. This alternation between manic episodes and depressive states is characteristic of bipolar disorder (see ▶ Sect. 7.3.3).

7.3.2.4 Pathogenesis

Manic states involve action-validation loops that generate positive reinforcement (reward circuits). Unlike dissociative states, the obsessive behaviors in manic states focus on real situations, but their significance is pathologically amplified. In other words, the pathological aspect lies in the animal's exaggerated importance given to certain elements rather than its reaction to those elements. The manic state *disrupts decision-making*, as a single idea becomes overwhelming, preventing the consideration of alternatives. This lack of flexibility results in the obsessive behaviors observed.

The dramatic nature of aggressive episodes often enables the cat to achieve its desired outcomes. These successes reinforce the importance of the episode and the necessity of its intensity.

Similar to depressive states, manic states are likely driven by widespread dysregulation of certain monoamine neurotransmitter receptors. In particular, dysregulations in dopaminergic actions within the nucleus accumbens and cortical regions contribute to disturbances in positive reinforcement systems and action-validation loops. Serotonergic deficits impair the inhibition of actions, further exacerbating the condition.

As a mood disorder, the individual's motivation is entirely directed toward a singular goal. This amplified motivation can be seen as the opposite of the depressive state.

- **Manic State and Dysthymic State**

What is now referred to as the manic state was previously termed the dysthymic state in the nosography proposed by Pageat (1998). However, to align terminology with human psychiatry, where dysthymia characterizes a mood tending toward a depressive state, we have chosen to refer to it as the manic state.

This change also allows for a clearer distinction between nosographic terms and the names of pathological states.

7.3.2.5 Symptomatology

- **Presenting Complaints**

In its early stages, the manic state manifests as brief episodes that are not easily forgotten due to their unpredictability and intensity. Frequently, the cat, displaying mydriasis, initiates an offensive attack. Initially, these attacks may appear triggered by the owners' movements, but over time, they occur even when the owners are still, reflecting mood swings originating from the cat rather than the environment.

Owners typically express two main concerns: the danger posed by the aggression and the obvious discomfort experienced by the animal.

When episodes persist, a palpable tension develops in the household, which becomes difficult to endure. Practitioners familiar with hospitalizing such cats during manic episodes are also aware of how the presence of an aggressive cat, even when confined, can heighten stress in the cattery.

If manic episodes are not accompanied by aggression, consultations are rare. In females, the onset of estrus may sometimes trigger the first episode.

Clinical Illustration

Forbie is a young European cat preparing for a trip to Norway, requiring a rabies vaccination. Two days after receiving the vaccine, Forbie begins vocalizing intensely while pacing through the house. Her cries are loud and seem aggressive.

A few days later, Forbie refuses to be petted and becomes threatening toward her owners, especially when they pass through the hallway leading to the bedroom. She isolates herself in the bedroom and attacks her owners with multiple bites and scratches, clinging tightly to their arms.

The owner, frustrated, believes this behavior is a side effect of the vaccine and suspects his cat is "rabid." The timing is suspicious, prompting a recommendation for hospitalization. A diagnosis of sexual dysthymia is quickly established.

The episode lasts a week, and once calm is restored (until the next episode), spaying is proposed as a solution.

- **Symptoms**

The alternation between periods of complete normalcy and episodes of crisis is a key factor in establishing the diagnosis.

In cats, these episodes are typically characterized by aggression. Unlike usual aggression, this behavior does not correspond to a readily identifiable trigger. Changes to the environment or interactions do not mitigate the episode, and the aggression persists. The response of the owners to this unusual behavior often exacerbates the situation.

The episode resolves spontaneously and tends to have a consistent duration each time. Some cats experience daily episodes lasting only a few minutes, while others have episodes only a few times per year, lasting several days.

7.3.2.6 Diagnosis

Table 7.5 summarizes the diagnostic criteria for the manic state.

The diagnosis is based on the presence of episodic obsessive behaviors, referred to as manic episodes, which occur without a clear contextual trigger and are only partially understandable. The inability to escape these episodes contrasts with the individual's normal behavior between episodes.

The diagnostic criteria for concluding a manic state are:

- The onset of obsessive manic episodes during puberty or early adulthood, alternating with periods of normalcy
- During manic phases, the animal's perception of reality is altered, leading to disproportionate reactions to external stimuli

These productive phases are characterized by:

- Unpredictable onset and end of episodes
- A change in gaze due to near-constant mydriasis
- Hypervigilance and motor agitation
- Idiopathic aggression: Often presents as high-intensity, uncontrolled irritation-based aggression from the first episodes. This abnormally elevated intensity is inconsistent with the triggering context
- Hyposomnia caused by a marked increase in vigilance.

7.3.2.7 Differential Diagnosis

It is essential to differentiate these obsessions from genuine needs. For example, hormonal imbalances causing overstimulation should always be considered.

The first condition that may be confused with the manic state is the compulsive state. In compulsive states, productive behaviors are expressed continuously, with no alternation between productive phases and normal phases.

The manic state must also be distinguished from the dissociative state. While there may be a resemblance due to the alternation of normal and productive phases, dissociative behaviors appear genuinely incoherent, as the animal responds to elements that are not visible to others.

Confusion is also possible with certain anxious states. In communication disorders, sudden, seemingly inexplicable aggressive outbursts can occur. Unlike manic episodes, these aggressions escalate gradually and are consistent within the same context, with no periods of normalcy.

7.3.2.8 Prognosis

Currently, few medications are effective in improving the manic state, and the prognosis is poor. Full recovery is not currently feasible. However, long-term treatment can provide reasonably good clinical results for several years. When treatment is initiated, it must be lifelong.

In severe cases, the intensity of the attacks may instill too much fear in people to allow time to stabilize the situation, leading to the decision for euthanasia. Rehoming can only be considered if the adopters are fully aware of the severity of the condition and are committed to providing lifelong treatment for the cat.

7.3.2.9 Treatments

Pharmacological Treatment

Fluoxetine, prescribed at a high dose (2 mg/kg), can help control the intensity of reactions during productive phases. This medication also reduces the severity of aggression but is rarely sufficient to fully stabilize the animal. Adding gabapentin, pregabalin, or trazodone may sometimes reduce the frequency of episodes.

Behavioral Therapy

Behavioral therapy should focus on two essential objectives: reducing danger and preventing the apparent triggering context. Limiting the often-significant danger is paramount. It is crucial that owners and others in the household avoid interacting with the cat during manic phases. The recommended approach is to ignore the cat while it is in a manic episode. When feasible, *isolating the cat* during productive phases helps to manage its danger effectively.

Clinical Illustration

Flytox, a three-year-old European cat, regularly experiences aggressive episodes. Initially, her owners were attacked when passing near a chair in the kitchen, leading them to believe she was frightened. Over time, the manic phases became more "predictable," occurring approximately once every 2 months and lasting 1–2 days. During these episodes, Flytox attacks anyone who comes near her. The episodes, which initially occurred only in the kitchen, later began happening in any room.

The veterinarian diagnosed a manic state and prescribed fluoxetine at 1.5 mg/kg, later increased to 2 mg/kg. The frequency of episodes decreased, and the owners learned to isolate Flytox in the garden shed during her manic phases. Flytox now voluntarily retreats to the shed when an episode occurs. No one attempts to enter the shed, and everything remains calm.

After about 2 days, Flytox returns to the house and becomes highly social, seeking interaction. In the weeks following an episode, her owners can pet her without triggering aggression, and she greatly enjoys the attention.

Summary of ▶ Sect. 7.3.2
The manic state is a mood disorder characterized by episodes or productive phases during which the cat typically exhibits significant instability and, consequently, aggression.

During episodes, mydriasis is often observed, accompanied by hypervigilance and motor agitation. Aggression in manic states, when present, consists of idiopathic attacks resembling intense irritation-based aggression, with a severity disproportionate to the triggering context.

This condition is distinct from dissociative states (triggered by internal, nonvisible factors) and compulsive states (marked by systematic obsessions).

The prognosis ranges from guarded to poor, as the condition is incurable. High-dose fluoxetine provides clinical improvement but requires lifelong administration. During episodes, it is essential to allow the cat to isolate itself.

Some forms of sexual dysthymia may be associated with the manic state. Early spaying often prevents recurrence, but intervention is necessary for the first episodes.

7.3.3 Bipolar State

As in human psychiatry, clinical cases exist where manic and depressive states alternate. This alternation of manic and depressive states is referred to as the *bipolar state.*

When a clinician encounters a manic state, it is important to investigate the presence of a depressive state between manic phases, as this may necessitate a different medical approach to address the depressive component.

The alternation can occur rapidly, with successive state changes within a single day, or much more slowly, with months of depression interspersed with occasional manic phases.

7.4 Cognitive Pathological States

Cognitive pathological states include states of self-control deficits, such as impulsive and compulsive states on one hand, and states of derealization, such as senile and dissociative states, on the other hand.

7.4.1 Self-Controls Pathological States

In the French psychopathological model (Beata et al. 2015; Pageat 1995), self-control deficit states characterize animals that are highly productive and unable to regulate their activity, particularly locomotor activity. This terminology is retained because it is widely used in France, but self-control deficit states more accurately correspond to a broader dysfunction of executive functions.

Clinical experience and psychopharmacological data have led to further distinctions within this category. Specifically, impulsive and compulsive states are separated to improve diagnostic precision. Previously, self-control deficit states encompassed all cats with executive function impairments. However, the description of this state overlapped significantly with a specific condition, the HSHA syndrome.

In clinical practice, it is common to encounter individuals with executive function dysfunctions, such as impulsivity or compulsivity, without exhibiting all the clinical signs of this neurodevelopmental disorder. Moreover, impulsive and compulsive states also occur in other conditions, such as addictive behaviors. Making this distinction enhances the coherence of the model and allows for more precise diagnostics.

7.4.1.1 Pathogenesis of Self-Controls Pathological States

The neural circuits whose dysfunction leads to pathological states of executive function are described in ▶ Chap. 2 (see ▶ Sect. 2.3.3.4). These include the motivational circuit and the action control circuit.

Advances in imaging have enhanced our understanding of the neurobiological mechanisms underlying impulsivity and compulsion in human medicine. In both states, the brain fails to effectively say "no." These pathological states represent cognitive rigidity and inflexibility caused by dysfunction in loops involving the prefrontal cortex and various subcortical structures, notably the striatum.

Simplified, impulsivity and compulsion can be seen as resulting from activating processes originating in the amygdala and hypothalamus, traveling from the striatum to the prefrontal cortex. This neurobiological pathway is referred to as "*bottom-up*" (from bottom to top). Impulsivity originates in the ventral striatum (mesolimbic dopaminergic loop), whereas compulsion begins in the dorsal striatum.

Conversely, the prefrontal cortex sends "*top-down*" (from top to bottom) information to the striatum. These cortical signals have the potential to inhibit initial desires. Impulsive actions and compulsive behaviors result from the balance between *bottom-up and top-down information*. A low level of prefrontal cortical inhibition or excessive motivation transmitted by the striatum will lead to impulsive or compulsive states, depending on the circuit involved (◘ Fig. 7.8).

The balance between these two systems allows the individual to avoid being overwhelmed by their emotions or desires and to consider alternative options before acting. The top-down system is specifically responsible for presenting these alternatives, emphasizing the short-, medium-, and long-term consequences of choosing one decision over another.

This enables the animal to be less dependent on the immediacy of internal or environmental stimuli and to take into account the consequences of its actions by integrating past experiences.

When this balance is absent or impaired, the *top-down effect* is reduced or nullified, leading to actions that are taken quickly and with high intensity: this is the impulsive state.

In addition to the diminished top-down action, pathologically impulsive individuals seem to assign an exaggerated value to rewards. The pleasure derived from these rewards is often extremely intense (e.g., such cats are "hyper-curious"). This heightened response applies to all emotions, often leading to generalized *emotional dysregulation*.

Over time and with repetition, the impulsivity circuit involving the ventral striatum begins to self-inhibit, giving way to another circuit located in the *dorsal striatum*. In normal individuals, this transition requires numerous repetitions. However, in individuals with a compulsive predisposition, only a few repetitions may suffice, leading to the automation of behaviors.

When this circuit becomes overactive, it results in the production of stereotyped behaviors through cascades of neuroadaptations (creation of synapses and neurons in the dorsal striatum), transforming an impulsive

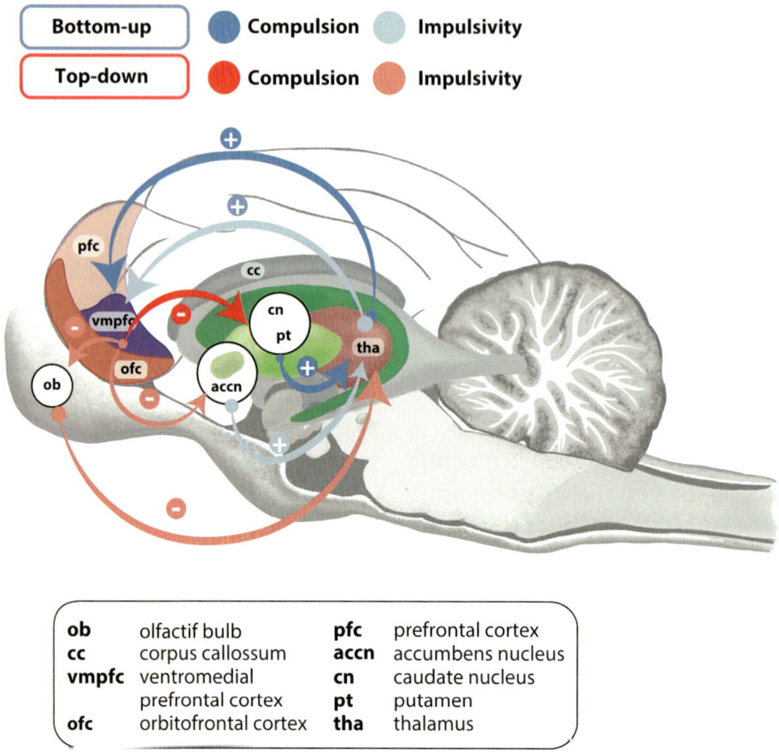

Fig. 7.8 Bottom-up and top-down loops. Dopaminergic pathways activate signals traveling from the striatum and thalamus to the cortex, forming the bottom-up pathway. The top-down pathway is the regulatory feedback loop from the cortex to the striatum. An excess of activation in the bottom-up pathway or a failure of inhibition by the top-down pathway results in impulsive and compulsive states

behavior into a compulsive one. The transition from one circuit to the other involves the amygdala and hippocampus, integrating emotions and memory into the process.

In other words, if the impulsivity circuit, initially leading to reward and pleasure, is overstimulated too strongly or too frequently, it eventually results in stereotyped behaviors that no longer bring pleasure but instead serve to alleviate the discomfort that becomes constant in their absence. This marks a shift from pleasure to addiction and withdrawal (see ► Chap. 9, ► Sect. 9.2).

Most addictive substances can rapidly facilitate the transition from the impulsive loop to the compulsive loop. This occurs either because the substance's properties enable it to quickly reach high concentrations in the striatum (e.g., addictive drugs like nicotine), due to a genetic vulnerability in certain individuals where the transition between loops happens more rapidly (individuals predisposed to addiction), or a combination of both.

In cats, overactivation of this circuit can occur during intense and regular reinforcement, as may happen during certain types of play with owners in animals that are abnormally motivated. Excessive behavior-reward associations can lead to a compulsive state in individuals predisposed to addiction.

At the cellular level, pleasure is triggered by dopamine, which is released by mesolimbic neurons onto the neurons of the nucleus accumbens. Through repeated behavior, the functioning of dopaminergic neurons adapts so that dopamine release becomes greater when the anticipation of reward is high. This is particularly evident at the beginning of learning, when the outcome of an action exceeds expectations and provides heightened satisfaction (Schultz 2011).

The modulation of dopamine release is influenced by GABA and glutamate neurons after every rewarded action, occurring following each dopamine neuron discharge. Opioids and cannabinoids are also involved in this reward circuit.

When this loop becomes imbalanced, either due to overstimulation or genetic vulnerability, the transition to compulsion occurs. Dopaminergic neurons in the ventral striatum stop responding to stimuli (e.g., play, food, social contact), and activity shifts to other dopaminergic neurons in the dorsal striatum. These dorsal neurons sustain the action without the association of pleasure, resulting in highly automated behavior.

This shift reduces the involvement of the prefrontal cortex in controlling the action, allowing the cortex to focus on other tasks, even as the automated behavior persists.

7.4.1.2 Impulsive State

Definition and Description

The impulsive state is a pathological mental condition characterized by an imbalance between activation and inhibition, or *bottom-up* versus *top-down* processes. This imbalance leads to behavioral responses where the appetitive phase is shortened, and the intensity of the response is potentially heightened (◘ Fig. 7.9).

Impulsivity is a pathological state characterized by the inability to refrain from initiating a behavior, often described as "acting before thinking." An impulsive individual cannot wait to obtain a reward, whether it be food, social contact, or play. They even demonstrate a preference for a smaller, immediate reward over a larger, delayed one.

An impulsive cat remains motivated even when faced with punishment.

Such an animal appears to no longer make decisions independently but instead reacts immediately to stimuli, with very rapid actions.

State Variations

The impulsive state can be specific to a particular context, referred to as a *contextual impulsive state*. This is often observed in hunting behaviors that are triggered systematically, regardless of their outcome or relevance.

It can also become pervasive, leading to hypermotricity and attention deficits caused by a "zapping"effect. This occurs when triggering contexts are numerous, making the cat appear unable to stop because it compulsively reacts to all stimuli in its environment.

It is essential to distinguish this from a compulsive state, where, despite the persistence or even absence of the same stimuli, the animal is unable to terminate its behavioral sequence. In the compulsive state, there is a failure in the *stopping signal*, accompanied by an *absence of the appeasement phase.*

Evolution

The impulsive state can become pervasive from the outset if *the sensory filter* is compromised. In such cases, the cat's *hypersensitivity*, behavioral responses to low-intensity stimuli, leads to successive impulsive reactions. This hypersensitivity creates a "zapping" effect, where behavioral sequences are shortened and interrupted by new reactions replacing the previous ones.

Additionally, in the impulsive state, behavioral responses may initially occur in a specific context but can generalize to other situations through learning. For example, an impulsive cat that bites its owner's hands during play may develop an impulsive response toward humans outside of play and begin biting when people walk past it.

The impulsive state can progress into the *compulsive state*. When impulsive behaviors are repeated and reinforced, the reward circuit is stimulated and can shift to involve the dorsal striatum, a circuit that excludes the nucleus accumbens (the pleasure center). The pleasure that originally initiated the action diminishes, yet the behavior is repeated despite the lack of positive outcomes (absence of reward) or even in the face of negative consequences (positive punishment).

Thus, an impulsive sequence can evolve into a compulsive sequence, which becomes increasingly less satisfying. However, this progression is not systematic and likely depends on an initial vulnerability. Additionally, some individuals may exhibit compulsive states from the outset.

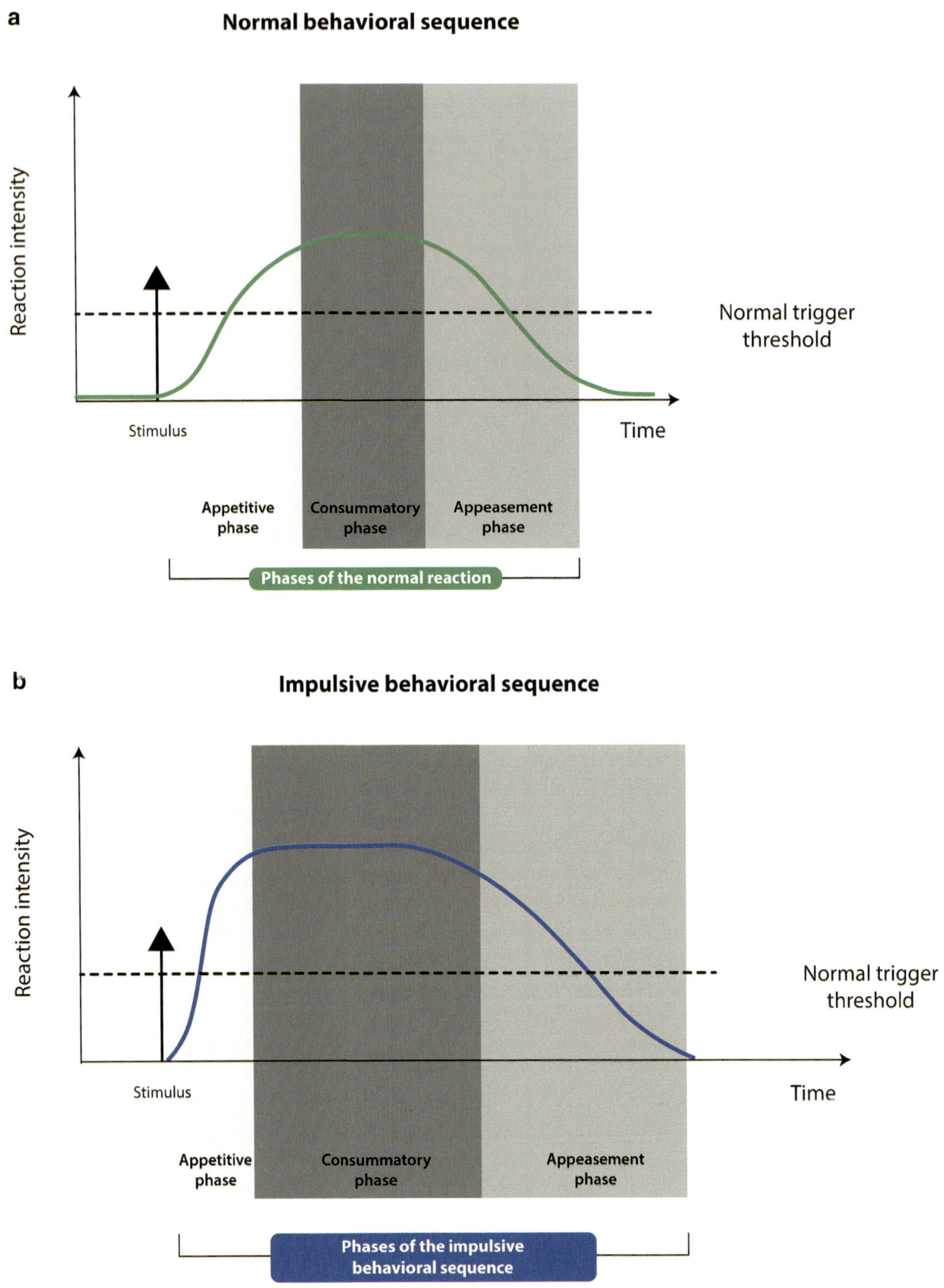

7

Fig. 7.9 Characteristics of reactions in the normal adaptive state (**a**) and the impulsive state (**b**). In a normal adaptive state, the stimulus must reach a sufficiently high intensity to trigger a behavioral response. Behavioral sequences can be divided into three phases: appetitive, consummatory, and calming. In the impulsive state, the reactions exhibit a shortened appetitive phase and an increased intensity. Consequently, the calming phase is prolonged

Clinical Illustration

Negus is an 18-month-old Abyssinian cat who is very lively and frequently runs around the house without any apparent reason. These bursts of activity are rather endearing, and the owner doesn't mind them. However, the owner wishes he could pet his cat without issues. Whenever he tries, Negus starts play-fighting, gets over-excited, and progresses from gentle nibbling to full-on biting. The bites become increasingly intense, and these interactions usually end poorly, as the owner becomes angry and chases the cat away while shouting.

This behavior occurs with everyone, though it escalates more quickly with the owner because Negus anticipates the reprimands but cannot stop himself. Despite this, Negus enjoys being petted and occasionally sits on his owner's lap, purring. However, if the owner reaches out to pet him, Negus pounces on the hand.

This is an example of uncontrolled impulsive behavior. Punishment will not inhibit the initiation of this behavior, and medical intervention is essential to help Negus learn to regulate himself.

Impulsive states frequently coexist with secondary pathological anxious states. A cat's impulsivity often leads owners to employ coercive methods to manage the behavior, especially when it results in biting or scratching. However, such methods are harmful to the animal and contribute to the development of anxiety. For example, a cat that frequently plays with its owner's hands and cannot resist biting and scratching may be punished during play. The inability to control its impulses renders these punishments counterproductive, and impulsivity will often escalate into aggressive behavior.

Additionally, impulsive states predispose cats to various somatic disorders. Osteoarticular issues may arise from injuries caused by a lack of attention during play or running. Digestive problems, such as the ingestion of non-food objects, can lead to obstructions or conditions like gastritis and enteritis caused by inflammation of the digestive system. Cardiac conditions may develop, potentially due to the heightened stress and excitement associated with impulsivity. Dermatological problems are also common, including excessive grooming or the ingestion of stinging insects.

▪ Symptomatology

Presenting complaints

Consultations are often prompted by behaviors that inconvenience owners. These are typically *productive symptoms* of impulsivity, such as destructive actions, the inability to touch the cat without triggering aggressive play, or the ingestion of non-food objects. One of the most frequent presenting complaints is the *exhausting* nature of the cat's behavior.

Owners rarely question the *cat's well-being* because the animal seems to derive pleasure from these actions. However, insights from humans with ADHD suggest that the well-being of animals with similar disorders may also be compromised. Owners of cats whose impulsivity has improved frequently and spontaneously report that their companions seem more peaceful and appear to enjoy life more fully.

Additionally, the relationship with the owner is often strained due to *incessant and ineffective reprimands* aimed at reducing impulsivity. In practice, these cats spend much of their time being punished while lacking the neurological capacity to comply with expectations. The escalation of punishments by owners attempting to defend themselves against the cat's impulsive actions often triggers phobic and anxious states in the animal.

The clinician must address the owner's immediate concerns while also focusing on reducing overall impulsivity to achieve meaningful progress. Restoring a better level of neurological self-control in areas beyond the specific behavior of concern will provide the animal with a stronger "braking system," aiding in improvement across all areas, including those targeted by the owner.

Symptoms

The symptoms indicative of an impulsive state arise from the animal's difficulty in refraining from reacting to environmental stimuli. To an observer, the response may appear deliberate, but the cat has no true free will in its actions. The cat's enthusiasm can be misleading, as its response is often inappropriate and causes the animal distress.

In *contextual forms*, the most commonly observed symptoms include:

- A heightened motivation to respond to certain stimuli or groups of stimuli, leading to behaviors such as exaggerated oral exploration, accidents during play (e.g., bites or scratches), or injuries—often scratches—during social interactions due to a lack of control over movement and claws
- Altered behavioral sequences with a shortened appetitive phase
- Sensory hypersensitivity caused by impaired sensory filtering, leading to reactions to inconspicuous environmental stimuli
- Motivational, emotional, or cognitive invasiveness restricted to specific contexts

In more severe, *pervasive forms,* the following symptoms are observed:

- Difficulty remaining idle, with incessant demands, including requests for petting, which may escalate into injurious play
- Hypermotricity and attention deficits in response to normal environmental stimuli
- Hyposomnia caused by difficulty resting during the day when triggers are present. These cats are exhausted but unable to sleep as long as there is any form of stimulation.

A number of consequences stemming from these symptoms are frequently observed:

- Poor-quality communication, often overly abrupt
- Secondary irritation-based aggression in contexts of frustration, fear, or pain, often following physical punishment.

7

Affected Conditions

The deficit in self-controls, whether impulsive or compulsive, is a necessary component for diagnosing HSHA syndrome, just as the dissociative state is essential for diagnosing dissociative syndrome. However, the impulsive state can also be present in most psychiatric disorders, where it acts as an additional factor of severity.

This state can be observed in conditions related to fear, territoriality, or social relationships, as well as in primary psychiatric disorders such as dysthymia or dissociative syndrome. Additionally, the impulsive state may occur in communication disorders or painful somatic conditions.

Diagnosis

Table 7.6 summarizes the diagnostic criteria for impulsive states.

The diagnosis of an impulsive state is based on the combination of multiple symptoms. No single symptom is pathognomonic; rather, it is their combination that establishes the diagnosis. The characterization of behavioral sequences is essential to this process (Fig. 7.10).

The diagnostic criteria for concluding an impulsive state are as follows:

- Modification of the behavioral sequence of one or more behaviors, with a shortened appetitive phase, increased reaction intensity, and consequently, a prolonged calming phase
- These sequence modifications persist despite attempts to alter the behavior
- When multiple stimuli are present, the sequences may end with the initiation of a new appetitive phase.

Differential Diagnosis

The main challenge is distinguishing the impulsive state from the *reactive state*, which involves determining whether the reactions are normal or pathological. We have already discussed the apparent joy of the impulsive cat. The practitioner must assess the possibility of choice (free will) and evaluate

Table 7.6 Diagnostic criteria for impulsive and compulsive states

Pathological state	Diagnostic criteria	State variations	Differential diagnosis
Impulsive	*The impulsive state* is a pathological mental state characterized by an imbalance between activation and inhibition, or bottom-up versus top-down processes, leading to behavioral responses with a shortened appetitive phase and potentially increased intensity	*Contextual to invasive form* The level of intrusion in the impulsive state depends on the number of behaviors involved: • Modification of the behavioral sequence of one or more behaviors, with a shortened appetitive phase, high reaction intensity, and prolonged calming phase • These sequence modifications persist despite attempts at behavioral modification • When multiple stimuli are present, sequences may stop, and a new appetitive phase may begin	Reactive state Environmental adjustments and meeting physiological needs lead to the rapid return to the normal adaptive state. Impulsive behaviors disappear completely Compulsive state The consummatory phase does not provide calming. Repetitive or stereotypical behaviors, often identifiable (e.g., play, frustration, eating) Phobic state Fear reactions (escape, aggression, inhibition, agitation) in the presence of a phobic stimulus. Absence of excitement or frustration. Motor agitation is linked to exaggerated fear response, not excess motivation Anxious state Ear reactions (escape, aggression, inhibition) even in the absence of a phobic stimulus. Absence of excitement or frustration. Motor agitation is related to hypervigilance, not excess motivation **Organic diseases** *Pain* may cause motor agitation but is rarely present from adoption. Once the organic issue is resolved, the impulsive behavior should disappear

(continued)

Table 7.6 (continued)

Pathological state	Diagnostic criteria	State variations	Differential diagnosis
Compulsive	*The compulsive state* is a pathological mental state characterized by the production of invariable behavioral responses, with a delayed, shortened, or absent calming phase. Compulsion can be defined as the inability to stop a behavioral sequence, regardless of the action's consequences. It is as though the consummatory phase, which normally leads to calming through action execution, remains ineffective	*Contextual to invasive form* The level of intrusion in the compulsive state depends on the time spent on compulsive behavior: • Modification of the behavioral sequence of one or more behaviors, with an elongation of the consummatory phase, leading to the absence of a calming phase • These sequence modifications persist despite attempts at behavioral modification • When the compulsive behavior is performed, there is indifference to other environmental stimuli • Transformation of compulsive activity into stereotypy in the most intrusive cases	Reactive state Environmental adjustments and meeting physiological needs lead to the rapid return to the normal adaptive state. Compulsive behaviors disappear completely **Maladies organiques** **Organic diseases** *Neurological disorder* partial temporal epilepsy, syringomyelia *Pain* may lead to stereotypical behaviors *Maladie dermatologique* atopy, abscesses, wounds Once the organic issue is resolved, compulsive behavior should disappear Impulsive state The consummatory phase provides calming Anxious state Fear reactions (escape, aggression, inhibition), even in the absence of a phobic stimulus. Absence of excitement or frustration. Motor agitation is related to hypervigilance, not excess motivation Manic state Manic phases are not permanent, and there are periods of normality during which repetitive behavior is not produced Dissociative state The trigger for compulsive behavior is not identifiable (internal stimulus)

whether the inappropriate behaviors result in suffering. Physical injuries to the owners, who sometimes perceive these vulnerable actions during play as aggression, are an important clue for the practitioner. It is important to note that by puberty, a cat should be able to control its biting, scratching, and locomotion.

In the case of the impulsive state, the presence of the symptoms mentioned is often associated with a decline in the animal's well-being. In response to disruptive impulsive behaviors, owners often reduce environmental enrichment to avoid triggering contexts. However, this only serves to reinforce the animal's suffering.

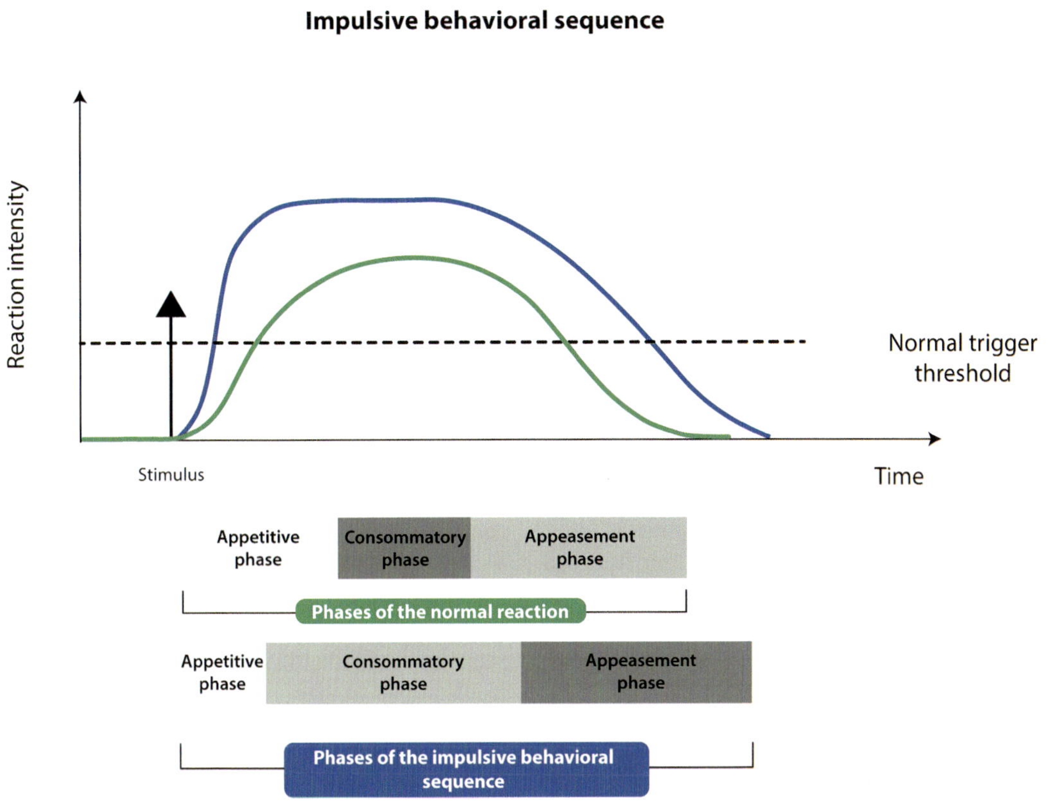

Fig. 7.10 Alteration of behavioral sequences in the impulsive state. When an animal exhibits an impulsive state, its reactions are marked by a shortened appetitive phase and increased intensity. As a result, the calming phase is prolonged

The impulsive state must also be distinguished from the *compulsive state*. The presence of a stopping signal is a key differentiating factor. In the impulsive state, the cat will eventually stop the behavior spontaneously and is still capable of responding to a disruption. In compulsive behavior, it seems as though the cat could continue the behavior indefinitely, and the calming phase is no longer present.

Furthermore, in the impulsive state, when the triggering stimuli are removed, the cat can quickly return to calm. In contrast, in the compulsive state, removing the stimulus increases frustration, and the cat actively seeks to resume the behavior.

Distinguishing between impulsive states and *phobic or anxious states* can be challenging. In both cases, the cat exhibits *context-dependent motor agitation*. Fear responses often become impulsive behaviors through learning. In a fear response, the amygdala circuit quickly activates a short pathway, allowing the subject to react quickly to potential threats. In both cases, the appetitive phase is shortened, but fear signs are clearly observable during the first executions of the behavior. In contrast, in the impulsive state, excitement and desire are visible from the outset. Therefore, the animal's intent is the most reliable way to differentiate between the two. This distinction becomes even more difficult when the owner's reactions trigger anticipatory defense behaviors in the cat.

As previously mentioned, the impulsive state can coexist with emotional pathological states (phobic and anxious), mood disorders (depressive, manic), or derealization states (senile, dissociative).

Prognosis

The prognosis ranges from moderate to excellent.

In general, the more pervasive the impulsivity, the worse the prognosis. However, certain factors can influence the chances of improvement:

- *Early intervention* improves the effectiveness of rehabilitation, simply because the cat has not been overly reinforced during the expression of impulsive behavior
- *Coercive methods* increase the risk of anxiety states and aggression, and deteriorate the human-animal relationship
- *The deterioration of the relationship* due to the exhausting nature of these cats may lead to abandonment or euthanasia. Thus, the level of urgency should be assessed during the first consultation. Delaying the implementation of medical treatment represents a lost opportunity for the animal, as owners may become demotivated
- *Comorbidity* with other pathological states is an aggravating factor.

Treatments

Pharmacological treatment

Monoamines play a role at various levels in the top-down and bottom-up circuits of impulsivity. In practice, it is important to remember that the *top-down* inhibitory effect is promoted by serotonin. Attention, which is adrenergic, is crucial for decision-making. The *bottom-up* part of the circuit is more influenced by dopamine (reward, motivation, learning).

Medications such as fluoxetine or other SSRIs or SNRIs, depending on the symptoms, should be prioritized. In practice, fluoxetine is the most commonly used drug, at a dose of 1–2 mg/kg once per day.

If the owner is unable to administer the medication to the cat, non-generic Effexor™ can be used at a dose of 1 to 2 mg/kg once per day, equivalent to 4–8 granules (contained in one capsule) per kilogram (Metz et al. 2021; Masson et al. 2024).

When the impulsive state is associated with a phobic or anxious state, an additional anxiolytic medication may be used to complement the anti-impulsive action. For example, gabapentin, pregabalin, or trazodone can be added. In depressive states, impulsivity is most often observed during the transition out of the depressive state, and adjustments to the treatment, either by changing medications or adjusting doses, are necessary.

Behavioral therapy.

Behavioral therapy will depend on the diagnosed condition (see ► Chap. 29, ► Sect. 29.3.2) and will include exercises focusing on the following elements:

- Positive training methods, meaning the cessation of all positive punishments
- Ensuring the fulfillment of physiological needs, particularly regarding feeding and play
- Avoid overexerting the cat physically in an attempt to tire it out for calmness. Additionally, do not eliminate triggers entirely, although reducing them may be a good idea at the beginning of therapy
- Working on attention and concentration
- Teaching calmness by stopping interactions or isolating the cat.

Prevention

First and foremost, *genetic prevention* must be considered by selecting non-impulsive mothers for breeding.

In terms of breeding, learning self-control is partially taught by adult individuals. Therefore, *early separation* from the mother or having mothers who intervene very little to regulate excitement are risk factors for developing an impulsive state. If the mother appears deficient in this learning process, interactions with adult cats who regulate excess excitement without being overly harsh should be encouraged. European cats that produce up to four litters per year are unable to provide a full education for their kittens. Cats raised entirely by humans are highly predisposed to impulsivity. For these animals, the presence of another cat, typically a neutered adult, who will take responsibility for their behavioral development, is crucial.

Teaching all cats, from the moment of adoption, calmness, attention, and tolerance to frustration are protective factors. These exercises are especially beneficial during the first vaccinations when impulsive behaviors are observed.

Meeting the *animal's physiological needs* is important, but excessive stimulation or play can lead to heightened excitement. For example, laser pointer games often lead to frustration in these cats because they can never catch their prey. Their use can generate compulsive states.

Early intervention is protective for two main reasons. First, the kitten does not learn harmful behaviors like shredding objects, scratching the owner's legs to climb, or biting during play. Second, and most importantly, early intervention prevents the deterioration of the relationship with the owner, which almost always happens over time.

7.4.1.3 Compulsive State

▪ Definition and Description

The compulsive state is a pathological mental condition characterized by the production of invariant behavioral responses, with a delayed, shortened, or absent calming phase.

Compulsion can be defined as the inability to stop a behavioral sequence regardless of the action's consequences. It is as if the consummatory phase, which normally leads to the calming phase through the execution of the action, has no effect (◘ Fig. 7.11).

In compulsivity, actions are no longer directed toward an objective based on the knowledge or desire to achieve positive consequences. Compulsive actions become automated behaviors, triggered mechanically regardless of environmental variations, with insensitivity to their positive or negative consequences.

Therefore, the compulsive state is more severe than the impulsive state. All cats in a compulsive state also exhibit impulsive behaviors with almost instantaneous reactions. However, when one or more behaviors no longer have a stopping signal, the individual transitions into the compulsive state.

This state may emerge immediately, in which case it is likely the result of significant genetic vulnerability. However, it can also develop through the excessive repetition of impulsive behaviors (pathogenesis of impulsive and compulsive states). The area responsible for this dysfunction appears to be the striatum, particularly the pallidum. The highly mechanical behaviors observed seem to escape the individual's control, similar to what is seen in Parkinson's disease.

▪ State Variations

The compulsive state, like the impulsive state, can be more or less pervasive in the cat's life and therefore varies in severity. It can affect a specific context, such as play or grooming. In this case, we would refer to it as a *contextual compulsive state.*

The compulsive state can also be *invasive.* Some cats may spend several hours engaged in behaviors such as spinning, fabric chewing, or repetitive play. It is important to note that the more pervasive the compulsive state, the more likely the behaviors will evolve into stereotypical actions without a stopping signal, meaning they *lose their original function*. For example, solitary play can remain a normal play sequence if its duration is not considered. However, in the case of spinning, for instance, there is a loss of function, and the behaviors become increasingly stereotypical.

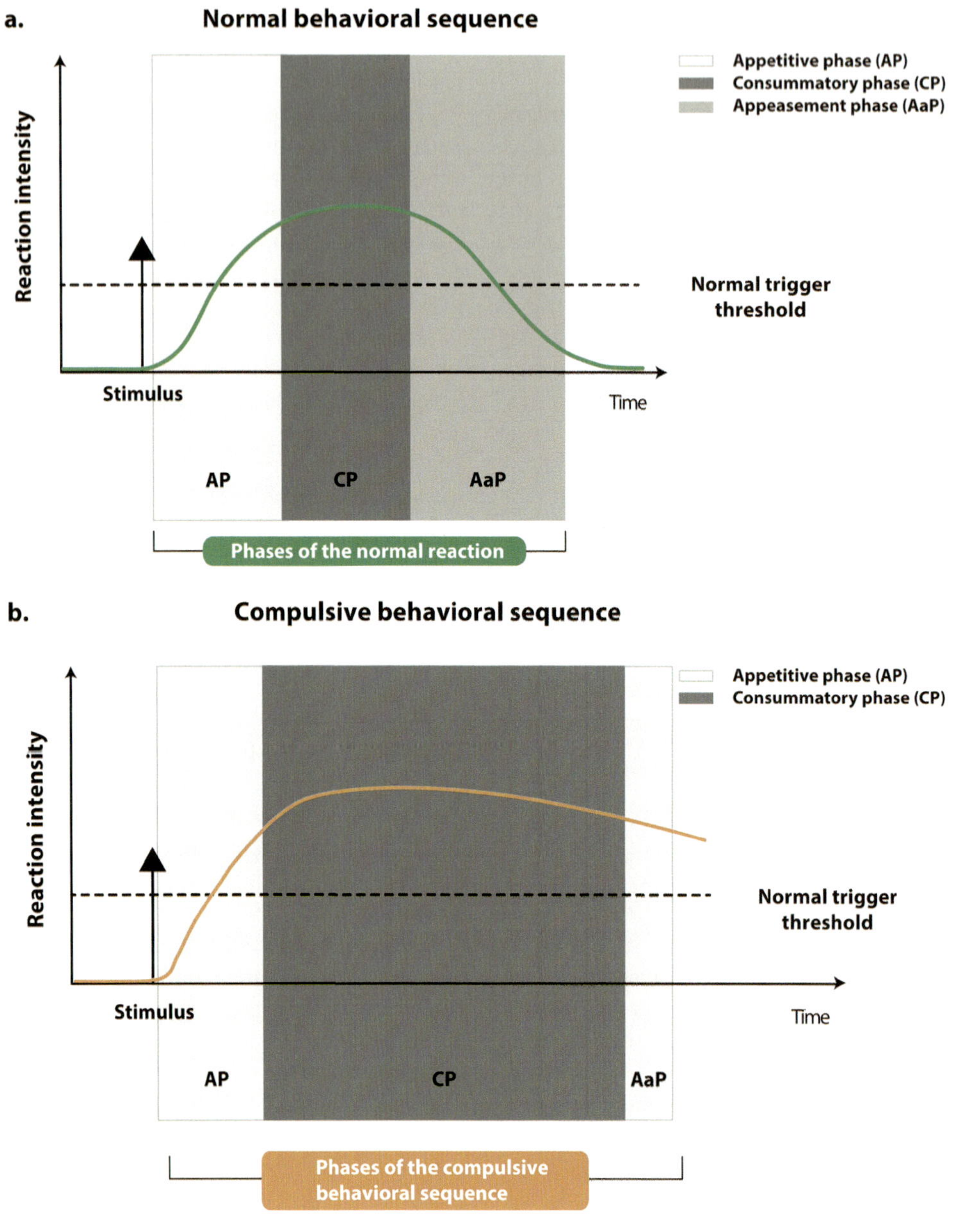

7

Fig. 7.11 Characteristics of reactions in the normal adaptive state (**a**) and the compulsive state (**b**)

Clinical Illustration

Mimie is a six-year-old, spayed Norwegian Forest Cat. This cat is obsessed with food, and all of her activities revolve around eating. She is capable of spending hours in front of the refrigerator door. As soon as she receives food, whether dry or wet, she rushes to it and eats ravenously while growling. When her meals have been offered ad libitum, as has been attempted several times, she only stops eating to vomit and eventually becomes exhausted from her excessive intake.

This behavior began after her spaying and has gradually intensified. Naturally, the cat has had numerous indigestions and even required surgery for an intestinal blockage after swallowing packaging.

With fluoxetine treatment, her appetite has decreased, and her kibble, placed in a food puzzle, allows her to have free access with limited amounts, spreading her eating throughout the day. This results in more normal feeding behavior sequences.

Important Point of View

Do OCDs exist in cats?

In human psychiatry, obsessive-compulsive disorder (OCD) refers to conditions that involve repetitive behaviors without apparent function, such as Tourette syndrome, autism, gambling addiction, bulimia, ADHD, drug addiction, trichotillomania, and other obsessions. The individual performs these ritualized behaviors despite being aware of their irrationality and uselessness. In veterinary medicine, the term "compulsive disorder" is preferred due to doubts about the individual's ability to have obsessive thoughts. However, this likely refers to the same group of disorders and the same neurobiological mechanisms.

Given the cognitive abilities of cats, the capacity to have obsessive thoughts seems quite realistic.

Clinical Illustration

Magritte is a three-year-old neutered Siamese cat. He appears normal and healthy, but his owners are considering giving him up because he is obsessed with licking fabrics. When he was around one year old, his kneading behavior during "cuddles" increased, and gradually, he began to suck on fabrics whenever he was petted. His owners responded by pushing him away, which initially proved effective. However, the behavior resurfaced when Magritte got a woolen cushion in his bed. This wool cushion triggered the return of the kneading and sucking behavior. Unlike with his owners, his reaction wasn't immediate (the discomfort was less apparent).

Over time, Magritte began sucking on all fabrics and more specifically started swallowing woolen materials. He is capable of creating holes in a sweater in just a few minutes. No form of punishment seems effective. This is compulsive behavior, and the fluoxetine treatment prescribed has shown limited effectiveness. It remains necessary to remove all access to wool. Since Magritte has been allowed outside, he damages fewer sweaters, but the habit persists.

Evolution

Helping a cat overcome its compulsive state remains a challenge for the clinician, as these automated behaviors respond poorly to medical treatments, and when they do improve, relapses are common. Associated treatments are often lifelong.

Like the impulsive state, the compulsive state can be linked to emotional pathological states, such as phobic and anxious states. A cat in a compulsive state performs an activity without deriving pleasure from it. The compulsive behavior becomes less and less soothing. Since stopping the behavior is nearly impossible, the compulsive cat is often physically blocked, leading to significant frustration. Aggression can then occur in the form of irritation-based attacks in a frustrated context, especially when efforts are made to prevent the compulsive behavior.

The aggression and inability to control the cat often result in hostility from the family environment, which, in turn, contributes to emotional disturbances.

Finally, as with impulsive states, the compulsive cat is at risk of somatic diseases, particularly osteoarticular and dermatological conditions, directly resulting from the compulsive behavior (e.g., arthritis, licking granuloma).

Clinical Illustration

Alison is a five-year-old neutered male Birman cat. He has been living for four years with Diamant, an approximately ten-year-old European cat who arrived as an adult. They get along well and sleep together. Alison excessively licks his upper lip, resulting in a large lesion that extends to his nostrils. This area is often bloody and painful.

When treated with Clomicalm at 2.5 mg per day, the licking decreases but does not completely stop. This behavior suggests a compulsive state. However, after Diamant's death, the lesion completely disappears. This suggests that the licking was more likely a substitute behavior related to anxiety. The seemingly good relationship between the cats likely wasn't as soothing for Alison as it appeared.

▪ Symptomatology

Presenting complaints

The primary reason for consultation almost always concerns *compulsive behavior*: licking lesions are common. This can also extend to play. For example, solitary play sessions may continue as long as the toy is not taken away from the animal. Feeding behavior, when the animal does not reach satiety, is expressed as an *obsession* with food, leading to constant searching for more. Ingestion of non-food materials is much rarer in cats than in dogs.

In these cases, stereotypic behaviors such as spinning without a stopping signal or instrumentalized licking, whether on the animal itself, objects, or individuals, are often observed.

Clinical Illustration

Solice is a two-year-old Chartreux cat. He constantly licks himself, and his fur has been reduced to a wide strip along his back. The skin examination is normal, and the fur on the areas he licks is merely abrased. The irritation is relatively mild considering the intensity of the grooming.

The behavior stops during meals and cuddle time. The cat seems to sleep normally. The history shows that Solice has never exhibited any fear reactions to items in the house. He does not appear stressed and shows neither hypervigilance nor Rolling Skin Syndrome (RSS). This appears to be a compulsive licking behavior.

▪ Symptoms

The symptoms that lead to the diagnosis of a compulsive pathological state stem from the observation of behaviors that have lost their *stop signal*. Contextual forms are associated with a specific context (play, eating), while more severe forms are more intrusive, without a clear triggering context.

In all affected behaviors, the behavioral sequence is altered with partial or total loss of the stop signal, leading to repetitive behaviors. The consummatory phases of the behavioral sequences lengthen, with no spontaneous refractory period appearing (**◘** Fig. 7.12**)**. Motivation is extremely high, even reaching obsession. In the most intrusive forms, the periods during which the animal does not perform the compulsive behavior are short. Hypersomnia may accompany the most intrusive compulsive behaviors.

Different behaviors can be observed depending on the contexts and the individual or racial predispositions. Hunting behaviors directed at any object or reflections of light may be seen.

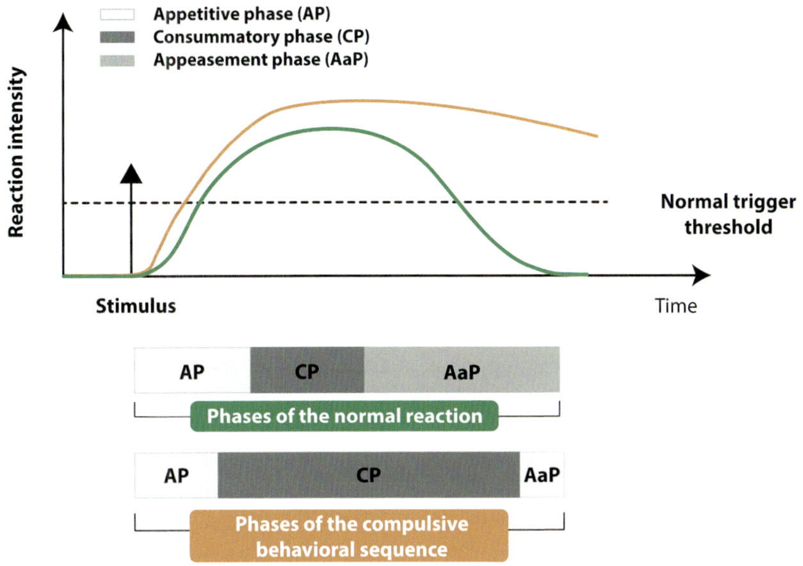

Fig. 7.12 Modification of the behavioral sequence in the compulsive state. When the animal is in a compulsive state, its reactions exhibit a shortened appetitive phase, but it is mainly the consummatory phase that is prolonged, as the calming phase is delayed or even nonexistent

These symptoms often lead to a decline in the quality of the relationship with the owners. The cat is perceived as "crazy," and owners often identify compulsive behaviors by saying that their cat is "touched."

Aggression is not part of the initial clinical picture of compulsions, but it is common as a result of attempts to interrupt the behavior. These are most often redirected aggression due to irritation, stemming from the frustration felt by the cat or from the owners' reactions that make the context anxiety-inducing.

Affected Conditions

In veterinary psychiatry, we do not refer to compulsive disease, as we consider compulsion to be a *pathological state* present in various diseases. Compulsive states can be encountered in the same behavioral disorders as those associated with impulsive states: HSHA syndrome or ADHD, fear-related conditions, as well as essential psychiatric disorders such as dysthymia or dissociative syndrome. Additionally, painful somatic diseases may also be accompanied by a compulsive state.

Diagnosis

Table 7.6 summarizes all the diagnostic criteria for compulsive states.

The diagnosis of a compulsive state is based on the identification of a behavior that is invariably repeated, without a refractory period. In the most severe cases, the compulsive behavior becomes stereotyped and loses its function.

The diagnostic criteria that allow for the conclusion of a compulsive state are as follows:

- Modification of the behavioral sequence of one or more behaviors, with an elongation of the consummatory phase, extending to the absence of the calming phase;
- These sequence modifications persist despite attempts at behavioral modification;
- when the compulsive behavior is performed, there is an indifference to other environmental stimuli;
- In the most severe cases, a transformation of the compulsive activity into a stereotype is observed.

Differential Diagnosis

The main difficulty lies in distinguishing this state from a *reactive state*, linked to poor environmental conditions. It should be noted that the environment must be of very poor quality to produce compulsive states in a healthy individual. In cats, such an environment more often leads to a depressive state with a lack of reactions, rather than a compulsive state.

The second difficulty is eliminating somatic causes:

- Neurological: partial temporal epilepsy, brain tumor, etc.
- Pain: may lead to stereotypy.
- Dermatological: atopy, abscess, wound, etc.

The compulsive state is difficult to distinguish from the *impulsive state* because, in both cases, the animal's behaviors reflect a high motivation for a specific action. A cat in a compulsive state almost always displays impulsive behaviors in addition to its compulsive behaviors. The criterion that helps the practitioner lean toward one diagnosis over the other is the presence of a stop signal in the behavioral sequences.

It is worth noting that the compulsive state may coexist with emotional pathological states (phobic and anxious states) or with manic or dissociative states. These multiple pathological states obviously complicate the clinician's work in reaching the final diagnosis.

Finally, it should be noted that in cats, this diagnosis is much rarer than in dogs, likely because fear is activated more quickly, and the selection process has not aimed to produce individuals capable of repeating a behavior in a highly persistent manner.

Prognosis

The prognosis is moderate, even reserved, because compulsive states respond much less favorably to medication than impulsive states. Relapses are frequent as soon as an attempt is made to taper off the medication. For this reason, it is not uncommon for these treatments to need to be maintained for life.

Treatments

Pharmacological treatment

The compulsive state is no longer under the same dopaminergic dependence, and the shift to the dorsal striatum marks a loss of response to treatments targeting the reward circuits. The most commonly used medications are SSRIs at high doses for their impact on the prefrontal cortex in a top-down manner, although they are not always effective. Fluoxetine at a dose of 2 mg/kg is the best first-line option, but, as often happens with cats, when tablet administration is difficult, the use of Effexor™ may be considered at a dosage of 1.5–2 mg/kg once a day (Metz et al. 2021; Masson et al. 2024). Fluoxetine can also be reconditioned into a liquid form by some specialized pharmacies.

Thus, it is common to need to add other medications to this molecule in an attempt to reduce the intrusion of compulsive behaviors.

In human medicine, numerous molecules are being studied for various types of addiction and would be interesting to test in cats with compulsive states, such as bupropion, which is used in alcohol and tobacco withdrawal.

Behavioral therapy

The measures will depend on the condition involved, meaning the specific nosographic diagnosis. However, it is possible to describe measures that are frequently prescribed in these pathological states.

First, access to the triggering stimulus—i.e., the object of the compulsion (such as a toy)—must be limited, if not eliminated. The trigger should be removed if compulsive behavior is observed.

Compulsive activity is often highly automated, and pleasure postures may not necessarily be visible during the activity. Therefore, once pharmacological treatment is initiated, it is important to work on less stimulating activities that the cat enjoys. This also helps restore pleasure in the relationship.

Finally, it is possible to try to trigger the compulsive behavior by teaching the cat to produce a stopping phase. For this, *relaxation protocols* are crucial to practice outside of the triggering context before using them in a triggering situation.

Prevention

As with impulsive states, there is strong evidence for the genetic component of this pathological state. Certain breeds are indeed predisposed to specific stereotypies, and the first preventive measure is to avoid breeding individuals that exhibit these behaviors. For example, Siamese and Burmese cats are known for ingesting wool (Borns-Weil et al. 2015).

In breeding, as with impulsive states, it is important to promote the presence of adult cats and the very early learning of the stop signal during excitement. Similarly, early educational measures during adoption that encourage calmness and frustration management should be promoted.

The living environment should be sufficiently rich and varied, as opposed to a very impoverished environment that could lead to stereotypical behaviors. However, it is important to remember that the lifestyle of most cat owners is not likely to be responsible for this type of state in an individual without underlying vulnerabilities.

Summary of "Impulsive and Compulsive States"

Impulsive states are more common than compulsive states in cats. Both accompany numerous diseases and may coexist with emotional, mood, or derealization disorders.

In impulsive states, the cat demonstrates an inability to delay or avoid initiating a behavior due to excess motivation or a lack of inhibition. It acts without thinking. The behavioral sequence is altered, with a marked shortening of the appetitive phase. The intensity of the reaction is often heightened.

The compulsive state is more severe, either present initially or evolving from an impulsive state, in which the cat is no longer able to stop its behavioral response, losing the stop phase. This response may manifest as repetitive behaviors, sometimes reaching the point of stereotypy, where the behavior loses its function.

Impulsive states, whether contextual or intrusive, are dopaminergic and serotonergic-dependent. They respond well to fluoxetine at doses ranging from 1 to 2 mg/kg. For compulsive states, the reference biological treatment is high-dose fluoxetine (2 mg/kg), but these states remain a therapeutic challenge.

Behavioral therapy is specific to the underlying condition, meaning it depends on the nosographic level of the diagnosis. When it concerns the axis of self-control, numerous exercises will teach the cat to better control the various phases of behavioral sequences. For example, relaxation protocols, frustration management, and learning calmness will be essential foundational elements.

7.4.2 Derealization Pathological States

7.4.2.1 Dissociative State

Definition and Description

The dissociative state is a pathological mental condition in which the animal exhibits behavioral responses to stimuli that are not objectively present in the environment. For an external observer, there is no environmental element that can explain why the animal is reacting in this way. This state is often intrusive, with the individual quickly becoming compulsive in response to these internal stimuli. Compulsive and dissociative states are frequently associated with one another.

State Variations

The *hallucinations* that prompt the animal to react can have various origins. They may *depend on the context*. Depending on whether they originate in the sensory organs (primary perception) or in the corresponding brain areas, even the associative cortex (interpreted, secondary perception), they can be more or less intrusive.

Some hallucinations can be *triggered*. This is particularly observed with visual hallucinations that resemble insect hunting behaviors. These sequences may follow point-like light stimuli (e.g., bright flashes, laser pointers) that initiate the cat's hunting activity. Initially, there is a trigger, but when the behavior intensifies while the stimulus has disappeared, this constitutes a *dissociative state*.

Similarly, the individual's reactivity can vary. A wide range of behavioral sequences can occur during these *episodes*. Hunting actions can manifest as hunting for invisible prey, stalking, pouncing, etc. It is also common to observe defensive reactions, such as threatening or attacking an imaginary enemy or even part of the body, such as hunting or attacking the tail, which needs to be distinguished from an anxious state associated with Rolling Skin Syndrome.

While most reactions fall within the usual behavioral spectrum, certain advanced forms of the state lead to compulsions that drive the animal into continuous activity. For example, a brief chasing of the tail triggered by tail movement may be observed occasionally in normal cats. However, it should not be confused with a prolonged spinning activity lasting several hours.

■ **Evolution**

Some perceptions may disappear with the loss of the perceptual organ that generates them. For example, visual hallucinations originating from the retina may cease with blindness, and auditory reactions may have a transient existence (Kompus et al. 2013). Other hallucinations may originate elsewhere in the brain, and their persistence can be permanent (Allen et al. 2008).

Habituation seems impossible, and one must then expect a progressive invasion and deregulation of the response. In very intrusive forms, the survival of the individual may be threatened. Some animals only stop their dissociative behavior to respond to other physiological needs. The behavioral response to the hallucination ceases only due to hunger or fatigue.

This clinical aspect prompts owners to question the *well-being of their animal*, which exhausts itself in a purposeless activity. Euthanasia may be requested and can be justified.

■ **Pathogenesis**

We are discussing perceptual origin constructs. The dissociative animal reacts to a perception that may be a *primary perception* resulting from a dysregulation of the sensory organs, or a *secondary perception* due to a dysregulation of the areas involved in processing sensory information. However, hallucinations are not only the result of dysfunction within the sensory or associative cortex. They are associated with a dysregulation of the CSTC (corticostriatal-thalamo-cortical) loop, characterized by a deficiency in the top-down processing and an excess of the bottom-up processing (Allen et al. 2008).

The dissociative state corresponds to the schizoid symptoms described in humans. At the neurotransmitter level, hallucinations are linked to dysfunction of D2 dopamine receptors, along with disruptions in serotonergic pathways and glutamate signaling (Rolland et al. 2014; Stahl and Muntner 2021). The dopaminergic circuits of the striatum would be the starting point for the dysregulations of perception and hallucinations (Cassidy et al. 2018; Schmack et al. 2021).

■ **Symptomatology**

Presenting complaints

Owners primarily seek consultation when the behavior becomes intrusive. More rarely, they may want to know if this behavior, often perceived as "strange," is physiological.

Many videos of dissociative animals are available on the Internet, illustrating the amusement these behaviors can cause in owners, who do not always recognize the underlying impairment of well-being associated with these atypical behaviors. The perception of discomfort usually only arises when the behavior becomes intrusive.

Two secondary reasons may prompt owners to seek help. The first is the perception of potential danger. Indeed, cats in a dissociative state may engage in threatening activities in response to their internal stimuli. These *aggressive actions* can be concerning. The second reason involves the compulsive activities that accompany these states, which can lead to physical injuries, prompting the consultation. Even before these injuries occur, the *perception of discomfort* in the animal is often a reason for owners to seek advice.

Symptoms

For the practitioner, the symptomatic study should address several important points:

Is there a danger? The cat's behavioral reaction is directed toward elements that only it perceives, making it difficult to anticipate its reactions. Although accidents appear to be rare with these conditions, these cats are hard to predict. These factors lead to the establishment of mechanical safety measures to reduce the danger, especially isolation during active phases.

Is there an alteration of well-being? This discomfort may result from the intrusiveness of the behavior, deterioration of the relationship with the owner, or even physical pain. This is the primary reason why medical treatment for these cats is essential.

Tickie

Tickie is a 2-year-old neutered male cat. He lives in a large apartment, and his development has been very harmonious. His appetite is regular, and he only eats dry food, which is available for free feeding. His owners are attentive. They believe Tickie's sleep is normal, although his room is closed at night. In the mornings, Tickie often comes to find them and meows at the bedroom door.

For the past 2 weeks, and this is the reason for the consultation, Tickie has been chasing "flies" on the walls. It's very amusing, but since this sometimes lasts for a long time (up to an hour), the owners are concerned. They also feel guilty because they have played a lot with a laser pointer and wonder if there could be a connection.

The behavior can be triggered by the smallest thing. Even a ray of sunlight can unsettle Tickie, who starts focusing on the wall and soon begins to attack it. He remains very gentle even during these episodes. However, even when held for a moment in an attempt to calm him down, he returns to the original spot and resumes his hunting. This is an example of imaginary hunting behavior. MRI and other complementary tests have all come back negative. It is likely that Tickie is experiencing a dissociative state.

It seems, as is often observed, that a change in context leads to a temporary calming of the disorder, but this is not sustainable, and the dissociative behavior resumes.

▪ Affected Conditions by Dissociative State

The dissociative state can occur in various nosographic conditions: *dissociative syndrome*, of course, for which the dissociative state is a mandatory criterion, partial temporal epilepsy, and the side effects of certain glutamate receptor antagonists such as ketamine. It is quite easy to notice that these reactions are triggered in various situations, independent of the environment.

The full diagnostic criteria for the dissociative state are presented in ◘ Table 7.7. The symptoms that allow for the conclusion of a dissociative state are as follows:

- Evidence of repetitive activity, unrelated to external stimuli.
- Progressive loss of usual activities in favor of this repetitive behavior.
- No organic cause can be identified through routine complementary tests (blood tests, MRI).

▪ Differential Diagnosis

The challenge is to clearly limit this diagnosis to reactions to imaginary perceptual stimuli. Phobias and anxiety must be excluded, as they correspond to reactions to non-existent dangers with a faithful perception of the environment.

On the other end, the dissociative state must be distinguished from the *compulsive state*. As we have seen, compulsions can arise from these imaginary perceptions, and these states can coexist. Despite the repetitive nature present in both states, it is the trigger of the behavioral sequence that allows for differentiation.

The diagnosis of *partial temporal epilepsy* is sometimes proposed for these dissociative states. The difference between this condition and an essential psychiatric disorder, such as dissociative syndrome, is not always clear in clinical practice, and sometimes only a therapeutic trial can help guide the preference toward one or the other of the conditions.

At the other end of the spectrum, it is important to distinguish the dissociative state from the compulsive state. However, we have seen that compulsions can emerge following these imaginary perceptions, and these states may coexist. Despite the repetitive nature present in both conditions, the key distinction lies in the trigger of the behavioral sequence.

A diagnosis of *partial temporal epilepsy* is sometimes proposed for dissociative states. Differentiating this condition from a primary psychiatric disorder, such as dissociative syndrome, is not always possible in clinical practice. In some cases, only a therapeutic trial can help guide the diagnosis toward one condition or the other.

Clinical Illustration

Terton is a cuddly cat. As soon as he has the chance to jump on someone, he rushes over and starts licking them. He generally licks the skin, but if he cannot, he settles for licking the clothes. This behavior causes tension because it quickly becomes unpleasant. When pushed away, he tries to continue on the couch before eventually leaving.

At first, his owners were able to control him and could pet him, but this has now become impossible. This is a compulsive state.

▪ Treatments

Pharmacological treatment

It is extremely limited. Currently, we can only work on reducing reactivity: high doses of fluoxetine provide temporary improvement. It is likely that antipsychotics could be beneficial in some cases, but these are second-generation antipsychotics, and clinical studies demonstrating their use in cats are lacking. One of the reasons for this is that this condition is still underdiagnosed due to limited knowledge. As a result, the cases observed are scattered, and with the prognosis being poor, few therapeutic trials have been conducted.

Antiepileptics are, to some extent, capable of reducing reactivity. They are sometimes added to fluoxetine to control the dissociative behavior.

Behavioral therapy

The therapy mainly consists of accident prevention measures. There are few measures capable of preventing compulsive behavior. Of course, if triggers are identified, it is wise to remove them from the environment. Regular evaluation of the animal's well-being should be carried out with the owners.

▪ Prevention

This state, when it is not iatrogenic, is difficult to prevent except through the exclusion of breeding affected individuals. A genetic etiology is highly probable.

Summary of ▶ Sect. 7.4.2.1

The dissociative state is a pathological derealization condition in which the animal exhibits behavioral responses to stimuli that are not objectively present in the environment. The symptomatology is dominated by hallucinations and repetitive behaviors that the cat cannot control. These behaviors may involve self-aggression or aggression toward others. When they occur, they are often spectacular and typically idiopathic in nature. The treatment is primarily medical. High doses of fluoxetine (up to 2 mg/kg) are the main treatment. It may be combined with certain antiepileptics (phenobarbital).

Behavioral therapy mainly includes safety measures designed to minimize danger during episodes. The prognosis remains very reserved.

7.4.2.2 Senile or Confusional State

▪ Definition and Description

The *senile state* is a pathological condition characterized by various behavioral abnormalities resulting from the involution process and its effects on the functioning of normal neural circuits. This abnormal aging is marked by a self-perpetuating worsening, which ultimately leads to a complete disorganization of brain functions.

State Variations

All brain functions are affected, and cognitive disturbances (learning, memory), emotional disorders, mood disorders, and endocrine disturbances are observed. Combinations of these various types of disruptions are most often seen and functionally add to the somatic disturbances that occur simultaneously.

The involution process does not manifest with the same intensity in all brain areas. As a result, there are dysfunctions of varying severity depending on the affected regions, and at the onset of this state, the pathological expression may take different forms depending on the most deficient area. Over time, as the degradation becomes self-perpetuating, the brain damage becomes global.

Evolution

There are two quite different types of evolution.

The *normal evolution*, linked to the aging process, which intensifies and leads to an involution process. Aging is an inevitable process that is part of species adaptation, allowing for the death of individuals and the renewal of populations. This phenomenon, related to *membrane stiffening*, the action of *telomerases*, and the destructive power of *free radicals*, leads to a gradual slowing of brain function (see ► Chap. 4, ► Sect. 4.2.8).

Clinically, this degradation manifests as a decrease in motivation and cognitive losses. It is often difficult to distinguish between capacity loss and brain loss. An older cat that no longer hunts mice might be deaf or arthritic. It might also be demotivated.

Similarly, it is often impossible to determine whether the loss of motivation is primary or secondary to the loss of physical abilities. The same cat that no longer reacts when a mouse passes might simply be unable to react quickly enough and knows it cannot capture prey due to its arthritis. Therefore, it no longer pays attention to what it cannot manage.

In the *pathological evolution* of neuronal death, vascular lesions and tumor processes occur. The degradation is more abrupt, and its characteristics depend on the affected area. These lesions most often combine with *global dysfunction*. As in other forms, local lesions lead to damage across the entire brain. The degradation becomes generalized.

Pathogenesis

Aging is a global degradation of all regulatory and protective systems.

A decrease in synaptic transmission is frequently described, accompanied by an increase in the release of free radicals. These free radicals are less effectively countered, leading to lesions that, in turn, produce more free radicals. These lesions result in a decline in synaptic transmission and synaptic reactivity. The phospholipid membranes stiffen, and all neurotransmitter receptors become less mobile. As a result, there is a global deficit in neuronal networks. Over time, this slowing leads to a decrease in motivation due to reduced reinforcement, memory loss, and cognitive deficits. There is also a reduction in synaptogenesis, particularly in the *hippocampus*.

Neuronal death is not present in all forms of the condition, but it can occur and worsen the overall picture. Apart from vascular causes, the exact cause of these cell losses is unknown.

Cognitive losses are notable because they represent a symptom *not observed in other states*. However, brain dysfunction may begin with emotional or mood disturbances. There is no reason why cognitive areas should be more vulnerable than others, and the onset of the senile state can therefore be confused with an anxious or depressive state.

Furthermore, the decline in abilities acts as negative punishment. Poorly executed behaviors no longer lead to a reinforcing result. This cognition-reinforcement loop helps explain that emotional disturbances will occur, even if the limbic system is not initially affected by the involution process.

Among the negative somatic or psychological influencing factors, the importance of *painful phenomena* should be emphasized. The overall impact of pain on the individual's functioning is well-known. It becomes increasingly significant with the onset of involution and can lead to significant degradation. Unfortunately, it is with age that *chronic pain* is most likely to develop.

Finally, on a *relational level,* the decrease in activity is accompanied by a dilution of emotional bonds. Additionally, it is common for episodes of inappropriate elimination to accompany this state, leading to severe reactions from the owners. Punishments worsen the disturbances previously described.

■ Symptomatology

Presenting complaints

Behaviors associated with aging are often well-known and tolerated by everyone. This tolerance leads to consultations being made *too late*.

Cognitive disorders that trigger a consultation are often related to *learning losses* and *inappropriate elimination*. Sometimes, mistakes in identifying individuals lead to fear reactions from the cat, which perceives its owners as strangers. These misunderstandings can become a reason for consultation.

It is primarily emotional disturbances that lead to consultations. Unexplained fears arise. The anxious cat meows, searching for its owners, whose presence provides reassurance. When this happens at night, the owners often have difficulty coping with it.

Symptoms

The onset of cognitive disorders associated with and triggering emotional disturbances is a major characteristic of this state. These cognitive disorders can be of various types, and among them, the most characteristic is the *loss of orientation*. These disorders can affect *temporal* or *spatial orientation*.

In the case of spatial disorientation, the animal appears to be disoriented in its usual environment. This disorientation can sometimes be difficult to identify.

Clinical Illustration

Petra is a 15-year-old cat who has always lived with her owners in the same place and has no medical history. Her owner visits the veterinarian because recently she has been urinating and defecating all over the house, and her meowing has become incessant. After further questioning, it is revealed that for several months, Petra has been meowing more frequently at night until her owner puts her back to bed. She barely goes outside anymore, has become very difficult with her food, and can beg for food even when her bowl is full. She hisses after some petting. Blood tests (including thyroid hormone levels) and urine tests show no abnormalities explaining the inappropriate elimination or behavioral disturbances. Signs of arthritis are present in the radiographs, but they are not sufficient to explain all the symptoms. Petra presents with pathological aging, along with cognitive and emotional disturbances.

Mild losses can be masked by compensations. The cat often simply calls its owner, who then seeks to satisfy its needs. The lost cat quickly finds the response it was looking for in this help, and no one notices the disorientation. However, there are errors that are difficult to hide: failure to recognize an individual, errors in direction (such as the cat getting lost in the garden or attempting to enter a house that is not its own…).

Temporal errors are often difficult to identify, except in cases of nocturnal waking.

Emotional and mood disturbances can also be the starting point for the senile state. In this case, they are typically either anxiety disorders or depressive disorders. Although not characteristic, these emotional disturbances, when they appear in an elderly animal without a clear explanation, are significant indicators of this state. Only the evolution of the condition, with the appearance of associated cognitive disorders, allows for an accurate diagnosis. While not an obligatory criterion, these emotional disturbances are therefore common in the senile state.

Finally, there is a somatic aspect to the symptoms. Self-amplification is a characteristic of this state. For example, a locomotor disorder may lead to a cascade of deterioration, such as a loss of motivation or cognitive decline in an elderly animal, while the same lesion would have little impact on a younger animal.

The combination of these degradations interacting negatively forces the diagnosis. The starting point of the condition is variable (emotional, humoral, cognitive, or somatic), but it will gradually lead to global damage, in which emotions and cognition mutually reinforce each other, worsening the clinical picture. This vicious circle is quickly compounded by relational degradation. It is this latter aspect that prematurely condemns the cat by demotivating the owners from acting (◘ Fig. 7.13).

Diagnosis

◘ Table 7.7 summarizes the full diagnostic criteria for the senile state. The symptoms that allow for the diagnosis of the senile state are as follows:

- The cat is over 8 years old;
- Exclusion and management of somatic diseases that do not allow for the resolution of symptoms;
- Loss of abilities, which occur in one or more of the following areas (cognitive, emotional, mood, somatic, relational) with functional self-aggravation.

Differential Diagnosis

Of course, there may be somatic conditions that cause similar dysfunctions. These must be excluded and treated as a priority, and it is when they do not resolve the symptoms that a

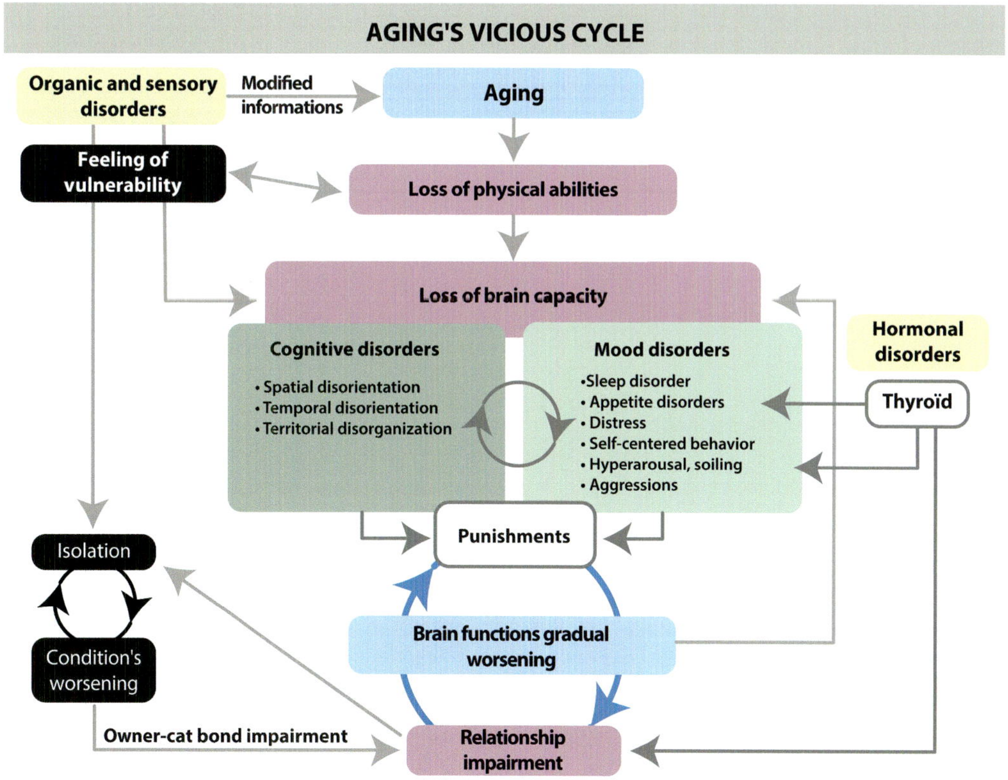

◘ Fig. 7.13 Cascades of physiological and pathophysiological events related to aging

Table 7.7 Diagnostic criteria for dissociative and senile states

Pathological state	Diagnostic criteria	State variations	Differential diagnosis
Dissociative	The *dissociative state* is a pathological mental condition in which the animal exhibits behavioral responses to stimuli that are not objectively present in the environment Common symptoms: • First symptoms appear in early adulthood • Evidence of repetitive activity (spinning, hallucinations, self-aggressions), unrelated to external stimuli, with no identifiable trigger • No organic cause can be identified through routine complementary tests (blood tests, MRI)	*Contextual to invasive form* The level of intrusiveness of the dissociative state depends on the time spent engaging in repetitive behavior. In parallel, a gradual loss of usual activities is observed The cat becomes less responsive to external interventions	**Maladies organiques** *Pain* managing pain allows for the immediate cessation of episodes Complementary blood tests may be abnormally indicative of conditions like insulinoma or porto-systemic shunt Imaging (MRI) helps exclude malformations, such as tumors Anxious state Repetitive behaviors occur as part of hypervigilance or behavioral inhibition. There is no alteration in the perception of reality. External interventions usually stop the repetitive behaviors, at least temporarily Manic state Obsessive behaviors are aggressive and occur in spontaneous or cat-provoked interactions. There is no alteration in the perception of reality Compulsive state One or more identifiable trigger stimuli are present. There is no alteration in the perception of reality Senile state Symptoms appear in senior cats. Stereotypies, while possible, are not central

(continued)

7

Table 7.7 (continued)

Pathological state	Diagnostic criteria	State variations	Differential diagnosis
Senile	The *senile state* is a pathological condition characterized by various behavioral abnormalities resulting from involution and its effects on normal neural circuit functioning. This abnormal aging is marked by a self-perpetuating worsening, ultimately leading to a complete disorganization of brain functions Common symptoms: • The cat is over 8 years old • Management of somatic diseases does not resolve the symptoms	Loss of abilities occurs in one or more areas, with functional self-aggravation The observed symptoms are cognitive, with spatial or temporal disorientation and loss of learning. They often extend to emotional and relational disorders, even mood disturbances	Trouble organique et sensoriel Blindness, deafness: To be excluded Chronic pain Intercurrent diseases In these organic disorders, there is no loss of spatial and temporal orientation during normal aging. Management of these conditions leads to improvement Dissociative state The first symptoms generally appear around adolescence or early adulthood Anxious state No spatial or temporal disorientation No loss of learning Depressive state No spatial or temporal disorientation No loss of learning

psychological or concomitant condition should be suspected.

Particular attention should be paid to endocrinology. Firstly, hormonal disorders may originate in the hypothalamus or pituitary gland, and thus correspond to a brain dysfunction. Furthermore, even though the relationships are still poorly understood, hormones play a role in mood and brain functions in general.

Older cats are often affected by hyperthyroidism, which is associated with severe behavioral changes. For older cats, this hormone should be routinely measured. Similarly, the search for hypertension is essential. These cardiovascular disorders can lead to blindness but also to brain lesions. Discomfort frequently accompanies them.

Additionally, any disease that occurs in an elderly animal can trigger a series of organic and behavioral reactions that may mark the onset of the senile state. These conditions act as triggers.

Clinical Illustration

Loustic is a very old cat, probably around 20 years old. Gradually, his territory has become smaller, and he now lives on a windowsill. His owners have arranged a quiet life for him. There is a small staircase that allows him to climb into his bed, as he can no longer jump. Recently, the owners have found him in various places around the house, completely lost. He often falls asleep at the foot of walls in improbable positions. Sometimes he meows to signal that he is lost, but he needs to be escorted back to his spot.

Since this happened too often, the owners decided to confine him to his windowsill. A small fence has been installed, and Loustic can no longer go outside. Twice a day, his owners give him a small outing in the room that lasts about 10 min, after which he requests to return to his bed.

The old cat is often unable to manage a large territory. As he ages, he loses his landmarks and gets lost in a territory he once knew very well. By imposing a restriction on his space, despite the constraint it represents, the owners are making life easier for the cat and ensuring his aging process is safe and without accidents.

■ **Prognosis**

The prognosis is inevitably poor. It is impossible to prevent aging. However, the implementation of treatments for each identified dysfunction and relational management can transform this vital stage. The complete elimination of discomfort is illusory, but by minimizing *suffering* and *pain*, it is possible to significantly slow down the process. Special attention must be given to painful phenomena.

■ **Treatments**

Pharmacological treatment

The treatment will be established at three levels: cellular, neuronal, and relational.

At the cellular level, the treatment will primarily be pharmacological. The most useful action will be to combat free radicals. Selegiline is likely the most suitable molecule currently. Traditionally, supplementation with selenium is recommended by the food industry.

Any attention given to the elderly animal improves its *quality of life* and performance. However, it seems that NSAIDs may prevent a number of pathological brain processes. Several publications have shown, in human medicine, their impact on the progression of senile diseases (Calvo-Rodríguez et al. 2015). In this context, supplementation with *unsaturated fatty acids*, such as omega-3, has been described as effective (Pifferi et al. 2020).

In many species, vascular disorders are responsible for brain lesions. At the neuronal level, it is important to maintain the functioning of circuits. It seems that activating the reward circuit is crucial to avoid the demotivation loop. Performing simple tasks leading to large rewards is a way to restart motivation. Overall, it is also likely that synaptic transmission is weakened by the involution process. Depending on the symptoms, the use of antidepressants may promote the resumption of transmission activity.

With aging, degradation is general. The functional treatment must address all the *functions* of the animal and focus on *pain management*. Maintaining physical activity appropriate to the animal's capabilities and *preserving the relationship* are essential.

Treating emotional disturbances is important and will involve the use of standard medications for these conditions (anxiolytics, anti-depressants). Special attention must be given to drug combinations, as elderly cats often take multiple medications. Gabapentinoids have both analgesic and anxiolytic effects, making them particularly useful for elderly animals. They can be used occasionally in cases of day/night reversal to help the animal calm down.

Finally, the *relationship with the owner* is an important factor to consider. Isolation and rejection are accelerators of the involution process, and maintaining soothing rituals helps preserve a high-quality relationship.

Behavioral therapy

It has already been mentioned above, but several principles can be emphasized. The activation of activity in general is crucial. In the cat, managing the territory while taking into account the physical abilities of the aging animal will help maintain intellectual stimulation.

Of course, punishments, which can sometimes be tempting with a senile cat that seems to be doing it on purpose, should be eliminated. However, simple requests that lead to rewards are a good therapeutic approach.

As cats age, they naturally restrict their territory. The owner must accept this and provide the necessary comfort within a confined area (litter box, food, resting places, etc.).

Summary of "Senile State"

The senile state is a pathological derealization condition that occurs in elderly animals. It is characterized by various behavioral abnormalities resulting from involution and its effects on the functioning of normal neural circuits. The main symptoms of this state include a loss of spatial and temporal orientation, as well as the disappearance of previously acquired learning. Hormonal anomalies and the presence of pain can contribute to the onset and maintenance of this state. The senile state is, by definition, irreversible, but the animal's well-being can be significantly improved. Selegiline at 1 mg/kg is the most suitable molecule for treating this state. Management of comorbid organic conditions and pain is essential. Therapy focuses on reactivating the cat's activities, strengthening the reward and motivation circuits as much as possible.

References

Allen P, Larøi F, McGuire PK, Aleman A (2008) The hallucinating brain: a review of structural and functional neuroimaging studies of hallucinations. Neurosci Biobehav Rev 32(1):175–191. https://doi.org/10.1016/j.neubiorev.2007.07.012

Beata C, Muller G, Arpaillange C, Graff E, Lachapèle D, Marion M, Marlois N, Massal N, Mège C (2015) Pathologie comportementale du chat (AFVAC). AFVAC

Bennett PC, Rutter NJ, Woodhead JK, Howell TJ (2017) Assessment of domestic cat personality, as perceived by 416 owners, suggests six dimensions. Behav Process 141:273–283. https://doi.org/10.1016/j.beproc.2017.02.020

Borns-Weil S, Emmanuel C, Longo J, Kini N, Barton B, Smith A, Dodman NH (2015) A case-control study of compulsive wool-sucking in Siamese and Birman cats (n = 204). J Vet Behav Clin Appl Res. https://doi.org/10.1016/j.jveb.2015.07.038

Calvo-Rodríguez M, Núñez L, Villalobos C (2015) Non-steroidal anti-inflammatory drugs (NSAIDs) and neuroprotection in the elderly: a view from the mitochondria. Neural Regen Res 10(9):1371. https://doi.org/10.4103/1673-5374.165219

Caspi A, Sugden K, Moffitt TE, Taylor A, Craig IW, Harrington H, McClay J, Mill J, Martin J, Braithwaite A, Poulton R (2003) Influence of life stress on depression: moderation by a polymorphism in the 5-HTT gene. Science (New York, NY) 301(5631):386–389. https://doi.org/10.1126/science.1083968

Cassidy CM, Balsam PD, Weinstein JJ, Rosengard RJ, Slifstein M, Daw ND, Abi-Dargham A, Horga G (2018) A perceptual inference mechanism for hallucinations linked to striatal dopamine. Curr Biol 28(4):503–514.e4. https://doi.org/10.1016/j.cub.2017.12.059

DePorter TL, Bledsoe DL, Beck A, Ollivier E (2019) Evaluation of the efficacy of an appeasing pheromone diffuser product vs placebo for management of feline aggression in multi-cat households: a pilot study. J Feline Med Surg 21(4):293–305. https://doi.org/10.1177/1098612X18774437

Duffy DL, de Moura RTD, Serpell JA (2017) Development and evaluation of the Fe-BARQ: a new survey instrument for measuring behavior in domestic cats (Felis s. Catus). Behav Process 141:329–341. https://doi.org/10.1016/j.beproc.2017.02.010

Elhwuegi AS (2004) Central monoamines and their role in major depression. Prog Neuro-Psychopharmacol Biol Psychiatry 28(3):435–451. https://doi.org/10.1016/j.pnpbp.2003.11.018

Hargrave C (2014) Pheromonatherapy and animal behaviour: providing a place of greater safety. Companion Anim 19(2):60–64. https://doi.org/10.12968/coan.2014.19.2.60

Haroon E, Miller AH, Sanacora G (2017) Inflammation, glutamate, and glia: a trio of trouble in mood disorders. Neuropsychopharmacology 42(1):193–215. https://doi.org/10.1038/npp.2016.199

Jesulola E, Micalos P, Baguley IJ (2018) Understanding the pathophysiology of depression: from monoamines to the neurogenesis hypothesis model – are we there yet? Behav Brain Res 341:79–90. https://doi.org/10.1016/j.bbr.2017.12.025

Kompus K, Falkenberg L, Bless J, Johnsen E, Kroken RA, Kråkvik B, Larøi F, Løberg E-M, Vedul-

Kjelsås E, Westerhausen R, Hugdahl K (2013) The role of the primary auditory cortex in the neural mechanism of auditory verbal hallucinations. Front Hum Neurosci 7. https://doi.org/10.3389/fnhum.2013.00144

Leuchter AF, McCracken JT, Hunter AM, Cook IA, Alpert JE (2009) Monoamine oxidase a and catechol-O-Methyltransferase functional polymorphisms and the placebo response in major depressive disorder. J Clin Psychopharmacol 29(4):372–377. https://doi.org/10.1097/JCP.0b013e3181ac4aaf

Martinowich K, Manji H, Lu B (2007) New insights into BDNF function in depression and anxiety. Nat Neurosci 10(9):1089–1093. https://doi.org/10.1038/nn1971

Masson S, Metz D, Bleuer-Elsner S, Schwobthaler F (2024) Retrospective study on the use of venlafaxine in 176 cats diagnosed with behavioral disorders. J Vet Behav Clin Appl Res. https://doi.org/10.2139/ssrn.4790331

McPeake K, Sparkes A, Billy C, Endersby S, Collin JF, De Jaeger X (2023) Development of a cat behaviour issues assessment scale (CABIAS) assessing problem behaviours in cats. Animals 13(18):2992. https://doi.org/10.3390/ani13182992

Mège C, Béata C, Beaumont-Graff E, Diaz C, Habran T, Marlois N, Muller G (2003) Pathologie comportementale du chien, Elsevier Masson edn. Masson-AFVAC

Metz D, Medam T, Masson S (2021) Double-blind, placebo-controlled trial of venlafaxine to treat behavioural disorders in cats: a pilot study. J Feline Med Surg. https://doi.org/10.1177/1098612X211036792

Pageat P (1995) Pathologie comportementale du chien. Editions du point vétérinaire Maison-Alfort

Pageat P (1998) Pathologie du comportement du chien. Editions du Point Veterinaire

Pifferi F, Cunnane SC, Guesnet P (2020) Evidence of the role of Omega-3 polyunsaturated fatty acids in brain glucose metabolism. Nutrients 12(5):1382. https://doi.org/10.3390/nu12051382

Rolland B, Jardri R, Amad A, Thomas P, Cottencin O, Bordet R (2014) Pharmacology of hallucinations: several mechanisms for one single symptom? Biomed Res Int 2014:307106. https://doi.org/10.1155/2014/307106

Schmack K, Bosc M, Ott T, Sturgill JF, Kepecs A (2021) Striatal dopamine mediates hallucination-like perception in mice. Science 372(6537):eabf4740. https://doi.org/10.1126/science.abf4740

Schroeter ML, Abdul-Khaliq H, Sacher J, Steiner J, Blasig IE, Mueller K (2010) Mood disorders are glial disorders: evidence from *in vivo* studies. Cardiovasc Psychiatry Neurol 2010:1–7. https://doi.org/10.1155/2010/780645

Schultz W (2011) Potential vulnerabilities of neuronal reward, risk, and decision mechanisms to addictive drugs. Neuron 69(4):603–617. https://doi.org/10.1016/j.neuron.2011.02.014

Stahl SM, Muntner N (2021) In: Grady MM (ed) Stahl's essential psychopharmacology: neuroscientific basis and practical applications, 5th edn. Cambridge University Press. https://doi.org/10.1017/9781108975292

Wong M-L, Licinio J (2004) From monoamines to genomic targets: a paradigm shift for drug discovery in depression. Nat Rev Drug Discov 3(2):136–151. https://doi.org/10.1038/nrd1303

Zhang L, Bian Z, Liu Q, Deng B (2022) Dealing with stress in cats: what is new about the olfactory strategy? Front Vet Sci 9:928943. https://doi.org/10.3389/fvets.2022.928943

Disorders Related to Dysregulation of the Fear Axis

Contents

8.1 Categorical Phobias – 290
8.1.1 Description – 290
8.1.2 Diagnosis – 298
8.1.3 Prognosis – 303
8.1.4 Treatments – 303
8.1.5 Therapeutic Follow-Up – 307
8.1.6 Prevention – 308
8.1.7 Clinical Vignettes – 308

8.2 Sensory Deprivation Syndrome – 311
8.2.1 Description – 311
8.2.2 Diagnosis – 314
8.2.3 Prognosis – 316
8.2.4 Treatments – 317
8.2.5 Therapeutic Follow-Up – 317
8.2.6 Prevention – 317
8.2.7 Clinical Vignettes – 318

8.3 Post-traumatic Phobias or PTSD – 320
8.3.1 Description – 320
8.3.2 Diagnosis – 321
8.3.3 Prognosis – 323
8.3.4 Treatments – 323
8.3.5 Therapeutic Follow-Up – 323
8.3.6 Prevention – 323
8.3.7 Clinical Vignettes – 323

References – 326

© The Author(s), under exclusive license to Springer Nature Switzerland AG 2026
S. Bleuer-Elsner et al., *Veterinary Psychiatry of the Cat*,
https://doi.org/10.1007/978-3-032-06260-4_8

In this chapter, we describe psychiatric disorders related to the dysregulation of the fear axis. We present sensory deprivation syndrome, categorical phobias, and post-traumatic phobias. The so-called categorical phobias include social phobias, healthcare phobia, noise phobia, and transport phobia, among others.

Due to its position in the food chain, being both a predator and prey, the cat is an animal quickly overwhelmed by fear. Its reactions are often highly amplified, and the clinician must be vigilant to distinguish between different disorders.

For each disorder, the clinical presentation, diagnosis, treatment, prognosis, and follow-up are detailed, and clinical vignettes illustrate the challenges and particularities of fear expression in cats.

Fear is a normal emotion, and the behaviors associated with it are often life-saving. Fear is considered normal whenever it leads to adaptive behaviors that allow the cat to find safety in a protective context. However, when fear becomes overwhelming, it no longer fades spontaneously and causes discomfort. In such cases, it is possible to speak of pathological fear. These are the situations in which we refer to the dysregulation of the fear axis.

In this chapter, we address various disorders labeled as "phobias" or "phobic disorders." These conditions can be associated with diverse pathological states, including phobic, anxious, or depressive states.

Important Point of View

Phobia: The Pathological State and Nosographic Disorder

The term "phobia," used generically to describe fear-related disorders, is highly misleading as it implies that the pathological state associated with the disorder is always a phobic state. However, the disorders discussed in this chapter are often linked to anxious or depressive pathological states. Despite this, we have retained the term "phobia" because this terminology is systematically used to describe the disorders we cover here.

In our approach, we distinguish between a phobic pathological state and various "phobic disorders," which may be accompanied by a phobic, anxious, or depressive pathological state.

8.1 Categorical Phobias

8.1.1 Description

8.1.1.1 Definition

Categorical phobia is a disorder in which the animal exhibits fear reactions that are inappropriate and exaggerated in their duration or intensity in response to one or more stimuli. This disorder results from the interaction between *genetic vulnerability* and a *triggering event*, which, at first glance, is not particularly traumatic. It can occur at any age. When associated with anxious or depressive states, it is not always possible to identify the triggering stimuli. To do so, it is necessary to reconstruct the *pathological pathway*. Indeed, the phobogenic stimulus, often identifiable during initial exposures, may no longer be recognizable over time. Through anticipation and generalization, the animal's fear reactions extend to all stimuli that have been associated with the initial phobogenic stimulus.

Unlike sensory deprivation syndrome, categorical phobia is observed in animals raised under *acceptable breeding conditions* and sometimes even in *very high-quality environments*.

The term "categorical" indicates that there are different categories of phobias based on the *triggering stimulus* (Table 8.1).

In practice, the diagnostic statement could be, for example, "intraspecific social phobia associated with an anxious state." Multiple phobias can coexist in the same individual, and the list helps specify the triggering elements.

Table 8.1 The different types of phobias included under the term "categorical phobia" and their corresponding triggering stimuli

Name of the phobic disorder	Target of the pathological fear reaction
Intraspecific social phobia	Cats
Interspecific social phobia	Another species, most often humans
Noise phobia	Firecrackers, fireworks, alarms
Storm phobia	Storms, but sometimes wind
Specific phobia	Vacuum cleaner or another specific isolated object
Healthcare phobia	Veterinary care, grooming
Travel phobia	Car rides or other means of transportation

8.1.1.2 Etiology and Pathogenesis

Categorical phobia manifests in *genetically vulnerable* cats (Montague et al. 2014; Tateo et al. 2021). Depending on the degree of vulnerability and the environment they are exposed to, symptoms may appear very gradually or more abruptly.

The age of onset varies, but puberty is a common time for consultations. As previously mentioned, this is a period of behavioral instability, during which some cats may experience a worsening of their condition. They may respond aggressively to stimuli that they previously avoided.

In some cases, the history reveals suboptimal yet standard environmental conditions. Certain events are identified as potential triggers but would not be considered traumatic for most individuals. Examples include a routine veterinary examination conducted with care, an elective surgery, a few hours of car travel, construction work near the owners' home, or an encounter with a human or another cat that provoked an initial aggressive reaction.

Cats are not social animals. Encounters with conspecifics are typically triggers for aggression, and such aggression alone cannot be considered a pathognomonic symptom of social phobia. For a situation to be classified as a phobia, the expression of fear must be out of control, which excludes simple territorial aggression or even irritation-induced aggression toward a conspecific.

Exposure to events that seem *minimally traumatic for most cats* triggers an exaggerated fear response in others. For instance, some cats, associating the carrier with other events (car rides, veterinary visits), refuse to enter it to the point of violently attacking their owner or displaying extreme reactions such as urination or defecation due to emotional stress. However, during the first visit, the history often reveals that the veterinary team took all necessary precautions regarding gentle handling and pain management. This indicates a *heightened vulnerability* and *sensitivity* in certain individuals, often reflected in owners' descriptions of their animal's behavior prior to surgery. These cats are frequently described as fearful, reserved, or timid, with a tendency to flee or inhibit themselves during their juvenile period.

Categorical phobias illustrate the animal's vulnerability to phobic disorders. These vulnerable animals respond poorly to unpleasant experiences, even when these experiences are of moderate intensity. In other words, they are fragile and less adaptable to environmental changes. Moderate adversity becomes a highly negative experience, triggering the disorder. Yet, unpleasant experiences are inevitable, regardless of one's living environment. Once the disorder is established, the triggering event is perceived as the cause of emotional suffering.

All species experience innate fears that are protective for their survival. For example, the fear of heights or the fear of silhouettes with eight legs is innate in humans (Hoehl et al. 2017). These so-called *atavistic fears* are represented in the brain by synaptic connections in the lateral external nucleus of the amygdala (LA).

When an individual encounters a new element in the presence of these existing fears, their LA associates the pathways from the novel stimulus with those of the initially known phobogenic stimulus. Through classical conditioning, these synaptic connections trigger a strong and rapid fear response to the stimulus, bypassing the prefrontal cortex. In vulnerable individuals, this system is overly sensitive, and synaptic connections strengthen more easily (Rumpel et al. 2005), while connections to the cortex are less effective (Lai et al. 2012). This is known as the *conditioned emotional response* (◘ Fig. 8.1). Glutamate, via NMDA and AMPA receptors, promotes the transmission within the LA (Migues et al. 2010), but more importantly, facilitates the memorization of this association. As a result, the presence of the new stimulus elicits an intense fear response. *Sensitization* has occurred instead of *habituation*. At the same time, this particular sensitivity promotes intense fear reactions, while the prefrontal cortex loses its ability to control the fear response and induce extinction (Mao et al. 2006).

In other words, if the brain has a tendency to sensitize easily, the number of stimuli triggering fear responses increases, and the likelihood of new associations grows even more. This exponential progression can plunge the individual into exaggerated fears in response to almost any stimulus. Anticipation exacerbates the condition in anxious states, and the animal becomes hypervigilant (◘ Fig. 8.2).

The genetics underlying fear behaviors are complex, with numerous candidate genes contributing to vulnerabilities. Interactions between these different genes (epigenetics) create almost infinite possibilities for clinical expression, explaining their diversity.

8.1.1.3 Epidemiology

The nosographic diagnosis of categorical phobia is common. In our veterinary psychiatry consultations, it represents a significant proportion of cats presented for evaluation (Yamada et al. 2020).

There are no prevalence studies in the general population, but the incidence of this condition is likely underestimated. When the phobia does not severely impair the cat, owners may opt to avoid the problematic situation. For example, many owners avoid visiting the veterinarian because their cat has a phobia of healthcare or simply of transport, and they feel it is preferable not to subject their pet to such stressful situations (Karn-Buehler and Kuhne 2022). These phobias often go untreated, particularly because veterinary visits are infrequent. By avoiding exposing the phobic animal, the condition does not improve; in fact, it often worsens.

8.1.1.4 Presenting Complaints

Most often, subtle signs are already observed in kittens at the time of adoption, during the socialization and juvenile periods. When these signs appear despite development under good conditions, they reflect an individual vulnerability related to the fear axis. It is common for consultation requests to arise at puberty for several reasons: the kitten establishes its territory and tolerates changes less well, it matures in its relationships and begins to reject direct contact with adult conspecifics, and neutering usually occurs during the pubertal period, combining these inconveniences with the stress of a veterinary visit. Consequently, difficulties placing the cat in a carrier or defensive behaviors during handling may prompt consultation when the problem worsens.

In fear reactions, cats may respond in one of four ways (4 F's): *freeze, flight, fight*, or *fidget*, corresponding to inhibition, escape, aggression, or agitation. Generally, noise phobias rarely lead to aggression and are more often associated with inhibition or escape behaviors. Conversely, social phobias fre-

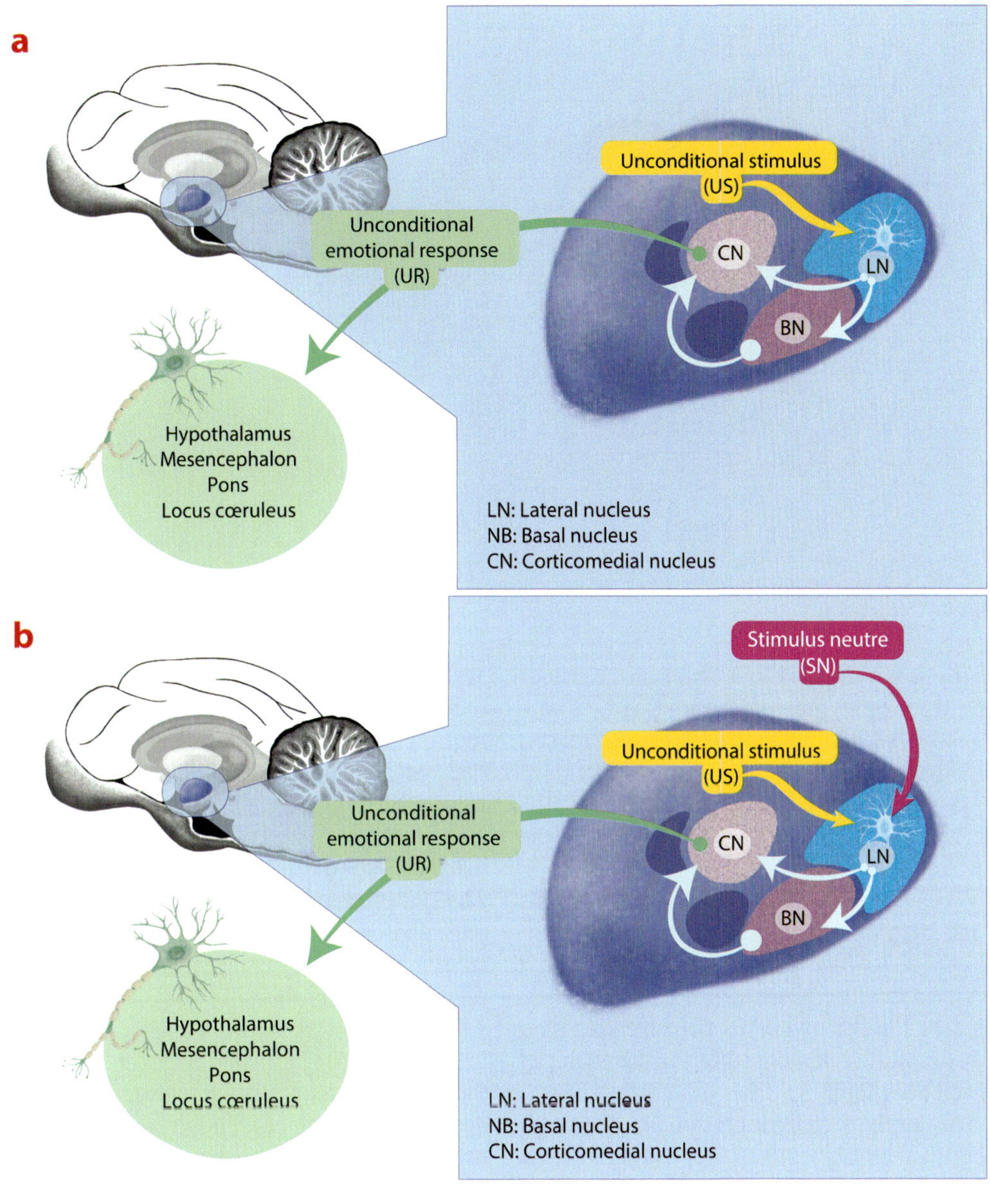

Fig. 8.1 Neurophysiology of classical fear conditioning. The association between a known phobogenic stimulus (**a**) and a neutral stimulus (**b**) strengthens certain synapses in the LA. This new pathway enables a rapid emotional reaction when exposed to the new stimulus, identical to the reaction triggered by the initial stimulus (**c**). The pathogenesis involved corresponds to that described for phobic and anxious pathological states, and the individual's genetic vulnerability activates these circuits. A depressive state may come into play when the individual gives up the struggle and is overwhelmed

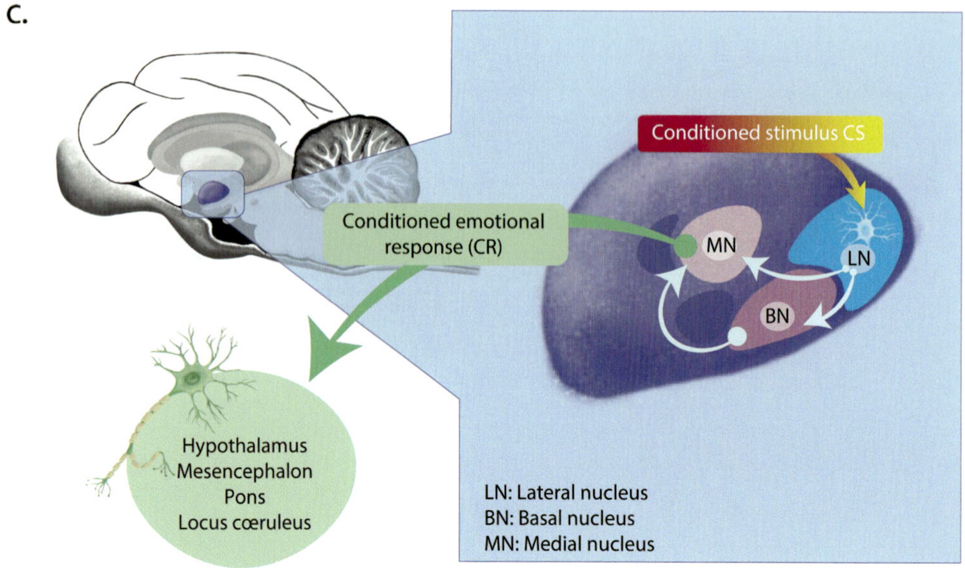

8

Fig. 8.1 (continued)

quently result in aggressive reactions, particularly when the animal is confined. These aggressive responses are the primary reason for consultations for this condition. The danger posed by such situations indicates the severity of the condition and influences the approach to management.

When the cat is not aggressive but exhibits escape behaviors, the phobia may go unnoticed by the owner, particularly if the animal has free access to the outdoors. In such cases, it is rare for the owner to perceive this as a problem or to seek a consultation for their pet (Powell et al. 2023).

The level of maladaptation in cats with categorical phobia varies depending on the frequency of exposure to triggering situations, but it quickly becomes debilitating for owners, who then avoid these situations, such as hosting visitors or attending veterinary appointments. This *handicap* often becomes a reason for consultation.

In cases of inhibition and escape, the cat's distress is not always apparent to the owners, and it is less likely to be brought for behavioral consultation than an aggressive counterpart. However, as previously noted, the level of suffering is not necessarily lower. When the *distress* is recognized, it becomes the owners' primary concern, prompting a request to restore a better quality of life for their pet, whom they perceive as isolated and suffering.

8.1.1.5 Symptomatology

The symptomatology varies depending on the pathological state associated and the type of strategy "chosen" by the cat to respond to the phobogenic stimulus.

Through anthropomorphism, we often interpret a cat approaching an individual during stress as a sign of emotional dependency. For the cat, however, this likely involves a different mechanism, more related to the concept of a safety zone. When the cat is near certain individuals, it feels protected in much the same way as it does within its isolation zone. This seeking of contact is, therefore, a way to find safety, which we likely misinterpret as a sign of attachment. Consequently, the presence or absence of the owner or another household cat can alter the animal's reactivity.

The following three sections describe the symptoms associated with the different forms of the disorder.

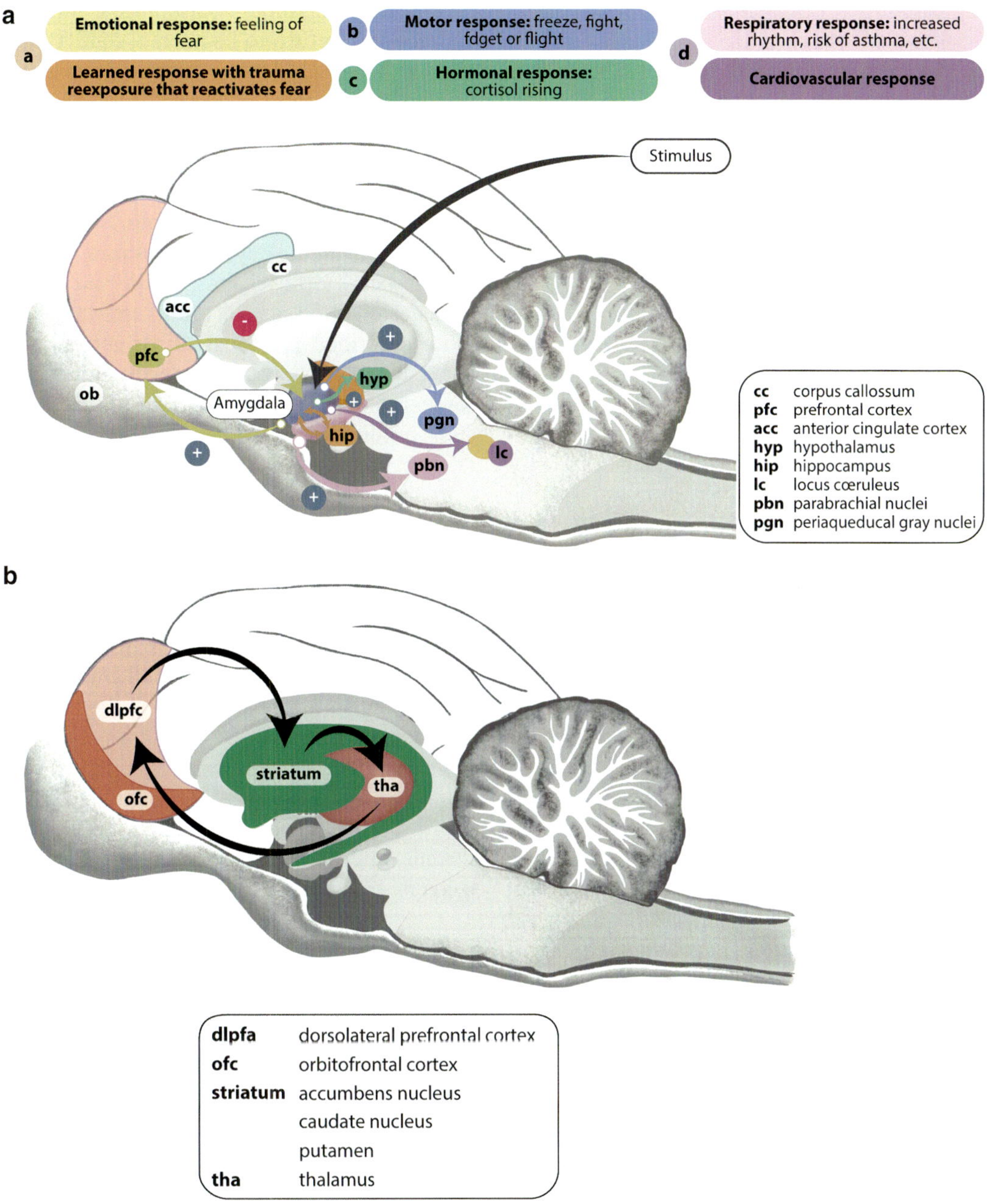

Fig. 8.2 Circuits of fear and anticipation. In categorical phobias, the fear circuits (**a**) and anticipation circuits (**b**) are dysregulated, reacting when they should not

- **Categorical Phobia Associated with a Phobic State**
  - In a phobic state, the cat exhibits fear symptoms in response to one or more identifiable stimuli, without progressing toward adaptation despite repeated exposures. When the stimulus is removed, the cat quickly returns to normal. In the absence of the phobogenic stimulus, it is impossible to detect the disorder.
  - These behavioral responses can take various forms, and in most cases, the cat adopts one of the following strategies:
  - inhibition (freeze): Complete immobilization of the cat, with a lack of exploration and withdrawal, often accompanied by salivation;
  - avoidance (flight): Escape responses, which can escalate into disappearance, with the cat hiding in a safe isolation zone;
  - aggression (fight): Irritation-driven aggression in a context of fear. These aggressions are often simple and occur before the cat seeks shelter. Uncontrolled emotional aggressions can occur in confined situations and are always associated with autonomic manifestations (emotional urination, anal gland emptying, etc.);
  - motor agitation (fidget): Disorganized movements with restlessness around the phobogenic stimulus.
  - Observable postures: Agitated tail, hissing, ears flattened backward.
  - Autonomic manifestations: Tachypnea, tachycardia, tremors, salivation, emotional urination and defecation, and anal gland emptying.

8

- **Categorical Phobia Associated with an Anxious State**

In an anxious state, fear reactions occur regularly, and the triggering factor is no longer always identifiable. The fear appears objectless. While the fear responses characteristic of the phobic state may still be observed, when the phobogenic stimulus disappears, the animal does not return to its normal state. It displays either productive or inhibited reactions that are intermittent or generalized, allowing the type of anxiety to be characterized (see ▶ Chap. 7, ▶ Sect. 7.2.2). In the most severe cases, hypervigilance leads to hyposomnia, with frequent startled awakenings. Sleep, particularly during the day, becomes rare.

The behavioral responses observed in the presence of phobogenic stimuli are similar to those described previously. Cats often favor one strategy but may switch depending on the context: for example, aggression indoors and flight outdoors.

The same symptoms as those associated with the phobic state are present, but additional *substitute activities* emerge, such as increased licking, food-seeking, and heightened skin sensitivity, often with RSS (see ▶ Chap. 15, ▶ Sect. 15.1.5.4). These behaviors are triggered by the anticipation of contexts signaling phobogenic stimuli and tend to increase in duration over time if the anxiety remains untreated.

Many organic conditions can be exacerbated by this anxious state, including idiopathic cystitis, feline pruritic alopecia, and chronic intestinal diseases (see ▶ Chap. 14, ▶ Sects. 14.2 and 14.5).

- **Categorical Phobia Associated with a Depressive State**

In the depressive state, the animal stops reacting to its environment and withdraws into itself. Clinically, this state is characterized by intense inhibition, social withdrawal, sleep and appetite disturbances, and anhedonia. Periods of agitation may occur, reflecting *intense discomfort*. During these phases, the animal, which typically isolates itself, may instead display aggressive behaviors. Identifying triggers for mood changes becomes increasingly difficult.

The observed behavioral responses correspond to a general decrease in *motivation*. Reactions typically seen in phobic and anxious states are less frequently observed. The clinical presentation fluctuates in a sawtooth pattern.

Some common behaviors include:
- refusal of social interactions, including rituals, play, and petting
- urine soiling due to withdrawal and territorial restriction
- inhibition and isolation at the slightest disturbance
- erratic appetite: alternating periods of hypersomnia and decreased appetite with episodes of motor agitation and increased appetite; irritation-driven aggression when interactions are imposed

This form is severe and is often associated with multiple phobias or highly stimulating contexts relative to the environment in which the animal was raised. It is more frequently observed in cases of *sensory deprivation syndrome*.

8.1.1.6 Evolution

A phobic state can remain stable over time, particularly when the animal is infrequently exposed to phobogenic stimuli and has the ability to avoid them. However, when exposure becomes more frequent, processes of anticipation and generalization exacerbate the situation, especially if the exposure occurs without the possibility of avoidance. In such cases, the CSTC loop is activated, leading to an anxious state. It is worth noting that this anxious state may also present from the outset. The animal attempts to cope with an environment that is not objectively hostile, but the body suffers the effects of chronic stress. Without resolution, the situation may progress to a depressive state, characterized by even more severe impairment and an abandonment of coping efforts. A depressive state may also appear directly.

Although these patterns of progression are observed across all types of phobic disorders, specific characteristics are unique to each type of phobia.

Veterinary healthcare phobias always arise following events perceived as unpleasant (ear cleaning, vaccination) but take on exaggerated proportions in vulnerable cats (Tateo et al. 2021). Without intervention, the spontaneous progression of veterinary healthcare phobias is poor. As veterinary visits are often associated with fear or pain, the animal increasingly anticipates these contexts and eventually exhibits fear as early as the announcement of care (e.g., the sight of the carrier, specific odors) (Kirby-Madden and Herron 2024). The examination table is a confined setting that promotes the emergence of aggressive reactions.

For these reasons, it is essential to recognize subtle signs of fear during consultations and to address the issue as early as possible. Behavioral training is often necessary for a favorable outcome once the initial signs of fear are detected. Unfortunately, the time investment required for therapy is significant compared to the short duration of veterinary visits. As a result, motivating owners can be challenging. It is likely that many cats miss out on appropriate care for this reason, as owners prefer to avoid subjecting their pets to unpleasant sessions that are also distressing for themselves (Mariti et al. 2016).

Transport phobia often begins with a first trip to the veterinary clinic. It is easy to understand how the unfavorable outcome of this journey, combined with the potential nausea from early trips, provides a straightforward mechanism for acquiring this phobia. It quickly generalizes, encompassing a phobia of the carrier, the trip itself, veterinary care, and the veterinarian. Aggressive reactions are also common, especially when physical restraints are applied.

In transport phobia, it is important to consider digestive issues. It is not uncommon for vomiting to be accompanied by diarrhea, which soils the cat and increases its discomfort, accelerating the development of the phobia. In some cases, a single transport event during which the cat panics is sufficient to trigger a severe phobia. Management should address both the digestive and behavioral components of this condition.

Phobias limited to new objects (e.g., vacuum cleaners, the noise of machines, fans) are rarely the primary reason for consultation and are seldom the only elements of the clinical presentation. Management is therefore often broader and more comprehensive.

Social phobias frequently evolve into aggression. *Interspecific social phobias*, involving fear of humans, are a common reason for consultation. Aggressive reactions may be observed early in life if escape is difficult. These phobias may target a specific subset of the population or extend to all humans from the outset.

Dog phobia is somewhat unique. Some cats experience genuinely dangerous situations in their lives, making the term "phobia" less appropriate when the danger is real. Additionally, these cats, even if they react to unfamiliar dogs, may become familiarized with the household dog. Therefore, the goal of treating dog phobia is often limited to familiarizing the cat with specific canine individuals.

Intraspecific social phobias are rare. These do not involve fights between two cats but rather an intense fear directed toward all cats. Among the animals requiring assistance, some exhibit a global deprivation syndrome, fearing everything external to their familiar environment.

Treatment must address both the fears related to the external environment and the fears of conspecifics. Since cats are a non-social species, it is normal for them to refuse interaction with conspecifics after puberty. Aggression toward other cats encountered in their territory should not automatically be interpreted as a sign of a phobia.

Conflicts between cats living in the same household are typically not related to phobia. Abnormal behavior in such cases is more often due to disruptions in territorial management or disorders affecting self-control, which result in poor communication.

8.1.2 Diagnosis

8.1.2.1 Diagnostic Criteria

Cats with a categorical phobia must present with a *phobic, anxious,* or *depressive* pathological state. The associated symptomatology is described in the previous section and corresponds to the characteristic symptoms of these pathological states. Depending on the type of phobia, certain responses are more frequently observed. The diagnostic criteria are therefore presented by type of categorical phobia.

Other cognitive pathological states may also be present (◘ Fig. 8.3), either due to the coexistence of another psychiatric disorder (e.g., comorbid HSHA syndrome with categorical phobia) or due to progression (e.g., the emergence of a senile disorder in old age).

▪ Noise Phobia

The diagnosis is based on identifying a phobic, anxious, or depressive pathological state associated with an intense and anticipatory fear reaction in the presence of storms, detonations, or specific noises such as alarms or metallic sounds. The severity of symptoms ranges from reactions occurring only in the presence of the phobogenic stimulus to anticipatory panic attacks, during which the animal may place itself in physical danger (Mills 2005).

Storm phobia often appears immediately upon the first exposure. It is difficult to treat as exposures are *uncontrolled* and *too infrequent* to allow for gradual habituation. The associated state is most often phobic. Using recordings of storm sounds does not prevent the onset of this phobia, likely because the sound level does not replicate that of a real storm, and the context is different (e.g., absence of wind or rain).

This phobic state is generally stable over time but can worsen into an anxious state through generalization to other sounds or due to environmental changes (e.g., moving, sensory decline with aging, or changes in household composition).

A depressive state may occur when this phobia is not the sole psychiatric disorder present or in cases of long-term progression with aging, although this remains an exception.

When the first exposure is severe and involves immersion in a confined environment, an acute depressive state followed by chronic depression linked to *learned helplessness* **(see box below)** may be observed. In such cases, the condition is more appropriately classified as post-traumatic phobia (see ► Sect. 8.3) rather than categorical phobia.

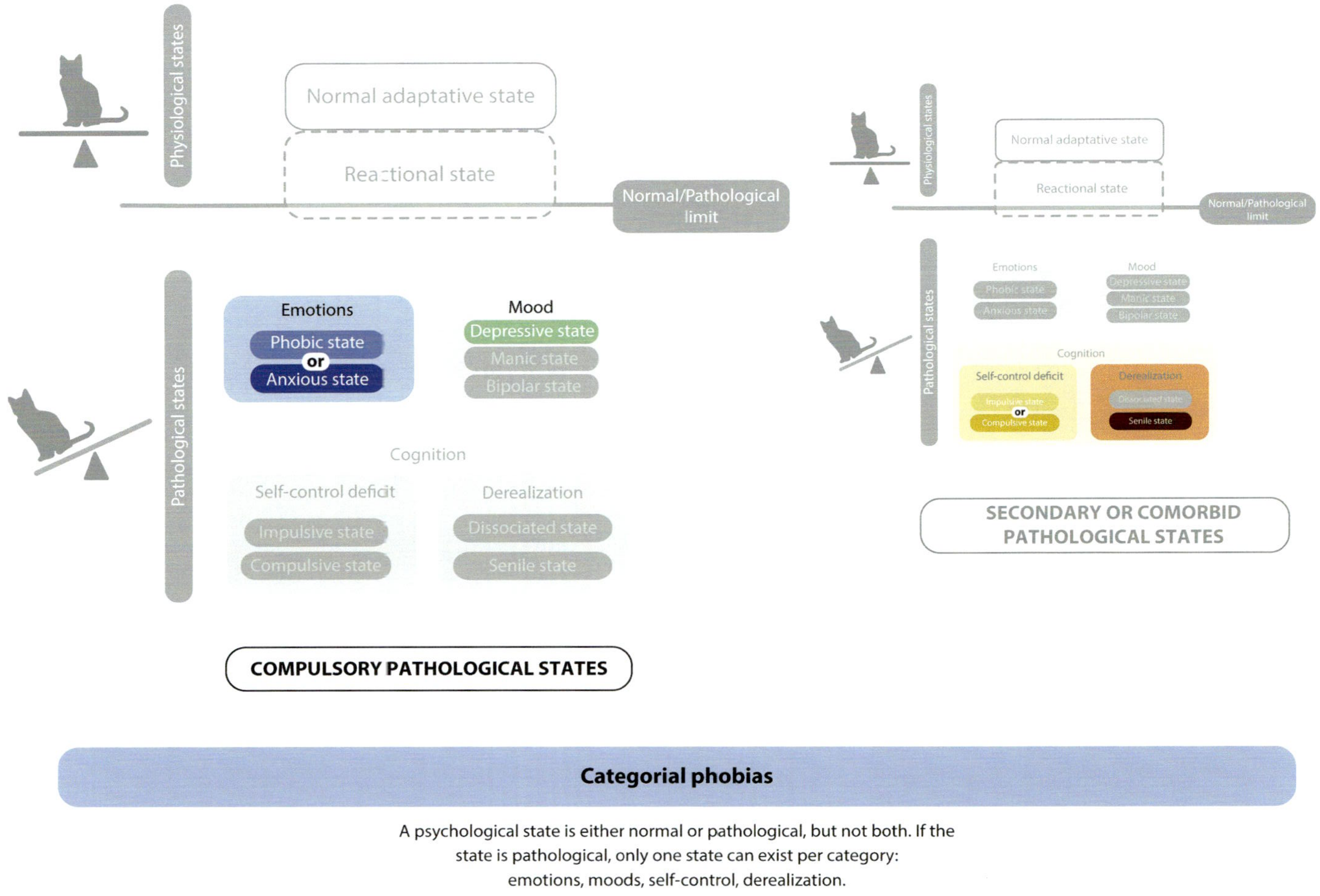

Fig. 8.3 Pathological states observed in categorical phobias. Mandatory pathological states: phobic **or** anxious state, **and/or** depressive state. Frequently comorbid pathological states: impulsive **or** compulsive state, **or** senile state

Important Point of View

Learned Helplessness

Martin Seligman, a psychologist, proposed the concept of learned helplessness in the 1970s (Maier and Seligman 1976). In his laboratory, he conducted an experiment that would no longer be ethically permissible today.

He placed dogs in harnesses from which they could not escape and subjected them to electric shocks paired with a visual signal. Dogs in the first group could press a pedal to stop the electric shock. Dogs in the second group could do nothing, but their shock stopped when a dog from the first group pressed the lever. Thus, both groups received shocks of the same duration at the same time, but only the first group had control over stopping them.

Shortly after, Seligman placed the dogs in a room where the floor delivered electric shocks following the same visual signal. The room was divided in two by a low barrier that the dogs could easily jump over. On the other side of the barrier, there were no electric shocks.

The dogs that had been subjected to inescapable shocks in the harness remained lying down and passively endured the shocks without attempting to cross the barrier, while the dogs that had been able to actively stop the shocks quickly jumped over the barrier to escape.

Seligman defined *learned helplessness* as a pervasive and general sense of powerlessness resulting from an individual's experience of being placed repeatedly or for extended periods in situations where their choices do not influence the outcome (lack of control).

Cases of uncontrolled immersion (exposure to a high-intensity phobogenic stimulus without the possibility of avoidance) lead to depressive states through this mechanism. For example, a cat forced to live in a house with a dog that terrifies it learns to isolate itself and restrict its territory without ever habituating to the dog. One might believe the cat has adapted, but it is important to emphasize that its behavior remains abnormal and its calmness is due to isolation rather than habituation. In general, violent immersion leads to a depressive state, which is why such immersion must be avoided.

8

▪ Healthcare Phobia

The diagnosis is based on identifying a phobic pathological state associated with a fear reaction of varying intensity, often anticipatory, in the context of care provided by a veterinarian, owner, or groomer. Symptom severity ranges from reactions occurring only in the presence of the phobogenic stimulus (e.g., the examination table) to anticipatory responses involving agitation, attempts to flee, or aggression at the first signs of upcoming care (e.g., the sight of the carrier). The inhibited form also exists but is almost never associated with an owner's request for consultation, even though it can be as severe as other forms. Practitioners can point this out to owners when they observe it to help them recognize the animal's distress, even in the absence of defensive reactions (Powell et al. 2023).

The healthcare phobia described here corresponds to the animal's *sensitization* to care-related experiences. There is no evident traumatic event, but unpleasant manipulations inherent to certain veterinary procedures generate exaggerated reactions and anticipation that compromise future care. These can include vaccination visits or a simple examination for a skin infection.

For adult cats, calming is achieved primarily through territorial organization and pheromones. Exposure to an aversive stimulus outside their territory, in a room where they have not left their soothing markings, adds another factor contributing to sensitization to care. For cats, the presence of the owner is not as reassuring as it is for dogs.

In some cases, merely being in the waiting room is enough to create an unpleasant situation, leading to a phobia. While the carrier provides protection, for some cats, being confined in it while dogs approach or even threaten them can be genuinely traumatic. In all cases, it is important to discourage owners from removing the cat from its carrier in the waiting room.

Sometimes, the animal experiences a true trauma, such as undergoing major orthopedic surgeries requiring prolonged immobilization and strict rest or enduring frequent bandage changes on extensive wounds with repeated painful procedures. If the cat subsequently develops healthcare phobia, the condition falls under post-traumatic phobias, described in ▶ Sect. 8.3 of this chapter. Forced handling, where the animal is physically restrained and struggles, is also a traumatic event for most cats. If this type of action is the only way to perform care, it must be done under sedation, as the impact on the overall well-being of a traumatized animal is far worse than the risks associated with routine sedation. Such harsh restraint practices must be avoided.

Unless the cat requires very frequent care, healthcare phobia rarely progresses to an anxious or depressive state. The use of a restraint cage, which immobilizes the cat by pressing it between two grates for injections, should be banned from veterinary practice. This type of method consistently results in a phobic state.

■ Transport Phobia

We previously described how transport phobia is often associated with healthcare phobia, as veterinary visits remain the primary reason for transporting cats. However, this phobia can also exist independently.
The diagnosis is based on identifying a phobic pathological state associated with a fear reaction of varying intensity, often anticipatory, during car transport, whether in a closed carrier or not. Symptom severity ranges from highly inhibited reactions to productive responses, such as vocalizations, urination, or defecation in the carrier. Most commonly, the inhibited form is observed in cats, where the animal hides under a seat or at the back of the carrier, often accompanied by autonomic manifestations.

As with healthcare phobia, it is usually possible to identify an initial unpleasant experience from the cat's perspective. Motion sickness often leads to anticipatory fear. When this experience is repeated without treatment, the cat reinforces its association between car transport and discomfort. Like humans, the maturation of the vestibular system reduces nausea with age, and true motion sickness in adults is uncommon. However, the fear associated with transport persists well beyond this point due to classical conditioning, and simply placing the cat in the car can trigger nausea.

Intense signs of fear are often observable during transport, such as trembling, hypersalivation, panting, or restlessness. Addressing this issue is often possible, but both behavioral and digestive aspects must be considered, as we will detail in the section on therapies. This phobic state rarely progresses to an anxious or depressive state.

■ Social Phobias

For cats, we have retained the term *social phobia*, although it could simply be referred to as *human phobia*. However, it is challenging to speak of intraspecific phobia, as cats naturally prefer to live alone within their territory. Friendly cats or those that truly accept territorial sharing are exceptions rather than the norm.

There can, however, be rare cases of intraspecific phobia, which manifest as an *out-of-proportion* fear reaction when another cat invades the territory. Aggressive behavior in such cases may be mistaken for simple territorial defense. To describe this as a phobia, the reaction must be disproportionate or pervasive.

Additionally, we have discussed dog phobia, which can develop in most cats but remains a normal reaction to what represents a real danger.

The diagnosis is based on identifying a phobic, anxious, or depressive pathological state associated with a fear reaction of varying intensity, often anticipatory, in the presence of humans. The severity of the condition depends on the type of reaction exhibited by the individual. As soon as the cat reacts aggressively, the level of danger changes, requiring more restrictive measures during therapy. Moreover, the animal learns that this reaction is effective, as most individuals who are attacked either freeze or withdraw from the interaction. This creates a reinforcing effect. A cat that primarily seeks to repel a potential danger learns that aggression is more effective than freezing or fleeing.

It is important to note that, in social phobias, aggressive reactions are quickly instrumentalized. The behavioral sequence evolves into an aggressive sequence with minimal or non-existent warning signals, and signs of fear may no longer be visible during the consultation. It is therefore essential to reconstruct the initial exposures to understand the emotion that motivated the *first aggressive reactions* and to map the pathological progression before concluding that the condition is a social phobia.

8.1.2.2 Direct Observation

Observing the cat allows for an assessment of the relevant stimuli and the extent to which the phobia impacts the cat, depending on the owners' lifestyle. For most phobias, it is possible to evaluate the animal's reaction during the consultation and determine which therapeutic strategies are effective and feasible. Regardless of the specific phobia, the practitioner must assess, during the consultation, the level of discomfort caused by the phobia and the relevance of initiating treatment. Owners' expectations are not always realistic, and direct observation, combined with professional experience, enables immediate reframing of objectives, available resources, and the veterinarian-owner relationship—essentially, the *therapeutic alliance.*

For example, a socially phobic cat at the anxious stage, whose strategy involves aggressive reactions when visitors attempt to approach, is unlikely to become friendly and enjoy petting in all circumstances. For such a cat, complete familiarization with all individuals cannot be envisioned.

8.1.2.3 Comorbidities

It is common for the developmental conditions of the cat to have been suboptimal or for the sensory filter to be of poor quality. Consequently, frequent comorbidity with *HSHA syndrome* or *feline ADHD* is not surprising. The learning capacity of hyperactive cats is also responsible for a severe limitation of sensory homeostasis thresholds.

At puberty, the presence of social phobias with aggressive reactions often results in frequent sequences of territorial aggression. Some breeders prefer to keep kittens for an extended period, only allowing adoption at around 4 months of age. This practice can sometimes lead to difficulties with socialization and subsequent phobias if the kittens have had limited exposure to humans in their birthplace. For cats, socialization with other species must begin before 2 months of age to allow generalization to the species (Campbell et al. 2024; Martínez-Byer et al. 2023).

Comorbidities with categorical phobias are therefore frequent. The practitioner must prioritize therapeutic interventions, either addressing the condition that aligns with the owners' primary concern to maintain a strong therapeutic alliance or focusing on the causal or most pervasive condition while explaining the reasoning behind this approach.

8.1.2.4 Differential Diagnosis

Identifying a phobic, anxious, or depressive pathological state leads the practitioner to consider *sensory deprivation syndrome* and *post-traumatic phobia* as part of the differential diagnosis. Early developmental conditions and the identification of a tangible traumatic event are key factors that help distinguish between these conditions.

Motor agitation observed in the presence of phobogenic stimuli can be mistaken for an *impulsive* or *compulsive state*. In the case of phobias, the animal reacts differently depending on the context. Unlike a hyperactive cat, a phobic cat behaves normally when it is not afraid, i.e., in its daily environment.

Finally, a sudden onset of phobia after puberty, without evidence of trauma, should prompt consideration of an *organic disease* underlying the observed pathological state. For example, joint pain may cause avoidance of contact, irritation-driven aggression during care, or avoidance of interaction with certain or all humans.

In such cases, it is essential to conduct *additional examinations* to identify the trigger for the categorical phobia while also addressing the pathological state. It is not uncommon for the organic cause to be treated while the phobic reaction persists. The response to medical treatment and therapy also aids in establishing the nosographic diagnosis.

8.1.3 Prognosis

The severity of the disorder is highly variable and depends on the pathological state, the *number of affected categories*, and the *frequency of exposure*. A cat suffering from multiple categorical phobias is frequently subjected to intense fear, increasing the likelihood of progressing to an anxious or even depressive state. The prognosis worsens under several conditions:

- if the disorder manifests early
- if the condition has progressed for a long time before the first consultation
- if aggressive reactions are present in the clinical presentation

Additionally, the inability to partially or fully shield the animal from phobogenic stimuli is an aggravating factor (◘ Table 8.2). For example, living in a household with numerous children for a socially phobic cat or near a shooting range for a cat phobic of detonations significantly hinders recovery. The living environments of some cats may be incompatible with their adaptive capacities, and their progress will remain limited until they are relocated to environments more suitable for their sensory homeostasis.

8.1.4 Treatments

The principle of managing phobic disorders is to use medication to reduce the area of extreme discomfort, allowing for the application of behavioral therapy, which most often involves a desensitization and counterconditioning technique (DSCC), as illustrated in ◘ Fig. 8.4.

◘ **Table 8.2** Factors influencing the prognosis of disorders related to the fear axis

Criteria considered	Protective factors	Aggravating factors
Pathological state	Phobic state	Anxious or depressive state
Frequency of exposure to phobogenic stimuli	Variable and controllable	Very frequent and unavoidable
Exposures contexts	Open situations possible	Confined situations
Age of onset	After puberty	Very young
Type of reaction	Inhibition and avoidance	Aggressive reactions
Duration since symptom onset	Managed at the onset of symptoms	Managed after several months or even years
High-quality exposures before puberty	Limited number of exposure attempts	Numerous exposures with slow improvements

8.1.4.1 Pharmacological Treatment

It is important to differentiate pharmacological treatments based on the state diagnosis, as the choice of medication depends on the specific neurophysiological pathway involved.

It should be noted that, in cases of reactive states, we are not dealing with categorical phobias but rather simple fear reactions. Returning to a normal state can be facilitated by the behavioral measures described below. The practitioner may also include nutraceuticals (alpha-lactalbumin, alpha-casozepine, hemp oil) (Beata et al. 2007; Landsberg et al. 2017) and pheromone diffusers to promote the return to a normal adaptive state (Frank et al. 2010).

▪ Phobic States

In phobic states where exposure to the stimulus is infrequent, such as healthcare phobias, noise phobias, transport phobias, or certain

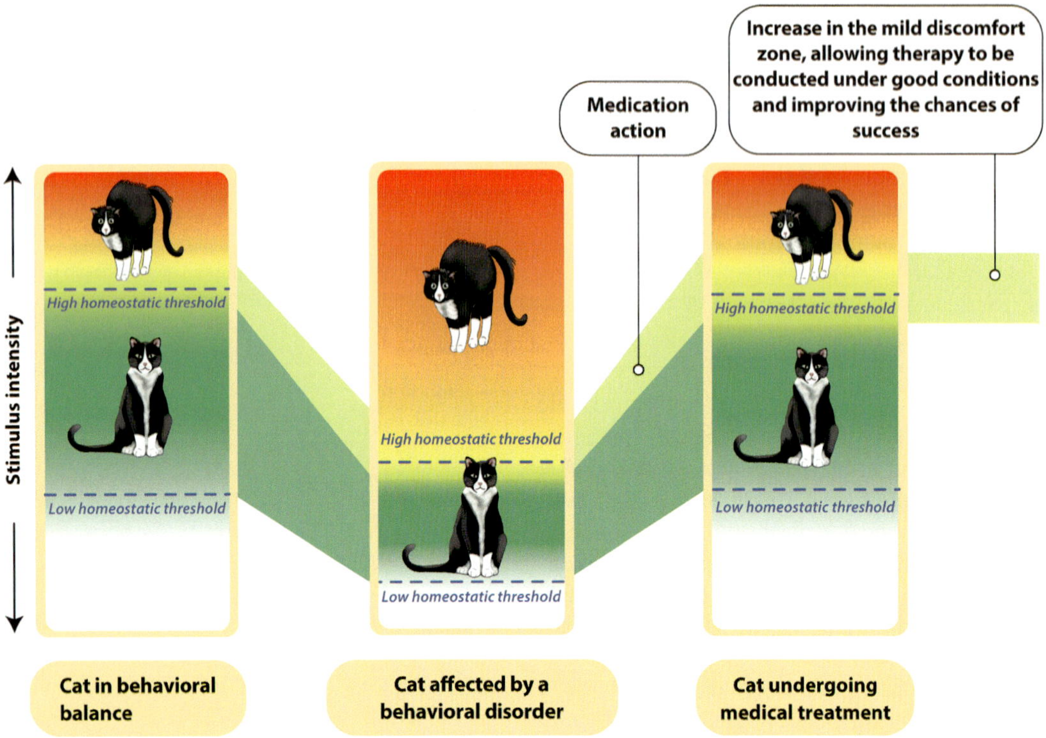

Fig. 8.4 Comfort zones in a cat with categorical phobia. To improve intense fear reactions, the work must be done within the cat's mild discomfort zone (yellow) without pushing it into a state of severe discomfort (red). In affected cats, the yellow zone is very narrow, and the transition to the red zone happens quickly. The role of pharmacological treatment is to expand the yellow zone, making behavioral therapy more effective

social phobias with limited contact opportunities, it is possible to prescribe medications solely for exposures or training sessions. *Gabapentin* is highly safe and can be used as needed (PRN) at doses ranging from 10 to 30 mg/kg 2 h before exposure (Kruszka et al. 2021). In cases of panic attacks where exposure could not be anticipated, gabapentin can be administered as soon as symptoms are observed, reducing the severity and consequences of the attack. *Pregabalin* may also be used for similar indications at a dose of 5 mg/kg per exposure (Lamminen et al. 2023) and has recently received regulatory approval for treating transport and healthcare phobias in cats. It is marketed under the name Bonqat™ in France.

Trazodone can also be used to provide an additional sedative effect, such as during veterinary visits (Orlando et al. 2015).

Dexmedetomidine oromucosal gel may be used occasionally at a dose of 4 mg/m^2, providing 2 h of efficacy. However, this dose almost invariably causes vomiting within minutes of administration and moderate sedation (Smith et al. 2020).

For the specific case of transport phobia, it is essential to include anti-nausea medications such as *metoclopramide* (0.5–1 mg/kg) during the initial stages of behavioral therapy while symptoms of vomiting persist. The animal should also be fasted before transport until the phobia is fully resolved and the vestibular system matures, typically by 1 year of age.

Sedative neuroleptics, such as *acepromazine*, are not a good choice for these indications, especially as monotherapy. This molecule is not anxiolytic and does not act on the circuits involved in phobia. It suppresses the visible manifestations of fear, thereby reducing the owners' discomfort, but leaves the cat powerless against its emotions, which it continues to experience at full intensity.

Benzodiazepines, such as alprazolam or lorazepam, can be used occasionally for transport in cases where gabapentin or pregabalin is insufficient (Kirby-Madden and Herron 2024).

When exposure is nearly daily, the use of medications such as clomipramine (Crowell-Davis et al. 2019) or venlafaxine (Metz et al. 2021; Masson et al. 2024) may be prescribed. These drugs limit the initiation of fear reactions through their action on adrenergic pathways. *Fluoxetine and other SSRIs* are more commonly used when there are comorbidities with an impulsive pathological state or when the animal's chosen reaction is intense aggressive behavior.

▪ Anxious States

In an anxious state, the CSTC circuit is involved in addition to the amygdala. This leads to pervasive anticipation and generalization for the animal. Medications can no longer be used intermittently, and treatment duration is rarely less than 6 months, even if improvements are observed within the first few weeks.

The choice of medication depends on the observed anxious state. SSRIs should be reserved for forms accompanied by productive and generalized anxiety, while medications with a stronger noradrenergic action, such as *clomipramine* and venlafaxine, are better suited for forms with intermittent or more inhibited anxiety. *Venlafaxine* is particularly useful due to its ease of administration, ensuring treatment adherence even when the cat refuses physical contact (Metz et al. 2021; Masson et al. 2024).

Pregabalin, gabapentin, and *trazodone* (Denenberg and Dubé 2018) can be added as second-line treatments when the above medications used as monotherapy are partially or wholly ineffective, reducing excessive fear responses through an alternative afferent pathway in the amygdala circuit.

Benzodiazepines should be avoided except in cases of panic attacks that cannot be controlled with the aforementioned treatments. Their amnesic effect hinders the effectiveness of behavioral therapies but allows for crisis management when primary treatments fail.

▪ Depressive States

Depressive states are extreme conditions that are difficult to overcome. Treatments must be maintained for more than 6 months to avoid a more or less rapid relapse after discontinuation.

Mirtazapine (Raekallio et al. 2024; Spano et al. 2023) and *mianserin* are first-choice medications as they help to restart exploratory behavior. They also have the advantage of *increasing appetite*, which facilitates behavioral therapy. Once behaviors are re-established and the depressive state has improved, it may be beneficial to switch to another medication from the anxiolytic class.

Venlafaxine, in the form of neutral granules contained in Effexor™, is particularly useful for cats that are difficult to medicate with tablets (Metz et al. 2021; Masson et al. 2024).

Pregabalin and *gabapentin* can be combined with the aforementioned medications when the initial prescription yields insufficient results.

8.1.4.2 Behavioral Therapy

The theoretical and general principle of fear therapies is always the same: to *gently push* the animal *out of its comfort zone* by *gradually* exposing it to increasingly phobogenic situations without crossing into a zone of extreme discomfort (◘ Fig. 8.4). In other words, it involves a straightforward process of "*active desensitization*" (*progressing to the first signs of fear*), often combined with *counter-conditioning* to enhance its effectiveness. Counter-conditioning involves using positive stimuli for the animal (play, food, petting) to induce an emotional state incompatible with the onset of panic and to accelerate the disappearance of the exaggerated fear response. This counter-conditioning aims to alter the emotion that arises uncontrollably.

In practice, overly frequent or abrupt exposure to phobogenic stimuli risks sensitizing the animal and worsening the condition. Similarly, complete avoidance of the phobic situation tends to exacerbate the issue. However, *partial avoidance* of the most phobogenic situations is protective and is systematically applied at the start of therapy to prevent triggering excessively intense fear reactions.

Additionally, implementing these therapies requires very close observation of the animal to determine when the discomfort becomes too intense or when it is possible to continue exposure. These therapies, while simple in theory, are therefore challenging to execute (Casey and Bradshaw 2008).

Without the aid of medication, the animal almost instantly shifts from its comfort zone to a zone of excessively intense emotional reactions, rendering behavioral therapy ineffective. The role of medication is thus to create a broader zone of mild discomfort, increasing the likelihood of successful desensitization combined with counter-conditioning (Fig. 8.5).

The success of therapy heavily depends on the owners' ability to finely observe the cat's reactions and adjust their strategy accordingly. When signs of intense fear are observed, it is necessary to revert to a less challenging level of exposure. Conversely, when the cat successfully overcomes a step, a more difficult situation must be introduced; otherwise, progress will stagnate.

In all cases, the animal must be allowed to explore the object of its fear to analyze the situation and make progress.

Environmental Modifications

Avoiding situations that trigger intense fear reactions in the animal is a measure that should be systematically applied. It is beneficial to designate a space (such as a quiet room) that provides the cat with protective isolation. Owners should be informed about the specific signs exhibited by their cat that indicate escalating emotions (e.g., autonomic manifestations, frequent appeasement signals).

The practitioner must observe these signs during the consultation or by reviewing videos when direct observation in context is not possible.

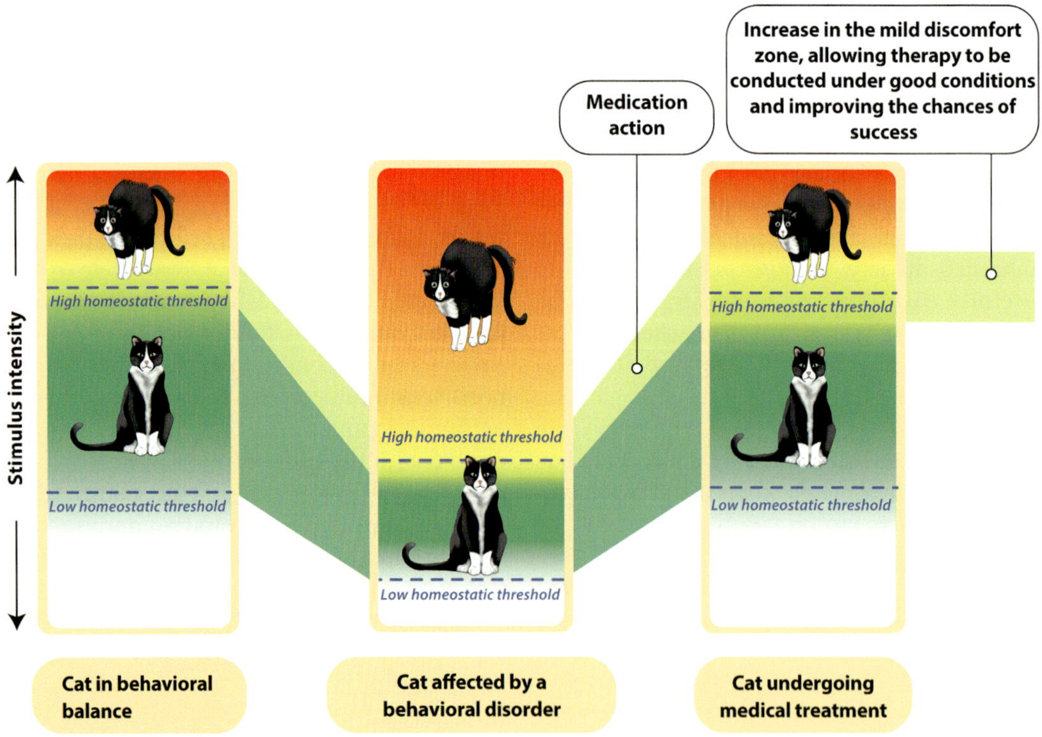

Fig. 8.5 Role of pharmacological treatment in expanding the comfort zones of cats. Medication helps widen the scope of work and improve the chances of success in behavioral therapy exercises

The use of a designated *room* or *hiding spot* is valuable for isolating the animal in situations to be avoided. For example, a socially phobic cat should have a safe refuge when there are large gatherings (e.g., family celebrations or big events) or when household children invite friends over. The practitioner should anticipate these high-risk situations. These measures serve a dual purpose: reducing the risk of injury by identifying high-risk contexts and improving the cat's comfort by avoiding situations that are emotionally overwhelming.

For healthcare phobias, tools such as *lick mats* can be used to desensitize the animal under suitable conditions. These mats can be combined with cooperative care techniques: when the animal stops licking, it indicates that it is distressed, and a pause is introduced until it resumes licking, signaling a return to a sufficiently positive emotional state. The idea is to make the animal an active participant in the care process, allowing it to interrupt the session when its emotional capacity is exceeded.

In practice, the practitioner selects from the available tools based on the individual cat, the overall clinical presentation, and discussions with the owners.

- **Behavioral Modifications**

For *noise phobia*, therapy is challenging because the cat can almost never come into direct contact with the source of its fear. For example, during storms, the cat cannot understand where the noise is coming from. There are many recordings of such noises available for download online, but direct recordings of the sounds using smartphones are often more effective as they correspond to the cat's actual environment. Regardless, the key to desensitization therapy is to play the recording on a loop for the cat, starting at a *low volume*, until the cat no longer exhibits hypervigilance. The cat should be as relaxed when the recording stops as it is when no sound is present. Day by day, the volume is increased. This exercise must be repeated *almost daily* to achieve good results. It is therefore very time-consuming and, incidentally, quite tedious for those involved. Achieving effective desensitization requires the ability to control the sound's intensity and reach its maximum intensity at the end of therapy, which is unrealistic for this type of phobia.

For *transport* and *healthcare phobias*, sessions should be conducted as frequently as possible, ideally *several times a week*. The procedures are detailed in the fourth part of this book. The general principle remains to respect the cat's emotional state and work in a triggering situation without reaching emotional overwhelm.

For *social phobias*, therapy always involves two steps. The first is to *achieve indifference in the cat*, meaning its ability to remain in the room when guests arrive. This stage is by no means socialization. Once this stage is successful, socialization can be considered. The goal is to progress to interactions that the cat finds pleasurable. For example, strangers can offer the cat small spoonfuls of wet food it particularly enjoys, extend a hand for the cat to make olfactory contact, and ideally, pet the cat. In most cases, familiarization rather than true socialization is achieved, as true generalization is unlikely. However, the primary goal is the cat's well-being, and ensuring it no longer experiences major emotional discomfort during visits is already a worthwhile objective.

Finally, it is important to remember that, in all the cases mentioned above, confined situations, coercive methods, and uncontrolled exposure to an excessive number of stimuli should be avoided (Gourkow and Fraser 2006).

8.1.5 Therapeutic Follow-Up

The practitioner must establish a *baseline* during the first consultation by characterizing the fear reaction. A good way to quantify this is to determine, out of ten exposures, how many resulted in an exaggerated fear reaction (frequency) and the type of reaction produced (intensity). This data is essential, as it allows the practitioner to assess whether the condition is *improving, stable, or worsening*.

During follow-up consultations, each phobia is reviewed and quantified again. Exercises are adjusted based on the observations made.

Whenever possible, real-life scenarios should be created during these follow-ups to observe the animal's actual reactions rather than relying solely on the owner's reports. However, for cats—a species for whom territory is highly significant—the clinical environment may not always be conducive to accurate observations. In such cases, videos showing the cat in its natural environment are invaluable tools for the veterinarian.

As previously emphasized, while fear therapies are simple in principle, they are challenging to implement and require significant expertise to maximize the chances of successful desensitization. Owners should be cautioned about the risk of relapse to prevent discouragement.

8.1.6 Prevention

The *selection of breeding animals* helps limit the transmission of vulnerabilities (Montague et al. 2014; Salonen et al. 2019). In addition, phobic mothers often designate numerous stimuli as potentially dangerous to their kittens, increasing the risk of the kittens adopting these maladaptive fear responses.

Before adoption, prospective owners should be encouraged to compare the *breeding environment* with their *own living environment* and choose a breeder whose conditions are similar. Factors to consider include the presence of children, other cats, dogs, and the level of ambient noise.

An effective way to prevent the development of phobias in kittens is to introduce them to new stimuli gradually while monitoring their emotional state (Stella et al. 2014). Owners should observe whether the kitten's vigilance decreases in the presence of new stimuli, allowing the kitten to explore at its own pace without forcing interactions. New experiences should lead to positive outcomes for the kitten. Progress should be based on the kitten's individual capacity to adapt, which may vary.

In practice, repeated exposure to the same stimulus approximately 10 times over a few weeks often results in *complete resolution of fear reactions*. If this is not the case, a psychiatric consultation can help determine whether the kitten is exhibiting a pathological state that requires medical treatment to support behavioral therapy.

8.1.7 Clinical Vignettes

8.1.7.1 Alia

Alia is a young cat whose owners often travel to their countryside house. They have noticed that Alia enjoys the freedom of the countryside with its large garden, but traveling there has become a challenge for everyone, to the point that they have considered placing her in a boarding facility.

Alia is 18 months old, spayed, and does not fear the veterinarian who lives nearby (she "walks" there, according to her owners). She is easy to care for.

When she sees the carrier, she becomes increasingly anxious, seeming to associate it with car travel. She begins salivating and meowing. Upon entering the vehicle, she howls and tries to force the door of her carrier. Shortly after the journey begins, she vomits and experiences episodes of diarrhea.

Initially, the veterinarian prescribed metoclopramide, but it had no effect. Following the lack of improvement, gabapentin at 20 mg/kg was added.

A progressive therapy plan was implemented. First, Alia was trained to enter the carrier on command, with gabapentin administered before each exercise (Eagan et al. 2023). The next step involved placing the carrier with Alia inside into the stationary car 3 times a week. Gabapentin and metoclopramide were administered 2 h beforehand. Vomiting stopped, and metoclopramide was discontinued.

By the third week, short car rides were introduced, with gabapentin given 2 h prior. Alia quickly adapted to these brief trips, showing improved comfort.

The first full car journey was largely uneventful, with minimal signs of discomfort. Gabapentin was continued 2 h before each trip, and breaks were taken if needed. Any vocalizations ceased quickly once the car stopped.

Over the course of ten trips, gabapentin was maintained and then discontinued. Eventually, Alia no longer showed signs of discomfort during car travel.

8.1.7.2 Pastis

Pastis is a young neutered male European cat. His initial vaccinations went smoothly, but since his neutering, he has become aggressive at the clinic. Once there, he refuses to come out of the carrier and becomes threatening. His vaccination booster was postponed because handling him was impossible. The veterinarian prescribed pregabalin to be administered 2 h before the rescheduled visit. With this medication, the visit became feasible.

The owner noted that Pastis seemed more comfortable with a female veterinarian and was fearful of men. Through discussion, it was discovered that Pastis was also phobic of the owner's new partner. At home, he hides and avoids being petted when the man is present. He sleeps in another room and only "reappears" once the man has left.

A prescription for venlafaxine (4 granules/kg, or 1 mg/kg) was initiated until the next visit, which would not involve any procedures, with the goal of making veterinary visits less distressing for Pastis. Pregabalin would still be administered 2 h before this visit.

During the follow-up visit, Pastis was more relaxed. He allowed the assistant to take him out of the carrier and was weighed, though he needed to be quickly returned to the carrier. The treatment was maintained as there was no progress at home.

The recommendation was for the man to remain entirely indifferent and avoid attempting to approach Pastis. A follow-up visit was scheduled every 2 weeks with the veterinary assistant for a simple weighing session.

After 3 weeks, the owner reported that Pastis began to appear in the man's presence. Although he kept his distance, he occupied the sofa while the owners ate. Week by week, Pastis became more comfortable with the man's presence.

The owner and her partner now live together full-time, and after 2 months of treatment, Pastis no longer hides when moving around the house. The man was then instructed to create a ritual with small spoonfuls of wet food offered to Pastis twice a day. Within a few weeks, Pastis became friendly with the man.

At the clinic, Pastis no longer causes problems. However, when a male veterinarian attempted to interact with him, Pastis became threatening again. It was agreed that only female members of the veterinary team would handle Pastis, and successful trials were conducted. An injection was even administered without incident.

After 6 months, venlafaxine was discontinued. Pastis is now fully comfortable with the man, who can pet him without any issues. Pastis even rubs against the man's legs, engaging in allomarking, a sign of high familiarity. Pregabalin is still used for veterinary visits.

8.1.7.3 Nino

Nino is a 5-year-old neutered male cat belonging to a young man who has recently moved in with a mother of two children, aged 2 and 4. Nino spends most of his time hiding, struggles when held, and may even scratch to escape and retreat to safety. His tail lashes constantly. He frequently licks his belly, which is completely bare and hairless. These symptoms, indicative of anxiety, appeared shortly after the move, even though the owners report no unpleasant interactions with the children.

Previously, Nino was a sociable cat, though he lived alone with his owner and primarily interacted with visiting adults. He would often hide at the start of visits, but if guests stayed for the evening and were patient, they could eventually pet him.

Two issues prompted therapeutic intervention: the woman feared Nino might scratch the children if cornered, and Nino's persistent urine soiling had not improved despite numerous treatments.

Nino suffers from generalized anxiety without the ability to escape or find shelter. His only moments of respite occur when the children are absent, but these periods are infrequent and too brief for him to regain balance.

In addition to pharmacological treatment with clomipramine (*Clomicalm ND*, 2.5 mg/day), it was decided that a room off-limits to the children would be designated, where Nino's litter box and food bowl would be placed.

Thanks to the treatment and environmental modifications, Nino's urine soiling resolved quickly, but his phobia of children persisted. When the children are at school, Nino ventures out into the house, but he immediately retreats to his room when he hears them returning. Over time, as the children became calmer, Nino gradually adapted to their presence.

The treatment was maintained for 5 years and then discontinued. Brief relapses occur when the couple entertains guests with children, particularly if Nino's designated room is occupied. These predictable relapses are managed with occasional gabapentin administration, which prevents the return of anxiety and limits the issue to short episodes of discomfort.

8.1.7.4 Rubson

Rubson is a 4-year-old spayed female cat. Her owner is seeking help as a last resort because, for the past 3 years, Rubson has been urinating exclusively on the sofa and curtains in the living room. The owner has not identified any specific event that triggered this behavior. She described an extensive list of treatments that have generally failed, though some temporary improvements were observed. Like the others, the initial prescription failed after a brief period of improvement, with only a slight reduction in the number of soiling incidents. Rubson's inappropriate elimination involves both feces and urine, which is uncommon. Otherwise, Rubson appears relatively calm when her owner does not harass her.

Following the therapeutic failure, Rubson was hospitalized for observation and to give her owner some respite. In the cage, the cat continued to eliminate inappropriately, refusing to use the litter box entirely. Even when the entire floor of her cage was covered with litter trays, Rubson pushed the trays aside to eliminate. The hypothesis of a substrate phobia was proposed. A litter box lined with newspaper was introduced, and Rubson immediately used it for elimination.

The cat was quickly returned to her owner with instructions to use newspaper as the elimination substrate. The inappropriate elimination ceased immediately.

The cause of substrate aversion is not always identifiable, but in Rubson's case, the owner recalled later that the cat had experienced an episode of cystitis and subsequent diarrhea at around 1 year of age. It is possible that the inappropriate elimination began during that time. Substrate aversion can occur due to any painful event experienced during elimination, resulting from classical conditioning in which the cat associates the litter box (neutral stimulus) with pain (unconditioned stimulus).

These aversions are true phobias and should always be considered in cases of inappropriate elimination. Different hypotheses should be tested systematically, altering only one element at a time, litter type, location, size, or whether the box is open or closed.

Summary of ▶ Sect. 8.1
Categorical phobia is a condition in which the animal exhibits maladaptive and exaggerated fear reactions in duration or intensity in response to one or more stimuli belonging to the same category. This condition arises from the interaction between genetic vulnerability and environmental factors and can occur at any age.

It is always associated with a phobic, anxious, or depressive pathological state.

The symptoms reflect the associated pathological states. The cat displays one of the four typical fear reactions: inhibition, flight, agitation, or aggression.

Different types of categorical phobias exist, including interspecific social phobia, intraspecific social phobia, noise phobia, healthcare phobia, and transport phobia, though this list is not exhaustive.

These phobias tend to worsen over time, progressing from a phobic state to an anxious state through anticipation and generalization. Puberty is a common period of worsening, often marked by the emergence of aggressive behaviors. Aggression becomes the cat's preferred strategy as it is more effective in repelling the perceived threat.

Pharmacological treatment involves using medications to reduce fear reactions and long-term anticipation, such as clomipramine, venlafaxine, or SSRIs, particularly when aggression is frequent. For shorter term, targeted action, pregabalin and gabapentin support therapy by reducing the intensity of fear reactions.

Environmental modifications include providing a refuge area for the cat to avoid uncontrolled exposure to triggers.

Behavioral therapy typically involves desensitization (gradual exposure to the phobogenic stimulus) combined with counter-conditioning (training a pleasurable response using a highly appealing reward or another source of motivation). These therapies are time-intensive, even when supported by pharmacological treatment.

Prevention involves selecting breeding animals to avoid promoting cats genetically predisposed to fear.

8.2 Sensory Deprivation Syndrome

8.2.1 Description

8.2.1.1 Definition

Sensory deprivation syndrome encompasses several conditions associated with dysregulation of the fear axis, manifesting as multiple phobias and an inability to adapt to the daily living environment.

This condition arises from the *cat's development in* poor environmental conditions or in environments too different from what their life demands. It is particularly observed in feral cats that are taken in to live as family pets. These cats grow up without human contact and learn to distrust their surroundings. The term "kennel syndrome," used for dogs, does not apply and is never used for cats.

Cats affected by sensory deprivation syndrome exhibit a phobic pathological state or, more commonly, an *anxious or depressive state,* with symptoms present *from the time of adoption.* Rather than categorizing the phobias in cats with sensory deprivation syndrome, it is more practical to evaluate the size of the area in which the cat displays normal behavior. For some cats, this area may be limited to a single room in the house, while for others, it extends to places they encounter for the first time.

In practice, the diagnosis will be stated as "sensory deprivation syndrome associated with a phobic, anxious, or depressive state."

8.2.1.2 Etiology and Pathogenesis

The etiology of sensory deprivation syndrome is partially distinct from that of categorical phobias. In categorical phobias, the condition results from a negative experience and specific vulnerabilities. In sensory deprivation syndrome, the disorder arises from a *lack of early experience.*

Synaptic pruning, which occurs between 14 and 60 days of age, provides these cats with a poor-quality database containing only the few environmental elements they were exposed to during the first 2 months of life (Campbell et al. 2024). Conversely, overly intense exposure increases the risk of sensitization in kittens, which can be avoided by learning to respect signs of fear and offering more gradual exposure to these kittens (Graham et al. 2024).

It is also crucial to advise owners to adopt a kitten that has been raised in an environment similar to the one it will experience after adoption. If the living conditions are more stimulating than the development environment, it is recommended to introduce the kitten to elements that pique its curiosity without provoking an overly intense fear reaction. Stimulations should not be excessive but should involve regular, brief exposures to new experiences. Early outdoor access to a garden should be established if the goal is for the cat to become skilled and well-adapted. However, the downside of this approach is the creation of a dependency that can quickly become a necessity.

Adopting an adult cat may carry greater risks, especially if the early living conditions and the cat's origins are unknown. The behavioral development period is a critical stage (see ▶ Chap. 4). Early learning experiences are often decisive, and contrary to many beliefs, providing more comfort and love is not always sufficient to make a cat happy if it feels trapped in an environment it fears (Reisner et al. 1994).

Most cats with sensory deprivation syndrome exhibit reactions similar to those observed in categorical phobias, but they often have multiple phobias, and the source of their fear is not easily identifiable once they are removed from their primary living area. Within the home, their main comfort zone, they may display completely normal behavior as long as there are no environmental changes, such as the presence of guests. Owners often feel as though they have two different cats: one that behaves normally with them and another that guests never see.

These animals require an emotional crutch to survive in an environment they perceive as highly hostile, which activates the attachment system. This can lead to the development of *compensatory anxious dependence*, though it is far less common in cats than in dogs. For cats, this is more likely a form of seeking shelter than seeking comfort. Generally, the cat chooses to retreat to an isolated location far from strangers, even if it means being away from its owner. In rare cases, the owner may serve as this calming refuge (Potter and Mills 2015).

The amygdala circuits and the CSTC are involved (◘ Fig. 8.2). It is as if the animal's emotional alarm system is malfunctioning. It triggers when it should not or fails to stop when the cortex should be able to inhibit the reaction after analyzing the situation.

The pathogenesis is therefore similar to that of categorical phobias, but in this case, the cat's neural networks are of poor quality. Neophobia is the norm, generalization to other non-phobogenic stimuli is almost impossible, and the overall number of neurons is reduced due to excessive synaptic pruning (◘ Fig. 8.6).

8.2.1.3 Epidemiology

Sensory deprivation syndrome is a common condition. Adopting kittens *before* 2 months of age increases the likelihood of developing the condition, especially if they display signs of fear at the time of adoption (Graham et al. 2024; Martínez-Byer et al. 2023). Adopting a kitten from a breeding environment where animals are highly isolated, with limited early experiences and few stimuli similar to their future living environment, is another factor that predisposes them to this syndrome (Campbell et al. 2024).

Adoption after 4 months might also seem disadvantageous, but cat breeders generally find it easier to create a calming family-like environment compared to dog breeders. Moreover, exposure to unfamiliar environments is less common in the life of domestic

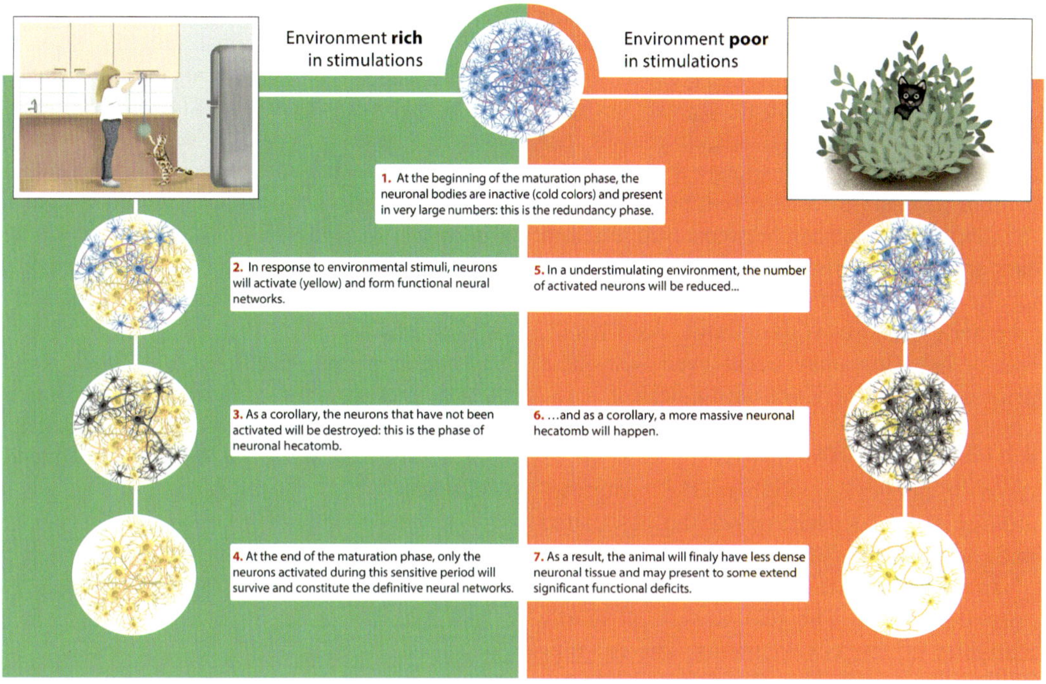

Fig. 8.6 Synaptic pruning in an environment with poor stimulation. In sensory deprivation syndrome, the animal's database is misaligned with its living environment. Neural networks are underdeveloped, and many neurons were prematurely destroyed during the socialization period. This seems to underlie the limited generalization capacities

cats, so the need to adapt to street stimuli, for example, is minimal.

Conversely, kittens born to feral mothers and rescued late in life, after the critical socialization period, often have limited social capacities. Their continuous exposure to a family environment can cause ongoing distress throughout their lives.

8.2.1.4 Presenting Complaints

Presenting complaints often involves *fear-based reactions,* particularly aggressive behavior when the cat is restrained. As with categorical phobias, the cat may respond with inhibition, flight, aggression, or agitation.

When the cat is not aggressive but displays avoidance behaviors, the consultation is often prompted by the owner's dissatisfaction with the cat's behavior, as they want to interact with their pet rather than triggering fear and avoidance.

The degree of maladaptation in cats with sensory deprivation syndrome varies depending on their living conditions, but it quickly becomes a burden for owners, who find themselves avoiding triggering situations and adjusting their lives to accommodate the cat's limitations.

In cases of inhibition, the cat's discomfort is not always recognized by the owners, who may describe their pet as "very calm," and such cats are less likely to be brought in for consultation (Powell et al. 2023). However, when the *cat's discomfort* is observed, it becomes a key reason for seeking help.

A frequent reason for consultation is the owner's *inability to understand the cat's behavior*, as it often seems random. For example, the cat may react aggressively toward specific types of people or situations without the owners being able to identify consistent triggers. Socially phobic cats often tolerate being near a person but refuse physical contact, sometimes aggressively. Men or individuals wearing certain items (hoods, hats, canes, etc.) are often perceived as more threatening. Children may also be more intimidating due to their different movements, unpredictable reactions, and loud noises. It is not uncommon for children to chase the cat for amusement, exacerbating the issue.

Being held can create strong aversions, as the cat experiences significant discomfort in a confined situation where escape is impossible.

It is also important to remind owners of the *sensory differences* between humans and cats, as these differences mean we lack access to the same information cats use to assess whether a stimulus is dangerous. As a result, fearful behaviors may occur even without a prior negative experience with a person.

8.2.1.5 Symptomatology

As with categorical phobias, the symptomatology varies depending on the associated pathological state (see ► Sect. 8.1.1.5).

In the *phobic state,* the cat appears perfectly normal at home. The phobic state primarily manifests in new environments. The cat may display star-patterned exploration, with familiar areas serving as the "center" of the star. Due to the etiopathogenesis of this syndrome, this pathological state is rarely observed.

In the *anxious state,* the cat becomes harder to manage when its living environment changes and remains hypervigilant even at home. While moments of relaxation are possible, anxious cats often react strongly to everyday environmental stimuli despite daily exposure. These cats are often described as wary, as their hypervigilance leads them to anticipate potential intrusions, disappearing well before they come into contact with new elements such as visitors.

With repeated exposure, these cats may accept certain objects or people and become accustomed to their presence, but any new object or person will still trigger fear. For example, the cat may tolerate the owner's presence but remain fearful of everyone else. This inability to *generalize is a hallmark of sensory deprivation syndrome* and is a source of significant challenges for both the cats and their owners.

It is not uncommon for these cats to develop *organic conditions* as a consequence of their anxious state. Idiopathic cystitis is one such example (see ► Chap. 14), and in these cases, inappropriate elimination is often the primary reason for consultation.

In the *depressive state,* cats exhibit abnormal behavior for their age and show low motivation. They isolate themselves, play infrequently, and are highly disturbed by changes. Territorial marking behaviors are absent or disorganized. Although the signs may appear less dramatic to owners than in anxious states, they reflect severe cerebral disruption and distress.

Inappropriate elimination is also frequently observed. This behavior often leads to repeated punishments, which worsen the cat's emotional state. It must be promptly addressed.

8.2.1.6 Evolution

Evolution is better when management is *initiated early*, as learning is easier when the cat is young. The pubertal period is often a phase of worsening symptoms. Kittens affected by deprivation syndrome transition from avoidance behaviors to aggressive reactions.

When the living environment is not significantly different from the developmental environment, progress is often possible. However, for the most severe cases (e.g., a cat raised on a farm, adopted at 6 months old, and now living in a city), the cat will struggle to improve unless placed in an environment closer to its developmental one.

The question of quality of life and well-being for these cats must be central to their management, as the chances of recovery are lower than for post-traumatic or isolated categorical phobias. These *animals' ability to generalize* is rarely sufficient to hope for complete recovery.

8.2.2 Diagnosis

8.2.2.1 Diagnostic Criteria

Cats affected by deprivation syndrome necessarily exhibit a pathological phobic, anxious, or depressive state (◘ Fig. 8.7). Symptoms appear as early as 2 months of age. Developmental conditions in an environment

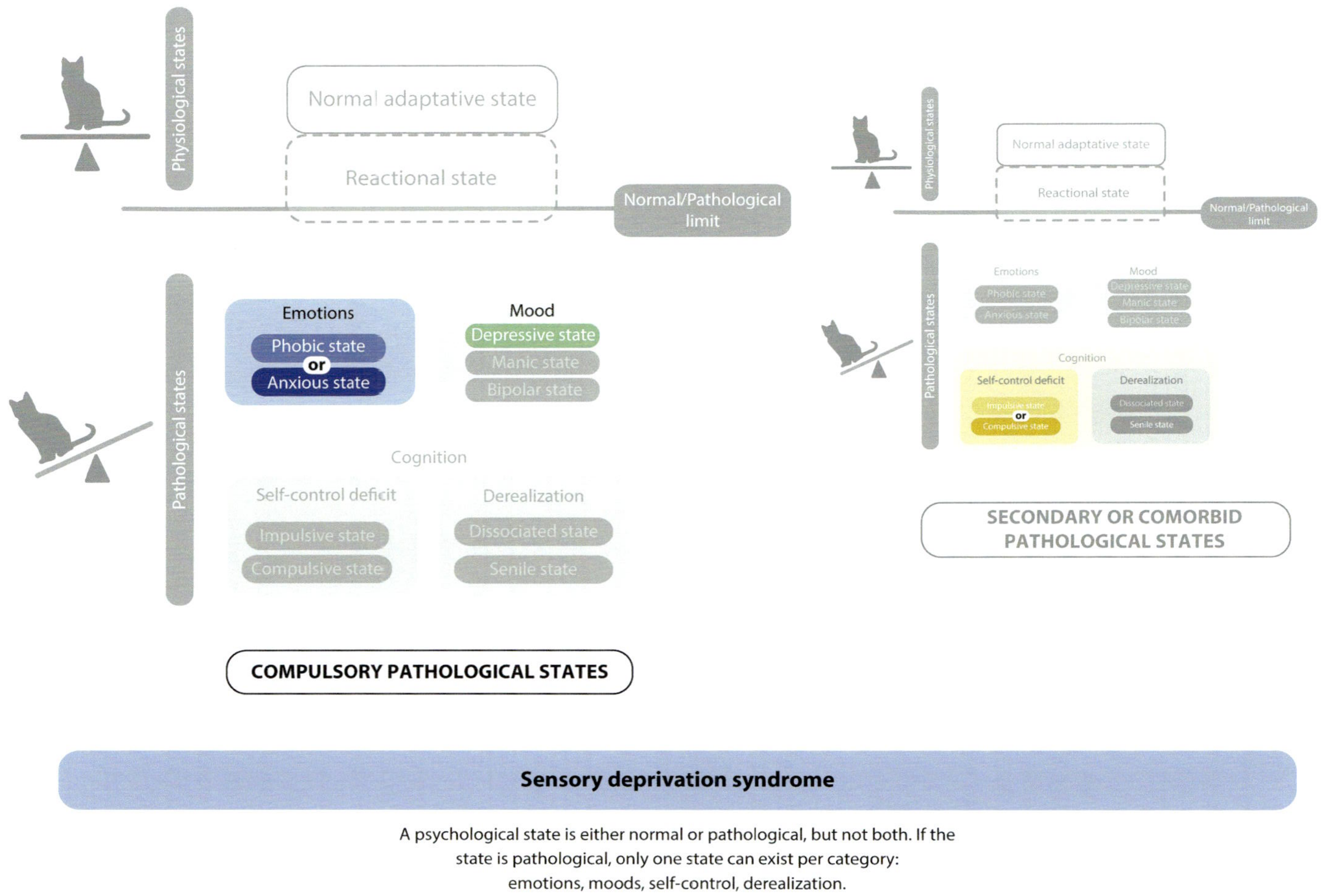

Fig. 8.7 Pathological states observed in sensory deprivation syndrome. Mandatory pathological states: phobic **or** anxious state, **and/or** depressive state. Frequently comorbid pathological states: impulsive **or** compulsive state

markedly different from their current living environment can be identified.

Symptoms consist of intense fear reactions in response to numerous unfamiliar stimuli. In some cases, it is impossible to identify the triggering stimulus, as any new element in the environment provokes an exaggerated fear response.

8.2.2.2 Direct Observation

Direct observation is highly valuable, as it allows fear reactions to be identified and quantified for many stimuli. This approach also helps demonstrate to owners the sometimes subtle signs of fear that they may not have been trained to recognize.

8.2.2.3 Comorbidities

Deprivation syndrome results from development in an *environment lacking sufficient stimulation* compared to the cat's future living environment. The socialization and juvenile periods do not unfold normally, leading to frequent comorbidities. In cats, motor control is largely influenced by maternal interactions between 2 and 4 months of age (Martínez-Byer et al. 2023). Disruptions that result in sensory homeostasis disorders are often accompanied by poor acquisition of self-control. This issue is particularly common in bottle-fed kittens that are kept isolated until adoption, as they are separated from their mothers early.

For this reason, *HSHA syndrome* is a very frequent comorbidity of deprivation syndrome. Practitioners must always anticipate this possibility, as the cat observed during consultation, outside its comfort environment, may only exhibit fear-related problems. In such cases, it can be difficult to suspect hyperactivity. However, this may be identified by questioning owners about the cat's behavior at home in the absence of fear. Owners may describe an animal that struggles with self-control during play, wakes them at night through predatory or playful behavior, and cannot tolerate frustration, especially when held.

Managing this dual condition is challenging. Comorbidity improves some cats, as their lowered sensitivity threshold gives them enough curiosity to engage with new elements and make faster progress. However, in others, the impulsivity linked to their HSHA syndrome causes them to react more quickly to fear, increasing the risk of aggressive reactions. Furthermore, the impulsivity of some cats prevents them from learning effectively, as their disorganized exploration perpetuates deprivation syndrome.

As with categorical phobias, the presence of social phobias accompanied by aggressive reactions frequently results in sequences of aggression resembling territoriality disorders (see ▶ Chap. 10).

Finally, in very young animals, it is important to rule out common *congenital conditions* that could be mistaken for anxious or depressive states. *Painful illnesses, cardiac diseases,* or abnormal growth should be excluded as potential explanations when a kitten exhibits marked behavioral inhibition.

Moreover, as with categorical phobias, the presence of social phobias with aggressive reactions leads to frequent aggression sequences resembling a territoriality disorder (see ▶ Chap. 10).

Finally, in very young animals, it is important to rule out common *congenital diseases* that could be mistaken for anxiety or depressive states. *Painful conditions, heart disease,* or abnormal growth must all be considered and excluded when a kitten exhibits pronounced behavioral inhibition.

8.2.3 Prognosis

The severity of the disorder varies greatly and depends not only on the pathological state but also on the cat's age at the time of the first consultation. The younger the cat, the greater its recovery potential. Additionally, the greater the difference between the environment of birth and the current living environment, the poorer the prognosis.

The prognosis is often refined during the second consultation based on the cat's ability to generalize under treatment. The rate of progress allows a better estimation of achievable goals.

The prognostic criteria in ◘ Table 8.2 are applicable to deprivation syndrome.

8.2.4 Treatments

8.2.4.1 Pharmacological Treatment

All pharmacological treatments and behavioral therapies are identical to those used for managing categorical phobias (see ▶ Sect. 8.1.4.1).

The duration of treatments is often long, ranging from 6 months to several years. As long as the cat continues to make progress, there is no reason to discontinue medication. Once a plateau is reached and the animal achieves an acceptable quality of life for both the practitioner and the owner, gradual withdrawal of medication can be initiated.

Relapses are common, especially during changes in living conditions (e.g., moving, the arrival of a baby in the household). In such cases, treatment should be resumed without hesitation and will often be shorter than the initial course.

8.2.4.2 Behavioral Therapy

Behavioral therapies are identical to those described for categorical phobias (see ▶ Sect. 8.1.4.2). However, if the cat's living environment is significantly different from its developmental environment, it is advisable to avoid highly phobogenic contexts, as the animal will quickly become emotionally overwhelmed and unable to learn.

Environmental modifications involve creating a more calming environment by organizing hiding spots and safe escape areas for the cat.

When feasible, using a therapeutic adult cat can be beneficial for young kittens. Not all adult cats take on the role of educator. In our experience, neutered male cats without excessive fear responses are more likely to engage in this role, although there is no publication to validate this observation.

Improvements can also occur through interaction with a household dog. Sometimes, the dog becomes a regular companion, providing opportunities for play and pleasure, which encourages the kitten to engage in more exploratory behaviors in the dog's presence.

8.2.5 Therapeutic Follow-Up

The practitioner must establish a baseline during the initial consultation, which involves *characterizing and quantifying the fear response*. A useful way to quantify it is to determine the size of the zone in which the cat behaves normally without exhibiting exaggerated fear responses, given the context.

The practitioner must also assess whether the condition is improving, stable, or worsening.

During follow-up consultations, the animal's comfort zone and the frequency of its interactions with novelty are re-evaluated and quantified again, with exercises adjusted accordingly. Videos showing the cat in specific situations provide valuable information for the veterinarian.

For cats that are highly uncomfortable outside their home, follow-up consultations can be conducted via teleconsultation. These sessions provide important insights into the environment in which the cat lives. Whenever possible, situational observations should be conducted during follow-ups to directly observe the animal's reactions, which may differ from those described by the owners.

As previously mentioned, fear-based therapies are theoretically simpler than they are in practice. For this reason, owners must have frequent opportunities to communicate with the practitioner. These exchanges partially compensate for the work performed by dog trainers, who serve as therapeutic intermediaries but are still uncommon in the feline domain.

8.2.6 Prevention

Prevention begins by creating *high-quality breeding conditions* in which kittens remain with their mother until 12 weeks of age (Martínez-Byer et al. 2023) and are progressively exposed to various stimuli, adjusted to their individual fear reactions. Excessive environmental stimuli at an age when kittens show signs of fear can hinder proper development (Graham et al. 2024).

Before adoption, prospective owners should be encouraged to compare the breeding environment to their own and to select a breeder whose living conditions are similar to theirs. Factors such as the presence of children, dogs, and the surrounding noise level should be considered.

Additionally, as with categorical phobias, *selecting breeding animals* can help limit the transmission of vulnerabilities.

Future owners should be cautioned about their expectations when adopting stray cats often offered by rescue organizations. While adopting a stray is commendable, they must be aware of the potential risks, including adopting a cat in significant distress and the possibility of being disappointed if the cat does not meet their idealized expectations.

During the first vaccination, practitioners who identify exaggerated fear responses or poor environmental development conditions in a kitten should emphasize the importance of gradually exposing the animal to stimuli it will encounter in its new environment, without waiting for the vaccination protocol to be completed. Adaptation and recovery abilities, especially after a poor start, are significantly better *before 6 months of age.* Early medical intervention is often recommended to increase the chances of successful desensitization therapy (◘ Fig. 8.5).

The advice given in such cases aligns with the recommendations for other conditions described in this chapter. Kittens should be introduced to *new stimuli gradually*, while their emotional state is closely monitored. Practitioners should observe a reduction in vigilance levels in the presence of new stimuli and allow the kitten to explore *at its own pace* without forcing interactions.

Using a therapeutic adult cat can be very beneficial, but this is a delicate choice, as not all cats are willing to share their territory. Finally, the lack of therapeutic intermediaries for behavioral rehabilitation necessitates closer collaboration between the owner and the practitioner.

8.2.7 Clinical Vignettes

8.2.7.1 Wild

Wild is a young cat adopted from a farm. Based on the clinical examination, he is estimated to be around 6 months old, and the adoption occurred 3 months ago. This means Wild left the farm at approximately 3 months of age. He was born in the countryside and grew up playing in the barn. He encountered dogs, cows, chickens, and primarily the farmer's wife. Upon arriving at his new home (a small city house without a garden but with a courtyard), he hid behind the kitchen furniture.
After 2–3 days, he began exploring the kitchen but did not allow his owners to touch him, which was surprising since he had accepted being petted on the farm.

Wild does not eat when people are around, nor does he play with his owners, though he often stays in the same room as them. Sometimes, he seems to want to go outside, but as soon as the door is opened, he runs to hide behind the furniture.

The owners brought him in for neutering but are considering returning him to his original environment. What makes them hesitate is that Wild seems very happy in the morning when they wake up. He climbs onto the bed, circles around them while purring, and even allows them to hold him briefly.

The veterinarian diagnosed Wild with deprivation syndrome combined with a generalized inhibited anxious state.

A treatment plan was proposed, which the owners found acceptable, and a prescription for Venlafaxine was initiated. Ten granules of Effexor™ hidden in a small amount of wet food were administered once daily. Additionally, Zylkène™was given in the evening in the same manner. Slight improvements were observed quickly. Wild became more relaxed, allowed himself to be touched (though not actively seeking petting), and began exploring the entire house. On several occasions, he managed to leave his hiding spot during the evening.

After 6 months of treatment, progress plateaued. Wild lived normally in the house and courtyard but would hide whenever new people arrived. One day, his owners took him on vacation to an apartment, where he hid for the entire week. Upon returning home, they noticed a regression in his behavior. He started hiding in certain closets again, something he hadn't done for months.

After a few days, he returned to his pre-vacation behavior. The treatment was completely stopped a few months later. There was no relapse, but only the owners could easily handle him. When visitors arrived, he preferred to hide but would re-emerge after a few minutes to return to the main room. However, he continued to avoid people who tried to approach him.

8.2.7.2 Nico

Nico is a Persian cat purchased at a feline exhibition at 5 months of age. His owners take great pride in showing off his impressive pedigree and openly express their hope that he will win beauty contests in the future. Nico's breeder is well-known for presenting young cats with already thick and silky coats.
To achieve these results, the breeder keeps her champions in cages or small spaces to prevent damage to their fur. Nico is indeed stunning, but the breeder decided to sell him to reduce the number of breeding cats in her care.

In his new home, Nico remains almost motionless all day. He does not play and is even afraid of his toys. He does not explore the house and spends most of his time perched on a shelf, observing what happens in the home. When frightened—which happens often—he runs to hide, sometimes disappearing for up to 24 h.

Although he tolerates being brushed and handled, Nico isolates himself when not being held. When placed on the grass in the garden, he walks as if the grass is prickling him. He hesitates to place his paws down and returns indoors as soon as possible.

However, Nico loves traveling in his carrier, running to it whenever he has the chance. Once inside, he refuses to come out.

The veterinarian diagnosed Nico with deprivation syndrome combined with a chronic depressive state. Due to the owners' reluctance to use psychotropic medication, the veterinarian initially recommended administering *Zylkène*®, emphasizing that if no improvement was observed, they would need to consider a treatment plan tailored to Nico's chronic depression.

The owners agreed to try the treatment, but it had no noticeable effect, and they eventually discontinued it.

One year later, Nico was brought back for a consultation. He was dirty because he was no longer being brushed and had stopped grooming himself. He slept all the time and appeared unwell, with a reduced appetite. He almost entirely refused to touch and retreated to his carrier on the shelf.

The cat's unpleasant odor helped convince the owners to accept a prescription. *Clomicalm*™(5 mg) was prescribed at half a tablet per day for his 4.5 kg weight. Since Nico's lack of appetite prevented hiding the tablet in food, it was administered directly. The owners also agreed that, after the necessary shaving, Nico would be brushed daily. It was decided that therapy would begin after 2 months of treatment.

At his follow-up, Nico appeared livelier. His coat was in better condition, and his odor had improved. His appetite had slightly increased. The decision was made to remove his carrier from the shelf and encourage him to leave the shelf several times a day.

The results were unexpected. Since it was difficult to get Nico down from his hiding spot, the shelf was modified. Nico then relocated to the desk where his owner worked from home or to the back of the couch. His owners were able to pet him, which he seemed to enjoy, and he quickly adopted the desk as his new preferred spot, staying near his owner.

Over several months, a bond developed. Nico began moving around the house and settling near his owner while he worked. After a year, the treatment was discontinued, and Nico maintained his progress. He will never be an overly expressive cat, but he appears content with his calm lifestyle.

This case highlights the disconnect between the motivations of a breeder and those of an individual owner, as well as the importance of behavioral development in achieving adaptability to various living environments.

Summary of ▶ Sect. 8.2
Deprivation syndrome arises from a cat's development in environmental conditions that differ significantly from its imposed daily living environment. It is invariably associated with a pathological phobic state or, more commonly, an anxious or depressive state.

Symptoms manifest as soon as the cat leaves its developmental environment, typically upon adoption, often at the end of the socialization period around 2 months of age.

Deprivation syndrome represents a more or less severe inability to adapt to any new environment, especially if it is vastly different from the environment in which the cat developed during the first 2 months of life. The symptoms reflect the associated pathological states.

For cats living in an environment drastically different from their developmental one, complete recovery is unrealistic. Severe cases cannot improve without transitioning to an environment closer to that experienced during their early weeks of life.

Pharmacological treatment involves the use of medications that reduce fear responses and anticipation. Treatments typically last several months to years.

Environmental modifications focus on providing hiding places for the cat while gradually exposing it to phobogenic stimuli.

Behavioral therapy consists of desensitization, often combined with counter-conditioning. Results are typically limited to the individuals living with the cat and its daily environment. Any changes represent significant stressors for these cats, who risk becoming resensitized very quickly.

8.3 Post-traumatic Phobias or PTSD

8.3.1 Description

8.3.1.1 Definition

Post-traumatic phobia is a condition in which the animal exhibits fear reactions that are inappropriate and exaggerated in duration or intensity in response to a stimulus that was previously traumatic for the animal.
This condition presents similarly to categorical phobias and can share the same forms in terms of symptomatology, but it follows an *objectively identifiable traumatic event.* Depending on the associations the animal makes between the traumatic event and its contextual cues, the triggering stimuli may either expand or remain stable.

As with categorical phobias, various types can occur, and the condition is named accordingly. For example, it might be referred to as "post-traumatic intraspecific social phobia associated with an anxious state" or "post-traumatic healthcare phobia associated with a phobic state."

8.3.1.2 Etiology and Pathogenesis

The circuits involved in post-traumatic phobia are the same as those implicated in categorical phobia, as the pathological states induced by both conditions are identical: phobic, anxious, or depressive states. However, in the case of post-traumatic phobias, individual vulnerability is not a prerequisite. The animal experiences an *objective trauma*, such as electrocution, a car accident, or very painful medical procedures performed without anesthesia.

Of course, the more vulnerable an individual is to fear, the more likely they are to respond severely to the traumatic event.

8.3.1.3 Epidemiology

This condition can occur in any individual.

8.3.1.4 Presenting Complaints

The presenting complaints always involve a sudden behavioral change, and the traumatic event has often already been identified by the owners.

In cases of phobic states, the cat exhibits signs of fear when exposed to the context or stimulus linked to the trauma. In cases of anxious states, the fear generalizes through anticipation to other stimuli. Occasionally, the cat is so overwhelmed by the event that it develops an acute depressive state, which may progress to chronic depression.

The observed reactions are the same as those seen in categorical phobias, including typical fear behaviors such as inhibition, flight, aggression, or agitation.

8.3.1.5 Symptomatology

The symptomatology, as with categorical phobias, varies depending on the associated pathological state (see ▶ Sect. 8.1.1.5).

In the phobic state, the animal reacts only in the presence of the stimulus that triggered the trauma. For example, the cat may react to the sight of its carrier in cases of travel phobia, to the proximity of the veterinary clinic if it experienced very painful procedures, or to the sight of large black dogs (simple phobic state) or dogs in general (complex phobic state) if it was attacked by a large black dog.

In the anxious state, the cat anticipates the possible occurrence of the stimulus, becoming hypervigilant, even at home, at least intermittently. Owners are often unable to identify the cause of the reaction but note that it did not occur before the trauma. For example, this could involve a cat startled by the slightest noise in its home after being exposed to an exceptionally violent thunderstorm.

In the depressive state, trauma-related depression is more often acute compared to categorical phobias, sometimes progressing to chronic depression. The animal exhibits learned helplessness, as seen in cases of uncontrolled immersion. For example, a cat might refuse to leave the house, eat, or play after experiencing a street accident.

8.3.1.6 Evolution

The evolution of post-traumatic phobia can be entirely stable. For example, a cat injured in a street accident may remain indoors for the rest of its life due to fear of the outside world without developing additional fears.
However, progression can also occur in the weeks following the trauma. Anticipation and generalization processes may gradually expand the range of triggering contexts, leading the animal to develop an anxious or depressive state. Early intervention is therefore critical after the incident.

For instance, a cat attacked by another cat in its garden may initially react only to the sight of the aggressor. Over time, it may begin reacting to all cats, then refuse to go into the garden where the attack occurred (*anticipation*), and eventually isolate itself in a single room in the house.

In general, if the cat does not return to normal within a few days following the trauma, despite repeated exposure to the phobogenic stimulus under appropriate conditions, it should be assumed that desensitization will not occur easily, and intervention is necessary.

8.3.2 Diagnosis

8.3.2.1 Diagnostic Criteria

Cats with post-traumatic phobia necessarily exhibit a pathological phobic, anxious, or depressive state (◘ Fig. 8.8). A clear correlation must be established between the traumatic event and the onset of symptoms.

For most phobias, it is relatively easy to assess the animal's reaction during the consultation and determine whether the proposed therapeutic strategies are both effective and feasible.

8.3.2.2 Direct Observation

When it is possible to expose the cat to the object of the trauma, signs of intense fear are readily apparent. In cases involving medical procedures, the veterinarian can easily observe this reaction.

8.3.2.3 Comorbidities

Post-traumatic phobia is environmentally induced and can be comorbid with any behavioral disorder.

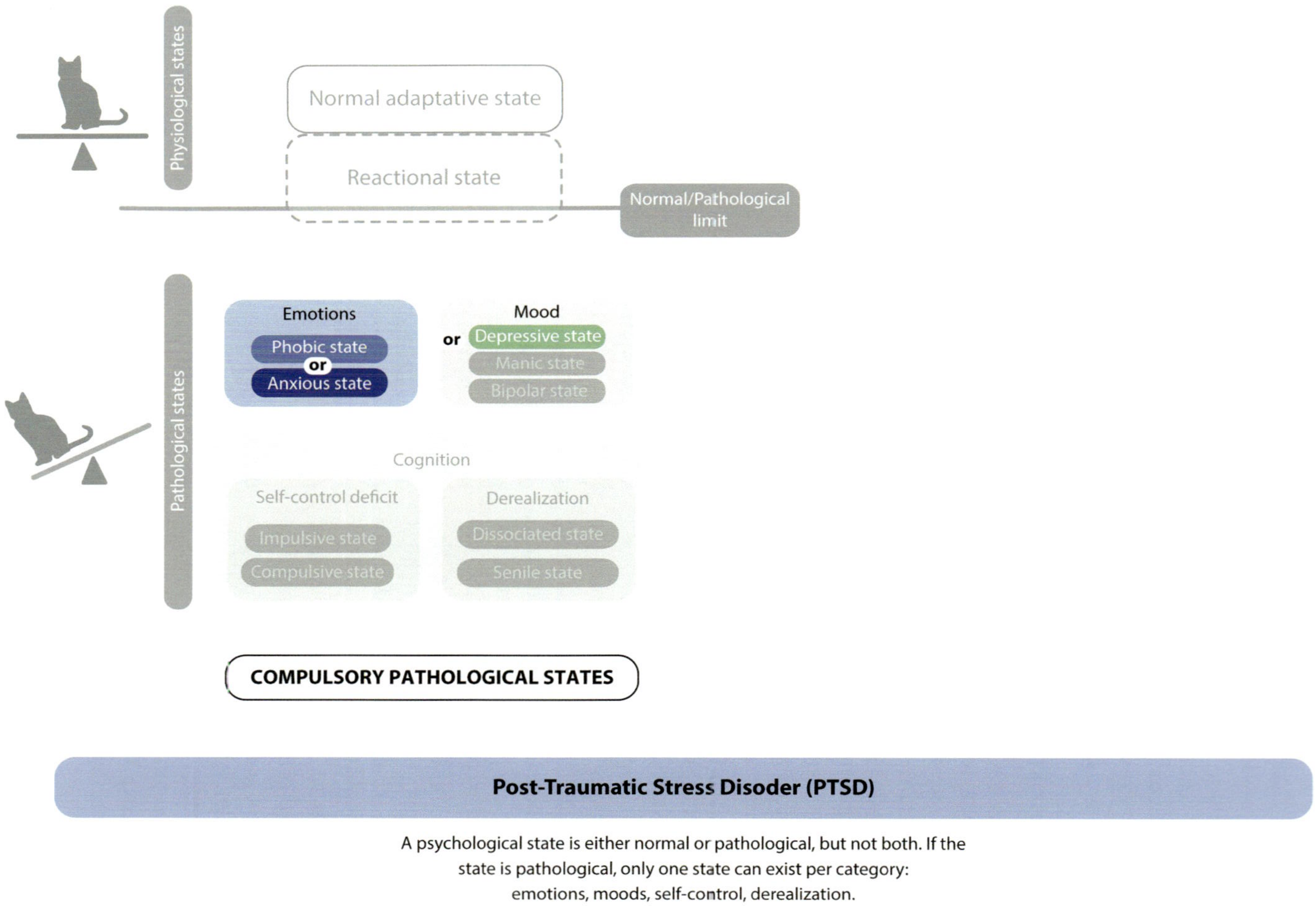

Fig. 8.8 Pathological states observed in post-traumatic phobias. Mandatory pathological states: phobic **or** anxious state, **and/or** depressive state

8.3.2.4 Differential Diagnosis

When identifying a phobic, anxious, or depressive pathological state, the practitioner must consider *categorical phobia* and *deprivation syndrome* as part of the differential diagnosis.

The identified traumatic event must be evaluated to distinguish post-traumatic phobia from categorical phobia. In post-traumatic phobia, the trauma is an objectively fear- or pain-inducing event for most individuals. In contrast, in categorical phobia, the animal demonstrates vulnerabilities from a young age and reacts excessively to events that are not typically traumatic for most individuals (e.g., an annual vaccination).

In cases of deprivation syndrome, the animal's developmental conditions are poor, and symptoms appear from an early age, without any connection to a specific traumatic event.

Finally, if no trauma can be identified and symptoms appear suddenly, the practitioner should investigate a somatic illness that might cause pain or physiological changes, explaining the abrupt behavioral shift. This is particularly important when symptoms arise in an adult or aging animal.

8.3.3 Prognosis

The severity of the condition, as with categorical phobias, depends on the animal's pathological state. The nature and intensity of the trauma are additional influencing factors correlated with the pathological state. Finally, the age at which the trauma occurred is another important consideration.

In the case of an adult cat with harmonious behavioral development before the trauma, recovery is possible, even though the memory of the trauma remains permanent. Conversely, trauma occurring in a young animal more frequently results in lasting effects and is harder to resolve.

◘ Table 8.2 outlines the factors influencing the prognosis of categorical phobias, which can be easily applied to post-traumatic phobias. It is important to note that, with identical symptoms, the prognosis for post-traumatic phobias is better than that for categorical phobias, and full recovery from post-traumatic phobia is achievable.

8.3.4 Treatments

All pharmacological treatments and behavioral therapies are similar to those used for managing categorical phobias (see ▶ Sects. 8.1.4.1 and 8.1.4.2). However, the duration of treatment required to restore emotional balance is generally shorter than in categorical phobias and can sometimes be very brief.

8.3.5 Therapeutic Follow-Up

The practitioner selects specific symptoms to evaluate in order to determine improvement or recovery and prescribes treatment for a limited duration. In the simplest cases, complete resolution is possible. Remote follow-up can validate recovery and allow for discontinuation of pharmacological treatment.

8.3.6 Prevention

Prevention is almost impossible since, by definition, post-traumatic phobias often result from accidents. Nevertheless, it is worth emphasizing that coercive methods and medical procedures performed under significant physical restraint should be avoided. Preventative measures primarily involve raising awareness about the major risks associated with these techniques (Anseeuw et al. 2006; Couture et al. 2022; Moody et al. 2020).

8.3.7 Clinical Vignettes

8.3.7.1 Boris

Boris is an adult neutered male cat. His owners sought consultation because, since burning himself on the stovetop, Boris has refused to enter the kitchen. He also fears high-

pitched chimes and hides while meowing if the oven or washing machine beeps.
Previously, Boris would often jump onto the kitchen counter. One day, he stepped on the still-hot stovetop and severely burned his paw pads. The stovetop immediately started beeping, and this sound coincided with the onset of pain. The association between the beeping and the pain created a clear phobia. Boris now has two fears: the kitchen and beeping sounds.

When he hears the oven or washing machine chime, he hides, salivates, and often experiences immediate diarrhea. His owners initially thought the problem would resolve on its own, but the situation has worsened over time. Gradually, the owners have developed their own aversion to beeping sounds, as they try to avoid triggering Boris's panic attacks.

A treatment with gabapentin was deemed sufficient for this recent trauma. However, because the beeping is difficult to anticipate, it was decided to administer a capsule systematically in the morning and evening (25 mg/kg twice daily). The beeping sounds were distinct, and Boris did not react to other types of noises.

An audio recording of the sound was made, and after two or three doses of the medication, desensitization was initiated. The chime was played in an adjacent room while Boris was being held and comforted. The results were rapid and lasting—Boris no longer reacts to the beeping sounds.

However, no treatment was implemented for the stovetop, which he continues to fear, and rightfully so. In this case, the phobic reaction is desirable as it is adaptive.

8

Summary of ▸ Sect. 8.3

Post-traumatic phobia is a condition in which the animal suddenly exhibits inappropriate and exaggerated fear reactions, either in duration or intensity, in response to one or more stimuli following an objectively identifiable traumatic event. It can occur at any age and may become apparent immediately or within a few days after the traumatic event.

The condition is invariably associated with a pathological phobic, anxious, or depressive state.

Recovery is possible, particularly if the animal exhibited normal behavior prior to the trauma and was an adult at the time of the incident.

Pharmacological treatment involves medications that target the amygdala circuit to reduce fear responses and anticipatory behaviors.

Behavioral therapies combine desensitization with counter-conditioning.

Disorders Associated with Dysregulation of the Fear Axis

Disorder's name	Subtype	State diagnosis	Diagnostic criteria	Differential diagnosis
Categorical phobias	Noise phobia	Phobic or Anxious Depressive	Pathological state linked to one or more noises (explosions, thunderstorms, household appliances, vehicles, etc.)	*Deprivation syndrome* Poor early developmental conditions *Phobie post-traumatique* Tangible and identifiable traumatic event *Self-controls axis disorders* Motor agitation not linked to fear but to excitement *Organic disease: Pain, hyperthyroidism, hypertension* these present with a truly sudden onset, and resolving the underlying condition eliminates the fear responses
	Healthcare phobia	Phobic	Pathological state linked to veterinary care or grooming	
	Transport phobia	Phobic	Pathological state linked to car travel	
	Interspecific social phobia	Phobic or Anxious Depressive	Pathological state linked to unfamiliar individuals (all or specific types)	
	Intraspecific social phobia	Phobic or Anxious Depressive	Pathological state linked to unfamiliar conspecifics (all or specific types)	
Deprivation syndrome		Phobic or Anxious Depressive	Fear reactions present from adoption Evidence of development in a hypostimulating environment Intense fear reactions to any environmental changes Learning capacities blocked by fear	*Self-controls axis disorders* Motor agitation not linked to fear but to excitement *Categorical phobias* Fear reactions limited to certain stimuli with good early developmental conditions *Post-traumatic phobias* Tangible and identifiable traumatic event Organic disease: pain, joint or cardiac malformations, hyperthyroidism
Post-traumatic phobia		Phobic or Anxious Depressive	Pathological state appears following an objectively identifiable and tangible trauma	Organic disease: pain, hyperthyroidism *Categorical phobias* No objectively traumatic event *Deprivation syndrome* Poor early developmental conditions

References

Anseeuw E, Apker C, Ayscue C, Barker L, Blair D, Brennan J, Brooks S, Case-Pall D, Caspersen H, Clark J, Colson L, Covill A, DeLong J, Dickey D, Harr K, Heine N, Krishna G, Lynch K, Maki J, Malamed R, McAuliffe M, McLucas B, Mengering C, Mulligan L, Nicholson M, Rodan I, Schobert-Nichols H, Steele A, Young R (2006) Handling cats humanely in the veterinary hospital. J Vet Behav Clin Appl Res 1. https://doi.org/10.1016/j.jveb.2006.06.003

Beata C, Beaumont-Graff E, Coll V, Cordel J, Marion M, Massal N, Marlois N, Tauzin J (2007) Effect of alpha-casozepine (Zylkene) on anxiety in cats. J Vet Behav Clin Appl Res 2. https://doi.org/10.1016/j.jveb.2007.02.002

Campbell GR, Arnott ER, Graham C, Niel L, Ward MP, Ma G (2024) Impact of early socialisation in foster care on kitten behaviour. Appl Anim Behav Sci:106306. https://doi.org/10.1016/j.applanim.2024.106306

Casey RA, Bradshaw JWS (2008) Owner compliance and clinical outcome measures for domestic cats undergoing clinical behavior therapy. J Vet Behav 3:114–124. https://doi.org/10.1016/j.jveb.2008.02.001

Couture M, Stellato AC, Moody CM, Niel L (2022) Owner perspectives of cat handling techniques used in the veterinary clinic. J Appl Anim Welf Sci:1–11. https://doi.org/10.1080/10888705.2022.2039144

Crowell-Davis SL, Murray T, de Dantas LMS (2019) Veterinary psychopharmacology, 2nd edn. Wiley-Blackwell, Hoboken

Denenberg S, Dubé MB (2018) Tools for managing feline problem behaviours: psychoactive medications. J Feline Med Surg 20:1034–1045. https://doi.org/10.1177/1098612X18806760

Eagan BH, Van Haaften K, Protopopova A (2023) Daily gabapentin improved behavior modification progress and decreased stress in shelter cats from hoarding environments in a double-blind randomized placebo-controlled clinical trial. J Am Vet Med Assoc:1–11. https://doi.org/10.2460/javma.23.01.0044

Frank D, Beauchamp G, Palestrini C (2010) Systematic review of the use of pheromones for treatment of undesirable behavior in cats and dogs. J Am Vet Med Assoc 236:1308–1316. https://doi.org/10.2460/javma.236.12.1308

Gourkow N, Fraser D (2006) The effect of housing and handling practices on the welfare, behaviour and selection of domestic cats (*Felis sylvestris catus*) by adopters in an animal shelter. Anim Welf 15:371–377. https://doi.org/10.1017/S0962728600030700

Graham C, Pearl DL, Niel L (2024) Too much too soon? Risk factors for fear behaviour in foster kittens prior to adoption. Appl Anim Behav Sci 270:106141. https://doi.org/10.1016/j.applanim.2023.106141

Hoehl S, Hellmer K, Johansson M, Gredebäck G (2017) Itsy bitsy spider…: infants react with increased arousal to spiders and snakes. Front Psychol 8:1710. https://doi.org/10.3389/fpsyg.2017.01710

Karn-Buehler J, Kuhne F (2022) Perception of stress in cats by German cat owners and influencing factors regarding veterinary care. J Feline Med Surg 24:700–708. https://doi.org/10.1177/1098612X211041307

Kirby-Madden T, Herron M (2024) Travel/carrier phobia and situational fear in a 1-year-old female spayed domestic shorthair cat: a case report. J Vet Behav 73:51–55. https://doi.org/10.1016/j.jveb.2024.03.003

Kruszka M, Graff E, Medam T, Masson S (2021) Clinical evaluation of the effects of a single oral dose of gabapentin on fear-based aggressive behaviors in cats during veterinary examinations. J Am Vet Med Assoc:1–7. https://doi.org/10.2460/javma.20.06.0307

Lai CSW, Franke TF, Gan W-B (2012) Opposite effects of fear conditioning and extinction on dendritic spine remodelling. Nature 483:87–91. https://doi.org/10.1038/nature10792

Lamminen T, Korpivaara M, Aspegrén J, Palestrini C, Overall KL (2023) Pregabalin alleviates anxiety and fear in cats during transportation and veterinary visits—a clinical field study. Animals 13:371. https://doi.org/10.3390/ani13030371

Landsberg G, Milgram B, Mougeot I, Kelly S, de Rivera C (2017) Therapeutic effects of an alpha-casozepine and L-tryptophan supplemented diet on fear and anxiety in the cat. J Feline Med Surg 19:594–602. https://doi.org/10.1177/1098612X16669399

Maier SF, Seligman ME (1976) Learned helplessness: theory and evidence. J Exp Psychol Gen. https://doi.org/10.1037/0096-3445.105.1.3

Mao S-C, Hsiao Y-H, Gean P-W (2006) Extinction training in conjunction with a partial agonist of the glycine site on the NMDA receptor erases memory trace. J Neurosci 26:8892–8899. https://doi.org/10.1523/JNEUROSCI.0365-06.2006

Mariti C, Bowen JE, Campa S, Grebe G, Sighieri C, Gazzano A (2016) Guardians' perceptions of cats' welfare and behavior regarding visiting veterinary clinics. J Appl Anim Welf Sci 19. https://doi.org/10.1080/10888705.2016.1173548

Martínez-Byer S, Hudson R, Bánszegi O, Szenczi P (2023) Effects of early social separation on the behaviour of kittens of the domestic cat. Appl Anim Behav Sci 259:105849. https://doi.org/10.1016/j.applanim.2023.105849

Masson S, Metz D, Bleuer-Elsner S, Schwobthaler F (2024) Retrospective study on the use of venlafaxine in 176 cats diagnosed with behavioral disorders. J Vet Behav. https://doi.org/10.2139/ssrn.4790331

Metz D, Medam T, Masson S (2021) Double-blind, placebo-controlled trial of venlafaxine to treat behavioural disorders in cats: a pilot study. J Feline Med Surg. https://doi.org/10.1177/1098612X211036792

8

Migues PV, Hardt O, Wu DC, Gamache K, Sacktor TC, Wang YT, Nader K (2010) PKMζ maintains memories by regulating GluR2-dependent AMPA receptor trafficking. Nat Neurosci 13:630–634. https://doi.org/10.1038/nn.2531

Mills D (2005) Management of noise fears and phobias in pets. In Pract 27:248–255. https://doi.org/10.1136/inpract.27.5.248

Montague MJ, Li G, Gandolfi B, Khan R, Aken BL, Searle SMJ, Minx P, Hillier LW, Koboldt DC, Davis BW, Driscoll CA, Barr CS, Blackistone K, Quilez J, Lorente-Galdos B, Marques-Bonet T, Alkan C, Thomas GWC, Hahn MW, Menotti-Raymond M, O'Brien SJ, Wilson RK, Lyons LA, Murphy WJ, Warren WC (2014) Comparative analysis of the domestic cat genome reveals genetic signatures underlying feline biology and domestication. Proc Natl Acad Sci 111:17230–17235. https://doi.org/10.1073/pnas.1410083111

Moody CM, Dewey CE, Niel L (2020) Cross-sectional survey of cat handling practices in veterinary clinics throughout Canada and the United States. J Am Vet Med Assoc 256:1020–1033. https://doi.org/10.2460/javma.256.9.1020

Orlando JM, Case BC, Thomson AE, Griffith E, Sherman BL (2015) Use of oral trazodone for sedation in cats: a pilot study. J Feline Med Surg 18. https://doi.org/10.1177/1098612X15587956

Potter A, Mills DS (2015) Domestic cats (felis silvestris catus) do not show signs of secure attachment to their owners. PLoS One 10:1–17. https://doi.org/10.1371/journal.pone.0135109

Powell L, Watson B, Serpell J (2023) Understanding feline feelings: an investigation of cat owners' perceptions of problematic cat behaviors. Appl Anim Behav Sci 266:106025. https://doi.org/10.1016/j.applanim.2023.106025

Raekallio M, Törmänen T, Kujala M, Vainio O (2024) Pharmacological treatment of canine and feline undesirable behaviors by Finnish veterinarians. J. Vet. Behav. 73:16–22. https://doi.org/10.1016/j.jveb.2024.04.005

Reisner IR, Houpt KA, Erb HN, Quimby FW (1994) Friendliness to humans and defensive aggression in cats: the influence of handling and paternity. Physiol Behav 55:1119–1124. https://doi.org/10.1016/0031-9384(94)90396-4

Rumpel S, LeDoux J, Zador A, Malinow R (2005) Postsynaptic receptor trafficking underlying a form of associative learning. Science 308:83–88. https://doi.org/10.1126/science.1103944

Salonen M, Vapalahti K, Tiira K, Mäki-Tanila A, Lohi H (2019) Breed differences of heritable behaviour traits in cats. Sci Rep 9:7949. https://doi.org/10.1038/s41598-019-44324-x

Smith P, Tolbert MK, Gould E, Taylor A, Knych H, Messenger K (2020) Pharmacokinetics, sedation and hemodynamic changes following the administration of oral transmucosal detomidine gel in cats. J Feline Med Surg 22:1184–1190. https://doi.org/10.1177/1098612X20917305

Spano V, Springer CM, Christensen E, Albright JD (2023) Effects of transdermal mirtazapine and oral gabapentin as pre-veterinary visit pharmaceuticals for shelter cats. J Vet Behav 64–65:47–53. https://doi.org/10.1016/j.jveb.2023.06.001

Stella J, Croney C, Buffington T (2014) Environmental factors that affect the behavior and welfare of domestic cats (Felis silvestris catus) housed in cages. Appl Anim Behav Sci 160:94–105. https://doi.org/10.1016/j.applanim.2014.08.006

Tateo A, Zappaterra M, Covella A, Padalino B (2021) Factors influencing stress and fear-related behaviour of cats during veterinary examinations. Ital J Anim Sci 20:46–58. https://doi.org/10.1080/1828051X.2020.1870175

Yamada R, Kuze-Arata S, Kiyokawa Y, Takeuchi Y (2020) Prevalence of 17 feline behavioral problems and relevant factors of each behavior in Japan. J Vet Med Sci 82:272–278. https://doi.org/10.1292/jvms.19-0519

Disorders Related to Dysregulation of the Self-Control Axis

Contents

9.1 Hypersensitivity-Hyperactivity Syndrome (HSHA) or Feline ADHD – 332
9.1.1 Description – 332
9.1.2 Diagnosis – 339
9.1.3 Prognosis – 345
9.1.4 Treatments – 346
9.1.5 Therapeutic Follow-Up – 350
9.1.6 Prevention – 351
9.1.7 Clinical Vignettes – 351

9.2 Addictions – 353
9.2.1 Description – 353
9.2.2 Diagnosis – 356
9.2.3 Prognosis – 358
9.2.4 Treatments – 358
9.2.5 Therapeutic Follow-Up – 359
9.2.6 Prevention – 359
9.2.7 Clinical Vignettes – 359

References – 361

© The Author(s), under exclusive license to Springer Nature Switzerland AG 2026
S. Bleuer-Elsner et al., *Veterinary Psychiatry of the Cat*,
https://doi.org/10.1007/978-3-032-06260-4_9

In this chapter, we describe psychiatric disorders associated with dysregulation of the self-control axis. We present the HSHA syndrome or feline ADHD and addictive behaviors.

HSHA syndrome is described with several subtypes: hypersensitive, hyperreactive, or mixed.

For each condition, the clinical presentation, diagnosis, treatment, prognosis, and follow-up are detailed, with clinical vignettes included to enhance the text.

This chapter focuses on disorders of the self-control axis. As explained in ► Chap. 2 (see ► Sect. 2.3.3), motor and emotional controls allow an animal to halt its behaviors and manage its emotions. The broader term, executive functions, includes attention and memory capacities, which are also affected by these disorders.

When the circuits governing impulse and compulsion become dysregulated, they result in pathological *impulsive* and *compulsive* states. *Impulsivity* refers to the inability to refrain from initiating a behavior, essentially acting without thinking. In contrast, *compulsion* is the inability to stop a behavior. In both cases, the brain fails to "say no."

The neurophysiological circuits involved include *bottom-up pathways,* which are activatory, and *top-down pathways,* which are inhibitory and enable decision-making based on experience. Impulsive and compulsive states stem from an imbalance between top-down and bottom-up pathways, favoring the latter.

Both activation and inhibition processes are impaired, often severely affecting learning. Motivation may appear excessive, but it can be challenging to distinguish between true excessive motivation, the effects of disproportionate reinforcement, or deficient inhibition. As with other behavioral disorders, these circuits may dysfunction at various levels, which current diagnostic tools cannot fully elucidate. However, these variations explain the differences observed in clinical presentations from one patient to another.

As discussed in ► Chaps. 2 and 4, self-control acquisition occurs from the socialization period to the end of puberty. Therefore, self-control disorders are often observable from the first months of life, such as upon adoption. By adulthood, spontaneous improvement is unlikely.

The conditions addressed in this chapter share a defining characteristic: the associated pathological state will invariably be either impulsive or compulsive. Comorbidities are frequently observed, particularly with anxious states, and more rarely with phobic or depressive states (◘ Fig. 9.3).

We describe two groups of disorders: *hypersensitivity-hyperactivity syndrome (HSHA),* also known as feline ADHD, and *addiction.* It is likely that these conditions encompass several distinct entities, which is why we refer to them as syndromes rather than specific diseases. For each, we outline different subtypes.

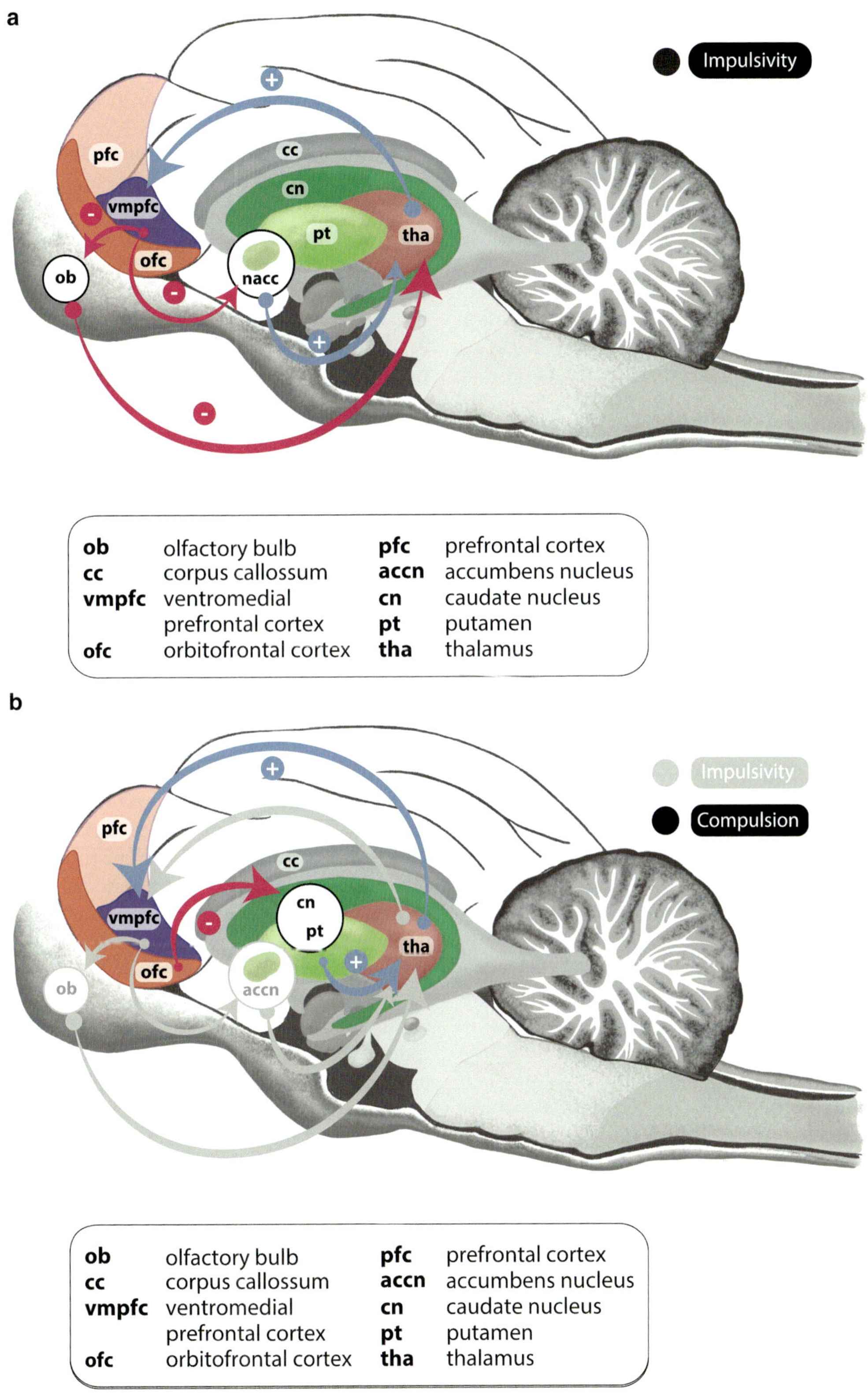

Fig. 9.1 Brain circuits of impulsivity and compulsion

9.1 Hypersensitivity-Hyperactivity Syndrome (HSHA) or Feline ADHD

9.1.1 Description

9.1.1.1 Definition

This syndrome is the feline equivalent of ADHD in humans. It is extensively described in dogs, with canine ADHD serving as a spontaneous model for the human condition (Hoppe et al. 2017; Stern and Schell 2018).

Hypersensitivity-Hyperactivity Syndrome (HSHA) is a syndrome encompassing multiple developmental disorders associated with a pathological, impulsive, or compulsive state. In these conditions, the cat produces amplified, uncontrolled, and therefore contextually inappropriate responses.

9

The responses may be exaggerated in intensity, as seen in a cat that bites the hands of people trying to play with it. This behavior represents **hyperreactivity**, where the reaction is qualitatively appropriate (in this case, play) but excessively intense and unsuitable for the context (◘ Fig. 9.5).

Responses may also occur in reaction to stimuli that are normal in the cat's environment and should not trigger a response. For example, a cat might react as soon as it sees a person move their foot, attacking as if it were a genuine invitation to play. These cats appear to "jump" from one stimulus to the next without completing a proper exploration sequence. This behavior represents *hypersensitivity,* defined as a behavioral response to stimuli that should not elicit a reaction (◘ Fig. 9.4).

All the behaviors of the animal can be affected by this malfunctioning "braking system" in the brain, characterized by pathological states: impulsive when the initiation is too rapid, or compulsive when the behavioral sequence lacks a stop signal. We will detail the forms this can take in ► Sect. 9.1.1.5. It is important to note that *emotions are also affected*. Most HSHA cats will appear "overjoyed," "overly sad," "overly frustrated," or "overly impatient." HSHA syndrome encompasses several types of dysfunctions (trigger thresholds, reaction intensity) and a wide spectrum of disorders. Some individuals are affected in a targeted way, only in specific contexts, whereas others exhibit a more severe condition that affects all behaviors. In its most *generalized forms*, the disorder is highly invasive and debilitating. These cats are challenging for owners to manage, and without appropriate care, the risk of abandonment or euthanasia is significant.

9.1.1.2 Etiology and Pathogenesis

In humans, it is established that the three symptoms of ADHD (*inattention, hyperactivity, and impulsivity*) originate from abnormalities in circuits involving the prefrontal cortex (Arnsten 2009). The dorsolateral and inferior regions regulate attention, cognitive control, and inhibition, while the orbital and ventromedial structures regulate motivation and affect (Arnsten and Rubia 2012).

In 2017, a longitudinal study conducted on more than 1700 children and adults diagnosed with ADHD, compared to 1500 non-affected individuals, revealed that the sizes of the amygdala, hippocampus, and nucleus accumbens also differ in ADHD subjects. These structures, like the prefrontal cortex, exhibit delayed maturation (Hoogman et al. 2017). It has also been demonstrated that alterations in these circuits are associated with changes in neurotransmitters, particularly dopamine and norepinephrine, in specific areas of the prefrontal cortex (Prince 2008) (◘ Fig. 9.2). More recently, researchers have highlighted the role of glutamate in this neurodevelopmental disorder, noting an increase in glutamate in the anterior cingulate cortex (Bauer et al. 2018). Lastly, familial studies have shown that ADHD is a disorder with a strong genetic component (Anita 2009; Zayats and Neale 2019).

In both cats and dogs, we lack imaging data that would allow for such studies, but the observed similarities in symptoms, progression, and treatments strongly support *mechanisms very similar to those seen in humans.*

One significant difference lies in early environmental conditions, which are more extreme

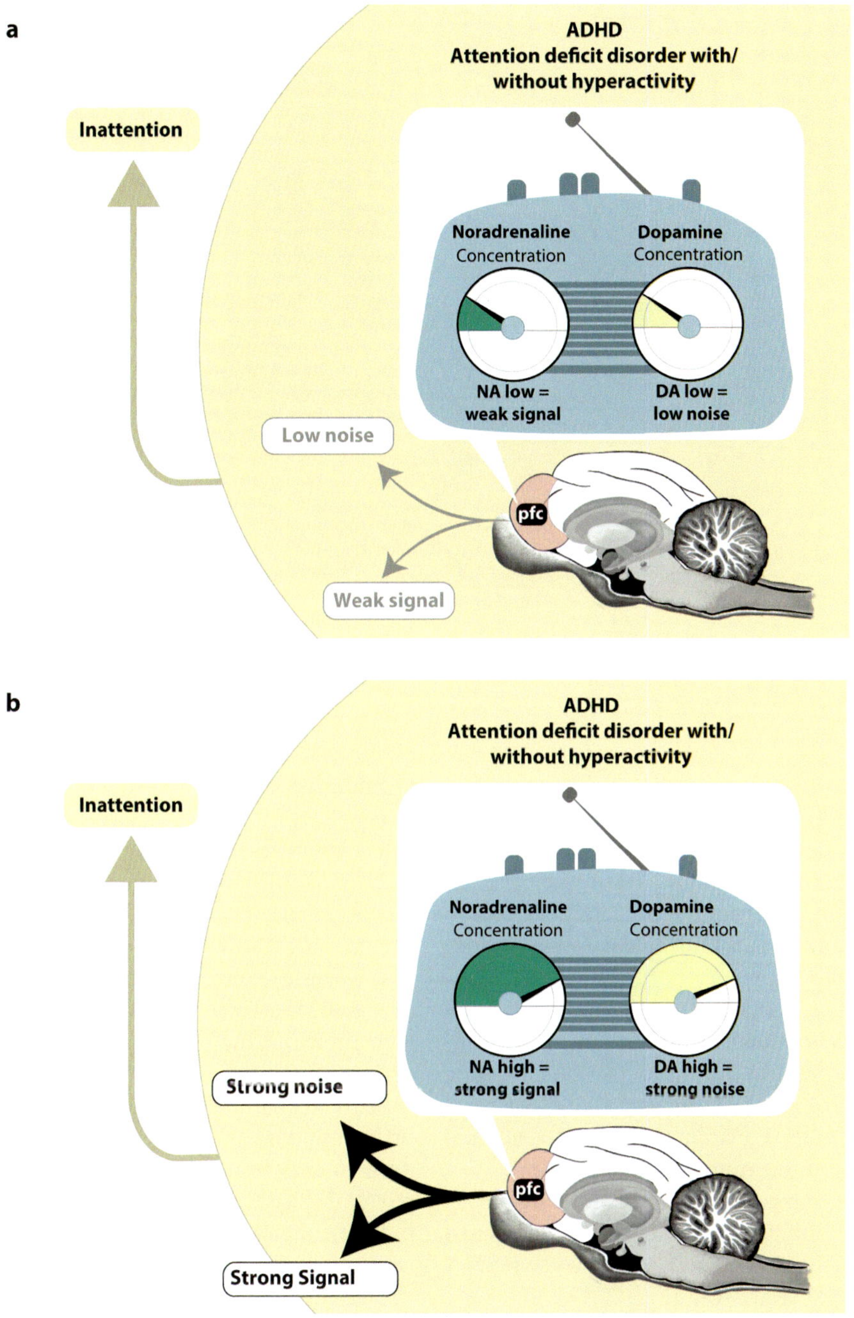

Fig. 9.2 Role of dopamine and norepinephrine in ADHD (Adapted from Stahl). (**a**) A low level of norepinephrine does not support concentration, despite a low level of dopamine, resulting in minimal disruption (or "noise"). (**b**) A high level of norepinephrine generates a strong signal, but concomitant high dopamine levels produce significant noise, impairing concentration. (**c**) Balanced levels of norepinephrine and dopamine optimize the signal while controlling noise, enabling concentration and calmness

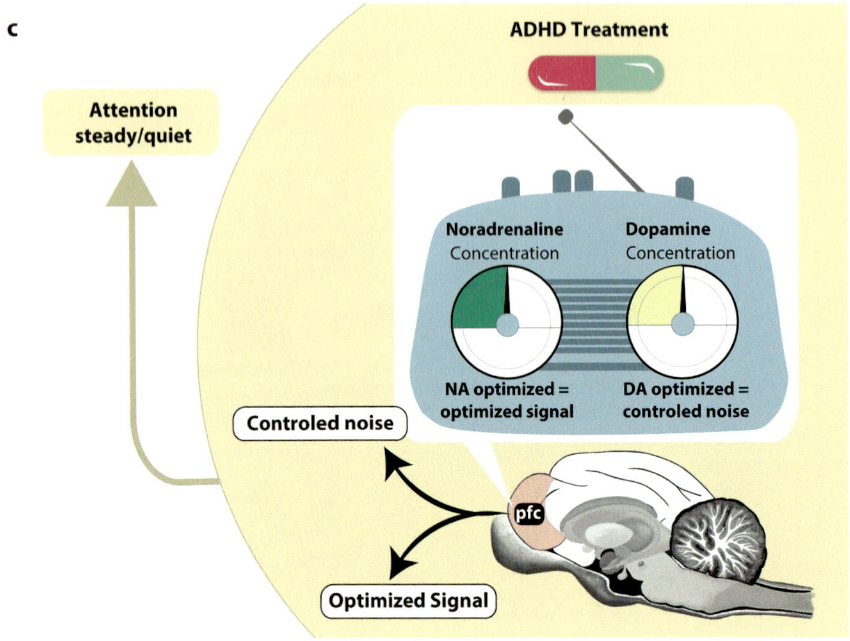

Fig. 9.2 (continued)

in animals than in humans, both genetically and developmentally. When these conditions involve a socialization period without sufficient maternal oversight until the age of 14 weeks, the disorder is more likely to develop (Ahola et al. 2017). During this same period, the acquisition of sensory filtering determines the baseline threshold for behavioral reactions. Development in understimulating conditions promotes the onset of the disorder: the triggering threshold for all reactions is lowered, and the animal reacts to normal environmental stimuli (hypersensitivity).

The strong genetic origin is undeniable, as the vulnerability genes identified in humans have also been found in dogs with HSHA (Hejjas et al. 2007; Lit et al. 2013), and they likely exist in cats as well.

9.1.1.3 Epidemiology

As with other animal psychiatric disorders, we lack large-scale studies to determine the prevalence of HSHA syndrome in cats.

All cats can potentially develop HSHA syndrome, particularly if raised in understimulating environments without the guidance of adult cats to acquire self-control. Early separation of mothers from their kittens is a contributing factor (Ahola et al. 2017). However, even when the mother is present, her interventions are not always appropriate. Some mothers are overly tolerant and ineffective educators, which can also promote the emergence of HSHA syndrome.

It is worth noting, however, that this disorder appears to be significantly less common in cats than in dogs. This difference is partly due to the distinct selection processes between the two species: in dogs, the pursuit of performance traits for work has led to the selection of individuals highly predisposed to this disorder (Fadel et al. 2016). In contrast, selection in cats has focused more on producing good companion animals. Based on clinical experience, the prevalence of this disorder in cats seems significantly lower than the 3–5% reported for dogs and humans (Faraone et al. 2003; Polanczyk et al. 2007).

9.1.1.4 Presenting Complaints

The timing of consultation varies depending on the severity of the disorder. Severe cases often seek consultation during the juvenile period, as the kitten is already very difficult to live with. However, for most cats, the first consultation tends to occur at *the end of puberty or the beginning of adulthood.* Symptoms are typically present from the time of adoption, but consultations happen later for several reasons. By the end of the juvenile period, tolerance toward the animal decreases. Additionally, injuries sustained during play become more significant. Owners often attempt to control these "attacks" through reprimands, which rapidly deteriorates the relationship.

Hyperactive cats often struggle to structure their territory and are frequently more attached to their owners because they do not find significant comfort in their environment. This often leads to the development of an intense bond, and attacks occurring in adulthood are perceived as unpredictable.

Similarly, the damage caused by the kitten during play, on furniture or curtains, becomes intolerable as it escalates with the animal's growth. Reprimands in such cases also typically lead to further deterioration in the relationship with the owner.

The presenting complaints are numerous but almost always revolve around the fact that these cats are *exhausting for their owners.*

The most common reasons for consultation include the damage and injuries caused by the cat, stemming from motor agitation and excessive reactions such as jumping, exaggerated play, or the inability to interact with the cat without chaos ensuing. Owners often express frustration at not being able to simply pet their cat. In this disorder, the behaviors observed are often normal for the species but are remarkable due to their intensity or frequency. As a result, it can be difficult for owners to initially recognize the "abnormal" nature of the observed behavior.

Aggression can also be a potential reason for consultation, as the *injurious sequences* caused during *excitement* and *play* are sometimes misinterpreted as aggression by the owners. As the cat grows, it becomes harder to control, and the injuries can be significant, even though the animal has no intention of causing harm. It is important to quickly differentiate these injurious sequences from genuine aggression.

However, HSHA cats can also exhibit true *irritation-related aggression*, often triggered by the *fear of reprimands.* These irritation-based aggressive behaviors frequently emerge as the cat gains weight with age.

In multi-pet households, play sequences with other animals can sometimes escalate into fights during this same period, due to the exuberant behaviors of the HSHA kitten, which eventually angers its companions. Fights with other cats in the household can, therefore, be another reason for consultation. In some cases, this can secondarily lead to intra- or interspecific relational disorders.

It is important to note that many HSHA cats, despite having brains with dysfunctional inhibitory and control circuits, are capable of *improvement and learning*. In fact, some are even advantaged in learning due to their high levels of motivation. They are tireless and try harder than average. Their exuberant enthusiasm often earns them the nickname "dog-cat" from their owners, as they are capable of behaviors uncommon in most cats, such as spontaneously retrieving toys or following their owners on walks.

Depending on their level of attention, they can learn quickly and effectively in *non-stimulating environments*, although they struggle significantly more in stimulating contexts.

9.1.1.5 Symptomatology

The symptomatology depends on the type of HSHA, and as this syndrome can manifest at varying levels of severity, the symptoms will also be more or less troubling for owners. The associated pathological states include the *impulsive state* and the *compulsive state,* which determine different levels of severity and distinct symptoms. The compulsive state, characterized by the absence of a stop signal, always indicates a poorer prognosis and should therefore raise concern promptly.

Due to the relational difficulties induced by the roughness associated with this disorder, which often leads to reprimands, and also because of the cat's significant genetic vulnerability related to the fear axis, an *anxious state* is frequently observed secondarily in adulthood.

It is important to note that not all behaviors are necessarily affected, and the symptomatology described here for each behavior may only be partially present in any given individual.

In kittens, their constant activity is often noted, and owners frequently feel as though no amount of play is enough to satisfy their pet.

In young cats, the acquisition of *control over biting, scratching,* and *locomotion* develops gradually from the socialization period to early adulthood, as discussed in ▶ Chap. 4, in parallel with the *maturation of the inhibitory system* in the prefrontal cortex. Logically, when the inhibitory system does not develop optimally, the consequences are apparent. Kittens tend to struggle with playing without biting or scratching forcefully. They have much greater difficulty acquiring control over their jaws and claws. These kittens exhibit either an *impulsive pathological state* or a *compulsive pathological state,* making it challenging for them to inhibit their actions. Their excitement during play results in *rough interactions.* Owners often describe a kitten that they are afraid to touch and are unable to pet. This can also manifest as a tendency to "attack people" as they move around the house. While this behavior is common in all kittens and typically limited to mock-charging, an affected kitten may escalate to biting. Consequently, it is not uncommon for these turbulent kittens to make the household's children cry, as their attempts to play are met with frequent biting or scratching.

Regarding *exploration* in kittens, it is worth noting that, unlike puppies, kittens do not explore their environment with their mouths but rather through olfactory investigation. However, HSHA kittens often exhibit *oral exploration*, which can persist into adulthood. The intensity of this oral exploration, combined with a compulsive state, may lead these animals to *ingest non-edible items* such as elastic bands, hair ribbons, threads, and strings (Bradshaw et al. 1997). Such ingestions are a severe symptom and warrant veterinary consultation.

It is important to differentiate between the ingestion of genuinely *non-edible items*, such as an elastic band, and objects that may have some food-related association, such as a sausage string or food packaging.

This abnormal oral exploration, when associated with an impulsive state, often results in destructive behavior. Periods of excitement frequently culminate in scratching various household fabrics. Many of these cats learn that scratching the sofa or curtains is an effective way to capture their owners' attention. These scratching behaviors are generally unrelated to marking.

Feeding behavior can also be affected. The impulsive state leads to gluttonous eating behaviors, with kittens swallowing kibble so quickly that it appears they are "inhaling" it. For these cats, owners may have already tried *anti-gobble bowls* or *"food puzzles."* They might have also switched to a schedule of two or three meals per day in an attempt to control the rapid intake. However, this approach exacerbates the issue, as a cat's natural feeding pattern requires the ability to nibble throughout the day.

One consequence of this meal-schedule approach in a cat with unregulated appetite is the emergence of attacks driven by frustration. A hunger-stimulated cat may position itself to attack its owner, often around feeding times. These attacks can sometimes escalate to violent behavior.

Clinical Illustration

Joddy is a very active 5-month-old kitten who constantly demands food. His owners eventually found a solution by placing his kibble in a plastic bottle with a narrow neck. To get a piece of kibble, Joddy jumps on the bottle, and the centrifugal force releases one or two pieces. However, when the bottle is empty, Joddy becomes restless and starts meowing. If his owners do not refill the bottle, his exaggerated frustration drives him to attack their legs, scratching and biting them. Despite the environmental enrichment provided, his impulsive state requires treatment.

Sometimes, the cat eats so quickly that it regurgitates immediately afterward and then consumes what it has just regurgitated. In compulsive states, a complete absence of a stop signal is observed, resulting in a *lack of satiety*. These cats can be given multiple meals and will eat as if they had not been fed before.

Sudden jumps or other *abrupt movements* during contact are also common presenting complaints. HSHA cats are more motivated than average, and their behavior reflects this excessive motivation in all interactions. If they have access to the outdoors, these kittens frequently put their lives at risk.

Sleep is often reduced in HSHA cats, with affected individuals sleeping less than 10 h a day, whereas the normal sleep duration for cats averages 15 h per 24-h period (Yamazaki et al. 2020). The quality of sleep is not necessarily altered, but when the number of sleep hours drops below eight, or even six, per day, the kittens become so exhausted that they often fall into a deep sleep. Sleep can overtake them suddenly in the middle of an activity, without giving them time to settle down.

Hyposomnia may become more apparent during weekends when the family is home, as the cat is unable to isolate itself to rest while there is movement in the house. These cats often wake their owners at night, attempting to play with their feet, and are quickly locked out of the bedroom because people struggle to sleep in their presence. Hyposomnia is always a high-severity symptom because it significantly interferes with the consolidation of information gathered throughout the day.

Interactions, whether with humans or other animals, can be highly motivated in hyperactive cats, often leading to interactions that escalate. These cats tend to be demanding, persistently seeking petting, play, and interaction, but these interactions frequently spiral into excessive behaviors involving biting or scratching. With their companions, they often continue playing even after being threatened. Like other behaviors, hyperactive cats give the impression that it takes a strong reprimand to get their attention. This symptom often leads to increasingly severe punishments from humans.

Adult cats do not typically react to these mock-aggressions during play until the hyperactive cat matures. The scolding owners administer to punish these play-related "attacks" often push the animal toward anxiety. This anxiety develops even more quickly in hyperactive cats because they struggle to structure their territory. Notably, they are unable to establish a soothing isolation zone. Unlike what is observed in normal young cats, HSHA cats do not establish calming rituals and do not live in a predictable, structured environment.

Various *productive behaviors* may also appear as part of the deficit in self-control, including spinning, licking people or objects, and seeking constant contact. While these behaviors may initially be well tolerated by the owner, they eventually seem abnormal.

All these behaviors must be considered within the broader clinical picture to determine if they stem from a regulation deficit. An isolated behavioral sequence does not allow for a conclusive diagnosis.

The HSHA cat *is not inherently aggressive.* Aggression may develop following punishments or in the presence of a comorbid condition, but initially, these cats have primarily positive intentions. They are perpetual kittens who view life as one big game. However, due to their roughness, they can unintentionally cause injuries and struggle with physical constraints, as such situations demand immobility. They often poorly tolerate frustration and have difficulty staying still when asked to wait, but their response tends to manifest as motor agitation or excessive meowing rather than threats.

On the other hand, when an *anxious state* is added to the initial clinical picture, irritation-related aggression may emerge. These episodes are uncontrolled and, therefore, severe and dangerous. Anxiety is most often induced by coercive communication that develops because the owners feel overwhelmed.

When anticipation is involved, the diagnosis becomes more challenging. The cat initiates a play sequence but immediately becomes aggressive because it anticipates punishment. As a result, it appears motivated by aggression.

9.1.1.6 Evolution

The nature of the disorder, which involves delayed maturation of the *prefrontal cortex*, *amygdala*, and *hippocampus*, explains the associated clinical signs. While maturation is delayed, it does occur, leading to improvements in the cat's ability to control and inhibit behaviors as it transitions from adoption to adulthood.

In humans, the notion that ADHD resolves spontaneously by adulthood was long perpetuated. However, advances in functional imaging have shown that ADHD still involves different brain functioning even in adults. Moreover, while motor hyperactivity often decreases to near-normal levels in adulthood, issues with inattention, concentration, and emotional sensitivity persist.

In cats, motor agitation, oral exploration, and motor control partially improve with educational efforts. However, without comprehensive management (medical intervention and re-education focused on self-control), the likelihood of developing an anxious state exceeds that of spontaneous recovery.

Two factors are critical in determining the course of the disorder: the *severity of the condition* and the use of *coercive methods* to "control" these cats.

The more severe and pervasive the disorder, the more difficult the cat is for owners to handle. Abnormal behaviors often lead to *punishment*, further *deteriorating the relationship*. These cats' territories will never be entirely calming or structured, as their pheromonal marking is often disorganized.

When pharmacological treatment is implemented, improvements are rapid and significant in the vast majority of cases. This treatment enhances sleep quality, which in turn improves memory and learning processes. For the most severe cases, medical treatment may exceptionally need to be maintained for life. As with all psychiatric or somatic conditions, it is essential to evaluate the quality of life of the cat (and its owners) with and without treatment to decide on the best course of action. However, unlike what is observed in dogs and humans, the small proportion of cats requiring lifelong treatment suggests that cortical maturation is sufficient to allow a normal life after discontinuing medication.

The second key factor in the evolution of the condition is the use of *coercive methods*. These hyperactive cats, who do not meet their owners' expectations of a typical cat, often push their owners to the brink. It is easy to understand how repeated failures and frequent injuries caused during rough play lead owners to try anything. As a result, HSHA cats are often punished, sometimes harshly.

However, these undesirable behaviors are not intentional. Despite reprimands, progress is slow, and owners often resort to avoiding situations that could escalate. It is common to see a cat engaging in misbehavior, anticipating punishment, and fleeing immediately after, even though avoiding the behavior in the first place would prevent the issue. Unfortunately, the cat is incapable of such restraint.

This punitive factor is a *major aggravating factor*, as it almost invariably leads to an *anxious state*. The animal is punished for an action it cannot help but perform. The emergence of aggressive sequences linked to the fear of punishment is common in this transition to an anxious state and significantly worsens the prognosis.

In contrast, when owners remain patient and kind, the outcome after treatment is generally positive, even though recovery may not be possible in severe cases. Achieving a highly acceptable quality of life for both the cat and the owner is entirely realistic.

The evolution of the condition is therefore highly variable: it ranges from mild cases, which recover with behavioral re-education focused on self-control and environmental enrichment, to severe cases exacerbated by physical punishment, with a high level of danger and much more challenging management.

For all these reasons, as with all neurodevelopmental disorders, it is recommended to establish the diagnosis as early as possible and to begin veterinary management promptly.

Important Point of View

Breed and HSHA

We know that there are genetic predispositions to this condition. It is likely that certain cat breeds are more vulnerable. However, unlike the selection process in dogs, which aimed to create working lines capable of near-limitless endurance, cats have not undergone such intense active selection based on behavioral traits. Consequently, these trends are far less pronounced in cats than in dogs.

Regardless, raising awareness among veterinarians, breeders, and trainers will give cats a better chance of achieving balance in adulthood. By doing so, we can promote early, comprehensive, and high-quality management of this condition.

9.1.2 Diagnosis

9.1.2.1 Diagnostic Criteria

Cats with HSHA syndrome necessarily exhibit an *impulsive pathological state* or a *compulsive pathological state* (◘ Fig. 9.3). Additionally, their symptoms are present from a very young age, as this is a neurodevelopmental disorder.

The diagnosis is based on the presence of several symptoms from the following list:

- delayed acquisition of bite inhibition and claw retraction control
- oral exploration, potentially leading to the ingestion of non-food items
- delayed acquisition of locomotion control, expressed as visible clumsiness
- absence or deficit of food satiety
- rough play, with difficulty controlling excitement, often escalating to scratching or biting
- exaggerated demands for interaction
- hyposomnia not associated with an anxious state
- disorganized exploration
- disordered pheromonal marking, with difficulty structuring territorial fields

The number of symptoms present from the list above indicates the severity of the disorder. An isolated criterion is never sufficient to establish the diagnosis. Therapeutic decisions guiding management are based on identifying irreversible suffering in the animal and its inability to adapt.

In human medicine, the symptoms of ADHD are classified into *hyperactivity, impulsivity*, and *inattention*. Hyperactivity describes uncontrolled and incessant motor restlessness. Impulsivity refers to the inability to control one's actions, think before acting, and regulate emotions. Finally, inattention describes the inability to focus on a task and the tendency to be easily distracted. This classification results in three types of ADHD in humans: Type 1 is primarily hyperactive-impulsive, Type 2 is primarily inattentive, and Type 3 is mixed.

In cats, we also observe different forms that deserve specific descriptions. These include the *hypersensitive form (HS)*, the *hyperreactive form (HA)*, and the *mixed form (HSHA)*.

During diagnosis, the practitioner will be able to identify the specific form of HSHA syndrome in their patient, which will lead to different approaches in both medical treat-

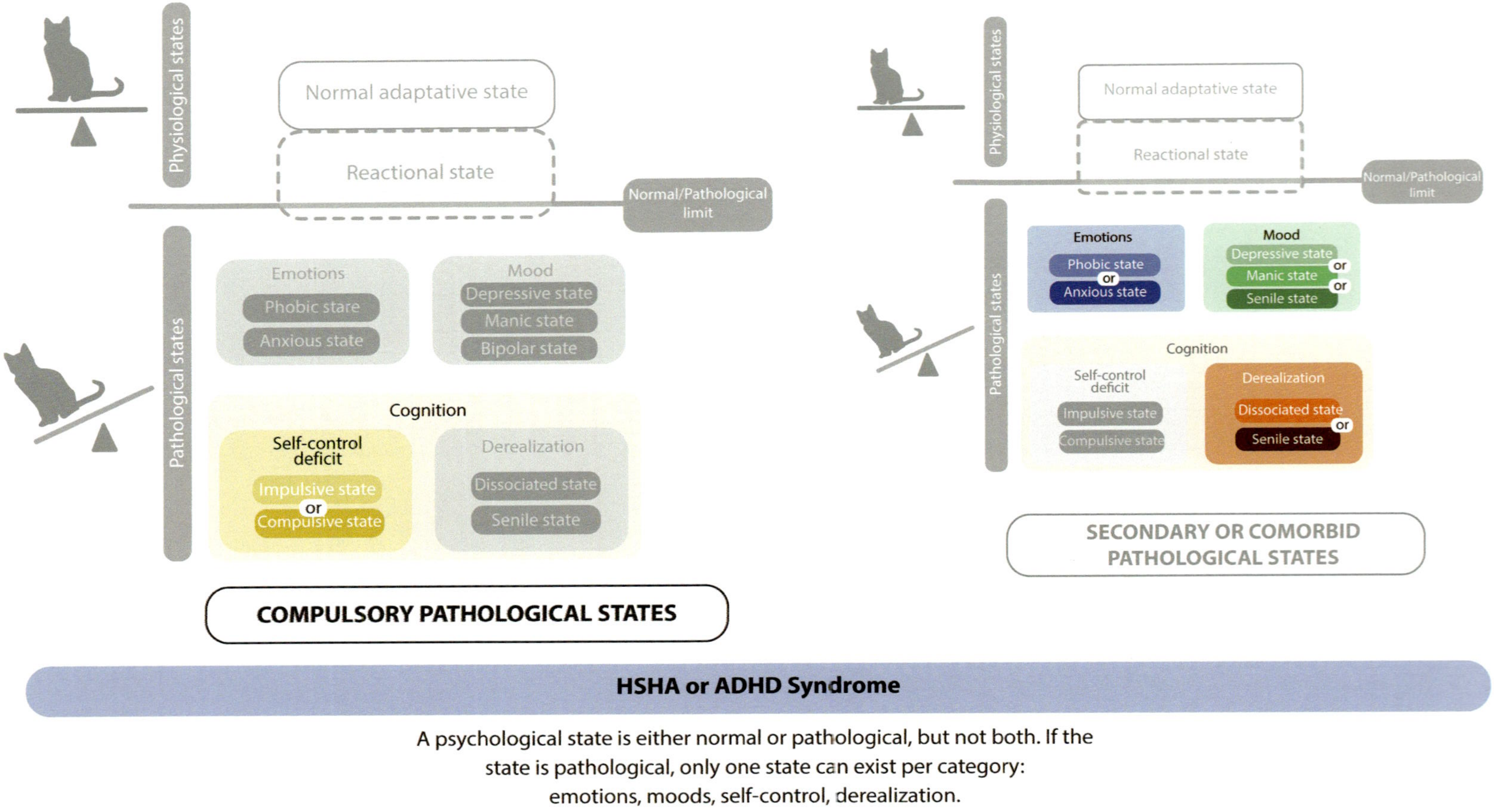

Fig. 9.3 Pathological states associated with HSHA syndrome. Mandatory pathological state: impulsive **or** compulsive state. Possible comorbid pathological states: phobic **or** anxious state, **and/or** depressive or manic state, bipolar state, **and/or** dissociative state **and/or** senile state

ment and *behavioral therapy*. A cat diagnosed with HSHA syndrome may therefore be classified as having the *mixed form (HSHA)*, while another may be classified as having the *HS form* or the *HA form*.

▪ Motor Agitation Related to HSHA Syndrome

Hypermotricity refers to exaggerated locomotor activity, meaning that the cat is almost constantly in motion.

This hypermotricity can be anxiety related and is often accompanied by neurovegetative manifestations such as tachypnea, hypervigilance, or appeasement signals. The animal moves because it is monitoring and anticipating the arrival of a potential danger. In such cases, we prefer to use the terms *hypervigilance* and *agitation*, which clarify the emotion underlying the hypermotricity.

Hypermotricity can also be observed in an HSHA cat reacting to all stimuli in a hypersensitive form. At the extreme, if no stimulus is present, its exploration becomes so intense and disorganized that it creates disorder, which in turn becomes a stimulus. In such cases, we prefer to use the term *motor agitation generated by hypersensitivity*.

Finally, hypermotricity can be observed in an HSHA cat exhibiting hyperreactivity. For example, a cat that enjoys playing hide-and-seek with its owner may initiate endless running episodes even when the owner is no longer participating in the interaction. In this case, we use the term *motor agitation generated by hyperreactivity*.

▪ The Hypersensitivity (HS) Form of HSHA Syndrome

In the hypersensitive form, the cat exhibits an excessively low reaction threshold, which may affect some or all of its behaviors. This leads to incomplete behavioral sequences, interrupted by the appetitive phase of the next sequence (◘ Fig. 9.4). The animal reacts to non-salient stimuli or stimuli that are continuously present

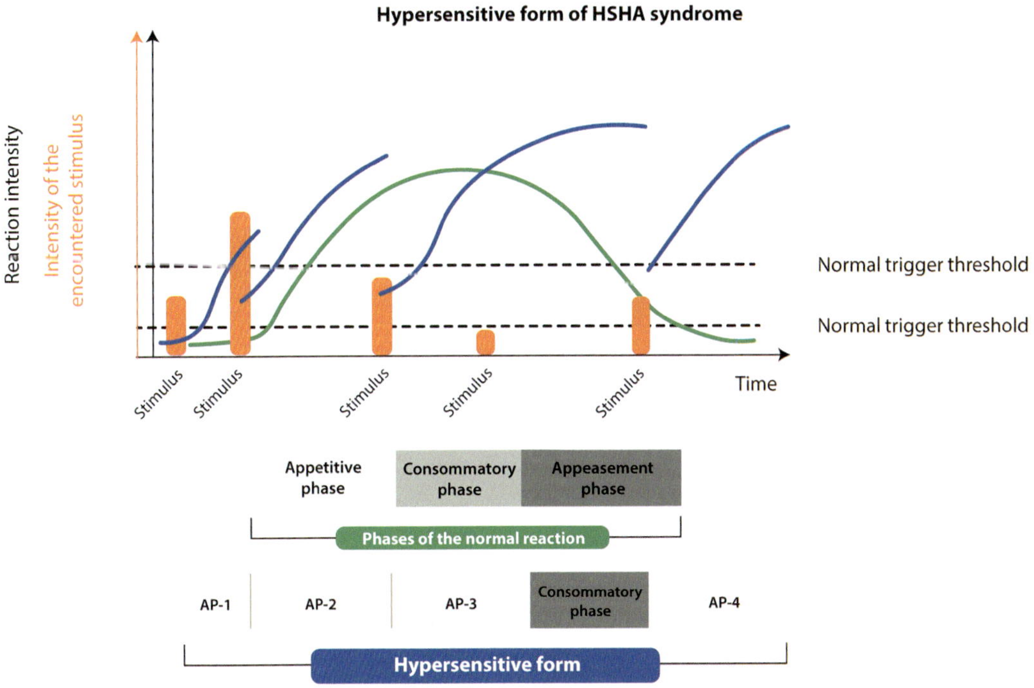

◘ **Fig. 9.4** Behavioral reactions in the hypersensitivity form of HSHA syndrome. Normal environmental stimuli trigger reactions. The consummatory phase is incomplete and leads to a new appetitive phase rather than a stopping phase. This results in a "zapping" effect in the cat's reactions, causing it to shift from one stimulus to another without completing behavioral sequences in its environment. There is also a reappearance of an appetitive phase before the completion of the previous behavioral sequence.

Typically, these are cats that attempt to create movement when none is present by pushing objects with their paws or knocking them over to provoke a reaction to the ensuing motion.

- **The Hyperreactivity (HA) Form of HSHA Syndrome**

In the hyperreactivity form, the cat exhibits a reaction that is appropriate to the context and the encountered stimulus, but with an intensity that is excessively high (Fig. 9.5). For example, these cats do not play more often than other cats, but when they do engage in play, they use their claws or bite, causing injury because the intensity they bring to the activity is excessive.

- **The Mixed Form of HSHA Syndrome**

The concurrent presence of the HS and HA forms is common, with the relative proportions of each form varying between individuals (Fig. 9.6).

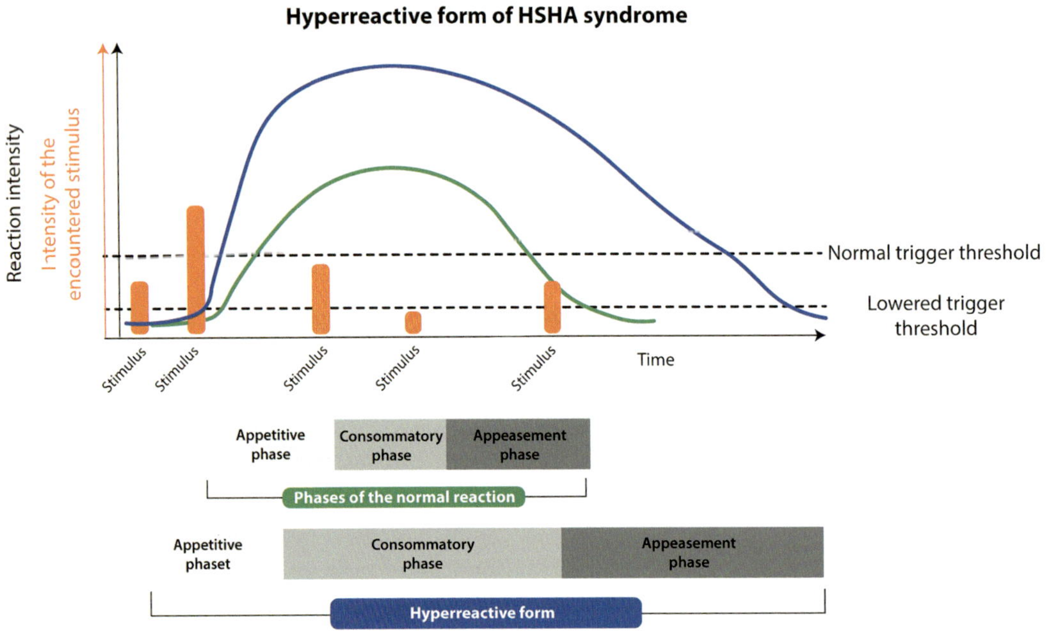

Fig. 9.5 Behavioral Reactions in the Hyperreactivity Form of HSHA Syndrome. Reactions are triggered by relevant stimuli, but the intensity of these reactions is significantly heightened. Consequently, both the consummatory phase and the calming phase are prolonged

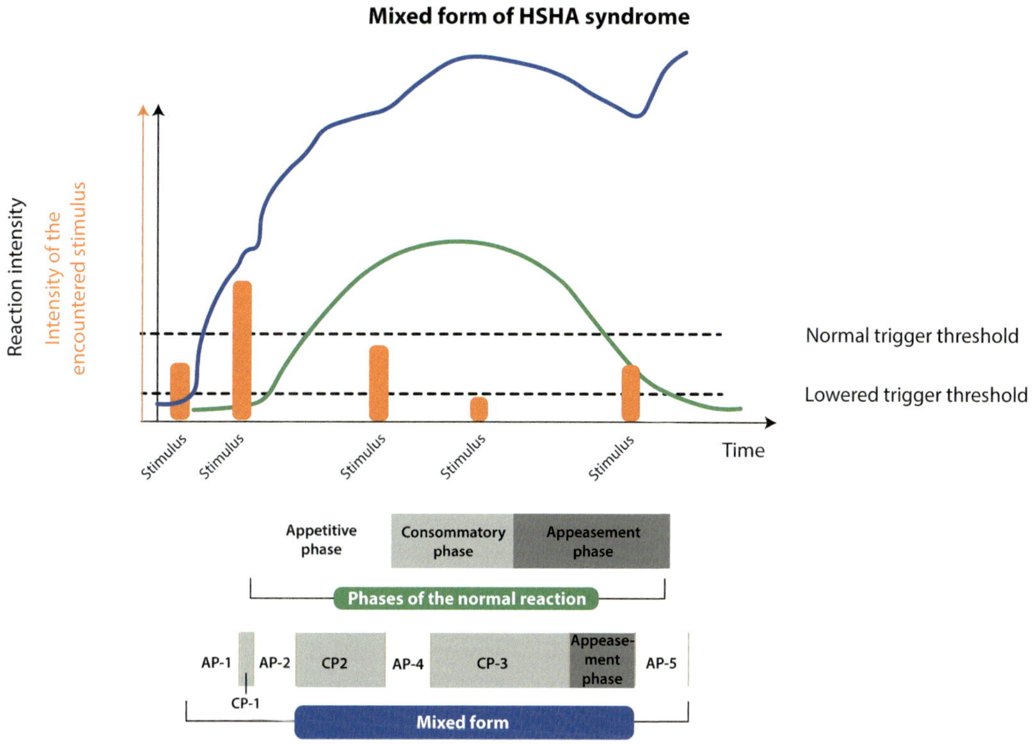

Fig. 9.6 Behavioral reactions in the mixed form of HSHA syndrome. Reactions are triggered by normal environmental stimuli (hypersensitivity), and the intensity of behavioral reactions is heightened (hyperreactivity). This results in significant behavioral agitation and markedly shortened calming phases

Important Point of View

Attention Tests in Humans

In human medicine, attention is measured using various tests that rely on language. These tests are used to detect and diagnose ADHD in children. The complex exercises involved are not easily translatable to animals.

The *Stroop test* measures selective attention. It involves asking participants to name the color in which a word is written, rather than reading the word itself. For example, the word "blue" might be written in orange, and the expected response would be "orange."

The *n-back test,* on the other hand, is a cognitive task that can be monitored through fMRI in the dorsolateral prefrontal cortex. It measures sustained attention, working memory, and problem-solving. During this test, participants are presented with a sequence of stimuli, and their task is to indicate when the current stimulus matches the one presented **n** steps earlier in the sequence. The **n** factor can be adjusted to make the task easier or more difficult.

Problems with activation in this part of the brain can be associated with ADHD, as well as schizophrenia, manic and depressive states, anxiety, pain, and sleep disorders.

In dogs, this attention deficit is also present and negatively impacts learning. In contrast, in cats, an attention deficit, if it exists, does not pose significant problems in daily life. For this reason, at this stage, we do not distinguish between cases with and without attention deficits, as the management approach remains the same.

9.1.2.2 Direct Observation

During consultation, *exploration* is the first notable feature. An HSHA cat tends to explore the room in a *disorganized* manner, venturing into areas that are not typically visited by other cats. These disorganized exploration sequences are far more characteristic than hypermotricity because their analysis reveals that they are incomplete. The cat may circle the room multiple times, sniffing numerous spots, and then return to them as if it had never investigated them, which is not typical exploratory behavior for a cat entering a new environment.

This type of cat is often excessively friendly and leaves its carrier much more readily than other cats. It appears abnormally comfortable in a space that should feel unfamiliar, as it lacks the cat's own pheromonal markings. The cat willingly jumps onto the practitioner's desk and seems unimpressed by the environmental change.

In hypersensitive forms, the cat reacts to the slightest noise or movement, such as a pencil falling, a chair moving, or a flying insect.

During the clinical examination, the cat is difficult to handle because it moves constantly. It is not aggressive (initially) but dislikes being immobilized or physically restrained, wriggling persistently to escape the physical constraint.

9.1.2.3 Comorbidities

The most common comorbidity is likely **categorical phobia** and its progression to **anxiety**. When the sensory filter is impaired, it often affects not only the self-control axis but also the fear axis. This comorbidity is especially frequent because the socialization period (2–8 weeks) and the period of acquiring self-control (2–16 weeks) are almost synchronous in cats. This overlap can lead to challenging clinical scenarios where, once the cat becomes accustomed to its owners, its exaggerated and overly rough play demands become intolerable.

When HSHA syndrome is associated with **deprivation syndrome,** the initial clinical presentation is dominated by the symptoms of deprivation syndrome, which are often severe. Once behavioral inhibition is reduced, impulsivity may emerge. The combination of fear-related disorders and HSHA syndrome increases the likelihood of aggressive reactions, particularly during episodes of social phobias.

However, this association can also have advantages. The innate curiosity of HSHA cats may counterbalance the behavioral inhibition associated with the fear response, helping the cat to explore its environment more quickly.

Conversely, this association can also be a handicap because the exploratory behavior of an HSHA cat is abnormal. It does not take the time to thoroughly explore and poorly analyzes its environment. It can be surprising to observe categorical phobias in a hyperactive cat, suggesting it has been subjected to sensory isolation, even though its owners have, on the contrary, provided a rich environment. Unfortunately, the affected cat is likely to fail at discovering its surroundings. Consequently, it also struggles to organize its territory effectively. Furthermore, frequent *hyposomnia* interferes with memory consolidation.

During encounters, the HSHA cat is often clumsy and poorly received by its peers. Its motor agitation leads to injuries and noisy events that make the environment hostile. This cat tends to focus on dangers, overlooking the finer details of its environment. It ends up in a situation similar to a cat experiencing sensory deprivation. Its phobias are caused by disorganized exploration, which prevents it from building a high-quality database of environmental information.

For such a cat, when HSHA syndrome is treated, the world suddenly becomes dangerous as it *encounters new stimuli* at a *late age.*

The practitioner should anticipate a shift in the balance between inhibition and agitation. In some cases, a cat treated for fear may become agitated and overly reactive. In others, a cat treated for hyperactivity may begin to exhibit fears as it starts discovering the world.

Territorial structuring is often highly incomplete in these cats, as they are not attentive to their environment. They frequently display a deficit in territorial marking, which diminishes their ability to establish a soothing isolation zone.

Relationship disorders of all types are very common, as these cats are challenging, and owners are often tempted to resort to physical punishment. The cat exhibits a duality: on the one hand, it actively seeks intense physical contact; on the other hand, it is often rough and bites the people it bonds with. These communication issues reflect the owners' *demotivation*, as they unsuccessfully attempt numerous strategies with their cat. This fatigue, or even exhaustion, must be identified to assess the *urgency of intervention.*

Hyperactivity is a possible reason for the cat being abandoned or, in severe cases, euthanized, especially when anxiety-related aggression worsens the clinical picture.

9.1.2.4 Differential Diagnosis

The first factor to rule out is an insufficient *fulfillment of the cat's physiological needs,* particularly regarding environmental enrichment. Cats may spend long periods without much activity, but they still require stimulation. Observation is key, as the value of a territory often depends more on its richness than its size. Additionally, if a cat has grown accustomed to outdoor activities filled with encounters and surprises, it may become unable to adapt to a single room or even a spacious apartment.

A cat lacking adequate activity may redirect its demands toward its owners and often toward food. An increase in food consumption is often a consequence of the owners' reaction to their cat's constant demands, as they attempt to satisfy it with more food. Naturally, the cat consumes these extra portions and quickly becomes dependent on this distribution. There are no universal rules, and each situation must be analyzed individually.

In assessing a cat's activities, it is important to consider interactions with its owners, the availability and quality of observation posts, human activity within the territory (e.g., visits, comings and goings), and the general richness of the environment. A cat deprived of outdoor access can still lead a harmonious and fulfilling life if its indoor environment is sufficiently enriched. Finally, it should be noted that a normal cat without sufficient exercise tends to become lazy rather than hyperactive.

Hunger or *pain* can also be potential causes of motor agitation and must be ruled out during the clinical examination.

Categorical phobias may also lead to motor agitation. However, unlike the persistent motor agitation of an HSHA cat, a phobic cat's behavior varies greatly depending on the circumstances. Additionally, HSHA cats exhibit motor agitation that is often linked to play or other generally positive emotional moments, whereas a cat with a categorical phobia—particularly in the early stages of the condition—displays retreat and avoidance behaviors when exposed to the phobic stimulus.

9.1.3 Prognosis

Several factors influence the prognosis of HSHA syndrome, as summarized in ◘ Table 9.1.

The *clinical severity* of HSHA syndrome is highly variable. The most clinically severe cases are also the most challenging to treat. Less severe cases can recover much of their lack of self-control within a few months, particularly if they have free access to the outdoors. For these cats, the disorder may not be noticeable to most people. In contrast, cats with the most severe symptoms may experience a highly disabling and pervasive condition, causing significant environmental damage or injuries during interactions, particularly in play or contact with strangers. These cats are so difficult to manage that their owners often isolate them to avoid problems.

The spontaneous progression of severe cases is poor, as the relationship with the owner deteriorates rapidly.

The severity of the disorder is assessed by the intensity of the behavioral responses and the *number of behaviors* in which the disorder manifests. When sleep is affected, the condition is often severe, as memory consolidation and learning are significantly disrupted.

Table 9.1 Factors influencing the prognosis of HSHA syndrome

Criteria considered	Protective factors	Aggravating factors
Severity of the condition	Contextual and/or minimally invasive	Severe and/or invasive
Pathological state	Impulsive state	Compulsive state
Environment	Calm	Highly stimulating
Communication	Gentle but clear	Brutal
Other conditions	Absence of comorbidities	Anxious state or other behavioral disorder
Age of intervention	Juvenile period	Adulthood
Progression since adoption	Low progress with possible learning	Symptom aggravation

The *compulsive pathological state* responds less effectively to *pharmacological treatment* compared to the *impulsive state.*

When coercive methods have been used, which is common due to the difficulty of controlling these cats, the risk of transitioning into an *anxious state* is significant, and this greatly worsens the prognosis. Similarly, the presence of comorbid behavioral disorders (e.g., categorical phobia, interspecific or intraspecific relational disorders) also negatively impacts the prognosis.

Management must therefore be initiated early, not because the chances of recovery are higher, but because the relationship with the owner has not yet deteriorated and punitive methods are less likely to have been employed. Additionally, all undesirable behaviors learned by the HSHA cat before treatment cannot be unlearned. These behaviors add complexity to the clinical picture and make *behavioral therapy* more challenging.

In practical terms, young cats with severe forms of the condition may require long-term treatment even if managed early. Conversely, less severely affected cats can often be rehabilitated within 4–6 months of treatment combined with therapy aimed at improving self-control.

9.1.4 Treatments

The complete cure of HSHA syndrome is unrealistic because these brains genuinely function differently, as shown by MRI studies in humans (Hoogman et al. 2017). However, achieving symptom improvement that allows for a high quality of life for both the cat and its owners is very often possible. In practice, clinical recovery is frequently achievable.

It is essential to distinguish between the management of an HSHA cat, even one severely affected but without *anxious comorbidities,* and that of an HSHA cat that has learned numerous undesirable behaviors, including aggression. If the animal is also anxious, management becomes more complex, and the prognosis less favorable.

The mildest cases involve young cats under 6 months of age with poor acquisition of self-control. These cats are noticeable during their first vaccinations because they squirm, nip, and struggle to control themselves. They enjoy contact but escalate it into play-fighting. Their reactivity to injections is often amplified. This state should be considered *reactive* and must be addressed promptly; otherwise, these cats may transition into impulsive or compulsive states, requiring medical treatment and re-education.

When this reactive state is observed, owners must be guided and their responses carefully directed. It is important to emphasize that the reactive state is transitory. If, after 1–2 months of appropriate intervention, the cat has not returned to normal or made significant progress, a comprehensive consultation should be scheduled to confirm or rule out the diagnosis of HSHA syndrome and provide more complete treatment.

9.1.4.1 Pharmacological Treatment

In human medicine, the reference medication is *methylphenidate,* known as Ritalin. This amphetamine derivative primarily acts on dopamine. While this treatment may be effective in cats, no data are currently available for this species.

In feline medicine, the most commonly used drug for HSHA syndrome is *fluoxetine* **(see box below),** administered at doses higher than the usual and traditionally described range (Carney et al. 2014; Hart et al. 2005; Pryor et al. 2001), specifically 1–2 mg/kg. For *compulsive states,* high-dose fluoxetine is also the treatment of choice.

As in human medicine, compulsive states respond poorly to medication. While high-dose SSRIs remain the recommendation, it is often necessary to combine them with other treatments, such as gabapentinoids, to achieve a satisfactory balance.

Venlafaxine is another possible treatment, particularly valued for its ease of use in cats. While it may be less effective than fluoxetine for managing extreme impulsivity or a lack of food satiety, it is particularly indicated when the cat exhibits a *phobic or anxious state* in addition to its impulsive or compulsive state. The recommended dosage is 1 to 2 mg/kg (equivalent to 4–8 granules/kg) (Metz et al. 2021; Masson et al. 2024).

Selegiline has also been described as potentially effective for treating HSHA syndrome. This molecule has a metabolite that is an amphetamine derivative, which has led to its consideration as potentially effective for *impulsive states* (Fadai et al. 2009; Mohammadi et al. 2004). While preliminary studies suggest its potential utility, the level of evidence supporting the effectiveness of selegiline in cats for this indication remains unproven (see ► Chap. 19, ► Sect. 19.2).

Important Point of View

Fluoxetine and HSHA

It is reasonable to question how and why this SSRI, which primarily acts on serotonin, can regulate a dysfunction primarily involving norepinephrine and dopamine in the prefrontal cortex.

First, like all SSRIs, fluoxetine acts on the SERT transporter. This transporter, along with all monoamine transporters, has a preferential affinity for serotonin but can also transport other monoamines. In the prefrontal cortex, there are very few dopamine and norepinephrine transporters. As a result, these monoamines remain free and readily occupy the SERT transporter.

Thus, any molecule capable of acting on the SERT transporter will influence dopamine and norepinephrine levels in the prefrontal cortex.

However, this does not fully explain why fluoxetine appears to be more effective than other SSRIs in regulating the impulsivity associated with HSHA syndrome.

There are several hypotheses to explain this. First, fluoxetine has an additional effect on norepinephrine transporters (NET), which becomes evident only at high doses. This effect may contribute to the noradrenergic action necessary to balance norepinephrine in the dysfunctional prefrontal cortex. This could explain why doses below 2 mg/kg fail to produce therapeutic effects in many dogs.

Second, fluoxetine possesses antagonistic activity on the *5HT2C receptor*. This postsynaptic receptor regulates the release of norepinephrine and dopamine, particularly in the prefrontal cortex. Blocking the serotonergic action of this receptor in the prefrontal cortex increases the release of norepinephrine and dopamine (◘ Fig. 9.7). This mechanism contributes to improved concentration.

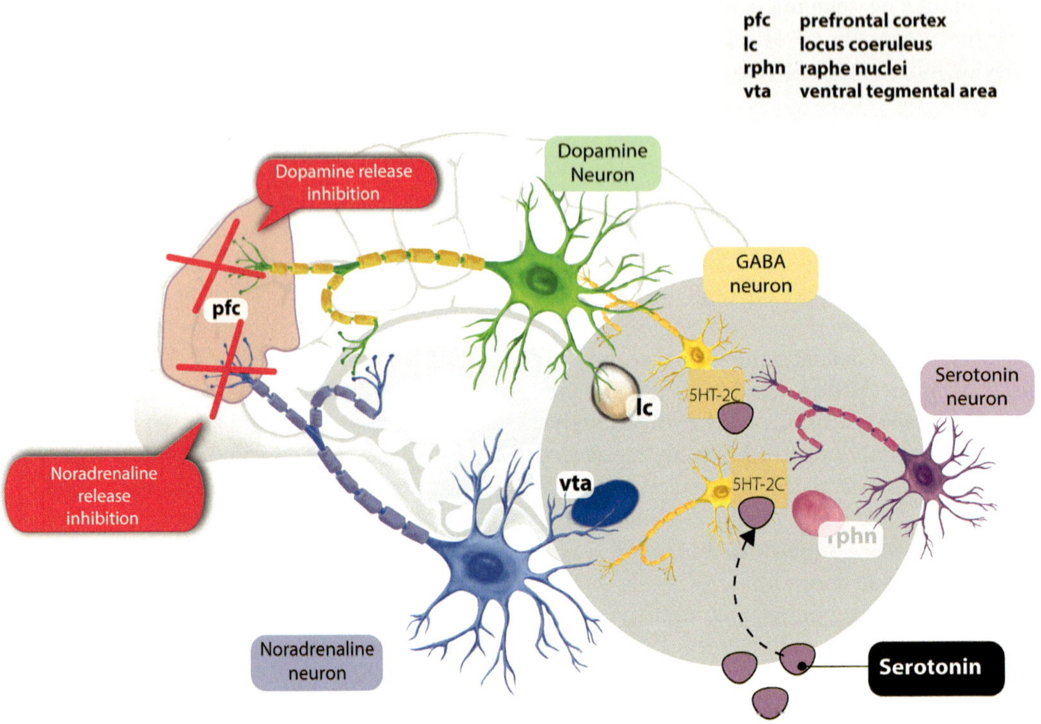

Fig. 9.7 The function of 5-HT2C receptors in the prefrontal cortex. The action of serotonin on 5-HT2C receptors stimulates GABAergic interneurons, which inhibit the release of norepinephrine and dopamine in the prefrontal cortex. 5-HT2C antagonists, such as fluoxetine, "release the brake," leading to an increase in norepinephrine and dopamine release in the prefrontal cortex

9

This receptor also plays a role in appetite regulation. Indeed, *5HT2C knock-out mice*, which lack the 5HT2C receptor, are consistently obese (Miller and Wacker 2010). A 5HT2C antagonist would therefore be expected to increase appetite, yet high-dose fluoxetine causes a decrease in appetite, sometimes even leading to anorexia. In fact, fluoxetine is the only SSRI approved for the treatment of bulimia.

These observations about receptors highlight how far we still are from fully understanding the regulatory effects of the molecules we use. A study in humans demonstrated that fluoxetine has opposite effects in the cortex when used in autistic individuals versus hyperactive individuals. Functional MRI shows hypo-activation of the medial prefrontal cortex (mPFC) in autistic individuals and hyper-activation of the same region in hyperactive individuals. High-dose fluoxetine induces *upregulation* of this region in autistic patients and *downregulation* in hyperactive subjects.

Although the initial dysfunction appears "opposite," fluoxetine leads to normalization in all cases, bringing both groups closer to the same functioning as non-affected individuals (Chantiluke et al. 2015).

For HSHA cats with comorbidities involving *phobic* or *anxious states* that complicate the clinical picture, it is common to combine *fluoxetine* or *venlafaxine* with *gabapentin* or *pregabalin*.

9.1.4.2 Behavioral Therapy

▪ Environmental Modifications

The choice of *toys* is crucial, and laser pointers should be avoided unless they are used in conjunction with a reward (e.g., a treat or toy) at the end of the session to prevent frustration. Despite this, laser pointers are often favored by owners because they are easy to use. However, a small ball, such as a ping-pong ball, is sufficiently motivating for most cats. Paper balls are also popular. These toys help redirect the cat's predatory instincts and are an excellent response when the cat becomes overly excited during a cuddle.

Food can be provided in the form of *food puzzles*, and standard bowls should be avoided for HSHA cats, even more so than for other cats, due to their limited food satiety and typically high motivation levels. It is especially important for HSHA cats to curb their tendency to eat too quickly and to encourage grazing throughout the day. Additionally, food puzzles enrich the cat's environment and provide activities aligned with their natural hunting behavior, which involves extensive observation.

The environment should also be enriched with numerous *hiding spots* and designed in three dimensions to provide a variety of stimuli that meet the high motivational needs of these individuals (see ► Chap. 29 for details on environmental enrichment strategies).

Special care should be taken to ensure the cat cannot ingest hair ties. Sewing threads (with or without needles) are also extremely dangerous.

In HSHA cats with the hypersensitive form, the level of *environmental stimulation* has a significant impact on their activity level. These cats are unable to ignore stimuli and tend to never rest if their environment is not calm. The use of an isolated room is a viable option to help them learn how to rest and stop their activity. This resting zone should be utilized whenever the cat's behavior becomes unmanageable. It can also be used to manage interactions with visitors by isolating the cat if it becomes overly excitable or during times when children are playing in the home and the cat has a tendency to chase them.

The possibility of free *outdoor access* often contributes positively to the normalization process for these cats. However, it comes with risks, as caution is not typically one of their strengths. Owners should be warned about this danger, even though exposure to the challenges of the outside world can be beneficial for their development.

▪ Behavioral Modifications

The exercises will vary depending on the specific form of HSHA encountered. Details on therapies targeting the self-control axis are presented in the fourth part (see ► Chap. 29, ► Sect. 29.3.2).

In all cases, the use of *gentle communication* is essential. Owners should be encouraged to speak softly and avoid sudden movements, as these tend to have the opposite effect, exciting the cat. Owners often resort to shouting in an attempt to get the attention of their hyperactive or distracted cat, but this only exacerbates the problem.

Preventing the cat from learning undesirable behaviors is a critical aspect of therapy. This is achieved by isolating the cat in low-stimulation environments when it is unable to calm itself.

Teaching *control over biting* and *scratching* during interactions is a central aspect of therapy. Initially, owners should be advised to provoke only very mild excitement and immediately break contact with the cat as soon as it oversteps by extending its claws or starting to bite. A similar approach can be applied to petting or encouraging tolerance of immobility, with owners gradually increasing their expectations each day, always maintaining a calm environment.

If the cat becomes overly excited, the owner can try to calm it by breaking contact or, for some cats, gently restraining it to limit physical agitation. If these measures prove ineffective, isolating the cat in a separate room remains the best solution to prevent injuries and teach the cat that such behavior does not allow it to continue its actions. In other words, if the cat wants to interact, it must learn to do so "politely."

Finally, as the cat improves in self-control, intellectual exercises of various kinds (e.g., scent work, learning new behaviors) can help improve concentration and tire the cat much more effectively than uncontrolled play. *Hyposomnia* should be addressed by encouraging the cat to take naps by placing it in a quiet area.

Regarding *play,* it is possible to redirect the cat's activity toward a ball when it begins to nip or scratch. Using a small paper ball, it is also possible to teach the cat to retrieve objects. This gradually enables the introduction of other types of training and periods of waiting between throws. Naturally, a gradual approach can be taken by introducing increasingly stimulating games for the cat, stopping the play session, or isolating the cat before it loses control of its biting and scratching due to excessive excitement.

The entire process revolves around helping the cat experience returning to a calm state more quickly, starting from increasingly intense levels of excitement.

In the *hypersensitive form,* it is essential to begin by limiting stimuli. Exercises should be conducted in the calmest environment possible, and the cat should be isolated at the first signs of excessive excitement.

Regarding *play*, we have mentioned that it is possible to redirect a cat's activity toward a ball when it starts to nibble or scratch. A small paper ball can also be used to teach the cat to retrieve objects. This exercise gradually facilitates other learning processes, including waiting periods between two throws. It is also possible to progressively increase the intensity of the games by using toys that are increasingly stimulating for the cat, while stopping play or isolating the cat before it loses control over its biting and scratching due to excessive excitement. The key objective is to help the cat experience a return to calm more and more quickly, even after highly stimulating activities.

In the *hypersensitive* form, stimulation should initially be limited. Exercises should take place in an extremely calm environment, and the cat should be placed in isolation as soon as signs of overexcitement appear.

In general, therapy for HSHA focuses on teaching the cat how to *"brake,"* whether *emotionally* or *physically.* The exercises are therefore primarily aimed at working on returning to a calm state or stopping behavioral sequences. Since the cat is impulsive and acts quickly, the responses of the person interacting with it must also be swift. The veterinarian's support throughout the therapy process is a crucial factor in its success.

9.1.5 Therapeutic Follow-Up

The therapeutic follow-up for an HSHA cat is always lengthy. Management spans from the beginning of treatment until the *brain reaches full adulthood,* which typically requires *6 months* to *2 years*, depending on the case. Throughout this period, the focus must remain on the consistent work of acquiring self-control. These therapies rely heavily on repetition and regular encouragement to ensure the cat learns to regulate its emotions in all circumstances.

Additionally, living with an HSHA cat can be exhausting for owners, and their motivation to perform the re-education exercises often fluctuates. Discouragement is common and understandable, making *motivational interviews* critical to success.

It is also important to remember that the maturation of the prefrontal cortex is delayed in cats with HSHA syndrome. The optimal time to discontinue treatment, when possible, appears to be the age at which the brain has reached maturity, usually around 18 months.

Finally, for cats whose treatment begins in adulthood, something that ideally should not occur but remains common, treatment durations are relatively short (6–10 months) when no comorbidities are present and the HSHA is not severe. However, treatment may become significantly longer when the cat has developed anxiety or presents with multiple behavioral disorders.

9.1.6 Prevention

HSHA syndrome has a strong hereditary component. The best way to prevent it is to *avoid breeding animals severely affected by the condition.*

It is also crucial to stop confusing weaning with the end of education. Many kittens are separated from their mothers as soon as they can feed themselves independently. These early adoptions ignore the educational role of mothers, and it is always preferable to wait at least 2 months, preferably 3 months, according to some studies (Ahola et al. 2017), before separating the queen from her kittens.

There are also overly tolerant mothers who fail to regulate their kittens or mothers unable to do so due to continuous breeding cycles. In such cases, the kittens should be placed in a home with an adult cat capable of taking over the educational role, though this can be challenging to arrange. It is common for neutered male cats to take on this educational role.

9.1.7 Clinical Vignettes

9.1.7.1 Mistigri

Mistigri is a 2-year-old neutered male Siamese-type cat. Since the age of 7 months, he has become overtly aggressive. Whenever his owners sit in a chair or on the sofa, he attacks them. In the evening, before dinner, when the lady of the house begins cooking, he lies in wait and pounces on her feet whenever she moves. To manage the situation, the owners have adopted two solutions: they close off the living room in the evening and the kitchen when she starts preparing meals. The consultation is prompted by their realization that they are rarely in the same room as their cat. Despite these precautions, the lady has been scratched on her arm and bitten on her heel, leaving deep marks.

Our questioning reveals a cat that was once a very lively and adorable kitten. Until the age of 6 months, he was focused solely on play and could enjoy long cuddles during which he would fall asleep purring. His primary flaw during this time was his appetite. The veterinarian had recommended free feeding, but it was never possible. As soon as he receives his kibble, he rushes at it, devouring it greedily while growling and shaking his head.

He sleeps very little, and his owners had to close their bedroom door to get some rest. Usually, around 6 a.m., he begins jumping on the door handle, waking his owners—this has now become their usual wake-up time.

Additionally, Mistigri is very clever and managed to open the pantry door where his kibble was stored. To do so, he slid the latch and jumped on the handle. This escapade resulted in a significant bout of indigestion. His owner successfully taught him to "give his paw" using his love for food as motivation. Mistigri retrieves balls and often carries a toy in his mouth, proudly parading around the living room.

During the consultation, he leapt out of his carrier and darted wildly around the examination room. He quickly located the stash of treats, which had to be hidden. Eventually, he abruptly fell asleep on the desk while playing with a pencil.

The diagnosis of hyperactivity is straightforward. In Mistigri's case, it is more of hyperreactivity than hypersensitivity. His ability to learn is a positive aspect, and the owners' strong attachment to him is remarkable.

A prescription for fluoxetine at a dose of 8 mg/day was initiated. Mistigri weighs 6 kg, and after 1 month, there was some improvement, though it remained insufficient. His appetite was still uncontrolled, and attacks persisted, albeit less intense, but the owners' fear was still evident. However, administration of the medication was easy. The dose was adjusted to 12 mg/day, and Mistigri's behavior improved significantly. He could stay on the sofa for extended periods, calmly watching his owner cook, and the bedroom door was now left open at night. A complex feeder made from a hollow brick, requiring him to eat kibble one piece at a time, nearly resolved the food-related issues. Gradually, the amount of kibble he consumed decreased, particularly after adopting a diet based on fiber-enriched kibble.

Guidelines were provided regarding cuddles. Whenever Mistigri began playing with his hands, his owner was instructed to stop the interaction and make him leave the room. She would then confine him to his designated room. Mistigri complied and now often goes to sleep in this room voluntarily.

The treatment was maintained for 1 year, then reduced to half the dose for 2 month before being discontinued. Mistigri stabilized at a weight of approximately 5.5 kg.

Summary of ▶ Sect. 9.1

HSHA syndrome is a neurodevelopmental disorder in cats. Many characteristic features of this condition are comparable to what is observed in dogs and humans. It is characterized by exaggerated excitement reactions or responses to non-salient environmental stimuli, resulting in hypermotricity and hyperemotivity. This condition encompasses a wide range of clinical presentations, with symptoms varying from mild and contextual to severe and highly invasive.

Cats affected by this syndrome can be clinically classified based on their predominant symptoms: hypersensitive form, hyperreactive form, or mixed form. In the predominantly hypersensitive form, the behavioral reaction threshold is lowered, and the animal reacts to normal environmental stimuli. This clinical presentation reflects an anomaly in the sensory filter. In the predominantly hyperreactive form, the reactions are contextually appropriate in nature, but the intensity of the reaction is excessively elevated.

The diagnosis is based on identifying an impulsive or compulsive pathological state, combined with the presence of characteristic symptoms.

The most commonly used medication for this condition is fluoxetine at a dose of 1 to 2 mg/kg.

Behavioral therapy aims to prevent phases of overexcitement and encourage a return to calm. Teaching controlled biting and scratching is an essential objective that must be achieved before considering discontinuing treatment.

The prognosis is generally good, particularly in the absence of comorbidities. It depends on the severity of the clinical presentation and the ability to begin management before complications arise.

9.2 Addictions

9.2.1 Description

9.2.1.1 Definition

Addiction is a psychological condition in which an individual persistently engages in an activity to obtain reinforcement that, due to neurobiological disruptions, no longer occurs. This compulsive and obsessive behavior is driven by a sense of lack that only the performed activity seems able to compensate for.

In addictive behaviors, a *compulsive state* is mandatory, with an absence of a stopping signal observable in one or more behavioral sequences. However, this condition is not merely a compulsion. The symptoms reflect an overactivation of the *reward system,* which becomes overstimulated either in frequency (training) or intensity (addictive drugs). In both cases, dopamine is massively released in the nucleus accumbens, leading to a decrease in dopaminergic sensitivity (Wise 2009).

Simultaneously, an imbalance in the brain's opioid system induces withdrawal symptoms (Le Merrer et al. 2009; Van Ree et al. 2000). What begins as a *pursuit of positive reinforcement* quickly transforms into a behavior aimed at *avoiding the sensation of withdrawal*, thus becoming a form of *negative reinforcement.*

This results in an active search for the stimulus at the center of the compulsion (withdrawal symptoms) and an absence of pleasure when the behavior is performed again. Cats exhibiting addictive behaviors struggle to initiate alternative behavioral sequences. Their actions become highly mechanical, with a noticeable repetition of the behavior and a lack of observable pleasure. The animal frequently seeks the object of its addiction, becoming disconnected from its environment when it is present, as its focus is entirely on the addictive stimulus, leading to *highly automated* responses.

Addictive behaviors are likely possible in both cats and dogs, though they are rarely observed in clinical practice.

9.2.1.2 Etiology and Pathogenesis

As with HSHA syndrome, addiction is likely to have a genetic basis. The most frequent cases observed in cats involve **addiction to fabrics**, with affected cats sucking on fabric to the point of creating holes.

In humans (Vermeire et al. 2012) and dogs, this genetic foundation is well-documented (Tang et al. 2014). In cats, such behaviors are often discussed in the context of **pica**, with studies concluding that the environment has no significant effect, although the genetic basis has not been clearly established (Demontigny-Bédard et al. 2016). However, certain breeds, such as Siamese and Birman cats, appear predisposed to fabric ingestion (Borns-Weil et al. 2015).

- **Dopamine and Addiction**

Dopamine plays a central role in the reward system, inducing pleasure through its release in the nucleus accumbens. This release increases the motivation to repeat the same behavior. It is followed by an inhibitory phase during which dopamine release in the nucleus accumbens is reduced.

However, with repeated experiences, the response in the nucleus accumbens changes. The dopamine released no longer triggers the activation responsible for pleasure, although the subsequent inhibitory phase remains unaffected, during which dopamine release in the nucleus is reduced. The receptors most involved in this habituation phenomenon are the D2 receptors. This change explains the absence of pleasure observed during the consummatory phase of compulsive behavior.

In addictions, dopaminergic neurons lose their sensitivity and stop responding to the primary reinforcer (such as food or play). However, anticipation sustains the conditioned response, and the announcement of the reinforcer (e.g., seeing the ball or hearing the refrigerator door) triggers stimulation.

With repeated exposure to the object of addiction and the development of habits, dopamine release shifts. Initially, it occurs in the nucleus accumbens of the ventral striatum (when the first behaviors are produced). Over time, it is released in the dorsal striatum during the execution of the conditioned response or upon seeing the stimulus (Stahl 2013).

This progressive shift, which involves both a decrease in sensitivity and a relocation of reinforcement to the dorsal striatum, characterizes the pathological nature of addiction.

Important Point of View

Compulsive Behavior and Stereotypy

In a book on stereotypic behaviors observed in various animals, G. Mason and J. Rushen cite detrimental environmental conditions and individual vulnerabilities as potential causes.

They define *stereotypic behavior* as a behavior that is: (i) frequently repeated,(ii) invariant in its motor sequence, interaction with the environment, or purpose, and(iii) without apparent function, inappropriate or traumatic for the individual, or highly inappropriate or "bizarre" (Mason and Rushen 2006).

Compulsive behaviors, in contrast to stereotypies (with which they share the absence of a stopping signal and repetitive nature), vary in their motor sequence and remain goal-directed. Compulsions can thus be viewed as behaviors with an inappropriate goal, whereas stereotypy appears devoid of any purpose. In both cases, the individual demonstrates a very high level of persistence in producing the behavior.

Compulsive behaviors are generated by a dysfunction of the cortico-striato-thalamo-cortical loop, often involving the cortical part, with a lack of inhibition in the top-down portion of this circuit.

Stereotypies, on the other hand, result from a dysfunction in the same circuit but specifically in the *basal ganglia*. Within this part of the circuit, the *direct pathway* (striatonigral) increases behaviors, while the *indirect pathway* (striatopallidal) inhibits behaviors. Numerous neurotransmitters, including dopamine, serotonin, opioids, GABA, and glutamate, are involved.

Stereotypies result from an imbalance between these direct and indirect pathways, either due to insufficient activation of the indirect pathway or excessive activation of the direct pathway (◘ Fig. 9.8). Drugs that enhance the activity of the direct circuit or inhibit the activity of the indirect circuit induce motor stereotypies.

Stereotypy is a symptom that can be found in conditions of very different origins, such as HSHA, addictive behaviors, autism, or extremely poor environmental conditions that induce an anxious state, often accompanied by stereotypies.

Finally, it should be noted that in the scientific literature, these terms are often confused, and these strict definitions are not universally adopted by all authors, which contributes significantly to the confusion surrounding *obsessive-compulsive disorders (OCD)*.

9

9.2.1.3 Epidemiology

No epidemiological data are available. Cases are rarely reported, and it is possible that if the addictive behavior does not directly disturb the owners, they simply do not seek help. The most frequently reported cases involve the ingestion of non-food items, often fabrics. However, many of these cases also present additional symptoms and are, in reality, cats affected by HSHA syndrome.

9.2.1.4 Presenting Complaints

The most common reason for consultation is the inconvenience caused by these repetitive behaviors. Wool-eating cats often suffer from health issues, and excessive licking can be unpleasant for owners. Cats with poor food regulation are perpetually demanding and quickly become obese.

A cat may become so insistent with its play partners or during other interactions that it leads the owner to seek help or to discontinue those interactions entirely. Withdrawal symptoms drive affected animals to constantly seek the object of their addiction. When access is denied, they become a nuisance to their owners.

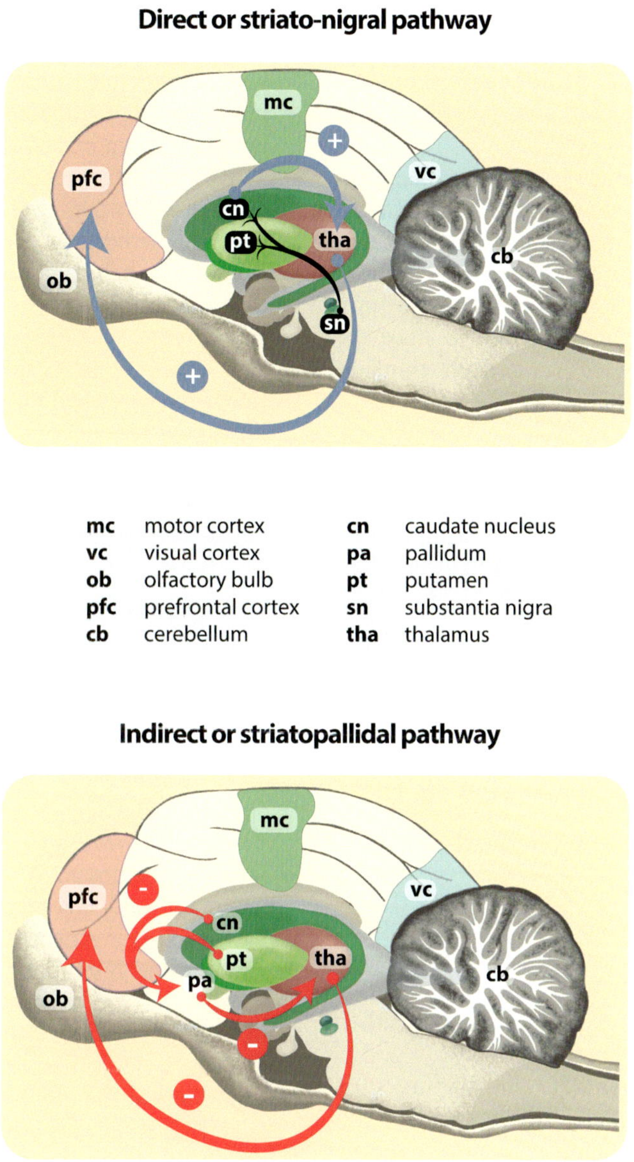

Fig. 9.8 Direct and indirect pathways in stereotypies

Regarding *food-related behaviors*, the lack of food satiety, which drives continuous food-seeking behavior, can be a reason for consultation. Some food-obsessed cats are affected by **HSHA syndrome** and exhibit additional clinical signs beyond the lack of food satiety. However, some do not meet any of these criteria and seem solely fixated on food.

These cats often have a poor quality of life, as they experience little pleasure and are frequently in a state of withdrawal. Medical treatment is necessary for these cases.

9.2.1.5 Symptomatology

Symptomatology can be divided into two parts: the symptoms observed while the compulsion is being performed and those observed during withdrawal phases.

▪ Symptoms Observed During the Addictive Phase

The *absence of pleasure* is noticeable. The animal's actions appear mechanical, with little regard for the outside world. It is highly focused and anticipates the next sequence, displaying very few moments of relaxation between repetitions. As behaviors become automated and repeated numerous times, the associated pleasure diminishes significantly.

In addictive behaviors, the sequences—though actively sought by the animal—do not seem to bring genuine pleasure. However, this alone is insufficient to diagnose addiction. Instead, it indicates the instrumentalization of the behavior, which is adaptive since the animal no longer needs to actively think to perform the behavior.

- **Symptoms Observed during the Withdrawal Phase**

In the absence of the object of compulsion, the animal shows little interest in other stimuli. It may even appear *apathetic* or *demotivated.* However, as soon as access to the object becomes possible, the animal expresses its frustration, and withdrawal symptoms persist. Other activities offered fail to capture its interest, and it frequently returns to seeking the object of its addiction.

9.2.1.6 Evolution

The boundary between normal and pathological behavior is not easy to define in this condition, as the behaviors themselves fall within the normal repertoire of the animal; it is their repetition that becomes problematic.

It is important to note that the ingestion of foreign objects, even fabrics, as well as excessive food consumption, can have significant pathological consequences. Conditions such as intestinal obstruction or obesity may necessitate intervention.

9.2.2 Diagnosis

9.2.2.1 Diagnostic Criteria

Cats with addictive behaviors necessarily exhibit a *compulsive pathological state* (◘ Fig. 9.9). Symptoms appear progressively and become noticeable in early adulthood. The diagnosis is based on identifying all or part of the following symptoms:

- behavior repeatedly performed by the cat in an automated manner, not associated with signs of pleasure;
- inability to spontaneously stop the behavior when it can be performed.
- active behavior seeking the object of the compulsion when it is absent, with evidence of withdrawal symptoms: meowing, stereotypic behaviors, and fixation if the object is visible.
- this intense seeking prevents the animal from engaging in its species' normal, routine behaviors when the compulsive behavior is possible.
- increased irritation-based aggression in contexts of frustration (these aggressions may be redirected).

9.2.2.2 Direct Observation

The seeking behavior is not necessarily observable during the consultation, but the mechanization of the compulsive behavior associated with the absence of pleasure is often easy to identify, particularly with current video recording methods.

There are three types of addictive behaviors that can be observed.

There are *wool ingestion behaviors* (Borns-Weil et al. 2015). Affected cats seek out any fabric surfaces resembling wool. They grab fibers and suck on them while manipulating them with their paws, similar to a kitten suckling. They quickly detach small strands of wool and repeat the process.

Likewise, some cats attempt to constantly *lick any individuals* who allow them to engage in a relationship. This behavior may disappear when treated and does not prevent other forms of contact.

Finally, some cats are *obsessed with food.* These deviant eating behaviors are perhaps the most suggestive of an addiction. The animal becomes entirely focused on food-seeking, and its entire life seems to revolve around obtaining food.

9.2.2.3 Comorbidities

Cats affected by *HSHA syndrome* are excellent candidates for developing addictive behaviors secondarily, as their motivation is

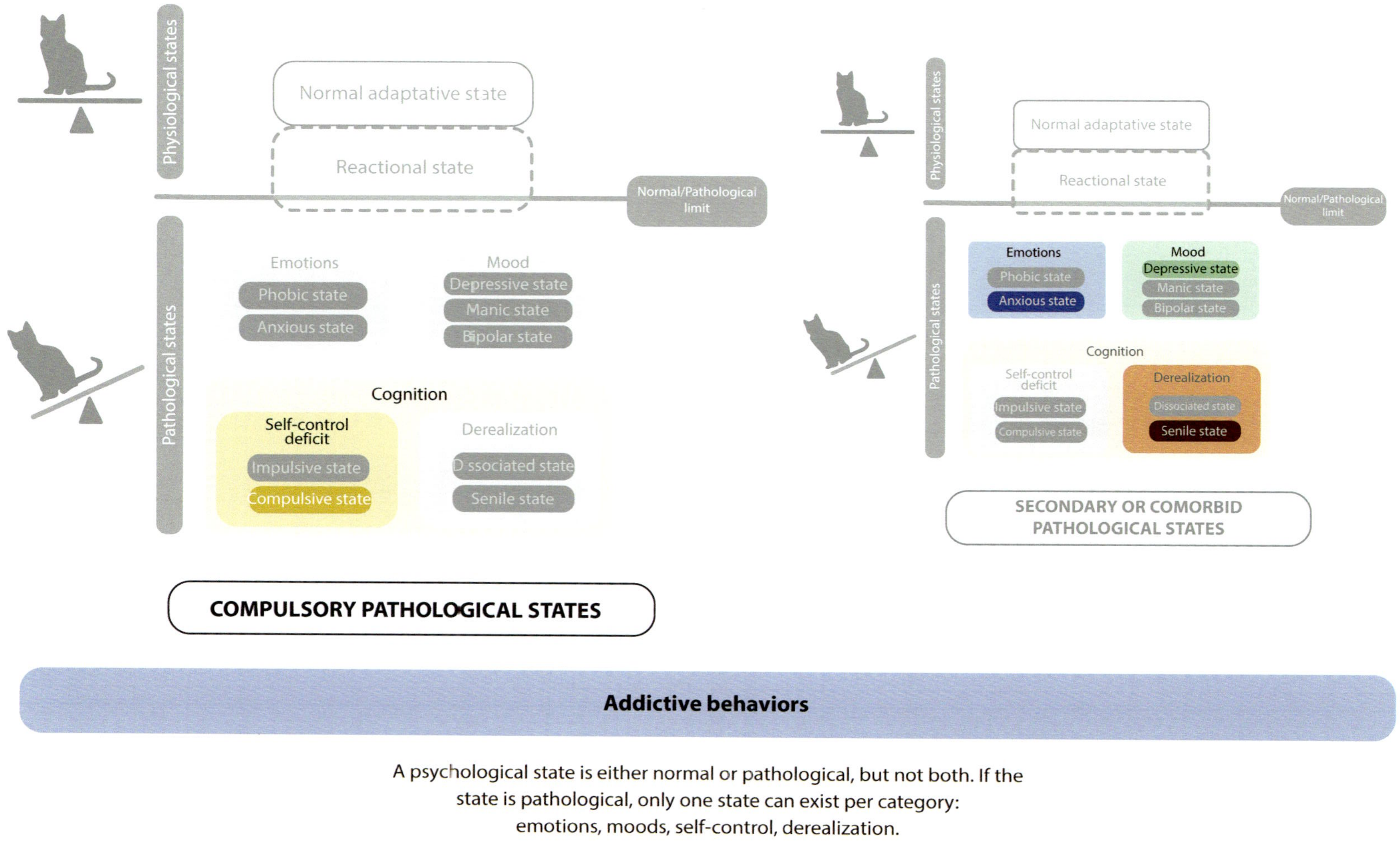

Fig. 9.9 Pathological states associated with addictions. Mandatory pathological state: compulsive state. Possible comorbid pathological states: anxious state, **and/or** depressive state **and/or** senile state

very high. It is possible that due to incomplete development, the inability to structure a calming environment promotes the search for relief through spontaneously pleasurable behaviors. Food intake, of course, is one such behavior, but the suckling behavior observed in wool-eating cats points to an infantile behavior that may highlight a behavioral maturation deficit.

All conditions affecting the *fear axis* can also coexist with these addictive behaviors. In the case of comorbidity with an *anxious state,* it can sometimes be difficult to determine whether the addiction was generated by frustration and the anxious state or whether, on the contrary, the anxious state is a consequence of the inability to perform the behavior. Indeed, in cases of anxiety, it is common to see substitute activities develop (licking of the flank or abdomen, fur-pulling), and in an impulsive or compulsive animal, this can quickly lead to a clinical picture in which this behavior becomes obsessive.

9.2.2.4 Differential Diagnosis

The first condition to consider in the differential diagnosis is HSHA syndrome. Both conditions can actually coexist. A compulsive state is possible in both conditions (mandatory in addictive behavior), but the rest of the clinical picture is different. In hyperactivity, discomfort during periods of non-performance of a behavior is not dominant in the clinical picture.

On the other hand, in addictive behaviors, if the object of the addiction and related behaviors are removed, it is not possible to identify the criteria for HSHA syndrome. Finally, these behaviors are not spontaneous: they are the result of repetition and tend to worsen over time. As such, they may appear in young adults, whereas HSHA syndrome is a neurodevelopmental disorder.

9.2.3 Prognosis

Complete recovery is illusory, and relapses are common, likely due to the highly genetic etiology of the condition. Medication treatments lead to improvement but rarely result in full clinical resolution of the symptoms. The goal is to control the disease and its consequences.

9.2.4 Treatments

9.2.4.1 Pharmacological Treatment

In human medicine, there is a wide range of psychotropic treatments tailored to specific addictions (e.g., tobacco, alcohol, binge eating). For obsessive-compulsive disorders, high-dose SSRIs are recommended. In cases of treatment failure, which is common, combining SSRIs with an atypical antipsychotic is advised.

In practice with cats, *fluoxetine* is the drug of choice for reducing appetite in cases of obsessive eating behaviors. For other types of obsessions, *clomipramine* at a dosage of 0.5–1 mg/kg (Seksel and Lindeman 1998) or *venlafaxine* at a dosage of 1.5–2 mg/kg (equivalent to 6–8 granules/kg) may be prescribed (Masson et al. 2024).

In the future, the use of molecules modulating the effects of endogenous opioids may potentially prove useful for this indication.

9.2.4.2 Behavioral Therapy

- **Environmental Modifications**

Depending on the type of addiction, the measures to be implemented will vary. For food-related addictions, access is typically monitored by the owner, who becomes accustomed to carefully storing all food items.

For addictions involving wool or fabrics, access to closets is often strictly controlled, and the same applies to the ingestion of non-food objects. It is essential, however, to ensure that the cat's environment is adequately managed. These behaviors are not necessarily more prevalent in cats living in unstimulating environments, and their resolution often requires medical intervention rather than environmental modifications. Nonetheless, practitioners should always aim to provide the cat with a stimulating environment that includes hiding spots, food presented as part of interactive games, varied toys, and observation posts.

Pharmacological treatment should aim to restore quality of life and even facilitate the organization of a self-managed, complex feeding routine. It is important to note that food bowls designed to make food retrieval more challenging do not address the underlying issue. On the contrary, these may exacerbate the problem by causing the cat to spend excessive time engaging in its addictive behavior, which does not support recovery.

Behavioral Modifications

The goal of therapy is to reintroduce *signs of pleasure* into addictive sequences and observe a significant reduction in *withdrawal signs* when the object of the addiction is removed. Ideally, the cat should transition more quickly to another activity, indicating progress in overcoming the addictive behavior.

9.2.5 Therapeutic Follow-Up

The key to success lies in rediscovering pleasure in activities. Often, once owners are educated about the significant difference between pleasure and addiction, they become more attentive to their animal's responses and, whenever possible, remove the object of addiction.

In some cases, treatments may need to be maintained for life, especially for food-related addictive behaviors.

9.2.6 Prevention

Addictions likely have a hereditary component. Therefore, it is advisable to avoid breeding affected animals.

9.2.7 Clinical Vignettes

Pito

Pito is a 4-year-old sterilized Persian cat. The issue was mentioned during a vaccination consultation, as the owner considers it to be under control. Whenever possible, Pito rushes to knead and suck on wool sweaters, especially cashmere. She consistently creates holes, indicating that she ingests the wool.

The owner is aware of this behavior and has become very meticulous about storing her sweaters. Pito no longer has access to wool, and the owner avoids petting her while wearing a sweater. Any lapse in this routine results in immediate damage, as the sweater becomes unusable.

This behavior began approximately two and a half years ago, when Pito was around 18 months old. Currently, the behavior is almost entirely absent due to the owner's vigilance. However, the obsession persists, and the addiction could resurface if the opportunity arises. The owner has expressed that she does not wish to pursue any intervention.

Summary of ► Sect. 9.2

Addictive behaviors are a group of behavioral disorders caused by a dysfunction in the reward system. They typically emerge during puberty and worsen over time and with repetition.

These behaviors are characterized by an active search for the object of addiction, accompanied by observable signs of distress (e.g., vocalizations, pacing), a lack of pleasure during the execution of the behavior, and a mechanical or sometimes stereotypical performance of the behavior.

Diagnosis is based on identifying a compulsive pathological state, along with withdrawal symptoms in the absence of the addictive object and an absence of pleasure when the object is present.

The most commonly used pharmacological treatment is fluoxetine, at a dosage of 1–2 mg/kg; however, no pharmacological approach guarantees a cure at this time.

Therapy primarily relies on avoiding the object of addiction, at least partially, as a strategy to achieve a behaviorally acceptable balance for both the cat and the owner.

Disorders associated with dysregulation of the self-controls axis

Disorder's name	Subtype	State diagnosis	Diagnostic criteria	Differential diagnosis
HSHA syndrome or feline ADHD	There are 3 forms: **Hypersensitive (HS) form** Lowered threshold for triggering motor and emotional responses **Hyperreactive (HA) form** Increased intensity of motor and emotional responses **Mixed form** Proportion of both HS and HA forms	Impul-sive or com-pulsive	Diagnosis relies on the presence of several symptoms from the following list: • Delayed acquisition of control over biting and clawing • Oral exploration that may lead to ingestion of non-food objects • Delayed acquisition of locomotion control, manifested as visible clumsiness • Absence or deficit in food satiety • Rough play, with difficulties controlling excitement, often resulting in scratches or bites • Excessive demand for interactions • Hyposomnia not associated with an anxious state • Disordered exploration • Disordered pheromonal marking with difficulty structuring territories	*Insufficient covering of physiological needs* Symptoms disappear with the restoration of normal food intake, play, and social contact *Pain* Motor agitation arises from the search for an antalgic context (e.g., specific positions or locations) *Categorical phobias* Motor agitation linked to fear occurs only in specific phobogenic contexts *Addictions* Emergence during early adulthood A compulsive state is mandatory *Relational disorders* Impulsive reactions vary depending on the person The cat is capable of controlling its behavior in many situations
Addictions		Com-pulsive	Symptoms gradually appear and become noticeable in early adulthood. Repetition of the addictive behavior may occur spontaneously through intensive self-reinforcement or through intentional training by the owner The diagnosis is based on identifying some or all of the following symptoms: • Behavior is performed repetitively and automatically, without signs of pleasure • Inability to spontaneously stop the behavior when it is possible to perform it • Active search for the object of compulsion when it is absent, with evident withdrawal symptoms, such as meowing, stereotypical behaviors, and fixation if the object is visible • Intense searching prevents the animal from engaging in normal species-specific behaviors when the compulsive behavior is possible • Increased aggression due to irritation in frustration contexts, with possible redirected aggression	*Insufficient covering of physiological needs* Symptoms disappear with the restoration of normal food intake, play, and social contact *Pain* Motor agitation arises from the search for an antalgic context (e.g., specific positions or locations) *Categorical phobias* Motor agitation linked to fear occurs only in specific phobogenic contexts *HSHA syndrome* Symptoms appear at a young age and tend to improve with appropriate intervention and management

References

Ahola MK, Vapalahti K, Lohi H (2017) Early weaning increases aggression and stereotypic behaviour in cats. Sci Rep 7(1):10412. https://doi.org/10.1038/s41598-017-11173-5

Anita T (2009) An overview on the genetics of ADHD: an overview on the genetics of ADHD. Acta Psychol Sin 40(10):1088–1098. https://doi.org/10.3724/SP.J.1041.2008.01088

Arnsten AFT (2009) Toward a new understanding of attention-deficit hyperactivity disorder pathophysiology: an important role for prefrontal cortex dysfunction. CNS Drugs 23(Supplement 1):33–41. https://doi.org/10.2165/00023210-200923000-00005

Arnsten AFT, Rubia K (2012) Neurobiological circuits regulating attention, cognitive control, motivation, and emotion: disruptions in neurodevelopmental psychiatric disorders. J Am Acad Child Adolesc Psychiatry 51(4):356–367. https://doi.org/10.1016/j.jaac.2012.01.008

Bauer J, Werner A, Kohl W, Kugel H, Shushakova A, Pedersen A, Ohrmann P (2018) Hyperactivity and impulsivity in adult attention-deficit/hyperactivity disorder is related to glutamatergic dysfunction in the anterior cingulate cortex. World J Biol Psychiatry 19(7):538–546. https://doi.org/10.1080/15622975.2016.1262060

Borns-Weil S, Emmanuel C, Longo J, Kini N, Barton B, Smith A, Dodman NH (2015) A case-control study of compulsive wool-sucking in Siamese and Birman cats (n = 204). J Vet Behav Clin Appl Res. https://doi.org/10.1016/j.jveb.2015.07.038

Bradshaw JWS, Neville PF, Sawyer D (1997) Factors affecting pica in the domestic cat. Appl Anim Behav Sci. https://doi.org/10.1016/S0168-1591(96)01136-7

Carney HC, Sadek TP, Curtis TM, Halls V, Heath S, Hutchison P, Mundschenk K, Westropp JL (2014) AAFP and ISFM guidelines for diagnosing and solving house-soiling behavior in cats. J Feline Med Surg 16(7):579–598. https://doi.org/10.1177/1098612X14539092

Chantiluke K, Barrett N, Giampietro V, Santosh P, Brammer M, Simmons A, Murphy DG, Rubia K (2015) Inverse fluoxetine effects on inhibitory brain activation in non-comorbid boys with ADHD and with ASD. Psychopharmacology 232:2071–2082. https://doi.org/10.1007/s00213-014-3837-2

Demontigny-Bédard I, Beauchamp G, Bélanger M-C, Frank D (2016) Characterization of pica and chewing behaviors in privately owned cats: A case-control study. J Feline Med Surg 18(8):652–657. https://doi.org/10.1177/1098612X15591589

Fadai F, Akhondzadeh S, Mohammadi M, Mohammadi MR (2009) Selegiline in comparison with methylphenidate in treatment of adults with attention deficit yperactivity disorder: a double-blind, randomized trial. Iran J Psychiatry 4(4):126–130

Fadel FR, Driscoll P, Pilot M, Wright H, Zulch H, Mills D (2016) Differences in trait impulsivity indicate diversification of dog breeds into working and show lines. Sci Rep 6(February). https://doi.org/10.1038/srep22162

Faraone SV, Sergeant J, Gillberg C, Biederman J (2003) The worldwide prevalence of ADHD: is it an American condition? World Psychiatry 2(2):104–113

Hart BL, Cliff KD, Tynes VV, Bergman L (2005) Control of urine marking by use of long-term treatment with fluoxetine or clomipramine in cats. J Am Vet Med Assoc 226(3):378–382. https://doi.org/10.2460/javma.2005.226.378

Hejjas K, Vas J, Topal J, Szantai E, Ronai Z, Szekely A, Kubinyi E, Horvath Z, Sasvari-Szekely M, Miklosi A (2007) Association of polymorphisms in the *dopamine D4 receptor* gene and the activity-impulsivity endophenotype in dogs. Animal Genetics 38:629–633. https://doi.org/10.1111/j.1365-2052.2007.01657.x

Hoogman M, Bralten J, Hibar DP, Mennes M, Zwiers MP, Schweren LSJ, van Hulzen KJE, Medland SE, Shumskaya E, Jahanshad N, de Zeeuw P, Szekely E, Sudre G, Wolfers T, Onnink AMH, Dammers JT, Mostert JC, Vives-Gilabert Y, Kohls G et al (2017) Subcortical brain volume differences in participants with attention deficit hyperactivity disorder in children and adults: A cross-sectional mega-analysis. Lancet Psychiatry. https://doi.org/10.1016/S2215--0366(17)30049-4

Hoppe N, Bininda-Emonds ORP, Gansloßer U (2017) Correlates of attention deficit hyperactivity disorder (ADHD)-like behavior in domestic dogs: first results from a questionnaire-based study. Vet Med—Open J 2(3):95–131. https://doi.org/10.17140/VMOJ-2-122

Le Merrer J, Becker JAJ, Befort K, Kieffer BL (2009) Reward processing by the opioid system in the brain. Physiol Rev 89(4):1379–1412. https://doi.org/10.1152/physrev.00005.2009

Lit L, Belanger JM, Boehm D, Lybarger N, Oberbauer AM (2013) Differences in Behavior and Activity Associated with a Poly(A) Expansion in the Dopamine Transporter in Belgian Malinois. PLoS ONE 8, e82948. https://doi.org/10.1371/journal.pone.0082948

Mason G, Rushen J (eds) (2006) Stereotypic animal behaviour: fundamentals and applications to welfare, 2nd edn. CABI. https://doi.org/10.1079/9780851990040.0000

Masson S, Metz D, Bleuer-Elsner S, Schwobthaler F (2024) Retrospective study on the use of venlafaxine in 176 cats diagnosed with behavioral disorders. J Vet Behav. https://doi.org/10.2139/ssrn.4790331

Metz D, Medam T, Masson S (2021) Double-blind, placebo-controlled trial of venlafaxine to treat behavioural disorders in cats: a pilot study. J Feline Med Surg. https://doi.org/10.1177/1098612X211036792

Miller KJ, Wacker DA (2010) Discovery and Development of 5-HT2C Receptor Agonists for Obesity: Is

There Light at the End of the Tunnel? Future Med. Chem. 2:1761–1775. https://doi.org/10.4155/fmc.10.261
Mohammadi MR, Ghanizadeh A, Alaghband-rad J, Tehranidoost M, Mesgarpour B, Soori H (2004) Selegiline in comparison with methylphenidate in attention deficit hyperactivity disorder children and adolescents in a double-blind, randomized clinical trial. J Child Adolesc Psychopharmacol 14(3):418–425. https://doi.org/10.1089/cap.2004.14.418
Polanczyk G, de Lima MS, Horta BL, Biederman J, Rohde LA (2007) The worldwide prevalence of ADHD: a systematic review and Metaregression analysis. Am J Psychiatry 164(6):942–948. https://doi.org/10.1176/ajp.2007.164.6.942
Prince J (2008) Catecholamine dysfunction in attention-deficit/hyperactivity disorder: an update. J Clin Psychopharmacol 28(3):S39–S45. https://doi.org/10.1097/JCP.0b013e318174f92a
Pryor PA, Hart BL, Cliff KD, Bain MJ (2001) Effects of a selective serotonin reuptake inhibitor on urine spraying behavior in cats. J Am Vet Med Assoc 219(11):1557–1561. https://doi.org/10.2460/javma.2001.219.1557
Seksel K, Lindeman M (1998) Use of clomipramine in the treatment of anxiety-related and obsessive-compulsive disorders in cats. Aust Vet J 76(5):317–321. https://doi.org/10.1111/j.1751-0813.1998.tb12353.x
Stahl S (2013) Stahl's essential psychopharmacology. Cambridge University Press
Stern L, Schell M (2018) Management of attention-deficit disorder and attention-deficit/hyperactivity disorder drug intoxication in dogs and cats: an update. Vet Clin North Am Small Anim Pract 48(6):959–968. https://doi.org/10.1016/j.cvsm.2018.07.007
Tang R, Noh HJ, Wang D, Sigurdsson S, Swofford R, Perloski M, Duxbury M, Patterson EE, Albright J, Castelhano M, Auton A, Boyko AR, Feng G, Lindblad-Toh K, Karlsson EK (2014) Candidate genes and functional noncoding variants identified in a canine model of obsessive-compulsive disorder. Genome Biol 15(3):R25. https://doi.org/10.1186/gb-2014-15-3-r25
Van Ree JM, Niesink RJM, Van Wolfswinkel L, Ramsey NF, Kornet MLMW, Van Furth WR, Vanderschuren LJMJ, Gerrits MAFM, Van den Berg CL (2000) Endogenous opioids and reward. Eur J Pharmacol 405(1):89–101. https://doi.org/10.1016/S0014-2999(00)00544-6
Vermeire S, Audenaert K, De Meester R, Vandermeulen E, Waelbers T, De Spiegeleer B, Eersels J, Dobbeleir A, Peremans K (2012) Serotonin 2A receptor, serotonin transporter and dopamine transporter alterations in dogs with compulsive behaviour as a promising model for human obsessive-compulsive disorder. Psychiatry Res Neuroimaging 201(1):78–87. https://doi.org/10.1016/j.pscychresns.2011.06.006
Wise RA (2009) Roles for nigrostriatal—not just mesocorticolimbic—dopamine in reward and addiction. Trends Neurosci 32(10):517–524. https://doi.org/10.1016/j.tins.2009.06.004
Yamazaki A, Edamura K, Tanegashima K, Tomo Y, Yamamoto M, Hirao H, Seki M, Asano K (2020) Utility of a novel activity monitor assessing physical activities and sleep quality in cats. PLoS One 15(7):e0236795. https://doi.org/10.1371/journal.pone.0236795
Zayats T, Neale BM (2019) Recent advances in understanding of attention deficit hyperactivity disorder (ADHD): how genetics are shaping our conceptualization of this disorder. F1000Res 8:2060. https://doi.org/10.12688/f1000research.18959.1

Disorders Related to Dysregulation of the Territoriality Axis

Contents

10.1 Disorders Resulting from Territorial Poverty – 365
10.1.1 Description – 365
10.1.2 Diagnosis – 367
10.1.3 Prognosis – 369
10.1.4 Treatments – 369
10.1.5 Therapeutic Follow-Up – 370
10.1.6 Prevention – 371
10.1.7 Clinical Vignette – 371

10.2 Disorders Resulting from Territorial Disrupting – 372
10.2.1 Description – 372
10.2.2 Diagnosis – 375
10.2.3 Prognosis – 377
10.2.4 Treatments – 377
10.2.5 Therapeutic Follow-Up – 378
10.2.6 Prevention – 378
10.2.7 Clinical Vignette – 379

10.3 Territoriality Disorders Secondary to Another Behavioral Disorder – 381
10.3.1 Description – 381
10.3.2 Diagnosis – 382
10.3.3 Prognosis – 382
10.3.4 Treatments – 384

© The Author(s), under exclusive license to Springer Nature Switzerland AG 2026
S. Bleuer-Elsner et al., *Veterinary Psychiatry of the Cat*,
https://doi.org/10.1007/978-3-032-06260-4_10

10.3.5 Therapeutic Follow-Up – 384
10.3.6 Prevention – 384
10.3.7 Clinical Vignette – 384
References – 388

In this chapter, we describe psychiatric disorders related to dysregulation of the territoriality axis.

We distinguish between different territoriality-related disorders: those caused by a lack of environmental stimuli, those resulting from excessive stimuli and environmental instability, and those secondary to another behavioral disorder. For each condition, the clinical presentation, diagnosis, treatment, prognosis, and follow-up are detailed, with clinical vignettes included to enrich the text.

10.1 Disorders Resulting from Territorial Poverty

10.1.1 Description

10.1.1.1 Definition

These disorders are observed in cats living in small spaces without sufficient stimulation or, more precisely, without the opportunity to express behaviors that are intrinsic to being a cat. To precisely describe this disorder, we would ideally refer to a detailed ethogram, which we do not currently have. However, it is possible to outline the common pathological evolution of behavior in cats living in environments lacking sensory stimulation.

This definition allows us to describe this disorder as resulting from the "poverty" of the environment.

10.1.1.2 Etiology and Pathogenesis

As described in previous chapters, a cat's territory is an organized space with activity zones that ideally align with the experiences the cat had during its developmental period. Within a framework determined by fixed genetic criteria, synaptic selection prepares each animal for life in a context that mirrors its early experiences. Through synaptic pruning (see ► Chap. 4), specific needs and limits are established that shape its future life.

Every cat, therefore, has an ideal territory. A cat's psychological balance depends on how closely its living environment aligns with this ideal. When the living conditions prevent or fail to allow the expression of certain behaviors, a deficiency can develop, leading to an anxious or depressive state.

10.1.1.3 Epidemiology

This disorder is closely linked to a cat's inability to engage in the normal behaviors of its species. Cats living in small apartments with no environmental enrichment and without outdoor access are at higher risk of developing this condition. With urbanization, more city dwellers have adopted cats, leading to a higher incidence of this disorder today compared to past decades.

However, the recent focus on environmental enrichment has raised awareness among owners about the importance of improving their pets' living conditions when outdoor access is unavailable. Advice on providing food through play, offering hiding spots, and enabling three-dimensional exploration has likely contributed to improvements.

The approach of providing animals with a protected environment designed for human comfort does not always ensure their well-being. It is necessary to go further to help cats achieve balance (see ► Chap. 5, ► Sect. 5.1.2).

10.1.1.4 Presenting Complaints

Whether in a reactive, anxious, or depressive state, cats affected by this disorder often exhibit symptoms that owners find problematic. Two primary symptoms serve as presenting complaints: frustration-related aggression and excessive marking behaviors as the cat attempts, unsuccessfully, to make its environment more comforting.

Aggression is closely linked to frustration. When a cat cannot express its normal behaviors due to an unappealing environment, *frustration* builds, often redirected as aggression. In other words, the environment is so impoverished that the cat responds to even the slightest stimulation, and the only moving objects are often its owners. These cats may attack legs by hiding behind doors or furniture, which quickly worsens as the owners' defensive reactions increase their fear of the cat. A common comorbidity in these cases is a secondary interspecific relational disorder.

Another way cats express that their environment is not comforting is through "overmarking" their surroundings to manifest frustration. While facial marking or scratching may be tolerated by owners, *urine marking* often becomes a significant reason for consultation.

In some rarer cases, owners seek help for *withdrawal behavior* or *substitute activities* such as excessive licking. These cats are often those who have "given up the fight," enduring their environment with resignation. They compensate for the lack of stimulation with substitute activities and may develop binge-eating behavior. Obesity, though rarely a primary reason for consultation, should alert practitioners as it is frequently associated with generalized inhibited anxious states.

10.1.1.5 Symptomatology

Urine soiling is a significant symptom for owners. The practitioner must distinguish *urine marking* from *elimination* behavior. In disorders related to the territoriality axis, an initial increase in marking sequences is almost always observed. This is because the cat is attempting to organize its environment to make it more comforting. At the onset, this represents a reactive state.

If living conditions remain unchanged and the environment continues to lack adequate stimulation, the cat's ineffective attempts to reorganize its territory through marking can lead to anxiety. Urine soiling involving full elimination sequences may also occur. In cats, chronic stress can easily result in bladder issues, and idiopathic cystitis of anxiety-related origin is common.

Other forms of *pheromonal marking* are likely also significantly disrupted, but owners rarely notice or report them spontaneously. By asking targeted questions, it is often possible to identify increased marking attempts. Scratching, however, is more likely to be noted by owners.

In these territorial disorders, it is common for cats to display inhibited forms of anxiety accompanied by significant substitute activities. This may include licking or binge eating. For cats, food consumption is the final act of hunting behavior. If all other phases of hunting are obstructed, it is natural for this activity to intensify as a substitute behavior.

For some cats, frustration caused by a lack of stimulation manifests as hunting behaviors directed at any moving object in their environment. Unsurprisingly, owners are the primary targets, and in households with children or vulnerable individuals, these become preferred targets.

These hunting actions are often poorly tolerated by owners, leading to frequent punishment. This quickly creates a vicious cycle in which both the cat and the owners increasingly fear each other, resulting in an escalation of aggression. Punishment is also common in response to urine marking.

Discussions about the measures taken with cats affected by these disorders often reveal a *poor or even non-existent relationship*. Many people decide to adopt a cat because of its reputation as an independent and low-maintenance animal, but modern living conditions often leave little time for meaningful interaction. As a result, comorbidities involving the relationship axis are frequent.

Additionally, to prevent damage to the home, some cats are confined to limited spaces when their owners are absent or asleep. This further reduces their territory and exacerbates territoriality disorders.

Important Point of View

Feeding Behavior of Cats

A significant portion of a cat's schedule is dedicated to hunting, which represents a major activity. This behavior is closely linked to feeding behavior because the beginning of a hunt is triggered by hunger, and in natural settings, the end of the hunting sequence culminates in ingestion. However, as an ambush predator, a cat's hunting behavior also includes extensive territorial monitoring and prey detection.

For a cat unable to engage in hunting or the preparatory behaviors of monitoring and detecting potential prey, compensatory behavior can manifest as food consumption. This activity is the closest available alternative to what motivates the animal. During this state of activation, eating acts as a reinforcer and can quickly become an effective substitute activity, especially if the food is palatable.

For this reason, it is crucial to provide indoor cats with observation points that can compensate for the absence of prey and help regulate their feeding behavior. Additionally, feeding systems that significantly slow down food consumption are essential to align eating behavior more closely with natural hunting behavior (see ► Chap. 29, ► Sect. 29.2.1.2).

10.1.1.6 Evolution

Spontaneous improvement is generally uncommon unless a change in the environment occurs. It is not unusual to see cats brought in for consultation due to urine soiling that has persisted for years or cats that habitually scratch or bite their owners when they pass through a hallway. Often, owners try to avoid the problem as much as possible, unaware that the situation can be improved.

Inhibition and resignation are common outcomes. The cat stops struggling and spends its time sleeping and eating without seeking other activities. Owners typically do not worry and describe the cat as a lazy glutton who only thinks about eating, without questioning what else the cat might be doing.

In depressive states, the cat may stop grooming altogether. Its coat appears tangled and dirty.

10.1.2 Diagnosis

10.1.2.1 Diagnostic Criteria

Cats affected by disorders resulting from territorial poverty must exhibit an anxious or depressive pathological state (◘ Fig. 10.1). Symptoms typically appear in early adulthood, a period when the territory should serve as the cat's primary source of comfort.

The identification of a living environment lacking stimulation, enclosed, and preventing the cat from structuring its various activity zones is a crucial diagnostic element.

The diagnosis is based on the presence of multiple symptoms indicating the cat's frustration with an environment that is too poor in stimulation, selected from the following list:

- Irritation-driven aggression in the context of frustration, targeting any moving object
- Altered facial pheromonal marking, often increased or sometimes absent
- Increased scratching around areas not associated with the cat's resting places
- Urine marking in the early stages, sometimes complicated by urine soiling
- Compensatory binge eating
- Licking leading to extensive alopecia, often on the ventral side
- Idiopathic cystitis
- Dirty, unkempt coat in cases of depressive states

10.1.2.2 Direct Observation

Aside from obesity and, potentially, the appearance of the coat when the cat is no longer grooming itself, the animal may appear normal during a consultation. When this type of disorder is suspected, a video of the cat's living environment or, at a minimum, a floor plan of the apartment is essential to confirm that the environment is unsuitable for the cat.

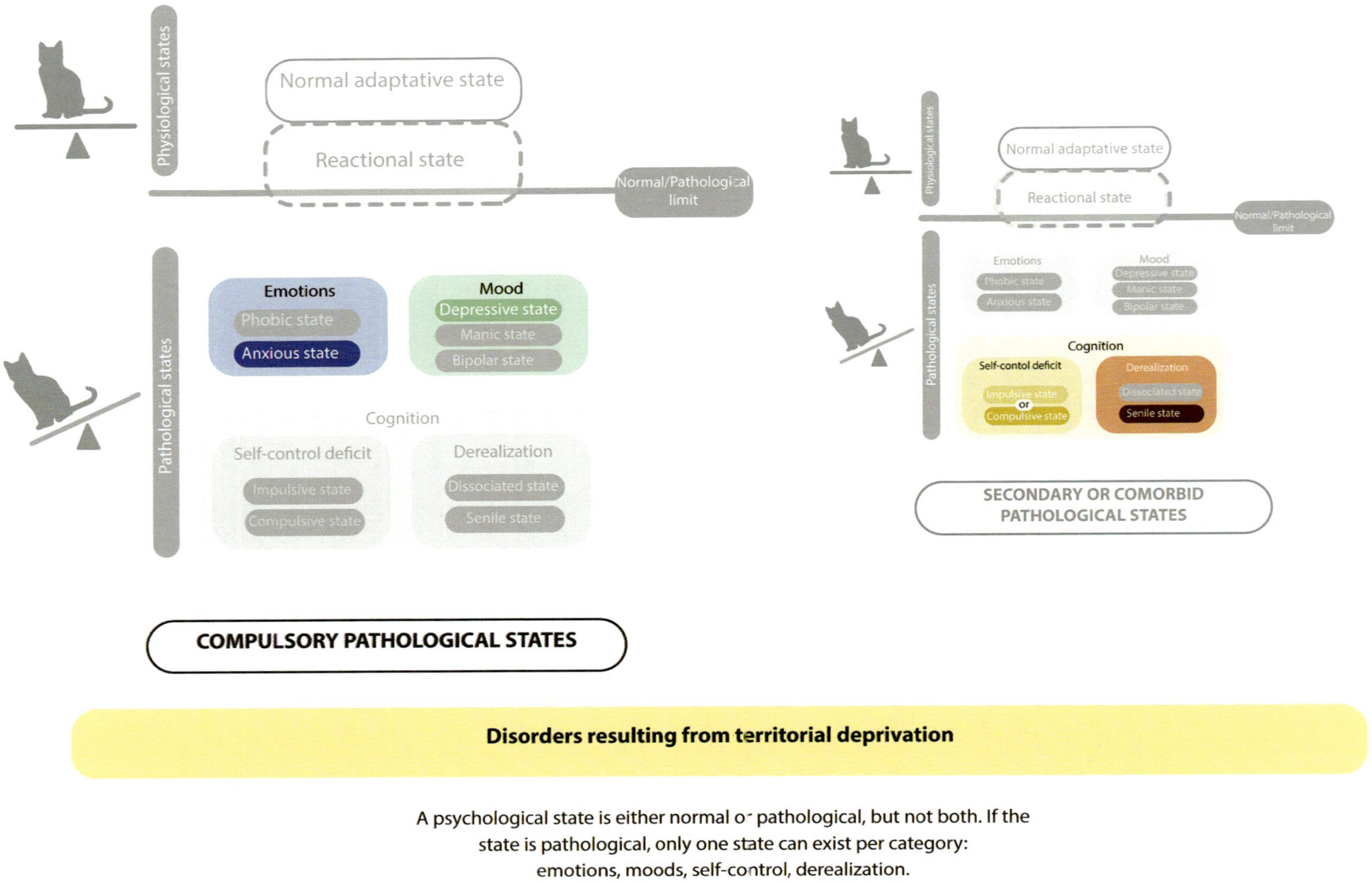

Fig. 10.1 Pathological states associated with disorders resulting from territorial poverty. Mandatory pathological states: anxious **and/or** depressive state. Possible comorbid pathological states: phobic state, and/or impulsive **or** compulsive states

10.1.2.3 Comorbidities

This disorder is highly environmental in nature, impacting all four other axes of behavior by exacerbating any existing conditions or triggering pathological shifts in an already vulnerable animal.

First, it can compound developmental disorders. In cases of *fear axis disorders*, a cat's phobias prevent it from exploring its entire territory, favoring this type of shift even if the environment is not exceedingly poor. Self-isolation and depressive states are more likely to develop when these two conditions coexist.

In cases of *self-control axis disorders*, the frustration typical of impulsive or compulsive states makes a poor environment even harder for these cats to tolerate. Symptoms related to disorders resulting from territorial poverty are often dramatic and may overlap with harmful sequences observed during play.

Conversely, this disorder can pave the way for other conditions, particularly *relational disorders*. As previously noted, symptoms like urine soiling or irritation-driven aggression in the context of frustration often lead to punishments, which in turn favor the development of interspecific relational disorders.

Finally, as the animal ages, if the disorder caused by a poor environment persists, a shift to a depressive state is common, compounded by the emergence of additional physical conditions.

10.1.2.4 Differential Diagnosis

The most challenging aspect is distinguishing between an environment that is genuinely too poor, causing the pathological state, and the heightened needs of a hyperactive animal that appears to lack stimulation. In both cases, the cat's frustration will be evident. However, in the case of *HSHA syndrome*, the frustration stems from the animal itself, even in an adequately enriched environment.

To differentiate between the two, the pathological pathway is often key. HSHA syndrome is a neurodevelopmental disorder, with symptoms appearing as early as adoption and persisting despite enrichment efforts. In contrast, disorders resulting from territorial poverty typically present with symptoms that develop progressively, often appearing at the end of puberty.

10.1.3 Prognosis

While the theoretical prognosis is good to excellent, as enriching the animal's environment is sufficient to address the issue, caution is warranted.

Enriching the environment is always possible, even in small spaces, by using interactive feeding bowls, leveraging three-dimensional spaces, and establishing routines. However, these measures require significant effort from the owners, which can be challenging to achieve.

Additionally, cats that are entirely inhibited do not cause disturbances, making owners less motivated to change their living conditions, as they may not perceive a problem.

Finally, even after improvements, not all issues can be fully resolved. Obesity that has developed remains difficult to eliminate.

10.1.4 Treatments

10.1.4.1 Pharmacological Treatment

The goal of pharmacological treatment is to restore a normal psychological state.

To address anxious states, *venlafaxine*, administered at a dose of 1–2 mg/kg once daily, is a preferred option (Hopfensperger 2016; Masson et al. 2024b; Metz et al. 2021). Its availability in granular form (Effexor™) makes it particularly convenient, which is always an asset when treating cats.

Clomipramine, at a dose of 0.25–1 mg/kg once daily, is another viable option (King et al. 2004). Its muscarinic side effects can help address symptoms of urine soiling and urine marking by strengthening the bladder sphincters. However, its strong sedative effects due to histaminergic action may be bothersome to many owners (Landsberg and Wilson 2005; Mertens et al. 2006; Seksel and Lindeman 1998).

In cases of aggression or binge eating, *fluoxetine* is an interesting option at a dose of 1–2 mg/kg once daily (Hart et al. 2005; Kaur et al. 2016). Monitoring for potential risks such as urinary retention or bladder distension (DiCiccio and McClosky 2022) is essential.

For depressive states, *mirtazapine,* at a dose of 1–2 mg/kg once daily, is the treatment of choice. However, in cases of obesity, it should be avoided or used only for a limited time, as it is highly orexigenic.

The transition back to a reactive state is not without challenges. Owners must be informed about the potential resurgence of marking behaviors and increased general activity levels.

10.1.4.2 Behavioral Therapy

- *Environmental Modifications*

The primary goal is to *enrich the cat's territory* (Herron and Buffington 2012; Houser and Vitale 2022; Kry and Casey 2007). While a complete relocation is not realistic, reconfiguring the living space is often sufficient to compensate for territorial poverty. Several elements must be addressed:

First, free up space and remove restrictions. If outdoor access is feasible, it should be organized. The objective is to ensure that all activity zones are identified and utilized (see ► Chap. 29, ► Sects. 29.2.1.1 and 29.2.1.2).

Once the cat has access to the entire apartment, create an *observation post*. Ideally, this involves a window setup that allows the cat to look down at the ground. Adding a small perch as an observation platform will enhance the experience.

Feeding routines also require adjustment. The cat should spend time locating and consuming its food. Complex *feeding bowls*, which owners can often create themselves, are highly effective (Dantas et al. 2016; Delgado and Dantas 2020; Naik et al. 2018). For instance, a piece of kibble hidden under a rug can keep a cat engaged for hours.

The play area must be enriched, as it helps reduce the cat's frustration and strengthens the relationship with the owner (Delgado et al. 2024, 2025). Play is an excellent activity for redirecting frustration, especially in cases where the cat regularly attacks its owners.

The owner's involvement is crucial. Establishing rituals is simple yet highly effective. For instance, medication administration can become part of a feeding ritual (see ► Chap. 29, ► Sect. 29.2.3.3). Rituals involving water or play sessions can also be created.

The creation (or utilization) of a "*small fear zone*" is particularly useful. This refers to a peripheral area of the territory that the cat rarely visits but perceives with a heightened sense of emotion. The area should allow the cat to retreat quickly to its secure territory. For example, the shared stairwell of an apartment building can serve as an enrichment zone. Once or twice a week, the cat can be allowed to explore the landing under supervision, with the apartment door left open for a quick retreat. Care must be taken to monitor for any urine marking behavior.

- *Behavioral Modifications*

The first prescribed modification is to stop the use of *positive punishment*, as it is counterproductive.

Active owner participation in territorial enrichment is typically the starting point for creating a new bond, which is often the most enriching change for the cat (Wojtaś et al. 2024). This also provides an opportunity to establish rituals, such as stroking or brushing, which can become part of the enrichment process.

Owners should be clearly warned about the risk of marking behavior as the cat transitions out of an inhibited anxious state. This risk can be particularly significant when activity levels resume, especially if areas that were previously off-limits are made accessible.

10.1.5 Therapeutic Follow-Up

Relapses are uncommon, and once the environment is truly enriched, the cat's psychological state can typically return to normal. However, partial successes do occur. Pharmacological treatment can increase the cat's tolerance, and while owners may make some efforts, the results may remain limited.

If the territory remains impoverished, it may be necessary to continue treatment.

In older cats, this condition often evolves into issues associated with aging. In such cases, the treatment should be maintained.

It is also important to emphasize that introducing a second cat as a solution for boredom is often a poor choice. Sharing a territory is frequently challenging for a cat, particularly if the territory is already impoverished.

10.1.6 Prevention

At the beginning of this subchapter, we discussed the alignment between the developmental environment and living conditions. It is crucial to consider this when acquiring a cat. Attempting to confine an adult cat rescued from the streets to a small apartment can be particularly challenging. Animal protection organizations are not always the best advisors in such cases. For these organizations, the primary goal is to provide a comfortable environment, but this may involve constraints that are difficult for the cat to endure.

Veterinarians also play an important role during changes in living conditions. Family separations are increasingly common and often lead to relocations to smaller homes. Advising owners to enrich their cat's environment before problems arise is a proactive way to support them, and it may be a service they value.

10.1.7 Clinical Vignette

- *Balty*

Balty is a 3-year-old male Abyssinian cat, neutered at 6 months. Acquired when his owner lived with a partner in a townhouse, he now resides in a two-room apartment on an upper floor. His owner works long hours and often returns home late, enjoying an active social life in the evenings.

The apartment is spacious enough, but when his owner is absent, Balty is confined to the hallway and kitchen. A small couch is present, and the owner places Balty's food bowl and litter box there when he is away. The owner dislikes Balty sleeping on his bed and worries about claw marks on the couch.

The consultation was prompted by nighttime conflicts. When the owner returns home tired, he wants to rest, but Balty demands attention, jumping on him for petting and playtime. After the owner's evening outings, Balty continues to seek interaction and insists on sleeping on the bed. If the owner shuts the door, Balty meows persistently and ends up confined in the kitchen.

The description of Balty's life is troubling, and the owner finds it difficult to recount. Awareness of the situation develops quickly during the discussion.

The diagnosis is a disorder resulting from territorial poverty associated with a reactive state.

The owner feels significant guilt. Balty is the last remnant of a previous life shared with his ex-partner, and the strict rules about cleanliness and order reflect a lingering sense of loyalty to that past (no fur, no mess, everything must remain pristine). Once this loyalty is acknowledged, it becomes easier to address.

The owner, who originally initiated Balty's adoption, recognizes his role in caring for the cat. It is agreed that Balty should be allowed on the bed, and a throw blanket is added to the couch to protect it. Balty is given full access to the apartment, significantly improving his living conditions.

A "small fear zone" is created for Balty, allowing him to explore the stairwell twice a week. Although timid, Balty ventures one floor up or down before quickly returning. Occasionally, he gets "lost" upstairs and meows until he finds his way back.

Balty is not anxious, so no pharmacological treatment is prescribed. A pheromone diffuser is placed in the apartment to promote relaxation.

The changes quickly yield positive results. The owner regains restful sleep, and during the day, Balty alternates between lounging on the couch (on his cushion) and watching the world from the window (in his basket).

The cleaning lady, who comes regularly, grows fond of Balty and allows him to run in the stairwell during her visits.

Summary of ▶ Sect. 10.1
Cats can live in confined spaces, but they require a range of activity zones. When their territory becomes impoverished and no longer provides the minimum activity they need, they may develop anxiety or even depression. The most notable symptoms are linked to the expression of frustration, often presenting as irritation-driven aggression in a context of frustration or urine marking.

If the anxious state persists without environmental changes, substitute activities like binge eating and depressive states may develop.

Treatment involves restoring a reactive psychological state that allows the cat to resume normal activities within a restructured territorial space. In addition to enriching the environment, rebuilding the relationship with the owners is an essential part of therapy. However, the cat's resumption of activity does not always equate to improvement for the owners, particularly when territorial marking behaviors return.

10.2 Disorders Resulting from Territorial Disrupting

10.2.1 Description

10.2.1.1 Definition

A cat's territory must be suited to its needs and remain relatively stable to fulfill its role in providing comfort. Variations can occur, but they must fall within the cat's tolerance range, which is determined by its behavioral development.

When faced with such variations, the cat may initially exhibit a *reactive state*. If it cannot adapt, it transitions into various pathological states. Consequently, we identify disorders originating from excessive variations in the territory, which may be perceived as either too rich or inconsistent, depending on the criteria applied.

10.2.1.2 Etiology and Pathogenesis

The key factor is the cat's ability to adapt. For example, a cat with a phobia of humans would find life in a bustling household with many family members to be an overwhelming and unmanageable territory. Conversely, for other cats, living with a single person who never leaves the apartment or receives visitors could represent territorial poverty, leading to disorders (see previous section).

Here, we refer to a cat that has undergone a high-quality developmental period and is well-suited to the environment it is offered.

A cat living in balance within its territory may be disrupted by various changes, which could include *geographical changes* such as moving to a new home, sudden access to the outdoors, or *familial changes* like a new baby, a divorce, cohabitation, or the adoption of another pet.

It is important to note that the environment is defined by the cat's perception of it. Organic changes in the cat, such as diminished sensory abilities or locomotor difficulties, can act as territorial disruptions. Aging-related illnesses may even begin with such disruptions.

This disorder is, therefore, the result of a cat's inability to adapt to changes in its territory, whether those changes are real or perceived.

10.2.1.3 Epidemiology

It is easy to imagine that this disorder is common. For cats, territorial disruptions are critical because they prevent them from finding comfort in the predictability of their environment. A parallel can be drawn with canine autonomopathies (Masson et al. 2024a), where a lack of predictability or abrupt changes affects the attachment bond in vulnerable individuals, leading to anxious symptoms in both the absence and presence of their owners **(see box: "Can autonomopathies be discussed in cats?")**.

Modern lifestyles, characterized by increasing mobility and frequent changes, contribute to the prevalence of these disorders.

Important Point of View

Can Autonomopathies be Discussed in Cats?

In dogs, problematic behaviors that occur when owners are absent form an important and frequent group of disorders known as autonomopathies (Amat et al. 2020; de Assis et al. 2020).

In cats, two primary reasons explain why autonomy-related disorders are less significant.

First, cats are territorial animals, and their sense of comfort in adulthood stems primarily from their territory in a broad sense, which includes relationships with the individuals within that territory. A fearful or anxious cat typically seeks isolation in a designated zone or isolation field that provides a sense of security.

Second, for kittens adopted very early, particularly before 5 weeks of age, a primary attachment bond with a human individual can sometimes form. Without experiencing maternal separation, such kittens may fail to structure a comforting territory at puberty, leaving them incapable of living independently when their owner is absent. In natural conditions, maternal separation is inevitable for cats, and all individuals eventually become autonomous. However, in human households, it is theoretically possible to create dependency in bottle-fed kittens whose autonomy is discouraged by humans. These cases are exceedingly rare because humans regularly leave the home, and cats are genetically programmed to structure themselves within their territory when alone.

For these reasons, we do not encounter cats in consultation displaying symptoms of distress during owner absences coupled with genuine emotional dependence in their presence. While cat behavior can vary depending on the owner's presence or absence, such absence is often perceived as a territorial change, potentially leading to anxiety as with other territorial disruptions. However, emotional dependency does not manifest in cats as it does in social species.

For cats, their territory appears to consist of both immobile and mobile objects, including humans. Cats even deposit different pheromones on these categories. When the absence of humans triggers anxiety, it is more consistent to classify such cases under disorders resulting from territorial disruptions (see ▶ Chap. 10).

The very existence of attachment as defined by Mary Ainsworth's Strange Situation Test (Main and Solomon 1990) is debated and lacks consensus, particularly regarding the experimental procedure's applicability to cats and the interpretation of observed behaviors (Chang et al. 2025; Potter and Mills 2015; Edwards et al. 2007).

Finally, very few studies address attachment-related disorders in cats (De Souza Machado et al. 2020; Schwartz 2002, 2003). Close examination of these studies reveals highly nonspecific symptoms, such as urine soiling and substitute activities, without identifying signs of distress or exaggerated contact-seeking behaviors in the owner's presence. This further supports the idea that the term "autonomopathy" is poorly suited to a non-social species like the cat.

10.2.1.4 Presenting Complaints

The reasons for consultation encompass a range of symptoms associated with anxious states that may disturb the owner: urine marking, scratching, excessive licking, idiopathic cystitis, or redirected aggression. However, what owners often mention early in the discussion is the identifiable cause linked to these symptoms. This condition typically arises from a change in the environment, which is often easily identifiable. Examples include the arrival of a newborn, the adoption of a new dog or cat, moving to a new home, or even the presence of a new cat in the neighborhood.

When a practitioner identifies symptoms consistent with an anxious state and no apparent cause is found, it is crucial to investigate any significant environmental changes that may explain the transition to an anxious state. A cat's territory is the foundation of its comfort, and any disruption to this environment can trigger a shift to a pathological psychological state.

If no environmental factors explain the anxious state, a potential organic cause should be investigated.

10.2.1.5 Symptomatology

The disorder always develops in the same way. Initially, the cat attempts to restore its preferred territory. It enters a *reactive state*, primarily expressed through *increased pheromonal marking* and the protection and defense of *isolation zones*. As described in ▶ Chap. 1 (see ▶ Sect. 1.3.2.1), various types of pheromonal marking serve specific functions.

Facial marking is primarily intended for the cat itself (except in cases of sexual overmarking), though the olfactory markers can be interpreted by other cats. Spot urine marking and scratching convey messages directed at other cats, even if the resident cat may derive some comfort from these actions. Increased marking should always be interpreted as a warning. It can signify an attempt to claim ownership of the space or to exclude or limit the presence of an external threat.

If the territory does not quickly regain its comforting properties, the cat will escalate its marking behaviors. Over time, these efforts lose their effectiveness, as the anxiety becomes entrenched. This diminishing effectiveness further exacerbates the anxiety.

It is as if you were trying to redecorate your home by adding more and more decorative objects, which progressively lose their personal significance and ultimately clutter your space. Alternatively, it resembles placing numerous "No Entry" signs in your garden, to the point where you can think only of potential intrusions, while the overwhelming signs detract from the space they were meant to protect.

The reactive state eventually collapses, and the cat becomes *anxious* or even *depressive*. In both cases, it fails to restructure its environment, leading to increased discomfort. The symptoms then evolve into the classic manifestations of anxiety or depression.

Aggression may occur, driven by the irritability that often accompanies anxiety. If the cat perceives an invasion (or experiences it as such) by an inaccessible intruder, such as another cat passing through the garden or an individual who is protected or too intimidating to attack, redirected aggression toward household members may arise. This adds complexity to the situation, as such aggression can sometimes be difficult to interpret.

Marking behavior, which initially escalates significantly, begins to decrease, often replaced by urine soiling, most commonly a consequence of idiopathic cystitis.

In inhibited forms of anxiety, substitute activities may emerge. The most frequent is a significant increase in food consumption, often accompanied by an obsession with food. This is exacerbated by owners' attempts to impose dietary restrictions to combat obesity.

Licking, particularly of the abdomen, is also commonly observed. While it may initially provide some relief for the cat, it eventually becomes a substitute activity that no longer alleviates the animal's discomfort.

> **Important Point of View**
>
> **Territorial Invasion**
>
> The invasion of a cat's territory by another cat is merely a specific instance of territorial variation. The discomfort is heightened because cats, as non-social animals, feel compelled to defend their territory. However, sharing is sometimes possible if the territory is sufficiently large and enriched.
>
> When a cat is forced to share its territory with another, the symptoms of territorial discomfort are compounded by those of intraspecific conflicts, which are detailed in the section on relational disorders (see ▶ Chap. 11).

In some cases, cats may leave their home to settle a few houses away in another household, or rather, in a different territory that better suits them. This change in territory can also occur when female cats are forced to coexist with their offspring once they have reached adulthood. Unable to expel the now-grown kitten, the mother may compensate by leaving, thereby facilitating the kitten's transition to independence.

10.2.1.6 Evolution

If the reactive state fails to restore balance, a favorable outcome is generally unlikely. Amplified marking, which does not achieve comfort, gives way to an anxious state characterized by urine soiling, hypervigilance, and substitute activities.

These symptoms are often poorly tolerated by owners, exacerbating the discomfort. In addition to the territorial disorder, conflicts may arise, often accompanied by punishments, further degrading the relationship between the cat and its owner.

In some cases, the source of the problem may disappear—for instance, the baby grows up, a new move prevents encounters with the neighborhood cat, or the animal invading the cat's space passes away. In such situations, a return to normalcy might seem possible. However, in most cases, the anxious state persists even after the cause has been removed.

10.2.2 Diagnosis

10.2.2.1 Diagnostic Criteria

Cats affected by a condition resulting from territorial changes inevitably exhibit an anxious or depressive pathological state (◘ Fig. 10.2). Symptoms typically appear in early adulthood, at a time when the territory should serve as the cat's primary source of reassurance.

The identification of a modification in the living environment is essential.

The diagnosis is based on the presence of several symptoms that reflect the cat's inability to make its territory comforting, selected from the following list:

- irritation-driven aggression, particularly in response to any moving object;
- altered facial pheromonal marking, often increased or sometimes absent;
- increased scratching around areas that are not associated with the cat's isolation zones;
- urine marking in the early stages, sometimes complicated by urine soiling;
- compensatory binge eating;
- abdominal licking;
- idiopathic cystitis.

10.2.2.2 Direct Observation

Direct observation is not specific to this disorder. The practitioner identifies anxiety or depression by observing the typical symptoms of these states. Obesity is common, and owners often describe their cat as unstable, irritable or grumpy, lazy, and overly focused on food.

In older cats, it may be necessary to investigate performance declines that could alter the structure of the territory. For example, an arthritic cat may lose access to its observation areas, or a visually impaired cat may struggle to manage its territory if changes are present.

10.2.2.3 Comorbidities

In theory, any comorbidity can exist. However, we have assumed that the cat was initially capable of structuring its territory normally, requiring a sufficiently harmonious developmental background for this ability.

Fear axis disorders often set the stage for this condition, particularly because these cats are well-adapted to impoverished territories. Cats that developed in low-stimulation environments are the primary victims of this disorder.

The introduction of a new individual can also trigger a phobia. For instance, consider a cat with a phobia of babies, whose owners welcome a child into the home. This phobia complicates territorial management and must be addressed alongside the territorial disorder.

Aging-related disorders also reduce adaptability to change, further predisposing cats to territorial disruption.

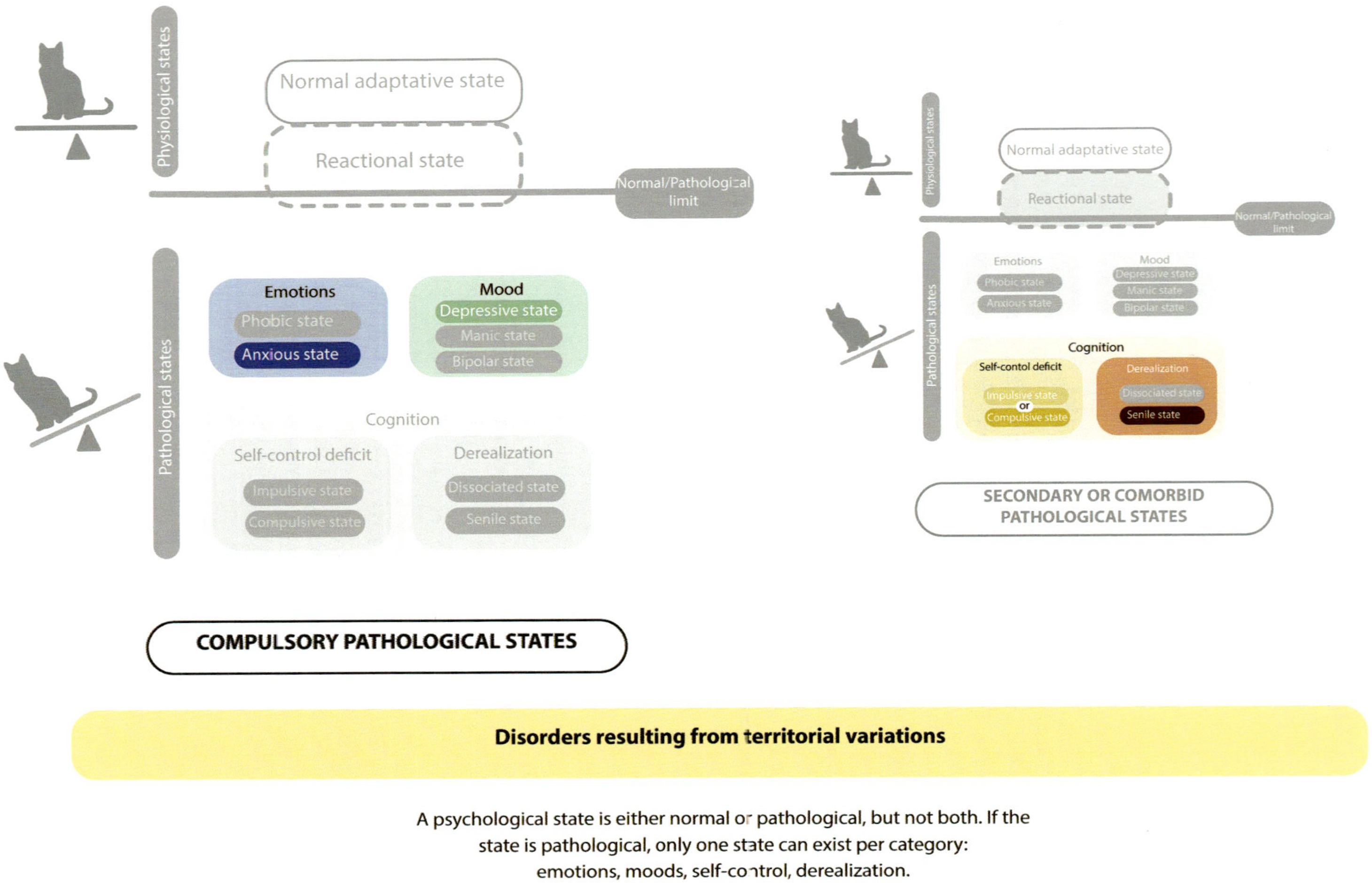

■ **Fig. 10.2** Pathological states associated with disorders resulting from territorial disruption. Mandatory pathological states: anxious **and/or** depressive state. Possible comorbid pathological states: phobic state, and/or impulsive **or** compulsive states

Self-control axis disorders do not directly contribute to the emergence of this condition. However, their presence prior to its onset can complicate and amplify the clinical picture, particularly in cases involving impulsive and compulsive states. Substitute activities, when present, are often highly pronounced due to the poorly regulated behavioral sequences in these cats.

Finally, *relational disorders* are a frequent, almost inevitable, complication of disorders resulting from territorial variations if left untreated. Living with an anxious cat exhibiting urine soiling strains owners, often leading to punishment or isolation of the cat, which worsens the disorder and degrades the human-animal relationship.

10.2.2.4 Differential Diagnosis

The practitioner must investigate the causes of anxiety. The challenge is to avoid automatically attributing pathological states to the territory and its variations as the primary cause. For a cat transitioning into an anxious state, there is always a phase of territorial destabilization.

The practitioner must identify a genuine environmental cause, such as the arrival of another animal, a move, or any other change, to diagnose a pathological state resulting from environmental variations.

The differential diagnosis should explore other possible causes of the anxious or depressive state, particularly by ruling out *sensory deprivation syndrome* and *other fear axis disorders*, where environmental changes alone do not sufficiently explain the rapid onset of anxiety. In such cases, the cat's vulnerabilities are more likely the cause than environmental modifications.

Relational disorders, whether interspecific or intraspecific, should also be excluded. In these cases, aggression toward a member of the group often precedes signs of pheromonal marking. Relationship degradation is typically observed early in the behavioral changes.

10.2.3 Prognosis

The prognosis depends on the changes the owners can implement. However, they do not always have unlimited alternatives to offer. Events such as the arrival of a child, the need to reorganize the living space, or the sudden obligation to work from home can disrupt the cat without any possibility of reversal. In most cases, these changes are permanent. For instance, adopting a puppy or even another cat represents irreversible modifications to the household.

The prognosis therefore hinges on the possibility of internal adjustments to the home and the cat's capacity to adapt. Despite these challenges, the prognosis is often good with appropriate treatment.

Deciding to discontinue pharmacological treatment is more complex because if the underlying cause persists, a return to the anxious state is likely. Thus, while treatment can provide a solution, the risk of relapse remains high.

10.2.4 Treatments

Pharmacological treatment aims to restore a reactive state that enhances the cat's adaptability. The choice of psychotropic medication will be guided by the specific pathological state and symptoms observed.

Behavioral therapy will focus on minimizing the impact of the disruptive environmental factor.

10.2.4.1 Pharmacological Treatment

The use of *pheromone diffusers* theoretically facilitates the restructuring of the territory (DePorter et al. 2019; Hargrave 2021), although the level of evidence for their efficacy remains low (Frank et al. 2010). Diffusers are preferable to sprays because they do not impose marks on the cat but instead provide a constant presence. Manual marking with a spray is often clumsy—too high, too intense, then too weak, or poorly positioned. The number of diffusers should be sufficient, with one per floor and for every 50 m^2 of space.

The objective of pharmacological treatment is to restore a normal psychological state. To treat anxious states, *venlafaxine*, administered at a dose of 1–2 mg/kg once daily, is a preferred option (Hopfensperger 2016; Masson et al. 2024a; Metz et al. 2021). It is also easy to administer, which is particularly beneficial for cats.

Clomipramine, at a dose of 0.25–1 mg/kg once daily, is another viable choice (King et al. 2004). Its muscarinic side effects help address urine soiling and marking by strengthening the bladder sphincters. However, the significant sedation caused by its antihistaminic effects can be problematic for many owners (Landsberg and Wilson 2005; Mertens et al. 2006; Seksel and Lindeman 1998).

In cases of aggression or binge eating, *fluoxetine* is an interesting option at a dose of 1–2 mg/kg once daily (Hart et al. 2005; Kaur et al. 2016). Monitoring for risks such as urinary retention or bladder distension is essential (DiCiccio and McClosky 2022).

For depressive states, *mirtazapine*, at a dose of 1–2 mg/kg once daily, is the drug of choice. However, in cases of obesity, its use should be avoided or limited over time, as it is highly orexigenic.

10.2.4.2 Behavioral Therapy

- *Environmental Modifications*

It is essential for the cat to regain stability in the organization of its territory. Owners must assist in reconstructing the lost environmental stability.

The primary need is an *isolation zone* that is calm and secure. Installing *interior cat doors* or *baby gates* allows the cat to access a refuge area while preventing disruptive elements from entering. This zone should always be quickly accessible, especially if the cat needs to escape from a new arrival (e.g., a dog, another cat, or a child invading its territory).

Similarly, owners must ensure that the cat's various activity zones are adequately protected to allow the intended activities to take place. For instance, an observation zone on a windowsill may no longer feel safe if the disruptive element is external, such as a new cat in the neighborhood. A play zone may become uncomfortable if the baby or dog causing the problem can access it every time the cat is about to engage in play.

- *Behavioral Modifications*

It is important to help owners understand the cat's discomfort so they can respond effectively. Punishments, even those of very low intensity, should be eliminated, as they can quickly lead to anticipatory hypersensitivity in the cat.

The environment must be stabilized to make it as unthreatening as possible from the cat's perspective. This may sometimes involve limiting the cat's space to create a protective zone while encouraging owners to compensate for this isolation by enhancing their relationship with the cat. Introducing new rituals can contribute significantly to reducing stress and promoting comfort.

Finally, a period without changes is necessary to allow the cat to readjust and stabilize within the newly created environment.

10.2.5 Therapeutic Follow-Up

Owners tend to quickly forget the problem, so it is important to monitor that, once the cat has adapted to one change, they do not impose another. A common scenario involves treating a cat after the arrival of a second cat, only for the owners to adopt a third once balance has been restored. A simple warning may suffice to prevent this.

Follow-up should focus on identifying events that could disrupt the cat, allowing owners to decide against introducing such changes. Owners often do not realize that significant life events, such as home renovations, the birth of a child, or the departure of a teenager, can also affect the cat.

10.2.6 Prevention

The key to prevention is educating owners about the ethological functioning of cats. A cat's olfactory and pheromonal world is inaccessible to humans, making it difficult for people to understand that adopting a new companion can be such a traumatic experience for their animal. Furthermore, the fact that cats are non-social animals that find comfort in their environment is a challenging concept for humans, who belong to a social species. Communicating with owners to help them better understand the feline world is likely the most effective preventative measure.

In line with this, the use of pheromone diffusers is much more effective as a preventative measure than as a treatment. Encouraging owners to set up diffusers at the start of a change, while the cat is still in a reactive state, is an excellent way to prevent the development of pathological states.

10.2.7 Clinical Vignette

▪ *Emily*

Emily is a 3-year-old spayed European cat. She lives comfortably with a young couple in a large two-story house that provides access to a city garden, which is shared with other gardens. The space is large, and this network of neighborhood gardens creates a calm and peaceful area. Emily rarely goes out, but she enjoys exploring her lawn and doing her business outside, provided it's not raining. She watches birds without actively hunting them. In the evening, she likes to stay close to her owners for cuddles while watching television.

Recently, two simultaneous events disrupted her world. Her owner gave birth to a very loud baby, which is completely out of place in Emily's environment. A few days before this unpleasant arrival, the neighbors moved in, bringing with them a new neutered male cat in one of the neighboring houses. This cat lives at least four gardens away, but it occasionally roams, even daring to mark Emily's garden.

Since then, Emily has been unable to look out the bedroom window, which is now closed (this is where the baby resides). Moreover, she no longer feels comfortable going into the garden because she fears encountering the other cat.

Emily has suddenly become incontinent: she has urinated in front of the garden door a few times. Since that didn't change anything, she has now begun urinating outside her litter box at least three times a week. She also attacked her owner quite aggressively as she passed by the door, with no apparent reason. Whenever she can, Emily enters the baby's room and urinates there. If the baby starts crying, she runs off and hides.

Emily meows a lot, circles around her owner, and demands more attention, but her owner is less available.

Thus, Emily's territory has experienced two simultaneous changes that both contribute to a restriction (more dangerous access to the outside, prohibition from the bedroom, and the observation post she previously had). It's also possible that Emily is exhibiting an abnormal fear of other cats, but it may simply be a natural fear of a larger, unfamiliar individual. She is encountering a newborn for the first time, and a shift toward a phobia is possible.

The diagnosis established is a territorial disorder resulting from territorial disruptions associated with an intermittent productive anxious state.

To assist Emily, venlafaxine at a dose of 1 mg/kg (approximately 15 granules per day for her weight of 4.2 kg) is prescribed. The ease of administration is a key factor in not overwhelming the owners, who are already managing their newborn. The owner is instructed to allow Emily access to the bedroom when the baby is not sleeping. Taking advantage of spring, the owner tries to help Emily return to the garden when he is outside himself (Emily enjoys staying with him when he gardens). These measures lead to quick results, and Emily is clean again by the first week of treatment, although she still hisses at the window when she sees the other cat on the garden wall. The owner was almost attacked, but Emily merely hissed when she entered the room.

After 4 months, discontinuing the treatment is being strongly considered, but without warning, the owner brings home a puppy "to grow up with their daughter." Emily then retreats upstairs (there are unoccupied rooms) and refuses to come down. However, the puppy seems to want to play…

Venlafaxine is maintained, and gabapentin is added 2 h before bringing Emily into the living room in the evening, while controlling the puppy's movements to prevent it from frightening her. The instruction is to place Emily high on her cat tree, where she used to rest before the puppy's arrival. The puppy is somewhat confined to the kitchen when it gets too excitable, allowing Emily to come and explore its scent.

Quickly, a first contact occurs, and the puppy receives a paw swipe to the nose. Since then, it has been much more cautious and approaches Emily more calmly: it gently approaches and freezes. It even dared to give Emily a lick, who hissed in response.

During a garden outing, Lou (the puppy) barked at the neighbor's cat on the wall. The cat fled, and now when Lou is outside, Emily dares to go out. The dog remains cautious when approaching Emily, but she tolerates him without issue.

Gabapentin is discontinued, and Emily continues to roam the house freely. Six months after the puppy's adoption, venlafaxine is stopped. Emily remains stable and even accepts affection from the child, who is now actively moving around.

10

Summary of ▶ Sect. 10.2

Cats organize their territory into different activity zones, and it is the consistency of this organization that allows them to find the comfort they need. Small modifications to their living space are acceptable if they have had a sufficiently rich behavioral development. However, too many changes quickly become a source of stress for the cat, and if adaptation is not possible, it can lead to a pathological anxious or depressive state, or both.

After a period of reactive state marked by intensified marking, particularly urine marking, anxiety sets in, sometimes expressed through aggression or, in inhibited forms, through substitute activities. Some cats transition directly into a depressive state.

Treatment is essential and most often involves venlafaxine or clomipramine to address the anxious state. In cases of substitute behaviors like binge eating, fluoxetine may be a good option to help reduce appetite.

The owner must assist the cat in rebuilding a stable territory by ensuring a certain degree of consistency and stability in the environment.

10.3 Territoriality Disorders Secondary to Another Behavioral Disorder

10.3.1 Description

10.3.1.1 Definition

This section describes certain cats that, despite ideal conditions, are unable to establish a comforting territory that meets their needs. These cats exhibit dysfunction in another behavioral axis, such as the self-control axis, the fear axis, or the relational axis. Their underlying condition prevents the effective construction and structuring of their living space.

10.3.1.2 Etiology and Pathogenesis

The etiology is straightforward. If a cat is affected by a deprivation syndrome and experiences multiple phobias in its living space, it is unable to produce the appropriate pheromones, and its phobic or anxious state often prevents it from structuring its environment effectively.

Cats with dysregulation on the self-control axis are often immature and disorganized. They are incapable of subdividing their living space and cannot find the necessary comfort for balance in their pheromonal markings. While they may recognize pheromonal traces, they do not stop long enough to calm themselves by reading them. Furthermore, these markings are often poorly distributed, and the marking itself is disorganized.

Cats with dysregulation on the relational axis tend to isolate themselves to cope with relational issues. By not utilizing the full extent of their territory, they are unable to structure their territory properly.

10.3.1.3 Epidemiology

These secondary disorders are common because disruptions in the fear or self-control axes are relatively frequent, and their management is often overlooked unless they cause significant disturbance for the owners. Moreover, primary disorders are rarely identified because owners may dismiss them as normal cat behavior or attribute them to a difficult early life, such as saying, "He was weaned too early, or abandoned by his mother and found on the street." While these explanations can shed light on the pathological process, they do not justify the abnormality of the behavior.

The lack of adaptability caused by these symptoms ultimately affects the cat's well-being, leading to a relational or territorial disorder in adulthood, as the cat remains unable to adapt.

10.3.1.4 Presenting Complaints

The most common reason for consultation is urine marking or soiling. If aggression is present, it can become the primary reason for consultation. While the symptoms of the initial disorder (fear axis or self-control axis) might not have significantly disturbed the owners, the complication into a secondary territorial disorder is often much more bothersome. There may be an accumulation effect, where the new symptoms, added to the previous ones, trigger the consultation.

10.3.1.5 Symptomatology

These animals, due to their primary disorder, quickly progress to pathological anxious and depressive states.

Hyperactive cats often also have significant relational issues. Their inability to remain calm frequently leads to punishments, and they are predisposed to becoming compulsive, making them prime candidates for compulsive eating with constant food-seeking behavior. If they develop other substitute behaviors, they will often be even more pronounced.

Cats that suffer from fear axis disorders tend to become depressed quickly and often present with highly inhibited behaviors.

Cats with relational disorders are often perceived as trying to get back at their owners. However, they simply need to express their distress in some way, and the idea that they are acting with intent toward their owner is, of course, a human construct. Rebuilding a calming relationship and assuring the owners that the cat is not acting "on purpose" becomes crucial.

10.3.1.6 Evolution

Abnormally fearful cats often end up retreating to an isolation zone and cease using the full space available to them. Their state is typically depressive. Some may become phobic and, outside of their isolation zone, they cannot tolerate a particular person or another animal. Aggression is possible in these cases.

Hyperactive cats are more likely to evolve into aggressive behaviors. In these cases, the demand for treatment is often more urgent, and they are treated more frequently than the previously mentioned cats, who do not disturb their owners as much.

In relational disorders and those involving territoriality, the combined worsening of both conditions is almost systematic: relational disorders often secondary lead to territoriality disorders, as the environment becomes impoverished, and conversely, territorial issues frequently lead to a deterioration in the relationship due to symptoms that are difficult for owners to tolerate. Therefore, if left untreated, it is common to see these disorders evolve into comorbidities.

10.3.2 Diagnosis

10.3.2.1 Diagnostic Criteria

Cats with territoriality disorders secondary to another behavioral condition must exhibit a pathological anxious or depressive state (◘ Fig. 10.3). Symptoms related to this condition typically appear in early adulthood, but the symptoms of the primary disorder are present earlier, with other possible pathological states before that.

The diagnosis is based on the inability to explain the entire clinical picture solely by the territoriality disorder. Behavioral signs clearly reveal a territory that the cat cannot structure, but other untreated symptoms are also noted, which constitute the primary cause or an aggravating factor of the clinical signs.

10.3.2.2 Direct Observation

Cats with these disorders do not present characteristics other than those of their primary condition. Depressive states may be highlighted by a neglected appearance of the coat.

In almost all cases, the cat's relationship with its owners is disturbed, either because the relational disorder is primary or due to the anxious symptoms of the secondary disorder, particularly urine soiling.

10.3.2.3 Comorbidities

Comorbidities are systemic, as the primary disorder is the cause of the territoriality disorder. This results in significant disruptions with severe clinical presentations, and the combination of these disorders makes treatment more challenging. Additionally, the concurrent presence of these disorders promotes premature aging.

10.3.2.4 Differential Diagnosis

The goal is to clearly differentiate the comorbidities in order to prioritize one treatment axis over another. This decision will be based on the owner's concerns and the symptoms that are most disruptive to them.

In general, primary conditions should be addressed first, as well as the pathological state that is the main reason for the consultation, which aligns with the owners' primary concerns.

10.3.3 Prognosis

The prognosis is guarded due to the combination of disorders, but also because owners often seek consultation late, when they are overwhelmed by a highly advanced condition. However, if the disorder is diagnosed early, it is possible to achieve a state of equilibrium. Most often, developmental issues are involved, compounded by maladaptation to the living environment, and it is crucial to adapt the environment to what the cat is capable of structuring.

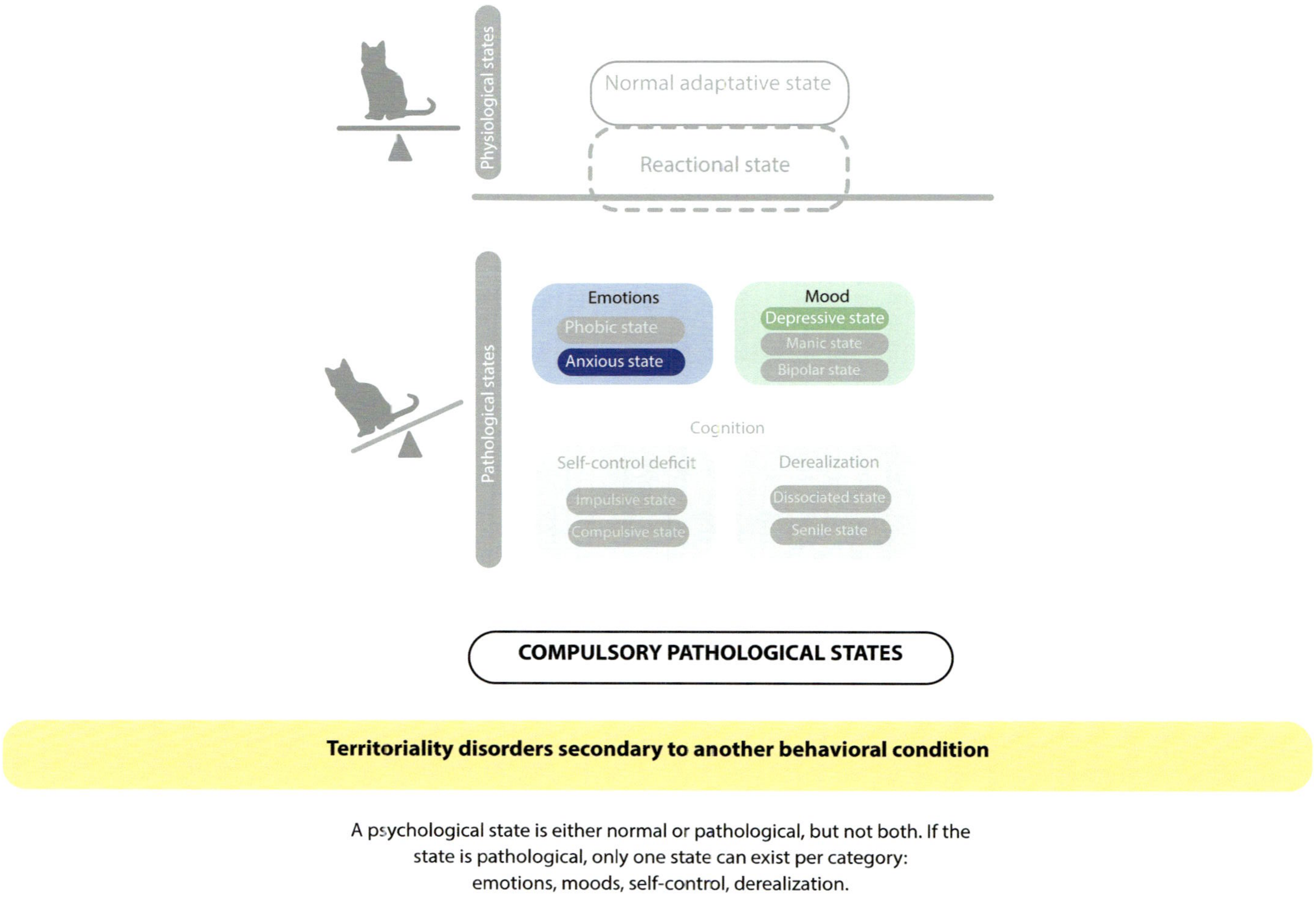

Fig. 10.3 Pathological states associated with territoriality disorders secondary to another behavioral condition. Mandatory pathological states: anxious **and/or** depressive state

10.3.4 Treatments

10.3.4.1 Pharmacological Treatment

The treatment will focus on the primary condition, and we refer to ▶ Chaps. 8 and 9, which address these disorders. The following chapter discusses relational disorders, and as we have already mentioned, there are numerous connections with territoriality disorders.

Due to the accumulation of disorders, it is not uncommon for the treatment to need to be maintained long term, with the possibility of resuming or continuing the treatment for life during relapses.

10.3.4.2 Behavioral Therapy

- *Environmental Modifications*

In addition to the measures taken for the primary disorder, specific territorial management measures must be added. The use of pheromone diffusers is often essential. Creating dedicated zones is important, and the owner can assist in creating a comforting territory by initially limiting it to part of the living space, gradually expanding it to the entire home.

For impulsive and compulsive cats, outings can be arranged to meet their high motivation levels. However, it is best to organize these only once the cat has made progress on the self-control axis to avoid accidents.

- *Behavioral Modifications*

Positive punishment must be eliminated. The owner should be educated that their animal is sick and can improve. This is not a matter of vengeance. The goal is to implement therapies for the other dysfunctional axes. It is sometimes surprising in practice to observe that owners are astonished to learn that even the initial condition they have always viewed as inevitable can be improved.

10.3.5 Therapeutic Follow-Up

As we have emphasized, these are often primary developmental disorders, and while improvement is possible, the cat will never be completely normal. Follow-up should anticipate potential problems caused by changes that may seem insignificant to the owner, who believes their cat is cured.

The balance achieved is often fragile, and relapses are common.

10.3.6 Prevention

It is essential to focus on developmental periods and challenge the idea that love alone can solve all problems. Addressing behavioral disorders as early as possible is the best way to prevent secondary disorders from developing, even if the initial symptoms are not very bothersome for the owner. Veterinarians play a crucial role in identifying these issues during early vaccinations or puberty consultations, and they are well-placed to explain that failing to address these concerns can lead to more troublesome secondary disorders later on (see ▶ Chap. 31).

Finally, highlighting the cat's distress, even when symptoms do not seem to disturb the owner, is often an effective way to raise awareness that a real illness exists that requires treatment.

10.3.7 Clinical Vignette

- *Tactic*

Tactic is a two-and-a-half-year-old neutered male cat. He was rescued after being found in a shoebox at the bottom of a trash bin when he was about 10 days old. He was bottle-fed and cared for by his owner, with whom he developed a strong bond. From his first vaccinations, he showed signs of being difficult. A handling test revealed severe inhibition issues, and the veterinarian warned his owner about the risk of unfavorable development.

Since the consultation, Tactic has become incontinent, restless, and increasingly aggressive.

At 6 months, Tactic began urinating around the house, marking everywhere. Due to his disruptive behavior, his right to sleep with his owner was taken away, and since then, he has become agitated at night to the point where he is now confined to the laundry room every evening. His incontinence has become systematic, as he never uses his litter box.

His owner has been in a relationship for about a year. Her partner tried to bond with Tactic, but from day one, the cat was aggressive. Now, when he tries to touch Tactic, the cat hides or attacks the owner, who has been bitten several times in these situations.

Tactic's latest issue is an obsessive behavior around food. Whenever he can, he steals food. His kibble is eaten in seconds, and he becomes threatening when his owner prepares meals. In response, she has to lock herself in the kitchen when she weighs his kibble, leaving him alone during mealtime.

The owners are exhausted, and euthanasia is being considered during the consultation.

The diagnosis established by the veterinarian is a generalized impulsive disorder associated with intermittent productive anxiety due to feline HSHA and deprivation syndrome. Additionally, Tactic suffers from a secondary interspecific relational disorder caused by both developmental disorders.

As a result, Tactic is unable to structure his territory. He also suffers from insufficient sleep and abruptly falls asleep at his owner's feet during the consultation.

After an ethological readjustment, fluoxetine 5 mg/day is prescribed to manage the impulsive state and aggression. An NSAID is also prescribed to treat an idiopathic cystitis, as blood was found in his urine.

The therapy involves creating a resting area different from the laundry room and imposing this space on Tactic at night once he is under control from the fluoxetine. A complex feeding bowl project to slow down his food intake and make it more natural is planned but postponed for later.

The relational disorder will be addressed later with her partner and will be handled after the follow-up consultation.

The effects of fluoxetine are rapid. His appetite is quickly regulated, and the aggression around meals disappears. He becomes clean again, despite two markings on the bedroom door. Tactic starts seeking solitude and no longer demands constant contact. He quickly adopts his new bed.

After a year, the situation seems stable. Tactic still does not want to establish a relationship with his owner's partner, but everyone has adjusted.

The situation remains stable, but it is decided not to stop the treatment, as Tactic remains irritable. His hyperactivity and impulsiveness eventually lead him to escape after a move, and he becomes a victim of a street accident, which proves fatal.

Summary of ▶ Sect. 10.3

This section describes cats that, despite ideal conditions, are unable to establish a comforting territory due to underlying behavioral disorders. These disorders may involve disorders from the self-controls axis, the fear axis, or the relational axis, preventing the effective structuring of their living space. These cats often exhibit anxious or depressive pathological states, and their inability to structure their territory leads to problematic behaviors such as urine marking, aggression, or eating disorders. Primary disorders are often overlooked by owners, as their symptoms do not immediately disrupt daily life. However, over time, these issues can evolve into complex secondary conditions that are difficult to treat. Treatment includes pharmacological approaches and behavioral therapies, including environmental modifications. However the treatment relies on treating the primary condition. Therapeutic follow-up is crucial to anticipate potential future issues, and prevention relies on early intervention for behavioral disorders.

Disorders associated with dysregulation of the territoriality axis

Disorder's name	State diagnosis	Diagnostic criteria	Differential diagnosis
Disorders resulting from territorial poverty	Anxious Depressive	Symptoms appear in early adulthood when the territory should be the cat's central point of comfort. The presence of an impoverished environment with a lack of stimulation, where the cat cannot structure its activity zones, is key. The diagnosis is based on several symptoms that express the cat's frustration in the face of an impoverished environment: Aggression due to irritation in response to any moving object Altered facial pheromonal marking, often increased, sometimes absent Increased scratching around non-isolation zones Early-stage urine marking, sometimes complicated by urine soiling Compensatory binge eating Abdominal licking Idiopathic cystitis	*Organic disorders* Loss of sensory abilities Any painful condition *HSHA syndrome* Environmental enrichment does not resolve the condition, as the environment is never stimulating enough for cats suffering from HSHA syndrome. *Territorial disorder secondary to another behavioral disorder* Environmental improvement does not significantly improve the clinical picture. Investigate another behavioral disorder that may explain the observed pathological state. *Aging disorders* The animal is over 8 years old. The disorder appears at this age in an environment that has not become poorer and was previously well tolerated.
Disorders resulting from territorial disruption	Anxious Depressive	Symptoms appear in early adulthood when the territory should be the central point of comfort for the cat. Identifying a living environment that has been modified, often enriched with new stimuli responsible for variations that disturb the cat, is essential. The diagnosis is based on the presence of several symptoms that express the cat's inability to make the territory comforting. These symptoms include: Aggression due to irritation Altered facial pheromonal marking, often increased, sometimes absent Increased scratching around areas that are not designated isolation zones for the cat Early-stage urine marking, sometimes complicated by urine soiling Compensatory binge eating Abdominal licking Idiopathic cystitis	*Organic disorders* Loss of sensory abilities Any painful condition *Social phobia or deprivation syndrome* Evidence of fear reactions before the suspected environmental change that is thought to be the cause of the disorder. *Interspecific relational disorder* Poor relationships existed prior to the environmental change, and symptoms of aggression due to irritation toward certain individuals are observed beforehand. *Intraspecific relational disorder* Poor relationships themselves constitute the environmental change, and symptoms of aggression due to irritation toward the other cat are observed beforehand. *Aging disorders* The animal is over 8 years old. The condition appears at this age in an environment that has not become poorer and was previously well tolerated.

Territoriality disorders secondary to another behavioral disorder	Anxious Depressive	Symptoms related to this condition appear in early adulthood, but symptoms of the primary condition are present earlier, with other possible pathological states emerging before. The diagnosis relies on the inability to explain the entire clinical picture solely by the territorial disorder. While there is a territory that the cat cannot structure, other untreated symptoms are noted during behavioral observation, indicating the presence of the primary condition as the underlying cause or an aggravating factor of the clinical signs.	*Organic disorders* Loss of sensory abilities Any painful condition *Disorders resulting from territorial disruption* Before the environmental modification, the cat exhibits no behavioral symptoms and has had normal development. *Aging disorders* The animal is over 8 years old. The disorder appears at this age in an environment that has remained stable and was well tolerated prior to aging.

References

Amat M, Le Brech S, Camps T, Manteca X (2020) Separation-related problems in dogs. Adv Small Anim Care 1:1–8. https://doi.org/10.1016/j.yasa.2020.07.001

Chang H, Zhang J, Huang H, Aviles-Rosa EO, Huang H, Yan G, Xiao Z, Liu Q, Deng B, Zhang L (2025) The effects of owner-cat interaction on oxytocin secretion in pet cats with different attachment styles. Appl Anim Behav Sci 106524. https://doi.org/10.1016/j.applanim.2025.106524

Dantas LM, Delgado MM, Johnson I, Buffington CT (2016) Food puzzles for cats: feeding for physical and emotional wellbeing. J Feline Med Surg 18(9):723–732. https://doi.org/10.1177/1098612X16643753

De Souza Machado D, Oliveira PMB, Machado JC, Ceballos MC, Sant'Anna, A. C. (2020) Identification of separation-related problems in domestic cats: a questionnaire survey. PLoS One 15(4):e0230999. https://doi.org/10.1371/journal.pone.0230999

de Assis LS, Matos R, Pike TW, Burman OHP, Mills DS (2020) Developing diagnostic frameworks in veterinary behavioral medicine: disambiguating separation related problems in dogs. Front Vet Sci 6:499. https://doi.org/10.3389/fvets.2019.00499

Delgado M, Dantas LMS (2020) Feeding cats for optimal mental and behavioral well-being. Vet Clin N Am Small Anim Pract 50(5):939–953. https://doi.org/10.1016/j.cvsm.2020.05.003

Delgado M, Marcinkiewicz E, Rhodes P, Ellis SL (2025) Assessing the impact of a human behaviour change intervention to reduce owner-related barriers to interactive cat play. Appl Anim Behav Sci 282:106479. https://doi.org/10.1016/j.applanim.2024.106479

Delgado M, Marcinkiewicz E, Rhodes P, Ellis SLH (2024) Identifying barriers to providing daily playtime for cats: a survey-based approach using COM-B analysis. Appl Anim Behav Sci 280:106420. https://doi.org/10.1016/j.applanim.2024.106420

DePorter TL, Bledsoe DL, Beck A, Ollivier E (2019) Evaluation of the efficacy of an appeasing pheromone diffuser product vs placebo for management of feline aggression in multi-cat households: a pilot study. J Feline Med Surg 21(4):293–305. https://doi.org/10.1177/1098612X18774437

DiCiccio VK, McClosky ME (2022) Fluoxetine-induced urinary retention in a cat. JFMS Open Rep 8(2):20551169221112065. https://doi.org/10.1177/20551169221112065

Edwards C, Heiblum M, Tejeda A, Galindo F (2007) Experimental evaluation of attachment behaviors in owned cats. J Vet Behav 2(4):119–125. https://doi.org/10.1016/j.jveb.2007.06.004

Frank D, Beauchamp G, Palestrini C (2010) Systematic review of the use of pheromones for treatment of undesirable behavior in cats and dogs. J Am Vet Med Assoc 236(12):1308–1316. https://doi.org/10.2460/javma.236.12.1308

Hargrave C (2021) Pheromones and 25 years of pheromonotherapy: what are they and how do they work? Vet Nurse 12(3):116–122. https://doi.org/10.12968/vetn.2021.12.3.116

Hart BL, Cliff KD, Tynes VV, Bergman L (2005) Control of urine marking by use of long-term treatment with fluoxetine or clomipramine in cats. J Am Vet Med Assoc 226(3):378–382. https://doi.org/10.2460/javma.2005.226.378

Herron ME, Buffington CAT (2012) Environmental enrichment for indoor cats: implementing enrichment. Compendium (Yardley, PA) 34(1):E3

Hopfensperger MJ (2016) Use of oral venlafaxine in cats with feline idiopathic cystitis or behavioral causes of periuria. Proceedings of the Veterinary Behavior Symposium

Houser B, Vitale KR (2022) Increasing shelter cat welfare through enrichment: a review. Appl Anim Behav Sci 248:105585. https://doi.org/10.1016/j.applanim.2022.105585

Kaur G, Voith VL, Schmidt PL (2016) The use of fluoxetine by veterinarians in dogs and cats: a preliminary survey. Vet Record Open. https://doi.org/10.1136/vetreco-2015-000146

King JN, Steffan J, Heath SE, Simpson BS, Crowell-Davis SL, Harrington LJM, Weiss A-B, Seewald W (2004) Determination of the dosage of clomipramine for the treatment of urine spraying in cats. J Am Vet Med Assoc 225(6):881–887. https://doi.org/10.2460/javma.2004.225.881

Kry K, Casey R (2007) The effect of hiding enrichment on stress levels and behaviour of domestic cats *(Felis sylvestris catus)* in a shelter setting and the implications for adoption potential. Anim Welf 16(3):375–383. https://doi.org/10.1017/S0962728600027196

Landsberg GM, Wilson AL (2005) Effects of clomipramine on cats presented for urine marking. J Am Anim Hosp Assoc 41(1):3–11. https://doi.org/10.5326/0410003

Main M, Solomon J (1990) Procedures for identifying infants as disorganized/disoriented during the Ainsworth Strange Situation. In: Attachment in the preschool years: theory, research, and intervention, vol 1, pp 121–160

Masson S, Bleuer-Elsner S, Muller G, Medam T, Chevallier J, Gaultier E (2024a) Veterinary psychiatry of the dog: diagnosis and treatment of behavioral disorders. Springer. https://doi.org/10.1007/978-3-031-53012-8

Masson S, Metz D, Bleuer-Elsner S, Schwobthaler F (2024b) Retrospective study on the use of venlafaxine in 176 cats diagnosed with behavioral disorders. J Vet Behav 77. https://doi.org/10.2139/ssrn.4790331

Mertens PA, Torres S, Jessen C (2006) The effects of clomipramine hydrochloride in cats with psychogenic alopecia: a prospective study. J Am Anim Hosp

Assoc 42(5):336–343. https://doi.org/10.5326/0420336

Metz D, Medam T, Masson S (2021) Double-blind, placebo-controlled trial of venlafaxine to treat behavioural disorders in cats: a pilot study. J Feline Med Surg 1098612X2110367. https://doi.org/10.1177/1098612X211036792

Naik R, Witzel A, Albright JD, Siegfried K, Gruen ME, Thomson A, Price J, Lascelles BDX (2018) Pilot study evaluating the effect of feeding method on overall activity of neutered indoor pet cats. J Vet Behav 25:9–13. https://doi.org/10.1016/j.jveb.2018.02.001

Potter A, Mills DS (2015) Domestic cats (felis silvestris catus) do not show signs of secure attachment to their owners. PLoS One 10(9):1–17. https://doi.org/10.1371/journal.pone.0135109

Schwartz S (2002) Separation anxiety syndrome in cats: 136 cases (1991–2000). J Am Vet Med Assoc 220(7):1028–1033. https://doi.org/10.2460/javma.2002.220.1028

Schwartz S (2003) Separation anxiety syndrome in dogs and cats. J Am Vet Med Assoc 222(11):1526–1532. https://doi.org/10.2460/javma.2003.222.1526

Seksel K, Lindeman M (1998) Use of clomipramine in the treatment of anxiety-related and obsessive-compulsive disorders in cats. Aust Vet J 76(5):317–321. https://doi.org/10.1111/j.1751-0813.1998.tb12353.x

Wojtaś J, Czyżowski P, Kaszycka K, Kaliszyk K, Karpiński M (2024) The impact of environmental enrichment on the cortisol level of shelter cats. Animals 14(9):1392. https://doi.org/10.3390/ani14091392

Disorders Related to Dysregulation of the Relationships Axis

Contents

11.1 Interspecific Relational Disorders – 392
11.1.1 Description – 393
11.1.2 Diagnosis – 395
11.1.3 Prognosis – 399
11.1.4 Treatments – 399
11.1.5 Therapeutic Follow-Up – 400
11.1.6 Prevention – 400
11.1.7 Clinical Vignettes – 400

11.2 Intraspecific Relational Disorders – 402
11.2.1 Description – 402
11.2.2 Diagnosis – 406
11.2.3 Prognosis – 406
11.2.4 Treatments – 409
11.2.5 Therapeutic Follow-Up – 410
11.2.6 Prevention – 410
11.2.7 Clinical Vignette – 411

References – 414

© The Author(s), under exclusive license to Springer Nature Switzerland AG 2026
S. Bleuer-Elsner et al., *Veterinary Psychiatry of the Cat*,
https://doi.org/10.1007/978-3-032-06260-4_11

This chapter describes psychiatric conditions related to dysfunctional relationships in cats. These disorders may involve interspecific relational difficulties, such as issues between the cat and humans or the family dog, as well as intraspecific relational difficulties among cats within the same household. For each disorder, the clinical picture, diagnosis, treatment, prognosis, and follow-up are discussed in detail, with clinical vignettes enhancing the content.

The cat is a solitary, territorial, and non-social species. Stable social rules, particularly a stable and predictable hierarchy, have not been demonstrated in this species (Bradshaw 2016). Therefore, unlike the dog (Masson et al. 2023), we cannot speak of social integration disorders since there are no true social relationships in the species. Furthermore, the existence of interspecific attachment is debated, and there is currently no consensus on the subject (Vitale et al. 2019; Potter and Mills 2015).

While the cat is considered by most authors to be a non-social species, the domestic cat forms relationships that can lead to shared pleasure. These positive relationships are often highlighted by behaviors such as allomarking, mutual grooming, and great tolerance in sharing territorial fields. However, as in all relationships, these can encounter difficulties in establishing limitations, or even deteriorate following events related to the individuals involved in the relationship.

Thus, these relational disorders are limited to communication errors that degrade the relationship between two individuals. These relational difficulties often lead to challenges in managing territory (see ▶ Chap. 10).

11.1 Interspecific Relational Disorders

In this section, we detail two types of relational disorders. On one hand, the interspecific relational disorder is *linked to dysfunctional communication*, where the human fails to communicate properly and is responsible for the deterioration of the relationship. On the other hand, the interspecific relational disorder *secondary to another behavioral condition*, where the cat behaves abnormally, leading to the deterioration of the relationship.

Important Point of View

Can autonomopathies be discussed in cats?

In dogs, problematic behaviors that occur when owners are absent form an important and frequent group of disorders known as autonomopathies (Amat et al. 2020; de Assis et al. 2020).

In cats, two primary reasons explain why autonomy-related disorders are less significant.

First, cats are territorial animals, and their sense of comfort in adulthood stems primarily from their territory in a broad sense, which includes relationships with the individuals within that territory. A fearful or anxious cat typically seeks isolation in a designated zone or isolation field that provides a sense of security.

Second, for kittens adopted very early, particularly before 5 weeks of age, a primary attachment bond with a human individual can sometimes form. Without experiencing maternal separation, such kittens may fail to structure a comforting territory at puberty, leaving them incapable of living independently when their owner is absent. In natural conditions, maternal separation is inevitable for cats, and all individuals eventually become autonomous. However, in human households, it is theoretically possible to create dependency in bottle-fed kittens whose autonomy is discouraged by humans. These cases are exceedingly rare because humans regularly leave the home, and cats are genetically programmed to structure themselves within their territory when alone.

For these reasons, we do not encounter cats in consultation displaying symptoms of distress during owner absences coupled with genuine emotional dependence in their presence. While cat behavior can vary depending on the owner's presence or absence, such absence is often perceived as a territorial change, potentially leading to anxiety as with other territorial disruptions. However, emotional dependency does not manifest in cats as it does in social species.

For cats, their territory appears to consist of both immobile and mobile objects, including humans. Cats even deposit different pheromones on these categories. When the absence of humans triggers anxiety, it is more consistent to classify such cases under disorders resulting from territorial disruptions (see ► Chap. 10).

The very existence of attachment as defined by Mary Ainsworth's Strange Situation Test (Main and Solomon 1990) is debated and lacks consensus, particularly regarding the experimental procedure's applicability to cats and the interpretation of observed behaviors (Chang et al. 2025; Potter and Mills 2015; Edwards et al. 2007).

Finally, very few studies address attachment-related disorders in cats (De Souza Machado et al. 2020; Schwartz 2002, 2003). Close examination of these studies reveals highly nonspecific symptoms, such as urine soiling and substitute activities, without identifying signs of distress or exaggerated contact-seeking behaviors in the owner's presence. This further supports the idea that the term "autonomopathy" is poorly suited to a non-social species like the cat.

11.1.1 Description

11.1.1.1 Definition

The communication of a cat is often delicate, and its understanding is not always intuitive for humans. For both humans and dogs, the communication model is a social one, and there is often a need to understand the functioning rules of the group.

It is therefore common for humans not to communicate in a way that is ethologically acceptable to the cat, leading to a *relational disorder linked to dysfunctional communication*. These communication errors can be classified into several types: *incoherent* or *violent* communication. Communication is *incoherent* when it relies on contradictory demands or inconsistent requests.

A *contradictory request* is a situation in which the verbal and paraverbal signals of the human have opposing meanings for the cat. This communication error between the cat and the human arises from the fact that the two species do not place the same importance on the different communication channels (see ► Chap. 1, ► Sect. 1.3.2): the cat primarily focuses on paraverbal signs (postures, intonations...), while the human gives more weight to words. In some cases, poor body language hinders the cat's interpretation of the human's request. For example, immobility is perceived as a form of aggression by the cat, while the owner is simply expressing patience. In other cases, the owner is upset and does not consider it necessary to control their emotions, so their paraverbal signals convey a threatening aspect that worries the cat, which prefers to distance itself.

An *inconsistent request* is a situation where the same behavior of the cat is sometimes tolerated, even encouraged, and other times punished. This communication error between the cat and the human arises from the fact that the rules imposed on the cat must be clear, but they too often depend on the fluctuating moods of the owners. Cats tolerate these variations in rules based on mood better than dogs, as long as the associations between mood and rule remain consistent, which is something we are not very capable of achieving.

Communication is *violent* when it is based on positive punishment or negative reinforcement (see ► Chap. 3). It is possible to set limits and prohibitions for a cat, but there is no reason to enforce them violently. For the cat, prohibitions are generally not about stopping a specific action, but rather about restricting access to certain places. For example, prohibiting the cat from begging at the table simply means forbidding the cat from jumping on the table, at least during mealtime. This is a spatial or temporal prohibition. Prohibitions should be clear and predictable, but the cat will always tend to disregard them if no one is present.

Conversely, it is also common for cats with behavioral disorders to fail to communicate normally. In this case, their behavior, often aggressive, will quickly lead to an interspecific *relational disorder secondary to another behavioral condition.*

The symptoms as well as the presenting complaints of these two interspecific relational disorders will often be very similar, but it is up to the practitioner to determine the origin of the disorder. Indeed, we will see that the treatment for each of these disorders differs depending on whether the animal or the owner is the source of the issue.

11.1.1.2 Etiology and Pathogenesis

The communication of cats, as described in ► Chap. 1 (see ► Sect. 1.3.2), is very different from that of humans. In most cases, communication errors reflect clumsiness rather than malice, but they can cause suffering for the animal.

The fact that owners do not naturally understand the functioning of a non-social, territorial species like the cat leads to many communication errors. This can simply involve carrying the cat and holding it close to oneself. In the case of children, carrying and physical contact can become overwhelming. The cat becomes sensitized very quickly, likely due to its dual role in the prey-predator food chain.

In cats with fear axis disorders, interspecific social phobia can promote the sensitization process in seemingly harmless situations. In cats with self-control axis disorders, their inability to control scratching and biting can injure the owners, who will naturally defend themselves... and the relational disorder develops through a few play incidents that result in injuries.

11.1.1.3 Epidemiology

This condition is based on violent communication, dysfunctional communication, and developmental disorders.

Violent communication is less frequent in cats compared to dogs, where educational demands are higher and methods can sometimes be coercive. However, since cats are much more sensitive, aversive stimuli that seem harmless can trigger phobias. It is not uncommon for cats to be traumatized by a water spray or a sharp noise originally intended to stop undesirable behaviors.

Relational disorders caused by dysfunctional communication, on the other hand, are more often due to a lack of understanding from humans, and thus can occur in any cat.

Bottle-fed or adopted cats before 8 weeks are more at risk of developing developmental disorders. As a result, they are also more likely to miscommunicate and develop secondary relational disorders.

11.1.1.4 Presenting Complaints

Relational disorders are common in cats, but many of them remain untreated. It is not uncommon for a cat to quickly avoid one of the adults in the family, and in response, the adult may also avoid trying to establish a relationship. The cat's flight or isolation response is often respected. In the case of children, the cat is more likely to be pursued, and the fear of scratching will prompt owners to seek consultation more quickly.

The main reason for consultation is aggression. Most often, this involves aggression due to irritation in the context of fear, but it can also involve uncontrollable emotional aggression.

The prey status of the feline species and its closeness to wild species, due to passive domestication (see ► Chap. 1), make the cat more prone to sensitization and anticipation. Aggression is therefore quickly instrumentalized, and owners genuinely fear their cat's reactions. They quickly learn not to approach the cat in certain contexts and begin to dread interactions.

When the problem arises from human communication, owners often do not understand how their actions may have caused a disorder, as they do not perceive the world the way the cat does.

Sometimes, the relational disorder is secondary to a territorial disorder, and punishments occur due to urine soiling or marking. In such cases, marking can be an additional reason for consultation, alongside aggression.

11.1.1.5 Symptomatology

The main symptom is often confused with the reason for consultation, namely aggression. Although a reactive state is theoretically possible if there have only been one or two altercations, the fear experienced by both individuals during the aggression quickly leads to a pathological *phobic state*, specifically in the presence of the individual, or even an *anxious state* due to the individual's frequent presence. Some cats may stop interacting and isolate themselves in a *depressive state*.

Here, we are talking about aggressive reactions to contact requests or simply the presence of the owner. In these aggressive interactions, a vicious circle often develops. The owner, fearing the cat's reactions, becomes increasingly clumsy and hesitant to touch the animal. They then resort to sudden movements that provoke sharp reactions from the cat. For both the owner and the cat, *anticipation exacerbates* the clumsiness.

In the case of secondary disorders, owners also describe altercations and situations where they defend themselves, but it is quite easy to see that it is the cat that has exaggerated. For example, in the case of a hyperactive kitten, owners may describe watching television without moving, and the kitten suddenly jumped and scratched their feet. As a reflex, the person defended themselves by pushing the kitten away, which then leads to escalating aggression due to anticipation and sensitization.

Collecting symptoms requires careful questioning, and the practitioner must consider whether such reactions are frequent when these scenes are described. Simply put, out of 100 cats who would experience this, how many would react by scratching or becoming aggressive? If the answer is very few, it's likely that the communication issue comes from the cat, meaning the relational disorder is secondary. The practitioner can also question the kindness and quality of communication from the owners, trying to rationalize and determine which party is exaggerating in the dysfunctional communication.

11.1.1.6 Evolution

The evolution is rapid, even very rapid. The state cannot remain reactive beyond a few confrontations. As the two parties are forced to coexist, the phobic state, if it exists, is only temporary, and the cat quickly becomes anxious or depressed. The initial symptoms are complicated by the typical symptoms of these states. There is usually an aggravation of the main symptom, with the cat becoming more aggressive, and frequently complications arise, including symptoms that suggest typical territoriality disorders. The poor relationship with one or more individuals in the cat's environment forces it to change its perception of its territory, which no longer provides comfort at certain times or in specific areas. For the owner, this worsening is sometimes interpreted as rebellious behavior, but more often, the fear of aggression or the sadness of being attacked by an animal they believe to be affectionate is deeply unsettling. The owner becomes frustrated and finds the cat ungrateful.

11.1.2 Diagnosis

11.1.2.1 Diagnostic Criteria

Cats with an interspecific relational disorder must present a pathological phobic, anxious, or depressive state (◘ Figs. 11.1 and 11.2). The symptoms can appear at any age and are indicative of an escalation in poor-quality communication.

The diagnosis is based on the presence of one or more of the following symptoms toward one or more members of the household:

- Identification of dysfunctional communication.

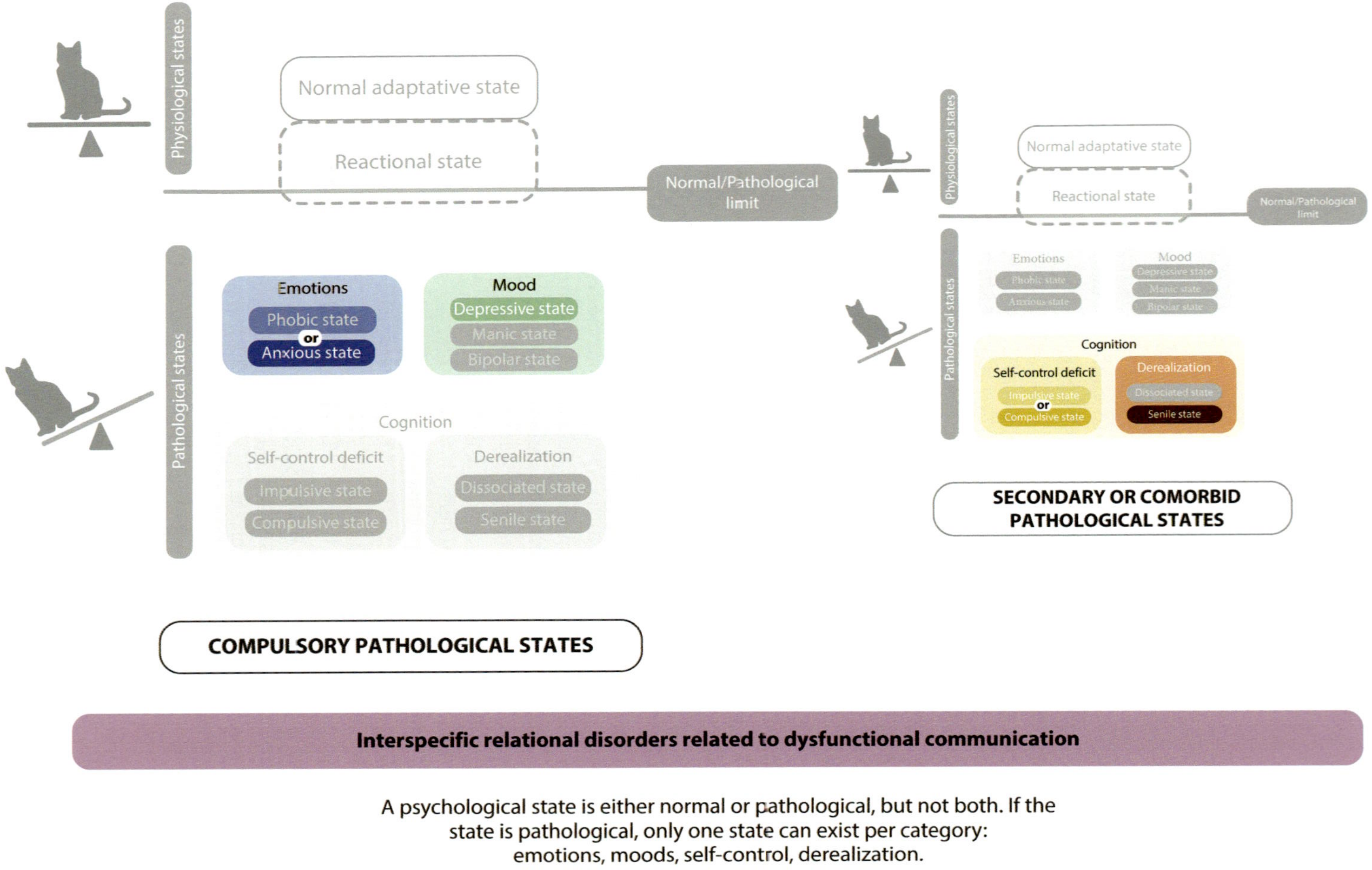

Fig. 11.1 Pathological states associated with interspecific relational disorders linked to dysfunctional communication. Mandatory pathological states: phobic **or** anxious **and/or** depressive state. Possible comorbid pathological states: impulsive **or** compulsive states **and/or** senile

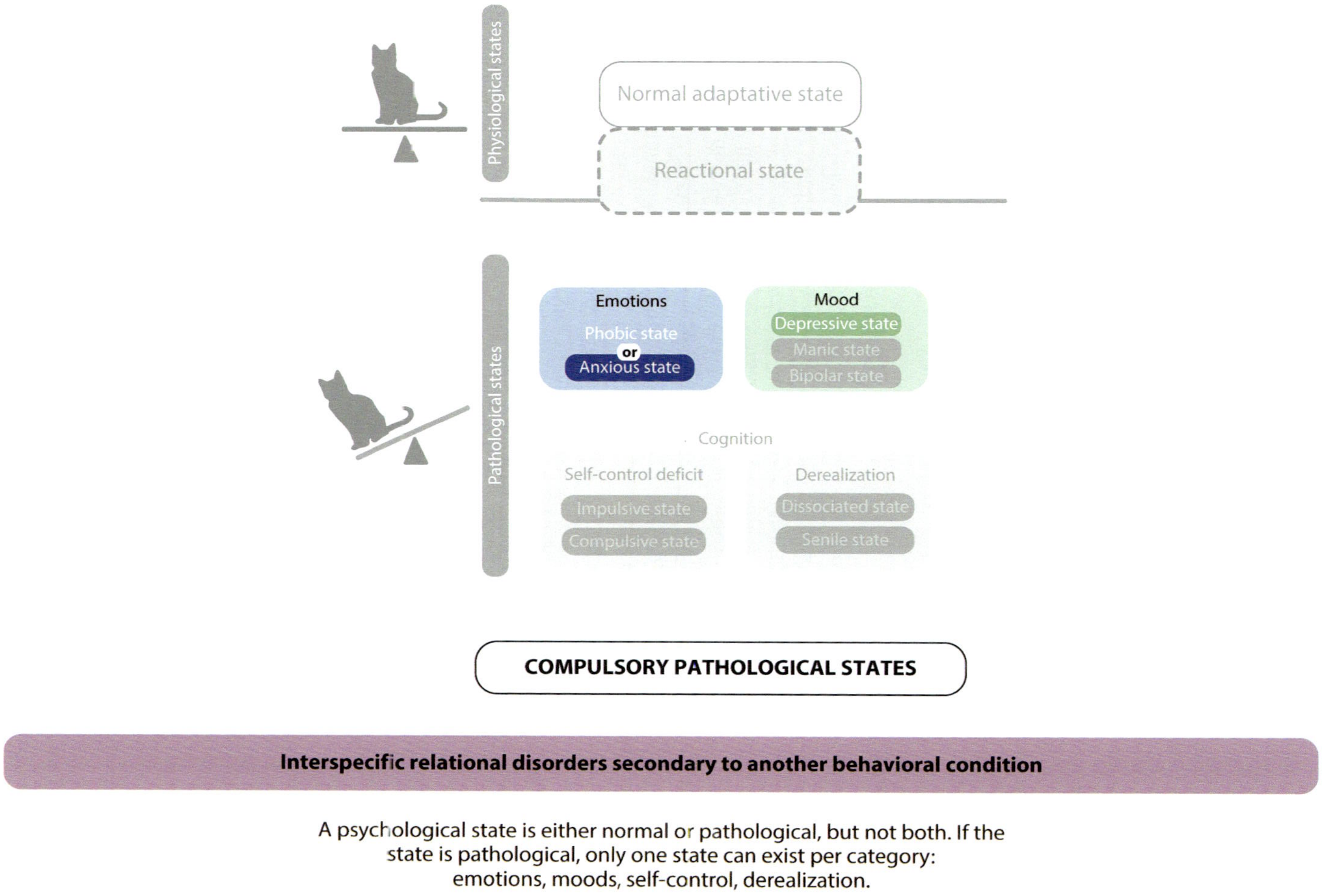

Fig. 11.2 Pathological states associated with interspecific relational disorders secondary to another behavioral disorder. Mandatory pathological states: phobic **or** anxious **and/or** depressive state

- Aggressions due to irritation in a fearful context, directed at one (or more) household members, becoming more pronounced and worsening over time.
- Anticipation of actions from this individual, with a tendency to flee or isolate.
- Hypervigilance in the presence of the individual(s) involved.

The origin of the conflicts is the key factor that allows the practitioner to establish whether the relational disorder is linked to dysfunctional communication or secondary to another behavioral disorder. In the latter case, the entire clinical picture cannot be explained solely by the symptoms listed above.

11.1.2.2 Direct Observation

If the person involved in the aggression is not present, the cat will appear perfectly normal during the consultation. When the person (or dog) is present, the cat may exhibit hypervigilance. However, the veterinary clinic environment tends to inhibit most cats' reactions. Deprived of the comfort of their territory, they may not express aggression, even in situations where it would typically occur in their home.

On the other hand, the owner's fear of the cat is noticeable, with a refusal to touch the cat, while the animal remains relatively non-reactive with the practitioner. These situations can sometimes highlight dysfunctional communication by observing how the person interacts with their cat (jerky or irritating petting), helping to validate the hypothesis of dysfunctional communication.

In the case of secondary disorders, self-control deficits can be observed, aiding in the understanding of the pathological pathway. The same applies to social phobias, which are the main differential diagnosis in the case of repeated aggression.

The practitioner's examination will be particularly attentive to signs of pain, as pain is a major factor in irritability and significantly lowers the threshold for aggression triggers.

11.1.2.3 Comorbidities

When the relational disorder is secondary, it can coexist with any of the conditions described in the axes of self-control, fear, territoriality, or aging. The process is easy to understand: the disturbances caused by these conditions lead to punishments, which quickly give rise to the relational disorder, adding anxiety to the initial clinical picture.

When dysfunctional communication is caused by the owners, the children in the household, or the dog, the disorder may be isolated. However, as we have pointed out, it quickly disrupts the cat's territorial organization and can lead to symptoms such as disrupted marking or even substitute activities (see ▶ Chap. 10, ▶ Sect. 10.3).

11.1.2.4 Differential Diagnosis

The primary differential diagnosis is with isolated *social phobia* or as part of a *sensory deprivation syndrome.* Indeed, a cat with a phobia of humans (or dogs) will react aggressively, and the relationship will deteriorate. The distinction is easy to make, as in the case of phobias, the owner is usually the least affected; often, the cat has become familiar with them. In contrast, in relational disorders, it is often the opposite: the cat has a history with a specific individual (usually one with whom it cohabits), and this individual will be more attacked than strangers.

It is also possible to differentiate this relational disorder from *dysthymia.* While aggression toward the owners occurs in both conditions, in dysthymia, the triggering context is hard to identify. Furthermore, in dysthymia, the aggressions are often of extreme intensity compared to what seems to have caused them. These are more likely to be idiopathic aggressions than aggressions due to irritation in a fearful context.

In *relational disorders*, there is a logic to the communication error, and the same contexts produce the same effects. For example, carrying the cat to cuddle it will always lead to defense in the case of a relational disorder.

HSHA syndrome can also cause injuries to the owners, but these vulnerable sequences are often produced in a context of excitement or frustration. They involve all individuals who attempt to interact with the cat and do not lead to anticipation in the presence of a specific person. However, it is worth noting that the aggression of these hyperactive cats quickly leads owners to defend themselves to protect themselves from injury, and thus, comorbidity with a secondary relational disorder is frequent.

11.1.3 Prognosis

In the case of interspecific relational disorders related to dysfunctional communication, the prognosis is generally favorable, especially if the cat is still in a reactive state at the time of consultation. When the observed state is already anxious or depressive, the prognosis is less favorable. If the practitioner is able to induce a change in the owner's attitude, recovery is entirely possible.

In the case of secondary relational disorders, the prognosis mainly depends on the prognosis of the primary disorder. It is therefore highly variable and largely depends on the time elapsed between the emergence of the symptoms and the consultation. If this time is very long, the owners' patience is limited, and the bond with the animal is significantly affected.

11.1.4 Treatments

The treatment combines pharmacological treatment, which should quickly calm the relationship, along with behavioral therapy, which will help reinforce positive interactions. This is done while ensuring that the environment allows the cat to retreat to isolation zones rather than opting for aggressive interactions.

11.1.4.1 Pharmacological Treatment

The pharmacological treatment aims to restore a normal psychological state. Since aggression is often the reason for consultation, *fluoxetine* will be an interesting option at a dose of 1–2 mg/kg once daily (Hart et al. 2005; Kaur et al. 2016), while monitoring the risk of urinary retention and bladder distention (DiCiccio and McClosky 2022). Indeed, restoring the relationship should be at the heart of concerns, and this medication helps by giving the cat emotional control, making its reactions less intense.

For treating anxious states, *venlafaxine* at a dose of 1–2 mg/kg once daily is the drug of choice (Hopfensperger 2016; Masson et al. 2024; Metz et al. 2021). It also has the advantage of being easily administered (granules inside Effexor™ capsules), which is always a great help in cats, especially if the cat is aggressive and the owner is hesitant to interact with it. *Clomipramine,* administered at a dose of 0.25–1 mg/kg once daily, is also possible (King et al. 2004), as its antihistaminergic side effects will make the cat sleep and thus be calmer (Landsberg and Wilson 2005; Mertens et al. 2006; Seksel and Lindeman 1998).

In the case of depressive states or when the animal adopts a strategy of fleeing and withdrawing, *mirtazapine* at a dose of 1–2 mg/kg once daily is the drug of choice. The increased appetite allows the use of tasty treats to reestablish a bond with the cat. However, in cases of obesity, it should be avoided or used temporarily, as it is highly orexigenic.

11.1.4.2 Behavioral Therapy

- **Environmental Modifications**

A part of our therapeutic approach also involves describing the cat's territorial organization and its functioning, including markings and pheromone deposits. The necessity of an isolation zone and its respect is important, as it should allow the cat to retreat without aggression, avoiding its owners when desired. Creating an isolated room where the cat can be placed if it begins to show signs of irritation is often a safety valve that helps prevent the escalation of aggression.

Conversely, if the cat has chosen to avoid certain individuals in the house by isolating itself in particular rooms or upstairs, it will be necessary to regain control by bringing the cat to a higher point in the living area, ensuring that the source of its fear (such as a child or dog) does not approach or force contact. To do this, it is essential to ensure there is a secure spot that is appreciated by the cat. If it is not, this spot should be made attractive, for example, by offering food treats there while the room is completely calm. The cat must realize it is safe and will not be disturbed if it settles in that spot.

Behavioral Modifications

The core of the therapy is based on changes in the relationship. For this, it is essential that the owner understands that the cat fears interactions just as much as they do, and on the other hand, that all their defensive gestures contribute to worsening the situation because they will be perceived as positive punishment by the cat. It is rare for owners to be unreceptive to advice, but getting them to suppress emotional reactions can sometimes be tricky. Indeed, the fear of being scratched triggers hesitant and sharp gestures, which activate the cat's response. Therefore, it is crucial to teach the owner to make slow, deliberate movements.

To overcome these difficulties, it is helpful to create new types of interaction by asking for the completion of sequences that should become rituals. In these new sequences, neither party anticipates the actions, and a positive exchange can develop, forming the basis for a new understanding.

When the animal is in an inhibited or avoidance strategy, these rituals are essential. They can be done by placing the cat in its secure spot and offering exceptional treats in the presence of the individual causing the problem. If it is a child, the owner can show the child how to pet the cat, doing it with them and holding their hand to ensure the gentleness of the touch. If it is a dog, the owner can place their hand between the cat and the dog to ensure the cat's safety and prevent overly rough nose bumps.

11.1.5 Therapeutic Follow-Up

Normally, once the correct behaviors are established and the cat's condition has returned to normal, there is no reason for relapse. However, sometimes a clumsy gesture may trigger the cat's reactivity. If the owner is aware of this risk, they can be prepared to intervene immediately by reinforcing the therapeutic elements used previously. A temporary resumption of medication may sometimes be helpful to prevent the cat from regressing into anxiety.

11.1.6 Prevention

Prevention will take place during the initial consultations. The veterinarian must take the time to explain how to interact positively with the cat, the importance of not forcing interactions, or excessively restricting the cat, as it can quickly become sensitized. These explanations are particularly important when adopting an adult cat. Relational issues are more common in these cases due to the owner's hesitations in their gestures, not knowing how the cat will react. In the presence of children, it is even more crucial to emphasize the necessary supervision, ensuring that the child's gestures are gentle enough for the cat to tolerate.

11.1.7 Clinical Vignettes

11.1.7.1 Boule

Boule is a European cat adopted from a shelter by a young couple of students at the age of 2 months. He is a perfectly normal and resilient cat until he turns 6. Despite the student lifestyle, frequent visits, and moving around, he is sociable, unafraid, and friendly with everyone, even seeking affection when guests arrive. The couple also has two dogs, with whom Boule gets along well, although they occasionally chase him for play. He knows how to keep himself safe by climbing to higher places, sometimes even teasing the dogs by watching them from his perch.

When Boule turns 6, the couple has their first child. As long as the child is immobile, Boule continues to be the cat they have always known, even being quite affectionate with their son, cuddling up with him during naps. However, as the child begins to grow and move, the parents notice unexpected interactions: their son starts petting Boule in the typical chaotic manner of a child, even chasing him to try and touch him. While Boule is typically tolerant, he starts to avoid these interactions with the increasingly invasive child.

After a few months, when the child is two and a half years old, Boule becomes so disturbed by the child that he tries to scratch him once or twice and increasingly retreats to the garage whenever the child is present in the house. He, who had always been with his owners, no longer comes up to the living room while the child is awake. With his owners, he remains perfectly normal but seems to suffer from the lack of relationship, asking for more affection when he can join them in their room at night.

The behavioral veterinarian diagnoses an interspecific relational disorder caused by dysfunctional communication. The communication is violent (the cat regularly has its tail and ears pulled). Boule shows signs of intermittent inhibited anxiety.

The veterinarian prescribes venlafaxine (mirtazapine is avoided due to Boule's moderate overweight of 5 kg) to be given once a day, with 12 granules mixed in his food. In addition, a new cat tree is installed, and it is decided that the feeding ritual will take place on this tree at a height where the child cannot reach Boule. Every evening, when the child is present, Boule is brought into the living room, and the child, with his mother, gives him small pieces of tuna.

Instead of scolding the child, it was chosen to have him interact with the cat, controlling his gestures (his mother holds his hand) to teach him how to gently pet the cat.

The progress is rapid, and the treatment is stopped after 4 months. Boule has regained his place in the living room, though he remains cautious around the child's movements. Over time, the vigilance decreases as the child's actions become more controlled.

11.1.7.2 Tama

Tama is a cat found on the street at just 3 weeks old and brought to the veterinarian. He was kept at the clinic for 2 weeks and displayed signs of poor behavioral development, such as hissing at every human approaching. When one of the veterinarians decided to adopt him, she quickly noticed that he had very poor self-control during play. The veterinarian has two children, ages 7 and 8. Tama was given venlafaxine, which helped him improve on the fear axis, and he became familiar with the family members in just a few days. After 2 months, the treatment was stopped because Tama had made significant progress; although he still hid during visits, he would return within an hour and accept being in the same room as guests.

At around 8 months old, self-control problems resurfaced, and Tama began chasing one of the children. Although he could do this with any family member, the youngest child would often scream and run when Tama tried to attack his calves. The more the child screamed, the more excited Tama became, playing more intensely with claws out. The child started crying and calling for help, eventually becoming fearful of Tama. The parents had to intervene repeatedly to stop Tama, and each attack became more violent. Tama's pupils would dilate, and even when calmly pushed away by one of the parents, he would immediately go after the child again.

The relationship between Tama and the child deteriorated, and it soon became impossible for them to be in the same room. The child was scared of the cat, and Tama was ready to pounce on him at the slightest movement. When Tama attacked, he would scratch the child's legs and then run away to hide, as he had learned the parents would intervene. He never attacked any other family members.

Tama was diagnosed with an interspecific relational disorder secondary to an impulsive state (HSHA syndrome). He also had intermittent, productive anxiety, which complicated his already moderate impulsive behavior from a young age. A treatment with fluoxetine was prescribed at 1 mg/kg once a day. The child was given instructions on how to protect himself without moving or exciting the cat: when he was alone at home, he always had access to a large box to use if Tama started flicking his tail at him. Initially, the child didn't need to use it, but knowing that the option was there reassured him.

Once this step was passed, positive interactions were established between the child and Tama. The child taught the cat to do "check" and "marmot," and if Tama started getting too excited, the child would ask him to perform these commands, which Tama easily did, helping to defuse the situation.

After 6 months, the relationship was calm, and the treatment was stopped. Tama no longer showed anxiety or attacked the child, though his impulsive behavior sometimes resurfaced during play. The situation remained stable for the following 6 years, and the child grew up. To date, no further treatment has been necessary.

Summary of ▶ Sect. 11.1

Disorders resulting from interspecific communication issues evolve rapidly and must be distinguished from interspecific phobias. These disorders involve a deterioration of the relationship between the cat and one or more individuals with whom it has had negative experiences. The two main causes of poor communication originating from humans are inconsistency and violence. Additionally, various behavioral disorders can impair the cat's communication and lead to negative experiences. Organic diseases, such as hearing impairments or pain, may also contribute to communication difficulties, and these should be ruled out if the cat seems to be the source of the communication problem.

In cases of secondary relational disorders, restoring quality communication can only occur after addressing the underlying issues. Pharmacological treatment aims to restore a normal adaptive emotional state, compatible with learning. The elimination of punishments, along with the intervention of a skilled professional to implement the therapy, often leads to recovery.

11.2 Intraspecific Relational Disorders

In this section, we detail two types of relational disorders. On the one hand, *intraspecific relational disorders are linked to dysfunctional communication*: since cats are non-social, they can easily become very intolerant toward their peers, leading to hostile communication. On the other hand, *intraspecific relational disorders are secondary to another behavioral disorder*: in this case, one of the cats is affected by a behavioral disorder that causes it to communicate abnormally, which leads to the deterioration of the relationship.

11.2.1 Description

11.2.1.1 Definition

Due to its non-social nature, a cat generally does not seek to form bonds with other members of its species. For a cat, encounters with others either involve distancing itself from individuals that should not share the same territory or organizing the sharing of this territory. There is therefore an inherent negative bias, and it is not surprising that the first interactions between two adult cats are often aggressive. Some cats will be more tolerant than others when it comes to accepting feline cohabitation, while others will be greatly disturbed by the presence of another cat and may communicate violently in an attempt to establish exclusivity over all or part of the territory.

As with interspecific communication, different communication anomalies can arise, as it will largely depend on the message sent and how the other cat receives it. Misunderstandings can quickly occur. If one of the cats is bolder in claiming the territory, an imbalance that leads to a pathological state in one of the cats is possible. These intraspecific relational disorders will be linked to *dysfunctional communication*.

Moreover, it is also common for cats with behavioral disorders to communicate abnormally. In this case, their abnormal behavior becomes even more disruptive for the other cats in the household. They may, for instance, become too intrusive if they are impulsive or compulsive, and these anomalies will quickly lead to an *intraspecific relational disorder secondary to another behavioral disorder.*

In both cases, the initial violent communications between the cats, even though they are normal for the species (hissing, distancing, etc.), can intimidate the owners, who may complicate the situation by intervening in the relationship between the cats in an attempt to "correct" things.

The symptoms and presenting complaints of these two intraspecific relational disorders will often be very similar, but just like with interspecific relational disorders, it is up to the practitioner to identify which cat is the source of the problem. Whether it stems from dysfunctional communication or an underlying behavioral pathology.

11.2.1.2 Etiology and Pathogenesis

Here, we must distinguish between several possible origins of this condition. This distinction is important for the prevention and prognosis of these diseases.

The communication rules of cats are described in ▶ Chap. 1 (see ▶ Sect. 1.3.2), and they are learned. If these learnings are not accomplished under good conditions, communication becomes awkward and difficult. These awkwardnesses, in turn, lead to inappropriate reactions from individuals during communication, and the situation quickly deteriorates. These early learnings are largely handled by the mother and other adults who interacted with the kitten from birth to puberty. When the number of these interactions is limited, communication is less likely to be of good quality due to this lack of practice. Kittens that are abandoned at a very young age, therefore, are more likely to communicate awkwardly and are at greater risk of being reprimanded by another cat without knowing how to react.

In some cases, initial encounters go well, and the cat's communication skills are good until the day it has a bad experience. In just a few negative confrontations, true intraspecific fear can develop. Often, this is not extreme, but communication is altered because the memory of the bad encounter remains present for both parties. Communication becomes hesitant and, in some cases, incoherent due to the emotions of fear that disrupt the messages. One of the cats gains the upper hand in the encounter, and the situation becomes unbalanced, with one cat monopolizing the space while the other avoids encounters by staying close to the walls. If the situation deteriorates further, one cat may begin to watch the other, preventing any intrusion into its space and attacking as soon as the other passes certain zones. In the most extreme cases, a cat may retreat to a room or the garage, or even go outside and refuse to return to the living space. These unsuccessful cohabitations often lead cats to change homes, as life in their original residence becomes so unpleasant.

When fleeing is not possible, the problem persists, and the emotional state of one or both cats deteriorates. The risk for the victimized cat is a shift into an anxious state, constantly monitoring the potential presence of the aggressor, or even into a depressive state if it seeks refuge in a location without daring to come out. For the persecuting cat, the risk of anxiety is real, as it remains on constant alert to ensure that the other cat does not threaten its space. For both cats, the construction of a territory is not possible. As long as the conflict is unresolved, each cat is only defending a living space that is in no way a place of peace.

Moreover, not all cats have the same communication codes. First, physical characteristics must be considered. Some cats (like Persians, Scottish) have reduced facial mobility, which diminishes their expressiveness and can hinder communication. Additionally, powerful cats, due to their strength, are inclined to respond to stress with aggression. The instrumentalization of their aggression is easier because it is immediately effective. Finally, the level of precision and tolerance varies between cats. Some individuals are highly tense, and the slightest communication flaw triggers a reaction, while others are more tolerant and able to share their living space with another cat.

In addition to these predispositions that can interfere with communication, behavioral disorders such as *sensory deprivation syndrome* and *HSHA syndrome* contribute to poor communication quality. These conditions predispose cats to have negative experiences and often, during puberty, to aggressively and instrumentally attack other cats. At this stage, the behavioral sequence does not allow for distinguishing the origin of the problem.

11.2.1.3 Epidemiology

Intraspecific relational disorders are common, and their frequency increases with population density. Two cats with adjacent territories may cross paths. This encounter is not a surprise, as it has been anticipated through pheromonal markings, and the individuals crossing each other are already informed about one another. The interactions are often brief and codified, and agonistic interactions are rare.

On the other hand, sharing a territory is more delicate. For cats living in the same space, adjustments and exchanges over territorial boundaries are inevitable. It is important to note that the sex of the animals, as long as they are all sterilized, does not influence this type of disorder (Lindell et al. 1997).

Predisposing factors for this disorder include the number of cats present in the household, as this increases the difficulties in sharing territory. The *age of the cats* is also a factor. Since pheromonal marking begins at puberty, it is easier for cohabiting cats to get along if both were adopted as kittens, or at least if one of the two was adopted as a kitten, as this prevents marking competition. The risk of developing intraspecific relational disorders is much higher when adopting an adult cat.

11.2.1.4 Presenting Complaints

The consultation is prompted by intraspecific aggressive behaviors of one or more cats. The perception of the owners plays a significant role in this situation. Their desire is always for harmony to prevail and for them not to have to manage injuries. For most owners, it would seem normal that two individuals who cross paths become friends, even though we know that, even for us as a social species, this is sometimes difficult **(see the box "From kitten to adult cat")**.

> **Important Point of View**
>
> **From kitten to adult cat**
>
> In normal upbringing, the mother gradually establishes her territory separate from her kittens, except in the case of a matriarchy with her daughters. It is therefore normal for a mother cat to reject her kitten once it reaches adulthood. Aggressive reactions (hissing, swatting) from a mother cat toward her sexually mature kittens often surprise owners, as do agonistic interactions between siblings who have grown up together. The concept of a nonsocial species poses real challenges for humans, who expect their cat to behave like a social species.

11.2.1.5 Symptomatology

The condition typically progresses in three stages, each increasingly severe, which can be described as follows: the first phase is called *distancing*, the second stage is known as the *skirmish phase*, and finally, the third stage is termed the *obsession phase*.

Depending on the progression of the condition, the practitioner intervenes at one of these stages.

The *distancing phase* is a normal phase when two adult cats encounter each other.

The cats are either in a normal or reactionary state, and this is a feline communication interaction. Each cat signals its presence and its intent to occupy the territory. At a variable but relatively short distance, usually just a few meters, the two cats hiss at each other and stare. The interaction should result in one of the cats retreating, though it may be pursued by the "winner." This establishes a territorial rule, and these "disputes" should occur repeatedly until the entire territory is shared. What owners find difficult to tolerate is that a scratch or bite may occur. For them, this is evidence of a negative progression, but their intervention may prevent the evolution toward a resolution, with one cat gaining part of the territory in the moment.

If the rule cannot be established, for whatever reason, the relationship evolves into the *skirmish stage*. These are real conflicts where the cats physically fight, potentially causing injuries. These fights still have the potential to resolve the conflict. It may require several fights to settle each part of the territory. At this stage, one or both cats may already be in a phobic state.

If territorial sharing is not established during the second phase, this is when the animals transition into pathological states. It may begin as a transient phobic state, but it quickly evolves into an anxious state because the confrontations are frequent. The third phase arrives, and its hallmark is that it cannot progress favorably. Communication is blocked, and the conflict becomes permanent. This phase is called the *obsession phase*. Both cats are fixated on the territorial dispute. One of the participants spends its time preventing any movement from the other, while the other withdraws and spends its time avoiding interaction.

In this situation, owners clearly perceive one cat as the aggressor and the other as the victim. The harassing cat is constantly threatening, while the other seems to want to avoid the conflict at all costs. However, this perception is not entirely accurate. In order for the territory to be shared, conflict and its resolution are essential. The cat that withdraws without leaving its refuge does not offer a solution and is ultimately as responsible for the conflict as the one that threatens it. It is important to note, however, that this situation only exists because cats are forced into sharing. Normally, one cat would flee, leaving the shared space, which resolves the conflict.

At this stage, progression leads to a pathological state for at least one of the animals, if not both. The persecuted cat often has self-inflicted grooming wounds and possibly injuries caused by the conflict. It will often be in an anxious or depressed state. In most cases, the owners have separated the two animals. The persecuting cat needs to monitor all of the other cat's actions, which inevitably leads to an anxious state.

11.2.1.6 Evolution

From this phase of obsession, the evolution is blocked. The deterioration of the health condition continues. In some cases, the "persecuted" cat may become depressed. In certain situations, if possible, the pursued cat may choose to move to another living space. If it has access to the outdoors, it may avoid returning to the house for increasingly long periods, isolating itself in the garden, a shelter, or a garage, rather than enduring this forced territorial sharing.

When territorial sharing involves many cats, which can be observed in some catteries or when owners accumulate animals in a restricted space, a stabilization may occur, and the conflicts disappear. Each cat resolves to live with its discomfort in its confined space. It is often difficult to make owners of these shared spaces, who are usually avid cat lovers, understand that the situation is not ideal and resembles more of a prison-like environment for a cat, which needs space it can structure on its own.

In the case of *intraspecific relational disorders secondary to another behavioral disorder*, the evolution depends on the management of the primary disorder. However, even when this primary disorder is resolved, the lasting conflict that has been established does not resolve automatically because the history of altercations is not erased. This presents an additional challenge to manage.

11.2.2 Diagnosis

11.2.2.1 Diagnostic Criteria

Cats with an intraspecific relational disorder present a pathological state of fear, anxiety, and/or depression (◘ Figs. 11.3 and 11.4). At the very beginning of the disorder, they may be in a reactive state.

Symptoms can appear at any age and are indicative of an escalation in poor communication quality.

The diagnosis is based on the presence of one or more of the following symptoms toward one or more members of the group:

- Repeated conflicts between two or more cats in the household, characterized by irritation-driven aggression in a fearful context and territorial aggression;
- Territorial aggression between two or more cats in the household;
- Anticipation of actions with a tendency toward fleeing for one cat and aggression for the other;
- Hypervigilance in the presence of the conspecific(s) involved.

The origin of the conflicts is the key element that will allow the practitioner to establish the diagnosis of a relational disorder due to dysfunctional communication or a disorder secondary to another behavioral condition.

11.2.2.2 Direct Observation

Aggressions between cats are sometimes visible during consultations, but more often through videos recorded by the owners. The fear of the owners, who have witnessed the fights, is palpable when attempting to introduce the cats to one another, and the tension between them is often visible, even though it is much less intense than what is observed in their home.

11.2.2.3 Comorbidities

This condition, even in its primary form, does not exclude any other behavioral disorder.

Many cats with a chaotic development in the fear or self-control axes will develop secondary communication disorders that contribute to these relational issues. In cases with comorbidities, determining the origin of the dysfunctional communication will be more challenging, and the practitioner will need to prioritize treatments. Separating the cats while treating all of their individual conditions will often be essential.

11.2.2.4 Differential Diagnosis

The main differential diagnosis will be made between the two forms of intraspecific relational disorders. However, identifying the origin of the communication disorder is the key challenge. It could involve normal cats whose owners prevent territorial sharing, it could involve normal cats forced to live in too small a space, or it could involve cats whose communication is disturbed by preexisting conditions.

Finally, it is important to ensure that the relational disorder is not systematic for one of the cats, in which case the question of a true *intraspecific social phobia* should be considered.

11.2.3 Prognosis

The prognosis is reserved if the cause is related to overpopulation and the owners are not willing to rehome one of the cats or provide a larger territory. In this case, treatment will act as a chemical restraint, improving the well-being of the animals but not resolving the issue as long as the cat density remains high.

It is important to note that a suitable territory for a cat is primarily one that is sufficiently enriched to compensate for the population density (see ► Chap. 10).

Owners must adjust their view of the species they house and accept that the cats are not "friends" but cohabitants who must share a space. In many cases, it is necessary to emphasize that once the conflict between the two cats is resolved, it is advisable not to introduce a third cat. This mistake is often made, and the challenge lies in the fact that once the cat is adopted, it is difficult to rehome them.

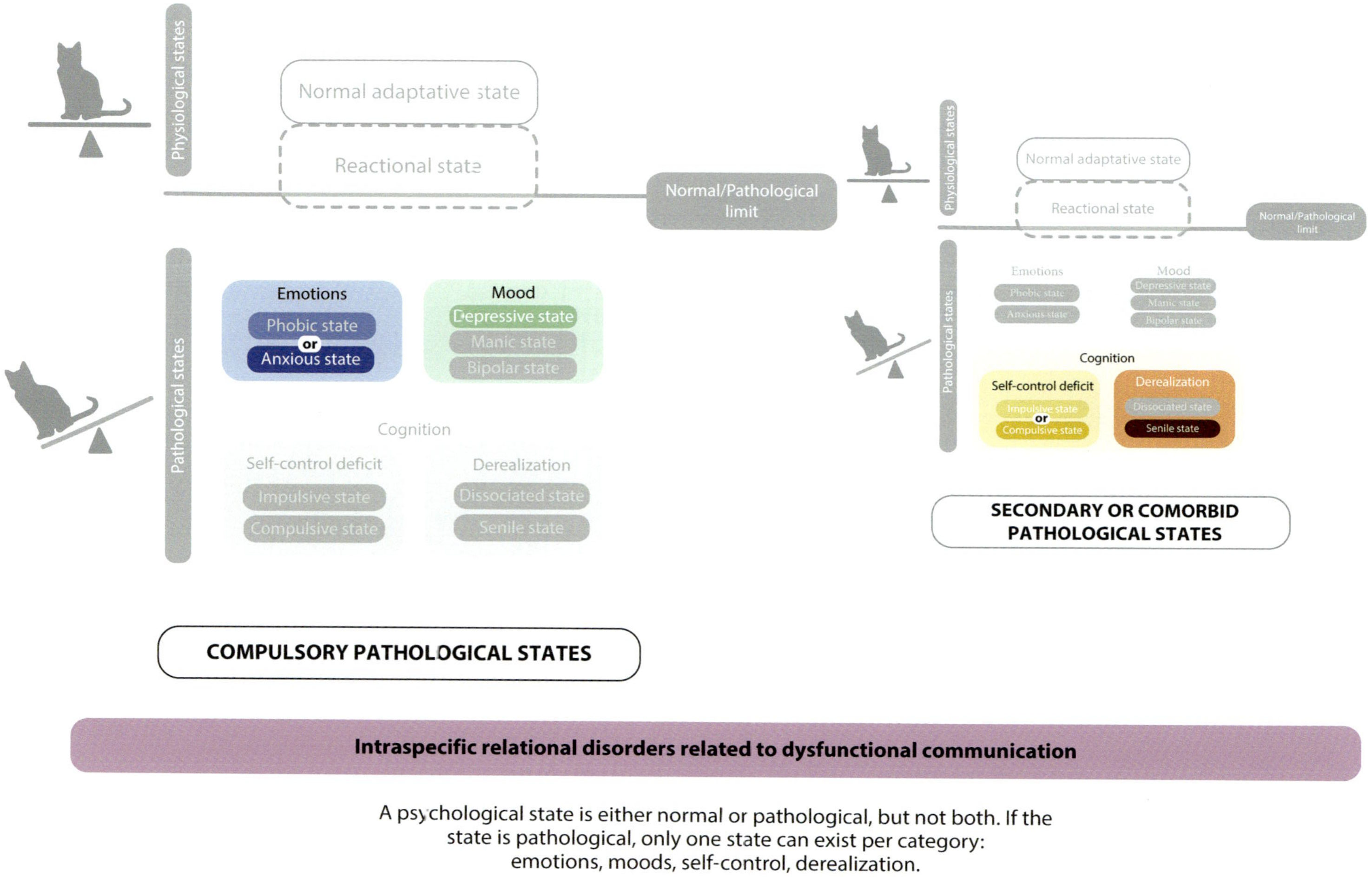

Fig. 11.3 Pathological states associated with intraspecific relational disorders linked to dysfunctional communication. Mandatory pathological states: phobic **or** anxious **and/or** depressive state. Possible comorbid pathological states: impulsive **or** compulsive states **and/or** senile

11

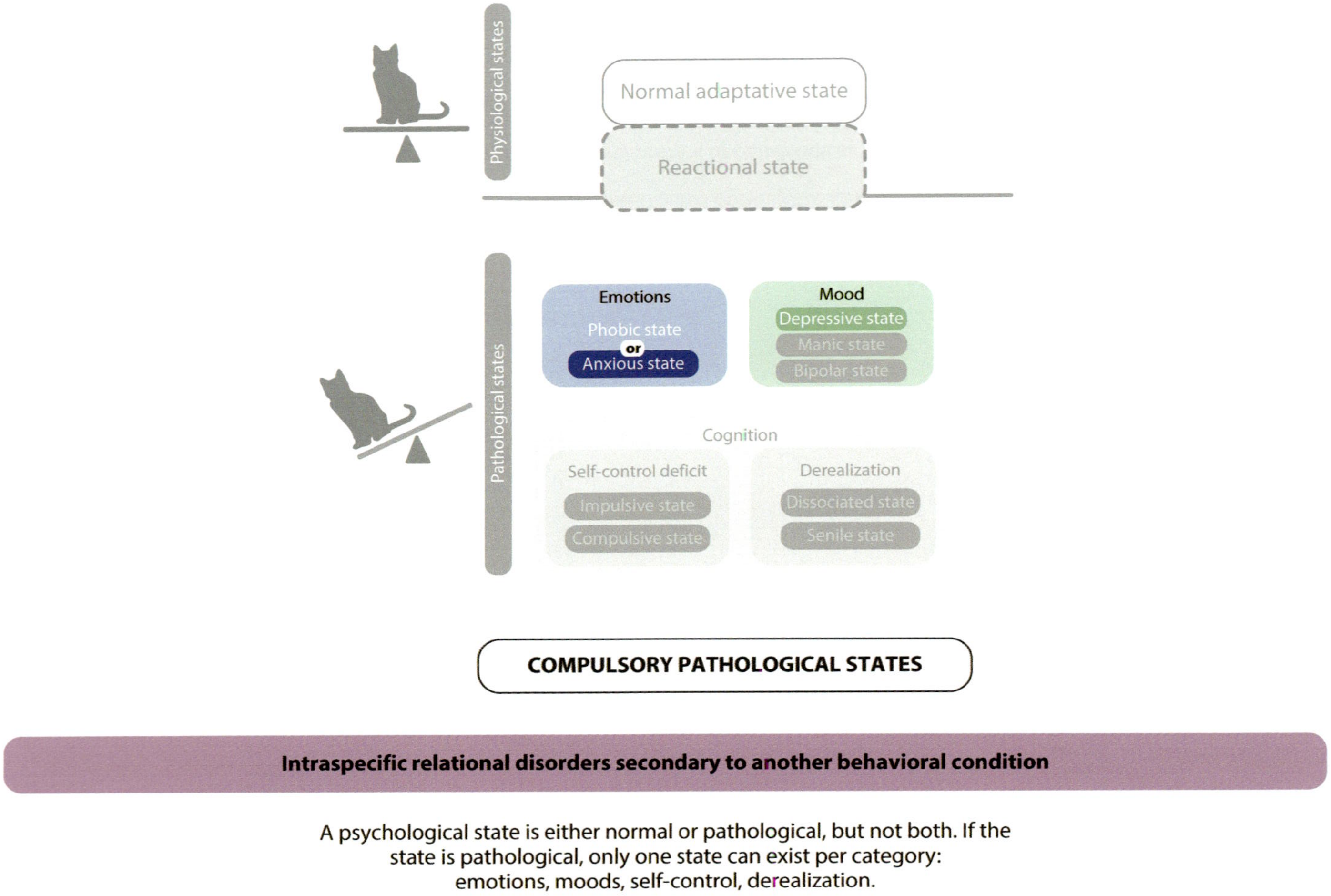

Fig. 11.4 Pathological states associated with interspecific relational disorders secondary to another behavioral disorder. Mandatory pathological states: phobic **or** anxious **and/or** depressive state

> **Important Point of View**
>
> **The harassing cat**
>
> It is easy to understand that in these conflicts, the harassing cat has an advantage over the other. This typically happens when a new cat arrives from outside to settle in a home that already has an animal.
>
> The newcomer is immediately in a reactive state, which favors adaptation, while the resident cat is already on the defensive. For the owners, the unfortunate cat from outside needs support, while the resident "could make an effort." This different perception places the cats in a predisposition to conflict.

11.2.4 Treatments

11.2.4.1 Pharmacological Treatment

At the beginning of the conflict, medication is not necessarily required. The use of pheromones can facilitate territorial sharing (Frank et al. 2010; Hargrave 2021). Additionally, the use of familiarization pheromone fractions (Cat Appeasing Pheromone or Felifriend™) can limit aggression, especially if used immediately upon the arrival of the second cat (DePorter et al. 2019).

Once the conflict reaches the skirmish stage, it is helpful to introduce anxiolytics. In any case, it is important to address the pathological states with a pharmacological treatment that restores a normal psychological state.

Fluoxetine is a good option for the aggressive cat at a dose of 1–2 mg/kg once a day (Hart et al. 2005; Kaur et al. 2016), while monitoring the risk of urinary retention and bladder distension (DiCiccio and McClosky 2022). Indeed, restoring the relationship should be the primary concern, and this medication helps by giving the cat emotional control, which leads to less intense reactions.

To treat anxious states, *venlafaxine* (Effexor ™) at a dose of 1–2 mg/kg once a day is a drug of choice (Hopfensperger 2016; Masson et al. 2024; Metz et al. 2021). It also has the advantage of being easy to administer, especially if one or both cats have difficulty taking medication. Clomipramine should be avoided for the inhibited cat due to its marked histaminergic effects, which could isolate it further (Landsberg and Wilson 2005; Mertens et al. 2006; Seksel and Lindeman 1998).

For depressive states in the more inhibited cat, which has adopted a strategy of flight and withdrawal, *mirtazapine* at a dose of 1–2 mg/kg once a day is the drug of choice.

Depending on the stage of the problem, especially if the cats have had to be completely separated, *gabapentin* and *pregabalin* can be used as supplements to support the more inhibited cat. This is often the case when the stage of obnubilation has been reached, which involves the most severe situations. These medications greatly increase the chances of the first interactions going smoothly, thus establishing a virtuous cycle for future contacts. Initially, they can be administered PRN 2 h before the cats are brought together at a dose of 20–40 mg/kg. In the first few days of full reintroduction, they can be given twice a day at a dose of 10–20 mg/kg to minimize the risk of sensitization during less controlled interactions by the owners. After about 15 days without issues, they can be discontinued.

> **Important Point of View**
>
> **Male cats and female cats: what about sexual status?**
>
> Territorial sharing cannot occur if the sexual status is not identical. Intact male cats cannot live alongside a sexually active female. In general, for a successful territorial sharing to occur, it is necessary that both parties are either neutered or in a sexual resting phase.

11.2.4.2 Behavioral Therapy

The therapy should aim, on one hand, to enrich the territory and, on the other hand, to allow conflicts to unfold in a healthy environment. The tension expressed by the owners is often extreme, and at the stage of obnubilation, it is preferable to separate or even hospitalize one or both cats to reduce this tension and allow the medication to start taking effect.

Environmental Modifications

The first step is to control the sexuality by ensuring that all individuals are neutered or chemically rendered sexually inactive.

At the stage of obnubilation, and sometimes at the stage of skirmishes, it is necessary to begin by separating the cats in conflict. This can be done by isolating one of the cats in a comfortable room, with the best option being to allow the more inhibited cat to have the most space, so it can reclaim this space. Initially, the space of each cat is limited, so they can structure it. After about 2 weeks, once the pharmacological treatment starts to take effect, it is beneficial to allow the cats to make olfactory contact with each other's space (it is possible to switch the spaces or let one of the cats outside while the other explores the area of the "competitor").

Then, the cats can be allowed to see and sniff each other through a door. The final stage will be the reintroduction into a shared space, initially under supervision to intervene if the tensions are too high and often with the help of temporary treatments.

Once these steps are completed, the treatment is continued for another month or two to ensure the stability of interactions and verify that each cat is structuring its territories well and that the territorial sharing is accepted.

In parallel, to facilitate the sharing of territory, which is often implicated in the communication disorder, the territory should be enriched, especially by providing each cat with a well-defined and comfortable isolation area (Kry and Casey 2007; Vinke et al. 2014). This area should be easily accessible and allow the cat to exit without being controlled by the other cat. Access to different rooms should be equitable and, in any case, constant.

To achieve a sufficient level of enrichment, it is often necessary to multiply the zones for elimination, feeding, and hiding places, offering more than the number of cats. If possible, access to the outdoors is desirable, as it allows the cats to artificially expand the territory they share, if not by size, then by the times each cat will occupy the indoor space.

Behavioral Modifications

Working with the owners is crucial. They must accept that a significant part of the issue between the cats is played out on an olfactory level, which is inaccessible to us. We should not intervene to establish what seems right to us, but instead manage the environment in advance to ensure it meets the needs of each cat. Occupying a territorial area does not grant ownership. Sharing depends on many factors that we do not fully understand, such as the timing of the first cat's arrival and the emotional state of each cat. Attempting to control ownership rights is an anthropomorphic idea that leads to failure.

Finally, a certain level of stability is necessary. Of course, this cannot be a prescription, but it serves as a warning about the risk of relapse in case of significant territorial changes (family changes, construction work in the habitat, or the arrival of a new individual).

11.2.5 Therapeutic Follow-Up

Relapses are possible, particularly due to territorial instability and changes in the protagonists caused by illness or aging. These relapses should be anticipated and treated as early as possible to prevent the emergence of new emotional outbursts.

11.2.6 Prevention

Cats are not meant to live in groups, so it is important to discuss this with owners when they express the intention to have multiple cats. Whenever possible, it is preferable to start with a litter. The arrival of another adult cat is always delicate, although counterexamples do exist.

One idea that needs to be addressed, and which is common enough to warrant mention, is that when a cat loses its companion, owners often think that it is wise to replace it. While this is a commendable intention, it fails to account for the personalities of the individuals involved. Not just anyone can replace a beloved companion without a long adaptation period. This should be emphasized to the owners.

Similarly, combating a cat's potential boredom by bringing in another cat is a poor idea. A cat's boredom is more often a sign of territorial poverty. Having to share this territory with a second cat will, in most cases, be more challenging. And with two cats, the environment will need to be even further enriched. Boredom is likely to turn into stress, and the result will be unfavorable.

11.2.7 Clinical Vignette

11.2.7.1 Caresse and Lloyd

Ms. X consulted because she was unable to make her two cats coexist. In fact, she had the intention of taking in a stray cat, Lloyd, who had been living alone in her garden under the rain. She has a large house and feels guilty seeing this cat wet outside while her Persian cat, Caresse, relaxed on the sofa. Sensing the potential risk, she was initially cautious and protected her delicate Caresse, who seemed frail compared to the large street cat. Her fears were confirmed as the situation quickly escalated. At first, there were some threats, but soon she had to separate Caresse and Lloyd as they started fighting.

During the consultation at her home, Caresse no longer dared to come down the stairs because Lloyd was guarding the staircase, ready to chase her. However, Lloyd is very friendly with them, and according to the owner, Caresse could have made more of an effort to tolerate him. Nevertheless, Lloyd had become openly aggressive toward Caresse. The diagnosis of intraspecific relational disorder due to dysfunctional communication was made, associated with a generalized inhibited anxiety state for Caresse and an intermittent productive anxiety state for Lloyd. The owner's perception was the main aggravating factor that allowed the problem to develop.

Treatment was implemented for both cats. Lloyd was given fluoxetine at a dose of around 2 mg/kg once daily. Caresse took venlafaxine (Effexor) granules in her food, 12 granules daily. Initially, the two cats were kept completely separate. Lloyd was confined to the garage, while Caresse was kept in the kitchen. This decision was made to avoid hospitalizing both animals. After 15 days, Lloyd was allowed inside under supervision, but only for short periods. Caresse was allowed to leave the kitchen when he was absent. Pheromone diffusers were installed on each floor. Gradually, Caresse started reoccupying the whole house. When Lloyd came inside, she was confined again to the kitchen. There were some hissing noises from both sides of the door.

After 3 weeks of hesitation, Ms. X dared to allow a meeting. She had been warned that a threat would likely happen, but it was limited to a mini charge and a hiss.

After 2 months, the situation had stabilized. Much to the owner's disappointment, the two cats still didn't get along. Every encounter was a threat, and at best, they shared the sofa, each at opposite ends. Lloyd now sleeps on the owners' bed, and Caresse sleeps in the children's room. As a precaution, treatment was continued, with weaning to be considered if the situation remained stable for at least 2 more months.

Lloyd uses the cat flap and often goes outside. Caresse only passes through the door if it's open, to lounge on the terrace, but she doesn't venture far. What truly pleases the owner is that recently, Caresse was threatened by a street cat that entered the garden, and Lloyd protected her.

11.2.7.2 Chouquette and Cacao

Ms. D. consulted because her two cats, Chouquette and Cacao, who are siblings, are no longer getting along. She adopted them at 3 months from an animal protection association, and for the first 6 months, they lived together peacefully, playing and sleeping together. However, their personalities have always been different: Cacao is more confident and enjoys being around guests, while Chouquette is shy and avoids social interactions, especially with strangers. Ms. D. considered this behavior normal, as many of her past cats were also less sociable. Chouquette has never been aggressive, never hurting anyone with her claws or teeth.

Chouquette's behavior started to change after a vaccination visit, where she urinated on herself during the injection and tried to flee. Later, after Cacao was hit by a car and hospitalized for a week, he had to be confined to a cage for 2 months to recover. When he returned, Chouquette hissed at him and ran away, refusing to return to the house. Afterward, Cacao tried to approach her, but she would always run away. Eventually, she stopped playing even with Ms. D., spending more time in the garage. Ms., feeling sad about Chouquette's isolation, occasionally brought her to the bedroom, where she would calm down with some petting but quickly return to the garage once the door was opened.

For a year, the veterinarian prescribed Zylkène and gabapentin (25 mg daily) in an attempt to improve the situation, but things continued to worsen. At the time of the consultation, Chouquette, now 2.5 years old, only came into the house for 5 min a day when Ms. served her food at 5 p.m. The veterinarian diagnosed an intraspecific relational disorder secondary to a phobic state (social phobia and fear of medical care). Chouquette was found to be chronically depressed and severely anxious, while Cacao remained normal and adapted, though his presence triggered Chouquette's fear and flight responses.

The veterinarian recommended a month-long separation, with Cacao sent to Madame's mother's house, while Chouquette underwent treatment. Chouquette was prescribed mirtazapine (2 mg/kg daily) and gabapentin (100 mg) to be given 2 h before Cacao's return and for any contact between the cats. During this time, Chouquette was isolated in Madame's bedroom, where she seemed to appreciate her owner's presence. After 2 days, the door was opened, allowing Chouquette to explore the house again. She began to sniff furniture and rub against it, behaviors she had stopped doing previously. Chouquette also resumed playing with Ms.

After a month, Cacao returned and was confined to the kitchen. Chouquette was still able to roam the house but frequently sniffed around the kitchen. After 3 days, Ms. brought Cacao into the living room while Chouquette was on the sofa. The two cats sniffed each other. Chouquette hissed but did not flee. Over the course of 10 days, their contact time increased, and they were no longer separated. Ms. continued to reserve her bedroom solely for Chouquette.

After a month of cohabitation, the cat flap was reopened, and Chouquette quickly went outside. The treatment was continued for another 3 months, and the cats seemed to be living together peacefully. After stopping the medication, there was no regression. Two years later, Ms. D. reports that there has been no recurrence of the issue, although she remains attentive to the pheromone markings of her cats and continues to regularly spend time alone with Chouquette in her bedroom.

Summary of ▶ Sect. 11.2

Communication between cats can only occur effectively if both individuals are healthy. If one of the cats is suffering from a pathological condition, it can lead to secondary relational issues, and the underlying condition must be treated before addressing the communication problem.

When two healthy cats are required to share the same space, rules for territorial sharing must be established. Conflicts are inevitable and typically start with mild threats. If these initial exchanges don't resolve the issue, skirmishes may occur, which usually facilitate territorial sharing. However, if the sharing remains impossible and the cats cannot separate from one another, the situation can become blocked, with one cat constantly monitoring the other, who seeks refuge in a confined space—this is the "obnubilation" phase.

Emotional states evolve from normal (distancing phase) to reactive, then phobic, and finally anxious. In some cases, a depressive state can emerge if the situation persists.

Cats should be treated according to their pathological state, and they must be temporarily separated to allow for gradual reintroduction. Even when the situation improves, it typically doesn't progress beyond a distant acceptance between the cats.

Disorders associated with dysregulation of the relationship axis

Disorder's name	Subtype	State diagnosis	Diagnostic criteria	Differential diagnosis
Interspecific relational disorder	• Linked to a dysfunctional communication • Secondary to another behavioral disorder	• Phobic or anxious • Depressive	The diagnosis is based on the presence of one or more of the following symptoms toward one or more members of the group: • Identification of dysfunctional communication • Aggression due to irritation in a fearful context directed at one or more individuals in the household, becoming more pronounced and worsening over time • Anticipation of the actions of the individual, with a tendency to flee • Hypervigilance in the presence of the individual(s) involved	*Organic disorders* Loss of sensory abilities Any painful condition *Interspecific social phobia* The aggressions concern unknown individuals, whereas in relational affection, there is a genuine poor relationship between two particular cats. *Dysthymia* The aggressions become independent of context over time and are much more unpredictable and violent. *HSHA syndrome* The triggering sequences occur during play or excitement. They do not lead to avoidance of a specific individual.

Disorder's name	Subtype	State diagnosis	Diagnostic criteria	Differential diagnosis
Intraspecific relational disorder	• Linked to a dysfunctional communication • Secondary to another behavioral disorder	• Phobic or anxious • Depressive	The diagnosis is based on the presence of one or more of the following symptoms toward one or more members of the group: • Repeated conflicts between two or more cats in the household, characterized by aggression due to irritation in a fearful context and territorial aggression • Territorial aggression between two or more cats in the household • Anticipation of the actions of this individual, with a tendency to flee for one and aggression for the other • Hypervigilance in the presence of the conspecific(s) involved	*Organic disorders* Loss of sensory abilities Any painful condition *Intraspecific social phobia* The aggressions involve all unknown cats, whereas in relational affection, there is a genuine poor relationship between two particular cats.

References

Amat M, Le Brech S, Camps T, Manteca X (2020) Separation-related problems in dogs. Adv Small Animal Care 1:1–8. https://doi.org/10.1016/j.yasa.2020.07.001

Bradshaw JW (2016) Sociality in cats: a comparative review. J Vet Behav 11:113–124

Chang H, Zhang J, Huang H, Aviles-Rosa EO, Huang H, Yan G, Xiao Z, Liu Q, Deng B, Zhang L (2025) The effects of owner-cat interaction on oxytocin secretion in pet cats with different attachment styles. Appl Anim Behav Sci 106524. https://doi.org/10.1016/j.applanim.2025.106524

De Souza Machado D, Oliveira PMB, Machado JC, Ceballos MC, Sant'Anna, A. C. (2020) Identification of separation-related problems in domestic cats: a questionnaire survey. PLoS One 15(4):e0230999. https://doi.org/10.1371/journal.pone.0230999

de Assis LS, Matos R, Pike TW, Burman OHP, Mills DS (2020) Developing diagnostic frameworks in veterinary behavioral medicine: disambiguating separation related problems in dogs. Front Vet Sci 6:499. https://doi.org/10.3389/fvets.2019.00499

DePorter TL, Bledsoe DL, Beck A, Ollivier E (2019) Evaluation of the efficacy of an appeasing pheromone diffuser product vs placebo for management of feline aggression in multi-cat households: a pilot study. J Feline Med Surg 21(4):293–305. https://doi.org/10.1177/1098612X18774437

DiCiccio VK, McClosky ME (2022) Fluoxetine-induced urinary retention in a cat. JFMS Open Rep 8(2):20551169221112065. https://doi.org/10.1177/20551169221112065

Edwards C, Heiblum M, Tejeda A, Galindo F (2007) Experimental evaluation of attachment behaviors in owned cats. J Vet Behav 2(4):119–125. https://doi.org/10.1016/j.jveb.2007.06.004

Frank D, Beauchamp G, Palestrini C (2010) Systematic review of the use of pheromones for treatment of undesirable behavior in cats and dogs. J Am Vet Med Assoc 236(12):1308–1316. https://doi.org/10.2460/javma.236.12.1308

Hargrave C (2021) Pheromones and 25 years of pheromonotherapy: what are they and how do they work? Vet Nurse 12(3):116–122. https://doi.org/10.12968/vetn.2021.12.3.116

Hart BL, Cliff KD, Tynes VV, Bergman L (2005) Control of urine marking by use of long-term treatment with fluoxetine or clomipramine in cats. J Am Vet Med Assoc 226(3):378–382. https://doi.org/10.2460/javma.2005.226.378

Hopfensperger MJ (2016) Use of oral venlafaxine in cats with feline idiopathic cystitis or behavioral causes of periuria. Proceedings of the Veterinary Behavior Symposium

Kaur G, Voith VL, Schmidt PL (2016) The use of fluoxetine by veterinarians in dogs and cats: a preliminary survey. Vet Record Open. https://doi.org/10.1136/vetreco-2015-000146

King JN, Steffan J, Heath SE, Simpson BS, Crowell-Davis SL, Harrington LJM, Weiss A-B, Seewald W (2004) Determination of the dosage of clomipramine for the treatment of urine spraying in cats. J Am Vet Med Assoc 225(6):881–887. https://doi.org/10.2460/javma.2004.225.881

Kry K, Casey R (2007) The effect of hiding enrichment on stress levels and behaviour of domestic cats *(Felis sylvestris catus)* in a shelter setting and the implications for adoption potential. Anim Welf 16(3):375–383. https://doi.org/10.1017/S0962728600027196

Landsberg GM, Wilson AL (2005) Effects of clomipramine on cats presented for urine marking. J Am Anim Hosp Assoc 41(1):3–11. https://doi.org/10.5326/0410003

Lindell EM, Erb HN, Houpt KA (1997) Intercat aggression: a retrospective study examining types of aggression, sexes of fighting pairs, and effectiveness of treatment. Appl Anim Behav Sci 55(1–2):153–162. https://doi.org/10.1016/S0168-1591(97)00032-4

Main M, Solomon J (1990) Procedures for identifying infants as disorganized/disoriented during the Ainsworth Strange Situation. In: Attachment in the preschool years: theory, research, and intervention, vol 1, pp 121–160

Masson S, Bleuer-Elsner S, Muller G, Medam T, Chevallier J, Gaultier E (2023) Psychiatrie vétérinaire du chien (No Ledge Editions, 1–1). No Ledge Editions

Masson S, Metz D, Bleuer-Elsner S, Schwobthaler F (2024) Retrospective study on the use of venlafaxine in 176 cats diagnosed with behavioral disorders. J Vet Behav 77. https://doi.org/10.2139/ssrn.4790331

Mertens PA, Torres S, Jessen C (2006) The effects of clomipramine hydrochloride in cats with psychogenic alopecia: a prospective study. J Am Anim Hosp Assoc 42(5):336–343. https://doi.org/10.5326/0420336

Metz D, Medam T, Masson S (2021) Double-blind, placebo-controlled trial of venlafaxine to treat behavioural disorders in cats: a pilot study. J Feline Med Surg 1098612X2110367. https://doi.org/10.1177/1098612X211036792

Potter A, Mills DS (2015) Domestic cats (felis silvestris catus) do not show signs of secure attachment to their owners. PLoS One 10(9):1–17. https://doi.org/10.1371/journal.pone.0135109

Schwartz S (2002) Separation anxiety syndrome in cats: 136 cases (1991–2000). J Am Vet Med Assoc 220(7):1028–1033. https://doi.org/10.2460/javma.2002.220.1028

Schwartz S (2003) Separation anxiety syndrome in dogs and cats. J Am Vet Med Assoc 222(11):1526–1532. https://doi.org/10.2460/javma.2003.222.1526

Seksel K, Lindeman M (1998) Use of clomipramine in the treatment of anxiety-related and obsessive-compulsive disorders in cats. Aust Vet J 76(5):317–321. https://doi.org/10.1111/j.1751-0813.1998.tb12353.x

Vinke CM, Godijn LM, van der Leij WJR (2014) Will a hiding box provide stress reduction for shelter cats? Appl Anim Behav Sci 160(1). https://doi.org/10.1016/j.applanim.2014.09.002

Vitale KR, Behnke AC, Udell MAR (2019) Attachment bonds between domestic cats and humans. Curr Biol 29(18):R864–R865. https://doi.org/10.1016/j.cub.2019.08.036

Essential Psychiatric Disorders

Contents

12.1 Dissociative Syndrome – 418
12.1.1 Description – 418
12.1.2 Diagnosis – 420
12.1.3 Prognosis – 422
12.1.4 Treatments – 422
12.1.5 Therapeutic Follow-Up – 423
12.1.6 Prevention – 423
12.1.7 Clinical Vignette – 423

12.2 Dysthymia – 424
12.2.1 Description – 424
12.2.2 Diagnosis – 426
12.2.3 Prognosis – 428
12.2.4 Treatments – 429
12.2.5 Therapeutic Follow-Up – 429
12.2.6 Prevention – 430
12.2.7 Clinical Vignette – 430

References – 432

© The Author(s), under exclusive license to Springer Nature Switzerland AG 2026
S. Bleuer-Elsner et al., *Veterinary Psychiatry of the Cat*,
https://doi.org/10.1007/978-3-032-06260-4_12

In this chapter, we describe the so-called essential psychiatric disorders. This chapter focuses on dissociative syndrome and dysthymia.

For each disorder, the clinical presentation, diagnosis, treatment, prognosis, and follow-up are detailed, complemented by clinical case studies that enrich the text.

In this chapter, we aim to describe specific psychiatric disorders that cannot be classified within the axes we outlined in previous chapters.

These disorders, which we refer to as essential, appear to have a strong genetic determinism. While no definitive cure exists for these conditions, their management can significantly reduce symptoms and provide the affected cat with a good quality of life despite the illness.

Notably, some of these disorders, particularly dissociative syndrome, lie at the intersection of neurology and psychiatry, placing them "apart" from the other conditions previously described. It is likely that this arbitrary classification choice will be reconsidered in the future.

Additionally, numerous orphan genetic disorders have been identified in human psychiatry, and it is evident that we have not yet reached this level of identification in cats (Bradshaw 2018; Montague et al. 2014; Robinson and Bergen 2021). This chapter is therefore intended to evolve, incorporating new entities as clinical research progresses.

12.1 Dissociative Syndrome

12.1.1 Description

12.1.1.1 Definition

Dissociative syndrome is one of the disorders associated with a state of derealization, known as the dissociative state. This condition is sometimes considered the feline equivalent of *schizophrenia*. Indeed, it is a disorder of varying severity in which the cat experiences hallucinations (dissociative state) that can entirely occupy its waking hours.

This condition can manifest in various ways. Currently, the diseases underpinned by a dissociative state remain poorly defined, and our nosography is far from complete. There are certainly other disorders associated with this state that have yet to be identified. For now, clinical presentation is the only available basis for differentiating these conditions. Consequently, the neurobiological distinction between the dissociative state and dissociative syndrome is not clearly established. However, as with other disorders, and to maintain coherence within this model, it seems important to distinguish between the *state level* and the *nosographic level* to allow for adjustments to the nosography when it becomes possible to differentiate the various entities grouped under dissociative syndrome.

Hallucinatory behaviors following ketamine anesthesia are regularly reported (Boudreau et al. 2012; Varì et al. 2022). Similarly, *temporal epilepsy* could be associated with hallucinations and trigger episodes typical of this group of disorders situated at the intersection of neurology, dermatology, and psychiatry (Gómez Álvarez and Soler Arias 2021).

In cats, this condition is often referred to as feline hyperesthesia (see ▶ Chap. 14, Sect. 14.2), where its symptoms are described. However, many cases of feline hyperesthesia are actually epilepsy or cutaneous hypersensitivity and are not necessarily associated with hallucinations, even though prominent symptoms, such as tail-chasing or tail attacks, may appear in all these cases (Amengual Batle et al. 2019).

12.1.1.2 Etiology and Pathogenesis

While the etiology remains incomplete, it is possible to describe vulnerability factors. In human schizophrenia, the genetic component is unquestionable, with numerous implicated genes (Schizophrenia Working Group of the Psychiatric Genomics Consortium 2014). Additionally, environmental factors such as substance abuse, particularly cannabis use (van Os et al. 2010), act as triggering factors.

In cats, no developmental conditions have been identified that increase this vulnerability. Therefore, we can consider *genetic vulnerabil-*

ity and describe *triggering events*. As in humans, these are often life events that are violent or, at the very least, emotionally intense (Misiak et al. 2018).

Cats affected by this condition often display poor self-regulation prior to experiencing derealization. The deficit in self-control, whether impulsive or compulsive, is likely the clinical expression of certain vulnerability genes. In schizophrenia, *progressive dilation of the cerebral ventricles* is described, which increases after the first episode (Kempton et al. 2010). This dilation has occasionally been observed in dogs, although it remains highly debatable, and it has not been documented in cats. Finally, the development of *addictions* is more frequently observed in individuals predisposed to this condition in human psychiatry (Reginsson et al. 2018), although this remains to be demonstrated in cats.

12.1.1.3 Epidemiology

The number of affected cats is low, and cases are typically seen in young adults. Currently, the lack of large-scale data prevents determining whether any specific breed is more frequently affected than others.

12.1.1.4 Presenting Complaints

Consultation is prompted by a clear sense of behavioral abnormality, accompanied by concerns about the *well-being* of the affected animals. The peculiar or sometimes stereotyped abnormal behaviors, including back rippling and, most notably, tail attacks, draw the owners' attention.

In the early stages of the condition, the disturbances are not invasive and may even appear amusing, but this is short-lived. Moreover, in cases where the cat engages in self-aggression (attacking its tail or hindquarters), owners may become worried about the animal's health.

12.1.1.5 Symptomatology

Dissociative syndrome can manifest in several forms.

Many cats compulsively and violently attack their tails. The onset of symptoms resembles what is observed in feline hyperesthesia, with back skin rippling (Rolling Skin syndrome), but the intensity and frequency of the attacks are distinct in this condition. In some cases, tail-chasing is accompanied by spinning behaviors, with the cat appearing to "fail" at catching its fleeing tail. As a predator, *the cat's behavior resembles a hunting action*.

Other forms involve sequences where the cat *chases imaginary objects*. In these cases, the abnormal behavior mimics the pursuit of small prey, with the cat leaping and extending its forelegs toward an imaginary target.

It is worth noting that the behaviors displayed during these episodes may belong to the cat's normal behavioral repertoire, particularly predatory behavior. What makes the behavior peculiar is not the sequence itself but the absence of an identifiable trigger. The animal is reacting to *a stimulus that does not exist in reality*.

12.1.1.6 Evolution

At the onset, the disease is minimally invasive.

Productive episodes are often triggered by various external phenomena with strong emotional consequences. These triggers may not always be objectively identifiable, but owners frequently provide detailed descriptions of the onset of the condition. Many owners believe that playing with laser pointers was a triggering factor. However, not all cats that play with these light pointers develop dissociative behavior.

For severely affected cats, a change in context often appears to be beneficial. Unfortunately, as soon as the novelty is assimilated, the dissociative behavior resurfaces. Between episodes, these cats can behave completely normally.

Finally, some cats experience frequent episodes without being severely affected. In these *partially affected cats*, there is often an initial contextual triggering factor, such as light flashes or sound signals. Over time, the episodes become independent of this trigger, even though they do not become more invasive.

Important Point of View

Human schizophrenia

The comparison between dissociative syndrome and human schizophrenia appears to be quite relevant. Schizophrenia typically affects young adults, although numerous predisposing factors can be identified during early life. The prodromal phase is non-specific before the first episode (Newton et al. 2018).

The condition is influenced by genetic predispositions combined with emotionally intense triggers (Misiak et al. 2018). It manifests through episodes of "positive symptoms," during which the individual acts in response to hallucinations (Ueda et al. 2018). These episodes become increasingly frequent and eventually dominate the individual's waking hours. Between episodes, the person is calm or even apathetic, characterized by "negative symptoms" (Correll and Schooler 2020). The individual becomes withdrawn and introspective, with diminished social interactions.

In dissociative syndrome, the various clinical expressions could similarly depend on the intensity of the positive and negative symptoms.

12

12.1.2 Diagnosis

12.1.2.1 Diagnostic Criteria

Cats affected by dissociative syndrome necessarily present a *pathological dissociative state* (◘ Fig. 12.1).

The diagnosis is based on the identification of the following symptoms:

- Onset of the first symptoms in early adulthood;
- Evidence of repetitive behaviors (hallucinations and self-aggression) unrelated to external stimuli, with no identifiable trigger;
- No organic cause identified through standard complementary tests (bloodwork and magnetic resonance imaging [MRI]);
- During episodes, the animal appears disconnected from its environment;
- Episodes become increasingly difficult to interrupt.

The presence of repetitive and identical behaviors across episodes, *in the absence of external triggers perceptible to all*, is pathognomonic of these disorders. The dissociative state is mandatory. Outside of episodes, the cat may appear normal. However, during the episode itself, it is almost impossible to capture the cat's attention.

The disorder most often begins in early adulthood.

Abnormal behaviors are not necessarily invasive but are at least repetitive and frequent. Currently, we also classify animals that exhibit episodes triggered by a specific context within this condition. At first glance, these behaviors may sometimes appear to stem from a compulsive state, making it tempting to classify these animals as being in a compulsive state. Indeed, outside of productive episodes, the cats sometimes seem "focused" on a stimulus, as if they were searching for the trigger of their compulsion. However, this compulsion is not consistent, and some animals are more likely to enter an episode when no other activity is provided.

It is important to emphasize that, unlike what occurs in the compulsive state, the object of the compulsion is neither identifiable nor present: the state is indeed derealized. There is, therefore, a *genuine loss of connection with reality*.

12.1.2.2 Direct Observation

Episodes are sometimes observed during the consultation. Additionally, it is common for owners to bring video recordings. Such observations help rule out generalized epilepsy and also allow for an assessment of the discomfort experienced by the affected animal.

The clinical examination may reveal subtle lesions resulting from spinning behaviors or repeated self-aggression. However, these lesions are not always present, and the clinical examination may appear completely normal.

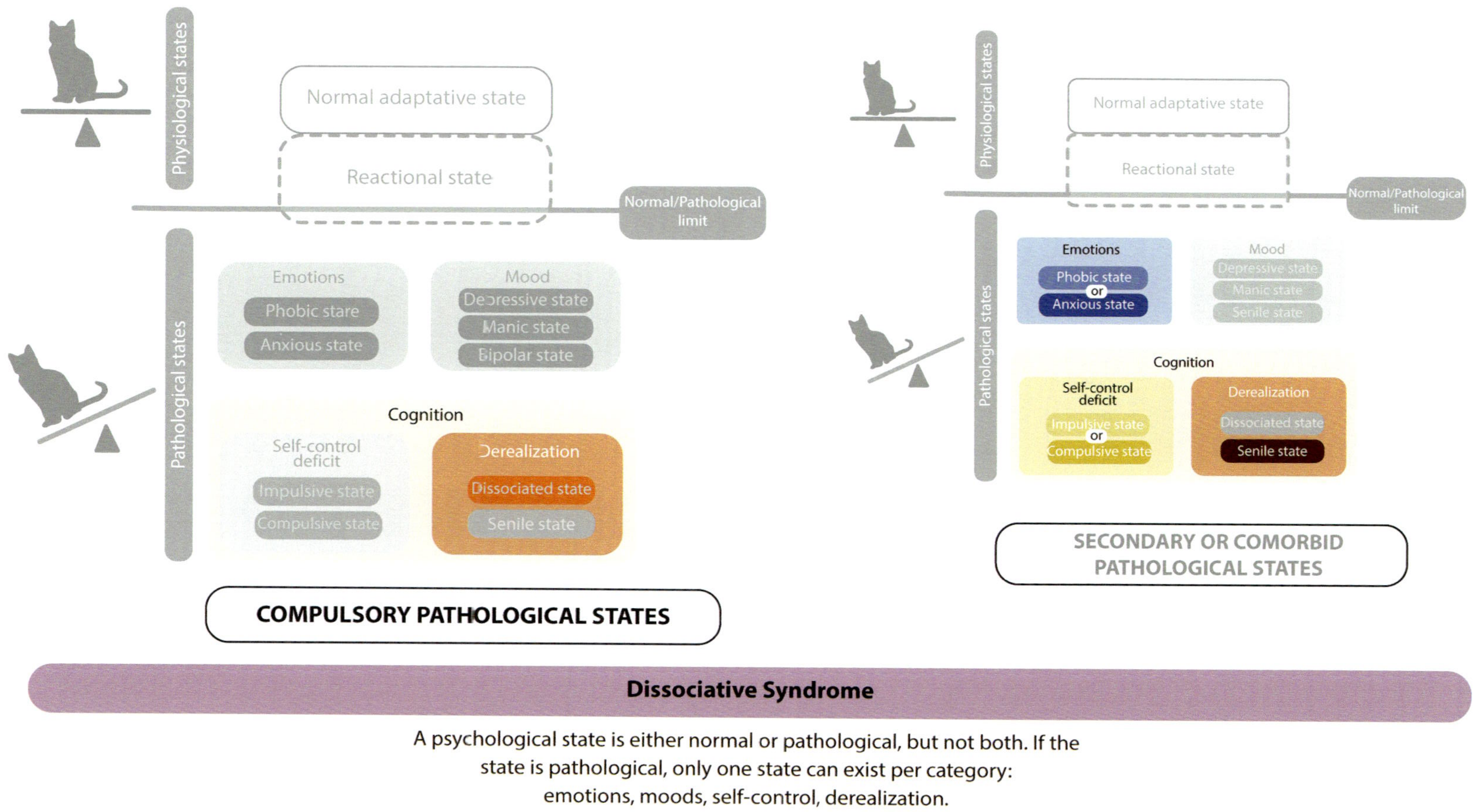

Fig. 12.1 Pathological states associated with dissociative syndrome. Mandatory pathological states: dissociative state. Possible comorbid pathological states: impulsive **or** compulsive states **and/or** senile. These states are often of lower priority compared to the symptoms associated with the dissociative state

12.1.2.3 Comorbidities

These conditions do not preclude the presence of other psychiatric disorders. However, their clinical expression is often so invasive that it obscures the symptoms of associated conditions. Managing dissociative syndrome is almost always the priority.

12.1.2.4 Differential Diagnosis

Behaviors resulting from hallucinations often belong to the cat's normal behavioral repertoire. The key to clinical differentiation lies in the *intensity of these behaviors* and, more importantly, the *environmental context* in which they occur. In some cases, cats are managed similarly to dogs with the same disorder, including tail amputation to address self-induced lesions or because their behavior suggests *pain*. Genuine pain can indeed provoke self-aggression, and this hypothesis must be included in the differential diagnosis. However, distinguishing between the two is generally straightforward. A cat in pain can localize the painful area precisely, often biting or licking the affected region. Additionally, a therapeutic trial using analgesics can help differentiate between the two. Pain-induced behaviors are also more easily triggered.

As previously noted, dermatological hypersensitivities, also referred to as feline hyperesthesia, can produce similar tail-related issues. However, the intensity of the aggression is usually less severe in such cases.

Similarly, *brain tumors* should be considered as a potential cause, although such behaviors are rarely associated with these lesions. In the early stages, their progression can produce similar symptoms, but imaging helps make the distinction. Cats with brain tumors are generally older.

Finally, *partial temporal epilepsy* can also explain the symptoms, although the episodes in such cases are typically shorter. Pharmacological treatment to stabilize the seizures is likely the best way to differentiate, as brain imaging will return negative results in both scenarios.

Cats with *hypersensitivity-hyperactivity (HSHA) syndrome* associated with an *impulsive or compulsive state* struggle to regulate themselves during periods of excitement. During play excitement, they may react to their tails or even chase them, but it is the movement that triggers their reaction. In moments of heightened excitement, they may attempt to climb walls or even feign hunting behaviors. The presence of an identifiable trigger and the non-invasive nature of these activities help differentiate them. Moreover, these sequences are brief and occur only in contexts of excitement or frustration.

The behavior of a dissociated animal can, in some cases, resemble an addiction, particularly when the behaviors are stereotyped. However, in the dissociative state, the triggers are not real, and the animal experiences few periods of normality. Finally, in addictions, the object of the addiction is not imaginary.

12.1.3 Prognosis

The prognosis is variable but often ranges *from poor to grim*. Some animals may stabilize with treatment or experience a non-progressive form that minimally affects their quality of life. However, in the most severe cases, treatment may prove entirely inadequate.

The time required to establish an effective treatment often feels too long for owners due to the discomfort caused by the episodes. It is not always possible to find a treatment quickly enough to reduce the impact of the condition. Owners frequently request euthanasia, unwilling to pursue what they perceive as unnecessary medical persistence.

12.1.4 Treatments

12.1.4.1 Pharmacological Treatment

To date, no medication ensures systematic recovery. High-dose fluoxetine (1–2 mg/kg) often increases cortical control and reduces the duration of crisis episodes (Epstein 2020). It appears beneficial to combine it with gabapentin due to its anti-epileptic and analgesic properties (Amengual Batle et al. 2019).

Given the similarity of these conditions to certain forms of schizophrenia, second-generation neuroleptics have been proposed, although their use in cats remains limited. Some authors report good results with clomipramine, paroxetine (Denenberg and Dubé 2018), or venlafaxine (Masson et al. 2024).

Phenobarbital and topiramate may also be considered, either as adjunctive treatments or monotherapies (Denenberg and Dubé 2018; Gómez Álvarez and Soler Arias 2021).

In general, the number of documented cases of dissociative syndrome remains insufficient to establish definitive conclusions regarding its etiology, symptoms, or treatment (Amengual Batle et al. 2019).

To date, no medication ensures systematic recovery. *High-dose fluoxetine* (1–2 mg/kg) often increases cortical control and reduces the duration of episodes (Epstein 2020). Combining fluoxetine with *gabapentin* appears beneficial due to its anti-epileptic and analgesic properties (Amengual Batle et al. 2019).

Given the similarity of these conditions to certain forms of schizophrenia, second-generation *neuroleptics* have been proposed, although their use in cats remains limited.

Some authors report good results with clomipramine, paroxetine (Denenberg and Dubé 2018), or *venlafaxine* (Masson et al. 2024).

Phenobarbital and *topiramate* may also be considered, either as complementary treatments or as standalone options (Denenberg and Dubé 2018; Gómez Álvarez and Soler Arias 2021).

Overall, the number of reported cases of dissociative syndrome remains insufficient to draw definitive conclusions regarding its etiology, symptoms, or treatment (Amengual Batle et al. 2019).

12.1.4.2 Behavioral Therapy

As expected with these disorders, which likely have a significant neurological and genetic basis, therapies often prove ineffective. It appears that anything that reduces the number of episodes slows the progression of the condition, as if the disease's expression worsens its own course.

▪ Environmental Modifications

It is important to keep affected animals in very calm and predictable environments. This measure can be effective if the disorder is not invasive.

▪ Behavioral Modifications

These primarily involve providing these cats with stable living conditions that respect their physiological needs and promote high-quality communication. This approach helps maintain a bond with the owners, who are often deeply affected by their pet's visible suffering and may become discouraged.

Of course, physical punishments must be eliminated, and the focus should be on quality of life, providing these animals with activities as diverse as their condition allows.

12.1.5 Therapeutic Follow-Up

The abnormal behaviors associated with these disorders are challenging to endure, and it is essential to maintain close contact with the owners in case of improvement, as they may find it extremely difficult to cope with a relapse. Additionally, treatments are typically lifelong, and follow-up is indispensable, at least to monitor potential side effects (e.g., regular biochemical assessments depending on the medications used, at least annually).

12.1.6 Prevention

To date, the etiology remains unknown. Our only tool for prevention is *excluding affected animals from reproduction*, which could potentially be extended to their family members. In any case, severely affected animals are no longer suitable for reproduction.

12.1.7 Clinical Vignette

▪ Pinchos

Pinchos is a 2-year-old neutered male cat who is very playful. His owner visits the clinic with a video recording. On the video, the cat is seen engaging in intense hunting behavior for imaginary prey. The scene takes place in the owner's bedroom, where Pinchos is on the bed. He suddenly leaps to catch an imaginary fly near the window, then turns and runs toward the opposite wall. If a real fly were present, this behavior would seem entirely normal, but there is no insect, and the actions appear completely disconnected from reality.

The owner seeks consultation because the cat injured himself during an episode and fears that 1 day Pinchos might jump out the window due to his intense focus on these imaginary flies, which occupy him for several hours each day. Complementary tests (a full blood panel and MRI) prescribed by the general practitioner revealed no abnormalities.

A treatment of fluoxetine at 2 mg/kg per day is prescribed, and as the episodes always occur in the bedroom on bright days, it is decided that the cat will no longer have access to the room except at night.

The results are quickly very positive and remain stable for several years. Pinchos relapses at 5 years old, after his imaginary hunting behavior had ceased. Gabapentin is added at a dose of 50 mg in the morning and evening, with a positive effect. Now 12 years old, Pinchos is still under treatment with no further relapses to date.

Summary of ▶ Sect. 12.1

Dissociative syndrome encompasses a group of conditions in which cats display compulsive activity in response to stimuli that are imperceptible to us. In these conditions, cats are in a dissociative state and often experience hallucinations, either intermittently or invasively.

These conditions are frequently compared to schizophrenia in human medicine and are primarily observed in young adult cats. Some affected cats experience episodic crises, often context-dependent. The behavioral reactions fall within the usual repertoire of feline behavior (e.g., tail-chasing and imaginary prey hunting), but their duration and intensity are abnormal.

The critical factor is the absence of an identifiable trigger for an external observer. Between episodes, the cat may appear normal, although it often seems fatigued. The etiology of the condition is largely unknown, but a strong genetic component is suspected.

Medical treatment is not always effective, but stabilization is achievable in some cases. Medications that reduce sensitivity and enhance cortical control can provide significant improvement. Fluoxetine at 1–2 mg/kg or clomipramine at 1 mg/kg once daily have been reported as effective. Gabapentin and antiepileptics can also be combined to reduce episodes and improve the quality of life for both the animal and its owner. Treatment must be continued throughout the cat's lifetime.

12.2 Dysthymia

12.2.1 Description

12.2.1.1 Definition

Dysthymia is a condition associated with a mood disorder in which the cat experiences cyclic episodes triggered by what appears to be an internal origin (◘ Fig. 12.2). Cats brought in for consultation are often those whose episodes include *aggressive sequences*, which seem to be *common*.

This condition could also suggest a form of derealization; however, the aggressive behavior in dysthymia is directed toward individuals, most often the owners, and the cat reacts, albeit in a *disproportionate and unjustified* manner, to perceptible stimuli.

12.2.1.2 Etiology and Pathogenesis

This condition corresponds to a mood disorder and typically affects young adult cats. At present, no contextual triggers or favorable developmental conditions have been identified.

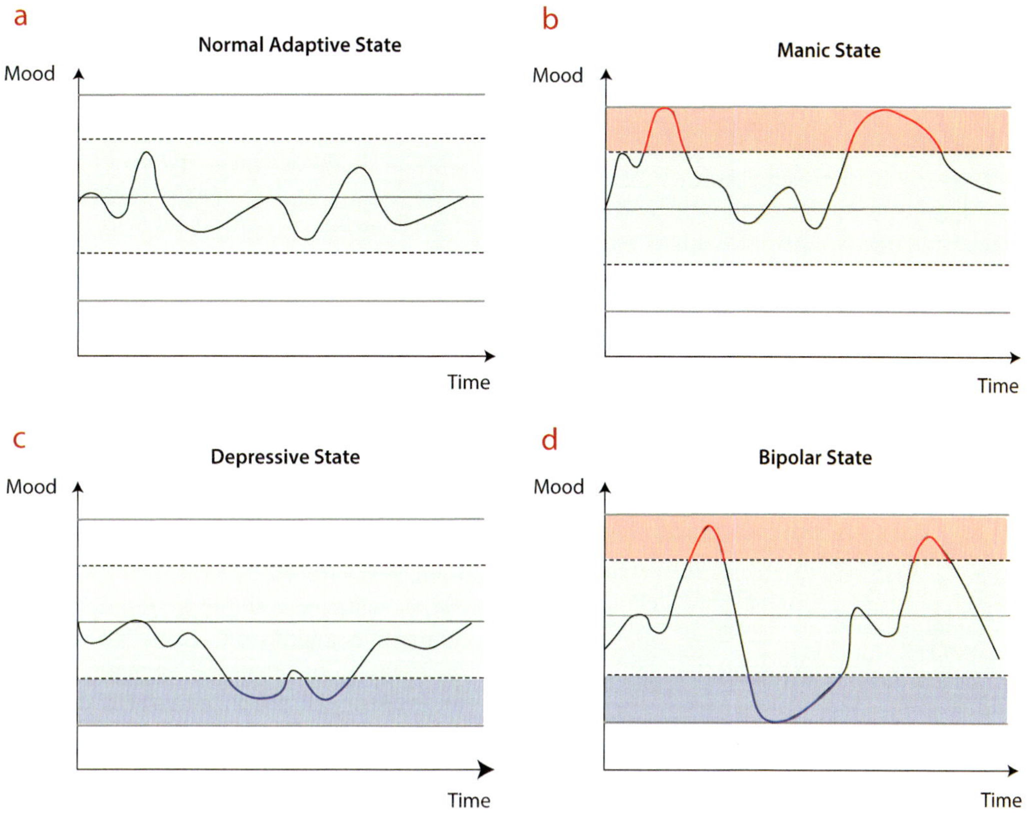

Fig. 12.2 Different mood variations leading to dysthymia. Mood fluctuates slowly around a central zone of balance **a**. Dysregulation can induce **b** an excessive elevation of mood leading to the pathological manic state, **c** an excessive decrease in mood leading to the pathological depressive state, or **d** an alternation between extremes or between extremes and normalcy, leading to the pathological bipolar state

Certain breeds appear to be predisposed, suggesting a probable genetic component, although no specific gene has been identified to date. This condition is comparable to *bipolar disorders* described in human medicine, where the etiological factors are also likely to be numerous. In humans, multiple genes are thought to contribute as vulnerability factors. Cortical gray matter volume appears to be significantly higher in patients at high genetic risk of developing bipolar disorders (Fusar-Poli et al. 2012).

In cats, comparable imaging data is not yet available.

12.2.1.3 Epidemiology

This condition is relatively rare, but despite the affected animals' limited ability to reproduce, it persists, and cases continue to be regularly observed.

12.2.1.4 Presenting Complaints

Overall, affected cats display *highly aggressive* behavior during episodes, and the fear of being bitten is the most common reason for consultation. It is likely that some affected animals exhibit non-aggressive forms of the condition, but these cases entirely escape medical attention.

12.2.1.5 Symptomatology

The most notable symptom is the occurrence of *episodes*. The cat is generally normal between these episodes.

By comparison with bipolar disorders in humans, it has been hypothesized that cats experience depressive periods between aggressive episodes. This appears true for some cats, which exhibit distinct withdrawal periods with signs of chronic depression (bipolar state) between episodes. However, for many dysthymic cats, the *manic phases* are separated by calm periods without depressive states.

The episodes are often so striking for the owners that, in cases where the cat is in a depressive state between episodes, the owners may simply feel relieved not to endure the attacks and may not perceive this as an abnormal decline in motivation.

Owners living with these cyclic cats recognize mood changes and describe them as sudden but easy to identify. They often report a change in the cat's expression, typically associated with pupil dilation (mydriasis). They become aware that the cat should not be disturbed during these times and quickly learn to leave it alone. However, many of these cats do not isolate themselves during episodes and instead display idiopathic aggression, unrelated to context and often offensive, without any identifiable trigger.

Finally, some forms of this condition are linked to the *hormonal cycle of female cats*. Estrus may coincide with the onset of the first manic phase.

12.2.1.6 Evolution

At the onset of the condition, the progression is rapid, sometimes abrupt, and heavily influenced by the owners' responses. The more they attempt to oppose the cat, the more the aggression intensifies. Very quickly, the situation stabilizes, and the episodes recur at relatively regular intervals.

Depending on the individual cat, episodes may occur daily and last a few hours or take place weekly or monthly, sometimes lasting several days. Owners who decide to keep the cat must then isolate it during these periods.

It is important to note that outside of these phases, the cat appears normal. Aggressive behaviors may still occur, but they are proportional and consistent with the context. The mood alteration characteristic of dysthymia leads to the aggressive responses associated with this condition.

12.2.2 Diagnosis

12.2.2.1 Diagnostic Criteria

Cats affected by dysthymic disorder necessarily present a *pathological manic or bipolar state* (◘ Fig. 12.3). It is worth noting that the depressive form may alternate with normal phases, but cats without manic episodes are rarely brought in for consultation. If they are, the diagnosis is unlikely to be dysthymia and is more often identified as a chronic depressive state.

The first manic episodes typically appear in early adulthood. The diagnostic criteria characterizing these episodes include some or all of the following symptoms:

- A change in expression due to almost continuous mydriasis;
- Hypervigilance and motor agitation;
- Idiopathic aggression, often manifesting as irritation or territorial aggression, with an intensity disproportionate to the triggering context;
- Changes in appetite: during these episodes, cats often stop eating;
- Sexual dysthymia: a sudden behavioral change occurring during estrus that does not resolve after the end of the estrus cycle.

Between episodes, the cat is either normal (manic state) or depressed (bipolar state).

The sudden onset of unfounded aggressive episodes in a young cat should raise concern. When observed, the aggression often resembles territorial aggression but is clearly disproportionate.

The dramatic nature of the episodes and the intensity of the aggressive responses should alert the practitioner, especially if the cat is in its pubertal phase or a young adult. Establishing a definitive diagnosis often requires multiple consultations and a therapeutic trial, but the presence of phases in a cat that reacts differently in the same context is a warning sign of a *mood disorder*. The fear experienced by owners who have been subjected to these attacks is also a frequent red flag for dysthymic disorder.

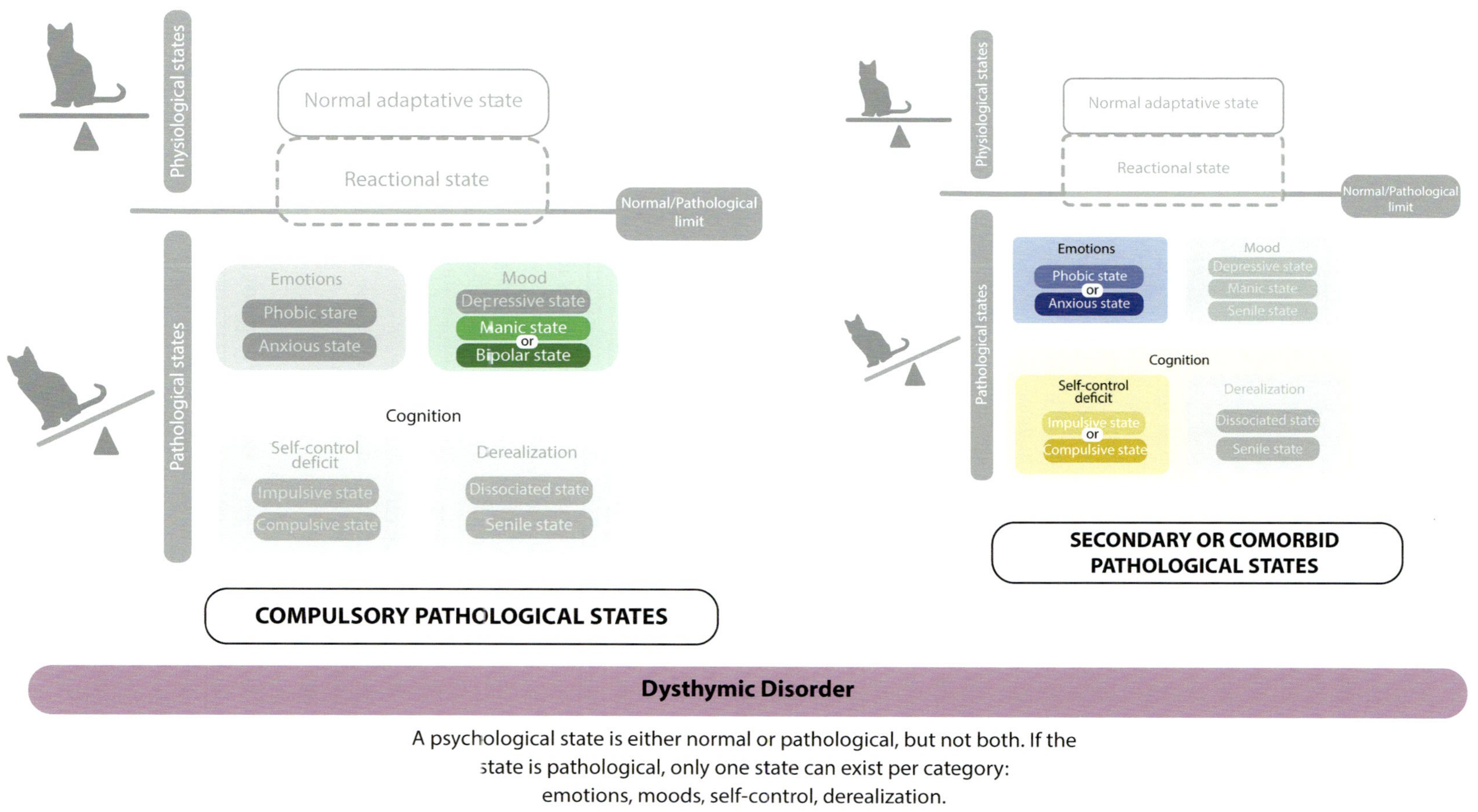

Fig. 12.3 Pathological states associated with dysthymia. Mandatory pathological states: manic **or** bipolar state. Possible comorbid pathological states: phobic **or** anxious states **and/or** impulsive **or** compulsive states **and/or** senile. These states are often of lower priority compared to the symptoms associated with the dissociative state

The severity of the episodes can be so extreme that it is not uncommon for owners to leave their home and call emergency veterinary services or firefighters before daring to return. Sometimes, they manage to isolate the cat in a part of the home, but when they crack open the door to check if the episode has ended, they often report that the cat is already attacking the door, forcing them to extend the isolation period.

12.2.2.2 Direct Observation

Unless the owner has recorded a video, the cat may appear normal during the consultation. In many cases, the episodes have led to strong reactions from the owners, resulting in a deteriorated relationship.

The owners are often on edge, closely monitoring the cat's movements out of fear of another attack. Meanwhile, the cat, which has indeed attacked but has also experienced defensive actions from the owners, often remains hypervigilant for several hours or even days following the episodes.

12.2.2.3 Comorbidities

Mood disorders are quite invasive but do not rule out the presence of other behavioral conditions. The coexistence of phobias cannot be excluded, and the conflicts inherent to dysthymia significantly impact the owner. Relationships with the owner are often strained, and it is not uncommon for the initial request to be for euthanasia. The practitioner must be persuasive if they hope to implement a therapeutic plan.

Due to the deterioration of the cat-owner relationship, an anxious state generally develops quickly alongside the manic state. With age, intercurrent illnesses associated with aging often emerge as well.

12.2.2.4 Differential Diagnosis

A *thorough differential diagnosis* must be conducted, starting with confirming the absence of organic diseases that could cause pain or sensory alterations.

In terms of psychiatric conditions, dysthymic disorder is often mistaken for a fear axis disorder, particularly one associated with an anxious state. However, a cat's anxious state is more commonly linked to a categorical or post-traumatic phobia, deprivation syndrome, or a territoriality disorder.

The sudden onset of aggressive episodes in an animal that was previously in a normal adaptive state should raise suspicion of dysthymic disorder.

It is also possible to confuse dysthymic disorder with *dissociative syndrome* associated with the dissociative state. In dissociative syndrome, however, the cat appears disconnected from reality, and episodes are not linked to any identifiable trigger. Additionally, the periodicity of derealization episodes is much higher.

Some redirected aggression episodes can also mimic a crisis, as the cause is not always easy to identify. A thorough anamnesis is necessary to uncover an external trigger. Typically, the cat quickly returns to normal after a redirected aggression, whereas it remains hypervigilant for an extended period following a manic episode.

Predatory sequences involving the owners' feet shortly before meal distribution—when meals are not free-fed—should also be part of the differential diagnosis. In such cases, the behavior is predatory, triggered by frustration associated with hunger and meal anticipation. Semi-darkness often exacerbates these attacks, but the sequences are short and much more controlled.

Finally, in unspayed female cats, sexual dysthymias may occur. In this form of dysthymia, lordosis postures and marking behaviors through urination, which are characteristic, can be easily observed. Aggressive episodes occur during estrus, and spaying often resolves the condition.

12.2.3 Prognosis

The prognosis is *poor* when considering the chances of complete recovery. However, in many cases, it is possible to manage the animal with pharmacological treatment, ensuring good quality of life for both the cat and its owners. The apparent normality of the cat under treatment often gives the impression of a cure, and it is essential to inform the owner not to discontinue the treatment.

Sexual dysthymias can be completely resolved through spaying, although some cases may persist even after surgery.

12.2.4 Treatments

12.2.4.1 Pharmacological Treatment

Treatment is primarily pharmacological, as the condition is neither triggered nor exacerbated by contextual factors. Currently, high-dose fluoxetine (1–2 mg/kg) appears to be the most effective treatment. Clomipramine can also be used at a dose of 1 mg/kg, although it tends to cause stronger sedation. Adding gabapentin may sometimes enhance the stabilization of episodes.

The use of mood stabilizers, such as valproic acid, or neuroleptics is standard in human medicine for bipolar disorders. However, these drugs are not approved for use in feline medicine. The lack of a significant number of case studies and documented research on this condition makes management far less advanced than that of bipolar disorders in human medicine. Additionally, most of these cats are quickly euthanized due to owner fear following a first episode, leading to limited therapeutic trials.

For the sexual form of dysthymia, *spaying* is the treatment of choice.

12.2.4.2 Behavioral Therapy

- **Environmental Modifications**

The primary goal is to ensure the safety of people who may come into contact with the cat during episodes. Isolating the animal in a quiet room, when feasible, is a simple and effective solution. In all cases, contact with the animal should be avoided during episodes.

- **Behavioral Modifications**

There is no established behavioral therapy for this condition. However, ensuring the safety of individuals during manic phases is essential. One focus during initial consultations is teaching owners to recognize the *early signs of an episode*.

It is worth noting that owners often become adept at identifying the onset of a crisis, largely due to the severity of the situations they have previously experienced. Early indicators of aggression may include pupil dilation as well as changes in tail movement, which progresses from subtle to more agitated as the episode unfolds.

Owners should also be made aware that a cat in crisis is *not in a state to learn* and will not respond to punishment or reward. The key priorities are, first, to prevent accidents involving the owners and, second, to minimize conflict. Under no circumstances should attempts be made to correct the cat, nor should people outside the family attempt to do so out of ignorance. Symptoms cannot be punished; attempting to punish aggression is as ineffective as trying to forbid fever or enteritis.

The best approach is to isolate the cat during the episode, which may last from a few hours to several days. Using "shields," such as wooden panels, to calmly guide the animal into a room that can be closed off is an effective method for ensuring safety. This technique protects against scratches and bites, which can be numerous and severe during these episodes. Owners often feel as though they are facing a wild animal and no longer recognize their cat.

12.2.5 Therapeutic Follow-Up

Therapeutic follow-up is crucial, as there is no hope for a cure, and the condition poses a danger. Additionally, the practitioner must regularly monitor for potential side effects of the treatment. A protocol for medication renewal should be established in advance, and the practitioner must actively ensure that improvement does not lead to discontinuation of the treatment.

12.2.6 Prevention

Currently, genetic prevention through testing is unrealistic. A large number of genes likely contribute to vulnerability, making selection overly complex. The simplest solution would be to exclude affected animals from reproduction. However, most of the cases we encounter are already sterilized, yet the condition continues to persist.

12.2.7 Clinical Vignette

▪ Romance

Romance is a 2-year-old spayed European cat. The initial consultation was highly emotional. The owners arrived distressed, requesting euthanasia for an animal they love but fear.

They struggled to place Romance in her carrier, and she arrived agitated, displaying signs of fear toward the veterinarian. After the owners expressed their emotions about the difficulties they faced, the owner explained her fear of the cat. She had recently suffered a violent attack, resulting in scratches and bites to her arm while protecting her face. The couple, planning to have a child, found the situation intolerable.

12

The description was telling that Romance had previously been normal and even affectionate. The first episode occurred without any apparent cause while the couple and Romance were watching television. Suddenly, Romance began growling and ran off. The owner, who had noticed morning agitation, attempted to speak to her. Romance then leapt at her face. For 2 days afterward, Romance roamed the house avoiding contact before returning to her usual behavior. Since then, the owner avoided sitting with her on the couch and began locking herself in the bedroom at night.

The owners' attachment to Romance was evident, and a trial treatment was initiated. Fluoxetine was prescribed at half a capsule per day (10 mg for 4 kg). However, 8 days later, the owners returned, reporting they could not get Romance to ingest the powdered contents of the capsules. Despite the challenging situation, they agreed to try venlafaxine instead. This medication was easier to administer, given at 8 granules per kilogram mixed into palatable wet food.

Approximately 10 days after starting treatment, Romance returned to her usual behavior, allowing the owner to relax. Gabapentin was prescribed as a backup for managing episodes despite the treatment, but the owners never needed to use it.

A year later, a relapse occurred when the owners stopped the treatment. After some persuasion, they resumed it, with successful results. Now 6 years old, Romance coexists well with the new baby. Annual check-ups monitor her condition, including bloodwork at each visit. The owners report occasional minor agitation, but these episodes are far less severe than before and are managed by leaving her alone.

Summary of ▶ Sect. 12.2

Dysthymia is a condition associated with the manic or bipolar state, making it a mood disorder. It can be mistaken for intermittent anxiety, but the absence of clinical signs supporting anxiety helps guide the diagnosis toward dysthymia.

Manic phases are often preceded by subtle behavioral changes, with the cat becoming more agitated without apparent cause. Dysthymic disorder is characterized by alternating manic phases and periods of normality. The aggression exhibited by the cat during manic phases frequently leads to a loss of trust from the owners and a request for euthanasia.

The treatment of dysthymic disorder involves fluoxetine at 1–2 mg/kg once daily. Venlafaxine at 2 mg/kg once daily also yields good results, particularly when administering medication is challenging. Combining these medications with gabapentin or pregabalin can sometimes improve symptom management.

The episodes do not always disappear but become less frequent and less intense.

Although the condition is associated with a high level of danger, isolating the cat during episodes is usually sufficient to prevent accidents.

Close collaboration between the practitioner and the owner is essential. Regular check-ups should ensure the animal's stability and adherence to treatment.

Essential psychiatric disorders

Disorder's name	State diagnosis	Diagnostic criteria	Differential diagnosis
Dissociative syndrome	Dissociative	First dissociative episodes appear in *adulthood*. The diagnosis is based on identifying the following symptoms: • Onset of symptoms in early adulthood. • Evidence of repetitive activity (hallucinations and self-aggressions) unrelated to external stimuli, with no identifiable trigger. • No organic cause identified through standard complementary tests (e.g., bloodwork and MRI). • During episodes, the animal appears disconnected from its environment. • Increasingly difficult to interrupt episodes.	*Somatic diseases* ***Pain*** The painful area can be identified, and pain management resolves the issue. ***Partial temporal epilepsy*** Episodes are shorter and less invasive. ***Dermatological hypersensitivity*** Pruritus treatment resolves the condition. ***Brain tumor*** Imaging can exclude this, although early evolution may appear similar. *Anxious states with self-centered behaviors* Behavior aligns with contextual triggers; anxiety management improves behavior. *HSHA syndrome or feline ADHD* Episodes are short and occur only in contexts of excitement or frustration. *Addictions* Triggers are real, and the focus of the addiction is not imaginary.

Disorder's name	State diagnosis	Diagnostic criteria	Differential diagnosis
Dysthymia	Manic or bipolar	First manic episodes appear in *early adulthood.* Diagnostic criteria include some or all of the following symptoms: • Altered gaze due to almost continuous mydriasis. • Hypervigilance and motor agitation. • Idiopathic aggression, often irritation based or territorial, with disproportionate intensity to the triggering context. • Appetite changes, often refusal to eat during episodes. • Sexual dysthymia: Abrupt behavioral change during estrus that does not resolve post-estrus. Between episodes, the cat is normal (manic state) or depressed (bipolar state).	*Interspecific relational disorder* Reestablishing communication resolves or diminishes episodes. The same actions consistently produce the same effects, unlike dysthymia, where mood changes lead to varying reactions despite identical contexts. *Dissociative syndrome* The cat appears disconnected from reality; episodes are not relational and are more invasive than those in dysthymia. *Sexual dysthymia* ***Gestation*** Diagnosed via ultrasound. Maternal aggression appears post-parturition, not during gestation ***Pyometra*** Excluded via bloodwork and ultrasound. Aggression is rare and less dramatic despite pain. Clinical signs include apathy and polyuria/polydipsia.

12

References

Amengual Batle P, Rusbridge C, Nuttall T, Heath S, Marioni-Henry K (2019) Feline hyperaesthesia syndrome with self-trauma to the tail: retrospective study of seven cases and proposal for an integrated multidisciplinary diagnostic approach. J Feline Med Surg. https://doi.org/10.1177/1098612X18764246

Boudreau AE, Bersenas AME, Kerr CL, Holowaychuk MK, Johnson RJ (2012) A comparison of 3 anesthetic protocols for 24 hours of mechanical ventilation in cats. J Vet Emerg Crit Care 22(2):239–252. https://doi.org/10.1111/j.1476-4431.2012.00722.x

Bradshaw J (2018) Normal feline behaviour: … and why problem behaviours develop. J Feline Med Surg 20(5):411–421. https://doi.org/10.1177/1098612X18771203

Correll CU, Schooler NR (2020) Negative symptoms in schizophrenia: a review and clinical guide for recognition, assessment, and treatment. Neuropsychiatr Dis Treat 16:519–534. https://doi.org/10.2147/NDT.S225643

Denenberg S, Dubé MB (2018) Tools for managing feline problem behaviours: psychoactive medications. J Feline Med Surg 20(11):1034–1045. https://doi.org/10.1177/1098612X18806760

Epstein ME (2020) Feline neuropathic pain. Vet Clin N Am Small Anim Pract 50(4):789–809. https://doi.org/10.1016/j.cvsm.2020.02.004

Fusar-Poli P, Howes O, Bechdolf A, Borgwardt S (2012) Mapping vulnerability to bipolar disorder: a systematic review and meta-analysis of neuroimaging studies. J Psychiatry Neurosci 37(3):170–184. https://doi.org/10.1503/jpn.110061

Gómez Álvarez CM, Soler Arias EA (2021) Feline hyperesthesia syndrome: epilepsy as possible aetiology in two cats. Vet Rec Case Rep 9(3). https://doi.org/10.1002/vrc2.132

Kempton MJ, Stahl D, Williams SCR, DeLisi LE (2010) Progressive lateral ventricular enlargement in schizophrenia: a meta-analysis of longitudinal MRI studies. Schizophr Res 120(1–3):54–62. https://doi.org/10.1016/j.schres.2010.03.036

Masson S, Metz D, Bleuer-Elsner S, Schwobthaler F (2024) Retrospective study on the use of venlafaxine in 176 cats diagnosed with behavioral disorders. J Vet Behav 77. https://doi.org/10.2139/ssrn.4790331

Misiak B, Stramecki F, Gawęda Ł, Prochwicz K, Sąsiadek MM, Moustafa AA, Frydecka D (2018) Interactions between variation in candidate genes and environmental factors in the etiology of schizophrenia and bipolar disorder: a systematic review. Mol Neurobiol 55(6):5075–5100. https://doi.org/10.1007/s12035-017-0708-y

Montague MJ, Li G, Gandolfi B, Khan R, Aken BL, Searle SMJ, Minx P, Hillier LW, Koboldt DC, Davis BW, Driscoll CA, Barr CS, Blackistone K, Quilez J, Lorente-Galdos B, Marques-Bonet T, Alkan C, Thomas GWC, Hahn MW et al (2014) Comparative analysis of the domestic cat genome reveals genetic signatures underlying feline biology and domestication. Proc Natl Acad Sci 111(48):17230–17235. https://doi.org/10.1073/pnas.1410083111

Newton R, Rouleau A, Nylander A-G, Loze J-Y, Resemann HK, Steeves S, Crespo-Facorro B (2018) Diverse definitions of the early course of schizophrenia—a targeted literature review. NPJ Schizophr 4(1):21. https://doi.org/10.1038/s41537-018-0063-7

Reginsson GW, Ingason A, Euesden J, Bjornsdottir G, Olafsson S, Sigurdsson E, Oskarsson H, Tyrfingsson T, Runarsdottir V, Hansdottir I, Steinberg S, Stefansson H, Gudbjartsson DF, Thorgeirsson TE, Stefansson K (2018) Polygenic risk scores for schizophrenia and bipolar disorder associate with addiction: polygenic risk and addiction. Addict Biol 23(1):485–492. https://doi.org/10.1111/adb.12496

Robinson N, Bergen SE (2021) Environmental risk factors for schizophrenia and bipolar disorder and their relationship to genetic risk: current knowledge and future directions. Front Genet 12:686666. https://doi.org/10.3389/fgene.2021.686666

Schizophrenia Working Group of the Psychiatric Genomics Consortium (2014) Biological insights from 108 schizophrenia-associated genetic loci. Nature 511(7510):421–427. https://doi.org/10.1038/nature13595

Ueda N, Maruo K, Sumiyoshi T (2018) Positive symptoms and time perception in schizophrenia: a meta-analysis. Schizophr Res Cogn 13:3–6. https://doi.org/10.1016/j.scog.2018.07.002

van Os J, Kenis G, Rutten BPF (2010) The environment and schizophrenia. Nature 468(7321):203–212. https://doi.org/10.1038/nature09563

Varì MR, Ricci G, Cavallo M, Pichini S, Sirignano A, Graziano S (2022) Ketamine: from prescription anaesthetic to a new psychoactive substance. Curr Pharm Des 28(15):1213–1220. https://doi.org/10.2174/1381612828666220510115209

Aging Disorders

Contents

13.1 Common Points to All Aging Disorders – 438
13.1.1 Description – 438
13.1.2 Diagnosis – 440
13.1.3 Prognosis – 441
13.1.4 Treatments – 441
13.1.5 Therapeutic Follow-Up – 441

13.2 Cognitive Dysregulation in Elderly Cats – 443
13.2.1 Description – 443
13.2.2 Diagnosis – 447
13.2.3 Prognosis – 449
13.2.4 Treatments – 449
13.2.5 Therapeutic Follow-Up – 450
13.2.6 Prevention – 450
13.2.7 Clinical Vignette – 450

13.3 Emotional Dysregulation in Elderly Cats – 451
13.3.1 Description – 451
13.3.2 Diagnosis – 453
13.3.3 Prognosis – 453
13.3.4 Treatments – 456
13.3.5 Therapeutic Follow-Up – 457
13.3.6 Prevention – 457
13.3.7 Clinical Vignette – 457

13.4 Mood Dysregulation in Elderly Cats – 458
13.4.1 Description – 458
13.4.2 Diagnosis – 459
13.4.3 Prognosis – 462
13.4.4 Treatments – 462
13.4.5 Therapeutic Follow-Up – 463
13.4.6 Prevention – 463
13.4.7 Clinical Vignette – 463

© The Author(s), under exclusive license to Springer Nature Switzerland AG 2026
S. Bleuer-Elsner et al., *Veterinary Psychiatry of the Cat*,
https://doi.org/10.1007/978-3-032-06260-4_13

13.5 Self-Controls Dysregulation in Elderly Cats – 464
13.5.1 Description – 464
13.5.2 Diagnosis – 466
13.5.3 Prognosis – 466
13.5.4 Treatments – 466
13.5.5 Therapeutic Follow-Up – 468
13.5.6 Prevention – 468
13.5.7 Clinical Vignette – 468

References – 471

In this chapter, we describe the psychiatric disorders associated with pathological aging. As with dogs, there are likely four forms of age-related cerebral decline: cognitive dysregulation in elderly cats, emotional dysregulation, mood dysregulation, and, finally, dysregulation of self-control. However, in cats, the clinical presentation is more subtle, and distinguishing between these different forms often proves challenging.

We begin by describing the commonalities among all these disorders, particularly their progression toward a generalized deterioration of all behavioral axes. In the second part, we attempt to provide clinical insights into approaching the aging process through the lens of these various mechanisms.

In the first part, we described the aging process at the cellular level and, more generally, at the cerebral level. The mechanisms of this progressive deterioration follow one another and reinforce each other. Somatic aging accelerates cognitive and emotional decline. All regulatory controls are challenged, and the major behavioral developments deteriorate. The evolution of these processes is inexorable, always leading to the same outcome after a degradation process of varying duration.

However, not all feline end-of-life scenarios are the same, and we can distinguish several types of disorders in elderly cats depending on which brain region deteriorates first. The relevance of this distinction, when possible, is not merely theoretical. Our interventions are only effective for these disorders if they are initiated early. It is therefore crucial to recognize the early signs of these conditions as soon as possible. The sooner we intervene, the better we can adapt our therapeutic approach.

We have identified four entry points into this aging spiral:

- The *fear and emotional axis*, which leads to the "emotional dysregulation of elderly cats" (Beata et al. 2015);
- The *mood axis*, which results in the "mood dysregulation of the elderly," described by Pageat in dogs (Pageat 1995) as "involutional depression";
- The *cognitive axis*, which leads to confusion syndrome or "cognitive dysregulation of elderly cats," referred to as "cognitive dysfunction syndrome" (CDS) in all Anglo-Saxon models (Sordo and Gunn-Moore 2021);
- The *self-controls axis*, which leads to the "self-control dysregulation of elderly cats." This fourth entry point into the aging spiral must exist, as this circuit can also malfunction (Masson et al. 2024a). However, it is rarely observed, likely because owners rarely seek consultation at the onset of this condition. As a result, early intervention is more challenging in this form of aging.

Before describing the specific characteristics of these different types of aging, it is essential to emphasize its *organic component* and its particular impact on disorders in elderly cats.

The defining characteristic of aging is the individual's *inability to adapt to change*. Whenever an organic or cerebral disorder occurs that requires a real adaptation from the organism, the aging process risks accelerating deterioration and triggering a runaway system. Take the example of simple arthritic pain. The affected animal may still be able to move, but it becomes slower and less resilient. It may still reach its elevated observation spot, for instance, but it needs a step to get there and sometimes even requires assistance. The risk is that the cat gives up: "I no longer go to the window, it's too high, and in any case, I can't catch prey." At this point, it enters a state of *demotivation* and may soon stop even noticing the bird in the garden in front of the door. If the animal manages to maintain its observation spot, it avoids this downward spiral. However, this adaptive choice is often too demanding for an elderly cat.

The practitioner must be vigilant for potential tipping points into aging and intervene in different ways to help the animal adapt. In treatment, the priority is not necessarily the absence of side effects or the complete restoration of function. The focus should be on how our intervention *influences the progression of aging* and on the quality of life of this cat in a state of varying degrees of deterioration.

For example, it may be useful for the cardiologist to minimize outdoor outings. However, contact with the outside world is also a source of pleasure and helps maintain proper brain function. Therapeutic intervention must take a holistic approach to the consequences on the life of the aging cat.

We previously mentioned pain, but the same applies to all somatic disorders. *Sensory loss* and arthritic pain are very common. Aging is also a period of *endocrine disruptions* that, in addition to their somatic impact, can alter the animal's reactivity and motivations.

For all these reasons, the management of aging must be *comprehensive*, and we must ensure that veterinary medicine remains integrative, avoiding as much as possible the hyperspecialization seen in human medicine.

Before addressing the four major cerebral aging disorders, we will first describe the main commonalities among all these conditions. We will then highlight their differences.

Clinical Illustration

Pantoufle is a somewhat timid cat, and she is afraid of the neighbor's dog, even though he is much smaller than she is. When she goes into the garden, she positions herself on the shared wall and keeps an eye on him. She takes pleasure in perching on this elevated spot.

Recently, she has been unable to jump onto the wall and no longer dares to approach the garden because of the dog. With a tablet of gabapentin, she manages to climb onto the wall and will return to the garden if the dog is not there. With an anti-inflammatory, she jumps onto the wall but does not stay. However, with a treatment combining non-steroidal anti-inflammatory drugs (NSAIDs) and an anxiolytic, she resumes her position there permanently.

13.1 Common Points to All Aging Disorders

13.1.1 Description

13.1.1.1 Definition

Diseases in aging cats ultimately lead to a common state of deterioration, characterized by a global decline in brain function. This decline combines varying degrees of *cognitive* impairment, *depressive states*, *anxious states*, and *loss of inhibitions*. Each deficit contributes to or exacerbates the others, creating a cycle of progressive decline. These losses in ability are further compounded by organic disorders, particularly a decrease in functionality, which accelerates this downward spiral.

All diseases of elderly cats ultimately lead to the same state of deterioration, characterized by a global decline in brain function, combining to varying degrees *cognitive impairment*, a *depressive state*, an *anxious state*, and *loss of self-controls*. Each loss triggers or exacerbates other forms of deterioration. These functional impairments are compounded by organic disorders, particularly a decline in physiological functionality, which further fuels this downward spiral.

13.1.1.2 Etiology and Pathogenesis

The processes responsible for this deterioration have been extensively described in ► Chap. 4. At the cellular level, three main phenomena are involved: the loss of membrane plasticity, the decreased efficiency of telomerases, and the increased production of free radicals due to inflammation.

In certain pathological forms, these involution processes are further aggravated by large-scale neuronal loss of unknown origin, selectively affecting specific brain regions. This phenomenon transforms a slow and harmonious aging process into a rapid cascade of deterioration (Chambers et al. 2015).

At the individual level, the first disorders correspond to a loss of adaptability, preventing an appropriate response to contextual changes. These changes are often of organic origin and are themselves linked to functional decline. This reduction in adaptability is accompanied by cognitive impairments, leading to mood disturbances, emotional dysregulation, and alterations in cortical control.

At the relational and territorial levels, the modified behavior of the aging cat is initially tolerated, as it is perceived as inevitable. However, cognitive decline soon leads to a deterioration in learned behaviors. This results in reprimands, followed by punishments, which further impair cognitive function and accelerate the rate of deterioration. Additionally, this decline in abilities makes the perception of the environment increasingly hostile, prompting social withdrawal.

This vicious cycle of aging inevitably leads to multiple disorders, both organic and psychological, which become difficult to manage once they are established (◘ Fig. 13.1).

13.1.1.3 Epidemiology

There is no escape from aging, but its onset varies in timing. A recent study conducted on 206 cats aged 7–10 years found that only 12% showed no abnormalities during a thorough clinical examination (Dowgray et al. 2022). Among the most common diseases, the authors cite obesity, arthritis, dental pain, azotemia, and hyperthyroidism.

Regarding CDS, studies report a prevalence of 36% in cats over 7 years old, which can rise to 88% in cats over 16 years old (Landsberg et al. 2010).

Anglo-Saxon publications do not distinguish between different types of behavioral disorders associated with aging, but it is reasonable to assume that the four forms of deterioration are represented within this 36%. However, it is likely that certain forms of aging lead to more veterinary consultations than others, and the exact distribution of these different forms remains unknown (Sordo et al. 2020; Sordo and Gunn-Moore 2021).

13.1.1.4 Presenting Complaints

Requests related to aging are rare and always come late, as owners often believe that what is attributed to aging does not require treatment. It is common during consultations for senior cats to hear statements such as *"He's*

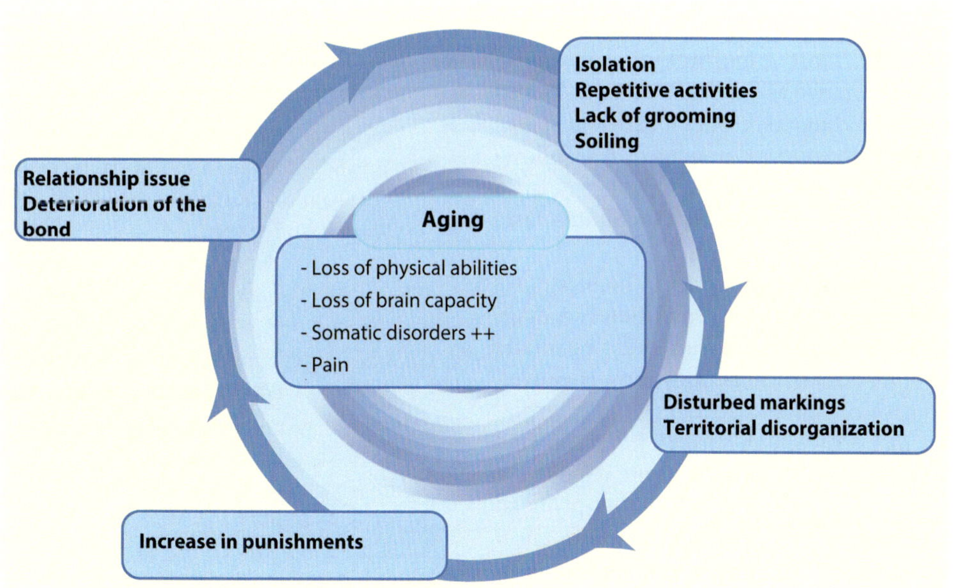

◘ **Fig. 13.1** The vicious cycle of pathological aging. Regardless of the entry point into aging, a vicious cycle sets in, characterized by declining performance, avoidance of effort, relational disturbances and loss of social bonds, reduced soothing behaviors, and an increase in punishments *deaf, but that's normal at his age"* or *"He doesn't go outside anymore, but that's not surprising, he's old now."*

The most frequent reasons for seeking veterinary care are disruptive behavioral changes, such as urine soiling, aggression, or excessive vocalization. However, these complaints usually arise too late in the progression of the condition.

13.1.1.5 Symptomatology

The specific symptoms of each type of disorder will be detailed in the following sections. However, the progression and final state of affected cats are always the same. This presentation is influenced by the accompanying somatic deterioration. The downward spiral leading the cat toward a senile state follows the same pattern, differing only in its point of entry, this is the focus of our symptomatological analysis.

Aging invariably results in a senile state, which may be accompanied, depending on the case, by a depressive, anxious, impulsive, or compulsive state.

13.1.2 Diagnosis

13

In an elderly animal over 8 years old, an unexplained and, most importantly, *non-adaptive* behavioral change that persists is sufficient to establish the diagnosis, once organic diseases have been ruled out and managed. The following sections highlight the importance of a more precise diagnosis, which facilitates *early intervention*.

In this book, we argue that the diagnosis of a behavioral disorder is not merely a diagnosis of exclusion that can only be made after ruling out all other conditions. However, in the case of aging disorders, it is essential to conduct a thorough series of tests to rule out potential organic diseases, which are numerous in elderly animals. Attempting to manage senile deterioration without carefully addressing all concurrent organic conditions is unrealistic.

The practitioner must pay particular attention to three key factors: *pain, sensory organ disorders, and hormonal imbalances. Pain* can rapidly trigger behavioral modifications in elderly cats, as it leads to a *loss of motivation* and an inability to respond according to learned behaviors. Similarly, *sensory loss*, particularly vision and hearing impairment, isolates the animal and prevents it from making appropriate responses to its environment. Renal insufficiency, hypertension, and dental health issues are also common conditions in elderly cats that contribute to *behavioral modifications.*

Finally, *hormonal disorders* can cause severe motivational disturbances, as seen in thyroid disorders, which should be systematically investigated in aging cats.

Important Point of View

Hyperthyroidism and Propranolol

The prevalence of hyperthyroidism in elderly cats is high, affecting approximately 10% of cats over 10 years old (Köhler et al. 2016; Stephens et al. 2014). The first signs of this condition are often noticed by owners, as it is accompanied by behavioral modifications.

The most common behavioral changes include:

- Changes in appetite, with an initial increase followed by a decrease.
- Episodes of agitation with excessive vocalization and demands.
- Sleep disturbances.

There may also be some vomiting, slight weight loss, and increased blood pressure. In some cases, this hypertension leads to cerebral lesions and, in unfavorable scenarios, blindness. The condition clearly causes discomfort for the animal.

Hypertension is often concurrent with hyperthyroidism and appears to be responsible for a portion of the discomfort and agitation (Stammeleer et al. 2024).

The treatment of severe cases involves anti-thyroid medications, but before initiating these drugs, it is often useful to manage the early stages of the disease. Based on our clinical experience, propranolol is an effective and elegant solution, addressing both hypertension and behavioral agitation. However, tablets are often difficult to administer. The treatment induces mild sedation, which is generally well tolerated by owners, especially when they are informed in advance about this side effect.

13.1.3 Prognosis

Once the aging spiral has begun, progression is rapid. This underscores the importance of early diagnosis, which is the key advantage of our classification by type of deterioration. Depending on the condition and its symptomatology, the degree of disruption for owners varies, potentially leading to requests for euthanasia at different stages, sometimes prematurely.

13.1.4 Treatments

13.1.4.1 Pharmacological Treatment

Treatment is generally similar across the different forms of aging disorders, once the specific therapies targeting the initial symptoms have been administered.

One of the main therapeutic goals is counteracting the effects of free radicals. Selegiline, at a dose of 1 mg/kg, contributes to this and is free of adverse effects in elderly cats. It does not interfere with age-related hormonal disorders and can be safely combined with commonly used anti-thyroid medications. However, while the manufacturer claims an anxiolytic effect, this has not been demonstrated at the 1 mg/kg dose.

Numerous dietary supplements containing selenium and vitamin E have been proposed to combat neuronal deterioration. These elements are sometimes directly incorporated into kibble, which facilitates compliance.

Fighting inflammation appears to have a protective effect against neuronal degradation and the production of free radicals (Pan et al. 2013). In humans, early administration of NSAIDs at limited doses has been shown to have positive effects on lifespan (Franceschi et al. 2007; He et al. 2014).

In addition to these general treatments, specific therapies are used to address depressive, anxious, impulsive, or compulsive states. Gabapentin and pregabalin, with their analgesic and anxiolytic effects, are particularly beneficial for elderly cats.

13.1.4.2 Behavioral Therapy

The primary therapy is to address each identified condition (▣ Fig. 13.2).

It consists of two key components. First, *environmental modifications* involve adapting the cat's environment to its capabilities. The priority is to pay particular attention to grooming and claw trimming. These aspects are often neglected because the animal is old. However, maintaining coat hygiene and checking claws are essential for its well-being, as they prevent physical deterioration and help counteract demotivation.

Second, particularly in the early stages of progression, *behavioral modifications* aim to support cognitive activity. It is essential to reinforce motivation and maintain learned behaviors. Overall, the goal is to help the cat develop compensatory strategies, replacing one lost learned behavior with another. This compensation is only possible if the cat remains engaged. Stimulation plays a crucial role, as does maintaining comforting rituals. For example, a disoriented cat can be encouraged to find its way by relying on its remaining functional senses, such as vision or smell.

13.1.5 Therapeutic Follow-Up

The treatment of elderly cats with aging disorders primarily falls within the scope of *general practice veterinarians*. Establishing continuous collaboration between the owner and the

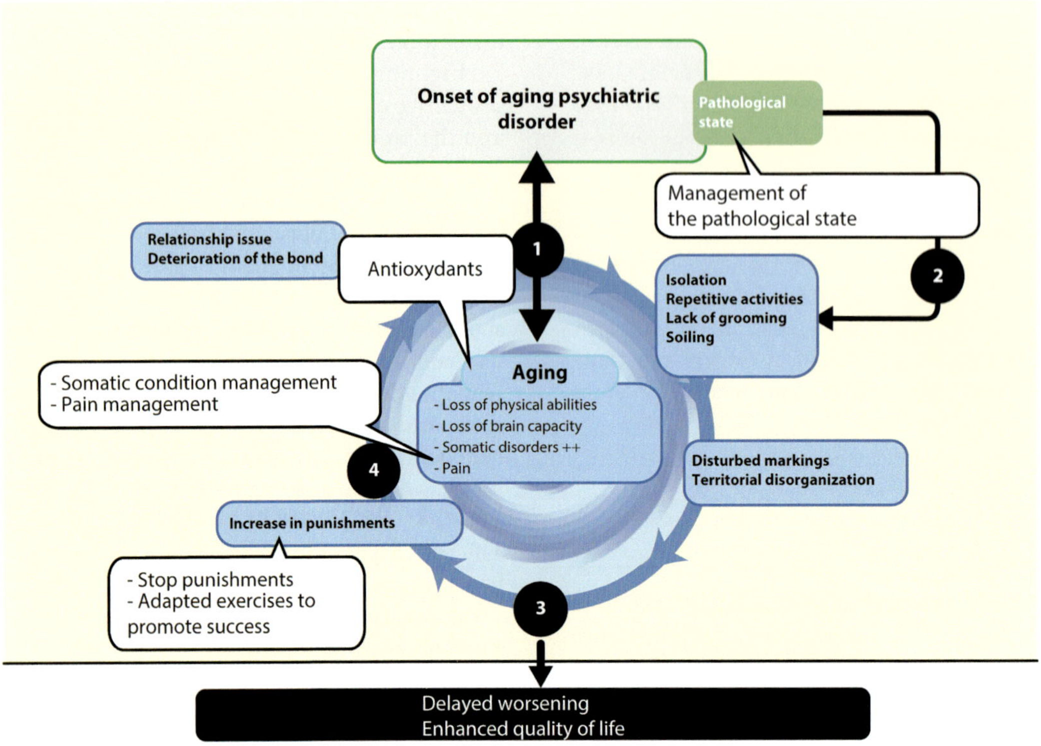

Fig. 13.2 Therapeutic pathways for aging disorders. Each identified condition should be addressed with therapeutic intervention as early as possible

13

veterinarian is essential. The owner must be aware that they can consult the practitioner whenever difficulties arise, while the veterinarian, in turn, relies on the owner to observe and report any changes. This partnership, between the owner, who knows the individual cat within its specific context, and the veterinarian, who specializes in aging, enables *a comprehensive and personalized approach.*

The goal is to systematically address *each aspect of deterioration*, even the most minor, to maintain an acceptable state of well-being. Mutual trust and respect between veterinarian and owner are fundamental. It may also be beneficial to conduct assessments that demonstrate how each discomfort can be counterbalanced by a source of pleasure. These assessments should be designed for both the cat and the owner to ensure that their relationship remains harmonious.

Summary of ▶ Sect. 13.1
Aging disorders all lead to a global decline in brain function. The initial trigger may be an emotional, mood, or cognitive disorder, but the pathological aging spiral inevitably results in an accumulation of these conditions, making it difficult to distinguish them in the final clinical presentation. This final stage typically involves a combination of cognitive decline, a depressive state, an anxious state, and loss of inhibitions.

Despite the significant impact of these disorders on the well-being of elderly cats, requests for veterinary care are rare and consistently delayed, as owners often view aging as an unavoidable and untreatable process.

The aging disease spiral most often culminates in a senile state, which may be accompanied by depression, loss of inhibition, or anxiety.

More than in other behavioral disorders, the practitioner must carefully consider concurrent physical and physiological conditions in elderly cats, particularly pain, sensory organ disorders, and hormonal imbalances. One of the primary therapeutic objectives is to counteract the effects of free radicals.

Behavioral modifications in therapy aim to support cognitive activity by reinforcing motivation and maintaining established routines.

13.2 Cognitive Dysregulation in Elderly Cats

13.2.1 Description

13.2.1.1 Definition

This condition in elderly animals, referred to as *confusion syndrome* in the initial model proposed by Pageat (1995) for dogs, is characterized by the loss of cognitive abilities. It manifests as errors in orientation, including the loss of spatial reference points and territorial disorganization, temporal confusion with a loss of time-related cues and soothing rituals, and relational errors leading to the deterioration of learned social behaviors.

Among aging disorders, *cognitive dysregulation* is the most common. It is often diagnosed quickly, as the cat's errors become disruptive for owners. This condition closely resembles *Alzheimer's disease* in humans.

Using the canine nomenclature for reference, this condition corresponds to *CDS* in the strict sense of the term in Anglo-Saxon models, even though it is likely that all four types of deterioration described here will eventually be grouped under CDS in Anglo-Saxon terminology (Gunn-Moore 2009).

13.2.1.2 Etiology and Pathogenesis

As described in ▶ Chap. 4, the mechanisms underlying *normal age-related deterioration* are relatively well understood. Several processes are involved, each amplifying the other. In terms of cognitive functions, one key factor is the decline in long-term potentiation, meaning a reduction in memory formation due to membrane rigidity. Overall, cerebral plasticity is significantly diminished. Synaptic responsiveness slows, and the defense against free radicals weakens. No new neural connections are formed, and new learning ceases, while existing memories progressively deteriorate.

In addition to these normal aging processes, somatic issues contribute primarily to a *decrease in motivation*. Sensory loss leads to a general decline in behavioral intensity. Reinforcements and soothing stimuli become delayed and less frequent, while the effort required to achieve the same outcomes increases. In other words, for the same level of exertion, the rewards are weaker. These gradual processes are observed in all aging cats but *do not involve neuronal loss*.

However, in certain individuals, without a currently identified explanation, *large-scale neuronal loss* can occur, most often localized to specific brain regions, particularly the hippocampus and the olfactory bulb, both of which are the most active areas for neurogenesis (Department of Comparative Medicine, School of Medicine, University of Washington, Seattle, WA, USA et al. 2020; Scheff and Price 2006; Telegina et al. 2019). These neuronal losses are characteristic of *pathological aging* and do not affect all aging cats. In these pathological forms, cognitive symptoms are significantly amplified.

Memory is particularly affected, especially topographic memory. Neuronal lesions and cell death generate large quantities of free radicals, causing a rapidly *exponential process* (Peña-Bautista et al. 2021). What begins as mild confusion quickly escalates into chaotic thinking, ultimately degrading the entire system.

In addition to this microscopic pathogenesis, there are also the social consequences of cognitive disorders. One of the key factors in demotivation is the reaction of those around the animal in response to its errors. Where positive reinforcement would be needed, the cat is often met with punishment or, at the very least, incomprehension. The anticipation of failure thus contributes to a decline in activity from the early stages of progression.

13

Important Point of View

Alzheimer's Disease and Free Radicals

Alzheimer-type dementia is associated with amyloid deposits in the affected brain regions. The presence of these aggregates appears to result from free radical production. In elderly dogs, plaques rather than deposits are observed (Torp et al. 2000; Urfer et al. 2020), and their quantity seems to correlate with the diagnosis of confusion syndrome.

Cats, however, appear to be one of the few species that exhibit senile plaques, amyloid deposits, and neurofibrillary tangles characteristic of Alzheimer's disease, making them a particularly valuable model for research (Chambers et al. 2015; Head et al. 2005).

13.2.1.3 Epidemiology

No studies have been conducted on the incidence of these dysfunctions, and even less on the prevalence of their pathological forms. Confusion and cognitive decline appear to be the norm, with few cats escaping them beyond a certain age. However, this form of the disease seems to be more common than the condition discussed in ▶ Sect. 13.4, in which the animal experiences a depressive state. The fact that cognitive dysregulation is more disabling for both the animal and its owners may explain why it more frequently leads to veterinary consultations.

13.2.1.4 Presenting Complaints

Three main themes emerge as presenting complaints: the cat is *urine soiling*, the cat *wakes* its owners during the night, or the cat exhibits episodes of *aggression*.

Urine soiling is particularly poorly tolerated by owners. The initial assumption is often that the cat is acting *"to annoy its owners."* However, caution is needed before diagnosing a cognitive disorder, as polyuria has numerous causes in elderly cats. For instance, cats with kidney disease or diabetes tend to drink more. Mobility issues due to pain may also prevent the cat from making the effort to reach the litter box. Additionally, an infection could be a contributing factor.

It is important to remember that these causes of urine soiling can coexist with cognitive dysfunction, which often amplifies the problem. Emotional disturbances can also lead to interstitial cystitis, a frequent cause of urine soiling in cats. Frequent punishment worsens the condition, reinforcing and aggravating the issue.

Owner complaints about *nocturnal agitation* typically arise late, often when they are already exhausted from sleep deprivation. *Spatial disorientation* is another common reason for consultation. For the cat, disorientation represents a significant emotional disturbance. As territorial structure deteriorates, all familiar soothing cues disappear, creating the distressing impression of living in an unfamiliar environment. This can quickly lead to an anxious state, further complicating the clinical picture.

In some cases, the cat seeks comfort from its owners when the environment no longer provides security. It becomes clingy and excessively demanding. While this change may initially seem endearing, it is often confusing and eventually leads to frustration.

More commonly, however, the opposite occurs. A cat that feels lost within its own territory may also fail to recognize its owners, becoming distant or even aggressive. These aggressive responses can be exacerbated by the anticipation of reprimands and by the presence of arthritic pain.

13.2.1.5 Symptomatology

The first defining symptom is the *cat's age*. This disorder does not begin in adulthood. Affected animals are always elderly, although not necessarily very old. While it is common to consider a cat as entering its senior phase at 8 years old, some cats do not begin to show signs of aging until much later.

Major life changes also appear to play a role. Relocations, for instance, require significant adaptation and can trigger the first signs of difficulty. However, the arrival of another animal is another critical factor. It is not uncommon for owners to acquire a younger pet in anticipation of replacing their aging companion. Beyond the territorial adjustments required, this transition is often challenging for the elderly cat. Additionally, attention from the owner is quickly divided, leading to a gradual but sometimes abrupt social withdrawal.

When a practitioner suspects this form of aging disorder, they must *actively look for all the mistakes the cat is making*. Most of these errors are harmless and may go unnoticed or unreported by the owner. The symptoms to identify always involve a contrast: something the cat previously did but can no longer do or no longer wants to do. In nearly all aging disorders, the cat's territorial range progressively shrinks, ultimately limiting its movement to only a small portion of its original territory.

In cases of *cognitive decline*, as previously mentioned, *spatial and temporal reference* points may be affected.

Over time, owners and their cats establish rituals that are both soothing and essential to their relationship. Initially, the cat may make mistakes in timing, requesting attention or food at inappropriate hours. The owner may misinterpret this as a change in needs, thinking "He's drinking more, he's hungrier, he needs more affection, and so forth."

This loss of temporal reference points can also lead to nocturnal awakenings or requests to go outside during the night. If the cat also has difficulty orienting itself, it may vocalize to find its bearings, making nighttime disturbances quickly unbearable for owners. Temporal disorientation may also manifest during mealtimes, with the cat failing to appear at its usual feeding time and later demanding food at an unusual hour. These confusions often go unnoticed by owners, who attribute them to normal fluctuations in appetite or expected effects of aging.

At home, the cat may start frequenting unusual rooms or meowing in specific corners. In reality, it is lost and seeking help. The meows are often distress calls that the owner struggles to interpret. In response, the owner may speak to the cat, which helps it regain its orientation. The cat then resumes its activity, seemingly ignoring the owner. This creates the impression that the cat meowed for no reason, increasing mutual misunderstanding.

The accumulation of these errors often leads to a *deterioration in the owner-cat relationship*, which in turn complicates symptom assessment. The practitioner must investigate the history of the disorder to determine how this deterioration began and whether the cat's mistakes were truly responsible for its worsening condition.

13.2.1.6 Evolution

The progression of the disease has two main characteristics (◘ Fig. 13.3). First, it is *inevitable*, as the condition, unfortunately, cannot be cured. However, it is possible to slow its progression by addressing each aggravating factor individually. As the deterioration follows an *exponential pattern*, intervention must occur as early as possible. This is why recognizing the *earliest signs* of the disorder is essential.

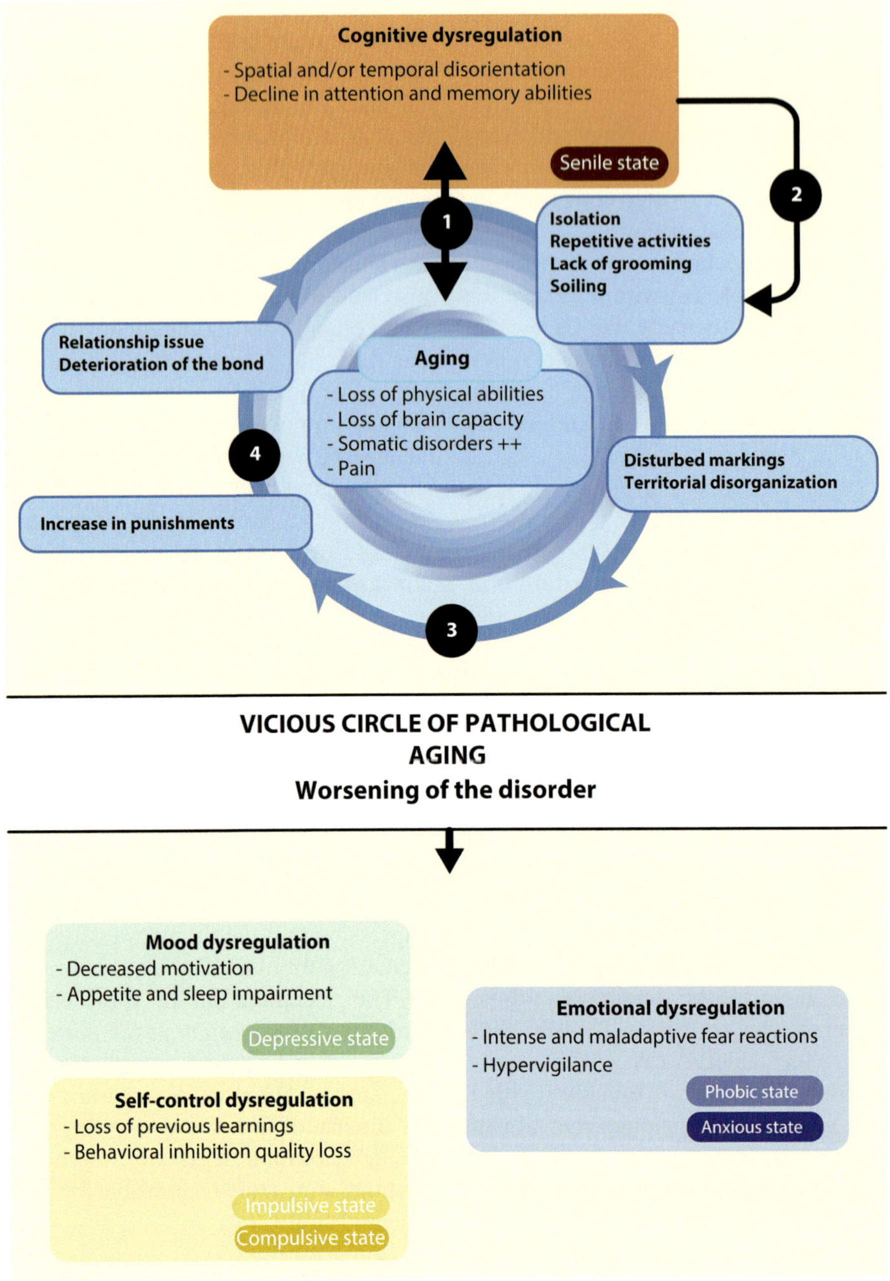

Fig. 13.3 Evolution of cognitive dysregulation in elderly cats. In the absence of targeted interventions against each factor contributing to pathological aging, a vicious cycle sets in, accelerating all forms of deterioration

The progression of the disease is also *variable,* as it depends on the cat's cognitive potential and external events. The more the animal has learned throughout its life, the less disruptive the loss of any one skill or behavior will be. Similarly, a large and enriched territory minimizes the perceptibility of territorial reduction. However, the downside is that a cat accustomed to a rich environment with frequent stimuli develops a greater need for engagement and may struggle more with the loss of its abilities.

The *compensation mechanisms* mentioned earlier, particularly regarding vocalizations, play a crucial role. For instance, a disoriented cat searching for the kitchen where its food bowl is located may rely on scent to navigate, compensating for its loss of spatial reference points. These compensatory strategies are often numerous and help prevent *isolation and demotivation,* making it essential to encourage them. However, because compensation occurs instinctively, it can sometimes *mask clinical signs*, delaying intervention if owners are not attentive to subtle behavioral changes.

Finally, *life events* significantly influence disease progression. A hallmark of aging cats is their diminished ability to adapt by forming new learning patterns. For them, change is particularly disruptive. This change can be *external,* such as a relocation or the loss of a family member, or *internal,* such as the onset of pain or a new somatic dysfunction.

13.2.2 Diagnosis

13.2.2.1 Diagnostic Criteria

Cats with cognitive dysregulation inevitably exhibit a *senile pathological state.* In the condition described here, the cat does not initially present with an anxious or depressive state (◘ Fig. 13.4). Symptoms emerge during the *senescence period.*

The diagnosis is based on identifying multiple *cognitive errors* in a cat over *8 years old.* These errors manifest through some or all of the following symptoms:

- loss of spatial reference points, leading to misplaced searches for familiar environmental elements;
- loss of temporal reference points, characterized by increased nocturnal activity compensated by excessive daytime rest, as well as inappropriate timing of requests relative to usual predictive stimuli;
- loss or incorrect execution of ritual behaviors;
- loss of urine soiling control.

To establish the diagnosis, multiple cognitive errors must be present. The cat remains cheerful and alert, without experiencing a change in personality—at least in the early stages of the disease.

If the practitioner intervenes late, the diagnosis concerns an older animal that has generally lost its cognitive abilities and has withdrawn from its environment. At this stage, distinguishing cognitive dysregulation from other aging disorders becomes difficult, if not impossible. Additional pathological states, most commonly anxious or depressive states, may further complicate the diagnosis.

13.2.2.2 Direct Observation

It is not always possible to observe loss of reference points during a consultation. In-clinic, cats often appear perfectly normal, requiring an active investigation through detailed discussions with the owner to identify cognitive errors.

13.2.2.3 Differential Diagnosis

Two key aspects must be examined carefully: *organic diseases* and *other aging disorders.*

The importance of somatic conditions, which are present in most cases, has already been emphasized. It is crucial to determine whether the predominant issue is a physical handicap or an inability to adapt to new circumstances.

In the late stages, aging disorders in cats become increasingly similar. Differentiating cognitive dysregulation from other behavioral aging disorders is only possible in the early stages of progression. Cognitive impairments in this disorder should emerge without emotional or mood disturbances and without loss of self-control. It is possible for elderly cats with cognitive decline to live their final days contentedly, without developing anxiety or depression.

13.2.2.4 Comorbidities

Organic comorbidities will not be discussed further here, as they are naturally numerous.

From a behavioral perspective, it is important to note that reaching old age does not erase pre-existing behavioral issues. In fact, aging may introduce an *additional vulnerability* in animals that have historically experi-

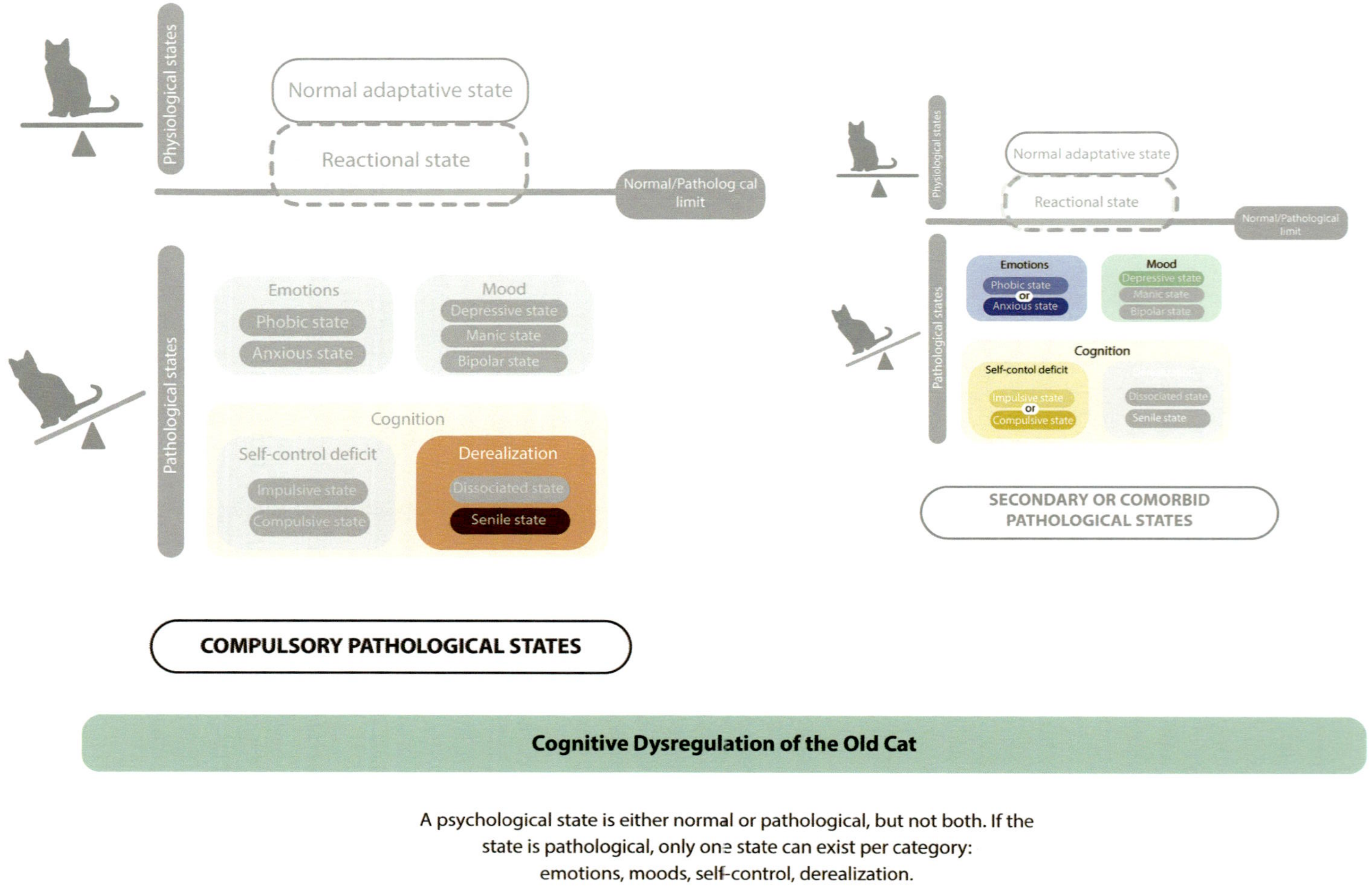

Fig. 13.4 Pathological states associated with cognitive dysregulation in elderly cats. Mandatory pathological state: senile state. Possible comorbid pathological states: phobic or anxious state, and/or impulsive or compulsive state, and/or depressive state

enced anxious states or other behavioral imbalances. However, these cats are more likely to develop emotional disturbances rather than cognitive deficits as they age.

13.2.3 Prognosis

The *prognosis* is inevitably poor, as aging is an irreversible process. The outcome of this disorder is ultimately fatal. However, this outlook should be tempered by emphasizing our ability to *slow progression* and *maintain an acceptable quality of life*.

The prognosis is thus comparable to that of certain terminal conditions, such as cancer or major endocrine disorders. While we cannot eliminate the disease, we can ensure that the animal maintains a good quality of life for as long as possible. Early intervention is crucial, along with establishing a care plan with the owners, who often require guidance and support. Ultimately, they remain the primary actors in therapeutic management.

13.2.4 Treatments

Three equally important pillars support treatment: pharmacological treatment, environmental modifications, and behavioral therapy aimed at stimulating cognitive function.

13.2.4.1 Pharmacological Treatment

At present, treatment options remain **limited**. The **drug of choice** is ***selegiline***, which has a well-documented effect on **free radicals** (Denenberg and Landsberg 2017; de Souza Dantas 2019). Additionally, it is **stimulating** and likely enhances **feeding motivation** in the presence of a reward, as observed in dogs (Mills and Ledger 2001). The recommended dosage is **1 mg/kg daily.**

Furthermore, all compounds capable of **counteracting free radicals** have therapeutic value. Certain **diets enriched with vitamin E, tocopherol, and selenium** have been reported to **reduce symptoms of aging** (Castillo and Hernández 2015; Pan et al. 2013).

Important Point of View

Free Radicals

Aerobic cellular metabolism, cellular oxidation, and apoptotic processes all lead to the release of free radicals. Synaptic function also involves the release of free radicals within the synaptic cleft. While free radicals can be harmful to cells, specific enzymatic pathways exist to degrade them and prevent toxicity.

The primary enzyme involved in this mechanism is superoxide dismutase (SOD). Selenium, vitamin C, and vitamin E also play a crucial role in this detoxification process. Aging leads to increased production and accumulation of free radicals, primarily due to a decline in SOD efficiency, as well as the rigidification of cell membranes and mitochondrial DNA (Beckman and Ames 1998; Castillo and Hernández 2015).

13.2.4.2 Behavioral Therapy

A strong **collaboration** between the veterinarian and the owner is essential. The owner, who knows the cat intimately, must take an **active role** in developing strategies to help the animal **compensate for its cognitive decline.** It is important to recognize that the **owner remains the expert on their cat,** having firsthand knowledge of its past behaviors and routines.

For this therapy to be effective, it must be built through **ongoing communication** between the owner and the veterinarian, allowing for **individualized adjustments** based on the cat's specific needs and capabilities.

- **Environmental Modifications**

Various environmental adjustments can be implemented to support the aging cat. For a cat that can no longer jump onto a stool to look out the window, providing a step to facilitate access is essential. Without such an adaptation, social isolation can develop rapidly.

Mechanical adaptations that compensate for physical limitations should be discussed with the owner based on the cat's living conditions.

Additionally, hygiene and maintenance play a crucial role in keeping the cat engaged within its social group. *Regular grooming, claw trimming, and the use of rugs or mats* to prevent slipping on tiled floors are all measures that help prevent isolation and demotivation, supporting the cat's overall well-being.

▪ Behavioral Modifications

In addition to environmental adaptations, *activity stimulation therapy* should be implemented. The goal is to reactivate functional synapses and restore motivation. The cat must experience its ability to influence its environment, reinforcing the connection between actions and rewards.

The owner should regularly engage the cat in interactive activities that encourage it to seek and request rewards. This approach helps prevent global demotivation and promotes a renewed willingness to act.

Another crucial aspect of treatment involves the owner's relationship with the cat. The owner must recognize the cat as a patient in need of care, avoiding punishment or scolding, which can lead to withdrawal and isolation. Instead, the focus should be on supporting the cat and *integrating it into family life,* ensuring that it remains engaged and emotionally connected.

13.2.5 Therapeutic Follow-Up

The therapy for cognitive disorders, like the care of elderly cats in general, falls within the domain of general practice. This therapy is only possible if the practitioner has long established a relationship of trust with the owner. The objectives must be presented with complete transparency, similar to *palliative care*. The owner and practitioner work together to maintain a shared *quality of life*. Problems are addressed individually, and availability is essential.

13.2.6 Prevention

Prevention is difficult to implement because owners tend to remain unmotivated until the first signs of decline appear. In humans, individuals with higher intellectual activity appear to be more resistant to neurodegeneration (Selkoe 2012).

The principle is simple: the greater the cognitive potential of the animal, the less noticeable the deterioration, and the *more effective compensatory mechanisms* will be.

As a preventive measure, it seems advisable to develop the intellectual capacities of cats by encouraging them to acquire various learned behaviors and maintaining mental activity throughout their lives. These exercises have the additional advantage of not relying on physical abilities, which often diminish with age.

13.2.7 Clinical Vignette

▪ Kytox

Kytox is a 12-year-old neutered European male cat who has always spent a lot of time outdoors hunting. Recently, his owners had their garden redesigned by a landscaper. During the renovations, he was kept indoors for several weeks, and now that he is allowed outside again, he refuses to go, having lost all his familiar reference points in the newly structured garden. Additionally, he has had some issues with urine soiling, as his litter box was typically kept outside. However, 4 years ago, when similar exterior work was done, Kytox adapted without issue to using an indoor litter box, leading the owners to believe he is now acting out deliberately.

The consultation was prompted by sleep disturbances, as Kytox had developed the habit of waking his owners at night to request that a door be opened or to drink from the faucet. His age, good health, and normal blood work suggest an early stage of age-related disorientation. His spatial and temporal reference points are disrupted, and the presence of sleep disturbances indicates that intervention is already somewhat overdue.

His highly motivated owners agreed to a dual approach: first, administering selegiline at 1 mg/kg (one tablet per day), and second, accompanying Kytox outdoors to help him reacquaint himself with the new garden. They were encouraged to engage him by drawing his attention to the fish in the pond and guiding him to find a pathway to the shed at the back of the garden.

The initial phase was difficult, with Mrs. making significant efforts to persuade Kytox to go outside and explore. After 3 weeks, Kytox managed to catch a fish in the newly renovated pond. Mr. quickly agreed that the fish would be replaced, and within a week, Kytox resumed going outside and stopped waking them at night. In response, Mr. placed stones in the pond, allowing the fish to learn to hide.

One episode of urine soiling occurred due to a malfunctioning cat flap, but Mrs. did not react negatively, and the issue was resolved swiftly by repairing the flap. However, some residual tension remained, as Kytox continued to meow frequently at the door. The veterinarian suggested leaving the door open temporarily to assess whether the issue was related to hesitation about pushing the flap.

After 3 months, all issues had been resolved, except Kytox still refused to push the cat flap to go outside. During the summer, this was manageable, but in winter, it would become problematic. The owners gradually transitioned the door by first replacing it with a veil, then a fabric barrier, and after a few days of training, Kytox accepted this new setup. Additionally, a second litter box was placed indoors.

With these two key adjustments, a balance was restored between outdoor access and comfort. However, the owners were informed of the risk of rapid progression of his condition.

Summary of ▶ Sect. 13.2

Cognitive aging is the most widely described form of aging in the literature, commonly referred to as CDS. This deterioration, which occurs with advancing age, may progress slowly but can also accelerate when normal cellular aging is accompanied by extensive neuronal loss. In such cases, the condition closely resembles Alzheimer's disease in humans.

The symptomatology is characterized by loss of spatial and temporal reference points, reflecting impairments in learned behaviors. A vicious cycle of aging quickly sets in, leading to demotivation and emotional disturbances. Without intervention, this process ultimately results in a senile state and rapid global deterioration.

13.3 Emotional Dysregulation in Elderly Cats

13.3.1 Description

13.3.1.1 Definition

Emotional dysregulation in elderly cats is a behavioral disorder in which emotional disturbances are the most prominent signs in the early stages of progression. It is important to emphasize that, like other aging disorders, this condition ultimately leads to global behavioral collapse, marked by a depressive state and varying degrees of cerebral dysfunction.

As with all aging disorders, the vicious cycle follows a common trajectory, although the initial trigger may differ. The final clinical presentation invariably includes cognitive, mood, emotional, and self-control impairments.

In this disorder, two dominant symptomatologies are frequently observed, often occurring together: sleep disturbances and loss of autonomy. As is often the case, the owner's concerns will determine which symptom requires the most immediate attention.

13.3.1.2 Etiology and Pathogenesis

It can be assumed that in these animals, due to pathological aging, localized abnormalities develop in the regions responsible for emotional regulation.

In clinical practice, it is easy to observe that significant aging emotional disturbances tend to affect cats that have always lived on the edge of anxiety. Throughout their lives, these animals exhibit emotional fragility, often presenting with phobias and a predisposition to anxious states. With age, the delicate balance they have maintained is disrupted by a change in their environment. The aging animal lacks the resources to re-establish a sense of security, becoming anxious and losing its autonomy. The environmental change itself is merely a trigger, and although owners tend to focus on this event, it is essential to explain that a return to the previous state is no longer possible.

At the cerebral level, the usual rigidification processes and loss of synaptic plasticity activate mechanisms of sensitization, anticipation, and generalization. Alongside these typical aging-related impairments, a deterioration in the owner-cat relationship further exacerbates symptoms and accelerates their progression.

As demotivation sets in, the cat withdraws and quickly enters a depressive state, further intensifying the vicious cycle of aging.

13.3.1.3 Epidemiology

This disorder is not necessarily more common than cognitive dysregulation, but due to its highly disruptive symptomatology, it leaves a strong impression on owners and likely leads to more frequent veterinary consultations (Sordo et al. 2020).

13.3.1.4 Presenting Complaints

The primary reason owners seek veterinary consultation is the distress caused by their cat's behavior. Common complaints include *sleep disturbances* affecting the owners and *episodes of urine soiling.*

These concerns are often accompanied by confusion and frustration, as the cat's behavior appears inexplicable compared to its previous patterns.

Cats that were previously phobic may experience a resurgence of their condition, often worsening and becoming more generalized. As the senile state progresses, affected cats tend to become increasingly isolated, sometimes to an extreme degree.

13.3.1.5 Symptomatology

The cats described in this condition are elderly and have typically lived with their owners for many years. As a result, the owners know their cat extremely well and have often attempted to resolve the initial behavioral changes on their own, assuming them to be minor relational or household issues. Regardless of the primary symptom, owners almost always identify a triggering event that they believe caused the disorder.

The most striking feature of these aging-related disorders is *change itself*, which should serve as a key warning sign.

Sleep disturbances often begin with occasional nocturnal awakenings that the owner initially dismisses as insignificant. The cat wakes up, sometimes enters the bedroom, or asks to go outside. Given that the cat is usually well-behaved, the owner assumes its requests indicate a legitimate need. However, these nighttime disturbances quickly become a daily occurrence, and owners, in need of sleep, eventually seek veterinary help. By the time they consult, they have already tried various solutions. Most can describe their cat as sleeping excessively during the day and evening, only for problems to arise at bedtime.

Owners often create their own temporary solutions. For example, they allow the cat to sleep in their bedroom, thinking this will resolve the issue. Unfortunately, the anxious cat's sleep is restless, meaning this solution does not necessarily improve the owner's rest, or does so only briefly.

Urine soiling frequently accompanies the anxiety experienced by aging cats. The defining characteristic of this form of aging disorder is loss of adaptability, leading to a resurgence of phobias and fears. The inability to adapt worsens the condition and fosters the development of an anxious state.

As the disease progresses, *cognitive impairments emerge,* affecting spatial, temporal, and social reference points, as well as previously learned behaviors. Early in the course of the disorder, the cat slows down significantly, exhibiting a dual pathological state of anxiety and senility. Anxiety in elderly cats is often dominated by inhibition and withdrawal, with frequent self-directed behaviors.

In the late stages, demotivation becomes prominent, and a depressive state frequently develops. The reduction in territorial range is often dramatic.

13.3.1.6 Evolution

The evolution of this disorder is comparable to that of other aging disorders (◘ Fig. 13.5). As the condition worsens, cognitive and mood impairments become pervasive. Some cats also exhibit reduced impulse control, leading to behavioral disinhibition. Regardless of the specific trajectory, the final clinical presentation remains the same, characterized by demotivation, isolation, and withdrawal.

As symptoms accumulate, they become increasingly difficult for owners to manage, who also witness their companion's suffering. Requests for sedatives are common. However, elderly cats often react poorly to sedation, especially with benzodiazepines, which further disrupt their ability to navigate their environment and worsen territorial disturbances.

Every treatment decision requires careful consideration, balancing the comfort of the animal with the well-being of the owners. The practitioner plays a crucial role in guiding this process.

Requests for euthanasia are frequent and not always unreasonable.

13.3.2 Diagnosis

13.3.2.1 Diagnostic Criteria

Cats affected by emotional dysregulation necessarily present with a *pathological phobic* or *anxious state*. They do not initially show signs of senility, depression, or deficits in self-control (◘ Fig. 13.6).

The diagnosis is based on identifying a worsening of emotional distress in a *cat over 8 years old*. The following symptoms may be observed: signs of distress during the day, such as vocalization and urine soiling;

- signs of distress at night, including pacing and vocalization
- panic attacks
- restless sleep with frequent awakenings
- evidence of pre-existing vulnerabilities related to the fear axis before the onset of senescence
- worsening of a previously stable phobic state, leading to generalization and transition into an anxious state

13.3.2.2 Direct Observation

There are no distinct visible characteristics of this disorder in the early stages of decline. These cats exhibit anxiety and display the typical signs associated with this state. It is common to observe episodes of polypnea, dyspeptic disorders, and self-directed activity lesions, such as excessive licking leading to wounds. Panic attacks initially occur sporadically but tend to increase in frequency over time.

13.3.2.3 Comorbidities

The presence or re-appearance of an anxious state alongside a senile state is characteristic of this disorder. In addition to these specific symptoms, elderly cats often suffer from various other medical conditions. Particular attention should be given to painful disorders and hormonal imbalances.

13.3.2.4 Differential Diagnosis

The most challenging aspect of diagnosis is distinguishing this disorder from *mood dysregulation*, where mood disturbances and motivational impairments are the initial triggers of pathological aging. In emotional dysregulation, the primary issue is phobia or anxiety, which arises from the *cat's inability to adapt to change.*

It is crucial to thoroughly investigate *all possible somatic conditions* that could produce similar symptoms. In many cases, the triggering event may actually stem from a decline in physical capacities, particularly sensory impairments. Thyroid disorders and hypertension can cause significant sleep disturbances, while chronic pain may also lead to severe sleep disruptions.

Additionally, hyperadrenocorticism, diabetes, and renal insufficiency can contribute to urine soiling, further complicating the diagnostic process.

13.3.3 Prognosis

The prognosis is less favorable than in the previous disorder. Sleep disturbances are extremely difficult to manage without com-

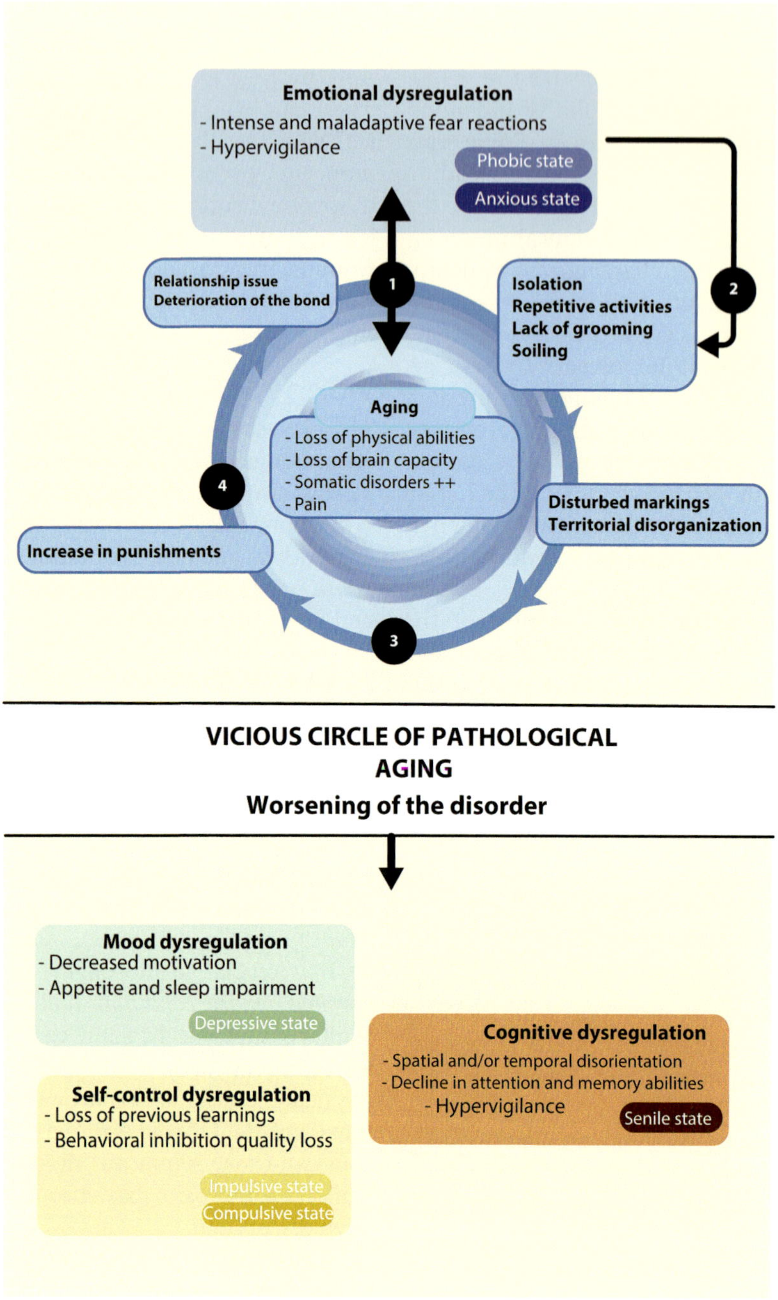

Fig. 13.5 Evolution of emotional dysregulation in elderly cats. Without targeted intervention for each factor contributing to pathological aging, a vicious cycle develops, accelerating all forms of deterioration

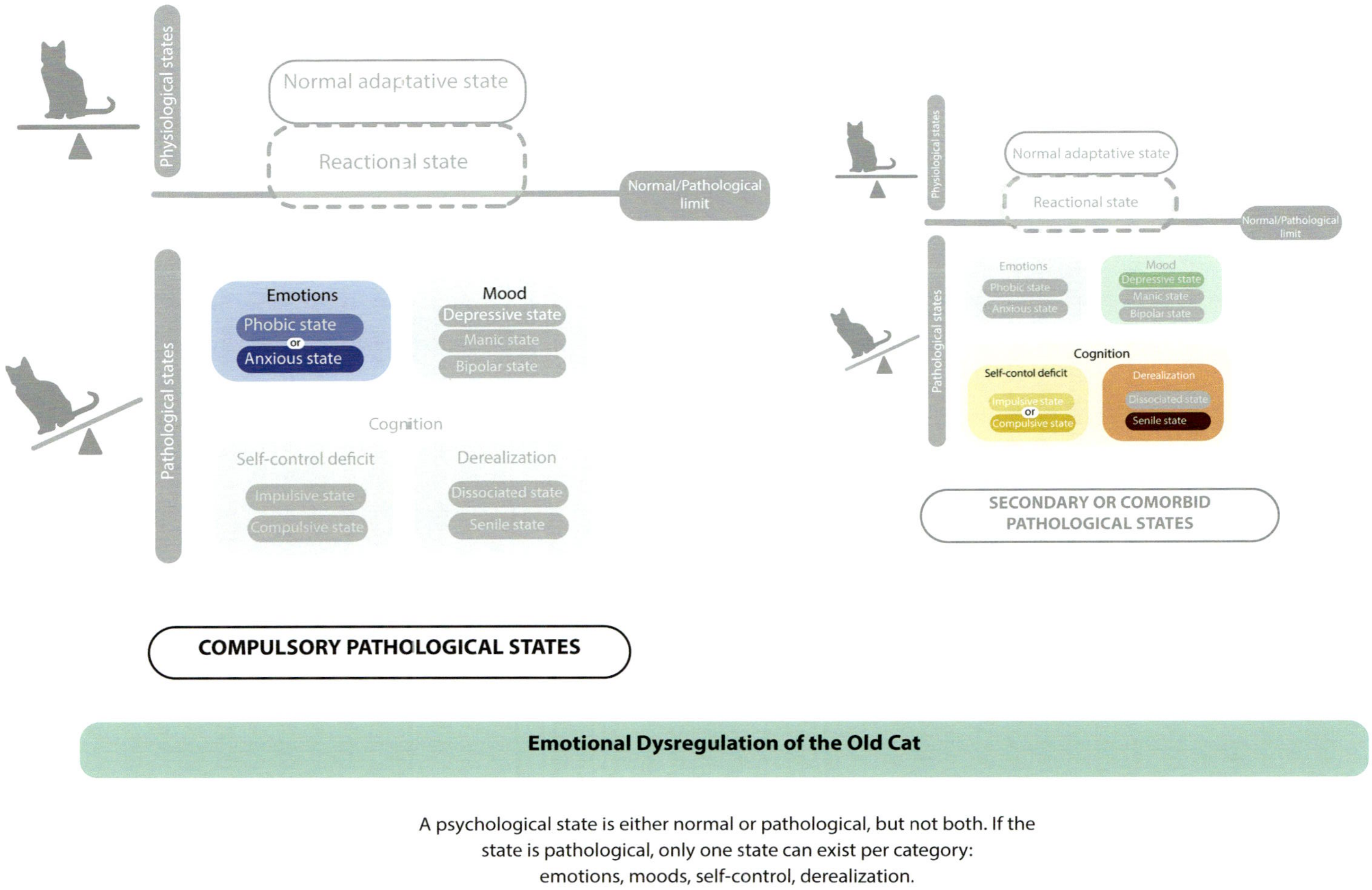

Fig. 13.6 Pathological states associated with emotional dysregulation in elderly cats. Mandatory pathological state: phobic **or** anxious state. Possible comorbid pathological states: impulsive **or** compulsive state, **and/or** depressive state, **and/or** senile state

pletely suppressing the cat's basic motivations. If the cat is unable to sleep, euthanasia is often requested quickly. Additionally, anxiety in elderly cats appears to be particularly persistent and resistant to treatment.

13.3.4 Treatments

13.3.4.1 Pharmacological Treatment

The priority is managing the anxious state. *Clomipramine*, administered at a dose of 0.25–1 mg/kg once daily, is recommended (King et al. 2004) because its histaminergic side effects induce drowsiness, making the cat calmer (Landsberg and Wilson 2005; Mertens et al. 2006; Seksel and Lindeman 1998).

Venlafaxine is highly effective and easy to administer (Metz et al. 2021), but it should be given at half the usual dose, or 0.5 mg/kg (two granules per kilogram), due to age-related reductions in metabolism and elimination capacity (Masson et al. 2024b).

Gabapentin or pregabalin, at doses of 10–30 mg/kg up to three times daily, can be added for their anxiolytic and analgesic effects. Renal function should be monitored when using these medications.

Trazodone, at a dose of 5–10 mg/kg as needed, is another option due to its hypnotic and anxiolytic properties.

Benzodiazepines can be used as a complementary treatment for managing treatment-resistant anxiety crises. However, all benzodiazepines interfere with short-term memory, so their use must be cautious, especially in cases of cognitive deficits. Despite these concerns, they remain excellent anxiolytics and can help establish a balance for some anxious cats, delaying disease progression.

Among herbal treatments, some useful combinations include *Eschscholzia* (California poppy) and *valerian*, where the former regulates sleep and the latter reduces anxiety.

13.3.4.2 Behavioral Therapy

▪ Environmental Measures

The first necessary change is a shift in the owner's perspective. It is essential that the cat be seen as a patient in need of support rather than an animal acting out of stubbornness. This change in approach helps restore the owner-cat relationship. Once this mindset is adopted, the owner becomes the primary caregiver, as they understand their cat's needs and want to ensure proper management of its condition.

The owner should provide assistance by modifying the environment to bypass obstacles and facilitate necessary compensations. The veterinarian should also guide the owner in maintaining the cat's coat, as poor grooming can lead to social rejection if the animal becomes unkempt or unpleasant to handle.

▪ Behavioral Modifications

The therapy consists of two main approaches. First, aggravating factors must be eliminated, particularly punishments, while encouraging the resumption of familiar rituals. As with the previous disorders, the goal is not to teach new behaviors but rather to motivate the cat by allowing it to regain positive interactions with its environment.

Creating a restricted but functional environment can sometimes be an effective solution. This involves limiting the cat's activity zone when it can no longer tolerate larger spaces due to fear. Keeping the cat in a single room can help, but this confinement must be balanced with regular but controlled outings. Short, brief excursions allow for some exercise and help prevent social withdrawal.

This strategy is primarily aimed at managing the worsening phase of the disorder and should not be implemented too early, but rather in the later stages of progression.

13.3.5 Therapeutic Follow-Up

The veterinarian must remain highly available and work closely with the owner to develop the necessary lifestyle adjustments. Maintaining open communication is essential, as discouragement is common in these cases.

13.3.6 Prevention

Although this condition is difficult to predict, vulnerable cats can often be identified. These are typically cats that have exhibited anxious tendencies throughout their lives. For these individuals, major life changes should be carefully monitored, and intervention should begin at the *first signs of loss of autonomy.*

13.3.7 Clinical Vignette

- **Gaston**

Gaston is a 9-year-old Persian cat who has always adapted to his owners' lifestyle, despite frequent relocations every 2 years due to their professional obligations. However, for the past month, he has refused to stay alone, and his continuous vocalizations have led his owners to give in. As a result, he is now allowed in the home office during telework sessions, which was previously off-limits.

Although Gaston had successfully adapted to all previous moves, he has been unable to cope with this latest relocation. His owners wonder if external factors such as vibrations or "harmful waves" could be affecting him. Following a complete blood workup and a thorough physical examination, the veterinarian rules out any organic disease and diagnoses emotional dysregulation in an elderly cat, accompanied by intermittent productive anxiety.

Treatment with clomipramine and gabapentin significantly reduces his anxiety-related behaviors within a few weeks, although his access to the office remains unchanged as part of the management plan.

Summary of ▸ Sect. 13.3

Emotional dysregulation in elderly cats is an aging-related disorder in which the earliest pathological signs indicate a disruption of the emotional axis. It is probably not the most common form of senility, but it leads to a loss of autonomy that quickly becomes difficult for the household to manage. As a result, it is a frequent reason for consultation, sometimes with early requests for euthanasia due to the severity of the symptoms.

As with other aging disorders in cats, emotional disturbances are merely an entry point into the broader process of global senility. Over time, the condition leads to significant cognitive, emotional, and mood-related impairments.

Treatment can only be effective if initiated early. The challenge is to restore a sense of calm to the animal without resorting to excessive sedation.

Therapeutic approaches should focus on eliminating aggravating factors such as punishment and rejection while enhancing the cat's motivation by introducing simple activities in fixed locations and at specific times of the day. This helps restore shared enjoyment and strengthens the bond between the cat and its owner. When signs of attachment dependence are observed, it is important to reinforce a soothing refuge and arrange a restricted, manageable space for the cat. Gradual access to a slightly larger area, such as an adjacent room or an entire floor, can give the cat the illusion of a well-controlled environment. Owners require significant support throughout the management process.

13.4 Mood Dysregulation in Elderly Cats

13.4.1 Description

13.4.1.1 Definition

Mood dysregulation in elderly cats, referred to in dogs by Pageat (1998) as *involutional depression*, is a common condition in aging cats, where demotivation and a depressive state dominate from the earliest stages.

As with other aging disorders, additional cerebral dysfunctions, including anxiety and senility, soon complicate the clinical picture, ultimately leading to global dysregulation of brain function.

This condition, like cognitive dysregulation, shares many similarities with Alzheimer-type dementia described in human medicine. As in this disorder, there is an association of cognitive decline and early onset depression (Neus Bosch et al. 2012; Schütt et al. 2016).

13.4.1.2 Etiology and Pathogenesis

In this particular form of cerebral aging, impairment of *motivational pathways* appears to develop early, often as a result of declining abilities. These losses are most commonly physical but can also be cognitive. In human medicine, significant memory loss is well documented, and it is likely that similar processes occur in animals.

13

A self-reinforcing cycle quickly takes hold: "I won't succeed, so there is no point in trying." This abandonment of tasks before even attempting them further exacerbates the decline in abilities, reinforcing a sense of helplessness and deepening *demotivation.* In this case, the depressive state acts more as a cause of capacity loss rather than simply a response to a deteriorating condition.

The progression then follows the typical pattern of other aging disorders, with cognitive decline becoming self-perpetuating. The deterioration in social interactions further contributes to global brain dysfunction, eventually leading to senile state as a comorbid condition.

13.4.1.3 Epidemiology

This form of disorder does not necessarily prompt owners to seek veterinary consultation. The disturbances it causes are relatively mild. The withdrawal and social isolation of an aging cat often seem acceptable to owners, making it difficult to estimate the prevalence of the condition.

13.4.1.4 Presenting Complaints

Consultations may be prompted by the consequences of secondary cognitive or emotional disorders, or by depressive manifestations that do not always present as simple silent withdrawal. Nocturnal restlessness and episodes of anxiety can lead to excessive activity or behavioral outbursts that are difficult to manage.

13.4.1.5 Symptomatology

Age is, of course, a key factor, as this is a disorder affecting elderly cats. The first symptoms are often subtle, with owners describing them as "anecdotal" or simply attributing them to aging. It is very common to hear statements such as *"he's going deaf with age,"* which should be considered a warning sign.

Hearing assessments are not commonly performed in cats, making it difficult to determine whether auditory capacity is truly declining. However, in this condition, it is frequently observed that cats are only partially or intermittently deaf. At certain times of the day, they appear to hear normally, while at other times they seem unresponsive to sounds, including familiar voices or household noises. They do not react when approached or spoken to and may be easily startled. It appears as though auditory stimuli no longer trigger a response. This perceptual non-integration phenomenon also affects other senses, but it is more challenging to detect.

Periods of deep sleep often accompany this decline in perception. The cat, disconnected from its surroundings, no longer reacts and sleeps profoundly. For owners, the lack of response to noise reinforces the belief that the cat is deaf, leading to a lack of concern and no request for veterinary care. As a result, the relationship between owner and cat gradually diminishes. A cat that does not seek attention receives none in return. It no longer appears for mealtimes, established rituals with the owner disappear, and the assumption is made that the cat simply prefers to be left alone.

Signs of depression appear quickly. This is a chronic depression accompanied by appetite and sleep disturbances. Sleep problems may be a reason for consultation. Before bedtime, the cat may experience restlessness resembling *hypnagogic anxiety*, similar to nightmares just before falling asleep. There may be sudden awakenings and, in some cases, nocturnal discomfort. Some cats vocalize their distress, but many do not even attempt to seek help and simply remain lying down. *Nocturnal urine soiling* can occur alongside these awakenings.

There are also moments of panic when the cat loses its bearings in an environment it can no longer mentally structure. During these anxiety episodes, the cat may call out to its owners, which can be particularly disruptive at night.

Cognitive decline follows as a consequence. A demotivated cat stops engaging in activities, leading to memory loss and skill deterioration. These losses of learned behaviors and spatial reference points, often associated with senility, are what first become noticeable to owners and typically prompt the initial veterinary consultations. The territory becomes disorganized and no longer serves its soothing function, contributing to the cycle of deterioration. At this stage, progression is rapid, leading to widespread disorganization.

The affected individual does not attempt to compensate but simply stops engaging in activities and becomes passive. Motor control loss, if present, is rarely reported, primarily because it is overshadowed by the generalized demotivation linked to the depressive state.

13.4.1.6 Evolution

In this condition, *motivational impairment* is the initial trigger for deterioration. Depression precedes cognitive decline, and the depressive state accelerates the effects of normal aging, often leading to senility.

As with other aging disorders, this depressive state is triggered by a change in the cat's environment or routine. There can be a significant delay, sometimes 1 or 2 years, between the first signs of demotivation and the onset of clinical depression.

The progression then follows a downward spiral, where cognitive decline, environmental changes, and relationship deterioration all contribute to the worsening of the condition (◘ Fig. 13.7).

13.4.2 Diagnosis

13.4.2.1 Diagnostic Criteria

Cats affected by mood dysregulation necessarily exhibit a *pathological depressive state.* At the onset, the cat does not present with anxious or senile state (◘ Fig. 13.8). Symptoms emerge *during the senescence period.*

The diagnosis is based on the identification of *chronic depressive state* in a cat *over 8 years old.* This generalized loss of motivation is expressed through some or all of the following symptoms:

- loss of interest in external events;
- inconsistent appetite;
- fluctuating sleep patterns, with episodes of very deep sleep;
- persistent demotivation despite treatment of any underlying medical conditions.

The earliest symptom is a reduced interest in external events. The cat stops engaging in certain activities despite still being physically capable of performing them. Naturally, this disorder can also affect a cat that has lost functional abilities, making diagnosis more complex. However, in these cases, functional improvement does not restore activity levels, confirming the presence of a mood disorder.

13.4.2.2 Direct Observation

Early forms are not noticeable, and it is through discussions with the owners that the first signs can be identified. *Deafness* is often mentioned. When the disease progresses fur-

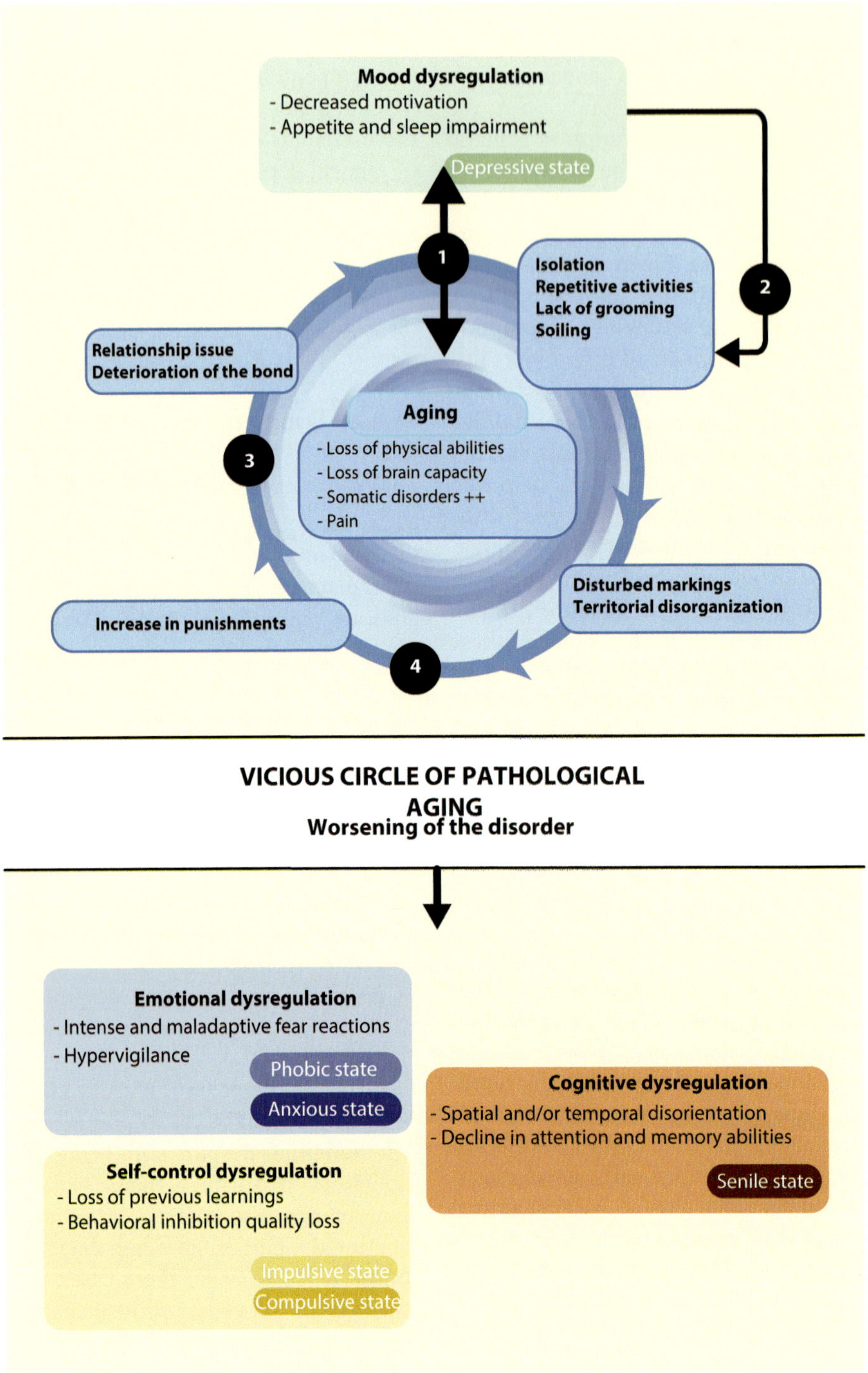

Fig. 13.7 Evolution of mood dysregulation in elderly cats. Without targeted intervention against each factor contributing to pathological aging, a vicious cycle develops, accelerating all forms of deterioration

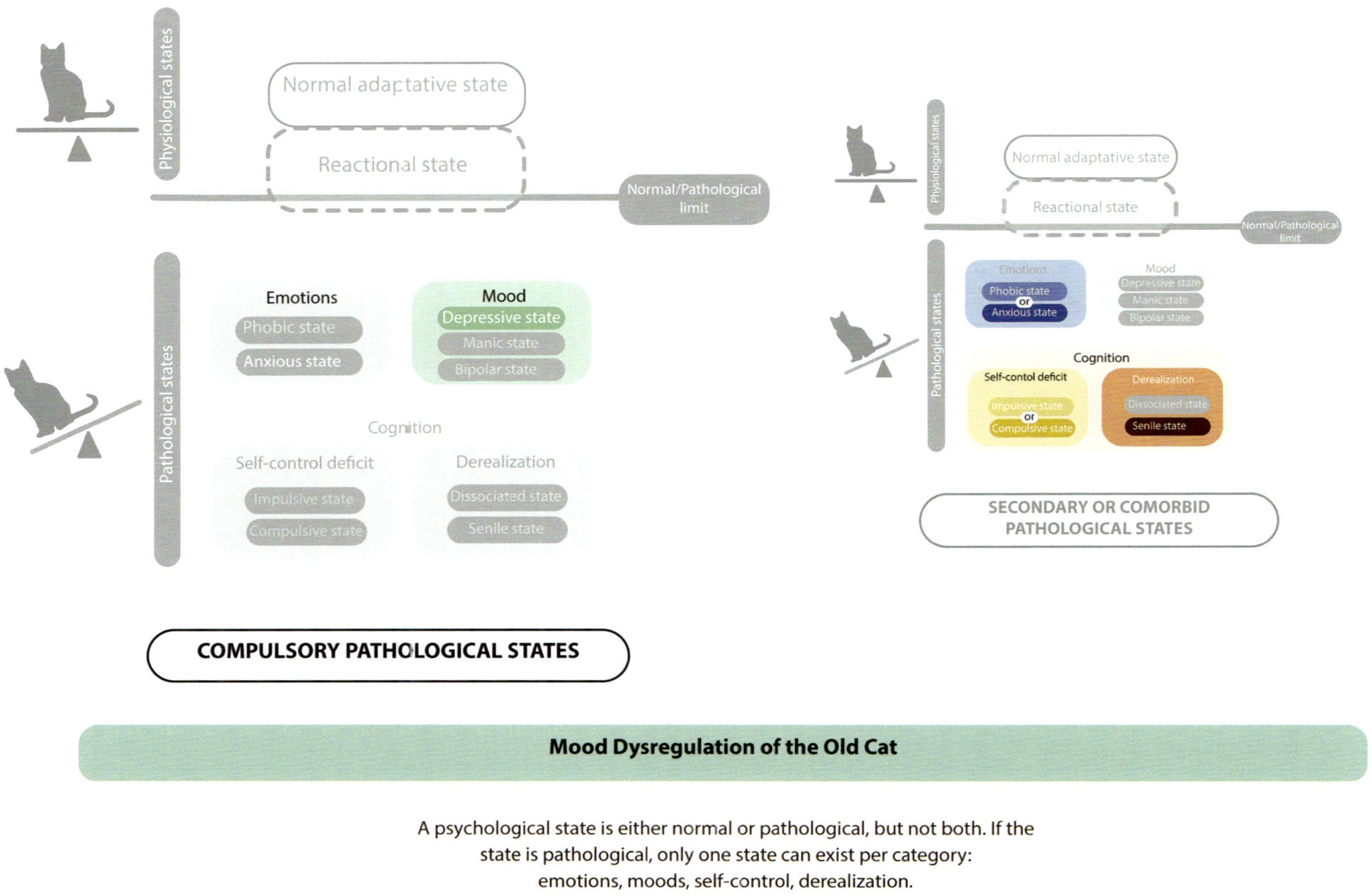

Fig. 13.8 Pathological states associated with mood dysregulation in elderly cats. Mandatory pathological state: depressive state. Possible comorbid pathological states: phobic **or** anxious state, **and/or** impulsive **or** compulsive state, **and/or** senile state

ther, a certain disinterest in the surroundings and environment can be observed. Cats do not seem to pay attention to what is happening around them.

13.4.2.3 Comorbidities

As always, aging animals accumulate disorders. It is important to assess hormonal functions, primarily the thyroid, as they can be responsible for depressive disorders. *Arthritic pain* associated with obesity is also a frequent trigger for demotivation. The management of this type of condition should focus on the animal's quality of life, and no aspect should be overlooked.

13.4.2.4 Differential Diagnosis

The differential diagnosis includes the three other aging disorders in cats. It can only be established at the early stages of the disease, as the terminal phases tend to resemble one another and accumulate all these conditions. Identifying the initial dysfunction within one pathway or another allows for distinction and leads to more effective management. In emotional dysregulation, emotional abnormalities should be investigated before cognitive, mood, or self-control disorders.

The differential diagnosis often includes *numerous organic diseases*, sometimes comorbid (such as deafness or arthritis), sometimes causing the behavioral disorder (hormonal imbalance, early oncological process, etc.), and sometimes resulting from the disorder (e.g., pressure sores due to decreased motivation). However, simply treating the somatic diseases that may explain part of the symptoms is not sufficient for ensuring well-being, and a *dual physical and psychological assessment is essential.*

13.4.3 Prognosis

The prognosis is the same for this condition as for other aging disorders. The added discomfort caused by the depressive state is significant.

13.4.4 Treatments

In the early stages of the condition, efforts should focus on *restoring motivation*. Antidepressants should be used, but they must not lead to a decrease in therapeutic vigilance. Therapy should align with this approach and make extensive use of rewards through the maintenance of rituals.

13.4.4.1 Pharmacological Treatment

Numerous compounds are available for aging disorders, including anti-oxidants and certain nutraceuticals (Benedetti et al. 2019; Pan et al. 2013). Their effectiveness remains debated and is difficult to distinguish from that of behavioral therapy, which restores motivation. However, in the early stages, they seem to provide an improvement from the owner's perspective.

The choice typically falls on an antidepressant, such as *mirtazapine*, at a dose of 1–2 mg/kg once per day. Its appetite-stimulating effect allows for the use of palatable treats to rebuild a bond with the cat. However, in cases of obesity, this medication should be avoided or used only for a limited time, as it has a strong orexigenic effect.

When administering medication is challenging, *venlafaxine* at a low initial dose (0.5 mg/kg, equivalent to 2 granules per kilogram) is a viable option even for cats that are difficult to medicate, without adding to their discomfort (Masson et al. 2024b).

Anxiety can sometimes manifest in strikingly productive ways, and in such cases, the use of *gabapentin* or *pregabalin* is recommended, with careful monitoring of renal function. Their analgesic effect is beneficial for arthritic pain in aging cats and, unlike tramadol, they can be combined with most prescriptions for older cats. Benzodiazepines, for the reasons previously mentioned in emotional disorders, are generally reserved for the most resistant cases, provided there are no signs of cognitive deficits.

13.4.4.2 Behavioral Therapy

Simply recognizing that this behavior in aging cats should be addressed already constitutes a therapeutic approach. As soon as owners understand that they can slow down the progression and that they must actively counter discouragement, the outlook becomes more favorable. It is essential to emphasize that mistakes should not be punished, as this would further increase demotivation.

From the outset, the goal is to provide *opportunities for reinforcement,* which can be *as simple as interactions with the owner.* Often, this involves re-establishing and strengthening routines as if they were complex exercises. The objective is not to create new learning experiences but to reinforce existing ones.

Introducing a young cat into the household is generally not advisable. Initially, an increase in activity may be observed in the senior cat, but there is a significant risk that the owners' attention will shift away, potentially leading to a decline in care and engagement.

13.4.5 Therapeutic Follow-Up

As with other aging disorders in cats, establishing a strong owner-therapist partnership is crucial. The veterinarian should rely on the owner's knowledge of their cat to suggest appropriate adjustments. The owner, in turn, should frequently consult the specialist for guidance and to adapt these modifications as needed. The success of the treatment and the cat's well-being depend on this collaboration. *Mutual trust* is essential.

13.4.6 Prevention

There is no way to prevent aging, but veterinarians are aware of the impact of changes on brain function. They can monitor for early signs of disorders and regularly encourage owners, during preventive or geriatric consultations, to describe any changes they observe in their cat.

13.4.7 Clinical Vignette

- **Féline**

Féline is a spayed Oriental cat over 12 years old. She is very attached to her owners and habitually requests that the faucets be turned on whenever one of them approaches the sink. However, in recent months, she has almost completely stopped doing so. Additionally, she often meows in her bed for no apparent reason, as if she were in pain. When spoken to, she no longer makes her usual small "pruit" sound of pleasure, and she hardly reacts to petting. Her owners suspect she is in pain and seek a consultation to obtain pain relief. A physical examination and blood tests reveal no abnormalities.

A diagnosis of chronic depressive state is made. Féline exhibits a capricious appetite, generalized demotivation, and a significant increase in sleep.

The veterinarian, who knows Féline well and recognizes that she is a shadow of her former self, prescribes mirtazapine to help restore her behaviors. The owners are instructed to place Féline near the sink and encourage the ritual that she once enjoyed. The medication takes effect quickly, leading to a resurgence of activity. After a few months, gabapentin is added to address occasional episodes of evening anxiety, and Féline soon returns to her usual habits.

Summary of ► Sect. 13.4

In this aging disorder in cats, the onset of the senility cycle occurs through a loss of motivation. This demotivation leads to a rapid decline in previously acquired skills, which further exacerbates the lack of motivation.

Cognitive impairments progress relatively quickly in this form, as indifference results in memory loss. This pattern of aging is somewhat similar to certain forms of Alzheimer's disease, which often begin with depression.

The treatment focuses on restoring activity. Mirtazapine is often beneficial, while venlafaxine is a viable alternative for facilitating medication administration in difficult-to-treat cats. Therapy should involve revisiting learned behaviors, followed by a reactivation of relational activities, particularly established rituals.

13.5 Self-Controls Dysregulation in Elderly Cats

13.5.1 Description

13.5.1.1 Definition

By symmetry, it is logical to describe a condition in which aging deteriorates the *axis of motor control*. This condition primarily manifests in cats that, in their youth, experienced difficulties with motor and emotional control and were impulsive or compulsive. The imbalance in this axis, which was previously manageable or tolerated, worsens as brain aging leads to a decline in cortical control. *Inhibition* becomes *insufficient*. This form of involution is the one that most resembles a "return to childhood."

13.5.1.2 Etiology and Pathogenesis

The impulsivity and compulsion loops are affected due to the aging process, marking the onset of the disorder. It is likely that a loss of inhibition is responsible for the resurgence of poorly controlled behaviors, particularly due to the weakened control of the prefrontal cortex over the striatum. Gradually, the cat loses the ability to properly execute stopping sequences and reverts to the inadequate control mechanisms of its early years.

13.5.1.3 Epidemiology

This condition is not mentioned in the model proposed by Pageat (1995) for dogs, nor is it distinguished from the "CDS" in Anglo-Saxon models (Landsberg et al. 2010). No prevalence studies are available. Our experience in general veterinary practice suggests that self-control dysregulation can be an initial trigger for aging disorders, but it only becomes a reason for consultation when it leads to aggression, which is not always the case.

13.5.1.4 Presenting Complaints

Early consultation should be prompted by a *loss of motor control* and difficulty calming down after moments of excitement, indicating a loss of cortical regulation.

13.5.1.5 Symptomatology

The most notable symptom is the resurgence of injurious behavior during play. A cat that had previously learned not to cause harm during interactions begins to do so again. Occasional hissing during contact or handling may escalate into forceful scratching or biting. Additionally, deficits in motor inhibition lead to a loss of abilities, which can initiate the downward spiral of aging.

A return to oral exploration may also be observed, and the ingestion of foreign objects at this age is a significant warning sign.

Finally, nocturnal vocalizations may emerge. While cats are naturally crepuscular and nocturnal, they typically learn to adjust their activity rhythm to match that of their owners. A decline in inhibitory processes can result in a resurgence of nocturnal behaviors, sometimes accompanied by disruptive vocalizations.

13.5.1.6 Evolution

The progression follows a typical aging pattern. The initial stage of the disorder involves a decline in inhibition and cortical control. This leads to impaired learning, followed by a deterioration in the cat's relationships. Senile, anxious, or depressive pathological states may then develop, further complicating the clinical picture. At this stage, management aligns with that of other advanced aging disorders in cats (◘ Fig. 13.9).

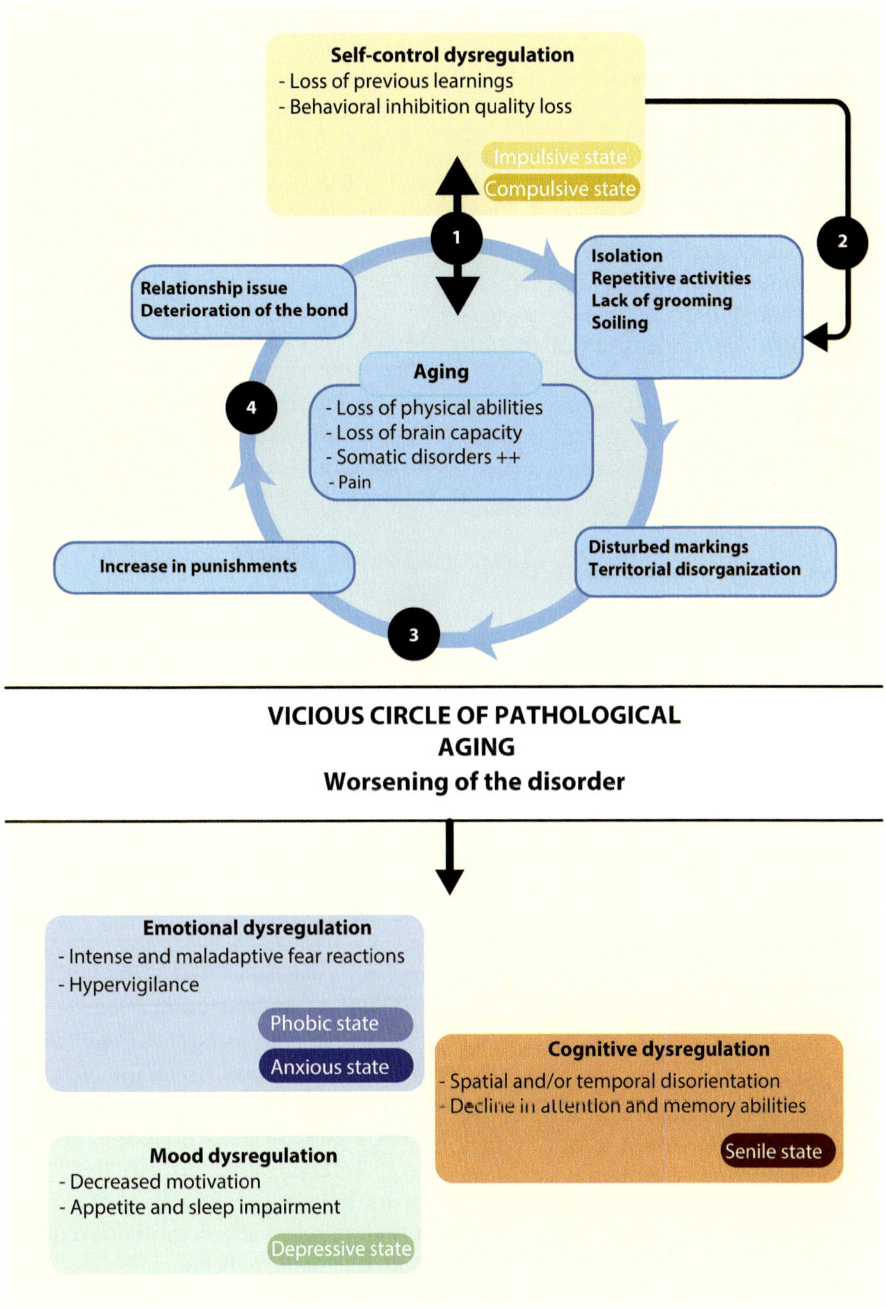

Fig. 13.9 Evolution of self-controls dysregulation in elderly cats. In the absence of intervention to address each contributing factor of pathological aging, a vicious cycle develops, accelerating overall deterioration

13.5.2 Diagnosis

13.5.2.1 Diagnostic Criteria

Cats affected by self-control dysregulation must present with an *impulsive or compulsive pathological state* (◘ Fig. 13.10). Symptoms emerge during the senescence period.

The diagnosis is based on the identification of motor and emotional control deficits in a cat *older than 8 years*. These dysregulations manifest through some or all of the following symptoms:

- Re-appearance of oral exploration;
- recurrence of impulsive or compulsive behaviors previously observed in youth;
- loss of control during interactions, with a return of rough exchanges;
- persistence of motor agitation despite treatment for organic conditions.

13.5.2.2 Direct Observation

This is likely the form of aging that least resembles a decline in its early stages. Elderly cats that seek play and exhibit juvenile behaviors may appear endearing. However, this change should be interpreted as an *abnormality*.

13.5.2.3 Comorbidities

Pain associated with this form of aging, combined with a loss of inhibition, can lead to aggression. Feeding behavior may also be disrupted, necessitating a thorough endocrine evaluation. *Hyperthyroidism* must be ruled out. A comprehensive blood panel should help eliminate most organic causes, as any sensation of discomfort can trigger agitation.

13.5.2.4 Differential Diagnosis

The differential diagnosis includes any painful somatic condition that may cause agitation.

In the early stages of progression, this loss of control should not be mistaken for agitation linked to emotional disorders.

Thyroid disorders must be ruled out due to the similarity of some symptoms with the condition described here.

13.5.3 Prognosis

The prognosis is better than that of other aging disorders in the early stages, as the discomfort is often moderate. However, if another pathological state is already present, the prognosis aligns with that of aging disorders as a whole.

13.5.4 Treatments

As the underlying mechanism of this cerebral imbalance is a loss of control, it is logical to propose a therapy aimed at restoring these controls.

13.5.4.1 Pharmacological Treatments

The clinical benefit of distinguishing this form from the other three aging disorders lies in the ability to intervene earlier in the first disrupted loop, which in this case is motor control.

The treatment of choice is *fluoxetine*, which helps reduce impulsivity and prevents deterioration of the human-animal relationship (Kaur et al. 2016). This medication is used at a lower dose than in very young cats and helps restore the missing inhibition.

13.5.4.2 Behavioral Therapy

▪ Environmental Modifications

The use of remote-play activities and the gradual reduction of play intensity at the end of each session helps the cat regulate its emotions, a process it can no longer manage on its own. Interactions and rituals, in all their forms, help maintain a level of control while stimulating the cat's cognitive circuits, making them particularly beneficial.

▪ Behavioral Modifications

Reintroducing exercises that promote self-control helps restore and maintain cortical inhibition at previously acquired levels. In this form of motor control loss, punishment may momentarily stop unwanted behavior, but it is even more strongly contraindicated. An aging cat that is punished may become highly sensi-

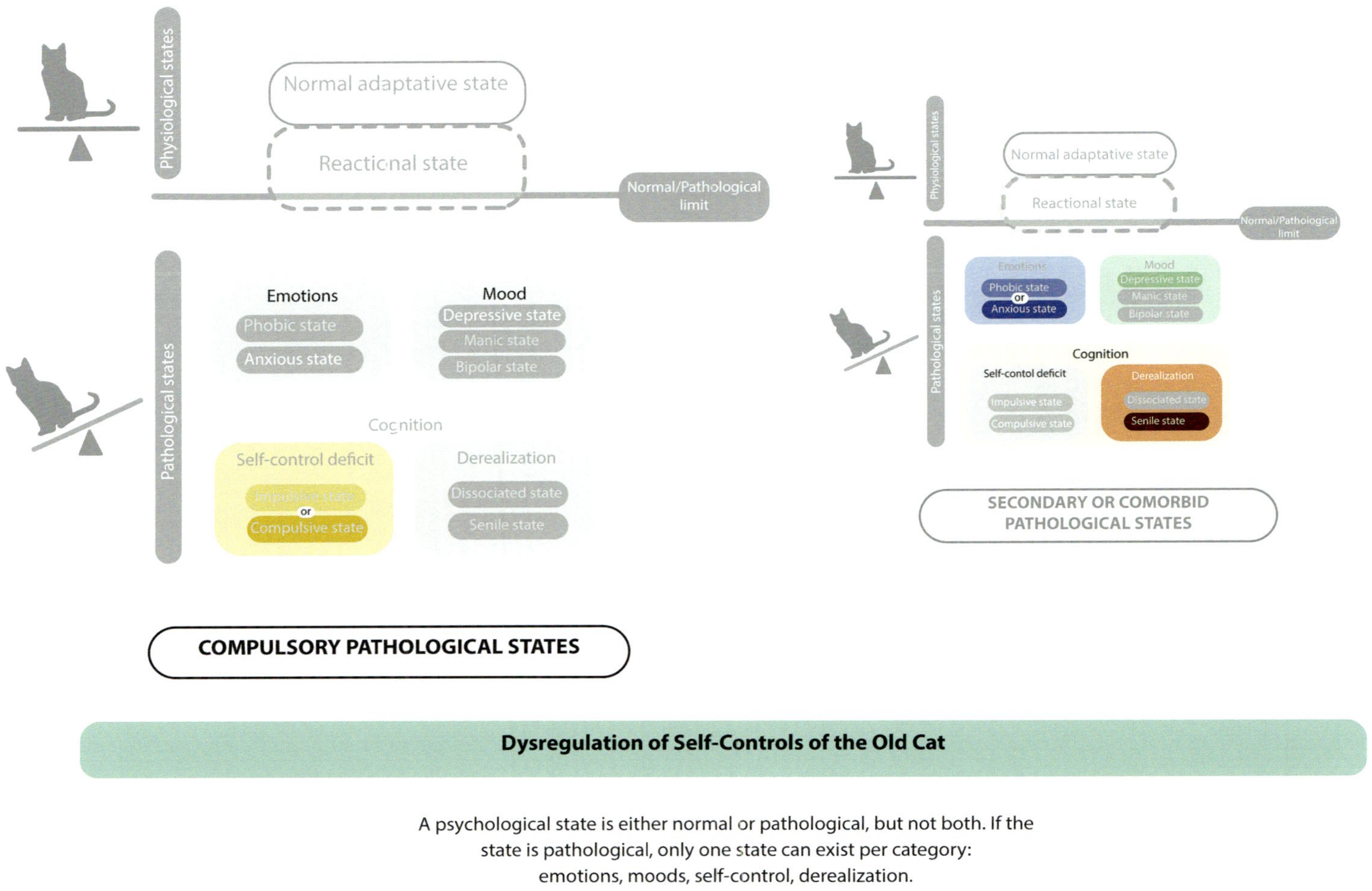

Fig. 13.10 Pathological states associated with self-controls dysregulation in elderly cats. Mandatory pathological state: impulsive **or** compulsive state. Possible comorbid pathological states: phobic **or** anxious state, **and/or** depressive state, **and/or** senile state

tive to reprimands, leading to a much faster shift toward anxiety and aggression than in a younger cat.

13.5.5 Therapeutic Follow-Up

As with other aging disorders in cats, it is essential to monitor the progressive decline in motor control and watch for the emergence of symptoms indicating cognitive, emotional, or mood impairment.

13.5.6 Prevention

To intervene at the earliest signs of deterioration, it is essential to question owners about all possible aspects of behavioral aging, including motor control, rather than focusing solely on emotional, mood, or cognitive decline.

13.5.7 Clinical Vignette

Modane is an Abyssinian cat who was difficult during her youth. She was impossible to pet because she would always attack hands. She was very restless and frequently attacked feet.

For the past 2 weeks, she has started biting feet again. She seems to be experiencing a second youth. However, her owners have vivid memories of the foot attacks that lasted for several years and are determined not to relive that experience.

Modane had recovered and had stopped exhibiting this behavior. They suspect that the behavior resumed with the switch to winter time. The possibility of an aging disorder is considered, and treatment with fluoxetine at a dose of 1 mg/kg is initiated. The owners are warned not to punish her.

The return to normal occurs quickly, but the treatment is maintained for life.

Summary of ▶ Sect. 13.5

Self-controls dysregulation in elderly cats is a condition in which the first sign of pathological aging is a dysfunction in the impulsivity and compulsion loops, accompanied by a decline in the effectiveness of cortical controls.

In this disorder, the cat appears to revert to kitten-like behavior, with the re-emergence of problematic habits from its youth. The initial symptoms may give the impression of a "return to youth," but they actually reflect a loss, particularly in inhibitory control. Because of the seemingly endearing nature of these early symptoms, this condition is not widely recognized, and owners are often reluctant to seek veterinary advice.

However, as the condition progresses, owners' patience is tested when the aging cycle accelerates, and anxiety compounds the clinical picture. Anxiety symptoms become even more striking due to the lack of control, prompting a rapid request for veterinary intervention. The treatment of choice in the early stages is fluoxetine, combined with behavioral work to restore inhibition.

13

Aging disorders

Disorder's name	State diagnosis	Diagnostic criteria	Differential diagnosis
Cognitive dysregulation in elderly cats	Senile	Cat older than 8 years. Evidence of multiple cognitive errors, expressed through some or all of the following symptoms: • Spatial disorientation: searching for familiar environmental elements in the wrong place • Temporal disorientation: day/night inversion, requests at unusual times relative to expected cues • Loss of learned behaviors: loss of litter box habits • Loss or incorrect performance of rituals	*Pain* Erratic behavioral responses related to pain *Organic disease* Polyuria/polydipsia (PUPD), hormonal disorders, decreased sensory acuity *Emotional dysregulation in elderly cats* Presence of worsening anxiety Loss of autonomy Absence of cognitive errors in the early stages *Mood dysregulation in elderly cats* Presence of depressive signs Absence of cognitive errors in the early stages *Self-controls dysregulation in elderly cats* Presence of signs of loss of inhibition Absence of cognitive errors in the early stages
Emotional dysregulation in elderly cats	Phobic or anxious	Cat older than 8 years. Evidence of a sudden worsening of emotional state and associated distress symptoms, including some or all of the following: • Daytime distress signs: vocalization, inappropriate elimination • Nocturnal distress signs: wandering and vocalization • Panic attacks • Restless sleep with frequent awakenings • Identification of pre-existing fear axis vulnerabilities before senescence • Worsening of a previously stable phobic state with generalization and progression into an anxious state • Loss of autonomy	*Pain* Behavioral responses related to pain *Organic disease* PUPD, hormonal disorders, decreased sensory acuity *Cognitive dysregulation in elderly cats* Presence of cognitive errors in the early stages Absence of anxiety or worsening emotional state initially *Mood dysregulation in elderly cats* Presence of depressive signs Absence of anxious agitation in the early stages *Self-controls dysregulation in elderly cats* Presence of signs of loss of inhibition Absence of anxious agitation in the early stages

Aging disorders

Disorder's name	State diagnosis	Diagnostic criteria	Differential diagnosis
Mood dysregulation in elderly cats	Depressive	Cat older than 8 years. Evidence of a general loss of motivation, expressed through some or all of the following symptoms: • Loss of interest in external events • Capricious appetite • Fluctuating sleep patterns • Persistent demotivation despite treatment of organic conditions	*Pain* Interaction avoidance due to pain *Organic disease* PUPD, hormonal disorders Deafness, and decreased sensory acuity *Cognitive dysregulation in elderly cats* Presence of cognitive errors in the early stages Absence of demotivation initially *Emotional dysregulation in elderly cats* Presence of worsening anxiety Loss of autonomy Absence of demotivation initially *Self-controls dysregulation in elderly cats* Presence of signs of loss of inhibition Absence of demotivation initially
Self-controls dysregulation in elderly cats	Impulsive or compulsive	Cat older than 8 years. Evidence of some or all of the following symptoms: • Re-appearance of oral exploration • Recurrence of impulsive or compulsive behaviors previously present in youth • Loss of control in interactions, with a return to rough exchanges • Persistent motor agitation despite treatment of organic conditions	*Pain* Agitation caused by pain *Organic disease* PUPD, hormonal disorders *Cognitive dysregulation in elderly cats* Presence of cognitive errors in the early stages Absence of controls deficit *Emotional dysregulation in elderly cats* Presence of worsening anxiety Loss of autonomy *Mood dysregulation in elderly cats* Presence of depressive signs Absence of anxious agitation in the early stages Presence of general demotivation

13

References

Beata C, Muller G, Arpaillange C, Graff E, Lachapèle D, Marion M, Marlois N, Massal N, Mège C (2015) Pathologie comportementale du chat. AFVAC

Beckman KB, Ames BN (1998) The free radical theory of aging matures. Physiol Rev 78(2):547–581. https://doi.org/10.1152/physrev.1998.78.2.547

Benedetti R, Marchegiani A, Tambella AM, Fruganti A, Serri E, Malfatti A, Spaterna A (2019) Effects of chronic supplementation of homotaurine on cognitive processes and spatial cognition in aged dogs: preliminary results. J Vet Behav 33:90–95. https://doi.org/10.1016/j.jveb.2019.07.003

Castillo C, Hernández J (2015) The role of oxidative stress in the development of cognitive dysfunction syndrome in cats. Importance of antioxidant prevention and therapy. In: The role of oxidative stress in the development of cognitive dysfunction syndrome in cats. Importance of Antioxidant Prevention and Therapy

Chambers JK, Tokuda T, Uchida K, Ishii R, Tatebe H, Takahashi E, Tomiyama T, Une Y, Nakayama H (2015) The domestic cat as a natural animal model of Alzheimer's disease. Acta Neuropathol Commun 3(1):78. https://doi.org/10.1186/s40478-015-0258-3

de Souza Dantas LM (2019) Monoamine oxidase inhibitors. Vet Psychopharmacol:185–199

Denenberg, S., & Landsberg, G. (2017). Current pharmacological and non-pharmacological approaches for therapy of feline and canine dementia. In G. Landsberg, A. Maďari, & N. Žilka (Éds.), Canine and feline dementia (p. 129–143). Springer International Publishing. https://doi.org/10.1007/978-3-319-53219-6_8

Department of Comparative Medicine, School of Medicine, University of Washington, Seattle, WA, USA, Klug J, Snyder MJ, Darvas MM, Imai D, Church M, Latimer C, Keene CD, Ladiges W (2020) Aging pet cats develop neuropathology similar to human Alzheimer's disease. Aging Pathobiol Ther 2(3):120–125. https://doi.org/10.31491/APT.2020.09.027

Dowgray N, Pinchbeck G, Eyre K, Biourge V, Comerford E, German AJ (2022) Aging in cats: owner observations and clinical finding in 206 mature cats at enrolment to the cat prospective aging and welfare study. Front Vet Sci 9:859041. https://doi.org/10.3389/fvets.2022.859041

Franceschi C, Capri M, Monti D, Giunta S, Olivieri F, Sevini F, Panourgia MP, Invidia L, Celani L, Scurti M, Cevenini E, Castellani GC, Salvioli S (2007) Inflammaging and anti-inflammaging: a systemic perspective on aging and longevity emerged from studies in humans. Mech Ageing Dev 128(1):92–105. https://doi.org/10.1016/j.mad.2006.11.016

Gunn-Moore D (2009) Feline cognitive dysfunction syndrome. In Small animal and exotics. Proceedings of the North American Veterinary Conference, Orlando, Florida, USA, 17–21 January, 2009. 2009. 1465–1468.

He C, Tsuchiyama SK, Nguyen QT, Plyusnina EN, Terrill SR, Sahibzada S, Patel B, Faulkner AR, Shaposhnikov MV, Tian R, Tsuchiya M, Kaeberlein M, Moskalev AA, Kennedy BK, Polymenis M (2014) Enhanced longevity by ibuprofen, conserved in multiple species, occurs in yeast through inhibition of tryptophan import. PLoS Genet 10(12):e1004860. https://doi.org/10.1371/journal.pgen.1004860

Head E, Moffat K, Das P, Sarsoza F, Poon WW, Landsberg G, Cotman CW, Murphy MP (2005) β-Amyloid deposition and tau phosphorylation in clinically characterized aged cats. Neurobiol Aging 26(5):749–763. https://doi.org/10.1016/j.neurobiolaging.2004.06.015

Kaur G, Voith VL, Schmidt PL (2016) The use of fluoxetine by veterinarians in dogs and cats: a preliminary survey. Vet Rec Open. https://doi.org/10.1136/vetreco-2015-000146

King JN, Steffan J, Heath SE, Simpson BS, Crowell-Davis SL, Harrington LJM, Weiss A-B, Seewald W (2004) Determination of the dosage of clomipramine for the treatment of urine spraying in cats. J Am Vet Med Assoc 225(6):881–887. https://doi.org/10.2460/javma.2004.225.881

Köhler I, Ballhausen BD, Stockhaus C, Hartmann K, Wehner A (2016) Prevalence of and risk factors for feline hyperthyroidism among a clinic population in Southern Germany. Tierarztl Prax Ausg K Kleintiere Heimtiere 44(3):149–157. https://doi.org/10.15654/TPK-150590

Landsberg GM, Wilson AL (2005) Effects of clomipramine on cats presented for urine marking. J Am Anim Hosp Assoc 41(1):3–11. https://doi.org/10.5326/0410003

Landsberg GM, Denenberg S, Araujo JA (2010) Cognitive dysfunction in cats. A syndrome we used to dismiss as « old age ». J Feline Med Surg. https://doi.org/10.1016/j.jfms.2010.09.004

Masson S, Bleuer-Elsner S, Muller G, Médam T, Chevallier J, Gaultier E (2024a) Aging disorders. In: Masson S, Bleuer-Elsner S, Muller G, Medam T, Chevallier J, Gaultier E (eds) Veterinary psychiatry of the dog. Springer Nature Switzerland, pp 503–538. https://doi.org/10.1007/978-3-031-53012-8_14

Masson S, Metz D, Bleuer-Elsner S, Schwobthaler F (2024b) Retrospective study on the use of venlafaxine in 176 cats diagnosed with behavioral disorders. J Vet Behav. https://doi.org/10.2139/ssrn.4790331

Mertens PA, Torres S, Jessen C (2006) The effects of clomipramine hydrochloride in cats with psychogenic alopecia: a prospective study. J Am Anim Hosp Assoc 42(5):336–343. https://doi.org/10.5326/0420336

Metz D, Medam T, Masson S (2021) Double-blind, placebo-controlled trial of venlafaxine to treat behavioural disorders in cats: a pilot study. J Feline

Med Surg. https://doi.org/10.1177/1098612X211036792
Mills D, Ledger R (2001) The effects of oral selegiline hydrochloride on learning and training in the dog: a psychobiological interpretation. Prog Neuro-Psychopharmacol Biol Psychiatry 25(8):1597–1613. https://doi.org/10.1016/S0278-5846(01)00205-6
Neus Bosch M, Pugliese M, Gimeno-Bayon J, Jose Rodriguez M, Mahy N (2012) Dogs with cognitive dysfunction syndrome: a natural model of Alzheimer's disease. Curr Alzheimer Res 9(3):298–314. https://doi.org/10.2174/156720512800107546
Pan Y, Araujo JA, Burrows J, De Rivera C, Gore A, Bhatnagar S, Milgram NW (2013) Cognitive enhancement in middle-aged and old cats with dietary supplementation with a nutrient blend containing fish oil, B vitamins, antioxidants and arginine. Br J Nutr 110(1):40–49. https://doi.org/10.1017/S0007114512004771
Pageat P (1995) Pathologie comportementale du chien. Editions du point vétérinaire Maison-Alfort
Pageat P (1998) Pathologie du comportement du chien. Editions du Point Veterinaire, Maisons-Alfort
Peña-Bautista C, Álvarez-Sánchez L, Ferrer I, López-Nogueroles M, Cañada-Martínez AJ, Oger C, Galano J-M, Durand T, Baquero M, Cháfer-Pericás C (2021) Lipid peroxidation assessment in preclinical Alzheimer disease diagnosis. Antioxidants 10(7):1043. https://doi.org/10.3390/antiox10071043
Scheff SW, Price DA (2006) Alzheimer's disease-related alterations in synaptic density: neocortex and hippocampus. J Alzheimer's Dis 9(s3):101–115. https://doi.org/10.3233/JAD-2006-9S312
Schütt T, Helboe L, Pedersen LØ, Waldemar G, Berendt M, Pedersen JT (2016) Dogs with cognitive dysfunction as a spontaneous model for early Alzheimer's disease: a translational study of neuropathological and inflammatory markers. J Alzheimer's Dis 52(2):433–449. https://doi.org/10.3233/JAD-151085
Seksel K, Lindeman M (1998) Use of clomipramine in the treatment of anxiety-related and obsessive-compulsive disorders in cats. Aust Vet J 76(5):317–321. https://doi.org/10.1111/j.1751-0813.1998.tb12353.x
Selkoe DJ (2012) Preventing Alzheimer's disease. Science 337(6101):1488–1492. https://doi.org/10.1126/science.1228541
Sordo L, Gunn-Moore DA (2021) Cognitive dysfunction in cats: update on neuropathological and Behavioural changes plus clinical management. Vet Rec 188(1):e3. https://doi.org/10.1002/vetr.3
Sordo L, Breheny C, Halls V, Cotter A, Tørnqvist-Johnsen C, Caney SMA, Gunn-Moore DA (2020) Prevalence of disease and age-related behavioural changes in cats: past and present. Vet Sci 7(3):85. https://doi.org/10.3390/vetsci7030085
Stammeleer L, Xifra P, Serrano SI, Rishniw M, Daminet S, Peterson ME (2024) Blood pressure in hyperthyroid cats before and after radioiodine treatment. J Vet Intern Med 38(3):1359–1369. https://doi.org/10.1111/jvim.17032
Stephens MJ, Neill DGO, Church DB, McGreevy PD, Thomson PC, Brodbelt DC (2014) Feline hyperthyroidism reported in primary-care veterinary practices in England: prevalence, associated factors and spatial distribution. Vet Rec 175(18):458–458. https://doi.org/10.1136/vr.102431
Telegina DV, Suvorov GK, Kozhevnikova OS, Kolosova NG (2019) Mechanisms of neuronal death in the cerebral cortex during aging and development of Alzheimer's disease-like pathology in rats. Int J Mol Sci 20(22):5632. https://doi.org/10.3390/ijms20225632
Torp R, Head E, Cotman CW (2000) Ultrastructural analyses of beta-amyloid in the aged dog brain: neuronal beta-amyloid is localized to the plasma membrane. Prog Neuro-Psychopharmacol Biol Psychiatry 24(5):801–810. https://doi.org/10.1016/s0278-5846(00)00107-x
Urfer S, Darvas M, Keene D, Czeibert K, Kubinyi E, Sándor S, Guscetti F, Kaeberlein M (2020) Amyloid beta-42 levels in companion dog brains correlate with age and cognitive Functio. Innov Aging 4(Suppl 1):887–887. https://europepmc.org/articles/PMC7742188

The Link Between Psychiatric Disorders and Somatic Disorders

Contents

14.1 Pain – 476

14.1.1 Psychiatric Disorders Secondary to a Painful Condition – 476

14.1.2 The Consequences of Psychiatric Disorders on Pain – 478

14.2 Dermatology – 479

14.3 Neurology – 481

14.4 Nutrition – 486

14.5 Internal Medicine – 489

14.5.1 Lower Urinary Tract Diseases – 489

14.5.2 Digestive Diseases – 490

14.5.3 Feline Hypertension – 491

14.5.4 Metabolic Diseases – 491

14.6 Reproduction – 492

14.7 Endocrinology – 493

14.7.1 Hyperthyroidism – 493

14.7.2 Diabetes – 494

14.7.3 Hypothyroidism – 494

14.7.4 Cushing's Syndrome or Hyperadrenocorticism – 495

14.7.5 Addison's Disease – 496

14.7.6 Sexual Dysendocrinopathies – 496

14.7.7 Acromegaly – 497

© The Author(s), under exclusive license to Springer Nature Switzerland AG 2026
S. Bleuer-Elsner et al., *Veterinary Psychiatry of the Cat*,
https://doi.org/10.1007/978-3-032-06260-4_14

14.8 Surgery – 497

14.9 Aging – 498

References – 498

In this chapter, we describe the bidirectional relationship between psychiatric disorders and somatic disorders. Various fields of veterinary medicine and their close ties with veterinary psychiatry are discussed: pain management, dermatology, neurology, nutrition, internal medicine, reproduction, surgery, and geriatrics.

For each category, we detail the possible behavioral consequences of these organic diseases on the animal's psychological state, as well as the impact of psychiatric disorders in these medical fields.

During a veterinary psychiatry consultation, part of the clinician's work is to ensure that the observed symptoms are not partially or entirely caused by an organic disease. Indeed, many organic diseases initially present with behavioral changes as their only visible symptoms. In some cases, these symptoms are exclusively behavioral, such as avoidance of interactions, irritability, aggression, urine soiling, or excessive licking. For this reason, psychiatry must be integrated with other medical disciplines.

It is not uncommon to hear that the approach to suspected psychiatric disorders should begin by ruling out every possible somatic disease. However, we find it inconsistent to treat veterinary psychiatry differently from other veterinary disciplines (◘ Fig. 14.1).

Our medical approach teaches us to perform targeted complementary examinations rather than conducting all possible tests on every cat during each visit. Moreover, insisting on ruling out all somatic disorders before addressing psychological aspects may lead to neglecting or delaying psychological care when an organic disease is detected. Yet, the links between somatic and psychological aspects of the body are numerous, and many cases cannot be treated without a comprehensive approach to the animal's health.

It is with this perspective that this chapter is organized. The connection between somatic and psychological aspects can manifest in different ways, which we will attempt to address in each section.

Any organic disease *can lead to psychological consequences*. A simple somatic disease, either because it is painful (see ▶ Chap. 3, ▶ Sect. 3.3) or because it is stressful (see ▶ Chap. 3, ▶ Sect. 3.1.2), can trigger a phobic, anxious, or depressive pathological state. Some diseases are particularly anxiety-inducing, such as pruritic dermatological conditions or diseases associated with chronic pain. The risk of developing a pathological psychological state can be minimized if this dimension is considered early.

A *somatic disease may also develop in an animal with a pre-existing but untreated psychological disorder*. In such cases, a combined approach is essential. For example, a cat suffering from extensive alopecia may have been anxious all its life or may have a compulsive state. The presence of psychiatric disorders in this case will hinder optimal treatment. Implementing a therapeutic plan must take these factors into account to avoid failure.

Thus, whether the consultation is prompted by a somatic or psychiatric concern, the clinician will encounter a clinical presentation that differs from the usual and must address both aspects.

Finally, *psychiatric disorders can have somatic consequences*. Recognizing these consequences allows veterinarians to refer several cases to a veterinary psychiatrist when psychiatric disorders are suspected.

It is often the lack of logical progression in the onset and evolution of the somatic condition that should alert the practitioner. The interplay between organic and psychiatric diseases disrupts the normal course of disease progression. Any irregularity in symptom development or appearance should prompt the clinician to investigate secondary or aggravating causes, including psychological factors.

A *somatic disease may develop in an animal with a pre-existing but untreated psychological disorder*. In such cases, a combined approach is essential. For example, a cat suffering from extensive alopecia may have been anxious all its life or may suffer from a compulsive state. The presence of psychiatric disorders in this case will hinder optimal treatment. Implementing a therapeutic plan must take these factors into account to avoid failure.

Thus, whether the consultation is prompted by a somatic or psychiatric con-

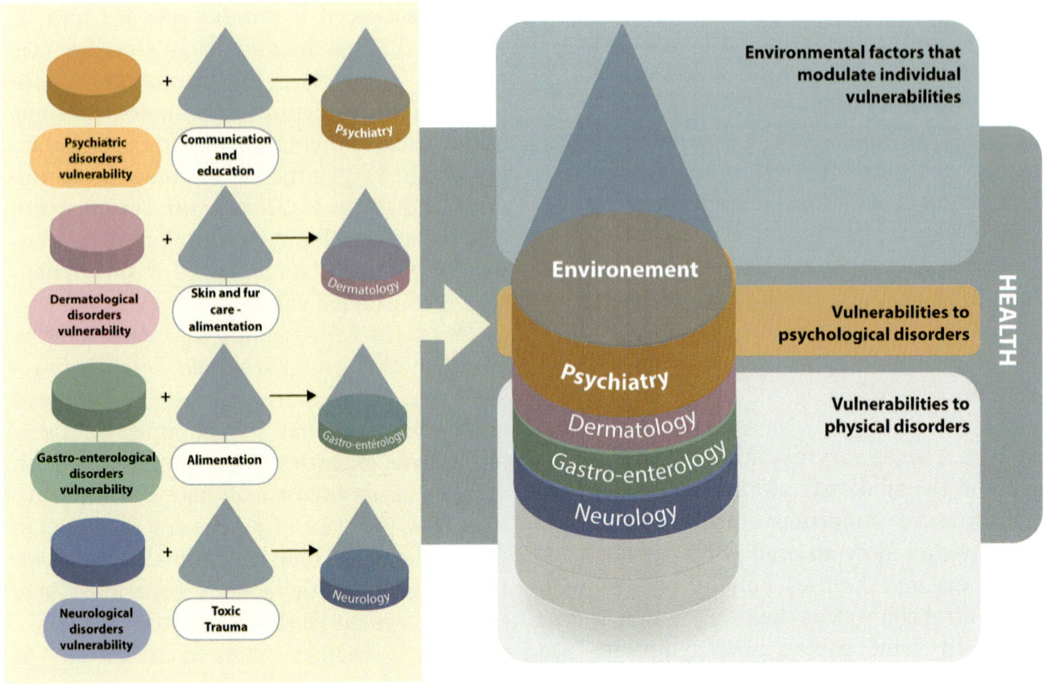

Fig. 14.1 A holistic approach to physical and mental health, illustrating that environmental factors combined with genetic vulnerabilities are identical across all disciplines

cern, the clinician will encounter a clinical presentation that differs from the usual and must address both aspects.

Finally, *psychiatric disorders can have somatic consequences.* Recognizing these consequences allows veterinarians to refer a number of cases to a veterinary psychiatrist when psychiatric disorders are suspected.

It is often the lack of logical progression in the onset and evolution of the somatic condition that should alert the practitioner. The interplay between organic and psychiatric diseases disrupts the normal course of disease progression. Any irregularity in symptom development or appearance should prompt the clinician to investigate secondary or aggravating causes, including psychological factors.

Before presenting an analysis tailored to the main disciplines linked to psychiatry, we dedicate the following section to pain, one of the sensations with the greatest impact on psychological functioning.

14.1 Pain

As discussed in ► Chap. 3 (► Sect. 3.3), pain, particularly *chronic pain*, induces cerebral changes that are very similar to those caused by *chronic stress*. Pain, whether nociceptive or neuropathic, is present in many conditions. It alters the animal's emotions and behavior and reduces its welfare.

14.1.1 Psychiatric Disorders Secondary to a Painful Condition

Pain creates a sense of vulnerability that makes the animal *hypervigilant* and *irritable*. As a result, any condition associated with pain can trigger aggressive reactions. These aggressions may occur in a cat that has never shown aggression before or become more frequent and intense in a cat that was already aggressive.

Acute pain is easily detected as it appears suddenly and causes visible symptoms such as lameness, vocalization at the moment of trauma, vomiting, or refusal to eat.

Chronic pain, on the other hand, is more subtle. Behavioral or postural changes associated with it often go unnoticed by owners. In some cases, they observe a change, but a routine examination does not reveal any lesions, and the animal lives with moderate, persistent, and almost constant pain.

In aging cats, pain becomes increasingly common, and practitioners are accustomed to seeing these animals become more irritable or less tolerant of certain constraints. Owners also tend to become more attentive. However, this progression is expected, understood, and more easily managed than when it occurs in a young animal.

The study of the *pathological pathway* often allows for the suspicion of chronic pain (■ Table 14.1). Indeed, the behavioral disorders described in our nosography follow a predictable progression. Take, for example, aggression due to irritation in the context of fear toward strangers. If such aggression appears at the age of 2 in a cat that was previously very friendly, without any identifiable trauma, further investigation becomes essential.

In all cases, the sudden onset of aggression at an age when behavioral stability is expected or within a condition that is not supposed to evolve in this way leads the practitioner to suspect an underlying somatic cause, with pain being the primary hypothesis. In some cases, pain may also manifest as an increased tendency to seek isolation. Here as well, it is the occurrence of a behavioral change at an unexpected age or in an unusual context that draws the attention of both owners and veterinarians.

A cat in pain tries different strategies to relieve its discomfort. This may include a simple change in posture or withdrawal from interaction. Certain *compulsive behaviors* may emerge as a means to alleviate pain, particularly licking, spinning, or self-directed aggression. Licking initially provides relief, mainly through the release of endorphins, but over time, endorphin secretion diminishes, and the behavior is no longer soothing (Willemse 1994). However, stopping the behavior leads to increased discomfort, which sustains it despite the absence of positive effects (Tynes and Sinn 2014; Virga 2003).

Chronic pain activates the same pathways as chronic stress. Indeed, it is impossible for an animal to experience pain without also experiencing fear, and repeated fear leads to an anxious state. Moreover, pain amplifies all of the animal's vulnerabilities. For example, an animal with a moderate noise phobia will become sensitized much more quickly if it also suffers from chronic pain. All stressors accumulate. Fear adds to pain, the animal's adaptive capacities are exceeded, and the shift to a pathological state becomes inevitable. Once this sensitization occurs and an anxious pathological state is established, eliminating the painful cause alone is no longer sufficient to restore a normal adaptive state.

The objective for the practitioner is not necessarily to determine the exact origin of the pain but rather to *identify the probable presence of a painful condition.*

The existence of such a condition should be considered not only when there is an inconsistency in the pathological pathway but also *when a proposed psychiatric treatment proves ineffective*. In such cases, the involvement of chronic pain hindering progress should be suspected, and additional examinations should be prescribed.

As soon as the practitioner suspects a painful process, a *therapeutic trial* with analgesics for 2–6 weeks can be conducted before introducing the medication indicated for the behavioral disorder (Mills et al. 2020). This approach remains valid even if complementary examinations yield negative results. However, it should be noted that pain can induce a pathological state that persists beyond the resolution of the painful condition.

Table 14.1 Criteria supporting pain involvement

Symptom considered	Criteria supporting pain involvement
Interspecific irritation-related aggression	The cat is well-socialized. Sudden onset or worsening No identifiable traumatic event The cat is aggressive toward people with whom it was previously friendly The sequence does not appear predictable in the described context The aggression occurs in the middle of what seems to be a positive interaction
Intraspecific irritation-related aggression	Sudden onset of aggression between cats in the same household, unrelated to puberty No identifiable traumatic event The sequence does not appear predictable in the described context The aggression occurs during play or a seemingly positive interaction The aggression is directed toward a cat with whom the level of familiarity was previously satisfactory
Increased isolation seeking	This sudden search for isolation appears in a cat that was previously well-acclimated to all household members
Compulsive behavior	Sudden onset of a compulsive behavior Identification of a painful organic condition, possibly acute, in the weeks preceding the onset of the compulsive behavior The behavior occurs irregularly and can sometimes be interrupted.
Fear behavior without an identifiable object or inappropriate to the context	Sudden onset The cat appears inhibited and seeks to escape or hide in situations that previously did not trigger similar behaviors The cat develops new or exacerbated phobias that seem spontaneous and unrelated to any identifiable pathological pathway The animal exhibits increasing emotional anticipation

14

14.1.2 The Consequences of Psychiatric Disorders on Pain

Anxious or compulsive states can lead to inflammation, such as idiopathic cystitis, chronic digestive diseases, or skin inflammation. These conditions cause pain, further complicating the management of the animal.

It is also important to consider cases in which a *healthcare phobia* has developed. Whether or not aggressive manifestations are present, this phobia complicates medical care and may even render it impossible. For this reason, practitioners treating behavioral disorders must anticipate potential future treatments and work proactively using medical training techniques (see ▶ Chap. 30, ▶ Sect. 30.2.2) to facilitate care under better conditions.

Cats in a *phobic or anxious state* tend to become *sensitized* and *anticipate* threats more quickly. As previously mentioned, the stress and pain pathways closely overlap, and these cats may experience *increased pain sensitivity,* known as *hyperalgesia*. When they sustain injuries that would typically cause mild pain, they often exhibit exaggerated withdrawal reactions or aggression, which seem disproportionate given the context.

Important Point of View

Clinical Pain Assessment in Cats

Cats are known for their lack of overt expression of signs related to organic diseases, particularly pain. A cat in pain exhibits altered *postures* and *facial expressions*, often in a subtle manner. To improve the detection of acute and chronic pain, numerous assessment tools have been developed. However, no consensus has yet been reached regarding their sensitivity and specificity (Hernandez-Avalos et al. 2019; Steagall and Monteiro 2019; Merola and Mills 2016; Holden et al. 2014).

Recent studies have attempted to develop measurement tools based on the automatic decoding of cat facial images (Martvel et al. 2024; Feighelstein et al. 2022, 2023).

Owners, who are highly familiar with their cats, often report that their pet does not have its "usual expression." It is important to trust these observations, as they are frequently accurate. Indeed, the cat's pain grimace is widely used to assess discomfort (Robinson and Steagall 2024; Evangelista and Steagall 2021; Steagall and Monteiro 2019; Evangelista et al. 2019). A Feline Grimace Scale (FGS) has been developed and tested under various conditions (Watanabe et al. 2020; Evangelista et al. 2020), including smartphone applications (Steagall 2019; Steagall et al. 2023).

Summary of ▶ Sect. 14.1

Acute pain causes sudden behavioral changes and is often noticed by owners. In contrast, chronic pain is more insidious, and the animal's behavioral adaptations to cope with it are subtle.

Any organic condition that causes chronic pain can induce anxious or compulsive pathological states. Other symptoms, such as the sudden onset or increase of irritation-related aggression in unexpected contexts, should prompt consideration of pain as a contributing factor in the clinical picture. In practice, behavioral semiological analysis can reveal inconsistencies in the pathological pathway of the behavioral disorder in question. In such cases, pain should be carefully investigated.

Pain and emotional dysregulation are closely linked, and the presence of one should prompt the practitioner to pay special attention to the other.

Pain assessment requires a trained eye, careful observation of postures, facial expressions, and even slight behavioral changes in the animal.

14.2 Dermatology

In human medicine, the connection between dermatology and psychiatry has long been established. Psoriasis is perhaps the best-known dermatological condition for its psychological impact (Moon et al. 2013), but any pruritic and chronic dermatological disorder has psychological consequences (Carr et al. 2014). Atopic dermatitis is also frequently associated with psychological effects, leading to a reduced quality of life (Wittkowski et al. 2004). Moreover, there is a significant relationship between stress levels and skin manifestations in patients suffering from atopic dermatitis or psoriasis (Arndt et al. 2008; Verhoeven et al. 2009).

The increase in endogenous glucocorticoids in any stressful situation in humans results in the inhibition of lipid synthesis. This inhibition leads to a reduction in the production and secretion of lamellar bodies from keratinocytes, directly causing a defective skin barrier (Choi et al. 2006).

In cats, as in humans, dermatological diseases can lead to behavioral dysregulation due to the discomfort and pain they cause. Conversely, certain psychiatric disorders can have dermatological manifestations.

If management does not consider both aspects, there is a real risk of self-aggravation. The psychological imbalances induced by dermatological disease can exacerbate skin manifestations, complicating the course of the condition. This bidirectional interaction reinforces both dermatological symptoms and psychological distress, creating a vicious cycle that is difficult to break.

Clinical manifestations such as miliary dermatitis, self-induced alopecia, head and neck pruritus, or the eosinophilic granuloma complex can all have psychological repercussions due to their chronic nature and associated discomfort. The differential diagnosis for these conditions includes flea allergy dermatitis, food allergy, and feline cutaneous atopic syndrome (Santoro et al. 2021).

Some licking behaviors may also reflect localized pain, such as cystitis or back pain, which can not only cause skin lesions but also worsen the discomfort associated with the underlying condition.

Cats suffering from inhibited anxious states may exhibit licking behavior. This is most often a *substitutive activity* that helps soothe the animal. Some *compulsive states* can lead directly to excessive licking or result from the progression of a substitutive activity into a *stereotypy*, meaning a repetitive behavior that has lost any adaptive function and occurs outside its original triggering context. However, some studies suggest that excessive licking of strictly behavioral origin appears to be relatively uncommon in cats (Waisglass et al. 2006).

Self-inflicted tail lesions require special attention. They may indicate a psychiatric disorder associated with a *compulsive* or *derealization state.* However, they may also be a clinical expression of *feline hyperesthesia syndrome*.

Important Point of View

Feline Hyperesthesia

Feline hyperesthesia is considered a rare behavioral and neurological disorder in cats, characterized by sudden episodes of cutaneous hypersensitivity and unusual behaviors. It is sometimes referred to as "rolling skin syndrome" due to the distinctive rippling of the skin observed on the cat's back during episodes. It is also sometimes called external neuritis or atypical neurodermatitis (Tuttle 1980).

Feline hyperesthesia manifests through compulsive behaviors such as tail chasing, excessive biting or licking of the back, flanks, and tail, muscle spasms, unusual vocalizations, uncontrolled jumping, and hallucinations (Amengual Batle et al. 2019). Involuntary rippling or contractions of the skin are observed, sometimes accompanied by twitching, pupil dilation, and increased sensitivity to touch. This behavior, which appears to correspond to cutaneous or deep dorsolumbar discomfort, intermittently occupies the cat and can become invasive. Between full episodes, subtle lumbar ripples may sometimes be observed without triggering a complete movement. Similarly, some episodes are limited to a few tail flicks without leading to an attack.

It should be noted that anxious cats often exhibit continuous tail flicking of varying intensity. This tail movement should be considered part of the signs of feline hyperesthesia.

The etiology of feline hyperesthesia remains unknown, and it may encompass various dermatological, behavioral, orthopedic, and neurological conditions. In severe cases, cats may mutilate their tails to the point of requiring surgical intervention.

Feline hyperesthesia thus includes various symptoms, and a thorough differential diagnosis must consider all possible causes. Psychiatric causes may include anxious states, impulsive states, compulsive states, or dissociative states. Neurological causes primarily involve focal epilepsy (Gómez Álvarez and Soler Arias 2021) or neuropathic pain. Dermatological causes include DHPP, food allergies, and feline atopic skin syndrome. Finally, pain, particularly orthopedic but also abdominal pain, must be considered.

In the most severe cases, cats affected by this syndrome may inflict serious injuries on themselves, even mutilating their tails, requiring immediate medical intervention. Amputation is almost never the right solution, and in any case, should never be performed without concurrent behavioral management. *Neurological disorders* other than feline hyperesthesia syndrome, as well as pain, must also be included in the differential diagnosis.

An individual's dermatological vulnerability can promote the development of a skin condition that impacts their psychological balance. Conversely, animals affected by psychiatric imbalances may inflict skin lesions on themselves, which in turn exacerbate their initial distress. Regardless of whether the initial cause is dermatological or psychiatric, a *cycle of self-aggravation* develops, making it difficult to resolve and requiring a comprehensive approach that integrates both dermatological and psychiatric aspects. A multimodal approach is therefore essential, considering both facets of the disease.

Finally, it should be noted that Elizabethan collars are *highly anxiety-inducing*. Their use should be limited and preferably accompanied by medication aimed at reducing the animal's stress and preventing behavioral deterioration.

Summary of ▶ Sect. 14.2

Dermatological conditions, due to the discomfort and pain they cause, frequently impact behavior. It is therefore essential to consider the psychological effects of skin lesions, particularly in the case of feline atopic syndrome.

Certain conditions, such as phobic, anxious, compulsive, or dissociative states, may manifest through excessive licking or self-inflicted skin injuries, further complicating the initial condition. Special attention should be given to self-inflicted injuries localized on a limb or the tail, as they may be associated with feline hyperesthesia syndrome.

Due to the close relationship between dermatology and psychological well-being, a phenomenon of self-aggravation is often observed. A combined dermatological and psychiatric approach is therefore often essential for effectively managing chronic or recurrent dermatological conditions.

The use of the Elizabethan collar should be minimized as much as possible. A behavioral approach incorporating anxiolytic medications, appropriate pain management, and care that is as gentle and stress-free as possible provides valuable support in managing chronic dermatological care.

14.3 Neurology

Neurology and psychiatry are closely related disciplines as they both focus on the same organ. Several conditions lie at the intersection of these two fields (Amadei et al. 2021). In cases of neurological disorders, behavioral changes are often the primary reason for consultation. Sometimes, these changes are intermittent, and the cat appears perfectly normal during the appointment. The use of video recordings to observe abnormal behavior, combined with clinical examination and imaging, are valuable tools for determining whether the condition has a neurological, psychiatric, or combined origin.

Many symptoms can serve as warning signs of neurological or psychiatric disorders, and the differential diagnosis often includes both aspects, as summarized in ◘ Table 14.3.

The relevant symptoms include *aggression due to irritation* or *idiopathic, generalized seizures, psychomotor seizures* or partial seizures (such as tail chasing or circling), *spatial* or *temporal disorientation*, and *hallucinations*. Similarly, the sudden onset of abnormal behaviors that cannot be linked to a pathological state or explained by a behavioral disorder should prompt the search for a neurological cause.

Whenever these symptoms are present, the practitioner's general examination should include a *neurological assessment* **(see box on the next page).** This can be performed without specialized equipment as part of a routine clinical examination. Genetic vulnerabilities are important considerations in both neurology and psychiatry. There are numerous breed predispositions. For example, Siamese cats are predisposed to hydrocephalus (Casal 2012; Silson and Robinson 1969), Birman cats to hypomyelination, which causes tremors in kittens (Dewey et al. 2016), Bengal cats to congenital neuropathies (Bensfield et al. 2011), Maine Coons to congenital cerebellar ataxia (Scott et al. 2022), and Burmese cats to feline orofacial pain syndrome (Rusbridge et al. 2010).

In young animals, malformations such as hydrocephalus, as well as metabolic and inflammatory conditions, are the most common, whereas in older animals, neoplastic and degenerative diseases are more prevalent.

14

Thus, the duration of symptoms, their onset, progression over time, and response to previous treatments are all important factors to consider. The presence of pain or systemic signs should be investigated. For all episodic symptoms occurring in the form of seizures, video recordings are highly useful.

Once the various examination findings have been gathered, it is possible to refine the localization of the lesion. In the context of a psychiatric consultation, the goal is not to establish a neurological diagnosis but to support the neurological hypothesis whenever the clinical neurological examination is abnormal. In such cases, this examination should lead to a referral for a specialized consultation or specific additional tests, such as a computed tomography (CT) scan or MRI.

Important Point of View

Neurological Examination

The examination of the nervous system consists of several steps:

- *Observation of the animal from a distance,* allowing it to move freely to assess its level of alertness (alert, disoriented, comatose, etc.), behavior (circling, sudden aggression, loss of spatial awareness, hemispatial neglect syndrome), posture (asymmetries, limb positioning on the ground, etc.), gait (limping, ataxia, paresis, etc.), vision (responses to invisible stimuli), and the presence of abnormal movements (myoclonus, tremors, etc.).
- *Postural reactions* to evaluate proprioception, which, when abnormal, indicates nervous system impairment.
- *Spinal reflexes,* including patellar, withdrawal, perineal, and cutaneous reflexes. The response can be normal, increased (indicating an upper motor neuron lesion), decreased, or absent (indicating a lower motor neuron lesion). A bilateral decrease suggests a spinal cord lesion, whereas a unilateral decrease indicates a lesion in a peripheral or highly lateralized spinal nerve.
- *Sensory evaluation* to detect pain, as stimulation of the affected area may trigger aggression, tachycardia, muscle contraction, vocalization, or mydriasis. This evaluation also includes assessing deep pain perception (nociception), which is tested on a limb that is already flexed to distinguish it from the withdrawal reflex.
- *Examination of cranial nerves* (see ◘ Table 14.2).

Knowing the typical progression of the most common neurological and psychiatric disorders allows the practitioner to *prioritize symptoms in order to establish a differential diagnosis* based on findings from the behavioral assessment and clinical examination (◘ Table 14.3).

Table 14.2 Cranial nerves, function, and clinical tests

Cranial nerve	Function	Test
Olfactory nerve (I)	Smell	Rarely performed in clinical practice due to its complexity and the lack of standardized tests.
Optic nerve (II)	Vision	Menace response, which involves rapidly approaching a hand toward the animal's eye without touching it, provoking a reflexive blink
Oculomotor (III), Trochlear (IV), and Abducens (VI) nerves	Eye movements	Observation of the position of the eyeballs, presence of strabismus, and pupillary response to light (direct and consensual pupillary reflex)
Trigeminal nerve (V)	Facial sensation and mastication	Palpation of the masticatory muscles to detect atrophy and testing facial sensitivity by touching innervated areas, eliciting a reflexive blink
Facial nerve (VII)	Control of facial expression muscles, taste perception on the part of the tongue, innervation of salivary and lacrimal glands	Observation of facial symmetry, mobility of the ears, eyelids, and lips, as well as the blink reflex in response to a menace
Vestibulocochlear nerve (VIII)	Vestibular function	Observation of posture and balance, presence of nystagmus, and reaction to sounds for the cochlear component
Glossopharyngeal (IX) and Vagus (X) nerves	Swallowing, phonation, and autonomic function control	Observation of swallowing and the gag reflex by gently stimulating the back of the throat
Accessory nerve (XI)	Control of neck and shoulder muscles	Clinical evaluation of this nerve is limited, but neck muscle atrophy may suggest dysfunction.
Hypoglossal nerve (XII)	Motor control of the tongue	Observation of the tongue for atrophy, fasciculations, or deviations during extrusion

Generalized seizures result from excessive synchronous activity of neurons in the cerebral cortex. These seizures are rarely associated with a psychiatric condition. They typically last less than 5 min and are accompanied by a prodrome that signals the onset of the seizure, followed by a postictal recovery phase during which abnormal behavior gradually resolves. This recovery phase can last from a few minutes to several hours. Additionally, during generalized seizures, rhythmic muscle movements are often observed, along with autonomic signs such as ptyalism, urination, and mydriasis. Finally, there is almost always a complete absence of response to external stimuli.

When seizures have a psychiatric origin, they do not follow the same strict pattern and may last longer. They tend to disappear as suddenly as they begin, with little to no recovery phase. These episodes are primarily observed during panic attacks in paroxysmal anxious states, where the cat loses locomotor control due to overwhelming fear. There are no prodromes, and recovery is rapid once the episode ends.

Partial seizures, also known as *psychomotor seizures,* are more challenging to differentiate because they do not present in a strictly identical manner each time and can resemble behaviors observed in various psychiatric disorders. Unlike generalized seizures, their cause remains unknown, but they appear to occur when the brain regions responsible for movement planning and regulation temporarily malfunction, likely due to highly localized

Table 14.3 Elements of differential diagnosis between neurology and psychiatry based on symptoms frequently observed in both disciplines

Symptom	Neurological or neurologically expressed conditions	Psychiatric disorder	Clinical tools for differentiation
Aggressions	Brain tumor Hepatic encephalopathy Hydrocephalus Pain	Anxious states Derealization states Phobic states Dysthymia in this cell	Consistency of sequences and pathological pathway with the overall clinical picture *Additional tests:* MRI, biochemical panel
Generalized seizures	Epilepsy Hypoglycemia Insulinoma Toxins	Paroxysmal anxious state	Age, onset, progression *Additional tests:* biochemical panel
Partial seizures (tail chasing or attacking the immediate environment)	Temporal lobe epilepsy Syringomyelia Hydrocephalus Brain tumor Cauda equina syndrome Phantom pain after amputation Hyperesthesia Toxins	Productive anxious state Impulsive state Compulsive state Dissociative state	Pathological pathway: onset, frequency, progression *Additional tests:* biochemical panel, MRI, specialized consultation
Spatial or temporal disorientation	Brain tumor Toxins	Cognitive dysregulation in elderly cats	*Additional tests:* MRI, biochemical panel
Hallucinations	Abnormal vision or blindness Toxins Brain tumor Encephalitis Anesthetic toxicity	Dissociative syndrome	Pathological pathway: onset, frequency, progression *Additional tests:* MRI, specialized consultation

epileptic activity. These seizures do not affect the cerebral cortex or the autonomic nervous system, resulting in *preserved environmental responsiveness* and an *absence of autonomic symptoms.*

In some cases, these seizures are triggered by identifiable external stimuli, making it even more difficult to distinguish between a neurological and a psychiatric etiology. The animal exhibits repetitive movements that appear difficult to control but remains aware of its surroundings. A wide range of behaviors may be observed, including tail chasing, attacking nearby objects, or, more generally, abnormal gait and postures that often prevent the cat from moving normally. These episodes can last from a few minutes to several hours, with a frequency ranging from monthly occurrences to multiple episodes per day. Each seizure resembles the previous one but may vary in intensity and duration.

The return to normal is immediate, with no postictal phase. During the episode, the cat remains conscious of its environment and may attempt to interact with it despite exhibiting abnormal movements.

Many psychiatric disorders can produce this type of psychomotor seizure. The most frequently described symptoms include tail chasing or attacking nearby objects without a clear target or apparent source of irritation. The cat may sometimes appear to be fleeing from pain while simultaneously displaying threatening behaviors, making pain assessment a crucial part of the differential diagnosis.

Regardless of the specific symptom, it is the pathological pathway identified during the behavioral assessment that will guide the practitioner toward one diagnosis or another. For example, tail chasing or attacking a limb is commonly reported but takes on very different meanings depending on the overall clinical picture. If clinical examination, including the neurological assessment, is normal, these self-directed attacks may result from an impulsive state associated with HSHA syndrome if they have been present since early life and are accompanied by other symptoms confirming impulsivity. They may also represent a compensatory (substitutive) behavior, occurring only occasionally in response to frustration (such as an unattainable prey behind glass) or during heightened excitement during play, suggesting a contextual impulsive state. However, if the behavior appears suddenly in an adult cat that seems disconnected from its surroundings or reacts aggressively when interrupted, other hypotheses, such as dissociative syndrome, should be considered.

The challenge of behavioral assessment lies precisely in moving from a non-specific clinical sign (such as tail chasing) to a meaningful symptom (see ► Chap. 15).

Partial seizures are typically positioned at the *intersection of neurology and psychiatry*, and it is likely that both forms exist, with clinical manifestations that are sometimes nearly indistinguishable. Anxious, impulsive, compulsive, and dissociative states should all be included in the differential diagnosis. The practitioner must rely on anamnesis, symptom progression, and the contexts in which the episodes occur to determine whether the condition is more likely psychiatric or neurological.

Some forms respond better to neurological treatments such as antiepileptic drugs, while others seem to be more effectively managed with SSRIs.

Hallucinations are interpretations of perceived reality. An animal experiencing these abnormal perceptions appears to react to a stimulus that we cannot perceive. From our perspective, its response is therefore inappropriate to the environment. The symptoms resemble those of partial seizures and the psychomotor seizures described earlier. It is often difficult to determine whether the episode is triggered by a possible hallucination, meaning the animal reacts to a stimulus that its sensory system perceives but does not actually exist, or whether the origin is internal, whether neurological or psychiatric.

Brain tumors cause a wide range of clinical signs but are more common in adult or elderly animals. Clinical signs vary significantly depending on the tumor's location, but their onset and progression are typically acute. The pathological pathway identified during behavioral assessment does not explain the timing of onset or the rapid evolution of symptoms. In such cases, the preferred diagnostic test is a CT scan.

Aggression can occur in any condition that causes pain. Neurological disorders can be highly painful, and the resulting aggression can sometimes be dramatic. As with psychomotor seizures, the overall clinical picture and the coherence of the pathological pathway will help the practitioner prioritize diagnostic hypotheses.

In some cases, there is *comorbidity* between a neurological and a psychiatric disorder. In such situations, it is essential to address both aspects, making collaboration between specialists crucial. The consulted practitioner may be a neurologist or a psychiatrist, but in either case, their hypotheses may not fully explain the entire clinical picture. Comorbidities may be suspected when neurological abnormalities are identified during the examination, but they do not account for all observed symptoms.

For example, a 4-year-old cat with severe social phobia may develop encephalitis, but the newly emerging symptoms would not explain the presence of long-standing aggression toward strangers due to irritation.

Summary of ▶ Sect. 14.3

Many symptoms commonly encountered in neurology and psychiatry overlap between the two disciplines. Among them, the most frequent are generalized seizures and, more notably, partial seizures, including hallucinations and self-directed aggression.

When these symptoms are observed during a psychiatric consultation, a full neurological examination should be performed. If abnormalities are detected, additional diagnostic tests (MRI, CT scan, biochemical panels, etc.) should be conducted, and referral to a neurology specialist is recommended.

Several conditions lie at the intersection of neurology and psychiatry, and their clinical manifestations do not always allow for clear differentiation. When complementary tests yield inconclusive results, a therapeutic trial may be necessary to rule out one hypothesis or the other. Finally, comorbidities can occur, and when neither the neurological nor the psychiatric evaluation fully explains the clinical picture, the practitioner must address both aspects through close collaboration with a neurologist.

14.4 Nutrition

Feeding behavior plays a crucial role in behavioral assessment. During consultations, discussions often lead to nutritional considerations, and practitioners must have a basic understanding of this discipline to provide informed guidance to pet owners.

First, it is important to recall that cats are predators of small prey and naturally consume *multiple meals throughout the day*. The palatability of these prey items is moderate, and capturing them requires effort. As motivation decreases when the animal is satiated, hunting efficiency declines, establishing a natural balance that prevents overconsumption. However, cats are physiologically capable of overeating while maintaining agility. This ability, which allows for fat storage, helps them cope with periods of food scarcity.

Imposing only one or two meals per day may lead to both behavioral and physiological disorders, such as *obesity*. A cat deprived of frequent meals may overconsume when food is available to compensate for its frustration. This significantly increases the risk of excess weight, and an unsatisfied cat may spend the entire day begging for food. Owners often find it difficult to ignore these demands and end up offering additional meals. These are typically small portions of kibble or commercial pet food, which cats rarely refuse due to their high palatability.

Pet food manufacturers invest heavily in creating highly appetizing products, and owners tend to choose food that their cats prefer. Additionally, marketing campaigns strongly encourage pet owners to express their love by offering increasingly flavorful meals.

It would be beneficial to find a type of kibble that is low in energy and low in palatability, so that the cat consumes it only out of *necessity*. However, this is not the goal of the pet food industry. Another option is to make food intake more challenging by using feeding toys or "complex" bowls that require the cat to "capture" its kibble. Diet kibble is often less palatable and lower in energy, making it more suitable for free-choice feeding.

Additionally, the issue of *hunting activity* must be considered. If a cat is unable to express its natural hunting behavior, such as stalking with prolonged periods of observation, it may compensate by engaging in the closest available behavior: food consumption. This explains why indoor cats have a *strong tendency* to overconsume food.

Beyond meeting energy requirements, the *qualitative balance* of the diet is also important. While a specialized nutrition consultation is not necessary, the practitioner should at least ensure that protein requirements are adequately met, as cats are obligate carnivores.

Cats require high hepatic enzymatic activity to break down proteins and maintain a constant energy supply from amino acids. Maintenance protein requirements for adult cats are approximately 5–6 g of protein per kg of body weight per day. Growing kittens, pregnant, or lactating females may require up to 9–10 g of protein per kg per day.

Additionally, dietary proteins must provide *essential amino acids* that cats cannot synthesize on their own: taurine for cardiac and retinal health, arginine for ammonia metabolism, methionine and cysteine for coat health, and lysine for growth and immune function.

Cats have a minimum protein requirement of 20–30% of total metabolic energy, which is two to three times higher than that of dogs. Unlike omnivores, they cannot adjust their metabolism to reduce protein breakdown when dietary intake is low. A diet rich in high-quality animal proteins is therefore essential for their health and well-being.

Cats are unique among mammals in their *fatty acid metabolism*. They can synthesize non-essential saturated and monounsaturated fatty acids de novo from glucose or amino acids through a common precursor, acetyl-CoA. The enzymes regulating these reactions are active when diets high in carbohydrates and low in lipids are consumed. However, to support the high concentrations of docosahexaenoic acid (DHA) required in retinal and neuronal tissues during development, kittens need an exogenous source of DHA (such as a DHA-enriched diet), as precursor conversion may be insufficient to meet these needs (Bauer 2006).

Regarding *starch digestion*, cats do not produce *salivary amylase* and, although they produce pancreatic amylase, their ability to digest carbohydrates remains limited due to the low production of this enzyme. This reflects their adaptation as obligate carnivores, requiring a diet primarily composed of animal proteins and fats.

In summary, a cat's diet should contain 20–30% protein, provide the essential amino acids that cats cannot synthesize, and be supplemented with unsaturated fatty acids, particularly during kitten development. Finally, energy requirements should be adjusted according to the animal's lifestyle (see box below). If needed, nutritional guidelines established by FEDIAF offer further analysis of kibble formulations (FEDIAF EuropeanPetFood 2024) (▣ Table 14.4).

Important Point of View

Calculation of Energy and Protein Requirements and Dietary Ration

Without delving into the detailed calculations of homemade rations as a nutritionist would, the practitioner should be able to quickly estimate a cat's energy and protein needs.

First, it is important to determine the animal's ideal weight and body condition score (BCS). The BCS is assessed on a scale of 1–9 or 1–5 **(see table on the next page).** This calculation depends on the nutritional score (Laflamme 1997), which is derived using the following formula established by the National Research Council (National Research Council 2006):

$$\text{Ideal Weight}\,(\text{IW}) = \text{Current Weight} \times 100\,/\,\left(100 + (\text{BCS} \times 10)\right)$$

Resting Energy Requirements (RER) are calculated for a maintenance-level animal using the formula:

$$\text{RER} = 70 \times \text{Ideal Weight}^{0.75}$$

From this formula, the Maintenance Energy Requirements (MER) of a cat can be determined based on activity level and physiological status:

- For an indoor or sedentary cat: MER = 1.2 × RER
- For a cat with outdoor access or moderate activity: MER = 1.4 × RER
- For a highly active cat: MER = 1.6 × RER
- For a growing kitten: MER = 2 × RER
- During gestation: MER = 1.3 × RER to 1.5 × RER (progressively increasing until the end of pregnancy)
- During lactation: MER = 2 × RER to 6 × RER (depending on the number of kittens)

Table 14.4 Body Condition Score (BCS) chart for cats: comparison of 9-point and 5-point scales

BCS 9-point scale	BCS 5-point scale	CHAT
1 	1	**Extremely emaciated with very poor body condition** Ribs, spine, and hips very visible. No fat, severe muscle loss. Severely tucked abdomen
2 		**Very thin** Ribs and spine easily palpable with very little fat. Prominent hips. Tucked abdomen
3 	2	**Thin** Ribs easily palpable with a thin fat layer. Spine and hips visible. Slightly tucked abdomen
4 		**Moderately thin** Ribs palpable with a very thin fat layer. Spine still noticeable. Less prominent hips. Slightly tucked abdomen
5 	3	**Ideal** Ribs palpable under a thin layer of fat. Spine smooth and not protruding. Slightly tucked abdomen. Visible waist when viewed from above
6 		**Slightly overweight** Ribs palpable under a thin fat layer. Smooth, non-prominent spine. Slightly tucked abdomen. Visible waist from above
7 	4	**Overweight** Ribs difficult to feel under a fat layer. Spine and hips barely palpable. Rounded abdomen. Waist barely visible or absent
8 		**Obesity** Ribs and spine difficult to feel under a thick fat layer. Well-rounded abdomen. Waist absent. Visible fat deposits
9 	5	**Severe obesity** Ribs impossible to feel. Spine and hips covered by fat. Large abdomen with prominent fat deposits. Waist completely absent

- For a neutered cat: MER = 1.0 × RER to 1.2 × RER

Neutered cats have significantly lower energy requirements, yet their needs for lipids, proteins, minerals, fiber, and vitamins remain unchanged. Ensuring adequate protein intake is a crucial first step. A protein deficiency may sometimes be identified before analyzing the amino acid composition of the diet, but this does not eliminate the need for further investigation into amino acid balance.

A mismatch between diet and external fat and muscle mass, whether an overweight or underweight animal despite an appropriate ration, should prompt the practitioner to investigate potential *internal causes*. These may include chronic digestive disorders or hormonal conditions such as hyperthyroidism or diabetes, as well as behavioral disorders.

However, appetite disturbances can have a *behavioral origin*. Cases of hyperphagia, or excessive food consumption, may be a symptom of *impulsive or compulsive states,* in which the cat struggles to achieve a sense of satiety. Certain *inhibited anxious states* can also lead to overeating, where food intake becomes a *substitution activity*. Pica can also be observed in these pathological states **(see box below),** and the need for digestive surgery following the ingestion of foreign objects should always be followed by a *postoperative behavioral assessment*.

Conversely, *anorexia* can be observed in the acute form of depressive states and is accompanied by a complete cessation of all behaviors (see ▶ Chap. 7, ▶ Sect. 7.3.1). *Dysorexia* may occur in chronic depressive states or generalized anxiety, particularly when associated with dyspepsia.

Important Point of View

Foreign Body Ingestion in Cats

Surgical interventions for obstruction due to foreign body ingestion are common in veterinary medicine. When they occur, particularly in cats, they should prompt further investigation. The ingestion of non-food objects, or pica, is an abnormal behavior. Consuming plastic, fabric, or elastic is always pathological.

These ingestions are frequently observed in cats with impulsive or compulsive states and are associated with disorders such as HSHA syndrome as well as addictive behaviors. Some of these behaviors appear to have a genetic basis, as seen in wool-eating cats, which are overrepresented in Asian breeds (Demontigny-Bédard et al. 2016; Borns-Weil et al. 2015; Bradshaw et al. 1997). In older cats, such behaviors may also emerge. Anxious states can also trigger these behaviors, as the cat attempts to find relief in some way. In such cases, they are substitution behaviors, most commonly characterized by fabric sucking or ingestion.

Summary of ▶ Sect. 14.4

Nutrition is closely linked to behavior.

On one hand, chronic nutritional imbalances—including inadequate energy intake, protein balance, amino acid composition, and various deficiencies—can have significant behavioral consequences.

On the other hand, behavior can strongly influence food intake. Acute depressive states may lead to anorexia, while impulsive, compulsive, and inhibited anxious states can result in a lack of satiety and potential pica. In chronic depressive states and generalized anxiety, appetite may become erratic, often associated with digestive disorders (see ▶ Sect. 14.5).

14.5 Internal Medicine

The objective of this section is not to provide an exhaustive list of all conditions with potential psychological repercussions, as the subject is vast. More generally, all organic diseases place stress on the body, and many cause pain. In this sense, any organic disease can have psychological consequences and lead to pathological anxious or depressive states.

In this chapter, we will focus on the diseases that seem most relevant to highlight in veterinary psychiatry: lower urinary tract diseases, digestive disorders, hypertension, and metabolic diseases.

Endocrinology, due to its importance in veterinary psychiatry, will be addressed in a dedicated section.

14.5.1 Lower Urinary Tract Diseases

The warning signs of lower urinary tract disease include *dysuria* and *pollakiuria*, often accompanied by *urine soiling*. When urine elimination occurs outside the litter box, the clinician must first rule out any infectious or inflammatory process. Pain or discomfort during urination may lead the cat to seek an alternative elimination surface, resulting in urine soiling.

Urinary tract infections, particularly those affecting the bladder and urethra, cause acute pain, which can lead to litter box aversion and inappropriate elimination.

From a psychiatric perspective, anxious cats are predisposed to developing cystitis **(see Feline idiopathic cystitis box), which**, in turn, can lead to urine elimination outside the litter box.

It is important to distinguish between urine elimination and territorial marking based on behavioral sequences and context (see ► Chap. 15, Sect. 15.1.4).

Important Point of View

Feline Idiopathic Cystitis

Feline idiopathic cystitis, also known as Pandora syndrome, encompasses urinary disorders and systemic manifestations associated with autonomic nervous system hyperreactivity (Jones et al. 2021).

This condition is not limited to the bladder but involves a multi-system dysregulation affecting the urinary tract, nervous system, immune system, and behavior.

Studies have shown that in animals suffering from this syndrome, the bladder wall is abnormally permeable (Parsons 2007; Lavelle et al. 2000), allowing irritating substances present in the urine to cause local inflammation.

Affected cats also exhibit autonomic nervous system hyperreactivity, which promotes stress emergence and increases bladder pain perception. Environmental stressors such as changes in the environment, low-stimulation settings, conflicts between cats, or disrupted routines—combined with low water intake, which can worsen urinary disorders—are the primary risk factors (Buffington et al. 2006a, b; Lekcharoensuk et al. 2001).

An anxious state is a major risk factor.

The symptoms of feline idiopathic cystitis include urinary signs—pollakiuria, hematuria, dysuria, inappropriate urination; signs of anxiety—aggression, tendency to isolate, and urine marking; as well as multi-systemic manifestations such as gastrointestinal issues, respiratory diseases, or skin conditions, often associated with stress periods.

Diagnosis is made by excluding other causes of urinary disorders (infections, urolithiasis, tumors) and requires, at a minimum, urinalysis and lower urinary tract ultrasound.

Management is primarily behavioral, with the use of psychotropic medication if a pathological anxious state is diagnosed. Litter box accessibility should be optimized at the start of treatment, especially if they have covers or doors. Environmental enrichment should be considered if the setting is too impoverished. The use of pheromones and behavioral therapy is recommended to eliminate positive punishment, establish stable routines, and promote positive interactions (Buffington et al. 2006a, b).

Additional measures include switching to wet food, which should be offered in multiple small meals to increase water intake and reduce urinary concentration. The use of analgesics (e.g., buprenorphine) and antispasmodics is also necessary for pain relief.

14

14.5.2 Digestive Diseases

Acute and chronic digestive diseases are the primary causes of fecal soiling and should be investigated first. A coproscopy can be performed to detect viral (FIP, FIV, coronavirus), bacterial (*E. coli*, *Salmonella*), or parasitic (coccidiosis, giardiasis) infectious agents. Additional blood tests and imaging may be required in cases of constipation or diarrhea to assess underlying conditions such as kidney disease, certain immune or oncological disorders such as digestive lymphomas, hyperthyroidism, or structural pelvic abnormalities (Carney et al. 2014).

However, digestive disorders can also be symptomatic of behavioral disorders. The link between anxiety and gastroenterology is well-established in human medicine (Byrne et al. 2017), and extensive research now focuses on the causes and consequences of the microbiota on anxious and depressive states (Foster and Neufeld 2013). Recent studies suggest that the microbiota may have an indirect influence on the development of behavioral disorders, although feline-specific research remains limited (Gorzelanna and Miszczak 2024; Hernandez et al. 2022).

Dyspepsia or chronic diarrhea can therefore be associated with anxiety disorders. One of the first signs of improvement in these cases is a return to more regular digestion, with a partial or complete resolution of gastroenteric disturbances and the associated fecal soiling. The use of a probiotic/prebiotic mixture could help stabilize the microbiota in affected cats, but the level of evidence remains limited.

Newly adopted kittens at 2 or 3 months of age are generally clean. Fecal soiling at this stage may result from early separation from the mother, as the kitten may not have learned litter box use through imitation (Bradshaw and Ellis 2016). In some cases, it can also be a sign of congenital neurological diseases or malformations, such as congenital hydrocephalus, cerebellar hypoplasia, or feline dysautonomia (Camps et al. 2015).

Chronic digestive inflammation is a good example of an organic disorder that can complicate pre-existing behavioral vulnerabilities: a cat may never regain proper elimination habits even after its digestive condition is controlled. Anticipatory pain at each elimination can lead to litter box aversion (see ► Chap. 29).

Behavioral disorders often act as a vulnerability, while organic diseases serve as triggers. A multimodal approach is therefore essential to achieve recovery.

14.5.3 Feline Hypertension

Feline hypertension is a common condition in older cats, often diagnosed in association with other diseases such as chronic kidney disease and hyperthyroidism (referred to as secondary hypertension), although some cases of primary hypertension have also been reported.

This disease can lead to visual disturbances or even vision loss due to hypertensive retinopathy. Visual impairment can have behavioral repercussions, such as disorientation or unexplained aggression. Cerebral functions are also highly sensitive to fluctuations in vascular pressure, and hypertension may be a contributing factor to certain depressive states. Cat owners frequently report an improvement in behavioral signs related to feline geriatric depression after antihypertensive treatment (Taylor et al. 2017).

Blood pressure measurement should be considered in cases of depressive signs or generalized inhibited anxiety in older cats, along with a retinal examination, as part of the differential diagnosis process.

14.5.4 Metabolic Diseases

Metabolic diseases can lead to behavioral changes.

Hepatic encephalopathy may cause isolated episodes with subtle behavioral changes. In cases of episodic and moderate behavioral alterations, such as tremors, aggression, or disorientation, where no clear cause is identified, a fasting blood glucose test and postprandial bile acid measurement should be performed.

Hypoglycemia or *uremia* can also lead to behavioral changes, including social withdrawal or even epileptiform seizures. These conditions should be distinguished from pathological depressive or anxious states through a comprehensive biochemical workup.

Polyuria-polydipsia (PUPD) may lead a cat to eliminate outside the litter box. It is essential to differentiate this type of urine soiling from feline idiopathic cystitis **(see *Feline idiopathic cystitis* box)**, anxiety-related marking (see ► Chap. 15, ► Sect. 15.1.4.4), or hormonal marking (see ► Sect. 14.7). The differential diagnosis for these conditions includes kidney failure, pyometra, diabetes, hypercalcemia, and diabetes insipidus.

A cat's daily water intake should not exceed 40–60 mL per kilogram of body weight for maintenance.

Summary of ▶ Sect. 14.5
Any organic disease can have repercussions on psychological state. Chronic illnesses, as they cause prolonged pain or stress, are particularly likely to induce anxious states, and their management should include a psychological component.

Lower urinary tract diseases can lead to litter box aversion due to anticipatory pain at the time of elimination. Feline idiopathic cystitis, on the other hand, has a significant behavioral etiology. Since inappropriate urination is one of the most common reasons for veterinary consultations, urinalysis and abdominal ultrasound are essential for differential diagnosis and treatment.

Intestinal and chronic gastric diseases are both a frequent cause and a consequence of pathological psychological states. Fecal soiling is often the primary reason for consultation. When inappropriate elimination occurs despite proper litter box setup and placement, digestive and urogenital conditions should be investigated. However, litter box aversion may persist even after the organic disease has been controlled or resolved.

Hypertension, particularly in older cats, should be considered in cases of sudden aggression, disorientation, or depressive signs.

Finally, metabolic diseases can cause significant behavioral changes, including loss of learned elimination habits. A routine blood test is recommended whenever psychiatric diagnostic hypotheses fail to fully explain all symptoms or their progression.

14.6 Reproduction

Sex hormones play a significant role in behavior (see ▶ Chap. 3, ▶ Sect. 3.2.3). In cats, the expression of sexual behaviors in males (vocalization, urine marking) and females (expressive heat cycles with vocalization, restlessness, and excessive seeking of contact) can quickly become disruptive. As a result, most cats not intended for breeding are neutered before the age of 1 year.

14

It is important to note that marking in intact males is a sexual behavior that emerges during adolescence, intended to attract females and deter other males (see ▶ Chap. 1, ▶ Sect. 1.3.2.1). This behavior is normal and driven by rising testosterone levels. Neutering eliminates *sexually* motivated *urine marking*; however, it will *not resolve anxiety-related urine marking*.

Urine marking is also a common behavior in females during heat. This behavior serves to signal their reproductive availability to nearby males.

During adolescence, cats establish their *territory* and may come into conflict with other cats in the household or neighborhood, even those they previously got along with as kittens. This is a normal stage in the maturation of the *territoriality* and *relationship axes* (see ▶ Chap. 4, ▶ Sect. 4.3.4). Depending on the context, the space available to cats, and individual vulnerabilities, this phase may be transient (a reactive state) with a quick return to balance and good social harmony, or it may progress into pathological states, leading to escalating aggression and a shift toward a pathological condition.

Some females, during their first heat cycles, may experience aggressive outbursts with sudden mood swings and manic episodes. This sexual dysthymia may persist even after sterilization, requiring control through polypharmaceutical treatment, most often combining fluoxetine or venlafaxine with a gabapentinoid (see ▶ Chap. 12, ▶ Sect. 12.2.4.1).

Important Point of View

Benefits of Sterilization

Beyond preventing sexually motivated urine marking in intact males and females in heat, sterilization also reduces roaming related to reproductive behavior, as well as the fights and injuries it causes. However, even a neutered cat remains a territorial animal. Access to the outdoors can still lead to confrontations and injuries due to territorial defense.

The placement of a deslorelin implant can be beneficial in intact breeding males to limit urine marking between reproductive periods or to avoid sterilization in individuals with high genetic potential.

Summary of ▶ Sect. 14.6

Sex hormones have a strong influence on feline behavior. Due to their often disruptive manifestations for owners, most companion cats are sterilized.

Adolescence and the rise of sex hormones coincide with the maturation of the territoriality and relationship behavior axes. Territorial or relational aggression can be influenced by hormones, but they also occur in neutered animals.

14.7 Endocrinology

Endocrine disorders have systemic repercussions, but their initial signs are often subtle and primarily behavioral.

The most common endocrine disorders in cats are *hyperthyroidism* and *diabetes*. Less frequently, *Cushing's syndrome*, *Addison's disease*, *hypothyroidism*, and *sexual endocrinopathies* may also be diagnosed. The cat's age and the speed of symptom onset should guide the clinician in investigating an endocrine disorder, especially when the pathological pathway observed in behavioral semiology is atypical.

In this section, we will discuss the primary symptoms of these endocrinopathies, with particular attention to hyperthyroidism, as it is the most frequent. Hyperthyroidism is included in the differential diagnosis of many feline behavioral disorders, especially in older cats.

14.7.1 Hyperthyroidism

Feline hyperthyroidism is a common endocrine disorder in older cats, characterized by excessive production of thyroid hormones T3 and T4. It is most often caused by benign hyperplasia of thyroid cells or, more rarely, by a thyroid adenoma (Mooney and Peterson 2023).

Hyperthyroidism significantly influences behavior due to the effects of excessive thyroid hormones on the central nervous system and metabolism. It creates a state in which the cat's metabolism and behaviors appear amplified.

Hyperthyroid cats are typically older than 7 years and exhibit increased appetite, often accompanied by weight loss. They may become obsessed with food, frequently begging or stealing food.

They are often more active or restless. Signs of hyperactivity and compulsive behaviors, such as excessive grooming, may be observed. Many hyperthyroid cats are also vocal, frequently and persistently meowing, especially at night, which is a common reason for veterinary consultation.

Excess thyroid hormones can cause irritability or sudden aggression toward other animals or humans, even in well-socialized or familiarized cats.

Affected cats may also experience a reduced normal sleep duration. Some hyperthyroid cats exhibit hypervigilance, reacting strongly to even minor environmental changes.

Social relationships can also be affected. Some cats become more clingy, seeking constant attention, while others may become distant, avoiding their usual interactions with their owner. In advanced stages, some cats may paradoxically become apathetic or lethargic (Mooney and Peterson 2023).

All the clinical signs described above can also suggest pathological anxious, impulsive, compulsive, depressive, or senile states. In any older cat presenting such a clinical picture, thyroid function should be checked.

A simple blood test measuring total T4 is sufficient for diagnosis (normal total T4 levels range from 10 to 50 nmol/L or 0.8 to 4.0 μg/dL). Careful palpation of the thyroid region may reveal a mass; however, as the feline thyroid gland can be diffusely affected, a negative palpation does not rule out the disease.

Medical treatment includes methimazole, thyroidectomy, or radioactive iodine therapy (I-131) (Carney et al. 2016). The management of any comorbid pathological condition should be part of the overall treatment plan.

14.7.2 Diabetes

Diabetes mellitus is a common endocrine disorder in cats, characterized by an absolute or relative insulin deficiency leading to chronic hyperglycemia. It is similar to type 2 diabetes in humans (Reusch 2014).

Cats over 7 years of age and neutered males are more vulnerable to this endocrinopathy. The etiology includes chronic pancreatitis, hyperthyroidism, acromegaly, and iatrogenic causes such as the administration of glucocorticoids or progestagens (Greco 2018).

Feline diabetes is primarily caused by peripheral insulin resistance, which leads to hyperglycemia and excessive stimulation of pancreatic beta cells. Over time, these cells become exhausted, reducing insulin secretion (Rand 2013).

The main causes of this resistance include:

- Obesity, a major risk factor that decreases insulin sensitivity.
- Sedentary lifestyle, which promotes fat accumulation.
- Genetic factors, as some breeds (Burmese, Siamese) are predisposed (Da Riz 2023).

Diabetes mellitus can have a significant impact on a *cat's behavior*.

This condition is accompanied by physical discomfort, often leading to lethargy or even apathy. In some cases, irritability may develop due to diabetic neuropathies or complications such as urinary tract infections.

Polyuria-polydipsia can result in elimination outside the litter box. Dehydration and discomfort may also lead to a decrease in grooming behaviors. During hypoglycemic episodes, cats may exhibit disorientation, aggression, or seizures. Additionally, a cat's social interactions may be affected during treatment due to frequent diagnostic and therapeutic interventions, which can ultimately lead to healthcare phobia.

Diagnosis is based on the observation of typical clinical signs, including polyuria-polydipsia, polyphagia, and weight loss, often accompanied by appetite disturbances. Biological tests reveal persistently elevated blood glucose levels above 2 g/L, glucosuria, and high serum fructosamine levels (Taylor et al. 2017). A single hyperglycemic episode is insufficient for diagnosis, as cats are prone to stress-induced hyperglycemia, and a routine veterinary visit can significantly elevate blood glucose levels.

Feline diabetes is managed with insulin injections, which are progressively adjusted based on glycemic monitoring, along with a specialized diet.

14.7.3 Hypothyroidism

Hypothyroidism is a rare endocrine disorder in cats, usually secondary to treatment for hyperthyroidism, such as thyroidectomy or radioactive iodine therapy. It results from insufficient production of thyroid hormones (T3 and T4), leading to a slowing of metabolism and associated behavioral changes.

Hypothyroid cats may appear lethargic, hypersomnolent, and show a lack of interest in play or social interactions (Peterson et al. 2018). Some individuals may exhibit signs of irritability, while others may become withdrawn.

Hypothyroidism can also cause a decline in grooming behaviors, leading to a dull or unkempt coat. Urine soiling may also occur. Although rare, memory disturbances or disorientation can be observed in hypothyroid cats, particularly older individuals. Conditions that lead to behavioral inhibition, such as chronic depressive states, generalized inhibited anxiety, and senile states, can present with similar clinical signs.

The differential diagnosis primarily relies on the presence of a history of hyperthyroidism treatment.

Diagnosis of feline hypothyroidism is based on clinical signs along with measurements of basal total T4 or free T4 levels and TSH concentrations (Greco 2006) (see ◘ Table 14.5).

14.7.4 Cushing's Syndrome or Hyperadrenocorticism

Cushing's disease is a rare endocrine disorder in cats, caused by excessive cortisol production by the adrenal glands. The associated behavioral changes often result from hormonal imbalances, muscle weakness, and secondary complications (Boland and Barrs 2017).

Affected cats exhibit *generalized muscle weakness*, which may lead to apathy, a lack of interest in play or exploration, and a tendency to remain immobile.

Excess cortisol affects the nervous system, increasing *stress sensitivity*. Although cortisol is a key hormone in stress adaptation, chronic overproduction disrupts normal physiological regulation, making affected cats more prone to irritability or social withdrawal, even from familiar household members.

Hyperphagia is a hallmark feature and can be striking. Food-seeking behaviors, including begging and food theft, are common and may sometimes be accompanied by aggression. Some cats become more distant and seek isolation, while others may become overly clingy, seemingly looking for comfort in response to their physical discomfort. Grooming behaviors may also be disrupted, leading to a dull or greasy coat. Skin fragility is a key characteristic of hyperadrenocorticism, and secondary wounds or infections caused by excessive scratching further exacerbate dermatological issues. Cushing's syndrome can also interfere with sleep-wake cycles, causing increased nocturnal activity, which may be a reason for veterinary consultation.

◘ Table 14.5 Reference values for thyroid hormone levels in cats

Hormone measured	Reference values *(may vary by laboratory)*
Total T4	10–50 nmol/L
Free T4	10–50 nmol/L
TSH	0.03–0.5 nmol/L

However, *endocrine tests* are essential for confirming a diagnosis of hyperadrenocorticism (HAC) in cats.

The *urinary cortisol-to-creatinine ratio (UCCR)* is used as an initial screening test due to its high sensitivity. In 13 cats diagnosed with HAC, median UCCR values ranged from 80 to 122 × 10^{-6} (Goossens et al. 1995). A negative result effectively rules out HAC, but a positive result requires further testing due to the test's low specificity. To avoid stress-induced elevations, urine samples should be collected at home.

The *low-dose dexamethasone suppression test (LDDST)* is the preferred initial diagnostic test for HAC. In cats, a dose ten times higher than that used in dogs (0.1 mg/kg IV) is required to improve specificity (Mellett Keith et al. 2013). In healthy cats, serum cortisol is suppressed to <40 nmol/L at 4 and 8 h. However, cats with HAC exhibit resistance to suppression. Some cases of HAC may still yield normal results, requiring additional testing, such as sex steroid analysis.

The ACTH stimulation test has low sensitivity (approximately one-third of HAC-positive cats test positive) and moderate specificity, making it less suitable as a primary screening test (Valentin et al. 2014).

Some pituitary tumors secrete α-melanocyte-stimulating hormone (α-MSH), which can be detected through specific tests. These tumors may lead to insulin resistance and clinical signs of HAC (Meij et al. 2005).

Finally, cytology of adrenal masses can help differentiate adrenocortical tumors from pheochromocytomas, but it does not distinguish between benign and malignant lesions or identify functional tumors.

14.7.5 Addison's Disease

Addison's disease is an extremely rare condition in cats, with only 48 documented cases in the literature since its first description in 1983 (Johnessee et al. 1983). It is an endocrine disorder caused by insufficient corticosteroid production by the adrenal glands. This hormonal deficiency leads to multisystemic effects, including behavioral changes, although these manifestations are often indirect and linked to physical symptoms.

Affected cats are typically adults between 4 and 9 years old and often show marked apathy and lethargy, resulting from general weakness and hypoglycemia associated with hormonal insufficiency. General fatigue is accompanied by anorexia, vomiting, constipation, and weight loss.

Cats with Addison's disease may become more withdrawn, avoiding interactions with their environment. They often seek quiet places to isolate themselves and appear less interested in play or social interactions.

Corticosteroid deficiency can also increase stress sensitivity, making cats more anxious or nervous, even in situations they previously managed well.

Anorexia is a common symptom, sometimes accompanied by food aversion or rejection of specific textures.

14

Due to hyponatremia and hyperkalemia induced by the disease, some cats may exhibit abnormal behaviors, such as disorientation, tremors, or exaggerated responses to sudden noises or movements.

The symptoms of Addison's disease can resemble generalized inhibited anxiety or depressive states.

The diagnosis is based on suggestive clinical signs, including gastrointestinal signs such as anorexia, vomiting, weight loss, and constipation. Some cats may present with adipsia, hyporexia, or polyphagia. Neurological signs may include generalized weakness, ataxia, dysphagia, empty swallowing motions, hind limb weakness, and difficulty jumping or climbing. Other symptoms often observed include marked lethargy, polyuria/polydipsia, hypothermia, and hair loss.

Diagnosis is confirmed with an ACTH stimulation test. A low resting cortisol concentration, combined with an inadequate (< 2 μg/dL) or absent response to synthetic ACTH, is consistent with a diagnosis of hypoadrenocorticism (Peterson 2012; Gunn-Moore 2005).

14.7.6 Sexual Dysendocrinopathies

Sexual dysendocrinopathies are related to disruptions in the production of sex hormones (estrogens, progesterone, testosterone) in intact cats. Although rare, these disorders can significantly affect the behavior of male cats due to testicular tumors or female cats due to ovarian tumors or persistent follicles. The sexual dysthymia of female cats has already been discussed (see ► Sect. 14.6).

Excess sex hormones in both sexes, regardless of the cause, can lead to an increase in *aggressive behaviors.* In males, testosterone overproduction may result in *hypersexuality,* characterized by frequent mounting behaviors and excessive urine marking. A plasma testosterone level above 5 ng/mL (17 nmol/L) indicates hypertestosteronemia. Treatment involves castration or the placement of a deslorelin implant.

In females, excessive estrogen production can lead to prolonged heat cycles, with persistent vocalizations often being the primary reason for consultation (Meler et al. 2011). Estrogen levels can be measured, with normal values as follows:

- anestrus (hormonal resting phase): <20 pg/mL (<73 pmol/L);
- proestrus / Estrus (heat cycle): 20–60 pg/mL (73–220 pmol/L);
- diestrus (post-ovulation phase): <20 pg/mL (<73 pmol/L).

A plasma estrogen level exceeding these values suggests a persistent follicle. Treatment consists of ovariectomy.

14.7.7 Acromegaly

Acromegaly, or *hypersomatotropism*, results from chronic excessive secretion of *GH* in adult animals. In cats, acromegaly is caused by a pituitary adenoma that produces excessive amounts of GH.

The characteristic effects of this hormone overproduction include the development of *diabetes mellitus* and the growth of acral body segments such as the jaw, extremities, and skull. Acromegaly primarily affects adult cats, predominantly males, and is often associated with diabetes mellitus. Hyperphagia is a consistent and striking symptom, sometimes leading to aggression due to frustration. Other clinical signs include stridor, jaw and extremity enlargement, increased lean body mass, and organomegaly (heart, liver, kidneys…).

Diagnosis is based on detecting elevated levels of GH or IGF-1 (or both) and identifying a pituitary mass using MRI or CT scan. To date, the most effective treatments are radiotherapy and surgery.

Many acromegalic cats exhibit lethargy or a decreased interest in usual activities, often related to poorly controlled secondary diabetes or general fatigue.

Bone, joint, and soft tissue enlargement can cause joint and muscle pain, leading to behavioral changes such as reluctance to jump, climb, or move, as well as increased irritability and aggression.

In advanced cases, pressure exerted by the pituitary tumor on the brain can cause neurological signs such as disorientation or abnormal behaviors.

Most commonly, affected cats have an abnormally high appetite and drink excessive amounts of water. They develop diabetes that is impossible to regulate due to their insatiable appetite.

Summary of ▶ Sect. 14.7
Diabetes is certainly the most common endocrine disorder in cats and should be considered in cases of symptoms suggestive of depression or anxiety.

Hyperthyroidism should be included in the differential diagnosis of all anxious, impulsive, compulsive, and senile states.

Other endocrine disorders are rarer, but it is important to keep them in mind, as hormonal imbalances can significantly influence an animal's behavior.

14.8 Surgery

Surgical procedures are generally painful and often require *intensive post-operative care.* Due to their sensitivity to phobias, cats may exhibit fear-related behaviors such as withdrawal, escape attempts, or, conversely, aggression during medical care. Post-operative care can be particularly challenging in cats that quickly become sensitized to handling or were already sensitive before surgery, with their condition further exacerbated by the surgical procedure.

Managing pain as effectively as possible during both the preoperative and post-operative phases is essential to prevent the development of *healthcare phobia.* To minimize sensitization, preoperative medication may be recommended. Gabapentin, at a dose of 15–30 mg/kg twice daily, can be administered, or alternatively, a single dose 2 h before handling. For a cat already sensitized to medical care, a long-term anxiolytic treatment may be considered throughout the recovery period, in addition to situational gabapentin use. *Venlafaxine*, in the granule form contained in Effexor™, is the preferred choice for this purpose.

It is important to note that among the available analgesic options, the concurrent use of tramadol and venlafaxine is strictly contraindicated.

Additionally, the use of an Elizabethan collar may be poorly tolerated by some cats. Assessing the cat's tolerance before hospital discharge allows for either implementing stress management strategies to help the cat cope with the collar or finding a more suitable alternative for the individual.

Summary of ▶ Sect. 14.8

Due to the feline species' rapid development of fear responses and its predisposition to phobic pathological states, effective pain management should begin in the preoperative phase and be maintained, particularly before potentially painful post-operative care procedures. In cats already sensitized to handling, a behavioral management plan before surgery is highly recommended.

14.9 Aging

Aging and its pathological consequences on feline behavior are detailed in ▶ Chap. 13. Here, we emphasize the importance of a *comprehensive approach* to managing cats over *8 years old.* Pain, endocrine disorders, and sensory degeneration can lead to various symptoms, from an unkempt cat due to insufficient grooming to one that vocalizes excessively, exhibits urine soiling, or becomes aggressive.

It is the responsibility of thc general practitioner to detect early signs of pathological aging before age-related conditions become too complex to manage and strain the bond between the owner and their cat. Educating owners about the *early signs of aging* during annual checkups is essential. Identifying and explaining these signs during consultations while offering *solutions to improve the cat's well-being* can significantly help both the animal and its owner.

Aging is inevitable, but with *proper support,* this phase can be experienced more comfortably. Ensuring a *strong dialogue between the owner and the veterinary team* is key to optimizing the cat's quality of life.

Summary of ▶ Sect. 14.9

The aging animal is often affected by multiple conditions, and psychological care is frequently delayed or overlooked due to the number of medical issues already requiring attention. However, addressing psychiatric disorders in older cats is essential, as they play a significant role in the loss of motivation and can contribute to the development of aggressive reactions or cognitive decline, potentially leading to premature euthanasia.

References

Amadei E, Cantile C, Gazzano A, Pierantoni L, Mariti C (2021) The link between neurology and behavior in veterinary medicine: a review. J Vet Behav 46:40–53. https://doi.org/10.1016/j.jveb.2021.07.003

Amengual Batle P, Rusbridge C, Nuttall T, Heath S, Marioni-Henry K (2019) Feline hyperaesthesia syndrome with self-trauma to the tail: retrospective study of seven cases and proposal for an integrated multidisciplinary diagnostic approach. J Feline Med Surg 21:178–185. https://doi.org/10.1177/1098612X18764246

Arndt J, Smith N, Tausk F (2008) Stress and atopic dermatitis. Curr Allergy Asthma Rep 8:312–317

Bauer JE (2006) Metabolic basis for the essential nature of fatty acids and the unique dietary fatty acid requirements of cats. J Am Vet Med Assoc 229:1729–1732. https://doi.org/10.2460/javma.229.11.1729

Bensfield AC, Evans J, Pesayco JP, Mizisin AP, Shelton GD (2011) Recurrent demyelination and Remyelination in 37 young Bengal cats with polyneuropathy. J Vet Intern Med 25:882–889. https://doi.org/10.1111/j.1939-1676.2011.0740.x

Boland LA, Barrs VR (2017) Peculiarities of feline hyperadrenocorticism: update on diagnosis and treatment. J Feline Med Surg 19:933–947. https://doi.org/10.1177/1098612X17723245

Borns-Weil S, Emmanuel C, Longo J, Kini N, Barton B, Smith A, Dodman NH (2015) A case-control study of compulsive wool-sucking in Siamese and Birman cats (n = 204). J Vet Behav 10:543–548. https://doi.org/10.1016/j.jveb.2015.07.038

Bradshaw J, Ellis S (2016) The trainable cat: a practical guide to making life happier for you and your cat. Basic Books, New York

Bradshaw JWS, Neville PF, Sawyer D (1997) Factors affecting pica in the domestic cat. Appl Anim Behav Sci 52:373–379. https://doi.org/10.1016/S0168-1591(96)01136-7

Buffington CAT, Westropp JL, Chew DJ, Bolus RR (2006a) Risk factors associated with clinical signs of lower urinary tract disease in indoor-housed cats. J Am Vet Med Assoc 228:722–725. https://doi.org/10.2460/javma.228.5.722

Buffington CAT, Westropp JL, Chew DJ, Bolus RR (2006b) Clinical evaluation of multimodal environmental modification (MEMO) in the management of cats with idiopathic cystitis. J Feline Med Surg 8:261–268. https://doi.org/10.1016/j.jfms.2006.02.002

Byrne G, Rosenfeld G, Leung Y, Qian H, Raudzus J, Nunez C, Bressler B (2017) Prevalence of anxiety and depression in patients with inflammatory bowel disease. Can J Gastroenterol Hepatol 2017:1–6. https://doi.org/10.1155/2017/6496727

Camps T, De La Fuente C, Pumarola M, Amat M, Le Brech S, Manteca X (2015) A case of spongiform polioencephalomyelopathy in a cat with a history of behavioural problems. J Feline Med Surg Open Rep 1:2055116915599172. https://doi.org/10.1177/2055116915599172

Carney HC, Sadek TP, Curtis TM, Halls V, Heath S, Hutchison P, Mundschenk K, Westropp JL (2014) AAFP and ISFM guidelines for diagnosing and solving house-soiling behavior in cats. J Feline Med Surg 16:579–598. https://doi.org/10.1177/1098612X14539092

Carney HC, Ward CR, Bailey SJ, Bruyette D, Dennis S, Ferguson D, Hinc A, Rucinsky AR (2016) 2016 AAFP guidelines for the Management of Feline Hyperthyroidism. J Feline Med Surg 18:400–416. https://doi.org/10.1177/1098612X16643252

Carr CW, Veledar E, Chen SC (2014) Factors mediating the impact of chronic pruritus on quality of life. JAMA Dermatol 150:613. https://doi.org/10.1001/jamadermatol.2013.7696

Casal ML (2012) Canine and feline congenital defects and genetic counseling. In: Lopate C (ed) Management of pregnant and neonatal dogs, cats, and exotic pets. Wiley, pp 169–205. https://doi.org/10.1002/9781118997215.ch11

Choi E-H, Demerjian M, Crumrine D, Brown BE, Mauro T, Elias PM, Feingold KR (2006) Glucocorticoid blockade reverses psychological stress-induced abnormalities in epidermal structure and function. Am J Physiol-Regul Integr Comp Physiol 291:R1657–R1662

Da Riz F (2023) Présentation du diabète sucré: similitudes et différences de la pathogénèse, de la présentation clinique et du diagnostic. Nouv Prat Vét Canine Féline 20:10–17. https://doi.org/10.1051/npvcafe/2024005

Demontigny-Bédard I, Beauchamp G, Bélanger M-C, Frank D (2016) Characterization of pica and chewing behaviors in privately owned cats: a case-control study. J Feline Med Surg 18:652–657. https://doi.org/10.1177/1098612X15591589

Dewey CW, da Costa RC, Da Costa RC (eds) (2016) Practical guide to canine and feline neurology, 3rd edn. Wiley Blackwell, Ames

Evangelista MC, Steagall PV (2021) Agreement and reliability of the Feline Grimace Scale among cat owners, veterinarians, veterinary students and nurses. Sci Rep 11:5262. https://doi.org/10.1038/s41598-021-84696-7

Evangelista MC, Watanabe R, Leung VSY, Monteiro BP, O'Toole E, Pang DSJ, Steagall PV (2019) Facial expressions of pain in cats: the development and validation of a feline grimace scale. Sci Rep 9:19128. https://doi.org/10.1038/s41598-019-55693-8

Evangelista MC, Benito J, Monteiro BP, Watanabe R, Doodnaught GM, Pang DSJ, Steagall PV (2020) Clinical applicability of the Feline Grimace Scale: real-time versus image scoring and the influence of sedation and surgery. PeerJ 8:e8967. https://doi.org/10.7717/peerj.8967

FEDIAF EuropeanPetFood [WWW Document] (2024) https://europeanpetfood.org/

Feighelstein M, Shimshoni I, Finka LR, Luna SPL, Mills DS, Zamansky A (2022) Automated recognition of pain in cats. Sci Rep 12:9575. https://doi.org/10.1038/s41598-022-13348-1

Feighelstein M, Henze L, Meller S, Shimshoni I, Hermoni B, Berko M, Twele F, Schütter A, Dorn N, Kästner S, Finka L, Luna SPL, Mills DS, Volk HA, Zamansky A (2023) Explainable automated pain recognition in cats. Sci Rep 13:8973. https://doi.org/10.1038/s41598-023-35846-6

Foster JA, Neufeld K-AM (2013) Gut–brain axis: how the microbiome influences anxiety and depression. Trends Neurosci 36:305–312

Gómez Álvarez CM, Soler Arias EA (2021) Feline hyperesthesia syndrome: epilepsy as possible aetiology in two cats. Vet Rec Case Rep 9:e132. https://doi.org/10.1002/vrc2.132

Goossens MMC, Meyer HP, Voorhout G, Sprang EPM (1995) Urinary excretion of glucocorticoids in the diagnosis of hyperadrenocorticism in cats. Domest Anim Endocrinol 12:355–362. https://doi.org/10.1016/0739-7240(95)00046-H

Gorzelanna Z, Miszczak M (2024) Through the intestines to the head? That is, how the gastrointestinal microbiota affects the behavior of companion animals. Pets 1:201–215. https://doi.org/10.3390/pets1030015

Greco DS (2006) Diagnosis of congenital and adult-onset hypothyroidism in cats. Clin Tech Small Anim Pract 21:40–44. https://doi.org/10.1053/j.ctsap.2005.12.007

Greco DS (2018) Diabetes mellitus in animals. In: Nutritional and therapeutic interventions for diabetes and metabolic syndrome. Elsevier, pp 507–517. https://doi.org/10.1016/B978-0-12-812019-4.00037-4

Gunn-Moore D (2005) Feline endocrinopathies. Vet Clin North Am Small Anim Pract 35:171–210, vii. https://doi.org/10.1016/j.cvsm.2004.09.002

Hernandez J, Rhimi S, Kriaa A, Mariaule V, Boudaya H, Drut A, Jablaoui A, Mkaouar H, Saidi A, Biourge V, Borgi MA, Rhimi M, Maguin E (2022) Domestic environment and gut microbiota: lessons from pet dogs. Microorganisms 10:949. https://doi.org/10.3390/microorganisms10050949

Hernandez-Avalos I, Mota-Rojas D, Mora-Medina P, Martínez-Burnes J, Casas Alvarado A, Verduzco-Mendoza A, Lezama-García K, Olmos-Hernandez A (2019) Review of different methods used for clinical recognition and assessment of pain in dogs and cats. Int J Vet Sci Med 7:43–54. https://doi.org/10.1080/23144599.2019.1680044

Holden E, Calvo G, Collins M, Bell A, Reid J, Scott EM, Nolan AM (2014) Evaluation of facial expression in acute pain in cats. J Small Anim Pract 55:615–621. https://doi.org/10.1111/jsap.12283

Johnessee JS, Peterson ME, Gilbertson SR (1983) Primary hypoadrenocorticism in a cat. J Am Vet Med Assoc 183:881–882

Jones E, Palmieri C, Thompson M, Jackson K, Allavena R (2021) Feline idiopathic cystitis: pathogenesis, histopathology and comparative potential. J Comp Pathol 185:18–29. https://doi.org/10.1016/j.jcpa.2021.03.006

Laflamme D (1997) Development and validation of a body condition score system for dogs. Canine Pr 22:10–15

Lavelle JP, Meyers SA, Ruiz WG, Buffington CAT, Zeidel ML, Apodaca G (2000) Urothelial pathophysiological changes in feline interstitial cystitis: a human model. Am J Physiol-Ren Physiol 278:F540–F553. https://doi.org/10.1152/ajprenal.2000.278.4.F540

Lekcharoensuk C, Osborne CA, Lulich JP (2001) Epidemiologic study of risk factors for lower urinary tract diseases in cats. J Am Vet Med Assoc 218:1429–1435. https://doi.org/10.2460/javma.2001.218.1429

Martvel G, Lazebnik T, Feighelstein M, Henze L, Meller S, Shimshoni I, Twele F, Schütter A, Foraita N, Kästner S, Finka L, Luna SPL, Mills DS, Volk HA, Zamansky A (2024) Automated video-based pain recognition in cats using facial landmarks. Sci Rep 14:28006. https://doi.org/10.1038/s41598-024-78406-2

Meij BP, Van Der Vlugt-Meijer RH, Van Den Ingh TSGAM, Flik G, Rijnberk A (2005) Melanotroph pituitary adenoma in a cat with diabetes mellitus. Vet Pathol 42:92–97. https://doi.org/10.1354/vp.42-1-92

Meler EN, Scott-Moncrieff JC, Peter AT, Bennett S, Ramos-Vara J, Salisbury SK, Naughton JF (2011) Cyclic estrous-like behavior in a spayed cat associated with excessive sex-hormone production by an adrenocortical carcinoma. J Feline Med Surg 13:473–478. https://doi.org/10.1016/j.jfms.2011.02.002

Mellett Keith AM, Bruyette D, Stanley S (2013) Trilostane therapy for treatment of spontaneous Hyperadrenocorticism in cats: 15 cases (2004–2012). J Vet Intern Med 27:1471–1477. https://doi.org/10.1111/jvim.12178

Merola I, Mills DS (2016) Systematic review of the behavioural assessment of pain in cats. J Feline Med Surg 18:60–76. https://doi.org/10.1177/1098612X15578725

Mills DS, Demontigny-Bédard I, Gruen M, Klinck MP, McPeake KJ, Barcelos AM, Hewison L, Van Haevermaet H, Denenberg S, Hauser H, Koch C, Ballantyne K, Wilson C, Mathkari CV, Pounder J, Garcia E, Darder P, Fatjó J, Levine E (2020) Pain and problem behavior in cats and dogs. Animals 10:318. https://doi.org/10.3390/ani10020318

Moon H-S, Mizara A, McBride SR (2013) Psoriasis and psycho-dermatology. Dermatol Ther 3:117–130. https://doi.org/10.1007/s13555-013-0031-0

Mooney CT, Peterson ME (2023) Feline hyperthyroidism. In: Mooney CT, Peterson ME, Shiel RE (eds) BSAVA manual of canine and feline endocrinology. British Small Animal Veterinary Association, pp 151–168. https://doi.org/10.22233/9781910443866.19

National Research Council (2006) Nutrient Requirements of Dogs and Cats. National Academies Press, Washington, DC. https://doi.org/10.17226/10668

Parsons CL (2007) The role of the urinary epithelium in the pathogenesis of interstitial cystitis/prostatitis/urethritis. Urology 69:S9–S16. https://doi.org/10.1016/j.urology.2006.03.084

Peterson ME (2012) Feline hypoadrenocorticism. In: Mooney CT, Peterson ME (eds) BSAVA manual of canine and feline endocrinology. British Small Animal Veterinary Association, pp 199–203. https://doi.org/10.22233/9781905319893.18

Peterson ME, Carothers MA, Gamble DA, Rishniw M (2018) Spontaneous primary hypothyroidism in 7 adult cats. J Vet Intern Med 32:1864–1873. https://doi.org/10.1111/jvim.15239

Rand JS (2013) Pathogenesis of feline diabetes. Vet Clin Small Anim Pract 43:221–231. https://doi.org/10.1016/j.cvsm.2013.01.003

Reusch CE (2014) Feline diabetes mellitus. In: Feldman EC, Nelson RW, Reusch CE, Scott-Moncrieff JCR, Behrend EN (eds) Canine and feline endocrinology, 4th edn, pp 258–314

Robinson AR, Steagall PV (2024) Effects of training on feline grimace scale scoring for acute pain assessment in cats. J Feline Med Surg 26:1098612X241275284. https://doi.org/10.1177/1098612X241275284

Rusbridge C, Heath S, Gunn-Moore DA, Knowler SP, Johnston N, McFadyen AK (2010) Feline orofacial

14

pain syndrome (FOPS): a retrospective study of 113 cases. J Feline Med Surg 12:498–508
Santoro D, Pucheu-Haston CM, Prost C, Mueller RS, Jackson H (2021) Clinical signs and diagnosis of feline atopic syndrome: detailed guidelines for a correct diagnosis. Vet Dermatol 32:26. https://doi.org/10.1111/vde.12935
Scott H, Marti J, Witte P (2022) Neuromuscular Disorders. In: Feline Orthopaedics. CRC Press, Boca Raton, pp 324–347. https://doi.org/10.1201/9780429091537-35
Silson M, Robinson R (1969) Hereditary hydrocephalus in the cat. Vet Rec 84(19):477
Steagall PV (2019) Feline grimace scale [WWW Document]. Feline Grimace Scale. https://www.felinegrimacescale.com/. Accessed 30 Dec 30
Steagall PV, Monteiro BP (2019) Acute pain in cats: recent advances in clinical assessment. J Feline Med Surg 21:25–34. https://doi.org/10.1177/1098612X18808103
Steagall PV, Monteiro BP, Marangoni S, Moussa M, Sautié M (2023) Fully automated deep learning models with smartphone applicability for prediction of pain using the Feline Grimace Scale. Sci Rep 13:21584. https://doi.org/10.1038/s41598-023-49031-2
Taylor SS, Sparkes AH, Briscoe K, Carter J, Sala SC, Jepson RE, Reynolds BS, Scansen BA (2017) ISFM consensus guidelines on the diagnosis and Management of Hypertension in cats. J Feline Med Surg 19:288–303. https://doi.org/10.1177/1098612X17693500
Tuttle J (1980) Feline hyperesthesia syndrome. J Am Vet Med Assoc 176:47
Tynes VV, Sinn L (2014) Abnormal repetitive behaviors in dogs and cats. Vet Clin North Am Small Anim Pract 44:543–564. https://doi.org/10.1016/j.cvsm.2014.01.011
Valentin SY, Cortright CC, Nelson RW, Pressler BM, Rosenberg D, Moore GE, Scott-Moncrieff JC (2014) Clinical findings, diagnostic test results, and treatment outcome in cats with spontaneous Hyperadrenocorticism: 30 cases. J Vet Intern Med 28:481–487. https://doi.org/10.1111/jvim.12298
Verhoeven EWM, Kraaimaat FW, De Jong E, Schalkwijk J, Van De Kerkhof PCM, Evers AWM (2009) Individual differences in the effect of daily stressors on psoriasis: a prospective study. Br J Dermatol 161:295–299
Virga V (2003) Behavioral dermatology. Vet Clin North Am Small Anim Pract 33:231–251. https://doi.org/10.1016/S0195-5616(02)00102-X
Waisglass SE, Landsberg GM, Yager JA, Hall JA (2006) Underlying medical conditions in cats with presumptive psychogenic alopecia. J Am Vet Med Assoc 228:1705–1709. https://doi.org/10.2460/javma.228.11.1705
Watanabe R, Doodnaught GM, Evangelista MC, Monteiro BP, Ruel HLM, Steagall PV (2020) Inter-rater reliability of the feline grimace scale in cats undergoing dental extractions. Front Vet Sci 7:302. https://doi.org/10.3389/fvets.2020.00302
Willemse T (1994) The effect of haloperidol and naloxone on excessive grooming behavior of cats. Eur Neuropsychopharmacol 4:39–45. https://doi.org/10.1016/0924-977X(94)90313-1
Wittkowski A, Richards HL, Griffiths CEM, Main CJ (2004) The impact of psychological and clinical factors on quality of life in individuals with atopic dermatitis. J Psychosom Res 57:195–200. https://doi.org/10.1016/S0022-3999(03)00572-5

Semiology in Feline Psychiatry

Contents

15.1 Centripetal Behaviors – 506
15.1.1 Semiology of Feeding Behavior – 506
15.1.2 Semiology of Drinking Behavior – 510
15.1.3 Semiology of Sleeping Behavior – 510
15.1.4 Semiology of Elimination Behavior – 511
15.1.5 Semiology of Repetitive Behaviors – 517

15.2 Centrifugal Behaviors – 523
15.2.1 Semiology of Injurious Sequences – 523
15.2.2 Semiology of Sexual and Maternal Behavior – 528
15.2.3 Semiology of Play and Mutual Grooming – 529
15.2.4 Semiology of Exploratory Behavior – 531

15.3 Other Elements of Semiology – 535
15.3.1 Semiology of Behavioral Development – 535
15.3.2 Semiology of Direct Observation – 537
15.3.3 Semiology of Behavior in Aging Cats – 539

15.4 Semiology by Behavioral Axis – 542
15.4.1 Disturbances Related to the Fear Axis – 542
15.4.2 Disturbances Related to the Self-Control Axis – 542
15.4.3 Disturbances Related to the Territoriality Axis – 543
15.4.4 Disturbances Related to the Relational Axis – 544

References – 544

© The Author(s), under exclusive license to Springer Nature Switzerland AG 2026
S. Bleuer-Elsner et al., *Veterinary Psychiatry of the Cat*,
https://doi.org/10.1007/978-3-032-06260-4_15

In veterinary psychiatry, the transformation of collected signs into symptoms is a key element. The semiological assessment is a complex technique that requires a precise interview with the owner. The details of this assessment, behavior by behavior, are described to highlight the complexity of this exercise and emphasize the importance of the accuracy of the collected information.

Medicine is gradually shifting away from the classical diagnosis, what could be called a "literary diagnosis," in favor of grouping measurable data that transform our observations into quantifiable certainties. In the past century, it was still common to hear discussions of the *art* of medicine, but slowly, this art is becoming a science. The complementary tests available in veterinary medicine now allow for a progressive validation of clinical findings.

Despite all efforts, psychiatry still largely escapes this numerical precision, even with the validation of several owner-directed evaluation scales (Duffy et al. 2017). Behaviors and their intentionality remain difficult to measure, and our diagnosis continues to rely primarily on clinically observed or owner-reported symptoms.

However, diagnosis is not based on mere impressions. Each disorder, each state, corresponds to a set of symptoms that must be carefully assembled to establish a diagnosis. This work must be rigorous if we are to transform our art into a precise and teachable discipline.

15

The semiological exercise consists of identifying a symptom by detecting signs. This constructivist approach, though debatable, remains the foundation of all scientific methodology.

Let us take the example of a cat urinating against a wall. This description constitutes a sign. When we state, "the cat is urine-marking," we are interpreting this behavior, assigning it intentionality, or at the very least categorizing it. The sign thus becomes a symptom **(see box below "*Signs and symptoms*").**

This approach integrates behavioral medicine within the broader field of medical disciplines. Grouping various symptoms to establish a classification is a methodological choice that aligns the discipline with psychiatry and solidifies its place within medicine. By transforming signs into symptoms, the consultation becomes a semiological assessment. The practitioner structures this process, and the synthesis of symptoms ultimately leads to the diagnosis.

Important Point of View

Signs and Symptoms

These two terms can be confusing due to the different meanings attributed to them. In Anglo-Saxon literature and human medicine, a *symptom* is subjectively described or experienced by the patient, while a *sign* or *clinical sign* is an objective element measurable by the healthcare professional. For example, the symptom might be *"Doctor, I am always tired,"* while the sign would be an abnormally low blood pressure of less than 90/60 mmHg.

In veterinary psychiatry, particularly within the so-called *French school* of veterinary psychiatry, a *sign* is what we observe, as we do not have direct access to the animal's emotions or cognitions. A sign is thus objective and measurable, as defined above. For instance, *"the cat attacks its tail three times a day for five minutes."*

A *symptom*, on the other hand, consists of transforming this sign into an emotional and cognitive interpretation. In this example, it may be interpreted as *"the cat is normal and playing," "the cat exhibits a lack of control over its scratching during play," "the aggression is pain-induced," "the aggression is induced by pruritic dermatitis,"* or *"the cat is experiencing focal epileptic seizures."*

A single sign can be linked to multiple symptoms. By selecting the most probable hypothesis from the differential diagnosis list, the practitioner interprets signs to define them as symptoms. In this case, the context, intensity, frequency, and the cat's ability to spontaneously stop the *tail-attacking* behavior, along with all other semiological elements collected, allow the practitioner to determine the most likely hypothesis. If additional signs indicate a lack of control over locomotion, biting, and scratching across various behaviors, the hypothesis of impaired self-control would be favored. Conversely, if the cat exhibits perfect control and these episodes have only appeared in recent weeks, with a visible wound upon clinical examination, the hypothesis of a behavior sequence induced by pain would be prioritized (see ◘ Table 15.1).

Throughout this book, we have attempted to approach behavior through four axes to facilitate clinical reasoning and symptom classification: fear, self-control, territoriality, and social relationships. These axes correspond to the animal's development, to which we must add disturbances related to communication and the environment. Our semiological approach follows this framework, and we propose that clinicians construct an oriented semiology: the identification of a characteristic disturbance within one of these axes should naturally lead them to examine *other symptoms within the same axis more closely*. The logic behind this reasoning is that if one axis is dysfunctional, signs of this dysfunction will be present in various behaviors, allowing the practitioner to interpret these signs as symptoms and validate the most probable hypotheses.

However, we will present this semiology by examining each behavior individually, beginning with *centripetal behaviors*, followed by *centrifugal* and *mixed* behaviors. In each case, we will attempt to indicate the signs described by owners, which will then become symptoms of a disturbance in a particular axis.

Semiology can be structured in different ways, and each practitioner will develop their own approach. However, in all cases, two pitfalls must be avoided. The first is overlooking an important aspect, thereby missing signs that could influence the diagnosis. The second is being overly exhaustive, even bookish, and failing to omit parts of the semiological process that are clearly uninformative because the relevant axis is within the normal range. Nothing is more frustrating for an owner than feeling that their time is being wasted, and this can have significant consequences on the *therapeutic alliance.*

We will successively examine the different elements of semiology in the following order.

◘ **Table 15.1** From sign to symptom

Owner reported sign	Clinician-retained symptom
He urinates against the kitchen wall at least twice a week while standing	Urine marking
He only eats at night	Exaggerated fear responses to environmental stimuli, to the extent that he only eats at night
He is rarely seen sleeping, and his sleep duration is estimated at less than 12 h per 24-h period	Hyposomnia
He does not chew his food and constantly begs for more	Deficit in control during food intake
He hides as soon as visitors arrive and does not come out until about an hour after they leave	Exaggerated fear responses to unfamiliar humans

The semiology of centripetal behaviors:
- feeding behavior
- drinking behavior
- sleeping behavior
- elimination behavior
- repetitive behavior

The semiology of centrifugal and mixed behaviors:
- injurious sequences
- exploratory behavior
- sexual and maternal behavior
- play and mutual grooming
- exploratory behavior

Finally, we will address specific aspects of the semiological assessment, such as developmental considerations, observation during consultation, and behaviors in aging cats.

15.1 Centripetal Behaviors

15.1.1 Semiology of Feeding Behavior

The study of feeding behavior, meal composition, and the mode of food distribution is one of the key aspects of the semiological assessment. For owners, feeding is often the primary (and unfortunately sometimes the only) act that reflects an emotional bond. The relationship established around feeding is not solely dictated by dietary considerations. Feeding is accompanied by numerous signs of affection, which may not always be understood by the animal but serve as markers of the owner's attachment to their pet (see ◘ Table 15.2).

15

◘ **Table 15.2** Reported signs in the semiology of feeding behavior and possible interpretations

Reported sign	Symptoms *In all cases, a detailed exploration of the frequency, intensity, and context in which the sign arises is necessary to qualify it as a symptom*	Underlying dysfunction
Aggression during feeding related to an inadequate food distribution method and/or highly palatable food	Excessive motivation, which may escalate to irritation-induced aggression in a context of frustration, particularly with overly palatable food	Imbalance in the self-control axis
	Fractionated food distribution causing intense anticipation, which may trigger irritation-induced aggression in a context of frustration	Imbalance in the communication or self-controls axis
Food fussiness	Decreased appetite in situations of intense fear	Imbalance in the fear axis
	Owner misinterprets an attention-seeking behavior as a food request	Imbalance in the communication axis
	Dysorexia with sometimes dramatic variations	Chronic depressive state
	Any organic condition causing reduced appetite	Organic disorder
	Medication-induced appetite suppression (e.g., fluoxetine, venlafaxine)	Iatrogenic

(continued)

Table 15.2 (continued)

Reported sign	Symptoms ***In all cases, a detailed exploration of the frequency, intensity, and context in which the sign arises is necessary to qualify it as a symptom***	Underlying dysfunction
Anorexia	Climatic adaptation (high external temperature) Suspicion of an odor linked to the introduction of a medication	Reactive state
	Decreased appetite in situations of intense fear	Imbalance in the fear axis
	Following intense stress or sensory deprivation (likely more frequent)	Acute depressive state
	Any organic condition affecting appetite. Self-perpetuated hepatic lipidosis due to prolonged anorexia	Organic disorder
	Medication affecting appetite (NSAIDs, certain antibiotics, and fluoxetine)	Iatrogenic

15.1.1.1 Mode of Food Distribution

Food can be distributed either continuously or in portions. It is well established that cats should spend a significant portion of their time engaged with their food. If we want this time to involve more than just ingestion, it is essential to discuss with owners what strategies they have implemented to slow down food consumption.

Ideally, the cat should have free access to food, but food intake should be structured in a way that the reinforcement gained from ingestion is relatively low compared to the effort required to obtain it.

There are two ways to achieve this: increasing the difficulty of obtaining the food or reducing its palatability to lower the cat's motivation.

In nature, cats hunt prey that are difficult to capture and likely far less palatable compared to commercial pet food. They also spend a great deal of time observing and waiting before hunting, as well as experiencing frequent failures in their hunting attempts.

It is therefore important to determine when the owner provides food and what type of feeding setup is used. Additionally, it is necessary to understand how the owner selects food, whether based on price, a desire to please the cat, dietary considerations, or other factors.

Feeding disorders are often the result of learned behaviors. A cat may vocalize to gain attention, and the owner may interpret this as a request for food. This owner response is reinforced by the fact that the cat quickly learns that the best place to receive attention is the area where its food bowl is located. If the owner repeatedly offers food in response to these requests, the cat learns to associate food with attention, developing an "attention-seeking feeding behavior" that can quickly become obsessive due to the dual reinforcement of both food and owner interaction.

It is easy to see that the more demanding the cat is, whether due to hyperactivity or living in a poor environment, the higher the risk of developing obsessive feeding behaviors. Additionally, the more the cat demands, the more the owner may feel guilty and reinforce the behavior by giving in to the requests.

Finally, it is important to highlight—and this calls for cautious reflection—that feeding disorders may result from a prescribed "weight-loss diet" by a colleague who aims to reduce the cat's food intake but merely specifies a fixed quantity without considering the animal's living conditions. This approach can lead to significant frustration and may pose long-term risks of hepatic or pancreatic issues due to excessive fasting periods.

Such a prescription may be excellent for the cat's physical health, but it must be adapted to the necessary psychological conditions by ensuring that the cat receives a meal distribution that remains physiologically acceptable for its needs (see ▶ Chap. 14, ▶ Sect. 14.4).

15.1.1.2 Aggression During Feeding

At the precise moment when a cat is about to receive a palatable food item, there is often significant anticipation. This is particularly common in cats fed a continuous dry diet who receive two wet meals per day, a feeding method frequently recommended by nutritionists.

This heightened activation in the context of food stimulation can easily trigger sequences of irritation-induced aggression due to frustration or excitement. Some owners may be attacked or even bitten during food distribution. The faster they rush to provide the food, the more the cat is reinforced and excited. This type of aggression has a high likelihood of becoming instrumentalized, as it leads to a reinforcing outcome.

A similar process occurs in cats that receive food immediately upon their owner's awakening. In such cases, it is natural for them to learn to provoke waking, even by biting if necessary, to obtain their reward.

Irritation-induced aggression can also be observed when an animal is eating highly palatable food. The cat eats while growling if approached. It may accelerate its food intake, and the intensity of the growling may be inversely proportional to the distance of the approaching person or animal.

15.1.1.3 Food Fussiness

Some cats are described as fussy in their food preferences. It is true that a cat's tastes remain very stable over time, and in general, cats do not seem to enjoy discovering new flavors. Furthermore, once a food is rejected, it becomes difficult to get the cat to accept it again.

However, learned behaviors can also play a role. A cat may circle around a full food bowl while appearing to demand something, but this does not necessarily indicate a dislike of the food. This behavior may be a request for attention, and it would be a mistake to respond by changing the food.

Such attention-seeking behavior should be interpreted as a symptom of territorial poverty, once dental and oral pain have been ruled out.

Chronic depressive states are accompanied by periods of dysorexia, while acute depression is marked by anorexia. Appetite also decreases in fearful situations, reflecting an imbalance in the fear axis.

It is important to remember that a typical cat does not immediately rush to its food bowl when eating. The label of *fussy* is often applied too quickly to cats that are, in fact, behaving normally. Owners tend to seek highly palatable food that their cat will eagerly consume, but a healthy cat often nibbles and eats without particular enthusiasm. The priority should therefore be an ethological reassessment of normal feline feeding behavior, with a focus on selecting kibble that will not be rapidly devoured.

15.1.1.4 Anorexia

Anorexia is most often a consequence of organic diseases and is frequently complicated by hepatic lipidosis, which perpetuates food refusal.

Stress, such as that associated with hospitalization, can also trigger a phase of anorexia. While a temporary loss of appetite lasting a few days may initially be a stress response, it can quickly lead to hepatic disorders, making recovery difficult. In all cases, anorexia should primarily be considered a symptom of an organic disorder, even though its origin may also be behavioral.

15.1.1.5 Hyperphagia and Licking or Ingestion of Non-edible Items

Some cats display excessive food-seeking behavior, which can escalate into food obsession. A bulimic pattern, characterized by rapid ingestion followed by regurgitation, is frequently described (Proux 2022).

Hyperphagia can result from a food obsession. If this obsession stems from insufficient food intake, the cat's body condition score will be abnormally low. However, hyperphagia may also be linked to an impulsive or compulsive state associated with a self-control disorder, in which case other semiological signs will be present. Hyperactive cats are rarely obese in the early stages of the condition.

Hyperphagia may also serve as a compensatory behavior in an anxious state, functioning as a substitute activity. In such cases, obesity is often severe.

Additionally, it can be a sign of territorial poverty, where the cat compensates by ingesting food. The underlying cause may be a simple pastime or a substitute activity linked to an anxious state.

The ingestion of non-edible items can be observed in certain cases. Cats with a deficit in self-control may ingest foreign objects, most often during episodes of excitement while playing.

Some substitute activities or food-related stereotypies may emerge in cases of generalized inhibited anxiety states and chronic depression. These behaviors can manifest as licking or sucking on fabrics, sometimes leading to their destruction and ingestion.

A compulsive wool- or fabric-licking behavior with a genetic predisposition has been identified in Siamese and Sacred Birman cats. In the study concerned, this behavior was linked to early weaning, and all the cats involved also exhibited compulsive food-related behaviors (Borns-Weil et al. 2015).

The ingestion of non-food foreign objects is always pathological, and its underlying cause must be investigated, whether it results from excessive licking and hair ingestion, impulsive behaviors, or compulsive behaviors (◘ Table 15.2).

Summary of ▸ Sect 15.1.1

Three key elements must be documented when studying a cat's feeding behavior.

The first concerns the *mode of food distribution*. Cats require multiple small meals throughout the day, and feeding should not become their sole activity, as this could lead to obsessive behaviors.

The second relates to *food intake* itself, which can be a symptom of an underlying condition. In anxious states, eating may become a substitute activity. In compulsive states following deficits in motor control, food intake may become obsessive. In depressive states, eating patterns tend to be irregular. Excitement surrounding food intake can also lead to aggressive behaviors.

The third aspect involves assessing the *communication* and *rituals* that develop around feeding between the cat and its owner. The feeding area is where the cat most easily captures its owner's attention, and there is often confusion between food-seeking behavior and attention-seeking behavior.

Three key elements must be considered when assessing a cat's feeding behavior.

The first concerns the *mode of food distribution*. Cats require multiple small meals, and feeding should not be their sole activity, as this could lead to obsessive behaviors.

The second pertains to the *act of food intake itself*, which can be a symptom of a pathological state. In anxious states, food intake becomes a substitute activity. In compulsive states linked to deficits in motor control, eating behaviors may become obsessive. In depressive states, food intake tends to be irregular. Excitement surrounding feeding can also trigger aggressive behaviors.

The third point involves evaluating the *communication and rituals established around feeding* with the owner. Feeding remains the easiest way for a cat to capture its owner's attention, and confusion between food-seeking and attention-seeking behaviors is common.

15.1.2 Semiology of Drinking Behavior

In cats, water intake does not become dysregulated for behavioral reasons as it does in dogs. An increase in water consumption should therefore lead the practitioner to investigate potential organic disorders.

However, it is common for cats to enjoy playing with water and to be particularly drawn to a continuously running faucet, interacting with the stream. *Rituals* surrounding drinking behavior often develop as part of communication with owners. Their presence generally indicates a strong bond and complex interactions that contribute to the cat's psychological well-being.

In some cases, specific drinking behaviors may be observed in cats with self-control axis disorders. However, drinking can often be a challenging task for cats, as they may struggle to clearly perceive the water's surface and adjust their mouth position accordingly. The act of dipping a paw into the water or preferring a flowing water source is therefore a logical and adaptive behavior.

Cats are also highly sensitive to the taste of water and often prefer rainwater puddles over fresh tap water. The container holding the water may have a particular odor that deters the cat from drinking. For these reasons, many cats prefer to drink outdoors.

A cat's preference for moving water may encourage owners to provide dynamic drinking systems such as cat water fountains. However, to date, no study has demonstrated a significant increase in water consumption or urine dilution following the introduction of a fountain (Grant 2010).

Summary of ▶ Sect. 15.1.2

Water intake should be documented quantitatively, as many organic diseases are associated with polydipsia. Cats are sensitive to the taste and smell of the water they are offered.

Rituals may develop around drinking behavior, contributing to environmental enrichment.

15.1.3 Semiology of Sleeping Behavior

Cats are heavy sleepers, capable of resting for more than 18 h per day (Panaman 2010). For this reason, excessive sleep disturbances can be difficult to assess. Inhibited anxious cats may sleep even more, only waking to eat or engage in self-directed activities.

Conversely, cats with self-control deficits tend to sleep very little. The amount of sleep should be carefully evaluated, taking into account that in multi-person households, hyposomnia may go unnoticed, as the cat may take advantage of different wakeful periods to play or interact.

Productively anxious cats exhibit hypervigilance, which can disrupt sleep quality, leading to abrupt awakenings and a reduction in overall sleep duration.

Depressed cats often experience disturbed sleep, with sudden awakenings occurring just after falling asleep. Owners can be asked whether they observe their cat dreaming—characterized by rapid eye movements behind partially closed eyelids—since this is typical of a healthy cat. However, the absence of observed dreaming does not necessarily mean it is absent; it may simply have gone unnoticed. Nonetheless, anxious or depressed cats will have altered sleep phases and frequent startle-induced awakenings, which can be difficult to document.

Aging cats may also experience sleep disturbances. They frequently exhibit a reversal of their sleep-wake cycle, as well as nocturnal awakenings accompanied by periods of distress.

Assessing sleeping locations provides limited information, except for verifying whether the cat identifies the area as a restful space by marking it visually with scratch marks in the vicinity.

In cats, sleep study and sleeping locations are generally less informative than in dogs. The primary focus should be on evaluating the total amount of sleep (see ◘ Table 15.3).

Table 15.3 Reported signs in the semiology of sleeping behavior and possible interpretations

Reported sign	Symptoms *In all cases, a detailed exploration of the frequency, intensity, and context in which the sign arises is necessary to qualify it as a symptom*	Underlying dysfunction
Hypersomnia (more than 20 h per day)	Behavioral inhibition due to overwhelming and persistent fear	Generalized inhibited anxious state
	Behavioral inhibition due to a general decrease in motivation	Chronic depressive state
	Chronic pain	Any organic disorder leading to chronic pain
Hyposomnia (less than 15 h per day)	Lowered behavioral response threshold, hypersensitivity	Impulsive state
	Behavioral agitation due to generalized fears and increased vigilance	Intermittent or generalized productive anxious state
	Daytime or nighttime agitation related to pathological aging in a cat over 8 years old	Senile state with emotional and/or cognitive dysregulation in aging cats
Startle-induced awakenings	Hypervigilance	Intermittent or generalized productive anxious state
	Hypersensitivity	Impulsive or compulsive state
	Agitation linked to an underlying somatic condition, such as a painful disorder	Organic disease causing discomfort

Summary of ► Sect. 15.1.3

Hyposomnia indicates a deficit in self-control or hypervigilance, which is observed in complex phobic states or intermittent anxiety. Hypersomnia is a symptom associated with mood disorders or generalized inhibited anxiety. These states are also characterized by disturbances in sleep quality, which may include startle-induced awakenings, although these can often be difficult to observe.

15.1.4 Semiology of Elimination Behavior

Elimination behavior and marking behavior are usually studied together, as owners are not necessarily able to distinguish between them. Traditionally, elimination is classified as a centripetal behavior, while marking should be categorized as a centrifugal behavior, directed outward.

15.1.4.1 Typical Elimination Behavior

The normal elimination sequence consists of three phases:

- appetitive phase: the cat searches for an area where it can bury its excreta. It may scratch the surface, and this phase can be intense. Once the cat has created a depression in the loose substrate, it assumes a crouching position with its tail held horizontally and eliminates.
- consummatory phase: urine and feces are eliminated with slightly different postures. The cat's back is more arched when defecating.

- appeasement phase: the cat turns around to explore the area by sniffing, then covers its excreta. The olfactory exploration sometimes interrupts the covering process, which then resumes. Some cats exhibit excitement after eliminating and run away quickly; this behavior can indicate either joy or discomfort and is not pathognomonic of any particular condition.

A study compared the elimination sequence of cats at home, using a large litter box, to their sequence in a hospital setting with a smaller litter box (McGowan et al. 2017). Hospitalized cats exhibited a different and significantly longer elimination sequence. They hesitated before entering the litter box, spent more time scratching surfaces other than the litter, and took longer to sniff their excreta after elimination. This suggests that a shorter elimination sequence is associated with a more positive litter box experience. Additionally, when given the opportunity, cats use a considerable amount of space during elimination. Contrary to the common belief that cats immediately eliminate outside the litter box when dissatisfied with their environment, this study indicates that they continue to use the litter box, even if their behavior suggests frustration.

This sequence is markedly different from what is observed in urine marking.

In marking behavior, the appetitive phase also involves olfactory exploration, but the cat focuses on a *vertical* surface rather than a horizontal one.

15

During the consummatory phase, the cat remains standing, turning its back to the marked surface. Unlike elimination, which occurs in a crouching position, urine marking is performed while the cat is upright. A distinctive tail movement, quivering at the tip, is systematically observed as the cat sprays a small quantity of urine in a fine jet. This behavior occurs in both sexes but varies depending on sexual status (intact or neutered) and, in intact females, the phase of the estrous cycle (Horwitz 2019).

The final phase consists of exploring the marked area. In the case of *sexual marking*, the cat may also deposit facial pheromones. The key distinguishing factors between elimination and marking, crouching versus standing position, and whether or not the excreta is covered allow for clear differentiation. If the cat is crouching, it is eliminating. A cat that scratches after eliminating is not marking.

These are objective indicators that assist the practitioner. Owners often mistakenly believe that a cat exhibiting inappropriate elimination is expressing discontent. They frequently misinterpret observed eliminations as marking behavior.

Marking behaviors may also conclude with "emphasis behaviors." Some cats vocalize after marking, particularly in the context of sexual marking.

In cases of idiopathic cystitis, it is not uncommon for a urine marking sequence to develop into true inappropriate elimination.

15.1.4.2 Elimination and Litter Box

One of the reasons for the cat's success as a companion animal is its natural ability to be "clean" without requiring training.

Unfortunately, this well-recognized aptitude makes any lapse in litter box use intolerable for many owners. Their first instinct when faced with inappropriate elimination is often to punish the cat.

Punishment can have two major consequences. First, it can disrupt the cat's territorial organization, creating a fear of the litter box—especially if the punishment includes forced placement in the litter box. Second, and more concerning, it can lead to anxiety, which in turn may trigger idiopathic cystitis as an initial consequence.

Punishment is generally ineffective, and most owners abandon it quickly. As a result, they often fail to mention it during consultations. Without making them feel guilty, it is important to identify whether punishment has occurred, as this can help explain the cat's behavior **(see box *Punishment and inappropriate elimination*).**

Important Point of View

Punishment and Inappropriate Elimination
For most owners, cleanliness is considered an innate trait in cats, and any failure in this regard leads to punishment. When seeking consultation for inappropriate elimination, owners often state that they have "tried everything." This typically means they first attempted to control the issue through punishment, then stopped punishing, and are now at a loss for what to do next.

Punishment for inappropriate elimination is both ineffective and harmful. It can quickly lead to relational axis dysregulation, fostering the development of phobias or anxiety, or exacerbating these conditions if they are already present. The first step in addressing the issue is the immediate cessation of all punishment, while helping the owner understand that inappropriate elimination is never intentional.

There are specific rules regarding litter box use (including number, size, substrate, cleanliness, and placement), which are detailed in ► Chap. 29, ► Sect. 29.2.1.2. If these conditions are not met, inappropriate elimination may result from an unsuitable elimination area. If the elimination area meets all appropriate criteria, a medical cause must be investigated, whether it is organic or behavioral.

15.1.4.3 Inappropriate Elimination of Organic Origin

Elimination behavior is highly sensitive and can be easily disrupted. Once inappropriate elimination begins, it can quickly become complicated by behavioral issues, even if the initial episodes had an organic etiology.

When a cat experiences pain or fear during elimination, it may become sensitized and, through classical conditioning, associate the pain with the litter box or its surrounding environment. A cat experiencing intense pain while urinating may "believe" that the pain was caused by eliminating in that particular spot. As a result, it may seek a different location to avoid the same unpleasant sensation in future eliminations. Regardless of the organic cause leading to inappropriate elimination, it is not uncommon to require a behavioral rehabilitation process to address this litter box phobia, often referred to as litter aversion.

In young cats, congenital malformations can occasionally be responsible for inappropriate elimination, although they are quite rare. When present, they are usually detected early, as the cat's elimination behavior appears abnormal from the start, often with signs of incontinence or pollakiuria (Foster et al. 1999; John et al. 1996).

In cases of complete incontinence, the elimination sequence is absent, as the cat has no control over its urination. The fur on the hind legs is often wet, and the cat engages in excessive licking. These cases, however, remain extremely rare.

Inappropriate elimination is most commonly observed in cases of *enteritis* and *cystitis*, affecting both adult and young cats. These conditions frequently lead to litter aversion, as elimination becomes associated with discomfort or pain. Owners, who may lack the patience to avoid punishing the cat, can inadvertently contribute to the development of relational issues.

Enteritis may sometimes cause fecal incontinence, though this is not systematic. Additionally, owners tend to notice changes in stool consistency and frequency more easily and are generally more tolerant in these cases.

Cystitis, on the other hand, can be much more challenging to diagnose. Urine may not appear noticeably different, and affected cats often eliminate outside the litter box, likely due to pain or pollakiuria. If the litter box is difficult to access or perceived as uncomfortable, the cat may choose an alternative location for elimination.

Practitioners should rely on the description of the elimination sequence to identify a possible case of cystitis. If the cat has been reprimanded, it may wait until the owner is out of sight before eliminating. In such cases, the owner may be unable to describe the behavior, having never witnessed it. The only remaining distinction between marking and inappropriate elimination will be the location of the urine deposit, vertical for marking, on the floor for elimination.

In *aging cats*, common organic conditions can be further complicated by reduced adaptability. For instance, an obstacle in the path to the litter box may lead to inappropriate elimination in an arthritic cat, despite the absence of urinary, digestive, or behavioral disorders. In this case, the issue stems from discomfort, prompting the cat to seek alternative elimination surfaces.

Once organic causes have been ruled out, we must consider behavioral disorders associated with inappropriate elimination. Although urinary and fecal elimination share many similarities, we will distinguish between their respective semiologies. Both may coexist, and it is important to remember that identifying a behavioral cause does not exclude the presence of an underlying organic condition. In complex cases, addressing both medical and behavioral aspects is essential.

15.1.4.4 Behavioral Urinary Inappropriate Elimination

This section addresses inappropriate elimination, meaning urination outside the litter box. Urine marking is discussed separately.

When faced with urinary inappropriate elimination, the first step for the practitioner should be a urinalysis to rule out infectious cystitis or urinary crystals. Cystocentesis is a simple procedure and should be followed by urine specific gravity measurement, centrifugation, and microscopic examination. Additionally, cytobacteriology and an ultrasound examination are recommended. It is important to note that a diagnosis of infectious cystitis or urinary crystals does not exclude behavioral causes.

The elimination area must also be thoroughly assessed as part of the semiological examination. This assessment should be performed systematically in all cases of inappropriate elimination, even when an infectious cause or urinary crystals have been identified. A comprehensive evaluation should include the number of litter boxes, their size, whether they are open or closed, the type of substrate used, the cleaning routine, and their location (see ► Chap. 29, ► Sect. 29.2.1.2).

Ethological mismanagement of the elimination area may lead to a reactive state, in which case environmental modifications alone may be sufficient to resolve the issue. However, if inappropriate elimination persists, the underlying condition is likely pathological and must be treated directly. In these cases, viewing inappropriate elimination as a clinical sign of an underlying disorder is essential. Relying solely on environmental adjustments without addressing the psychological pathology constitutes a serious professional oversight.

Dysfunctions leading to inappropriate urination may involve the relational axis, the fear axis, or the territoriality axis.

In summary, the behavioral causes of inappropriate urinary elimination outside the litter box include reactive states to changes in the elimination area, dysregulations in the fear, self-control, relational, or territoriality axes, and aging-related issues. In older cats, inappropriate elimination may result from cognitive impairment or simply reduced mobility, making the litter box inaccessible given the cat's declining locomotor abilities (see ◘ Table 15.4).

Table 15.4 Differential diagnosis of urinary and fecal inappropriate elimination and their associated characteristics

Reported sign	Symptoms ***In all cases, a detailed exploration of the frequency, intensity, and context in which the sign arises is necessary to qualify it as a symptom***	Underlying dysfunction
Urine elimination outside the litter box	Complete elimination sequence. Dysuria, stranguria, and hematuria may be observed, and the elimination sequence may become simplified due to pain or the presence of owners and associated punishment **Presence of crystals and/or bacteria**	Infectious cystitis and/or related to urinary crystals.
	Complete elimination sequence. Dysuria, stranguria, and hematuria may be observed, and the elimination sequence may become simplified due to pain or the presence of owners and associated punishment **Sterile urine without crystals**	Idiopathic cystitis associated with an anxious state.
	Complete elimination sequence. No signs of pain or inflammation at the onset	Litter aversion due to elements that have repelled or frightened the cat during elimination. Most often associated with dysfunction in the fear or relational axis, or age-related impairments
Urine marking	Urine spray on vertical surfaces without burying behavior	Sexual marking in intact male cats at puberty and in females in heat Persistent territorial competition Sudden environmental change associated with dysregulation of the self-control axis, territoriality axis, or cognitive dysregulation in aging cats
Feces outside the litter box	Normal stools and a complete elimination sequence with an attempt to bury. If pain is present, the cat often runs away immediately after defecation and may vocalize during elimination (this can also occur with urination)	Litter aversion due to elements that have repelled or frightened the cat during elimination. Most often associated with dysfunction in the fear or relational axis, or age-related impairments
	Soft or diarrheic stools, either continuous or intermittent	Intestinal disorder Phobic or anxious state
	Normal stools not buried, usually found outdoors and at the edges of the territory	Fecal marking (rare)

Important Point of View

Urine Marking

Urine marking is a normal behavior in cats, serving to indicate their presence in a specific area, define their territory, and communicate status, particularly sexual status. The distinguishing features of marking behavior, as opposed to elimination, have been previously described. This distinction is crucial, as the management approach differs entirely between the two behaviors.

Urine marking reflects territorial competition and is never pathological in itself. However, it can rapidly evolve into a relational disorder associated with anxiety, especially due to punitive owner responses, persistent territorial competition, or intraspecific aggression resulting from these conflicts. This anxiety state can subsequently trigger idiopathic cystitis, which in turn leads to inappropriate elimination. Thus, a non-pathological urine-marking behavior can progress to pathological inappropriate urinary elimination (see ▸ Chap. 14, ▸ Sect. 14.5.1).

Urine marking is more common in multi-cat households or in environments where stray cats are present nearby (Pryor et al. 2001).

Major environmental changes, such as moving to a new home, can also trigger urine marking. This behavior is typically transient and should eventually be replaced by facial marking. If this does not occur, it is essential to investigate underlying factors that indicate territorial instability, including dysregulation of the territoriality axis, self-control axis, fear axis, or cognitive impairments in aging cats, which prevent them from properly structuring their territory.

In most cases, urine marking becomes complicated by inappropriate urinary elimination due to the associated anxiety, which further exacerbates the clinical picture.

15

15.1.4.5 Fecal Inappropriate Elimination

Fecal inappropriate elimination requires an evaluation of the elimination area similar to that conducted for urinary inappropriate elimination (see ▸ Sect. 15.1.4.2).

Regarding fecal elimination, a study found that cats prefer covered litter boxes available on the market over open ones, provided that they offer sufficient space (Beugnet and Beugnet 2020).

Fecal elimination outside the litter box is rarely linked to a psychopathological condition. In most cases, it results from an inadequate elimination area due to its design, size, location, cleanliness, or, more commonly, from pain or intense fear associated with defecation.

When assessing inappropriate fecal elimination, the first step for the practitioner should be to investigate pain-related causes. If the litter box has been associated with aversive or highly unpleasant experiences, inappropriate elimination tends to involve both urine and feces.

A stool examination (coproscopy) should be performed, along with a review of the cat's history for episodes of diarrhea or recurrent constipation (see ▸ Chap. 14, ▸ Sect. 14.5.2). Discomfort or pain during defecation, particularly during periods of intestinal disorders, often leads the cat to seek an alternative elimination site.

Additionally, chronic diarrhea is sometimes associated with phobic states and anxiety disorders, which may progress to inappropriate fecal elimination. In aging cats, age-related disorders can also lead to inappropriate fecal elimination (see ◘ Table 15.4).

Important Point of View

Fecal Marking

Cats are well known for their meticulous habit of covering their excrement. However, in certain situations, particularly outside their primary living area, they may leave their feces exposed. This behavior could represent a form of fecal marking. Cats likely use the scent of their feces and anal gland secretions as a means of communication. When two unfamiliar cats encounter each other, they often spend a significant amount of time circling and attempting to sniff each other's perianal region. If they are not overly antagonistic, they will eventually allow mutual sniffing (Houpt 2018).

Summary of ▸ Sect. 15.1.4

In most cases, the deposition of urine or feces in an inappropriate location, one not chosen by the owner, is perceived as inappropriate elimination. Since cats are widely regarded as exceptionally clean animals, this behavioral issue is often poorly tolerated by owners and quickly leads to relational conflicts. It is also one of the primary reasons for veterinary psychiatry consultations in cats.

When addressing both urinary and fecal inappropriate elimination, a thorough examination of the elimination area and a systematic investigation of potential organic disorders are essential for an accurate diagnosis. Idiopathic cystitis is a particularly important cause of inappropriate urinary elimination.

Dysfunctions involving the fear axis and relational axis are more commonly associated with inappropriate urinary elimination, whereas self-control and territoriality axis disorders are more frequently linked to urine marking, especially in the early stages.

Inappropriate fecal elimination is most often digestive in origin, but digestive disturbances can also result from an underlying phobic or anxious state.

15.1.5 Semiology of Repetitive Behaviors

This section describes the repetitive behaviors of cats, which we have chosen to classify as centripetal behaviors. These behaviors stem from the expression of a natural behavior that becomes reinforced or even automated due to internal or external reinforcement.

Cats exhibit numerous self-directed activities, meaning behaviors in which the animal redirects its actions onto a part of its own body. Additionally, many activities acquire a repetitive nature and become automatisms, losing their original function, which may have initially been relational (rituals) or soothing (substitutive activities). Licking, nibbling, suckling, or sucking behaviors—often considered signs of anxiety—are the most common.

Unusual behaviors, such as self-directed aggression or imaginary hunting, are rarer in this species. However, they can still occur, particularly in dissociative states. The veterinarian must carefully examine these signs, as they may be symptoms belonging to various states or disorders (see ◘ Table 15.5).

15.1.5.1 Self-Directed Behaviors

This section describes feline behaviors directed toward their own body, including grooming behaviors and self-aggressive behaviors.

▪ Grooming Behaviors

Cats spend approximately 6% of their time grooming (Houpt 2018; Eckstein and Hart 2000a; Swenson and Randall 1977). One of the primary functions of grooming is parasite control, particularly flea removal (Eckstein and Hart 2000b). Licking triggers the release of endorphins, which is why grooming behaviors may also occur in response to stress following a conflict (Van Den Bos 1998).

Grooming can also be a reaction to frustration; it is common to observe cats grooming themselves after failing to catch prey. In such cases, it serves as a substitutive activity that helps them alleviate their frustration.

Table 15.5 Reported signs in the semiology of repetitive behaviors and possible interpretations

Reported sign	Symptoms ***In all cases, a detailed exploration of the frequency, intensity, and context in which the sign arises is necessary to qualify it as a symptom***	Underlying dysfunction
Ritualized activity	Repetitive behaviors diverted from their initial purpose, becoming a means of communication. The cat or its ritual partner can stop the behavior spontaneously	Positive reinforcement by the owner Contextual compulsive state
Substitutive activity	Redirection of behaviors toward repetitive activities providing a coping strategy or relief from internal or external stress that cannot be resolved: Licking, sucking, grooming. These behaviors are contextual and intermittent	Pain Inhibited anxious state Chronic depressive state
Stereotypy	An invasive, repetitive behavioral sequence that cannot be stopped spontaneously and has lost its function	Invasive compulsive state Chronic depressive state Senile state Derealized state

The clinician must therefore assess grooming behavior, considering its intensity, frequency, and context, to determine whether it is a normal and adaptive behavior or a compensatory response that may occur in reactive or pathological states.

The first step is the direct observation of the cat, which provides immediate insight in cases of grooming deficiency. Grooming deficits in cats can have several causes:

- pain, preventing the cat from reaching certain areas of its body
- obesity, which limits the cat's ability to groom certain parts of its body
- chronic depressive state or senile state, which may lead the cat to neglect or stop grooming altogether

Additionally, poorly maintained, matted fur may also result from the texture of the coat, particularly in long-haired cats with fine fur, making grooming through licking insufficient. Sterilization alters the structure of the fur, making grooming more difficult.

For cats that cannot groom themselves, the presence of tangles and mats often reflects grooming intolerance rather than owner neglect. This intolerance may stem from pain, which can escalate into a phobia of brushing. These cats are frequently brought to the veterinarian for sedation and full shaving because their owners are unable to maintain their coats. Unfortunately, behavioral management of brushing phobia is rarely considered, despite being the only approach that fully addresses the problem.

Excessive grooming can have various underlying causes:

- Pruritus: most often due to the presence of fleas or other skin diseases (see ▶ Chap. 14, ▶ Sect. 14.2).
- Pain, which may lead the cat to excessively lick the skin covering a painful area in an attempt to relieve discomfort. This can occur in response to abdominal, urinary, or osteoarticular pain.
- Grooming as an adaptive strategy in response to stress, regardless of its origin. Licking triggers the release of central opioids, providing the cat with a sense of relief (Tynes and Sinn 2014; Willemse 1994). This licking can become excessive, leading to changes in coat quality, hairless patches, abrasions, and skin infections.

Excessive grooming can occur as an occasional response to transient stress but may also become more pronounced, developing into a true substitutive activity or even a compulsive state. In the case of substitutive behavior, it is still possible to interrupt the licking by redirecting the cat to other activities, although the cat may quickly return to it. This type of licking is particularly observed in intermittent anxious states and in the early stages of chronic depression.

Over time, the licking loses its soothing function and becomes compulsive. What initially started as a substitutive activity evolves into a stereotypy. Unlike substitutive behavior, stereotypic licking is difficult to interrupt. Stereotypies are particularly observed in generalized inhibited anxiety states and in advanced stages of chronic depression.

Clinical Illustration

Ms. X brings her cat, Atax, to the consultation. Atax has been living with an Elizabethan collar, which he wears due to an anxious behavior that drives him to attack his tail compulsively and incessantly. On the day of the consultation, the cat has been wearing the collar for 2 years and is also receiving clomipramine at a dose of 1 mg/kg. The owner is satisfied because the fur on his back has grown back.

The only time the collar can be removed is when Atax is in the countryside. There, he is calm and scratches very little.

The consultation is lengthy but reveals a crucial detail: in the countryside, the cat receives flea treatment, which he does not receive in the city.

The lesions observed in photos taken during a "crisis" show a hairless triangular area at the base of the tail, a typical pattern of flea allergy dermatitis. A detailed review of the medical record also shows that long-acting corticosteroid injections had improved his condition.

Atax suffers from an allergy, which explains why Clomicalm™ alone was insufficient. A rigorous flea control treatment allowed him to stop wearing the collar. The antidepressant treatment had to be maintained for 1 year.

This case highlights the necessity of integrating both physical and psychological management in treatment. A comprehensive approach is essential for resolving many chronic conditions.

▪ Self-Aggressive Behaviors

Self-aggressive behaviors refer to aggressive sequences in which the cat attacks a part of its own body, most commonly the tail. Some cats may also attack one of their limbs.

Tail-directed self-aggressive behaviors are often improperly described under the term feline hyperesthesia syndrome (see ▶ Chap. 14, ▶ Sect. 14.2, feline hyperesthesia box). It is important to remember that this term merely refers to one or several poorly defined symptoms, including Rolling Skin Syndrome (RSS), tail attacks, and reactions suggesting dorsal, lumbar, or sacral pain leading to the observed behaviors. Thus, beyond the fact that these cats attack their tails, it is crucial to understand the underlying cause of the behavior. At this stage, feline hyperesthesia only describes a symptom with various potential origins, and the term falsely implies a well-defined clinical entity. It is therefore imperative to abandon its use when describing these cases.

These self-aggressive behaviors can be highly dramatic and may be symptomatic of pathological states, including impulsive or compulsive states, productive intermittent anxiety states, and dissociative states. The key to differentiation lies in the context in which the aggression occurs. Self-aggressive behaviors can result from redirected frustration, acting as a compensatory behavior that may evolve into a substitutive activity, similar to the licking behavior described above. In some cases, they manifest as self-aggression episodes with no apparent or identifiable cause, which may be linked to a dissociative pathological state. Finally, organic causes such as pain, including vertebral bone tumors or muscle contractures, or focal epilepsy must always be considered.

15.1.5.2 Kneading and Licking

Some cats enjoy "kneading" during petting, a movement in which they rhythmically push their front paws back and forth, mimicking the kneading of the mother's teats during nursing. In kittens, this movement involves extending the claws, which are barely developed at that stage.

While this behavior is charming in a young kitten, it becomes problematic in an adult cat, especially if the claws are long and if the behavior is accompanied by suckling. In adult cats, excitement may also lead to slight salivation.

In a natural environment, the mother discourages this behavior at puberty, and the kitten quickly learns to stop, except on rare occasions when kneading soft surfaces before settling down comfortably. If owners tolerate it, the cat has no reason to stop, especially since this action is often associated with petting and, therefore, pleasure.

If owners find the behavior bothersome, the first recommendation is to move the cat away as soon as it begins kneading to indicate that it is unwelcome. However, if the owners have tried to stop the behavior and it persists, it is reasonable to question the cat's ability to control itself and to suspect an impulsive or compulsive state. In such cases, the cat often struggles to stop and may replace the initial behavior with fabric sucking or suckling on the owner's clothing.

15.1.5.3 Ritualized Activities

Rituals are activities that have been diverted from their original purpose to become a means of communication within a group. While they are discussed here, they are actually centrifugal behaviors.

Any behavior can become a way to gain attention and be integrated into a cat's communication strategies. When this happens, they are classified as "attention-seeking behaviors." In cats, these attention-seeking behaviors develop based on the owners' concerns. Many of them are related to food intake, but numerous cats also establish rituals around drinking water, while others demand a ritual involving the cleaning of their litter box. Some rituals involve contact behaviors, including licking.

In most cases, cats enjoy engaging in repetitive activities with their owners at the same time of day and under the same conditions. These behaviors may remain simple relational activities, but they can also take on a compulsive or even stereotypic nature. Even in their simplest form, these relational activities provide comfort to the cat and often to the owner as well. Their suppression is perceived as a change in the environment and can lead to frustration, which most often manifests as repetitive meowing, a behavior owners frequently find difficult to tolerate.

Rituals are therefore an important source of enrichment in a cat's life, but they are very difficult to eliminate. However, they can be used therapeutically to redirect undesirable behaviors by offering an acceptable alternative that benefits both the cat and the owner. In this way, they provide an effective and elegant solution for enriching the cat's environment.

Clinical Illustration

Luna has established a well-structured ritual with her owner. Every morning, when he goes to the sink to brush his teeth, she asks him to let a thin stream of water run. She then pretends to drink, licking the stream but not swallowing. She waits for petting until the end of the brushing routine. The ritual ends with a facial contact, after which she disappears and does not allow herself to be touched for the rest of the day.

Clinical Illustration

Mr. D has a diabetic cat. The veterinarian struggles to regulate the cat's diabetes because of a ritual involving food rewards. In addition to the prescribed diet, Mr. D gives his cat two generous 100 g portions of fish, which disrupts blood sugar control. While the veterinarian expresses concern and tries to reason with his client, Mr. D responds, "But if I don't give it to him, Doctor, he bites me!"

15.1.5.4 Substitutive Activities

A wide range of behaviors, from simple to complex, can become substitutive activities. A substitutive activity is a behavioral sequence that emerges to replace another sequence that cannot be carried out in response to a given stimulus. In practical terms, this sequence provides emotional relief that the normal consummatory phase failed to achieve, allowing the animal to return to an emotionally balanced state—at least initially.

A classic example is licking behavior that appears after a failed hunting attempt and the resulting frustration. The likelihood of a substitutive behavior emerging increases if it is related to the original behavioral sequence. For instance, a cat that fails to catch a bird may attack and shake a toy violently. While this response is not entirely appropriate, it helps reduce tension and provides temporary relief. The cat may also rush to its food bowl and eat a few kibbles.

When frustration occurs repeatedly, self-directed behaviors become a systematic response. Initially, they serve to soothe the animal and help maintain emotional balance. However, over time, these sequences become less effective, as they no longer lead to the release of endorphins. As a result, the cat tends to repeat them more frequently without experiencing the expected relief, increasing the risk of transitioning into a pathological state.

Excessive grooming often leads to hairless patches of varying extent. The most commonly affected area is the belly between the thighs, but it can spread across the entire abdomen and down the legs. In most cases, excessive grooming corresponds to a self-directed activity indicative of an *inhibited anxious state.*

Some cats that develop anxiety due to a poor-quality environment compensate for their boredom by eating. This behavior can become a substitutive activity, eventually leading to severe obesity.

Attacks on the tail can be observed in certain intermittent anxiety states. They are associated with the well-known RSS, which manifests as a wave-like skin twitch running along the spine, sometimes ending in a sudden tail flick followed by an attack. This symptom, often referred to as feline hyperesthesia, can be particularly pronounced at certain times of the day. In its milder form, it may be limited to a slight shiver or occasional tail movement. The frequency of these episodes reflects the degree of emotional intrusion affecting the cat. A cat in a completely calm state does not exhibit skin twitching or tail agitation. Assessing these signs and their frequency is therefore essential during the consultation. Observing these behaviors on video recordings is highly beneficial for the clinician.

The clinician must be cautious when interpreting skin-related signs. While some may indicate anxiety or depressive disorders, it is essential to rule out dermatological conditions, whether parasitic or non-behavioral. Organic imbalances often disrupt psychological stability and vice versa, making a comprehensive psycho-dermatological approach the only effective solution for complete resolution.

As symptoms progress and worsen, the practitioner may observe behaviors that appear aimless and occur repeatedly, sometimes without any apparent trigger. In cases of anxiety, derealization, or compulsions, the stimulation is internally driven. These behaviors may take up a significant portion of the cat's waking time. While initially serving as a form of relief, they become progressively less effective. In depressive states, these substitutive activities tend to be less frequent. The animal becomes hypoactive and no longer seeks comfort. As is often the case in behavioral medicine, examining the cat's overall behavioral profile is key to guiding the clinician toward the most plausible hypothesis.

15.1.5.5 Stereotypies

Stereotypies are *repetitive behaviors that have lost all function*. They may arise from a *ritualized activity* that no longer serves a communicative purpose or from a *substitutive activity*. In cats, behavioral sequences become automated quickly, and many ritualized behaviors are so stable that they nearly qualify as stereotypies. The same applies to certain persistent substitutive activities, which likely no longer provide any soothing effect. In *dissociative states*, some behaviors may also be classified as stereotypies.

The most frequently observed stereotypy in cats is licking of the upper lip. This behavior likely originates from an appeasement-related communication gesture. When repeated excessively, it causes irritation, which in turn leads to increased licking. This vicious cycle rapidly transforms the behavior into a compulsive and then a stereotypic act. Affected cats develop lip deformities, which can split all the way to the nostril. The resulting lesions may be mistaken for eosinophilic dermatitis or neoplastic processes, as they produce similar lesions. These injuries are particularly difficult to treat, as there is no effective way to protect the affected area.

Clinical Illustration

Allison is a male Sacred Birman cat, 5 years old, with a very gentle temperament. He presents with an extensive facial licking lesion. His upper lip is thickened, and the area between the lip and the nostril is raw and bleeding. This symptom is classified as an anxiety-related sign, and treatment with clomipramine at 1 mg/kg/day leads to a slight improvement, though without complete resolution.

Allison lives with an older cat, and they spend a lot of time curled up together, which appears to provide Allison with psychological stability. The treatment is maintained for 3 years until the older cat passes away. Shortly after, Allison's lip lesion improves and then disappears entirely. He becomes more active, and clomipramine treatment is eventually discontinued. The licking does not return.

15

- **Normal Behaviors in Derealization States**

Derealization states, particularly *dissociative states*, can take various forms. It is common to observe hunting behaviors directed at imaginary objects in these states. During these episodes, the animal is completely disconnected from its surroundings, and its attention is difficult to capture. Unlike stereotypies, where the cat appears to act mechanically, in derealization sequences, the cat seems to be responding to a stimulus with intent.

As with other abnormal behaviors associated with dissociative states, these behaviors gradually evolve from occasional occurrences to highly intrusive patterns. Derealization-related behaviors appear to be less frequent in cats than in dogs.

Summary of ► Sect. 15.1.5

Repetitive behaviors in cats can take various forms, including licking, sucking, self-aggression, kneading, and meowing. The semiological analysis of these behaviors should help the clinician determine whether they correspond to a ritualized activity, a substitutive activity, or a stereotypy. The pathological states and behavioral disorders associated with these different forms of repetitive behaviors vary and will influence the choice of treatment approach.

15.2 Centrifugal Behaviors

15.2.1 Semiology of Injurious Sequences

Bites and injuries caused by cats can be a reason for consultation, often leading to emotionally charged discussions with owners. The veterinarian must calmly address and dispel certain emotional biases while maintaining a strong sense of empathy.

To achieve this, it is paradoxically necessary to first accept and validate these emotions. This approach allows the veterinarian to use metacommunication, a technique that highlights their awareness and understanding of the owner's emotions. Recognizing and verbalizing these feelings, such as stating, *"I noticed that you are feeling emotional, and I understand,"* helps to soothe the owner and create a calm, constructive dialogue. Expressing empathy in this way fosters a strong therapeutic alliance, which is essential for collecting accurate information during the consultation.

Clinical Illustration

Imagine a scenario where parents consult you because their cat has scratched their child's face. The minimum requirement is to acknowledge the internal conflict these parents are experiencing. Of course, they love their cat, but any harm to their child is unacceptable. Even though you need technical information, the first step is to address these emotions and express empathy. Without this initial, seemingly non-essential step from a semiological perspective, it will be impossible to conduct an effective consultation.

Our goal here is to situate injurious sequences within behavioral semiology. The semiological approach in this context aims to *determine the intention* behind the cat's aggressive behavior. This allows us to answer key questions: "Why did this cat bite or scratch? How can we prevent it from happening again? Is it related to a pathological state?" It is essential to remember that aggression is a normal part of the feline behavioral repertoire, meaning that not every cat that has bitten or scratched is necessarily exhibiting a pathological condition.

To understand the cat's intent, it is helpful to refer to the *modified Moyer classification* (see ► Chap. 1, ► Sect. 1.3.2.3), which categorizes different types of aggression based on the objective pursued by the individual displaying it (see ◘ Table 15.6):

- maternal aggression is used to protect the nesting area and the litter
- territorial aggression is used to defend the territory and maintain its integrity
- irritation aggression is used to protect the individual when exposed to pain or discomfort
- uncontrolled emotional or survival aggression occurs when the individual's life is threatened or perceived as such
- idiopathic aggression refers to injurious behaviors displayed by cats suffering from a dysthymic and sometimes dissociative state, where the intent is difficult to define

Additionally, not all injurious sequences qualify as aggression, particularly those occurring during poorly controlled play or predation, which is aimed at obtaining food.

Some cats may be inadvertently injured during play or movement. These sequences often indicate a lack of control over locomotion, excitement, biting, or scratching, leading to a significant number of injuries.

It is therefore essential to analyze the behavioral sequence and context to identify the type of injurious behavior and, if applicable, classify the aggression.

15.2.1.1 Semiology of Maternal Aggression

Maternal aggression is characterized by a sudden charge from a queen to defend her litter. The sequence may be limited to warning signs, in which case it resembles territorial aggres-

Table 15.6 Classification of injurious sequences in cats based on the modified Moyer classification

Name of the sequence	Subtypes	Offensive/ defensive	Target	Trigger	Intention	Level of control
Non-aggressive injurious sequences						
Predation		Offensive	Prey	Prey	Kill to consume	
Vulnerant sequence during play		Offensive	Play partner	Excitement	Play	
	Aggressions					
Maternal aggression		Defensive	Outsider to the group	Approche des chatons	Defend her kittens + repel the threat	
Territorial aggression		Defensive	Outsider to the group	Intrusion du territoire	Defend against intrusions into the territory + repel the threat	
Irritation aggression	Frustration excitement Fear Pain	Defensive	Indifferent but greater tolerance toward familiars	Approach of the individual	Repel the threat	
Uncontrolled emotional aggression		Defensive	More frequent with strangers but possible with anyone	Approach of the individual	Repel the threat	
Idiopathic aggression		Defensive or offensive	Indifferent	No specific rule	Unknown	

sion. In some cases, the queen is highly reactive and attacks any individual passing nearby who she perceives as a potential threat. The trigger for aggression is not the actual danger but the perceived threat, based on the queen's fears and past experiences.

This type of aggression often startles the victim and provokes flight. For someone unaware of the presence of a litter, the attack may appear to be an inexplicable fit of aggression.

Maternal aggression is hormonally driven and can sometimes occur due to ovarian dysfunction, even in the absence of a litter. This type of attack can be observed in certain forms of dysthymia (see ▶ Chap. 12, ▶ Sect. 12.2).

15.2.1.2 Semiology of Territorial Aggression

In cats, this is the most ritualized form of attack. The most subtle form of threat is immobilization with a fixed gaze directed at the intruder. The cat's tension is intense, and its tail is often agitated. Brief charges may occur, accompanied by short hissing sounds that resemble sneezes. To increase its body size, the cat may present its side and advance sideways like a crab toward its target. The tail is often puffed up, and the back is arched. If the distance between the two individuals is sufficient, the threatened cat may retreat without being attacked.

The threat phase can last a long time, alternating between moments of calm and renewed aggression depending on the movements of the other individual. Fleeing inevitably provokes an attack if the distance allows it. Remaining motionless inhibits the aggression but does not change the situation. These sequences are typically observed between neighborhood cats when they encounter each other.

This pattern of interrupted sequences can also be seen in certain play behaviors. If, during play, the cat jumps onto a hand, arm, or foot and then freezes in a biting and scratching position around its imagined prey, the aggression may stop if the captured object remains still. Any attempt to withdraw, however, will trigger the aggression or continue the play sequence.

15.2.1.3 Semiology of Irritation Aggression

This type of aggression occurs in response to irritation, triggered by an emotion such as pain, fear, or frustration that accumulates until it manifests as aggression. The level of control in this aggression varies depending on the intensity of the underlying emotional state.

The attack is typically preceded by growling and hissing. The cat's ears are flattened against the back of its head, and its pupils are dilated. The attack itself may involve biting or scratching. In less controlled cases, the cat may launch itself entirely at the opponent, holding the bite while simultaneously scratching with all four paws.

In a confined space, such as during a clinical examination where retreat or escape is impossible, irritation aggression can quickly escalate into uncontrolled emotional aggression, which is particularly injurious and accompanied by numerous neurovegetative signs (see next section).

Important Point of View

Redirected Aggression

Redirected aggression is a form of irritation aggression that is not directed at the individual who initially triggered the irritation, either because that individual is inaccessible or because the cat fears retaliation. Instead, the cat attacks another individual, most often the owner, who happens to be nearby.

Some cats, after becoming agitated by watching another cat outside, may even run across the entire house to attack their owner.

This type of aggression is often completely misunderstood by the victim, who, by defending themselves, may inadvertently trigger another attack.

15.2.1.4 Semiology of Uncontrolled Emotional Aggression

This type of aggression follows the same sequence as irritation aggression, but cortical control is completely ineffective. Biting and scratching continue until the cat is able to escape or until the victim withdraws.

There is no true warning phase, as the attack is immediate. However, a sudden shift onto the side with front legs extended and claws out signals that an attack will occur if the aggressor enters the cat's defensive zone. The key distinction from irritation aggression lies in the presence of autonomic signs, although the idea that uncontrolled emotional aggression is simply an extreme form of irritation aggression remains consistent.

This form of aggression is directed at any individual whom the cat perceives as a life-threatening danger. The notion of a "deadly threat" is often a subjective impression arising from panic fear.

A defining characteristic of this aggression is the absence of a complete behavioral sequence. Bites and scratches occur immediately, sometimes accompanied by simultaneous growling. The attacks are forceful and multiple and serve only to create an escape route. The aggression may appear excessively violent, but its sole purpose is to secure an exit rather than destroy the opponent. In such cases, attacks to the victim's face are common.

The trigger for these attacks is not necessarily a confined environment but rather the *cat's perception of being trapped* in a *closed situation*.

The cat's overall behavior expresses extreme fear, including *pupil dilation*, *uncontrolled urination*, *defecation*, and *anal gland emptying*. These autonomic manifestations are a central feature of this type of aggression.

15.2.1.5 Semiology of Idiopathic Aggression

Certain mood disturbances and some derealization states in cats can lead to *unexplained violent attacks*. Unlike the aggressive sequences described in the previous sections, it is not always possible to identify a clear progression of events. The cat launches an attack in response to a perceived threat that has no objective basis in reality.

In dysthymic states, the cat experiences periods of heightened aggression, often accompanied by pupil dilation and growling, even in the absence of a visible threat. During these episodes, crossing an invisible aggression threshold can trigger an attack in such cases, isolating the cat until it calms down is the safest approach.

Derealization states are often accompanied by predatory sequences directed at imaginary or unidentifiable prey.

The description of these abnormal forms of aggression must be approached with caution. While similarities to other types of aggression can be noted, the primary goal of this classification, determining the cat's intent, becomes unattainable; hence the term idiopathic aggression. Differential diagnoses must also be considered, including pain, central neurological disorders such as focal epilepsy, and hormonal imbalances.

In addition to the aggression sequences described in the previous sections, it is important to consider the process of instrumentalization, which alters the behavioral sequence. Certain sequences may appear impossible to interpret. When a sequence occurs under the same circumstances, meaning the same causes and the same context, it gradually becomes an automatic or *instrumentalized* act. With anticipation, its occurrence becomes progressively earlier. Over time, the associated emotions fade, the initial warning signs disappear, and the return-to-balance phase is no longer necessary.

This phenomenon makes it difficult to identify the aggression sequence. Therapeutic intervention is also more complex because modifying the emotions associated with an instrumentalized bite becomes impossible. Instrumentalization functions as a *cognitive economy mechanism,* preventing the activation of the emotional system. This process occurs more quickly in cats than in dogs.

Recognizing the instrumentalization of a sequence allows the practitioner to establish a prognosis, which in this case is unfavorable.

15.2.1.6 Semiology of Vulnerant Sequences During Play

Not all injuries inflicted by a cat are intentional. Some occur during play. Under normal circumstances, the cat controls its movement, the strength of its jaw, and the extension of its claws. Play must remain within certain limits, which are taught from an early age by the mother. She intervenes in kitten play to moderate the level of excitement and prevent excessive aggression. By the age of 6 weeks, a kitten is capable of moderating the intensity of its bite and claw use, but this ability must be refined until the age of 12–14 weeks, which is the ideal adoption age for a kitten (see ▶ Chap. 4, ▶ Sect. 4.3.2).

Some accidental injuries may occur in adulthood, but they remain rare and should not be interpreted as a lack of control. However, a kitten older than 3 months that frequently injures its play partner should be suspected of having poor emotional regulation. An adult cat should never cause injury without intent. It is capable of controlling both its claws and the intensity of its bite. Under normal conditions, there is a strong inhibition against facial scratching, and any facial injury should raise concern for the clinician. Injuries inflicted during play are almost always associated with impulsive states or even compulsive states.

The learning process for play moderation is highly refined. A cat is capable of adapting to its play partner. It is natural for a kitten to be gentler with a more vulnerable partner. In kittens that have not yet fully developed this control, play sessions should be closely monitored to prevent early learning from being disrupted by overly rough interactions. The intensity of play also applies to interspecies interactions. Rough children can cause a kitten in development to lose control, and conversely, rough kittens can trigger excessive reactions from children. It is the responsibility of the children's parents to supervise these interactions. Play can be intense, but intervention should occur before excitement escalates. If children no longer want to play with the cat and begin avoiding its presence, it is necessary to assess the cat's self-control. If the cat avoids the child and seeks refuge when the child is playing, it is important to check its socialization, investigate possible pain, or determine if the child has made a movement that caused discomfort.

The specific case of play interactions with owners should also be noted. Owners frequently complain about biting or scratching during play. In all cases, a well-balanced adult cat does not cause injury during playful interactions. It is important to emphasize that offering one's hand for biting or scratching is not an educational practice. Many owners keep their kittens occupied by allowing them to nibble on their hands while they engage in other activities. This strategy may be convenient, but it reinforces the behavior and promotes dysregulation.

Additionally, if the owner overstimulates the kitten, injuries may occur. A common pattern involves physical punishment during play. At first, the kitten perceives that the owner is intentionally escalating the intensity and responds accordingly. Later, it realizes that the owner is angry and begins to feel fear. This may lead to a fear-induced bite, which is a form of irritation aggression occurring in a fear context. If play gets out of control, the owner should not punish but rather interrupt the session and, most importantly, help the cat calm down.

15.2.1.7 Semiology of Predation

This is not, strictly speaking, an aggression sequence, except from the victim's perspective. The cat is a solitary hunter that relies on ambush. During this phase, the cat adopts a highly attentive observational posture. Its body remains tense, and its ears are pointed toward the prey, tracking its movements. The capture leap may be preceded by small hindquarter movements and intense tail flicking, indicating the cat's excitement.

Once the prey is immobilized, the cat often remains motionless over it. Sequences of releasing and recapturing the prey are common, resembling the interrupted immobilization observed in other attack behaviors, particularly in play. It is well established that a cat may play with its prey and even derive pleasure from it. During these playful sequences, the cat may purr or roll over its dying prey.

Similar sequences are observed in kitten play and interactions with owners. A typical example is a game of "hide and seek," where the cat waits in ambush behind a piece of furniture or at the edge of a room before initiating a pseudo-predation sequence when a person walks by, often targeting their legs. However, this behavior is more closely related to play than to true predatory behavior.

In these playful interactions, bites and scratches should be controlled and non-injurious. A normal cat is capable of playing without extending its claws and while moderating its bite force. The presence of wounds or scratch marks on an owner's arms or legs suggests a deficit in self-control.

Summary of ▶ Sect. 15.2.1

An injurious sequence is a clinical sign related to an emotion, an intention, and sometimes a mood disorder. The description of the sequence, the identification or absence of a trigger, and the context will determine whether the injurious behavior should be classified as aggression, play, or a predation sequence.

In the case of aggression, it is necessary to specify the type using information provided by the owner and, in some cases, video support. This classification will then guide appropriate medical and behavioral measures.

Aggression is never insignificant. Even though there is no legal framework, at least in France, for assessing the dangerousness of a cat, semiology should help measure the risk the cat presents to its environment, the owners' awareness of this risk, and their willingness to implement measures to reduce it.

15.2.2 Semiology of Sexual and Maternal Behavior

This aspect of semiology is optional in consultation since the vast majority of cats are sterilized.

15.2.2.1 Estrus in the Female Cat

The female cat belongs to a seasonal and polyestrous species (see ▶ Chap. 3, ▶ Sect. 3.2.3.2). Estrus manifests through behaviors that are often disruptive for owners. If the cat is free to roam, she will call for males in a highly demonstrative manner for one or two nights. Mating induces ovulation, ending estrus and allowing the female to return to normal behavior.

If the cat is kept indoors, the vocalizations persist for several days, interrupted by brief periods of rest. The female's demands become very intense, and her irritability increases. She also engages in frequent urine marking.

This estrous behavior can also result from a persistent ovarian cyst, in which case the cat remains in estrus (a condition known as medical nymphomania). This condition may lead to aggressive behaviors, described under the term sexual dysthymia (see ▶ Chap. 12, ▶ Sect. 12.2).

Although variations exist between breeds, the average interval between estrous cycles is primarily influenced by environmental factors, such as photoperiod, rather than genetics (Romagnoli et al. 2019).

15.2.2.2 Male Sexual Behavior

Unneutered male cats become sexually active between 1 and 2 years of age. Depending on competition, the first matings may be delayed, and in some contexts, reproduction does not begin immediately. However, the male cat is fertile and capable of reproduction as early as 1 year old.

Male cats gather around a female in estrus and take turns mating with her. Fights may occur to establish the order of mating. A single female may mate with up to 15 different males (Dards 1983).

Some prepubescent or sterilized males are tolerated by females and may participate in kitten rearing and protection. Observations of feral cat groups show that young males leave their birth and development area before the age of three (Liberg 1983).

In some wild feline species, a male may kill a female's kittens to bring her back into estrus more quickly and mate with her. However, this behavior is rare in domestic cats (*Felis catus*).

15.2.2.3 Maternal Behavior

From a semiological perspective, it is important to note that some queens will not tolerate being disturbed during the first days after giving birth. If they have had kittens removed from previous litters, they may develop a tendency to hide with their offspring until the end of nursing or even until the kittens are 2 months old.

Kitten care can be shared among multiple queens, typically those with a familial bond (MacDonald 1981). An interesting variation of this behavior was observed in a recently spayed female who stole the kittens of another cat in the household. The issue was easily resolved by isolating the biological mother in a room with her kittens. This case suggests that maternal behavior is independent of ovarian hormones and/or is stimulated by a sudden drop in estrogen and progesterone levels, as occurs after spaying or parturition (Houpt 2018).

Owners should also be warned that estrus often returns quickly after giving birth. These postpartum heats are both highly fertile and discreet. To prevent back-to-back litters, early outdoor access should be restricted. A free-roaming female can technically have three litters per year, but this quickly leads to exhaustion.

Finally, in France, sterilization has become increasingly common, largely due to awareness campaigns led by animal welfare organizations, which have educated the public on the importance of neutering cats with outdoor access.

Summary of ▸ Sect. 15.2.2

The normal sexual activity of cats is highly intense, which is why most domestic cats are sterilized. This activity is loud and accompanied by urine marking. During sexual encounters, frequent fights occur, leading to injuries and the transmission of diseases. In natural conditions, multiple males mate with the same female.

Female cats may display aggression when they have a litter but can also engage in co-nursing, simultaneously nursing multiple litters.

15.2.3 Semiology of Play and Mutual Grooming

This section covers play behaviors and intraspecific mutual grooming, extending the discussion to petting and interspecific physical contact. A summary of the semiology of these behaviors is presented in ◘ Table 15.7.

15.2.3.1 Play

Previous sections have described certain play sequences involving owners. In general, feline play mimics predatory behavior. Anything that moves is perceived as potential prey in motion. Predatory behaviors, which are normally partially inhibited and controlled, are commonly observed.

Motor control mechanisms must be present. While mild injuries from kittens are acceptable, they should quickly learn to moderate their attacks. By 8 weeks of age, and no later than 12–14 weeks, a kitten should be able to play without causing injury.

Some cats engage in object retrieval games, which often surprise owners since this behavior is typically associated with dogs. However, like dogs, cats can learn that returning a ball or a crumpled paper allows the owner to throw it again. This game serves as an excellent territorial enrichment activity.

Play involving laser pointers can be highly stimulating for cats but may also lead to frustration since the "prey" can never be caught. To avoid this, such play sessions should end with the capture of a tangible object, such as a tossed kibble or a fabric ball.

Another favorite feline game is facial hide-and-seek with owners. This is symbolic since only the eyes must be hidden, while the rest of the body can remain visible.

Regardless of the type of play, these interactions indicate a positive relationship. Their regularity is particularly important from a semiological perspective.

Table 15.7 Semiology of play and mutual grooming

Reported sign	Symptoms *In all cases, a detailed exploration of the frequency, intensity, and context in which the sign arises is necessary to qualify it as a symptom*	Underlying dysfunction
Excessive and/or injurious play	The cat spends a large part of its time playing at the expense of sleep (less than 12 h)	Dysregulation of the self-control axis (feline hypersensitivity-hyperactivity).
	The cat has a seemingly normal play duration, but play sequences with a partner frequently result in injurious interactions	Dysregulation of the self-control axis (feline hypersensitivity-hyperactivity)
	The movements of household members are the only stimulation triggers, leading to aggressive play	Dysregulation of the territoriality axis due to environmental impoverishment.
Absence of play	The cat plays very little or not at all	Dysregulation of the fear axis (sensory deprivation syndrome) Aging Relationship disorders Organic pathology
Aggression during contact	The cat tolerates contact but quickly turns to scratch or bite. Often described as "petting-induced aggression syndrome."	Dysregulation of the self-control axis (feline hypersensitivity-hyperactivity) Dysregulation of the inter- or intraspecific relational axis Dysregulation of the fear axis (social or contextual phobia due to past punishments). Pain

15

A cat that plays excessively or too aggressively may be suspected of having a self-control or territoriality disorder, as these conditions often generate frustration. Conversely, a cat that rarely or never plays may suffer from a fear-axis disorder, senile pathological state, or depression.

15.2.3.2 Mutual Grooming

▪ Intraspecific Mutual Grooming

In domestic cats, mutual grooming is a cooperative activity that strengthens social bonds (Bradshaw 2016). However, Van den Bos (1998) questioned the purely affiliative nature of this behavior, suggesting that it may also serve as a mechanism to defuse aggression.

Findings from a recent study (Van Belle et al. 2024) indicate that mutual grooming plays a role in playful, agonistic, or intermediate interactions, as it is often associated with behaviors such as paw scratching and biting, as well as a focus on grooming the cranial region. These observations suggest that mutual grooming has a complex function within feline social dynamics.

▪ Interspecific Contacts

Not all cats experience the same level of pleasure from physical contact. Some enjoy proximity but do not tolerate direct touch on their body. These cats may solicit petting or simple physical contact, only to bite the person touching them. This behavior is most commonly observed in cats suffering from sensory deprivation syndrome, who no longer fear their owners but are unable to tolerate prolonged or invasive contact. A similar behavior can also be seen in cats with deficits in motor control, who may request affection but become overly excited, interpreting the interaction as the beginning of a play session, leading to a sudden wrestling attempt or loss of control over their excitement.

These signs are often grouped under the term "petting-induced aggression syndrome." However, organic causes, particularly pain, should also be considered as a possible underlying factor.

The owner's reaction plays a key role in shaping the evolution of these interactions. Cats quickly instrumentalize their responses. If the sequence has, on one or two occasions, ended in punishment, such as a shout or physical reprimand, the cat may begin to associate contact-seeking behavior as a trigger for irritation aggression. Understanding the initial interactions is essential to tracing the origin of this aggressive response.

These reactions to human contact can also extend to interactions with other species, such as dogs. In these cases, the contact is not in the form of petting but rather licking or mouthing, which the cat may tolerate to varying degrees.

> **Summary of ► Sect. 15.2.3**
> Play helps identify dysfunctions in the self-control axis when a cat fails to regulate its use of claws or teeth during interactions, as well as intense inhibition states when play behaviors have completely disappeared.
>
> Mutual grooming, like interspecific touch, provides insight into a cat's tolerance for physical contact. A lack of tolerance may indicate dysfunctions related to the fear axis, self-control deficits, or relational disturbances.

15.2.4 Semiology of Exploratory Behavior

This section examines the semiological aspects of exploratory behavior and marking in cats (see ◘ Table 15.8).

◘ **Table 15.8** Semiology of exploratory behavior and marking

Reported sign	Symptoms ***In all cases, a detailed exploration of the frequency, intensity, and context in which the sign arises is necessary to qualify it as a symptom.***	Underlying dysfunction
Inhibited exploration	During consultation, the cat quickly seeks to flee or hide. In a confined situation, it may become aggressive. At home, it occupies only a limited area of the territory due to its inhibited behavior	Dysregulation of the fear axis (sensory deprivation syndrome, social phobia, healthcare phobia). Chronic inhibited anxious state Chronic depressive state Senile state
Disorganized and excessive exploration	The cat is highly agitated during consultation, reacting to all stimuli. It struggles with frustration, and restraint may be difficult, as its excitement rapidly escalates, potentially leading to aggression	Dysregulation of the self-control axis (feline hypersensitivity-hyperactivity)

(continued)

Table 15.8 (continued)

Reported sign	Symptoms ***In all cases, a detailed exploration of the frequency, intensity, and context in which the sign arises is necessary to qualify it as a symptom.***	Underlying dysfunction
Facial marking	Excessive: – The cat faces significant territorial changes and attempts to restructure its environment; – The cat fails to structure its territory through facial marking due to impulsivity Deficient: – The cat's behavior is inhibited, including territorial marking	Reactive state. Dysregulation of the self-control axis (feline hypersensitivity-hyperactivity) Chronic inhibited anxious state Chronic depressive state Senile state
Lack of allomarking	The cat does not recognize one or more individuals as part of its territory or as a source of reassurance The cat is unable to structure its territory globally	Dysregulation of the fear axis (sensory deprivation syndrome, social phobia). Dysregulation of the inter- or intraspecific relational axis Dysregulation of the self-control axis (feline hypersensitivity-hyperactivity)
Excessive scratching	The cat experiences significant territorial disruptions and scratches vertical surfaces unrelated to its isolation area, marking multiple locations in its territory. The cat fails to structure its territory effectively	Dysregulation of the territoriality axis due to changes in the territory Intermittent productive anxiety state Dysregulation of the self-control axis (feline hypersensitivity-hyperactivity)
Urine marking	Except for sexual marking in intact male cats and females in heat, urine marking inside the territory is always pathological. The cat faces significant territorial upheavals and is unable to reorganize its territorial markers through facial marking or scratching	Dysregulation of the territoriality axis due to significant territorial changes

15

15.2.4.1 Observation During Consultation

The cat's behavior during a veterinary consultation should be carefully observed. Cats are generally unaccustomed to leaving their territory, and their behavior in an unfamiliar examination room is always influenced to some degree by fear.

More pronounced fear behaviors may be the result of previous veterinary experiences, and their intensity can escalate to healthcare phobia, which is particularly common in this species. These are behavioral disorders that can be managed, but unless the appointment is specifically dedicated to this issue, they are not usually the primary reason for consultation.

However, if a cat exhibits a phobic reaction toward handling or the veterinarian, this may indicate sensory deprivation syndrome or interspecific social phobia, both of which significantly increase the likelihood of such responses.

On the other hand, some cats are much more comfortable in an unfamiliar environment. While inhibited behavior during consultation is not always informative from a semiological perspective, a calm attitude in a new setting reflects good emotional balance and enriched early development. A well-adjusted cat should explore the room calmly, showing moderate caution.

Seeking a hiding place under furniture is a sign of fear, while choosing an elevated refuge does not indicate full confidence. Every veterinarian becomes familiar with the hiding spots in their office, and with experience, they can effectively assess a cat's level of comfort when allowed to move freely in the consultation room.

A cat may also explore with abnormally high curiosity. This type of exploration is often rapid and interspersed with startle reactions triggered by unexpected sounds or movements. Objects may be knocked over during this exploration, and clumsiness is often a sign of motor control deficits. Even in cats affected by hypersensitivity-hyperactivity syndrome (HSHA), exploration may eventually lead to seeking refuge, as self-control deficits do not exclude the presence of fear-axis disorders.

During this exploratory phase, the veterinarian should observe how the cat initiates contact. Following ethological principles, the veterinarian should establish an olfactory interaction before attempting to touch the cat, allowing the animal to sniff them first. This preparation for handling is also an effective way to prevent scratches, which can sometimes be anticipated from the cat's initial visual interactions. Refusals of contact should be interpreted cautiously, as they may indicate healthcare phobia rather than a general aversion to interaction.

A normal cat tolerates being touched and, in most cases, handled outside of its carrier. However, some cats refuse specific manipulations, such as touching the tail, being placed on their back, or enduring prolonged petting, any of which may trigger defensive behavior, even in a cat generally tolerant of contact.

Finally, as previously mentioned, play-related reactions involving biting and scratching may be triggered simply by an initial touch. This behavior is indicative of self-control deficits. However, it is relatively normal in kittens, provided that it remains moderate and can be voluntarily interrupted.

The practitioner must pay close attention to interruptions in contact. A fearful cat may initially accept being picked up and carried but then refuse further contact once it has been released. Therefore, it is preferable not to let go of the cat once contact has been established, especially if there is noticeable tension.

Observation should also focus on pupil dilation and any particularities in facial expression. Pupil dilation often accompanies states of fear. Some cats display permanently dilated pupils. Apart from conditions such as blindness or hypertension, this symptom may indicate an anxious or dysthymic state.

The practitioner may also observe flehmen behavior, during which the cat inhales and swallows air with its mouth slightly open. This is an exploratory behavior used to detect pheromone traces.

Excessive salivation is a sign of acute fear, commonly seen in phobic states.

Sweating from the paw pads can also provide information about the cat's immediate emotional state. A cat with moist paw pads is often highly anxious yet inhibited. Such a cat typically tolerates contact despite its large, dilated eyes. Conversely, the same pupil dilation combined with dry paw pads is more likely a warning sign of imminent aggression.

15.2.4.2 Description at Home

In addition to observing the cat during the consultation, the practitioner must ask the owners about the cat's behavior at home. Owners' observations are often detailed and provide valuable information. Several semiological aspects should be documented: the cat's daily routine and its marking behaviors, including facial marking, allomarking, urine marking, and scratching.

Daily Routine

Cats often have a very consistent routine. It is not uncommon to hear owners describe their habits in detail, such as *"in the morning, he's on my bed, then around 1 PM he walks in the garden and later naps on the couch."*

This description provides insight into the cat's different activities, feeding rhythm, and its interactions with the owners. While cats sleep extensively, they should also engage in hunting behaviors or, at the very least, spend time observing their surroundings. The less time a cat spends outdoors, hunting, or watching its environment, the more important its interactions with the owners become.

The cat must also find stability in its living environment. During these discussions, the practitioner can assess the richness of the cat's environment. This evaluation can be complemented by a *floor plan of the home*, where owners indicate different activity zones. Owners may also mark areas where the cat deposits pheromonal signals.

It is useful to identify *relational rituals*, which are part of the cat's routine. These short, repetitive interactions occur between the owner and the cat at the same place, at the same time of day, and in the same manner. These rituals provide comfort to the cat, strengthen its bond with the owner, and enrich its territory on an emotional level.

Facial Marking

15

There are two types of facial marking. The first type, allomarking, is directed toward living beings and serves to identify individuals within the cat's territory. This behavior is often highly demonstrative and is perceived by owners as a form of affection or "cuddling." Allomarking has a calming effect, as the marked individual is recognized as a friend or a source of comfort and pleasure. This behavior is an important factor when evaluating the cat's relational axis. A lack of allomarking toward all or some household members should prompt the practitioner to further investigate the cat's social relationships within the group.

The second type of facial marking is intended for the cat itself. It consists of pheromonal deposits on objects, which help structure the territory and provide reassurance as the cat moves through its environment. This marking is often subtle, and owners may need guidance to recognize and observe it. It is important to determine whether this behavior is present, increasing, or disappearing.

A stable and content cat will exhibit consistent but discreet facial marking. Minor territorial changes typically cause a temporary increase in this behavior, indicating a reactive state. This is commonly observed when new furniture is introduced or when objects are rearranged.

A significant increase in facial marking often indicates that the cat is struggling to structure its environment or is experiencing difficulty adapting to major changes, which may suggest an anxious state. Conversely, a lack of facial marking may be associated with self-control deficits, where the cat is unable to organize its territory, or with an inhibited anxious state or chronic depression.

Urine Marking

Urine marking is a distancing signal intended to restrict or limit access to a specific area within the cat's living space. Occasional marking at territorial boundaries is normal and must be periodically renewed. Since these marks are meant to be detected by other occupants, they are unnecessary if the territory is stable and unthreatened. Conversely, urine marking increases in response to perceived invasion threats.

For cats, urine spots are not calming signals but protective markers. When urine marking becomes more frequent, it can become anxiety-inducing, as its increased occurrence reflects its ineffectiveness in maintaining territorial boundaries.

The appearance of urine marking in a previously unmarked area indicates that the cat feels its space is being invaded. The source of this perceived invasion must be identified—whether it is another cat, an animal, or a new human in the household.

When urine marks become excessively frequent, this strongly suggests that the cat is experiencing severe emotional distress. At this stage, the cat's psychological state is pathological, often indicative of intermittent or generalized productive anxiety.

In unneutered cats, urine marking may also serve a sexual function, predominantly observed in male cats. After creating a urine spot, the cat often reinforces it with facial pheromone deposits. This marking is a normal consequence of sexual activity, intended to inform potential mates of the cat's hormonal status. In females, urine marking may also occur physiologically during estrus.

▪ Scratching

Like urine marking, scratching is often quickly reported by owners, who find the resulting damage to their environment distressing. However, scratching also serves as a marking behavior, intended to be read by other individuals entering or inhabiting the same territory. These visual and olfactory marks indicate an isolation zone that is not meant to be shared. Scratches are most commonly found around the cat's resting area. In some cases, for cats with both indoor and outdoor access, markings may also be present near entryways.

Scratching consists of vertical visual marks and should not be confused with stretching-related scratch marks, which are usually made on horizontal surfaces as part of claw maintenance.

While an increase in scratching marks may suggest territorial disorganization, it is not always a significant indicator of distress and can sometimes be excessive in well-balanced cats. However, particularly excessive scratching, which does not align with the cat's isolation areas and persists despite the presence of appropriate scratching posts, should raise concerns about a pathological condition related to dysregulation of the territoriality axis or self-control axis.

Summary of ► Sect. 15.2.4

The semiology of exploratory behavior begins with assessing exaggerated fears in response to various stimuli such as noises, humans, or other cats, as well as determining the extent of these reactions—whether they are contextual, limited to certain individuals, or generalized. This requires detailed descriptions of the cat's initial reactions and postures during first-time exposures, as behavioral evolution may lead to instrumentalized responses that obscure the original cause of the behavior.

Observing the cat during consultation provides valuable semiological insights that can be linked to fear-axis disorders, self-control deficits, territorial dysregulation, or relational issues. Therefore, allowing the cat to move freely during consultation is crucial. When possible, observing the cat at home, either directly or through videos provided by the owner, can further clarify its behavior.

A thorough evaluation of the cat's behavioral fields, along with its facial, scratching, and urine markings, is an essential part of the exploratory behavior semiology.

15.3 Other Elements of Semiology

15.3.1 Semiology of Behavioral Development

For a complete semiological evaluation, it is essential to gather information about developmental conditions.

The first weeks of development are crucial in shaping an individual's personality. Understanding early life conditions and major events that influenced the cat's formative period provides valuable insight into the origins of behavioral disorders. Conversely, when early development appears optimal, yet numerous symptoms are present, this suggests significant inherent vulnerabilities. In such cases, it is useful to investigate the behavior of other individuals from the same family to assess the potential role of *genetic predisposition.*

This information helps practitioners understand the *pathological pathway* and may guide treatment decisions.

However, there is a risk in placing too much emphasis on these early-life factors. Regardless of the origins of a disorder, therapeutic intervention must continue. A common pitfall is to abandon treatment efforts under the assumption that past events cannot be changed. If a behavioral disorder has genetic roots, it may seem that the only option is to accept the condition, rather than seeking ways to manage or modify it.

Owners often seek explanations in past events to justify their cat's current behavior. This mirrors the concerns raised in discussions on temperament classification. Common statements include: *"He is fearful because he was separated from his mother too early; there is nothing we can do." "It's the breed, these cats are always fearful."*

Such assumptions can lead to fatalistic attitudes, which may hinder effective treatment and management strategies.

We can establish a number of facts regarding this development and focus on informing our semiology based on the following points:

- *age at adoption:* the ideal adoption age is 3 months to allow the self-control axis to mature under the mother's care (Ahola et al. 2017). Earlier adoption may promote the development of impulsive or compulsive states. Conversely, later adoption can also be problematic for the development of the fear axis, particularly if the kitten's living conditions before adoption differ significantly from those provided by the new owners;
- *adoption environment:* the closer the kitten's early environment is to its future living conditions, the lower the risk of fear axis dysregulation. Kittens born to responsible breeders in a family setting and exposed to various stimuli from an early age are more likely to develop balanced behaviors;
- *Family history of behavioral disorders:* a mother exhibiting dysregulation of a behavioral axis, particularly the fear or self-control axes, is likely to pass this imbalance on to her kittens, either genetically or through imitation and emotional contagion. Kittens may either model their mother's behavior or fail to develop proper self-regulation. Determining the respective influence of genetics and learning is complex, but both contribute to the kitten's behavioral maturation;
- *special case of street kittens*: kittens adopted at a very young age and bottle-fed are particularly susceptible to developing dysregulation of the self-control axis. It is essential to look for suggestive signs during the first consultations in order to implement an appropriate management plan, including medication, from an early age. Additionally, street-born kittens and those from stray mothers are often left to fend for themselves very early in their development. This occurs because the mother cat quickly returns to estrus after giving birth and is ready to have a new litter well before the previous one reaches 3 months of age. This situation also fosters the development of self-control axis dysregulation;

  Moreover, if these kittens are born in the street but adopted later, after 6 weeks, they may develop severe phobias of humans. Despite owners believing that patience will improve the situation, these cases are often very severe and require early pharmacological intervention to facilitate socialization with humans. Finally, street cats have been subject to very little genetic selection pressure from humans. Stray females reproduce with males that may be highly unsocial (McCune 1995). This does not favor the selection of cats exhibiting behaviors better suited to a pet lifestyle and human interactions.
- age of onset and duration of disorders: adolescence is the period when a cat fully establishes its territory. However, in multi-cat households, puberty is often a time of tension, particularly leading to dysregulation of the territoriality axis.

In practice, the clinician must gather this information to validate the coherence of their observations with the rest of the semiology. However, our treatments are based on the observation of present facts, here and now. Past elements are not essential for helping the animal progress. In many cases, the past remains unknown, yet this does not prevent us from providing effective treatment.

That being said, certain past events can help confirm a diagnosis and clarify the prognosis. For example, a cat that suddenly develops a phobia toward a specific human is more likely to suffer from post-traumatic phobia rather than deprivation syndrome—especially if there is sufficient evidence to conclude that the early environment and developmental period were adequate and if a traumatic event can be identified as the trigger for this phobia. In this example, a post-traumatic social phobia has a much better prognosis than deprivation syndrome.

15.3.2 Semiology of Direct Observation

15.3.2.1 Direct Observation Before Clinical Examination

Cats usually arrive at the consultation in a transport carrier. Compared to dogs, this means the veterinarian loses some information regarding the cat's behavior in the waiting room and its communication with the owner. However, some particularly well-socialized individuals may be brought in on a leash and harness. Certain veterinary clinics and hospitals have developed enclosed cat-only waiting areas where cats can be allowed to roam freely. However, given the possibility of encountering conspecifics, this is not necessarily an optimal solution in terms of feline well-being.

Once inside the consultation room, it is preferable to wait a few moments before releasing the cat. From within its transport carrier, the cat engages in a remote assessment of the room and its occupants. This period allows the veterinarian to focus on the owner.

Following this initial phase, and unless otherwise advised by the owner, the transport carrier can be opened for most cats. The carrier should be placed on the floor, offering the cat control and the choice to explore the room at its own pace. Meanwhile, the discussion with the owner can continue.

For initial contact with the cat, it is recommended to extend a hand, allowing the cat to initiate olfactory contact with the practitioner.

Whenever possible, it is best to use the same consultation room consistently. This enables the veterinarian to establish a reference for typical feline behaviors under these conditions and compare observations accordingly.

The cat's behavior in the consultation room provides valuable insights into the four behavioral axes. In this section, we present the significant attitudes associated with each axis.

15.3.2.2 Fear Axis

To assess feline behavior along the fear axis, the practitioner must identify the four expressions of this emotion: immobility, escape, aggression, and agitation.

Some cats remain completely immobile and will often stay inside their carrier.

In other cases, the cat appears agitated. It quickly exits the carrier and attempts to escape from the consultation room. These cats often crawl rapidly across the floor before seeking refuge in a corner or under a piece of furniture.

Some cats may also exhibit aggression. This aggression is rarely offensive; it typically emerges when the veterinarian approaches within a certain distance. The cat, which usually remains in its carrier or has fled to a corner of the room, hisses, swats, and may bite. This type of aggression is classified as irritation-induced aggression in a fear-related context or, in some cases, as uncontrolled emotional aggression, which can be highly injurious. Such responses may be accompanied by defecation or urination. These autonomic manifestations provide insight into the cat's emotional state. Fear may also trigger panting, salivation, pupil dilation, and sweating of the paw pads.

The cat and its behavior should first be evaluated from a distance and then reassessed during handling and the clinical examination, if possible.

Fear is a normal and adaptive emotion. Its intensity, frequency, and contexts of occurrence help differentiate normal fear from pathological fear, such as a *phobic state* or *anxious state*.

The observations gathered during consultation pertain to the cat's fear response within this specific context. However, it is essential to obtain additional information regarding the expression, intensity, and frequency of the cat's fear in other settings, such as at home with familiar household members, at home in the presence of unfamiliar people or animals, or in outdoor environments. While a cat's reactions may vary significantly between the clinic and its home environment, it is exceedingly rare for a cat that is completely unapproachable in a clinical setting to be perfectly well-balanced elsewhere.

15.3.2.3 Self-Controls Axis

A cat lacking self-control is often restless and intrusive. Its exploration is disorganized and repetitive because it remains incomplete. If the room allows, exploration may quickly take place at higher elevations. Affected cats are often clumsy and may knock over objects.

These hypersensitive cats exhibit intense reactions to handling and to pain induced by medical procedures. Their aggression is poorly controlled and can sometimes be very sudden.

Kittens exhibiting self-control axis dysregulation are relatively easy to identify. The clinical examination can be challenging, as these kittens or cats tolerate frustration poorly. They quickly become agitated during restraint and may turn aggressive if they cannot free themselves. Some kittens display strong extension and defensive reactions when lifted by the scruff. However, the absence of a scruff-carrying reflex should be interpreted with caution, as it is not pathognomonic (Moody et al. 2020).

Self-control axis dysregulation is often indicative of inadequate developmental conditions and is frequently accompanied by fear axis dysregulation. In cats, impulse control disorders are therefore commonly associated with fear-axis disorders. Depending on the context and severity of the dysregulation, affected cats may present as inhibited or as immediately aggressive due to being rapidly overwhelmed by fear. In such cases, the classic signs of impulse control deficits may be harder to identify during consultation.

15.3.2.4 Territoriality Axis

Dysregulation of the territoriality axis is not typically observable outside the cat's usual territory. However, some cats may begin to engage in facial marking or allomarking during the consultation. This provides an opportunity not only to demonstrate to owners what these different types of marking entail but also to assess the cat's emotional and cognitive balance during the consultation.

Facial marking during the consultation is generally indicative of good emotional balance. However, some cats with self-control axis dysregulation may engage in marking certain objects in the room in a manner that appears repetitive or obsessive.

Allomarking on the owners or veterinary staff during the consultation is usually a sign of good socialization to humans. However, as with facial marking, excessive frequency and intensity are commonly observed in cats with poor impulse control.

15.3.2.5 Relational Axis

As mentioned in the previous section, allomarking during the consultation is a sign of good socialization and strong familiarization abilities.

During the consultation, various types of communication between the owner and their cat can be observed, such as rituals, play, or conversely, fears exhibited by either party. It is important to assess whether the owner is able to correctly interpret their cat's behavioral and emotional signals in order to evaluate the quality of their relationship. Misunderstandings and misinterpretations are common on both sides.

The most frequent issue is the owner's failure to respect the signals sent by the cat, leading to forced interactions that are often poorly tolerated by the animal.

15.3.2.6 Clinical Examination

We have discussed observation and the signs to look for according to different axes. However, the *organic aspect* of this assessment must not be overlooked. A thorough clinical examination, from nose to tail, is essential.

The veterinarian must focus on identifying and evaluating the following:

- the presence of articular or abdominal pain
- the quality of sensory organs
- the presence of abnormalities in the neurological examination
- the condition of the skin and coat
- the detection of any other abnormalities

Veterinary psychiatry is a medical field like any other, with links to all other areas of medicine, as discussed in ► Chap. 14.

Diagnosis must always go through the differential diagnosis stage, which should include both organic and behavioral causes in order to comprehensively explain the clinical picture.

Finally, a cat's behavior during the clinical examination provides valuable information. Behavioral inhibition is more commonly associated with a fear-axis disorder, frustration intolerance is more indicative of a self-control axis disorder, and aggression or attempted aggression should also be carefully assessed.

15.3.2.7 Communication and Environment

Direct examination must be complemented by an assessment of the cat's communication and environment.

Throughout the consultation, the words and attitudes of the owners and their companions can help support a diagnosis. These observations should be noted, and sometimes it is best to simply retain them without making immediate comments. Therapy will take these elements into account, and ethological adjustments can be introduced either during the consultation or at the moment of explaining the therapeutic approach.

Regarding the environment, the information provided by owners about the enrichment of the cat's living space, or the lack thereof, should also be recorded by the clinician. Enhancing the cat's stimulating environment may be part of the treatment plan.

Summary of ► Sect. 15.3.2

Direct observation provides valuable insights into communication, environment, and the four behavioral axes. However, the clinician must interpret these observations with nuance, considering the unique context of the consultation room. The displacement of a cat from its usual environment masks many of its natural behaviors.

A clinical examination is an integral part of this observation and remains essential in every case.

15.3.3 Semiology of Behavior in Aging Cats

We have established that aging is an inevitable process. Whether or not it is accompanied by pathological mechanisms, it leads to behavioral changes. Slowing down this deterioration is sometimes possible if intervention occurs early, although therapeutic options remain limited. No treatment can reverse the aging process.

A key aspect of this semiological assessment is that it requires comparing the cat's current state to its previous state. What matters here are changes, initially subtle but requiring intervention before they become overwhelming. As with aggression, owners often seek help only when they no longer understand their cat. If they can find an explanation for the observed behavior, this understanding provides reassurance, which in turn reduces their demand for therapeutic intervention. When an aging cat loses certain abilities, owners may simply conclude, "He's getting old!" It is therefore the veterinarian's role to actively gather relevant information that may indicate abnormal deterioration.

As cats age, their habits gradually change. Senile disorders tend to follow similar patterns, but the earliest symptoms often affect one specific behavioral axis first. Identifying these initial changes is crucial, as early intervention may help slow the progression.

Special semiological attention must be given to somatic conditions, as illnesses in elderly cats often have subtle early signs and may be mistakenly attributed to psychological disorders.

The senile elderly cat exhibits consistent behavioral characteristics. It is an animal that reduces its activity and occupies only a small portion of its former territory. Often, it settles near an observation area and establishes its isolation zone there. When it leaves this zone for broader exploration, it experiences moments of anxiety and struggles to cope with these excursions, which seem to become real adventures. It gets lost and becomes distressed until it returns to its isolation zone.

For these reasons, it is advisable to organize a very small territory where the cat has access to an elimination area, a feeding area, a bed, and, if possible, some observation points. It is useful to prevent the cat from leaving this mini-territory and to arrange for short, supervised outings that end with the cat being gently returned to its bed.

The first signs of senility are primarily a decrease in activity, which is most evident in a restriction of territorial occupation.

An attentive practitioner monitors weight trends, but more importantly, the loss of muscle mass, which is the main indicator of entry into advanced old age. This muscle loss is associated with various organic disorders, including dental issues, tumors, kidney failure, hypertension, hyperthyroidism, and diabetes. Given the reduction in the cat's occupied space, it is essential to monitor food intake and its consistency.

It is also important to ensure that the cat's ability to move within its territory aligns with its physical capabilities. The onset of senility can sometimes result from physical impairments.

Finally, in the early stages of aging, certain behavioral issues may emerge, or rather, re-emerge, after having been managed in the cat's youth. This may include the recurrence of fears and phobias in cats with a fragile fear axis, the return of poorly controlled behaviors such as nibbling or scratching in cats with self-control difficulties, or the reappearance of inconsistent marking behaviors in cats that previously struggled with territorial structuring.

Important Point of View

Urine Soiling in Elderly Cats

In anxious cats, and particularly in elderly cats, significant episodes of urine soiling may occur. It is important to remember that cats are highly susceptible to idiopathic cystitis of emotional origin. This non-infectious bladder inflammation can lead to persistent urine soiling if left untreated.

Beyond this primary cause, several age-specific factors can contribute to urine soiling in elderly cats. Polydipsia associated with kidney failure and diabetes can lead to elimination issues due to an increased frequency of urination. Additionally, chronic joint pain may make it difficult for the cat to reach and use its usual elimination area, as mobility issues can hinder its ability to climb into the litter box. Finally, in the early stages of senile pathological states, the cat may struggle to locate the litter box or may be unable to hold its urine long enough to reach it.

15

Four types of disturbances occur with aging: *cognitive, mood, emotional, and self-control disorders* (◘ Table 15.9.), along with somatic difficulties. This section does not cover signs associated with tumors or other brain lesions.

The practitioner should focus on identifying the earliest signs of deterioration. As the condition worsens, dysfunctions accumulate, making it increasingly difficult to distinguish between the different clinical presentations.

15.3.3.1 Cognitive Disorders

■ **Loss of Learned Behaviors**

Learning disorders in cats are often subtle and more difficult to detect compared to dogs. Unlike in dogs, a decline in sensitivity and reactivity, particularly auditory reactivity, is not commonly observed.

Table 15.9 Key warning signs in the semiology of aging to determine the starting point of the disorder

Starting point of behavioral change	Most frequently observed signs
Cognitive dysregulation	Loss of spatial or temporal orientation Loss of learned behaviors Nocturnal and daytime vocalizations
Mood dysregulation	Decreased motivation Social withdrawal
Emotional dysregulation	Intensification or emergence of severe fears Nocturnal and daytime vocalizations
Self-control dysregulation	Loss of control in interactions, sometimes with the reappearance of uncontrolled scratching and biting

Some cats may change their territorial occupation, settling in places they did not frequent before. This shift may indicate a loss of spatial reference points and the onset of cognitive disorders.

The resulting *spatial disorientation* is often subtle. Affected cats may meow for help, and owners typically respond by calling them, allowing the cat to use their voice as a navigational cue.

As spatial orientation deteriorates, the cat progressively restricts its territory to a very small area.

A reversal of sleep cycles may also occur without an actual sleep disturbance. This is a simple temporal disorientation, but it can be accompanied by nocturnal or daytime vocalizations, which often become a primary reason for consultation in elderly cats.

15.3.3.2 Mood Disorders

Aging can begin with mood disturbances, which are primarily characterized by sleep disorders. A depressive cat often experiences nocturnal awakenings, sometimes accompanied by episodes of panic or intense fear. After waking, the cat wanders around, calling for its owners. These nocturnal awakenings, which may also result from confusion between day and night cycles, are often associated with loss of territorial reference points. In many cases, the fear accompanying these distressing episodes leads to urination and defecation.

Mood disorders are also marked by a loss of motivation, which manifests as a disappearance of the cat's preferred activities. In the depressive course of aging, sleep and appetite disturbances are very common. This pattern of nocturnal activity quickly becomes unbearable for owners. The emotional deterioration progresses rapidly, fueled by the owners' frustration and the resulting breakdown in their relationship with the cat.

15.3.3.3 Emotional Disorders

As cats age, they may become anxious due to the reappearance of old fears that they had previously learned to manage. This state can sometimes lead to aggression, prompting owners to seek veterinary consultation. The sudden and unexplained nature of this change often leads owners to look for an explanation.

The first hypothesis to consider is that the aggression is triggered by pain or the anticipation of pain. Pain itself lowers the threshold for reactivity, making irritation-induced aggression occur more quickly.

Anxiety episodes can also manifest through vocalizations, which may become highly disruptive for owners.

15.3.3.4 Self-Control Disorders

In some cases, the onset of dysregulation stems from a loss of motor control. Cats may appear to experience a second youth, responding to petting with poorly controlled play-fighting. This behavior can quickly strain the relationship with the owner. As the relationship deteriorates, it can lead to further emotional, mood, and cognitive disturbances.

Summary of ▶ Sect. 15.3.3

The symptoms of aging are often subtle at first and must be actively investigated, as owners often do not consider them worth mentioning. The aging spiral represents the typical pattern of worsening, regardless of the initial trigger.

The most common clinical signs include intense vocalizations of emotional or cognitive origin, elimination disorders, and sometimes a loss of control in interactions, leading to potentially harmful play or contact behaviors. A thorough examination is essential to assess all possible organic causes that may explain some or all of the observed clinical signs.

The loss of abilities in aging cats must be addressed early, as they can quickly become overwhelming and impossible to manage if left untreated.

15.4 Semiology by Behavioral Axis

The semiological assessment can be conducted using the approach described above, analyzing centrifugal behaviors first, followed by centripetal behaviors, then developmental conditions, and finally direct observation. It is also possible to proceed by examining each axis individually, focusing on the specific characteristics of a given disturbance.

15

When a symptom suggests dysfunction in a particular axis, it is relevant to explore other potential disturbances within the same axis.

However, this approach carries a risk: the lack of a systematic method may lead to interrupting the semiological assessment as soon as a disorder is identified, even though comorbidities are common. On the other hand, its advantage is that it avoids redundant and tedious analyses of strictly normal behaviors, which could unnecessarily complicate the examination and cause owners to lose focus.

The ideal approach is to gather this information throughout the consultation, but this requires experience to ensure nothing is overlooked. Although this method is more fluid and pleasant, it may paradoxically give the impression of being less professional, as owners do not always perceive the reasoning behind the diagnosis.

Thus, it is often wise to combine these different methods. With this in mind, we present here a semiological assessment structured by axis.

15.4.1 Disturbances Related to the Fear Axis

As mentioned earlier, fear is a normal and adaptive emotion (see ▶ Sect. 15.3.2.2). It becomes pathological when its intensity and frequency are no longer appropriate to the context. The veterinary psychiatrist must therefore assess both the frequency and intensity of the cat's fears and determine whether they are contextually appropriate.

Dysregulation of fear underlies phobic and anxious states, as well as various fear axis disorders. Once it is established that the cat's fear responses are not adaptive, the next step is to use the semiological assessment to differentiate between the various pathological states and fear-related disorders, which are discussed in ▶ Chaps. 7 and 8 of this book.

Some clinical signs observed during consultation may guide the practitioner, but these findings must be compared with the owner's observations in the cat's home environment, the cat's developmental conditions (if known), and its history since adoption.

15.4.2 Disturbances Related to the Self-Control Axis

Owners who seek consultation for self-control axis dysregulation most often describe a cat that is restless and scratches and bites during contact and play. It is often useful to observe or ask owners to show their hands and forearms, as they frequently bear multiple scratch or bite scars.

The key objective of the semiological assessment is to evaluate the quality of the cat's motor and emotional control, particularly its tolerance to frustration.

Direct observation quickly provides relevant information. The difficulty of the clinical examination and intolerance to certain types of handling can immediately guide the clinician.

The persistence of oral exploration, and even more so, the ingestion of non-food foreign objects, should strongly suggest a diagnosis of impulsive or compulsive state and HSHA syndrome.

Motor control disturbances are among the easiest to identify during a consultation. If these disturbances are not associated with deprivation syndrome or other fear-axis disorders, the cat tends to be intrusive compared to other patients. It actively explores the environment, jumping onto furniture in an attempt to investigate further. This excessive exploration may result in objects being knocked over, as the cat often lacks awareness of its own body size. Such exploratory behavior is interspersed with brief fear reactions, indicative of hypersensitivity.

If the cat is offered petting, it may sometimes respond with playful attacks, which can escalate into violent reactions. Its reactivity to medical handling or injections is generally high. Through questioning, the veterinarian should assess whether there has been a decrease in sleep duration. Additionally, food intake is often difficult to regulate.

The absence of directly observable signs does not rule out a self-control axis disturbance. Fear and its associated disorders may completely mask motor control deficits. These conditions are often coexisting, reflecting poor early developmental conditions.

Important Point of View

Carrying Test in Kittens

The carrying reflex in kittens begins at around 2 weeks of age when the mother starts transporting them. During this process, the kitten adopts a reflexive posture, with its tail curled against its abdomen and complete immobility, facilitating transport. This response, known as the carrying reflex, can persist throughout the cat's life and is generally considered an indicator of good self-controls regulation. However, its presence does not rule out dysregulation of this axis.

During a consultation, it is possible to assess the presence of this reflex, provided it is performed under appropriate conditions.

When lifting a cat by the scruff, it is essential to support its body weight with the other hand to avoid excessive tension on the skin and eliminate any risk of pain.

In mother cats, carrying behavior naturally stops around 6 weeks of age, likely due to the increased weight of the kittens.

15.4.3 Disturbances Related to the Territoriality Axis

Disorders affecting the territoriality axis are generally not directly observable in a clinical setting. Instead, it is through discussions with owners that relevant information is gathered.

It is essential to collect details regarding facial marking, scratching behavior, and urine marking, especially if owners do not spontaneously report it (see ► Sect. 15.2.4).

Feeding behavior disorders, particularly overconsumption leading to obesity, can be an important sign of territorial deprivation or at least an underlying anxiety state that warrants further investigation, even if it is not directly related to territoriality.

The practitioner must be cautious not to stop the assessment after identifying a territoriality-related disorder. These disturbances are common, and comorbidities are frequent. Improving the cat's territory and environment is often a necessary first step in treating behavioral disorders, but it may not always be sufficient to fully restore the cat's psychological well-being.

15.4.4 Disturbances Related to the Relational Axis

The relationship between the owner and their cat is often observable during a consultation, although owners may be reluctant to disclose their use of punishment.

Observing how the owner handles their cat can be very revealing, as it often reflects their level of trust or fear toward the animal.

It is important to note that some owners do not realize that a panicked cat may scratch them. In such cases, it may be necessary to advise them not to place their hands near the cat, even if they appear confident in doing so.

The consultation should help determine whether the cat has ever been punished. In cases of urine soiling, punishment is common and often has long-lasting effects. Owners frequently believe that simply stopping punishment is enough to reassure the cat. However, a single punishment can instill lasting fear, significantly impacting the progression of therapy.

Summary of ▸ Sect. 15.4

Behavioral semiology can be approached using a systematic method or by analyzing different behavioral axes. Examining a specific axis helps identify related disorders but carries the risk of overlooking comorbidities if the evaluation stops too soon. The ideal approach is to combine multiple methods to ensure a comprehensive analysis while avoiding redundancy and owner fatigue. in all cases, a rigorous and nuanced approach is necessary to establish an accurate diagnosis and prevent misinterpretation.

15

Summary of This Chapter

During the consultation, regardless of the method used by the veterinarian, certain signs will be identified and validated as symptoms. After reviewing all relevant domains, the practitioner must compile a list of symptoms leading to a differential diagnosis, followed by the most probable diagnosis.

Once this first step is completed, it is essential to define the behavioral disorder being observed.

The proposed diagnosis must account for all observed symptoms. The diagnostic process must be structured and logical, and by explaining the pathological pathway of the cat using the semiological assessment, the practitioner builds a therapeutic alliance with the owner.

Naming the disorder affecting the patient demonstrates that we understand what is happening. This reassures the owner, as we can describe the progression of the condition and provide a prognosis. Once the disease has been identified, treatment can begin with a clear direction, rather than relying on uncertain approaches.

References

Ahola MK, Vapalahti K, Lohi H (2017) Early weaning increases aggression and stereotypic behaviour in cats. Sci Rep 7:10412. https://doi.org/10.1038/s41598-017-11173-5

Beugnet VV, Beugnet F (2020) Field assessment in single-housed cats of litter box type (covered/uncovered) preferences for defecation. J Vet Behav 36:65–69. https://doi.org/10.1016/j.jveb.2019.05.002

Borns-Weil S, Emmanuel C, Longo J, Kini N, Barton B, Smith A, Dodman NH (2015) A case-control study of compulsive wool-sucking in Siamese and Birman cats (n = 204). Journal of Veterinary Behavior 10, 543–548. https://doi.org/10.1016/j.jveb.2015.07.038

Bradshaw JWS (2016) Sociality in cats: A comparative review. Journal of Veterinary Behavior 11, 113–124. https://doi.org/10.1016/j.jveb.2015.09.004

Dards JL (1983) The behaviour of dockyard cats: interactions of adult males. Appl Anim Ethol 10:133–153. https://doi.org/10.1016/0304-3762(83)90117-7

Duffy DL, De Moura RTD, Serpell JA (2017) Development and evaluation of the Fe-BARQ: a new survey instrument for measuring behavior in domestic cats (Felis s. catus). Behav Process 141:329–341. https://doi.org/10.1016/j.beproc.2017.02.010

Eckstein RA, Hart BL (2000a) The organization and control of grooming in cats. Appl Anim Behav Sci 68:131–140. https://doi.org/10.1016/S0168--1591(00)00094-0

Eckstein RA, Hart BL (2000b) Grooming and control of fleas in cats. Appl Anim Behav Sci 68:141–150. https://doi.org/10.1016/S0168-1591(00)00095-2

Foster SF, Hunt GB, Malik R (1999) Congenital urethral anomaly in a kitten. J Feline Med Surg 1:61–64. https://doi.org/10.1016/S1098--612X(99)90011-6

Grant DC (2010) Effect of water source on intake and urine concentration in healthy cats. J Feline Med Surg 12:431–434. https://doi.org/10.1016/j.jfms.2009.10.008

Horwitz DF (2019) Common feline problem behaviors: urine spraying. J Feline Med Surg 21:209–219. https://doi.org/10.1177/1098612X19831203

Houpt KA (2018) Domestic animal behavior for veterinarians and animal scientists, 6th edn. Wiley, Hoboken

John MK, Cari AO, Jody PL, Rodney EO (1996) Inherited and congenital diseases of the feline lower urinary tract. Vet Clin Small Anim Pract 26:265–279. https://doi.org/10.1016/S0195-5616(96)50207-X

Liberg O (1983) Courtship behaviour and sexual selection in the domestic cat. Appl Anim Ethol 10:117–132. https://doi.org/10.1016/0304--3762(83)90116-5

MacDonald D (1981) The ecology and control of feral cats: proceedings of a Symposium Held at Royal Holloway College, University of London, 23rd and 24th September, 1980. Universities Federation for Animal Welfare

McCune S (1995) The impact of paternity and early socialisation on the development of cats' behaviour to people and novel objects. Appl Anim Behav Sci 45:109–124. https://doi.org/10.1016/0168-1591(95)00603-P

McGowan RTS, Ellis JJ, Bensky MK, Martin F (2017) The ins and outs of the litter box: a detailed ethogram of cat elimination behavior in two contrasting environments. Appl Anim Behav Sci 194:67–78. https://doi.org/10.1016/j.applanim.2017.05.009

Moody CM, Mason GJ, Dewey CE, Niel L (2020) Getting a grip: cats respond negatively to scruffing and clips. Vet Rec 186:385–385. https://doi.org/10.1136/vr.105261

Panaman R (2010) Behaviour and ecology of free-ranging female farm cats (Felis catus L.). Z Für Tierpsychol 56:59–73. https://doi.org/10.1111/j.1439-0310.1981.tb01284.x

Proux M-A (2022) Étude du comportement boulimique du chat domestique. Oniris- École nationale vétérinaire, agroalimentaire et de l'alimentation Nantes-Atlantique

Pryor PA, Hart BL, Bain MJ, Cliff KD (2001) Causes of urine marking in cats and effects of environmental management on frequency of marking. J Am Vet Med Assoc 219:1709–1713. https://doi.org/10.2460/javma.2001.219.1709

Romagnoli S, Bensaia C, Ferré-Dolcet L, Sontas HB, Stelletta C (2019) Fertility parameters and reproductive management of Norwegian Forest cats, Maine coon, Persian and Bengal cats raised in Italy: a questionnaire-based study. J Feline Med Surg 21:1188–1197. https://doi.org/10.1177/1098612X18824181

Swenson RM, Randall W (1977) Grooming behavior in cats with pontile lesions and cats with tectal lesions. J Comp Physiol Psychol 91:313–326. https://doi.org/10.1037/h0077331

Tynes VV, Sinn L (2014) Abnormal repetitive behaviors in dogs and cats. Vet Clin North Am Small Anim Pract 44:543–564. https://doi.org/10.1016/j.cvsm.2014.01.011

Van Belle M, Moons C, Mills D (2024) Unravelling feline social dynamics: a video-based observational study on allogrooming in domestic cats. In: European veterinary Congress of behavioural medicine and animal welfare, 6th annual meeting, Abstracts

Van Den Bos R (1998) Post-conflict stress-response in confined group-living cats (Felis silvestris catus). Appl Anim Behav Sci 59:323–330. https://doi.org/10.1016/S0168-1591(98)00147-6

Willemse T (1994) The effect of haloperidol and naloxone on excessive grooming behavior of cats. Eur Neuropsychopharmacol 4:39–45. https://doi.org/10.1016/0924-977X(94)90313-1

Psychopharmacology

Contents

Chapter 16 Introduction to psychopharmacology – 549

Chapter 17 Selective serotonin reuptake inhibitors (SSRIs) – 565

Chapter 18 Tricyclic antidepressants – 585

Chapter 19 Monoamine Oxidase Inhibitors (MAOIs) – 597

Chapter 20 Benzodiazepines – 603

Chapter 21 Sympatholytics – 611

Chapter 22 Antiepileptic drugs – 619

Chapter 23 Tetracyclic antidepressants or NaSSAs – 637

Chapter 24 Neuroleptics or antipsychotics – 645

Chapter 25 Other psychotropic drugs – 657

Chapter 26 Hormonal therapies – 679

Chapter 27 Over-the-counter treatments – 687

Introduction to Psychopharmacology

Contents

16.1 The Mechanism of Action of Psychotropic Drugs – 550
16.1.1 Actions on Ionotropic Receptors – 550
16.1.2 Actions on Metabotropic Receptors – 553
16.1.3 Actions on Reuptake Transporters – 553
16.1.4 Actions on Voltage-Dependent Ion Channels – 556
16.1.5 Actions on Enzymes – 556

16.2 Prescription of Psychotropic Drugs – 559
16.2.1 Basics of Prescription – 559
16.2.2 Monotherapy and Polytherapy – 560
16.2.3 Duration of Treatment and Withdrawal – 561

References – 562

© The Author(s), under exclusive license to Springer Nature Switzerland AG 2026
S. Bleuer-Elsner et al., *Veterinary Psychiatry of the Cat*,
https://doi.org/10.1007/978-3-032-06260-4_16

Psychopharmacology is the study of chemical substances that affect the psyche. In this section, we will review the active compounds that can be used in feline veterinary psychiatry. They are categorized by family and mechanism of action. Thus, chapter by chapter, we will describe selective serotonin reuptake inhibitors, tricyclic antidepressants, benzodiazepines, sympatholytics, antiepileptics, tetracyclic antidepressants or Noradrenergic and Specific Serotonergic Antidepressants. (NaSSAs), neuroleptics, monoamine oxidase inhibitors, and other molecules that do not fit into the aforementioned families. This section concludes with treatments using over-the-counter molecules.

The available data on feline psychopharmacology remain limited. We have aimed to be as comprehensive as possible despite the lack of information. Whenever feasible, the text refers to relevant studies, and any uncertainties are clearly indicated.

Before considering the numerous molecules available in veterinary psychiatry, it is important to revisit the general mechanisms of action of psychotropic drugs and the fundamental principles of their prescription.

16.1 The Mechanism of Action of Psychotropic Drugs

A psychotropic drug is an active substance that alters brain function. This substance consists of a molecule or its active metabolites, which must meet two essential criteria. They must be able to cross the blood-brain barrier to reach the brain and have a certain affinity for one or more sites of action. Their effects result from their binding to these targets and the subsequent modifications.

There are five sites of action for neurotransmitters: ionotropic receptors, metabotropic receptors, reuptake transporters, voltage-dependent ion channels, and enzymes.

16.1.1 Actions on Ionotropic Receptors

Ionotropic receptors were described in the first section (see ▸ Chap. 2, Sect. 2.3). They are composed of protein subunits that span the cell membrane and form a channel through which ionic exchanges occur (◘ Fig. 16.1).

Each of the protein subunits contains binding sites for various natural ligands, primarily neurotransmitters but also ions or hormones. Psychotropic drugs exhibit varying levels of affinity for these binding sites, competing with other ligands.

Thus, psychotropics can be categorized as agonists, which significantly increase the frequency of ion channel opening compared to the baseline frequency; partial agonists, which act similarly but with a lesser effect than a full agonist; antagonists, which maintain the baseline frequency of channel opening while competing with the natural agonist ligand or neurotransmitter; and inverse agonists, which completely close the ion channel.

It is important to note that a partial agonist competes with the natural ligand while reducing the ion channel's opening frequency, thereby diminishing the effects of the natural ligand (◘ Fig. 16.2).

Psychotropics can also act on other binding sites distinct from the natural ligand's site. These do not involve competition but act in conjunction. These sites are called allosteric modulatory sites. In the case of a positive allosteric modulator, the combined binding of the natural ligand and the psychotropic drug enhances the opening of the channel compared to the action of the ligand alone. Conversely, in the case of a negative allosteric modulator, the combined binding of the natural ligand and the psychotropic drug reduces the channel's opening compared to the ligand's isolated action (◘ Fig. 16.3).

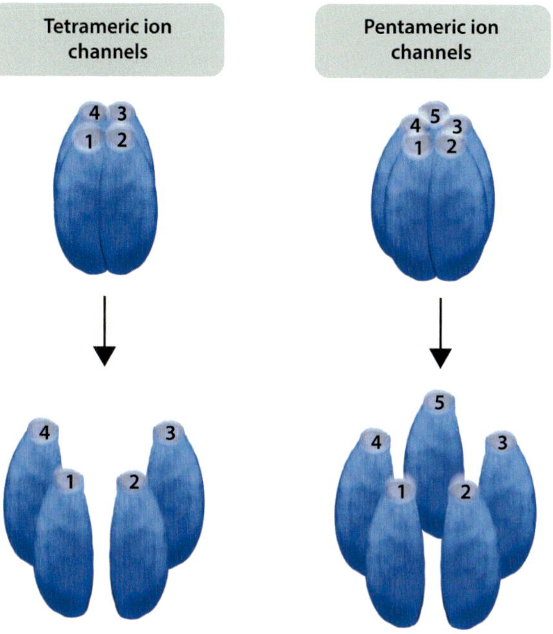

Fig. 16.1 Subunits of the ionotropic receptor. Ionotropic receptors are composed of protein subunits that assemble to form a transmembrane ion channel. The regulation of ion entry into the cell controls the transmission of the cell's action potential. There are tetrameric ion channels, consisting of four subunits (e.g., NMDA and AMPA), and pentameric ion channels, consisting of five subunits (e.g., GABA, 5-HT3, and nicotinic cholinergic receptors)

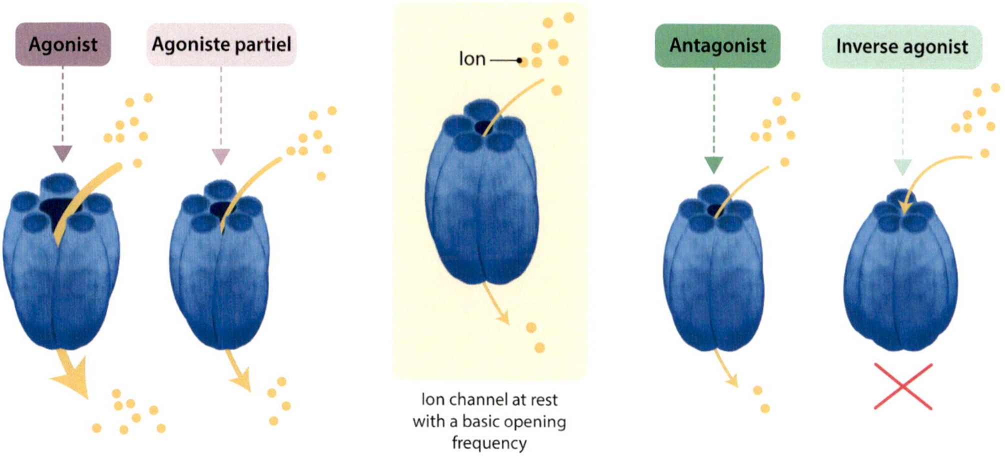

Fig. 16.2 Mechanism of action of psychotropics on ionotropic receptors. In the center, the ion channel at rest with a baseline frequency of opening. (**a**) The agonist significantly increases the frequency of opening. (**b**) The partial agonist increases the frequency of opening but to a lesser extent. As it competes with the natural ligand, it reduces its effects. (**c**) The antagonist maintains the baseline frequency of opening. (**d**) The inverse agonist closes the channel

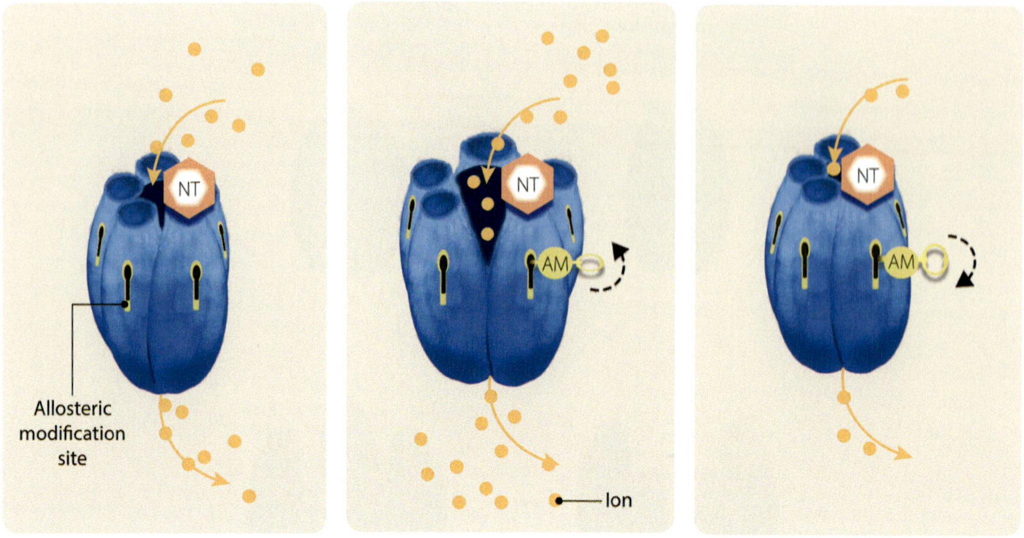

Fig. 16.3 Allosteric modulators. (**a**) The binding site of the allosteric modulator is distinct from that of the natural ligand. The allosteric modulator acts in the presence of the neurotransmitter. (**b**) The positive allosteric modulator increases the channel opening compared to the action of the neurotransmitter alone. (**c**) The negative allosteric modulator decreases the channel opening compared to the action of the neurotransmitter alone

Important Point of View

Action of psychotropic drugs on autoreceptors and heteroreceptors

α2-adrenergic, serotonergic 5-HT1A, and dopaminergic D2 receptors are autoreceptors of adrenergic, serotonergic, and dopaminergic neurons, respectively. They are located on the presynaptic membrane, and their binding to neurotransmitters released by the presynaptic neuron leads to a reduction in the release of the same neurotransmitter. This process is a form of autoregulation (see ► Chap. 2, Sect. 2.4).

These same receptors are also found on neurons that release a neurotransmitter different from the one to which they have an affinity. In such cases, they are referred to as heteroreceptors. Their binding to their ligand results in a decrease in the release of the neurotransmitter specific to the neuron on whose membrane the heteroreceptor is expressed. For example, the binding of norepinephrine to α2-adrenergic heteroreceptors on the membrane of a serotonergic neuron induces a reduction in serotonin release by that neuron.

Some psychotropic drugs act as agonists of autoreceptors, promoting a decrease in neurotransmitter release (e.g., clonidine acts on α2-adrenergic receptors and reduces noradrenergic transmission). Other psychotropics act as antagonists of autoreceptors, favoring an increase in neurotransmitter release (e.g., mirtazapine acts on α2-adrenergic receptors and increases noradrenergic transmission). These same psychotropic drugs can also act on heteroreceptors, leading to more or less desirable side effects (e.g., mirtazapine's action on α2-adrenergic heteroreceptors of serotonergic neurons and the resulting increase in serotonergic transmission).

16.1.2 Actions on Metabotropic Receptors

Metabotropic receptors, as described in the first part of this book (see ► Chap. 2, Sect. 2.3), are complex proteins whose activation leads to a cascade of events: regulation of the opening of voltage-dependent ion channels, regulation of the opening of ionotropic receptors, activation of genetic transcription, and activation of intracellular enzymatic processes (◘ Fig. 16.4).

The metabotropic receptor has a baseline activity that is modulated by its natural ligand or psychotropic drugs that compete with the latter. Thus, as with ionotropic receptors, there are agonists, partial agonists, antagonists, and inverse agonists (◘ Fig. 16.5).

Important Point of View

Up-regulation and down-regulation

There is a balance between the gene expression of membrane receptors and the extracellular environment. This applies not only to neurons but also to hormonal and immune cells.

The presence of numerous ligands or molecules such as psychotropics in the extracellular space induces an increase in the gene expression of the receptor. This phenomenon is called "upregulation" or increased regulation.

Conversely, if many receptors are occupied by their natural ligand or a psychotropic drug, the gene expression of the receptor decreases. This phenomenon is referred to as "downregulation" or decreased regulation. The downregulation of receptors can lead to desensitization of the neuron for that receptor.

Indeed, for a certain period, no new receptors are produced, and all existing receptors are occupied (see ◘ Fig. 16.6).

16.1.3 Actions on Reuptake Transporters

Transporters are assemblies of 12 transmembrane proteins with a high affinity for a neu-

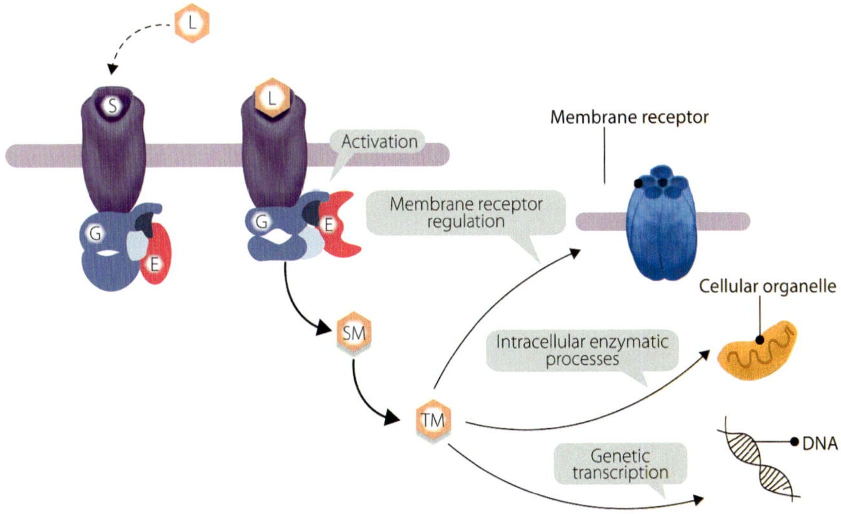

◘ **Fig. 16.4** Baseline activity of a metabotropic receptor. The receptor is composed of three subunits: a transmembrane protein with a binding site (S) for the ligand (L), which binds to an intracellular protein known as the G-protein. This G-protein, in turn, activates an enzyme (E) that releases a second messenger (SM). The second messenger subsequently activates a third enzymatic messenger (TM), which, through the addition or removal of phosphate ions, regulates the opening of voltage-dependent ion channels, the opening of ionotropic receptors, the activation of genetic transcription, and the activation of intracellular enzymatic processes

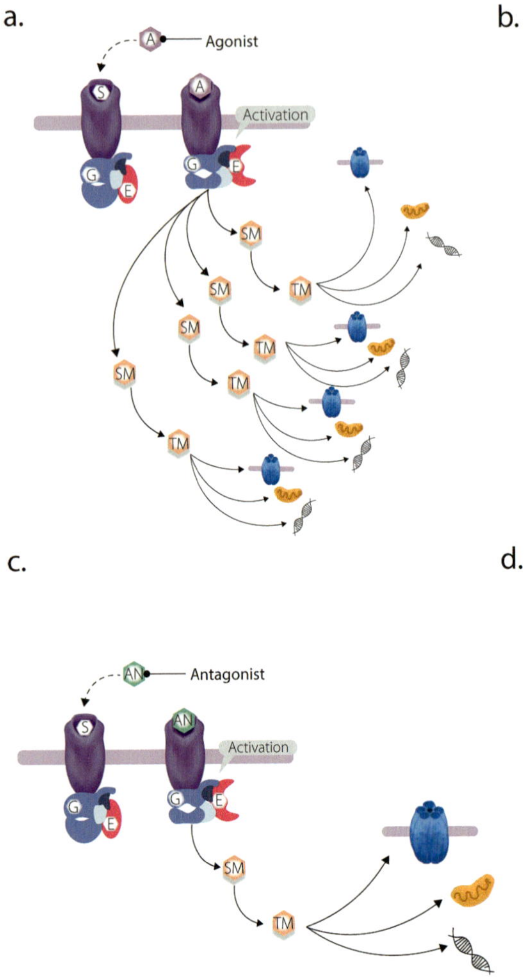

Fig. 16.5 Mechanism of action of psychotropics on metabotropic receptors. (**a**) The agonist significantly increases the cascade of events produced by the activation of the metabotropic receptor. (**b**) The partial agonist has a similar but less pronounced effect than the agonist. (**c**) The antagonist maintains the baseline cascade of events. (**d**) The inverse agonist completely inhibits the cascade of events

16

rotransmitter. Monoamine transporters are particularly significant in pharmacology and are named according to the neurotransmitter they transport: SERT for serotonin transporters, NET for norepinephrine transporters, and DAT for dopamine transporters. Transporters for glycine, gamma-aminobutyric acid (GABA), and glutamate have also been identified.

Transporters have a preferential affinity for one neurotransmitter but can accommodate other molecules. For instance, NET also accepts dopamine and amphetamines, DAT accepts norepinephrine, and SERT accepts 3,4-methylenedioxymethamphetamine (MDMA) (or ecstasy) (Stahl and Muntner 2021).

Transporters are coupled to ATPases, which act as sodium pumps. By expelling sodium from the cell, they create a gradient that activates the transporter, facilitating the reuptake of neurotransmitters from the synaptic cleft (Fig. 16.7).

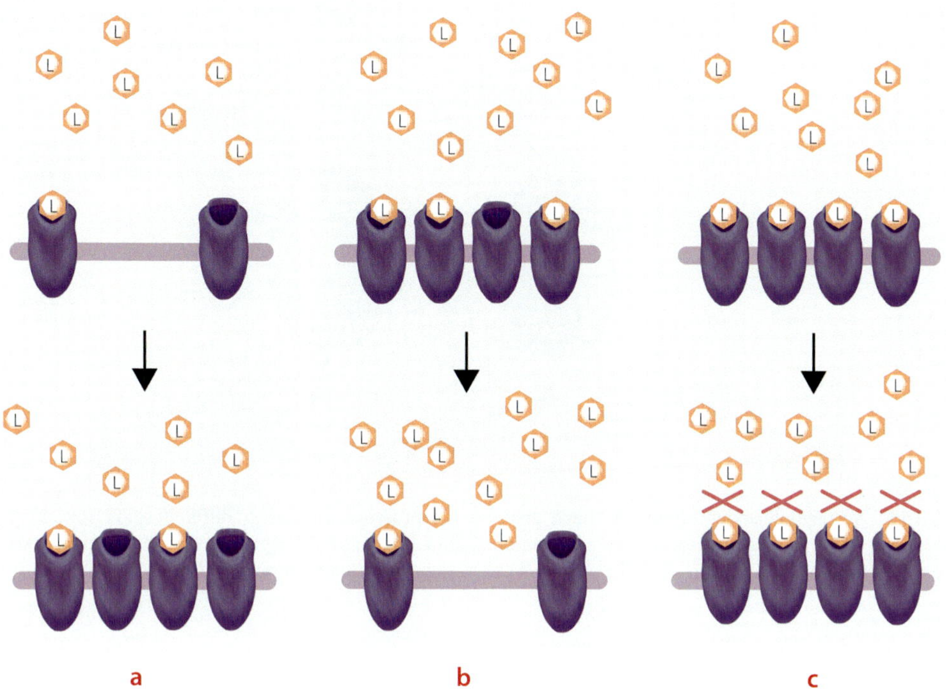

Fig. 16.6 Upregulation and downregulation. (**a**) *Upregulation*: The presence of numerous ligands (or psychotropics) in the membrane environment leads to increased gene expression and a higher number of membrane receptors. (**b**) Downregulation: The occupation of many receptors by their ligand (or psychotropic) results in a decrease in the gene expression of the receptors. (**c**) When all receptors are occupied and receptor gene expression is inhibited, the neuron is considered desensitized for the receptor in question

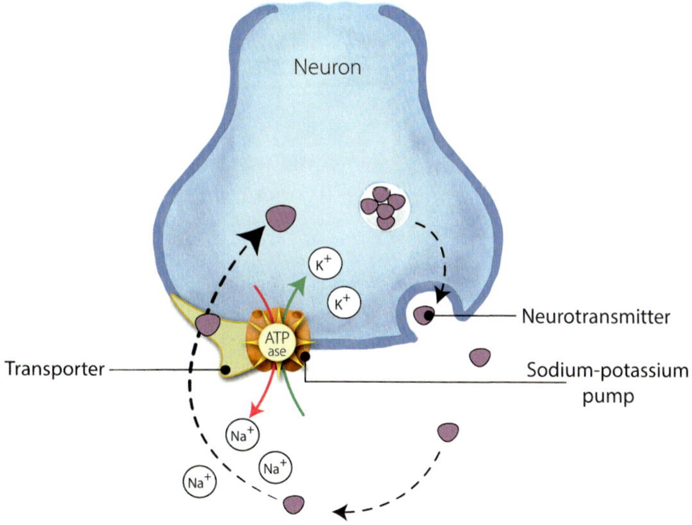

Fig. 16.7 Sodium pump and transporter. The ATPase-Na + transporter coupling enables neurotransmitter reuptake by the presynaptic membrane. Psychotropics with an affinity for transporters are allosteric modulators. By altering the transporter's shape, they prevent the neurotransmitter from binding and being reintroduced into the intracellular environment

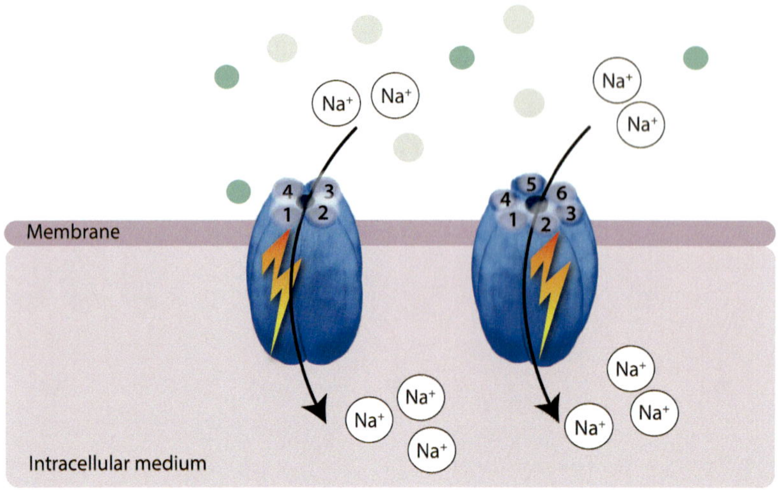

Fig. 16.8 Voltage-dependent channels. Each channel is composed of four sets of six subunits that combine to form an ion pore. Within each set of subunits, one protein is sensitive to differences in electrical charges and promotes the opening of the ion pore upon the arrival of an action potential

16.1.4 Actions on Voltage-Dependent Ion Channels

Voltage-dependent channels, described in the first section (see ▶ Chap. 2, Sect. 2.3), are composed of an assembly of four subunits surrounding an ion pore, with each subunit itself consisting of six segments. One of these segments is sensitive to differences in charge between the intracellular and extracellular environments and regulates the opening or closing of the ion pore (Fig. 16.8).

There are voltage-dependent sodium ion channels and voltage-dependent calcium ion channels. The former facilitates the propagation of the action potential (AP) along the neuron to the terminal bouton. Calcium channels, on the other hand, are closely associated with the vesicles of the presynaptic neuron. The opening of calcium channels enables the release of synaptic vesicles, their fusion with the cell membrane, and the release of the neurotransmitter they contain (Fig. 16.9).

When the AP reaches the synaptic bouton, the final sodium channels activate the calcium channels through their opening and induce depolarization. The release of calcium ultimately triggers the release of the neurotransmitter.

Some psychotropics are inhibitors of voltage-dependent channels. By interacting with these channels, they prevent ionic flow, leading to the inhibition of action potential (AP) transmission along the neuron when acting on sodium channels, and the inhibition of synaptic vesicle release when acting on calcium channels.

16.1.5 Actions on Enzymes

Enzymes play a crucial role in regulating synaptic transmission (see ▶ Chap. 2, Sect. 2.3). By acting on enzymes, psychotropics can modulate the information transmitted from one neuron to another. The three main enzyme targets of psychotropics are monoamine oxidases (MAOs), acetylcholinesterase, and glycogen synthase kinase.

Psychotropics function as competitive inhibitors of the target enzyme. Their binding can be either irreversible—in which case the enzyme permanently loses its function—or reversible, meaning the natural neurotransmitter can compete with the psychotropic. In the latter case, the enzyme regains its function once the psychotropic is released (Fig. 16.10).

It is worth noting that the vast majority of psychotropics are metabolized in the liver by

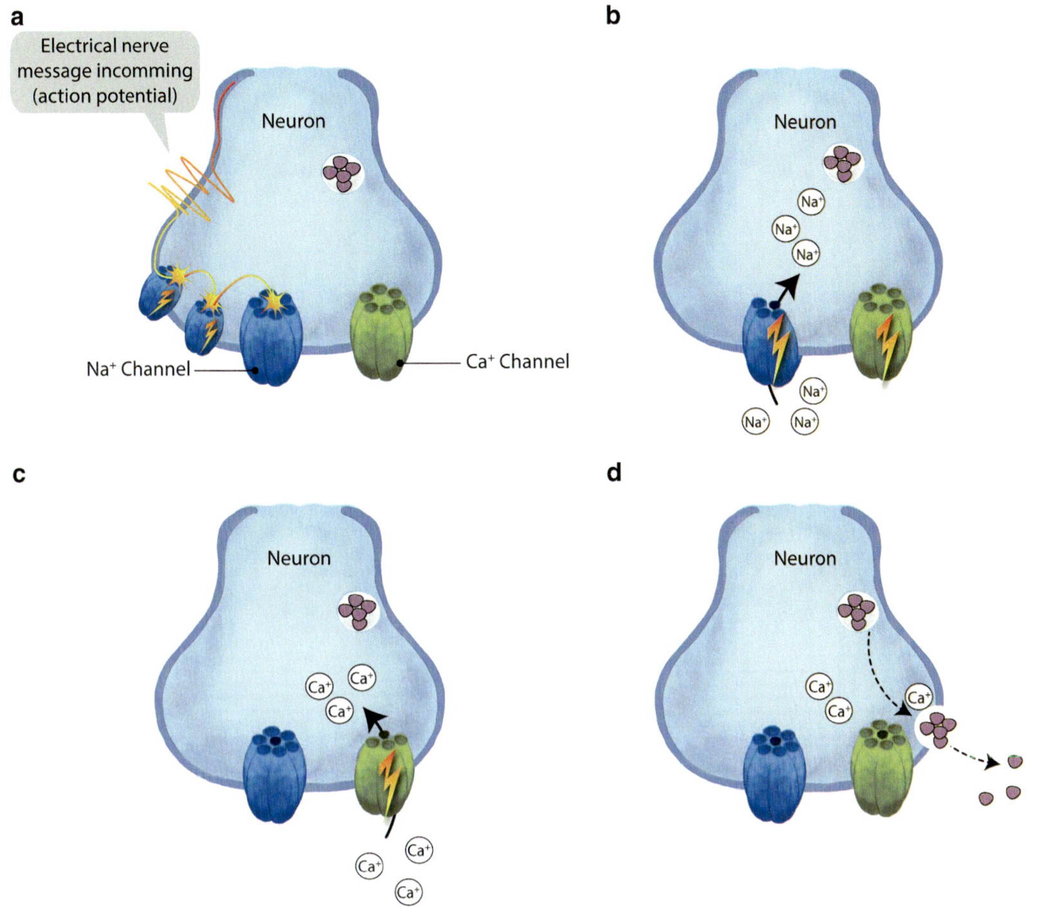

Fig. 16.9 Voltage-dependent sodium and calcium ion channels. (**a**) Voltage-dependent sodium ion channels propagate the action potential (AP) to the synaptic bouton. (**b**) Under the influence of the AP, the final sodium channels open. (**c**) The potential difference is detected by the calcium channels. (**d**) The calcium channels then open, allowing calcium ions to enter the cell. The presence of calcium ions within the cell triggers the release of the synaptic vesicle, which fuses with the membrane and releases its contents

CYP enzymes, a family of cytochrome P450 enzymes (see Sect. 2.2, box "Cytochrome P450 or CYP450"). By occupying the action sites of certain CYP enzymes during their metabolism, psychotropics block these enzymes, which are also involved in other metabolic pathways. This competition within metabolic pathways can lead to an increase or decrease in plasma levels of molecules administered simultaneously.

This explains certain contraindications for combining molecules that rely on the same metabolic pathways. For this reason, whenever known, the metabolic pathways and cytochromes involved will be detailed for each molecule.

The action of psychotropics on voltage-dependent ion channels or ionotropic receptors is very rapid. Their binding immediately alters intracellular ion fluxes, either promoting or inhibiting the membrane action potential.

The action on membrane transporters is immediate; however, the desired effect takes several days to several weeks. This is because the sudden increase in neurotransmitter concentration in the synaptic space initially triggers upregulation, followed by downregulation or even desensitization of the postsynaptic membrane for that neurotransmitter (see Sect. 1.2, box « Upregulation et downregulation »). It takes time for a new equilibrium to be established.

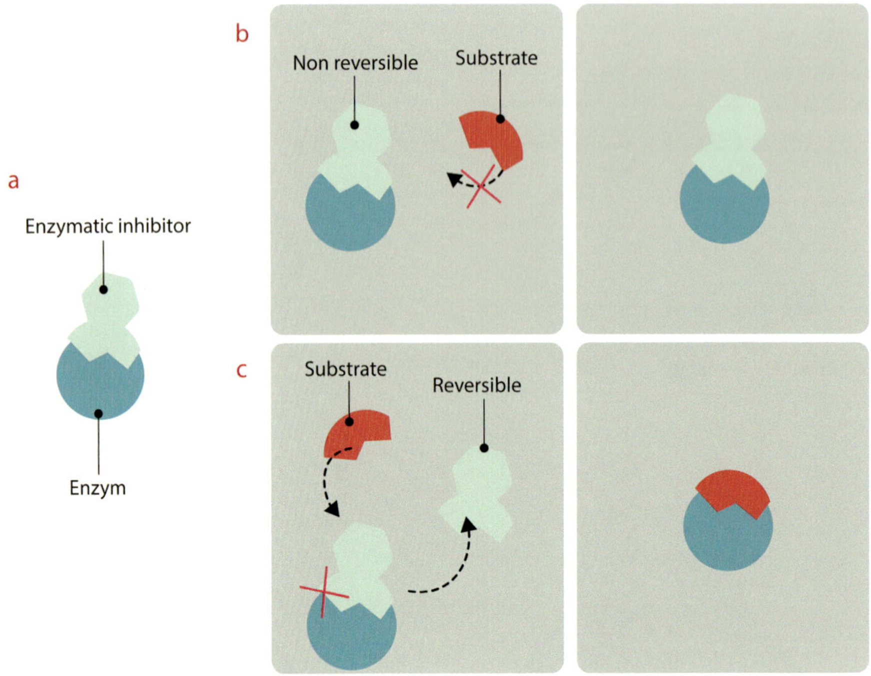

Fig. 16.10 Enzymes inhibitors. (**a**) The enzyme inhibitor binds to the enzyme in competition with the natural substrate (most often a neurotransmitter), (**b**) either irreversibly, preventing the substrate from binding, (**c**) or reversibly, allowing the substrate to compete with the enzyme inhibitor

In contrast, the action on metabotropic receptors is complex. The cascade of events resulting from receptor activation can take several hours to several days. In particular, the influence on gene transcription may take several weeks.

The action on enzymes is also relatively slow, as it requires neutralizing a sufficient number of enzymes despite their constant renewal.

Finally, it is worth noting that even in animals, any therapeutic intervention involves a placebo effect. The success of the treatment or the occurrence of side effects is not solely attributable to the pharmacological action of the administered molecule (see adjacent box).

16

Important Point of View

The placebo effect

The placebo is defined as any broad medical intervention (medication, vaccination, surgery, physical manipulations, rituals, food, or dietary supplements) that has a psychological or psychophysiological therapeutic effect without having specific activity on the condition being treated (McMillan 1999; Shapiro and Shapiro 1997). The placebo effect is, therefore, the nonspecific psychological or psychophysiological therapeutic effect induced by a placebo.

This effect can be either favorable or unfavorable to the individual; in the latter case, it is referred to as a nocebo effect. The placebo effect is reduced or even nullified by the administration of naloxone, which blocks opioid receptors (Levine et al. 1979), and rimonabant, which blocks endocannabinoid receptors (Benedetti et al. 2011). Opioid and cannabinoid pathways thus play a significant role in the placebo effect, along with dopamine, by acting within reward circuits, conditioning, expectations (optimism or pessimism), and pain modulation.

The production of these neurotransmitters and their receptors is genetically regulated, suggesting that the placebo effect is also modulated by an individual's genes and, therefore, varies between subjects. In humans, 11 brain regions involved in the placebo effect have been identified (Jubb and Bensing 2013; Lidstone and Stoessl 2007), including the nucleus accumbens and the dorsal striatum.

Classical conditioning plays a significant role in the placebo effect, including in animals. The scientific literature is filled with experimental protocols demonstrating the involvement of this form of learning in the placebo effect. In Pavlov's historical experiment, dogs repeatedly injected with morphine in an experimental cage exhibited symptoms of morphine action as soon as they entered the cage, even before the product was injected (Pavlov 1927).

The placebo effect can be measured biologically. For instance, guinea pigs repeatedly exposed to a specific odor along with an allergenic substance showed elevated histamine levels in their blood upon mere exposure to the odor, even in the absence of any allergenic substance (Russell et al. 1984).

The expectations of pet owners play a crucial role in the placebo effect in animals. Several recent studies have shown that the owner's attitude, stress level, and expectations regarding the effectiveness of the treatment can influence the perception of results and interactions with the animal (Carlino and Benedetti 2014). Indeed, an owner convinced of the efficacy of a treatment may unconsciously alter their behavior toward the animal, for instance, by increasing attention or reducing stressful stimuli. This change in attitude can then improve the animal's well-being, not due to the pharmacological effect of the treatment, but because of the more positive environment and reduced stress. This is an effect often observed after a consultation, especially when the therapeutic alliance is of high quality.

Treatment expectations play a role in both humans and animals. It is an anticipatory process that allows for better, or conversely, worse, adaptation to a situation (Kirsch 1985; Van Weel 2001). Studies in humans have shown that the same brain regions are activated by real or imagined changes. In companion animals, the therapeutic environment, such as a veterinary clinic, if associated with relief or even pleasure, can itself induce a placebo effect (or a nocebo effect under opposing conditions).

16.2 Prescription of Psychotropic Drugs

16.2.1 Basics of Prescription

The prescription of psychotropic drugs depends on various parameters.

The first is the patient's state diagnosis. As described in the second part of this book, there are numerous pathological states associated with emotional, mood, or cognitive dysfunctions (◘ Table 16.1).

Each pathological state corresponds to a dysregulation of a brain circuit involving several neurotransmitters. For example, a phobic state involves a disruption of the fear circuits, engaging the amygdala and prefrontal cortex. This circuit involves norepinephrine, serotonin, GABA, and glutamate. To restore the balance of brain function, the practitioner will select one or more molecules that act on one or more of these neurotransmitters.

The first step is to establish as precise a state diagnosis as possible to identify the dysfunctional circuits and their associated neurotransmitters. The choice of psychotropic drugs can then be adjusted based on the needs and preferences of the owner. For example, in treating a senile pathological state with sleep disturbances and nocturnal incontinence, it may be beneficial to use a molecule with antihistaminic properties that

Table 16.1

Emotional pathological states		**Mood pathological states**	
Phobic	Simple /complex	Depressive	Acute/chronic
Anxious	Productive/inhibited	Manic	Contextual/invasive
	Intermittent/generalized	Bipolar	
Pathological states of cognition			
Pathological states of lack of self-control		**Pathological states of lack of derealization**	
Impulsive	Contextual/invasive	Dissociated Senile	
Compulsive	Contextual/invasive	Place Senile	

promote sleep, combined with peripheral muscarinic cholinergic properties that promote urinary retention.

Similarly, in the case of treating a simple phobia that occurs only in a specific context, it would be preferable to choose an anxiolytic with very rapid action, administered on a PRN (pro re nata) basis (pro re nata = as needed), rather than a daily treatment.

Important Point of View

International nomenclature

In France, prescriptions typically specify the number of doses per day in full words. The international nomenclature uses Latin acronyms, summarized as follows:

- **SID (semel in die)** = once per day
- **BID (bis in die)** = twice per day
- **TID (ter in die)** = three times per day
- **PNR (pro re nata)** = as needed

We will use the international nomenclature in the tables in Part III of this book.

16

Finally, there is the prescriber's experience, and the specific characteristics of the pathologies encountered in their clientele. Over time, the veterinary psychiatrist may develop the ability to match a psychopathological profile to a pharmacological profile, thereby refining their treatment choices.

16.2.2 Monotherapy and Polytherapy

The more we understand the neurophysiological and neuropathological mechanisms underlying behavior, the more we discover their complexity. Psychotropics rarely have a single, specific action; they are likely to affect multiple neurotransmitters and circuits.

Each state diagnosis category corresponds to a distinct brain circuit. It is possible to encounter dysfunctions in multiple circuits, leading to a multiple state diagnosis that includes emotional, mood, and cognitive disorders. For example, a cat may exhibit both a simple phobic state and an impulsive state. The practitioner prioritizes the dysfunctions as much as possible, focusing treatment on the primary dysfunction. However, there are cases where multiple circuits must be addressed simultaneously, requiring the prescription of combined drug treatments.

While monotherapy should be preferred whenever possible, the practitioner may need to use a combination of psychotropics or psychotropics and hormones to address multiple brain circuit dysfunctions and enhance treatment efficacy.

In cases of *polytherapy*, certain rules should be followed:

- Do not combine two molecules from the same family or sharing the same primary mechanism of action.

- Do not combine molecules that modify monoamines with a monoamine oxidase inhibitor (MAOI), such as combining selective serotonin reuptake inhibitors (SSRIs), tricyclics, or tetracyclics with an MAOI.
- Reduce doses when combining two molecules from different families that share certain mechanisms of action (e.g., SSRI + tricyclic).
- Reduce doses when combining two molecules with hepatic metabolic pathways that share the same cytochromes (see box below).

Important Point of View

Cytochromes P450 or CYP450

Cytochrome P450 enzymes form a family of hepatic hemoprotein isoenzymes. These are proteins that use a heme molecule (also found in hemoglobin, for example) as a cofactor. CYP450 enzymes are responsible for the redox metabolism of a variety of natural substrates involved in endogenous metabolism as well as xenobiotics.

They have been extensively studied in humans but have also been identified in all living organisms, from bacteria to mammals. They are numbered according to the gene encoding the specific cytochrome. For example, CYP3A4, which is involved in the metabolism of fluoxetine, corresponds to the CYP3A4 gene that encodes this cytochrome.

Many drugs can either increase or decrease the activity of CYP isoenzymes. This can occur either by inducing the biosynthesis of a specific isoenzyme, referred to as enzyme inducers (such as phenobarbital on CYP1A, CYP2C, CYP2D, and CYP3A), or by occupying part of the available CYP pool, leading to inhibition of CYP activity for other molecules.

Changes in CYP enzyme activity can affect the metabolism and elimination of various drugs. For instance, administering two drugs that rely on the same CYP enzyme(s) for metabolism may lead to potentially toxic accumulation. Similarly, one drug can reduce the production of the active metabolite of another drug by occupying the same CYP enzyme.

In the same way, combining a drug with a second enzyme inducer can lower the plasma levels of the first drug by accelerating its metabolism. As a result, drug interactions may require dosage adjustments or, in some cases, contraindications for coadministration.

Finally, it should be noted that variations exist across species, sexes, and even individuals for a given CYP enzyme (Honda et al. 2011; Shah et al. 2007; Sugiyama et al. 2019). Unlike humans, for whom extensive data are available, comprehensive knowledge of metabolic pathways and the CYP enzymes involved for all drugs in companion animals remains incomplete (Fink-Gremmels 2008). In the remainder of this book, when a CYP enzyme has been identified in cats, it will be indicated. Other references remain extrapolations from human medical data.

16.2.3 Duration of Treatment and Withdrawal

A cat undergoing pharmacological and psychotherapeutic treatment should be regularly re-evaluated. This can be done using a score provided by the owner and veterinarian on a scale from 0 to 10, where 0 represents the starting point at the first consultation and 10 corresponds to normalized behavior indistinguishable from that of a nonmedicated cat. An average score of at least 7 or 8 suggests that therapeutic objectives have been achieved. Once the cat's condition has stabilized at this level for 2 or 3 months, the question of discontinuing pharmacological treatment arises.

There are relatively few molecules that cause true dependence in brain function, with the exceptions being opioids, benzodiazepines (after more than 1 week of use), alcohol, and all stimulants.

Regardless of the molecule used, a gradual reduction in dosage is recommended. The primary goal is to confirm the cat's emotional stability without medication and to ensure that the learned behaviors acquired during treatment are retained.

Summary of This Chapter

Psychotropic drugs are chemical substances that affect the psyche through their potential actions on five different sites:

- Ionotropic receptors
- Metabotropic receptors
- Reuptake transporters
- Voltage-dependent ion channels
- Enzymes

By binding to a receptor with which it has an affinity, a psychotropic drug can exert six types of actions:

- *Agonist*: producing the same result as the natural ligand
- *Partial agonist*: producing a weaker result than the natural ligand
- *Antagonist*: maintaining a resting state (to be differentiated from a lack of receptor activity)
- *Inverse agonist*: inhibiting all receptor activity, rendering it inactive
- *Desensitization*: resulting from prolonged receptor occupancy (regardless of the action) and downregulation
- *Enzyme inhibition*: reversible or irreversible inhibition of one or more enzymes

Psychotropic drugs are prescribed following a rigorous diagnostic process, based on the state diagnosis and taking into account the owner's needs. While monotherapy is preferred, many cases require polytherapy. In such cases, the following rules of combination must be observed:

- Avoid combining two molecules with the same primary mechanism of action
- Avoid combining a monoamine modulator with an MAOI
- Halve the doses when two psychotropics share certain mechanisms of action
- Verify common metabolic pathways and the cytochromes involved

16

References

Benedetti F, Amanzio M, Rosato R, Blanchard C (2011) Nonopioid placebo analgesia is mediated by CB1 cannabinoid receptors. Nat Med 17(10):1228–1230. https://doi.org/10.1038/nm.2435

Carlino E, Benedetti F (2014) Placebo and nocebo effects. In: Mostofsky DI (ed) The handbook of behavioral medicine, 1st edn. Wiley, pp 36–57. https://doi.org/10.1002/9781118453940.ch3

Fink-Gremmels J (2008) Implications of hepatic cytochrome P450-related biotransformation processes in veterinary sciences. Eur J Pharmacol 585(2–3):502–509. https://doi.org/10.1016/j.ejphar.2008.03.013

Honda K, Komatsu T, Koyama F, Kubota A, Kawakami K, Asakura H, Uno Y, Kitazawa T, Hiraga T, Teraoka H (2011) Expression of two novel cytochrome P450 3A131 and 3A132 in liver and small intestine of domestic cats. J Vet Med Sci 73(11):1489–1492. https://doi.org/10.1292/jvms.11-0098

Jubb J, Bensing JM (2013) The sweetest pill to swallow: how patient neurobiology can be harnessed to maximise placebo effects. Neurosci Biobehav Rev 37(10):2709–2720. https://doi.org/10.1016/j.neubiorev.2013.09.006

Kirsch I (1985) Response expectancy as a determinant of experience and behavior. Am Psychol 40(11):1189

Levine JD, Gordon NC, Bornstein JC, Fields HL (1979) Role of pain in placebo analgesia. Proc Natl Acad Sci 76(7):3528–3531. https://doi.org/10.1073/pnas.76.7.3528

Lidstone SCC, Stoessl AJ (2007) Understanding the placebo effect: contributions from neuroimaging. Mol Imaging Biol 9(4):176–185. https://doi.org/10.1007/s11307-007-0086-3

McMillan J (1999) Cognitive psychology and hermeneutics: two irreconcilable approaches? Philos Psychiatr Psychol 6(4):255–258

Pavlov I (1927) Conditioned reflexes: an investigation of the physiological activity of the cerebral cortex. Oxford Univ Press

Russell M, Dark KA, Cummins RW, Ellman G, Callaway E, Peeke HVS (1984) Learned histamine release. Science 225(4663):733–734. https://doi.org/10.1126/science.6205449

Shah SS, Sanda S, Regmi NL, Sasaki K, Shimoda M (2007) Characterization of cytochrome P450-mediated drug metabolism in cats. J Vet Pharmacol Ther 30(5):422–428. https://doi.org/10.1111/j.1365-2885.2007.00902.x

Shapiro AK, Shapiro E (1997) The powerful placebo: from ancient priest to modern physician. Johns Hopkins University Press

Stahl SM, Muntner N (2021) In: Grady MM (ed) Stahl's essential psychopharmacology: neuroscientific basis and practical applications, 5th edn. Cambridge University Press. https://doi.org/10.1017/9781108975292

Sugiyama S, Uno Y, Amano T, Kitazawa T, Teraoka H (2019) Genetic diversity of cytochrome P450 3A with different metabolic activity in domestic cats. J Vet Med Sci 81(4):598–600. https://doi.org/10.1292/jvms.18-0692

Van Weel C (2001) Examination of context of medicine. Lancet 357(9258):733–734. https://doi.org/10.1016/S0140-6736(00)04176-3

Selective Serotonin Reuptake Inhibitors (SSRIs)

Contents

17.1 General Characteristics of All SSRIs – 567
17.1.1 General Pharmacology of SSRIs – 567
17.1.2 Common Side Effects and Usage Precautions for All SSRIs – 568
17.1.3 Clinical Use and Dosage – 569

17.2 Fluoxetine – 569
17.2.1 Description, Chemical Formula, and Brand Names – 569
17.2.2 Pharmacology – 569
17.2.3 Common Side Effects and Usage Precautions – 570
17.2.4 Clinical Use and Dosage – 572

17.3 Fluvoxamine – 573
17.3.1 Description, Chemical Formula, and Brand Names – 573
17.3.2 Pharmacology – 573
17.3.3 Common Side Effects and Usage Precautions – 574
17.3.4 Clinical Use and Dosage – 575

17.4 Paroxetine – 576
17.4.1 Description, Chemical Formula, and Brand Names – 576
17.4.2 Pharmacology – 576
17.4.3 Common Side Effects and Usage Precautions – 577
17.4.4 Clinical Use and Dosage – 578

© The Author(s), under exclusive license to Springer Nature Switzerland AG 2026
S. Bleuer-Elsner et al., *Veterinary Psychiatry of the Cat*,
https://doi.org/10.1007/978-3-032-06260-4_17

17.5 Sertraline – 578
17.5.1 Description, Chemical Formula, and Brand Names – 578
17.5.2 Pharmacology – 578
17.5.3 Common Side Effects and Usage Precautions – 579
17.5.4 Clinical Use and Dosage – 580

References – 582

Selective serotonin reuptake inhibitors (SSRIs) share the common mechanism of inhibiting presynaptic serotonin reuptake pumps, thereby increasing the availability of this neurotransmitter in the synaptic space.

Currently, no SSRI has marketing authorization for use in cats.

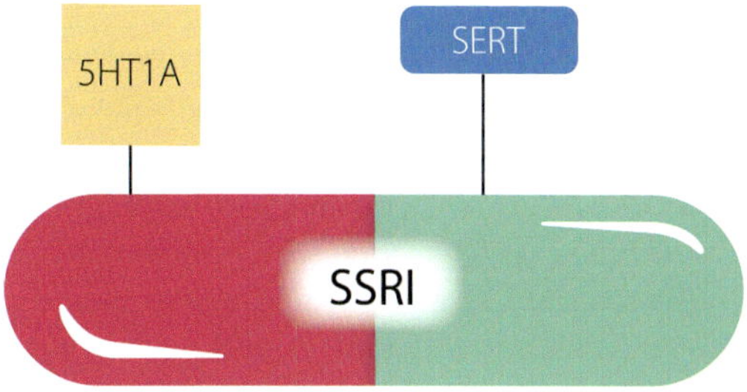

Fig. 17.1 General structure of SSRIs. All SSRIs inhibit the SERT receptor and desensitize the 5-HT1A receptor

17.1 General Characteristics of All SSRIs

Selective serotonin reuptake inhibitors (SSRIs) are the most commonly used class of psychotropics in canine veterinary psychiatry. In cats, the challenge of administering medication often limits their prescription. Serotonin plays a central role in numerous circuits, and the action of SSRIs on serotonin makes this class of drugs valuable for treating a wide range of pathological states.

After describing the common characteristics of all SSRIs, we will outline the specific features and indications of several of them: fluoxetine, fluvoxamine, paroxetine, and sertraline. Escitalopram can be prescribed for dogs but is not used in feline veterinary medicine due to a lack of information on its potential effects.

17.1.1 General Pharmacology of SSRIs

17.1.1.1 Mechanism of Action

Medications in the SSRI family share the dual property of inhibiting the presynaptic serotonin transporter (SERT) and desensitizing presynaptic serotonin 5-HT1A autoreceptors (Fig. 17.1). This increases serotonin concentrations in synapses.

In addition to these shared properties, each SSRI has specific affinities for additional serotonergic receptors (e.g., 5-HT2C for fluoxetine), other monoamine receptors (norepinephrine and dopamine), or other receptors (muscarinic cholinergic, histaminic). These differences among SSRIs help explain their distinct mechanisms of action, specific indications, and unique side effects.

17.1.1.2 Metabolism and Involved Cytochromes

All SSRIs are metabolized in the liver by various types of cytochrome P450 enzymes. The metabolites of SSRIs are highly diverse, ranging from minimally present or inactive metabolites to those playing a significant role (e.g., norfluoxetine). Some SSRIs are inhibitors of multiples cytochromes.

17.1.1.3 Pharmacokinetics

Pharmacokinetics vary greatly between SSRIs. Half-lives (see box below) are highly variable and, in some cases, unknown in cats. The elimination pathways in felines are partially known, varying between biliary, urinary, or mixed elimination depending on the molecules.

Important Point of View

The Half-Life of a Drug

In pharmacology, the half-life refers to the time required for the plasma concentration of an active molecule to decrease by half of its initial value. It is influenced by metabolic pathways, the mode of transport, and elimination routes.

The half-life varies significantly between molecules, ranging from a few hours (e.g., diazepam, methylphenidate) to several days (e.g., norfluoxetine). It determines the number of daily doses required to achieve the desired effect. A short half-life molecule can be particularly useful for as-needed administration, a method not feasible for long half-life molecules.

In general, as-needed administrations correspond to molecules with a half-life of no more than 8 h. Prolonged use of these short half-life molecules requires readministration every 8 h. Molecules with a half-life of up to 12 h are typically administered every 12 h, and those with longer half-lives are given once a day.

Finally, it is important to note that half-lives can vary significantly between individuals. The times reported are averages or ranges derived from clinical studies.

17.1.2 Common Side Effects and Usage Precautions for All SSRIs

17.1.2.1 Side Effects

In humans, the most commonly reported side effects include sexual dysfunction, drowsiness, weight gain, dry mouth, insomnia, fatigue, nausea, dizziness, and tremors (Cascade et al. 2009). Due to the limited data available, it is not possible to generalize the side effects of SSRIs in cats at this time.

17.1.2.2 Precautions and Contraindications

Severe hepatic insufficiency should lead to a reduction in SSRI doses due to their hepatic metabolism.

In cases of renal insufficiency, only the dosage of paroxetine needs to be adjusted. Other SSRIs are unaffected.

In cases of renal insufficiency, caution is advised, and the onset of side effects should be closely monitored. The elimination pathways in cats are still poorly understood, and the accumulation of the molecule cannot be ruled out.

Studies on toxicity during pregnancy and lactation are incomplete. While no major risks have been demonstrated, prescribing remains at the discretion of the practitioner, who must assess the risk-benefit balance.

17.1.2.3 Drug Interactions

The following drug combinations are *prohibited:*

- Amitraz and selegiline (monoamine oxidase inhibitors or MAOIs): for all drugs in the SSRI family, the major contraindication is the combination with monoamine oxidase inhibitors. This association carries a significant risk of serotonin syndrome
- Tramadol (analgesic): SSRIs may inhibit the hepatic metabolism of tramadol into active metabolites. This combination increases the risk of serotonin syndrome.
- Ketoconazole (antifungal): this combination increases the risk of serotonin syndrome.
- Venlafaxine: combining an SSRI with venlafaxine may increase plasma levels of both molecules and promote the onset of serotonin syndrome.
- Clomipramine, amitriptyline (tricyclic antidepressants): combining an SSRI with tricyclic antidepressants may elevate plasma levels of the tricyclics and increase the risk of serotonin syndrome.

The following drug interactions have been reported in humans or animals receiving an SSRI (combined use is not necessarily contraindicated, but the risks must be assessed by the practitioner):

- Propranolol, metoprolol (beta-blockers): SSRIs may increase the blood levels of beta-blockers
- Nonsteroidal anti-inflammatory drugs (NSAIDs): combining NSAIDs with SSRIs may increase the risk of gastrointestinal ulcers and bleeding.
- Metoclopramide (antiemetic): this combination increases the risk of serotonin syndrome.
- Clopidogrel, heparin, warfarin (anticoagulants): their use with SSRIs is associated with an increased risk of bleeding in humans.

17.1.2.4 Toxicity

The LD_{50} values vary between different SSRIs. Major toxicity results from the onset of serotonin syndrome (see box below). No cases of fatal poisoning have been reported in the literature (Pugh et al. 2013).

> **The Half-Life of a Drug**
>
> **Serotonin Syndrome and Its Treatment**
> This syndrome is primarily described in humans and is believed to result from excessive stimulation of presynaptic 5-HT1A receptors. The observed symptoms can be classified into three categories: mental, neuromuscular, and autonomic (Kuisma 1995; Prakash et al. 2021).
>
> Mental symptoms include agitation, confusion, delirium, and irritability. Neuromuscular symptoms include tremors, hyperreflexia, myoclonus (involuntary muscle contractions), and muscle rigidity. Autonomic symptoms include hyperthermia, tachycardia, vomiting, diarrhea, and hypersalivation.
>
> Serotonin syndrome is rarely observed in veterinary medicine, but it is important to recognize and manage it if it occurs. A study by Pugh (2013) involving 33 cats accidentally intoxicated by ingesting SSRIs showed that clinical signs resolved within 12–14 hours and that none of the cats died. Reported symptoms included mental signs (agitation), motor symptoms (tremors), autonomic signs (vomiting, diarrhea, hypersalivation), and general symptoms (hyperthermia).

Treatment is symptomatic, involving the administration of anticonvulsants, management of hyperthermia, and shock treatment through fluid therapy. Cyproheptadine, an antihistamine, can be used as a specific treatment. In human medicine, it is administered at an initial dose of 12 mg, followed by 2 mg every 2 hours until symptoms resolve. In veterinary medicine, a dose of 2 mg intravenously or 8 mg orally is recommended, with readministrations every 8 to 12 hours until symptoms resolve (Norris et al. 1998).

17.1.3 Clinical Use and Dosage

SSRIs are used to treat all pathological states, as serotonin's action can modulate all other neurotransmitters. The choice of SSRI is guided by the pathological state being treated and potential side effects.

17.2 Fluoxetine

17.2.1 Description, Chemical Formula, and Brand Names

Fluoxetine, chemically named *N-methyl-3-phenyl-3[4-(trifluoromethyl)phenoxy]propan-1-amine* (International Union of Pure and Applied Chemistry, IUPAC nomenclature), belongs to the family of antidepressants known as selective serotonin reuptake inhibitors (SSRIs).

A veterinary formulation with marketing authorization (MA) exists for use in dogs: Reconcile™, available in chewable tablets of 8, 16, 32, and 64 mg. In cats, fluoxetine is used off-label. It is primarily known under its original brand name, Prozac™, but is available in various dosages and generic names depending on the country. Fluoxetine is found in tablets, capsules, delayed-release capsules, and oral solutions.

A transdermal gel formulation has been proposed, but current clinical studies conclude that this form is ineffective and can cause skin irritation in cats (Ciribassi et al. 2003; Eichstadt et al. 2017).

17.2.2 Pharmacology

17.2.2.1 Mechanism of Action

Fluoxetine increases serotonin concentration in the synapse through mechanisms shared by all SSRIs:

- Inhibition of presynaptic serotonin transporters (SERT) increases the amount of serotonin in the synaptic cleft
- Desensitization of presynaptic serotonergic 5-HT1A autoreceptors promotes serotonin release.

Fluoxetine also enhances the action of norepinephrine and dopamine through the following mechanisms:

- Antagonism of postsynaptic 5-HT2C receptors, which increases dopamine and norepinephrine release in the prefrontal cortex, resulting in procognitive and especially antidepressant effects (Stahl and Muntner 2021).
- Inhibition of presynaptic norepinephrine reuptake pumps (NET), which raises norepinephrine levels in the synaptic space.

Fluoxetine also moderately enhances cholinergic and histaminergic activity due to its antagonistic action on muscarinic and histaminergic receptors (◘ Fig. 17.2).

17.2.2.2 Metabolism and Involved Cytochromes

There are no specific studies on the cytochromes involved in the metabolism of fluoxetine in cats. Therefore, we refer to human data regarding fluoxetine metabolism.

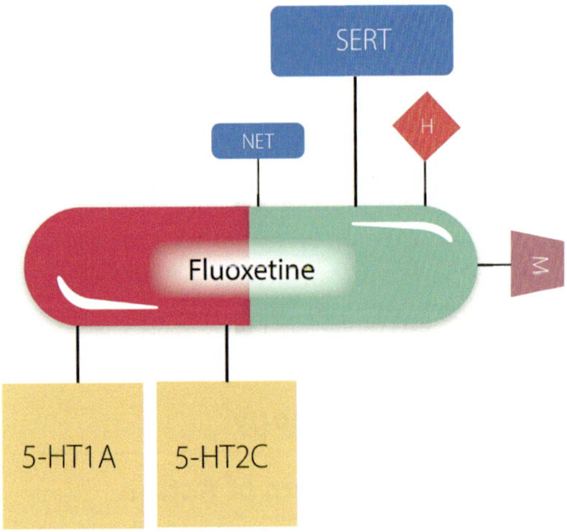

◘ **Fig. 17.2** General structure of fluoxetine. SERT inhibitor—5-HT1A desensitizer—5-HT2C antagonist—NET inhibitor—histaminergic antagonist—muscarinic antagonist

Fluoxetine is metabolized by the cytochrome CYP2D6 into norfluoxetine, its active metabolite. Fluoxetine inhibits the cytochromes CYP3A4, CYP2C19, and CYP2C9.

17.2.2.3 Pharmacokinetics

Fluoxetine is absorbed at nearly 100% after oral administration (Papich, 2016). It is distributed throughout all organs, with the highest concentrations found in the liver and lungs. Fluoxetine is extensively metabolized into various metabolites in the liver, primarily producing its active metabolite, norfluoxetine.

After administering an oral dose of 1 mg/kg of fluoxetine, the maximum plasma concentration of fluoxetine (95 ng/mL) is reached in 6.5 h, and the maximum concentration of norfluoxetine (21 ng/mL) is reached in 72 h. The average half-life in this study was 47 h for fluoxetine and 55 h for norfluoxetine (Ciribassi et al. 2003).

Fluoxetine, when used in transdermal gel form, is absorbed through the skin to reach systemic circulation. However, its relative bioavailability is only about 10% compared to oral administration (Ciribassi et al. 2003).

The molecule and its metabolites are primarily eliminated via the biliary route, though some elimination may occur through the urinary route. However, no studies have yet determined the exact fraction excreted renally.

Therefore, impaired liver function may slow down the elimination of fluoxetine and norfluoxetine. Renal function appears to have little impact on the pharmacokinetics of these molecules, but available data are limited.

17.2.3 Common Side Effects and Usage Precautions

17.2.3.1 Side Effects

Reported side effects of fluoxetine include anxiety, irritability, sleep disturbances, tremors, vomiting, diarrhea, hypersalivation, hyperthermia, and urinary retention (DiCiccio and McClosky 2022; Pugh et al. 2013). These effects are typically observed in cases of overdose.

Depending on the severity of the side effects, either a reduction in dosage or simply waiting for approximately 10 days can resolve the issue. It is worth noting that if the dose is reduced and the side effects are resolved, it is often possible to return to the original dose without a recurrence of the initial side effects.

These effects are primarily due to the regulation of 5-HT1A receptors in synaptic membranes. This regulation explains why such side effects are observed only at the beginning of treatment.

17.2.3.2 Precautions and Contraindications

In humans, fluoxetine is contraindicated in cases of diabetes mellitus as it alters glucose metabolism and plasma concentration (Breum et al. 1995; Santos et al. 2001).

For cardiac disease, SSRIs should be prescribed with caution, as they may have vasodilatory and cardiodepressive effects (Pacher et al. 1998). While no such studies exist in cats, caution is advised in cases of diabetes or heart disease.

In humans, isolated cases of bleeding during childbirth and fetal cardiovascular changes have been reported in patients taking fluoxetine during pregnancy. Fluoxetine is excreted in breast milk at 20–30% of its plasma level, but lactation while taking fluoxetine is not a major contraindication. The recommended approach is monitoring the newborn, with treatment discontinuation if prolonged sedation or agitation occurs.

There are no studies in cats regarding fluoxetine's effects on the fetus during pregnancy or on kittens during lactation.

In cases of severe liver failure, fluoxetine doses should be reduced. In renal insufficiency, due to a lack of data, caution is required, and potential side effects should be closely monitored in case of impaired drug elimination.

In both human epileptic cases (Ribot et al. 2017) and animal models (Cardamone et al. 2013; Igelström and Heyward 2012), fluoxetine does not increase seizure frequency but may actually help reduce seizures. However, fluoxetine inhibits cytochrome enzymes, altering the pharmacokinetics of certain drugs, including phenobarbital.

When combining fluoxetine and phenobarbital, treatment should begin at half the normal dose for both drugs, following a precautionary principle. Dose adjustments should then be made gradually, based on observed clinical effects.

17.2.3.3 Drug Interactions

The following drug combinations are *prohibited:*

- Selegiline (monoamine oxidase inhibitors or MAOIs): as with all medications in the SSRI family, the major contraindication is the combination of fluoxetine with monoamine oxidase inhibitors. This combination carries a significant risk of serotonin syndrome. In human medicine, the recommendation is a six-week washout period when transitioning from fluoxetine to selegiline, and a 2-week period for the reverse transition.
- Tramadol (analgesic): fluoxetine may inhibit the hepatic metabolism of tramadol into its active metabolites. The combination of these two molecules increases the risk of serotonin syndrome.
- Clomipramine, amitriptyline (tricyclic antidepressants): combining fluoxetine with tricyclic antidepressants may elevate plasma levels of the tricyclics, thereby increasing the likelihood of serotonin syndrome.
- Venlafaxine: combining venlafaxine with an SSRI may increase the plasma levels of both drugs, thereby promoting serotonin syndrome.
- Ketoconazole (antifungal): this combination increases the risk of serotonin syndrome.

The following drug interactions have been reported in humans or animals receiving fluoxetine. The combination is not necessarily contraindicated, but the risks must be assessed by the practitioner.

- Diazepam, alprazolam (benzodiazepines): fluoxetine, through its effects on hepatic cytochromes, may increase their blood levels.

- Propranolol, metoprolol (beta-blockers): fluoxetine may elevate blood levels of beta-blockers.
- Trazodone (serotonin antagonist and reuptake inhibitor, SARI): combining trazodone with fluoxetine may increase trazodone plasma levels and promote serotonin syndrome.
- Hypericum perforatum (St. John's Wort): combining it with fluoxetine increases the risk of serotonin syndrome.
- Clopidogrel, heparin, warfarin (anticoagulants): their use with fluoxetine is associated with an increased risk of bleeding in humans.
- Buspirone (anxiolytic): combining buspirone with fluoxetine increases the risk of serotonin syndrome.
- Cyproheptadine (antihistamine): cyproheptadine may diminish or nullify the effects of fluoxetine. It is used to treat serotonin syndrome.
- Diuretics: when combined with fluoxetine, diuretics may increase the risk of hyponatremia.
- Insulin: fluoxetine may alter the required doses needed to maintain diabetes control.
- Isoniazid (antibiotic used in tuberculosis treatment): combining it with fluoxetine increases the risk of serotonin syndrome.
- Methadone (synthetic opioid): in dogs, fluoxetine increases the peak plasma concentration and half-life area under the curve (KuKanich et al. 2011) of methadone.
- Nonsteroidal anti-inflammatory drugs (NSAIDs): combining NSAIDs with SSRIs may increase the risk of gastrointestinal ulcers and bleeding.
- Pentazocine (analgesic): when combined with fluoxetine, a pseudoserotonin syndrome may occur.
- Phenytoin (antiepileptic): fluoxetine may increase plasma levels of phenytoin.

17

17.2.3.4 Toxicity

The LD_{50} (lethal dose 50) is 452 mg/kg in rats, 100 mg/kg in dogs, and is unknown in cats. However, a median toxic dose has been reported, with signs of serotonin syndrome occurring at doses as low as 3.7 mg/kg (Pugh et al. 2013).

17.2.4 Clinical Use and Dosage

17.2.4.1 According to the Scientific Literature

Fluoxetine is recommended for the treatment of urinary and fecal incontinence, urine marking (Pryor et al. 2001), aggression, stereotypic and compulsive behaviors, and anxiety (Kaur et al. 2016).

The recommended doses range from 0.5 to 1 mg/kg orally, once daily for all these indications (Denenberg & Dubé, 2018).

A treatment period of 4–8 weeks is required before its efficacy can be evaluated ("Drug Monographs Plumb's Veterinary Drugs," 2024).

17.2.4.2 Experience Feedback

Fluoxetine is primarily a serotonergic enhancer. However, it also acts, albeit less effectively, by enhancing noradrenergic and dopaminergic transmission. It is useful for stimulating behavior as well as for controlling aggression and high levels of impulsivity.

Depending on the psychopathological state to be treated, the dosage is adjusted as follows:

- Chronic depressive state: 0.25 mg/kg to 0.5 mg/kg orally once daily
- Phobic and anxious states: 0.5 mg/kg to 1 mg/kg orally once daily
- Impulsive and compulsive states: 1 mg/kg to 2 mg/kg orally once daily

Administering the molecule with progressive dosing, with adjustments every 2 weeks, helps to avoid significant side effects. The practitioner will start with a dosage close to the lower end of the range based on the state being treated and then increase or maintain the dose depending on the severity of side effects.

From a nosographic perspective, fluoxetine is one of the drugs of choice for treating

HSHA Hypersensitivity-Hyperactivity Syndrome (HSHA), with a dosage of 1 mg/kg to 2 mg/kg orally once daily.

Clinical stabilization is observed after 2–4 weeks of treatment. This corresponds to the onset of therapeutic effects following the resolution of side effects. Indeed, side effects are present during the first 2 weeks of treatment and also contribute to reduced impulsivity (e.g., lethargy, fatigue). Therefore, improvements in behavior may be observed within the first few days of treatment, but these are due to the downregulation effects linked to the modification and redistribution of synaptic receptors (see ► Chap. 17, Sect. 1.2, Box "Up-regulation and Down-regulation").

It is not uncommon for impulsivity to increase once the side effects subside, requiring a dosage increase to achieve the desired therapeutic effect.

17.2.4.3 Possible Combinations

- Fluoxetine-gabapentin combination: helps regulate both impulsivity and intense fear reactions in phobic and impulsive states. This combination is also beneficial for better control of anxiety disorders.
- Fluoxetine-benzodiazepine combination: provides better management of phobic or anxious states.
- Fluoxetine-trazodone combination: aids in better sleep regulation in cases of anxiety, depression, and senile disorders.

17.2.4.4 Withdrawal

Given the very long half-life of norfluoxetine, a natural withdrawal process occurs even after abrupt discontinuation.

Withdrawal remains clinically relevant after 6 months of treatment, though not due to any potential addiction. Typically, withdrawal is performed gradually by first reducing the dose to 0.25 mg/kg per day, then transitioning to alternate-day dosing. This method establishes withdrawal stages to ensure that symptoms do not reappear. Additionally, it allows verification that the behaviors learned during treatment remain intact.

If withdrawal fails—meaning that symptoms partially or completely reappear—this procedure helps determine the minimum effective dose for the cat at the end of treatment, allowing it to be maintained over time while periodically attempting withdrawal. However, it is important to note that returning to the previous effective dose does not always restore the initial therapeutic effects.

The pace of dose reduction varies. In general, complete withdrawal is carried out over 1 week per month of treatment. For example, for a cat treated for 1 year, withdrawal would be completed over 12 weeks.

17.3 Fluvoxamine

17.3.1 Description, Chemical Formula, and Brand Names

Fluvoxamine, chemically named *2-[(E)-[5-methoxy-1-[4-trifluoromethyl)phenyl]pentylidene]amino]oxy]ethanamine*, belongs to the family of antidepressants known as selective serotonin reuptake inhibitors (SSRIs).

There is no veterinary-specific formulation for fluvoxamine, so it is used off-label. It is available in tablet form and under various human brand names depending on the country (e.g., Luvox™, Favoxyl™, Floxyfral™). A prolonged-release tablet form is also available (Luvox™ CR).

17.3.2 Pharmacology

The pharmacology of fluvoxamine has been scarcely or not at all studied in the scientific literature. In this section, we present the available data in humans, supplemented by the authors' personal experience.

17.3.2.1 Mechanism of Action

Fluvoxamine increases serotonin concentration in the synapse through mechanisms common to all SSRIs:

- Inhibition of presynaptic serotonin transporters (SERT) increases the amount of serotonin in the synaptic cleft.

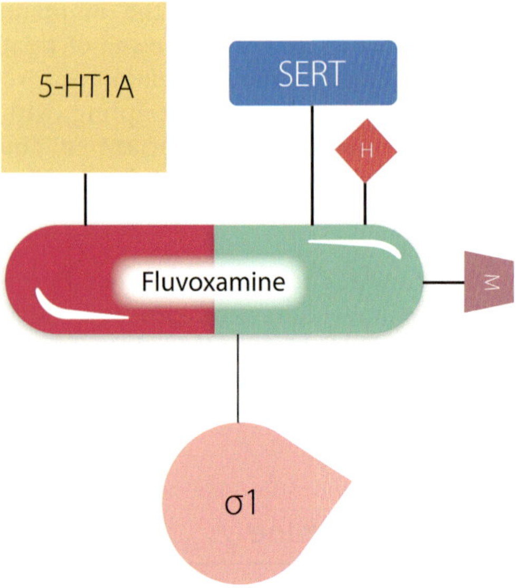

Fig. 17.3 Structure of fluvoxamine. SERT inhibitor—5-HT1A desensitizer—σ1 agonist—histaminergic antagonist—muscarinic antagonist

- Desensitization of presynaptic serotonergic 5-HT1A autoreceptors promotes serotonin release.

Fluvoxamine has a unique affinity for receptors, which is not yet fully understood but may explain its anxiolytic effects.

It also exhibits low affinity for muscarinic and histaminergic receptors (Fig. 17.3).

17.3.2.2 Metabolism and Involved Cytochromes

Nine hepatic metabolites have been identified, with little to no effect on serotonergic transporters. The majority of the administered dose is excreted in the urine as metabolites in humans (Van Harten 1995).

17

Fluvoxamine inhibits the cytochromes P450 CYP3A4, CYP1A1, CYP2C9, and CYP2C19 in humans. Its interaction with hepatic enzymes has implications for drug interactions (see ▶ Chap. 16, Sect. 2.2, Box « Cytochromes p450 or cyp450 »).

17.3.2.3 Pharmacokinetics

The molecule is believed to be fully absorbed and then metabolized by the liver. Its half-life in humans is 15 h (Van Harten 1995).

17.3.3 Common Side Effects and Usage Precautions

17.3.3.1 Side Effects

The primary side effects observed by the authors of this book in their daily practice include fatigue during the first few days of treatment. Decreased appetite is common but less pronounced than in dogs, likely due to the cat's fractioned feeding habits. It is also less severe than what is observed with fluoxetine. No occurrence or worsening of aggressive behavior has been reported at standard doses (see Sect. 3.4).

17.3.3.2 Precautions and Contraindications

Severe hepatic insufficiency should lead to a reduction in the administered dose of fluvoxamine. In cases of renal insufficiency, no dose adjustment is necessary.

There are no studies on the effects of fluvoxamine during gestation and lactation in cats. In rats, fluvoxamine has been reported to increase pup mortality at birth (Glazova et al. 2014). In humans, it is excreted in milk without significant consequences (Wright et al. 1991).

17.3.3.3 Drug Interactions

The following drug combinations are *prohibited:*

- Selegiline (monoamine oxidase inhibitors or MAOIs): as with all medications in the SSRI family, the major contraindication is the combination of fluvoxamine with monoamine oxidase inhibitors. This combination poses a significant risk of serotonin syndrome.
- Tramadol (analgesic): fluvoxamine may inhibit the hepatic metabolism of tramadol into active metabolites. The combination of these two drugs increases the risk of serotonin syndrome.
- Clomipramine, amitriptyline (tricyclic antidepressants): the combination may increase plasma levels of tricyclic antidepressants, promoting the onset of serotonin syndrome.

- Venlafaxine: combining venlafaxine with an SSRI may elevate plasma levels of both drugs, increasing the risk of serotonin syndrome.
- Ketoconazole (antifungal): this combination increases the risk of serotonin syndrome.

The following drug interactions have been reported in humans or animals receiving fluvoxamine. The combination is not necessarily contraindicated, but the risks must be assessed by the practitioner.

- Buspirone (anxiolytic): combining buspirone with fluvoxamine reduces the clinical efficacy of buspirone.
- Cisapride (gastrointestinal motility stimulant): fluvoxamine may increase blood levels of cisapride, leading to cardiac toxicity.
- Cyproheptadine (antihistamine): its use may reduce or nullify the effects of fluvoxamine. Cyproheptadine is used in the treatment of serotonin syndrome.
- Diazepam, alprazolam (benzodiazepines): fluvoxamine, through its effects on hepatic cytochromes, may increase their blood levels.
- Diltiazem (calcium channel blocker): its effects are enhanced when combined with fluvoxamine. In humans, this may increase the risk of bradycardia.
- Methadone (synthetic opioid): in dogs, fluvoxamine increases the peak plasma concentration of methadone, which may lead to toxic doses.
- Metoclopramide (antiemetic): combining fluvoxamine with metoclopramide increases the risk of serotonin syndrome.
- Mexiletine (antiarrhythmic): fluvoxamine may slow the metabolism of mexiletine.
- Phenytoin (antiepileptic): fluvoxamine may increase plasma levels of phenytoin.
- Propranolol, metoprolol (beta-blockers): Fluvoxamine may increase blood levels of beta-blockers.
- Theophylline (bronchodilator): fluvoxamine may increase blood levels of theophylline.
- Warfarin (anticoagulant): its use with fluvoxamine is associated with an increased risk of bleeding.

17.3.3.4 Toxicity

The LD_{50} (lethal dose 50) is unknown in cats. In dogs, it exceeds 464 mg/kg, and doses above 10 mg/kg are likely to cause tremors and lethargy.

17.3.4 Clinical Use and Dosage

17.3.4.1 According to the Scientific Literature

In human psychiatry, the primary indications for fluvoxamine are obsessive-compulsive disorders and social anxiety.

In cats, fluvoxamine appears to be useful for cases of productive anxiety and aggression.

The doses range from 0.25 to 2 mg/kg orally, once daily. Three to eight weeks may be required before the effectiveness of the treatment can be evaluated.

17.3.4.2 Experience Feedback

Fluvoxamine is primarily a serotonergic enhancer.

Its main indication is the control of impulsive or compulsive states. The anxiolytic potential of the molecule is likely underutilized in practice.

The recommended dosage is 0.25 to 0.5 mg/kg orally once daily (Crowell-Davis et al. 2019).

In practice, prescribing between 1/8 and 1/4 of a 25 mg tablet per day is possible, which corresponds to a high dose and may not be easy for owners to administer. Therefore, fluvoxamine is generally reserved for cases refractory to fluoxetine or other psychotropics and is rarely prescribed as a first-line treatment.

17.3.4.3 Possible Combinations

Fluvoxamine can be combined with gabapentinoids and benzodiazepines in cases of anxious and impulsive states.

17.3.4.4 Withdrawal

A withdrawal period spanning 3 weeks is recommended after a treatment duration of at least 3 weeks (Stahl and Muntner 2021).

17.4 Paroxetine

17.4.1 Description, Chemical Formula, and Brand Names

Paroxetine, chemically named *(3S-trans)-3-[(1,3-benzodioxol-5-yloxy)methyl]-4-(4-fluorophenyl)piperidine hydrochloride hemihydrate*, belongs to the family of antidepressants known as selective serotonin reuptake inhibitors (SSRIs).

There is no veterinary-specific formulation for paroxetine, so it is used off-label. This medication is available under various brand names and in generic forms, including Paxil™ and Deroxat™ (France). Available dosages vary by country, and paroxetine is available in tablets, extended-release tablets, and oral solution forms.

17.4.2 Pharmacology

The pharmacology of paroxetine has been scarcely or not at all studied in the scientific literature. In this section, we present the available data in humans, supplemented by the authors' personal experience.

17.4.2.1 Mechanism of Action

Paroxetine increases serotonin concentration in the synapse through mechanisms common to all SSRIs:

- Inhibition of presynaptic serotonin transporters (SERT) increases the amount of serotonin in the synaptic cleft.
- Desensitization of presynaptic serotonergic 5-HT1A autoreceptors promotes serotonin release.

17

Paroxetine has limited noradrenergic activity due to its low potential for inhibiting presynaptic norepinephrine transporters (NET). This property slightly increases the amount of norepinephrine in the synaptic space.

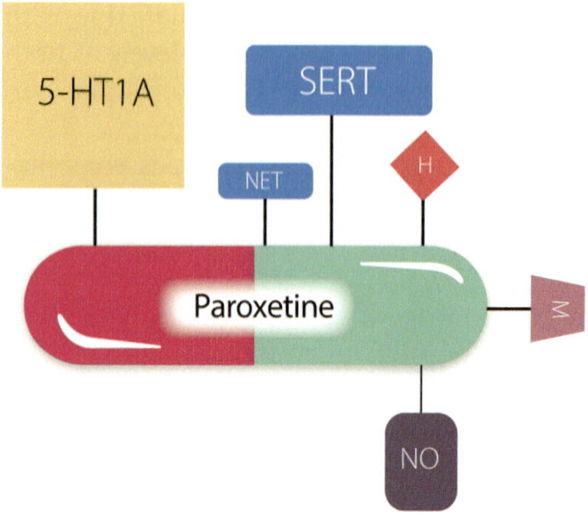

Fig. 17.4 Structure of paroxetine. SERT inhibitor—5-HT1A desensitizer—NO inhibitor—NET inhibitor—histaminergic antagonist—muscarinic antagonist

Paroxetine also modestly enhances cholinergic activity through its agonistic action on muscarinic receptors.

It is an inhibitor of nitric oxide synthase, the enzyme responsible for producing nitric oxide (NO). NO acts as a cerebral and immune neuromodulator and also has vasodilatory effects. The inhibitory action on nitric oxide synthase is associated with the sexual dysfunction caused by paroxetine in humans (Fig. 17.4).

17.4.2.2 Metabolism and Involved Cytochromes

Paroxetine is extensively metabolized by the liver. All its metabolites are inactive (Stahl and Muntner 2021). Paroxetine inhibits the P450 cytochromes CYP2C9 and CYP2D6 in humans.

17.4.2.3 Pharmacokinetics

In humans, the half-life is approximately 24 h. Elimination is 64% urinary and 36% fecal. Only 2% is excreted as paroxetine, with the remainder excreted as inactive metabolites (Kaye et al. 1989).

17.4.3 Common Side Effects and Usage Precautions

17.4.3.1 Side Effects

There are no specific studies available for cats. Therefore, we will refer to the common side effects of all SSRIs.

17.4.3.2 Precautions and Contraindications

Severe hepatic insufficiency should lead to a reduction in the administered dose of paroxetine. In cases of renal insufficiency, it is also recommended to reduce the administered doses.

There are no studies on the effects of paroxetine during gestation and lactation in cats. Paroxetine crosses the placental barrier and is excreted in milk, but at low levels. Since its safety during gestation and lactation has not been evaluated, the prescriber must base their decision on a benefit-risk assessment.

17.4.3.3 Drug Interactions

The following drug combinations are *prohibited:*

- Selegiline (monoamine oxidase inhibitors or MAOIs): as with all medications in the SSRI family, the major contraindication is the combination of paroxetine with monoamine oxidase inhibitors. This combination poses a significant risk of serotonin syndrome.
- Clomipramine, amitriptyline (tricyclic antidepressants): combining paroxetine with tricyclic antidepressants may increase their plasma levels and promote the onset of serotonin syndrome.
- Venlafaxine: combining venlafaxine with an SSRI may elevate plasma levels of both drugs, increasing the risk of serotonin syndrome.
- Tramadol (analgesic): paroxetine may inhibit the hepatic metabolism of tramadol into active metabolites. The combination of these two drugs increases the risk of serotonin syndrome.

The following drug interactions have been reported in humans or animals receiving paroxetine. The combination is not necessarily contraindicated, but the risks must be assessed by the practitioner.

- Methylene blue (dye and treatment for methemoglobinemia): combining methylene blue with paroxetine may induce serotonin syndrome.
- Clopidogrel, heparin, warfarin (anticoagulants): their use with paroxetine is associated with an increased risk of bleeding in humans.
- Buspirone (anxiolytic): combining buspirone with paroxetine increases the risk of serotonin syndrome.
- Codeine (morphinic alkaloid): paroxetine may reduce the conversion of codeine into morphine.
- Cyproheptadine (antihistamine): the combination may reduce or nullify the effects of paroxetine. Cyproheptadine is used to treat serotonin syndrome.
- Diuretics: When combined with paroxetine, diuretics may increase the risk of hyponatremia.
- Doxorubicin (anticancer drug): paroxetine may increase plasma concentrations of doxorubicin.
- Insulin: paroxetine may alter the doses required to maintain diabetes control.
- Isoniazid (antibiotic used in tuberculosis treatment): combining isoniazid with paroxetine may increase the risk of serotonin syndrome.
- Methadone (synthetic opioid): in dogs, paroxetine increases the peak plasma concentration and the area under the half-life curve (KuKanich et al. 2011) of methadone. It also increases the risk of serotonin syndrome.
- Metoclopramide (antiemetic): combining metoclopramide with paroxetine may induce serotonin syndrome.
- Mexiletine (antiarrhythmic): paroxetine may decrease the metabolism of mexiletine.

- Mirtazapine (antidepressant): combining mirtazapine with paroxetine may induce serotonin syndrome.
- Nonsteroidal anti-inflammatory drugs (NSAIDs): combining NSAIDs with an SSRI increases the risk of gastrointestinal ulcers and bleeding.
- Ondansetron (anti-nausea drug): combining ondansetron with paroxetine may induce serotonin syndrome.
- Pentazocine (analgesic): when combined with paroxetine, a pseudo-serotonin syndrome may occur.
- Phenobarbital (anticonvulsant): combining phenobarbital with paroxetine may decrease paroxetine concentrations.
- Phenytoin (antiepileptic): paroxetine may increase plasma levels of phenytoin.
- Propranolol, metoprolol (beta-blockers): paroxetine may increase blood levels of beta-blockers.
- Hypericum perforatum (St. John's Wort): combining St. John's Wort with paroxetine may induce serotonin syndrome.
- Trazodone (serotonin antagonist and reuptake inhibitor, SARI): combining trazodone with paroxetine may increase plasma levels of trazodone and promote the onset of serotonin syndrome.

17.4.3.4 Toxicity

There are no reliable data to evaluate an LD_{50} in cats. The symptoms of toxicity are believed to be those common to all SSRIs.

17.4.4 Clinical Use and Dosage

17

17.4.4.1 According to the Scientific Literature

Paroxetine has been used to treat cats exhibiting urine marking and aggression towards humans and other cats (Pachel 2014; Pryor 2003; Frank and Dehasse 2003).

The recommended dose ranges from 0.25 to 1 mg/kg orally once daily (Crowell-Davis et al. 2019). It is advised to start treatment at the lowest dose and adjust it every 2 months based on results.

17.4.4.2 Experience Feedback

Our limited clinical experience with paroxetine does not allow us to further expand on the data presented above.

17.4.4.3 Possible Combinations

Possible combinations are the same as those reported for other SSRIs. Due to our lack of clinical data, no specific combination can be recommended.

17.4.4.4 Withdrawal

In human psychiatry, paroxetine is the SSRI most commonly reported to induce withdrawal syndrome (Price et al. 1996). It is recommended to reduce the dose by 25% once per week and return to the previous dose for 1–2 months if withdrawal symptoms occur.

17.5 Sertraline

17.5.1 Description, Chemical Formula, and Brand Names

Sertraline, chemically named *(1S,4S)-4-(3,4-dichlorophenyl)-N-methyl-1,2,3,4-tetrahydronaphthalen-1-amine*, belongs to the family of antidepressants known as selective serotonin reuptake inhibitors (SSRIs).

There is no veterinary-specific formulation for sertraline, so it is used off-label. It is widely marketed under its brand name Zoloft™ and is also available in various generic forms. The available dosages vary by country, and the drug is found in tablet, capsule, and oral solution forms.

17.5.2 Pharmacology

The pharmacology of sertraline has been scarcely or not at all studied in the scientific literature. In this section, we present the available data in humans, supplemented by the authors' personal experience.

17.5.2.1 Mechanism of Action

Sertraline increases serotonin concentration in the synapse through mechanisms common to all SSRIs:

- Inhibition of presynaptic serotonin transporters (SERT) increases the amount of serotonin in the synaptic cleft.
- Desensitization of presynaptic serotonergic 5-HT1A autoreceptors promotes serotonin release.

It also increases dopamine levels by inhibiting presynaptic dopamine transporters (DAT). This secondary affinity for dopaminergic transporters (DAT) is weaker than its affinity for serotonergic transporters (SERT).

Like fluvoxamine, it has an affinity for certain receptors whose mechanisms are not yet fully understood, which may explain its anxiolytic effects.

Unlike most other SSRIs, sertraline does not have muscarinic or histaminergic activity. Therefore, cholinergic or sedative side effects are more limited (◘ Fig. 17.5).

17.5.2.2 Metabolism and Involved Cytochromes

Sertraline is extensively metabolized by the liver. It is converted into N-desmethylsertraline, a metabolite with weaker SERT inhibition than the parent molecule. In humans, sertraline inhibits the P450 cytochromes CYP2C9 and CYP2D6.

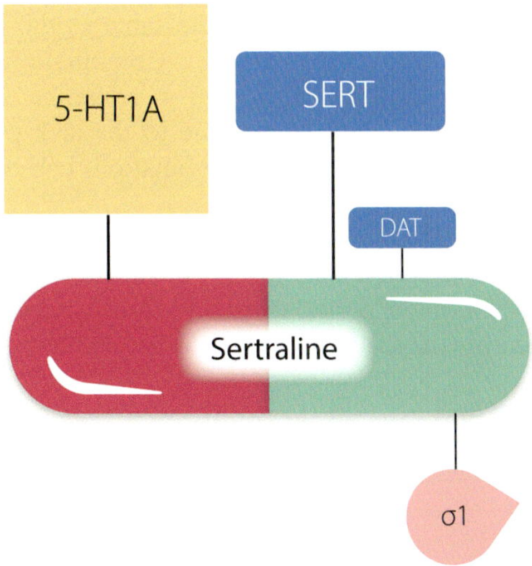

◘ **Fig. 17.5** Structure of sertraline. SERT inhibitor—5-HT1A desensitizer—σ agonist—DAT inhibitor

17.5.2.3 Pharmacokinetics

There is limited data available for cats. In humans, the molecule is 98% plasma protein-bound. The primary elimination pathway is biliary. The half-life is 26 h for sertraline and 80 h for N-desmethylsertraline, its active metabolite.

17.5.3 Common Side Effects and Usage Precautions

17.5.3.1 Side Effects

In the absence of specific studies on sertraline's side effects in cats, reference should be made to the common side effects of all SSRIs. In humans, sertraline is the most well-tolerated SSRI.

17.5.3.2 Precautions and Contraindications

In cases of severe liver failure, sertraline doses should be reduced. In cases of renal insufficiency, caution is advised in cats due to the lack of available data. Since the safety of sertraline during pregnancy and lactation has not been evaluated in cats, the benefit-risk balance should guide the prescriber's decision.

17.5.3.3 Drug Interactions

The following drug combinations are *prohibited:*

- Selegiline (monoamine oxidase inhibitors or MAOIs): as with all medications in the SSRI family, the major contraindication is the combination of sertraline with monoamine oxidase inhibitors. This combination poses a significant risk of serotonin syndrome.
- Clomipramine, amitriptyline (tricyclic antidepressants): combining sertraline with tricyclic antidepressants may increase their plasma levels and promote the onset of serotonin syndrome.
- Venlafaxine: combining venlafaxine with an SSRI may elevate plasma levels of both

drugs, increasing the risk of serotonin syndrome.
- Tramadol (analgesic): sertraline may inhibit the hepatic metabolism of tramadol into active metabolites. Combining these two drugs increases the risk of serotonin syndrome.

The following drug interactions have been reported in humans or animals receiving sertraline. The combination is not necessarily contraindicated, but the risks must be assessed by the practitioner.
- Clopidogrel, heparin, warfarin (anticoagulants): their use with sertraline is associated with an increased risk of bleeding in humans.
- Buspirone (anxiolytic): combining buspirone with sertraline increases the risk of serotonin syndrome.
- Cyproheptadine (antihistamine): combining cyproheptadine with sertraline may reduce or nullify sertraline's effects. Cyproheptadine is used in the treatment of serotonin syndrome.
- Diazepam, alprazolam (benzodiazepines): sertraline may increase their blood levels through its effects on hepatic cytochromes.
- Metoclopramide (antiemetic): combining metoclopramide with sertraline increases the risk of serotonin syndrome.
- Mexiletine (antiarrhythmic): sertraline may decrease the metabolism of mexiletine.
- Pentazocine (analgesic): when combined with sertraline, a pseudo-serotonin syndrome may occur.
- Hypericum perforatum (St. John's Wort): combining St. John's Wort with sertraline increases the risk of serotonin syndrome.

17.5.3.4 Toxicity

In cats, no LD_{50} has been established. In dogs, for whom the recommended daily dose is higher than in cats, a daily dose exceeding 40 mg/kg induces mydriasis, hind limb weakness, hyperactivity, and anorexia. A daily dose exceeding 80 mg/kg causes elevated alkaline phosphatase levels. Seizures have been observed at doses exceeding 90 mg/kg. A daily dose above 160 mg/kg results in elevated hepatic transaminases (ALT) (Crowell-Davis et al. 2019).

17.5.4 Clinical Use and Dosage

17.5.4.1 According to the Scientific Literature

Sertraline is indicated for obsessive-compulsive behaviors, inappropriate elimination, and cases of anxiety. Doses range from 0.5 to 1.5 mg/kg orally once daily (Crowell-Davis et al. 2019). It is recommended to start treatment at the lowest dose and adjust it every 2 months based on results (Stahl and Muntner 2021).

17.5.4.2 Experience Feedback

Our experience does not allow us to add further information to the above.

17.5.4.3 Possible Combinations

Possible combinations are the same as those reported for other SSRIs. Combinations should always be considered in reference to cytochrome interactions. Due to the lack of clinical data, no specific combinations can be recommended.

17.5.4.4 Withdrawal

A withdrawal period of two to three weeks is recommended (Stahl and Muntner 2021).

Summary of This Chapter

Summary table of SSRIs

Substance	Mechanism of action	Pathological states for which the molecule is indicated	Dosage	Cytochromes
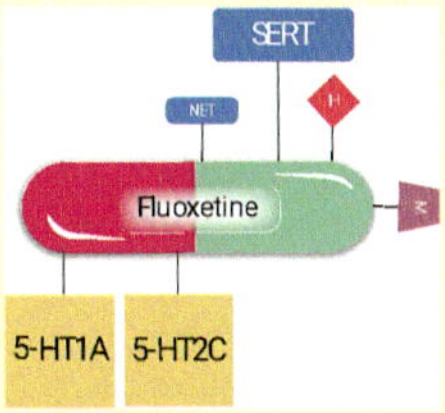	SERT inhibitor	Chronic depressive	0.25–1 mg/kg SID	CYP3A4
	5-HT1A desensitizer 5-HT2C antagonist	Phobic or anxious		CYP2C19 CYP2C9
	NET inhibitor Histaminic antagonist	Impulsive or compulsive		CYP2D6
	Muscarinic antagonist			
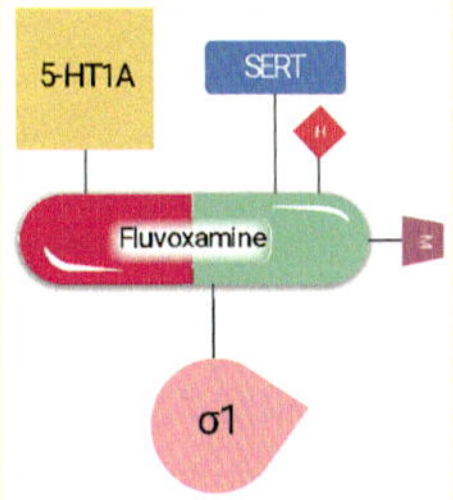	SERT inhibitor 5-HT1A desensitizer	Chronic depressive Phobic or anxious	0.25–0.5 mg/kg SID	CYP3A4 CYP1A1
	σ1 Agonist Histaminic antagonist	Impulsive or compulsive		CYP2C9 CYP2C19
	Muscarinic antagonist			
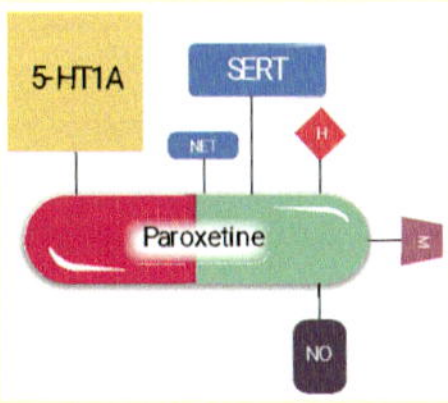	SERT inhibitor	Chronic depressive	0.25–1 mg/kg	CYP2C9
	5-HT1A desensitizer	Phobic or anxious	SID	CYP2D6
	NO inhibitor	Impulsive or compulsive		
	NET inhibitor			
	Histaminic antagonist			
	Muscarinic antagonist			
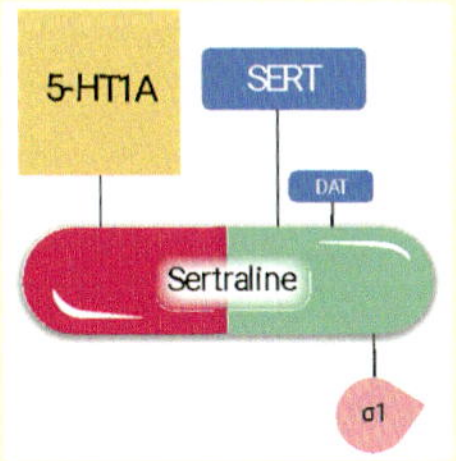	SERT inhibitor	Chronic depressive	0.5–4 mg/kg	CYP2C9
	5-HT1A desensitizer	Phobic or anxious	SID	CYP2D6
	DAT inhibitor	Impulsive or compulsive		
	σ1 Agonist			

References

Breum L, Bjerre U, Bak JF, Jacobsen S, Astrup A (1995) Long-term effects of fluoxetine on glycemic control in obese patients with non-insulin-dependent diabetes mellitus or glucose intolerance: Influence on muscle glycogen synthase and insulin receptor kinase activity. Metabolism 44(12):1570–1576. https://doi.org/10.1016/0026-0495(95)90077-2

Cardamone L, Salzberg M, O'Brien T, Jones N (2013) Antidepressant therapy in epilepsy: Can treating the comorbidities affect the underlying disorder? Brit J Pharmacol 168(7):1531–1554. https://doi.org/10.1111/bph.12052

Cascade E, Kalali AH, Kennedy SH (2009) Real-World Data on SSRI Antidepressant Side Effects. Psychiatry (Edgmont) 6, 16–18

Ciribassi J, Luescher A, Pasloske KS, Robertson-Plouch C, Zimmerman A, Kaloostian-Whittymore L (2003) Comparative bioavailability of fluoxetine after transdermal and oral administration to healthy cats. Am J Veterinary Res 64(8):994–998. https://doi.org/10.2460/ajvr.2003.64.994

Crowell-Davis SL, Murray TF, Mattos de Souza, Dantas L (2019) Veterinary psychopharmacology, 2nd edn. Wiley Blackwell

Denenberg S, Dubé MB (2018) Tools for managing feline problem behaviours: Psychoactive medications. J Feline Med Surg 20(11):1034–1045. https://doi.org/10.1177/1098612X18806760

DiCiccio VK, McClosky ME (2022) Fluoxetine-induced urinary retention in a cat. J Feline Med Surg Open Report 8(2):205511692211120. https://doi.org/10.1177/20551169221112065

Eichstadt LR, Corriveau LA, Moore GE, Knipp GT, Cooper BR, Gwin WE (2017) Absorption of transdermal fluoxetine compounded in a lipoderm base compared to oral fluoxetine in client-owned cats. Int J Pharm Compound 21(3):242–246

Frank D, Dehasse J (2003) Differential diagnosis and management of human-directed aggression in cats. Vet Clin Small Anim Pract 33:269–286. https://doi.org/10.1016/S0195-5616(02)00131-6

Glazova NY, Merchieva SA, Volodina MA, Sebentsova EA, Manchenko DM, Kudrun VS, Levitskaya NG (2014) Effects of neonatal fluvoxamine administration on the physical development and activity of the serotoninergic system in white rats. Acta Naturae 6(3):98–105. https://doi.org/10.32607/20758251-2014-6-3-98-105

Igelström KM, Heyward PM (2012) The antidepressant drug fluoxetine inhibits persistent sodium currents and seizure-like events. Epilepsy Res 101(1-2):174–181. https://doi.org/10.1016/j.eplepsyres.2012.03.019

Kaur G, Voith VL, Schmidt PL (2016) The use of fluoxetine by veterinarians in dogs and cats: A preliminary survey. Veterinary Record Open 3(1):e000146. https://doi.org/10.1136/vetreco-2015-000146

Kaye CM, Haddock RE, Langley PF, Mellows G, Tasker TCG, Zussman BD, Greb WH (1989) A review of the metabolism and pharmacokinetics of paroxetine in man. Acta Psychiatrica Scandinavica 80(S350):60–75. https://doi.org/10.1111/j.1600-0447.1989.tb07176.x

Kuisma M (1995) Fatal serotonin syndrome with trismus. Ann Emerg Med 26(1):108. https://doi.org/10.1016/S0196-0644(95)70250-4

KuKanich B, KuKanich KS, Rodriguez JR (2011) The effects of concurrent administration of cytochrome P-450 inhibitors on the pharmacokinetics of oral methadone in healthy dogs. Veterinary Anaesthesia and Analgesia 38, 224–230. https://doi.org/10.1111/j.1467-2995.2011.00602.x

Norris CR, Boothe DM, Esparza T, Gray C, Ragsdale M (1998) Disposition of cyproheptadine in cats after intravenous or oral administration of a single dose. Am J Veterinary Res 59(1):79–81. https://doi.org/10.2460/ajvr.1998.59.01.79

Pacher P, Ungvari Z, Kecskemeti V, Furst S (1998) Review of cardiovascular effects of fluoxetine, a selective serotonin reuptake inhibitor, compared to tricyclic antidepressants. Curr Med Chem 5(5):381–390

Pachel CL (2014) Intercat Aggression: Restoring Harmony in the Home. Vet Clin North Am Small Anim Pract 44:565–579. https://doi.org/10.1016/j.cvsm.2014.01.007

Papich MG (2016) Saunders handbook of veterinary drugs: Small and large animal, 4th edn. Elsevier

Prakash S, Rathore C, Rana K, Prakash A (2021) Fatal serotonin syndrome: a systematic review of 56 cases in the literature. Clin Toxicol 59(2):89–100. https://doi.org/10.1080/15563650.2020.1839662

Price JS, Waller PC, Wood SM, Mackay AVP (1996) A comparison of the post-marketing safety of four selective serotonin re-uptake inhibitors including the investigation of symptoms occurring on withdrawal. Brit J Clin Pharmacol 42(6):757–763. https://doi.org/10.1046/j.1365-2125.1996.00498.x

Pryor P (2003) Animal behavior case of the month. J Am Vet Med Assoc 223:790–792. https://doi.org/10.2460/javma.2003.223.790

Pryor PA, Hart BL, Cliff KD, Bain MJ (2001) Effects of a selective serotonin reuptake inhibitor on urine spraying behavior in cats. J Am Vet Med Assoc 219(11):1557–1561. https://doi.org/10.2460/javma.2001.219.1557

17

Pugh CM, Sweeney JT, Bloch CP, Lee JA, Johnson JA, Hovda LR (2013) Selective serotonin reuptake inhibitor (SSRI) toxicosis in cats: 33 cases (2004–2010). J Vet Emerg Critic Care 23(5):565–570. https://doi.org/10.1111/vec.12091

Ribot R, Ouyang B, Kanner AM (2017) The impact of antidepressants on seizure frequency and depressive and anxiety disorders of patients with epilepsy: Is it worth investigating? Epilepsy & Behavior 70:5–9. https://doi.org/10.1016/j.yebeh.2017.02.032

Santos EA, Keller R, Rodriguez E, Lopez L (2001) Effects of serotonin and fluoxetine on blood glucose regulation in two decapod species. Brazil J Med Biol Res 34(1):75–80. https://doi.org/10.1590/S0100-879X2001000100009

Stahl SM, Muntner N (2021) In: Grady MM (ed) Stahl's essential psychopharmacology: neuroscientific basis and practical applications, 5th edn. Cambridge University Press. https://doi.org/10.1017/9781108975292

Van Harten J (1995) Overview of the pharmacokinetics of fluvoxamine. Clin Pharm 29(Supplement 1):1–9. https://doi.org/10.2165/00003088-199500291-00003

Wright S, Dawling S, Ashford J (1991) Excretion of fluvoxamine in breast milk [letter]. Brit J Clin Pharmacol 31(2):209–196. https://doi.org/10.1111/j.1365-2125.1991.tb05519.x

Tricyclic Antidepressants

Contents

18.1 General Characteristics of All Tricyclics – 586
18.1.1 General Pharmacology of TCAs – 586
18.1.2 Common Side Effects and Usage Precautions for All TCAs – 587
18.1.3 Clinical Use and Dosage – 588

18.2 Clomipramine – 588
18.2.1 Description, Chemical Formula, and Brand Names – 588
18.2.2 Pharmacology – 588
18.2.3 Side Effects and Usage Precautions – 589
18.2.4 Clinical Use and Dosage – 590

18.3 Amitriptyline – 591
18.3.1 Description, Chemical Formula, and Brand Names – 591
18.3.2 Pharmacology – 591
18.3.3 Side Effects and Usage Precautions – 592
18.3.4 Clinical Use and Dosage – 593

References – 594

© The Author(s), under exclusive license to Springer Nature Switzerland AG 2026
S. Bleuer-Elsner et al., *Veterinary Psychiatry of the Cat*,
https://doi.org/10.1007/978-3-032-06260-4_18

Tricyclic antidepressants (TCAs) share the common mechanism of inhibiting presynaptic serotonin and norepinephrine reuptake pumps, thereby increasing the presence of these neurotransmitters in the synaptic space. No tricyclic antidepressant has been approved for veterinary use in cats. Below, we detail the two tricyclic antidepressants used in cats: clomipramine and amitriptyline. Clomipramine, marketed under the name Clomicalm™ in France, holds veterinary approval but only for dogs.

18.1 General Characteristics of All Tricyclics

Tricyclics are named after their three-ring molecular structure. Veterinary psychiatry makes limited use of different tricyclic antidepressants (TCAs). After describing the common characteristics of all the molecules in this family, we will detail clomipramine and amitriptyline. Other molecules, used only occasionally or anecdotally in dogs, are generally unsatisfactory and ineffective in cats. These include doxepin, which acts as a sedative in cats, and imipramine and amitriptyline, which cause hypersalivation in cats (Rodan and Heath 2016).

18.1.1 General Pharmacology of TCAs

18.1.1.1 Mechanism of Action

All tricyclic antidepressants share the common property of inhibiting presynaptic serotonin (SERT) and norepinephrine (NET) transporters (or reuptake pumps). They are also all antagonists of histaminergic H1 receptors, α1-adrenergic receptors, and muscarinic cholinergic receptors, as well as sodium (Na+) ion channels. The affinity of tricyclics for these receptors varies significantly between molecules, and some also inhibit β-adrenergic receptors and serotonergic 5-HT1A, 5-HT2A, and 5-HT2C receptors (◘ Fig. 18.1).

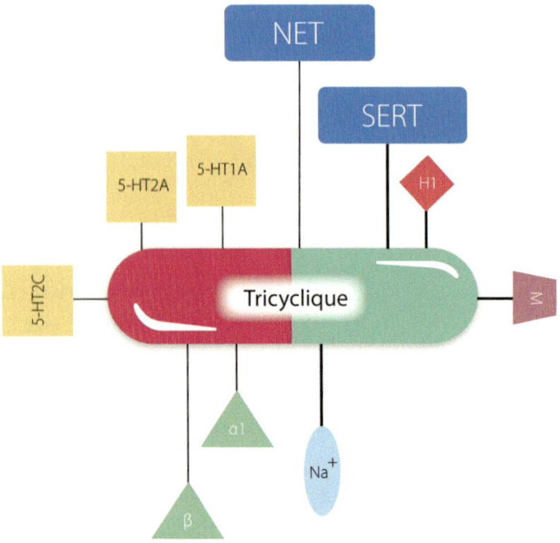

◘ Fig. 18.1 General structure of tricyclic antidepressants. SERT inhibitor—NET inhibitor—H1 histamine antagonist—M1 muscarinic antagonist—α1-adrenergic antagonist—Na+ channel blocker—β-adrenergic inhibitor—5-HT1A antagonist—5-HT2A antagonist—5-HT2C antagonist

18.1.1.2 Metabolism and Involved Cytochromes

Tricyclic antidepressants are metabolized in the liver through demethylation and aromatic hydroxylation, followed by glucuronidation of the aromatic hydroxyl group. Various P450 cytochromes are involved in these transformations. In humans, demethylation primarily involves CYP2C19, as well as CYP1A2 and CYP3A4, while hydroxylation is mainly mediated by CYP2D6 (Rudorfer and Potter 1999).

Cats exhibit a deficiency in glucuronyl transferase. In this species, glucuronidation and hydroxylation reactions occur more slowly than in dogs or humans (Baggot 2001; Lainesse et al. 2006). As a result, the prescribed doses and intervals between doses of TCAs are adjusted and extended compared to those for dogs.

18.1.1.3 Pharmacokinetics

All tricyclics are readily absorbed through the gastrointestinal tract. Peak plasma concentrations (2–6 h) and half-lives (up to 24 h) vary between molecules. Elimination occurs through both biliary and renal pathways, with varying proportions depending on the molecule.

18

18.1.2 Common Side Effects and Usage Precautions for All TCAs

18.1.2.1 Side Effects

Tricyclics share common side effects due to their inhibitory effects on histaminergic, adrenergic, cholinergic receptors, and sodium channels:

- Blocking histaminergic H1 receptors causes fatigue and a tendency toward weight gain (especially reported in humans).
- Blocking muscarinic M1 receptors results in dry mouth and dry eyes, as well as severe urinary and fecal retention.
- Blocking α1-adrenergic receptors leads to hypotension and dizziness.
- Partial sodium channel blockade at high doses can cause seizures and cardiac arrhythmias.

18.1.2.2 Precautions and Contraindications

Severe hepatic insufficiency should lead to a reduction in the administered doses of tricyclics due to their hepatic metabolism. It is also recommended to reduce doses in cases of renal insufficiency, given the partial renal elimination of the molecule. Finally, precautions should be taken with epileptic subjects due to the partial sodium channel blockade.

No major risks have been demonstrated during pregnancy and lactation. The practitioner should therefore base their decision on a benefit-risk assessment.

18.1.2.3 Drug Interactions

The following interactions are based on reports of toxicity in humans or animals, as well as theoretical considerations. The main interactions are listed here. For more specific information, refer to the section for each molecule.

- Acetazolamide/dichlorphenamide (antiglaucoma agents): the combination may increase blood levels of tricyclic antidepressants (TCAs).
- Detomidine, dexmedetomidine, medetomidine (α2-adrenergic agonists): the combination may increase the risks of respiratory depression, hypotension, and bradycardia.
- Amantadine (antiviral, NMDA antagonist, dopaminergic enhancer): the combination may increase the risks of urinary retention, constipation, ileus, hyperthermia, and QT interval prolongation.
- Ammonium chloride (expectorant): the combination may decrease blood levels of TCAs.
- Isoflurane (anesthetic agent): the combination may lead to hypotension, central nervous system (CNS) depression, QT interval prolongation, and death.
- Ketamine (anesthetic agent): the combination may lead to hypotension, CNS depression, QT interval prolongation, and death.
- Anthracyclines (anticancer agents): the combination may lead to QT interval prolongation.
- Antiarrhythmics: the combination may lead to QT interval prolongation, cardiac arrhythmias, and death.
- Apomorphine: the combination may lead to QT interval prolongation, cardiac arrhythmias, and death.
- Azole antifungals: the combination may cause constipation, ileus, urinary retention, and QT interval prolongation.
- Barbiturates: the combination may lead to respiratory system depression.
- Benzodiazepines: the combination may lead to central and respiratory system depression.
- Bromocriptine (dopaminergic agonist): the combination may cause central and respiratory system depression and serotonin syndrome.
- Buspirone: the combination may result in central nervous system depression and serotonin syndrome.
- Carbamazepine (anticonvulsant): the combination may decrease blood levels of tricyclic antidepressants (TCAs) and increase those of carbamazepine.

- Cimetidine (antacid): the combination may increase the risk of tricyclic antidepressant toxicity.
- Clonidine (α2-agonist): the combination may lead to fatal hypertension.
- Erythromycin (antibiotic): The combination may cause QT interval prolongation and cardiac arrhythmias.
- Fluoroquinolones: the combination may lead to potentially fatal QT interval prolongation.
- Levetiracetam (antiepileptic): the combination may result in central and respiratory system depression.
- Lithium: the combination may cause central nervous system depression and serotonin syndrome.
- Macrolides: the combination may result in potentially fatal QT interval prolongation.
- Methylphenidate (amphetamine): the combination may cause hypertension and overstimulation of the central nervous system.
- Metoclopramide (antiemetic): the combination may result in malignant dopaminergic syndrome. This combination is strongly contraindicated.
- Metronidazole (antiprotozoal antibiotic): the combination may lead to QT interval prolongation. Data in animals remain limited.
- Mirtazapine (antidepressant): the combination may cause constipation, urinary retention, central depression, and serotonin syndrome.
- Monoamine Oxidase Inhibitors (MAOIs) (Selegiline): the combination may lead to potentially fatal serotonin syndrome.
- Nonsteroidal anti-inflammatory drugs (NSAIDs): the combination may increase the risk of bleeding.
- Phenothiazines: the combination may increase TCA concentrations.
- Pregabalin: the combination may increase the risk of central depression.
- SSRIs: the combination may result in serotonin syndrome.
- Venlafaxine: the combination with a tricyclic antidepressant may increase plasma levels of both drugs and favor the occurrence of serotonin syndrome.
- Levothyroxine: the combination may pose risks of tachycardia, hypothyroidism, and thyrotoxicosis.
- Tramadol: the combination may lead to serotonin syndrome.
- Trazodone: the combination may result in constipation, ileus, urinary retention, central depression, and serotonin syndrome.

18

18.1.2.4 Toxicity

The LD_{50} varies between molecules. Toxicity manifests as serotonin syndrome, seizures, or cardiovascular disorders.

18.1.3 Clinical Use and Dosage

The diverse effects of tricyclic antidepressants make them highly versatile drugs that can be used for nearly all pathological states.

18.2 Clomipramine

18.2.1 Description, Chemical Formula, and Brand Names

Clomipramine, with the chemical name 3-(2-chloro-5,6-dihydrobenzo[b][1]benzazepin-11-yl)-N,N-dimethylpropan-1-amine, belongs to the tricyclic antidepressant family.Clomipramine is available for human use in pharmacies worldwide under its brand name Anafranil™ in capsule or tablet form.

18.2.2 Pharmacology

18.2.2.1 Mechanism of Action

Clomipramine increases the concentration of serotonin and norepinephrine in the synapse through the inhibition of presynaptic serotonin transporters (SERT), which increases the amount of serotonin in the synaptic space, and the inhibition of presynaptic norepinephrine transporters (NET), which increases the amount of norepinephrine in the synaptic space.

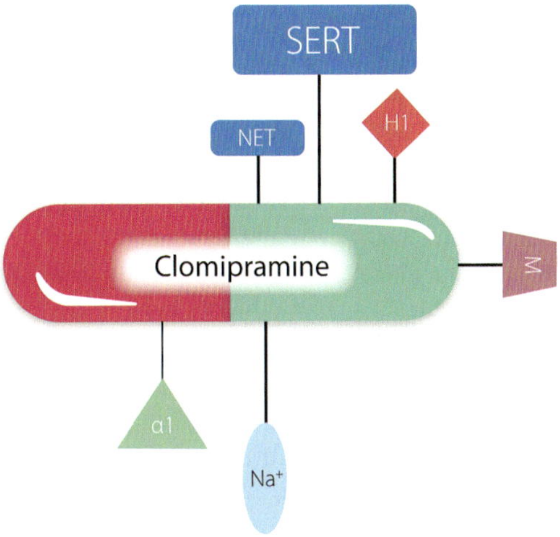

Fig. 18.2 Structure of clomipramine. SERT inhibitor—NET inhibitor—H1 histamine antagonist—muscarinic antagonist—α1-adrenergic antagonist—sodium channel blocker

Clomipramine has a greater affinity for SERT receptors, making it more effective at inhibiting serotonin reuptake than norepinephrine reuptake. It is considered the most "serotonergic" of the tricyclic antidepressants.

Like all molecules in its family, clomipramine possesses anticholinergic, antimuscarinic, antihistaminic, α1-blocking, and sodium channel-blocking properties, which are responsible for the drug's side effects or adverse reactions (Fig. 18.2).

18.2.2.2 Metabolism and Involved Cytochromes

Clomipramine is metabolized primarily through demethylation catalyzed by cytochrome P450 enzymes CYP3A4, CYP2C19, CYP2C9, and CYP1A2, producing its active metabolite desmethylclomipramine (Lainesse et al. 2007). Both clomipramine and desmethylclomipramine are subsequently metabolized via hydroxylation by CYP2D6 (Nielsen et al. 1996).

18.2.2.3 Pharmacokinetics

Clomipramine is highly lipophilic and crosses membrane barriers easily. Its bioavailability is 90% and increases when taken with food. In cats, the plasma concentration peak of clomipramine is reached in six hours, and its half-life is 12 h. The metabolism of clomipramine appears to be slower in males than in females (Lainesse et al. 2006).

Let me know if there are any additional details or corrections required.

18.2.3 Side Effects and Usage Precautions

18.2.3.1 Side Effects

Cats appear more susceptible than dogs to developing side effects due to the slower elimination of secondary metabolites in this species. The main reported side effects include mydriasis, urinary retention, constipation, diarrhea, dry mouth, and lethargy (Pfeiffer et al. 1999).

18.2.3.2 Precautions and Contraindications

Due to its action on ion channels, clomipramine is contraindicated in cats with a history of seizures or when used in combination with other drugs that may lower the seizure threshold (Koella et al. 1979).

Cats that received 10 mg/day of clomipramine for 28 days showed a decrease in total thyroxine (T4), triiodothyronine (T3), and free thyroxine (fT4) levels by 25%, 24%, and 16%, respectively. This effect could lead to a misdiagnosis of euthyroidism in cats with subclinical hyperthyroidism (Martin 2010).

Because of its anticholinergic effects, clomipramine should be used cautiously in cases of impaired intestinal motility, heart disease, urinary retention, or glaucoma. It is recommended to discontinue treatment before the use of anesthetics to minimize cardiac risk.

Let me know if additional adjustments are needed.

18.2.3.3 Drug Interactions

The following drug combinations are *prohibited:*

- Selective serotonin reuptake inhibitors (SSRIs): combining SSRIs with clomipramine increases the risk of serotonin syndrome. A washout period of four days is recommended when transitioning from an SSRI to clomipramine or vice versa (Trepanier 2013).

- Venlafaxine: combining venlafaxine with a tricyclic antidepressant may raise plasma levels of both drugs, increasing the risk of serotonin syndrome.
- Metoclopramide (antiemetic): combining metoclopramide with clomipramine may induce extrapyramidal malignant syndrome. This combination is contraindicated in humans and should be used with extreme caution in dogs.
- Selegiline (MAOI): combining selegiline with clomipramine significantly increases the risk of serotonin syndrome. This combination is contraindicated. A washout period of 14 days is required when transitioning between these drugs, regardless of the direction.
- Tramadol (opioid): combining tramadol with clomipramine increases the risk of serotonin syndrome (KuKanich 2013).

The following drug interactions have been reported in humans or animals receiving clomipramine. The combination is not necessarily contraindicated, but the risks must be assessed by the practitioner.

- Nonsteroidal anti-inflammatory drugs (NSAIDs): combining NSAIDs with clomipramine increases the risk of bleeding.
- Atropine, chlorpheniramine, glycopyrrolate (anticholinergic agents): use with caution when combining these agents with clomipramine.
- Buspirone (anxiolytic): combining buspirone with clomipramine increases the risk of serotonin syndrome.
- Cimetidine (antihistamine): combining cimetidine with clomipramine reduces the metabolism of tricyclic antidepressants, increasing their toxicity.
- Clonidine (α2-agonist): combining clonidine with clomipramine raises blood pressure.
- Cyproheptadine (antihistamine): cyproheptadine antagonizes the effects of clomipramine.
- Dextromethorphan (morphinan antitussive): combining dextromethorphan with clomipramine increases the risk of serotonin syndrome.
- Enalapril (angiotensin-converting enzyme inhibitor): combining enalapril with clomipramine increases toxicity risks.
- Levothyroxine (T4 thyroid hormone): combining levothyroxine with clomipramine increases the toxicity risks of both drugs.
- Linezolid (antibiotic): combining linezolid with clomipramine increases the risk of serotonin syndrome. This combination is contraindicated in humans.
- Meperidine (opioid): combining meperidine with clomipramine increases the risk of serotonin syndrome.
- Mirtazapine (antidepressant Noradrenergic and Specific Serotonergic Antidepressant (NaSSA)): combining mirtazapine with clomipramine increases the risk of serotonin syndrome.

18.2.3.4 Toxicity

In cats, the LD_{50} is unknown but is likely significantly lower than in dogs, where it ranges between 50 and 100 mg/kg/day orally. Common signs of intoxication in cats include mydriasis, lethargy, tachycardia, ataxia, agitation, and vocalization.

18.2.4 Clinical Use and Dosage

18.2.4.1 According to the Scientific Literature

In an open-label study on anxiety and compulsive disorders in cats, clomipramine (0.2–0.55 mg/kg/day) combined with behavioral modification improved six cases of urine marking, three cases of excessive licking, and one case of excessive vocalization. At higher doses, some cats exhibited reversible sedation, which resolved upon dose reduction (Seksel and Lindeman 1998).

In a double-blind clinical trial, clomipramine (0.5 mg/kg/day) was found to be as effective as fluoxetine (1 mg/kg/day) in treating urine marking (Hart et al. 2005). A single-blind trial reported a 75% reduction in marking in 80% of cats treated with clomipramine (5 mg/day) compared to a placebo (Dehasse 1997).

Other studies showed improvement in over 80% of cats receiving 0.5 mg/kg/day for 1 month, with side effects including drowsiness, decreased appetite, and anticholinergic effects (Landsberg and Wilson 2005). A meta-analysis confirmed a significant association between clomipramine and reduced urine marking (Mills et al. 2011).

Additionally, clomipramine has been effective in treating psychogenic alopecia in some studies (King et al. 2004; Mertens et al. 2006)

Clomipramine is used to treat anxiety and compulsive behaviors (Sawyer et al. 1999; Seksel and Lindeman 1998). It is also employed either alone (King et al. 2004) or in combination with fluoxetine (Hart et al. 2005) to address urine marking and urinary incontinence (Landsberg and Wilson 2005).

18.2.4.2 Experience Feedback

Clomipramine enhances noradrenergic and serotonergic transmission, making it useful for stimulating behaviors, managing fears, and, to a lesser extent, reducing impulsivity.

The recommended initial dose is 0.25–0.5 mg/kg once daily, with the possibility of increasing it to 1 mg/kg once daily, in combination with behavioral therapy. Taking the medication with food increases its bioavailability.

The dosage is adjusted according to the psychopathological condition being treated:

- For chronic depressive states: 0.25–0.5 mg/kg orally once daily.
- For phobic and anxious states: 0.25–1 mg/kg orally once daily.
- For impulsive or compulsive states: fluoxetine is generally preferred over clomipramine.

Due to its pronounced muscarinic actions, clomipramine is particularly beneficial in cases of inappropriate elimination, as its muscarinic side effects help improve cat-owner relationships, especially at the start of treatment. However, the marked histaminic effects can lead to lethargy, which is more pronounced than with SSRIs or drugs like venlafaxine.

In practice, a dose of 1/4 to 1 tablet of Clomicalm® 5 mg per cat per day (equivalent to 1.25–5 mg) is commonly administered.

18.2.4.3 Possible Combinations

For phobic states such as deprivation syndromes, categorical phobias, or post-traumatic phobias, combining clomipramine with gabapentin or alprazolam is recommended when clomipramine alone does not sufficiently mitigate the animal's exaggerated reactions.

18.2.4.4 Withdrawal

It is advisable to gradually reduce the dose over a period of 2 weeks to 1 month, depending on the treatment duration, or to taper the dose by reducing it over 1 week for every month of treatment.

18.3 Amitriptyline

18.3.1 Description, Chemical Formula, and Brand Names

Amitriptyline, chemically known as N,N-dimethyl-3-(2-tricyclo[9.4.0.03,8]pentadeca-1(15),3,5,7,11,13-hexaenylidene) propan-1-amine, belongs to the family of tricyclic antidepressants.

Various brand names and dosages are available depending on the country. Examples include *Elavil*™ in 25 mg tablets, *Laroxyl*™ in 25 mg and 50 mg tablets, and *Elatrolet*™ in 10 mg tablets. There are also formulations available as oral solutions.

18.3.2 Pharmacology

18.3.2.1 Mechanism of Action

As with all tricyclic antidepressants, amitriptyline increases the concentration of serotonin and norepinephrine in the synapse through the following mechanisms:

- The inhibition of presynaptic serotonin transporters (SERT), which increases the amount of serotonin in the synaptic space.

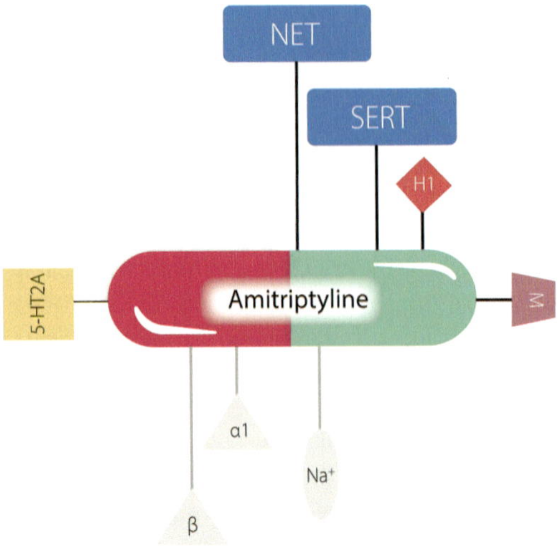

Fig. 18.3 Structure of amytriptiline. SERT inhibitor—NET inhibitor—Histaminergic H1 antagonist—muscarinic antagonist—5-HT2A antagonist. Other potential sites of action (in gray) are likely but remain poorly understood

- The inhibition of presynaptic norepinephrine transporters (NET), which increases the amount of norepinephrine in the synaptic space.

The molecule has additional actions common to all tricyclics, particularly an antagonistic effect on muscarinic, histaminergic, and 5-HT2A receptors. Other sites of action are likely but remain poorly understood (Fig. 18.3).

18.3.2.2 Metabolism and Involved Cytochromes

Amitriptyline is metabolized in the liver, primarily through hydroxylation, glucuronoconjugation, methylhydroxylation, and N-demethylation in dogs (Lee et al. 2015). Its metabolism has not been described in cats. The primary active metabolite is believed to be nortriptyline. In humans, the cytochromes involved are CYP2C19, CYP1A2, CYP3A4, and CYP2D6.

18

18.3.2.3 Pharmacokinetics

In cats, the plasma concentration peak of amitriptyline is reached 1–2 h after absorption. A transdermal gel exists but has very low absorption and is not recommended for use in cats (Mealey et al. 2004).

The effects of food intake at the time of absorption have not been demonstrated in dogs (Norkus et al. 2015). The half-life in cats is unknown but is believed to be longer than in dogs, where it is eight hours for amitriptyline and twelve hours for its active metabolite (Lautz et al. 2021). The drug is eliminated via the kidneys.

18.3.3 Side Effects and Usage Precautions

18.3.3.1 Side Effects

As with all tricyclic compounds, amitriptyline can cause sedation, ataxia, disorientation, miosis, hypersalivation, anorexia, urinary retention, deterioration of coat quality, diarrhea, vomiting, and constipation (Chew et al. 1998; Kruger et al. 2003). Cardiovascular, gastrointestinal, hematological (thrombocytopenia, neutropenia), and endocrine side effects have been reported in humans.

18.3.3.2 Precautions and Contraindications

Due to hepatic metabolism and renal elimination, doses should be reduced in cases of hepatic or renal insufficiency.

The drug is excreted in milk and has been shown to be teratogenic in many rodent species, as well as in humans. Use during gestation and lactation should be evaluated based on the benefit-risk balance (Briggs et al. 2011).

As with clomipramine, precautions are advised for animals with cardiac conditions, hyperthyroidism, keratoconjunctivitis, glaucoma, intestinal motility disorders, or epilepsy (Koella et al. 1979). It is also recommended to discontinue the treatment before administering anesthetics.

18.3.3.3 Drug Interactions

The main interactions are those common to all tricyclics (see Sect. 18.1.2).

18.3.3.4 Toxicity

The toxicity threshold has not been established in cats. In mice, the oral LD_{50} ranges from 100 to 216 mg/kg and is likely lower in cats (Ribbentrop and Schaumann 1965).

18.3.4 Clinical Use and Dosage

L'amitriptyline was the first tricyclic antidepressant used for anxiety treatment in dogs and cats, initially for economic reasons, followed by practitioner familiarity and experience, particularly in Anglo-Saxon countries. However, compared to other molecules like clomipramine, amitriptyline demonstrates lower clinical efficacy and more side effects (Crowell-Davis et al. 2019; Overall and Dunham 2002; Virga et al. 2001).

18.3.4.1 According to the Scientific Literature

Amitriptyline is used to treat anxiety and phobias, though selective serotonin reuptake inhibitors (SSRIs) are preferred for their higher safety profile (Overall and Dunham 2002; Sawyer et al. 1999). It is also prescribed as an adjunctive treatment for pruritus, chronic neuropathic pain, feline lower urinary tract disease, and idiopathic cystitis, with inconsistent results for the latter two conditions (Chew et al. 1998; Kraijer et al. 2003; Kruger et al. 2003).

Recommended doses are 0.5–1.5 mg/kg once daily for adjunctive treatment of pruritus, neuropathic pain, anxiety, and phobias. For pain management, the dose may be increased up to 4 mg/kg once or twice daily (Monteiro 2020). For urinary disorders, a dose of 0.5–1 mg/kg once daily for 7 days is advised.

18.3.4.2 Experience Feedback

Our clinical experience with amitriptyline does not provide additional insights beyond the data presented above. Its use in France remains anecdotal compared to clomipramine.

18.3.4.3 Withdrawal

A 3-week tapering regimen with gradually decreasing doses is recommended following prolonged use of the molecule.

Summary of This Chapter

Summary table of tricyclic antidepressants

Substance	Mechanism of actions	Pathological states for which the molecule is indicated	Dosage	Cytochromes
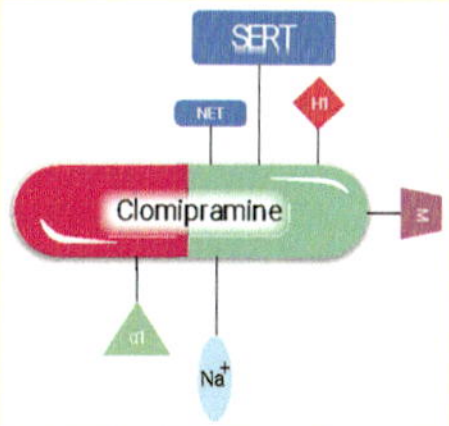	SERT inhibitor NET inhibitor H1 histamine receptor antagonist M1 muscarinic receptor antagonist α1-adrenergic receptor antagonist Voltage-gated sodium channel blocker	Chronic depressive Phobic or anxious Impulsive or compulsive (Consider using an SSRI instead)	0.25–1 mg/kg SID	CYP3A4 CYP2C19 CYP2C9 CYP1A2 CYP2D6
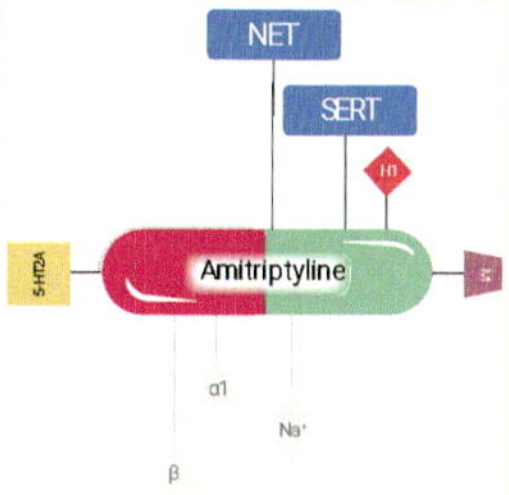	SERT inhibitor NET inhibitor H1 histamine receptor antagonist M1 muscarinic receptor antagonist α1-adrenergic receptor antagonist Na^{+} sodium channel blocker	Anxious Pain (Consider using an SSRI instead)	0.5–1.5 mg/kg SID	CYP2C19 CYP1A2 CYP3A4 CYP2D6

References

Baggot JD (2001) The physiological basis of veterinary clinical pharmacology. Blackwell Science

Briggs GG, Freeman RK, Yaffe SJ (2011) Drugs in pregnancy and lactation: a reference guide to fetal and neonatal risk, 9th edn. Wolters Kluwer Health—Lippincott Williams & Wilkins

Chew DJ, Buffington CA, Kendall MS, DiBartola SP, Woodworth BE (1998) Amitriptyline treatment for severe recurrent idiopathic cystitis in cats. J Am Vet Med Assoc 213(9):1282–1286

Crowell-Davis SL, Murray TF, Mattos de Souza, Dantas L (2019) Veterinary psychopharmacology, 2nd edn. Wiley Blackwell

Dehasse J (1997) Feline urine spraying. Appl Anim Behav Sci 52:365–371. https://doi.org/10.1016/S0168-1591(96)01135-5

Hart BL, Cliff KD, Tynes VV, Bergman L (2005) Control of urine marking by use of long-term treatment with fluoxetine or clomipramine in cats. J Am Vet Med Assoc 226(3):378–382. https://doi.org/10.2460/javma.2005.226.378

King JN, Steffan J, Heath SE, Simpson BS, Crowell-Davis SL, Harrington LJM, Weiss A-B, Seewald W (2004) Determination of the dosage of clomipramine for the treatment of urine spraying in cats. J Am Vet Med Assoc 225(6):881–887. https://doi.org/10.2460/javma.2004.225.881

Koella WP, Glatt A, Klebs K, Dürst T (1979) Epileptic phenomena induced in the cat by the antidepressants maprotiline, imipramine, clomipramine, and amitriptyline. Biol Psychiatr 14(3):485–497

Kraijer M, Fink-Gremmels J, Nickel R (2003) The short-term clinical efficacy of amitriptyline in the management of idiopathic feline lower urinary tract disease: A controlled clinical study. J Feline Med Surg 5(3):191–196. https://doi.org/10.1016/S1098-612X(03)00004-4

Kruger JM, Conway TS, Kaneene JB, Perry RL, Hagenlocker E, Golombek A, Stuhler J (2003) Randomized controlled trial of the efficacy of short-term amitriptyline administration for treatment of acute, nonobstructive, idiopathic lower urinary tract disease in cats. J Am Vet Med Assoc 222(6):749–758. https://doi.org/10.2460/javma.2003.222.749

KuKanich B (2013) Outpatient oral analgesics in dogs and cats beyond nonsteroidal antiinflammatory drugs: an evidence-based approach. Veterinary Clinics: Small Animal Practice 43, 1109–1125

Lainesse C, Frank D, Beaudry F, Doucet M (2007) Comparative oxidative metabolic profiles of clomipramine in cats, rats and dogs: Preliminary results from an in vitro study. J Vet Pharmacol Therap 30(5):387–393. https://doi.org/10.1111/j.1365-2885.2007.00893.x

Lainesse C, Frank D, Meucci V, Intorre L, Soldani G, Doucet M (2006) Pharmacokinetics of clomipramine and desmethylclomipramine after single-dose intravenous and oral administrations in cats. J Vet Pharmacol Therap 29(4):271–278. https://doi.org/10.1111/j.1365-2885.2006.00742.x

Landsberg GM, Wilson AL (2005) Effects of clomipramine on cats presented for urine marking. J Am Anim Hosp Assoc 41(1):3–11. https://doi.org/10.5326/0410003

Lautz LS, Jeddi MZ, Girolami F, Nebbia C, Dorne JLCM (2021) Metabolism and pharmacokinetics of pharmaceuticals in cats (Felix sylvestris catus) and implications for the risk assessment of feed additives and contaminants. Toxicol Letters 338:114–127. https://doi.org/10.1016/j.toxlet.2020.11.014

Lee J-Y, Lee SY, Lee K, Oh SJ, Kim SK (2015) Determination of species-difference in microsomal metabolism of amitriptyline using a predictive MRM-IDA-EPI method. Chemico-Biol Interact 229:109–118. https://doi.org/10.1016/j.cbi.2015.01.024

Martin KM (2010) Effect of clomipramine on the electrocardiogram and serum thyroid concentrations of healthy cats. J Vet Behav 5:123–129. https://doi.org/10.1016/j.jveb.2009.12.019

Mills DS, Redgate SE, Landsberg GM (2011) A meta-analysis of studies of treatments for feline urine spraying. PLoS One 6:1–10. https://doi.org/10.1371/journal.pone.0018448

Mealey KL, Peck KE, Bennett BS, Sellon RK, Swinney GR, Melzer K, Gokhale SA, Krone TM (2004) Systemic absorption of amitriptyline and buspirone after oral and transdermal administration to healthy cats. J Vet Intern Med 18(1):43–46. https://doi.org/10.1111/j.1939-1676.2004.tb00133.x

Mertens PA, Torres S, Jessen C (2006) The effects of clomipramine hydrochloride in cats with psychogenic alopecia: a prospective study. J Am Anim Hosp Assoc 42:336–343. https://doi.org/10.5326/0420336

Monteiro BP (2020) Feline Chronic Pain and Osteoarthritis. Vet Clin North Am Small Anim Pract 50(4):769–788. https://doi.org/10.1016/j.cvsm.2020.02.003

Nielsen KK, Flinois JP, Beaune P, Brøsen K (1996) The biotransformation of clomipramine in vitro, identification of the cytochrome P450s responsible for the separate metabolic pathways. J Pharmacol Exp Therap 277(3):1659–1664

Norkus C, Rankin D, KuKanich B (2015) Evaluation of the pharmacokinetics of oral amitriptyline and its active metabolite nortriptyline in fed and fasted

Greyhound dogs. J Vet Pharmacol Therap 38(6):619–622. https://doi.org/10.1111/jvp.12237
Overall KL, Dunham AE (2002) Clinical features and outcome in dogs and cats with obsessive-compulsive disorder: 126 cases (1989–2000). J Am Vet Med Assoc 221(10):1445–1452. https://doi.org/10.2460/javma.2002.221.1445
Pfeiffer E, Guy N, Cribb A (1999) Clomipramine-induced urinary retention in a cat. Canad Vet J La Revue Veterinaire Canadienne 40(4):265–267
Ribbentrop AU, Schaumann W (1965) Pharmakologische Untersuchungen mit Doxepin einem Antidepressivum mit zentral anticholinerger und sedierender Wirkung. Arzneimittel-Forschung 15(8):863
Rodan I, Heath S (2016) Feline behavioral health and welfare. Elsevier
Rudorfer MV, Potter WZ (1999) Metabolism of tricyclic antidepressants. Cell Mol Neurobiol 19(3):373–409. https://doi.org/10.1023/A:1006949816036
Sawyer LS, Moon-Fanelli AA, Dodman NH (1999) Psychogenic alopecia in cats: 11 cases (1993–1996). J Am Vet Med Assoc 214(1):71–74
Seksel K, Lindeman M (1998) Use of clomipramine in the treatment of anxiety-related and obsessive-compulsive disorders in cats. Aus Vet J 76(5):317–321. https://doi.org/10.1111/j.1751-0813.1998.tb12353.x
Virga V, Houpt K, Scarlett J (2001) Efficacy of amitriptyline as a pharmacological adjunct to behavioral modification in the management of aggressive behaviors in dogs. J Am Anim Hosp Assoc 37(4):325–330. https://doi.org/10.5326/15473317-37-4-325

Monoamine Oxidase Inhibitors (MAOIs)

Contents

19.1 General Characteristics of All IMAO – 598

19.2 Selegiline – 599
19.2.1 Description, Chemical Formula, and Brand Names – 599
19.2.2 Pharmacology – 599
19.2.3 Side Effects and Usage Precautions – 600
19.2.4 Clinical Use and Dosage – 600

References – 601

© The Author(s), under exclusive license to Springer Nature Switzerland AG 2026
S. Bleuer-Elsner et al., *Veterinary Psychiatry of the Cat*,
https://doi.org/10.1007/978-3-032-06260-4_19

Monoamine oxidase inhibitors (MAOIs) prevent the breakdown of monoamine neurotransmitters (norepinephrine, serotonin, and dopamine) in the synaptic space, thereby increasing their availability for membrane receptors. There are two types of MAOIs, classified as MAO-A and MAO-B. Each type is characterized by its specific action on monoamine neurotransmitters.

Selegiline, marketed under the name Selgian™, has veterinary authorization (MA) for use in dogs, but no MAOI holds an MA for use in cats.

19.1 General Characteristics of All IMAO

Monoamine oxidases (MAOs) are enzymes responsible for metabolizing monoamine neurotransmitters (norepinephrine, dopamine, and serotonin) released into the synaptic cleft. This enzymatic activity plays a crucial role in regulating neurotransmission (see ▶ Chap. 2, ▶ Sect. 2.3).

Monoamine oxidase inhibitors (MAOIs), which inhibit MAO activity, were the first class of antidepressants used in human psychiatry. They were discovered serendipitously when an antitubercular drug, iproniazid, was found to also have antidepressant effects. This drug exhibited MAOI activity alongside its other properties. Beyond their use as antidepressants, MAOIs also possess significant anxiolytic properties.

It is essential to distinguish between two types of MAOs, MAO-A and MAO-B (◘ Table 19.1). MAO-A metabolizes monoamines related to depression, such as serotonin and norepinephrine, while MAO-B does not affect these neurotransmitters. Instead, MAO-B metabolizes phenylethylamine, a regulator of catecholamines.

Both MAO-A and MAO-B metabolize dopamine. However, inhibiting only MAO-B does not result in a sufficiently stable elevation of dopamine because MAO-A also contributes to dopamine metabolism (and vice versa).

Both types of MAO metabolize tyramine, and their inhibition can lead to an increase in this molecule, which, in turn, raises norepinephrine levels and blood pressure in humans. This cascade, known as the "cheese effect," explains dietary restrictions (avoiding tyramine-rich foods such as certain cheeses, fish, avocado, or grapes) that can accompany the use of MAOIs. Notably, at low oral doses, selegiline does not alter tyramine metabolism, meaning no specific dietary restrictions are needed. This effect is even less concerning in cats, as the doses of selegiline used in this species are sufficiently low to avoid significant tyramine accumulation. Additionally, tyramine metabolism in cats seems to mitigate the impact of this side effect.

Inhibiting MAO-B offers no benefits in treating depression. Theoretical considerations also suggest that it is ineffective for anxiety because MAO-B does not act on serotonin or norepinephrine. MAO-B inhibitors are primarily used in humans to increase L-dopa production in patients with Parkinson's disease (Stahl and Muntner 2021).

In veterinary medicine, only MAO-B inhibitors are used, particularly selegiline. Other MAO-B inhibitors, such as rasagiline

◘ **Table 19.1** Subtypes of MAO

	MAO-A	MAO-B
Substrates	Serotonine Norepinephrine Dopamine Tyramine	Dopamine Tyramine Phenylethylamine
Tissue distribution	Brain, intestine, liver, placenta, skin	Brain, platelets, lymphocytes

19

and safinamide, have been developed for Parkinson's disease but have no reported veterinary applications.

As a result, the section on monoamine oxidase inhibitors will focus exclusively on selegiline.

19.2 Selegiline

19.2.1 Description, Chemical Formula, and Brand Names

Selegiline, with the chemical name (2R)-N-methyl-1-phenyl-N-prop-2-ynylpropan-2-amine, is an irreversible MAO-B inhibitor at the recommended doses. Available dosages for *Selgian*™ are 4, 10, and 20 mg. In the United States, *Anipryl*™ is available in doses of 2, 5, 10, 15, and 30 mg. In human medicine, *Deprenyl*™ is available as 5 mg tablets, and transdermal formulations also exist.

19.2.2 Pharmacology

19.2.2.1 Mechanism of Action

Selegiline is an irreversible inhibitor of MAO-B, showing minimal affinity for MAO-A at recommended doses (Fig. 19.1). This selectivity is not absolute, and rare cases of MAO-A inhibition have been reported in human patients. In humans, selectivity is lost at doses of 30–40 mg per day (Somerset Pharmaceuticals, Inc. 2003). Similar data is unavailable for cats.

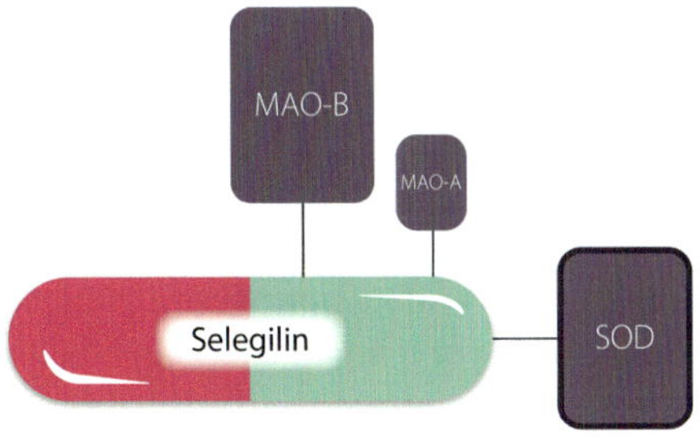

Fig. 19.1 Structure of seleginine. Irreversible MAO-B inhibitor—SOD activator—(irreversible MAO-A inhibitor)

Selegiline also enhances the activity of superoxide dismutase (SOD) and catalase enzymes, which is believed to contribute to its anti-aging effects by improving free radical elimination, particularly in the brain.

19.2.2.2 Metabolism and Involved Cytochromes

Selegiline is metabolized in the liver. While no specific data exists for cats, studies in humans indicate that the primary cytochrome involved is CYP2B6, with CYP2A6 and CYP3A4 playing secondary roles. Selegiline is converted into three main metabolites: L-amphetamine, L-methamphetamine, and N-desmethylselegiline (Benetton et al. 2007).

These findings are based on human studies, and extrapolations to feline veterinary medicine must be interpreted cautiously, as feline metabolism may differ significantly from that of humans or dogs.

The metabolites are believed to contribute to some of the pharmacological effects of selegiline (Fozard et al. 1985). L-amphetamine and L-methamphetamine act as central nervous system stimulants without MAO inhibitory activity, while N-desmethylselegiline retains MAO-B inhibitory properties.

Additionally, phenylethylamine, a modulator of catecholaminergic neurotransmissions and a substrate of MAO-B, increases in concentration following selegiline administration (Durden and Davis 1993).

19.2.2.3 Pharmacokinetics

No pharmacokinetic data is available for selegiline in cats. In dogs, following the oral administration of 1 mg/kg of selegiline, the plasma peak concentration is reached within 20 to 30 minutes. The molecule remains detectable in plasma for up to 3 hours postadministration. The half-life is approximately 78 minutes (Mahmood et al. 1994). Urinary excretion occurs in both conjugated and unconjugated metabolite forms.

19.2.3 Side Effects and Usage Precautions

19.2.3.1 Side Effects

Reported side effects include vomiting, diarrhea, agitation, disorientation, and a reduction in auditory acuity (Denenberg and Dubé 2018).

19.2.3.2 Precautions and Contraindications

The safety of selegiline in pregnant or lactating females has not been established. Studies in rats have not demonstrated teratogenic effects.

19.2.3.3 Drug Interactions

The challenges in combining selegiline with other drugs make its use particularly delicate. The mandatory therapeutic washout periods required when switching treatments are also a significant obstacle to its use. Any treatment involving selegiline requires the practitioner to remain vigilant whenever other medications are introduced.

The following drug combinations are *prohibited:*

- SSRIs: the combination may induce serotonin syndrome. Transitioning from an SSRI to selegiline requires a 5-week washout period, while a 2-week washout period is needed in the opposite direction (*Drug Monographs Plumb's Veterinary Drugs* 2024).
- Venlafaxine: the combination may induce serotonin syndrome.
- Tramadol: the combination may cause serotonin syndrome and cardiac arrest.
- Tricyclic antidepressants: the combination may induce serotonin syndrome. A 15-day washout period is recommended when transitioning between these drugs.
- Amitraz: concurrent use with selegiline is prohibited.
- Amitriptyline: the combination may cause neurotoxicity, epileptiform seizures, or serotonin syndrome.
- Bupropion: the combination may induce serotonin syndrome.
- Cyclobenzaprine (muscle relaxant): the combination may induce serotonin syndrome.
- Ephedrine: concurrent use with selegiline is prohibited.
- Meperidine and opioids in general: the combination may cause severe agitation, hallucinations, and death in human patients. Concurrent use of opioids should be avoided.

The following drug interactions have been reported in humans or animals receiving selegiline. These combinations are not necessarily contraindicated, but the risks should be assessed by the practitioner:

- Phenylpropanolamine, pseudoephedrine: this combination may result in hypertension and hyperthermia.
- Metoclopramide: the combination may induce hypertensive crises and reduce the efficacy of selegiline.

19.2.3.4 Toxicity

In laboratory animals, the LD_{50} ranges between 200 and 445 mg/kg. In one study, dogs receiving three times the recommended dose exhibited symptoms such as weight loss, salivation, decreased pupillary reflexes, shortness of breath, stereotypic behaviors, and dehydration. Cats treated with a dose up to 10 times the therapeutic dose showed no signs of toxicity (Ruhel 1996).

19.2.4 Clinical Use and Dosage

19.2.4.1 According to the Scientific Literature

Generally, selegiline is described for its effects in two main therapeutic domains: first, its impact on aging, and second, its purported anxiolytic effects, which are minimally documented (Sordo and Gunn-Moore 2021). The primary use of the drug in cats is for the treatment of feline cognitive dysfunction syndrome (Crowell-Davis et al. 2019). The recommended dose for the treatment of feline cognitive dysfunction syndrome is 0.25 to 1 mg/kg once daily (Ray et al. 2021). Administration is advised in the morning, as the drug may have a stimulating effect (Denenberg and Dubé 2018).

19.2.4.2 Experience Feedback

Scientific and pharmacological evidence provides little support for the use of selegiline in emotional or mood-related disorders. Selegiline requires monotherapy and careful precautions with almost all psychotropics. Any treatment change necessitates therapeutic washout periods of 15 days, which can be challenging for owners when their cat exhibits problematic behaviors. Based on our experience, selegiline is a valuable option to reserve for senile states at the recommended dose of 0.25 to 1 mg/kg. Even for this indication, attention must be paid to drug interactions.

19.2.4.3 Possible Combinations

Any combination with psychotropics affecting monoamines is potentially very dangerous. In senile states with sleep disturbances, combining selegiline with gabapentin at the start of treatment can be helpful to quickly enhance the cat's well-being and restore a harmonious relationship with its owner. Combinations with benzodiazepines are also possible, though these can have detrimental effects on senile states (see ▶ Chap. 21).

19.2.4.4 Withdrawal

Tapering is unnecessary. Discontinuing selegiline results in a gradual reappearance of MAO-B without requiring a progressive dose reduction.

Summary Table of MAOIs

Substance	Mechanism of action	Pathological states for which the molecule is indicated	Dosage	Cytochromes
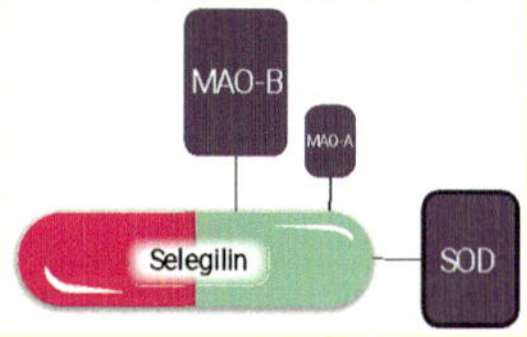	MAO-B inhibitor SOD activator (MAO-A inhibitor)	Senile	0.25–1 mg/kg SID	CYP2B6 CYP2A6 CYP3A4

References

Benetton SA, Fang C, Yang Y, Alok R, Year M, Lin C-C, Yeh L-T (2007) P450 phenotyping of the metabolism of selegiline to desmethylselegiline and methamphetamine. Drug Metab Pharmacokinet 22(2):78–87. https://doi.org/10.2133/dmpk.22.78

Crowell-Davis SL, Murray TF, Mattos de Souza Dantas L (2019) Veterinary psychopharmacology (Second edition). Wiley Blackwell

Denenberg S, Dubé MB (2018) Tools for managing feline problem behaviours: psychoactive medications. J Feline Med Surg 20(11):1034–1045. https://doi.org/10.1177/1098612X18806760

Durden DA, Davis BA (1993) Determination of regional distributions of phenylethylamine andmeta-andpara-tyramine in rat brain regions and presence in human and dog plasma by an ultra-sensitive negative chemical ion gas chromatography-mass spectrometric (NCI-GC-MS) method. Neurochem Res 18(9):995–1002. https://doi.org/10.1007/BF00966759

Fozard JR, Zreika M, Robin M, Palfreyman MG (1985) The functional consequences of inhibition of monoamine oxidase type B: comparison of the pharmacological properties ofl-deprenyl and MDL 72145. Naunyn Schmiedeberg's Arch Pharmacol 331(2–3):186–193. https://doi.org/10.1007/BF00634237

Mahmood I, Peters DK, Mason WD (1994) The pharmacokinetics and absolute bioavailability of selegiline in the dog. Biopharm Drug Dispos 15(8):653–664. https://doi.org/10.1002/bdd.2510150804

Ray M, Carney HC, Boynton B, Quimby J, Robertson S, St Denis K, Tuzio H, Wright B (2021) 2021 AAFP Feline Senior Care Guidelines. J Feline Med Surg 23(7):613–638. https://doi.org/10.1177/1098612X211021538
Ruhel W (1996) Effects of L-deprenyl in cats in a one-month dose escalation study. Vet Pathol 33:621
Sordo L, Gunn-Moore DA (2021) Cognitive dysfunction in cats: update on neuropathological and behavioural changes plus clinical management. Vet Rec 188(1):e3. https://doi.org/10.1002/vetr.3
Stahl SM, Muntner N (2021) In: Grady MM (ed) Stahl's essential psychopharmacology: neuroscientific basis and practical applications, 5th edn. Cambridge University Press. https://doi.org/10.1017/9781108975292

Benzodiazepines

Contents

20.1 General Characteristics of All Benzodiazepines – 604
20.1.1 Pharmacology – 604
20.1.2 Side Effects and Usage Precautions – 605
20.1.3 Clinical Use and Dosage – 606

20.2 Alprazolam – 607
20.2.1 Description, Chemical Formula, and Brand Names – 607
20.2.2 Pharmacology – 607
20.2.3 Side Effects and Usage Precautions – 607
20.2.4 Clinical Use and Dosage – 607

20.3 Other Benzodiazepines – 608
20.3.1 Pharmacology – 608
20.3.2 Side Effects and Usage Precautions – 608
20.3.3 Clinical Use and Dosage – 608

References – 609

© The Author(s), under exclusive license to Springer Nature Switzerland AG 2026
S. Bleuer-Elsner et al., *Veterinary Psychiatry of the Cat*,
https://doi.org/10.1007/978-3-032-06260-4_20

Benzodiazepines are all positive allosteric modulators of ionotropic Gamma-Aminobutyric Acid (GABA) receptors. Their presence enhances GABAergic transmission, which plays an inhibitory role in many cerebral circuits. Molecules in this family are used as anticonvulsants and anxiolytics. There is no oral benzodiazepine with veterinary marketing authorization (VMA). It is essential to note that, while no benzodiazepine has VMA for cats, their off-label use can be beneficial in certain clinical contexts. However, rigorous monitoring is required, particularly due to hepatic risks and the amnesic effect of benzodiazepines.

20.1 General Characteristics of All Benzodiazepines

Benzodiazepines are historically significant anxiolytic molecules in human psychiatry. In veterinary psychiatry, they are used to manage specific phobic or anxious pathological states, especially during panic attacks not controllable by other psychotropics. However, the fact that they impair memory and reduce the effectiveness of behavioral therapies leads to the preference for molecules from other chemical groups in many indications, especially as a first-line treatment.

After summarizing the common characteristics of all benzodiazepines, this chapter focuses on the most commonly used one: alprazolam. Diazepam is not used in cats due to the risk of severe hepatic failure (Center et al. 1996). Alprazolam is the only benzodiazepine not metabolized by glucuronidation in cats, making it the safest benzodiazepine for this species (Denenberg and Dubé 2018).

Other benzodiazepines, such as clonazepam, clorazepate dipotassium, lorazepam, and oxazepam, will be discussed briefly.

20.1.1 Pharmacology

20.1.1.1 Mechanism of Action

Benzodiazepines act as facilitators of GABA's action on Cl^- ion channels (◘ Fig. 20.1).

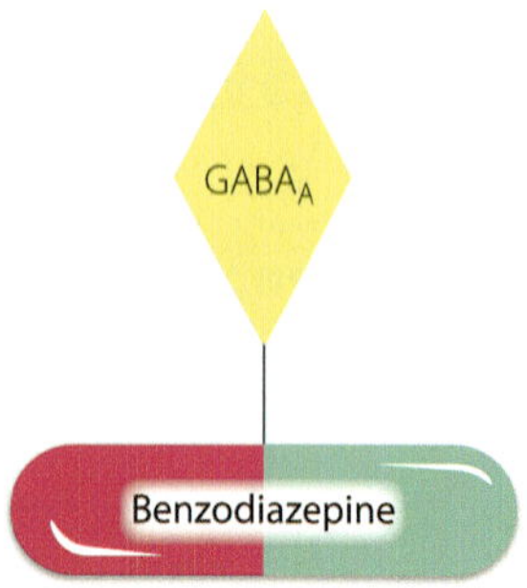

◘ **Fig. 20.1** General structure of benzodiazepines. Positive allosteric modulator of GABA$_A$

The presence of a benzodiazepine at the allosteric modulation site of GABA$_A$ receptors increases the frequency of Cl^- ion channel opening in the presence of GABA. This results in a greater intracellular ionic flux. Benzodiazepines function as positive allosteric modulators (◘ Fig. 20.2). The influx of chloride ions into the postsynaptic neuron prevents cell repolarization, producing an inhibitory effect. By acting in this way, particularly on neurons in the amygdala or the cortico-striato-thalamo-cortical loop, benzodiazepines reduce the intensity of fear responses in phobic and anxious states.

20.1.1.2 Metabolism and Involved Cytochromes

In humans, benzodiazepines are metabolized by the liver and excreted via the kidneys. Therefore, the functions of these two organs should be monitored during treatment. Although detailed data on the metabolism of these molecules in cats are limited, their use relies on extrapolations from human or canine data. Practitioners must exercise caution and adjust doses carefully. The primary cytochrome involved is P450 CYP3A4 for all benzodiazepines, although other cytochromes may also be involved depending on the specific molecule (Mandrioli et al. 2008).

20.1.1.3 Pharmacokinetics

Benzodiazepines act synergistically with GABA on ionotropic receptors of Cl^- ion channels. As a result, their action begins rapidly and lasts from a few minutes to several hours, depending on the molecule used.

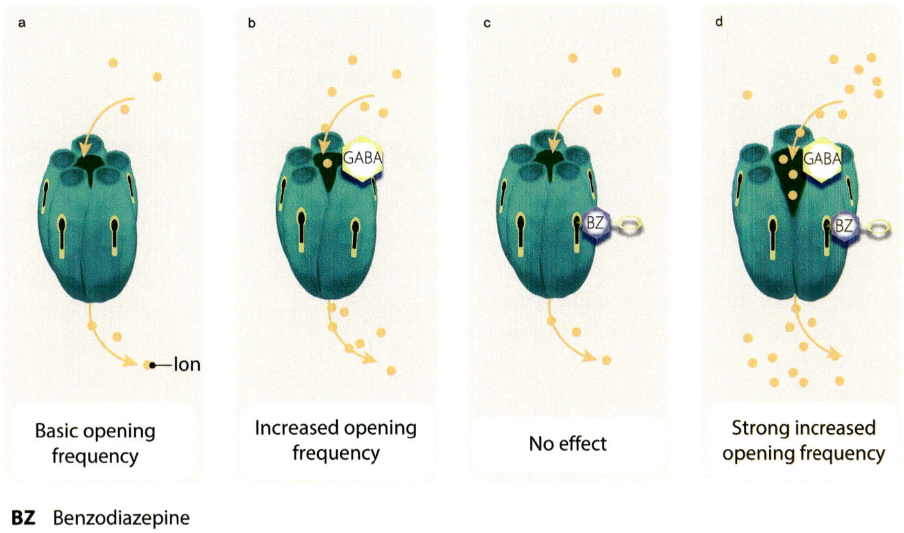

Fig. 20.2 Positive allosteric modulation by benzodiazepines. (**a**) In the absence of any ligand, the Cl^- ion channel exhibits a baseline frequency of opening. (**b**) The binding of GABA to the Cl^- ion channel increases the opening frequency, allowing significant Cl^- influx into the cell. (**c**) The benzodiazepine alone has no effect. (**d**) The simultaneous binding of GABA (at the natural ligand site) and a benzodiazepine (at the receptor's allosteric site) leads to a substantial increase in the channel's opening frequency, greater than that resulting from GABA binding alone

20.1.2 Side Effects and Usage Precautions

20.1.2.1 Side Effects

Side effects include sedation, ataxia, muscular hypotension, increased appetite, and paradoxical excitation. In humans, additional effects such as anxiety, hallucinations, aggression due to reduced social inhibitions, and insomnia have been observed.

Benzodiazepines also have amnesic effects, which can hinder the implementation of behavioral therapies.

20.1.2.2 Precautions and Contraindications

Benzodiazepines cross the placental barrier and are excreted in milk. Thus, their use should be highly cautious in pregnant or lactating females.

In human patients with cognitive disorders or dementia, such as Alzheimer's disease, benzodiazepines can be detrimental and worsen the patient's condition (Defrancesco et al. 2015). Although similar studies have not yet been conducted in veterinary medicine, caution is advised when prescribing these drugs to elderly cats with cognitive impairments.

20.1.2.3 Drug Interactions

Benzodiazepines are relatively safe in combination with other molecules. However, the following drug interactions have been reported in humans or animals receiving benzodiazepines. While the association is not necessarily contraindicated, the risks should be evaluated by the practitioner:

- Valproic acid (antiepileptic): the combination is teratogenic and prolongs the effects of benzodiazepines. Significant sedation may occur.
- Amiodarone (antiarrhythmic): increases the bioavailability and effects of benzodiazepines. Monitoring and dose adjustment are required.
- Antihypertensive agents (amlodipine, atenolol, lisinopril): risk of respiratory hypotension.
- Antacids: decrease the absorption time of benzodiazepines. A 2-hour interval between the administration of the two drugs can prevent this interaction.

- Tricyclic antidepressants (amitriptyline, clomipramine): risk of central nervous system depression and respiratory depression.
- Antiglaucoma agents (dichlorphenamide, methazolamide): risk of respiratory depression.
- Barbiturates: risk of central nervous system depression and respiratory depression.
- Digoxin: risk of increased plasma concentrations of digoxin. Blood monitoring is recommended.
- Fluoxetine and fluvoxamine: risk of increased plasma concentrations of benzodiazepines. Monitoring is recommended.
- Hepatic enzyme inducers (e.g., glucocorticoids, mitotane, rifampin, phenobarbital): risk of decreased plasma concentrations of benzodiazepines. Dose adjustment may be necessary.
- Hepatic enzyme inhibitors (e.g., cimetidine, erythromycin, isoniazid, ketoconazole, itraconazole): risk of increased plasma concentrations of benzodiazepines and significant sedation.
- Mirtazapine: increases the risk of central nervous system depression and respiratory depression.
- Opioids: increases the risk of central nervous system depression and respiratory depression.
- Propranolol: may slow the metabolism of benzodiazepines.
- Theophylline, aminophylline: reduces the effects of benzodiazepines.

20.1.2.4 Toxicity

Signs of benzodiazepine toxicity in cats include ataxia, prostration, agitation, vomiting, hyperesthesia, muscle tremors, coma, hypersalivation, aggression, and paresis.

Following oral administration of diazepam, cats have been reported to exhibit signs of idiosyncratic hepatic insufficiency. This side effect is dose-independent and occurs only with oral ingestion of diazepam, not with injection (Center et al. 1996). Although this adverse reaction has not been reported with alprazolam, the prescription of benzodiazepines in cats requires extreme caution **(see box below).**

Treatment of Benzodiazepine Intoxication

Administering activated charcoal is recommended if intervention occurs within 2 to 3 hours after ingestion. In dogs, flumazenil, a benzodiazepine receptor antagonist, can partially or completely reverse the effects. It must be administered intravenously (Crowell-Davis et al. 2019; Unkel et al. 2006).

If flumazenil is unavailable, strict symptomatic management with intensive monitoring is advised. This includes intravenous fluids to accelerate excretion and close observation of vital signs to prevent complications.

Beyond 3 hours postingestion, benzodiazepines are fully absorbed, making gastric lavage ineffective and potentially hazardous due to the risk of profound sedation or seizures.

Hypothermia should be corrected. Intravenous fluids help increase the renal excretion rate of benzodiazepines. The prognosis is generally excellent.

20.1.3 Clinical Use and Dosage

Benzodiazepines can be used in the treatment of phobic and anxiety disorders.

Benzodiazepines administered at low doses increase appetite, which can be useful for conducting therapy in cases of fear in cats (Denenberg and Dubé 2018).

The effects of benzodiazepines vary between individuals. Strong side effects or paradoxical reactions, opposite to the expected outcomes, may occur with doses that have proven effective in other patients. It is recommended to test and adjust dosages for each patient: excessive side effects should prompt a dose reduction, while paradoxical effects should lead to an increase.

Additionally, tolerance to benzodiazepines can develop after prolonged use. In such cases, the practitioner should adjust the daily doses by increasing them accordingly (Crowell-Davis et al. 2019).

20.1.3.1 Possible Combinations

Benzodiazepines can be combined with most other psychotropics. However, combinations with mirtazapine, venlafaxine, or Selective Serotonin Reuptake Inhibitor (SSRIs) require liver and kidney monitoring.

20.1.3.2 Withdrawal

There are no specific data available for cats; however, tapering is recommended if usage exceeds one continuous week. The main signs of withdrawal syndrome include tremors, muscle rigidity, reduced appetite, and tonic-clonic seizures (Salzman 1998).

The general rule for withdrawal is to reduce the dose by 25% to 33% each week, over a minimum duration of 4 weeks.

20.2 Alprazolam

20.2.1 Description, Chemical Formula, and Brand Names

Alprazolam, with the chemical name *8-chloro-1-methyl-6-phenyl-4H-[1,2,4]triazolo[4,3-a][1,4]benzodiazepine*, is the most commonly used and well-documented benzodiazepine.

Known under the brand name *Xanax*™, alprazolam is marketed under various names. All available forms are administered orally. Tablet strengths range from 0.25 to 2 mg, depending on the country and legal regulations.

20.2.2 Pharmacology

20.2.2.1 Mechanism of Action

Like all benzodiazepines, alprazolam facilitates GABA action on chloride ion channels.

20.2.2.2 Metabolism and Involved Cytochromes

Alprazolam, like other benzodiazepines, is metabolized in the liver. No specific studies have been conducted on its metabolism in cats.

In humans, alprazolam is metabolized into two main metabolites: alpha-hydroxy-alprazolam and benzophenone. Benzophenone is inactive, while alpha-hydroxy-alprazolam retains half the activity of the parent compound. The primary cytochromes involved in its metabolism are CYP3A4 and CYP3A5 (Mandrioli et al. 2008).

Monitoring liver function is recommended for cats undergoing daily treatment.

20.2.2.3 Pharmacokinetics

There is limited pharmacokinetic data available for cats. The plasma peak is reached 2 to 3 hours after administration (Denenberg and Dubé 2018).

20.2.3 Side Effects and Usage Precautions

All side effects and precautions are common to all benzodiazepines and are detailed in the introductory chapter on this class of psychotropic drugs.

20.2.4 Clinical Use and Dosage

20.2.4.1 According to the Scientific Literature

Alprazolam is commonly used as an adjunct treatment for anxiety and phobias. The prescribed dose is 0.125 to 0.25 mg per cat every 8 to 12 hours or on an as-needed basis, 30 to 60 minutes before exposure to a phobogenic stimulus (Denenberg and Dubé 2018; Nivy et al. 2019; Seibert 2004).

For phobias related to handling or transport, the dose can be increased to 0.5 to 1 mg per cat, administered 60 minutes before exposure (Erickson et al. 2021).

20.2.4.2 Experience Feedback

Alprazolam is rarely used alone but is typically combined with a first-line treatment when the latter's effect is insufficient.

Alprazolam can be combined with an SSRI, gabapentin or pregabalin, comipramine or venlafaxine. Its rapid onset of action allows for managing panic episodes at the start of treatment while waiting for other psychotropics with slower therapeutic effects to take over.

Alprazolam is also effective, either alone or in combination with sympatholytics or gabapentinoids, prior to exposure to a phobogenic stimulus.

The proposed dose is 0.125 to 1 mg per cat every 8 hours or 1 hour before exposure to the stimulus.

Paradoxical effects, including worsening of symptoms, are frequently observed regardless of dose. Close monitoring at the start of treatment and dose adjustments are essential.

20.3 Other Benzodiazepines

This section will review clonazepam, potassium clorazepate, lorazepam, and oxazepam.

20.3.1 Pharmacology

20.3.1.1 Mechanism of Action

Like all benzodiazepines, these compounds act as facilitators of GABA activity on chloride ion channels.

20.3.1.2 Metabolism and Involved Cytochromes

All benzodiazepines are metabolized by the liver, but data in cats are lacking. Here, we present human data.

- Clonazepam (*Rivotril*™) is metabolized into various inactive metabolites.
- Potassium clorazepate (*Tranxène*™) is rapidly converted into nordiazepam, an active metabolite, which is further hydroxylated into 3-hydroxynordiazepam and p-hydroxynordiazepam.
- Lorazepam (*Temestat*™ or *Ativan*™) undergoes glucuronidation, transforming into lorazepam glucuronide, a molecule with no central nervous system effects.
- Oxazepam (*Seresta*™) is predominantly converted into oxazepam glucuronide, an inactive metabolite.

The cytochromes involved in the metabolism of these molecules mainly belong to the CYP3A4 family.

20.3.1.3 Pharmacokinetics

The half-life of lorazepam ranges from 4 to 6 hours, while oxazepam and clonazepam have half-lives of 10 to 12 hours. Elimination occurs through both urinary and fecal pathways.

Lorazepam, oxazepam, and clonazepam are considered safer than potassium clorazepate, as they are not converted into directly active metabolites in cats (Center et al. 1996; Crowell-Davis et al. 2019; Denenberg and Dubé 2018).

20.3.2 Side Effects and Usage Precautions

All side effects and usage precautions are common to all benzodiazepines and are detailed in the introductory chapter for this psychotropic family. Rare cases of hepatic necrosis have been reported with the use of clonazepam ("Drug Monographs Plumb's Veterinary Drugs," 2024).

20.3.3 Clinical Use and Dosage

20.3.3.1 According to the Scientific Literature

- **Clonazepam**: The prescribed dose varies greatly, ranging from 0.02 to 0.25 mg/kg once or twice daily.
- **Potassium clorazepate**: Due to its long half-life, this molecule is beneficial for managing prolonged crises. The recommended dose is 0.2 to 0.5 mg/kg once or twice daily.
- **Lorazepam**: Depending on studies, the prescribed dose ranges from 0.025 to 0.08 mg/kg once or twice daily or as needed, or 0.125 to 0.25 mg per cat once or twice daily.
- **Oxazepam**: Primarily used for appetite stimulation, the prescribed dose varies widely, from 0.2 to 0.5 mg/kg once or twice daily.

20.3.3.2 Experience Feedback

Our clinical experience with these benzodiazepines does not allow us to expand upon the data above. These molecules remain rarely used in France, except as adjuncts in refractory cases unresponsive to conventional treatments or for seemingly uncontrollable panic episodes.

Summary Table of Benzodiazepines

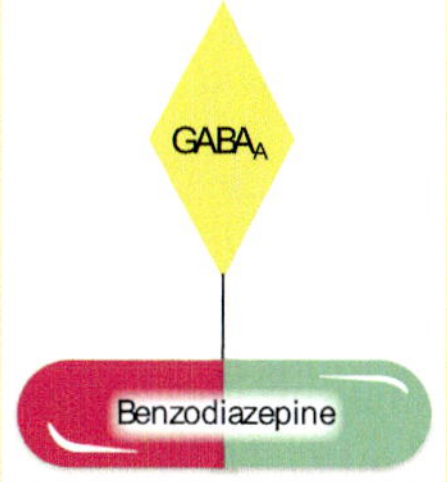	Mechanism of action	Pathological states for which the molecule is indicated	Dosage	Cytochromes
Alprazolam	Positive allosteric modulator of $GABA_A$	Phobic or anxious	0.125–1 mg per cat PRN or TID	CYP3A4 CYP3A5
Clonazepam	Positive allosteric modulator of $GABA_A$	Phobic (noise)	0.02–0.25 mg/kg PRN	CYP3A4
Clorazepate	Positive allosteric modulator of $GABA_A$	Phobic or anxious	0.2–0.5 mg/kg SID, BID, or TID	CYP3A4
Lorazepam	Positive allosteric modulator of $GABA_A$	Phobic or anxious	0.025–0.08 mg/kg PRN	CYP3A4
Oxazepam	Positive allosteric modulator of $GABA_A$	Phobic or anxious	0.2–0.5 mg/kg SID, BID, or TID	CYP3A4

References

Center SA, Elston TH, Rowland PH, Rosen DK, Reitz BL, Brunt JE, Rodan I, House J, Bank, S, Lynch LR, Dring LA, Levy JK (1996) Fulminant hepatic failure associated with oral administration of diazepam in 11 cats. J Am Vet Med Assoc 209(3):618–625

Crowell-Davis SL, Murray TF, Mattos de Souza Dantas L (2019) Veterinary psychopharmacology (Second edition). Wiley Blackwell

Defrancesco M, Marksteiner J, Fleischhacker WW, Blasko I (2015) Use of benzodiazepines in Alzheimer's disease: a systematic review of literature. Int J Neuropsychopharmacol 18(10):pyv055. https://doi.org/10.1093/ijnp/pyv055

Denenberg S, Dubé MB (2018) Tools for managing feline problem behaviours: psychoactive medications. J Feline Med Surg 20(11):1034–1045. https://doi.org/10.1177/1098612X18806760

Erickson A, Harbin K, MacPherson J, Rundle K, Overall KL (2021) A review of pre-appointment medications to reduce fear and anxiety in dogs and cats at veterinary visits. Can Vet J 62(9):952–960

Mandrioli R, Mercolini L, Raggi M (2008) Benzodiazepine metabolism: an analytical perspective. Curr Drug Metab 9(8):827–844. https://doi.org/10.2174/138920008786049258

Nivy R, Segev G, Rimer D, Bruchim Y, Aroch I, Mazaki-Tovi M (2019) A prospective randomized study of efficacy of 2 treatment protocols in preventing recurrence of clinical signs in 51 male cats with obstructive idiopathic cystitis. J Vet Intern Med 33(5):2117–2123. https://doi.org/10.1111/jvim.15594

Salzman C (1998) Addiction to Benzodiazepines. Psychiatry Q 69(4):251–261. https://doi.org/10.1023/A:1022125929946

Seibert LM (2004) Animal behavior case of the month. J Am Vet Med Assoc 224(10):1594–1596. https://doi.org/10.2460/javma.2004.224.1594

Unkel JH, Brickhouse TH, Sweatman TW, Scarbecz M, Tompkins WP, Eslinger CS (2006) A comparison of 3 routes of flumazenil administration to reverse benzodiazepine-induced desaturation in an animal model. Pediatr Dent 28(4):357–362

Sympatholytics

Contents

21.1 Propranolol – 612
21.1.1 Description, Chemical Formula, and Brand Names – 612
21.1.2 Pharmacology – 612
21.1.3 Side Effects and Usage Precautions – 613
21.1.4 Clinical Use and Dosage – 614

21.2 Clonidine – 614
21.2.1 Description, Chemical Formula, and Brand Names – 614
21.2.2 Pharmacology – 614
21.2.3 Side Effects and Usage Precautions – 615
21.2.4 Clinical Use and Dosage – 616

References – 617

© The Author(s), under exclusive license to Springer Nature Switzerland AG 2026
S. Bleuer-Elsner et al., *Veterinary Psychiatry of the Cat*,
https://doi.org/10.1007/978-3-032-06260-4_21

Sympatholytics encompass molecules that reduce noradrenergic transmission through two possible mechanisms: blocking postsynaptic noradrenergic receptors, these are the β-blockers—and decreasing norepinephrine release by the presynaptic neuron, these are the α2-agonists. This section will detail a β-blocker, propranolol, and an α2-agonist, clonidine. Dexmedetomidine is an α2-agonist used in dogs for its anxiolytic properties (Korpivaara et al. 2021), but no studies have been conducted on its use in cats. In this species, it is employed for its sedative and emetic properties (McSweeney et al. 2012; Thawley and Drobatz 2015).

It is important to note that the sympatholytics presented in this chapter are used off-label in cats. Their use must therefore be accompanied by close monitoring and appropriate precautions.

21.1 Propranolol

21.1.1 Description, Chemical Formula, and Brand Names

Propranolol, chemically known as 1-naphthalen-1-yloxy-3-(propan-2-ylamino)propan-2-ol, is a β-blocker primarily used in cardiology as an antiarrhythmic. Its action on noradrenergic transmission is theoretically of interest in managing certain phobias.

There is no veterinary-specific formulation. Propranolol is available in various generic forms: 40 mg tablets, extended-release tablets of 80 and 160 mg, and oral solutions of 5, 10, and 50 mg/5 mL.

21.1.2 Pharmacology

21.1.2.1 Mechanism of Action

Propranolol is a nonselective antagonist of β-adrenergic receptors (Fig. 21.1). Blocking these receptors results in vasoconstriction, inhibition of the synthesis of angiogenic factors such as vascular endothelial growth factor and basic fibroblast growth factor, induction of endothelial cell apoptosis, and downregulation of the renin-angiotensin-aldosterone system.

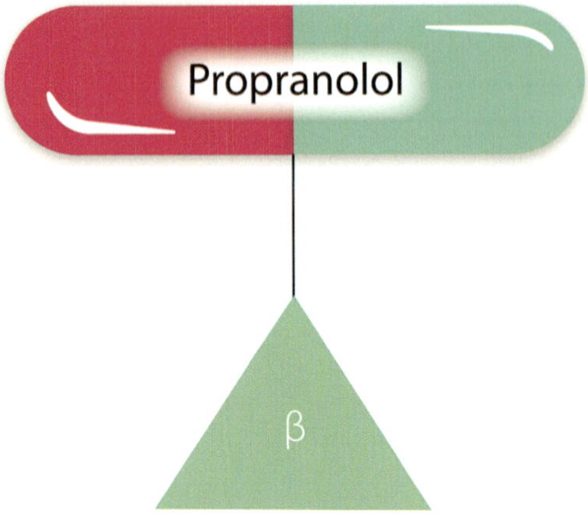

Fig. 21.1 Structure of propranolol. β-adrenergic antagonist

Propranolol may therefore modulate the autonomic expression of fear.

21.1.2.2 Metabolism and Involved Cytochromes

Although studies on the pharmacokinetics of propranolol in cats exist (Jacobs et al. 1997), detailed information on its metabolism, particularly the cytochromes involved, remains limited. In dogs, it is primarily metabolized by the liver through cytochrome CYP2D15. An active metabolite, 4-hydroxypropranolol, has been identified after oral administration in humans (*Drug Monographs Plumb's Veterinary Drugs* 2024).

21.1.2.3 Pharmacokinetics

Excretion is primarily urinary, with less than 1% excreted unchanged. Hyperthyroidism increases the bioavailability and reduces the clearance of propranolol in affected cats compared to healthy ones (Jacobs et al. 1997).

The reported half-life in cats is 58 minutes (Jacobs et al. 1997). Propranolol crosses the placenta and is excreted in milk in very small amounts (*Drug Monographs Plumb's Veterinary Drugs* 2024).

21.1.3 Side Effects and Usage Precautions

21.1.3.1 Side Effects

The side effects of propranolol in cats are similar to those observed in humans, though their exact frequency in this species has not been clearly established. Regular monitoring for signs of bradycardia and hypotension is recommended. These effects are most commonly reported in elderly animals or those with decompensated heart disease. They include bradycardia, lethargy, depression, atrioventricular conduction disturbances, onset of chronic heart failure or worsening of preexisting heart failure, hypotension, hypoglycemia, and bronchoconstriction. Syncope and diarrhea have also been reported.

21.1.3.2 Precautions and Contraindications

Propranolol is contraindicated in cats with heart failure unless it is secondary to a tachyarrhythmia that responds to β-blocker treatment. It is also contraindicated in cases of bundle branch block or sinus bradycardia. It is relatively contraindicated in patients with bronchospastic pulmonary disease.

Propranolol should be used with caution in cats with significant renal or hepatic insufficiency or sinus node dysfunction. In humans, it has been suggested to reduce the dosage by 50% (or more) in patients with hepatic insufficiency.

Propranolol can cause either hypoglycemia or hyperglycemia (*Drug Monographs Plumb's Veterinary Drugs* 2022) and should therefore be used cautiously in diabetic patients.

21.1.3.3 Drug Interactions

The following drug interactions have been reported in humans or animals receiving propranolol. While the combination is not necessarily contraindicated, the risks should be evaluated by the practitioner.

- α2-agonists (e.g., clonidine, romifidine): increased risk of sedation and severe bradycardia. Simultaneous use is not recommended (Cenani et al. 2017).
- Fluoxetine: may decrease the metabolism of propranolol.
- Tricyclic antidepressants (e.g., amitriptyline, clomipramine): concomitant use may increase the risk of orthostatic hypotension. Amitriptyline may potentiate propranolol's action.
- Thyroid hormones: may decrease the effects of β-blockers.
- General anesthetics: additive myocardial depression may occur when propranolol is used concurrently with myocardial depressant anesthetics.
- Bupivacaine (local anesthetic): concomitant use may increase bupivacaine concentrations.
- Lidocaine: clearance may be altered by propranolol.
- Phenobarbital: may increase the metabolism of propranolol.
- Antacids: may reduce the oral absorption of propranolol. Doses should be spaced at least one hour apart.
- Amiodarone (antiarrhythmic): concomitant use may increase the risk of bradycardia, hypotension, sinus arrest, and atrioventricular block.
- Anticholinergics: may negate the cardiac effects of β-blockers.
- Bupropion (atypical antidepressant): concomitant use may potentiate the effects of propranolol.
- Calcium channel blockers: concomitant use of β-blockers and calcium channel blockers (or other negative inotropes) should be approached with caution.
- Cimetidine: may decrease propranolol metabolism and increase blood levels.
- Digoxin: concomitant use may increase the risk of bradycardia and digitalis toxicity.
- Furosemide: concomitant use may increase the risk of hypotension and bradycardia.
- Hydrochlorothiazide (diuretic): concomitant use may increase the risk of hypotension, hyperglycemia, and hypertriglyceridemia.
- Insulin and other antidiabetic medications: propranolol may prolong the hypoglycemic effects of insulin therapy.

21.1.3.4 Toxicity

The most predominant signs of toxicity are hypotension and bradycardia. Other possible signs include central depression, seizures, bronchospasm, hypoglycemia, hyperkalemia, respiratory depression, pulmonary edema, arrhythmias, and asystole.

21.1.4 Clinical Use and Dosage

21.1.4.1 According to the Scientific Literature

The limited number of studies in the scientific literature does not allow conclusions to be drawn about the efficacy of propranolol in addressing phobic or anxious states. In cardiology, the prescribed doses range from 0.2 to 0.8 mg/kg every 8 to 12 hours ("Drug Monographs Plumb's Veterinary Drugs" 2024).

21.1.4.2 Experience Feedback

Propranolol demonstrates some activity in early-stage phobic reactions and has significant effects on behavioral agitation of emotional origin. It may be used to manage uncontrolled agitation in new situations that could be described as a deficit in noradrenergic control. However, in practice, this molecule is rarely used. Its use in this species should be approached cautiously, with careful dose adjustments and close monitoring of clinical responses.

When anticipation is present—an occurrence that quickly develops in all phobic reactions—propranolol appears ineffective. While this molecule blocks the physical fear response, it does not seem to reduce emotional anticipation.

21.1.4.3 Possible Combinations

Since propranolol's mechanism of action only reduces noradrenergic transmission, it can be combined with psychotropic drugs that have little to no effect on adrenergic pathways.

Its combination with venlafaxine or fluoxetine is theoretically possible as an adjunct treatment for phobias with strong adrenergic symptoms, such as tachycardia, tachypnea, sweating, and tremors.

21

21.1.4.4 Withdrawal

Exacerbations of clinical signs have been reported after the abrupt discontinuation of β-blockers in humans. It is recommended to gradually discontinue treatment in patients receiving the medication continuously. This approach is also justified due to its cardiac effects.

21.2 Clonidine

21.2.1 Description, Chemical Formula, and Brand Names

Clonidine, chemically known as N-(2,6-dichlorophenyl)-4,5-dihydro-1H-imidazol-2-amine, is an α2-agonist approved in human medicine for the treatment of hypertension. It is also used off-label for a variety of conditions, including substance addiction, attention-deficit hyperactivity disorder (ADHD), and as a sedative and analgesic agent. In perioperative or intensive care patients, clonidine is commonly used to reduce anxiety (Cao et al. 2009).

Clonidine does not have a veterinary-specific approval. It is available in pharmacies under various trade names, such as *Duraclon*™, *Catapressan*™, *Catapes*™, *Normopresan*™, and *Clonnirit*™, depending on the country. The most common pharmaceutical form is tablets dosed at 0.15 mg.

21.2.2 Pharmacology

21.2.2.1 Mechanism of Action

Clonidine is primarily an α2-agonist of adrenergic receptors (◘ Fig. 21.2). The molecule binds to α-2A, α-2B, and α-2C receptors, with the latter being responsible for its central hypotensive and antiarrhythmic effects. Clonidine also binds to imidazoline receptors, which contribute to its sedative and hypotensive effects (Head and Mayorov 2006).

The α2-adrenergic receptor is a presynaptic G-protein-coupled metabotropic receptor that inhibits adenylyl cyclase and activates

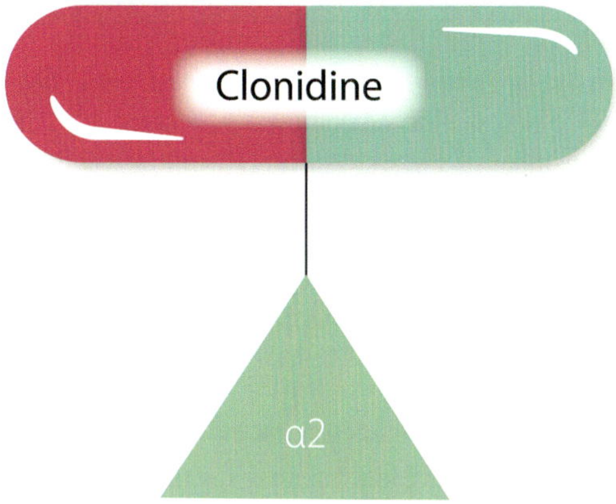

Fig. 21.2 Structure of clonidine. α2-adrenergic agonist

potassium channels. These presynaptic membrane modifications cause hyperpolarization of the presynaptic neuron, thereby inhibiting neurotransmitter release.

Stimulation of α2-adrenergic receptors in the locus coeruleus, a brain region involved in regulating wakefulness, may account for the hypnotic effects of clonidine.

Clonidine reduces norepinephrine release from presynaptic neurons in the prefrontal cortex, thereby alleviating fear or anxiety in rodents and primates (Söderpalm and Engel 1988). It also improves impulsive and hyperactive behavior observed in attention-deficit hyperactivity disorder (ADHD) in humans (Nguyen et al. 2014).

Additionally, clonidine can decrease the transmission of pain signals in the spinal cord and regulate blood pressure.

21.2.2.2 Metabolism and Involved Cytochromes

There is limited information on the metabolism of clonidine in cats. Even in humans, details about the metabolism of this substance remain incomplete. Clonidine is metabolized less than 50% by the liver into 4-hydroxyclonidine, an inactive form, through cytochromes CYP2D6, CYP1A1, CYP1A2, CYP3A4, and CYP3A5 (Claessens et al. 2010).

21.2.2.3 Pharmacokinetics

In cats, approximately 50% of clonidine is excreted unchanged in the urine. Tissue penetration is rapid, with brain concentrations being 1.7 times higher than those in the blood. The molecule's half-life is 80 minutes. Excretion is predominantly urinary. However, pharmacokinetics vary depending on the administered dose, partly because clonidine causes vasoconstriction that alters its tissue penetration (Paalzow and Edlund 1979).

21.2.3 Side Effects and Usage Precautions

21.2.3.1 Side Effects

Limited data exist on the side effects of orally administered clonidine in cats. It may induce vomiting, and bradycardia and hypotension have been observed after intraventricular administration (Ally 1997).

21.2.3.2 Precautions and Contraindications

Clonidine should be used with caution in patients with severe cardiovascular diseases, particularly conduction disorders or congestive heart failure. Lower doses are advised for patients with renal or hepatic insufficiency.

21.2.3.3 Drug Interactions

The following drug interactions have been reported in humans or animals receiving clonidine. While the combination is not necessarily contraindicated, the risks should be carefully assessed by the practitioner.

- Tricyclic antidepressants (e.g., amitriptyline, clomipramine): they increase the risk of severe hypertension. The combination should be avoided if possible.
- Antihypertensive medications (e.g., amlodipine, enalapril, hydralazine, sildenafil, telmisartan): additive hypotensive effects are possible.
- β-adrenergic antagonists (e.g., propranolol): they may increase the risk of bradycardia when administered with clonidine.

- Calcium channel blockers (e.g., diltiazem, verapamil, excluding amlodipine): there is an increased risk of sinus bradycardia and complete atrioventricular block. Blood pressure and heart rate should be closely monitored.
- CNS depressants (e.g., barbiturates, opioids): clonidine may exacerbate the effects of other CNS depressants.
- Digoxin: additional bradycardia may occur. Blood pressure and heart rate should be closely monitored.
- Stimulants (e.g., doxapram): concurrent use may increase the risk of hypertension and tachycardia. Pulse and blood pressure should be monitored.

21.2.3.4 Toxicity

Common clinical signs associated with overdose in cats include lethargy, sedation, bradycardia, ataxia, vomiting, and hypotension. In humans, seizures or respiratory depression are rarely observed. The reported LD50 for oral clonidine is 465 mg/kg in rats and 206 mg/kg in mice (*Drug Monographs Plumb's Veterinary Drugs* 2024).

21.2.4 Clinical Use and Dosage

21.2.4.1 According to the Scientific Literature

No studies describe the use of clonidine in the treatment of behavioral disorders in cats. It is used in the treatment of inflammatory bowel disease at a dose of 0.005 to 0.01 μg/kg two to three times daily, either subcutaneously or orally (PO) (Washabau and Day 2013). It is also experimentally used for the treatment of hypertension (Rodan and Heath 2016).

The recommended dose is intentionally low due to the sensitivity of adrenergic receptors in cats and should be adjusted based on the observed clinical response.

21.2.4.2 Experience Feedback

We do not have sufficient experience to provide feedback on the use of clonidine in cats. Its use in this species should therefore be approached with caution, adjusting doses and closely monitoring clinical responses.

21.2.4.3 Possible Combinations

The venlafaxine-clonidine and fluoxetine-clonidine combinations are commonly used to treat phobias in dogs and are theoretically possible in cats.

21.2.4.4 Withdrawal

In humans, abrupt discontinuation of orally administered clonidine can lead to a withdrawal syndrome characterized by agitation, nervousness, tremors, and hypertension.

Although veterinary information is lacking, and due to its effect on norepinephrine release, gradual withdrawal of clonidine is essential to avoid a resurgence of anxiety or cardiovascular clinical signs. A gradual taper over several days is recommended.

Summary Table of Sympatholytics

Substance	Mechanism of action	Pathological states for which the molecule is indicated	Dosage	Cytochromes
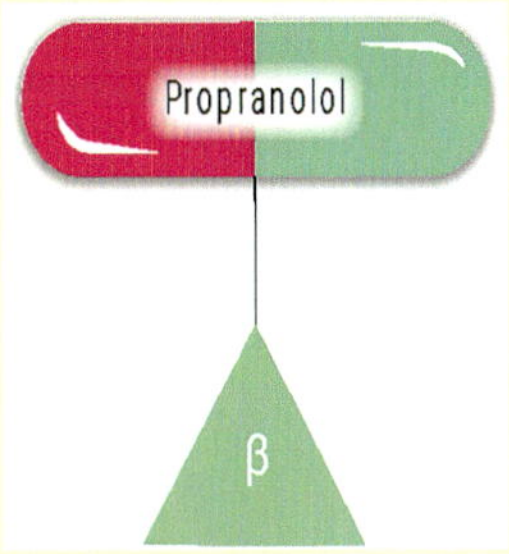	β-adrenergic antagonist	Phobic	0.2 to 0.8 mg/kg BID, TID	CPY2D15
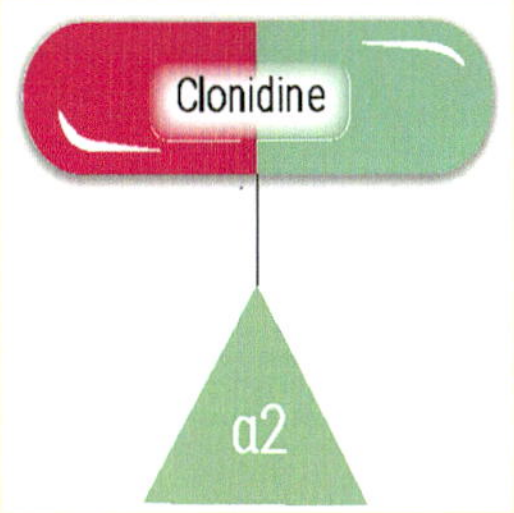	α_2-adrenergic agonist	Phobic	0.005 to 0.01 µg/kg SID, BID, or TID	CYP2D6 CYP1A2
				CYP3A4
				CYP1A1 CYP3A5

References

Ally A (1997) Cardiovascular effects of central administration of clonidine in conscious cats. Brain Res 761(2):283–289. https://doi.org/10.1016/S0006-8993(97)00353-3

Cao J, Shi X, Miao X, Xu J (2009) Effects of premedication of midazolam or clonidine on perioperative anxiety and pain in children. Biosci Trends 3(3):115–118

Cenani A, Brosnan RJ, Madigan S, Knych HK, Madigan JE (2017) Pharmacokinetics and pharmacodynamics of intravenous romifidine and propranolol administered alone or in combination for equine sedation. Vet Anaesth Analg 44(1):86–97. https://doi.org/10.1111/vaa.12410

Claessens AJ, Risler LJ, Eyal S, Shen DD, Easterling TR, Hebert MF (2010) CYP2D6 Mediates 4-Hydroxylation of Clonidine In Vitro: Implication for Pregnancy-Induced Changes in Clonidine Clearance. Drug Metab Dispos 38, 1393–1396. https://doi.org/10.1124/dmd.110.033878

Drug Monograph - Plumb's Veterinary Drugs [WWW Document], (2024). URL https://app.plumbs.com/drug-monograph/hVa7Ct9JYiPROD (accessed 7.19.21)

Head G, Mayorov D (2006) Imidazoline receptors, novel agents and therapeutic potential. Cardiovasc Hematol Agents Med Chem 4(1):17–32. https://doi.org/10.2174/187152506775268758

Jacobs G, Whittem T, Sams R, Calvert C, Ferguson D (1997) Pharmacokinetics of propranolol in healthy cats during euthyroid and hyperthyroid states. Am J Vet Res 58(4):398–403

Korpivaara M, Huhtinen M, Aspegrén J, Overall K (2021) Dexmedetomidine oromucosal gel reduces fear and anxiety in dogs during veterinary visits: a randomised, double-blind, placebo-controlled clinical pilot study. Vet Rec 189(12):e832. https://doi.org/10.1002/vetr.832

McSweeney PM, Martin DD, Ramsey DS, McKusick BC (2012) Clinical efficacy and safety of dexmedetomidine used as a preanesthetic prior to general anesthesia in cats. J Am Vet Med Assoc 240(4):404–412. https://doi.org/10.2460/javma.240.4.404

Nguyen M, Tharani S, Rahmani M, Shapiro M (2014) A review of the use of clonidine as a sleep aid in the child and adolescent population. Clin Pediatr 53(3):211–216. https://doi.org/10.1177/0009922813502123

Paalzow LK, Edlund PO (1979) Pharmacokinetics of clonidine in the rat and cat. J Pharmacokinet Biopharm 7(5):481–494. https://doi.org/10.1007/BF01062390

Rodan I, Heath S (eds) (2016) Feline behavioral health and welfare. Elsevier

Söderpalm B, Engel JA (1988) Biphasic effects of clonidine on conflict behavior: involvement of different alpha-adrenoceptors. Pharmacol Biochem Behav 30(2):471–477. https://doi.org/10.1016/0091-3057(88)90482-0

Thawley VJ, Drobatz KJ (2015) Assessment of dexmedetomidine and other agents for emesis induction in cats: 43 cases (2009–2014). J Am Vet Med Assoc 247(12):1415–1418. https://doi.org/10.2460/javma.247.12.1415

Washabau RJ, Day MJ (2013) Canine & feline gastroenterology. Elsevier Saunders

Antiepileptic Drugs

Contents

22.1 General Characteristics of All Antiepileptic Drugs – 621

22.2 Gabapentin – 621
22.2.1 Description, Chemical Formula, and Brand Names – 621
22.2.2 Pharmacology – 621
22.2.3 Side Effects and Usage Precautions – 622
22.2.4 Clinical Use and Dosage – 623

22.3 Pregabalin – 624
22.3.1 Description, Chemical Formula, and Brand Names – 624
22.3.2 Pharmacology – 624
22.3.3 Side Effects and Usage Precautions – 625
22.3.4 Clinical Use and Dosage – 625

22.4 Imepitoin – 626
22.4.1 Description, Chemical Formula, and Brand Names – 626
22.4.2 Pharmacology – 626
22.4.3 Side Effects and Usage Precautions – 627
22.4.4 Clinical Use and Dosage – 627

22.5 Phenobarbital – 628
22.5.1 Description, Chemical Formula, and Brand Names – 628
22.5.2 Pharmacology – 628
22.5.3 Side Effects and Usage Precautions – 629
22.5.4 Clinical Use and Dosage – 630

22.6 Levetiracetam – 631
22.6.1 Description, Chemical Formula, and Brand Names – 631

© The Author(s), under exclusive license to Springer Nature Switzerland AG 2026
S. Bleuer-Elsner et al., *Veterinary Psychiatry of the Cat*,
https://doi.org/10.1007/978-3-032-06260-4_22

22.6.2 Pharmacology – 631
22.6.3 Side Effects and Usage Precautions – 631
22.6.4 Clinical Use and Dosage – 632

References – 634

This chapter covers molecules with potential action on seizures and, for some, also on pain. These include gabapentinoids (gabapentin and pregabalin), imepitoin, phenobarbital, and levetiracetam.

Pregabalin has veterinary approval for use in cats and is available under the brand name *Bonqat*™. Phenobarbital also has veterinary approval for cats when combined with other antiepileptic drugs, as found in *Crisax*™.

22.1 General Characteristics of All Antiepileptic Drugs

The molecules discussed in this chapter are distinguished by their mechanisms of action.

Gabapentinoids have minimal direct influence on the gamma-aminobutyric acid (GABA) or glutamate systems but inhibit calcium voltage-dependent ion channels, indirectly affecting the release of various neurotransmitters, particularly glutamate. They play a role in controlling seizures and pain by inhibiting voltage-dependent calcium channels. Their use is beneficial in the management of phobic and anxious states.

Phenobarbital is a GABAergic drug that facilitates the inhibition of epileptic foci.

Levetiracetam, on the other hand, acts by binding to presynaptic vesicles, thereby reducing neurotransmitter release. Both are important molecules to be familiar with due to their relevance in behavioral disorders with a neurological component, as well as for treating anxious or phobic states in epileptic cats.

22.2 Gabapentin

22.2.1 Description, Chemical Formula, and Brand Names

Gabapentin, or 2-(1-(aminomethyl)cyclohexyl) acetic acid, is a molecule derived from GABA (gamma-aminobutyric acid). It is primarily known for its antiepileptic properties and its effects in managing chronic pain, particularly neuropathic in nature. Only recently have its anxiolytic properties been studied in cats.

Gabapentin is available under the brand name *Neurontin*™ or as the generic gabapentin, in capsules of 100, 300, and 400 mg, tablets of 600 and 800 mg, and an oral solution of 250 mg/5 mL.

22.2.2 Pharmacology

22.2.2.1 Mechanism of Action

The mechanism of action of gabapentin is not fully understood. Although gabapentinoids do not have a direct affinity for GABA receptors, they may have an indirect effect by slightly increasing GABA production, as suggested by some studies (Errante et al. 2002), but primarily by reducing glutamate production (Errante and Petroff 2003). Its action mainly depends on its affinity for the α2-δ-1 subunit of voltage-dependent calcium ion channels (Sills 2006) (Fig. 22.1).

Voltage-dependent calcium ion channels are distributed throughout the brain and serve various functions. They regulate neuronal excitability and gene expression, trigger neurotransmitter release, and initiate rhythmic action potential discharges (Benarroch 2010). The role of presynaptic voltage-dependent calcium channels is primarily to mediate the

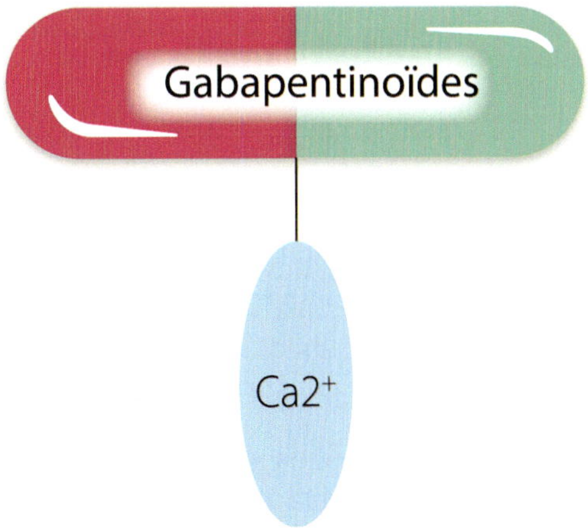

Fig. 22.1 Structure of gabapentinoids (gabapentin and pregabalin). Calcium channel Inhibitors

release of neurotransmitter vesicle contents in response to action potentials (see ▶ Chap. 17).

Bikbaev et al. (2020) demonstrated that α2-δ subunits of voltage-dependent calcium channels are linked to glutamatergic transmission, while α2-δ-2 and α2-δ-3 subunits are associated with GABAergic transmission.

By binding to the α2-δ-1 subunit of calcium ion channels, gabapentin reduces glutamate release from the presynaptic neuron, thereby attenuating postsynaptic excitation.

22.2.2.2 Metabolism

Gabapentin is not metabolized by the liver in cats, unlike in dogs. It is excreted unchanged (Di Cesare et al. 2023). In cats with renal insufficiency, gabapentin excretion may be slowed, warranting dosage adjustments, typically to half the usual dose.

22.2.2.3 Pharmacokinetics

Gabapentin is primarily eliminated via the urine. Its bioavailability is 90–95% following oral administration. The half-life is estimated, depending on the source, to range between 2.5 and 4 h. Plasma clearance is 2.7–3 mL/min/kg, indicating rapid elimination. Peak plasma concentration is reached between 65 and 100 min after a single administration and within 45 min with repeated administrations (Adrian et al. 2018; Di Cesare et al. 2023; Siao et al. 2010).

22.2.3 Side Effects and Usage Precautions

22.2.3.1 Side Effects

The primary side effects observed in cats are sedation and ataxia. Rare cases of vomiting and hypersalivation have been reported. These side effects resolved spontaneously within 8 h of administration (Van Haaften et al. 2017).

22.2.3.2 Precautions and Contraindications

Due to its urinary elimination, doses should be reduced in cases of renal insufficiency or hypertension (Quimby et al. 2022). The oral solution marketed as Neurontin™ contains *xylitol*, which does not appear toxic at the doses used in cats (Jerzsele et al. 2018). Gabapentin crosses the placental barrier and may be teratogenic and fetotoxic in rats (Petrere and Anderson 1994). It is also excreted in cow's milk (Malreddy et al. 2013). Therefore, its use is not recommended during pregnancy or lactation.

22.2.3.3 Drug Interactions

The following drug interactions have been reported in humans or animals receiving gabapentin. While the combination is not necessarily contraindicated, the risks should be evaluated by the practitioner.

- Antacids (containing aluminum, magnesium, or calcium): may reduce the oral bioavailability of gabapentin by 20%. If antacids are necessary, administration should be separated from gabapentin by at least 2 h.
- Buprenorphine: Increased risk of sedation when combined.
- Cannabidiol: increased risk of sedation when combined.
- Hydroxyzine: increased risk of sedation when combined.
- Minocycline: increased risk of sedation when combined.
- Mirtazapine: increased risk of sedation when combined.
- Opioids: increased risk of sedation when combined. Concurrent administration of gabapentin and hydrocodone may increase gabapentin's area under the curve (AUC), enhancing its efficacy or side effects. Gabapentin may reduce the AUC of hydrocodone, potentially diminishing its efficacy. Morphine may increase gabapentin levels.
- Venlafaxine: reduces the antinociceptive action of gabapentin in rats (Rode et al. 2006).

22.2.3.4 Toxicity

The LD_{50} in rats is greater than 8000 mg/kg.

Common signs of intoxication include ataxia, lethargy, sedation, and vomiting. No fatalities have been reported.

22.2.4 Clinical Use and Dosage

22.2.4.1 According to the Scientific Literature

Gabapentin is used to manage seizures resistant to other treatments at doses of 5–20 mg/kg every 8–12 h (Papich 2016).

It is also employed as an analgesic at doses of 5–10 mg/kg every 12 h. It is particularly recommended for the treatment of neuropathic pain (Steagall and Monteiro-Steagall 2013), as well as post-surgical pain, such as after ovariohysterectomy (Steagall et al. 2018), trauma and fractures (Vettorato and Corletto 2011), or dental pain (Di Cesare et al. 2023). While effective for treating arthritic pain, its sedative effects may impact a cat's quality of life (Guedes et al. 2018).

Studies using transdermal gels at doses of 5–10 mg/kg indicate good skin permeability and show that administration reduces pain scores in cats (Adrian et al. 2018; Slovak and Costa 2021).

Gabapentin can also be used as a sedative and for premedication in anesthesia at a dose of 10 mg/kg (Madan et al. 2023).

Two double-blind, randomized, placebo-controlled studies evaluated gabapentin in cats for preventing phobic reactions.

The first study showed a significant reduction in fear after administering 50–100 mg before spaying/neutering (Pankratz et al. 2018). The second study demonstrated an improvement in stress and cooperation after 100 mg was given before transport or veterinary examination, with no severe adverse effects (Van Haaften et al. 2017).

In psychiatry, gabapentin is utilized to treat phobias and anxiety. The recommended dose is 20–30 mg/kg or 100 mg per cat, administered 90 min before exposure to the stimulus or twice daily for continuous treatment. It is particularly effective for handling care-related phobias and managing stress during transport or veterinary visits (Kruszka et al. 2021). A single administration before the visit reduces the frequency of aggression, respiratory rate, and stress scores based on body posture and facial expressions during the visit (Buckley 2019; Erickson et al. 2021). Additionally, handling is facilitated, and cats are more cooperative following gabapentin administration (Gurney and Gower 2022).

Gabapentin is recommended for cardiac examinations as it facilitates the procedure without altering cardiovascular or echocardiographic parameters, such as blood pressure or heart rate (Allen et al. 2021; Ruviaro Tuleski et al. 2022). However, according to Hudec (Hudec and Griffin 2020), cortisol and glucose levels remain similar during veterinary examinations regardless of whether the cat received gabapentin beforehand. Cats that received gabapentin appeared less stressed, attributed to the sedative effect of the drug.

Finally, gabapentin may have orexigenic properties. In a study by Fantinati et al. (2020), it promoted faster appetite recovery compared to controls and was as effective as mirtazapine during the postoperative period in cats.

22.2.4.2 Experience Feedback

Gabapentin is particularly effective for treating phobic, anxious, impulsive, and compulsive states. By reducing glutamatergic activation, it can modulate most circuits involved in these conditions.

Depending on the severity and complexity of symptoms, doses of 10–30 mg/kg can be administered. Due to its very short half-life, the medication must be repeated every 8–12 h. It can also be given as a one-time dose before exposure to a phobic stimulus.

The therapeutic strategy is as follows:

- For isolated phobias (e.g., visitors, travel, veterinary visits): a dose of 20–30 mg/kg is recommended 90–120 min before exposure. This dose can be increased to 50 mg/kg for one-time use.
- For more frequent exposure to phobic stimuli (e.g., humans or other household animals) or for anxious, impulsive, or compulsive states that do not sufficiently respond to primary treatment: a dose of 10–20 mg/kg can be administered every 8–12 h over extended periods.

22.2.4.3 Possible Combinations

Gabapentin can theoretically be combined with all other psychotropic medications. It is a simple molecule to integrate into polytherapy for certain behavioral disorders that are difficult to manage.

Notably, gabapentin can be combined with venlafaxine or fluoxetine for the treatment of phobic and anxious states. Similarly, it can be used for impulsive or compulsive states poorly controlled by fluoxetine alone.

22.2.4.4 Withdrawal

There are no studies on the prolonged use of gabapentin. It appears that tapering doses before discontinuation is unnecessary. However, gradual dose reduction can help determine whether the cat's behavioral improvements persist after treatment cessation.

22.3 Pregabalin

22.3.1 Description, Chemical Formula, and Brand Names

Pregabalin, or (3S)-3-(aminomethyl)-5-methylhexanoic acid, is a derivative of GABA (gamma-aminobutyric acid). It is primarily known for its antiepileptic properties and its effects in managing chronic pain, particularly neuropathic pain. Pregabalin is pharmacologically very similar to gabapentin, sharing most of its properties.

Bonqat™ is approved for use in cats. Off-label, it is available under the brand name *Lyrica*™ or as generic pregabalin in capsules of 25, 50, 150, 200, and 300 mg, as well as in an oral solution of 20 mg/mL.

In France, pregabalin has required a secure prescription since May 2021, likely due to its recreational use, which appears to be a growing trend (Schjerning et al. 2016).

22.3.2 Pharmacology

22.3.2.1 Mechanism of Action

The mechanism of action of pregabalin is not fully understood. Despite its chemical similarity to GABA, pregabalin has no direct affinity for GABAergic receptors. However, it may slightly increase GABA production (Whitworth and Quick 2001) and decrease glutamate production (Errante and Petroff 2003). Its primary action relies on its affinity for the α2-δ-1 subunit of voltage-dependent calcium ion channels (Sills 2006) (◘ Fig. 22.1).

Voltage-dependent calcium ion channels are distributed throughout the brain and serve various functions. They regulate neuronal excitability and gene expression, trigger neurotransmitter release, and initiate rhythmic action potential discharges (Benarroch 2010). The α2-δ-1 subunits of calcium ion channels are linked to glutamatergic transmission (Bikbaev et al. 2020).

By binding to the α2-δ-1 subunit of presynaptic calcium ion channels, pregabalin inhibits intracellular calcium influx and reduces the release of several excitatory neurotransmitters, including glutamate, substance P, and norepinephrine. It also dampens postsynaptic excitability.

22.3.2.2 Metabolism and Involved Cytochromes

As in humans, pregabalin does not undergo significant hepatic transformation.

22.3.2.3 Pharmacokinetics

The bioavailability of pregabalin is 94% and is unaffected by food intake. Its elimination is primarily urinary. The maximum plasma concentration is reached within 0.5–3 h, and its half-life is 14.6 h after an oral dose of 5 mg/kg. Plasma clearance is 30 mL/h/kg (Lamminen et al. 2022; Orion Corporation Orion Pharma 2021). Despite its long half-life, more frequent dosing may be necessary to maintain optimal clinical efficacy due to interindividual variability and specific pharmacokinetic characteristics.

22.3.3 Side Effects and Usage Precautions

22.3.3.1 Side Effects

In humans, regardless of the gabapentinoid used, the most frequently reported side effects include drowsiness, dizziness, ataxia, difficulty concentrating, attention or memory impairment, blurred vision, dry mouth, peripheral edema, constipation, and weight gain. Syncope and congestive heart failure have been reported. Rarely, reversible renal failure and rhabdomyolysis have been observed. Hypersensitivity reactions have included angioedema, rash, blisters, and wheezing. Pregabalin treatment has been associated with decreased platelet production or elevated creatine kinase levels in some patients.

In cats, the main side effects of pregabalin, observed in fewer than 10% of cases, are sedation, vomiting, and ataxia. Less than 1% of cases may present with tremors, hypothermia, salivation, mydriasis, anorexia, weight loss, leukopenia, and neutropenia (Esteban et al. 2018; Lamminen et al. 2022; Orion Corporation Orion Pharma 2021). A case of superficial dermatitis has also been reported (Clark et al. 2017).

22.3.3.2 Precautions and Contraindications

Due to its urinary elimination, pregabalin doses should be adjusted in cases of renal insufficiency or hypertension. Pregabalin may cause bradycardia and reflex hypertension due to its sedative properties; it is therefore not recommended in cases of heart disease.

Pregabalin is excreted in milk and may cause malformations during gestation in rats (Drug Monographs Plumb's Veterinary Drugs, 2024). Its use is thus contraindicated during pregnancy or lactation.

22.3.3.3 Drug Interactions

The following drug interactions have been reported in humans or animals receiving pregabalin. While the combination is not necessarily contraindicated, the risks should be evaluated by the practitioner:

- Buprenorphine: increased risk of sedation.
- Cannabidiol: increased risk of sedation.
- Hydroxyzine: increased risk of sedation.
- Minocycline: increased risk of sedation.
- Mirtazapine: increased risk of sedation.
- Opioids: increased risk of sedation and respiratory depression.
- Benzodiazepines (e.g., alprazolam, diazepam, midazolam): increased risk of sedation and respiratory depression.
- Angiotensin-converting enzyme inhibitors (e.g., benazepril, enalapril): increased risk of edema and urticaria.
- Central nervous system depressants (e.g., acepromazine, dexmedetomidine, phenobarbital): increased risk of sedation and respiratory depression.

22.3.3.4 Toxicity

In cats, signs of toxicity have been observed following pregabalin administration at doses between 15 and 25 mg/kg. These include abnormal gait, ataxia, drowsiness, apathy, hypothermia, mydriasis, vomiting, hypersalivation, and loss of consciousness (Orion Corporation Orion Pharma 2021).

22.3.4 Clinical Use and Dosage

22.3.4.1 According to the Scientific Literature

Bonqat™ is approved for the "attenuation of acute anxiety and fear associated with transport and veterinary visits." It is therefore indicated for transport and care-related phobias caused by veterinary visits. The recommended dose is 5 mg/kg (0.1 ml/kg) administered approximately 1.5 h before transport or the veterinary visit. The liquid should be given directly into the cat's mouth or mixed with a small amount of food (Lamminen et al. 2023; Orion Corporation Orion Pharma 2021).

Pregabalin can also be used for other phobias and anxiety. The recommended dose is 5–10 mg/kg administered 90 min before exposure to a phobic stimulus or twice daily for continuous treatment.

Pregabalin is rarely used for controlling seizures resistant to other treatments. The recommended dose is 1–2 mg/kg every 12 h (Esteban et al. 2018).

It can be used as a sedative and premedication for anesthesia at doses of 4–10 mg/kg administered orally 90 min before anesthesia (El-Sherif and Mahmoud 2024; Madan et al. 2023). Unlike gabapentin, pregabalin can cause hypotension and is less recommended for echocardiographic examinations (Li et al. 2024).

Finally, pregabalin is used to manage certain chronic pain conditions, particularly neuropathic pain. The recommended dose is 1–3 mg/kg every 8–12 h (Goich et al. 2019; Korff and Williamson 2020).

22.3.4.2 Experience Feedback

Based on our experience, pregabalin appears promising in managing certain phobias, though the results remain incomplete. Its clinical indications seem to align with those of gabapentin. It would be interesting to determine whether the variability in effects between the two drugs depends on individual responses or if there are specific indications for which one is more effective than the other. The doses used are 5–10 mg/kg, administered twice daily or as a single dose.

22.3.4.3 Possible Combinations

Pregabalin can theoretically be combined with all other psychotropic drugs. It is a straightforward molecule to integrate into polytherapy for certain behavioral disorders that are difficult to manage. Notably, pregabalin can be combined with venlafaxine or fluoxetine for treating phobic and anxious states.

22.3.4.4 Withdrawal

There are no studies on the prolonged use of pregabalin. It appears unnecessary to taper doses before discontinuing the drug. However, gradual dose reduction can help determine whether the cat's behavioral improvements persist after treatment cessation.

22.4 Imepitoin

22.4.1 Description, Chemical Formula, and Brand Names

Imepitoin, chemically known as 3-(4-chlorophenyl)-5-morpholin-4-yl-4H--imidazol-2-one, is recognized for its anticonvulsant properties.

The molecule is available exclusively in the veterinary product *Pexion*™, which has European approval for controlling seizures in dogs. The same product has also received approval in the United States for the treatment of noise phobias in dogs. *Pexion*™ is available in 100 mg and 400 mg tablets. Imepitoin does not have veterinary approval for use in cats, and its use is based on data extrapolated from other species.

22.4.2 Pharmacology

22.4.2.1 Mechanism of Action

Imepitoin is a low-affinity partial GABAergic agonist at $GABA_A$ receptors. Its binding affinity is approximately 600 times lower than that of diazepam. The inhibitory effects mediated by $GABA_A$ receptors on neurons are enhanced, thereby preventing the onset of epileptic foci.

Additionally, imepitoin has a mild ability to inhibit calcium channels, which contributes to its anticonvulsant activity (◘ Fig. 22.2).

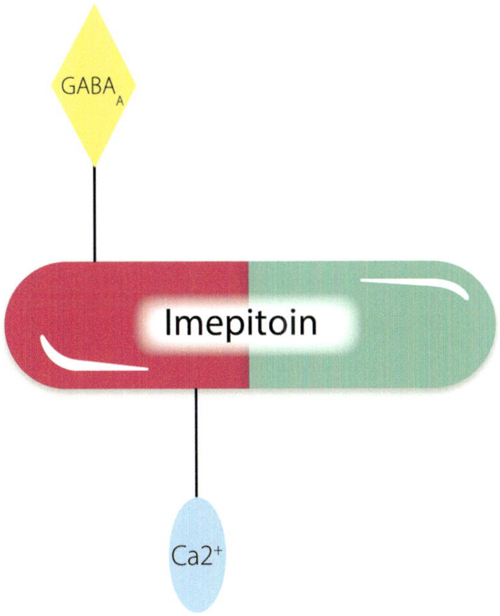

Fig. 22.2 Structure of imepitoin. Positive allosteric modulator of GABA-A—calcium channel inhibitor

22.4.2.2 Metabolism and Involved Cytochromes

Imepitoin is extensively metabolized through oxidative mechanisms into four main inactive metabolites. The specific cytochrome P450 enzymes involved have not been clearly identified.

22.4.2.3 Pharmacokinetics

In dogs, the half-life of imepitoin is approximately 1.5–2 h. Elimination is primarily biliary and fecal. No pharmacokinetic data are available for cats.

22.4.3 Side Effects and Usage Precautions

22.4.3.1 Side Effects

Reported side effects in cats include lethargy, vomiting, and decreased appetite (Engel et al. 2017).

In dogs, where studies are more extensive, side effects include ataxia, lethargy, increased appetite, vomiting, hyperactivity, drowsiness, polyuria, polydipsia, salivation, diarrhea, prolapse of the third eyelid, decreased vision, and reduced sensitivity to sound (Boehringer Ingelheim Vetmedica, Inc. 2018).

22.4.3.2 Precautions and Contraindications

Imepitoin is contraindicated in patients with severe hepatic insufficiency, renal disorders, or cardiovascular diseases.

22.4.3.3 Drug Interactions

Flumazenil (a GABA antagonist) has been reported to potentially suppress the anxiolytic and anticonvulsant effects of imepitoin (Bialer et al. 2013). Caution is advised when using imepitoin in patients receiving other medications.

22.4.3.4 Toxicity

The acute toxicity of oral imepitoin has been evaluated in rats and mice, with no mortality observed up to 2105 mg/kg (Woodward 2018). Flumazenil, a benzodiazepine antagonist, may potentially be used to reverse the effects of imepitoin toxicity.

22.4.4 Clinical Use and Dosage

22.4.4.1 According to the Scientific Literature

In dogs, *Pexion*™ has U.S. approval for the control of noise phobias (Boehringer Ingelheim Vetmedica, Inc. 2018).

In cats, there are no studies on the psychotropic effects of imepitoin. It is used as a second-line treatment for seizures after phenobarbital (Wyatt 2019). The recommended dose is 30 mg/kg every 12 h (Engel et al. 2017).

22.4.4.2 Experience Feedback

Our experience is very limited, and we do not have sufficient data to supplement current scientific publications.

22.4.4.3 Possible Combinations

Imepitoin can theoretically be combined with most psychotropics. Caution is advised when combining it with other GABAergic type A drugs, such as benzodiazepines or phenobarbital.

22.4.4.4 Withdrawal

A gradual tapering protocol has been proposed for epileptic dogs controlled with phenobarbital and potassium bromide, starting with 20 mg/kg twice daily for 1 month, followed by 10 mg/kg twice daily for 1 month, and then 10 mg/kg once daily for 1 month (Stee et al. 2017). The same precautions are recommended for cats. For short-term or occasional use, tapering is not required.

22.5 Phenobarbital

22.5.1 Description, Chemical Formula, and Brand Names

Phenobarbital, chemically known as 5-ethyl-5-phenyl-1,3-diazinane-2,4,6-trione, is a barbiturate used as a first-line treatment for epilepsy and is essential in veterinary medicine.

It is available under the brand name *Gardénal*™ and in veterinary-specific formulations as pure forms, *Épityl*™, *Phénoleptil*™, *Soliphen*™, or combined with bromide salts, *Antirutol*™, *Crisax*™, *Nervicanis*™. Among these, *Crisax*™ and *Nervicanis*™ are the only formulations with veterinary approval for use in cats. The others are intended for dogs. The phenobarbital concentrations in these commercial products range from 10 to 60 mg.

22.5.2 Pharmacology

22.5.2.1 Mechanism of Action

Phenobarbital acts on $GABA_A$ receptors, enhancing synaptic inhibition. This increases the seizure threshold and reduces propagation from a seizure focus. Phenobarbital can also inhibit calcium channels, leading to decreased release of excitatory neurotransmitters (◻ Fig. 22.3). The sedative-hypnotic effects of phenobarbital are likely due to its impact on the mesencephalic reticular formation, which controls central nervous system activation.

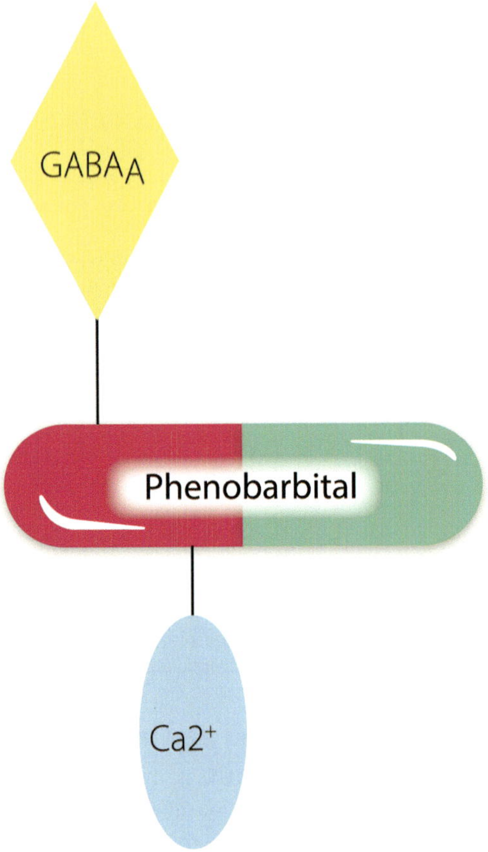

◻ **Fig. 22.3** Structure of phenobarbital. Positive allosteric modulator of GABA-A—calcium channel inhibitor

22.5.2.2 Metabolism and Involved Cytochromes

Phenobarbital is metabolized in the liver primarily through hydroxylation into p-hydroxyphenobarbital, along with other minor metabolites. All metabolites are inactive. In humans, the main cytochrome P450 enzyme involved is CYP2C19.

22.5.2.3 Pharmacokinetics

Phenobarbital is fully absorbed after oral administration, with a half-life of 34–43 h. Excretion is primarily renal (Cochrane et al. 1990).

Prolonged administration of phenobarbital can induce enzymatic activity, potentially decreasing its elimination half-life and serum levels over time.

22.5.3 Side Effects and Usage Precautions

22.5.3.1 Side Effects

A study by Marsh et al. (2021) involving 77 epileptic cats documented the side effects observed during phenobarbital treatment. Nearly half of the cats (47%) exhibited side effects, including sedation (89%), ataxia (53%), polyphagia (22%), polyuria (6%), and anorexia (6%).

Sedation and ataxia tend to occur when serum levels approach the upper limits of the therapeutic range. Excessive sedation caused by phenobarbital may sometimes be interpreted by owners as an improvement in symptoms, making it essential to regularly monitor the cat's clinical condition to accurately evaluate treatment effects.

Polyuria and polyphagia generally appear after 2–3 weeks of treatment.

An increase in liver enzymes is possible, though it is less common than in dogs. This increase is not necessarily indicative of hepatic dysfunction but should be monitored, and the treatment should be reassessed, as with dogs (Hermans et al. 2022; Thomas 2010).

More rarely, autoimmune cytopenias such as anemia, thrombocytopenia, or neutropenia have been observed (Dohány et al. 2023; Walton-Clark et al. 2022). These conditions are reversible if detected early. Coagulopathies have been reported after high-dose phenobarbital administration (10–40 mg/kg per day) (Peterson et al. 1995).

Phenobarbital, like other drugs stimulating $GABA_A$ receptors, such as benzodiazepines, may increase agitation and anxiety in cats.

22.5.3.2 Precautions and Contraindications

Barbiturates are contraindicated in patients with severe hepatic disease or those with a history of hypersensitivity reactions to the medication.

High doses are contraindicated in patients with respiratory dysfunction. Cats, in particular, are more sensitive to the respiratory depressant effects of barbiturates compared to dogs.

Barbiturates should be used cautiously in patients with hypovolemia, anemia, borderline adrenal function, nephritis, or cardiac or respiratory disease. Dose reductions may be necessary in patients with significant renal insufficiency.

Barbiturates cross the placental barrier and are excreted in human milk; therefore, their use is not recommended during pregnancy or lactation.

22.5.3.3 Drug Interactions

As phenobarbital is an enzymatic inducer, the risks of concurrent use with any other medication must be carefully evaluated.

We will limit this section to psychotropic drugs:

- Benzodiazepines: increased risk of additional respiratory depression.
- Mirtazapine: increased risk of additional respiratory depression.
- Trazodone: increased risk of additional respiratory depression.
- MAO (Monoamine Oxidase) inhibitors (MAOIs): risk of prolonged effects of phenobarbital.
- Levetiracetam: reduced half-life when combined with phenobarbital in dogs (Moore et al. 2011).
- Tricyclic antidepressants: their efficacy is reduced when combined with phenobarbital.
- Venlafaxine: increases brain concentration of phenobarbital in mice (Borowicz et al. 2011).

Important Point of View

Phenobarbital, Epilepsy, and Psychotropics
A practitioner may need to manage one or more pathological states in an epileptic cat. Additionally, an increase in seizures is often reported in epileptic cats suffering from phobic or anxious states. Combining a psychotropic with phenobarbital often raises questions.

Among the most commonly used psychotropics, fluoxetine, venlafaxine, and/or gabapentin can be safely combined with phenobarbital, as no aggravating effects have been described.

Note that the use of clomipramine is contraindicated.

22.5.3.4 Toxicity

The oral LD_{50} in rats is approximately 162 mg/kg.

Cats intoxicated with phenobarbital may exhibit ataxia, sedation with recumbency, hypothermia, and mydriasis.

Treatment for phenobarbital overdose involves neutralizing the ingested product in the intestine and providing cardiopulmonary support if necessary. Activated charcoal significantly enhances the elimination of phenobarbital, even when the drug has been administered parenterally. Activated charcoal acts as a trap, causing diffusion from the vascular system to the intestine.

Forced alkaline diuresis is also indicated to facilitate phenobarbital elimination in patients with normal renal function. Peritoneal dialysis or hemodialysis may be beneficial in cases of severe intoxication or anuric patients.

22.5.4 Clinical Use and Dosage

22.5.4.1 According to the Scientific Literature

There are very few publications describing the effects of phenobarbital in the treatment of behavioral or psychiatric disorders. Phenobarbital is primarily regarded as the first-choice medication for the treatment of essential epilepsy (Charalambous et al. 2018; Dewey 2006).

The recommended doses are 1–3 mg/kg twice daily or half a 15 mg tablet per cat twice daily. In cases of very frequent seizures, cats can be hospitalized for a loading dose of 12–24 mg/kg intravenously, followed by 3–4 mg/kg every 2–6 h. For *status epilepticus*, a bolus of 10 mg/kg can be administered, followed by 3 mg/kg intravenously, repeated every 20 min until the condition resolves.

Phenobarbital is also prescribed by neurologists for compulsive behaviors that may resemble partial epilepsy, such as shadow or light chasing, tail attacks, and hypersexuality (Liatis and Cherubini 2021). In such cases, the recommended doses are 0.5–1.5 mg/kg twice daily (Denenberg 2020).

Phenobarbital is used as an analgesic for chronic feline gingivostomatitis or feline hyperesthesia syndrome at doses of 1–3 mg/kg twice daily (Rodan and Heath 2016; Rusbridge 2024; Rusbridge et al. 2010).

Finally, phenobarbital may be used as an adjunct treatment for neuropathic pain (Tremont-Lukats et al. 2000).

22.5.4.2 Experience Feedback

The use of phenobarbital is typically reserved for cases where a neurological component is suspected, often in compulsive, manic, or derealized states. If a neurologist evaluates the cat first and suspects an epileptic origin for the disorder, they typically prescribe antiepileptic medications and may refer the case to a veterinary psychiatrist if the treatment does not adequately control symptoms.

Conversely, when faced with compulsive, manic, or derealized states, a veterinary psychiatrist may initially test a psychotropic such as fluoxetine but could later complement the treatment with phenobarbital.

The recommended doses are 1–3 mg/kg every 12 h (Drug Monographs Plumb's Veterinary Drugs, 2024). Plasma level monitoring is advised every 3 weeks to adjust the dosage based on the results.

22.5.4.3 Possible Combinations

Combining phenobarbital with venlafaxine or fluoxetine is possible. Following a precautionary approach, both drugs should be started at half the usual dose. Gradual adjustments to standard doses can be made based on observed results.

22.5.4.4 Withdrawal

A gradual reduction of 25% of the dose every 2 weeks is recommended.

22.6 Levetiracetam

22.6.1 Description, Chemical Formula, and Brand Names

Levetiracetam, chemically known as (2S)-2-(2-oxopyrrolidin-1-yl)butanamide, is an anticonvulsant molecule most commonly used in combination with phenobarbital for refractory epilepsy or when phenobarbital is not well tolerated by the patient.

The molecule is available under the brand name *Keppra*™ or in generic forms. Commercially, it is offered in tablets of 250, 500, and 1000 mg; sachets of gel containing 250, 500, 750, and 1000 mg; and an oral solution of 100 mg/mL.

22.6.2 Pharmacology

22.6.2.1 Mechanism of Action

The exact mechanism of levetiracetam's antiepileptic activity is not yet fully understood. Levetiracetam appears to affect neurotransmitter release in the central nervous system by selectively binding to the presynaptic protein Synaptic Vesicle Glycoprotein 2A (SV2A) on synaptic vesicles (◘ Fig. 22.4). It is thought to prevent the hypersynchronization of epileptiform foci and the propagation of epileptic activity, without affecting normal neuronal excitability (Drug Monographs Plumb's Veterinary Drugs, 2024).

◘ **Fig. 22.4** Structure of levetiracetam. SV2A inhibitor by suppressing the release of vesicular contents from the presynaptic neuron

22.6.2.2 Metabolism and Involved Cytochromes

Levetiracetam undergoes minimal metabolism in the body. The primary metabolic pathway is enzymatic hydrolysis of its acetamide group, producing an inactive carboxylic acid metabolite, L057, which accounts for approximately one-quarter of the total administered dose. The specific enzymes responsible for this reaction have not been clearly identified, but this pathway is known to be independent of hepatic Cytochrome P450 (CYP) enzymes.

22.6.2.3 Pharmacokinetics

Following oral administration, levetiracetam has an elimination half-life of approximately 3 h, 100% bioavailability, a clearance rate of 2 mL/kg/min, and 90% renal excretion (Carnes et al. 2011).

22.6.3 Side Effects and Usage Precautions

22.6.3.1 Side Effects

Levetiracetam appears to be well tolerated in cats. The most commonly reported side effects include hypersalivation, ataxia, and reduced appetite (Lowrie et al. 2017). These effects are generally more transient than those seen with phenobarbital administration.

Levetiracetam is excreted in milk and may cause fetal deaths in rats (Drug Monographs Plumb's Veterinary Drugs, 2024). Its use is therefore not recommended during pregnancy or lactation.

22.6.3.2 Precautions and Contraindications

Levetiracetam should be used cautiously in patients with renal insufficiency. Dosages or administration frequency should be adjusted accordingly.

22.6.3.3 Drug Interactions

- Central Nervous System (CNS) depressants (e.g., benzodiazepines, opioids): may increase the risk of central nervous system depression.
- Methotrexate (immunosuppressant): concurrent use with levetiracetam may increase methotrexate levels and the risk of toxicity.
- Phenobarbital: in cats, continuous use of phenobarbital does not appear to alter the pharmacokinetics of levetiracetam (Bailey et al. 2008).

22.6.3.4 Toxicity

Levetiracetam is considered a relatively safe agent. The LD_{50} in rats exceeds 5000 mg/kg (Drug Monographs Plumb's Veterinary Drugs, 2024). Clinical signs of intoxication include lethargy, hyperthermia, recumbency, and vomiting.

22.6.4 Clinical Use and Dosage

22.6.4.1 According to the Scientific Literature

Levetiracetam is used as a first-line treatment for epilepsy or as an adjunct therapy with phenobarbital and potassium bromide in cases of refractory epilepsy. The recommended dose is 20–25 mg/kg every 8 h (Bailey et al. 2008; Carnes et al. 2011; Lowrie et al. 2017).

22.6.4.2 Experience Feedback

Levetiracetam, like phenobarbital, may theoretically be beneficial in cats whose symptoms suggest a neurological component. Epileptic cats on levetiracetam may also exhibit phobic or anxious states, making it important to be familiar with the drug and its characteristics if polytherapy is considered. The recommended doses are 20–25 mg/kg every 8 h (Drug Monographs Plumb's Veterinary Drugs, 2024).

22.6.4.3 Possible Combinations

Gabapentinoids, venlafaxine, and fluoxetine can be combined with levetiracetam.

For fluoxetine, a precautionary approach is recommended, starting both drugs at half the usual dose. Gradual adjustments to the standard doses are made based on observed outcomes until the usual dose is reached for both drugs if needed.

22.6.4.4 Withdrawal

A withdrawal protocol does not appear to be necessary given the molecule's half-life.

Summary Table for Antiepileptics

Substance	Mechanism of action	Pathological states for which the molecule is indicated	Dosages	Cytochromes
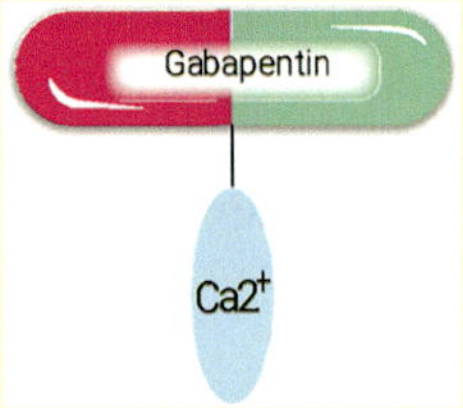	Voltage-dependent Ca^{2+} channel inhibitor	Phobic Anxious	10–20 mg/kg BID or TID 10–30 mg/kg PRN	No hepatic metabolism

Substance	Mechanism of action	Pathological states for which the molecule is indicated	Dosages	Cytochromes
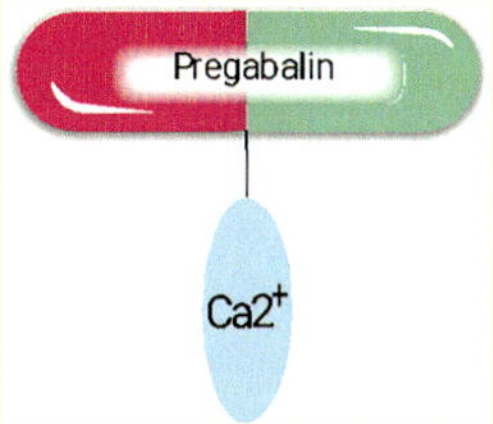	Voltage-dependent Ca^{2+} channel inhibitor	Phobic Anxious	5–10 mg/kg BID	No hepatic metabolism
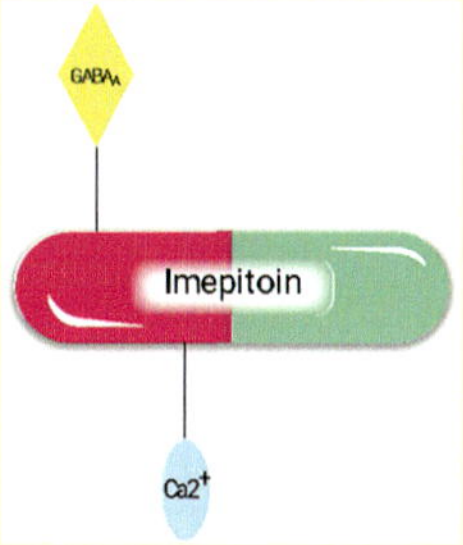	Voltage-dependent Ca^{2+} channel inhibitor Partial $GABA_A$ agonist	No current indications	30 mg/kg BID	–
	Voltage-dependent Ca^{2+} channel inhibitor	Manic	1–3 mg/kg BID	CYP2C19
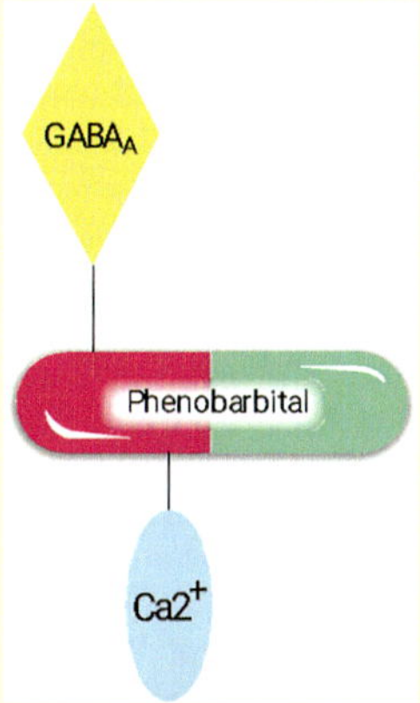	$GABA_A$ agonist	Compulsive Dissociated Neurological		
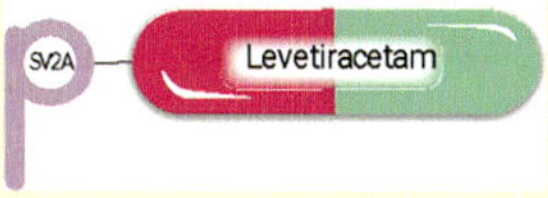	Inhibitor of synaptic vesicle release via SV2A	Neurological	20–25 mg/kg TID	No metabolism via cytochromes

References

Adrian D, Papich MG, Baynes R, Stafford E, Lascelles BDX (2018) The pharmacokinetics of gabapentin in cats. J Vet Intern Med 32(6):1996–2002. https://doi.org/10.1111/jvim.15313

Allen ME, LeBlanc NL, Scollan KF (2021) Hemodynamic, echocardiographic, and sedative effects of oral gabapentin in healthy cats. J Am Anim Hosp Assoc 57(6):278–284. https://doi.org/10.5326/JAAHA-MS-7081

Bailey KS, Dewey CW, Boothe DM, Barone G, Kortz GD (2008) Levetiracetam as an adjunct to phenobarbital treatment in cats with suspected idiopathic epilepsy. J Am Vet Med Assoc 232(6):867–872. https://doi.org/10.2460/javma.232.6.867

Benarroch EE (2010) Neuronal voltage-gated calcium channels: brief overview of their function and clinical implications in neurology. Neurology 74(16):1310–1315. https://doi.org/10.1212/WNL.0b013e3181da364b

Bialer M, Johannessen SI, Levy RH, Perucca E, Tomson T, White HS (2013) Progress report on new antiepileptic drugs: a summary of the eleventh EILAT conference (EILAT XI). Epilepsy Res 103(1):2–30. https://doi.org/10.1016/j.eplepsyres.2012.10.001

Bikbaev A, Ciuraszkiewicz-Wojciech A, Heck J, Klatt O, Freund R, Mitlöhner J, Enrile Lacalle S, Sun M, Repetto D, Frischknecht R, Ablinger C, Rohlmann A, Missler M, Obermair GJ, Di Biase V, Heine M (2020) Auxiliary α2δ1 and α2δ3 subunits of calcium channels drive excitatory and inhibitory neuronal network development. J Neurosci 40(25):4824–4841. https://doi.org/10.1523/JNEUROSCI.1707-19.2020

Boehringer Ingelheim Vetmedica, Inc. (2018) PexionTM (imepitoin tablets) Dogs

Borowicz KK, Gołyska D, Luszczki JJ, Czuczwar SJ (2011) Effect of acutely and chronically administered venlafaxine on the anticonvulsant action of classical antiepileptic drugs in the mouse maximal electroshock model. Eur J Pharmacol 670(1):114–120. https://doi.org/10.1016/j.ejphar.2011.08.042

Buckley LA (2019) Cats that get stressed when visiting the veterinary practice: can gabapentin help improve their welfare? Vet Evid 4(4):1–12. https://doi.org/10.18849/ve.v4i4.227

Carnes MB, Axlund TW, Boothe DM (2011) Pharmacokinetics of levetiracetam after oral and intravenous administration of a single dose to clinically normal cats. Am J Vet Res 72(9):1247–1252. https://doi.org/10.2460/ajvr.72.9.1247

Charalambous M, Pakozdy A, Bhatti SFM, Volk HA (2018) Systematic review of antiepileptic drugs' safety and effectiveness in feline epilepsy. BMC Vet Res 14(1):64. https://doi.org/10.1186/s12917-018-1386-3

Clark L, Doyle RS, Shroeder S (2017) Skin lesions following pregabalin administration in a cat. Vet Anaesth Analg 44(2):383–385. https://doi.org/10.1016/j.vaa.2016.03.008

Cochrane SM, Parent JM, Black WD, Allen DG, Lumsden JH (1990) Pharmacokinetics of phenobarbital in the cat following multiple oral administration. Can J Vet Res 54(3):309–312

Denenberg S (2020) Small animal veterinary psychiatry, 1st edn. CAB International

Dewey CW (2006) Anticonvulsant therapy in dogs and cats. Vet Clin N Am Small Anim Pract 36(5):1107–1127. https://doi.org/10.1016/j.cvsm.2006.05.005

Di Cesare F, Negro V, Ravasio G, Villa R, Draghi S, Cagnardi P (2023) Gabapentin: clinical use and pharmacokinetics in dogs, cats, and horses. Animals 13(12):2045. https://doi.org/10.3390/ani13122045

Dohány A, Guija-de-Arespacochaga A, Fux D, Silberbauer C, Pákozdy Á (2023) A retrospective evaluation of phenobarbital-induced hematologic changes in 69 cats. Vet Clin Pathol 52(4):601–606. https://doi.org/10.1111/vcp.13259

El-Sherif MW, Mahmoud MA (2024) Clinical anesthetic evaluation of Pregabalin as premedication for ketamine-induced general anesthesia in dogs and cats. SVU Int J Vet Sci 7(2):12–24. https://doi.org/10.21608/svu.2024.279269.1316

Engel O, Von Klopmann T, Maiolini A, Freundt-Revilla J, Tipold A (2017) Imepitoin is well tolerated in healthy and epileptic cats. BMC Vet Res 13(1):172. https://doi.org/10.1186/s12917-017-1087-3

Erickson A, Harbin K, MacPherson J, Rundle K, Overall KL (2021) A review of pre-appointment medications to reduce fear and anxiety in dogs and cats at veterinary visits. Can Vet 62(9):952–960

Errante LD, Petroff OAC (2003) Acute effects of gabapentin and pregabalin on rat forebrain cellular GABA, glutamate, and glutamine concentrations. Seizure 12(5):300–306. https://doi.org/10.1016/S1059-1311(02)00295-9

Errante LD, Williamson A, Spencer DD, Petroff OAC (2002) Gabapentin and vigabatrin increase GABA in the human neocortical slice. Epilepsy Res 49(3):203–210. https://doi.org/10.1016/S0920-1211(02)00034-7

Esteban MA, Dewey CW, Schwark WS, Rishniw M, Boothe DM (2018) Pharmacokinetics of single-dose oral pregabalin administration in normal cats. Front Vet Sci 5:136. https://doi.org/10.3389/fvets.2018.00136

Fantinati M, Trnka J, Signor A, Dumond S, Jourdan G, Verwaerde P, Priymenko N (2020) Appetite-stimulating effect of gabapentin vs mirtazapine in healthy cats post-ovariectomy. J Feline Med Surg 22(12):1176–1183. https://doi.org/10.1177/1098612X20916391

Goich M, Bascuñán A, Faúndez P, Valdés A (2019) Multimodal analgesia for treatment of allodynia and hyperalgesia after major trauma in a cat. J Feline Med Surg Open Rep 5(1):205511691985580. https://doi.org/10.1177/2055116919855809

Guedes AGP, Meadows JM, Pypendop BH, Johnson EG, Zaffarano B (2018) Assessment of the effects of gabapentin on activity levels and owner-perceived mobility impairment and quality of life in osteoarthritic geriatric cats. J Am Vet Med Assoc 253(5):579–585. https://doi.org/10.2460/javma.253.5.579

Gurney M, Gower L (2022) Randomised clinical trial evaluating the effect of a single preappointment dose of gabapentin on signs of stress in hyperthyroid cats. J Feline Med Surg 24(6):e85–e89. https://doi.org/10.1177/1098612X221091736

Hermans M, Charalambous M, Pakozdy A, Eisl-Glantschnigg U, Neßler J, Van Meervenne SA, Serrano G, Cornelis I, Van Ham L, Paepe D, Broeckx BJ, Bhatti SF (2022) Evaluation of the effect of phenobarbital administration on the biochemistry profile, with a focus on serum liver values, in epileptic cats. J Feline Med Surg 24(6):530–538. https://doi.org/10.1177/1098612X211037431

Hudec CP, Griffin CE (2020) Changes in the stress markers cortisol and glucose before and during intradermal testing in cats after single administration of pre-appointment gabapentin. J Feline Med Surg 22(2):138–145. https://doi.org/10.1177/1098612X19830501

Jerzsele Á, Karancsi Z, Pászti-Gere E, Sterczer Á, Bersényi A, Fodor K, Szabó D, Vajdovich P (2018) Effects of p.o. administered xylitol in cats. J Vet Pharmacol Ther 41(3):409 414. https://doi.org/10.1111/jvp.12479

Korff CP, Williamson BG (2020) Clinical presentation of Chiari-like malformation in 2 Persian cats. Top Companion Anim Med 41:100460. https://doi.org/10.1016/j.tcam.2020.100460

Kruszka M, Graff E, Medam T, Masson S (2021) Clinical evaluation of the effects of a single oral dose of gabapentin on fear-based aggressive behaviors in cats during veterinary examinations. J Am Vet Med Assoc 259(11):1285–1291. https://doi.org/10.2460/javma.20.06.0307

Lamminen T, Doedée A, Hyttilä-Hopponen M, Kaskinoro J (2022) Pharmacokinetics of single and repeated oral doses of pregabalin oral solution formulation in cats. J Vet Pharmacol Ther 45(4):385–391. https://doi.org/10.1111/jvp.13061

Lamminen T, Korpivaara M, Aspegrén J, Palestrini C, Overall KL (2023) Pregabalin alleviates anxiety and fear in cats during transportation and veterinary visits-a clinical field study. Animals (Basel) 13(3):371. https://doi.org/10.3390/ani13030371

Li M, Wu Y, Chen H, Xu X, Peng H, Wei B, Zhu Y, Yang Z (2024) Effect of oral administration of pregabalin on physiological and echocardiographic variables in healthy cats. J Feline Med Surg 26(7):1098612X241250245. https://doi.org/10.1177/1098612X241250245

Liatis T, Cherubini GB (2021) Hypersexuality responsive to phenobarbital in a male neutered domestic shorthair cat. J Vet Behav 41:7–11. https://doi.org/10.1016/j.jveb.2020.10.006

Lowrie M, Thomson S, Bessant C, Sparkes A, Harvey RJ, Garosi L (2017) Levetiracetam in the management of feline audiogenic reflex seizures: a randomised, controlled, open-label study. J Feline Med Surg 19(2):200–206. https://doi.org/10.1177/1098612X15622806

Madan RD, Cenani A, Montgomery E, Azevedo T, Vernau KM, Brosnan RJ (2023) Pregabalin produces similar effects as gabapentin for preanesthetic sedation in cats. J Am Vet Med Assoc 262(3):359–363. https://doi.org/10.2460/javma.23.09.0493

Malreddy PR, Coetzee JF, KuKANICH B, Gehring R (2013) Pharmacokinetics and milk secretion of gabapentin and meloxicam co-administered orally in Holstein-Friesian cows. J Vet Pharmacol Ther 36(1):14–20. https://doi.org/10.1111/j.1365-2885.2012.01384.x

Marsh O, Corsini G, Van Dijk J, Gutierrez-Quintana R, De Risio L (2021) Prevalence and clinical characteristics of phenobarbitone-associated adverse effects in epileptic cats. J Feline Med Surg 23(2):59–66. https://doi.org/10.1177/1098612X20924925

Moore SA, Muñana KR, Papich MG, Nettifee-Osborne JA (2011) The pharmacokinetics of levetiracetam in healthy dogs concurrently receiving phenobarbital. J Vet Pharmacol Ther 34(1):31–34. https://doi.org/10.1111/j.1365-2885.2010.01188.x

Orion Corporation Orion Pharma (2021) BONQAT-RCP

Pankratz KE, Ferris KK, Griffith EH, Sherman BL (2018) Use of single-dose oral gabapentin to attenuate fear responses in cage-trap confined community cats: a double-blind, placebo-controlled field trial. J Feline Med Surg 20:535–543

Papich MG (2016) Saunders handbook of veterinary drugs: small and large animal, 4th edn. Elsevier

Peterson JL, Couto CG, Wellman ML (1995) Hemostatic disorders in cats: a retrospective study and review of the literature. J Vet Intern Med 9(5):298–303. https://doi.org/10.1111/j.1939-1676.1995.tb01088.x

Petrere JA, Anderson JA (1994) Developmental toxicity studies in mice, rats, and rabbits with the anticonvulsant gabapentin. Toxicol Sci 23(4):585–589. https://doi.org/10.1093/toxsci/23.4.585

Quimby JM, Lorbach SK, Saffire A, Kennedy A, Wittenburg LA, Aarnes TK, Creighton KJ, Jones SE, Paschall RE, King EM, Bruner CE, Wallinger JN, Van Haaften KA (2022) Serum concentrations of gabapentin in cats with chronic kidney disease. J Feline Med Surg 24(12):1260–1266. https://doi.org/10.1177/1098612X221077017

Rodan I, Heath S (2016) Feline behavioral health and welfare. Elsevier

Rode F, Broløs T, Blackburn-Munro G, Bjerrum OJ (2006) Venlafaxine compromises the antinociceptive actions of gabapentin in rat models of neuropathic and persistent pain. Psychopharmacology

187(3):364–375. https://doi.org/10.1007/s00213-006-0430-3
Rusbridge C (2024) Neuropathic pain in cats: mechanisms and multimodal management. J Feline Med Surg 26(5):1098612X241246518. https://doi.org/10.1177/1098612X241246518
Rusbridge C, Heath S, Gunn-Moore DA, Knowler SP, Johnston N, McFadyen AK (2010) Feline orofacial pain syndrome (FOPS): a retrospective study of 113 cases. J Feline Med Surg 12(6):498–508. https://doi.org/10.1016/j.jfms.2010.03.005
Ruviaro Tuleski GL, Silveira MF, Bastos RF, Pscheidt MJGR, Prieto WDS, Sousa MG (2022) Behavioral and cardiovascular effects of a single dose of gabapentin or melatonin in cats: a randomized, double-blind, placebo-controlled trial. J Feline Med Surg 24(12):e524–e534. https://doi.org/10.1177/1098612X221124359
Schjerning O, Rosenzweig M, Pottegård A, Damkier P, Nielsen J (2016) Abuse potential of Pregabalin: a systematic review. CNS Drugs 30(1):9–25. https://doi.org/10.1007/s40263-015-0303-6
Siao KT, Pypendop BH, Ilkiw JE (2010) Pharmacokinetics of gabapentin in cats. Am J Vet Res 71(7):817–821. https://doi.org/10.2460/ajvr.71.7.817
Sills G (2006) The mechanisms of action of gabapentin and pregabalin. Curr Opin Pharmacol 6(1):108–113. https://doi.org/10.1016/j.coph.2005.11.003
Slovak JE, Costa AP (2021) A pilot study of transdermal gabapentin in cats. J Vet Intern Med 35(4):1981–1987. https://doi.org/10.1111/jvim.16137
Steagall PV, Benito J, Monteiro BP, Doodnaught GM, Beauchamp G, Evangelista MC (2018) Analgesic effects of gabapentin and buprenorphine in cats undergoing ovariohysterectomy using two pain-scoring systems: a randomized clinical trial. J Feline Med Surg 20(8):741–748. https://doi.org/10.1177/1098612X17730173
Steagall PV, Monteiro-Steagall BP (2013) Multimodal analgesia for perioperative pain in three cats. J Feline Med Surg 15(8):737–743. https://doi.org/10.1177/1098612X13476033
Stee K, Martlé V, Broeckx BJG, Royaux E, Van Ham L, Bhatti SFM (2017) Imepitoin withdrawal in dogs with idiopathic epilepsy well-controlled with imepitoin and phenobarbital and/or potassium bromide does not increase seizure frequency. Vet J 230:1–5. https://doi.org/10.1016/j.tvjl.2017.10.003
Thomas WB (2010) Idiopathic epilepsy in dogs and cats. Vet Clin N Am Small Anim Pract 40(1):161–179. https://doi.org/10.1016/j.cvsm.2009.09.004
Tremont-Lukats IW, Megeff C, Backonja M-M (2000) Anticonvulsants for neuropathic pain syndromes: mechanisms of action and place in therapy. Drugs 60(5):1029–1052. https://doi.org/10.2165/00003495-200060050-00005
Van Haaften KA, Forsythe LRE, Stelow EA, Bain MJ (2017) Effects of a single preappointment dose of gabapentin on signs of stress in cats during transportation and veterinary examination. J Am Vet Med Assoc 251(10):1175–1181. https://doi.org/10.2460/javma.251.10.1175
Vettorato E, Corletto F (2011) Gabapentin as part of multi-modal analgesia in two cats suffering multiple injuries. Vet Anaesth Analg 38(5):518–520. https://doi.org/10.1111/j.1467-2995.2011.00638.x
Walton-Clark M, Travail V, Best M (2022) Phenobarbital-induced autoimmune haemolytic anaemia, thrombocytopenia and peripheral lymphadenomegaly due to reactive lymphoid hyperplasia in a cat. JFMS Open Rep 8(1):20551169221079238. https://doi.org/10.1177/20551169221079238
Whitworth TL, Quick MW (2001) Upregulation of γ-aminobutyric acid transporter expression: role of alkylated γ-aminobutyric acid derivatives. Biochem Soc Trans 29(6):736–741. https://doi.org/10.1042/bst0290736
Woodward KN (2018) Pharmacovigilance for veterinary medicinal products. In: Marrs T, Woodward K (eds) Regulatory toxicology in the European Union. The Royal Society of Chemistry, pp 243–354. https://doi.org/10.1039/9781782622222-00243
Wyatt SE (2019) Should Levetiracetam or Imepitoin be used in preference as second-line treatment in Pharmacoresistent epileptic cats? Vet Evid 4(2). https://doi.org/10.18849/ve.v4i2.226

Tetracyclic Antidepressants or NaSSAs

Contents

23.1 Mianserin – 638

23.1.1 Description, Chemical Formula, and Brand Names – 638

23.1.2 Pharmacology – 638

23.1.3 Side Effects and Usage Precautions – 638

23.1.4 Clinical Use and Dosage – 639

23.2 Mirtazapine – 639

23.2.1 Description, Chemical Formula, and Brand Names – 639

23.2.2 Pharmacology – 639

23.2.3 Side Effects and Usage Precautions – 640

23.2.4 Clinical Use and Dosage – 642

References – 643

© The Author(s), under exclusive license to Springer Nature Switzerland AG 2026
S. Bleuer-Elsner et al., *Veterinary Psychiatry of the Cat*,
https://doi.org/10.1007/978-3-032-06260-4_23

Tetracyclic antidepressants, also known as noradrenergic and specific serotonergic antidepressants (NaSSA), share the ability to inhibit adrenergic autoreceptors, thereby promoting the release of norepinephrine into the synaptic cleft. Unlike Selective Serotonin Reuptake Inhibitor (SSRIs), NaSSA indirectly increase serotonin release by inhibiting α2-adrenergic autoreceptors rather than directly inhibiting serotonin reuptake. Indeed, the activation of adrenergic heteroreceptors on serotonergic neurons leads to a decrease in serotonin release into the synaptic cleft.

The two primary molecules in this class of antidepressants used in veterinary medicine are mianserin and mirtazapine. Mianserin is prescribed off-label. A transdermal gel containing mirtazapine, *Mirataz*™, has received veterinary approval for the treatment of anorexia in cats.

23.1 Mianserin

23.1.1 Description, Chemical Formula, and Brand Names

Mianserin, chemically known as 5-methyl-2,5-diazatricyclo[13.4.0.02,7.08,13]nonadeca-1(19),8,10,12,15,17-hexaene, is a tetracyclic antidepressant or noradrenergic and specific serotonergic antidepressant (NaSSA). No veterinary-specific formulation is available. Mianserin is available in pharmacies under its generic name in tablet forms of 10, 30, and 60 mg.

23.1.2 Pharmacology

23.1.2.1 Mechanism of Action

Mianserin acts as an antagonist of both α1- and α2-adrenergic receptors, enhancing noradrenergic transmission in the locus coeruleus and cortex. By affecting α2-adrenergic heteroreceptors on serotonergic neurons, it also increases serotonin release (see ► Chap. 16, ► Sect. 16.1.1, Box: "Action of Psychotropics on Autoreceptors and Heteroreceptors,"). However, mianserin's α1-antagonist activity minimizes this effect, making its primary influence the enhancement of noradrenergic transmission (Stahl and Muntner 2021).

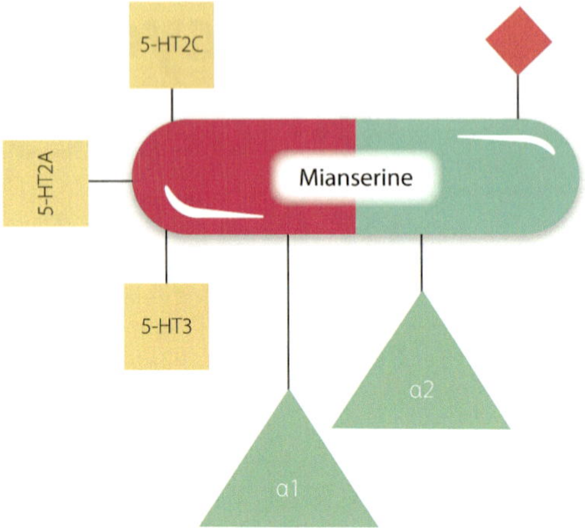

Fig. 23.1 Structure of mianserin. α1- and α2-adrenergic antagonist—5-HT2A, 5-HT2C, and 5-HT3 antagonist—histaminergic H1 antagonist

Additionally, mianserin blocks three serotonergic receptors: 5-HT2A, 5-HT2C, and 5-HT3. Finally, it also antagonizes histaminergic H1 receptors (Fig. 23.1).

23.1.2.2 Metabolism and Involved Cytochromes

The primary hepatic metabolite in rats is 8-hydroxydesmethylmianserin. The cytochromes involved have not been identified.

23.1.2.3 Pharmacokinetics

There is no information available on the pharmacokinetics of oral absorption in cats.

In humans, the half-life has been estimated to range between 38 and 87 h.

23.1.3 Side Effects and Usage Precautions

23.1.3.1 Side Effects

No specific information is available for cats. The main side effects reported in humans are drowsiness and dizziness. In humans, mianserin may lower the epileptogenic threshold, necessitating caution in cases of renal or hepatic insufficiency.

In cats, the molecule acts very quickly and can be used for acute depressive states in kittens. Although detailed pharmacokinetic studies in cats are lacking, clinical observations suggest rapid efficacy, with visible effects observed in less than 24 h.

23.1.3.2 Precautions and Contraindications

Mianserin induces a strong reduction of inhibition and a resurgence of activity. It is highly orexigenic, and prolonged use may lead to weight gain. Caution is advised when using it in older cats due to its disinhibitory effects.

No known teratogenic effects have been reported. However, the molecule is excreted in milk and is therefore not recommended during lactation.

23.1.3.3 Drug Interactions

In the absence of specific data on drug interactions, caution is advised when combining mianserin with other psychotropics or medications that affect the central nervous system.

23.1.3.4 Toxicity

The LD_{50} in rats has been determined to be 780 mg/kg (Drugbank Online, "Mianserin," 2024). No toxicity data are currently available for cats.

23.1.4 Clinical Use and Dosage

Mianserin is an older molecule and is less commonly used than mirtazapine. It is indicated for the treatment of acute depressive states and persistent anxiety, particularly when inhibition is severe. Mianserin can enhance exploratory behavior and counteract anorexia.

The recommended dose is 1–5 mg/kg, administered once or twice daily.

23.2 Mirtazapine

23.2.1 Description, Chemical Formula, and Brand Names

Mirtazapine, chemically known as *5-methyl-2,5,19-triazatricyclo[13.4.0.02,7.08,13] nonadeca-1(15),8,10,12,16,18-hexaene*, is a tetracyclic antidepressant or noradrenergic and specific serotonergic antidepressant (NaSSA).

The transdermal gel *Mirataz*™ has been approved for the treatment of anorexia.

In human medicine, the molecule is marketed under the names *Remeron*™, *Norset*™, and as a generic in 15, 30, and 45 mg tablets, depending on the country.

23.2.2 Pharmacology

23.2.2.1 Mechanism of Action

Mirtazapine acts as an antagonist at presynaptic α2-adrenergic inhibitory autoreceptors, leading to increased noradrenergic activity.

It also antagonizes heteroreceptors in the central nervous system, thereby enhancing serotonergic activity (see ▶ Chap. 16, ▶ Sect. 16.1.1, Box: "Action of Psychotropics on Autoreceptors and Heteroreceptors"). These effects are known to improve depressive symptoms in humans.

Mirtazapine is also a potent antagonist of serotonin receptors 5-HT2A, 5-HT2C, and 5-HT3.

In addition, mirtazapine acts as a peripheral α1-adrenergic antagonist, which may explain the episodes of orthostatic hypotension reported after its use.

It is a strong antagonist of histaminergic H1 receptors, contributing to its sedative effects. Finally, mirtazapine also has analgesic properties, possibly explained by its effects on opioid receptors (◘ Fig. 23.2).

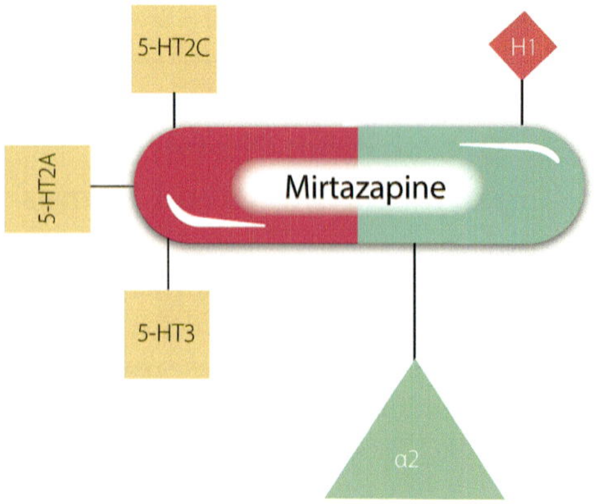

Fig. 23.2 Structure of mirtazapine. Antagonist of α2-adrenergic, 5-HT2A, 5-HT2C, 5-HT3, and H1 receptors

23.2.2.2 Metabolism and Involved Cytochromes

Mirtazapine is metabolized through multiple pathways, which vary between species. In cats, three primary hepatic metabolites have been identified: 8-OH mirtazapine, dimethyl-mirtazapine, and 8-OH mirtazapine-glucuronide (Fitzpatrick et al. 2018). In humans, the main cytochromes involved are CYP2D6, CYP1A2, and CYP3A4.

23.2.2.3 Pharmacokinetics

The half-life of orally administered mirtazapine in cats has been estimated at 10–16 h for doses between 1.88 and 3.75 mg (Quimby et al. 2011b). The clearance rate is 10–18 mL/kg/min, meaning the drug could theoretically be administered every 24 h. No accumulation was detected after daily administration for 6 days (Quimby et al. 2011b).

For transdermal administration, the half-life is 21–27 h. In cases of licking, the half-life is approximately 10 h, similar to oral administration. The bioavailability of mirtazapine is 65%, with the plasma concentration peak reached at 16 h for a single dose and 2 h for daily administration over 14 days (Buhles et al. 2018).

Data on the elimination pathways of mirtazapine in cats are lacking. In humans, 75% of the drug is excreted in urine and 15% in feces. Renal and hepatic insufficiencies significantly reduce the clearance of the molecule and prolong its half-life (Fitzpatrick et al. 2018). In cases of chronic kidney disease, it is recommended to extend the interval between oral administrations to every 48 h (Quimby et al. 2011a).

23.2.3 Side Effects and Usage Precautions

23.2.3.1 Side Effects

The most frequently observed side effects include vocalization, agitation, vomiting, and apathy (Quimby et al. 2011b). A reversible increase in alanine aminotransferase may occur, resolving after discontinuation of treatment (Fitzpatrick et al. 2018).

Due to the risk of serotonin syndrome, it is critical to monitor symptoms such as tremors, fever, or agitation, particularly when mirtazapine is combined with other serotonergic agents.

For transdermal administration, local reactions such as crusting, scaling, alopecia, dermatitis, irritation, and erythema are reported in 10% of cats (Ferguson et al. 2016).

Long-term use of the molecule may lead to significant weight gain. Owners should be informed about potential dietary adjustments to mitigate this effect.

23.2.3.2 Precautions and Contraindications

Treatment of patients with renal or hepatic insufficiency may require lower doses of mirtazapine and close monitoring during its administration, including transdermal application.

In humans, very rare cases of blood dyscrasias have been reported in patients with pre-existing hematological diseases, making mirtazapine contraindicated in cases of hematological abnormalities (Stübner et al. 2004).

Caution is advised when using mirtazapine in older cats due to its disinhibitory effects.

While the effects of antidepressants on glucose metabolism are well documented in humans and may affect blood glucose levels through indirect effects on the endocrine system, specific data for cats are limited.

Monitoring glucose levels in diabetic cats receiving mirtazapine is recommended.

Particularly for transdermal administration, wearing gloves is mandatory, and the animal must not be touched by people or other animals for 12 h after application. The product should not be used in cats intended for reproduction, pregnant or lactating queens, kittens under 7.5 months, or cats weighing less than 2 kg.

No teratogenic effects have been reported. In humans, mirtazapine is excreted in very small amounts into breast milk, so its oral use is not recommended in lactating queens (Kristensen et al. 2007).

23.2.3.3 Drug Interactions

The following drug combinations are *prohibited:*

- Monoamine oxidase inhibitors (MAOIs) (e.g., selegiline, amitraz): concurrent use with mirtazapine increases the risk of serotonin syndrome. Simultaneous administration or use within 14 days after discontinuing an MAOI is contraindicated.
- SSRIs (e.g., fluoxetine, fluvoxamine): concurrent use with mirtazapine increases the risk of serotonin syndrome.
- Venlafaxine: concurrent use with mirtazapine increases the risk of serotonin syndrome.
- Tricyclic antidepressants (e.g., amitriptyline, clomipramine): concurrent use with mirtazapine increases the risk of serotonin syndrome.
- Tramadol: concurrent use with mirtazapine increases the risk of serotonin syndrome.

The following drug interactions have been reported in humans or animals receiving mirtazapine. While coadministration is not necessarily contraindicated, potential risks should be assessed, and additional monitoring implemented as needed.

- Clonidine: concomitant use with mirtazapine may increase blood pressure.
- Benzodiazepines (e.g., diazepam, midazolam): concomitant use has minimal effects on mirtazapine blood levels but may further impair motor abilities.
- Buprenorphine: concomitant use with mirtazapine increases the risk of serotonin syndrome.
- Butorphanol: concomitant use with mirtazapine increases the risk of serotonin syndrome.
- Cimetidine: may increase plasma levels of mirtazapine.
- CNS depressants (e.g., antihistamines, opioids): concomitant use with mirtazapine may increase the risk of sedation and CNS depression.
- Cyproheptadine: may negate the effects of mirtazapine.
- Erythromycin: may increase plasma levels of mirtazapine.
- Ketoconazole: may increase plasma levels of mirtazapine.
- Various serotonergic agents (e.g., buspirone, lithium, metoclopramide, ondansetron): concomitant use with mirtazapine increases the risk of serotonin syndrome.
- Warfarin: may prolong prothrombin time when used with mirtazapine.

23.2.3.4 Toxicity

The LD50 in mice has been estimated at 830 mg/kg, but it remains unknown for cats.

Ferguson et al. (2016) studied the effects of mirtazapine intoxication in cats, reporting the following signs in order of frequency: vocalizations (56%), agitation (31%), vomiting (26%), abnormal gait or ataxia (17%), insomnia (14%), tremors (14%), hypersalivation (13%), tachypnea (12%), tachycardia (11%), and lethargy (11%). Less commonly observed were anorexia, disorientation, dyspnea, hypothermia, mydriasis, hypertension, discomfort, seizures, and weakness. The ingested doses ranged from 1.88 mg (1 cat) to 15 mg (40 cats).

In addition, serotonin syndrome is a possible complication. Activated charcoal is recommended alongside other standard monitoring procedures in cases of overdose (Drug Monographs Plumb's Veterinary Drugs, 2024).

23.2.4 Clinical Use and Dosage

23.2.4.1 According to the Scientific Literature

Mirtazapine is widely used in cats. *Mirataz™*, a transdermal ointment, is approved for use in cats as an appetite stimulant (Williams 2020). It is also used as an antiemetic (Quimby and Lunn 2013). The recommended dose is 2 mg of ointment per cat once daily, corresponding to 0.1 g of ointment or a 3.8 cm line applied to the inner pinna of one ear. Alternating ears is advised to prevent skin reactions, including necrosis.

In cats with stage 2 or 3 chronic kidney disease, mirtazapine promotes appetite recovery and weight gain. In such cases, it is administered every other day via the transdermal route at a dose of 2 mg per cat or orally at a dose of 0.5–1 mg/kg once daily.

In veterinary psychiatry, mirtazapine is more commonly used than mianserin due to its stronger anxiolytic and antidepressant properties. It facilitates disinhibition, increased activity and exploratory behavior, enhanced appetite, and improved sleep. Mirtazapine is used for acute depression and persistent anxiety. The recommended dose is 0.5–1 mg/kg or 1.88 mg per cat once daily (Overall 2015; Rodan and Heath 2016; Stahl and Muntner 2021).

23.2.4.2 Experience Feedback

Mirtazapine is effective in managing debilitating complex phobias, inhibited forms of generalized anxiety, and acute or chronic depression. It helps to reinitiate behaviors, reduce fears, and significantly improve food motivation—improvements that can subsequently support behavioral therapies. The recommended dose is 0.5–1 mg/kg once daily, reduced to 0.5–1 mg/kg every 2–3 days in cases of chronic kidney disease.

23.2.4.3 Possible Combinations

Combining mirtazapine with gabapentinoids can be beneficial for controlling phobic and anxious states where behavioral inhibition is significant.

23.2.4.4 Withdrawal

In humans, abrupt discontinuation of mirtazapine after long-term administration has led to withdrawal symptoms such as nausea, headaches, and malaise.

In cats, gradual tapering of the medication is advised to minimize potential withdrawal effects.

Summary Table of Tetracyclic NaSSAs

Substance	Mechanism of action	Pathological states for which the molecule is indicated	Dosage	Cytochromes
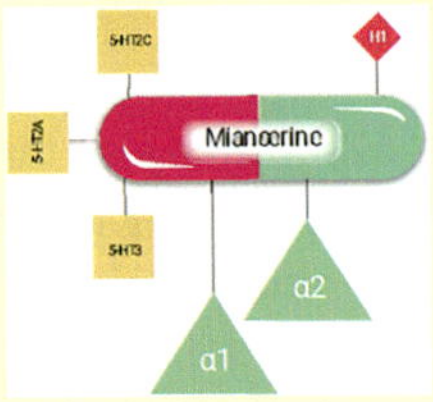	α1-adrenergic antagonist	Acute depressive Inhibited anxious	1–5 mg/kg SID to BID	–
	α2-adrenergic antagonist			
	5-HT2A antagonist			
	5-HT2C antagonist			
	5-HT3 antagonist			
	Histaminergic H1 antagonist			
	α2-adrenergic antagonist	Acute or chronic depressive inhibited anxious	1,88 mg/cat BID	CYP2D6 CYP1A2

Substance	Mechanism of action	Pathological states for which the molecule is indicated	Dosage	Cytochromes
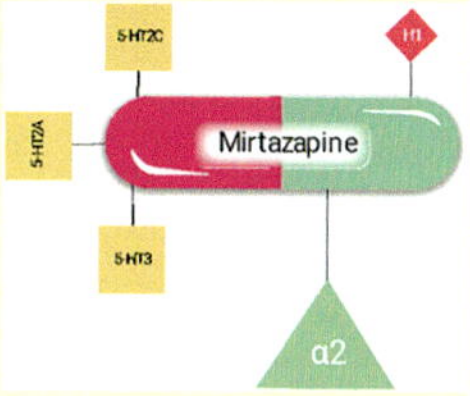	5-HT2A antagonist			CYP3A4
	5-HT2C antagonist			
	5-HT3 antagonist			
	Histaminergic H1 antagonist			

References

Buhles W, Quimby JM, Labelle D, Williams VS (2018) Single and multiple dose pharmacokinetics of a novel mirtazapine transdermal ointment in cats. J Vet Pharmacol Ther 41(5):644–651. https://doi.org/10.1111/jvp.12691

Ferguson LE, McLean MK, Bates JA, Quimby JM (2016) Mirtazapine toxicity in cats : retrospective study of 84 cases (2006–2011). J Feline Med Surg 18(11):868–874. https://doi.org/10.1177/1098612X15599026

Fitzpatrick RL, Quimby JM, Benson KK, Ramirez D, Sieberg LG, Wittenburg LA, Gustafson DL (2018) In vivo and in vitro assessment of mirtazapine pharmacokinetics in cats with liver disease. J Vet Intern Med 32(6):1951–1957. https://doi.org/10.1111/jvim.15237

Kristensen JH, Ilett KF, Rampono J, Kohan R, Hackett LP (2007) Transfer of the antidepressant mirtazapine into breast milk. Br J Clin Pharmacol 63(3):322–327. https://doi.org/10.1111/j.1365-2125.2006.02773.x

Overall KL (2015) Manual of clinical behavioral medicine for dogs and cats ([Enhanced Credo edition]). Elsevier, Credo Reference

Quimby JM, Gustafson DL, Lunn KF (2011a) The pharmacokinetics of mirtazapine in cats with chronic kidney disease and in age-matched control cats. J Vet Intern Med 25(5):985–989. https://doi.org/10.1111/j.1939-1676.2011.00780.x

Quimby JM, Gustafson DL, Samber BJ, Lunn KF (2011b) Studies on the pharmacokinetics and pharmacodynamics of mirtazapine in healthy young cats. J Vet Pharmacol Ther 34(4):388–396. https://doi.org/10.1111/j.1365-2885.2010.01244.x

Quimby JM, Lunn KF (2013) Mirtazapine as an appetite stimulant and anti-emetic in cats with chronic kidney disease: a masked placebo-controlled crossover clinical trial. Vet J 197(3):651–655. https://doi.org/10.1016/j.tvjl.2013.05.048

Rodan I, Heath S (2016) Feline behavioral health and welfare. Elsevier

Stahl SM, Muntner N (2021) In: Grady MM (ed) Stahl's essential psychopharmacology: neuroscientific basis and practical applications, 5th edn. Cambridge University Press. https://doi.org/10.1017/9781108975292

Stübner S, Grohmann R, Engel R, Bandelow B, Ludwig W-D, Wagner G, Müller-Oerlinghausen B, Möller H-J, Hippius H, Rüther (2004) Blood Dyscrasias induced by psychotropic drugs. Pharmacopsychiatry 37:70–78. https://doi.org/10.1055/s-2004-815513

Williams VS (2020) Mirtazapine pharmacokinetic data and daily dosing

Neuroleptics or Antipsychotics

Contents

24.1 General Characteristics of All Neuroleptics – 646

24.1.1 Description – 646

24.1.2 General Pharmacology of Neuroleptics – 646

24.1.3 Side Effects and Usage Precautions – 652

24.1.4 Clinical Use and Dosage – 654

References – 656

© The Author(s), under exclusive license to Springer Nature Switzerland AG 2026
S. Bleuer-Elsner et al., *Veterinary Psychiatry of the Cat*,
https://doi.org/10.1007/978-3-032-06260-4_24

24

Neuroleptics, also known as antipsychotics, encompass a group of molecules with a broad range of pharmacological effects and a wide variety of side effects. In humans, they are used to treat most forms of psychosis, such as schizophrenia or bipolar disorders.

In veterinary medicine, all neuroleptics are prescribed off-label. While their use in cats is very rare, they could theoretically be considered for treating severe disorders, such as compulsive states resistant to conventional treatments, dissociative syndrome, or dysthymia. It is important to note that their use remains off-label in veterinary medicine.

24.1 General Characteristics of All Neuroleptics

24.1.1 Description

In veterinary medicine, neuroleptics could theoretically be used for various psychopathological states requiring motor activity control. They have the potential to suppress conditioned responses in animals (Crowell-Davis et al. 2019).

24.1.2 General Pharmacology of Neuroleptics

All neuroleptics share the same primary action of blocking D2 dopaminergic receptors. However, two classes of neuroleptics can be distinguished: classical and atypical. Atypical neuroleptics are more recent, exhibit a slightly different mode of action, and, most importantly, cause fewer side effects (Stahl and Muntner 2021).

The dopaminergic properties of neuroleptics suggest a strong theoretical indication for treating compulsive, dissociative, manic, and bipolar states, as well as addictive behaviors.

24.1.2.1 Mechanism of Action

▪ **Classical Neuroleptics**

Classical neuroleptics were the first to be discovered in the 1950s, with the identification of the antipsychotic effects of chlorpromazine. Their key characteristic is their primary function as antagonists of dopamine D2 receptors. This blockade affects all four dopaminergic pathways (see ► Chap. 2, ► Sect. 2.2.4.3).

Dopaminergic suppression in the mesolimbic and mesocortical pathways underlies the improvement of positive symptoms of schizophrenia, such as hallucinations and behavioral disorganization. Patients become calmer and less responsive to their environment (◘ Fig. 24.1).

However, dopaminergic suppression in the nigrostriatal and tuberoinfundibular pathways contributes to the side effects of classical neuroleptics. Suppression in the nigrostriatal pathway leads to extrapyramidal symptoms, characterized by involuntary movements and, in severe cases, irreversible tardive dyskinesia. Suppression in the tuberoinfundibular pathway causes increased prolactin levels, which may result in gynecomastia and galactorrhea, even in males (◘ Fig. 24.2).

All classical neuroleptics also possess antimuscarinic, antihistaminic, and α1-adrenergic antagonist properties.

The antihistaminic properties of classical neuroleptics may cause reduced alertness and weight gain. Their α1-adrenergic antagonist properties can lead to hypotension and fatigue.

Neuroleptics with strong antimuscarinic properties are less likely to cause extrapyramidal symptoms. This is because the dopaminergic nigrostriatal pathway inhibits cholinergic pathways. The increased cholinergic activity caused by dopaminergic suppression from neuroleptics is partially offset by their antimuscarinic action.

Each classical neuroleptic has specific affinities for muscarinic, histaminic, and adrenergic receptors. As a result, they may vary in the intensity of side effects they induce while providing similar therapeutic effects (◘ Fig. 24.3).

An outdated classification of classical neuroleptics categorized them into two groups: low-potency neuroleptics, which require high doses to be effective and are associated with more side effects, particularly sedation (e.g., acepromazine, chlorpromazine); and high-

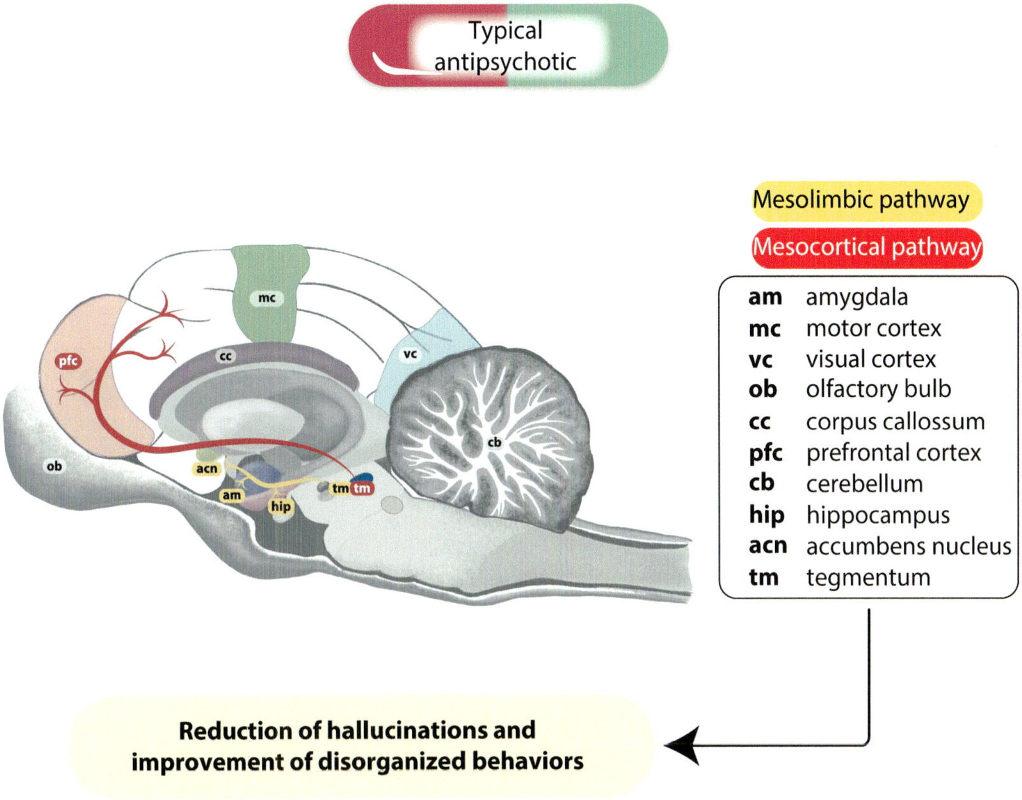

Fig. 24.1 Beneficial effects of neuroleptics on the mesolimbic and mesocortical pathways. By modulating dopaminergic transmission in these pathways, classical neuroleptics reduce hallucinations and improve disorganized behaviors

potency neuroleptics, which require lower doses and cause fewer side effects (e.g., fluphenazine).

Atypical Neuroleptics

Atypical neuroleptics were discovered later than classical neuroleptics, starting with clozapine in the 1960s. Their distinguishing feature is that they achieve the same therapeutic effects as classical neuroleptics but with significantly fewer extrapyramidal and hormonal side effects. This difference is attributed to their antagonistic action on serotonergic receptors, particularly 5-HT2A, and in some cases, 5-HT1A (Fig. 24.4).

Serotonin reduces dopamine release in the striatum through its action on 5-HT2A receptors. Atypical neuroleptics, by blocking 5-HT2A receptors, increase dopamine release while simultaneously blocking its action. The released dopamine and the neuroleptic then compete for D2 dopaminergic receptors, modulating the neuroleptic's effect. It is as though the atypical neuroleptic creates its own competitor (Fig. 24.5).

The occupancy of D2 receptors in the nigrostriatal pathway decreases from 80% with a classical neuroleptic to approximately 60% with an atypical neuroleptic. This lower occupancy is sufficient to eliminate extrapyramidal symptoms.

In the tuberoinfundibular pathway, the mechanism is slightly different. Serotonin, by stimulating 5-HT2A receptors, increases prolactin secretion, whereas dopamine inhibits prolactin secretion by stimulating D2 receptors.

Atypical neuroleptics simultaneously increase prolactin secretion by blocking D2 receptors and decrease it by blocking 5-HT2A receptors. This dual action modulates prolac-

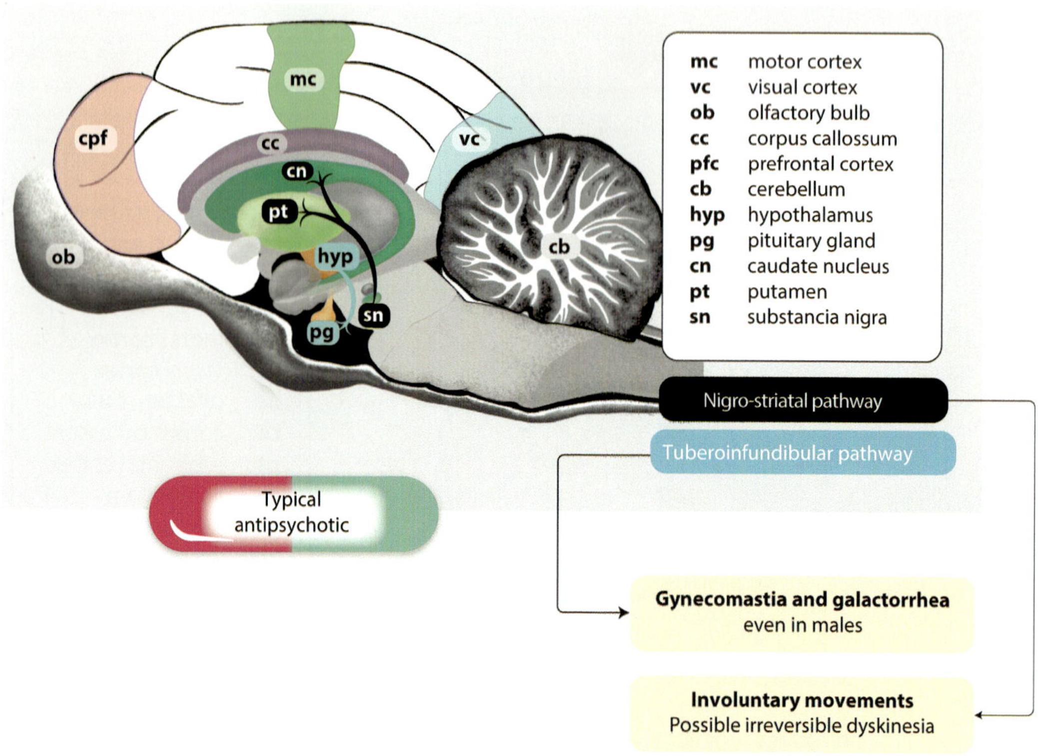

Fig. 24.2 Side effects of neuroleptics on the nigrostriatal and tuberoinfundibular pathways. Action on the nigrostriatal pathway leads to involuntary movements and, in severe cases, irreversible tardive dyskinesia. Action on the tuberoinfundibular pathway can result in gynecomastia and galactorrhea, even in males

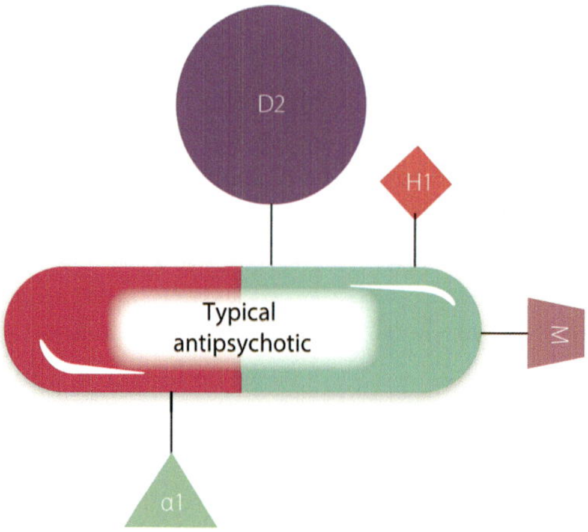

Fig. 24.3 Structure of a typical neuroleptic: D2 antagonist—muscarinic antagonist—histaminic H1 antagonist—α1-adrenergic antagonist

tin secretion, with varying levels of success depending on the specific neuroleptic (Fig. 24.6).

The 5-HT2A blocking action of atypical neuroleptics has less influence on the mesolimbic and mesocortical pathways, likely due to the reduced presence of 5-HT2A heteroreceptors in these areas. Consequently, atypical neuroleptics exhibit the same effects as classical neuroleptics on these pathways.

Atypical neuroleptics also act as partial agonists at 5-HT1A receptors. Their action on postsynaptic 5-HT1A receptors indirectly increases dopamine release. Activation of these postsynaptic 5-HT1A receptors inhibits glutamate neurons, which in turn cease stimulating Gamma-Aminobutyric Acid (GABA) neurons that normally inhibit dopaminergic neurons. This cascade of events results in increased dopamine release (Fig. 24.7).

Their action on presynaptic 5-HT1A receptors reduces serotonin release (via inhibitory autoreceptors), which in turn alleviates

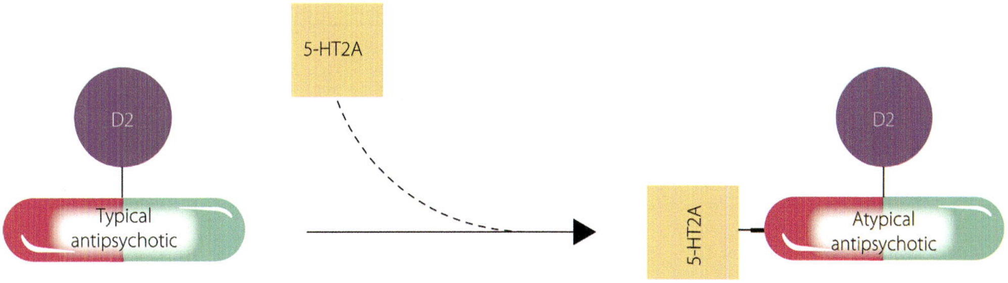

Fig. 24.4 Comparison of the structures of classical neuroleptics and atypical neuroleptics. The key distinction between the two types of neuroleptics lies in their antagonistic or inverse agonistic valence at 5-HT2A serotonergic receptors

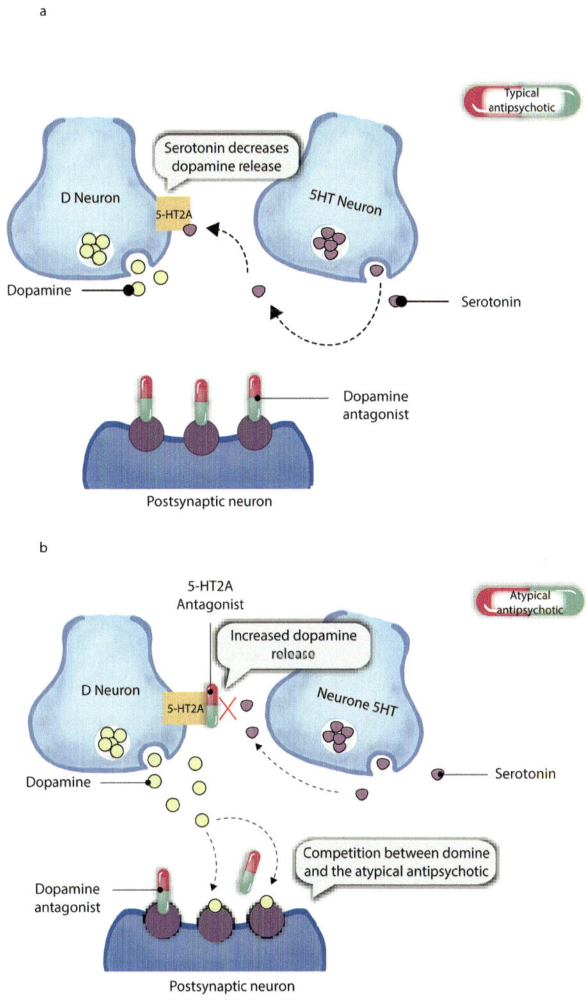

Fig. 24.5 Blocking 5-HT2A receptors in the nigrostriatal pathway by neuroleptics. **a** Classical neuroleptics do not act on 5-HT2A receptors. Serotonin reduces dopamine release through 5-HT2A receptors, and the neuroleptic occupies most of the action sites (80% or more), leading to undesirable extrapyramidal symptoms. **b** Atypical neuroleptics block 5-HT2A receptors, increasing dopamine release, which then competes with the atypical neuroleptic for D2 receptors. The neuroleptic occupies a smaller proportion of receptors (around 60%), effectively eliminating extrapyramidal symptoms

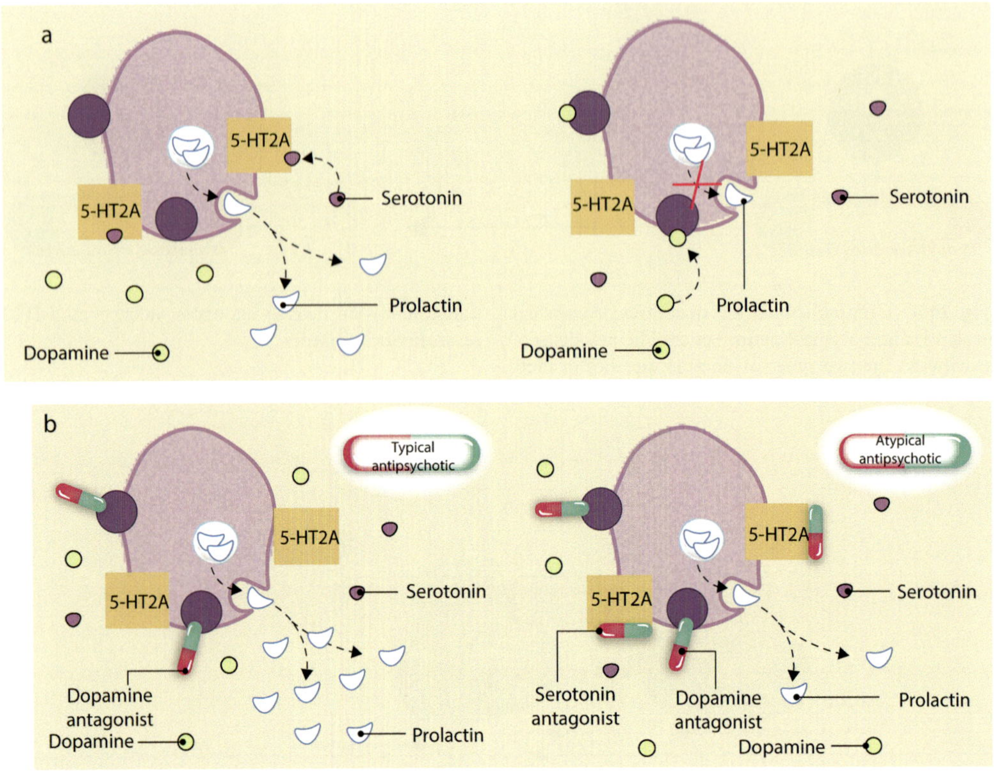

Fig. 24.6 Blocking 5-HT2A receptors in the tuberoinfundibular pathway. **a** Activation of 5-HT2A receptors increases prolactin secretion, while activation of D2 receptors inhibits it. **b** Classical neuroleptics, through their antidopaminergic action alone, increase prolactin secretion. In contrast, atypical neuroleptics, by acting on both receptor types simultaneously, limit prolactin elevation and associated side effects

the dopaminergic suppression mediated by 5-HT2A receptors (Fig. 24.7). By "cutting the brake" associated with 5-HT2A effects, atypical neuroleptics ultimately lead to increased dopamine release.

Atypical neuroleptics also interact with other serotonergic receptors and, like classical neuroleptics, with histaminic, muscarinic, and adrenergic receptors.

The differences observed among various atypical neuroleptics arise from the relative affinity of each molecule for these receptors. This gives each neuroleptic a unique therapeutic profile: some are more sedative, others more anxiolytic, while some exhibit antidepressant properties. However, all share the ability to reduce the positive symptoms of schizophrenia (hallucinations, behavioral disorganization) with relatively few side effects (Fig. 24.8).

24.1.2.2 Metabolism and Involved Cytochromes

The hepatic metabolism rate of neuroleptics is high, and their metabolites are generally inactive. The cytochrome P450 enzymes involved vary depending on the specific molecule.

24.1.2.3 Pharmacokinetics

The effects of neuroleptics manifest within an hour and last between 4 and 24 h, depending on the molecule. Half-lives range from 10 to 30 h in humans. Excretion is primarily urinary.

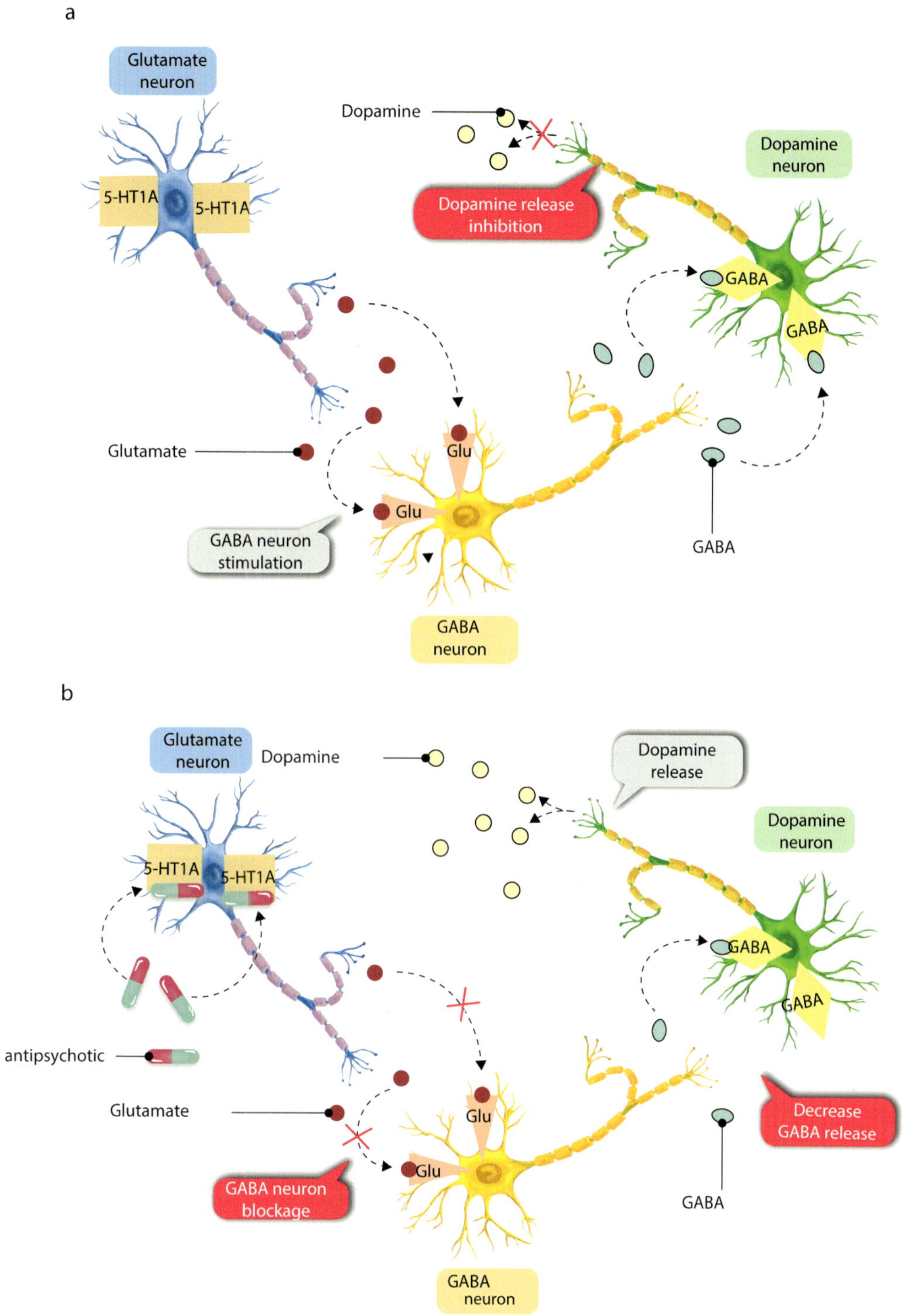

Fig. 24.7 Agonist action on 5-HT1A receptors. **a** The chain of glutamate activator neurons and GABA inhibitor neurons forms a braking system on dopamine release. **b** Neuroleptics, through their partial agonist action on postsynaptic somatic 5-HT1A heteroreceptors, reduce glutamate production, thereby lifting the brake and leading to increased dopamine release

24

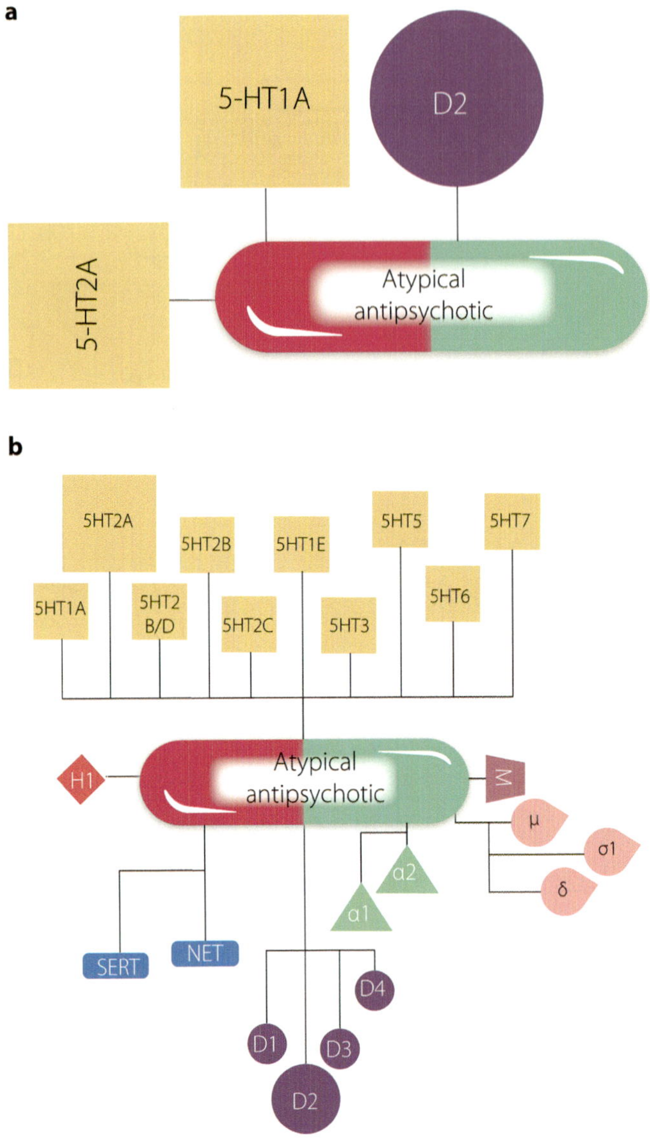

Fig. 24.8 Structure of atypical neuroleptics. D2 antagonist or inverse agonist—5-HT2A antagonist—partial 5-HT1A agonist. These neuroleptics also exhibit affinity for various other receptors. The specific therapeutic profile of each atypical neuroleptic is determined by its relative affinity for each receptor type

24.1.3 Side Effects and Usage Precautions

24.1.3.1 Side Effects

The use of neuroleptics can lead to extrapyramidal symptoms, such as difficulty initiating movement, muscle spasms, and agitation. Their administration also induces environmental indifference and reduced emotional expression. Irreversible dyskinesia is reported in 10–20% of human patients undergoing continuous treatment with classical neuroleptics. Gynecomastia and galactorrhea are also documented but are rarely observed with atypical neuroleptics.

Neuroleptics may cause hypotension and syncope, particularly in brachycephalic animals. They are also known to lower the seizure threshold, induce thrombocytopenia, hyperglycemia, and cause electrocardiogram abnormalities.

24.1.3.2 Precautions and Contraindications

Cats have unique metabolic characteristics, including limited glucuronidation, necessitating extreme caution when considering the use of neuroleptics. This is particularly important for cats with epilepsy, hepatic, renal, or cardiac insufficiency. Neuroleptic use is also discouraged in very young animals, geriatric patients, pregnant or lactating females. Close monitoring is essential to prevent severe side effects.

24.1.3.3 Drug Interactions

The following indications are theoretical or have been reported in humans and animals. This list may vary depending on the specific neuroleptic. Critical drug interactions include:

- Anesthetic agents (e.g., alfaxalone, isoflurane, ketamine): combining neuroleptics with anesthetics increases the risk of central depression.
- Cisapride: using neuroleptics with cisapride increases the risk of QT interval prolongation, cardiac arrhythmias, torsades de pointes, and death. Cisapride should not be used with acepromazine.
- Opioids (e.g., buprenorphine, fentanyl, morphine): neuroleptic dosages should be reduced when used in combination with opioids.

The following drug interactions have been reported in humans or animals receiving neuroleptics. While these combinations are not necessarily contraindicated, the associated risks should be carefully evaluated by the practitioner.

- α-adrenergic agonists (dexmedetomidine, xylazine): combining with neuroleptics increases the risk of depression
- Antipyretics (e.g., acetaminophen, dipyrone, NSAIDs): combining with neuroleptics may increase the risk of hypothermia
- Anticholinergic agents (e.g., atropine, glycopyrrolate, hyoscyamine): combining with neuroleptics increases the risk of anticholinergic effects (e.g., dry mucous membranes, constipation, urinary retention)
- Barbiturates (e.g., pentobarbital, phenobarbital): combining with neuroleptics increases the risk of central depression
- Benzodiazepines (e.g., diazepam, midazolam): combining with neuroleptics increases the risk of central depression
- Cannabinoids: combining with neuroleptics increases the risk of central depression
- Cimetidine: combining with neuroleptics increases the risk of sedation and hypotension
- Dopaminergic agonists (bromocriptine, cabergoline, pergolide): concurrent use reduces the efficacy of both agents
- Emetics (apomorphine, ropinirole): acepromazine may reduce the efficacy of emetics, and combining increases the risk of sedation and hypotension
- Epinephrine: combining with neuroleptics may cause severe hypotension
- Hypotensive agents (amlodipine, benazepril, enalapril, telmisartan): combining with neuroleptics may cause severe hypotension
- Methocarbamol: combining with neuroleptics increases the risk of central depression
- Metoclopramide: combining with neuroleptics increases the risk of extrapyramidal symptoms, such as tremors, catalepsy, rigidity, agitation, sweating, or seizures
- Organophosphates: neuroleptics enhance the toxicity of organophosphate antiparasitics and should not be administered within 1 month of their use
- Propranolol: the concentrations of both neuroleptics and propranolol may increase when administered together
- Quinidine: combining with neuroleptics may cause cardiac depression

24

24.1.3.4 Toxicity

Neuroleptic malignant syndrome, a rare but potentially fatal condition, characterizes neuroleptic toxicity. This syndrome is characterized by muscular rigidity, hyperthermia, tachycardia, cardiac rhythm disturbances, altered consciousness, coma, elevated liver enzymes and creatine kinase, and leukocytosis. Mortality rates can reach 20–30% in human patients affected by this syndrome. Treatment is symptomatic.

24.1.4 Clinical Use and Dosage

24.1.4.1 According to the Scientific Literature

The table below (◘ Table 24.1) summarizes the theoretical possible uses and associated dosages, if known, for neuroleptics in cats (Boehringer 2022, Crowell-Davis et al. 2019, Verstegen et al. 1996).

◘ **Table 24.1** Summary table of neuroleptics

Substance	Mechanism of action	Usage and dosage	Experience feedback	Cytochromes
Classical neuroleptics				
Acepromazine 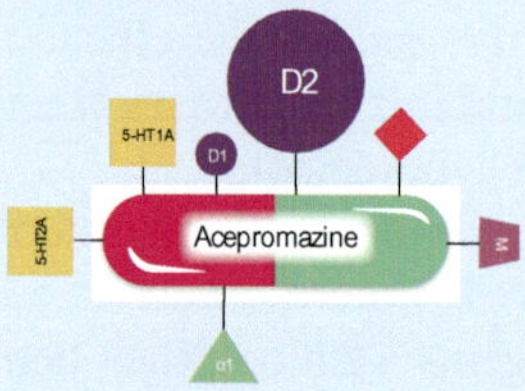 (Calmivet™, Vetranquil™, PromAce™)	**D2 antagonist** D1 antagonist Muscarinic antagonist α1-adrenergic antagonist Histamine antagonist 5-HT1A and 5-HT2A receptor antagonist	Sedation, tranquilization Pruritus Nausea Urethral obstruction Dose: 0.5–1 mg/kg PO 1.1–2.2 mg/kg IV, IM, or SC PRN or TID	Use discouraged	–
Azaperone 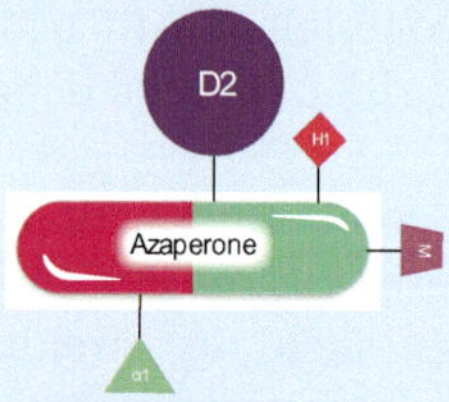 (*Stresnil*™)	**D2 antagonist** α1-adrenergic antagonist Histamine antagonist Muscarinic antagonist	Sedation, tranquilization	Use discouraged	–
Haloperidol 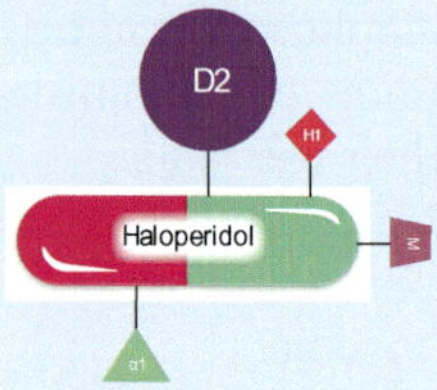 (*Haldol*™)	**D2 antagonist** α1-adrenergic antagonist Histamine antagonist Muscarinic antagonist	Impulsive, compulsive manic, bipolar dissociated	Not currently used	–

(continued)

Table 24.1 (continued)

<table>
<tr><th>Substance</th><th>Mechanism of action</th><th>Usage and dosage</th><th>Experience feedback</th><th>Cytochromes</th></tr>
<tr><td>Pimamperone
(Dipiperon™)

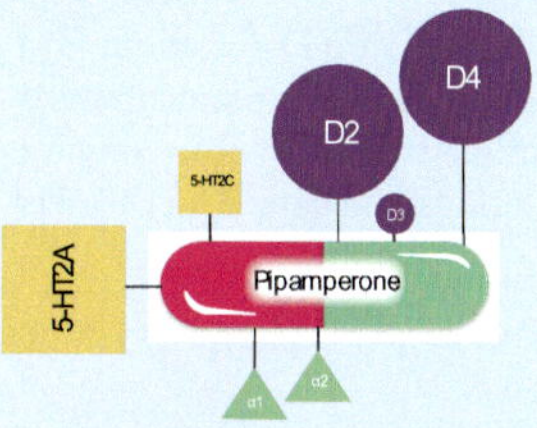

</td><td>D2 antagonist
D3 antagonist
D4 antagonist
5-HT2A antagonist
5-HT2C antagonist
α1- and α2-adrenergic antagonist</td><td>Impulsive, compulsive manic, bipolar dissociated</td><td>Not currently used</td><td>–</td></tr>
<tr><td>Sulpiride

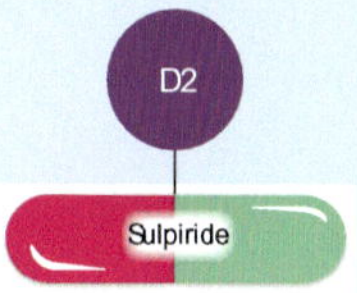

(Dogmatil™)</td><td>D2 antagonist (post and presynaptic)</td><td>Phobic, anxious</td><td>Not currently used</td><td>–</td></tr>
<tr><td>Atypical neuroleptics</td><td></td><td></td><td></td><td></td></tr>
<tr><td>
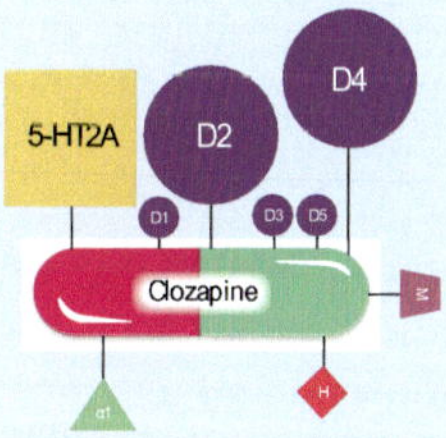

Clozapine (Léponex[ND] ou Clozaril[ND])</td><td>D2 antagonist
5-HT2A antagonist
D4 antagonist
D1, D3, D5 antagonist
α1-adrenergic antagonist
Histamine antagonist
Muscarinic antagonist</td><td>Impulsive, compulsive manic, bipolar dissociated</td><td>Not currently used</td><td>–</td></tr>
<tr><td>Risperidone

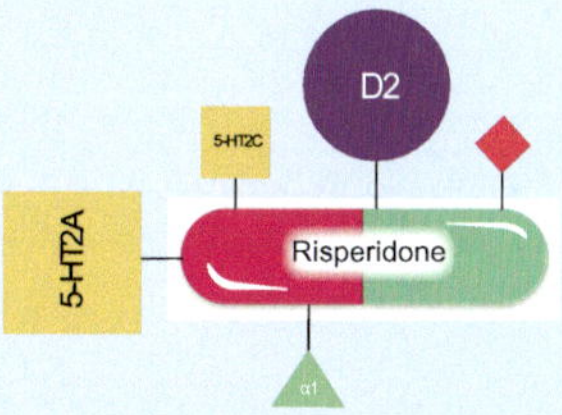

Risperdal™)</td><td>D2 antagonist
5-HT2A antagonist
5-HT2C antagonist
α1-adrenergic antagonist
Histamine antagonist</td><td>Impulsive, compulsive manic, bipolar dissociated</td><td>Not currently used</td><td>CYP2D6</td></tr>
<tr><td>Aripiprazole

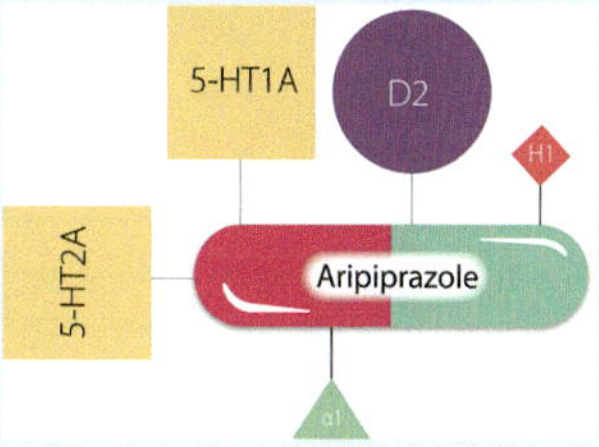

(Abilify™)</td><td>D2 antagonist
5-HT2A antagonist
5-HT1A antagonist
α1-adrenergic antagonist
Histamine antagonist</td><td>Impulsive, compulsive manic, bipolar dissociated</td><td>Not currently used</td><td>CYP2D6
CYP3A4</td></tr>
</table>

24.1.4.2 Experience Feedback

Our limited clinical experience does not allow us to expand on the data presented above.

References

Boehringer Ingelheim Animal Health USA Inc. (2022) PROMACE- acepromazine maleate injection

Crowell-Davis SL, Murray TF, de Souza M, Dantas L (2019) Veterinary psychopharmacology, 2nd edn. Wiley

Stahl SM, Muntner N (2021) In: Grady MM (ed) Stahl's essential psychopharmacology: neuroscientific basis and practical applications, 5th edn. Cambridge University Press. https://doi.org/10.1017/9781108975292

Verstegen J, Deleforge J, Dernblon D, Rosillont D (1996) Non-linear relationship between bioavailability and dose after oral administration of four single doses of acepromazine in dogs and cats. 23(2):47–51

Other Psychotropic Drugs

Contents

25.1 **Venlafaxine – 658**
25.1.1 Description, Chemical Formula, and Brand Names – 658
25.1.2 Pharmacology – 658
25.1.3 Side Effects and Usage Precautions – 659
25.1.4 Clinical Use and Dosage – 660

25.2 **Trazodone – 661**
25.2.1 Description, Chemical Formula, and Brand Names – 661
25.2.2 Pharmacology – 661
25.2.3 Side Effects and Usage Precautions – 662
25.2.4 Clinical Use and Dosage – 663

25.3 **Buspirone – 663**
25.3.1 Description, Chemical Formula, and Brand Names – 663
25.3.2 Pharmacology – 664
25.3.3 Side Effects and Usage Precautions – 664
25.3.4 Clinical Use and Dosage – 665

25.4 **Cannabinoids – 666**
25.4.1 Cannabidiol – 669
25.4.2 Dronabinol – 672

References – 675

© The Author(s), under exclusive license to Springer Nature Switzerland AG 2026
S. Bleuer-Elsner et al., *Veterinary Psychiatry of the Cat*,
https://doi.org/10.1007/978-3-032-06260-4_25

25

This chapter addresses psychotropic drugs used in veterinary psychiatry that cannot be categorized into the previously discussed families.

It covers venlafaxine (anxiolytic), trazodone (anxiolytic, antidepressant, and hypnotic), buspirone (anxiolytic), and cannabinoids (analgesic and anxiolytic in humans). None of these drugs have received marketing authorization for veterinary use.

25.1 Venlafaxine

25.1.1 Description, Chemical Formula, and Brand Names

Venlafaxine, chemically known as *1-[2-(dimethylamino)-1-(4-methoxyphenyl) ethyl]cyclohexan-1-ol*, is an antidepressant that functions by inhibiting the reuptake of serotonin and norepinephrine (SNRI). It does not have a veterinary-specific formulation. In human medicine, venlafaxine is available under the brand name Effexor or the generic name Venlafaxine, in extended-release capsules available in 37.5 mg, 75 mg, and 150 mg depending on the country.

25.1.2 Pharmacology

25.1.2.1 Mechanism of Action

The mechanism of action of venlafaxine is only partially understood. Venlafaxine and its active metabolite, O-desmethylvenlafaxine (ODV), inhibit the reuptake of serotonin and norepinephrine, with a stronger affinity for the serotonin transporter (SERT) than for the norepinephrine transporter (NET).

The serotonergic actions of venlafaxine manifest at low doses, whereas its noradrenergic effects gradually appear at higher doses. The noradrenergic activity of the ODV metabolite on the NET transporter is stronger than that of venlafaxine itself.

Both venlafaxine and ODV have mild inhibitory effects on dopamine reuptake. Unlike tricyclic antidepressants and selective serotonin reuptake inhibitors (SSRI), venlafaxine has no affinity for histaminergic, muscarinic, or α1-adrenergic receptors (◘ Fig. 25.1).

25.1.2.2 Metabolism and Involved Cytochromes

In humans, the hepatic metabolism of venlafaxine produces one major active metabolite, O-desmethylvenlafaxine (ODV), and two less active minor metabolites, N-desmethylvenlafaxine and N,O-didesmethylvenlafaxine. The formation of ODV is catalyzed by the cytochrome CYP2D6, while N-demethylation is catalyzed by CYP3A4, CYP2C19, and CYP2C9. Genetic variations among individuals influence the relative quantities of these metabolites. The ODV metabolite functions as a standalone SNRI and has also been developed as a primary psychotropic agent under the name desvenlafaxine (Stahl and Muntner 2021). Data for cats are currently unavailable.

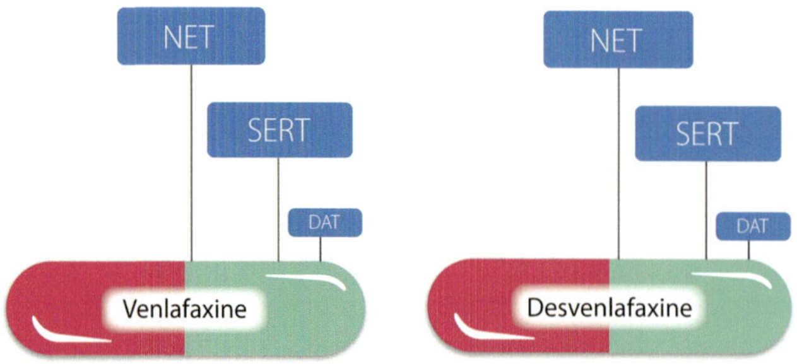

◘ **Fig. 25.1** Structure of venlafaxine and desvenlafaxine. SERT inhibitor—NET inhibitor—DAT inhibitor

25.1.2.3 Pharmacokinetics

In humans, venlafaxine has a half-life of 5 h. Both venlafaxine and its metabolites are primarily excreted through the urine. There are no pharmacokinetic data available for cats.

25.1.3 Side Effects and Usage Precautions

25.1.3.1 Side Effects

In humans, the most common side effects of venlafaxine are dose dependent. Frequently reported side effects include nausea, diarrhea, constipation, decreased appetite, headaches, dizziness, dry mouth, nervousness and psychomotor agitation, insomnia, sedation, sweating, and fatigue (Stahl and Muntner 2021). Less frequently, abdominal pain, vomiting, cutaneous bleeding and rectal bleeding, tremors, mydriasis and accommodation disorders, urinary retention, hypotension, skin rashes, alopecia, and seizures have been observed. Regular blood tests are recommended due to possible cholesterol elevation and hyponatremia (Egger et al. 2006). Increased blood pressure and intraocular pressure have also been reported, and it is advised to monitor blood pressure in patients undergoing treatment (Calvi et al. 2021). Other side effects, such as suicidal ideation and sexual dysfunction, exist but are less relevant in our context.

In cats, reported side effects, listed by frequency, include sedation, loss of appetite, mydriasis, vomiting, anuria, diarrhea, constipation, rectal bleeding, polydipsia, salivation, twitching during sleep, and excitement (Masson et al. 2024).

Depending on the severity of side effects, a dose reduction or a waiting period of about 10 days usually resolves the issue. Notably, if side effects resolve after a dose reduction, it is often possible to return to the initial dose without the reappearance of the initial adverse effects.

25.1.3.2 Precautions and Contraindications

In humans, venlafaxine is contraindicated in cases of ocular or cardiac issues, hypertension, hyponatremia, and bleeding disorders. For diabetic patients, adjustments to antidiabetic treatment doses may be required.

In cats, hepatic or renal insufficiency necessitates a reduction in the administered venlafaxine dose.

Due to its inhibitory action on cytochromes, venlafaxine can alter the pharmacokinetics of certain drugs. The initiation of a combined treatment should follow a precautionary principle, starting both drugs at half the usual dose. When introducing an antiepileptic treatment, enhanced monitoring of liver function is recommended. Dosage adjustments to normal levels should be made gradually based on observed results.

There are no studies on the effects of venlafaxine on feline fetuses during gestation or on kittens during lactation. Venlafaxine is excreted in variable proportions in milk, but lactation is not a major contraindication for venlafaxine use in humans (Newport et al. 2009). In humans, it also does not increase the risk of congenital malformations or miscarriages (Lassen et al. 2016).

25.1.3.3 Drug Interactions

The following drug combinations are *prohibited:*

- Amitraz and selegiline (monoamine oxidase inhibitors or MAOIs): this association carries a significant risk of serotonin syndrome.
- Tramadol (analgesic): SSRIs may inhibit the hepatic metabolism of tramadol into active metabolites. This combination increases the risk of serotonin syndrome.
- SSRI: combining an SSRI with venlafaxine may increase plasma levels of both molecules and promote the onset of serotonin syndrome.
- Clomipramine, amitriptyline (tricyclic antidepressants): combining venlafaxine with tricyclic antidepressants may elevate plasma levels of the tricyclics and increase the risk of serotonin syndrome.
- Ketoconazole (antifungal): this combination increases the risk of serotonin syndrome.

The following drug interactions have been reported in humans or animals receiving venlafaxine. The combination is not necessarily contraindicated, but risks should be evaluated by the practitioner.

- Diazepam, alprazolam (benzodiazepines): venlafaxine, through its effects on hepatic cytochromes, may increase their blood levels.
- Propranolol, metoprolol (beta-blockers): venlafaxine may increase blood levels of beta-blockers.
- Trazodone (serotonin antagonist and reuptake inhibitor, SARI): the combination with venlafaxine may increase trazodone plasma levels and promote the occurrence of serotonin syndrome.
- Hypericum perforatum (St. John's Wort): the combination with venlafaxine increases the risk of serotonin syndrome.
- Clopidogrel, heparin, warfarin (anticoagulants): their use with venlafaxine is associated with an increased risk of bleeding in humans.
- Buspirone (anxiolytic): the combination with venlafaxine increases the risk of serotonin syndrome.
- Diuretics: when associated with venlafaxine, they may increase the risk of hyponatremia.
- Insulin: venlafaxine may alter the doses required to maintain diabetes control.
- Isoniazid (antibiotic used in tuberculosis treatment): the combination with venlafaxine increases the risk of serotonin syndrome.
- Nonsteroidal anti-inflammatory drugs (NSAIDs): the combination with venlafaxine may increase the risk of gastrointestinal ulceration and bleeding.
- Pentazocine (painkiller): associated with venlafaxine, a pseudo-serotonin syndrome is possible.
- Phenytoin (antiepileptic): venlafaxine may increase plasma levels of phenytoin.

25.1.3.4 Toxicity

The LD_{50} is 350 mg/kg in rats and unknown in cats (Paulis et al. 2018), but it is greater than 50 mg/kg (Pugh et al. 2013).

In humans, signs of intoxication include electrocardiogram changes, tachycardia or bradycardia, hypotension, dizziness, altered consciousness, seizures, and digestive disturbances. Treatment is symptomatic and involves intravenous fluid therapy, induction of vomiting if appropriate, and administration of activated charcoal.

In cats, a retrospective study by Pugh et al. (2013) on 12 cats that ingested between 5 and 30 mg/kg of venlafaxine reported 6 symptomatic and 6 asymptomatic cases. Symptoms included sedation, depression, digestive disturbances, tachycardia, and hypertension. Additionally, complete inhibition of dorsal raphe neurons only occurs at doses starting from 5 mg/kg administered intravenously (Bjorvatn et al. 2000).

25.1.4 Clinical Use and Dosage

25.1.4.1 According to the Scientific Literature

Venlafaxine has been successfully used in the treatment of idiopathic cystitis and urinary marking at doses ranging from 1 to 2 mg/kg once daily (Hopfensperger 2016; Metz et al. 2022; Sinn 2018). It has also been used at the same dose for the treatment of fear and anxiety (Denenberg and Dubé 2018; Metz et al. 2022).

In a study by Pflaum and Bennett (2021), venlafaxine was prescribed at a dose of 1.1 mg/kg once daily and was effective in resolving aggression caused by impulsive behavior. Metz et al. (2022) highlighted the ease of administering venlafaxine compared to other therapeutic options. The medication is available as capsules containing microgranules, which are easily ingested by cats when mixed with food.

Masson et al. (2024) conducted a retrospective study involving 176 cats diagnosed with phobic, anxious, depressive, manic, senile, impulsive, compulsive, or dissociative states. The study reported significant improvement in 53% of cats, resolution of symptoms in 38%, and successful weaning without recurrence of symptoms or illness in 20% of cases.

The initial prescribed dose was 4 microgranules per kilogram, approximately 1 mg/kg. The study found that the dose needed adjustment based on the cat's age, treatment response, and diagnosed condition:

- For cats older than 8 years or in a depressive state, the dose was reduced to 0.5 mg/kg (2 microgranules per kilogram).
- For impulsive states, the dose was increased to 1.5 mg/kg (6 micro-granules per kilogram).
- For manic states, a starting dose of 2 mg/kg (8 micro-granules per kilogram) was recommended.

The study again emphasized the ease of administering venlafaxine due to its formulation.

25.1.4.2 Experience Feedback

Venlafaxine proves beneficial in conditions where serotonin and norepinephrine play a significant role, such as phobic, anxious, impulsive, compulsive, or depressive states. It can be used as a first-line treatment for these conditions and as a second-line option when initial treatments fail. Its formulation makes it particularly appealing for cats in whom oral administration of medication is challenging.

25.1.4.3 Possible Combinations

Combining venlafaxine with gabapentinoids (gabapentin or pregabalin) might enhance its anxiolytic effects, particularly in conditions like phobias or generalized anxiety, while mitigating some excitatory side effects.

25.1.4.4 Withdrawal

Given the short half-life of the molecule, withdrawal does not seem to be necessary. However, it is often carried out gradually to observe whether the symptoms reappear.

25.2 Trazodone

25.2.1 Description, Chemical Formula, and Brand Names

Trazodone, with the chemical name *2-[3-[4-(3-chlorophenyl)piperazin-1-yl]propyl]-[1,2,4]triazolo[4,3-a]pyridin-3-one*, is a serotonin antagonist and reuptake inhibitor (SARI). It belongs to the class of antidepressants and has particular utility in treating certain phobias and anxiety disorders. Additionally, it possesses hypnotic properties.

There is no veterinary-specific formulation of trazodone. In human medicine, it is marketed under the names *Desyrel*™ or *Trazodil*™, *Trittico*™, available in tablets of 50, 100, 150, and 300 mg depending on the country. It is not available in France.

25.2.2 Pharmacology

25.2.2.1 Mechanism of Action

The action of trazodone is *dose-dependent:*

At high doses (150–600 mg in humans):

- It inhibits the reuptake of serotonin, enhancing serotonergic transmission and producing an antidepressant effect.
- It antagonizes 5-HT2A auto- and heteroreceptors, promoting the release of monoamines, contributing to its antidepressant effects.
- It antagonizes 5-HT2C receptors, stimulating the release of dopamine and norepinephrine in the prefrontal cortex, enhancing antidepressant and procognitive effects.
- It antagonizes α1-adrenergic receptors and histamine H1 receptors, reducing alertness.

At lower doses (25–150 mg in humans), trazodone no longer inhibits serotonin reuptake. As a result, it loses its major antidepressant action while retaining its antagonistic effects on 5-HT2A, α1-adrenergic, and H1 receptors (Stahl and Muntner 2021). At these doses, it is used as a hypnotic and sedative (◘ Fig. 25.2). This property makes it particularly advantageous for inducing sleep at low doses, which corresponds to 50 mg in cats.

25.2.2.2 Metabolism and Involved Cytochromes

There is no available information regarding trazodone metabolism in cats. In humans, its full metabolic pathway has not been com-

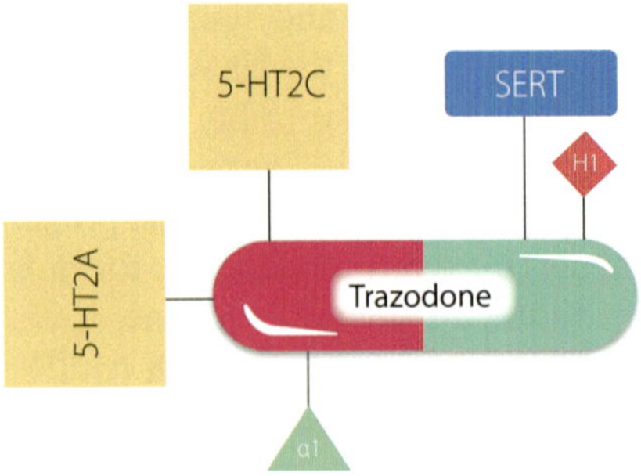

Fig. 25.2 Structure of trazodone. Inhibitor SERT—antagonist 5-HT2A—antagonist 5-HT2C—antagonist α1-adrenergic—antagonist histamine H1 (at low doses, the SERT inhibitory action is absent)

pletely elucidated. However, it undergoes significant hepatic metabolism, primarily via cytochromes CYP3A4 and CYP2D6, producing its active metabolite, m-chlorophenylpiperazine. Additional metabolites have been identified, including a dihydrodiol and a carboxylic acid (Rotzinger et al. 1998; Wen et al. 2008).

25.2.2.3 Pharmacokinetics

In cats, the half-life of trazodone is 5.12 h, and the peak plasma concentration is reached approximately 1.67 h after administration (Tucker et al. 2023). In humans, the oral bioavailability is 65%, with an elimination half-life of approximately 7 h. Excretion is primarily renal (70–75%) and partially fecal (20%) (Nilsen and Dale 1992).

25.2.3 Side Effects and Usage Precautions

25.2.3.1 Side Effects

In humans, the most frequently encountered side effects include fatigue, drowsiness, vision disturbances, dry mouth, hypotension, hyponatremia, sweating, and cardiac rhythm disorders. In dogs, common side effects are gastrointestinal disturbances such as nausea, vomiting, diarrhea, or colitis. Ataxia, sedation, tachycardia, increased anxiety, disinhibition, and aggression have also been reported (Gruen and Sherman 2008; Jay et al. 2013). Additionally, a transient case of priapism has been documented in a dog (Murphy et al. 2017).

In cats, at doses of 50–100 mg, or a maximum of 33.33 mg/kg, only sedation has been reported (Orlando et al. 2016; Stevens et al. 2016).

25.2.3.2 Precautions and Contraindications

Trazodone should be used with caution in patients with severe cardiac, hepatic, or renal insufficiency. While it is unlikely for trazodone alone to induce serotonin syndrome at clinically relevant doses, caution is advised when it is administered alongside other serotonergic medications.

25.2.3.3 Drug Interactions

The following drug interactions have been reported in humans or animals receiving trazodone. Unless otherwise indicated, the association is not necessarily contraindicated, but risks should be assessed, and additional monitoring performed if necessary.

- SSRIs (fluoxetine): the combination increases the risk of serotonin syndrome. Trazodone is commonly used with SSRIs, but patients should be monitored for signs of serotonin syndrome. Fluoxetine may also inhibit trazodone metabolism.
- Tramadol: the combination increases the risk of serotonin syndrome.
- Diuretics: the combination may increase the risk of hyponatremia.
- Fluoroquinolones (ciprofloxacin): there is an increased risk of QT interval prolongation.
- Isoflurane: premedication with trazodone reduced the minimum alveolar concentration of isoflurane required in cats (Brosnan et al. 2024).
- Monoamine oxidase inhibitors (amitraz, selegiline): the combination increases the risk of serotonin syndrome. Concomitant use is contraindicated in humans.
- Venlafaxine: the combination increases the risk of serotonin syndrome.
- Metoclopramide: the combination increases the risk of serotonin syndrome.

- Antihypertensive drugs: trazodone may potentiate their effects and cause hypotension.
- Azole antifungals (ketoconazole, fluconazole): they may increase trazodone blood levels. Concomitant use is contraindicated in humans.
- Cisapride: there is an increased risk of QT interval prolongation. Concomitant use is contraindicated in humans.
- Central nervous system (CNS) depressants: the combination with trazodone enhances CNS depressant effects.
- Digoxin: trazodone may increase digoxin levels.
- Linezolid: the combination with trazodone increases the risk of serotonin syndrome. Concomitant use is contraindicated in humans.
- Macrolide antibiotics (erythromycin, clarithromycin): they may increase trazodone blood levels.
- Methylene blue: the combination with trazodone increases the risk of serotonin syndrome. Concurrent use is discouraged in humans.
- Phenothiazines: they may increase trazodone blood levels.

25.2.3.4 Toxicity

The LD_{50} determined in rats is 690 mg/kg (Drugbank Online, "Trazodone," 2024). In cats, clinical signs of trazodone toxicity include sedation, lethargy, ataxia, and vomiting. A serotonin syndrome is also possible.

25.2.4 Clinical Use and Dosage

25.2.4.1 According to the Scientific Literature

In cats, trazodone is valuable as an *anxiolytic* and *sedative*. For cats with a fear of veterinary care and transportation, it has been shown to reduce manifestations of fear during transit and veterinary handling. The recommended dose is 50 mg per cat, administered 60–90 min before the veterinary visit (Stevens et al. 2016). Similarly, during hospitalization, administering 50 mg per cat reduces stress signals, largely due to its sedative effect (Delgado 2023).

Trazodone can also be used as a sedative at doses of 50–100 mg per cat. The peak effect is observed 2 h after administration and lasts up to 8 h (Orlando et al. 2016). Concurrent administration with gabapentin does not enhance sedation in cats and does not appear to offer any particular benefit (Tucker et al. 2023).

For anesthetic premedication, the recommended dose is 25 mg per cat. The maximum effect seems to occur after 1 h (Delgado 2023).

25.2.4.2 Experience Feedback

Trazodone is particularly useful for its antihistaminic and sedative effects in cats that are restless at night (anxious or senile meowing, or frustration in impulsive or compulsive cats). The dosage ranges from 5 to 10 mg/kg in a single dose at bedtime.

25.2.4.3 Possible Combinations

The most common combination is with fluoxetine. Combining trazodone with venlafaxine or tricyclic antidepressants should be done cautiously and under close monitoring, starting with the lowest doses of both psychotropics.

25.2.4.4 Withdrawal

Given the short half-life of the molecule, withdrawal does not appear to be necessary.

25.3 Buspirone

25.3.1 Description, Chemical Formula, and Brand Names

Buspirone, chemically named *8-[4-(4-pyrimidin-2-ylpiperazin-1-yl)butyl]-8-azaspiro[4.5]decane-7,9-dione*, is a nonbenzodiazepine anxiolytic belonging to the azapirone class. It is not available as a veterinary-specific product. The molecule can be obtained from pharmacies under the brand name Buspar® and in generic forms as 10 mg tablets.

25.3.2 Pharmacology

25.3.2.1 Mechanism of Action

Buspirone acts as an agonist at presynaptic 5-HT1A autoreceptors and as a partial agonist at postsynaptic 5-HT1A receptors, reducing serotonergic transmission. Over time, its action results in desensitization of the autoreceptors, leading to increased serotonin release in the synaptic cleft, which is responsible for its anxiolytic effects (Rang et al. 2012).

The molecule also exhibits low affinity for 5-HT2 serotonin receptors and functions as a partial antagonist of dopamine D2 autoreceptors. Additionally, it acts as an antagonist at presynaptic dopamine D3 and D4 receptors and as a partial agonist at α1-adrenergic receptors (Howland 2015; Drugbank Online, "Buspirone," 2024) (◘ Fig. 25.3).

25.3.2.2 Metabolism and Involved Cytochromes

The hepatic metabolism of buspirone primarily involves the cytochrome CYP3A4. This oxidation process produces four metabolites, including one pharmacologically active metabolite, 1-pyrimidinylpiperazine (1-PP). In animal studies, 1-PP exhibits approximately one-quarter of the pharmacological activity of buspirone (Drugbank Online, "Buspirone," 2024).

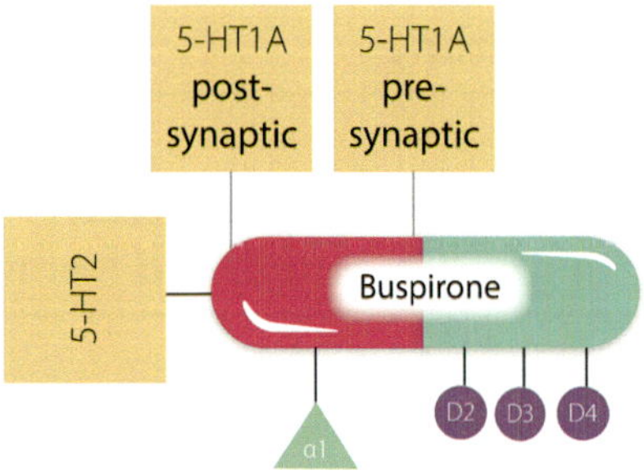

◘ **Fig. 25.3** Structure of buspirone. Presynaptic 5-HT1A agonist—postsynaptic 5-HT1A partial agonist—5-HT2 receptor antagonist—D2 receptor partial antagonist—D3 and D4 receptor antagonist—α1-adrenergic partial agonist

25.3.2.3 Pharmacokinetics

In humans, 29–63% of the administered dose is excreted in urine and 18–38% in feces, mainly as metabolites. In cats, the maximum concentration of the drug in the blood is reached 1.4 h after oral administration. Bioavailability appears to be lower in cats compared to humans (Mealey et al. 2004).

25.3.3 Side Effects and Usage Precautions

25.3.3.1 Side Effects

In humans, the most common side effects include dizziness, headaches, nausea, anorexia, motor agitation, and sedation. Less frequently, cardiovascular signs such as tachycardia may occur. Buspirone is generally well-tolerated in cats. Potential side effects include bradycardia, gastrointestinal issues such as transient anorexia or vomiting, and stereotypical behaviors like obsessive licking or circling (Drug Monographs–Plumb's Veterinary Drugs 2024).

25.3.3.2 Precautions and Contraindications

Use in animals with severe renal or hepatic disease is not recommended. Buspirone should be used with caution in aggressive individuals, particularly at the beginning of treatment, due to the initial decrease in serotonergic transmission.

The effects of buspirone typically take a week or longer to manifest; therefore, the drug is not suitable as monotherapy for the immediate treatment of phobic or anxious states.

25.3.3.3 Drug Interactions

The following drug interactions have been reported in humans or animals receiving buspirone. Coadministration is not necessarily contraindicated, but the risks should be assessed, and additional monitoring performed when necessary.

- MAOIs (amitraz, linezolid, selegiline): use of buspirone in direct combination or

within 14 days of discontinuing an MAOI is contraindicated. Severe hypertension and serotonin syndrome may occur.
- SSRIs (fluoxetine, paroxetine): combination increases the risk of serotonin syndrome. SSRIs may reduce the metabolism of buspirone.
- Venlafaxine: combination increases the risk of serotonin syndrome.
- Trazodone: use with buspirone may elevate Alanine Aminotransferase (ALT) levels and increase the risk of serotonin syndrome.
- Tricyclic antidepressants (amitriptyline, clomipramine): combination increases the risk of serotonin syndrome. Dose reduction may be necessary to minimize side effects.
- Mirtazapine: combination may increase the risk of serotonin syndrome.
- α2-adrenergic agonists (dexmedetomidine, xylazine, clonidine): combination increases the risk of respiratory depression and CNS depression.
- Anesthetic agents (e.g., isoflurane, ketamine, propofol): combination increases the risk of respiratory depression and CNS depression.
- Angiotensin-Converting Enzyme (ACE) inhibitors (e.g., enalapril): combination increases the risk of hypotension.
- Antihistamines (cetirizine, chlorpheniramine): combination increases the risk of respiratory depression and CNS depression.
- Apomorphine: combination may increase the risk of CNS depression, drowsiness, and dizziness.
- Azole antifungals (fluconazole, ketoconazole): may slow the metabolism of buspirone, potentially increasing concentrations and side effects.
- Barbiturates (methohexital, phenobarbital): combination increases the risk of respiratory depression and CNS depression.
- Benzodiazepines (alprazolam, midazolam): combination increases the risk of CNS depression.
- Carbamazepine: may lower plasma levels of buspirone.
- Cimetidine: may increase plasma levels of buspirone.
- Gabapentin and pregabalin: combination increases the risk of respiratory depression and CNS depression.
- Glucocorticoids (dexamethasone, prednisolone): may increase the metabolism of buspirone.
- Levetiracetam: combination increases the risk of respiratory depression and CNS depression.
- Metoclopramide: combination increases the risk of respiratory depression and CNS depression and may induce neuroleptic malignant syndrome.
- Opioids (fentanyl, morphine): combination increases the risk of severe respiratory depression, coma, and death.
- Tramadol: combination may increase the risk of serotonin syndrome and CNS and respiratory depression.
- Phenothiazines (acepromazine, chlorpromazine): combination increases the risk of respiratory depression and CNS depression.

25.3.3.4 Toxicity

The oral LD_{50} in dogs is established at 586 mg/kg (Drugbank Online, "Buspirone," 2024). Overdose incidents may result in vomiting, dizziness, drowsiness, miosis, and gastric distension. To date, no studies have been conducted in cats.

25.3.4 Clinical Use and Dosage

25.3.4.1 According to the Scientific Literature

Buspirone is indicated for the treatment of phobic or anxious behaviors such as urine marking and inappropriate elimination (Chávez et al. 2016; Hart et al. 1993), inter-cat aggression (Crowell-Davis et al. 2019; Denenberg and Dubé 2018), and excessive grooming (Sawyer et al. 1999; Selk Ghaffari and Sabzevari 2010). In cases of generalized inhibited anxiety or chronic depressive states, it may reduce the inhibition observed in affected cats (Crowell-Davis et al. 2019; Denenberg and Dubé 2018). The recommended dose is 1 mg/kg or 2.5–7.5 mg per cat once or twice daily.

25.3.4.2 Experience Feedback

Our experience does not allow us to expand upon the information above. Field trials have been inconclusive, and we rarely use this molecule.

25.3.4.3 Withdrawal

In human medicine, withdrawal is not necessary.

25.4 Cannabinoids

The term *cannabinoids* encompasses all synthetic or natural, external or internal molecules capable of interacting with the receptors of the endocannabinoid system.

The endocannabinoid system consists of a network of receptors, ligands, and enzymes:

- Two main subtypes of metabotropic receptors: The CB1 subtype and the CB2 subtype. *CB1 receptors* are the most abundant metabotropic receptors in the brain and are widely distributed. They are predominantly presynaptic and are particularly present in circuits responsible for pain, movement, and memory processes. Notably, they are widely distributed in brain regions associated with fear behaviors, the limbic system, and the prefrontal cortex (Tsou et al. 1998). Outside the central nervous system (CNS), CB1 receptors are also found in the pituitary gland, immune cells, reproductive cells, gastrointestinal cells, lymph nodes, the heart, lungs, bladder, and adrenal glands. *CB2 receptors* have been less studied and were long considered peripheral, with high density in immune cells, particularly B lymphocytes and natural killer (NK) cells. CB2 receptors are found in large quantities in the spleen and tonsils. Recent studies have shown that CB2 receptors are expressed on the membranes of postsynaptic neurons in the CNS (Gong et al. 2006; Roche and Finn 2010), particularly in the limbic system and the prefrontal cortex (Ivy et al. 2020). CB2 receptors are therefore thought to play a role in fear and anxiety behaviors (García-Gutiérrez et al. 2012).
- Other cannabinoid receptors are designated as G protein-coupled receptors (GPR), followed by an identifier number corresponding to their genetic coding (e.g., GPR55, GPR18, GPR3, GPR6, GPR12, GPR119). Their role is not yet fully understood.
- TRPV1 receptors (transient receptor potential cation channels subfamily V member 1) are involved in pain perception and thermoregulation.
- Ligands of the above receptors, termed endocannabinoids, include five identified to date: N-arachidonoylethanolamine or anandamide (Devane et al. 1992), 2-arachidonoylglycerol (2-AG), 2-arachidonyl glyceryl ether (Hanuš et al. 2001), O-arachidonoylethanolamine or virodhamine (Porter et al. 2002), and N-arachidonoyldopamine or NADA (Bisogno et al. 2000; Gaffuri et al. 2012; Mechoulam et al. 2014).
- Palmitoylethanolamide (PEA) is an endogenous fatty acid that increases anandamide activity via an entourage effect (Ho et al. 2008) and has a particular affinity for GPR55 and GPR119 receptors (Godlewski et al. 2009). It is not considered a classical endocannabinoid due to its weak affinity for CB1 and CB2 receptors.
- A system of pre- and postsynaptic enzymes is involved in the synthesis and synaptic metabolism of endocannabinoids. FAAH (fatty acid amide hydrolase) is postsynaptic, and MGL (monoacylglycerol lipase) is presynaptic (Muccioli et al. 2010; Pertwee 2014).

The endocannabinoid system is a lipid-based signaling system, unlike other neurotransmitter systems, which are protein-based. It is a phylogenetically ancient system present in all vertebrates.

The signaling mechanism works as follows: endocannabinoids are synthesized by FAAH enzymes from the postsynaptic cell membrane's phospholipids in response to intracellular calcium influx (Miller and Devi 2011). They are released immediately (rather than stored in vesicles like most other neurotransmitters) and diffuse to other neurons or glial cells via retrograde signaling.

This signaling occurs in the opposite direction of all previously described neurotransmission mechanisms in this book. Additionally, there is no neuronal specialization; all neurons are capable of producing endocannabinoids since they are derived from cell membranes.

Depolarization of the postsynaptic membrane triggers the production of endocannabinoids. These molecules then follow a retrograde pathway to regulate neurotransmitter release at the presynaptic neuron (◘ Fig. 25.4).

The binding of the ligand to presynaptic CB1 receptors modulates the release of neurotransmitters such as glutamate, GABA, acetylcholine, dopamine, histamine, serotonin, cholecystokinin, D-aspartate, glycine, and noradrenaline (Iffland and Grotenhermen 2017; Heifets and Castillo 2009). CB1 receptor stimulation in astrocytes enhances glial glutamatergic signaling (Castillo et al. 2012).

The stimulation of CB2 receptors on postsynaptic membranes in the central nervous system (CNS) is less studied. It is thought to inhibit the polarization of the postsynaptic neuron (Den Boon et al. 2012; Zhang et al. 2014). This suggests that the endocannabinoid system acts as a "supermodulator" of all other neurotransmitter systems.

Beyond endocannabinoids, cannabinoids include phytocannabinoids and synthetic cannabinoids.

Phytocannabinoids are primarily extracted from the resin of female *Cannabis sativa* plants. More than a hundred phytocannabinoids have been identified. The most studied include δ-9-tetrahydrocannabinol (THC), cannabidiol (CBD), tetrahydrocannabivarin, tetrahydrocannabiorcol, cannabichromene, and cannabigerol.

Synthetic cannabinoids were initially developed in academic or pharmaceutical research laboratories. Some have entered the illegal market and are known to drug users as "K2" or "Spice." These substances are particularly dangerous. Two synthetic cannabinoids have been developed for therapeutic purposes: nabilone and dronabinol.

Exogenous cannabinoids do not act solely on the natural ligand receptors CB1, CB2, GPR, or TRPV1. Some also interact with serotonergic 5-HT1A receptors (CBD) and opioid μ and κ receptors (CBD and THC).

From a pharmaceutical perspective, there are four approved molecules for human use, depending on national regulations, that act on CB1 and CB2 receptors. Two of these are synthetic: nabilone, a CB1/CB2 agonist, and dronabinol, a synthetic formulation of THC. The other two are phytocannabinoids derived from *Cannabis sativa*: Sativex®, which contains both THC and CBD, and EpidiolexND, which contains only CBD.

Palmitoylethanolamide is found in a veterinary dietary supplement, Palmidol PEA®.

As of now, there are no established veterinary guidelines for the use of cannabinoids. This section will detail cannabidiol (CBD), which is emerging in the veterinary product market, and briefly address dronabinol (THC) to better understand the differences between these two compounds.

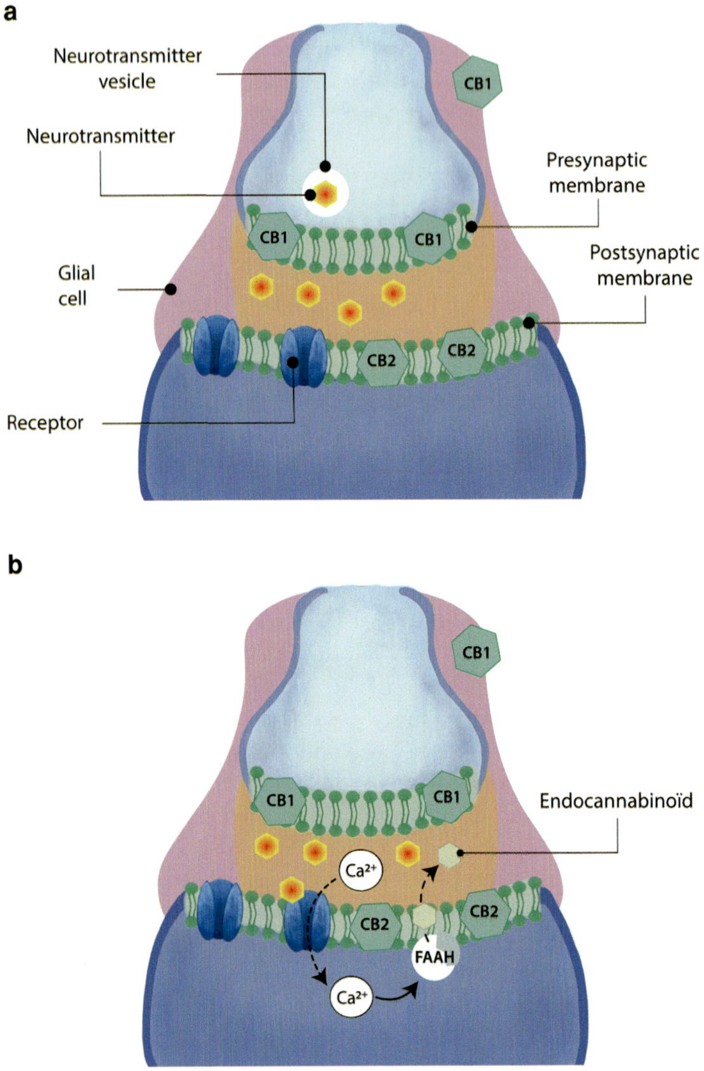

Fig. 25.4 CB1 and CB2 receptors. **a** and **b** The binding of an excitatory neurotransmitter to its receptor induces a Ca^{2+} influx, which subsequently activates the FAAH enzyme. FAAH converts molecules from the postsynaptic membrane into endocannabinoids, which are immediately released into the synaptic cleft. **c** These endocannabinoids travel retrogradely to CB1 receptors located on the presynaptic membrane and glial cells. The binding of endocannabinoids to presynaptic CB1 receptors reduces neurotransmitter release from the presynaptic neuron and stimulates glutamate release from glial cells. **d** FAAH and MGL enzymes metabolize the endocannabinoids in the synaptic cleft, terminating their actions. The binding of endocannabinoids to CB2 receptors on the postsynaptic membrane inhibits membrane polarization

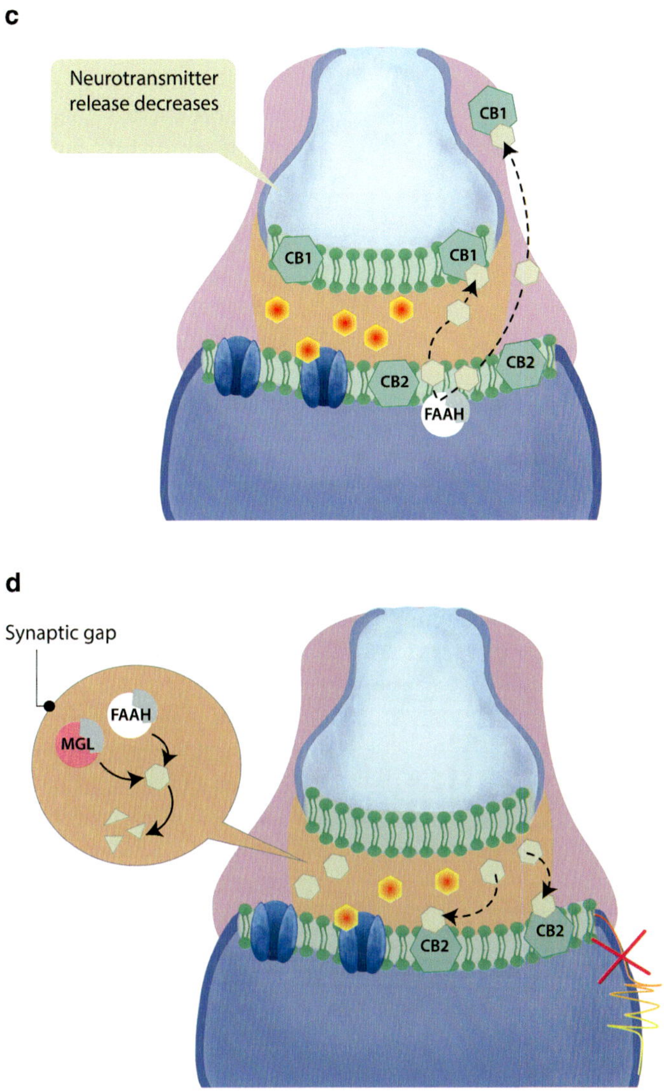

Fig. 25.4 (continued)

25.4.1 Cannabidiol

25.4.1.1 Description, Chemical Formula, and Brand Names

Cannabidiol (CBD), chemically named *2-[(1R,6R)-3-methyl-6-prop-1-en-2-ylcyclohex-2-en-1-yl]-5-pentylbenzene-1,3-diol,* is a phytocannabinoid. It is available as a veterinary dietary supplement under the name SativaVet® (oil form) and as a human pharmaceutical under the name Epidiolex™ (oral solution, 100 mg/mL).

25.4.1.2 Pharmacologie

- Mechanism of Action

CBD acts as a *negative allosteric modulator of CB1 receptors*, an inverse agonist of CB2 receptors, a desensitizer of TRPV1 receptors, a partial agonist of serotonergic 5-HT1A receptors, and an allosteric modulator of μ and δ opioid receptors (Murkar et al. 2021) (Fig. 25.5).

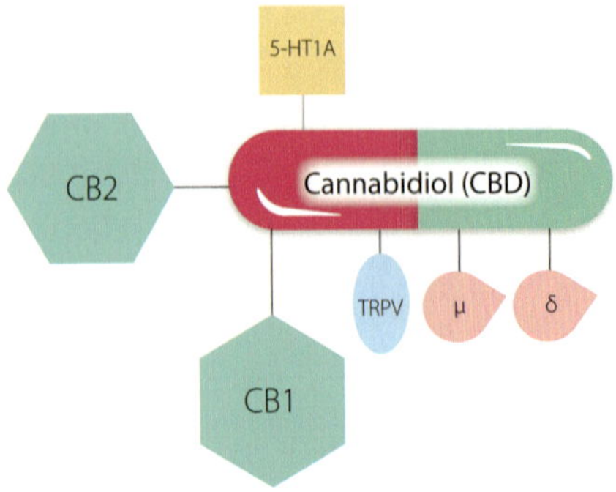

Fig. 25.5 Structure of cannabidiol (CBD). Negative allosteric modulator of CB1—inverse agonist of CB2—TRPV1 desensitizer—partial agonist of 5-HT1A—negative allosteric modulator of μ and δ opioid receptors

▪ **Metabolism and Involved Cytochromes**

CBD is metabolized in the liver by several cytochrome P450 isoenzymes, including CYP2C9, CYP2C19, CYP2D6, and CYP3A4. CBD can be stored in adipose tissues for up to 4 weeks, from where it is slowly released into the bloodstream at subtherapeutic doses. It is eliminated via renal and biliary routes. The primary active metabolite of CBD is 7-hydroxy-cannabidiol.

▪ **Pharmacokinetics**

The half-life of CBD in humans ranges from 1.44 to 1.81 h depending on the formulation used. In cats, the elimination half-life of CBD is between 1 and 3 h after administration (Deabold et al. 2019).

CBD is excreted through urine and feces. Elimination is faster in cats than in dogs, and cats appear to tolerate CBD better (Kulpa et al. 2021). The bioavailability of CBD is increased when consumed with a meal (Jukier et al. 2023).

CBD accumulates in tissues, particularly adipose tissue. A cumulative effect can occur, altering the pharmacokinetics of the molecule. With repeated administration, plasma concentrations of CBD increase over time (Child and Tallon 2022).

25.4.1.3 Side Effects and Usage Precautions

▪ **Side Effects**

Administered at a dose of 4 mg/kg/day orally, the most commonly used dosage in studies involving cats, CBD appears to be well tolerated.

The most frequent side effects include sedation, increased appetite, and mild gastrointestinal disturbances (Corsato Alvarenga et al. 2023; Deabold et al. 2019).

When administered at standard doses, an increase in alkaline phosphatase levels has been observed in healthy cats, without associated clinical symptoms (Coltherd et al. 2024; Gutierre et al. 2023). Additionally, no increase in urea or creatinine levels has been reported even at a dose of 80 mg/kg (Rozental et al. 2023).

▪ **Precautions and Contraindications**

CBD should be used cautiously in cases of hepatic or renal insufficiency. Its elimination is partially renal, and anandamide increases renal blood flow while decreasing the glomerular filtration rate, potentially affecting kidney function (Larrinaga et al. 2010).

▪ **Drug Interactions**

The following drug interactions have been reported in humans or dogs. Unless otherwise specified, these combinations are not necessarily contraindicated; however, risks should be assessed, and additional monitoring should be implemented as needed.

- Amiodarone (antiarrhythmic): the combination may lead to increased plasma concentrations of CBD.
- Azole antifungals (fluconazole, itraconazole, ketoconazole, voriconazole): the combination increases the risk of hepatotoxicity.
- Benzodiazepines: the combination may increase plasma concentrations of certain benzodiazepines such as alprazolam or diazepam, but not midazolam. It increases the risk of sedation and respiratory depression.
- Central nervous system depressants (acepromazine, gabapentin, pregabalin, tramadol, trazodone): the combination increases the risk of sedation.
- Chloramphenicol: the combination may lead to increased plasma concentrations of CBD.
- Cimetidine: the combination may lead to increased plasma concentrations of CBD.

- Clopidogrel (anticoagulant): CBD may inhibit the formation of clopidogrel's active metabolite. The combination may increase plasma concentrations of CBD.
- Dexamethasone: the combination may decrease plasma concentrations of CBD.
- Diltiazem (antiarrhythmic): the combination may lead to increased plasma concentrations of CBD.
- Fluoxetine: the combination may lead to increased plasma concentrations of CBD.
- Venlafaxine: the combination may lead to increased plasma concentrations of CBD.
- Loperamide: the combination may lead to increased plasma concentrations of loperamide.
- Non-steroidal anti-inflammatory drugs (carprofen, meloxicam): a 4-week combination did not result in any apparent side effects or significant laboratory value changes in dogs.
- Opioids (fentanyl, methadone, morphine, tramadol): there is an increased risk of CNS depression.
- Phenobarbital: the combination may lead to decreased plasma concentrations of CBD and increase the risk of hepatotoxicity.
- Propofol: the combination may lead to increased plasma concentrations of propofol.
- Proton pump inhibitors (omeprazole, pantoprazole): the combination may lead to increased plasma concentrations of CBD.
- Rifampin: the combination may lead to decreased plasma concentrations of CBD.
- Theophylline/aminophylline: the combination may lead to decreased plasma concentrations of theophylline.
- Tricyclic antidepressants (amitriptyline, clomipramine): the combination increases cardiovascular risks (hypotension, tachycardia).

Toxicity

Limited recent information is available on CBD toxicity.

Clinical signs of CBD intoxication may include hypersalivation, vomiting, hyperesthesia, urinary incontinence, prolapse of the third eyelid, mydriasis, tremors, agitation, altered consciousness, lateral recumbency, lethargy, ataxia, paroxysmal seizures, respiratory depression, bradycardia, hyperthermia, or hypothermia (Brutlag and Hommerding 2018).

Treatment involves inducing vomiting if ingestion is recent, administration of activated charcoal, and supportive care. In cases of agitation or seizures, butorphanol and benzodiazepines (diazepam, midazolam) are recommended. For doses of 30 mg/kg, clinical signs are transient and self-resolving (Kulpa et al. 2021).

25.4.1.4 Utilization Clinique et Posologie

According to the Scientific Literature

The use of cannabinoids, including CBD, in cats is poorly documented in veterinary medicine (Yu and Rupasinghe 2021).

In a study by Masataka (2024), CBD was administered at a dose of 4 mg/kg once daily as an anxiolytic and to facilitate cats' attachment to humans. The study involved 10 cats who received CBD for 2 weeks and then a placebo for the same duration, with a 30-day washout period between treatments. During the CBD phase, cats showed reduced avoidance of caregivers and spent more time near them compared to the placebo phase. However, physical interactions with caregivers were longer during the placebo phase than during the CBD phase. This is currently the only study of its kind.

For pain management, however, there is more extensive research, and CBD appears more effective than it is for phobic and anxious states. Studies in cats have demonstrated CBD's efficacy in treating chronic gingivostomatitis (Coelho et al. 2023). The recommended dose is 2–4 mg per cat, administered twice daily.

Additional studies have explored CBD's use in managing arthritic pain (Gutierre et al. 2023), perioperative pain (Zanelli et al. 2024), and oncologic pain (Banach and Ferrero 2023; Buranakarn 2020), with positive outcomes. For pain management, prescribed doses range from 0.2 to 0.25 mg/kg twice daily. However, protocols and products vary widely, and the results are inconsistent across studies.

In humans, the meta-analysis by Blessing et al. (2015) highlights the potential of CBD in treating anxiety in laboratory animals. Subsequent case studies suggest promising effects in treating generalized anxiety (Elms et al. 2019) and insomnia (Shannon et al. 2019). Many synthetic cannabinoids are under evaluation for various psychiatric disorders; however, no evidence currently supports a positive effect of these molecules.

- **Experience Feedback**

The use of CBD for reactive states could be considered. However, the lack of proof of efficacy, its cost, and the difficulty of administration in cats prevent us from recommending its prescription. If prescribed, it is advised to use pure products containing only CBD, such as *Epidiolex*™, which has regulatory approval. However, these products are often expensive and difficult to obtain.

- **Possible Combinations**

Refer to the section "Drug interactions" above.

- **Withdrawal**

Due to limited information, a cautious withdrawal is advised, especially if the treatment duration was extended because of the molecule's cumulative effects in tissues.

25.4.2 Dronabinol

The importance of discussing this molecule lies in better understanding the effects of THC alone and deriving guidelines for the use of THC or components containing it. Information about cats is nonexistent, apart from certain toxicological reports (Brutlag and Hommerding 2018). The following information pertains to humans, except for specific toxicological data.

25.4.2.1 Description, Chemical Formula, and Brand Names

Dronabinol, with the chemical name (6aR,10aR)-6,6,9-trimethyl-3-pentyl-6a,7,8,10a-tetrahydrobenzo[c] chromen-1-ol, is a synthetic δ-9-tetrahydrocannabinol (or THC). No veterinary formulation exists. Dronabinol is available in human pharmacies in countries where its use is permitted under various trade names, such as *Marinol*™ (capsules of 2.5 mg), *Syndros*™ (oral solution of 5 mg/ml), *Reduvo*™, and *Adversa*™.

25.4.2.2 Pharmacology

- **Mechanism of Action**

THC acts as a partial agonist of CB1 and CB2 receptors, with greater affinity for CB1 receptors. It is also a negative allosteric modulator of μ and δ opioid receptors (Murkar et al. 2021) (■ Fig. 25.6).

- **Metabolism and Involved Cytochromes**

The hepatic metabolism of THC is particularly complex, with 80 metabolites identified. The primary active metabolite, 11-hydroxy-δ-9-tetrahydrocannabinol (11-OH-δ- 9-THC), can produce psychological and behavioral effects (Fabritius et al. 2012).

- **Pharmacokinetics**

Dronabinol and its primary active metabolite, 11-OH-δ-9-THC, are present in approximately equal concentrations in the plasma.

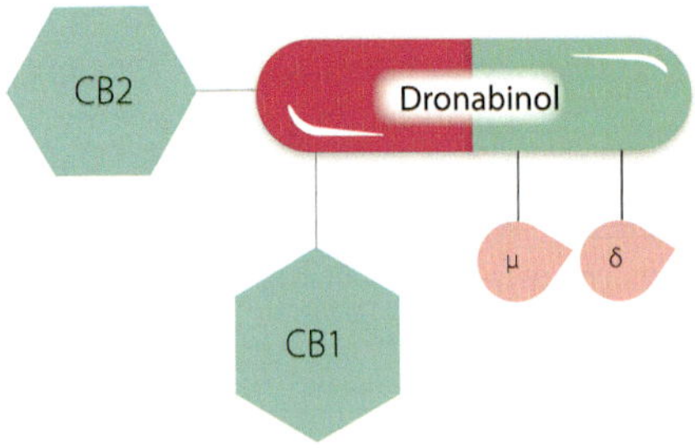

■ **Fig. 25.6** Structure of dronabinol. Partial agonist CB1—partial agonist CB2—allosteric modulator of μ and δ opioid receptors

Their plasma concentrations peak about 30 min to 4 h after oral administration. Their elimination occurs via the urine (15%) and feces (50%) over several days.

There is tissue accumulation of THC, particularly in adipose tissue, leading to a potential cumulative effect that can alter the molecule's pharmacokinetics. With repeated administration, the plasma concentration of THC increases over time (Child and Tallon 2022).

25.4.2.3 Side Effects and Usage Precautions

Side Effects

Published studies indicate no major toxicity and relatively good tolerance for dronabinol (Schimrigk et al. 2017). Dependence is rare and typically observed only with prolonged use at high doses (U.S. Food and Drug Administration, 2004). Common side effects include headaches, dizziness, fatigue, myalgia, and muscle weakness (Schimrigk et al. 2017). These side effects are dose-dependent and rarely require discontinuation of treatment (Beal et al. 1995; Plasse et al. 1991).

In cats, a study using THC and CBD at doses of up to 41.5 mg/kg and 30.5 mg/kg, respectively, reported transient and self-resolving side effects, including lethargy, hypothermia, ataxia, and prolapse of the third eyelid (Kulpa et al. 2021).

Precautions and Contraindications

Dronabinol should be used cautiously in elderly patients and those with hepatic insufficiency.

Drug Interactions

Reported interactions in humans (Narouze and MacCallum 2021) include the following.

- Opioids, barbiturates, benzodiazepines, lithium, ethanol, antihistamines, muscle relaxants, other CNS depressants: coadministration with THC may exacerbate additive drowsiness and CNS depression.
- Amphetamines, cocaine, other sympathomimetics: coadministration with THC may exacerbate hypertension, tachycardia, and potentially cause cardiotoxicity.
- Atropine, scopolamine, antihistamines, other anticholinergic agents: coadministration with THC may exacerbate tachycardia and drowsiness.
- Amitriptyline, amoxapine, desipramine, other tricyclic antidepressants: coadministration with THC may exacerbate tachycardia, hypertension, and drowsiness.
- Fluoxetine, disulfiram: coadministration with THC may induce a hypomanic reaction.
- Barbiturates, antipyrine: coadministration with THC may reduce the elimination of these drugs.
- Theophylline: coadministration with THC may increase the metabolism of theophylline.

Toxicity

In cases of intoxication in cats, symptoms include altered consciousness, paroxysmal seizures, ataxia, vocalizations, behavioral changes, hypersalivation, diarrhea, vomiting, bradycardia or tachycardia, hypothermia or hyperthermia, tremors, hyperesthesia, and mydriasis (Brutlag and Hommerding 2018; Janeczek et al. 2018). No toxic dose has been established for cats.

25.4.2.4 Clinical Use and Dosage

According to the Scientific Literature

Dronabinol is used in human medicine as an orexigenic, an analgesic, an antiemetic, in the treatment of certain sleep apneas, and for cannabis addiction (Bar-Sela et al. 2019).

Experience Feedback

As noted at the beginning of this section, there is no information on the use of dronabinol and THC in cats, as no product has marketing authorization for this purpose. Moreover, depending on the dosage, products containing these active principles are prohibited for sale in France.

The key points to retain include the partial agonist effect on CB1 and CB2 receptors, the potential for a very long half-life, hepatic metabolism, and biliary excretion as the primary elimination pathway.

The LD50 is high, ensuring a relative safety margin for the product. However, the therapeutic potential of cannabinoids in the treatment of certain psychopathological conditions remains unvalidated.

Extreme caution should be exercised regarding the expected outcomes, and standardized, placebo-controlled clinical studies are necessary. The therapeutic margins are broad, and these products appear to have low toxicity.

Finally, it is important to note that the effects of natural extracts of *Cannabis sativa* are not limited to THC and CBD, whose concentrations vary from extract to extract. These extracts also contain numerous other phytocannabinoids with complex, highly nonselective pharmacological effects that are largely understudied. It is therefore recommended to opt for pharmaceutical-grade products with a known and controlled composition, as permitted by national legislation.

Summary Table of Other Psychotropic Medications

Substance	Mechanism of action	Pathological states for which the molecule is indicated	Dosage	Cytochromes
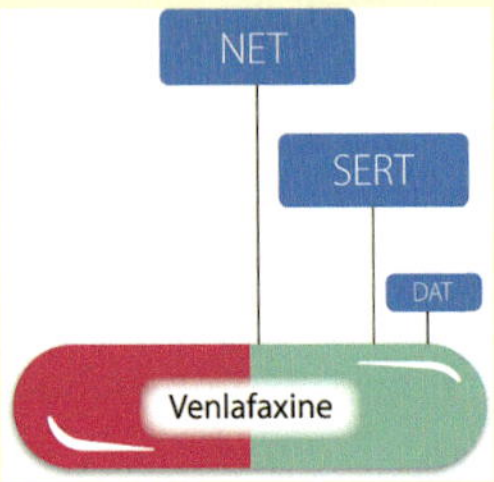	Inhibitor of SERT Inhibitor of NET Inhibitor of DAT	Anxious impulsive or compulsive states	1–2 mg/kg BID	CYP2D6 CYP3A4 CYP2C19 CYP2C9
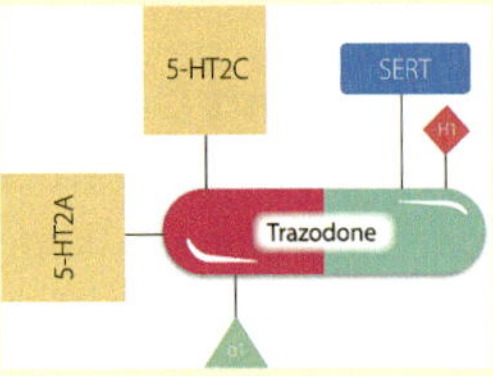	Inhibitor of SERT Antagonist of 5-HT2A Antagonist of 5-HT2C α1-adrenergic antagonist Histamine antagonist	Phobic Anxious Impulsive or compulsive states	7–10 mg/kg PRN (every 8 h) in combination with fluoxetine, start at 5 mg/kg	CYP3A4 CYP2D6
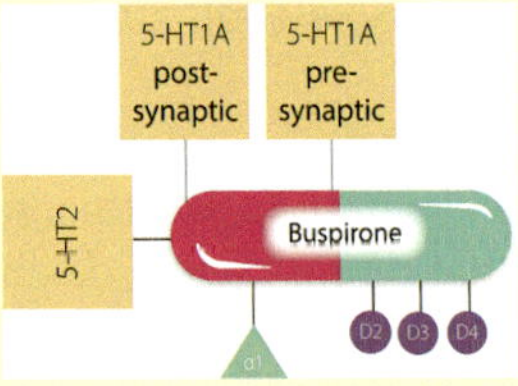	Partial agonist of 5-HT1A (presynaptic) Partial agonist of 5-HT1A (postsynaptic) Antagonist of 5-HT2 Partial antagonist of D2 D3, D4 antagonist Partial α1-adrenergic agonist	Phobic or anxious	0.5–1 mg/kg BID or TID	CYP3A4

Substance	Mechanism of action	Pathological states for which the molecule is indicated	Dosage	Cytochromes
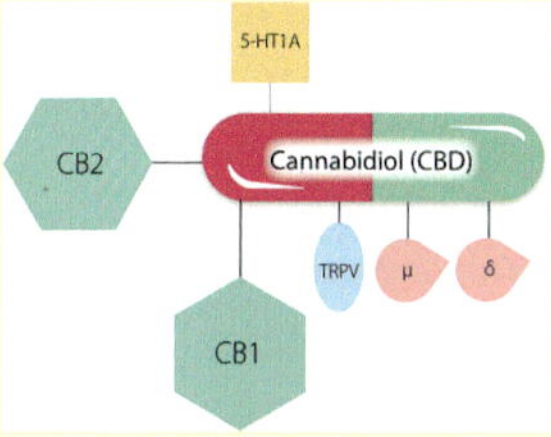	**Negative allosteric modulator of CB1** **Inverse agonist of CB2** TRPV desensitizer Partial agonist of 5-HT1A Negative allosteric modulator of µ-opioids Negative allosteric modulator of δ-opioids	Phobic or anxious	1.5–2 mg/kg BID increase as needed up to 10 mg/kg BID	CYP2C9 CYP2C19 CYP2D6 CYP3A4
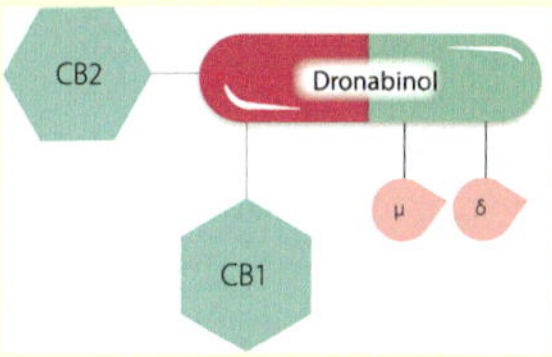	**Partial agonist of CB1** **Partial agonist of CB2** Negative allosteric modulator of µ-opioids Negative allosteric modulator of δ-opioids	Phobic or anxious Impulsive or compulsive	2.5–5 mg per individual (in humans)	

References

Banach D, Ferrero P (2023) Cannabis and pathologies in dogs and cats: first survey of phytocannabinoid use in veterinary medicine in Argentina. J Cannabis Res 5(1):39. https://doi.org/10.1186/s42238-023-00209-5

Bar-Sela G, Zalman D, Semenysty V, Ballan E (2019) The effects of dosage-controlled cannabis capsules on cancer-related cachexia and anorexia syndrome in advanced cancer patients: pilot study. Integr Cancer Ther 18:153473541988149. https://doi.org/10.1177/1534735419881498

Beal JE, Olson R, Laubenstein L, Morales JO, Bellman P, Yangco B, Lefkowitz L, Plasse TF, Shepard KV (1995) Dronabinol as a treatment for anorexia associated with weight loss in patients with AIDS. J Pain Symptom Manag 10(2):89–97. https://doi.org/10.1016/0885-3924(94)00117-4

Bisogno T, Melck D, Bobrov MY, Gretskaya NM, Bezuglov VV, De Petrocellis L, Di Marzo V (2000) N-acyl-dopamines: novel synthetic CB1 cannabinoid-receptor ligands and inhibitors of anandamide inactivation with cannabimimetic activity in vitro and in vivo. Biochem J 351(3):817–824. https://doi.org/10.1042/bj3510817

Bjorvatn B, Fornal CA, Martín FJ, Metzler CW, Jacobs BL (2000) Venlafaxine and its interaction with WAY 100635: effects on serotonergic unit activity and behavior in cats. Eur J Pharmacol 404(1-2):121–132. https://doi.org/10.1016/S0014-2999(00)00622-1

Blessing EM, Steenkamp MM, Manzanares J, Marmar CR (2015) Cannabidiol as a potential treatment for anxiety disorders. Neurotherapeutics 12(4):825–836. https://doi.org/10.1007/s13311-015-0387-1

Brosnan RJ, Pypendop BH, Cenani A (2024) Effects of trazodone and dexmedetomidine on fentanyl-mediated reduction of isoflurane minimum alveolar concentration in cats. Vet Anaesth Analg 51(1):80–89. https://doi.org/10.1016/j.vaa.2023.09.130

Brutlag A, Hommerding H (2018) Toxicology of marijuana, synthetic cannabinoids, and cannabidiol in dogs and cats. Vet Clin N Am Small Anim Pract 48(6):1087–1102. https://doi.org/10.1016/j.cvsm.2018.07.008

Buranakarn V (2020) Sarcoma cancer treatment using extracted cannabis oil in cat. Int J Sci Innov Technol 3(1):35–40

Calvi A, Fischetti I, Verzicco I, Belvederi Murri M, Zanetidou S, Volpi R, Coghi P, Tedeschi S, Amore M, Cabassi A (2021) Antidepressant drugs effects on blood pressure. Front Cardiovasc Med 8:704281. https://doi.org/10.3389/fcvm.2021.704281

Castillo PE, Younts TJ, Chávez AE, Hashimotodani Y (2012) Endocannabinoid signaling and synaptic function. Neuron 76(1):70–81. https://doi.org/10.1016/j.neuron.2012.09.020

Chávez G, Pardo P, Ubilla MJ, Marín MP (2016) Effects on behavioural variables of oral versus transdermal buspirone administration in cats displaying urine marking. J Appl Anim Res 44(1):454–457. https://doi.org/10.1080/09712119.2015.1091335

Child RB, Tallon MJ (2022) Cannabidiol (CBD) dosing: plasma pharmacokinetics and effects on accumulation in skeletal muscle, liver and adipose tissue. Nutrients 14(10):2101. https://doi.org/10.3390/nu14102101

Coelho JC, Duarte N, Bento Da Silva A, Bronze MDR, Mestrinho LA (2023) Placebo-controlled trial of daily oral cannabidiol as adjunctive treatment for cats with chronic gingivostomatitis. Animals 13(17):2716. https://doi.org/10.3390/ani13172716

Coltherd JC, Bednall R, Bakke AM, Ellerby Z, Newman C, Watson P, Logan DW, Holcombe LJ (2024) Healthy cats tolerate long-term daily feeding of cannabidiol. Front Vet Sci 10:1324622. https://doi.org/10.3389/fvets.2023.1324622

Corsato Alvarenga I, Panickar KS, Hess H, McGrath S (2023) Scientific validation of cannabidiol for management of dog and cat diseases. Annu Rev Anim Biosci 11(1):227–246. https://doi.org/10.1146/annurev-animal-081122-070236

Crowell-Davis SL, Murray TF, de Souza M, Dantas L (2019) Veterinary psychopharmacology, 2nd edn. Wiley Blackwell

Deabold KA, Schwark WS, Wolf L, Wakshlag JJ (2019) Single-dose pharmacokinetics and preliminary safety assessment with use of CBD-rich hemp nutraceutical in healthy dogs and cats. Animals 9(10):832. https://doi.org/10.3390/ani9100832

Delgado CV (2023) Managing stress using trazodone-effects in hospitalized cats

Den Boon FS, Chameau P, Schaafsma-Zhao Q, Van Aken W, Bari M, Oddi S, Kruse CG, Maccarrone M, Wadman WJ, Werkman TR (2012) Excitability of prefrontal cortical pyramidal neurons is modulated by activation of intracellular type-2 cannabinoid receptors. Proc Natl Acad Sci 109(9):3534–3539. https://doi.org/10.1073/pnas.1118167109

Denenberg S, Dubé MB (2018) Tools for managing feline problem behaviours: psychoactive medications. J Feline Med Surg 20(11):1034–1045. https://doi.org/10.1177/1098612X18806760

Devane WA, Hanuš L, Breuer A, Pertwee RG, Stevenson LA, Griffin G, Gibson D, Mandelbaum A, Etinger A, Mechoulam R (1992) Isolation and structure of a brain constituent that binds to the cannabinoid receptor. Science 258(5090):1946–1949. https://doi.org/10.1126/science.1470919

Egger C, Muehlbacher M, Nickel M, Geretsegger C, Stuppaeck C (2006) A review on hyponatremia associated with SSRIs, reboxetine and venlafaxine. Int J Psychiatry Clin Pract 10(1):17–26. https://doi.org/10.1080/13651500500410216

Elms L, Shannon S, Hughes S, Lewis N (2019) Cannabidiol in the treatment of post-traumatic stress disorder: a case series. J Altern Complement Med 25(4):392–397. https://doi.org/10.1089/acm.2018.0437

Fabritius M, Staub C, Mangin P, Giroud C (2012) Distribution of free and conjugated cannabinoids in human bile samples. Forensic Sci Int 223(1–3):114–118. https://doi.org/10.1016/j.forsciint.2012.08.013

Gaffuri A-L, Ladarre D, Lenkei Z (2012) Type-1 cannabinoid receptor signaling in neuronal development. Pharmacology 90(1–2):19–39. https://doi.org/10.1159/000339075

García-Gutiérrez MS, García-Bueno B, Zoppi S, Leza JC, Manzanares J (2012) Chronic blockade of cannabinoid CB_2 receptors induces anxiolytic-like actions associated with alterations in $GABA_A$ receptors. Br J Pharmacol 165(4):951–964. https://doi.org/10.1111/j.1476-5381.2011.01625.x

Godlewski G, Offertáler L, Wagner JA, Kunos G (2009) Receptors for acylethanolamides—GPR55 and GPR119. Prostaglandins Other Lipid Mediat 89(3–4):105–111. https://doi.org/10.1016/j.prostaglandins.2009.07.001

Gong J-P, Onaivi ES, Ishiguro H, Liu Q-R, Tagliaferro PA, Brusco A, Uhl GR (2006) Cannabinoid CB2 receptors: immunohistochemical localization in rat brain. Brain Res 1071(1):10–23. https://doi.org/10.1016/j.brainres.2005.11.035

Gruen ME, Sherman BL (2008) Use of trazodone as an adjunctive agent in the treatment of canine anxiety disorders: 56 cases (1995–2007). Journal of the American Veterinary Medical Association 233, 1902–1907. https://doi.org/10.2460/javma.233.12.1902

Gutierre E, Crosignani N, García-Carnelli C, Di Mateo A, Recchi L (2023) A case report of CBD and THC as analgesic therapy in a cat with chronic osteoarthritic pain. Vet Med Sci 9(3):1021–1025. https://doi.org/10.1002/vms3.1057

Hanuš L, Abu-Lafi S, Fride E, Breuer A, Vogel Z, Shalev DE, Kustanovich I, Mechoulam R (2001) 2-Arachidonyl glyceryl ether, an endogenous agonist of the cannabinoid CB_1 receptor. Proc Natl Acad Sci 98(7):3662–3665. https://doi.org/10.1073/pnas.061029898

Hart BL, Eckstein RA, Powell KL, Dodman NH (1993) Effectiveness of buspirone on urine spraying and inappropriate urination in cats. J Am Vet Med Assoc 203(2):254–258. https://doi.org/10.2460/javma.1993.203.02.254

Heifets BD, Castillo PE (2009) Endocannabinoid Signaling and Long-Term Synaptic Plasticity. Annu. Rev. Physiol. 71, 283–306. https://doi.org/10.1146/annurev.physiol.010908.163149

Ho W, Barrett DA, Randall MD (2008) 'Entourage' effects of *N*-palmitoylethanolamide and *N*-oleoylethanolamide on vasorelaxation to anandamide occur through TRPV1 receptors. Br J Pharmacol

155(6):837–846. https://doi.org/10.1038/bjp.2008.324
Hopfensperger MJ (2016) Use of oral venlafaxine in cats with feline idiopathic cystitis or behavioral causes of periuria
Howland RH (2015) Buspirone: Back to the Future. J Psychosoc Nurs Ment Health Serv 53, 21–24. https://doi.org/10.3928/02793695-20151022-01
Iffland K, Grotenhermen F (2017) An update on safety and side effects of cannabidiol: a review of clinical data and relevant animal studies. Cannabis and cannabinoid research 2, 139–154
Ivy D, Palese F, Vozella V, Fotio Y, Yalcin A, Ramirez G, Mears D, Wynn G, Piomelli D (2020) Cannabinoid CB2 receptors mediate the anxiolytic-like effects of monoacylglycerol lipase inhibition in a rat model of predator-induced fear. Neuropsychopharmacology 45(8):1330–1338. https://doi.org/10.1038/s41386-020-0696-x
Janeczek A, Zawadzki M, Szpot P, Niedzwiedz A (2018) Marijuana intoxication in a cat. Acta Vet Scand 60(1):44. https://doi.org/10.1186/s13028-018-0398-0
Jay AR, Krotscheck U, Parsley E, Benson L, Kravitz A, Mulligan A, Silva J, Mohammed H, Schwark WS (2013) Pharmacokinetics, bioavailability, and hemodynamic effects of trazodone after intravenous and oral administration of a single dose to dogs. American Journal of Veterinary Research 74, 1450–1456. https://doi.org/10.2460/ajvr.74.11.1450
Jukier T, Cruz-Espindola C, Martin D, Boothe DM (2023) Disposition of a single oral dose of a cannabidiol medication in healthy cats. Front Vet Sci 10:1181517. https://doi.org/10.3389/fvets.2023.1181517
Kulpa JE, Paulionis LJ, Eglit GM, Vaughn DM (2021) Safety and tolerability of escalating cannabinoid doses in healthy cats. J Feline Med Surg 23(12):1162–1175. https://doi.org/10.1177/1098612X211004215
Larrinaga G, Varona A, Pérez I, Sanz B, Ulgade A, Cándenas ML, Pinto FM, Gil J, López JI (2010) Expression of cannabinoid receptors in human kidney. Histol Histopathol 25:1133–1138. https://doi.org/10.14670/HH-25.1133
Lassen D, Ennis ZN, Damkier P (2016) First-trimester pregnancy exposure to venlafaxine or duloxetine and risk of major congenital malformations : a systematic review. Basic Clin Pharmacol Toxicol 118(1):32–36. https://doi.org/10.1111/bcpt.12497
Masataka N (2024) Is cannabidiol (CBD) effective to ease separation anxiety? Heliyon 10(3):e25851. https://doi.org/10.1016/j.heliyon.2024.e25851
Masson S, Metz D, Bleuer-Elsner S, Schwobthaler F (2024) Retrospective study on the use of venlafaxine in 176 cats diagnosed with behavioral disorders. Preprint article
Mealey KL, Peck KE, Bennett BS, Sellon RK, Swinney GR, Melzer K, Gokhale SA, Krone TM (2004) Systemic absorption of amitriptyline and Buspirone after oral and transdermal administration to healthy cats. J Vet Intern Med 18(1):43–46. https://doi.org/10.1111/j.1939-1676.2004.tb00133.x
Mechoulam R, Hanuš LO, Pertwee R, Howlett AC (2014) Early phytocannabinoid chemistry to endocannabinoids and beyond. Nat Rev Neurosci 15(11):757–764. https://doi.org/10.1038/nrn3811
Metz D, Medam T, Masson S (2022) Double-blind, placebo-controlled trial of venlafaxine to treat behavioural disorders in cats: a pilot study. J Feline Med Surg 24(6):539–549. https://doi.org/10.1177/1098612X211036792
Miller LK, Devi LA (2011) The highs and lows of cannabinoid receptor expression in disease: mechanisms and their therapeutic implications. Pharmacol Rev 63(3):461–470. https://doi.org/10.1124/pr.110.003491
Muccioli GG, Naslain D, Bäckhed F, Reigstad CS, Lambert DM, Delzenne NM, Cani PD (2010) The endocannabinoid system links gut microbiota to adipogenesis. Mol Syst Biol 6(1):392. https://doi.org/10.1038/msb.2010.46
Murkar A, De Koninck J, Merali Z (2021) Cannabinoids: revealing their complexity and role in central networks of fear and anxiety. Neurosci Biobehav Rev 131:30–46. https://doi.org/10.1016/j.neubiorev.2021.09.002
Murphy LA, Barletta M, Graham LF, Reichl LJ, Duxbury MM, Quandt JE (2017) Effects of acepromazine and trazodone on anesthetic induction dose of propofol and cardiovascular variables in dogs undergoing general anesthesia for orthopedic surgery. javma 250, 408–416. https://doi.org/10.2460/javma.250.4.408
Narouze SN, MacCallum CA (2021) Cannabinoids and pain. Springer
Newport DJ, Ritchie JC, Knight BT, Glover BA, Zach EB, Stowe ZN (2009) Venlafaxine in human breast milk and nursing infant plasma: determination of exposure. J Clin Psychiatry 70(9):1304–1310. https://doi.org/10.4088/JCP.08m05001
Nilsen OG, Dale O (1992) Single dose pharmacokinetics of trazodone in healthy subjects. Pharmacol Toxicol 71(2):150–153. https://doi.org/10.1111/j.1600-0773.1992.tb00535.x
Orlando JM, Case BC, Thomson AE, Griffith E, Sherman BL (2016) Use of oral trazodone for sedation in cats: a pilot study. J Feline Med Surg 18(6):476–482. https://doi.org/10.1177/1098612X15587956
Paulis MG, Hafez EM, El-Tahawy NF, Aly MKM (2018) Toxicological assessment of venlafaxine: acute and subchronic toxicity study in rats. Int J Toxicol 37(4):327–334. https://doi.org/10.1177/1091581818777470
Pertwee RG (2014) Elevating endocannabinoid levels: pharmacological strategies and potential therapeutic applications. Proc Nutr Soc 73(1):96–105. https://doi.org/10.1017/S0029665113003649
Pflaum K, Bennett S (2021) Investigation of the use of venlafaxine for treatment of refractory misdirected

play and impulse-control aggression in a cat: a case report. J Vet Behav 42:22–25. https://doi.org/10.1016/j.jveb.2020.10.008

Plasse TF, Gorter RW, Krasnow SH, Lane M, Shepard KV, Wadleigh RG (1991) Recent clinical experience with dronabinol. Pharmacol Biochem Behav 40(3):695–700. https://doi.org/10.1016/0091-3057(91)90385-F

Porter AC, Sauer J-M, Knierman MD, Becker GW, Berna MJ, Bao J, Nomikos GG, Carter P, Bymaster FP, Leese AB, Felder CC (2002) Characterization of a novel endocannabinoid, virodhamine, with antagonist activity at the CB1 receptor. J Pharmacol Exp Ther 301(3):1020–1024. https://doi.org/10.1124/jpet.301.3.1020

Pugh CM, Sweeney JT, Bloch CP, Lee JA, Johnson JA, Hovda LR (2013) Selective serotonin reuptake inhibitor (SSRI) toxicosis in cats: 33 cases (2004-2010). J Vet Emerg Crit Care 23(5):565–570. https://doi.org/10.1111/vec.12091

Rang HP, Dale MM, Ritter JM, Flower RJ, Henderson G (2012) Rang and Dale's pharmacology, Seventh edition, main edition. ed, Student consult. Elsevier, Churchill Livingstone, Edinburgh London New York Oxford Philadelphia St. Louis Sydney Toronto

Roche M, Finn DP (2010) Brain CB2 receptors: implications for neuropsychiatric disorders. Pharmaceuticals (Basel) 3(8):2517–2553. https://doi.org/10.3390/ph3082517

Rotzinger S, Fang J, Baker GB (1998) Trazodone is metabolized to m-chlorophenylpiperazine by CYP3A4 from human sources. Drug Metab Dispos 26(6):572–575

Rozental AJ, Gustafson DL, Kusick BR, Bartner LR, Castro SC, McGrath S (2023) Pharmacokinetics of escalating single-dose administration of cannabidiol to cats. J Vet Pharmacol Ther 46(1):25–33. https://doi.org/10.1111/jvp.13100

Sawyer LS, Moon-Fanelli AA, Dodman NH (1999) Psychogenic alopecia in cats: 11 cases (1993-1996). J Am Vet Med Assoc 214(1):71–74

Schimrigk S, Marziniak M, Neubauer C, Kugler EM, Werner G, Abramov-Sommariva D (2017) Dronabinol is a safe long-term treatment option for neuropathic pain patients. Eur Neurol 78(5–6):320–329. https://doi.org/10.1159/000481089

Selk Ghaffari M, Sabzevari A (2010) Successful management of psychogenic alopecia with buspirone in a crossbreed cat. Comp Clin Pathol 19(3):317–319. https://doi.org/10.1007/s00580-009-0878-8

Shannon S, Lewis N, Lee H, Hughes S (2019) Cannabidiol in anxiety and sleep: a large case series. Perm J 23(1):18–041. https://doi.org/10.7812/TPP/18-041

Sinn L (2018) Advances in behavioral psychopharmacology. Vet Clin N Am Small Anim Pract 48(3):457–471. https://doi.org/10.1016/j.cvsm.2017.12.011

Stahl SM, Muntner N (2021) In: Grady MM (ed) Stahl's essential psychopharmacology: neuroscientific basis and practical applications, 5th edn. Cambridge University Press. https://doi.org/10.1017/9781108975292

Stevens BJ, Frantz EM, Orlando JM, Griffith E, Harden LB, Gruen ME, Sherman BL (2016) Efficacy of a single dose of trazodone hydrochloride given to cats prior to veterinary visits to reduce signs of transport- and examination-related anxiety. J Am Vet Med Assoc 249(2):202–207. https://doi.org/10.2460/javma.249.2.202

Tsou K, Brown S, Sañudo-Peña MC, Mackie K, Walker JM (1998) Immunohistochemical distribution of cannabinoid CB1 receptors in the rat central nervous system. Neuroscience 83(2):393–411. https://doi.org/10.1016/S0306-4522(97)00436-3

Tucker LE, Sanchez A, Valverde A, Blois S, Uccello O, Rutherford A, Monteith G, Reinhart JM, Keating S, Gu Y, Johnson R (2023) Pharmacokinetic, sedative, and physiological effects of oral compounded formulations of trazodone alone or in combination with gabapentin in male cats. J Vet Pharmacol Ther 46(5):300–310. https://doi.org/10.1111/jvp.13384

U.S. Food and Drug Administration (FDA). (2004). Guidance for Industry: Assessment of Abuse Potential of Drugs. U.S. Department of Health and Human Services

Wen B, Ma L, Rodrigues AD, Zhu M (2008) Detection of novel reactive metabolites of trazodone: evidence for CYP2D6-mediated bioactivation of m-chlorophenylpiperazine. Drug Metab Dispos 36(5):841–850. https://doi.org/10.1124/dmd.107.019471

Yu CHJ, Rupasinghe HPV (2021) Cannabidiol-based natural health products for companion animals: recent advances in the management of anxiety, pain, and inflammation. Res Vet Sci 140:38–46. https://doi.org/10.1016/j.rvsc.2021.08.001

Zanelli GR, Vieira GBM, Souza RVM, Aguiar AJDA, Cassu RN (2024) Perioperative analgesic and sedative effects of cannabidiol in cats undergoing ovariohysterectomy. Animals 14(16):2286. https://doi.org/10.3390/ani14162286

Zhang H-Y, Gao M, Liu Q-R, Bi G-H, Li X, Yang H-J, Gardner EL, Wu J, Xi Z-X (2014) Cannabinoid CB_2 receptors modulate midbrain dopamine neuronal activity and dopamine-related behavior in mice. Proc Natl Acad Sci 111(46):E5007–E5015. https://doi.org/10.1073/pnas.1413210111

Hormonal Therapies

Contents

26.1 Thyroid Hormones – 680
26.1.1 Description, Chemical Formula, and Brand Names – 680
26.1.2 Pharmacology – 680
26.1.3 Side Effects and Usage Precautions – 681
26.1.4 Clinical Use and Dosage – 681

26.2 Sexual Hormones – 681
26.2.1 Megestrol Acetate – 681
26.2.2 Oxytocin – 681
26.2.3 Deslorelin – 682

References – 684

© The Author(s), under exclusive license to Springer Nature Switzerland AG 2026
S. Bleuer-Elsner et al., *Veterinary Psychiatry of the Cat*,
https://doi.org/10.1007/978-3-032-06260-4_26

26

Due to their effects on brain function, hormonal therapies hold significant interest in veterinary psychiatry. This chapter covers the use of sex and thyroid hormones.

Thyroid hormones are key regulators of gene expression. Any imbalance must be identified and corrected. Several veterinary medications based on thiamazole (an inhibitor of iodine incorporation) have received marketing authorization.

Therapies involving sex hormones include megestrol acetate, which is discussed from a historical perspective, oxytocin for its role in social behaviors, and deslorelin for its antiandrogenic properties. Among these molecules, only deslorelin has veterinary marketing authorization.

26.1 Thyroid Hormones

As previously discussed in the first part of this book (see ▶ Chap. 3, ▶ Sect. 3.1.2.2) and in the second part (see ▶ Chap. 14, ▶ Sect. 14.7.1), primary hyperthyroidism is a common disorder in aging cats with behavioral repercussions. A comprehensive blood panel is recommended during a behavioral consultation for elderly cats, and this factor must be assessed before initiating any other treatment.

Depending on the medical diagnosis (e.g., thyroid adenoma, thyroid adenocarcinoma), various treatments are available. Options include an iodine-restricted diet, surgical mass excision, radiotherapy, or medical treatment. The medical approach involves administering an antithyroid agent, methimazole (Trepanier 2007).

26.1.1 Description, Chemical Formula, and Brand Names

Methimazole, also known as thiamazole, has the chemical name 1-methyl-3H-imidazole-2-thione. It limits thyroid hormone production by inhibiting the enzyme thyroperoxidase, which is responsible for iodinating tyrosine. Tyrosine is a precursor to the hormones thyroxine (T4) and triiodothyronine (T3).

Several veterinary formulations exist depending on national regulations: Felimazole™, tablets of 1.25, 2.5, and 5 mg; Apelka™, oral solution at 5 mg/ml; Thiafeline™, tablets of 2.5 and 5 mg; Thiamacare™, oral solution at 10 mg/mL.

In human medicine, methimazole is available as Thyrozol™ and generic formulations. The methimazole precursor, carbimazole, is also found in human medicine under the brand Neo-mercazole™ and generic forms.

26.1.2 Pharmacology

26.1.2.1 Mechanism of Action

Methimazole works by inhibiting the enzyme thyroperoxidase. Thyroperoxidase is responsible for iodinating tyrosyl residues on tyrosine, a precursor of the thyroid hormones T3 and T4. Methimazole also prevents already iodinated tyrosyl residues from combining, thereby inhibiting the formation of thyroxine. It does not affect circulating thyroid hormones, so its effect is not immediate. A reduction in serum hormone concentrations is typically observed 2 to 3 weeks after the initiation of treatment (Trepanier 2007).

26.1.2.2 Metabolism

Methimazole is metabolized in the liver with the assistance of cytochrome P450 enzymes, particularly CYP3A4, CYP2C9, and CYP2C19, and is excreted in the urine (Guo et al. 1997; Jansson et al. 1985).

26.1.2.3 Pharmacokinetics

In cats, orally administered methimazole has a bioavailability of 80% and an average elimination half-life of 5 to 6.5 hours, with considerable individual variability. Administering methimazole with food impairs its absorption and reduces its efficacy; it is therefore recommended to delay meals by at least 1 hour. Peak serum concentrations after oral administration are achieved 30 to 60 minutes postadministration, and the half-life is approximately 10 to 14 hours.

26.1.3 Side Effects and Usage Precautions

Common side effects of methimazole include anorexia, diarrhea, vomiting, vocalization, and lethargy. Less frequently, eosinophilia, leukopenia, lymphocytosis, thrombocytopenia, and prolonged coagulation times may occur. Rare side effects include hepatopathy, blood dyscrasias, hemolytic anemia, facial excoriations, and pruritus.

Methimazole does not halt the progression of thyroid disease. It may exacerbate underlying renal disease by reducing the glomerular filtration rate. Methimazole is contraindicated in cats with systemic conditions such as autoimmune diseases, hepatic, renal, hematological, or coagulation disorders (*Drug Monographs Plumb's Veterinary Drugs* 2024).

26.1.4 Clinical Use and Dosage

The recommended starting dose for treatment is 2.5 mg per cat twice daily. After 3 weeks of treatment, the dose can be adjusted based on blood test results, in increments of 2.5 mg per cat twice daily. The maximum recommended dose is 10 mg per cat twice daily (*Drug Monographs Plumb's Veterinary Drugs* 2024).

26.2 Sexual Hormones

Sex hormones and their actions have been detailed in the first part (see ▶ Chap. 3, ▶ Sect. 3.2.3) and the second part (see ▶ Chap. 14, ▶ Sect. 14.7.6) of this book. This section focuses on their relevance in the treatment of behavioral disorders. Specifically, we will discuss megestrol acetate, oxytocin, cyproterone acetate, and deslorelin.

26.2.1 Megestrol Acetate

Megestrol acetate was historically widely used as a contraceptive for intact female cats. Its use has been associated with the development of diabetes (Peterson 1987), fibroadenomatosis, and mammary adenocarcinoma (Hayden et al. 1989). Additionally, it has been employed in veterinary psychiatry for managing various behavioral disorders in cats, particularly aggression and urine marking (Norsworthy and Romagnoli 2016; Rodan and Heath 2016).

As a hormone, megestrol acetate may seem promising for mood disorder management. However, its numerous side effects and inconsistent outcomes have led to its abandonment in veterinary psychiatry. If considered for use, it should be reserved as a last resort when other options have failed, and only to avoid euthanasia in conditions such as dysthymia or dissociative syndromes that are unresponsive to conventional treatments.

26.2.2 Oxytocin

Beyond its role in lactation, oxytocin appears to be a hormonal factor supporting the human-cat bond (Koyasu et al. 2022; Nagasawa et al. 2022).

In theory, oxytocin may have potential therapeutic applications in addressing interspecies and intraspecies relational disorders, as well as psychopathological conditions involving fear and anxiety with a social component. For example, in humans, intranasal administration of oxytocin has been studied in patients suffering from social anxiety and post-traumatic stress disorder (PTSD). In the case of social anxiety, intranasal oxytocin has been shown to enhance patient confidence during public speaking (Alvares et al. 2012; Guastella et al. 2009). Functional MRI studies have demonstrated that intranasal oxytocin decreases activation in the amygdala, rostral anterior cingulate cortex, and medial prefrontal cortex, key brain regions involved in fear and anxiety, in patients with social anxiety (Tully et al. 2018). In PTSD patients, intranasal oxytocin has shown some efficacy in strengthening extinction processes during therapy (Acheson et al. 2013).

In cats, studies in shelter environments suggest potential benefits. Intranasal administration of oxytocin increased the frequency of human-directed approaches (Sakaguchi 2019). Another study by Hattori et al. (2024) found that intranasal oxytocin increased the duration of cats' gaze towards humans, regardless of familiarity.

However, genetic variability in oxytocin receptors can modulate individual responses to the hormone, as seen in humans (Inoue et al. 2010; Rodrigues et al. 2009; Wu et al. 2012) and cats (Arahori et al. 2016, 2017). This variability contributes to inconsistent results following intranasal administration of oxytocin. Additionally, intranasal administration presents practical challenges in cats, making widespread application difficult.

In summary, while oxytocin may hold promise in managing relationship-related disorders in cats, further studies are required to identify the profiles of cats and specific conditions that could benefit from such treatment. Given the challenges of administration, achieving good compliance in cats appears unlikely at present.

26.2.3 Deslorelin

26.2.3.1 Description, Chemical Formula, and Brand Names

Deslorelin is a gonadotropin-releasing hormone (GnRH) agonist. In veterinary medicine, it is available as a long-acting subcutaneous implant containing 4.7 mg or 9.4 mg of deslorelin, marketed under the brand name *Suprelorin*™. This implant is authorized for temporary infertility induction in sexually mature, intact, and healthy male dogs, male cats, and ferrets. In feline veterinary medicine, it is used off-label as a reversible chemical sterilization agent for both male and female cats, though its use in females is not officially approved.

26.2.3.2 Pharmacology

Deslorelin has two primary effects: stimulation of GnRH receptors followed by desensitization of the same receptors. GnRH receptor stimulation increases follicle-stimulating hormone and luteinizing hormone levels, leading to an elevated release of sex hormones. This results in increased testosterone levels in males and elevated estrogen levels in females (Goericke-Pesch et al. 2011; Toydemir et al. 2012). The initial rise in gonadotropins and sex hormones is referred to as the "flare-up."

An increase in plasma testosterone is observed within 20 minutes after implant placement and gradually decreases over an average of 12 days until reaching undetectable levels. In clinical practice, estrus in females and exacerbated sexual behaviors in both males and females may occur shortly after implant placement and can persist for up to 3 weeks.

The subsequent desensitization of GnRH receptors leads to a fully reversible downregulation of testicular and ovarian endocrine function in male and female cats (Fontaine and Fontbonne 2011).

26.2.3.3 Side Effects and Usage Precautions

The side effects associated with deslorelin implants stem from two different mechanisms.

The first set of side effects is short-lived and directly results from the flare-up phase. Symptoms related to increased sex hormones, such as estrus, urine marking, aggression, and mounting behaviors, may intensify within the first 10 days after implant placement. Estrus induced by the flare-up effect can be fertile, presenting potential challenges for pet owners. Although these effects are temporary, they can be significant and require careful management.

The second set of side effects aligns with those typically observed after surgical castration. These include increased appetite and weight gain. Notably, the side effects caused by the implant mirror those of surgical castration but are reversible, which is a critical distinction.

Additionally, a significant reduction in testicular volume is observed, which usually reverses shortly before the behavioral effects of the treatment subside.

26.2.3.4 Experience Feedback and Posology

Deslorelin is primarily used in physiological conditions by pet owners who wish to avoid permanent surgical castration while still preventing their pets from reproducing. It allows observation of the animal's behavior as if sterilized and provides insight into potential effects, such as weight gain, associated with this status.

It is also indicated when owners want to delay breeding. By mitigating behaviors associated with estrus in females and urine marking in intact males, deslorelin can offer a temporary solution. Additionally, deslorelin is advised for females deemed too young or small for breeding. Implanting deslorelin in prepubescent females delays the onset of estrus by approximately 3 months (Risso et al. 2012).

During the flare-up effect, estrus may still be induced. In dogs, early placement of a deslorelin implant delays the closure of growth plates, requiring caution in larger breeds. Such studies have not been conducted in cats (Kaya et al. 2015).

Similarly, in prepubescent males, the placement of a deslorelin implant delays puberty until at least 11 months of age. Some males may never become fertile, even after the removal or exhaustion of the implant (Mehl et al. 2016).

To date, deslorelin is used far less frequently in cats than in dogs. The implant's duration of action varies significantly in cats, ranging from 6 to 37 months in females and 5 months to over 48 months in males. The implant's efficacy is not absolute in cats, with 1% to 3% of both males and females not achieving complete sterility (Goericke-Pesch 2017; Toydemir et al. 2012).

The implant can be placed near the umbilicus, allowing for its removal under anesthesia before it is fully exhausted. Over time, the implant degrades, and multiple fragments may form, sometimes making extraction challenging during removal (Fontaine and Fontbonne 2011). After removal, fertility is typically restored within 10 to 45 weeks in male cats and within a few weeks in female cats.

Summary Table of Complementary Hormonal Treatments

Substance	Mechanism of action	Pathological states for which the molecule is indicated	Dosage	Cyto-chromes
Thiamazole	Prevents iodination of T3 and T4 precursors	Hyperthyroidism	Methimazole 2.5–10 mg/cat BID	CYP450
Megestrol acetate	Testosterone inhibitor GnRH inhibitor	Not indicated		CYP3A4
Deslorelin	GnRH agonist	Behavioral issues related to reproduction ("physiological tat")	Implant PRN	–

26

References

Acheson D, Feifel D, De Wilde S, Mckinney R, Lohr J, Risbrough V (2013) The effect of intranasal oxytocin treatment on conditioned fear extinction and recall in a healthy human sample. Psychopharmacology 229(1):199–208. https://doi.org/10.1007/s00213-013-3099-4

Alvares GA, Chen NTM, Balleine BW, Hickie IB, Guastella AJ (2012) Oxytocin selectively moderates negative cognitive appraisals in high trait anxious males. Psychoneuroendocrinology 37(12):2022–2031. https://doi.org/10.1016/j.psyneuen.2012.04.018

Arahori M, Hori Y, Saito A, Chijiiwa H, Takagi S, Ito Y, Watanabe A, Inoue-Murayama M, Fujita K (2016) The oxytocin receptor gene (OXTR) polymorphism in cats (Felis catus) is associated with "Roughness" assessed by owners. J Vet Behav 11:109–112. https://doi.org/10.1016/j.jveb.2015.07.039

Arahori M, Chijiiwa H, Takagi S, Bucher B, Abe H, Inoue-Murayama M, Fujita K (2017) Microsatellite polymorphisms adjacent to the oxytocin receptor gene in domestic cats: association with personality? Front Psychol 8:2165. https://doi.org/10.3389/fpsyg.2017.02165

Drug Monograph - Plumb's Veterinary Drugs [WWW Document], (2024) URL https://app.plumbs.com/drug-monograph/hVa7Ct9JYiPROD (accessed 7.19.21)

Fontaine E, Fontbonne A (2011) Clinical use of GnRH agonists in canine and feline species. Reprod Domest Anim 46(2):344–353. https://doi.org/10.1111/j.1439-0531.2010.01705.x

Goericke-Pesch S (2017) Long-term effects of GnRH agonists on fertility and behaviour. Reprod Domest Anim 52(S2):336–347. https://doi.org/10.1111/rda.12898

Goericke-Pesch S, Georgiev P, Antonov A, Albouy M, Wehrend A (2011) Clinical efficacy of a GnRH-agonist implant containing 4.7 mg deslorelin, Suprelorin®, regarding suppression of reproductive function in tomcats. Theriogenology 75(5):803–810. https://doi.org/10.1016/j.theriogenology.2010.10.020

Guastella AJ, Howard AL, Dadds MR, Mitchell P, Carson DS (2009) A randomized controlled trial of intranasal oxytocin as an adjunct to exposure therapy for social anxiety disorder. Psychoneuroendocrinology 34(6):917–923. https://doi.org/10.1016/j.psyneuen.2009.01.005

Guo Z, Raeissi S, White RB, Stevens JC (1997) Orphenadrine and methimazole inhibit multiple cytochrome P450 enzymes in human liver microsomes. Drug Metab Dispos 25(3):390–393

Hattori M, Kinoshita K, Saito A, Yamamoto S (2024) Exogenous oxytocin increases gaze to humans in male cats. Sci Rep 14(1):8953. https://doi.org/10.1038/s41598-024-59161-w

Hayden DW, Barnes DM, Johnson KH (1989) Morphologic changes in the mammary gland of megestrol acetate-treated and untreated cats: a retrospective study. Vet Pathol 26(2):104–113. https://doi.org/10.1177/030098588902600202

Inoue H, Yamasue H, Tochigi M, Abe O, Liu X, Kawamura Y, Takei K, Suga M, Yamada H, Rogers MA, Aoki S, Sasaki T, Kasai K (2010) Association between the oxytocin receptor gene and amygdalar volume in healthy adults. Biol Psychiatry 68(11):1066–1072. https://doi.org/10.1016/j.biopsych.2010.07.019

Jansson R, Lindström B, Dahlberg PA (1985) Pharmacokinetic properties and bioavailability of methimazole. Clin Pharmacokinet 10(5):443–450. https://doi.org/10.2165/00003088-198510050-00006

Kaya D, Schäfer-Somi S, Kurt B, Kuru M, Kaya S, Kaçar C, Aksoy Ö, Aslan S (2015) Clinical use of deslorelin implants for the long-term contraception in prepubertal bitches: effects on epiphyseal closure, body development, and time to puberty. Theriogenology 83(7):1147–1153. https://doi.org/10.1016/j.theriogenology.2014.12.015

Koyasu H, Takahashi H, Sasao I, Takagi S, Nagasawa M, Kikusui T (2022) Sociality of cats toward humans can be influenced by hormonal and socio-environmental factors: pilot study. Animals 13(1):146. https://doi.org/10.3390/ani13010146

Mehl NS, Khalid M, Srisuwatanasagul S, Swangchan-uthai T, Sirivaidyapong S (2016) GnRH-agonist implantation of prepubertal male cats affects their reproductive performance and testicular LH receptor and FSH receptor expression. Theriogenology 85, 841–848. https://doi.org/10.1016/j.theriogenology.2015.10.031

Nagasawa T, Kimura Y, Masuda K, Uchiyama H (2022) Physiological assessment of the health and welfare of domestic cats—an exploration of factors affecting urinary cortisol and oxytocin. Animals 12(23):3330. https://doi.org/10.3390/ani12233330

Norsworthy GD, Romagnoli S (2016) Use of megestrol in cats. J Feline Med Surg 18(3):248–249. https://doi.org/10.1177/1098612X16631235

Peterson ME (1987) Effects of megestrol acetate on glucose tolerance and growth hormone secretion in the cat. Res Vet Sci 42(3):354–357

Risso A, Corrada Y, Barbeito C, Diaz J, Gobello C (2012) Long-term-release GnRH agonists postpone puberty in domestic cats. Reprod Domest Anim 47(6):936–938. https://doi.org/10.1111/j.1439-0531.2012.01994.x

Rodan I, Heath S (eds) (2016) Feline behavioral health and welfare. Elsevier

Rodrigues SM, Saslow LR, Garcia N, John OP, Keltner D (2009) Oxytocin receptor genetic variation relates to empathy and stress reactivity in humans. Proc Natl Acad Sci 106(50):21437–21441. https://doi.org/10.1073/pnas.0909579106

Sakaguchi, N. K. (2019). *The effects of oxytocin on proximity-seeking behavior and vocalizations in shelter cats* (Honors thesis, Oregon State University). Oregon State University ScholarArchive. Retrieved from https://ir.library.oregonstate.edu/concern/honors_college_theses/tm70n170z

Toydemir TSF, Kılıçarslan MR, Olgaç V (2012) Effects of the GnRH analogue deslorelin implants on reproduction in female domestic cats. Theriogenology 77(3):662–674. https://doi.org/10.1016/j.theriogenology.2011.07.046

Trepanier LA (2007) Pharmacologic management of feline hyperthyroidism. Vet Clin N Am Small Anim Pract 37(4):775–788. https://doi.org/10.1016/j.cvsm.2007.03.004

Tully J, Gabay AS, Brown D, Murphy DGM, Blackwood N (2018) The effect of intranasal oxytocin on neural response to facial emotions in healthy adults as measured by functional MRI: a systematic review. Psychiatry Res Neuroimaging 272:17–29. https://doi.org/10.1016/j.pscychresns.2017.11.017

Wu N, Li Z, Su Y (2012) The association between oxytocin receptor gene polymorphism (OXTR) and trait empathy. J Affect Disord 138(3):468–472. https://doi.org/10.1016/j.jad.2012.01.009

Over-the-Counter Treatments

Contents

27.1 Pheromone-Based Treatments – 688
27.1.1 Maternal Pheromone – 688
27.1.2 Facial Pheromones – 688

27.2 Nutraceuticals and Dietary Supplements – 689
27.2.1 Alpha-Casozepine – 689
27.2.2 α-Lactalbumine and Magnesium Combination – 690
27.2.3 L-Tryptophan – 691
27.2.4 Fish Hydrolysate and Superoxide Dismutase Extracted from Melon Combination – 691
27.2.5 Other Dietary Supplements – 691

27.3 Phytotherapy – 692
27.3.1 L-Theanine – 692
27.3.2 Crataegus Monogyna—Passiflora Incarnata—Griffonia Simplicifolia Combination – 692
27.3.3 Aromatherapy – 692
27.3.4 Bach Flowers and Homeopathy – 693

References – 694

© The Author(s), under exclusive license to Springer Nature Switzerland AG 2026
S. Bleuer-Elsner et al., *Veterinary Psychiatry of the Cat*,
https://doi.org/10.1007/978-3-032-06260-4_27

This chapter discusses treatments that do not require a prescription and are most often sold over the counter. Over-the-counter treatments can be effective in preventing the worsening of reactive states due to their impact on mild and transient symptoms. However, their efficacy is limited when addressing pathological conditions that require medical management. This chapter will cover pheromone-based treatments, nutraceuticals, and various dietary supplements.

Additionally, phytotherapy, aromatherapy, Bach flower remedies, and homeopathy will be discussed. The chapter concludes with a summary table providing the scientific evidence level for each over-the-counter treatment, offering practitioners a useful guide for clinical decisions.

27.1 Pheromone-Based Treatments

Pheromones are chemical substances emitted to act as messages for individuals of the same species (Ghaemmaghami 2019). They influence physiology and behavior (see ▶ Chap. 1, ▶ Sect. 1.3.2.1, and ▶ Chap. 3, ▶ Sect. 3.2.4.2).

27.1.1 Maternal Pheromone

In dogs, a pheromone known as canine appeasing pheromone (CAP) is produced within 3 days after parturition in the intermammary sulcus of lactating bitches. This pheromone is vaporized into the immediate atmosphere through association with saprophytic bacteria. It stimulates the receptors of the vomeronasal organ located in the nasal cavity of puppies. This organ transmits the pheromonal signal to the accessory olfactory bulb, which is connected to other brain regions, particularly the amygdala and hypothalamus (Pageat and Gaultier 2003). CAP is believed to be one of the neurophysiological elements of attachment in dogs.

In cats, a maternal pheromone, also referred to as CAP (cat appeasing pheromone), has been identified. Although few studies have been conducted on feline CAP, it has been synthesized into Feliway Friends® (CEVA). This product is available as a plug-in diffuser, effective for 30 days and covering up to 70 m^2. It is recommended to place the diffuser in the room most frequented by the cats.

The pheromone appears promising for strengthening intraspecific behaviors, fostering attachment, and potentially alleviating symptoms related to phobic and anxious states. It has shown potential in improving relationships between cats (DePorter et al. 2019) and between cats and dogs (Prior and Mills 2020) within the home (see ▶ Chap. 3, ▶ Sect. 3.2.4.2).

In DePorter's 2019 study, the statistical significance of the results was limited (P-value ranging between 0.03 and 0.99). In the study by Prior and Mills (2020), no control group of animals without pheromone exposure was included, and the authors recommended using canine appeasing pheromone instead.

We recommend using *Feliway Friends®* (CEVA) when introducing a new animal into the home. In such cases, it can help transition the animal from a reactive to a physiological state (see ▶ Chap. 1, ▶ Sect. 1.1.2.3). However, if the cat is already in a pathological state, this product alone will not suffice, and medical intervention is advised.

27.1.2 Facial Pheromones

Facial pheromones are chemical signals released during facial marking by cats, playing a role in structuring their territory (see ▶ Chap. 1, ▶ Sect 1.3.2.1). Fractions F2, F3, and F4 have been identified, with distinct functions: F2 is involved in sexual marking, F3 in object marking, and F4 in interspecific or intraspecific social marking (Pageat and Gaultier 2003).

27.1.2.1 The F3 Fraction of Facial Pheromones

The synthesized F3 fraction is incorporated into *Feliway®*, *Feliway Help®*, *Feliway Optimum®* (CEVA), and *Zenifel®* (Virbac).

Feliway® (CEVA) is available as a spray and a diffuser. The diffuser works for 30 days and covers an area of up to 70 m^2. It is recommended to place the diffuser in the room most frequented by the cat. The spray should be applied directly to the cat's cage, sleeping area, or transport carrier about 15 min before the cat is introduced to these spaces. Its effects are reported to last up to 5 h.

A double-blind study by Pereira et al. (2016) demonstrated a reduction in stress levels in cats during veterinary consultations when *Feliway*® (CEVA) was used. However, the study's methodology has been criticized, for example, due to inconsistencies in the procedures performed on the cats. Doonan (2018) concluded that *Feliway*® (CEVA) alone is insufficient to significantly reduce stress during veterinary examinations.

A meta-analysis by Mills et al. (2011) revealed a decrease in urine marking following the use of *Feliway*® (CEVA). However, this reduction was less pronounced than that achieved with fluoxetine or clomipramine and was not always distinguishable from a placebo effect. The evidence level is low.

A double-blind study by Pereira et al. (2023) reported only minor differences in the intensity and frequency of scratching on undesirable surfaces when *Feliway*® (CEVA) was used.

The *Zenifel*® (Virbac) product also contains nepetalactone, a catnip extract. It is available as a spray to be applied 15 min before introducing the cat to the treated area. A single study (Bernachon et al. 2015) suggests that *Zenifel*® is more effective at reducing stress compared to *Feliway*® (CEVA). However, the effects are short-lived, with no significant difference observed compared to the control group after a few hours.

The purpose of these products is to restore facial marking behavior after environmental disruptions, such as moving to a new home, furniture rearrangements, returning from hospitalization, or transport. Additionally, they encourage the cat to perceive a space as familiar, such as a veterinary clinic or a transport cage. These products are therefore recommended for managing reactive states to help the cat return to a physiological state. However, they are insufficient for addressing pathological states.

27.1.2.2 The F4 Fraction of Facial Pheromones

The F4 fraction of the facial complex was marketed under the name *Felifriend*™ (CEVA) as a spray. As of today, the product is no longer available for purchase. A published study found no significant effect of the product on the ease of handling cats during a veterinary consultation compared to a control group (Bonnafous et al. 2005).

27.2 Nutraceuticals and Dietary Supplements

Nutraceuticals are dietary supplements that may have health benefits (Sachdeva et al. 2020).

In dogs, it has been demonstrated that a specific diet can significantly and positively affect neuroendocrine parameters and the metabolites resulting from cellular oxidation (Sechi et al. 2017). Thus, the use of diet and nutraceuticals as part of the treatment for behavioral disorders appears to be of interest.

27.2.1 Alpha-Casozepine

Alpha-casozepine corresponds to the 91–100 fragment of bovine αS1-casein and is extracted from milk. The molecule shares part of the biochemical structure of an endogenous ligand of central and peripheral GABAA_AA receptors, giving it a "benzodiazepine-like" affinity for the GABAA_AA receptor, although this affinity differs from that of diazepam (Violle et al. 2006).

While alpha-casozepine demonstrates an affinity for GABAA_AA receptors that is 10,000 times weaker than diazepam in vitro, it is 10 times more effective in the conditioned defensive burying paradigm in rats (a reflexive burying response to danger) (Miclo et al. 2001).

Alpha-casozepine is therefore theoretically of interest for the treatment of anxiety.

It is available in over-the-counter veterinary products such as *Zylkene*™ (Vetoquinol). It is also included as an additive in some commercial diets, such as *Calm*™ (Royal Canin and Concept for Life), *Urinary Stress*™ (Hill's), or treats like *Feliway Happy Snack*™ (CEVA). In such products, alpha-casozepine is often combined with tryptophan, which may interfere with its efficacy. Additionally, the effective dose may not always be reached through the consumption of these diets alone.

A study by Beata et al. (2007) demonstrated a positive effect of *Zylkene*™ (Vetoquinol) in managing social phobias and anxiety in cats. However, a recent meta-analysis in dogs (Buckley 2017) concluded that there is currently no robust evidence supporting the efficacy of alpha-casozepine when administered minutes to days before exposure to an anxiogenic stimulus.

A study on *Urinary Stress*™ kibble from Hill's showed a reduction in fears and greater tolerance of contact (Meyer and Bečvářová 2016). However, this study involved only 10 cats, allowed the use of calming pheromones during the study, showed no reduction in aggression, and lacked a control group. Another study on *Calm*™ kibble from Royal Canin reported a slight decrease in fear in a new environment, but no significant effect in the presence of an unfamiliar person (Landsberg et al. 2017).

While alpha-casozepine may have a role in reducing anxiety in cats over the medium and long term, the level of evidence supporting its efficacy remains weak.

In our experience, higher doses than those recommended by the manufacturer may be necessary to observe a reduction in fear-related symptoms, particularly for acute events like transport, moving, or veterinary visits. In such cases, doses of 2–4 times the manufacturer's recommendation (15–30 mg/kg) were used, and administration began 1–2 days before the event.

Alpha-casozepine can also be recommended for longer-term use, such as during the adoption of an animal or the introduction of a new pet into the household, to facilitate adaptation and reduce anxiety-related behaviors.

27.2.2 α-Lactalbumine and Magnesium Combination

The alpha-lactalbumin and magnesium combination is available as a nonprescription veterinary supplement under the brand name *Alphazium*™ (TVM). Alpha-lactalbumin is a protein derived from whey and is extensively converted into tryptophan during digestion (Verschoor et al. 2010). Tryptophan, an essential amino acid, is a precursor of serotonin, which crosses the blood-brain barrier. Serotonin is involved in regulating mood, appetite control, sleep, alertness, and pain (see ► Chap. 2, ► Sect. 2.2.4.3). Magnesium acts as a negative allosteric modulator of NMDA glutamate receptors. Neuronal membrane depolarization dislodges magnesium ions (Mg^{+}), allowing calcium channels to open under the influence of glutamate and glycine. Magnesium thereby functions as an N-méthyl-D-aspartate (NMDA) receptor antagonist and plays an antidepressant role (Poleszak et al. 2007). Additionally, magnesium has anxiolytic properties by potentiating GABAergic receptors, either alone or in combination with benzodiazepines (Poleszak 2008). Given these mechanisms of action, the alpha-lactalbumin and magnesium combination appears promising for treating depression and anxiety. However, only one study has been published, which found no significant differences in fear- or anxiety-related symptoms between the placebo group and the group receiving *Alphazium*™ (TVM) (Beata et al. 2018).

Alphazium™ (TVM) is available as palatable 5 mg chews. The recommended dose is 1 mg/kg administered 1 3 days before an anxiety-inducing event such as transport, veterinary visits, or thunderstorms. It may also be given once daily for a month during continuous exposure scenarios such as boarding, hospitalization, moving, or the introduction of a new animal. We advise using this supplement during reactive states to facilitate a return to a physiological state. Its use is less appropriate for pathological conditions where medical intervention may be necessary.

27.2.3 L-Tryptophan

Tryptophan is an essential amino acid that is a precursor to serotonin. It is converted into 5-hydroxytryptophan and then into serotonin, a neurotransmitter essential for the regulation of appetite, sleep, mood, and pain. Tryptophan is naturally found in dairy products, meats, brown rice, fish, and soy. It would therefore theoretically be interesting to supplement a cat's diet with tryptophan to increase brain serotonin levels.

L-tryptophan is the levorotatory enantiomer of tryptophan. The molecule is found in many dietary supplements and as additives in some commercial cat foods.

Studies on the effects of L-tryptophan on cat behavior are limited and focus on supplemented kibble (Beynen 2016; Landsberg et al. 2017; Pereira et al. 2010). These studies show a limited effect on fear, and the kibble also contains alpha-casozepine."

27.2.4 Fish Hydrolysate and Superoxide Dismutase Extracted from Melon Combination

Fish hydrolysate PC60, derived from cod and mackerel extracts, exhibits effects similar to benzodiazepines on the hypothalamic-pituitary-adrenal axis, sympathoadrenal activity, and GABA levels in the hypothalamus (Bernet et al. 2000). Stabilium 200, a derivative of PC60, has demonstrated anxiolytic properties in humans (Dorman et al. 1995) and facilitated learning in both rats and humans (Le poncin, M. 1996; Le Poncin and Lamproglou 1996).

Concentrated melon juice is rich in superoxide dismutase, an enzyme that prevents cellular oxidation and may positively affect various stress and fatigue symptoms in humans (Milesi et al. 2009).

The combination of fish hydrolysate and superoxide dismutase extracted from melon is available as Anxivet®. The recommended dosage for cats is 50 mg/kg once daily, either administered short-term before a stressful situation or long-term during continuous exposure. As with similar products, we recommend its use during reactive states to help restore physiological balance.

No studies on its efficacy in cats have been published.

27.2.5 Other Dietary Supplements

Various other dietary supplements aim to alleviate symptoms associated with cognitive disorders in aging cats:

- Aktivait® (VetPlus) is a blend of N-acetylcysteine, alpha-lipoic acid, vitamins C and E, L-carnitine, coenzyme Q10, docosahexaenoic acid (DHA), eicosapentaenoic acid, and phosphatidylserine. A single study on aging dogs demonstrated improvements in signs of disorientation, altered interactions, and inappropriate elimination (Heath et al. 2007). The recommended dosage is one capsule once daily.
- Sitalan® (TVM), a mixture of cystine, glutamic acid, glycine, vitamin E, vitamin C, and organic selenium, is marketed by its manufacturer as supporting the physiological functions of elderly cats and acting as an antioxidant. No published studies are available. The recommended dosage is half a capsule once daily.
- Novifit® (Virbac) and Zentonil® (Vetoquinol) contain S-Adenosyl-L-Methionine (SAMe) as their primary component. SAMe has shown antidepressant properties in humans (Papakostas 2009). In cats, studies on SAMe primarily focus on its role in supporting hepatic function (Center et al. 2005; Webb et al. 2003). The recommended dosage is 10–20 mg/kg once daily.

27.3 Phytotherapy

27.3.1 L-Theanine

L-theanine, or N-ethyl-L-glutamine, is an amino acid found exclusively in green tea and historically reported as a relaxing agent. L-theanine may increase cerebral serotonin, dopamine, and acide γ-aminobutyrique (GABA) levels and has affinities for acide α-amino-3-hydroxy-5-méthyl-4-isoxazolepropionique (AMPA), , kainate, and NMDA glutamate receptors. Additionally, it has demonstrated neuroprotective effects due to its antagonistic action on metabotropic glutamate receptors (Nathan et al. 2006). While the theoretical mechanisms of action of L-theanine and the Crataegus-Passiflora-Griffonia combination are promising, their efficacy in cats has not been substantiated by robust clinical studies. Peak concentrations in the blood and liver are reached 1 h after administration and in the central nervous system after 5 h. Both concentrations then decline over a 24-h period (Nathan et al. 2006).

L-theanine is marketed under the names Anxitane® (Virbac) and Solliquin® (Nutramax).

A study involving 33 cats reported a slight reduction in anxiety, as evaluated by a veterinarian, after 1 month of treatment with Anxitane® (Virbac) (Dramard et al. 2018). However, the study lacked a control group.

Anxitane® S (Virbac) is available as palatable tablets. The recommended dosage is ½ tablet of Anxitane® (Virbac) per cat, morning and evening, for up to 2 months.

27.3.2 Crataegus Monogyna—Passiflora Incarnata—Griffonia Simplicifolia Combination

Extracts of *Crataegus monogyna*, or hawthorn, are traditionally used as complementary treatments for hypertension and heart disease (Cloud et al. 2020). Proanthocyanidins extracted from hawthorn have antioxidant activity that strengthens cardiomyocytes and arteries and increases blood oxygenation (Bernatonienė et al. 2008). More recently, they have been explored for the treatment of anxiety, depression, and pain in a rat study (Zanganenejad et al. 2018). However, the underlying mechanisms of action remain unknown.

Extracts of *Passiflora incarnata*, or passionflower, are employed for the treatment of anxiety, insomnia, addictions, seizures, neuropathic pain, and attention deficit disorders (Miroddi et al. 2013). The mechanisms of action of these extracts are not fully understood, but they act as antagonists of $GABA_A$ and $GABA_B$ receptors and limit the reuptake of GABA in the synaptic cleft (Appel et al. 2011; Janda et al. 2020).

Extracts of *Griffonia simplicifolia*, or griffonia, are used to treat anxiety, depression, migraines, epilepsy, insomnia, and hyperphagia (Mannino et al. 2023). The plant's extracts contain 5-hydroxytryptophan, a precursor of melatonin and serotonin (Subbaraju et al. 2005), leading to increased serotonin levels in the brain (Guerra Ribas et al. 2021).

The combination of *Crataegus monogyna–Passiflora incarnata–Griffonia simplicifolia* is marketed under the name Sérénité® (Wamine). No published studies are available on the product's efficacy. The recommended dose is one tablet per day, with no specified duration of treatment.

27.3.3 Aromatherapy

Aromatherapy is a branch of phytotherapy that involves using the properties of essential oils or aromatic essences. Products are available as diffusers, sprays, and wipes (Petscool® Anidev). There is limited published evidence on the effectiveness of essential oils used in aromatherapy. A meta-analysis in human medicine concluded that the efficacy of aromatherapy is not supported by rigorous clinical trials (Cooke and Ernst 2000).

In cats, one study demonstrated that aromatherapy in the form of a lavender oil diffuser reduced signs of stress during hospitalization (Goodwin and Reynolds 2018). Another study on Pet Remedy® spray, which contains a combination of valerian, vetiver, basil, and sage essential oils, reported behavioral changes after exposure to the product without detailing their interpretation (Vaghela 2022).

27.3.4 Bach Flowers and Homeopathy

Bach flowers or Bach remedies are alcoholic macerations of plants, or "floral elixirs," made from the flowers of 37 plant species, with the addition of rock water, resulting in 38 "elixirs" in total. This practice was conceptualized between 1928 and 1936 by Edward Bach, an English homeopathic physician. Studies in humans have shown no differences between the effects of Bach flowers and a placebo (Armstrong and Ernst 2001; Pintov et al. 2005; Walach et al. 2001). No studies exist for cats. Based on our experience, the active ingredient concentrations in various preparations are highly variable and uncontrolled, and their efficacy does not exceed that of a placebo.

Homeopathy involves administering infinitesimal doses of substances capable, at higher doses, of producing symptoms similar to those of the condition being treated. A meta-analysis published in 2005 demonstrated that the effects of homeopathic remedies are not superior to placebo in humans (Shang et al. 2005). In cats, one study suggested a slight effect of homeopathy in reducing aggression and fear (Mathie et al. 2010). Overall and Dunham (2009) emphasized the need for studies on homeopathic preparations to follow rigorous scientific methods and concluded that, to date, there is no evidence supporting the efficacy of homeopathy in veterinary psychiatry.

It is important to note that, at present, no rigorous studies demonstrate the efficacy of either homeopathy or Bach flowers in cats, and their use relies on principles that lack scientific validation.

Summary of This Chapter

The level of evidence regarding the effectiveness of over-the-counter products varies from one product to another (see table below). In most cases, these products are not intended to treat pathological conditions but rather to prevent the worsening of a reactive state into a pathological state. They can be useful at the counter or during preventive medicine consultations if accompanied by guidance that informs owners and, when necessary, encourages them to seek a psychiatric consultation.

Summary table of over-the-counter products

Substance	Proposed mode of action	Reactional and Pathological states for which the molecule is indicated	Level of evidence
Feliway Friend®	Synthetic analogue of CAP	Kitten habituation Introduction of a new animal Phobic Anxious	One study published for cat interactions, low evidence level. One study published for dog-cat interactions, no control group.
Feliway®	Synthetic analogue of F3 pheromonal fraction	Disruption of facial marking and scratching	Numerous studies, low evidence level.
Zenifel®	Synthetic analogue of F3 pheromonal fraction	Disruption of facial marking and scratching	One published study, low evidence level.

27

Substance	Proposed mode of action	Reactional and Pathological states for which the molecule is indicated	Level of evidence
Zylkene®	Positive allosteric modulator of $GABA_A$ (10,000 times weaker than diazepam)	Phobic	One published study, low evidence level.
Alp-hazium®	Tryptophan + NMDA antagonist GABA agonist	Phobic Anxious Depressive	One study published, no significant difference between treated and placebo groups
Anxitane®	L-theanine: purportedly increases monoamine levels	Anxious Impulsive or compulsive	Highly inconsistent results across studies.
Anxivet®	Benzodiazepine-like	Phobic	No studies published.
Aktivait®	DHA + EPA + phospha-tidylserine	Senile	One study in dogs. Multicentric double-blind placebo-controlled trial showed effects on disorientation, interac-tions, and inappropriate elimina-tion.
Sitalan®	Antioxydant	Senile	No published studies.
Novifit® Zentonil®	SAMe	Senile	No studies published on anxiolytic properties in cats.
L-Theanine	Serotonin + Dopamine + GABA NMDA inhibitor AMPA inhibitor	Phobic or anxious	One study in cats, no control group. Anxiety scores improved.
Serenité®	Serotonin + GABA	Phobic	No published studies.
Aroma-therapy	No data	Transport phobia	Study against placebo.
Homeopa-thy	No data		One study in cats, low evidence level.

References

Appel K, Rose T, Fiebich B, Kammler T, Hoffmann C, Weiss G (2011) Modulation of the γ-aminobutyric acid (GABA) system by *Passiflora incarnata* L. Phytother Res 25(6):838–843. https://doi.org/10.1002/ptr.3352

Armstrong NC, Ernst E (2001) A randomized, double-blind, placebo-controlled trial of a Bach Flower Remedy. Complement Ther Nurs Midwifery 7(4):215–221. https://doi.org/10.1054/ctnm.2001.0525

Beata C, Beaumont-Graff E, Coll V, Cordel J, Marion M, Massal N, Marlois N, Tauzin J (2007) Effect of alpha-casozepine (Zylkene) on anxiety in cats. J Vet Behav 2(2):40–46. https://doi.org/10.1016/j.jveb.2007.02.002

Beata C, Gérard-Champod M, Lachapele D, Saunier E (2018) Alphazium: Premiers pas pour un nouveau nutraceutique à visée bien-être et équilibre comportemental. Revue Vétérinaire Clinique 53(2):31–38. https://doi.org/10.1016/j.anicom.2018.04.001

Bernachon N, Beata C, Crastes N, Monginoux P, Gatto H, McGahie D (2015) Response to acute stress in domestic cats using synthetic analogues of natural appeasing pheromones with Nepeta cataria extract rich in Nepetalactone: a double-blinded, randomized, positive controlled cross-over study. Int J Appl Res Vet Med 13(2):125–134

Bernatonienė J, Masteikova R, Majienė D, Savickas A, Kėvelaitis E, Bernatonienė R, Dvoráčková K, Civinskienė G, Lekas R, Vitkevičius K, Pečiūra R (2008) Free radical-scavenging activities of Crataegus monogyna extracts. Medicina 44(9):706. https://doi.org/10.3390/medicina44090091

Bernet F, Montel V, Noël B, Dupouy JP (2000) Diazepam-like effects of a fish protein hydrolysate (Gabolysat PC60) on stress responsiveness of the rat pituitary-adrenal system and sympathoadrenal activity. Psychopharmacology 149(1):34–40. https://doi.org/10.1007/s002139900338

Beynen AC (2016) Therapeutic de-stressing foods for cats. Unpublished https://doi.org/10.13140/RG.2.2.19746.02241

Bonnafous L, Lafont C, Gaultier E, Falewée C, Pageat P (2005) Interest in the use of a new galenic form of the Feline Allomarking Pheromone (F4) analog (Felifriend) during medical examination

Buckley LA (2017) Is alpha-casozepine efficacious at reducing anxiety in dogs? Vet Evid 2(3). https://doi.org/10.18849/ve.v2i3.67

Center SA, Randolph JF, Warner KL, McCabe-McClelland J, Foureman P, Hoffmann WE, Erb HN (2005) The effects of S-adenosylmethionine on clinical pathology and redox potential in the red blood cell, liver, and bile of clinically Normal cats. J Vet Intern Med 19(3):303–314. https://doi.org/10.1111/j.1939-1676.2005.tb02699.x

Cloud A, Vilcins D, McEwen B (2020) The effect of hawthorn (Crataegus spp.) on blood pressure: a systematic review. Adv Integr Med 7(3):167–175. https://doi.org/10.1016/j.aimed.2019.09.002

Cooke B, Ernst E (2000) Aromatherapy: a systematic review. Br J Gen Pract 50(455):493–496

DePorter TL, Bledsoe DL, Beck A, Ollivier E (2019) Evaluation of the efficacy of an appeasing pheromone diffuser product vs placebo for management of feline aggression in multi-cat households: a pilot study. J Feline Med Surg 21(4):293–305. https://doi.org/10.1177/1098612X18774437

Doonan C (2018) The effects of Feliway on the stress of cats during veterinary examination. Steigerwald, Griffith

Dorman T, Bernard L, Glaze P, Hogan J, Skinner R, Nelson D, Bowker L, Head D (1995) The effectiveness of Garum armoricum (stabilium) in reducing anxiety in college students. J Adv Med 8:193–200

Dramard V, Kern L, Hofmans J, Rème CA, Nicolas CS, Chala V, Navarro C (2018) Effect of l-theanine tablets in reducing stress-related emotional signs in cats: an open-label field study. Ir Vet J 71(1):21. https://doi.org/10.1186/s13620-018-0130-4

Ghaemmaghami F (2019) Phéromones: La sémiochimie inconsciente les phéromones putatives humaines. Éditions Hugo Stern

Goodwin S, Reynolds H (2018) Can aromatherapy be used to reduce anxiety in hospitalised felines. Vet Nurs 9(3):167–171. https://doi.org/10.12968/vetn.2018.9.3.167

Guerra Ribas RDM, De Oliveira DCL, Da Silva PC, Lima Martins HAD, De Moraes JG, Da Silva MP, Ribas VR, Mendes Calafange CDO, Silva Castro AETMD, De Castro RM (2021) The different roles of Griffonia simplicifolia in the treatment of depression: a narrative review. Int J Complement Altern Med 14(6):167–172. https://doi.org/10.15406/ijcam.2021.14.00553

Heath SE, Barabas S, Craze PG (2007) Nutritional supplementation in cases of canine cognitive dysfunction—a clinical trial. Appl Anim Behav Sci 105(4):284–296. https://doi.org/10.1016/j.applanim.2006.11.008

Janda K, Wojtkowska K, Jakubczyk K, Antoniewicz J, Skonieczna-Żydecka K (2020) Passiflora incarnata in neuropsychiatric disorders—a systematic review. Nutrients 12(12):3894. https://doi.org/10.3390/nu12123894

Landsberg G, Milgram B, Mougeot I, Kelly S, De Rivera C (2017) Therapeutic effects of an alpha-casozepine and L-tryptophan supplemented diet on fear and anxiety in the cat. J Feline Med Surg 19(6):594–602. https://doi.org/10.1177/1098612X16669399

Le Poncin M, Lamproglou I (1996) Experimental study: stress and memory. Eur Neuropsychopharmacol 6:110

Le poncin, M. (1996) Nutrient presentation of cognitive and memory performances. Eur Neuropsychopharmacol 6:187

Mannino G, Serio G, Gaglio R, Maffei ME, Settanni L, Di Stefano V, Gentile C (2023) Biological activity and metabolomics of Griffonia simplicifolia seeds extracted with different methodologies. Antioxidants 12(9):1709. https://doi.org/10.3390/antiox12091709

Mathie RT, Baitson ES, Hansen L, Elliott MF, Hoare J (2010) Homeopathic prescribing for chronic conditions in feline and canine veterinary practice. Homeopathy 99(4):243–248. https://doi.org/10.1016/j.homp.2010.05.010

Meyer HP, Bečvářová I (2016) Effects of a urinary food supplemented with milk protein hydrolysate and L-tryptophan on feline idiopathic cystitis—results of a case series in 10 cats. Int J Appl Res Vet Med 14(1):7

Miclo L, Perrin E, Driou A, Papadopoulos V, Boujrad N, Vanderesse R, Boudier J-F, Desor D, Linden G, Gaillard J-L (2001) Characterization of α-casozepine, a tryptic peptide from bovine α_{s1}-casein with benzodiazepine-like activity. FASEB J 15(10):1780–1782. https://doi.org/10.1096/fj.00-0685fje

Milesi M-A, Lacan D, Brosse H, Desor D, Notin C (2009) Effect of an oral supplementation with a proprietary melon juice concentrate (Extramel®) on stress and fatigue in healthy people: a pilot, double-blind, placebo-controlled clinical trial. Nutr J 8(1):40. https://doi.org/10.1186/1475-2891-8-40

Mills DS, Redgate SE, Landsberg GM (2011) A meta-analysis of studies of treatments for feline urine spraying. PLoS One 6(4):e18448. https://doi.org/10.1371/journal.pone.0018448

Miroddi M, Calapai G, Navarra M, Minciullo PL, Gangemi S (2013) Passiflora incarnata L.: ethnopharmacology, clinical application, safety and evaluation of clinical trials. J Ethnopharmacol 150(3):791–804. https://doi.org/10.1016/j.jep.2013.09.047

Nathan P, Lu K, Gray M, Oliver C (2006) The neuropharmacology of L-theanine(N-ethyl-L-glutamine). J Herb Pharmacother 6(2):21–30. https://doi.org/10.1300/J157v06n02_02

Overall KL, Dunham AE (2009) Homeopathy and the curse of the scientific method. Vet J 180(2):141–148. https://doi.org/10.1016/j.tvjl.2008.10.006

Pageat P, Gaultier E (2003) Current research in canine and feline pheromones. Vet Clin N Am Small Anim Pract 33(2):187–211. https://doi.org/10.1016/S0195-5616(02)00128-6

Papakostas GI (2009) Evidence for S-Adenosyl-L -methionine (SAM-e) for the treatment of major depressive disorder. J Clin Psychiatry 70(suppl 5):18–22. https://doi.org/10.4088/JCP.8157su1c.04

Pereira GD, Fragoso S, Pires E (2010) Effect of dietary intake of L-tryptophan supplementation on multi-housed cats presenting stress related behaviours. BSAVA 53rd Annual Congress 2010, Scientific Proceedings, Veterinary Programme

Pereira JS, Fragoso S, Beck A, Lavigne S, Varejão AS, Da Graça Pereira G (2016) Improving the feline veterinary consultation: the usefulness of Feliway spray in reducing cats' stress. J Feline Med Surg 18(12):959–964. https://doi.org/10.1177/1098612X15599420

Pereira JS, Salgirli Demirbas Y, Meppiel L, Endersby S, Da Graça Pereira G, De Jaeger X (2023) Efficacy of the Feliway® Classic Diffuser in reducing undesirable scratching in cats: a randomised, triple-blind, placebo-controlled study. PLoS One 18(10):e0292188. https://doi.org/10.1371/journal.pone.0292188

Pintov S, Hochman M, Livne A, Heyman E, Lahat E (2005) Bach flower remedies used for attention deficit hyperactivity disorder in children—a prospective double blind controlled study. Eur J Paediatr Neurol 9(6):395–398. https://doi.org/10.1016/j.ejpn.2005.08.001

Poleszak E (2008) Benzodiazepine/GABAA receptors are involved in magnesium-induced anxiolytic-like behavior in mice. Pharmacol Rep 60:996

Poleszak E, Wlaź P, Kędzierska E, Nieoczym D, Wróbel A, Fidecka S, Pilc A, Nowak G (2007) NMDA/glutamate mechanism of antidepressant-like action of magnesium in forced swim test in mice. Pharmacol Biochem Behav 88(2):158–164. https://doi.org/10.1016/j.pbb.2007.07.018

Prior MR, Mills DS (2020) Cats vs. dogs: the efficacy of feliway friendsTM and AdaptilTM products in multispecies homes. Front Vet Sci 7:399. https://doi.org/10.3389/fvets.2020.00399

Sachdeva V, Roy A, Bharadvaja N (2020) Current prospects of nutraceuticals: a review. Curr Pharm Biotechnol 21(10):884–896. https://doi.org/10.2174/1389201021666200130113441

Sechi S, Di Cerbo A, Canello S, Guidetti G, Chiavolelli F, Fiore F, Cocco R (2017) Effects in dogs with behavioural disorders of a commercial nutraceutical diet on stress and neuroendocrine parameters. Vet Rec 180(1):18–18. https://doi.org/10.1136/vr.103865

Shang A, Huwiler-Müntener K, Nartey L, Jüni P, Dörig S, Sterne JA, Pewsner D, Egger M (2005) Are the clinical effects of homoeopathy placebo effects? Comparative study of placebo-controlled trials of homoeopathy and allopathy. Lancet 366(9487):726–732. https://doi.org/10.1016/S0140-6736(05)67177-2

Subbaraju GV, Kannababu S, Vijayakumar K, Murthy P, Vanisree M, Tsay H-S (2005) Spectrophotometric estimation of L- 5-hydroxytryptophan in Griffonia simplicifolia extracts and dosage forms. Int J Appl Sci Eng 3(2):111–116

Vaghela V (2022) Can the use of pet remedy's calming spray elicit a calming response in domestic cats (Felis catus)? https://doi.org/10.21203/rs.3.rs-1318478/v1

Verschoor E, Finlayson G, Blundell J, Markus CR, King NA (2010) Effects of an acute α-lactalbumin manipulation on mood and food hedonics in high- and low-trait anxiety individuals. Br J Nutr 104(4):595–602. https://doi.org/10.1017/S0007114510000838

Violle N, Messaoudi M, Lefranc-Millot C, Desor D, Nejdi A, Demagny B, Schroeder H (2006) Ethological comparison of the effects of a bovine αs1-casein tryptic hydrolysate and diazepam on the behaviour of rats in two models of anxiety. Pharmacol Biochem Behav 84(3):517–523. https://doi.org/10.1016/j.pbb.2006.06.017

Walach H, Rilling C, Engelke U (2001) Efficacy of Bach-flower remedies in test anxiety: a double-blind, placebo-controlled, randomized trial with partial crossover. J Anxiety Disord 15(4):359–366. https://doi.org/10.1016/S0887-6185(01)00069-X

Webb C, Twedt D, Fettman M, Mason G (2003) S-adenosylmethionine (SAMe) in a feline acetaminophen model of oxidative injury. J Feline Med Surg 5(2):69–75. https://doi.org/10.1016/S1098-612X(02)00017-7

Zanganenejad Z, Ahmadynasab M, Setorki M (2018) Effect of Crataegus Monogyna extract on anxiety, depression, and pain in Streptozotocin-induced diabetic rats. J Diabetes Nurs 6(3):539–549

Psychotherapy

Contents

Chapter 28 Introduction to veterinary psychotherapies – 699

Chapter 29 Practical application of veterinary psychotherapie – 723

Chapter 30 The causes and consequences of veterinary care – 797

Chapter 31 Veterinary psychiatry in the daily life of the gen – 841

Introduction to Veterinary Psychotherapies

Contents

28.1 Types of Therapies in Human Medicine – 700

28.1.1 Historical Context – 700

28.1.2 Cognitive-Behavioral Therapies – 701

28.1.3 Systemic Therapies – 701

28.2 Practical Tools for the Veterinary Practitioner: Consultation Techniques – 709

28.2.1 The System in Veterinary Medicine – 710

28.2.2 Alliance and Resistances – 710

28.2.3 The Therapist Positioning – 711

28.2.4 Consultation Techniques – 713

28.2.5 Constructing the Therapy – 718

References – 721

© The Author(s), under exclusive license to Springer Nature Switzerland AG 2026
S. Bleuer-Elsner et al., *Veterinary Psychiatry of the Cat*,
https://doi.org/10.1007/978-3-032-06260-4_28

This chapter, after a brief overview of the types of psychotherapy used in human medicine, lists the techniques applicable in veterinary consultations. Practical tools and consultation techniques are detailed to help the practitioner gather essential signs and symptoms for diagnosis, as well as to construct the key elements of behavioral therapy.

Psychotherapies encompass all nonpharmacological interventions designed to treat behavioral disorders in cats. In this chapter, we will present the fundamental principles of psychotherapies used in human medicine (see ▶ Sect. 28.1), followed by practical tools that can be adapted for use in veterinary consultations (see ▶ Sect. 28.2).

In humans, *psychotherapy* aims to help the patient overcome the factors underlying their condition—namely, a mental configuration that restricts their ability to make choices, causes suffering, and leaves them unable to adapt to their environment. The goal of psychotherapy is not only to alleviate *symptoms* but also to enhance the patient's *quality of life* solely through therapist-patient interaction. No single psychotherapy provides universal solutions. The therapist aids the patient in discovering personal solutions using methods specific to each type of therapy.

Psychologists and psychiatrists often specialize in one or more types of therapy and, consequently, in certain conditions. Becoming a therapist typically requires several years of training in each form of therapy. Therefore, it is unrealistic to claim expertise in systemic, strategic, or cognitive-behavioral therapies (CBTs) after simply reading this chapter. However, understanding their applications, fundamental principles, and contributions is essential for incorporating practical tools from these therapies into behavioral consultations (see ▶ Sect. 28.2).

In veterinary psychiatry, psychotherapy as we define it involves a set of exercises prescribed by the therapist to the animal or a member of its system, aiming at recovery. It only makes sense if a diagnosis has been established, as it is a contractual form of care. Moreover, for psychotherapy to be effective, it must be complemented by *medical treatment* when a pathological psychological state is identified. Without this combination, the animal's brain is not in a state conducive to learning. Even a coherent and symptom-appropriate psychotherapy will be slowed if the cat is unable to engage successfully in the prescribed exercises.

Most of our therapeutic prescriptions fall under the domain of cognitive-behavioral therapies, which will be detailed in ▶ Chap. 30. Numerous other therapies, initially developed for human medicine, exist and can provide valuable support for certain conditions or serve as communication tools in consultations (see ▶ Sect. 28.1).

28.1 Types of Therapies in Human Medicine

28.1.1 Historical Context

The care of the psyche was already described by the Greeks. However, it was only at the end of the nineteenth century that advancements in medicine and psychology allowed for a transition from rudimentary therapies to more rigorous psychotherapeutic techniques.

Historically, Sigmund Freud pioneered the first psychotherapies with psychoanalysis, of which he was the main theorist. The term "psychoanalysis" appeared in a French text for the first time in 1896 (Weinberg 2012). *Psychoanalysis* flourished throughout the twentieth century, with numerous theorists such as Melanie Klein and Jacques Lacan contributing to its development. It posits that the psychological problems experienced by an individual arise from unconscious internal conflicts. The principle of psychoanalytic therapy is to guide the patient through the process of resolving these conflicts, enabling them *to find their own solutions.*

Apart from psychoanalysis, two major therapeutic approaches emerged during the twentieth century: cognitive-behavioral therapies (CBT) and systemic therapies.

28.1.2 Cognitive-Behavioral Therapies

Cognitive-behavioral therapies (CBT) encompass a range of treatments for psychiatric disorders based on scientific psychology. Historically, CBT emerged from the convergence of two schools of thought: behaviorism, which focuses on modifying behavior, and cognitive psychology, which aims to alter an individual's representations and thought patterns.

The behaviorist approach, from which CBT originates, began with Pavlov and classical conditioning in the late nineteenth century. Skinner later expanded these theories with operant conditioning in the mid-twentieth century. Concurrently, Wolpe introduced the first clinical applications of these theories through counterconditioning in cats with experimental neuroses (Wolpe 1954).

The cognitive approach, on the other hand, developed in the 1970s, particularly through the works of Ellis (1962) and Beck (Nichols and Beck 1960). Until the 1980s, there was significant opposition between behaviorists, who believed that cognition could evolve simply by implementing adaptive behaviors, and cognitivists, who maintained that changing cognition alone was sufficient to alter behavior.

Bandura emphasized the interactions between individuals and their environment. He highlighted the role of cognitive phenomena, such as self-regulation, in behavior control and underscored the close relationship between emotions and behaviors. His work bridged the gap between behaviorists and cognitivists, influencing the development of behavioral therapies (Bandura 1978).

Hayes pioneered the new wave of CBT (Friman et al. 1998), which remains in practice today. According to Hayes, modifying behavior requires manipulating contextual variables while also addressing the individual's emotions (Hayes 1987; Hayes and Hofmann 2017).

Since the 1980s, the divide between behaviorist and cognitive approaches has gradually diminished. The term *Cognitive Behavioral Therapy* (CBT) now encompasses the goals of both schools of thought, which involve changing an individual's representations by modifying specific behaviors. The patient's difficulties, expressed through their symptoms, are addressed by the therapist "here and now," using targeted exercises designed to alter behavior. Subsequently, changing specific behaviors leads the individual to adjust their perceptions, resulting in lasting change and improved adaptability.

CBTs are generally highly standardized and involve monitoring the patient's progress throughout therapy. They align with the principles of evidence-based medicine and represent the dominant therapeutic approach in the United States. Numerous studies demonstrate their effectiveness, beginning with their positive impact on anxiety disorders (Öst 2008). These therapies are also effective in treating phobias and addictions (Hofmann et al. 2012). Certain studies highlight the importance of combining medication with therapy to achieve optimal outcomes (Bandelow et al. 2015).

CBTs form the primary foundation of behavioral therapies in veterinary psychiatry. For this reason, a dedicated chapter explores them in detail (see ▶ Chap. 29).

28.1.3 Systemic Therapies

The systemic approach stems from general systems theory (see **box** below).

Important Point of View

General System Theory

General Systems Theory was established in 1937 by Karl Ludwig von Bertalanffy with the concept of the "open system," which posits that a "system" is a collection of interrelated elements where any change in one element results in changes to all the others.

This theory has influenced a wide range of fields, including psychology. In systemic approaches, an individual both influences and is influenced by the systems to which they belong through interactions. This represents a circular interpretation of relationships rather than a linear one, confined to cause and effect.

General Systems Theory is based on four key principles:

- *Principle of totality:* A change in one element results in changes to the entire system.
- *Principle of non-summativity:* A system is more than the sum of its parts.
- *Principle of equifinality:* The same outcome can have different causes.
- *Principle of homeostasis or self-regulation:* The system always seeks to return to equilibrium.

Each system is organized around explicit and implicit rules that provide its structure. A system naturally strives to maintain equilibrium. When a change occurs in an individual, it affects the entire system. Even a seemingly minor change can lead to a crisis and disrupt the previously maintained balance. In such cases, either a new equilibrium is established, often unconsciously, or the members of the system, such as a family, may struggle to create a satisfactory new dynamic. If this inability persists, the system can become disorganized, and the intervention of a therapist may become necessary.

Systemic therapies originated at the Palo Alto School in California in 1952, under the leadership of Gregory Bateson and his team. Bateson brought together students from various disciplines, such as communication studies (e.g., Jay Haley), psychiatry (e.g., William Fry), and anthropology (e.g., John Weakland and Edward Hall). This interdisciplinary collaboration was further enriched by contributions from other prominent scientists, including Milton Erickson, the founding father of Ericksonian hypnosis.

Bateson combined the diverse skills of his team members with his own background, which stemmed from work in an interdisciplinary group of mathematicians, logicians, anthropologists, psychologists, and economists. This group aimed to establish a general science of mental processes, and Bateson's project assembled an impressive array of leading scientists of the time.

Initially, Bateson's research focused on schizophrenia, leading to the formulation of the "double bind" theory. This groundbreaking concept (Bateson et al. 1956), often cited directly using the English term, refers to situations in which a person is trapped in a context where, no matter what they do, they "cannot win".

In 1959, American psychiatrist Don Jackson founded the Mental Research Institute (MRI) in Palo Alto and invited Bateson to join. Although Bateson declined, most of his team eventually became part of the MRI. The contributions of the Palo Alto School extend beyond the double bind theory, giving rise to all systemic therapies, including *family therapy*, *brief therapy*, and *strategic therapy*.

Thus, in the Americas, researchers who left the MRI or those influenced by it developed their own approaches derived from systemic theory: Virginia Satir established her school of communication (Woods and Martin 1984), which later evolved into Neuro-Linguistic Programming (NLP), while Jay Haley developed his strategic therapy approach (Madanes and Haley 1977). In Europe, systemic ideas were introduced notably through the founding of the Center for the Study of the Family in Milan in 1967 by Italian psychiatrist Mara Selvini Palazzoli.

Systemic approaches gradually expanded beyond family therapy to address conflict resolution in educational or professional settings, as well as international relations. In private or institutional medicine, systemic therapy enables a holistic understanding of the patient by incorporating their relationships into the therapeutic process.

Systemic therapies are also essential in our field, as our interaction with the animal is indirect and inherently involves the owner. We are always addressing a system, and it is crucial to incorporate this reality into therapeutic prescriptions. If the owner does not approve of the proposed solutions, even if they are technically sound, the therapy cannot be implemented.

This book does not aim to train readers in systemic therapies but rather to spark curiosity and provide references for those who wish to explore further.

The following sections introduce specific types of therapies developed by disciples of the Palo Alto school. These schools of thought have evolved in parallel, allowing family or systemic therapists, after completing their general training, to adopt their own "style." They can draw from these diverse approaches, all of which share the foundational principles of *general systems theory* on the one hand, and/or communication techniques derived from Milton Erickson's practices on the other.

Important Point of View

Evaluating the Effectiveness of Therapies

Assessing the effectiveness of various therapies remains challenging for several reasons. Firstly, the therapist's influence is significant, regardless of the psychotherapy employed. Secondly, evaluations often rely on patient satisfaction rather than actual recovery. Since measuring psychological distress is difficult to objectify, evaluating therapeutic efficacy is inherently complex.

In 2004, Inserm published a study evaluating the effectiveness of psychoanalysis, cognitive-behavioral therapy (CBT), and systemic therapy. The study concluded that CBT and systemic therapy were effective, while psychoanalysis demonstrated only marginal effectiveness (Canceil et al. 2004). In 2016, a meta-analysis compared the effectiveness of seven psychotherapies for depression. It found that all therapies appeared to be "potentially effective," but the levels of evidence and robustness varied significantly between therapies (Barth et al. 2016).

In veterinary psychiatry, the additional factor of the animal further complicates assessment. Since the animal cannot express its opinion, its perspective is represented—albeit indirectly—by the owner. This adds another layer of interpretation and subjectivity to the evaluation.

This lack of validation for the relative effectiveness of different therapies based on specific indications remains a challenge in both human and veterinary psychiatry. Many studies attempt to synthesize available therapies, but the recurring conclusion is that in feline veterinary medicine, research is still too limited to establish the definitive effectiveness of behavioral therapies (Houser and Vitale 2022). However, most authors agree on the critical role of environmental modifications in feline behavioral management (Ramos 2019).

28.1.3.1 Ericksonian Hypnosis

Ericksonian hypnosis is the most widely practiced form of therapeutic hypnosis in the world. It is named after its primary founder, Milton Erickson, an American psychiatrist and psychologist. His work forms the basis of Ericksonian hypnosis but also brief therapy, strategic therapy, and neurolinguistic programming (NLP) (see ► Sect. 28.1.3.4).

For readers interested in Erickson's work, *The Collected Papers of Milton H. Erickson on Hypnosis*, in three volumes, offers a deeper understanding of his research.

Erickson suffered from color blindness, dyslexia, and poliomyelitis. At the age of 17, following a coma that left him paralyzed, he was confined to a wheelchair. Erickson was convinced that he could overcome his paralysis by mentally stimulating his limbs. Through months of self-hypnosis, he managed to regain mobility. He then became a psychiatrist and revolutionized psychotherapy by applying this personal approach to his patients.

Many therapists attended Milton Erickson's sessions and described him as a magician for lost causes. They all seemed fascinated and admiring of his work, while acknowledging that his expertise was an art not necessarily easy to replicate. It is likely for this reason that many of them tried to theorize what they had learned from Erickson, creating movements such as brief therapy and strategic therapy.

The hypnotic state is defined as an altered state of consciousness. It is neither a state of sleep nor one of wakefulness but involves focused attention, reduced sensitivity to the environment, and an enhanced ability to respond to suggestions. Hypnosis therapists refer to this particular state as a trance. Every person experiences these trance states quite regularly without even realizing it. For example, this occurs when you drive from point A to point B and your mind wanders to a completely different subject than the road you are following. Upon arrival, you are unable to recall precisely what happened on the road, even though you didn't have an accident. You were in a trance.

During a hypnosis session, the therapist actively listens to the patient and adapts their speech style and approach to the individual they are treating. They rely on metaphors (see ► Sect. 28.2.4.9, "Metaphors"), anecdotes, and indirect suggestions to overcome resistance to change, allowing the patient to access their own resources. The goal is to facilitate the desired change.

The applications of therapeutic hypnosis in humans are numerous. It is commonly used for pain management, dermatological disorders, addictions, sports preparation, stress management, and emotional regulation (Häuser et al. 2016; Palsson 2015; Pinnell and Covino 2000; Shenefelt 2000). It is essential to keep in mind that, while hypnosis can have a significant impact on a patient's recovery, this practice does not heal. Hypnosis enables the patient to connect with their natural resources and improve their own recovery abilities. It does not replace appropriate medical care.

In cats, hypnosis is not currently used directly. However, the communication techniques employed by therapists (metaphors, active listening, reformulation, etc.) are widely utilized to establish a strong therapeutic alliance. Moreover, although it is less commonly applied than with dogs, relaxation protocol exercises are frequently used with cats. These exercises aim to induce a state of relaxation and calmness in the animal. These protocols are reminiscent of trance states. They are first taught to the cat outside of the problematic context, then gradually introduced into contexts increasingly closer to the challenging situation (see ► Chap. 29, "relaxation protocol").

28.1.3.2 Brief Therapies

Paul Watzlawick details his thought process in his works *How to Make Yourself Miserable* and *The Tactics of Change: Doing Brief Therapy*. Watzlawick theorized and synthesized Erickson's systemic approach and hypnotherapy to transform what was once considered an art or a form of magic into a reproducible clinical procedure. Paul Watzlawick is also known for his famous statement, "One cannot not communicate," a phrase often quoted by communication professionals.

Like all systemic therapies, brief therapies are based on the idea that each individual interacts with themselves, others, and their environment by creating their own interpretation of the world. They thus adopt a specific perspective and construct their reality. The direct consequence is that there is no single "true" reality, but rather as many possible realities as there are interactions between the individual and their environment. Furthermore, once this reality is constructed, the individual becomes subject to it.

From this perspective, psychiatric disorders result from a dysfunctional way of perceiving and reacting to this "constructed reality." The corollary is that if the clinician can alter the patient's perception of the problem, they will also induce changes in the patient's reactions, as they will then be confronted with a different reality.

This logic forms the foundation of brief therapies, and in clinical practice, it is embodied through the use of suggestions and refined manipulation that subtly modify the patient's perception of reality.

The goal of brief therapy is clear: to eliminate the symptoms presented by the patient and resolve their problem. However, it is neither a behavioral nor a symptomatic therapy. In fact, the alteration of the patient's perception of reality creates a subtle change, the result of which is the resolution of the patient's symptoms by the patient themselves. In other words, this aligns with the principle of totality from general systems theory: if a small change is introduced, the entire system is altered.

To achieve this goal, the therapist must break the vicious cycle within this circular system (homeostasis) that traps the patient in their dysfunctional reaction, leading them to repeatedly engage in the same behaviors because the response they receive from the system is unsatisfactory.

To do this, the therapist focuses very little on the past, unlike traditional forms of therapy. The critical point in brief therapy is the persistence of the problem rather than its origin. In other words, the practitioner focuses on the "how" rather than the "why." Thus, they position themselves in the "here and now" (hic et nunc), as the present is the only element that can be worked on.

In practice, the therapist seeks to answer three questions:

- What are the relationships between the patient and themselves, the patient and others, and the patient and the world?
- How does the reported problem function within this relational system?
- How can this problematic situation be changed in the fastest and most effective way?

At the Mental Research Institute (MRI), Watzlawick reported in 1974 a resolution rate of 72% of cases, with an average of seven sessions. At the Brief Family Therapy Center in Milwaukee, similar resolution rates were reported, but the therapists worked on reducing the number of sessions, achieving an average of five by 1984.

Regarding the tools used by the strategic therapist during a consultation, several key steps can be highlighted.

At the beginning of the session, the practitioner *reformulates the problem* and *identifies exceptions* (see ▸ Sect. 28.2.4.4, "Reformulating"). They then establish objectives and ask the patient to describe what will be different for them once the problem is resolved. Based on this, the therapist builds a schematic representation of the problem and the factors sustaining it.

The next step involves *constructing the strategic therapy*. This process relies on several possible options, such as transforming the exception into the rule, altering the structure of the schema itself, modifying its sequence or the individuals involved, adjusting the frequency or duration of the sequences within the schema, or introducing a random beginning or end. These strategies must be tailored to the patient, meaning they must use variables that resonate with that specific individual so they are willing to implement them. This aspect is perhaps the most challenging to master but is also central to strategic feedback.

Indeed, before presenting their strategy, the therapist begins by *complimenting* the patient and acknowledging what they are already doing that works. The therapist then provides cues to guide the patient toward constructing a therapy suited to their mode of functioning, highlighting specific actions the patient is already taking that align with the chosen strategy. These are highly effective therapeutic suggestions, as the patient selects from solutions that all orient them toward resolving their problem.

As part of these cues, the patient is often asked to observe any highly positive experiences they have during the week following the initial session. At the MRI, this prescription to note positive occurrences yielded remarkable results: most patients complied (90%) and reported elements they perceived as new or different, indicating that change was already underway.

The practitioner then proposes their therapeutic strategy to the patient and agrees with them on the objectives (symptom reduction), the means (therapy explained simply), and the relationship (between therapist and patient). This process builds a therapeutic alliance (see ▸ Sect. 28.2.2.1). If the proposed strategy is effective, symptoms decrease rapidly, and a gradual shift in the perception of the new reality takes place. The initial system, rigidified by the patient's dysfunctional reactions, becomes increasingly flexible. Over time, the patient realizes that resolving the system is possible and gains autonomy and self-esteem.

Follow-up sessions differ from the first, as they no longer focus in detail on the problem but instead aim to amplify what is already working. The therapist revisits initial techniques, such as offering compliments on what is effective, and spaces out the sessions as they observe substantial improvements in the patient's life. Most of the time, the intervals between sessions increase progressively, following a pattern of one week, then two, then three, then four, and so on.

This approach has achieved considerable success and is now being refined by researchers and clinicians, either by synthesizing various studies conducted by other researchers or by tailoring its application to specific clinical contexts. It is noteworthy that the founders of this approach took the time to demonstrate the effectiveness of strategic therapy, both in the short duration required to achieve clinical success and in the sustainability of these results after therapy concludes. The pursuit of clinical efficacy, supported by systematic evaluation of the therapy's lasting outcomes, has led therapists to regularly evolve this therapeutic model. Numerous trials of alternative or refined models have been tested and gradually adopted. Researchers working on this topic report that the difficulty others encountered in learning and applying their models often guided them toward greater *simplicity*, making the methods more *teachable*.

The paradox of this approach lies in its ability to resolve complex relational problems with very simple measures, even though these seemingly simple solutions are underpinned by a complex theory.

28.1.3.3 Strategic Therapies

Strategic therapies are directly derived from the work of Milton Erickson. They were primarily developed by Jay Haley and Cloé Madanes.

Jay Haley was neither a psychiatrist nor a clinical psychologist. He was a communication practitioner who founded and directed the Family Therapy Institute of Washington in 1975. He worked with Gregory Bateson and Milton Erickson at the first Palo Alto center. Haley was particularly interested in the need for a pragmatic approach to emotional problems and for a descriptive system that accounted for all parties, directly or indirectly involved, in a pathological system. An eclectic researcher, he worked in fields ranging from anthropology to psychiatry, including the development of brief therapies for schizophrenic patients. A prominent speaker at hypnosis and psychiatry conferences, he was a central figure in the field of systemic family therapies. In his book *Uncommon Therapy*, published in 1973, he introduced the term "strategic therapy" for the first time. Haley coined several phrases that are widely referenced in current veterinary practices, such as "to speak of hierarchy is to speak of protection and love."

Cloé Madanes, on the other hand, was a psychologist and also a student of Erickson. Alongside her husband Jay Haley, she contributed to the founding of the Washington Institute. Among her best-known works, readers can deepen their understanding of strategic therapies through *Strategic Family Therapy*, *Behind the One-Way Mirror*, or *The Secret Meaning of Love*. Her books have been translated into more than 20 languages, and she is also considered a pioneer in the field of strategic therapies.

In France, Jacques-Antoine Malarewicz redefines the core principles of strategic therapies, bringing his own perspective in his book *Fourteen Lessons in Strategic Therapy*. Like Haley, Malarewicz places significant emphasis on communication techniques. According to him, clinicians are increasingly well-trained in applying the concept of systems to the problems presented to them but are often poor communicators, which leads to failure. Indeed, "it is not enough to have good ideas; you must also know how to convey them." The strategic approach he employs combines a systemic perspective with communication techniques derived from Ericksonian hypnosis to succeed in clinical consultations. For him, therapy has two components: the first aims to understand the patient's problematic situation, while the second seeks to modify it. To succeed in this complex task, he believes it is essential for the therapist to have tools that are as simple as possible, enabling them to induce change effectively.

Malarewicz presents, in a highly didactic manner and through clinical cases, each of the tools he uses during his therapy sessions. He reviews how therapeutic purposes can be served by *paradoxes*, *circular questioning*, and *metaphors*.

These tools, developed for psychotherapies in humans, are valuable resources that we can adopt in veterinary consultations to improve communication with clients and enhance the effectiveness of therapeutic prescriptions for animals (see ▶ Sect. 28.2).

28.1.3.4 NLP (Neurolinguistic Programming)

In communication, the message is conveyed through a combination of verbal and nonverbal elements. Verbal cues include words, syntax, intonation, use of irony, humor, and so on. Nonverbal cues, on the other hand, consist of gestures, posture, accent, and voice timbre. In practice, during communication between two individuals, the message as received by the listener is decoded using this combination of verbal and non-verbal elements.

Unconsciously, we have learned to detect messages that are incongruent between *verbal* and *nonverbal* communication. For example, someone who says, "I'm sorry about what happened to you," while struggling to suppress laughter does not create the same effect as someone who says the same words with tears in their eyes.

Neuro-linguistic programming (NLP) focuses on individuals' verbal and nonverbal communication rather than the content of the message being conveyed. These communication techniques were inspired by Erickson's work, which already emphasized the importance of communication techniques during his consultations.

By studying the elements of a message beyond its content, NLP offers techniques to reduce information loss and improve communication. These observation skills are invaluable and not naturally acquired by everyone. Some individuals intuitively decode nonverbal cues, while others perceive them much less easily. A study conducted with final-year medical students showed that, during patient interviews, students demonstrated highly variable communication skills (Vogel et al. 2018).

The ability to detect these signals enhances *empathy* with the patient during consultations. Moreover, in certain conditions, such as autism spectrum disorders, it has been shown that individuals struggle in social interactions partly because they misinterpret nonverbal signals and also because they themselves exhibit atypical nonverbal cues that are misunderstood by others (Schaller et al. 2021).

When it comes to *verbal communication*, word choice can be adjusted to synchronize with the conversation partner.

The *level of language* is chosen to match that of the patient. However, it is possible to go further by identifying lexical fields that resonate most with them or by mirroring their paraverbal parameters, such as speech rhythm or volume. A classification system for *communication channels* has been developed: the Vakog system (◘ Fig. 28.1). This acronym refers to different sensory systems: visual, auditory, kinesthetic, olfactory, and gustatory. The principle is to identify which communication channel appears to be most dominant for the conversation partner and then use words that correspond to it. For example, a visual person might often use expressions like, "I see what you mean"; an olfactory person might say, "I can't stand the

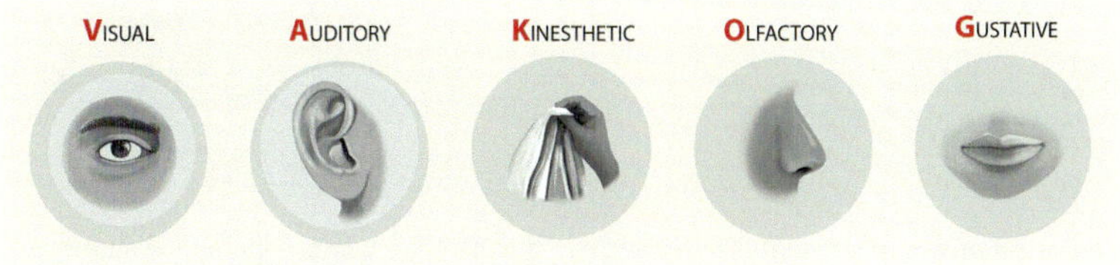

◘ **Fig. 28.1** Communication channels and the Vakog System

smell of him"; whereas an auditory person might say, "I hear what you're saying, but it doesn't sound right."

The practical application of this concept is to prioritize a specific channel to establish communication effectively. For instance, a patient who seems highly visual will benefit from therapy explanations presented with diagrams, while an auditory-oriented patient will respond better to verbal explanations. For kinesthetic patients, having them perform the exercise will yield the best results.

Regarding the *nonverbal aspect of communication,* it involves deciphering a person's posture, position, and micromovements, such as eye movements or specific facial expressions. Some of these micromovements can be consciously controlled by the individual, but many are not and instead reflect the person's emotional state. These subtle signals are somewhat equivalent to the calming signals we described in cats in ▶ Chap. 1.

Learning to detect these elements enables the therapist to verify whether the conversation partner truly means what they are saying. It can also sometimes help identify unspoken emotional feelings. For instance, a person who shifts their position, crosses their legs, leans back in their chair, and folds their arms indicates a negative reaction to what has just been said. Such a change in posture allows the therapist to recognize that they may have upset their interlocutor and provides an opportunity to address the message that was poorly received.

Pupil dilation is uncontrollable but also provides valuable emotional cues. A dilated pupil indicates activation of the sympathetic nervous system, signifying alertness, whereas constricted pupils (miosis) reflect a more relaxed emotional state.

Finally, eye movements are also uncontrollable, and rapid movements in certain directions can indicate what the individual is thinking (◘ Fig. 28.2). However, these movements are reversed between right-handed and left-handed individuals, making it impractical to memorize such patterns.

For example, when a right-handed person visually recalls a genuine memory, they look upward and to the left. In practice, the observation of such movements should not isolate a single action as an absolute truth. The goal is to learn to identify changes during a conversation that might suggest the person is not entirely sincere.

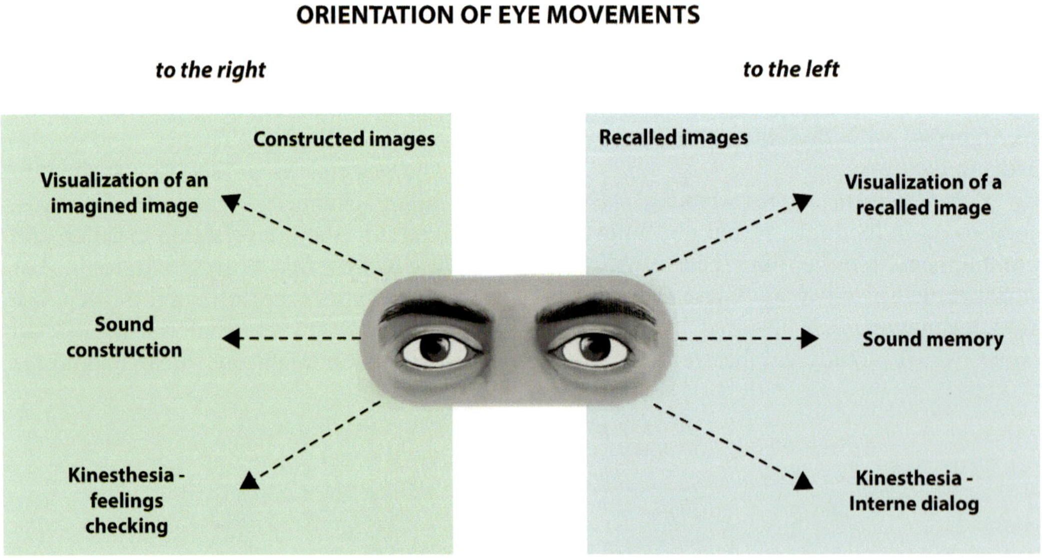

◘ **Fig. 28.2** Orientation of eye movements for recalled or constructed images. This diagram should be interpreted from the observer's perspective. In practice, this means that when a person visually recalls a memory, they look up and to the left, but the observer sees their eyes moving to the right

Summary of ▶ Sect. 28.1

In human medicine, psychotherapies have existed since Antiquity, but they have evolved significantly over the past century. Psychoanalysis was historically the first psychotherapy, though its effectiveness has not been proven.

The most commonly used psychotherapies are cognitive-behavioral therapies (CBT), which use specific exercises to alter an individual's perceptions and, consequently, their behaviors, and systemic therapies, which focus on the system in which the patient operates and are based on the idea that changing one element of the system will transform the system as a whole.

Systemic therapies were developed in Palo Alto by a group of pioneers who conducted intensive research in the second half of the twentieth century. The initial work done in Palo Alto includes that of Milton Erickson, the father of therapeutic hypnosis. The dynamic environment of the Palo Alto school laid the foundation for various psychotherapeutic approaches, all derived from systemic theories and the communication techniques used in Ericksonian hypnosis. From this foundation emerged brief therapies, strategic therapies, and NLP, all of which offer interesting variations that can be utilized by veterinary practitioners to enhance their consultation techniques.

This nonexhaustive list of major human psychotherapies aims solely to spark the reader's curiosity about this entire field of therapeutic tools and provide references to works that will allow them to delve deeper into the topic. This inventory can serve as a menu for the therapeutic practitioner. By knowing what exists and understanding the strengths and weaknesses of each type of therapy, practitioners can make choices based on the condition they are addressing, their client, and their own personality. Together, these choices define their "therapeutic style."

28.2 Practical Tools for the Veterinary Practitioner: Consultation Techniques

It is important to remember that our primary responsibility as veterinarians is to care for the animal, not its owner. However, in veterinary medicine, and even more so in the field of veterinary psychiatry, successful treatment requires the involvement of the owner.

Behavioral treatments are lengthy because they always include therapy and new learning processes that must be repeated over time to be fully integrated. The owners' commitment is therefore essential, but it often represents a significant obstacle, as important as financial considerations. Furthermore, the use of psychotropic medications, which accompanies most treatments, is a source of considerable concern for many owners.

Without the *owners' adherence* to the prescribed medical and behavioral treatment plan, recovery is impossible. To achieve this, the therapist must have strong communication skills, both to gather relevant data and to convince owners to commit to these long-term therapies.

It should also be noted that not all therapies can treat all conditions. In practice, as mentioned in ▶ Chap. 6, veterinary psychiatry prescriptions always include behavioral therapy, which involves performing one or more specific exercises. Whether these exercises take on a more traditional tone, such as CBT, or a more systemic approach depends on the individual therapist.

However, in general, when the behavioral condition involves the fear axis or the self-control axis, CBT is predominant. Conversely, when the condition pertains to the territoriality axis or the relational axis, tools derived from systemic therapies are often more appropriate.

Thus, training in *communication techniques* is a real advantage for the veterinary therapist. The following paragraphs describe various tools drawn from communication techniques used by psychotherapists practicing the therapies outlined in the previous section. These tools are numerous and varied, and the practitioner will need to learn to adopt some over others, depending on their personality, which will shape their "style."

The goal here is not to describe these tools in detail, as there are many books specifically dedicated to them, but rather to provide some guidance on what can be done to improve the quality of consultations, thereby maximizing the chances of successful treatment. Some might view this as a form of *manipulation*, as mastering these tools allows the veterinarian to convince owners effectively. However, it is primarily a way for the veterinarian to conduct the consultation calmly, manage tensions, and better support the owners for *optimal care* of their animal.

28.2.1 The System in Veterinary Medicine

Illness in general, and behavioral disorders in particular, have a significant impact on the entire family. A cat with inappropriate elimination creates genuine exhaustion for the owners, who, day after day, exhaust every possible strategy to try to improve the situation. Most owners of cats with behavioral disorders are judged by those around them and made to feel guilty for supposedly not providing their cat with everything it needs. This is also true for parents of children with disabilities or psychiatric illnesses. Not long ago, autism was still described as a consequence of insufficient parental care for affected children.

Behavioral consultations are therefore at the heart of this often arduous journey, providing an opportunity to relieve owners of the guilt they feel about their perceived responsibility for their animal's condition.

It is worth noting that, in many therapies, the goal is to achieve *lasting change* that allows for a better balance. This change must be *initiated by the therapist*.

In veterinary medicine, the simplest system we encounter consists of the cat and its owner. In some cases, the systems are more complex, such as in large families or blended families. It might be tempting to believe that the simpler the system, the easier it is to modify, but the opposite is true.

In a simple system, such as a single person living with their cat, changing the rules of daily life is complex because each individual in the system represents a significant component of that system. Conversely, in a large family, the inertia of each individual carries less weight within the overall system. Furthermore, in a family, even a small change by one member can influence all the others and significantly alter the system as a whole.

Thus, it is important to remember that the simpler a system is, the greater the inertia of each individual within it. Conversely, the more complex a system, the easier it is to modify.

28.2.2 Alliance and Resistances

28.2.2.1 Therapeutic Alliance

The *therapeutic alliance* is defined by a threefold agreement on the relationship, the treatment goals, and the methods implemented to achieve them.

In practice, the alliance is essential in our veterinary psychiatry consultations due to the extended duration of follow-ups. If the owner becomes discouraged during treatment, the process will not be completed and will ultimately fail.

The quality of the alliance can be assessed by asking the client about their feelings regarding the consultation, though the therapist can only form an impression of the strength of the alliance they have created. In chronic conditions, follow-up consultations help maintain a strong alliance, and the practitioner must pay close attention to ensuring agreement on the goals, methods, and relationship.

For example, during a follow-up visit, the therapist might discover that the owner has changed the medication dosage because a psychologist friend advised that it seemed too high. This realization allows the roles to be clarified by reminding the owner that the veterinarian is the one responsible for medication management. Statements like, "I understand that this concerned you, and I really believe you should have called or emailed me so we could decide together about this dosage change," open the discussion about the issue at hand.

The practitioner can then propose a solution to define the framework and the relationship, such as: "Can we agree that if you have concerns about the treatment in the future, you will notify me immediately so we can work together to find the best solution?"

28.2.2.2 Resistance

Resistance is a normal aspect of any veterinary therapy. It may appear to hinder the therapeutic process, manifesting as *attempts to avoid change*. One of the most common forms of resistance encountered in our field is opposition to medicalization through the use of psychotropic drugs.

This resistance is not directed at the therapist personally but rather at the *risk of change they represent*. It reflects a motivation to consider change, even though it also shows that the change is not desired. The therapist must identify these resistances, take them into account, and find ways to resolve them. Often, resistance centers around a key point of change, and one way to address it is to anticipate it. By artificially creating resistance around secondary issues, the practitioner can prevent its emergence on more critical points.

For example, if the therapist's goal is to transition the cat from having meals twice a day to ad libitum feeding, they might prescribe that a new blue food bowl with tubes should be placed in the living room. By emphasizing the color and location, the owners may focus their attention on this detail and possibly resist by negotiating a different type of bowl or location. However, they are less likely to negotiate against the critical point: the introduction of ad libitum feeding with food puzzles.

The *precision of the prescription* (e.g., "a blue bowl with tubes") in this case helps direct the owners' focus to a secondary point, making it easier for the practitioner to achieve acceptance of a more critical aspect of the treatment plan.

28.2.3 The Therapist Positioning

28.2.3.1 Establishing the Framework of the Consultation

The veterinarian is the professional responsible for setting the *framework* of the consultation. It is their responsibility to manage how the consultation unfolds. They are free to arrange the furniture as they see fit, and these choices will already have an impact, as they create the client's first impression. Owners may need to discuss intimate aspects of their lives, emotions, and relationships with their cat, and sometimes even with each other, as a problematic cat can be a source of conflict. It is therefore essential that they feel at ease.

Increasingly, specialists in the field recommend using a room with comfortable chairs and no table, which might act as a barrier between the practitioner and the owner, particularly for long veterinary psychiatry consultations. At the very least, each practitioner should choose a room where they feel comfortable. Decorations, such as portraits or images of cats, can also facilitate discussion. For example: "Yes, I completely understand. I've had very large cats myself—see the Maine Coon behind you? She was my first cat." Such conversations humanize the professional and help establish a trusting environment.

The framework also includes boundaries that the practitioner wants to see respected. This might include requiring a transport crate to bring the cat to the clinic and not allowing the cat to leave the crate until the practitioner has assessed the safety of letting it out. If this framework is not respected, it is the practitioner's responsibility to remind the client of the rule.

This approach situates the therapist-client relationship within an atmosphere of trust and kindness, while also emphasizing the need to respect boundaries. It also clarifies the roles of each party in the relationship during the treatment process that will follow.

28.2.3.2 Therapeutic Listening

The practitioner must be exceedingly polite. This means ensuring that clients feel comfortable. Of course, there are explicit rules of politeness that must be followed, but that alone is not enough. Being "professionally" polite also involves creating a sense of ease by adapting one's demeanor and vocabulary.

Using Nonverbal Language

Depending on the client, the therapist must be able to adapt. As discussed in ▶ Sect. 28.1.3.4, communication channels are crucial. The practitioner should aim to find the right tone and appropriate words for their client from the very beginning of the consultation. In this effort, observing the client's non-verbal language is a valuable tool.

Learning to interpret non-verbal language should be part of the foundational knowledge for all therapists. The goal is not to detect lies or decode a person's message based solely on their non-verbal cues. Instead, the aim is to identify *incongruences* between verbal and non-verbal communication. When the gap between the "container" (non-verbal) and the "content" (verbal) becomes too wide, it becomes challenging to grasp the underlying message.

There are three main types of gestures that it is important to identify in an interlocutor:

- *Self-focused gestures:* these involve the patient making contact with their own body. Such gestures indicate discomfort and a tendency to withdraw.
- *Regulatory gestures:* these punctuate interactions and can include head movements or leaning back in a chair. They provide insight into the quality of the exchange and the listener's level of attention and agreement.
- *Emotion-related gestures:* these are involuntary and often subtle but must be quickly identified by the therapist to consider them in their subsequent intervention. Examples include flushed cheeks or tear sacs filling with liquid.

Of course, the isolated presence of one of these elements should not be overinterpreted by the practitioner. However, their repeated occurrence, sudden appearance, or increased frequency is information that should not be overlooked.

Active Listening

In all cases, it is essential to practice high-quality listening, often referred to as active listening. This is not merely the practitioner remaining silent but a type of listening that *ensures the interlocutor feels they have the professional's full attention.* The practitioner's role is to understand not only what the client is saying but also their life context, emotions, hesitations, fears, and doubts. Active listening is conveyed through eye contact, nodding, and prompting the conversation forward.

Body posture is part of non-verbal communication and reinforces the client's impression. For instance, without speaking, the practitioner can reassure the client simply by sitting upright, looking at them, and resting their hands on the desk. Conversely, avoiding eye contact while the client speaks, leaning back in the chair, and crossing their arms may create a sense of discomfort.

The beginning of the consultation is a key moment for listening. Clients often feel the need to explain what brought them to the consultation. If they feel comfortable, they will spontaneously provide valuable information. The practitioner can also use this time to identify the most appropriate communication channels or lexical fields to later ensure effective communication with the client.

For example, if the owners talk about their children, the practitioner will know that the people in front of them already have experience with education and have likely adopted implicit family rules. This allows the use of *metaphors* related to parenting to help explain concepts more effectively.

Compassionate Listening

The therapist's role is to help both the cat and its owner. At times, they may encounter inadequate practices or a lack of respect for the cat's physiological needs. However, the therapist must maintain a *compassionate stance* and *refrain from judging the owner*. Any form of judgment would undermine the therapeutic alliance. Furthermore, most owners have already faced judgment from family, neighbors, or others and often arrive at the consultation with a deep sense of guilt.

It is also important to emphasize that when a behavioral disorder is present, educating the cat becomes a genuine challenge. Criticism of the owner's communication methods is often more a reflection of the difficulties posed by the cat than an inherent incompetence on the owner's part. The same applies to the use of coercive methods. These are often *symptoms* of the challenges owners face with an unmanageable cat rather than an intentional desire to harm or cause suffering.

- *Empathy*

Empathy is the ability to understand the emotions of others without feeling them yourself. Empathy is bounded by antipathy and sympathy. *Antipathy* is the experience of feeling an emotion opposite to that of another person, whereas *sympathy* involves feeling the same emotion. Sympathy and antipathy trap individuals in their emotions, which are uncontrollable and detrimental to the working relationship.

The practitioner must be empathetic but should avoid both sympathy and antipathy in the therapeutic context.

A physiotherapist illustrates this balance well. Rehabilitation often involves pain. If the physiotherapist is overly sympathetic, they may not push hard enough, and progress stalls. If they are too harsh, the pain becomes unbearable, and movements are inhibited, again halting progress. Instead, they must induce mild discomfort, fully aware of the patient's pain but without experiencing it themselves.

Thus, to ensure the success of therapy, the practitioner must skillfully navigate between empathy, sympathy, kindness, and rapport-building while remaining strictly within the *therapeutic framework*.

28.2.3.3 Low-Position Approach

Consultations can sometimes be delicate and may escalate into conflict. It is crucial that the veterinarian remains the director of the exchange.

Communication specialists explain that the one who leads the relationship is the person who adopts a low position. Being in a low position means accepting help and even asking for it. This indirect method of encouraging change by asking the owner to take responsibility for part of the prescription forces them to make suggestions they feel capable of implementing. This is an indirect yet effective way to foster change, particularly when the therapist encounters *resistance*.

This approach may seem counterintuitive since clients are ostensibly coming to seek help. However, to mobilize their engagement, it is important to demonstrate that your expertise is nothing without their active participation. Clients must understand that they remain central to the situation. Simple prompts such as "I don't understand," "Explain this to me," or "I don't know" are examples of low-position statements. The low position can also be taken further with statements like: "I need you to tell me what I should do," or "I don't feel capable of managing the details of your environment on my own; I need your help."

In contrast, a high-position approach is risky. Some clients may immediately place the practitioner in a high position by assuming a low position themselves. For instance, they might say, "Everyone told me to come to you because you're supposed to be the best." While these comments may seem flattering, the practitioner should take note. It is common for such clients to have already seen several therapists, who have likely failed. Furthermore, if the therapy does not meet their expectations, it becomes easy for them to say, "You were supposed to be the best, but I see that's not true. It's been two weeks, and the problem isn't solved yet."

Of course, these are only examples. Sometimes, the therapist must assume a high position of authority, particularly when explaining technical points. However, recognizing the dynamic of positional roles during the consultation is crucial to avoid many pitfalls in the exchange.

28.2.4 Consultation Techniques

Consultation techniques alone could easily be the subject of a substantial book.

It is important to emphasize that these are often learned skills, and anyone can work toward mastering them. Some individuals are naturally talented, while others will need to make sustained efforts over a long period to succeed. However, everyone is capable of acquiring these techniques.

These methods might be perceived as manipulative since they provide those who use them with significant persuasive power. However, *therapeutic manipulation* is only harmful when it is not used in the service of the patient.

28

The following section presents a few tools that can be applied in all consultations.

28.2.4.1 Inducing Change

In 1984, Prochaska and DiClemente, two psychologists, explored the different stages that precede change (Prochaska and DiClemente 1984). According to them, individuals struggling with addiction go through a series of stages of change: precontemplation, contemplation, preparation, action, maintenance, relapse (◘ Fig. 28.3).

This model of change is very useful for identifying the stage at which owners are regarding the problem presented by their cat. These stages, adapted to our field, can be described as follows:

- *Precontemplation*: the owner does not believe their cat has a real problem and, as a result, does not consider seeking consultation. This is a form of denial.
- *Contemplation*: at this stage, ambivalence begins to emerge. The owner considers a change in behavior but hesitates to give up the perceived benefits of the current situation. They weigh the pros and cons of change versus the current state until deciding in favor of one or the other.
- *Preparation*: the owner feels ready to begin behavioral treatment for their cat in the near future. They make decisions and start implementing them gradually.
- *Action*: the change is underway through behavioral modifications, regular exercises, and more. Support and encouragement are essential during this stage.
- Maintenance: during this consolidation phase, caution is required, as there are many temptations to revert to the problematic behavior.
- *Relapse*: relapse is possible and part of the normal process of change. It is not pathological but may require additional time to achieve the final success of the process.

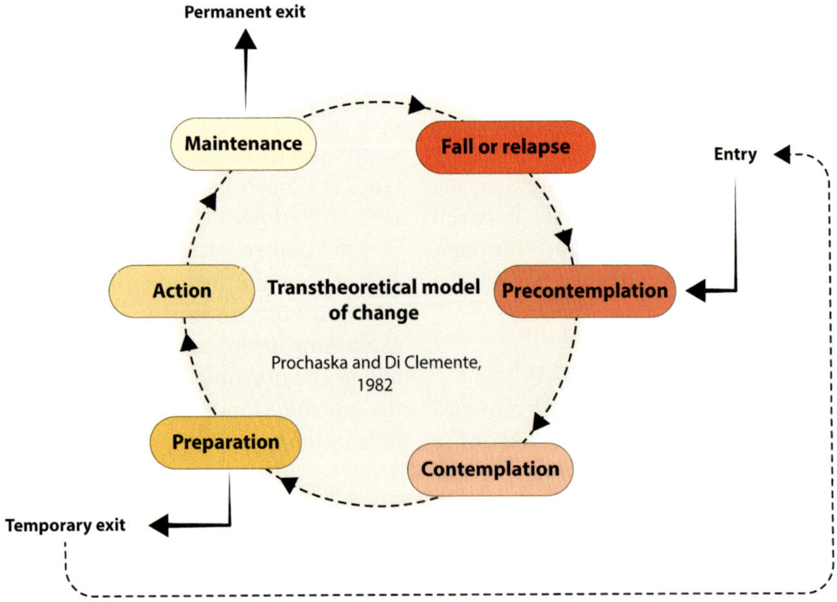

◘ **Fig. 28.3** Stages of change according to Prochaska and Di Clemente

- *Permanent exit or recovery*: this stage marks the final success of the process, during which the owner consolidates the maintenance phase.

This approach, initially developed for addiction treatment, is an interesting model for representing the stages necessary for change. Depending on the stage of change identified by the practitioner, the consultation will take a different direction. The primary goal is to guide the owner toward the next stage.

For instance, if the practitioner identifies the precontemplation stage, which is common in developmental disorders in young animals, they conduct a *motivational interview*. In this case, the focus is on increasing the owner's motivation by highlighting the benefits of the final outcome, rather than pushing directly for behavioral treatment.

28.2.4.2 Validating Emotions

Depending on their personality, some individuals communicate more on a rational level, while others prefer an emotional level. However, under certain circumstances, the emotional level will surface regardless of personality. In our consultations, it is common for owners to recount difficult events that have deeply affected them, such as a bite or a panic episode involving their cat. As a result, the beginning of a consultation is often filled with strong emotions.

The practitioner must identify these elements and *validate the owner's emotions* before proceeding to a rational discussion. If the practitioner responds rationally right away, they will not be heard. On the contrary, the owner may become defensive, feeling that the professional does not grasp the emotional difficulty they are facing.

For example, if your client is crying heavily while describing how her 8-month-old son was scratched by the family cat, you should begin by listening and acknowledging what you observe. "Madam, I can see that this is still a very emotional situation for you, and I understand. I know that in a similar situation, I would feel completely overwhelmed (or angry or terrified…)."

Only by validating this emotion can the practitioner help the owner move past the emotional outburst and begin discussing the facts.

28.2.4.3 Metacommunication

Metacommunication is a digression within an exchange, an interruption that allows the participants to discuss the communication process itself. This tool is particularly useful for addressing feelings that arise and disrupt the quality of intellectual discourse.

It is especially valuable when the practitioner senses discomfort. This may occur because the practitioner feels the clients are withholding some information or notices incongruent nonverbal cues. To reduce the tension that gradually builds in the exchange, the practitioner engages in metacommunication by expressing their discomfort. For example, they might say: "I apologize, but I feel like you disagree with what I just said," or, "Sir, I feel uneasy. I don't sense that I have your trust, and we need to clarify this situation."

These simple statements often help to dissipate discomfort. On one hand, the therapist can continue with the consultation without questioning the validity of their sense of unease. On the other hand, if the owner genuinely disagrees, they now have an opportunity to voice their concerns.

28.2.4.4 Guiding the Consultation to Gather Relevant Information

- ***Open-Ended Verses Closed-Ended Questions***

The practitioner leads the consultation, and the way they ask questions significantly influences the responses they receive. *Closed-ended questions* are typically used to clarify a specific answer, while *open-ended questions* elicit broader information that cannot be obtained through closed-ended questions.

For example, if the practitioner asks, "How does your cat eat?" the client might discuss the location, the type of kibble, the ritual they use when feeding their cat, and might even spontaneously mention the time the cat tried to scratch them while being served a meal. On the other hand, asking, "Does your cat have a bowl of kibble available in your kitchen?" will yield a very precise answer but is unlikely to provide additional context.

Thus, when the client talks extensively and tends to provide too many details in their account, the practitioner can use closed-ended questions to obtain the specific answers they need. Conversely, a client who speaks very little should be asked open-ended questions to encourage them to express themselves freely.

- *Silences*

Silences are very useful for giving the owner time to express themselves. They align with the use of open-ended questions and often help gather valuable information. For instance, pausing after a client mentions their dissatisfaction with the behaviorist who recently visited their home is likely to prompt them to elaborate and provide details about what led to this feeling.

- *Rephrasing*

Rephrasing is a technique that fosters the therapeutic alliance, as it demonstrates good listening skills. It involves the practitioner restating, in their own words, what the client has just said. This rephrasing reflects the client's worldview as perceived by the practitioner, encouraging further discussion.

For example, the practitioner might say: "If I understand correctly, you were sitting completely still on your couch when your cat approached with dilated pupils and attacked your leg?"

- *Summarizing*

During the consultation, it may be helpful to pause for various reasons: to refocus and redirect, clarify, or highlight progress or a digression in the conversation. *Summarizing* is an effective way to achieve this. It also serves as an opportunity to rephrase after an initial discussion. Additionally, it allows the practitioner to emphasize the evolution of the client's request.

For example, after a long description of a cat's aggressive behavior, the practitioner might summarize as follows: "If I understand correctly, your cat has urinated outside its litter box an average of five times a week over the past three months, and you feel this is getting worse?"

- *Circular Questioning*

Circular questioning involves asking questions about the relationship between two individuals to a third person, in the presence of the two individuals. The goal is not to gather precise information but rather to give everyone access to relational insights that exist within the system.

For example, when a couple, Mr. and Mrs., who disagree on how to discipline their cat, come in for a consultation, the therapist might ask: "Mrs., how does your husband react when your cat climbs onto the kitchen table, which is off-limits?" Or: "Mr., the last time Gino tried to scratch you, did your wife come to help when she heard the hissing?"

Circular questioning provides a wealth of information about the relationships within the system and is also an excellent tool when certain individuals are not present at the consultation.

28.2.4.5 Using Humor

Humor is a very useful tool. The stories shared by clients are often sad, heavy with consequences, and sometimes difficult to hear. Humor can help lighten the atmosphere in certain cases, but the therapist must use it appropriately and in connection with the work being undertaken. Any humorous remark should serve the consultation.

For example, if Mr. criticizes Mrs. for her lack of diligence in cleaning the cat's litter box, the therapist might look at Mrs. and say: "I think Mr. is trying to tell you that he does it much better than you. It's almost a request to have a monopoly on cleaning the litter box, wouldn't you agree?" Mrs., who was about to respond defensively to her husband, offended by the unfair remark, bursts into laughter and says to the therapist, "Excellent idea." In this case, the therapist defends Mrs. and creates an alliance with her. Of course, care must be taken not to discredit Mr. in the process.

28.2.4.6 Speaking to the Cat

In the same vein of lightening the mood after difficult news from the owners or emotional tension (e.g., Mr. and Mrs. arguing about the cat), it can be effective to deliver a message by *speaking to the cat*. For instance, in the case of a budding argument, the practitioner might address the animal and say: "You see, you're going to have to make an effort, because they're going to end up fighting over your antics."

28.2.4.7 Avoiding Taking Sizes

A consultation involving more than one owner carries the risk of the therapist being drawn into taking sides. For example, one owner might say: "Doctor, could you please explain to my wife that we shouldn't let the cat play with us while we're watching a movie?"

If the therapist agrees, they align with Mr., creating an alliance with him but alienating Mrs. If they say nothing, they risk upsetting Mr. An elegant way to navigate this situation could be to respond:

> You're right, Mr., playing while you would prefer your cat to stay calm isn't ideal. However, seeing how intensely your cat plays no matter the timing, I think I wouldn't be able to manage it better than Mrs. does.

28.2.4.8 Making the Absent Speak

For behavioral consultations, it is ideal for all individuals involved in the cat's relationship to be present. Unfortunately, scheduling conflicts often make this difficult, requiring some flexibility.

Two main reasons support the need for everyone's presence. First, the information provided by one individual is often less complete when they are the only one describing the situation. Many details emerge from differing perspectives, which the therapist risks missing if everyone is not present. It is therefore important to engage absent parties by contacting them during the consultation if possible or postponing part of the discussion if necessary.

The second reason is the risk of resistance. Therapy always involves change, and this change can only occur if everyone agrees to it. An absent person may not understand the need for change and could oppose it. This possibility should be addressed openly, which means bringing it up and metacommunicating about it.

In some situations, the absence itself is a form of communication and may indicate principled opposition. For instance: "My husband stayed in the car..." By default, one way to gather the perspective of absent individuals and evaluate their role in the system is through *circular questioning*. For example, the therapist might ask: "How does your husband react when your cat urinates outside the litter box?"

28.2.4.9 Use Analogies

An *analogy* is a thought process that identifies a similarity in structure between two things that are otherwise of different natures or categories. In conversation, an explicit analogy is a *comparison*, while an implicit analogy is a *metaphor.*

- ***Comparison***

Comparison helps the owner accept an idea because it seems evident in a different context from the one being discussed. Once they recognize that the situation they are experiencing is comparable to another, acceptance often follows naturally.

For example, if an owner expresses concern about the potentially lifelong medicalization proposed for their cat, the practitioner might respond: "Would you give me the same answer if I had told you your cat had a heart condition or an allergy?" The owner would likely respond negatively. The practitioner continues: "Yet, the brain is an organ, just like the heart or the skin. It is a complex organ, and when it malfunctions, my duty as a veterinarian is to try to restore it to proper functioning. If this malfunction is due to a biological anomaly, I use medications that target this anomaly, just as I would if your cat had a heart condition or an allergy." The owner then replies: "Oh, I'd never thought of it that way, but what you're saying makes sense."

- ***Metaphor***

A *metaphor* represents observed reality within a different world or context. This approach allows facts and their intensity to be described without harshness. To be effective, the metaphor must provide a simple image of what is happening and should not require effort to understand. It can also serve as a gentler form of rephrasing, and even incorporate humor!

For example: "Sir, you're describing that your cat urinates on the bed intentionally to get revenge. You're strict with him and punish him by rubbing his nose in his mess, but the more you try to correct him, the less cooperative he becomes! Do you remember your math classes? Were you ever scolded for mistakes you deliberately made to get a bad grade?"

The metaphor allows the listener to construct their own mental image from the one provided by the practitioner. It is an indirect way of encouraging change.

For a metaphor to be relevant and effective, it must meet certain criteria. First, it should align with the owner's sensory orientation (Vakog system). It should also reference the owner's favorite activities and not shy away from repetition, so they can fully grasp the concept contained within the metaphor. In a way, it should perfectly fit the owner's reality to maximize its impact.

Additionally, the metaphor should not be followed by an explanation, nor should it interrupt the flow of the conversation.

28.2.4.10 Using Paradox

A *paradox* describes an absence of logic hidden beneath an illusion of logic. Most often, it arises from the encounter of multiple logical levels. This encounter can be intentional or unintentional. Paradoxes can be either harmful or beneficial.

Paradoxes are *harmful* when they prevent the recipient from responding, trapping them in what is known as a "double bind." A double bind is a paradoxical situation in which two propositions represent contradictory logical levels that cancel each other out. For example, telling someone "Be spontaneous!" is an example of a harmful paradox.

Conversely, paradoxes can be *beneficial* when used in a therapeutic context. This also means there is always an exit for the owner if they wish to opt out. The main characteristic of this type of therapeutic paradox is that it forces change. This obligation is theoretical, as the owner always retains the right to refuse and step away from the therapeutic framework.

This paradox often involves a disorienting question, such as: "Tell me what I should do to help you?" This phrasing aims to reposition the practitioner in a low position while encouraging the owner to take a stance.

28.2.5 Constructing the Therapy

The construction of the therapy occurs throughout the consultation, not just at the end. Owners will often provide useful details during the process. These might include preferences, such as the cat being more motivated by play than by treats, or mentioning that they enjoy DIY projects and creating spaces for their cat, which could be relevant to the environmental aspects of the therapy. These details, shared over the course of the discussion, should be noted by the practitioner and incorporated when building the treatment plan.

28.2.5.1 Reason for Consultation, Request, Hidden Request

The practitioner's treatment plan should not aim solely to target a symptom but rather to address the owner's request while improving the animal's health and quality of life.

Owners often arrive at a consultation with a stated reason, such as: "Our cat urinates outside the litter box to get revenge whenever he's left alone." During the discussion, the practitioner must clarify the owner's actual request, which may differ significantly from the stated reason. In this case, the request might be: "I want to be able to leave him alone," or "I'd like to go away for the weekend more often without worrying," or even "I want him to stop this behavior because my partner refuses to move in with me as long as this problem persists."

By clarifying the owner's expectations, the practitioner avoids falling into the trap of addressing only the *hidden request*. Many owners present a reason for consultation that is quite distant from what is truly bothering them.

28.2.5.2 Exception

Clients seek consultations because their pet exhibits behavioral symptoms of varying severity. As a result, they often focus on describing all the problems and difficult moments their cat causes. While it is essential to understand these details to identify precise symptoms, it is equally important to identify exceptions—those moments, contexts, or situations where the problematic behavior does not occur.

These exceptions provide an easy starting point for therapy, as they represent approaches that have already worked for the owner and align with their existing practices.

Identifying a successful exception is an excellent way to gather personalized exercises tailored to the clinical case. The practitioner should then develop and propose a version that transforms this exception into a rule.

28.2.5.3 Compliments

No matter the owner, there are always actions or insights they provide that can be highlighted for their relevance. This is not about lying or fabricating positive elements but rather about recognizing and acknowledging the positive steps the owner has already taken. At the very least, they have made the effort to seek consultation and have shown patience in dealing with a challenging cat without giving up or abandoning it.

Compliments help *alleviate the owner's guilt*, focus the discussion on the *hic et nunc* (here and now), and strengthen the therapeutic alliance to find a solution. The practitioner must deliver these compliments sincerely; insincerity will be perceived by the owner and will undermine the alliance.

28.2.5.4 Presenting the Therapy

Consultations in veterinary psychiatry are inherently lengthy due to the detailed collection of semiological data. The presentation of the therapy must be clear and comprehensible so that owners can effectively implement the proposed treatment.

Based on the communication channels identified during the consultation, the practitioner should choose the most appropriate way to present the therapy. For example, they might use diagrams for a visually oriented client or a story for an auditory-oriented one. The practitioner should regularly confirm that the owners understand the explanations and, if necessary, invite them to rephrase in their own words.

In cats, environmental adjustments play a significant role in therapy, often requiring less owner effort than the re-education exercises typically prescribed for dogs. This is likely one reason why outcomes are often excellent in feline cases.

When possible, using practical demonstrations followed by having the owners replicate them is a valuable technique. It allows the practitioner to assess the feasibility of the exercises for both the cat and its owner. However, this approach may sometimes result in failure, which the practitioner can use as an opportunity to empathize with the owners, acknowledging how challenging the cat is and validating their struggles. This strengthens the therapeutic alliance and ensures the prescribed therapies are realistic and achievable.

The number of exercises prescribed should be limited to avoid overwhelming the owner and discouraging them. Additionally, since the cat in consultation is not as comfortable as it would be at home, many exercises cannot be demonstrated during the session. These exercises should instead be explained as clearly as possible, often with visual aids, so that the owner fully understands the requirements. Demonstration videos can be a valuable tool for helping owners grasp the prescribed exercises.

Finally, the therapy plan should be prioritized to focus on the most relevant exercises. It will evolve during follow-up consultations, based on the progress observed since the initial session.

Summary of ▶ Sect. 28.2

Therapeutic tools borrowed from human psychotherapies are numerous and allow practitioners to be more effective during consultations. The system encountered in veterinary medicine involves the animal and its owner and follows the same principles as systemic interviews. Training in these techniques is therefore a valuable asset for any professional conducting consultations.

The goal of the practitioner examining a cat for behavioral issues is to maintain a strong therapeutic alliance, as treatment often spans several months to allow the implementation and assimilation of various learning processes. This alliance is built on a threefold agreement: on the goals, the means, and the relationship between the practitioner and the owner.

Resistance expressed by the owner represents an opportunity for change and should not be perceived negatively by the therapist.

The therapist must remain in control of the consultation's framework while practicing therapeutic listening—both active and compassionate. They should demonstrate empathy without falling into sympathy or antipathy and aim to adopt a low position whenever possible to indirectly encourage change.

There are numerous techniques that can be used during consultations, such as validating emotions, metacommunicating, using humor, talking to the cat, engaging absent parties, or employing analogies and paradoxes. These techniques also contribute to developing and presenting the therapy plan, enabling practitioners who master them to significantly increase the likelihood of successful outcomes for the animal's treatment.

Summary of This Chapter

In the previous section, we discussed how medications serve as valuable tools for providing animals with greater mental flexibility, enabling them to adopt new behaviors that are more conducive to their well-being and health, as well as to the safety and comfort of those around them.

A sine qua non condition for the successful implementation of behavioral therapy, as well as medication prescription, lies in a particular aspect of feline psychiatry: the patient, the cat exhibiting undesirable behaviors, is not the one carrying out the treatment. It is the owner who administers the medication and follows the therapy instructions, placing them in a central role in the care process. Their voluntary and effective involvement constitutes one of the three fundamental pillars for successful treatment, alongside medication and behavioral therapy. Neglecting this factor represents a significant loss of opportunity for the animal. History shows that in human medicine, applying techniques aimed at convincing and motivating the family has sometimes produced remarkably rapid and consistent results.

From a theoretical perspective, these techniques broaden the scope of both the problem and its solution to include the entire family, rather than focusing solely on the symptomatic individual. These so-called systemic or family therapies address all members of the system and their interactions within the group. Agreement among all members on the goals, the means to achieve them, and the role of each participant forms the foundation of the therapeutic alliance.

While the systemic approach allows the therapist to understand relationships that are either helpful or resistant to change, it is the use of communication techniques tailored to the individuals and the situation that equips the practitioner with the persuasive power needed to maintain the therapeutic alliance throughout the treatment process.

References

Bandelow B, Reitt M, Röver C, Michaelis S, Görlich Y, Wedekind D (2015) Efficacy of treatments for anxiety disorders: a meta-analysis. Int Clin Psychopharmacol 30(4):183–192. https://doi.org/10.1097/YIC.0000000000000078

Bandura A (1978) On paradigms and recycled ideologies. Cogn Ther Res 2(1):79–103. https://doi.org/10.1007/BF01172518

Barth J, Munder T, Gerger H, Nüesch E, Trelle S, Znoj H, Jüni P, Cuijpers P (2016) Comparative efficacy of seven psychotherapeutic interventions for patients with depression: a network meta-analysis. Focus 14(2):229–243. https://doi.org/10.1176/appi.focus.140201

Bateson G, Jackson DD, Haley J, Weakland J (1956) Toward a theory of schizophrenia. Behav Sci 1(4):251–264. https://doi.org/10.1002/bs.3830010402

Canceil O, Cottraux J, Falissard B, Flament M, Miermont J, Swendsen J, Teherani M, Thurin JM (2004) Psychothérapie: trois approches évaluées (p. 565 pages, figures, tableaux). [Report, Institut national de la santé et de la recherche médicale(INSERM)]. https://hal-lara.archives-ouvertes.fr/hal-01570642

Ellis A (1962) Reason and emotion in psychotherapy. Lyle Stuart, New Jersey, p 442

Friman PC, Hayes SC, Wilson KG (1998) Why behavior analysts should study emotions: the example of anxiety. J Appl Behav Anal 31(1):137–156. https://doi.org/10.1901/jaba.1998.31-137

Häuser W, Hagl M, Schmierer A, Hansen E (2016) The efficacy, safety and applications of medical hypnosis: a systematic review of meta-analyses. Dtsch Arztebl Int 113(17):289–296. https://doi.org/10.3238/arztebl.2016.0289

Hayes SC (1987) A contextual approach to therapeutic change. In: Psychotherapists in clinical practice: cognitive and behavioral perspectives. Guilford Press, New York, pp 327–387

Hayes SC, Hofmann SG (2017) The third wave of cognitive behavioral therapy and the rise of process-based care. World Psychiatry 16(3):245–246. https://doi.org/10.1002/wps.20442

Hofmann SG, Asnaani A, Vonk IJJ, Sawyer AT, Fang A (2012) The efficacy of cognitive behavioral therapy: a review of meta-analyses. Cogn Ther Res 36(5):427–440. https://doi.org/10.1007/s10608-012-9476-1

Houser B, Vitale KR (2022) Increasing shelter cat welfare through enrichment: a review. Appl Anim Behav Sci 248:105585. https://doi.org/10.1016/j.applanim.2022.105585

Madanes C, Haley J (1977) Dimensions of familial therapy. J Nerv Ment Dis 165(2):88–98. https://doi.org/10.1097/00005053-197708000-00002

Nichols RC, Beck KW (1960) Factors in psychotherapy change. J Consult Psychol 24(5):388–399. https://doi.org/10.1037/h0044678

Öst L-G (2008) Cognitive behavior therapy for anxiety disorders: 40 years of progress. Nord J Psychiatry 62(sup47):5–10. https://doi.org/10.1080/08039480802315590

Palsson OS (2015) Hypnosis treatment of gastrointestinal disorders: a comprehensive review of the empirical evidence. Am J Clin Hypn 58(2):134–158. https://doi.org/10.1080/00029157.2015.1039114

Pinnell CM, Covino NA (2000) Empirical findings on the use of hypnosis in medicine: a critical review. Int J Clin Exp Hypn 48(2):170–194. https://doi.org/10.1080/00207140008410047

Prochaska JO, DiClemente CC (1984) The transtheoretical approach: crossing traditional boundaries of therapy. Dow Jones-Irwin, Homewood

Ramos D (2019) Common feline problem behaviors: aggression in multi-cat households. J Feline Med Surg 21(3):221–233. https://doi.org/10.1177/1098612X19831204

Schaller UM, Biscaldi M, Burkhardt A, Fleischhaker C, Herbert M, Isringhausen A, Tebartz van Elst L, Rauh R (2021) ADOS-eye-tracking: the archimedean point of view and its absence in autism spectrum conditions. Front Psychol 12:584537. https://www.frontiersin.org/article/10.3389/fpsyg.2021.584537

Shenefelt PD (2000) Hypnosis in dermatology. Arch Dermatol 136(3):393–399. https://doi.org/10.1001/archderm.136.3.393

Vogel D, Meyer M, Harendza S (2018) Verbal and non-verbal communication skills including empathy during history taking of undergraduate medical students. BMC Med Educ 18(1):157. https://doi.org/10.1186/s12909-018-1260-9

Weinberg A (2012) Sigmund Freud. Éditions Sciences Humaines. https://www.cairn.info/load_pdf.php?ID_ARTICLE=SH_MARMI_2012_01_0069&download=1

Wolpe J (1954) Reciprocal inhibition as the main basis of psychotherapeutic effects. Arch Neurol Psychiatr 72(2):205. https://doi.org/10.1001/archneurpsyc.1954.02330020073007

Woods MD, Martin D (1984) The work of Virginia satir: understanding her theory and technique. Am J Fam Ther 12(4):3–11. https://doi.org/10.1080/01926188408250192

Practical Application of Veterinary Psychotherapies

Contents

29.1 General Information – 724
29.1.1 General Principles of Therapies – 724
29.1.2 General Principles of Conditioning – 731
29.1.3 How to Modify the Probability of a Behavior Occurring? – 733

29.2 Organizing the Cat's Environment and Respectful Communication: Prevention Tools for All Cats – 740
29.2.1 Organizing the Environment – 740
29.2.2 Respectful Communication – 757
29.2.3 Learning and Useful Exercises for All Cats – 759

29.3 Behavioral Axis Therapies – 768
29.3.1 Disorders Related to a Dysregulation of the Fear Axis – 768
29.3.2 Disorders Related to a Dysregulation of the Self-Controls Axis – 774
29.3.3 Disorders Related to a Dysregulation of the Territoriality Axis – 782
29.3.4 Disorders Related to a Dysregulation of the Relationships Axis – 786
29.3.5 Other Psychiatric Disorders – 789

References – 792

© The Author(s), under exclusive license to Springer Nature Switzerland AG 2026
S. Bleuer-Elsner et al., *Veterinary Psychiatry of the Cat*,
https://doi.org/10.1007/978-3-032-06260-4_29

After a general overview of the objectives of therapy, this chapter describes the different types of learning and their specificities. General methods for generating new behaviors or, conversely, eliminating behaviors are detailed, along with communication guidelines and the use of appropriate tools.

Therapies for specific disorders are then addressed axis by axis, each time presenting general principles, environmental modifications, and the most suitable behavioral adjustments.

It is worth noting that for cats, a nonsocial species, the environmental modification aspect plays a much larger role than for dogs. The objectives differ. Cats are not expected to be walked in urban areas, nor are they required to heel or respond reliably to recall commands. These differing needs lead to simpler and often more easily implemented behavioral therapies. This is likely one of the reasons why treating cats for behavioral disorders is generally easier than it is for dogs.

29.1 General Information

29.1.1 General Principles of Therapies

29.1.1.1 Objectives of Therapies

▪ Education and Rehabilitation

The objectives, benefits, and roles of education and behavioral rehabilitation for cats must be clearly distinguished.

Education aims to teach cats the basic rules that allow them to coexist with humans, such as not scratching or biting people, as well as tolerating grooming or medical care. Education, in the sense of habituation, involves making the cat adaptable to everyday situations, thereby preventing the emergence of behavioral issues related to the environment.

Rehabilitation, on the other hand, seeks to address behavioral disorders in cats, regardless of their etiology. It aims to correct faulty or poor-quality learning to restore appropriate behavioral responses to various situations.

Feline trainers are becoming increasingly numerous, and this profession is becoming more accessible. Their role is to explain to owners how to provide their cats with an environment that meets their needs. For instance, by visiting the home, they can clarify the rules around litter box use, ensure that activity zones are clearly defined and recognized by the cat, and teach owners to communicate kindly with their pet, eliminating any form of physical punishment, which would be anxiety-inducing for the animal.

The intervention of a feline trainer, or a veterinary assistant who often assumes this advisory role as a first point of contact, is necessary and sufficient when the cat's problematic behaviors stem from poor environmental arrangements or inadequate communication by its owners. However, if the cat is dysregulated in one of the behavioral axes, correcting these deficits and communication issues is insufficient to resolve the cat's pathological psychological state (▪ Fig. 29.1).

A competent feline trainer improves communication and adjusts the environment appropriately, which, even in cases of behavioral disorders, helps to *minimize observed symptoms* or *slow their progression.* However, they must consider the normality of the behavioral axes, as without joint management involving a veterinarian, progress will be slower or even impossible if the pathological condition is not addressed.

The intervention of a veterinary behaviorist becomes essential as soon as the animal's psychological state is no longer normal, adaptive, or reactive, but pathological. Therefore, the trainer must consider collaborating and referring cases when communication or environmental adjustments alone fail to explain the cat's problematic behaviors, or more simply, whenever these behaviors persist or worsen.

▪ Environmental and Behavioral Modification

Behavioral education and rehabilitation aim to modify, either by decreasing or increasing, the likelihood of specific behavioral responses to certain stimuli. This modification involves both the motor and emotional aspects of the animal's behavior. It can be applied directly to the cat's behavior itself

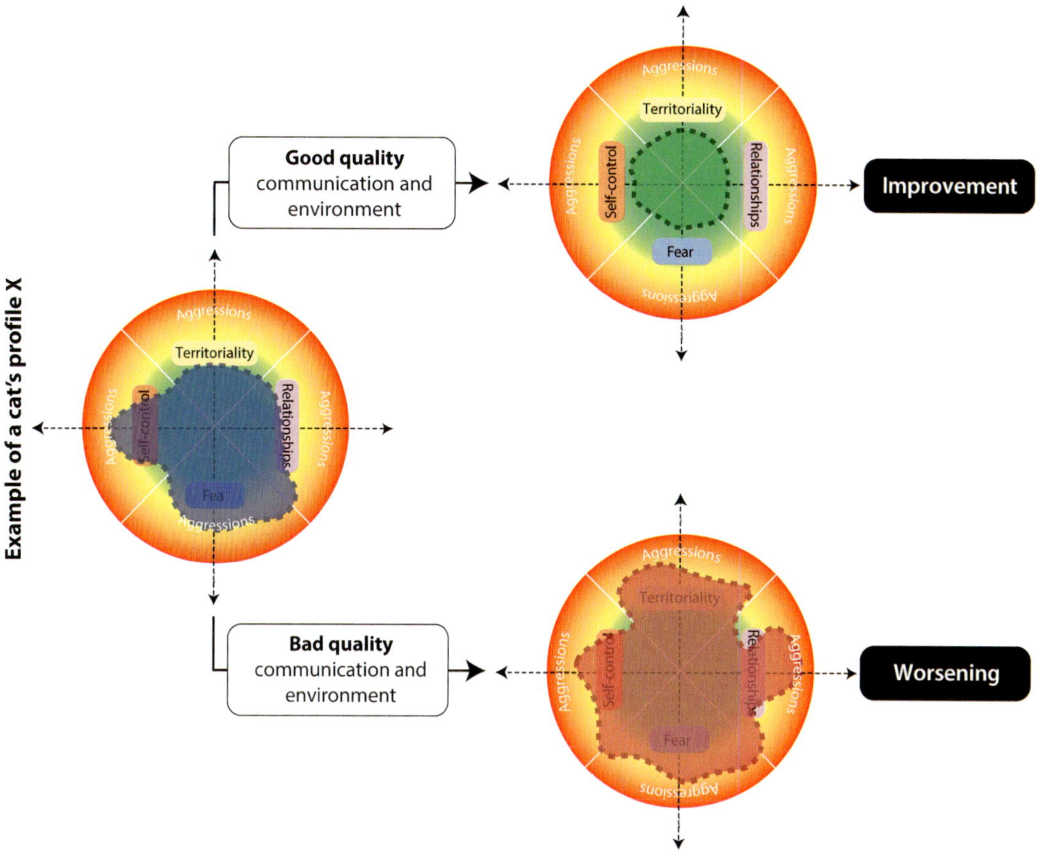

Fig. 29.1 Influence of communication and environmental adjustments on the cat's progress

(behavioral modification) or indirectly by acting on the cat's environment (environmental modification).

Environmental modification involves changes to the type or location of the cat's food, sleeping area, or activities, as well as the organization of spaces to coherently arrange litter boxes, feeding stations, and observation posts, in line with feline species-specific needs. These modifications can be implemented as a preventive measure or as remediation (rehabilitation) for the cat's behavioral disorders. They are applied once and for all and do not require ongoing exercises, apart from potentially habituating the cat to the new accessories being used.

Modifying the animal's environment also means altering that of the owners. Practitioners must master appropriate communication techniques during consultations to avoid resistance and facilitate change (see ▶ Chap. 28, ▶ Sect. 28.2.2.2).

Behavioral modification relies on various protocols to change the cat's mental representations and behavioral responses to different stimuli: *habituation* and *desensitization* protocols, based on the cat's experiences with environmental stimuli; *classical* and *operant conditioning* protocols, based on learning relationships between events; and *cognitive therapies*, based on learning *communication rituals* within the household where the cat lives.

For each therapeutic prescription, practitioners will select the appropriate elements of *environmental* or *behavioral modification* based on the presented clinical case. For cats, the environmental component of therapy is both significant and essential.

29.1.1.2 Different Methods of Behavioral Modification

■ **Habituation and Desensitization**

Habituation refers to the process of progressively reducing an unlearned alert behavioral response to an unfamiliar stimulus following repeated or prolonged exposure to the stimulus, provided the exposure has no negative consequences for the animal. Thus, habituation reduces the animal's emotional reactions to a novel stimulus.

As a fundamental process, habituation is the opposite of *sensitization*, which corresponds to the progressive increase of the same alert behavioral response. The process of habituation also differs from extinction in the nature of the behavioral response: the response is innate in habituation, whereas it is acquired in extinction.

As a behavioral modification protocol, habituation differs from *flooding* in terms of the intensity of the phobic stimulus to which the animal is exposed. In flooding, the intensity is not controlled, which carries risks of traumatizing the animal, whereas in habituation, the intensity is very gradually increased to prevent the animal from becoming sensitized. The habituation protocol also differs from the *desensitization* protocol in the nature of the concerning stimulus: in habituation, the stimulus is unfamiliar, whereas in desensitization, the stimulus is familiar because the animal has already been sensitized to it.

Thus, desensitization protocols are *habituation* protocols with added complexities. Habituation occurs quickly, as soon as the animal understands that the stimulus is not dangerous, since there are no strong neural connections between the stimulus and the animal's alert response. In contrast, desensitization is more challenging because it requires creating an alternative neural pathway without being able to completely undo the one the animal has already learned (see ◘ Fig. 29.2).

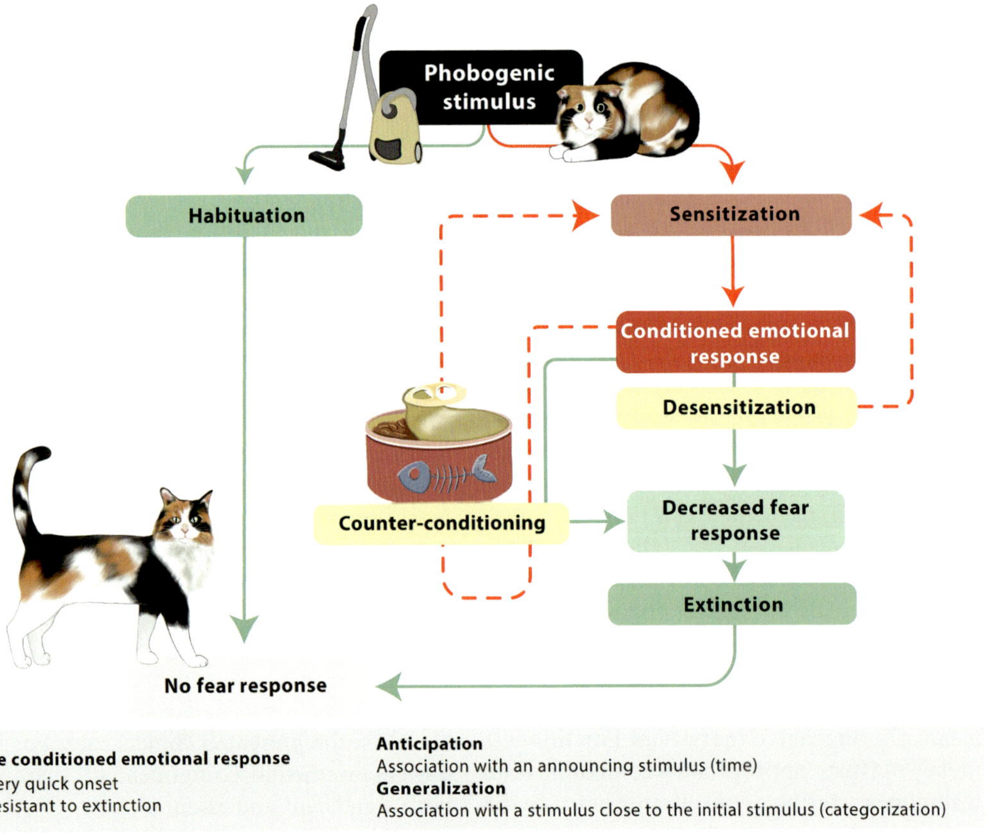

◘ **Fig. 29.2** Stimuli and responses in habituation, sensitization, desensitization, and counterconditioning

To promote habituation rather than sensitization, a habituation or desensitization protocol must meet several conditions: the animal must be *repeatedly exposed* to a stimulus of *minimal intensity* at the start, in an *open environment* (◘ Table 29.1).

In practice, adhering to the last two conditions requires training cat owners to recognize their pet's fear signals so they can adjust the intensity of the stimulus the cat is exposed to, without ever pushing it beyond its zone of mild discomfort (◘ Fig. 29.3).

Classical and Operant Conditioning

Conditioning refers to the learning of a relationship between a stimulus and a response. When this stimulus has previously been associated with another response, the term *counterconditioning* is used.

A conditioning protocol is therefore always faster and easier to implement than a counterconditioning protocol, as the latter requires counteracting associations, i.e., already established neural connections.

As explained in the first part of this book, conditioning can be divided into two main types: classical and operant.

Classical conditioning establishes an association between a neutral stimulus (NS), which initially elicits no response from the animal, and an unconditioned stimulus (US), which innately elicits a response. The neutral

◘ **Table 29.1** Conditions favoring habituation or sensitization

Criteria favoring habituation	Criteria favoring sensitization
Open environment Exposure to a stimulus of low intensity Repeated exposure	Closed environment Immediate exposure to a stimulus of maximum intensity Isolated exposure

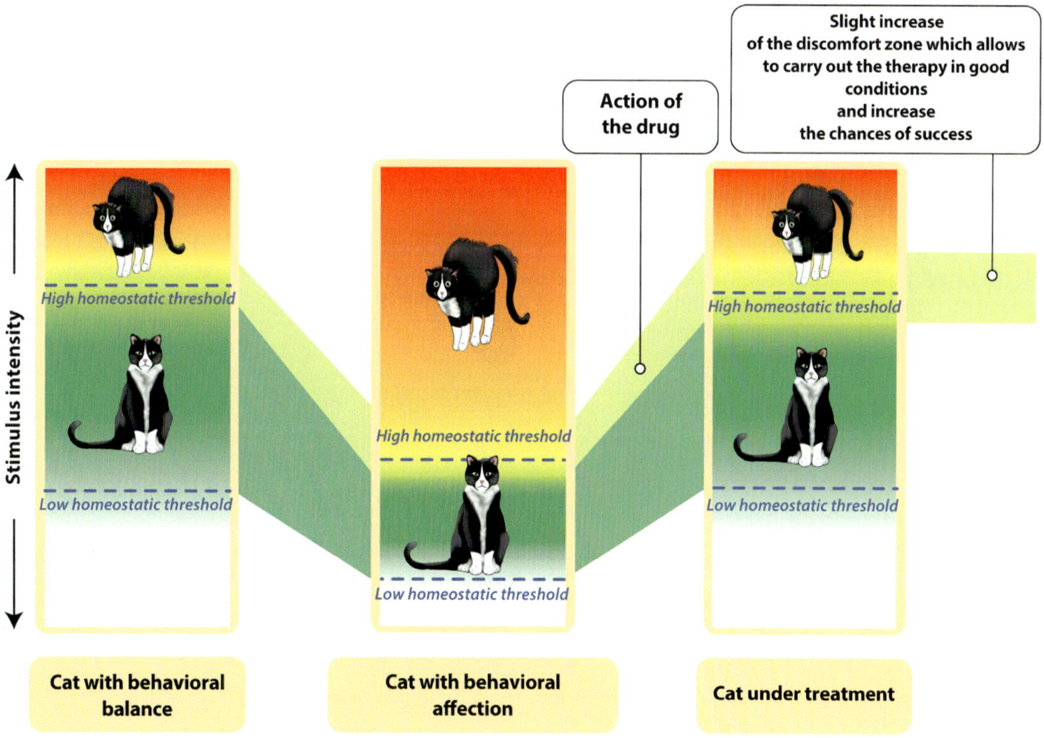

◘ **Fig. 29.3** Cat comfort zones and habituation

stimulus then becomes a conditioned stimulus (CS). The conditioned stimulus triggers the response previously associated solely with the unconditioned stimulus, even in the absence of the latter.

Thus, classical conditioning expands the range of stimuli that provoke emotional reactions in the animal (◘ Fig. 29.2).

For example, using a clicker as a training tool (see ▶ Sect. 29.3.2.1, "Using the Clicker,") requires classical conditioning to associate the sound of the clicker (*NS*) with the delivery of a food reward (*US*), so that the sound of the clicker (*CS*) alone elicits in the animal a representation of the food reward (*US*).

Operant conditioning establishes an association between a response and an *appetitive stimulus* (i.e., pleasant) or an *aversive stimulus* (i.e., unpleasant). The appearance or disappearance of the stimulus serves as the consequence of this response.

A *reinforcement* is a stimulus that appears or disappears following the execution of a behavior, resulting in an increased likelihood of that behavior occurring. A *punishment* is a stimulus that appears or disappears after a behavior, resulting in a decreased likelihood of that behavior occurring.

Thus, reinforcement enriches the animal's behavioral repertoire, while punishment reduces it.

There are four types of operant conditioning, each of which can be associated with a primary emotion to better understand what the animal might feel (◘ Fig. 29.4).

Positive reinforcement relies on pleasure. It involves adding an *appetitive stimulus* following a behavior to increase the likelihood of its recurrence. For example, a spoonful of wet food given to a cat that comes back inside the house when called is a positive reinforcement, as the pleasure associated with the food increases the probability that the cat will return the next time it is called.

Negative reinforcement relies on relief. It involves removing an *aversive stimulus* following a behavior to increase the likelihood of its recurrence. For example, the release of the veterinarian's hands after holding the cat in an uncomfortable position is a negative reinforcement, as the relief felt when the hands are removed increases the probability that the cat will try to avoid the handling the next time a similar restraint is attempted.

Positive punishment relies on fear or pain. It involves adding an *aversive stimulus* following a behavior to decrease the likelihood of its

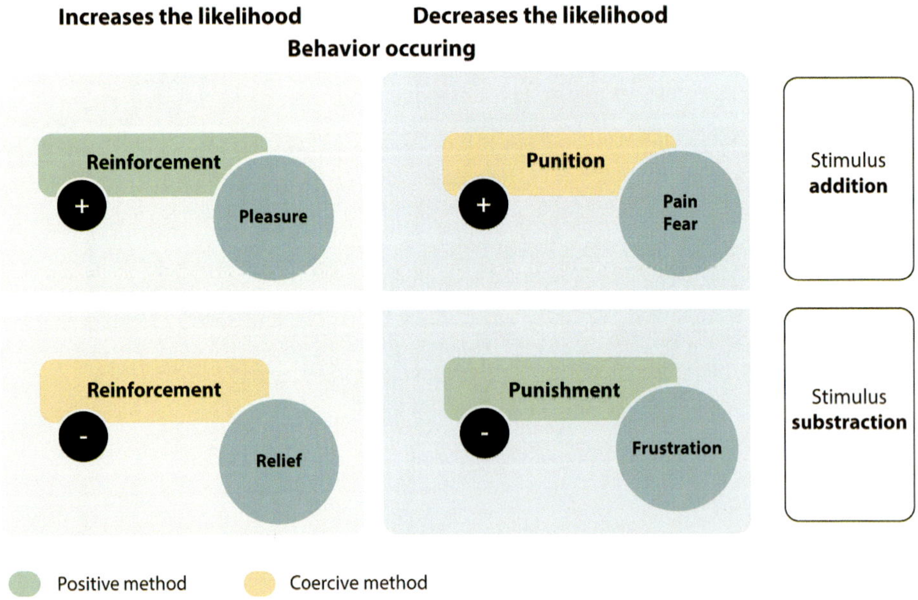

◘ **Fig. 29.4** Characteristics of positive and negative reinforcements used in operant conditioning

recurrence. For example, a slap on a cat's nose after it jumps on the table is a positive punishment, as the fear or pain experienced decreases the likelihood that the cat will repeat the behavior.

Negative punishment relies on frustration. It involves removing an *appetitive stimulus* following a behavior to decrease the likelihood of its recurrence. For example, stopping all interaction with a kitten that nips or scratches during play with its owner is a negative punishment, as the resulting frustration decreases the likelihood that the kitten will nip during future interactions, seeking to avoid the frustration caused by the cessation of interaction.

It is important to note that in cats, due to their predator-prey status, the delays between a behavior and the stimulus must be very short, much shorter than in dogs. Whether using a reward or positive punishments, the owner must anticipate, as the association may not be properly established if a delay occurs between the stimulus and the response.

▪ Cognitive Therapies

Cognitive therapies aim to modify the cat's perception of itself, others, and the world around it. These therapies include establishing *consistent rituals* and *communication* tailored to the cat, as well as *verbal* and especially *non-verbal communication* methods adapted to the cat.

Establishing consistent rituals and communication with the cat, based on the principle of *mutual respect* between the cat and its owners, relies on the following general rules:

- The cat must communicate "politely" with its owners, meaning it should control its biting and scratching during interactions. In return, the owners must communicate verbally with their cat in a calm and clear manner, avoiding shouting or violence.
- The cat should not be disturbed during its *rest periods*, but it must also respect its owners' moments of rest, accepting that they may refuse interaction or restrict access to certain rooms.
- The cat should respect its owners' belongings, refraining from destroying or damaging them. Conversely, the owners should understand that any modification to the furniture significantly affects the cat's perception of its environment. Therefore, they should organize the living space so that the cat's activity zones are well-defined, providing an environment that meets the ethological needs of the feline species.

Establishing an organized environment and consistent communication with the cat helps *prevent* and *address* the emergence of communication disorders, relational disorders, or Territoriality axis disorders.

29.1.1.3 Choice of Training Methods and Exercises

▪ Choice of Training Methods

Positive education or reeducation methods differ from coercive education or reeducation methods based on the types of conditioning they employ (◘ Fig. 29.4). On one hand, positive education methods use *appetitive* unconditioned stimuli to establish *classical conditioning* and rely on positive reinforcement or negative punishment to establish *operant conditioning*. On the other hand, coercive education methods use *aversive* unconditioned stimuli, such as fear or pain, to establish *classical conditioning* and employ negative reinforcement or positive punishment to establish *operant conditioning*.

In other words, positive methods utilize stimuli that are pleasant for the animal, which are given when the animal displays a desirable behavior but are removed when the animal exhibits undesirable behavior. Conversely, coercive methods rely on stimuli that are unpleasant for the animal, which are applied when the animal misbehaves but cease when the animal behaves correctly.

t has been extensively demonstrated in dogs that coercive methods, with the use of electric collars being the most notable example, are no more effective than positive methods (Cooper et al. 2014; Hiby et al. 2004; Ziv 2017) and are highly detrimental to the dog's welfare and its relationship with its owners (Masson et al. 2018a, 2018b; Schilder and Van Der Borg 2004).

In cats, publications studying the effects of coercive methods are far fewer, but the conclusions regarding recommendations and best practices are identical (Riemer et al. 2021). For example, the sound signal from an electric collar interrupts an undesirable behavior in cats solely because the animal associates the sound with the same negative emotions it experienced during the initial shock (Kasbaoui et al. 2016). However, methods that evoke intense negative sensations (fear, pain) are generally used as a last resort by incompetent trainers or overwhelmed owners.

More often, this sense of helplessness reflects the existence of a pathological condition in the animal. Coercive methods should therefore be addressed and replaced rather than outright condemned by veterinarians. This requires presenting the owners with the consequences of these practices and their alternatives, helping them find cat trainers trained in *respectful methods*. It is also crucial to explain that their cat experiences emotions and that its disobedience may stem from an emotional or cognitive incapacity to respond correctly to their requests, rather than from their incompetence or the cat's ill will.

However, positive methods are not incompatible with establishing a structured environment and setting boundaries for the cat, or even occasionally punishing it (see ► Sect. 29.1.1.2). That said, since the cat does not belong to a social species, it cannot adhere to a rule in the same way and tends to respond to requests in a more contextual manner. For instance, a cat can easily learn not to climb onto the countertop while its owners are cooking, but when no one is present, it will act based on what seems appropriate at the moment. Positive methods are therefore compassionate rather than permissive. We prefer to use the term *respectful education* to describe our approach, which employs positive methods while still allowing for the establishment of boundaries. Through verbal disruptions or calmly moving the cat away, owners can communicate that certain actions are bothersome and therefore prohibited. Patience is essential, given the cat's nonsocial nature, but respectful coexistence is entirely achievable.

▪ Choice of Exercises

The selection of exercises prescribed as part of a cat's behavioral therapy directly depends on the origin of its behavioral issues.

If there is an environmental deficiency, particularly in the organization of the cat's activity zones, the practitioner must address this as a priority, as such a problem can only exacerbate the cat's dysregulation across behavioral axes (◘ Fig. 29.1). Similarly, if the owners are communicating inappropriately with the cat, re-establishing respectful communication with the animal will be a prerequisite before initiating behavioral therapy. The principles of consistent communication with the cat are detailed in ► Sect. 29.2.2.

If the cat exhibits dysregulation on one or more behavioral axes, the practitioner must propose a limited number of exercises to the owners, ensuring the following points are addressed:

- Exercises aimed at *improving the cat's motivation and attention* should be prioritized above all others; without these, no learning is possible (see ► Sect. 29.1.3.3. « Attention et motivation »).
- No exercise should begin without implementing necessary safety measures when required. For instance, prescribing exercises involving interactions with visitors for an aggressive cat is both dangerous and counterproductive if isolation is not possible.
- Exercises designed to address situations that are particularly challenging for the cat's owners should take precedence over exercises targeting more frequent but less critical problems. The quality of the therapeutic alliance between the practitioner and the cat's owners, as well as the owners' relationship with their cat, depends on this prioritization.
- Exercises that address multiple behavioral axes simultaneously, provided they are relevant for the cat, should be preferred over exercises specific to a single axis. This approach minimizes the number of exercises prescribed to the owners.

- All exercises should be broken down into successive steps to maintain both the cat's motivation and that of its owners, while also facilitating the practitioner's ability to track the cat's progress.

Reeducation exercises, although prescribed for cats as well as dogs, constitute a less significant part of therapy compared to *environmental modification*. Nevertheless, the guidelines outlined above must be followed for any exercises that are to be carried out.

In the case of a cat in a *normal adaptive* or reactive state with vulnerabilities in one of the axes, behavioral prescriptions alone will suffice to resolve the issue. This is commonly observed in many cats with mild disorders who receive guidance from competent feline trainers or general veterinarians.

For a cat in a *pathological state*, these exercises must be combined with a *medication prescription*. The medication's role is to enable the animal to perform the exercises or adapt to environmental changes under favorable conditions, thereby ensuring the success of the therapy. This success is the key to accelerating recovery.

29.1.2 General Principles of Conditioning

29.1.2.1 Laws of Conditioning

Several laws govern the functioning of classical and operant conditioning (◘ Table 29.2), with the exception of the "law of effect," which is specific to operant conditioning.

According to the law of effect, "any act that produces satisfaction in a given situation is more likely to occur again if a similar situation arises," while "any act that produces dissatisfaction will tend to diminish if that situation occurs again" (Thorndike 1898). In other words, a behavior increases if rewarded and decreases if punished.

The *law of temporal contiguity* states that establishing an association between a neutral stimulus and an unconditioned stimulus requires the first to appear in close temporal proximity to the second. Thus, the interval between the two must not exceed a few seconds. As a result, a reward should be given as quickly as possible after the execution of a behavior during the learning process.

However, when maintaining an already learned behavior, reinforcement can be

◘ **Table 29.2** Laws of conditioning

Name of the law	Classical conditioning	Operant conditioning
Effect		An act that brings satisfaction will reoccur; an act that brings dissatisfaction will disappear (in similar contexts)
Temporal contiguity	The NS and US must be contiguous for an association to form (a few seconds at most)	Same for establishing a new behavior, but reinforcement should be delayed during the maintenance phase
Repetition	The more the US-CS association is repeated, the better it is memorized	The more the US-CS association is repeated, the better it is memorized
Extinction	A behavior that is neither reinforced nor punished will disappear	A behavior that is neither reinforced nor punished will disappear
Generalization	Conditioning to a stimulus extends to a similar but nonidentical stimulus	Conditioning to a stimulus extends to a similar but nonidentical stimulus
Discrimination	Conditioning to a stimulus does not occur in response to a similar but nonidentical stimulus	Conditioning to a stimulus does not occur in response to a similar but nonidentical stimulus

NS neutral stimulus, *CS* conditioned stimulus, *US* unconditioned stimulus delayed to reduce the behavior's dependence on the reward, making it more resistant to extinction. This effect is similar to that produced by intermittent reinforcement schedules (see ▶ Sect. 29.1.2.3).

According to the *law of repetition*, a conditioned response is better learned when the association between the conditioned stimulus (CS) and the unconditioned stimulus (US) is frequently repeated. If this repetition is insufficient, forgetting may occur.

According to the *law of extinction*, a conditioned response disappears or "extinguishes" when the unconditioned stimulus is no longer presented. It is important to note that extinction is preceded by what is called the "extinction burst," a temporary and paradoxical increase in the frequency of the behavior that is no longer being reinforced.

According to the *law of generalization*, a conditioned response to a specific stimulus also occurs in response to a similar but nonidentical stimulus. This law explains why cats can respond to commands given with different voices, intonations, or postures.

According to the *law of discrimination*, a conditioned response to a specific stimulus does not occur in response to a similar but nonidentical stimulus. This law, complementary to the law of generalization, highlights its limitations, without which it would be impossible to distinguish between different stimuli. This also explains why cats stop responding when the form of the request differs too much from the usual one. For example, it is common for the family cat not to come when called by children but to respond immediately when called by the lady of the house.

29.1.2.2 Different Reinforcers and Punishments

A *reinforcer* is a stimulus that possesses or acquires a rewarding value from the perspective of the recipient. It depends directly on the animal's motivation, which varies between individuals, as well as according to context and time. There are different types of reinforcers:

- *Primary reinforcers*, such as water, food, sleep, or avoidance of pain, correspond to a vital need for the animal. *Secondary or conditioned reinforcers*, such as verbal praise, petting, or the sound of a clicker, have no initial reinforcement value but acquire it through classical conditioning by being associated with a primary reinforcer.
- *Natural reinforcers*, such as the ability to hide in another room or receive petting, are directly present in the animal's environment. *Artificial reinforcers*, such as food or toys, are not naturally present.
- *Imperative reinforcers* are those that correspond to a strong need or desire of the animal, whereas *attenuated reinforcers* correspond to a weaker motivation, particularly in cases of habituation to the reward. Habituation manifests as a loss of interest in a reward when it remains the same over time. All food rewards can provoke this type of reaction, which is why it is essential to vary the rewards used.

Maintaining a cat's motivation requires carefully selecting reinforcers. Since cats are not a social species, their need for contact with their owner is less significant than that of dogs, and high-value food rewards are often necessary to capture their attention. For some cats, petting or play may hold greater reward value, and individual variations are significant (Bentosela et al. 2008; Willson et al. 2017).

The process begins with observing what the cat values most. For instance, a cat may prefer contact with its owner over play or favor food rewards over playing with its owner.

Primary reinforcers are more prone to *attenuation* than secondary reinforcers, and natural reinforcers are more potent and practical than artificial ones. However, using *imperative reinforcers* is the key to sustaining a cat's motivation. It is therefore essential to identify an individual cat's preferences for rewards and to ensure that reinforcers are varied according to context and time to prevent habituation.

Having a variety of available reinforcers allows owners to move away from *food* rewards if a suitable and motivating alternative exists for the cat. However, it is important to note that, unlike in dog training, where food rewards are typically phased out once a behavior is learned, food rewards are often maintained for life in cats and form an integral part of the request made to the cat. For example, in medical training, receiving food rewards during care is part of the care process itself.

A *punishment* is either the removal of a stimulus with rewarding value or the introduction of an aversive stimulus to the animal. In the first case, its effectiveness depends directly on the animal's motivation, while in the second case, it partially depends on the animal's sensitivity.

However, the effectiveness of *positive punishment* is directly proportional to the intensity of the aversive stimulus, which carries the risk of causing phobic or anxious states, aggression, or learned helplessness in the cat (Maier and Seligman 1976; Seligman and Maier 1967). As previously mentioned, learned helplessness is a depressive state induced by the inability to control a situation.

Consequently, *negative punishment* should be preferred over positive punishment to preserve the cat's emotional well-being and its relationship with its owners.

29.1.2.3 Different Reinforcement and Punishment Schedules

A *continuous reinforcement schedule,* in which every correct response is reinforced, allows for quick and easy learning. This approach is used at the beginning of the acquisition of a new behavior. Reinforcement must be applied at the end of the behavioral sequence, adhering to the law of temporal contiguity (see ► Sect. 29.1.2.1). For cats, this type of reinforcement is generally sufficient, as it is for zoo animals (Fernandez and Martin 2021).

An *intermittent reinforcement schedule,* in which only some correct responses are reinforced, allows for long-lasting learning. This approach is used to ensure that a learned behavior is resistant to extinction.

In other words, it is essential to systematically reinforce a desired behavior during the initial learning phase but to switch to intermittent reinforcement once the behavior is understood. By progressively *spacing out rewards*, the learning is consolidated, making it highly resistant to extinction.

These conclusions are derived from the different types of reinforcement schedules (◘ Fig. 29.5).

In contrast, a *continuous punishment schedule* is more effective than an intermittent punishment schedule. Therefore, positive punishment should be applied every time an undesirable behavior occurs, if it cannot be eliminated through other means (see ► Sect. 29.1.3.2). Additionally, for positive punishment to be effective, it should be applied as early as possible in the behavioral sequence.

However, as previously mentioned, in a consistent relationship and communication with a cat, the use of positive punishment should simply be excluded.

29.1.3 How to Modify the Probability of a Behavior Occurring?

29.1.3.1 How to Teach a New Behavior?

The process of learning any new behavior always follows the same five steps **(see box below)**, which correspond to different stages in the cat's acquisition of the behavior. These steps allow the practitioner to assess the cat's progress.

In practice, while our way of living with dogs and taking them everywhere often necessitates completing all five steps, this approach is rarely applied to cats. Therefore, these steps and the accompanying theory are summarized in the following section. However, for cats, we generally stop at the first step: generating the behavior and systematically rewarding it.

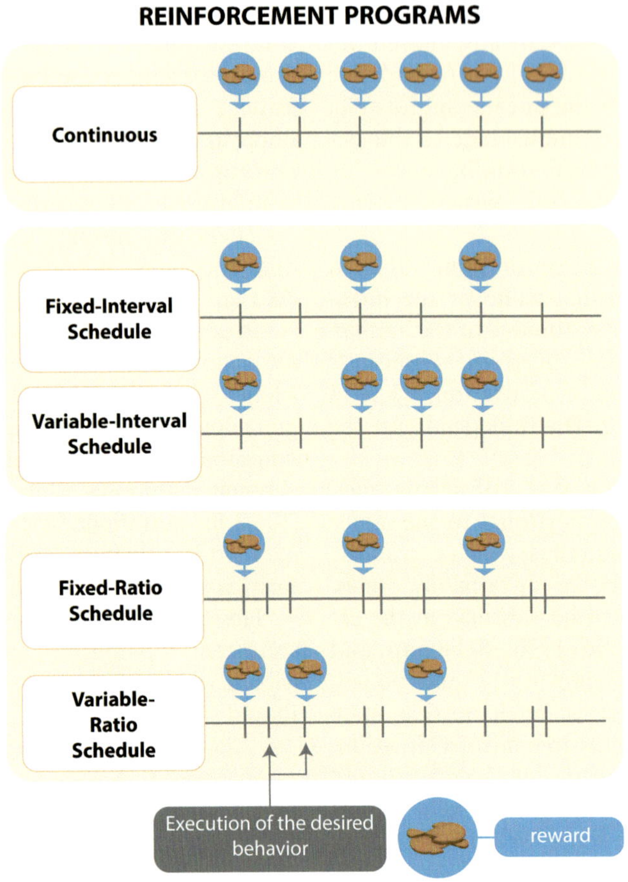

Fig. 29.5 Different programs of reinforcement

29

> **Important Point of View**
>
> **The Five Steps of Learning**
>
> Teaching a cat any behavior follows the same five steps. The first two steps and the fourth should be conducted in a calm, distraction-free environment.
>
> The first step involves *generating the behavior* by using one of three techniques: capturing, guiding, or shaping, to produce the desired behavior.
>
> The second step focuses on *eliminating unnecessary or extraneous movements* that may have been introduced during the first step, ensuring the cat responds only to the intended verbal or gestural command.
>
> The third step involves *repeating the first two steps in various contexts* to confirm that the cat can generalize its learning to these new situations.
>
> The fourth step is to gradually *space out and eventually* remove rewards, ensuring that the cat can maintain the learned behavior with only praise or natural reinforcers, such as petting.
>
> Finally, the fifth step involves *repeating the exercise in different contexts* to confirm that the cat can consistently perform the learned behavior without reinforcement beyond praise or natural reinforcers.

Teaching a new behavior to a cat begins by generating the behavior. There are three methods to achieve this: *capturing, guiding,* and *shaping.* Each method has its own advantages and disadvantages and should be selected based on the context.

Capturing

Capturing a behavior involves reinforcing a spontaneous behavior of the animal, meaning a behavior produced without any external intervention. Praising a cat that has just sat on a seat is an example of capturing this behavior. Similarly, giving wet food to a cat meowing in front of the refrigerator captures the meowing behavior and encourages its repetition.

The *advantage* of this method is that it reinforces behaviors already present in the cat's behavioral repertoire, making them "natural" and likely enjoyable for the cat. The cat's intrinsic motivation to perform these behaviors already exists, making reinforcement straightforward.

The *disadvantage* is the need to wait for the desired behavior to occur, which might never happen if it is not part of the cat's behavioral repertoire or if the cat's emotions interfere with the behavior's execution. Capturing requires careful and frequent observation of the cat to identify and reinforce any spontaneous behavior that aligns with the desired outcome.

Guiding

Guiding a behavior involves inducing a behavior in the animal using verbal and, more importantly, gestural cues, followed by reinforcement. This step allows the cat to understand what is expected of it in response to a verbal command. Most of the time, this is achieved by using a lure, often a treat, to direct the animal's attention or movements (◘ Fig. 29.6). For example, placing a treat in front of the cat, then moving it above its nose and between its ears to encourage it to sit is a way of guiding the "sit" behavior. Slowly raising your hand vertically above the cat's muzzle, rather than moving it backward between its ears, guides the cat to stand upright and perform the "prairie dog" trick.

It is important to note that the lure also serves as a reward when the desired behavior is produced. In theory, the lure should be eliminated once the behavior is almost learned to avoid creating "unnecessary gestures."

The *advantage* of this method is that it allows reinforcement of behaviors without waiting for their spontaneous occurrence, as

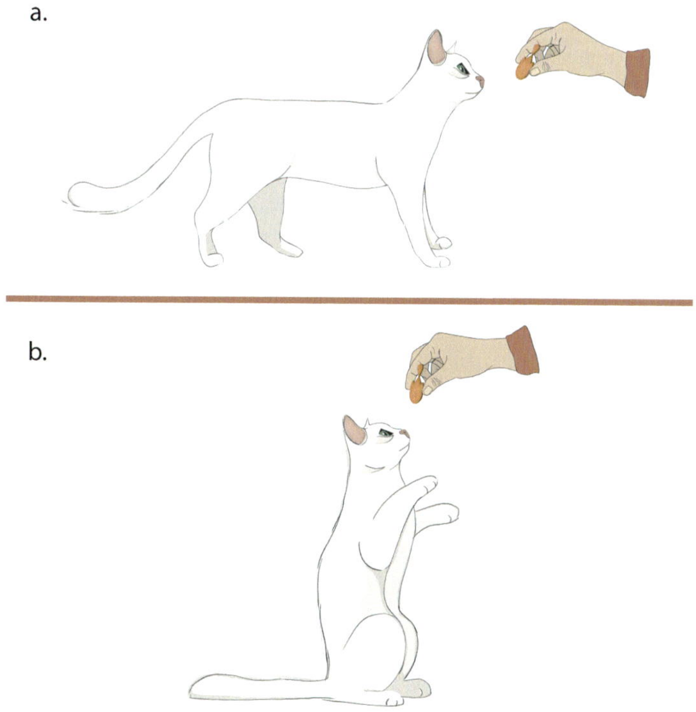

◘ **Fig. 29.6** Teaching the "prairie dog" behavior through guidance

the capturing method requires, thereby speeding up the learning process.

The *disadvantage* is that it requires some practice to perform the correct technical gesture, which a feline trainer can teach. Additionally, many tutorials are now available online and on social media, making this type of training much more accessible.

Furthermore, guidance often leads to the emergence of unnecessary gestures, which is more problematic for dogs than for cats. In practice, this technique is widely used to teach desired behaviors to cats.

Important Point of View

Lure and Reward

A *lure* is an appetitive stimulus, such as a treat, used to direct the animal's attention or movements, while a reward is an appetitive stimulus used to *reinforce* a desirable behavior. In practice, when teaching a new behavior to a cat, the lure used to generate the behavior often also serves as a reward once the cat performs the desired action.

However, the use of the lure should be discontinued as soon as the cat understands what is expected, if the goal is to achieve the behavior in response to a simple gesture or verbal command. If this step is not taken, the lure will become essential, and the cat will not perform the behavior without it. From the cat's perspective, the gesture with the treat becomes an integral part of the command, just like the verbal or gestural cues.

It is important to note that *physical guidance* involving forced contact, such as pressing on a cat's back to make it lie down, should be avoided. These actions often cause pain and almost always result in discomfort or fear, making them coercive methods. This does not mean that physical contact with the cat is forbidden; on the contrary, the cat must also learn to be *handled* and *touched*, but in a gentle and respectful manner. However, if the contact causes the cat to show signs of discomfort or fear, it should be avoided, and another form of guidance should be considered.

Cat behaviors suited to guidance include those involving adopting a position or performing a movement. This technique is also widely used in nonrestrictive handling during medical training, as treats can serve as a lure to guide the cat into remaining still during the procedure.

- **Shaping**

Shaping a behavior involves reinforcing *successive approximations* of the desired final behavior. Typically, the final verbal cue for the command is introduced only once the behavior is fully realized, which occurs at the end of the shaping process. For example, reinforcing a series of steps, first lying down, then lying on the side, then lying on the back, and finally rolling over, can be used to shape the "roll over" behavior. The word "roll" is only associated with the sequence once the cat performs a full roll, ensuring the cat understands the entire action as a single gesture linked to one word, avoiding confusion with distinct behaviors.

The *advantage* of this method is that it allows for the reinforcement of complex behaviors or behaviors for which the *cat's motivation is low*, as it provides frequent reinforcement at each completed step. Even when the behavior is incomplete or imperfect, the cat receives a reward that validates the step, signaling that it is on the "right track."

The *disadvantage* is that it requires carefully breaking down the desired behavior into simple steps to maintain the cat's attention and preserve its motivation. While this method is widely used with dogs to break down complex behaviors for canine sports, cat training is generally less extensive, and few owners require it for everyday interactions. However, it is worth noting that agility trials are now being organized for cats in the United States, and in such cases, this technique can be effectively applied.

29.1.3.2 How to Eliminate a Behavior?

While positive punishment theoretically allows for the elimination of a behavior, it has no place in the respectful education we advocate. The harmful consequences it almost invariably produces prevent us from endorsing its use for both scientific and ethical reasons (Riemer et al. 2021).

Of course, this does not mean blaming owners who have physically punished their cat. Most cats with behavioral issues are difficult to live with, leading owners to try anything they are advised, including drastic measures like sprays or devices that create loud noises to scare the cat. In cases of aggression, it is not uncommon for owners to hit the cat reflexively, simply in self-defense. However, it is exceedingly rare for such actions to stem from malicious intent.

The discussion around *positive punishment* should, therefore, serve as a foundation for building a *therapeutic alliance*, agreeing that since the cat will now be managed holistically, such punishments will no longer be used. Conveying the message that violence is *always a sign of failure* and that positive punishments should be limited to a rare few instances in a cat's lifetime helps avoid making owners feel guilty about their actions before the consultation.

An undesirable behavior can be eliminated using three methods: *extinction*, *negative punishment*, and *counterconditioning*. Each method has its advantages and disadvantages, which we will detail.

Logically, extinction should be the first choice, as it has minimal impact on the animal and, when feasible, is highly effective. Negative punishment is often possible, at least in its avoidance form, but it is frequently supplemented by counterconditioning, as it does not teach the cat a more appropriate solution to the situation. By systematically reviewing, in this order, the feasibility of using each method, the practitioner can develop a structured approach to determining the exercise they will recommend.

▪ Extinction

Extinction is the first strategy to use when attempting to eliminate an undesirable behavior. The goal is to remove any form of reward. Most often, this involves ignoring the cat each time it exhibits the behavior to be eliminated. For example, a cat that meows when its owners approach the refrigerator or kitchen can learn to stop meowing if the owners consistently refrain from giving it food in the minutes following this contextual meowing.

For extinction to be effective, the cat must not receive any reinforcers. External reinforcers are theoretically eliminated through the owner's neutral attitude. However, many reinforcers cannot be removed, particularly internal reinforcers. A cat in a pathological state, such as an impulsive state, has abnormal motivation and may find pleasure in actions that would not provide the same satisfaction to a cat in a normal state.

Thus, ignoring a cat's meowing in front of the refrigerator is usually effective for a normal cat but ineffective for an impulsive cat. The internal motivation of such a cat is strong enough for it to derive pleasure from meowing, even when faced with an unresponsive person. The advantages and disadvantages of extinction stem directly from this observation.

The *advantage* of this method is that it can eliminate an undesirable behavior effortlessly.

The *disadvantage* is that it is only applicable when the targeted behavior can be ignored without causing significant consequences. For example, ignoring a cat scratching at the bedroom door during the night or vocalizing for several hours may be challenging.

This technique must be explained clearly to owners for two reasons. First, undesirable behaviors must never be reinforced. If they are reinforced "occasionally," due to the highly effective intermittent reinforcement schedule (see ► Sect. 29.1.2.3), the rare reinforcements will have the opposite effect, preventing the behavior from disappearing.

Second, due to the phenomenon of the "extinction burst" (see ► Sect. 29.1.3.2), the undesirable behavior may initially increase in frequency and intensity before quickly disappearing. If owners are not forewarned, they may assume that the advice given is ineffective or even that the professional offering it is incompetent.

Behaviors that are well-suited to extinction include all *attention-seeking behaviors*, such as meowing to obtain treats or food or to have a door opened. Results should appear quickly; if they do not, it is important to verify that the procedure has been implemented correctly or to consider the possibility that the cat is suffering from a pathological state.

- **Negative Punishment**

Negative punishment is the strategy to employ for directly eliminating an undesirable behavior when extinction is not feasible. It involves removing an appetitive stimulus as soon as the undesirable behavior occurs. When the problematic behavior arises during an interaction with the cat (e.g., petting or playing), negative punishment consists of *immediately interrupting the interaction*. For example, if a cat starts scratching or biting during play, the play session should stop until the cat calms down.

When the undesirable behavior occurs outside of a direct interaction with the cat (e.g., meowing at the bedroom door at night to gain attention), negative punishment involves *isolating the cat* so it cannot continue the behavior. If isolation is not possible, an alternative is to leave the room and prevent the cat from following. The duration of the isolation should be proportional to the situation. As with counterconditioning, negative punishment by isolation can be preceded by the word "no," giving the cat a chance to stop the behavior. Over time, this can allow the undesired behavior to be interrupted with the verbal cue alone, without needing to isolate the cat each time. The ineffectiveness of negative punishment, like that of extinction, often indicates *abnormally high motivation* in the cat, which in itself justifies behavioral management.

The *advantage* of this method is that it effectively eliminates many problematic behaviors. In most cases, it involves avoiding the situation when it arises.

The drawback is that it requires the owner to be present. For this reason, it is often replaced by anticipatory avoidance of the problematic situation. For example, an owner might isolate their cat in a room every night to prevent it from meowing at the bedroom door or close the bedroom door because they know the cat tends to "play" with their feet during the night and has scratched them on previous occasions.

While this avoidance strategy is useful in preventing the cat from learning undesirable behaviors, it does not teach the animal an appropriate response in the presence of the stimulus. For instance, closing the bedroom door to prevent the cat from playing with feet during the night does not help the cat learn to control its scratching when it jumps at its owner's feet moving slightly under the blanket.

Thus, avoidance often provides a *partial solution* but should be supplemented with counterconditioning in problematic situations.

On the other hand, negative punishment is still a *form of punishment*: it generates frustration in the cat, a negative emotion, and can limit its behavioral repertoire (see ► Sect. 29.1.1.2). Therefore, it should be used while monitoring the level of frustration it causes in the cat and, more importantly, should be combined with learning opportunities in problematic contexts.

For instance, when the owner returns home, they can move their hands under a blanket to observe the cat's reaction. If the cat becomes overexcited and starts scratching, the owner can disrupt the behavior and stop the interaction, or even temporarily isolate the cat until it plays gently or remains calm in response to slight hand movements (counterconditioning). Cat behaviors suited to negative punishment are typically those related to *excessive excitement*.

- **Counterconditioning**

Counterconditioning is the third strategy used to eliminate an undesirable behavior. It involves *teaching the cat a new behavior that is incompatible with the problematic behavior*, using the techniques described in ► Sect. 29.1.2.1.

This strategy requires interrupting the cat's undesirable behavior through *disruption*, which involves redirecting its attention away from the object of the unwanted behavior. This can be achieved with a verbal cue like "no," followed by asking the cat to perform a more acceptable behavior that it has preferably already learned in a different context.

For counterconditioning to be effective, it is crucial to use a behavior the cat already knows. This is a limitation in cats compared to dogs, as most dogs are familiar with simple commands like "sit," whereas this is far less common with cats.

The *advantage* of this method is that it eliminates the cat's undesirable behavior while enriching its behavioral repertoire and avoiding the induction of negative emotions.

The *disadvantage* is that, if misused, it can inadvertently encourage the cat to first exhibit the problematic behavior before performing the expected one.

In theory, many cat behaviors lend themselves to counterconditioning. However, in practice, negative punishment is more frequently used because cats generally have a less developed behavioral repertoire in response to verbal commands.

29.1.3.3 Golden Rules

▪ Attention and Motivation

As mentioned earlier, a cat's attention and motivation are prerequisites for any learning process, yet they are often overlooked in the training or reeducation of cats.

Gaining the cat's attention before addressing it is essential for any exercise. There are two strategies to enhance attention:

- teaching the cat to focus on its owner on command (e.g., using the cat's name followed by "look" or "watch").
- capturing spontaneous moments of attention by rewarding the cat whenever it pays attention to its owner.

Similarly, *managing the cat's motivation* is essential for successful learning. Excessive motivation can make the cat unable to focus on anything else, leading to overexcitement or frustration (e.g., when trying to obtain a piece of tuna). Conversely, insufficient motivation prevents learning entirely, as the cat will not even attempt the desired behavior. Cats with self-control deficits have the advantage of being more motivated than the average cat, which, in this species, is beneficial as it makes them more eager to engage in learning activities (Kogan et al. 2017).

Several strategies can be used to adjust motivation.

To *decrease* motivation, the *quality or quantity of rewards can be reduced.* Another option would be to use positive punishment, but this is excluded within the framework of respectful education that we advocate (see ► Sect. 29.1.1.3).

To *increase* motivation, the *quality or quantity of rewards can be enhanced.* Another approach is to use negative punishment, which induces frustration. However, it should be applied sparingly to avoid generating excessive frustration (see ► Sect. 29.1.1.2).

The cat's attention and motivation can also be influenced by an altered psychological state. For this reason, any pathological state must be stabilized with treatment, if necessary, before implementing any behavioral therapy.

▪ Managing Exercises

The learning process for cats, especially kittens or cats with self-control deficits and limited concentration abilities, is facilitated by short training sessions. A few minutes of practice once or twice a day is more effective than a 1-h session per week, even though the literature on this topic currently only exists for dogs (Demant et al. 2011; Meyer and Ladewig 2008). Ideally, exercises should be integrated into the daily routines of both the cat and its owners.

Each exercise should be introduced in successive steps so that both the cat and its owners can master it without becoming discouraged. The difficulty levels should gradually increase as the cat progresses. For example, when teaching a cat to tolerate claw trimming, the process begins by teaching the cat to give its paw. The duration for which the paw is held without the cat moving is gradually increased, without introducing the nail clipper at this stage. Finally, the ability to perform the task in the presence of background noise can be added as a further step. Difficulties should never be combined in a single stage.

This step-by-step approach, while essential, should not lead the practitioner to prolong exercises unnecessarily. If a cat is unable to progress as quickly as others, it is crucial to investigate the underlying cause of this difficulty.

- **What to Do in Case of Failure or Slow Progress?**

As previously mentioned, the speed of a cat's progress is a valuable indicator for veterinarians, their assistants, or feline trainers in identifying a psychopathological dysregulation in one or more of the major behavioral axes, or a communication issue between the cat and its owners.

The main causes of learning failure are as follows:

- The cat's attention and motivation are not optimal because they were overlooked.
- Some successive steps in the learning process were omitted or rushed through too quickly.
- The exercise intended to help the cat progress was poorly designed, poorly explained, or poorly understood, and therefore improperly implemented.
- The cat is unable to learn normally due to its physical or mental health condition.

If the first three causes of failure can be ruled out, owners should be promptly referred to a behavioral consultation to obtain a diagnosis of the cat's emotional state.

29

Summary of ► Sect. 29.1

Behavioral therapies encompass a range of techniques aimed at introducing a new behavior or eliminating an undesirable one. These techniques apply to both training and behavioral reeducation.

Therapeutic methods draw on the principles and theories of learning, such as habituation, desensitization, classical conditioning, and operant conditioning. They also consider the perceptions of both the cat and its owner, as well as consistent communication rules.

Positive methods that respect the cat's well-being are the only ethically acceptable and scientifically validated approaches. Techniques involving fear or pain must be strictly avoided.

Behavioral therapies are tailored solutions designed primarily to restore coherent interspecific communication. Once necessary safety measures are established, these therapies focus on improving the cat's motivation and concentration, followed by addressing the most challenging behaviors for the owners.

Three methods are recommended for generating a new behavior: capturing, guiding, and shaping. To eliminate an undesirable behavior, three techniques can be employed: extinction, negative punishment, and counterconditioning.

The practitioner must ensure that the therapy is practical and propose steps that are appropriate for the patient, taking into account the resources available within the system.

29.2 Organizing the Cat's Environment and Respectful Communication: Prevention Tools for All Cats

This section, which applies to all cats, outlines the organization of the cat's environment, the rules for respectful communication, and useful exercises to practice with cats. Together, these elements help prevent the development of behavioral issues.

If adhering to these guidelines does not prevent the emergence of problematic behaviors or eliminate them within a few weeks, a behavioral consultation is necessary. This step ensures that the cat's psychological state is not the underlying cause of these undesirable responses.

29.2.1 Organizing the Environment

29.2.1.1 Organizing the Cat's Living Space

- **Indoor Cats Versus Outdoor Cats**

The territorial nature of cats requires ensuring that their living space adheres to feline territorial organization principles. The importance of the environment is so significant that in feline behavioral medicine, a recurring question is: "Would the cat be happier if it could go outside?"

From our clinical experience, we frequently encounter perfectly balanced indoor cats that, upon moving to a home where they can go outside, suddenly show significant behavioral deterioration. Going outdoors introduces the cat to potentially frightening stimuli, such as cars, the neighbor's dog, and various unfamiliar encounters.

Conversely, we also often see outdoor cats in behavioral consultations that display pathological psychological states. Furthermore, a Brazilian study found that owners of exclusively indoor cats tended to care for their animals better and provided a higher level of medical attention compared to owners of cats with outdoor access (Oliveira et al. 2023).

The question of whether to let a cat go outdoors is complex and raises many ethical debates. In North America, allowing a cat to roam outside is often considered dangerous due to risks such as traffic, predators, and disease exposure. A survey indicates that older owners and those living in rural areas are more likely to allow their cats outdoor access (Foreman-Worsley et al. 2021).

Currently, there is a trend toward offering controlled outdoor access using systems such as inward-angled mesh installed at the top of fences, preventing the cat from leaving the property. These systems are increasingly adopted by owners concerned about letting their cats roam freely, with positive impacts reported for both the cat and the owner (De Assis and Mills 2021).

That said, it is clear that a cat without outdoor access has identical needs to those of an outdoor cat: the need for predation, play, and frequent nibbling throughout the day. The goal is to provide an environment that meets these physiological needs.

To compensate for the lack of outdoor access, adjustments must be made. Without fields full of rodents to hunt, the cat's environment should include elements that fulfill its needs for observation and predation. Meal distribution must also be adapted, allowing the cat to nibble throughout the day, and a litter box must be provided since there is no soil in the home for elimination.

▪ Environmental Enrichment

The term *environmental enrichment* refers to measures aimed at making the cat's living space more stimulating and satisfying for its natural needs. These techniques have been significantly developed in recent years and appear to play a key role in ensuring the well-being of indoor cats. For example, a study showed that providing opportunities for enrichment engaging the cat in play, cognitive, social, and hunting behaviors can reduce undesirable behaviors such as inappropriate urination or aggression toward owners (Strickler and Shull 2014).

However, a recent systematic review of the literature examining the impact of environmental enrichment on cats' well-being highlighted that little research exists on the effectiveness of most of these recommendations. The authors concluded that more research is needed, particularly studies exploring the impact of complex environments involving multiple types of enrichment that may interact, such as the presence of children, dogs, and a combination of various artificial enrichments (Foreman-Worsley and Farnworth 2019).

These environmental measures are necessary to compensate for the shortcomings of indoor life compared to having access to a garden, but they are far from sufficient to ensure a cat's well-being. Additionally, some cats with outdoor access encounter unpleasant stimuli and dangers, making it essential to also adapt their indoor environment. For example, it may be better to keep a cat indoors if it is frequently attacked by neighborhood cats, to avoid injuries. This can be offset by providing additional toys or observation posts within the home. In fact, some cats choose to stay indoors after experiencing one or more negative outdoor incidents.

What seems most important is ensuring the cat feels comfortable, whether living exclusively indoors or with outdoor access, rather than stigmatizing individuals who cannot or choose not to let their cats roam outdoors. After all, having outdoor access does not necessarily guarantee perfect psychological balance.

Normal Pathological Limit

The environment is only one part of the equation for ensuring a cat's psychological well-being. As highlighted by Zubin and Spring's stress vulnerability model (Zubin and Spring 1977) (▶ Chap. 1, stress vulnerability model box), once an animal develops a pathological psychological state, an enriched environment alone cannot normalize its condition. Furthermore, animals vulnerable to behavioral disorders may develop them even in a perfect environment (◘ Fig. 29.7).

In the following section, we will review the various activity fields of a cat and how to adapt them. It is important to remember that if such adaptations do not prevent or eliminate undesirable behaviors, a behavioral consultation will be essential to determine the underlying causes.

29

29.2.1.2 The Different Activity Fields

As explained in part I (▶ Chap. 1, ▶ Sect. 1.3.1.4), cats organize their habitat into distinct activity fields, which they delineate using their pheromones. One of the primary principles in environmental adaptation is to identify and differentiate each of these fields (◘ Fig. 29.8).

Before detailing each activity field and the appropriate adaptations, it is essential to ensure that the spatial and temporal organization of the cat's environment encourages the use of each field at specific times of the day. Facial markings, allomarkings, and scratch marks around the isolation field should be identifiable by the owners, and the various fields should be as distinct as possible.

In smaller apartments, it may not be feasible to dedicate a separate room for each activity. However, the three-dimensional space can

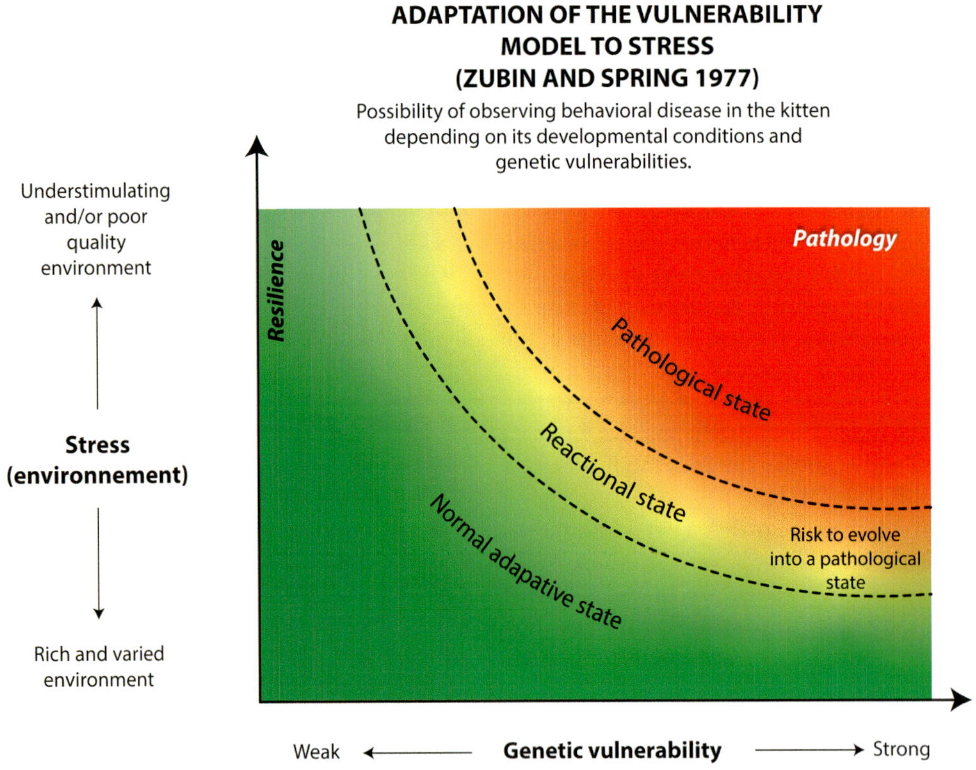

◘ **Fig. 29.7** At the two extremes: (a) very low genetic vulnerability protects the individual despite poor environmental conditions; (b) high genetic vulnerability predisposes the individual to pathology despite excellent environmental conditions

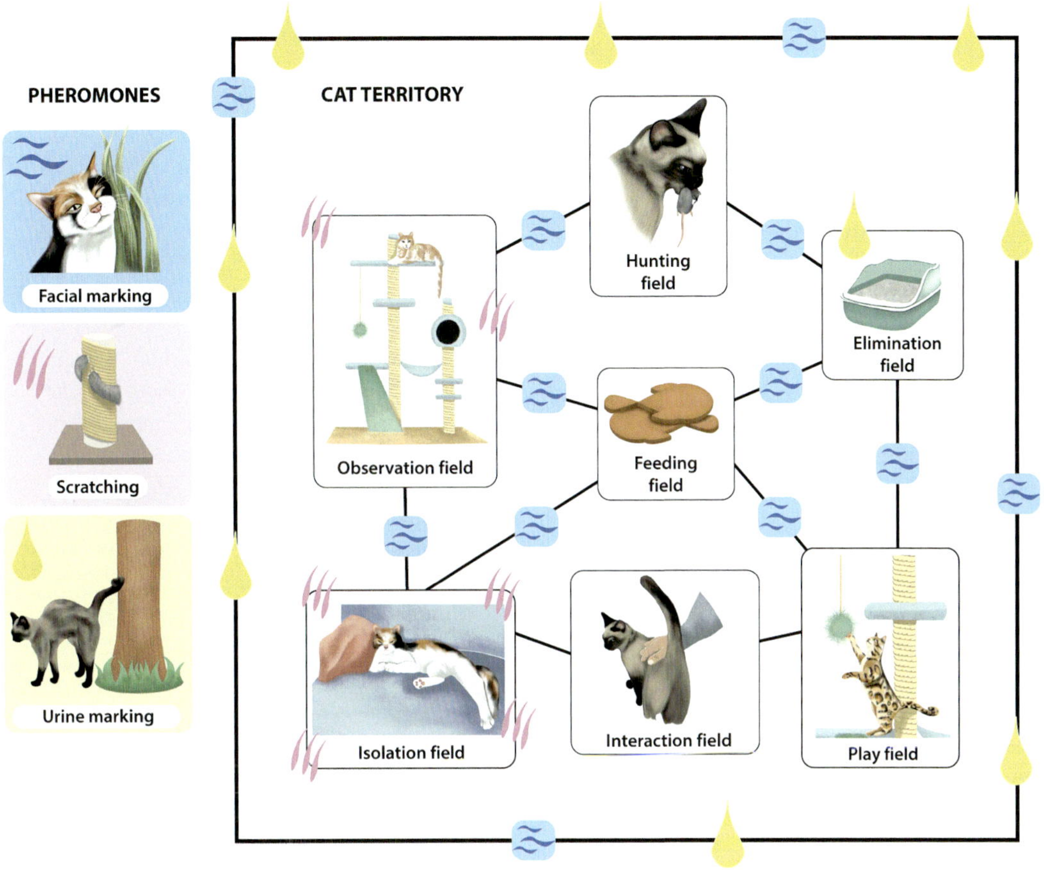

■ **Fig. 29.8** Organization of the cat's activity fields using pheromonal markings

be utilized to allow the cat to establish boundaries between zones. For instance, the litter box can be placed on the floor under a table, while the feeding area can be positioned on the table's upper level.

It is important to recognize that each activity field is associated with a specific emotional and physiological state. For example, a tired cat will go to its resting zone, where occupying that space promotes sleep. When hungry, the cat will move to its hunting zone, where the environment itself stimulates motivation for this activity. Temporal organization also plays a role, as each zone corresponds to specific periods of the day.

Although this concept may seem complex, it is intuitive. Humans, for example, can organize an optimal workspace that they use only at certain times of the day, fostering concentration. It would seem incongruous to eat or sleep in that same space, as deviating from such an organization often leads to stress.

Let us now examine in detail the key considerations for adapting each of the cat's activity fields.

■ Feeding Field

Role of the Feeding Field

In the wild, a free-roaming domestic cat, as a strict carnivore, dedicates a significant portion of its time to hunting. Much of this time is spent watching and waiting for prey, with a success rate that is often modest. Picture the scene: the cat, patiently stationed, observes a section of the garden, waiting for a small prey to appear. This activity can last several hours—sometimes up to 8 h a day—in order to catch enough prey to meet its energy needs.

A cat might miss a grasshopper, then catch a fly, fail to catch a mouse, miss another grasshopper, and finally capture a shrew. In other words, the cat eats small quantities but frequently. It is a grazer. Its prime hunting times are dawn and dusk, when its vision—optimized for low light and moving objects—gives it a significant advantage.

As veterinarians, it is crucial to understand how owners feed their cats. Typically, a bowl filled with kibble—often very rich—is provided once or twice a day. This practice transforms skilled hunters, accustomed to intense activity to acquire food, into sedentary consumers of abundant, rich, and easily accessible meals. This shift largely explains why so many domestic cats are overweight.

This is especially true because, for cats, stalking prey is an activity closely associated with eating. This activity does not provide calories, but it involves significant energy expenditure. By removing this "hunting" phase and offering a bowl of kibble instead, the cat loses this energy-burning step, replacing it with the mere act of consuming kibble. Additionally, the high palatability of kibble acts as a strong reinforcer.

Another concerning aspect is how cats regulate their food intake. In natural environments, cats rarely face an overabundance of food. As a result, they have numerous biological signals that trigger food-seeking behavior but very few mechanisms to limit consumption. For indoor cats, every activation of food-seeking behavior directly results in food consumption. There is no failure, no stalking phase, and no waiting period, which further disrupts their natural regulation of food intake.

Important Point of View

Hepatic Lipidosis

Cats are "ethologically programmed" for food-seeking behavior, and a cat that does not eat for more than 2 or 3 days is at risk of developing hepatic lipidosis.

Feline metabolism relies heavily on proteins and fats, making cats particularly vulnerable to fasting (Plantinga et al. 2011). When a cat fasts, its body quickly begins mobilizing fat reserves to provide energy. However, a cat's liver is not as efficient as that of other animals. During fasting, the influx of fat overwhelms the liver, which starts storing it in its cells. These fat-laden cells lose their ability to function normally, and critical detoxification activities normally carried out by the liver are impaired.

Additionally, cats have a high protein requirement, and when food intake decreases, they begin using muscle proteins to maintain vital functions, further exacerbating hepatic stress. At this stage, the cat enters hepatic lipidosis, a potentially life-threatening condition if left untreated. In summary, it is crucial to avoid fasting in cats, even as a strategy for inducing weight loss.

Organization of the Feeding Field

It is essential to respect the cat's natural tendency as a grazer. The idea of feeding a cat in two or three meals per day should be abandoned, as such a regimen is unsuitable for their physiology and can cause distress.

Some cats we see in consultations are capable of nibbling responsibly from a simple bowl of always-full kibble, but this is not the general rule. For these cats, changing their feeding method is not a priority.

However, for all others, it is necessary to completely change the feeding method by replacing traditional food bowls with food enrichment devices (Ellis 2009). Standard bowls must be entirely removed, as cats will naturally gravitate toward the easiest option, feeding primarily from the simpler, more accessible bowl. Of course, some cats that spend significant time outdoors hunting insects, birds, and mice have less need for these systems. However, for indoor cats, it is rare for feeding activities to take more than a few minutes daily when using traditional bowls (Naik et al. 2018).

Food enrichment systems are therefore recommended from a young age and for all cats, particularly those prone to overeating (Dantas et al. 2016).

The food enrichment device industry has expanded considerably in recent years. Historically, one of the first systems was the Pipolino®, a large plastic tube with adjustable holes to control difficulty. Nowadays, a wide variety of food puzzle toys are available: some require the cat to roll an object, others involve retrieving kibble with a paw placed inside a tube, and some take the form of mazes. One of the most challenging options is the *Turn Around*® by Trixie®, where the cat must rotate tubes around an axis to release kibble. Once released, the kibble falls into a labyrinth with small plastic pegs, requiring further paw work for extraction.

These systems significantly slow down the speed of ingestion and mimic feeding behaviors more akin to hunting in the wild, thereby improving the cat's feeding experience.

For owners who do not wish or are unable to purchase these types of devices, it is entirely possible to create food puzzles using everyday household items. For example, plastic cups or glass jars taped together can be partially filled with kibble at the bottom to encourage the cat to retrieve the food with its paw. Empty paper towel or toilet paper rolls can be perforated with small holes and sealed at the ends to create a homemade version of a Pipolino®. More creative owners can design even more ingenious systems that rival commercially available options.

There are many online resources and social media platforms offering ideas for homemade food puzzles, including articles with photos of these systems (Ellis 2009).

The more gluttonous and food-motivated the cat, the more challenging the system should be to make obtaining kibble more effortful. Additionally, multiple feeding stations can be set up to provide variety in the type of effort required. Placing some of these stations at elevated locations, such as on a windowsill, encourages physical exercise.

This approach is particularly important in multicat households to prevent tension around a single feeding point.

This free-access feeding system also avoids creating a "refrigerator ritual," where the cat meows whenever its owner approaches the area where food is kept. Food puzzles can be refilled every few days, disconnected from any specific behavior by the cat, which further discourages such associations.

To complement these systems, owners can be advised to hide a few kibbles in the home each morning so that the cat uses its sense of smell to locate them throughout the day.

This approach does not preclude offering wet food to the cat, adhering to the current principles of bi-nutrition. This practice is recommended to increase water intake and reduce the risk of lower urinary tract disease (Black 2018). Additionally, wet food contains fewer calories per volume than dry food and helps promote satiety. Similarly, adding cooked zucchini to the diet can reduce calorie density, though not all cats will eat it, and some owners may find the preparation too inconvenient. Zucchini is also recommended to lower the risk of calcium oxalates (Blanchard et al. 2022).

Attention should be given to the composition of wet foods. Manufacturers prioritize palatability to ensure the cat enjoys the food and consumes it in large quantities. Palatability is often driven by high fat content, which is abundant in many products. Like a chef, food manufacturers prioritize pleasure over nutritional balance. Many marketed products are far from what is optimal for a cat's health, often heavily promoted through intense advertising.

As veterinarians, we are particularly concerned that a majority of cats today are overweight or even obese (Arena et al. 2021). It is well known that preventing obesity is far easier than helping an obese cat lose weight. For cats prone to being overweight, a tendency often noticed around the time of sterilization, these feeding systems are essential.

If these measures are insufficient, it becomes necessary to adjust the type of kibble. Since the number of meals cannot be reduced and kibble must remain available at all times, the only lever to pull is the caloric density of the kibble. Switching to diet kibble, which is high in protein and fiber, allows calorie reduction while maintaining similar portion sizes.

It may also be necessary to help owners understand that the food their cat prefers is not always the most suitable for its health. Owners should be reassured that it is normal for a cat not to rush to its food unless it is hungry. This perspective often contrasts with messages promoted in advertising, so it is important to *repeat these explanations* to convince owners.

Despite these recommendations, cats in impulsive or compulsive states may continue to consume far more calories than they need. For these cats, dietary adjustments must be complemented with pharmacological management, such as fluoxetine, to address excessive food intake.

▪ Drinking Field

Role of the Drinking Field

Felines originally evolved as desert animals with highly efficient renal systems capable of concentrating urine to a significant degree, allowing them to survive with minimal water intake. In the wild, cats obtain most of their hydration from their prey, which consists of approximately 70% water. Occasionally, licking dew-covered grass in the morning is sufficient to meet their water needs. However, this remarkable ability to concentrate urine also makes them more vulnerable to urinary and renal conditions.

As domestic animals, particularly those fed dry kibble, it is crucial to compensate for their reduced water intake by providing additional hydration sources. Cats have a lower thirst drive compared to other species, making them more susceptible to medical complications when their water consumption is insufficient.

Additionally, many cats are highly selective about their water source, often preferring fresh, moving water. It is not uncommon for cats to attempt drinking directly from a faucet rather than a bowl. At home, this natural preference must be considered to ensure proper hydration (Zanghi et al. 2018) for domestic cats.

This is why it is also recommended to regularly include wet food in a cat's diet. Cats consuming wet food tend to drink more water as well, which helps increase their overall water intake and prevent conditions related to insufficient hydration (Zoran and Buffington 2011).

Organization of the Drinking Field

To encourage hydration in cats, it is advisable to place multiple water bowls in various areas of the home. This strategy promotes more frequent drinking. It is crucial to position these bowls away from the litter box and feeding area to prevent potential contamination or discomfort.

Water should be changed regularly, as cats are particularly sensitive to the cleanliness of their water and prefer fresh water. Using bowls made of stainless steel, ceramic, or glass is recommended, as these materials do not retain odors like plastic, which can deter cats from drinking. Observing the cat's preferences can help identify the type of bowl that suits it best.

Cats often struggle to visually adjust to the surface of the water and tend to prefer shallow containers, such as plates or plant saucers. Including a partially submerged object in the water can make drinking easier for them.

Water fountains are also highly attractive to cats, as they provide moving water, which can increase water consumption. However, it is essential to clean the fountain regularly to prevent the buildup of bacteria, which could not only harm the cat's health but also make it reluctant to drink.

For cats that tend to drink little, adding a small amount of unsalted chicken broth to their water can make it more appealing. Adding ice cubes to the water can also help keep it fresh, especially during hot months.

Incorporating wet food into a cat's daily diet is another effective way to increase hydration. Wet food contains a high water content, compensating for low drinking habits. Additionally, these feeding rituals can facilitate medication administration when necessary.

Some cats enjoy playing with water. Introducing floating toys in a large bowl of water can stimulate their interest and encourage them to drink more while having fun.

Many cats with outdoor access, even if limited, develop the habit of drinking rainwater collected in various outdoor containers. This naturally distilled water may sometimes be mixed with leaves or soil residues, but there is no need for concern. Replacing it with tap water could deter the cat from drinking. However, owners should ensure the cat has access to such water (e.g., keeping doors open in winter) and that freezing temperatures or drought do not prevent the cat from drinking.

It is vital to monitor each cat's drinking habits. Any sudden changes, whether excessive water consumption or a drastic reduction in intake, should be investigated. Educating owners on this point during routine visits is essential.

By following these recommendations, owners can ensure their cat remains well-hydrated, a key factor in preventing renal and urinary diseases, while respecting the cat's natural drinking behaviors.

Elimination Field

Role of the Elimination Field

The elimination field refers to the areas a cat selects for its toileting needs. Interestingly, many cat owners proudly note that their kitten was already "clean" upon arrival, as if this were an extraordinary skill. This contrasts with the active training required for puppies. In reality, kittens quickly learn to use a litter box, often before adoption, by observing their mother's behavior.

Clinical Illustration

Looky, a 3-year-old Bengal cat, contracted a virus that caused diarrhea when her kittens were still very young. Due to her condition, she preferred to eliminate outdoors, preventing her kittens from observing her use of the litter box. Normally, kittens spontaneously develop the habit of using a litter box by 4 weeks of age without specific intervention. However, in Looky's litter, the kittens began toileting on blankets or directly on the floor. They only became fully litter-trained between 6 and 7 weeks, requiring regular placement in the litter box to encourage its use.

This case highlights how imitation of the mother can accelerate the development of litter box habits in kittens, although it is not strictly necessary for achieving cleanliness.

Interestingly, even bottle-fed kittens will instinctively use a litter box. It is naturally the most appealing place for a cat to eliminate in a household, a soft area where they can easily scratch to cover their waste.

Regarding *posture*, cats typically adopt a squatting position to urinate, with their tail held horizontally. This position is distinct from the posture used for urine marking, where they may slightly raise their hindquarters, hold their tail vertically, and spray urine onto a vertical surface.

A study compared the elimination sequence of cats in their homes, where large litter boxes were available, to their behavior in a clinical setting, such as a cage with a small litter box and limited space to move around (McGowan et al. 2017). Researchers observed that hospitalized cats displayed a different and much longer elimination sequence, particularly regarding interactions with the litter box after elimination. Specifically, these cats hesitated to enter the litter box, spent significant time scratching surfaces other than the litter, and took more time sniffing their waste after elimination. They also urinated less frequently. Interestingly, a shorter elimination sequence appears to indicate a more positive litter box experience. Moreover, when given the option, cats prefer a spacious area during their elimination sequence. Contrary to the common belief that cats immediately eliminate outside the litter box if dissatisfied with their environment, this study shows that cats will continue to use the litter box, even if their behavior signals frustration.

Research on cat behavior has also identified specific preferences for *elimination sites*. Using plastic mock feces and water to simulate waste, researchers tested whether the presence of these elements or the odor of previous waste, especially from another cat, bothers cats more. Surprisingly, cats preferred litter boxes that were the most "empty" of prior waste, showing no preference between real or fake waste, even if it was not their own (Ellis et al. 2017). The key factor appears to be the absence of urine or feces in the litter box, rather than whether another cat has used it previously.

The *size* of the litter box is another crucial factor: it must be large enough to allow the cat to turn around and scratch freely, with low edges to ensure easy access, particularly for kittens or elderly cats (Guy et al. 2014). Cats generally prefer litter with a fine grain size, as it feels more comfortable on their paws (Villeneuve-Beugnet and Beugnet 2018). Additionally, many cats prefer covered litter boxes for defecation and open ones for urination (Beugnet and Beugnet 2020).

However, as is often the case with cats, *individual preferences* override general guidelines, and there are no absolute rules. Some cats may even show incompatibility with certain types of litter due to aversive associations learned through experience.

Care should be taken with complex litter systems designed to blend with sophisticated interiors, which often prioritize owner satisfaction over the cat's comfort. Small litter boxes that cannot ventilate properly can retain odors, which may deter sensitive cats, even if the box is relatively clean.

It is essential to prioritize the cat's comfort over aesthetics and experiment to determine what suits each individual cat best.

Organizing the Elimination Field

To ensure a cat consistently and correctly uses the litter box, a few essential technical guidelines should be followed.

First, it is recommended to have one litter box per cat, plus an additional one, all placed in quiet areas, away from resting and feeding zones. It is important to note that two litter boxes placed side by side count as only one in the cat's perspective. In small living spaces, spatial constraints can make this configuration difficult, but using vertical space can help. For instance, the litter box can be positioned on the floor under a piece of furniture, while water bowls and food puzzle toys can be placed on top of the same furniture, creating distinct zones for the cat.

Regarding the choice of litter, materials that meet the cat's needs should be prioritized over those designed to appeal to the owner. Scented litters or aesthetically pleasing designs for humans may not suit felines. Studies have shown that cats have preferences for specific substrates, but it is essential to test different types of litter and remain mindful of the individual preferences of the cat (Frayne et al. 2019).

The litter box should be placed in a quiet location, away from noisy appliances such as washing machines. A sudden noise while the cat is using the litter box can create a lasting aversion to that area. Additionally, digestive issues like diarrhea or constipation can cause the cat to associate the litter box with pain, leading to elimination outside the box. Interactions between cats should also be monitored, as aggression following litter box use may result in litter box avoidance.

The litter box must be large enough for the cat to turn around and scratch freely. Ideally, its length should be one and a half times the length of the cat, with a width at least as long as the cat (excluding the tail). The edges should be low enough for easy access, especially for kittens or senior cats. Commercial litter boxes are often too small, particularly for large breeds like Maine Coons, which require oversized models typically available only from specialty retailers.

Many cats prefer to eliminate in a private area. Covered litter boxes can provide this privacy, but they must be regularly ventilated. Placing a litter box at the bottom of an open cardboard box can be a great solution for younger, agile cats that can easily jump into the box.

The type of substrate is also crucial: fine-grained litter is generally more comfortable for a cat's paws, and strongly scented litters should be avoided as they can be off-putting. It is important to observe the cat's behavior during elimination. The cat should enter the litter box confidently and cover its waste. A litter depth of 4–6 cm is recommended to allow the cat to scratch without reaching the bottom of the box.

To determine your cat's preferred litter, consider testing two different types of litter placed near each other and observing its preferences. Only one variable should be changed at a time to ensure clear conclusions.

Litter box hygiene is critical for a cat's comfort and continued use. Waste should be removed daily, and the litter should be completely replaced at least once a week. The litter box should be cleaned with hot water and mild soap, and using bleach is recommended, as its scent is attractive to cats. The litter box itself should be replaced annually, as the plastic can eventually absorb urine, making it difficult to clean effectively.

If a cat urinates outside the litter box, white vinegar or sparkling water are the best options for neutralizing odors. Products containing ammonia or bleach should be avoided in these cases, as they can encourage the cat to eliminate in the same spot again.

It is essential not to punish the cat for such incidents, as this can increase its stress and lead it to avoid its owner rather than learn to use the litter box properly.

To discourage a cat from eliminating in inappropriate areas, these spots can be made less appealing by using aluminum foil, artificial grass, or emergency blankets, which create unpleasant textures. At the same time, the litter box should be made more attractive by regular cleaning and the use of bleach, whose scent draws cats. Rewarding the cat when it uses the litter box can also be effective, though this requires close monitoring.

Finally, if inappropriate elimination persists, even occasionally, a thorough evaluation including urinalysis, bloodwork, and a behavioral consultation should be undertaken promptly.

▪ Observation Field (Cat Trees)

Role of the Observation Field

The observation field is a crucial element for a cat's well-being. As sit-and-wait predators, cats spend a significant amount of time observing their environment, either in search of potential prey or to monitor their territory. This behavior, deeply rooted in their instincts, persists even in domestic settings. Cats require observation zones to satisfy their natural vigilance and curiosity. An environment rich in observation opportunities helps reduce stress and boredom, particularly for indoor cats without outdoor access.

Studies have shown that environmental enrichment in animal shelters, specifically through the addition of elevated structures and observation zones, significantly improves the well-being of cats and reduces stress-related problem behaviors (Wojtaś et al. 2024). Cats in enriched environments are more active, explore their surroundings more, and exhibit fewer signs of stress compared to those without such features.

These findings highlight the importance of vertical enrichment and opportunities for surveillance in ensuring cats' well-being, particularly in confined spaces or when outdoor access is restricted.

Observation is also a hunting-related activity and should be considered part of the feeding time. Cats with satiety disorders can benefit from increased observation time, which helps replace constant food intake with this engaging, instinctual behavior.

Organizing the Observation Field

To meet cats' natural need for observation, it is essential to create dedicated zones in their environment. Cat trees are an excellent way to provide elevated observation areas. Available in a wide variety of sizes and styles, some cat trees are designed more for the owner's aesthetic than the cat's comfort. It is advisable to choose a cat tree with multiple levels and stable platforms, ideally placed near a window, allowing the cat to observe the outside world. This placement enables the cat to watch outdoor activities such as birds, cars, or passersby.

Cats particularly enjoy elevated spots where they feel safe and can oversee their territory. In addition to cat trees, installing wall-mounted shelves, boards, or furniture can provide high spaces for the cat to explore. These structures are especially useful in small apartments or households where the cat cohabits with a dog or young children who might chase it.

Creating elevated pathways using wall-mounted boards offers the cat a safe and enjoyable way to move around the home. These pathways should be sturdy and wide enough to allow the cat to move comfortably. They can connect to different observation zones or lead to elevated refuges where the cat can relax and observe its surroundings in peace.

Placing perches or beds near windows is also an excellent idea. Owners can purchase window-specific perches or simply position a stable piece of furniture nearby to allow the cat to climb up. However, it is crucial to ensure that windows are secure to prevent accidents.

By providing a cat with suitable observation zones, the owner not only supports the cat's physical well-being by enabling climbing and movement but also enhances its mental well-being by offering visual stimulation and a sense of security. This is particularly important for indoor cats without access to a garden or balcony. Increasing the number of observation areas further contributes to the cat's overall well-being.

If, despite these accommodations, a cat shows little interest in observation and prefers to hide, a routine veterinary consultation is advised. If no physical issues are identified, a behavioral consultation should be considered. A healthy cat typically spends significant time observing its environment, and changes in this behavior may indicate a physical or mental health issue.

▪ Interaction Field

Role of the Interaction Field

The interaction field is essential for the social and emotional well-being of cats. These are the areas where cats interact with humans and other household animals. As previously mentioned, incorporating vertical space into the environment is crucial, and this also applies to interaction zones. Cats enjoy being in elevated areas not only for observation but also because it facilitates interactions with humans, such as when they seek attention, are spoken to, or try to reach a faucet. Elevated areas also provide a safe haven for cats to retreat from dogs or other interactions they wish to avoid.

Cats are domestic animals that require regular interactions to thrive. Interaction zones may include spaces such as sofas, beds, tables, and countertops. However, it is important to manage these areas so that cats can access them without causing conflicts with human activities, such as meals or meal preparation.

It is advisable to designate specific zones where the cat can engage with humans without disrupting daily activities. These spaces should balance accessibility, comfort, and the cat's need for safety and security.

Organizing the Interaction Field

To foster meaningful interactions between a cat and family members, it is essential to organize the space in a way that allows the cat to access various areas while managing its specific behaviors.

First, as with the observation field, utilizing three-dimensional space is critical. Adding elevated pathways and suitable furniture not only meets the cat's observation needs but also facilitates interactions with humans. Cats prefer elevated spaces as they allow for interactions at eye level, aligning with their natural mode of communication.

Regarding tables and countertops, it is possible to train a cat to avoid these surfaces during meals or meal preparation by consistently redirecting it to an appropriate location whenever it tries to climb up. For instance, owners can establish a routine where the cat is gently placed back on the floor while a specific command, such as *"no, not here,"* is repeated until the cat associates these areas with specific restrictions.

However, it is important to acknowledge that it is unrealistic to expect a cat to avoid all surfaces entirely, especially when owners are not present. Setting boundaries and providing alternative spaces that are both acceptable to the owners and appealing to the cat can help maintain a balance between human activities and feline behavior.

To manage this type of behavior, a more effective strategy is to redirect the cat to a more appealing area. For instance, placing objects on surfaces you want to protect can make these zones less accessible or less comfortable for the cat. At the same time, providing more attractive alternatives, such as a cat tree or a perch near a window, can offer the cat a comfortable spot to settle while continuing to interact with family members.

Interaction zones should also be welcoming and secure. It is essential to ensure there are enough comfortable spots where the cat can settle near its owners while they read, watch television, or work. These moments of proximity are crucial for strengthening the bond between the cat and family members.

If, despite these accommodations, a cat shows little interest in interacting with family members, it is important to investigate the underlying reasons. A lack of interaction may indicate physical or mental health issues that require attention.

▪ Isolation Field

Role of the Isolation Field

The isolation field refers to the space where cats can retreat to rest, hide, and feel secure. Cats are solitary animals by nature and require periods of isolation to relax, away from interactions and external stimuli. This need is particularly significant in environments with high activity levels or frequent changes.

For example, during the COVID-19 lockdowns, while pet owners found their pets helpful in coping with isolation (Hargrave 2020; Jezierski et al. 2021; Riggio et al. 2022), some cats may have experienced this period more negatively. The constant presence of humans in their habitat might have disrupted their territory and daily routines, leading to stress.

Scratching vertical surfaces around isolation zones is a natural territorial marking behavior in cats. By scratching, they deposit pheromones that create a familiar and calming environment, symbolically marking the entry to their resting space.

Providing a cat with spaces where it feels secure enough to rest is essential for its well-being, especially considering the multiple daily naps that are integral to its natural rhythm. These areas should be designed to meet the cat's need for isolation, offering a safe and quiet refuge.

Organization of the Isolation Field

When designing a cat's environment, one fundamental rule applies: *"the owner proposes, the cat disposes."* Structuring an isolation field involves offering multiple hiding spots and perches, allowing the cat to choose the most suitable place to rest and feel secure.

To create these hiding places, everyday items such as delivery boxes can be highly effective. Many owners leave boxes in living spaces for a few days, giving cats the opportunity to use them as temporary hideouts for play, rest, or observation. Once the cat loses interest, these boxes can be removed and replaced. Other options include using drawers filled with clothing, blankets placed under beds or behind furniture, or draped chairs to create secure spaces.

Elevated spaces are especially popular with cats. In addition to ground-level hiding spots, wall-mounted shelves or cat trees with elevated enclosures are highly recommended. These areas not only provide a secure refuge but also offer a commanding view of the living space, which is crucial for a cat's sense of security.

It is advisable to position these hiding spots in quiet areas of the home, away from noise and heavy foot traffic, to allow the cat to retreat in peace (Foreman-Worsley and Farnworth 2019). Encouraging scratching behavior around isolation zones by placing scratching posts nearby can also be beneficial. This allows the cat to mark its isolation field and enhances its comfort and security.

Observing each cat's preferences and adjusting hiding spots accordingly is essential. Every cat has its own preferences for rest and isolation areas. Providing multiple options and allowing the cat to decide which suits it best is recommended (Kry and Casey 2007).

If a cat shows a tendency to hide excessively at the expense of interacting or observing its environment, it is important to schedule a consultation to rule out any underlying health or welfare issues.

▪ Play Field

Role of the Play Field

Play is a vital component of a cat's well-being, as it mimics natural predatory behaviors and provides essential mental and physical stimulation. Play allows cats to fulfill their need for physical exertion, develop coordination, and refine hunting and combat techniques. It also helps prevent boredom, especially in indoor cats. Research by Bradshaw and Ellis highlights that interactive play reduces frustration levels in cats and contributes to emotional balance (Turner and Bateson 2014).

Games involving stalking, chasing, and capturing are particularly beneficial as they replicate the predation sequences cats would perform in the wild. These activities also strengthen the bond between cats and their owners, fostering a harmonious and fulfilling relationship (S. L. Ellis 2009; Henning et al. 2022). Play also serves as a medium for learning motor and emotional control, including the moderation of biting and scratching behaviors. Teaching cats to regulate these actions is essential for ensuring safe interactions with humans and other animals.

Playtime is especially crucial for kittens, peaking between 12 and 14 weeks of age, when it can account for nearly 10% of their day (Martin 1984). For adult cats, it is abnormal not to play at all. Play is a normal behavior reflecting well-being, typically occupying between 5% and 8% of an adult cat's day. Significant deviations from these values may indicate pathology and warrant attention.

For instance, a cat experiencing physical pain may gradually lose interest in playing. Similarly, an anxious cat might be too inhibited to engage in play, while an impulsive or compulsive cat could spend up to 30% of its time playing without finding it sufficient. These behavioral variations should be carefully considered when evaluating a cat's health and well-being.

Organizing the Play Field

To encourage play in cats, it is essential to provide a variety of toys and activities that cater to their natural instincts. Interactive toys, such as dangling feathers, balls, and mechanical toys, are ideal for stimulating a cat's hunting behaviors. Puzzle games that hide treats are also effective for keeping a cat engaged while rewarding their efforts. For example, cats in shelters show a strong preference for play activities, especially those involving interactions with caregivers, likely due to their limited opportunities for interaction compared to cats in homes (Houser and Vitale 2022).

Using laser pointers can be highly engaging for cats, but it is crucial to end these play sessions with a tangible reward, such as a treat or a toy the cat can "catch." This practice prevents excessive frustration and satisfies the cat's hunting instinct while offering a positive conclusion to the play experience.

Play also provides an opportunity for learning, enabling cats to develop specific skills and behaviors. For instance, some cats can be trained to fetch a ball or play hide-and-seek with their owner, further enriching their playtime and strengthening the bond with their human companions.

Clinical Illustration

A father had a habit of teaching the household cats a ritual where they would jump through a hoop formed by his arms before receiving their food. This ritual, learned in just a few days using shaping techniques, demonstrates that cats are capable of learning complex behaviors when properly stimulated.

It is important to emphasize that all cats have the ability to learn if provided with the right opportunities and incentives. Often, owners do not attempt to train their cats due to the belief that cats cannot learn like dogs. However, these limitations are more often imposed by humans than by the cats themselves. The key to success lies in patience and consistency during training.

Shared play sessions also strengthen the emotional bond between a cat and its owner, offering opportunities for positive communication and interaction. They help owners better understand their cat's needs and behaviors, fostering a harmonious cohabitation. To keep the cat engaged, it is important to vary toys and regularly introduce new activities. Studies have shown that interactive play can reduce problematic behaviors and improve cats' mental health (Vitale Shreve et al. 2017).

By providing an enriched and diverse environment, owners enable their cats to fulfill their play needs, contributing to their overall well-being. If, despite these measures, a cat shows signs of aggression or frustration during play sessions, a behavioral consultation is recommended to assess and address these excessive behaviors.

- **Hunting Field**

Role of the Hunting Field

The hunting field is a crucial element for satisfying a cat's natural predatory instincts. When a cat has outdoor access, its hunting territory often lies outside, determined by the availability of prey. Cats develop and adapt their hunting techniques depending on the types of prey in their environment, a behavior widely confirmed by numerous behavioral studies (Dickman and Newsome 2015). These techniques vary based on the prey: hunting insects, birds, or small mammals requires specific skills, highlighting the cat's adaptability as a predator.

Indoors, this innate need to hunt must be compensated for with a rich and stimulating environment, particularly through the play field. Frustration from a lack of stimulation can lead some cats, especially those with impulsive tendencies, to display aggressive behaviors. These behaviors often manifest as attacks on anything that moves within the home, including their owners.

Hunting behaviors occupy a significant portion of a cat's time and are essential not only for food acquisition but also for mental and physical stimulation. These activities allow cats to expend energy, keep their instincts sharp, and prevent boredom—a factor that can contribute to undesirable behaviors.

For indoor cats, creating an environment that satisfies these needs is imperative to ensure an optimal quality of life. This involves providing opportunities to mimic hunting behaviors, thereby promoting their well-being and reducing the risk of behavioral problems.

Organizing the Hunting Field

For cats with outdoor access, it is essential to secure their environment to minimize risks while allowing them to hunt naturally. When a cat ventures outside, it will instinctively explore prey-rich areas and use its specialized hunting techniques to capture prey. Adding small platforms, hiding spots, and elevated observation zones in outdoor spaces can stimulate hunting behaviors and provide observation opportunities that enhance their well-being.

For indoor cats, it is vital to offer alternatives to real hunting. Interactive toys, such as dangling feathers, balls, and mechanical toys, can mimic prey movements and stimulate a cat's hunting instincts. Games that encourage stalking, chasing, and capturing are particularly beneficial for fulfilling these innate needs.

The use of puzzle toys that conceal treats can simulate hunting while also providing a food reward, satisfying predatory drives. It is crucial to vary toys and activities to maintain the cat's interest and prevent monotony, which is a key factor in ensuring psychological well-being.

Enriching the environment with elevated pathways, hiding spots, and observation zones can also channel the cat's energy and address its hunting instinct. Cats have an innate need to climb and survey their territory, which can be encouraged by installing wall-mounted shelves, cat trees, and platforms, as previously discussed.

If a cat shows signs of frustration or aggression despite an enriched environment, a behavioral consultation is essential to assess and address any underlying issues.

29.2.1.3 Practical Implications

Naturally, the calming effects of organizing a cat's territory have numerous practical consequences on the cat's life. We will discuss some of these in the following section, as they represent common situations where owners often seek advice. Additionally, offering such guidance during routine consultations can serve as a preventive measure against the development of behavioral issues.

- **Moving a Cat**

Relocating a cat, whether for vacations, a veterinary visit, breeding, or a permanent move, is invariably more challenging than relocating a dog. While both animals are removed from their familiar environment, a dog primarily relies on its owner for comfort. In contrast, a cat's well-being is strongly tied to its territory and the soothing pheromonal markings in its surroundings. Taking this environment along is, of course, nearly impossible.

To mitigate stress during such transitions, it is essential to prepare in advance, taking steps to make the new or temporary environment as familiar and comfortable as possible for the cat.

Due to this, cats experience stress whenever they are relocated, and this stress is logically more pronounced after puberty, as pheromonal markings become their primary source of comfort at that stage. This likely explains why kittens during vaccination visits readily accept treats and interact almost as much as puppies with veterinary staff, whereas post-puberty, they are generally far more inhibited. It also explains why sterilization performed before puberty is often less behaviorally challenging for cats compared to procedures done at 7–8 months or later.

To minimize the stress associated with relocation, several strategies can be recommended. Providing a comfortable carrier and leaving it open at home allows the cat to familiarize itself with the space. We will discuss how to train a cat to enter a carrier on command in ▶ Sect. 29.2.3.3 of this chapter (Pratsch et al. 2018). Using calming pheromones in the form of sprays or wipes to line the carrier can significantly reduce the cat's stress during travel (Shu and Gu 2022).

Finally, it is recommended to always secure the carrier in the car to prevent sudden movements that might frighten the cat. Placing the carrier on the floor often helps the cat feel less affected by the movements of the vehicle, but the best option is to choose the location where the cat seems most at ease. For some cats, covering the carrier with a blanket can provide additional comfort and reduce stress.

Relocating a cat is never a trivial act and requires consideration of the stress it might cause versus the benefits. While owners sometimes have no choice, in other situations they may hesitate, and this is where advice can be valuable. Should the cat stay home or be boarded? Is it better to take the cat on vacation or leave it at home with a neighbor visiting to provide food and care?

Each cat is unique, but their territorial nature generally tips the scale toward staying in their home environment, more so than for dogs. For short trips lasting a few days, it is usually preferable to leave the cat at home with someone visiting regularly to check on their well-being and provide food. Similarly, when feasible, it is better to recommend that a caretaker visits the home rather than placing the cat in a boarding facility.

However, it is important to remember that a cat's environment includes the presence of its owners; it is not merely the house as an empty structure. For prolonged absences, it is essential to carefully evaluate different options before leaving a cat alone at home.

Most cats are capable of adapting to new territories, even when relocated for an extended period. In the absence of dangers, they generally need only a few days to familiarize themselves with and claim a new space. Thus, taking a cat along on vacation can sometimes be a perfectly viable option.

Finally, abandoning a cat during a move under the pretext that "it is attached to its territory" is a cruel and irresponsible act, characteristic of individuals merely seeking to rid themselves of their pet.

▪ Living with a Cat

Cohabitation Between Multiple Cats

A cat's need to establish comfort through territorial organization helps explain why cohabitation with another cat after puberty is often more challenging than cohabitation with humans or dogs, who do not mark objects with feline pheromones.

While some cats may be more sociable than others, cohabitation between multiple cats requires meticulous territorial organization to minimize conflicts and ensure the well-being of each cat. It is common to adopt multiple kittens from the same litter for cohabitation. If cohabitation issues arise, they typically appear after puberty, as territorial development is not fully established until this stage. Owners may struggle to understand why two kittens that cohabit peacefully might later fail to get along.

The size of the territory is a critical factor to consider. A domestic cat with outdoor access spends the vast majority of its time within 50 meters of its home. This freely chosen territory size provides insight into their ideal living conditions (Horn et al. 2011).

Naturally, it is easier to accommodate two cats in 200 m^2 than in 20 m^2. Moreover, when cats have outdoor access, they can distance themselves to avoid conflicts during periods of tension. In such cases, the size of the indoor space becomes less significant as the outdoors expands their territory.

Given that a cat's organization of space is spatiotemporal, two cats can use the same zones at different times, making shared spaces easier to manage. Thus, it is essential not only to consider the number of square meters available but also to evaluate how the space is utilized and the richness it offers to the cats.

Each area must be identified by each cat, requiring enough space and distinct zones or rooms for every cat to rest, hide, and observe. Thus, housing four cats in a 20 m^2 space without outdoor access is unrealistic and almost inevitably causes them distress. Conversely, a large space does not guarantee successful cohabitation. While there is no strict standard in square meters, a small, well-designed space for cats can be better than a large space lacking engaging resources. It is therefore essential to assess whether it is possible to provide everything a cat needs in the available space before considering adding another cat.

It is essential to multiply *hiding places* and resting areas. Providing diverse hiding spots, both on the ground and in elevated areas, allows each cat to find a refuge where they feel secure. Elevated spaces, such as wall-mounted shelves and cat trees, are particularly favored and help reduce tensions by offering retreat options

The distribution of food and *water points* is also crucial. Each cat must have access to these resources without being disturbed by another cat's presence. It is recommended to multiply feeding stations using food puzzles to stimulate mental activity while reducing tensions related to accessing food.

Litter management is another key consideration. It is generally advised to provide one litter box per cat, plus an additional one. These litter boxes should be spread across different areas of the home, ideally in quiet, low-traffic locations, to avoid conflicts during their use. Regular cleaning of the litter boxes is indispensable to maintain a hygienic and pleasant environment.

Finally, enriching the environment with play areas is vital for keeping the cats occupied and reducing frustration that could easily be redirected toward another cat. Interactive toys, food puzzles, and scratching posts placed throughout the home allow cats to expend their energy constructively. These setups help maintain emotional balance and prevent undesirable behaviors related to cohabitation.

If signs of frustration or aggression persist despite these measures, it is strongly recommended to schedule a behavior consultation to identify the root of the problem before one of the cats transitions into a pathological state.

Introducing a New Cat to One or More Existing Cat

Introducing a new cat into a household with one or more existing cats can be challenging due to the territorial and often solitary nature of the species. Cats are highly attached to their environment and routine, making the arrival of a new companion a potential source of stress and conflict if not managed carefully. Observing the challenges faced when introducing a new feline into a zoo enclosure provides valuable insights on this topic. Wild felines are strictly territorial, and their introduction to a new environment or conspecifics requires an extremely gradual approach.

Zookeepers typically employ phased introduction methods, often using pheromones like Feliway® to ease tensions and encourage peaceful interactions. This process can take weeks or even months, as each step must be meticulously planned to avoid potentially fatal conflicts. Initially, the animals are only allowed to detect each other's scent, followed by visual exposure, and finally, a controlled attempt to bring them together, often with keepers armed with anesthetic syringes to neutralize the animals if the introduction goes poorly (Gaultier et al. 2005).

Domestic cats, while less strictly territorial than their wild counterparts, retain similar instincts. The distinction between introducing a kitten and introducing an adult cat is critical. Kittens, especially when introduced to other kittens, generally adapt more easily due to their flexible and curious nature. Their ability to integrate into a new environment is much quicker, and they are often accepted more readily by adult cats because of their status as "noncompetitive" in terms of phero-monal signaling.

Conversely, introducing an adult cat into a home that already houses one or more adult cats requires increased attention and patience. As noted in ▶ Chap. 4, once they reach puberty, cats struggle to integrate into an already established group, even in outdoor living conditions (Crowell-Davis et al. 2004).

Adult cats are more territorial and less receptive to change. Moreover, during temporary separations of two adult cats, such as vacations or hospitalization, it is also recommended to reintroduce the cats gradually upon their return. This approach allows the gradual re-establishment of territorial markings and reduces the risk of fighting.

Progressive reintroduction involves initially separating the cats and then reintroducing them in a controlled manner, gradually increasing the duration of direct contact.

- *Isolate the new cat* in a separate room equipped with its own resources (food, water, litter box, toys). This allows the newcomer to acclimate to the environment without feeling threatened.
- After a few hours or days, depending on the reactions of each cat, *exchange objects* between the two areas, such as blankets or toys, so each cat can get used to the other's scent. Advising the use of synthetic pheromones in a diffuser can facilitate this process.
- Next, allow the cats to *see each other through a secure barrier,* such as a baby gate or a mesh screen. Observe their reactions and note any curiosity or aggression. Using pheromones like *Feliway Friends®* can help limit initial conflicts (DePorter et al. 2019).
- Once the cats seem comfortable with the scent exchanges and visual introductions, *organize supervised meetings.* Ensure that each cat has access to multiple hiding spots and escape routes if needed.
- Gradually *increase the duration of supervised interactions,* carefully monitoring for signs of fear or aggression. Reward calm and positive behaviors with treats or gentle petting.

It is crucial not to rush the process. Each cat adapts at its own pace, and forcing interactions can lead to lasting conflicts. In the case of a temporary separation between two adult cats (e.g., due to hospitalization), it is advisable to reintroduce the cats gradually, as if one were a new arrival. This helps reestablish territorial markings and reduces the risk of fights.

Finally, for many cats, cohabitation may never be perfectly harmonious, and hissing or growling may occur regularly. In such cases, it is important to verify that the relationship is not asymmetrical to the point where one cat is being mistreated by another. It is also crucial to ensure that these conflicts are limited to vocal intimidations and do not escalate into repeated fights or injuries. If conflicts persist, a behavioral consultation should be scheduled to stabilize the situation, as ongoing tension could push one of the cats into a pathological state.

Introducing a Puppy, a Dog or a Baby to a Cat

Introducing a puppy or an adult dog into a household where a cat already resides is generally easier than introducing another cat, but it remains a delicate process requiring careful management. This process can be facilitated by following specific steps and respecting the pace of each animal.

It is essential to prepare the environment before the dog's arrival. Ensure that the cat has secure areas where it can retreat, such as elevated hideouts or separate rooms inaccessible to the dog. Installing barriers to limit the dog's access to certain parts of the home, such as upstairs areas, or using a cat flap leading to a dog-free room, can create a safe retreat space for the cat.

Begin by isolating the dog in a separate area with its own resources to allow the cat to acclimate to the dog's scent without direct confrontation. Then, let the animals see each other through a secure barrier, such as a baby gate, while closely observing their reactions. Initial direct meetings should be supervised, with the dog kept on a leash to allow the cat the option to flee and retreat to a secure space.

Once both animals show signs of calmness or indifference through the barrier, the barrier can be gradually removed under supervision. Using a muzzle for the dog might be an option to prevent bites, but it's also essential to monitor the cat, which may display aggressive behavior and scratch the dog. Trimming the cat's claws or using soft claw covers can help reduce potential harm during initial interactions.

Gradually, the duration of supervised interactions should be increased, rewarding calm and positive behaviors from both animals with treats and gentle petting. Patience is crucial, as each animal adapts at its own pace. That said, if progress stalls beyond 2 or 3 weeks, it is advisable to schedule a consultation to check whether either animal might be exhibiting pathological behavior.

Finally, if one of the animals shows signs of intense fear, such as a cat retreating to a specific area of the house to avoid the dog, prompt intervention is necessary to prevent more severe issues from developing, such as anxiety or depressive disorders.

29.2.2 Respectful Communication

To establish a harmonious relationship with a cat, it is crucial to build a foundation of trust in the interactions between the cat and its owner. This involves fostering communication that respects feline behavioral codes. Achieving this requires first understanding the cat's nature and its methods of communication.

29.2.2.1 Understanding the Cat's Reaction

Cats are particularly sensitive to fear and trauma, a sensitivity rooted in their nature as small-sized animals often vulnerable to predation. Their survival instincts drive them to remain constantly alert to potential threats, which can heighten their reactivity to stressful or unfamiliar situations. As a nonsocial species, cats rely solely on themselves for protection. This has led to an increased sensitivity to environmental stimuli as a survival mechanism.

The domestication of cats occurred passively; they adapted to living alongside humans without undergoing active selective breeding. While this process has somewhat reduced their generalized fear responses, they have retained aspects of their primitive instincts. This is evident in their rapid and intense reactions to situations perceived as dangerous.

The cat's nervous system is particularly responsive to external stimuli, with a highly sensitive fear axis. This means that negative experiences can quickly lead to sensitization, making cats more prone to developing phobias. Their tolerance for inappropriate communication from their owners is much more limited than that of dogs, who are often more forgiving of human clumsiness.

▪ Evaluating the Emotional State of the Cat

Owners, who are often not trained to interpret their cat's signals, may not realize that certain behaviors are indicators of fear or stress. For example, a twitching tail, dilated pupils, or flattened ears signal that the cat is in a state of alert or distress. An untrained owner may misinterpret these signals, which can lead to misunderstandings and exacerbate the cat's emotional state.

The postures and key signals that owners need to know and learn to observe include:

- **Tail**: a rapidly flicking tail is often a sign of frustration or annoyance. In contrast, a tail held high with the tip curved often indicates a friendly greeting.
- **Pupils**: dilated pupils can signal fear, excitement, or aggression. They may also indicate that the cat is about to attack or flee.
- **Ears**: ears turned forward indicate a relaxed and curious cat, while ears flattened backward show fear or aggression.
- **Vocalizations**: high-pitched, short meows often indicate a greeting or a request, while low-pitched meows can indicate irritation. Purring, although often associated with contentment, can also occur in response to stress or pain.

It is essential to explain to owners the importance of taking the time to observe their cat in order to detect these signals. By learning to read and respect these signs, they can avoid worsening a stressful situation for their pet and, on the contrary, contribute to its well-being. Poor communication, on the other hand, will only exacerbate misunderstandings. It is also possible to teach them to recognize their cat's calming signals, enabling them to identify increasingly subtle changes in emotional states.

29.2.2.2 Owner's Communication

Verbal Communication

As detailed in the first part of this work (see ▶ Chap. 1 ▶ Sect. 1.3.2.2), communication with a cat must follow specific rules in order to be understood. The verbal message must be clear, congruent, and consistent (Mouzon et al. 2022; Saito and Shinozuka 2013).

Requests directed to the cat should be brief, and, if possible, consist of a single word. Repeating them, especially louder, is generally not helpful. In fact, when the cat does not respond to a request, it is much more likely because it is not sufficiently motivated or focused, rather than because it hasn't heard the command. The agitation of the owner tends to worsen these problems: shouting only reduces the cat's motivation, and increasing verbal cues only hinders its understanding. On the contrary, to increase comprehension, requests should be directed to the cat as calmly as possible, keeping the volume low, ideally to a whisper, and often accompanied by gestural guidance or visual signals (see next paragraph).

We have already emphasized the non-social nature of the cat, but one consequence of this is that there is no hierarchy within this species. It is still common to hear owners in consultation claim that one of their cats is dominant or submissive. In reality, this does not exist: in a given situation, there is no established order. It is the individual motivation that prevails, and the same situation can lead to different outcomes.

Clinical Illustration

Mr. X has the habit of giving milk to his cats when he makes coffee. They love it and often three or four of them come. When he fills two bowls with milk, it is not always the same cats that serve themselves first. There are no fights, it's more like "first come, first served." Sometimes they jostle slightly to access the milk, and regularly, while one cat drinks, another pushes it to take its place. We can't say which cat drinks more or first, because there is no established social rule among them.

Therefore, reprimanding a cat involves interrupting the undesirable behavior, often through disruption, and then redirecting to an alternative behavior to calm the animal (see the box "Disruption and Redirection"). In all cases, anything painful or frightening for the cat must be avoided as it represents positive punishment, which can be anxiety-inducing and may lead to aggressive reactions.

Important Point of View

Disruption and Redirection

The word "no" should only be used to indicate to the cat that its behavior *at the moment the word is spoken* is unacceptable. Therefore, this word should never be used *retroactively*.

Similarly, the word "no" cannot be used to indicate to the cat what behavior is expected of it at the moment the word is spoken, so it should never be used *alone*. On the contrary, it should be followed by a suggestion to do something else (for example, throwing a toy or gently lifting the cat off the table to indicate that it should get down), offering the cat an acceptable *alternative* that it might otherwise struggle to guess. The execution of this alternative behavior should be *rewarded* in order to increase the likelihood of this desired behavior appearing in the future and in the same context, rather than the initial undesirable behavior. This approach reinforces the appropriate response positively and gradually replaces the problematic behavior. It can be done by speaking gently: "Yes, you see, you play with your toys, or, you see the table, you can't climb on it, but you can stay on the floor."

The use of "no," which aims to redirect the cat's undesirable behaviors to desirable ones, is a *disruption*, not a *punishment*. It allows for an acceptable *counterconditioning* process and responds to the problematic behavior without altering the relationship or the emotional well-being of the cat.

Nonverbal Communication

Nonverbal communication is far more important for cats than verbal communication (see ► Chap. 1, ► Sect. 1.3.2.2). In fact, it has been shown that a vocal request is better understood by the cat when accompanied by a visual signal (De Mouzon and Leboucher 2023). In the same section of ► Chap. 1, we already detailed the importance of our body positioning relative to the space around the cat, that is, the concept of proxemics and its significance in this species (Čabraja 2021). The flight and aggression distance surrounding a cat is variable depending on the individual, and for the same individual, it changes over time based on experiences, especially its socialization, and its emotional and physiological states (see ► Fig. 1.15). For example, for most cats, the head is not part of the aggression zone, and touching this area is tolerated even by veterinarians. On the other hand, touching the belly is often never tolerated, as it is part of the cat's permanent aggression zone, or it may be tolerated but not by the veterinarian or during a stressful situation.

The respective meanings of our various gestures, postures, and movements for cats must be taken into account to communicate with them in a coherent and reassuring manner in different situations. To avoid aggression, it is recommended to establish visual contact first, followed by olfactory contact, before tactile contact, allowing the cat to sniff a finger, for example. It is also important to approach from a non-frontal position, use calming signals such as blinking or averting one's gaze, give the cat time to take in smells, and pet it in its preferred contact zones (see ◘ Fig. 29.9). Finally, a sustained contact is preferred over multiple discontinuous touches.

29.2.3 Learning and Useful Exercises for All Cats

We have seen in ► Sect. 29.1.3 how to teach a new behavior to a cat or, conversely, how to reduce the likelihood of an undesired behavior appearing.

A cat is not a dog, and it is not about teaching all cats to walk on a leash or perform sit and down commands. However, some types of learning are very useful for all cats and are often overlooked, particularly because of the widespread belief that "a cat can't be trained."

However, being examined, touched, allowing claw trimming, being transported in a carrier, or even accepting medication intake are essential lessons that significantly improve the well-being of all cats. Additionally, having cats that have been accustomed to these experiences also enhances the well-being of the veterinary team (see ► Chap. 30, ► Sect. 30.2.2).

◘ **Fig. 29.9** Preferred contact zones in cats for interactions

Having a cat with whom the owner can play without the risk of injury, but also a cat capable of not hiding under a bed at the sight of an unfamiliar person, is also an important learning goal that enhances the quality of life for both the cat and its owner.

Outside of this, there are *optional learning* exercises that, while contributing to improving the human-cat relationship, are not essential for living with a cat. For example, teaching a cat to fetch a ball or perform a roll. These exercises often allow the owner and cat to spend enjoyable moments together but can be achieved simply by following the steps outlined in ▶ Sect. 29.1.3. We will not dwell on these in this section.

Anything medical-related will be discussed in detail in ▶ Chap. 30, from the caregiver's perspective. However, here, let us focus on the exercises that should be recommended for all cats during vaccination consultations. These exercises are genuinely useful for the cat's lifelong well-being, help prevent the development of behavioral issues, and allow for quick intervention if the expected progress is not made.

29.2.3.1 Socialization

▪ General Principles

As we have mentioned several times, kittens that have not had any contact with humans before 2 months old have almost no chance of being socialized with humans later in life (Ahola et al. 2017; Lowe and Bradshaw 2002; Martin 1986) due to synaptic pruning that occurs at the end of the socialization period. Furthermore, poor-quality socialization predisposes the cat to behavioral issues later on (Mikkola et al. 2022).

Additionally, it is important to distinguish between familiarization and socialization of a cat with humans or dogs: the former refers to the cat's ability to interact with a very limited number of individuals, known as familiar ones, while the latter refers to the cat's ability to interact with all individuals, including strangers, of a species in general. A cat living with a dog or child will thus be familiarized with these particular individuals, but may not necessarily be socialized with unknown individuals of the same species.

From an early age, it is also important to remember that a critical aspect of a cat's development concerns the absence of intense fears related to daily sounds and humans. It is essential to gradually introduce the cat to different stimuli that will be part of its daily environment, without forcing it. A well-balanced companion cat is not supposed to have exaggerated fear reactions toward humans, household machinery noises, or the neighbor's doors opening and closing. A cat does not have to interact with every human who visits its home, but it should not exhibit exaggerated flight or aggression toward these strangers. These reactions can have negative effects primarily on the cat itself, generating stress and anxiety. It is therefore crucial to address these disproportionate emotional responses to preserve its well-being (see remediation).

▪ Prevention

As explained in ▶ Chap. 4, the *socialization period* during which a kitten is most capable of learning to communicate with its own species and other species, including humans, spans from 2 weeks to 2 months of age. It continues during the juvenile period, up to 6 months, with a high capacity for habituation, as long as the experiences are positive. Therefore, at the usual time of adoption, between 2 to 4 months, socialization can and should be continued by the new owners.

This work is particularly important if the kitten is expected to interact in the future with categories of humans or other animals it was not exposed to at the breeder's. For example, it is necessary to socialize the kitten with humans who present specific characteristics (age, gender, disabilities...), or with animals from species it has never encountered in its original environment, such as dogs, chickens, or others. This work must be done with care, ensuring that familiarization and socialization are not confused. Regularly introducing the kitten to the same individuals of other species does not guarantee that the animal will generalize its behavior to other members of the same species. For example, some cats living with dogs or other cats at home may still exhibit aggressive behavior toward other cats or dogs encountered outside. It is therefore crucial to understand that socialization involves a broader, more generalized adaptation, while familiarization remains limited to the specific individuals the cat already knows.

Finally, this work must be carried out while respecting the principles of habituation (see ▶ Sect. 29.1.1.2, "Habituation and desensitization"), meaning that the kitten must never be placed in an intensely uncomfortable situation. Furthermore, kitten owners must ensure that when humans interact with their pet, the contact is never forced or intrusive. In households with children, it is essential to teach them to respect the cat and understand its boundaries. For example, if the cat decides to leave when they want to pet it, they should not chase it or physically block its escape. However, it should be noted that many cats tend to be more tolerant with children than with adults.

Moreover, a cat's interspecies socialization should be maintained throughout its life, especially since certain periods or events can alter it. On the one hand, the juvenile period (from 2 months to puberty) is governed by fear of unknown stimuli, and the pubertal period marks the beginning of pheromone marking on objects. Visits to the veterinarian will therefore become more difficult, and if pain is associated with the visit, sensitization may occur. To counter this, regular positive and pleasant interactions with strangers are crucial for maintaining this socialization.

▪ Remediation

Despite these precautions, and due to poor early developmental conditions or traumatic experiences, an adult cat may exhibit imperfect sociability. This is the case in social phobias (either intra- or interspecific), but also in sensory deprivation syndrome and relational disorders (either intra- or interspecific).

Behavioral therapies, usually combined with medication, must then be implemented to desensitize the cat (in the first case) or to socialize it (in the second case) with the individuals toward whom it has inappropriate reactions. In both cases, the safety of the individuals to whom the fearful cat will be exposed must be ensured. In most situations, it will be necessary to use exceptionally high-value rewards to enhance the cat's interest in these people or pets, who initially trigger significant fear (see ▶ Sect. 29.3.1).

29.2.3.2 Self-Controls

▪ General Principles

As we discussed in the behavioral development of the cat in ▶ Chap. 4, a pet cat must quickly be able to control its biting, scratching, and locomotion. All of these behaviors are acquired progressively as the prefrontal cortex matures, from the socialization period until the end of puberty. The goal that should serve as a benchmark for the practitioner is that the cat should no longer cause injury with its teeth or claws during excitement or play by the beginning of puberty.

We also emphasized that cats most at risk of developing this type of problem related to delayed acquisition of self-control are those adopted too early, that is, before 14 weeks of age, particularly hand-raised kittens, who cannot benefit from early learning through interaction with an adult cat. For hand-raised kittens, the absence of siblings, even in the presence of adult cats, significantly reduces the acquisition of communication and controlled play skills (Guyot et al. 1980). These kittens should be identified so that emphasis can be placed on the exercises described below for their owners, with the aim of minimizing the risk of transitioning to an impulsive or compulsive pathological state. If a pathological state develops, these exercises will be part of the therapeutic management.

▪ Prevention

The principle of prevention involves playing with the kitten from the moment of adoption, using negative punishment as soon as the kitten gets too excited, that is, when it starts to nibble or extend its claws. The interaction should be stopped as soon as the behavior becomes overwhelming, and the game should be interrupted. The owner can use a sound signal like "no" or "ouch" to let the kitten know that its action is too intense. The goal is for the kitten to stop its behavior at the mere sound of this verbal signal.

It is important to explain to owners that movement is a powerful trigger of excitement in cats, as it also activates their predatory sequence. Therefore, when they stop playing, complete stillness must follow. Shouting, quick hand movements, or any other form of rapid movement should be avoided, which can be challenging for young children or people who are afraid of cats.

If the kitten does not seem to understand and continues to play aggressively by scratching or biting, owners can isolate it in a quiet space for a few minutes, such as a bedroom, bathroom, or its transport crate. The purpose of this isolation is to allow the kitten, whose cortex is still immature, to calm down gradually and experience a return to calmness. Owners should quickly see the effects of this exercise, and by 6 months, they should be able to play with their cat without risking these outbursts. Moreover, when they simply stop the play, the kitten should be able to do something else and manage its excitement.

▪ Remediation

Despite these precautions and initial advice, some kittens struggle to execute this exercise, either because the owners' movements are too stimulating for the kitten or because they fail to stop communication at the right moment, or because the kitten has vulnerabilities in the area of self-control.

In such cases, the kitten should be referred for consultation. It is likely experiencing a delay in acquiring self-control. For the most severe cases, the described technique will not work because the kitten's internal motivation is too high, and it may continue its aggressive play behavior on the owner's legs or hands, even as they attempt to remain stoic despite the injuries. In all cases, if the owners report vulnerable injuries and have marks on their arms or legs while trying to implement these tips, and the kitten is still under 6 months old, the severity of the situation requires full intervention.

29.2.3.3 Habituation to Routine of Veterinary Care, Transport Carrier and Accessories

In the essential learnings for all cats, it is crucial that the owner can handle their cat enough to observe injuries or groom it when its fur becomes tangled. Of course, a kitten is often more willing to be handled, and it is always possible to force things when the kitten is around 3 months old. However, such constraints should be avoided, as the cat might become sensitized and soon become untouchable, anticipating negative experiences during future handling. Additionally, if handling is only addressed when necessary, the cat may react poorly to the novelty, making care difficult without coercion.

Therefore, it is essential to begin habituating the kitten to these manipulations from an early age, and this should apply to all kittens. In this section, we will review the accessories and exercises useful for providing routine care.

▪ Feeding Ritual

Creating a *feeding ritual* is a simple exercise that serves almost as a prerequisite for all the following exercises, as it will significantly facilitate them. Like all the exercises presented in this section, it is very easy to establish for all cats in just a few days. However, it cannot be implemented at the last minute, the day the owner realizes that they need to administer medication to their cat and wants to hide it in the food. If the treatment is urgent, there will be a lack of time, and the owner or practitioner will be forced to administer it directly, which could make the cat more reluctant to future care.

Establishing such a ritual also has a great advantage for outdoor cats, as, if timed correctly, it can become an effective way to bring them inside for the night.

To implement an effective feeding ritual for the cat, it is recommended to encourage owners to introduce a small daily portion of wet food in addition to the usual dry food. This practice creates a positive association with a specific time of day, making it easier to administer future medication treatments. The choice of food is crucial: it should not only be appetizing for the cat, but also be compatible with the administration of medications, such as pills, which can be hidden within it.

It is recommended to offer small portions, such as the equivalent of a teaspoon, to ensure that the cat consumes the entire portion, which is particularly important when it contains medication. Owners should be encouraged to introduce this feeding habit well before the administration of medications becomes necessary, so that the cat associates the experience with a daily pleasure, rather than with a chore.

The practitioner can also emphasize another benefit of this routine: it makes it easier to bring indoor-outdoor cats inside by attracting them with this reward. Many owners want to know how to recall their cat effectively when they are outside, and such a ritual fully meets their needs. Additionally, for particularly cautious or delicate cats, it is essential to remind owners that the implementation of the ritual must be gradual and consistent, avoiding any attempts to hide medications during the first sessions. Once the ritual is well established and the cat is accustomed to receiving its treat, the addition of medications, such as generally appetizing dewormers, can be done without disrupting the animal.

It is also relevant to advise owners to choose a convenient and easily accessible treat, considering the longevity of this routine, which may extend over several years. By establishing this ritual carefully and in advance, they significantly increase the chances of a successful, stress-free medication administration for the cat, thus enhancing therapeutic compliance and the quality of at-home care.

▪ Habituation to the Transport Carrier

This exercise is crucial, not only to allow owners to easily transport their cat if they go on vacation and want to bring their pet along, but also to ensure minimal stress during veterinary visits.

A study conducted by Pratsch et al. (2018) highlights the significant impact of gradual training on reducing stress in cats during travel and veterinary visits. Two groups of cats were observed in this research: the first group received specific training for using the transport carrier, while the second, the control group, received no training.

The cats in the first group were gradually accustomed to spontaneously entering their transport carrier over a 6-week period, with regular sessions of 15 min, spaced every other day. The results were striking: the cats that were trained showed significantly lower levels of stress compared to those in the untrained group. Stress was measured using the "Cat Stress Score," an indicator based on observing the animal's behavioral responses. The cats trained to the carrier were more willing to accept food rewards during the veterinary visit and displayed fewer exaggerated fear postures. Additionally, the veterinary examination was much faster for these cats, as they allowed handling more easily. Another sign of reduced anxiety: the temperature variations between the departure and arrival were much less pronounced.

These marked differences between the two groups highlight the importance of such training to calm the animal during veterinary visits. It then becomes evident to owners that investing 5–6 h, spread over 1–2 months, to help their cat comfortably enter the transport carrier is a significant effort but one that is incredibly beneficial. It is not about waiting until the day of the appointment to worry about this but rather anticipating the moment to avoid unnecessary stress for the animal.

First, let us clarify that if the transport carrier has already been used for unpleasant trips for the cat, it must be thoroughly cleaned to eliminate any stress-related pheromones the cat may have left inside. On the other hand, Feliway™ spray in the carrier is specifically designed to provide a calming environment for the cat inside the carrier (Shu and Gu 2022).

Second important point: there is an impressive variety of transport carriers, which can make it difficult for the owner to choose. However, some criteria can help guide the decision: it is better to opt for a sturdy carrier rather than a bag, which easily deforms when pressed against a rigid surface, making the transport less comfortable for the cat. It is also recommended to choose a carrier with a removable top, making it easier to access the cat if it does not spontaneously exit, allowing the examination to be performed while keeping the animal in a reassuring environment. The carrier should preferably be opaque for the cat. A few openings should allow the cat to see outside, but it should not be easily seen. The carrier should be designed like a mashrabiya. Finally, the size of the carrier should be large enough for the cat, once fully grown, to stand and turn freely inside. The bottom can be lined with fabric or absorbent cellulose to avoid the cat sitting in its urine in case of an accident.

To habituate a cat to the transport carrier, the following progressive protocol is recommended:

- **Habituation to the "transport carrier" object**: place the carrier in a familiar location for the cat, with the door open and comfortable cushions inside. If the cat shows reluctance, remove the top of the carrier to make exploration easier. The carrier should remain accessible at all times, allowing the cat to familiarize itself at its own pace.
- **Encouragement to enter the carrier**: use tasty treats (or the food ritual mentioned in the previous section) or toys to encourage the cat to enter the carrier. Rewards should first be placed at the entrance, and then gradually placed inside the carrier, with positive reinforcement at each successful step.
- **Gradual closure of the carrier**: once the cat enters the carrier regularly, start closing the door for short periods, gradually increasing the duration the door remains closed. Treats can be offered while the door is closed to reinforce the positive experience.
- **Acclimatization to transport**: after the cat is comfortable with the closed carrier, gently begin moving the carrier around the house, and then gradually introduce it to the car for short trips, always offering rewards.

In the case of a difficult veterinary visit or transport that seems to have been a negative experience for the cat, it may be necessary to reintroduce regular training sessions to compensate for any negative experiences. By adopting this approach, we can not only improve the cat's experience during clinic visits but also strengthen the trust-based relationship between the animal and its owner, while facilitating long-term healthcare. Adding pheromones to the carrier also helps ease the journey by reducing stress during the transport phase (Frank et al. 2010; Shu and Gu 2022).

▪ Habituation to Accessories

Accessories refer to all the tools used on the cat: nail clippers, thermometers, ear liquid bottles, brushes, syringes, stethoscopes, etc.

If these veterinary tools are shown to a kitten a few weeks old, it will not be long before it stops being afraid of them. It has no prior negative associations and quickly realizes that the object is not dangerous. However, if you wait until it has already developed a fear due to a bad experience, everything will become more complicated, even difficult to resolve.

Habituation to accessories requires rewarding the cat each time it approaches, sniffs, and then touches the accessory. The simplest way is to present the object in your hand or, if the cat seems very scared, place it on the ground. For these exercises, it is important to use highly appetitive treats to create a positive association between unfamiliar accessories and a high-quality reward.

Harness and Leash

The *harness* and *leash* allow a cat to explore the outside world while remaining under the control of its owner, thereby reducing risks associated with complete freedom, such as traffic accidents, fights with other animals, or getting lost. Additionally, these outings provide valuable mental and physical stimulation, helping to prevent boredom and obesity in indoor cats. For cats living in apartments or urban environments, where access to the outdoors is limited or dangerous, the leash can be an excellent solution for safely enriching their environment.

When an owner wants to use a leash and harness for their cat, it is important to emphasize that they will likely need to maintain it for the cat's entire life if the cat enjoys the walks; otherwise, it may lead to great frustration. In such cases, it may be better not to start.

First and foremost, a properly fitting harness is one from which the cat cannot escape. There are many models specially designed for cats, which can be tested to ensure that they are comfortable while being snug enough to prevent the cat from getting out.

To accustom the cat to the harness, it is recommended to begin by familiarizing the cat with the harness by leaving it in an area the cat frequents so it can sniff and get used to it. After this exploration phase, you can encourage the cat to interact voluntarily with the harness, for example, by placing a treat near the harness and gradually moving the treat further away so that the cat is encouraged to approach it.

Once the cat seems comfortable with the presence of the harness, it is important to handle it without attempting to put it on immediately. The owner can handle the straps, buckles, and other fastening mechanisms of the accessory near the cat so it can get used to the sounds and movements associated with them. Then, the harness can be gently placed on the cat, without securing it, for short periods, gradually increasing the duration (5, then 10, then 15 s), rewarding the cat with treats or petting afterward.

When the cat is comfortable with the harness, it is time to introduce the leash. It is helpful to let the cat wander freely inside the house with the leash attached, without intervening, so it can get used to the sensation of the leash without feeling any pressure. The owner can then start holding the leash and gently guiding the cat, while continuing to use rewards to encourage the desired behavior.

Once the cat is ready, short walks can be introduced. It is recommended that owners start in a calm and secure environment before progressing to more stimulating locations. Using a retractable leash can be useful in a larger space, allowing the cat to explore while remaining safe.

The length of this learning process should be emphasized. Just like walking a dog on a leash, it takes many attempts before success is achieved. Rushing the process can lead to aversion, making future leash walks impossible.

Brushes

Brushing is an essential aspect of maintaining coat health, especially for long-haired cats, to prevent the formation of hairballs and dermatological problems. As veterinarians, it is crucial to educate owners about the importance of this care while providing practical advice on how to help their cat accept, and even enjoy, the brushing process.

Before starting the first brushing sessions, it is helpful to first introduce the brush, as described with all accessories: the goal is not necessarily to brush the animal right away but to associate the brush with a pleasant experience, during which tasty rewards are given.

It is important to highlight that for activities that require prolonged durations, such as brushing, using a licking mat smeared with pate is particularly useful. This device allows the owner to use both hands for the task while offering the cat a continuous reward throughout the grooming session, making the experience more enjoyable and less stressful for the animal.

Once the brush is accepted by the cat, short sessions should be carried out, starting with areas that the cat tolerates the best, such as the back. For particularly sensitive cats, it may be useful to have a few "empty" sessions, meaning handling the brush near the cat without actually brushing, allowing the cat to get used to the tool without pressure. It is also important to emphasize gentleness, especially on sensitive areas such as the belly and paws, where cats may be more reluctant.

For long-haired cats, it is essential to highlight the importance of more frequent grooming, while reassuring owners that, although this task may seem demanding, it can become an enjoyable routine for the cat if introduced properly from a young age. It is also worth noting that many cats eventually enjoy these grooming sessions, especially if the early experiences have been positive.

In cases of persistent difficulties where the cat systematically refuses grooming despite an appropriate approach, it is advisable to take the time for a behavioral consultation. This type of behavior may indicate underlying issues, whether physical (pain) or behavioral, and it is important to address them as soon as possible to prevent further complications. Additionally, it is highly likely that a cat who refuses to be groomed will present numerous other behaviors that will cause problems for future care.

Nail-Clippers, Blockers, and Accessories

Without going into detail, the procedure for familiarizing the cat with all the care items it is likely to encounter in its life, such as nail clippers, thermometers, syringes, and cleaning bottles, should follow the same protocols outlined earlier. This may seem like a tedious task, but doing it with a kitten typically takes one or two sessions per item and will be a great help throughout the cat's life.

Habituation to Healthcare

Tolerance to Handling

Habituation to healthcare in cats is a delicate process that requires a gradual and respectful approach to the animal's emotional state. To ensure easy handling during daily care or veterinary exams, it is crucial to follow a series of steps aimed at familiarizing the cat with contact and manipulation of various parts of its body, while ensuring that it shows no signs of fear or defense.

The use of *licking mats* is highly relevant in this type of training. Given the challenges posed by the cat's territorial nature, which makes trips to the veterinary clinic more difficult, especially after puberty, familiarizing the cat with being handled at home using a licking mat will allow the owner to bring the cat to the consultation with greater peace of mind. The licking mat helps familiarize the cat with the actions and treatments that will be carried out, while also calming it. Moreover, the cat will be more likely to consume the usual food used during these sessions, which will facilitate the interaction and reduce stress. The goal is to proceed step by step, starting with brief, noninvasive contact, and then gradually increasing the duration and intensity of these interactions. For example, handling a paw might begin with simply touching it, followed by holding it, and then gently bending it. It is often helpful to pair a soothing verbal command, such as "calm down," to reinforce the positive association between the handling and the desired behavior.

Daily petting sessions are an effective method for starting the habituation process. The veterinarian should recommend that owners begin with areas the cat enjoys, such as the head or neck, before gradually moving to the paws, belly, or other more sensitive areas. It is crucial that the cat is rewarded with treats or petting at each successful step to reinforce its comfort and tolerance. Additionally, the cat should never be forced; it must be able to withdraw at any time for these sessions to remain positive and stress-free.

If a cat shows persistent signs of refusal or discomfort despite a gradual and respectful approach, it is advisable to suggest that the owners consult a veterinary behaviorist. The behaviorist can assess if adjustments are needed in the way the handling is carried out or if the animal has underlying anxiety that requires specific intervention. Furthermore, it is recommended to encourage owners to replicate these exercises, if possible, at the veterinary clinic to reinforce the habituation process. When the clinic setup allows, these sessions can also be conducted by veterinary assistants.

Claw Trimming

When it comes to managing a cat's claws, it is important to note that most outdoor cats rarely require claw trimming. Their daily activities, such as climbing, scratching rough surfaces, or hunting, naturally help wear down their claws, keeping them at an appropriate length without human intervention. However, as cats age, especially those that go outside less or not at all, their claws may grow more quickly, requiring regular trimming to prevent complications such as claws curving or growing into the paw pad, which can be painful and prone to infection. For indoor cats, it is common to observe that they partially maintain their claws by using scratching posts or other rough surfaces, such as carpets or furniture. These scratching behaviors are not only essential for claw maintenance but also for pheromone marking, particularly during vertical scratching. Nonetheless, regular trimming may still be necessary. Therefore, starting this habituation at a young age is beneficial for all cats.

Introducing this practice early makes it easier and less stressful for both the cat and the owner in the long term. It is recommended to start at a young age. By regularly and gently handling the kitten's paws, without necessarily trimming the claws initially, the owner can help the animal get used to this contact. Early familiarization is essential for making claw trimming a routine the cat will accept.

The steps should be gradual. The owner can first touch and gently manipulate the paws, using rewards like treats, *a licking mat,* or petting to make the session pleasant for the cat. The cat should never be forced or restrained too tightly, nor should the claws be trimmed under strong restraint. Once the kitten is comfortable with this handling, it is advisable to gently press on the pads to extend the claws, continuing to positively reinforce this step.

Positive reinforcement is crucial to associate claw trimming with a pleasant experience. It is also important to remind owners not to rush the steps and to respect their cat's pace of adaptation. If the cat shows signs of stress, it is better to take a break and try again later.

For claw trimming, it is essential to use a nail clipper that is suitable for the size of the cat's claws. Owners must be trained to trim only the tip of the claw, avoiding the pink area that contains blood vessels, to prevent any pain or bleeding. Initially, the clipper should be presented to the animal, followed by contact with its paw, and eventually, a claw can be trimmed.

Even in adulthood, it is important to maintain this routine. If a cat is reluctant, the claw trimming process can be reintroduced gradually using the same techniques as for a kitten. In more delicate situations, the veterinarian may intervene to demonstrate the correct technique or perform the trimming during routine visits. However, these interventions often require some level of restraint, as cats are usually less willing to cooperate. In such cases, using a clipnose clipper is preferred over full restraint. This tool causes less stress for the cat, making the procedure easier and providing a better experience for the animal (Moody et al. 2020; Nuti et al. 2016; Pozza et al. 2008).

It is important to note that declawing is a painful procedure and is banned in many countries, including France. A less invasive alternative is the use of "Soft Paws®". These are plastic caps that are glued to the cat's claws. They usually fall off within 3 to 6 weeks, after which the claws are trimmed and the process can be repeated as necessary. In practice, these caps are mainly used to prevent damage caused by scratching in apartments, and in this case, they are applied continuously. However, they can also be used on a more occasional basis, such as to prevent a cat from scratching a wound and worsening it, or in the case of a kitten that, due to poor control over its scratching, may cause serious injuries. By applying these protective caps, the cat can no longer climb with its claws and must find other ways to adapt instead of scratching. The owner, no longer being scratched during interactions, finds it much easier to stay calm when moving the kitten and asking it to find another way of doing things. This can be very soothing for both the cat and its companions, and usually, a single application is enough for the cat to learn to find alternative solutions and gain better control of its scratching. Applying these caps requires the same tolerance for handling as trimming the claws, so it is crucial to habituate the cat to such handling in order to carry it out properly.

Summary of ▸ Sect. 29.2

The organization of a cat's environment is a key component of its psychological well-being and the prevention of behavioral disorders. Unlike dogs, cats have specific territorial and spatiotemporal needs that must be respected. Whether indoors or outdoors, a cat's environment should be arranged to provide distinct zones for each important activity: feeding, resting, playing, and eliminating.

An enriched environment is particularly crucial for indoor cats, as it compensates for the natural stimuli that the outside world would otherwise provide. Adding elements such as perches, hiding spots, and scratching areas helps create a stimulating and secure living space for the cat. However, even for cats with access to the outdoors, it is sometimes better to keep them indoors if the external environment presents sources of stress or danger.

Understanding and respecting the communication signals of cats are essential to establish a harmonious relationship between the animal and its owner. Cats communicate through subtle signals, such as tail movements, pupil dilation, and various body postures, which reflect their emotional state. A lack of understanding of these signals can lead to misunderstandings and stress for the cat. By carefully observing these indicators and adapting interactions accordingly, it is possible to foster coherent communication with the animal, thereby reducing undesirable behaviors related to misinterpretation of its needs and emotions.

The fundamental learnings for all cats, which are crucial tools for preventing behavioral issues, include socialization, acquiring self-control, and habituation to accessories and care. Exercises, though often underestimated, are essential for the cat's well-being and daily management. Socialization, initiated during the first few weeks of life, allows the cat to develop positive interactions with humans and other animals. Training in motor and emotional control helps the cat manage its impulses in various situations. Finally, gradual habituation to accessories and care, such as grooming or nail trimming, is crucial to avoid stress and facilitate necessary handling.

29.3 Behavioral Axis Therapies

In this section, we will review more specific therapies corresponding to disorders along a particular axis. We will study them successively, one behavioral axis after another. As we have stated, the state diagnosis helps determine the medication to be prescribed, while the nosological diagnosis, based on the disrupted axes, helps choose a behavioral therapy strategy.

For each axis and each disorder, we propose therapeutic actions. Of course, these are not strict formulas but suggestions that should be *tailored to each case*. This process of adaptation is what truly defines the quality of the therapist.

29.3.1 Disorders Related to a Dysregulation of the Fear Axis

29.3.1.1 General Principles

The therapy for a cat suffering from dysregulation of the fear axis generally involves implementing a *desensitization protocol* to phobic stimuli. This work is often complemented by a *counterconditioning protocol* aimed at associating these phobic stimuli with appetitive stimuli. Together, this is referred to as the desensitization-counterconditioning protocol or DS-CC protocol (see ▸ Chap. 30, ▸ Sect. 30.2.2.2).

This goal can only be achieved by teaching the cat's owners to closely observe its *communication signals* to assess in real time the intensity of its discomfort. This learning is necessary to find a balance between the systematic avoidance of phobic situations and the uncontrolled immersion of the cat in these situations.

It is important to remember that phobic or anxious cats trigger exaggerated fear responses in one or more contexts that should not be alarming (see ◘ Fig. 29.2). Without appropriate medical treatment to reduce the intensity of the animal's emotional reactions, rehabilitation is nearly impossible, as it does not allow the cat to progress as it should, keeping it in a state of heightened emotional suffering.

▪ Desensitization and Counterconditioning

The *desensitization protocol* involves exposing the cat to phobic situations in a way that takes it out of its comfort zone without ever bringing it into a state of excessive discomfort. Indeed, complete avoidance of these situations prevents learning. On the other hand, intense exposure to these situations, or immersion, may sensitize the animal further and lead it to anticipate and generalize to other contexts, thereby worsening its imbalance.

In practice, this means gradually increasing either the intensity of the stimuli presented or the duration of the cat's exposure to these stimuli, ensuring never to push the cat into a state of panic. The conditions that promote desensitization rather than sensitization are the same as those that favor habituation (◘ Table 29.1): *a stimulus of increasing intensity*, initially weak, should be presented to the cat. The presentation is initially brief but *frequent*. The situation should be open, meaning the cat must be able to distance itself if it wishes. For example, if the cat is phobic of the vacuum cleaner, one should start by exposing it to the object when it is stationary, before introducing the vacuum cleaner in motion and, later, with it turned on. Similarly, one would begin by showing the vacuum cleaner to the cat from several meters away and stationary before trying to bring the cat closer to it.

A desensitization protocol is, by definition, impossible to implement when the duration and intensity of phobic stimuli cannot be controlled, and it is often long and tedious. It is therefore a better strategy to prefer or combine it with a counterconditioning protocol.

The *counterconditioning protocol* consists of pairing the introduction of pleasant stimuli for the cat (play, treats, petting, etc.) with the appearance of stimuli it fears, in order to induce pleasure in the cat, creating an emotional state incompatible with fear.

Whether using a desensitization protocol or counterconditioning, the therapy requires *very fine observation skills* to determine whether it is possible to continue exposing the animal to distressing stimuli (◘ Fig. 29.10). These therapies are thus simple in theory but difficult in practice. They are long and tedious, but combining them with medication can accelerate the rehabilitation process.

▪ Protective Isolation and Open Situation

Since a cat with a dysregulation of the fear axis, when exposed to a phobic stimulus that is too intense or lasts too long, risks becoming more sensitized to it, it is essential to avoid situations that would be too emotionally complex for it to handle. It is important to explain to owners that the most stressful situations should be avoided, as they are counterproductive and dangerous. A frightened cat might injure itself or cause an accident if it tries to flee and may also attempt to bite or scratch to escape the source of its fear if it feels cornered. There will be other opportunities to help the cat progress in better conditions.

Negative reactions of flight and aggression can be avoided in two different but complementary ways. The first involves isolating the cat in particularly problematic contexts, such as the arrival of visitors for an animal suffering from an interspecific social phobia. The second involves always providing the cat with the option to escape such situations, for example, by allowing it to pass behind a barrier or a cat door to reach an area it considers secure. Indeed, the cat should never be blocked or forced to confront the object of its fear, as this could trigger an aggressive reaction. To make progress with its phobias, the cat must come

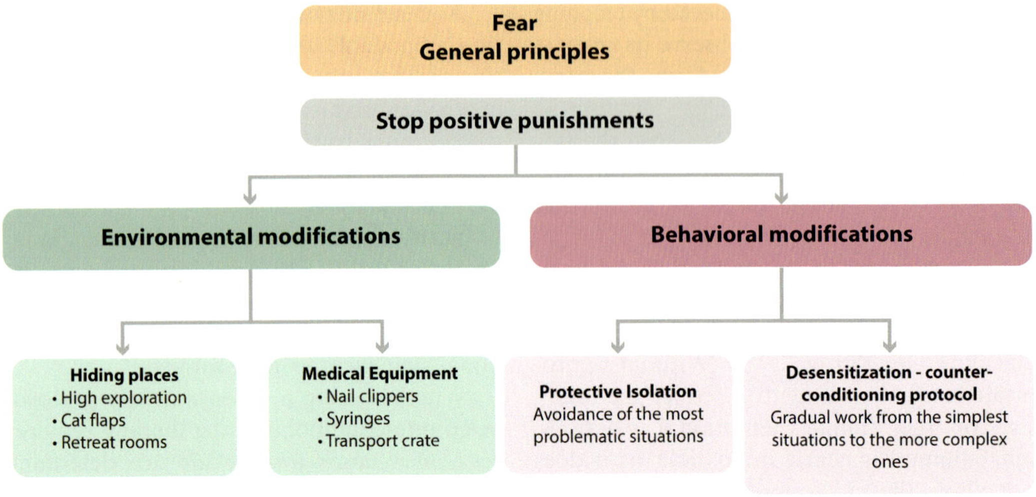

29

Fig. 29.10 General principles of fear axis therapies. This protocol is divided into environmental modifications (green) and behavioral modifications (pink). The latter includes avoiding aggravating situations and a tailored DSCC protocol for working on specific problematic situations

into contact with the object of its fear *on its own initiative and at its own pace.*

Thus, the implementation of desensitization and counterconditioning protocols in *open situations* and the *isolation of the cat* in closed situations plays a dual role: increasing the cat's comfort by avoiding situations that are emotionally too uncomfortable for it and reducing the risk of injury to both the cat and its owners.

▪ Reassuring Communication

Cat owners must be properly trained to detect *calming signals* and the *organo-vegetative manifestations* indicating their pet's emotional discomfort (see ► Sect. 29.2.2.1). For this, the practitioner should draw the owners' attention to these signs when they appear during the consultation. Since the cat's reactions differ greatly between its home environment and the consultation room, the practitioner can also ask the owners to film the cat in its home environment to analyze its reactions from videos captured there.

Owners should encourage the cat to approach the object of its fear by positioning themselves near it and scattering treats along the path between the cat and them. Using high-value treats, i.e., those that the cat doesn't typically receive but greatly enjoys, is an invaluable aid in advancing the therapy. As soon as the cat progresses toward the feared object, it should be verbally encouraged and praised.

In general, picking up the cat to bring it closer is a bad idea, as it makes the situation feel closed from the cat's perspective. Unless the cat is blocked and cornered, this should be avoided.

29.3.1.2 Environmental Modifications

▪ Isolation Rooms, Barriers, and Cat Flaps

Since a cat suffering from a fear axis disorder must be removed from stimuli that trigger its fear reactions when they are too intense, the use of a *cage* or an *isolated room* is helpful. A room is generally the simplest option for the cat, and it is possible to install a cat flap at the entrance so the cat can enter freely without being followed, and when it wishes. When the refuge is on another floor, it is common to use child barriers to prevent dogs, for example, from following the cat. Knowing that it is the only one capable of crossing a certain boundary helps reassure the cat about the security of the area where it chooses to take refuge. Of course, this area should include an elimination space, food and water, and possibly play areas if the cat is expected to spend time there.

Hiding spots such as closets, empty boxes, or laundry baskets are good refuges for the cat. However, if the cat constantly chooses to take shelter in these places without confronting the object of its fear, it may be wise to encourage it to approach the room where the stimulus is located. To do so, it is recommended to place the cat in a high place, such as a piece of furniture or its cat tree, so that it can observe from a distance and gradually get used to the object that scares it. If the cat chooses to flee, it should not be forced, but instead, the exercise should be regularly offered to help it overcome its fear without pressure.

If the cat suffers from phobias related to sounds, such as thunderstorms or fireworks, it is important to try to protect it in a *sound-proof room*. While Ford®'s active noise reduction kennel with anti-vibration blocks is currently just a prototype, a similar alternative can be created by covering a cage with acoustic panels to soundproof it.

In any case, the isolation area for a cat with a dysregulation of the fear axis should be carefully chosen to prevent escape, but also to avoid injuries the cat might inflict on itself and the damage it might cause in a panic crisis. Even with the use of psychotropic treatments that limit the intensity of the reactions, they do not completely eliminate them, and securing the cat during these crises is essential.

▪ Accessories

To limit panic reactions, it can be helpful to dress the cat in a *Thundershirt*® type garment, which applies light, constant pressure to the cat's body and has a calming effect. The effectiveness of these shirts varies from one individual to another (Lloyd 2017), but when such an accessory is soothing for an individual, it seems to be consistently so. However, the tolerance of cats for these vests varies, and they should only be used after ensuring that the cat is genuinely more relaxed and calm when wearing it.

In cases of care-related phobias, accessories like *lick mats* can be used as part of a desensitization-counterconditioning protocol to ensure that discomfort is minimized at every stage of care. If the cat stops licking or eating the food, it is a sign that the animal is feeling discomfort. At this point, it is important to pause the care until the cat returns to a sufficiently positive emotional state to resume its activity. This approach respects the cat's pace and helps prevent exacerbating stress. Therefore, the use of such accessories falls under *cooperative care*, where the animal's *consent* is prioritized.

▪ Work Sessions and Rewards

Desensitization and counterconditioning sessions should be *short* to avoid overwhelming the cat's emotional capacity, but they should be *frequent*. Regardless of the cat's fears, it will require *dozens*, if not *hundreds*, of *positive experiences* to change its perception of the feared object, considering it as, if not pleasant, at least harmless. Generalization takes longer once the socialization period has passed.

Additionally, it is important to remind owners of a cat with a fear-related condition that their animal will always remain very sensitive to negative experiences, and *vigilance must be maintained to prevent relapses*. Even though significant progress is possible through these therapies, the cat's vulnerability and history make it more fragile in this regard. As the cat ages, it may also become more susceptible to *emotional dysregulations*.

The *rewards* used in a counterconditioning protocol should be particularly appealing to the cat. It is important to reserve the cat's favorite rewards for these work sessions. For example, in the case of interspecific social phobia, giving exceptionally appetitive treats to the unfamiliar humans the cat interacts with is an excellent idea. This strategy allows for changing *the cat's cognition* by transforming its view of strangers from being dangerous and unpleasant to being associated with exceptional rewards, and therefore, pleasure.

29.3.1.3 Behavioral Modifications

The food ritual is beneficial regardless of the phobias, as all affected cats will need to receive medication in addition to behavioral therapy. Therefore, it is an essential prerequisite (see ▸ Sect. 29.2.3.3).

- **Noise Phobia**

Noise phobia therapy is particularly challenging because the cat does not necessarily understand where these disturbing stimuli come from and usually cannot interact with them. For instance, during a storm or fireworks, it is impossible to show the cat the source of the noise. Additionally, the intensity and duration of these stimuli are difficult to control, so the risk of exacerbating the cat's emotional state is high. However, isolated phobia of thunderstorms is relatively rare in cats, and often noise phobia is a consequence of the cat's generalized anxiety, which causes it to react to nearly all sounds. Prescribing psychotropic medication is therefore the key point that addresses the anxiety and leads to improvement, even without any specific exercises.

A desensitization or counterconditioning protocol can be implemented using recordings of the phobic noises, provided the cat reacts to them as it would to the actual sounds (which is rarely the case). For this reason, although many *recordings* are available for free online, it is often more effective to record them in the cat's real environment, which is simple today with smartphones by making a video of the scenes where the cat shows significant signs of fear. This recording should be played for the cat almost daily, gradually increasing the volume once the cat no longer reacts. Once the cat can listen to the sound of the object that initially triggered the fear at a normal volume, it is possible to confront the cat with the actual sound, waiting for it to be relaxed before ending the session. If, on the contrary, the cat shows strong fear reactions again, the session should be stopped. In many cases, the cat can be desensitized to the sound but not to the context that accompanies it. A second DS-CC (desensitization-counterconditioning) process involving both the noise and the context will be necessary, which can present an impossible challenge when the context is difficult to present progressively.

- **Healthcare Phobia**

Licking mats and foraging mats can be used as part of a habituation or desensitization protocol for medical care, serving two functions. The first is to associate a positive stimulus, food, with the care procedure within a *conditioning* or *counterconditioning framework*. The second is to establish *cooperative care* by seeking the animal's *consent*. The cat's interest in the mat reflects its emotional state: if the cat interrupts foraging or licking, it indicates discomfort, signaling the need to pause the procedure until the cat is relaxed enough to resume the activity (see ► Chap. 30, medical training and consent to care).

The treatment of healthcare phobia involves implementing habituation protocols, used here as desensitization protocols, as described in ► Sect. 30.2.2.2.

The cat must be desensitized to *routine care, including veterinary procedures. Care tools* (comb, brush, scissors, clippers, tick remover, flea treatment pipettes, compresses, etc.) should be selected to cause as little discomfort or pain as possible. For example, the brush should be chosen according to the cat's fur characteristics and tested on the cat to ensure it does not cause pain, observing the animal's behavior during brushing.

Accessories such as *licking mats* or *foraging mats* can be used in a care habituation or desensitization protocol, where they serve two purposes: the first is, of course, to associate a positive stimulus (food) with the care in the context of *conditioning* or *counterconditioning;* the second is to establish *cooperative care*, seeking the animal's *consent*, because the cat's interest in the mat reflects its emotional state. If the cat interrupts its search or licking of the food, it indicates discomfort, and a pause in care should be taken until the cat is calm enough to resume the activity (see ► Chap. 30, "Medical training and consent to care").

Additionally, habituation to the equipment used by the veterinarian can be carried out at home with a bit of creativity: the cat can be familiarized with the microchip reader using any device that makes a sound; with the syringe, using a ballpoint pen with the tip extended, etc.

These therapies are tedious and much more complex than simply habituating a non-phobic cat to care, especially since the veterinary context necessarily involves removing the cat from its soothing territory. For these cats, home visits or remote follow-up consultations via video can be helpful, as bringing them to a point where care can be easily provided is often unrealistic.

Of course, for these cats, learning to accept the carrier is one of the prerequisites for improving the quality of visits, and the gradual work with the various care accessories largely depends on the owners' motivation to implement the exercises at home.

▪ Social Phobias

The therapy for a social phobia, whether intra- or interspecific, always involves the same two steps.

The *first step* in the therapy for social phobias is to achieve the cat's *indifference* towards humans, dogs, or another cat. This step is considered successful when the cat is able to *move around these individuals and pass by them without reacting*, at progressively closer distances. Of course, this step first applies to individuals within the household and is only extended to outsiders once the household members have become familiar to the cat.

Indeed, for some cats, social phobia is so severe that they hide from the humans in the household, not just from unfamiliar people. For these cats, the principle of therapy remains the same, but they have very few refuge areas because they are exposed to the object of their phobia in a very intense way. It is not uncommon for owners to report that it took several weeks, or even months, for the cat they just adopted to allow itself to be touched. With the use of psychotropic medication, these delays are significantly reduced, and it is rare that progress is not spectacular within just a few days. This type of issue is also very common when owners acquire a dog or a new cat.

This step is a prerequisite but can also be the final goal of social phobia therapy, particularly for intraspecific phobias or interspecific phobias involving the household dog. This should be discussed with the owners during the consultation and is part of the therapeutic alliance, which includes agreement on the relationship, the objectives, and the means to achieve them.

The *second step* is to gain the cat's *interest* in humans, the dog, or its fellow cats, and this step is considered successful when the cat shows *pleasure in interacting with these individuals*. It is important to note that this step aims at the *sociabilization* of the cat, meaning the development of a positive attitude toward all unfamiliar humans and cats, and not just the *familiarization* of the cat with certain known individuals. For dogs, this step is often more of a goal than it is for cats because the need to take the dog outside results in more interactions with other individuals than is typically the case for cats. However, improving the cat's contact with unfamiliar individuals who cause it anxiety allows the cat to adapt more easily in many stressful situations, such as veterinary visits, having guests stay for several days, or gatherings with many guests.

To achieve these goals, the most effective strategy is the combination of a desensitization procedure and a counterconditioning procedure.

The *desensitization procedure* involves exposing the cat to unfamiliar humans or other cats at gradually decreasing distances, without ever placing it in a situation of intense discomfort. To achieve this, providing the cat with an observation post, such as a high piece of furniture or a cat tree, as a starting point, allows it to observe the feared object from a distance. For example, if the cat is able to eat a spoonful of wet food while visitors are present, it has already made progress. For this to work, it is important to create the association between the observation post and the reward, initially in the absence of the phobic stimulus, and then only in its presence. The situation must remain open, meaning the cat should have the option to escape if it wishes. It is important to note that if the protocol is properly followed, this intense fleeing response should never occur.

Similarly, the duration of exposure and the number of unfamiliar humans, dogs, or other cats the cat is exposed to simultaneously must only be increased very gradually. It is important to ask guests not to attempt to engage with the cat but to let the cat approach at its own pace. Of course, this is not possible when the phobia concerns a dog or another cat, but it is possible to control the dog by keeping it on a leash to prevent it from approaching the cat too closely, and interactions can be made through artificial barriers when dealing with another cat. The carrier, if the cat is accustomed to it, is an interesting way to introduce cats to each other, as the cat feels protected in it and can observe without being noticed. This allows for contact while ensuring the safety of both cats (see the protocol for introducing cats in ▶ Sect. 29.3.4.3).

The *counterconditioning procedure* involves reinforcing the cat's relaxation in the presence of unfamiliar individuals with an exceptional reward. Specifically, during the entire exposure to the cat, highly appetitive treats are used to encourage the cat to remain relaxed in the presence of the unknown, and gradually to approach the unknown, whether it be a human, a dog, or another cat.

In this approach, it is possible to teach the cat the "search" exercise to encourage it to engage in olfactory activity. In a quiet room, without the presence of strangers and without stimulation, it is enough to throw tasty treats on the floor while saying "search" to the cat, so that it learns that the word "search" is always associated with these jackpot rewards. Gradually, this procedure will be repeated in increasingly stimulating conditions. This training helps the cat focus on the treats and facilitates the desensitization therapy.

When the cat approaches strangers to ask for petting, it is obviously a significant advancement in therapy, but it is important to first ask them to present their hand and allow the cat to sniff it. They can then try to pet the cat very slowly, starting with a gentle touch, for example, near the corners of the mouth or on the back with their finger. A pause should be marked, and the therapist should wait for the cat to request attention again, as the contact stage can easily startle the cat. Of course, the situation should remain open, and attempts to block the cat should not be made.

29.3.2 Disorders Related to a Dysregulation of the Self-Controls Axis

29.3.2.1 General Principles

The therapy for a cat suffering from a disorder of the self-control axis aims to teach it how to control its *emotional* and *motor responses.* In general, the goal is to teach the cat to interact without biting or scratching its owners.

- **Respectful Communication**

Calm and consistent communication is important for all cats, but especially for those suffering from self-control issues. These cats often have difficulty managing their emotions and motor skills. This issue is made worse if the humans they interact with also have trouble controlling their own emotions and motor skills. Shouting and physical restraint, which reflect the owners' sense of helplessness, are particularly counterproductive and anxiety-inducing for these cats, who are often very emotionally sensitive.

Moreover, due to the impulsivity of the cats we are discussing, the individuals in charge of their rehabilitation must be particularly quick and precise in punishing inappropriate behavior or reinforcing appropriate behavior at the moment it occurs. *Anticipating* the cat's reactions is therefore essential in order to act at the right time. Any inaccuracy may trigger aggression, which is difficult to suppress and can quickly become ingrained.

Maintaining calm and consistent communication with a cat suffering from a self-control disorder is difficult but crucial for the success of the therapy. Support from a feline educator, although not as widespread as with dogs, can greatly help owners succeed. For instance, they could train the cat's owners in using a clicker® when their verbal or gestural communication lacks precision or sobriety (see the next section "Using the clicker").

Respectful communication is also expected from the cat. A cat with this type of disorder is usually intrusive and may, for example, disturb the owners at night or while they are working in a very insistent manner. Due to its impulsivity, it is naturally incapable of waiting before acting. Improving the relationship necessarily involves establishing a structured living environment and setting up communication rules that are appropriate.

Using the Clicker

Cats with a self-control disorder exhibit an excess of motivation, whether it be in play, food, or interactions. As such, they are excellent candidates for learning much more than the average cat. For this reason, we provide details here on the use of the clicker, which is more often used in this group of therapies. However, note that this tool can be used for any of the exercises presented in this chapter.

A *clicker* is an object that, when pressed, produces a distinct metallic sound. This click can be paired, through classical conditioning, with a food reward.

When the clicker is used in what is then a *passive learning* context, it has the advantage of producing a standard and unique sound that neutralizes the emotions conveyed by the voice. It also helps standardize responses from different people who are training the cat. In this case, it serves no other purpose than to praise the cat for executing a command correctly. In other words, it simply replaces verbal signals such as "yes!", "good!", or "well done!"

However, the clicker (see box on clicker training) can also be used in *proactive learning* to encourage the cat to spontaneously offer behaviors. In this approach, the cat understands that it has the power to trigger the click and therefore *actively* seeks to understand what is expected of it in order to receive a reward. This proactive use of the clicker has the advantage of giving the cat more control over its environment and much more *freedom of choice* than traditional learning methods. It is well suited for shaping because the cat learns to quickly offer behaviors. At this point, approximations of the desired behavior can easily be reinforced until the final version is reached, at which point a verbal command can be added. In this proactive approach to learning, the innovation lies in getting the cat to *think differently*, not in the tool itself. Thus, working proactively and using the sound of the voice to validate each desired behavior is also effective. It is important to note that there is no magic in the sound of the clicker (Grant and Warrior 2019; Kogan et al. 2017).

Important Point of View

Clicker Training

1. "Charging" the cat, meaning teaching the cat the association between the sound of the clicker (a neutral stimulus) and the arrival of a food reward (a primary reinforcer). To do this, reward the cat with a treat (without speaking) every time you click. This phase consists of a series of "click + reward" repetitions until the cat responds joyfully to the sound of the clicker, which has then become a secondary reinforcer.
2. Generate a behavior from the cat using capture, guidance, or shaping. Whichever option is chosen, click and then give the cat a treat when it spontaneously performs the desired behavior or at least an approximation of it. It is important to remain silent during this step. Every click must be followed by a reward, even if it was an accidental click, to maintain the click-reward association.
3. Let the cat spontaneously produce the desired behavior or an approximation of it before clicking and rewarding. Repeat the exercise until the behavior is perfectly executed.
4. Associate a command with the final behavior.

Motivation

Working on motivation is particularly important for cats with a self-control deficit. If motivation can be channeled into productive work without leading the cat into addictive behaviors, it will foster learning. However, if motivation is expressed through disruptive behaviors, it becomes particularly invasive and difficult to interrupt. *The regulation of motivation* in cats with a self-control disorder relies on the principles and exercises applicable to all cats. However, these need to be adapted to avoid the specific pitfalls associated with these conditions.

Managing the cat's *motivation* relies on two complementary strategies. On one hand, using *negative punishment* helps decrease the cat's motivation to perform undesirable behaviors. On the other hand, *using reinforcers* that are suited to the individual within a specific problematic context helps increase motivation to perform desirable behaviors.

However, *managing the motivation* of a cat in the context of strengthening self-control requires additional precautions.

First of all, it is important to highlight that behaviors associated with the cat's excitement (such as biting and scratching during play, meowing, etc.) can provide pleasure to the cat and thus become internal reinforcers. To avoid this mechanism and allow the cat to stay within an acceptable level of excitement, it is preferable to remove it from stimuli that trigger excessive excitation, which can no longer be modulated. Thus, as with the fear axis, *the most extreme situations should be avoided at first*. The cat's ability to control itself even in the most difficult situations will mark the end of the therapy.

Furthermore, it is important to avoid using *excessively* attractive reinforcers for the cat, such as a particularly stimulating toy or a highly appetitive treat. These may generate too much excitement or frustration, which can focus the cat's attention and prevent it from performing the behavior that would allow it to obtain the reinforcer. At its extreme, these reinforcers can foster or lead the cat into addiction, especially if a compulsive pathological state is present. Finally, when the cat's motivation regarding play or food remains excessive, it is useful to implement exercises that teach the cat to temporarily or permanently give up these resources.

■ **Acquisition of Self-Controls**

Self-control refers to the processes that underlie an individual's ability to *spontaneously* regulate their emotions and behaviors, and to produce an appropriate response to a given situation, without external assistance. By definition, self-control is not fully acquired as long as external intervention is needed to manage the emotional or behavioral responses of the cat. However, this *external control,* which must eventually be eliminated, is essential in the initial stages, as the cat cannot independently find the correct response, especially due to its impulsivity.

Strengthening self-control, therefore, involves two successive stages of increasing difficulty.

The first stage is that of *external control,* where the goal is to get the cat to stop producing undesirable behaviors in problematic situations, or even to produce desirable behaviors instead.

The second stage, that of *internal control,* involves getting the cat to refrain from producing undesirable behaviors on its own in these problematic situations, and to spontaneously offer the desirable behaviors that were taught during the first stage.

The reinforcement of *external control* over the cat relies on the use of two parallel strategies. On one hand, the *undesirable behavior of the cat must be suppressed,* and on the other, *an acceptable behavior* must be generated in the same context (see ► Sect. 29.1.3.2). At first, the desirable behavior can be obtained through guidance. However, over time, it should be achieved on command, just like any other learning. For example, if the cat nibbles or scratches during play, both the behavior should be discouraged by ignoring it or, when this is not possible, by using the word "no," but at the same time, an alternative behavior should be encouraged, by redirecting the cat to another game, for instance.

The reinforcement of *internal control* involves waiting for the cat to spontaneously produce a desirable behavior previously learned in problematic situations. For instance, in the case of overstimulation during play, the cat should be able to walk away and play by itself as soon as the owner stops the interaction. To achieve this, one must regularly wait after interrupting the activity to see if what was previously offered to the cat occurs spontaneously. Through repetition, this will become a habit for the cat, and it will naturally behave as expected. As with any new behavior learning, the desirable alternative behavior must be continuously reinforced until it is fully automated by the cat, after which it should only be reinforced intermittently.

The choice between a strategy of reinforcing *external control* and a strategy of reinforcing *internal control* of behavior should be guided by the *severity of the disorder* diagnosed during the consultation. As mentioned, external control precedes internal control. The severity of conditions related to the axis of self-control can vary significantly, and in the most severe cases, only external control is prescribed at the start of therapy. In milder or more contextual cases, it is possible to quickly achieve internal control.

Depending on the cat, the nature of the most exciting elements can vary, and the therapy must be adapted to the problematic contexts. For some, it is food; for others, it is play; and for some, it is friendly relationships.

It is noteworthy that the severity of these conditions is less pronounced in cats than in dogs, likely due to the active selection carried out in dogs to create working lines. Therefore, it is very rare to need lifelong treatments in cats, and it is also exceptional for them not to reach the stage of internal control once their brain has reached adult maturity.

If, in a given problematic context, the internal control of the cat is, although imperfect, sufficient to allow it to spontaneously produce the desired behavior (even if only occasionally and after some time), the implementation of external control over the cat becomes unnecessary. In this case, the cat's spontaneous desirable behavior should be reinforced using the capture method rather than the guidance method (see ► Sect. 29.1.3.1).

For this reason, the first response to a disruptive behavior in a cat is *indifference.* This strategy not only discourages the behavior but also allows the cat the opportunity to spontaneously propose another, more acceptable behavior, which can then be reinforced.

If the cat's self-control quality in a particular context is poor, it is essential to begin by reducing the intensity, distance, and duration of the triggering stimulus. This is why working on reinforcing a cat's self-control should not only be carried out during *random encounters* with the problematic stimuli for the cat but also within *structured exercises,* where all parameters (duration, distance, intensity) of these stimuli can be carefully controlled (■ Fig. 29.11).

29.3.2.2 Environmental Modifications

The environmental modifications that can be made in cases of self-control axis disorders are numerous. Indeed, most of these cats, by definition, have an abnormally high level of motivation for everything, which gives the impression that their environment should be endlessly enriched. They are highly motivated to learn, and for this reason, their owners often find that they behave like dogs. They are capable of following their owners on walks, retrieving toys... Most of the environmental enrichment strategies have been detailed in ► Sect. 29.2.1.2., so we will simply mention them in this section.

▪ Food Games

Impulsive and compulsive cats tend to lack satiety when it comes to food. The use of *food games* is therefore almost indispensable for them. As detailed in ► Sect. 29.2.1.2, complex feeding bowls have the dual advantage of preventing obesity in these at-risk cats while providing them with an activity that reduces their frustration by responding to their higher-than-average motivation. Another advantage of these systems is that they prevent these highly persistent cats from using meowing to demand food throughout the day.

For these cats, the most complex devices are preferred to offset their excess motivation and perseverance. For example, the "Turn Around Activity®" system from Trixie® combines several challenges and may be too complex for most cats who risk losing interest but is well-suited for a cat with a self-control axis disorder. Of course, these feeding bowls can easily be made by the owners, which is also a strategy that helps involve them in the therapy.

It is important to note that using a food dispenser in the form of a game slows down *eating speed* and limits the risk of regurgita-

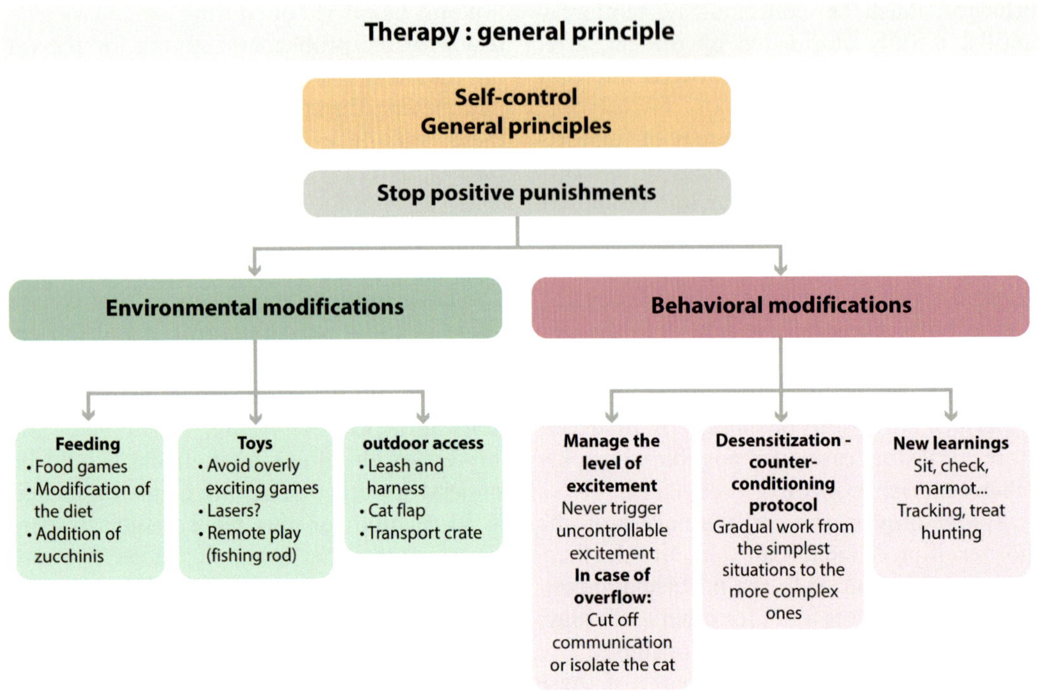

Fig. 29.11 General principles of therapies for the axis of self-control. This protocol is divided into environmental changes (green) and behavioral changes (pink). The latter includes, on one hand, the avoidance of aggravating situations and, on the other hand, a DSCC protocol adapted to the problematic situations that can be worked on

tion but has little effect on food satiety. However, adding a vegetable rich in water and insoluble fiber, with low-calorie content, to the cat's meals, such as cooked zucchini, can enhance its sense of fullness. In many cases of compulsive eating in cats, even with the addition of psychotropic drugs that help restore satiety, the use of high-fiber and high-protein diet kibble is essential.

Water Fountains and Other Moving Stimuli

In the same vein as keeping an overly motivated cat occupied, the use of *water fountains* is beneficial for cats with a self-control deficit. For some cats, watching moving water is genuinely interesting, so it's important to choose models that feature substantial water jets to entertain the cat.

It is also possible to use an aquarium or a cage with other moving animals, which can be stimulating for the cat, but this raises ethical concerns due to the potential lack of well-being for the animal inside the cage. To address this issue, there are now apps for cats that can be easily used on a tablet, allowing the cat to interact with moving elements on the screen (such as fish or a fake laser) using its paw.

Outdoor Access

Allowing a cat to have access to the outdoors for hunting in a natural environment can be very tempting to help these cats. Owners or practitioners, unaware of the cat's condition, often believe that the problem will be solved if the cat can go outside, given the visible frustration. However, this is not necessarily the ideal solution. It is important to remember that these conditions are caused by an excess of motivation, rather than a lack of stimulation. Increasing stimulation can often lead to more frustration in its absence, without providing the "satiation" that should result from the behavioral sequences. Additionally, these cats are often so impulsive that they lose all sense of caution, which increases the risk of accidents.

Therefore, before allowing outdoor access, it is essential to treat the cat with the appropriate medication so that these outings can actually be calming.

There are now many devices available to help secure cats' outdoor activities. For example, fences can be fitted with roller systems to prevent climbing or extended with inward-facing extensions (Kasbaoui et al. 2016). This allows the cat to access a secure garden safely.

For cats living in apartments, the temptation to arrange outdoor walks on a leash and harness is strong for the same reasons mentioned above—that the cat seems very frustrated and would be happy to frolic outside. In practice, as described in ► Sect. 29.2.3.3, it is essential to start by acclimating the cat to these accessories, which can be a challenge for these cats who struggle with immobility and constraints. Additionally, if the owner decides to initiate these walks, they must ensure that the chosen harness and leash will keep the cat secure, as it might try to escape if it manages to slip out of the harness.

- **Hiding Spots and 3D Exploration**

The ability to hide in various places is an effective way to engage highly motivated cats (Ellis 2009; Herron and Buffington 2012; Vinke et al. 2014). Once again, creating these varied and elevated hiding spots allows owners to get involved in the therapy, and it is particularly interesting because it encourages them to be creative when setting up their cats' territory (see ► Sect. 29.2.1.2.).

- **Toys**

The choice of *toys* for a cat with a deficit in self-control must meet several requirements. These toys should not be overly stimulating so that they can be used as reinforcers without triggering addictive behaviors or uncontrollable excitement. For example, the use of laser toys is often discouraged, partly because they generate strong excitement followed by intense frustration as the cat can never "capture" its prey. If owners still wish to use such toys, it is advisable to recommend throwing a treat or another toy after a few seconds to allow the cat to complete the predation sequence it started. Additionally, when the cat exhibits oral exploration and damages its toys, it is important to ensure their durability to prevent the ingestion of fabric or plastic pieces (Demontigny-Bédard et al. 2016).

Finally, especially at the beginning of therapy, it is better to recommend toys that maintain some distance from the cat, such as fishing rods or balls, to avoid direct interactions with the owners' moving hands or feet.

- **Soft-Paws®**

Cats with a self-control axis disorder often struggle to control their scratching. When taken in at a young age, owners may get injured because any interaction with the kitten leads to scratching. These kittens get used to climbing their owner's legs, scaling curtains, etc.

A simple way to encourage a different behavior is to use Soft Paws, which are small plastic caps placed on the claws with a specific glue. These claw covers have a foam tip that prevents the cat from digging its claws into surfaces. They fall off naturally after 3 to 6 weeks, allowing the claws to be trimmed if new ones are to be applied. For a kitten that cannot control its claws, this immediately soothes the situation because the owners are no longer injured, and their fear of interacting with their kitten decreases. They are also less likely to scream or move when the cat uses its claws, protected by the plastic caps. This is especially important when there are children in the household, as it provides safety for them against the mischievous kitten.

For the cat, it does not prevent it from extending and retracting its claws, but its daring actions become failures: if it tries to climb its owner's calves, it slips. Through extinction, it will quickly find other solutions. As an added benefit, this significantly reduces damage to furniture for cats that have started damaging curtains, chair backs, or sofas.

In many cases, this measure is so effective that a single application is enough to redirect the behaviors.

29.3.2.3 Behavioral Modifications

The feeding ritual is beneficial for cats with self-control deficits, but it is often very easy to implement for them (see ► Sect. 29.2.3.3). However, it is important to be cautious not to use food that is too appetizing, as they might go overboard to obtain it, for example, by bothering the owner as soon as they eat yogurt if yogurt has been used.

Next, several types of exercises may be prescribed depending on the contexts that trigger the cat's overflow behaviors. In all cases, these exercises will always be paired with the strict application of negative punishment, which involves interrupting the cat (disruption) and redirecting it to a toy, for example, or temporarily excluding it by calmly carrying it to its isolation room until the excessive excitement subsides. In practice, if the therapy is well-executed and because these cats receive appropriate treatment, the periods of isolation should become increasingly rare. However, before treatment, even a correctly applied version of the exercises described below often leads to overflow behaviors, where the play or interaction escalates to biting or scratching.

▪ Controlling Excitement in Interactions

The goal is to teach the cat to tolerate moderate constraints in order to increase its tolerance, allowing it to be treated when necessary. In fact, for cats with self-control deficits, even a simple general examination can be difficult because immobility generates significant frustration. To address this, two exercises can be prescribed: the relaxation protocol and teaching calm after excitement.

Relaxation Protocol

The *relaxation protocol,* or *relaxing massage*, involves gently petting the cat in a calm environment while saying "calm down" or "gently" until the cat is completely relaxed. You can associate this calming moment with a specific blanket or music to link relaxation with the object and promote tranquility session after session.

Relaxation is marked by a complete resting position, with closed or partially closed eyes. Often, the cat will purr and visibly enjoy this interaction. The goal is to find a way of petting the cat that almost leads it to fall asleep. When the contact is broken, the cat should be relaxed enough not to get up. The contact with the cat should be slow and continuous to avoid triggering excitement. To progress with the exercise, once the cat can relax in less than 30 s, it can be practiced in a livelier environment or after triggering the cat's excitement.

Learning Calm After Excitement

This exercise consists of two levels of increasing difficulty, with the respective goals of achieving *external control* first, followed by *internal control* over the cat's behavior.

The first level of the exercise involves *guiding* the cat back to calmness. The second level focuses on waiting for the cat to return to calmness *spontaneously.*

In both levels, the exercise starts by provoking a moderate level of excitement in the cat, such as through playing with a toy, running, or interacting with the cat.

As soon as the excitement becomes visible, the owner begins a gentle contact by petting the cat but preventing it from moving too much, allowing the cessation of movement to help the cat find calmness. Of course, the goal is not to forcefully constrain the cat, but to provoke a calming contact, similar to how we would soothe a young child having a "tantrum" by holding them gently to help them regain control over their emotions. This guidance should, however, gradually be reduced and eventually eliminated before progressing to the second level of the exercise.

The second level of the exercise aims to passively achieve calmness by simply stopping all interaction with the cat, or even calmly walking away.

In both levels, the cat should be praised and rewarded with a natural reinforcer when it complies. Practically speaking, this means that once the cat accepts to calm down, playtime can resume. This exercise uses the cat's heightened motivation to achieve calmness.

To progress with the exercise, the intensity of the excitement provoked in the cat will gradually be increased.

▪ Kibble Hunt

Hiding kibbles around the living space every day is also a great way to stimulate the cat olfactorily, and owners are often very impressed with how their cat manages to find all the hidden food. This also provides an easy way to involve children in the therapy by asking them to hide 6 or 8 kibbles once a day in different spots around the house. This practice helps to bring positivity back into the relationship, which may have been strained due to the cat's impulsive behaviors.

▪ Self-Control in Play

Learning to control oneself during play is often a central exercise for cats with deficits in self-control, as these are situations in which they easily bite and scratch. It is an excellent opportunity to teach them self-regulation while also controlling the type of play to initially provoke only mild excitement.

The principle of this exercise is to condition the cat not to scratch or bite when excited. To do this, the owner should start a play session with the animal. As soon as the cat shows signs of intense excitement, suggesting that it may cause harm, the activity should be immediately stopped. It is recommended to say a short phrase like "that's enough, no more play" and then calmly walk away if needed. It is important to wait until the cat regains calmness before resuming play, as this will act as a reward for the calm behavior. If the cat is unable to calm down, it is advised to isolate them for a few minutes before starting the next session with a lower level of excitement. With each session, the intensity of the excitement can be gradually increased, teaching the cat to manage their reactions over time.

▪ Intelligent Games

"Intelligent" games are often prescribed for these cats because they develop their reasoning and concentration abilities. These types of games require control rather than overflow. When it comes to play, the focus is not on the type of game, but rather on the way it is played. In these exercises, it is important to help the owner understand that no game is off-limits, but that the games should lead to controlling excitement rather than overwhelming it.

What matters in prescribing these various exercises is to provide the owners with ideas for what can be done, as well as the tools and steps to implement them. Once the principle of the steps is understood, their creativity will be the only limit.

Simple Cues

It is possible to teach a cat new commands such as "sit," "lie down," "high five," "roll over," "play dead," or "jump." The box "High Five!" shows how to build the steps, and can be adapted for all commands. It simply follows the rules outlined in ▶ Sect. 29.1.3.1 and can be replicated by analogy with other commands. The initial physical guidance is generally the only point that differs from one exercise to another.

Important Point of View

Teaching the "High Five!"
The goal is to teach your cat to tap their paw with your hand when you present it to them.
Here's how to proceed step by step, knowing that you should only move on to the next step once your cat is comfortable with the previous one:

1. Present your hand *very close to your cat's paw* and say "High Five!"
   - If the cat does not react, gently touch their paw.
   - As soon as the cat touches your hand, say "Good job!" and reward with their favorite treat.
   - Repeat the exercise about 10 times.
2. Gradually move your hand further away from the cat's paw when you ask for the "High Five!"
   - Reward them *as soon as they touch your hand.*
3. Once your cat is easily doing the "High Five!" start *spacing out the rewards.*

More Complex Exercises

It is possible to multiply exercises in order to stimulate the cat while strengthening the bond with its owner through these enjoyable learning moments. For example, the cat can learn to fetch a ball (see the box), or even play a game like the "bento" by first learning to knock over a cup, and then learning to knock over the specific cup under which there is a treat (see the box). Many commercially available toys also follow this principle of interaction, where the cat manipulates an object to open a compartment containing a treat.

Important Point of View

"Fetch the Ball"

The goal of this exercise is to help your cat control itself during play.

At first, it's helpful to use *multiple toys* to encourage your cat to bring back the first one more easily. Proceed step by step, ensuring that you only move on to the next step once your cat is comfortable with the current one:

1. Start by throwing the toy to begin the game. As soon as your cat catches it:
   - Encourage them to return to you to bring it back.
   - Once they drop it, throw another toy to restart the game.
2. Gradually, you should be able to stop using the second toy.

Make the exercise more challenging by asking your cat to place the toy in your hand.

Important Point of View

Shell Game

The goal of this exercise is to improve your cat's attention skills.

Proceed step by step, ensuring that you only move on to the next step once your cat is comfortable with the current one:

1. Hide a treat under a plastic cup, ensuring your cat sees you do it.
   - As soon as your cat sniffs or touches the cup with its paw, help them knock it over.
   - Your cat is rewarded with the treat hidden underneath.
   - Repeat this game 5 to 10 times.
   - Your cat should learn to knock over the cup on its own to get the reward.
2. Make the exercise more challenging by increasing the number of cups:
   - Your cat is only rewarded when it knocks over the correct cup.

29.3.3 Disorders Related to a Dysregulation of the Territoriality Axis

29.3.3.1 General Principles

Therapies related to the *territoriality axis* concern a discomfort caused by the cat's inability to feel calm in its environment. This can be due to a lack of stimulation (affection resulting from a poor territory), significant changes (invasions, disturbances), or, conversely, an environment that is too stimulating (conditions resulting from variations in the territory). Obviously, if this discomfort is secondary to another behavioral pathology, the therapy must first target the primary disease. Without this, it will be impossible to treat the territorial condition, and relapses will be inevitable. Environmental modifications will logically be crucial for this category of conditions, as shown in ◘ Fig. 29.12.

The use of pheromones makes perfect sense in this type of condition because they help cats structure their territory to make it calming. They should therefore be systematically added to other measures in the case of conditions related to the territoriality axis (Frank et al. 2010; Griffith et al. 2000).

29.3.3.2 Environmental Modifications

The strategy for modifying the environment will depend on the type of condition. In all cases, the analysis of the cat's pheromone markings must be systematic, as healing will be reflected through a return of these markings to normal. This return signifies that the cat is once again able to make its territory calming.

However, some aspects of the elimination area setup are common to all these conditions. As soon as symptoms include urine marking or incontinence, the *rules regarding the litter box* must be strictly reminded to the owner, as outlined in ► Sect. 29.2.1.2. It should also be noted that all these cases must first undergo a

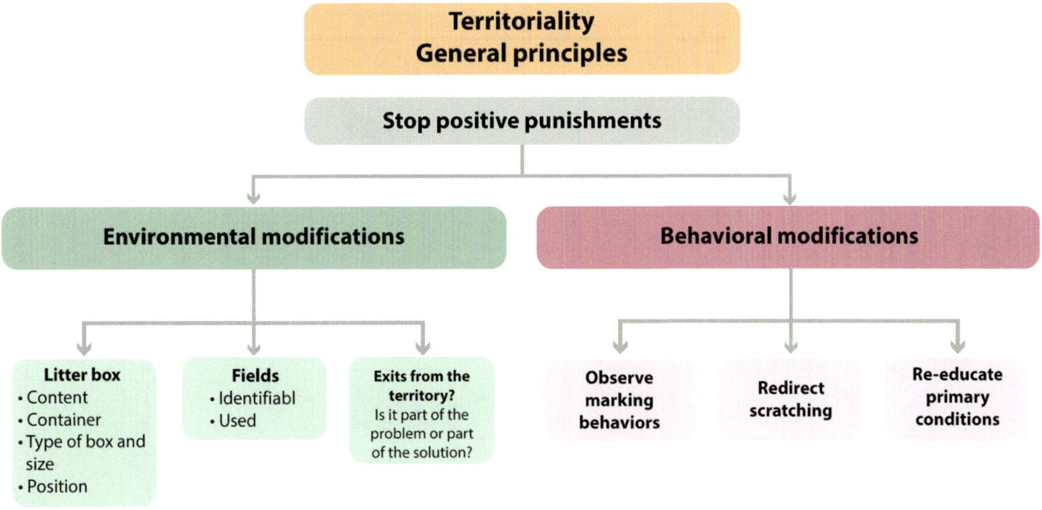

Fig. 29.12 General principles of Territoriality axis therapies. This protocol is divided into environmental modifications (in green) and behavioral modifications (in pink)

urinary examination and a complete blood panel, as urinary and renal problems are common in cats.

However, cats affected by these territorial conditions are generally anxious, and setting the rules around the litter box, when not combined with anxiolytic psychotropic treatment, does not resolve the issue. In fact, owners often claim to have tried everything in this area, yet the incontinence persists. Thus, these rules should be reviewed, but if consistency is maintained, it should also be considered that, over time, the cat either has a real aversion to the litter box or has developed new elimination habits. In such cases, it will be necessary to make the inappropriate elimination areas less attractive, for example, by using *survival blankets, synthetic grass*, or even *aluminum foil* on soiled areas, after cleaning. Additionally, efforts should be made to make the litter box more appealing, for example, by increasing its surface area or depth, or by rewarding the cat when it eliminates in the litter box.

In extreme cases, if the aversion is complete and the cat no longer eliminates in its litter box at all, it is necessary to create an extended zone, almost entirely within a small room (or a large cage), to encourage elimination in the litter box. However, it is important to note that it is very rare to have to resort to this approach once the pathological condition is addressed with appropriate medication.

Disorders Due to Territorial Poverty

This condition is likely overdiagnosed, as environmental enrichment has become the go-to response to all behavioral issues in cats and zoo animals. However, in the case of a genuine diagnosis of a condition caused by territorial poverty, the simple and effective solution is to enrich the cat's environment by organizing its territory in a way that significantly increases the available stimuli.

In an environment where the owners are rarely present and there is little interaction with the cat, it is important to provide the cat with hiding spots, food puzzles, a cat tree with an observation post, a clean litter box, etc., and ensure that the cat actually uses these various areas as shown in Fig. 29.13.

The question of outdoor access often arises when it comes to indoor cats. However, as discussed in ► Sect. 29.2.1.1., this is not essential if the environment is sufficiently enriched for a cat that is not in a pathological state. Therefore, solutions are entirely feasible, even if owners cannot allow their cat to go outside. It is crucial to reassure them on this point and avoid making them feel guilty by presenting alternatives, as they are often

29

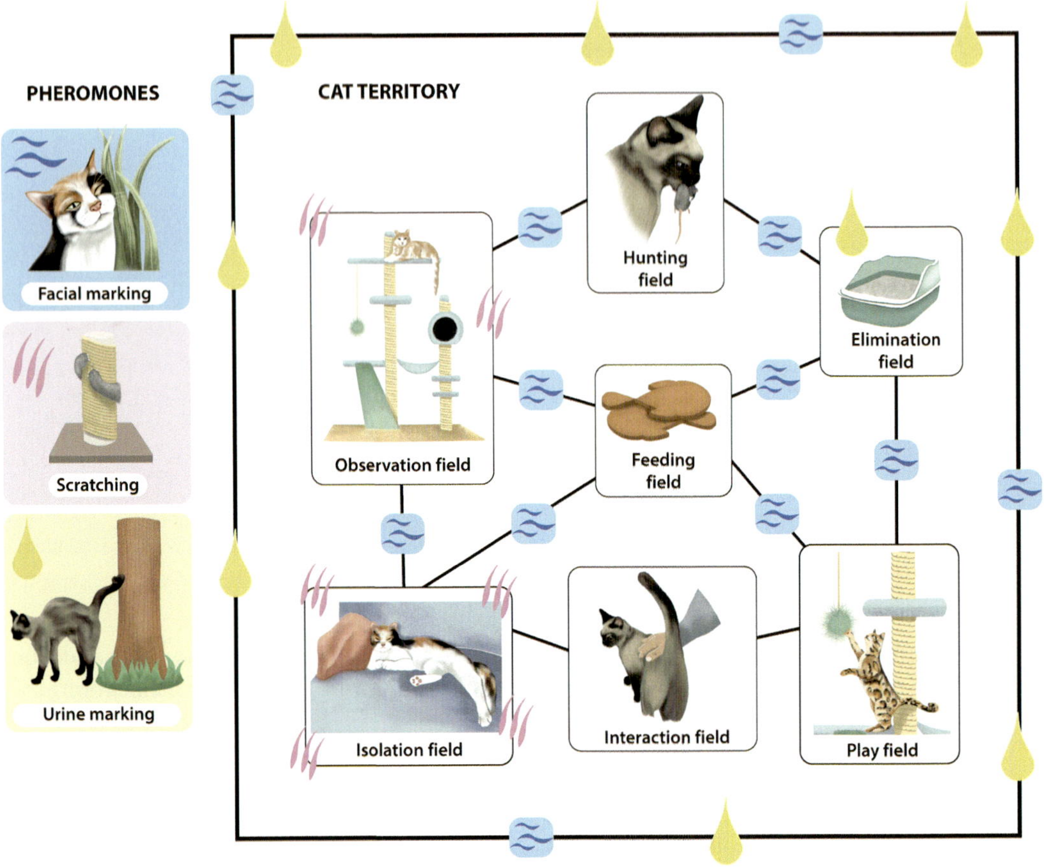

Fig. 29.13 The different activity zones for the cat

already overwhelmed with guilt, having been stigmatized in the past.

Disorders Due to Territorial Variations

In these conditions, it is essential to identify the type of stimulus that has led to the anxious state by disrupting the territory. This may include the arrival of a baby, a new cat, a dog, a move, or neighborhood cats attacking the affected cat. Therapeutic measures for conflicts between cats (e.g., adopting a second cat) will be addressed separately.

Depending on the identified element, the therapeutic strategy will be adapted to remove the animal from the problematic situation by offering an easily accessible isolation area, facilitated by the installation of cat doors or barriers, for example.

It is important to ensure that, despite the presence of this problematic stimulus, the cat's activity zones are all structured and used, and to reorganize them if the cat restricts access to certain areas. For instance, a cat that is attacked by other neighborhood cats and has decided not to go outside will need larger play or interaction activity zones than before to compensate for the absence of outdoor time, which it had previously.

The areas described in ▶ Sect. 29.2.1.2 must therefore be reviewed to ensure the cat can easily access them, even with the constraints imposed by the new stimuli.

Territorial Disorders Secondary to Another Behavioral Condition

It is important to primarily address the underlying behavioral condition responsible, and to draw from the environmental therapies for that condition while paying special attention to ensuring that the activity areas are reidentified and used by the animal.

29.3.3.3 Behavioral Modifications

The feeding ritual is beneficial regardless of the territorial condition, as the cats involved need to receive medication alongside behavioral therapy. It is therefore an essential prerequisite (see ▶ Sect. 29.2.3.3).

In cases of soiling or marking, it is very common for physical punishments to have been used afterward, as owners are often at their wit's end. It is important to remind owners that these soiling issues are a major cause of abandonment and euthanasia (Casey et al. 2009; Patronek et al. 1996). Therefore, without guilt-tripping, it must be explained to owners that these punishments are not only ineffective but, more importantly, counterproductive as they are anxiety-inducing.

Behavioral modifications are somewhat limited in comparison to other conditions, but general principles can still be drawn depending on the specific condition.

- **Disorders Due to Territorial Poverty**

Redirecting Undesirable Scratching Behaviors

Cats with these conditions often express a great deal of frustration due to an inability to express their normal behaviors. Therefore, it is common to need to redirect excessive scratching to appropriate locations. This can be done by protecting the areas being scratched accidentally and placing a new scratching post near these spots, previously enriched with Feliway Scratch®, to redirect the scratching to the desired surface. Disruption and redirection are possible if the owners observe the cat scratching an unwanted surface, keeping in mind that this should not involve a disruption that scares the cat, as this would then become a positive punishment, which we want to avoid because it would be anxiety-inducing and worsen the problem. It is important to help the owners understand and accept that a behavior cannot simply be eliminated. It must be replaced or redirected.

Increasing Social Interactions to Enrich the Environment

The individuals living with the cat are mobile but nonetheless integral parts of its environment. Encouraging owners to increase positive interactions with their cat when it suffers from territorial distress due to a lack of stimulation is essential.

Creating *new rituals* can be an interesting option. This can be achieved by using treats that the cat is asked to search for (see the "kibble hunt" exercise in ▶ Sect. 29.3.2.3) or through intellectual games that help reduce frustration caused by insufficient stimulation.

In this context, some owners consider adopting a second cat. This reasoning is logical, as a bored cat might enjoy having a playmate. However, introducing another adult cat or even a kitten may be poorly tolerated by the resident cat, potentially leading both animals to develop relational pathologies.

- **Disorders Due to Territorial Variations**

For such cases, it is essential to begin by providing the cat with the ability to isolate itself from the problematic stimuli. For instance, if the cat is disturbed by the arrival of a puppy, it should be given access to a retreat space, such as a room with an electronic cat flap that the puppy cannot pass through, or by installing a barrier that prevents the dog from reaching areas like an upstairs floor.

Once this has been achieved, a gradual approach can be taken to help the cat accept the problematic stimulus by employing a protocol of desensitization and counterconditioning.

Desensitization involves controlling the problematic stimulus, such as keeping the newborn or puppy as still as possible, while providing the cat with an observation point a few meters away.

Counter-conditioning involves pairing these controlled exposures with highly enticing treats to gradually change the cat's perception of the new and unpleasant stimulus. The goal is to shift the perception from negative to neutral, and eventually, to positive.

As therapy progresses, the distance between the cat and the stimulus is reduced, and the duration of exposure is incrementally increased.

- **Territorial Disorders Secondary to Another Behavioral Condition**

These conditions are common in cats with HSHA (Hypersensitivity-Hyperactivity Syndrome or Feline ADHD) syndrome, which are unable to produce structured pheromonal markings. Remediation involves addressing the primary condition (see ► Sect. 29.3.2). They can also be secondary to conditions related to the fear axis (see ► Sect. 29.3.1) or a relational disorder (see ► Sect. 29.3.4).

29.3.4 Disorders Related to a Dysregulation of the Relationships Axis

29.3.4.1 General Principles

In these disorders, we differentiate between interspecific relational disorders and intraspecific relational disorders.

The general environmental adjustment strategy in both cases involves providing the cat with a retreat area and hiding places to isolate itself from the individual with whom the relationship is problematic. However, depending on the type of disorder, the behavioral therapy to be applied will differ (◘ Fig. 29.14).

29.3.4.2 Environmental Modifications

As with territorial disorders where a specific stimulus disturbs the cat to the point of causing anxiety, it is essential to start by providing the cat with the means to isolate itself from the problematic stimuli. For example, if the cat is disturbed by a family member due to past physical punishment, it is advisable to provide the cat with a room or a high refuge where it is clearly explained to the owners that the cat will never be chased or disturbed.

In cases of fights between cats, the use of electronic cat doors that recognize each cat's microchip can limit access to a particular room for one of the cats. Additionally, in the most severe cases, whether intraspecific or interspecific, these measures allow for the complete and necessary separation of individuals while waiting for medication to take effect and for therapy to progress.

29.3.4.3 Behavioral Modifications

The feeding ritual is particularly useful in cases of interspecific relational disorders involving humans, as it helps to establish a calming routine. Additionally, cats affected by this group of disorders must receive medication alongside behavioral therapy, making this an essential prerequisite (see ► Sect. 29.2.3.3).

▪ Interspecific Relational Disorders

Cats affected by this disorder exhibit aggressive or avoidance behaviors toward one or more household members. In turn, the tar-

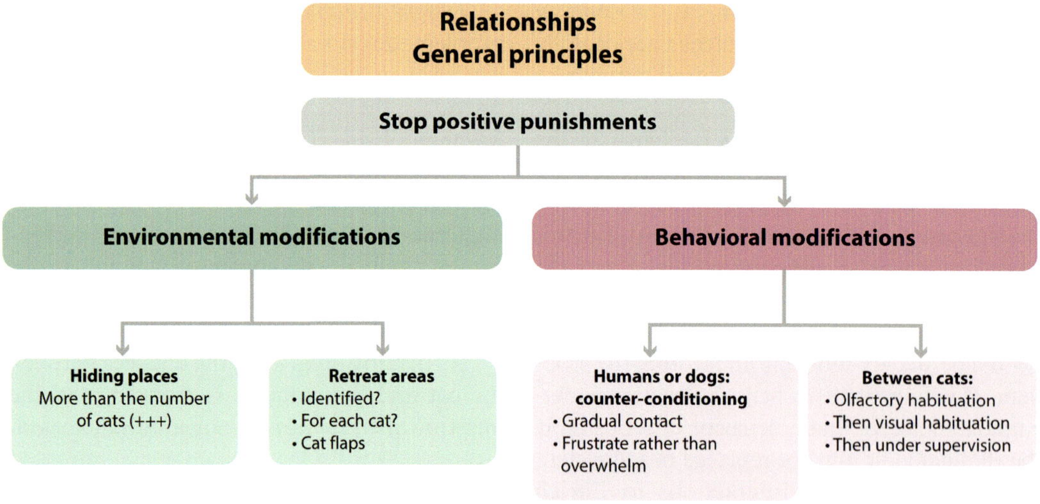

◘ **Fig. 29.14** General principles of therapies for the relationships axis. This protocol includes environmental modifications (in green) and behavioral modifications (in pink). The latter involves, on the one hand, avoiding aggravating situations and, on the other hand, implementing a DSCC protocol tailored to the problematic situations that may arise in interspecific or intraspecific relationships

geted individuals become fearful of interactions due to previous injuries, and both parties begin to anticipate negative outcomes. This creates a vicious cycle that further deteriorates the relationship.

For issues involving one specific human in the household, it is essential to begin by reviewing proper feline communication practices. This includes halting any form of punishment, as detailed in ► Sect. 29.2.2.2, and respecting preferential petting zones.

Often, an *ethological reframing* is necessary, particularly when owners describe their cat's behavior using terms like "revenge" or claim the cat is trying to "dominate" them. As explained, the cat does not belong to a social species, and its behavior is more difficult to interpret compared to that of a dog. While cats can find significant comfort in their human companions, they do not seek contact as intensely as some owners might desire. For example, a cat may enjoy being "near" its owner but may not want direct contact or petting, especially if restrained like a rag doll.

One of the initial steps is to prescribe *observation exercises* for the owners, helping them become more attuned to subtle postures that indicate moderate discomfort in their cat. This allows them to adjust their own behavior accordingly. For instance, many owners fail to notice subtle signs such as dilated pupils or slight tail flicking. Others may persist in forced contact, ignoring the calming signals the cat uses to express its desire to end the interaction. When the cat eventually resorts to irritation-induced aggression, which can sometimes be violent, the owners perceive the aggression as coming "out of nowhere."

One strategy to rebuild a positive relationship is to avoid forcing the cat into contact and instead use high-value reinforcers, such as treats or toys, to create *new rituals*. For example, owners can be instructed to schedule short sessions of a few minutes during which they "invite" their cat to play or offer wet food. The shared enjoyment of these moments should be the priority, and choosing a consistent time each day can help the cat by making interactions more predictable.

When owners express frustration about not being able to pet their cat or complain about having a cat that is "not affectionate enough," it can be beneficial to teach them how to play with the cat's sense of frustration. This involves offering brief, isolated petting sessions paired with a reward, followed by stopping and waiting for the cat to request more interaction. The goal of such an approach is to end the interaction well before any potential aggression might occur (see box chase me and I flee, flee me and I chase).

Important Point of View

Chase Me and I Flee, Flee Me and I Chase

The goal is to encourage your cat to spend more time in close contact with you.

- For various reasons, your cat may not have developed a *tolerance for physical contact.*
- Do not confuse wanting to be "near you" with wanting to be "touched."
- When your cat comes close to you and appears to want to interact, never pet them more than once. Then, wait for them to request further contact before petting them again.
- Over time, the cat will increasingly seek out interactions.

When the problematic relationship involves a child being attacked because they unknowingly exceeded the cat's tolerance limits, interactions must be supervised to gradually teach the child to be gentler and avoid forcing contact with the cat. It is best to give the child clear and specific instructions on how to pet the cat, such as: "Gently pet its head, then its back, then its tail, and stop!" The more precise the recommendation, the more the child can focus on it and follow it attentively.

When the issue involves a dog, it is unrealistic to expect the dog to remain calm either when the cat flees or when the cat attacks. Therefore, therapy requires adding progressive steps to reintroduce the cat and dog. This involves ensuring the cat has a retreat area and controlling the dog's movements during the initial reintroductions, as detailed in ► Sect. 29.2.1.3.

In cases where interactions escalate, such as when a cat is about to or has already started attacking, it is important to have a strategy to reassure the owners. This typically involves prescribing a disruption exercise (owners often yell spontaneously) followed by redirection to play. Owners should always have a toy, like a ball or a crumpled piece of paper, on hand to throw and create movement that catches the cat's attention, diverting it from its aggressive intent.

Of course, in all these situations, therapy can only be effective if the pathological psychological states of all individuals involved are addressed through biological treatment.

▪ Intraspecific Relational Disorders

Once the cats have been isolated from each other, the goal is to gradually reintroduce them using a very progressive exposure protocol. Naturally, depending on the severity of the situation, the cats may already be cohabiting, with fights occurring only occasionally. In such cases, the protocol can proceed more quickly, as the initial steps will have already been completed. To provide a comprehensive approach, however, we describe here the protocol to be followed when the relationship between the cats has led to a complete separation, sometimes lasting a significant amount of time. It is also essential to note that this protocol should only be applied after the pathological psychological states of each cat have been addressed with appropriate medication; otherwise, the behavioral therapy is likely to fail.

Since cats are a territorial species, it is important to explain to owners that distancing threats is not abnormal behaviors, even if they may seem alarming. The key to observing the cats lies in determining whether one cat exaggerates its intolerance toward the other—persistently "persecuting" the other without valid reason, chasing and injuring it even when it retreats—or whether the exchange of threats is simply a way to establish coexistence. This coexistence may never be perfectly harmonious but it is not necessarily abnormal between two cats.

Owners often assign blame to one of the cats, labeling it as "mean" or "violent." It is important for them to understand that the issue involves both cats. The one being attacked often displays behavior that is intolerable in feline communication. For example, remaining immobile and failing to actively occupy the territory is not a structuring behavior in the feline world.

As described in ▶ Sect. 29.2.1.3, the procedure involves following the same steps as for introducing a new cat. However, in this case, not only are one or more of the cats likely to exhibit pathological states, but the history of aggression between them has also altered their relationship. Their preconceived notions are, therefore, far from neutral, and careful progression through each step is the key to success. While the introduction of a kitten can often be successfully completed within a few days, in this scenario, each step will typically take several days to accomplish.

The steps to follow are as follows:

1. **Initial isolation**: isolate each cat in a separate room with its own resources (food, water, litter box, toys). This step allows each cat to calm down and feel secure in its own space.
2. **Scent exchange**: after a few days, once both cats appear more relaxed, begin exchanging objects between their areas, such as blankets or toys. This helps them get used to each other's scent without direct confrontation. Using synthetic pheromone diffusers, like Feliway Friends®, can facilitate this process by reducing tension.
3. **Visual introduction**: when the cats seem less reactive to each other's scents, allow visual interactions through a secure barrier, such as a baby gate or a screen door. Carefully monitor their reactions, noting any signs of curiosity or aggression. Pheromone use is also recommended during this stage to help ease tensions.
4. **Supervised meetings**: if visual interactions proceed smoothly, arrange supervised meetings. During these encounters, ensure each cat has multiple hiding spots and escape routes available. Constant supervision is crucial to intervene if aggression seems imminent.
5. **Gradual increase in interaction**: gradually extend the duration of supervised interac-

tions while closely observing for signs of fear or aggression. Consistently reward calm and positive behaviors with treats or gentle petting to reinforce them.

In the event of a fight between the cats, it is important to intervene using a disruption technique and physically separate the cats by placing a "shield" between them to avoid injury (e.g., a large piece of cardboard or a wooden board). If the protocol is followed gradually, it should not escalate to this point.

If a fight does occur, it is crucial to review what triggered the conflict and consider restarting the steps of the protocol, progressing more slowly to the next stage.

▪ **Relational Disorders Secondary to Another Behavioral Disorder**

As with territorial disorders, the underlying condition must be treated as a priority. These disorders are common in cats with HSHA syndrome, as their overactive behavior often leads to punishments when they bite or scratch their owners (or the household dog). In such cases, the punishment becomes the cause of the relational deterioration. However, addressing the HSHA syndrome is essential to restore harmonious interactions, as the overactive behaviors will not cease spontaneously (see ▸ Sect. 29.3.2).

These disorders can also be secondary to fear axis disorders (see ▸ Sect. 29.3.1). In such cases, the cat's heightened sensitivity is the primary issue, and the owner's initial reaction is not the causal factor.

Finally, in territorial disorders, relational deterioration may stem from punishments following behaviors such as urine marking or from frustration caused by an impoverished environment. These factors often lead to punishments that exacerbate relational issues (see ▸ Sect. 29.3.4).

29.3.5 Other Psychiatric Disorders

29.3.5.1 Aging-Related Disorders

▪ **General Principles**

Regardless of the initial cause, aging progresses through a worsening spiral: loss of capacity, loss of motivation, withdrawal, and functional decline. The goal of all interventions is to delay the onset of this spiral and slow its progression.

Therapies for aging-related disorders always begin by shifting the owners' perspective on their pet: they must understand that the problematic behaviors stem from the disease and not from the animal's unwillingness. Unlike therapies for other conditions, in these cases, the owners are the only ones who can implement the necessary constant support. Their acceptance of this role is therefore essential.

▪ **Environmental Modifications**

Special attention must be given to the comfort and hygiene of the elderly cat to improve its well-being and reduce both discouragement and the risk of irritation-induced aggression, particularly in the context of pain.

The *environment* of an elderly cat should be adapted to help it maintain its activities as much as possible. For example, *ramps* or *steps* can facilitate access to windows, sofas, or beds, preserving a semblance of autonomy. *Scent markers*, such as hydrosols, can be used to help a blind cat navigate by signaling obstacles to avoid or paths to follow, making it easier for them to move around the house. For *sleeping arrangements*, flat, *rimless beds* that the cat can easily access should be prioritized, along with *memory foam models* to relieve joint pain.

Maintaining the *hygiene* of an elderly cat is crucial. If pain prevents the cat from keeping its coat clean, it will experience discomfort without the ability to resolve it, grooming being a vital behavior for cats. Regular dental cleaning and grooming will not only enhance physical interactions between the cat and its owners but also improve the cat's quality of life. Claw trimming is particularly beneficial as it enhances mobility by preventing the cat's claws from snagging on various surfaces.

The quality of communication between owners and their pet is also fundamental to the progression of the disease. Owners of a cat that is beginning to lose its learned behaviors or experience mood changes may be tempted to ignore or isolate it, which can lead to anxiety and withdrawal in the cat. In cases of renal disease, polyuria may cause inappropriate elimination, often resulting in punishment that further deteriorates the relationship between the cat and its owners.

Here again, the practitioner must emphasize that the cat's problematic behaviors are not intentional. Owners should be encouraged to provide an environment that supports the cat's remaining skills and learned behaviors.

▪ **Behavioral Modifications**

The feeding ritual is particularly valuable for elderly cats, as it helps maintain a positive routine and strengthens the bond between the cat and its owner. Additionally, these cats often require multiple treatments, and ensuring that their administration does not cause stress is essential for providing optimal care (see ▶ Sect. 29.2.3.3).

To support the cognitive activity of an elderly cat, it is important to rekindle its motivation and stimulate its cognitive circuits. The goal is not to teach new behaviors but to maintain existing ones.

Owners should start by *reintroducing simple rituals* they used to share with their cat, whether through play—such as rolling a ball just a few centimeters to stimulate the cat—or by using a wand toy and moving it slowly, or even by carrying the cat to a faucet to let it drink from a running tap. During consultations, practitioners should focus on helping the owner recall positive memories by asking them to share stories of their past habits with the cat. This can reignite the owner's motivation to recreate these routines in a simplified form, preventing *demotivation* and encouraging engagement.

29.3.5.2 Essential Psychiatric Disorders

▪ **General Principles**

In cases of essential psychiatric disorders, behavioral modifications are often ineffective, and therapy is generally limited to *environmental adjustments* aimed primarily at *ensuring the safety of the cat and its surroundings.*

First and foremost, precautions must be taken. Keeping a "crazy and dangerous" cat in a household may seem entirely unreasonable to many people, especially given that during episodes of crisis, the attacks of cats suffering from these conditions can be particularly alarming. Owners often become very frightened, and experiencing another episode is rarely tolerated, often leading to the cat's euthanasia.

For this reason, presenting safety measures to the owners is a crucial argument, emphasizing that it is ultimately the cat who will pay the highest price if these measures are not followed.

▪ **Environmental Modifications**

Regardless of the specific essential psychiatric disorder, the first and most important environmental adjustment is the implementation of *strict safety measures to protect household members* from the cat. This involves a dual approach: monitoring for early warning signs of a crisis to isolate the cat in time and having a procedure in place to isolate the cat safely without risking injury. A designated room should be prepared for crisis episodes, and owners should have protective tools such as gloves, a piece of cardboard, or a wooden board to use as a shield, calmly guiding the cat into the isolation room.

These safety measures are complemented by *steps to ensure the cat's own safety and well-being*, tailored to its specific condition.

In the case of *dissociative syndrome*, the cat should be kept in a *calm and predictable environment*, with any identifiable *triggers* of crises removed whenever possible.

For *dysthymia*, the cat needs to be isolated during manic episodes. This isolation must be organized according to the practitioner's recommendations. Owners are typically trained to detect early warning signs of a crisis, such as sudden pupil dilation or progressive tail agitation, so they can isolate the cat before the crisis begins (see box « safety first »).

Important Point of View

Safety First

As soon as you notice a change in your cat's gaze (pupil dilation) or tail flicking:

1. Isolate the cat in the designated closed room.
2. Do not attempt to physically restrain the cat during these crisis moments; prioritize your safety.
3. Wait until the crisis subsides (when the cat has calmed down) before bringing it back to you.

Regardless of the cat's psychiatric condition, its owners must be thoroughly trained to detect changes in the cat's behavior (such as gaze or posture) that signal an impending crisis. This anticipation is critical, as the safety measures implemented for the cat will depend directly on the owners' ability to recognize these signs.

The practitioner must remain available throughout the cat's life to provide guidance and support. It may become necessary to reassess the conditions for keeping the cat in the household if significant family changes occur (e.g., birth, divorce, or family restructuring), as these situations may sometimes lead to decisions about euthanasia for safety reasons.

Behavioral Modifications

The feeding ritual is essential for administering lifelong medication to the cat without the need for physical restraint. Given the dramatic nature of these cats' aggressive reactions, forcing medication administration during a crisis is not an option (see ▶ Sect. 29.2.3.3).

Behavioral modifications are largely ineffective for most essential psychiatric disorders. However, maintaining rituals and creating moments of shared enjoyment, as in relational disorders, can be beneficial. In these cases, it is often the loss of the therapeutic alliance or the breakdown of the bond following one too many episodes that ultimately seals the cat's fate.

Summary of ▶ Sect. 29.3

Therapies can be approached based on the four main axes of behavioral disorders, complemented by treatments for aging-related disorders and essential psychiatric conditions.

Therapies for the *fear axis* are primarily based on the principles of desensitization and counterconditioning (DS-CC). The practitioner aims to find the therapeutic balance between avoidance, which prevents progress, and immersion, which can result in traumatic worsening (◘ **Fig. 29.10**).

Therapies for the *self-control axis* focus on transitioning from external to internal control of the animal. Given the cat's limited emotional and motor control, addressing excessive motivation is crucial. This transition involves various techniques, including environmental enrichment to satisfy heightened motivation, and behavioral modifications aimed at teaching the cat to regulate its excitement and develop control and focus (◘ **Fig. 29.11**).

Therapies for the *territoriality axis* involve modifying the cat's environment, either by enriching it in cases of stimulation deficits or by allowing the cat to isolate itself from disturbing stimuli, at least partially and when desired. Behavioral therapy then focuses on gradually working toward the acceptance of problematic stimuli (◘ **Fig. 29.12**).

Therapies for the *relationships axis* aim to restore harmonious relationships, whether between the cat and its owners or between cohabiting cats. This involves reinforcing the principles of proper feline communication in interspecific issues, avoiding any owner guilt, and creating new rituals and positive shared moments between the cat and the individuals involved in the relational disorder. For intraspecific issues, reintroducing cats must follow a highly gradual contact protocol while educating owners on distinguishing normal feline communication from problematic behaviors (◘ **Fig. 29.14**).

Therapies for *aging-related disorders* focus on slowing the self-aggravating spiral that leads to cognitive and emotional imbalances in elderly cats. This involves managing any pain, maintaining the cat's comfort and hygiene, and ensuring appropriate communication. Simple exercises that help sustain a sufficient level of interest and motivation for social interactions are essential.

Therapies for *essential psychiatric disorders* are limited to environmental modifications aimed primarily at ensuring the safety of the cat and its surroundings. Key elements include isolating the animal, detecting crises, and avoiding known triggers.

Summary of This Chapter

The combination of environmental modifications and cognitive therapies enables comprehensive management of all psychiatric disorders in cats. These approaches are grounded in learning theories, aiming to induce a shift in the perception of the world for both the animal and its owner through cognitive strategies. By employing educational or re-educational techniques tailored to each cat, it is possible to develop a specific therapeutic protocol that addresses the dysregulation of one or more behavioral axes.

Throughout this process, it is essential to respect the cat and maintain consistent communication to create an environment that is both predictable and conducive to the adoption of strategies suited to its species.

References

Ahola MK, Vapalahti K, Lohi H (2017) Early weaning increases aggression and stereotypic behaviour in cats. Sci Rep 7(1):10412. https://doi.org/10.1038/s41598-017-11173-5

Arena L, Menchetti L, Diverio S, Guardini G, Gazzano A, Mariti C (2021) Overweight in domestic cats living in urban areas of Italy: risk factors for an emerging welfare issue. Animals 11(8):2246. https://doi.org/10.3390/ani11082246

Bentosela M, Barrera G, Jakovcevic A, Elgier AM, Mustaca AE (2008) Effect of reinforcement, reinforcer omission and extinction on a communicative response in domestic dogs (Canis familiaris). Behav Process 78(3):464–469. https://doi.org/10.1016/j.beproc.2008.03.004

Beugnet VV, Beugnet F (2020) Field assessment in single-housed cats of litter box type (covered/uncovered) preferences for defecation. J Vet Behav 36:65–69. https://doi.org/10.1016/j.jveb.2019.05.002

Black V (2018) Approach to feline lower urinary tract disease. Compan Animal 23(7):388–394. https://doi.org/10.12968/coan.2018.23.7.388

Blanchard G, Amato C, André A, Bleis P, Ninet S, Zentek J, Nguyen P (2022) Beneficial effects of a prescription home-prepared diet and of zucchini on urine calcium oxalate supersaturation and urinary parameters in adult cats. J Feline Med Surg 24(12):1203–1211. https://doi.org/10.1177/1098612X211067931

Čabraja J (2021) Nonverbal communication: animals and humans. Sveučilište Josipa Jurja Strossmayera u Osijeku, Filozofski fakultet. https://urn.nsk.hr/urn:nbn:hr:142:147736

Casey RA, Vandenbussche S, Bradshaw JWS, Roberts MA (2009) Reasons for relinquishment and return of domestic cats (*Felis Silvestris Catus*) to rescue shelters in the UK. Anthrozoös 22(4):347–358. https://doi.org/10.2752/089279309X12538695316185

Cooper JJ, Cracknell N, Hardiman J, Wright H, Mills D (2014) The welfare consequences and efficacy of training pet dogs with remote electronic training collars in comparison to reward based training. PLoS One 9(9). https://doi.org/10.1371/journal.pone.0102722

Crowell-Davis SL, Curtis TM, Knowles RJ (2004) Social organization in the cat: a modern understanding. J Feline Med Surg 6(1):19–28. https://doi.org/10.1016/j.jfms.2003.09.013

Dantas LM, Delgado MM, Johnson I, Buffington CT (2016) Food puzzles for cats: feeding for physical and emotional wellbeing. J Feline Med Surg 18(9):723–732. https://doi.org/10.1177/1098612X16643753

De Assis LS, Mills DS (2021) Introducing a controlled outdoor environment impacts positively in cat welfare and owner concerns: the use of a New Feline

Welfare assessment tool. Front Vet Sci 7:599284. https://doi.org/10.3389/fvets.2020.599284
De Mouzon C, Leboucher G (2023) Multimodal communication in the human–cat relationship: a pilot study. Animals 13(9):1528. https://doi.org/10.3390/ani13091528
Demant H, Ladewig J, Balsby TJS, Dabelsteen T (2011) The effect of frequency and duration of training sessions on acquisition and long-term memory in dogs. Appl Anim Behav Sci 133(3):228–234. https://doi.org/10.1016/j.applanim.2011.05.010
Demontigny-Bédard I, Beauchamp G, Bélanger M-C, Frank D (2016) Characterization of pica and chewing behaviors in privately owned cats: a case-control study. J Feline Med Surg 18(8):652–657. https://doi.org/10.1177/1098612X15591589
DePorter TL, Bledsoe DL, Beck A, Ollivier E (2019) Evaluation of the efficacy of an appeasing pheromone diffuser product vs placebo for management of feline aggression in multi-cat households: a pilot study. J Feline Med Surg 21(4):293–305. https://doi.org/10.1177/1098612X18774437
Dickman CR, Newsome TM (2015) Individual hunting behaviour and prey specialisation in the house cat Felis catus: implications for conservation and management. Appl Anim Behav Sci 173:76–87. https://doi.org/10.1016/j.applanim.2014.09.021
Ellis SL (2009) Environmental enrichment: practical strategies for improving feline welfare. J Feline Med Surg 11(11):901–912. https://doi.org/10.1016/j.jfms.2009.09.011
Ellis JJ, McGowan RTS, Martin F (2017) Does previous use affect litter box appeal in multi-cat households? Behav Process 141:284–290. https://doi.org/10.1016/j.beproc.2017.02.008
Fernandez EJ, Martin AL (2021) Animal training, environmental enrichment, and animal welfare: a history of behavior analysis in zoos. J Zoolog Botan Gardens 2(4):531–543. https://doi.org/10.3390/jzbg2040038
Foreman-Worsley R, Farnworth MJ (2019) A systematic review of social and environmental factors and their implications for indoor cat welfare. Appl Anim Behav Sci 220:104841. https://doi.org/10.1016/j.applanim.2019.104841
Foreman-Worsley R, Finka LR, Ward SJ, Farnworth MJ (2021) Indoors or outdoors? An international exploration of owner demographics and decision making associated with lifestyle of pet cats. Animals 11(2):253. https://doi.org/10.3390/ani11020253
Frank D, Beauchamp G, Palestrini C (2010) Systematic review of the use of pheromones for treatment of undesirable behavior in cats and dogs. J Am Vet Med Assoc 236(12):1308–1316. https://doi.org/10.2460/javma.236.12.1308
Frayne J, Murray SM, Croney C, Flickinger E, Edwards M, Shoveller AK (2019) The behavioural effects of innovative litter developed to attract cats. Animals 9(9):683. https://doi.org/10.3390/ani9090683
Gaultier E, Falewee C, Bougrat L, Pageat P (2005) The introduction of a female tiger (Panthera tigris) in a pre-established group of two neutered males: a case study. Purdue University Press, West Lafayette
Grant RA, Warrior JR (2019) Clicker training increases exploratory behaviour and time spent at the front of the enclosure in shelter cats; implications for welfare and adoption rates. Appl Anim Behav Sci 211:77–83. https://doi.org/10.1016/j.applanim.2018.12.002
Griffith CA, Steigerwald ES, Buffington CT (2000) Effects of a synthetic facial pheromone on behavior of cats. J Am Vet Med Assoc 217(8):1154–1156
Guy NC, Hopson M, Vanderstichel R (2014) Litterbox size preference in domestic cats (Felis catus). J Vet Behav 9(2):78–82. https://doi.org/10.1016/j.jveb.2013.11.001
Guyot GW, Bennett TL, Cross HA (1980) The effects of social isolation on the behavior of juvenile domestic cats. Dev Psychobiol 13(3):317–329. https://doi.org/10.1002/dev.420130307
Hargrave C (2020) COVID-19: implications of self-isolation and social distancing for the emotional and behavioural health of cats. Compan Anim 25(5):138–143. https://doi.org/10.12968/coan.2020.0036
Henning JSL, Nielsen T, Fernandez E, Hazel S (2022) Factors associated with play behavior in human-cat dyads. J Vet Behav 52-53:21–30. https://doi.org/10.1016/j.jveb.2022.05.007
Herron ME, Buffington CAT (2012) Environmental enrichment for indoor cats: implementing enrichment. Compendium (Yardley, PA) 34(1):E3. https://www.ncbi.nlm.nih.gov/pmc/articles/PMC3933199/
Hiby EF, Rooney NJ, Bradshaw JWS (2004) Dog training methods: their use, effectiveness and interaction with behaviour and welfare. Anim Welf 13(1):63–69
Horn JA, Mateus-Pinilla N, Warner RE, Heske EJ (2011) Home range, habitat use, and activity patterns of free-roaming domestic cats. J Wildl Manag 75(5):1177–1185. https://doi.org/10.1002/jwmg.145
Houser B, Vitale KR (2022) Increasing shelter cat welfare through enrichment: a review. Appl Anim Behav Sci 248:105585. https://doi.org/10.1016/j.applanim.2022.105585
Jezierski T, Camerlink I, Peden RSE, Chou J-Y, Sztandarski P, Marchewka J (2021) Cat owners' perception on having a pet cat during the COVID-19 pandemic. PLoS One 16(10):e0257671. https://doi.org/10.1371/journal.pone.0257671
Kasbaoui N, Cooper J, Mills DS, Burman O (2016) Effects of long-term exposure to an electronic containment system on the behaviour and welfare of domestic cats. PLoS One 11(9):e0162073. https://doi.org/10.1371/journal.pone.0162073
Kogan L, Kolus C, Schoenfeld-Tacher R (2017) Assessment of clicker training for shelter cats. Animals 7(10):73
Kry K, Casey R (2007) The effect of hiding enrichment on stress levels and behaviour of domestic cats *(Felis sylvestris catus)* in a shelter setting and the implications for adoption potential. Anim Welf 16(3):375–383. https://doi.org/10.1017/S0962728600027196

Lloyd JKF (2017) Minimising stress for patients in the veterinary hospital: why it is important and what can be done about it. Vet Sci 4(2). https://doi.org/10.3390/vetsci4020022

Lowe SE, Bradshaw JWS (2002) Responses of pet cats to being held by an unfamiliar person, from weaning to three years of age. Anthrozoös 15(1):69–79. https://doi.org/10.2752/089279302786992702

Maier SF, Seligman ME (1976) Learned helplessness: theory and evidence. J Exp Psychol Gen. https://doi.org/10.1037/0096-3445.105.1.3

Martin P (1984) The time and energy costs of play behaviour in the cat. Z Tierpsychol 64(3–4):298–312. https://doi.org/10.1111/j.1439-0310.1984.tb00365.x

Martin P (1986) An experimental study of weaning in the domestic cat. Behaviour 3:221–249

Masson S, Nigron I, Gaultier E (2018a) Questionnaire survey on the use of different E-Collar types in France in everyday life with a view to providing recommendations for possible future regulations. J Vet Behav 26. https://doi.org/10.1016/j.jveb.2018.05.004

Masson S, Vega S, Gazzano A, Mariti C, Pereira G, Halsberghe C, Leyvraz A, Mcpeake K, Schoening B (2018b) Electronic training devices: discussion on the pros and cons of their use in dogs as a basis for the position statement of the European Society of Veterinary Clinical Ethology (ESVCE). J Vet Behav 25. https://doi.org/10.1016/j.jveb.2018.02.006

McGowan RTS, Ellis JJ, Bensky MK, Martin F (2017) The ins and outs of the litter box: a detailed ethogram of cat elimination behavior in two contrasting environments. Appl Anim Behav Sci 194:67–78. https://doi.org/10.1016/j.applanim.2017.05.009

Meyer I, Ladewig J (2008) The relationship between number of training sessions per week and learning in dogs. Appl Anim Behav Sci 111(3–4):311–320. https://doi.org/10.1016/j.applanim.2007.06.016

Mikkola S, Salonen M, Hakanen E, Lohi H (2022) Fearfulness associates with problematic behaviors and poor socialization in cats. iScience 25(10):105265. https://doi.org/10.1016/j.isci.2022.105265

Moody CM, Mason GJ, Dewey CE, Niel L (2020) Getting a grip: cats respond negatively to scruffing and clips. Vet Rec 186(12):385–385. https://doi.org/10.1136/vr.105261

Mouzon CD, Gilbert C, Di-Stasi R, Leboucher G (2022) How's my kitty? Acoustic parameters of cat-directed speech in human-cat interactions. Behav Process 203:104755. https://doi.org/10.1016/j.beproc.2022.104755

Naik R, Witzel A, Albright JD, Siegfried K, Gruen ME, Thomson A, Price J, Lascelles BDX (2018) Pilot study evaluating the effect of feeding method on overall activity of neutered indoor pet cats. J Vet Behav 25:9–13. https://doi.org/10.1016/j.jveb.2018.02.001

Nuti V, Cantile C, Gazzano A, Sighieri C, Mariti C (2016) Pinch-induced behavioural inhibition (clipthesia) as a restraint method for cats during veterinary examinations: preliminary results on cat susceptibility and welfare. Anim Welf 25(1):115–123. https://doi.org/10.7120/09627286.25.1.115

Oliveira IDA, Viana-Junior AB, De Azevedo CS (2023) Indoor and outdoor management for cats: inferences about the welfare and cat-caretaker relationship. J Vet Behav 60:70–78. https://doi.org/10.1016/j.jveb.2022.12.010

Patronek GJ, Glickman LT, Beck AM, McCabe GP, Ecker C (1996) Risk factors for relinquishment of cats to an animal shelter. J Am Vet Med Assoc 209(3):582–588

Plantinga EA, Bosch G, Hendriks WH (2011) Estimation of the dietary nutrient profile of free-roaming feral cats: possible implications for nutrition of domestic cats. Br J Nutr 106(S1):S35–S48. Cambridge Core. https://doi.org/10.1017/S0007114511002285

Pozza ME, Stella JL, Chappuis-Gagnon A-C, Wagner SO, Buffington CAT (2008) Pinch-induced behavioral inhibition ('clipnosis') in domestic cats. J Feline Med Surg 10(1):82–87. https://doi.org/10.1016/j.jfms.2007.10.008

Pratsch L, Mohr N, Palme R, Rost J, Troxler J, Arhant C (2018) Carrier training cats reduces stress on transport to a veterinary practice. Appl Anim Behav Sci 206:64–74. https://doi.org/10.1016/j.applanim.2018.05.025

Riemer S, Heritier C, Windschnurer I, Pratsch L, Arhant C, Affenzeller N (2021) A review on mitigating fear and aggression in dogs and cats in a veterinary setting. Animals 11(1):158. https://doi.org/10.3390/ani11010158

Riggio G, Borrelli C, Piotti P, Grondona A, Gazzano A, Di Iacovo FP, Fatjó J, Bowen JE, Mota-Rojas D, Pirrone F, Mariti C (2022) Cat–owner relationship and cat behaviour: effects of the COVID-19 confinement and implications for Feline Management. Vet Sci 9(7):369. https://doi.org/10.3390/vetsci9070369

Saito A, Shinozuka K (2013) Vocal recognition of owners by domestic cats (Felis catus). Anim Cogn 16(4):685–690

Schilder MBH, Van Der Borg JAM (2004) Training dogs with help of the shock collar: short and long term behavioural effects. Appl Anim Behav Sci 85(3–4):319–334. https://doi.org/10.1016/j.applanim.2003.10.004

Seligman M, Maier S (1967) Failure to escape traumatic shock. J Exp Psychol. https://doi.org/10.1037/h0024514

Shu H, Gu X (2022) Effect of a synthetic feline facial pheromone product on stress during transport in domestic cats: a randomised controlled pilot study. J Feline Med Surg 24(8):691–699. https://doi.org/10.1177/1098612X211041305

Strickler BL, Shull EA (2014) An owner survey of toys, activities, and behavior problems in indoor cats. J Vet Behav 9(5):207–214. https://doi.org/10.1016/j.jveb.2014.06.005

Thorndike EL (1898) Animal intelligence: an experimental study of the associative processes in animals.

29

Psychol Rev Monogr Suppl. https://doi.org/10.1037/h0092987
Turner DC, Bateson P (2014) The domestic cat: the biology of its behaviour, 3rd edn. Cambridge University Press
Villeneuve-Beugnet V, Beugnet F (2018) Field assessment of cats' litter box substrate preferences. J Vet Behav 25:65–70. https://doi.org/10.1016/j.jveb.2018.03.002
Vinke CM, Godijn LM, van der Leij WJR (2014) Will a hiding box provide stress reduction for shelter cats? Appl Anim Behav Sci 160(1). https://doi.org/10.1016/j.applanim.2014.09.002
Vitale Shreve KR, Mehrkam LR, Udell MAR (2017) Social interaction, food, scent or toys? A formal assessment of domestic pet and shelter cat (Felis silvestris catus) preferences. Behav Process 141:322–328. https://doi.org/10.1016/j.beproc.2017.03.016
Willson EK, Stratton RB, Bolwell CF, Stafford KJ (2017) Comparison of positive reinforcement training in cats: a pilot study. J Vet Behav 21:64–70. https://doi.org/10.1016/j.jveb.2017.07.007
Wojtaś J, Czyżowski P, Kaszycka K, Kaliszyk K, Karpiński M (2024) The impact of environmental enrichment on the Cortisol level of shelter cats. Animals 14(9):1392. https://doi.org/10.3390/ani14091392
Zanghi BM, Gerheart L, Gardner CL (2018) Effects of a nutrient-enriched water on water intake and indices of hydration in healthy domestic cats fed a dry kibble diet. Am J Vet Res 79(7):733–744. https://doi.org/10.2460/ajvr.79.7.733
Ziv G (2017) The effects of using aversive training methods in dogs—a review. J Vet Behav Clin Appl Res 19:50–60. https://doi.org/10.1016/j.jveb.2017.02.004
Zoran DL, Buffington CAT (2011) Effects of nutrition choices and lifestyle changes on the well-being of cats, a carnivore that has moved indoors. J Am Vet Med Assoc 239(5):596–606. https://doi.org/10.2460/javma.239.5.596
Zubin J, Spring B (1977) Vulnerability: a new view of schizophrenia. J Abnorm Psychol. https://doi.org/10.1037/0021-843X.86.2.103

The Causes and Consequences of Veterinary Care

Contents

30.1 Veterinary Care – 798
30.1.1 The Paradox of Veterinary Care – 798
30.1.2 Healthcare Phobia – 803
30.1.3 Consequences of Healthcare Phobia – 810

30.2 Practical Solutions in Veterinary Clinics – 813
30.2.1 Available Tools – 814
30.2.2 Medical Training in Veterinary Practices – 825

30.3 The Care Journey in Practice – 829
30.3.1 Reception and Waiting Area – 830
30.3.2 The Consultation Room – 831
30.3.3 Hospitalization – 833

References – 837

© The Author(s), under exclusive license to Springer Nature Switzerland AG 2026
S. Bleuer-Elsner et al., *Veterinary Psychiatry of the Cat*,
https://doi.org/10.1007/978-3-032-06260-4_30

This chapter first describes the causes and consequences of care in veterinary settings, as well as the mechanisms underlying the development of care-related phobias in cats. It then provides a series of environmental, behavioral, and pharmacological tools to manage or prevent the onset of such phobias. Finally, it outlines an ideal care pathway, offering adaptable elements that practitioners and veterinary facilities can incorporate into their own practices.

30.1 Veterinary Care

30.1.1 The Paradox of Veterinary Care

For pet owners, access to high-quality veterinary care is a key factor in improving their animal's well-being (Dawson et al. 2016). However, this mission carries inherent risks, particularly for cats, which are highly sensitive to their environment and human interactions.

Veterinary care requires close physical proximity to the animal, physical manipulations that may sometimes be painful, and restraint, which is often poorly tolerated by cats. These necessary elements for diagnosing and treating animals can paradoxically cause pain, fear, and frustration, thereby compromising the animal's short-term well-being, even though the ultimate goal is to improve its health. The unique characteristics of cats and their heightened sensitivity to their environment and uncontrolled handling make these interactions especially stressful (Taylor et al. 2022).

Humans face a similar paradox. Many people dread visits to the dentist or doctor, often due to past painful or unpleasant experiences. Dental care phobia, for example, is classified among post-traumatic stress disorders (Bracha et al. 2006). While humans can rationalize these experiences by weighing the long-term health benefits, animals cannot. Cats, in particular, are vulnerable when undergoing veterinary care. Thus, it is the responsibility of veterinarians to adapt care practices to minimize trauma, taking into account cats' specific behavioral needs, such as their need for control, predictability, and a secure environment (Rodan et al. 2022).

The following paragraphs will outline the negative factors a cat may encounter during veterinary care—pain, fear, frustration, and a perceived hostile environment, while introducing recommendations to minimize these factors. These include implementing cat-specific environmental modifications, adjusting handling techniques, and using feline-appropriate pain management strategies (Taylor et al. 2022).

30.1.1.1 Pain

- **Pain as the Origin of Care**

Pain, as defined earlier in this book and redefined in 2020 by the International Association for the Study of Pain (see ▶ Chap. 3, ▶ Sect. 3.3), is the unpleasant sensory and emotional experience associated with or resembling that associated with actual or potential tissue damage (Raja et al. 2020). The primary function of pain is to alert the animal to a threat to its physical integrity. The resulting allostatic load drives the animal to adopt behaviors aimed at alleviating the pain. For instance, a cat experiencing pain in its paw may adopt an abnormal posture or avoid placing weight on the affected limb, much like a dog. However, signs of pain are often more subtle in cats (Rodan et al. 2022).

In addition to its sensory pathways, pain also involves an emotional component (◘ Fig. 30.1). A fear emotion may become associated with the context in which the pain occurred. This mechanism has significant adaptive value: it enables the organism to prepare for a response by activating the release of cortisol and noradrenaline when the pain-associated context is recognized (see ▶ Chap. 3, ▶ Sect. 3.3.1.2). For example, a cat may avoid specific movements or locations if it anticipates pain, helping to prevent further injury or the triggering of more intense pain.

In cats, it is critical to recognize these subtle avoidance or withdrawal behaviors to effectively assess pain (Brondani et al. 2013).

Sick animals arriving at veterinary facilities often present with a level of pain that predisposes them to heightened sensitization. Despite all the precautions taken by the veterinary team, an animal's preexisting pain upon entering the clinic creates a context con-

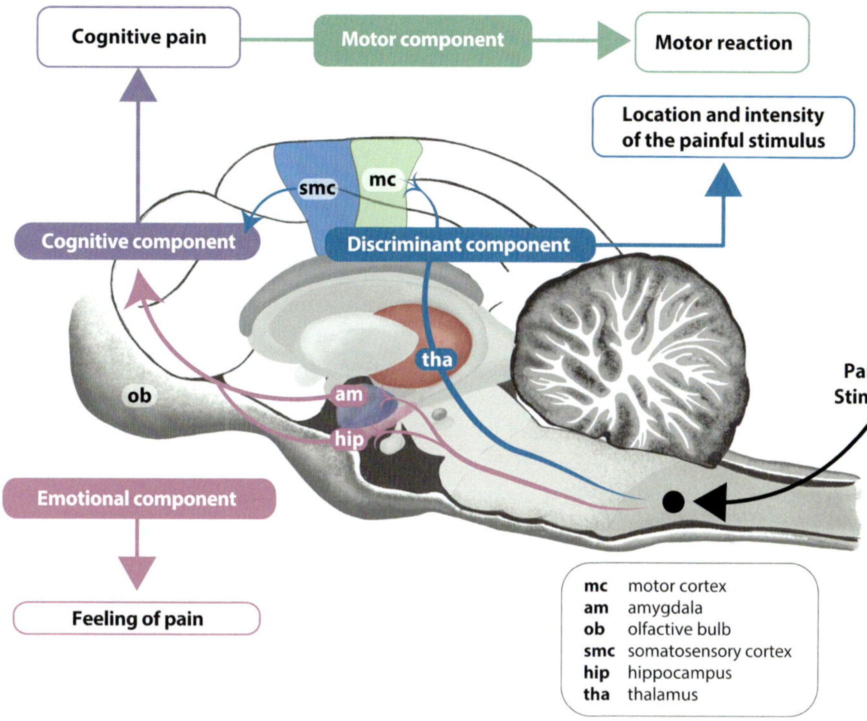

Fig. 30.1 Ascending pain pathways. The ascending pain pathways pass through the dorsal horn of the spinal cord to reach the somatosensory cortex (Sensory discriminative component) and the limbic system (emotional component). The information is then transmitted to the cortex (cognitive component), resulting in a motor response (motor component)

ducive to sensitization to care. This is particularly true for cats, which can exhibit marked reactions to sudden environmental changes, further lowering their pain tolerance threshold (Taylor et al. 2022).

As veterinarians, our responsibility is to manage patients in these challenging circumstances, sometimes under emergency conditions, where delaying care is not an option. It is essential to carefully assess the initial level of pain, as animals experiencing severe pain are at significantly increased risk of developing psychological trauma related to care.

In cases of intense pain, to prevent further sensitization, sedation is often preferable to performing procedures on a fully conscious cat. As with any therapeutic decision, this choice must be guided by a benefit-risk evaluation. Whenever sedation is feasible and does not carry disproportionate risks, it is less harmful than subjecting a severely painful animal to forced care. Sedation produces only temporary side effects, whereas a care-related phobia can permanently impair a cat's ability to undergo future interventions, making them extremely difficult, if not impossible (Rodan 2011).

Pain Induced by Veterinary Care

Every cat will, at some point in its life, face procedures that are potentially painful and stressful. Common diagnostic and treatment actions, such as injections, cystocentesis for urine collection, or rectal temperature measurements, are often unpleasant and sometimes painful. As Carlson and Birkett (2017) explain, any stimuli associated with a painful or stressful experience can form new connections in the amygdala, leading to a conditioned emotional response. This is especially true for cats, which are more prone to associating environmental elements with negative experiences (Steagall et al. 2022). The more a cat suffers during a procedure, the more likely it is to associate contextual elements (e.g., stethoscope, lab coat, examination table) with that pain and to avoid or display signs of fear during future visits.

Pain perception in cats, as in other species, is subjective, making it difficult to standardize pain assessment for a specific procedure. Some cats may exhibit heightened sensitivity to something as simple as a subcutaneous injection, while others may appear to tolerate more invasive procedures, such as blood draws or ear examinations, with apparent resilience (Brondani et al. 2013).

It is therefore crucial for practitioners to evaluate the pain and discomfort caused by each medical act on an individualized basis for every cat.

It is important to note that the more a cat displays signs of pain or stress during a medical procedure (such as crying out or attempting to flee during an injection), the more likely it is to anticipate those sensations during future visits. If these reactions are not addressed and managed, the cat will increasingly associate elements of the clinic—such as the examination table, instruments, smells, or staff—with pain or stress. This heightened association leads to escalating fear and anticipation with each subsequent visit (Rodan 2010). Initially, the cat may only avoid the examination table, but over time, it could start showing signs of stress upon entering the clinic, or even earlier, during transport to the clinic.

Whether it is the reason for the visit or caused by the procedures performed, pain has a significant impact on future care. The veterinary team must, therefore, pay close attention to detecting and evaluating pain in cats to uphold the fundamental principle of veterinary medicine: *primum non nocere*, first, do no harm. By correctly assessing pain and anticipating its consequences, we preserve our feline patients' ability to tolerate medical care and approach future visits with less stress.

▪ Neurophysiology of Pain and Links with Fear and Memory

Pain circuits involve the amygdala, which also functions as the center of fear. In cats, exposure to painful stimuli can quickly lead to durable, and sometimes irreversible, amygdala connections (see ▶ Chaps. 2 and 7). This process of sensitization is an essential adaptive mechanism, allowing the animal to associate pain with contextual elements present during its occurrence and thereby avoid future exposure to painful or pain-associated stimuli. This phenomenon, known as anticipation, occurs rapidly following an irreversible classical conditioning process. In cats, re-exposure to the same context immediately triggers the fear of the associated pain, thereby reinforcing stress during repeated veterinary visits (Vlaeyen et al. 2016).

The only way to inhibit this learned response is to teach the cat to control its pain or fear reaction via the prefrontal cortex, a process known as desensitization (Johnson and Casey 2015). However, it is much easier to prevent sensitization by gradually habituating the cat to its environment than to implement desensitization in an animal already phobic of veterinary care. Once established, care-related phobia is particularly challenging to eliminate, as the prefrontal cortex must learn to inhibit a deeply ingrained emotional response often linked to pain (Steagall et al. 2022; Stellato et al. 2019).

30.1.1.2 Fear

Fear is often associated with pain, and their combined consequences can lead to conditioned emotional responses, as explained in the previous section. In cats, the risk of sensitization to veterinary care increases when they experience initial pain, but this risk also exists if the animal is frightened. A cat that experiences fear during veterinary procedures may refuse to cooperate during future visits, associating the clinic with a hostile environment (Rodan 2011).

Non-painful but unfamiliar elements, such as reading the microchip, using a stethoscope, or employing an ophthalmoscope, can be particularly alarming for some cats, especially those with pre-existing behavioral disorders. Cats most vulnerable to such fear responses include those with fear-axis disorders, such as socially phobic cats (Furgala et al. 2022). Veterinary care often requires close physical proximity to unfamiliar people, which can provoke exaggerated fear in these animals. Any contextual elements associated with these individuals, such as sounds, objects, or smells, can quickly become linked to a phobic reaction.

Veterinarians aiming to deliver care without traumatizing the animal must continuously evaluate each patient for signs of fear, even if they are subtle. This ongoing assessment is crucial to prevent sensitization, even when the intervention is neither invasive nor painful. In cats, fear signals may include crouched postures, ears flattened backward, dilated pupils, or unusual vocalizations (Rodan et al. 2022).

Studies conducted in veterinary environments indicate that fear is highly prevalent among animals in clinics. While research specifically focused on cats is less common, it is well-established that the majority of cats display signs of fear or anxiety during clinical examinations. Additionally, approximately 50% of cats exhibit signs of fear or stress as soon as they arrive in the waiting area, triggered by noise, unfamiliar odors, or the presence of other animals, particularly dogs (Mariti et al. 2016).

Two approaches can help improve this situation. First, trauma can be mitigated by accustoming cats to veterinary care from an early age, reducing their likelihood of becoming sensitized to these stressful situations. Second, it is crucial to enhance the detection of cats vulnerable to fear or phobias related to veterinary care, enabling proactive management before sensitization occurs. These solutions are further discussed in ▶ Sect. 30.2 of this chapter.

30.1.1.3 Frustration

Frustration is defined as the inability to achieve a goal or satisfy a desire. In veterinary settings, care procedures are structured around constraints of time, space, personnel, and feasibility. Cats face various constraints during their care journey, such as waiting, being placed on a table or scale, remaining still during examinations, or being held in unnatural positions. While some cats tolerate these conditions, many show signs of discomfort or agitation, often attempting to escape the situation.

Frustration in cats can manifest as repeated attempts to evade or avoid, plaintive meowing, or, in more extreme cases, aggressive behaviors such as biting or scratching, particularly when the perceived constraint becomes excessive (Couture et al. 2022; Rodan 2010). Unlike dogs, signs of frustration in cats can be more subtle, including behaviors like excessive licking, pupil dilation, or sudden tail movements, which often precede more overt aggression (Steagall et al. 2022).

The primary causes predisposing cats to these reactions are vulnerabilities or disorders related to the self-control axis. When a cat lacking adequate control is restrained against its will, frustration quickly escalates, and the cat may struggle to escape. If the restraint persists or the veterinary team uses overly strong physical handling, this can lead to aggressive behaviors driven not by fear but by a desire to regain freedom of movement (Rodan et al. 2011).

For some cats, intolerance to restraint stems not solely from fear but also from anger or frustration at being immobilized. This is particularly true for cats unaccustomed to human handling or those who have had negative experiences during prior care. Forced immobility is often perceived by these cats as a direct threat to their safety and their need for predictability in their environment (Rodan et al. 2011).

To ensure the well-being of the patient, veterinarians must carefully observe and address signs of frustration as soon as they appear to prevent escalation. If a cat shows even moderate signs of intolerance to restraint during initial visits, it is essential to inform the owner and propose appropriate solutions (see ▶ Sect. 30.2). These may include a comprehensive behavioral evaluation based on the four behavioral axes by a specialized veterinarian to determine the best approach for managing such behaviors in a care setting.

30.1.1.4 The Veterinary Environment

The environment of a veterinary facility encompasses a wide range of stimuli that can be perceived as highly stressful for cats.

Unfamiliar odors, proximity to people and other animals, unusual objects, and handling that involves some level of restraint are all environmental factors likely to provoke intense emotional reactions in felines (Dawson et al. 2016). Cats are particularly sensitive to changes in their surroundings and often perceive these unfamiliar elements as threatening (Taylor et al. 2022).

For example, the vocalizations of other animals, especially dogs, a species that can be perceived as a potential predator by cats, or the sounds of animals expressing fear or pain can heighten a cat's vigilance and fear responses. Additionally, the tension of the owners themselves, along with their perception of the veterinary care and environment, significantly influences their cat's well-being during the visit (Couture et al. 2022; Mariti et al. 2016). Cats are highly attuned to their owners' emotions, which can further exacerbate their stress levels (Rodan et al. 2022).

Important Point of View

Benefits and Limitations of the « Fear Free® » Approach

The practices and standards promoted by the Fear Free® method are indeed useful and necessary for reducing stress associated with the veterinary environment. Initially, these practices focused primarily on environmental elements, such as the size and layout of rooms, as well as standardized handling procedures. While these aspects are important, they do not always account for the individual behavioral vulnerabilities of cats, their pre-existing conditions, or their emotional state at the time of the consultation. These factors are critical in determining how a cat perceives and responds to a care situation (Rodan 2010).

Thus, while the strict application of Fear Free® methods is necessary, it is not always sufficient to ensure that cats tolerate care. The risk of inducing care-related phobias or exacerbating pre-existing behavioral disorders remains, particularly for animals vulnerable to fear. The only reliable way to prevent these outcomes is through continuous, careful observation of the cat's behaviors and emotions throughout the care process and the real-time adaptation of veterinary interventions (Rodan et al. 2011).

The publication of guidelines by the International Society of Feline Medicine (ISFM) in 2022, addressing both the veterinary clinic environment (Taylor et al. 2022) and interactions with feline patients (Rodan et al. 2022), supports an individualized approach. These guidelines align closely with the recommendations presented in this book, promoting a tailored and responsive strategy for feline care.

Fear, pain, frustration, and other stressful stimuli can accumulate in cats and, through sensitization, result in a traumatic experience. A cat's individual tolerance threshold is influenced by its emotional vulnerability, which may be heightened due to behavioral or genetic predispositions. The stimuli frequently encountered in veterinary clinics represent an allostatic load for the cat. When the frequency, intensity, or accumulation of these negative experiences persists, this load can evolve into an allostatic overload, leading to the development of care-related phobia (Masson et al. 2024a). This care-related phobia, detailed in ► Sect. 30.1.2, arises from a detrimental combination of genetic vulnerabilities and environmental factors (◘ Fig. 30.2). It can result in intense defensive behaviors and significant suffering for the animal.

Veterinarians can mitigate this phobia in two ways: by minimizing stressful elements in the clinical environment and by promptly identifying vulnerable cats to provide early intervention (see ► Sect. 30.2.1.1) (Rodan et al. 2022). Both approaches are essential to prevent the situation from worsening into a pathological state (red zone).

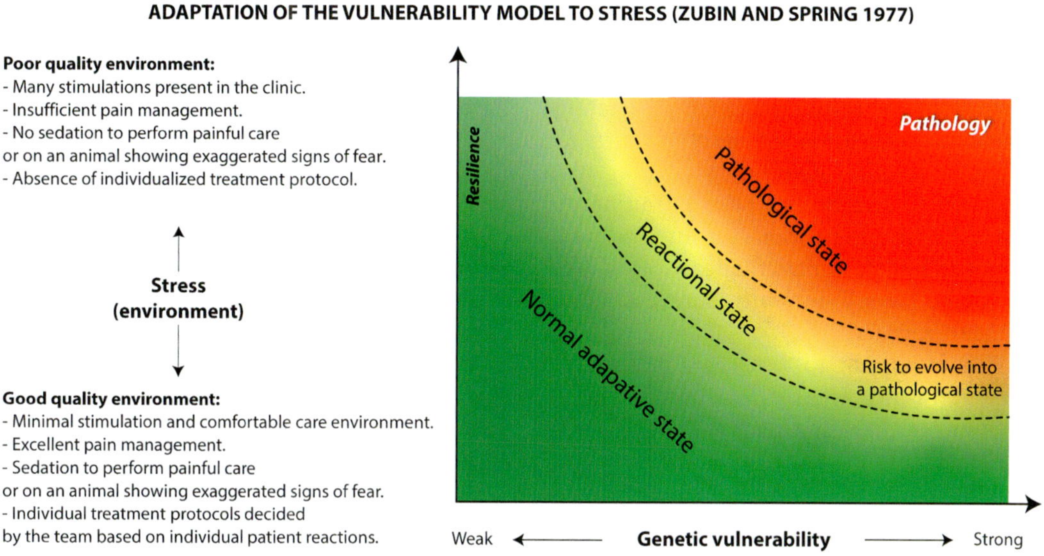

Fig. 30.2 Zubin model adapted for veterinary care

30.1.2 Healthcare Phobia

30.1.2.1 Genesis of Healthcare Phobia

As mentioned earlier, healthcare phobia in cats arises from the inherent paradox of veterinary care. While veterinary procedures are essential for improving long-term well-being, they can cause short-term discomfort, pain, and sometimes fear. Certain veterinary actions, even those deemed minor for most individuals, can be traumatic for more sensitive cats.

Healthcare phobia can develop under two distinct circumstances.

In some cases, the animal experiences a clear trauma, such as major surgery that is painful and disabling, or prolonged hospitalization involving frequent treatments. This type of phobia is best understood as a form of post-traumatic phobia (see ▶ Chap. 8, ▶ Sect. 8.3). Forced handling, during which a cat is physically restrained and struggles, is another frequent source of trauma. Such coercive handling practices are particularly harmful and should be avoided. They account for a significant number of the healthcare phobia cases encountered in daily veterinary practice.

When interviewing owners, a highly coercive restraint episode is often identifiable. This might involve a procedure where the restraint occurred, such as during an ocular treatment, the administration of a painful injection, or the management of an abscessed wound. Similar scenarios can also occur during handling by the owners.

In such situations, sedation is essential, as the negative consequences of such an experience for the cat are often far more severe than the risks associated with routine sedation.

In other cases, healthcare phobia develops gradually, without an identifiable traumatic event. Repeated handling may be perceived as unpleasant, leading to exaggerated fear and then anxious anticipation, which compromises subsequent visits. As described in ▶ Sect. 30.1.1, this phenomenon is exacerbated by pain, fear, and frustration experienced in the veterinary environment. A nontraumatic phobia can develop as early as the first vaccination visits or during a routine physical exam. Over time, as visits continue, the cat generalizes this fear to an increasing number of stimuli: first the examination table, then the veterinarian, followed by all staff wearing lab coats, and eventually any clinical environment (Furgala et al. 2022).

In both cases, the trauma is very real for the cat. In the first case, it is objective (e.g., painful postoperative care), whereas in the second case, it results from an exaggerated emotional perception of harmless stimuli. The cat's reaction determines what is traumatic for it, and both types of trauma can lead to healthcare phobia. It is therefore critical not to underestimate the reaction of a cat that defends itself violently during a simple physical examination.

It is important to remember that certain factors influence whether a cat becomes habituated or sensitized to care (see Table 30.1). In the context of veterinary care or during handling by owners, conditions that promote sensitization are often present.

The distinction between post-traumatic healthcare phobia and non-traumatic healthcare phobia is crucial for assessing a cat's vulnerability. The prognosis is generally more favorable for the post-traumatic form, where the environment is the primary factor, compared to the nontraumatic form, which often reveals deep behavioral vulnerabilities. In the latter case, a comprehensive behavioral diagnosis is essential for a holistic approach. This approach should not be limited to addressing healthcare phobia alone but must also include underlying behavioral disorders (Rodan et al. 2022).

30.1.2.2 Diagnosis of Healthcare Phobia

Clinical Description

Except in specific cases requiring regular and intensive care, isolated healthcare phobia in cats rarely progresses to generalized anxiety or depression. Instead, it is often accompanied by a simple or complex phobic state.

In a phobic state, the cat exhibits exaggerated fear responses to one or more identifiable stimuli, without adaptation despite repeated exposures. When the stimulus is removed, the cat's condition quickly returns to normal. In the absence of the phobic stimulus, the condition is often undetectable.

Cats' behavioral responses to healthcare phobia can vary, and they generally adopt one of the following strategies:

- **Inhibition (*freeze*)**: Complete immobility, lack of exploration, and withdrawal, often accompanied by dilated pupils and muscle tension.
- **Avoidance (*flight*)**: Rapid and sometimes dangerous escape responses, such as sudden jumps or attempts to flee the examination table.
- **Motor agitation (*fidget*)**: Disorganized movements and restlessness around the phobic stimulus, often accompanied by plaintive meowing or vocalizations.
- **Aggression (*fight*)**: Irritation-based aggression in the context of fear or pain, often involving scratching or biting. These aggressive responses are typically brief and not sustained, followed by an attempt to retreat. Uncontrolled emotional aggression can occur in closed situations, often accompanied by autonomic signs such as excessive salivation, urination, or defecation due to emotional stress.

The signs of fear observed in cats include:

- **Appeasement signals**: yawning, lip licking, avoiding eye contact, turning the head away, or moving ears backward.
- **Fear postures**: ears flattened against the head, tail curled under the body or held still, contracted muscles, and a fixed gaze. A sudden aggressive charge may occur, often followed by a quick retreat.
- **Autonomic manifestations**: rapid breathing (tachypnea), rapid heart rate (tachycar-

Table 30.1 Conditions favoring habituation or sensitization

Criteria favoring habituation	Criteria favoring sensitization
Open situation Exposure to a low-intensity stimulus Repeated exposure	Closed situation Exposure to a low-intensity stimulus Isolated exposure

dia), trembling, excessive salivation, emotional urination or defecation, and anal gland expression.

The intensity of these signs correlates with the level of fear experienced. Recognizing the early signs of fear can help the practitioner assess the cat's stress level and decide whether to continue or temporarily halt care.

- **Diagnostic Criteria**

The diagnosis of healthcare phobia in cats is based on identifying an exaggerated and anticipatory fear reaction in the context of care provided by a veterinarian or owner. The severity of symptoms varies, ranging from reactions triggered solely by the presence of the phobic stimulus, such as the examination table, to anxious anticipation associated with restlessness, attempts to flee, or aggression at the earliest signs of impending care, such as seeing the transport carrier.

The inhibited form of the phobia also exists but is rarely reported by owners, despite being equally concerning for the cat's well-being. In such cases, the cat may appear compliant during care, often showing signs like sweaty paw pads, while enduring significant distress. The practitioner can raise awareness among owners by pointing out these signs of intense fear, helping them recognize their pet's discomfort even in the absence of obvious defensive reactions (Powell et al. 2023).

- **Typical Progression**

In the absence of intervention, the progression of healthcare phobia in cats is often unfavorable. Repeated and painful handling reinforces the cat's perception and sensitization to the veterinary environment, increasing its reluctance to accept care. Owners of cats that are rarely ill and visit the clinic only once a year for vaccinations are generally less motivated to pursue behavioral therapy. Between visits, these cats recover sufficiently for their well-being to be affected only during veterinary appointments.

However, for other cats, particularly those that adopt an aggressive strategy, the progression can be rapid and problematic. When a cat scratches or bites during care, the immediate cessation of handling acts as a powerful negative reinforcement. This reinforcement accelerates the instrumentalization of aggressive behavior, which will then be repeated during each visit. This process can make handling impossible, severely compromising the ability to provide care, even in life-threatening emergencies.

In all cases, this unfavorable progression highlights the importance of preventing the development of healthcare phobia rather than managing a cat that has already become phobic.

30.1.2.3 General Considerations Before Addressing Healthcare Phobia

We distinguish between two types of cats with healthcare phobia based on their emotional state before the phobia develops. On one hand, there are cats in a normal adaptive or reactive state who experience an objective trauma during their care journey (e.g., painful or chronic illness, or rough handling). On the other hand, there are cats with a pathological emotional state before entering the care pathway.

The latter are more likely to develop healthcare phobia due to significant emotional vulnerabilities caused by underlying behavioral disorders. In these cases, managing healthcare phobia will be less effective if the comorbid behavioral disorder is not treated simultaneously.

- **Cats Without Behavioral Disorders at the Start of the Care Journey or with Isolated Healthcare Phobia**

Cats in a normal psychological state generally pose few challenges during the care journey. By definition, these cats are adaptable. When brought in for care, novelty does not trigger fear, and they tend to be compliant patients. Naturally, if they undergo very painful procedures or experience violent punitive actions, they may become sensitized.

However, when sensitization occurs, standard medical training techniques are usually sufficient to help them regain balance. This is particularly true if these techniques are anticipated and implemented early in the care process (habituation) or in the early stages of phobia development (desensitization).

Clinical Illustration

Luna is a 5-year-old European cat who has always been independent but relatively docile during veterinary visits. Since her adoption at 8 weeks old, Luna has attended her annual check-ups without notable issues. At home, she is generally calm, enjoys hiding in quiet spots, and does not display signs of excessive fear. She tolerates handling during routine visits, although she is not particularly fond of new situations.

Last year, Luna developed severe gingivitis requiring multiple dental procedures. During her most recent visit, the pain was particularly intense, and two veterinary team members had to restrain her for a thorough examination. This handling was traumatic for Luna, and since that episode, she refuses to leave her carrier at the clinic, showing marked signs of fear.

Today, when her owner brings her to the clinic, Luna freezes in her carrier and growls when attempts are made to remove her. On the examination table, she remains immobile, with ears flattened back, dilated pupils, and slight trembling. When a stethoscope is brought near, she hisses and tries to hide under any available object. Luna exhibits post-traumatic healthcare phobia linked to the painful dental intervention.

If precautions, such as a more gradual approach or preemptive sedation, had been taken during her earlier sessions, Luna would likely have tolerated the care better. Administering medications like gabapentin before each visit, combined with a gradual desensitization protocol during consultations, can significantly help rebuild her trust. For future dental procedures, anesthesia should be planned from the outset to manage her healthcare phobia effectively and avoid exacerbating the trauma.

30

Cats with Behavioral Disorders at the Start of the Care Journey or Comorbid Healthcare Phobia

Cats with a pathological emotional state before beginning their care journey often pose greater challenges for the veterinary team. By definition, cats with behavioral disorders struggle to adapt to certain contexts. When these difficulties are combined with the stress of veterinary care, these cats are much more likely to develop healthcare phobia compared to those in a stable emotional state.

A key indicator is that a cat whose healthcare phobia develops very quickly or becomes disproportionately severe in response to minor medical procedures likely has an underlying behavioral disorder in addition to healthcare phobia. Clinical observations show that the least cooperative and most stressed patients often have another behavioral condition. Their management must differ from that of cats with isolated healthcare phobia.

Cats with a *fear axis disorder* exhibit exaggerated fear in various contexts. Their likelihood of forming a conditioned emotional response during care, becoming sensitized, anticipating, and generalizing is *heightened* due to their initial condition. These cats often present as "shy" at first: they take time to approach, display low postures, and show inhibition or avoidance even during friendly interactions.

When young, these cats may comply with handling, but the experience is often uncomfortable or even unpleasant due to their vulnerabilities. The closed nature of clinical examinations, combined with pain, fear, or the repetitive nature of these visits, which they perceive as negative events, increases the risk of irritation-driven aggression during a visit. Over time, through learning mechanisms such as negative reinforcement and instrumentalization, these cats can quickly become unmanageable without sedation.

Even the use of inhalation anesthesia does not prevent the progression to healthcare phobia, likely because the induction process itself is highly unpleasant for the cat. Managing these untouchable cats is complex and must include treatment of their overarching fear-related condition. Identifying these cats before they develop healthcare phobia significantly improves their prognosis.

Cats with *a self-control axis disorder* present quite differently. They often navigate the consultation room confidently, as though they were at home. They readily exit their transport carrier and may even approach the practitioner entering data on the computer. If they have received rewards during the care process, they may appear excessively friendly, making the examination seem enjoyable.

These cats are often more tolerant than average, reducing the risk of associating certain stimuli with exaggerated fear responses. Consequently, they are less vulnerable than most cats to developing healthcare phobia. However, they are intolerant of immobility and often impatient.

When prolonged or repeated care is required, or if the veterinary team becomes too coercive, the constraint of immobility can sensitize them or even cause genuine trauma. Their responses then reflect their impulsive or compulsive state. These cats tend to be explosive, quick, and poorly controlled if they exhibit aggression. Identifying these cats before they develop healthcare phobia is therefore equally critical.

Moreover, the *territorial nature* of cats means that, from puberty onward, visiting a veterinary clinic presents a major challenge that does not typically exist for dogs. Cats are removed from their familiar environment, which is soothing to them due to their pheromonal marking. This simple fact means that cats with a pathological condition related to a disorder on the territoriality axis are often more affected by visits than other cats.

In such cases, their primary difficulty is not social phobia or being handled by strangers but rather the presence of markings from other individuals or the odors of other species. These factors significantly lower their tolerance threshold, increasing the likelihood of aggressive reactions because they perceive themselves as being in a "hostile" environment. Aggression exhibited at the veterinarian, regardless of the underlying behavioral disorder, is largely irritation-based and occurs within a context of fear in cats (Riemer et al. 2021).

Thus, for all cats with a pre-existing behavioral disorder before the onset of healthcare phobia, it is crucial to adopt a comprehensive approach. Establishing a behavioral diagnosis is essential to offer an appropriate and holistic management plan.

Veterinary Care and Managing Dangerous Behaviors

Unlike other manifestations of fear, which primarily affect the well-being of the animal itself, aggressive behaviors have a direct impact on others. While the perceived danger posed by a cat during veterinary care is generally lower than that of a dog, feline scratches and bites can be serious. They may expose veterinary staff and owners to injuries and infections, such as cat scratch disease (*Bartonella henselae*). Ensuring the safety of caregivers and owners must therefore remain a priority when managing an aggressive cat.

Although managing feline aggression does not involve the same legal implications as dog bites, preventing injuries is essential. Risk mitigation becomes the primary factor in determining care strategies for a potentially dangerous cat. These strategies rely on an appropriate assessment of both the cat and the situation.

As previously emphasized, the clinical examination inherently creates an anxiety-inducing context for the animal: a closed environment, unfamiliar objects, pain, and fear of pain. This combination of factors strongly predisposes the patient to exhibit aggressive responses, such as growling, attempted biting, or actual biting.

Additionally, when a cat becomes so difficult to treat that the entire team dreads its arrival at the clinic, it is often an indication of a comorbid psychiatric condition rather than a simple healthcare phobia. Identifying such cases early is crucial so that the veterinary team can recommend a comprehensive management plan.

Important Point of View

Safety Considerations

While veterinarians are often less concerned about cat aggression compared to dog aggression, cat bites and scratches can be serious and may result in severe infections. Studies suggest that nearly all veterinarians have been victims of cat bites or scratches at some point in their careers (Drobatz and Smith 2003). Preventing these incidents is therefore critical for ensuring the safety of both veterinary staff and pet owners.

Unlike dog bites, cat aggression does not impose strict legal responsibilities on veterinarians in France, such as mandatory reporting or behavioral follow-up. However, it remains essential to implement preventive measures to avoid accidents. These measures include using appropriate restraint techniques, administering sedation when necessary, and educating owners on how to manage their cat during veterinary care.

The semiology of aggression is detailed in the second part of this book (see ▶ Chap. 15, ▶ Sect. 15.2.1). Here, we focus only on the most common injurious sequences observed during veterinary care. It is important to remember that the goal of aggressive behaviors during care, whether growling, attempting to bite, or biting, is to stop an interaction that the animal finds intolerable.

Irritation-Based Aggression

Most aggressive incidents occurring during veterinary care are irritation-based and arise in a context of fear, pain, anticipation of pain, or frustration (Riemer et al. 2021). These aggressions are typically preceded by subtle threat signals, which follow postural cues that the veterinary team must be trained to recognize (see ▶ Sect. 30.2.1.2).

Irritation-based aggression can be more or less controlled, and thus more or less injurious. Such aggression is most often triggered by an *accumulation of stressful stimuli* (◘ Fig. 30.3).

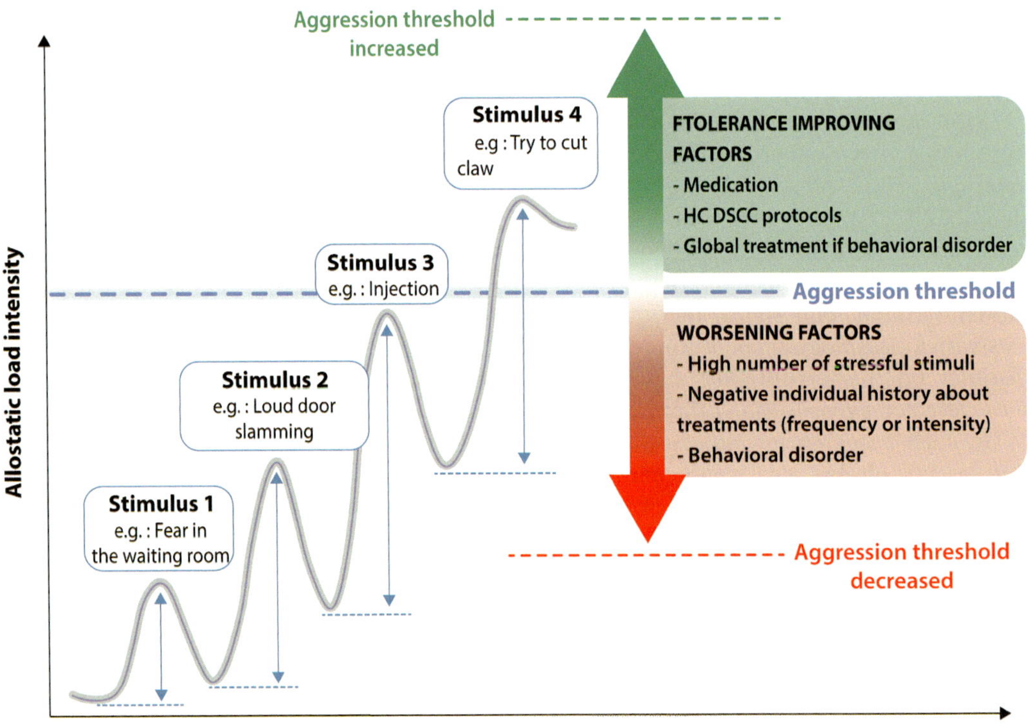

◘ **Fig. 30.3** Factors contributing to the onset of aggression during veterinary care

The likelihood of aggression during care is determined by two factors: the intensity of the allostatic load created by unpleasant stimuli encountered during the visit, and the *animal's aggression threshold* (Edwards et al. 2019).

The aggression threshold decreases in the following circumstances:

The threshold for triggering aggression lowers in the following situations:

- When the animal has negative memories of unpleasant experiences from previous care episodes
- In cases of pre-existing healthcare phobia
- In the presence of other behavioral disorders
- When the cat has a history of positive punishment

Conversely, the aggression threshold increases under the following conditions:

- Administration of anxiolytic medication (see ► Sect. 30.2.1.3)
- Use of habituation and counter-conditioning protocols (HC, see ► Sect. 30.2.2.1) or desensitization and counter-conditioning protocols (DS-CC, see ► Sect. 30.2.2.2)
- Management of the isolated healthcare phobia or any broader behavioral disorder if it exists

The *intensity of the allostatic load* depends on the sum of stimuli perceived as stressful by the individual during their visit:

- Stressful stimuli related to the care pathway: environment, equipment, individuals, etc.
- Events during the care pathway: handling, restraint, painful procedures
- Negative events recalled and acting as stressful anticipatory stimuli
- Absence of the owner during the clinical examination (Karn-Buehler and Kuhne 2022)

When irritation-based aggression occurs during veterinary care, it results from the cat's individual aggression threshold being exceeded. Several stressful stimuli accumulate, leading to an allostatic overload that triggers the aggression. The surprising nature of aggression during handling often stems from the fact that the negative stimulus perceived to have triggered the aggression may appear rela-

Clinical Illustration

Salem is a 3-year-old European cat described as calm but somewhat wary of strangers. She has never caused significant issues at home, but her owners note that she prefers quiet, solitary spaces. Salem arrives at the veterinary clinic for a routine visit, accompanied by her owner. Upon arrival, she remains frozen in her carrier, with her head lowered and ears pinned back. Her owner briefly reassures her, but she does not respond to his petting.

After a few minutes in the waiting room, the veterinarian calls Salem in for her appointment. Salem is removed from her carrier with difficulty, her pupils dilated and her muscles tense. She shows no apparent aggression but remains completely still on the examination table. The veterinarian performs a full examination without any major incident.

At the end of the exam, the owner, having forgotten to mention claw trimming earlier, requests that the veterinarian perform it. As soon as the veterinarian attempts to handle her paws, Salem startles and suddenly turns to bite his hand without warning. The veterinarian quickly steps back, surprised by the sudden reaction.

Salem accumulated stress from the moment she arrived at the clinic. Her fear went unnoticed because she displayed signs of inhibition rather than overt aggression. The claw trimming, perceived as intrusive and potentially painful in her heightened state of stress, abruptly expanded her aggression threshold and triggered irritation-based aggression.

tively mild in intensity. Frequently, the cat would not have reacted to that single stimulus if it had been presented in isolation. Instead, the aggressive response results from the cumulative effect of prior stressful stimuli, which expand the aggression threshold to the breaking point (◘ Fig. 30.3).

Avoiding aggression requires real-time evaluation of the cumulative effect of various stimuli on the patient during the care process. Additionally, when a cat becomes aggressive, it is essential to retrospectively analyze all the stressors experienced by the animal, not just the final triggering event. This analysis is crucial for identifying areas for improvement in future interactions with this particular patient, helping to prevent recurrence and potential accidents.

The analysis begins with the owner's actions, such as placing the cat in the carrier, a process that can already be stressful and influence the entire care pathway (Pratsch et al. 2018).

- **Uncontrolled Emotional Aggressions**

In a closed environment, or one perceived as such by the animal, and in response to intense fear, a cat may trigger *uncontrolled emotional aggression*. This type of aggression is characterized by an incomplete sequence: warning signals are absent, and biting is repeated and unrestrained as long as the triggering situation persists. Autonomic manifestations are always present, including mydriasis (pupil dilation), emotional urination, defecation, and anal gland expression.

The bites in these cases are potentially very injurious, and the loss of control is severe. Cats exhibiting such aggression provide clear evidence of having experienced trauma in a single incident, whether the trauma is objectively identifiable or not.

The onset of uncontrolled emotional aggression may result from the accumulation of stressful stimuli, as previously described for irritation-based aggression.

- **Other Injurious Sequences**

An imbalance in the self-control axis can lead to sudden, forceful, and potentially injurious behaviors from the cat, even though these are not strictly classified as aggression. The incidents resulting from such behaviors are outside the scope of this chapter, as they often involve owners describing being "attacked" by their cat when, in reality, these are sequences of play or uncontrolled excitement.

These types of incidents are rarely observed during veterinary consultations, likely because the cat is removed from its territory, and its size does not cause the same degree of injury through uncontrolled movement as might occur with a dog. However, as noted earlier, an impulsive cat in a clinical setting can have its tolerance for handling exceeded quickly, leading to irritation-based aggression.

Owners may sometimes struggle to understand that their cat, which they handle easily at home, could pose a risk of aggression. In certain cases, the practitioner must warn them against attempting to restrain their cat, as they risk being violently scratched or bitten. The veterinarian is responsible for controlling the animal, and protecting the owner is part of this obligation.

30.1.3 Consequences of Healthcare Phobia

Healthcare phobia may seem trivial to some owners, as it only occurs during the annual preventive visit, provided the animal is otherwise healthy. This perception can make it tempting to avoid addressing the issue, especially when weighing the low frequency of the problem against the time and financial investment required for therapy.

However, healthcare phobia significantly impacts the well-being of all parties involved in the care process: the cat, the veterinary team, and the owner. Moreover, it is wise to anticipate potential illnesses or accidents that may require extensive care, where treatment could be complicated by this pre-existing phobia.

30.1.3.1 For the Cat

First, the intense stress experienced during veterinary visits can alter biological parameters such as heart and respiratory rates, glucose levels, and cortisol levels, affecting the quality of clinical examinations and complementary tests (Rand et al. 2002; Siragusi et al. 2017; Willemse et al. 1993). Stress thus directly influences the reliability of results, complicating diagnosis and treatment.

Additionally, healthcare phobia indirectly reduces the quality of care for several reasons. Intense fear reactions may lead the veterinary team to delay or avoid performing certain complementary exams that are more stressful or painful, knowing that such procedures will be challenging with a phobic cat. Similarly, owners, aware of how poorly their cat tolerates consultations, may hesitate to schedule appointments promptly or to agree to necessary complementary exams (Karn-Buehler and Kuhne 2022; Mariti et al. 2016).

In severe cases, when the patient is noncompliant, many diagnostic or treatment procedures become inaccessible, leaving sedation as the only immediate option to perform these interventions. While sedation can make certain procedures possible, it also poses challenges. Owners may be reluctant to sedate their cat, or the frequency of required sedations may discourage the completion of necessary treatments. In some cases, the act of sedation itself becomes complicated, especially with extremely aggressive cats, further increasing the difficulty of care.

Finally, hospitalization of phobic patients is particularly problematic. It poses safety risks for both staff and the cat while exacerbating the animal's fear and pain, which increases the likelihood of further sensitization.

Thus, while healthcare phobia may seem trivial as long as the cat is healthy, it quickly becomes a significant issue when more complex or invasive care is required. This phobia becomes particularly debilitating when transitioning from preventive care to treatment. For example, some cats cannot be placed on an IV drip unless they are so lethargic that they remain compliant. Once they begin to recover physically, they must often be discharged and sent home because continued care at the clinic becomes impossible.

Moreover, for cats with broader behavioral disorders, often the least cooperative, failing to address the overall behavioral issue means neglecting a significant aspect of their psychological suffering. This lack of care is ethically unacceptable, as these cats endure psychological distress that warrants intervention.

Failing to treat healthcare phobia therefore represents a missed opportunity for the cat. On one hand, its psychological suffering, whether limited to the context of veterinary care or more generalized, is not addressed. On the other hand, the quality of care provided is compromised, as procedures cannot be performed under optimal conditions, and some may be entirely unfeasible.

30.1.3.2 For the Veterinary Team

Beyond the frustration of being unable to provide high-quality care to a phobic or aggressive cat, managing these patients represents a significant cost for veterinary teams on several levels.

First, handling noncompliant cats incurs a *time cost*. These patients require specific organization, including extended consultations and, at times, breaks to allow the animal to calm down or to wait for the sedative to take effect. This type of management is much more complex and demands additional time for procedures that would normally be simple and quick.

The *financial cost* of managing these cats is also considerable. The extra time spent handling these difficult patients is not always billable, making some consultations less profitable for the veterinary clinic. Furthermore, in some cases, costs increase due to the use of sedatives or additional equipment needed to handle difficult cats (Zhang et al. 2022).

Finally, managing these cats carries a significant psychological cost for the veterinary team. The stress of dealing with dangerous patients, where the safety of staff, owners, and other clients is at risk, imposes a mental burden that can greatly affect the well-being of caregivers (Moir and Van Den Brink 2020). Veterinarians report that handling aggressive cats can provoke heightened anxiety, worsened by the fear of being bitten or scratched during examinations or restraint (Beaver 2004).

A cat perceived as "untouchable" is often a source of intense stress for practitioners, especially general veterinarians who are on the front lines of such cases. These veterinarians know well the feeling of relief when a consultation with a potentially dangerous patient concludes without incident, but also the frustration of having to abandon certain procedures because of the risks involved. This constant balance between the benefits of care and the risks to the team creates a stressful dynamic, where every decision must be carefully weighed.

- **Prioritizing Actions When Handling a Dangerous Patient**

From both an ethical and legal standpoint, the veterinarian's primary priority is to ensure the safety of everyone present in the clinic, including staff, owners, and other clients. Managing aggressive cats must begin with a clear assessment of the risk they pose to their surroundings, even before addressing the issue of care. Ensuring this safety is not only a legal obligation but also a practical necessity to prevent serious incidents.

The second priority is to establish an appropriate protocol to treat the cat while minimizing its stress and avoiding exacerbation of its emotional state. In cases of life-threatening emergencies, it is essential to quickly tranquilize the animal to perform the necessary treatments. Even when there is no immediate urgency, the veterinary team must always adopt an approach that preserves the cat's emotional state by carefully selecting interventions and avoiding reinforcement of aggressive behaviors. Sedation protocols, often required for such challenging patients, must be used with caution and in accordance with the animal's overall health condition.

In terms of priority management, profitability comes after considerations of safety and patient care. While difficult cats do not generally jeopardize the financial sustainability of a clinic, they do represent a significant cost in terms of time and energy. This time is often undervalued, and every minute spent managing an aggressive patient reduces the time available for other interventions.

For example, when an adult cat presents for a simple vaccination but displays marked aggression, a prioritization of tasks is necessary. In this case, the vaccination itself is not a life-threatening urgency, and the risks involved do not justify the time and effort required to perform it immediately. The veterinarian can legitimately postpone the consultation to a later date to prepare a management plan for the cat's aggression. This protocol may include sedation methods for urgent cases and, in the long term, behavioral intervention to improve the cat's tolerance for non-urgent care.

Forced restraint, particularly for sedation, should only be used in life-threatening emergencies. Even in such cases, pre-established protocols should be in place to minimize stress and risks associated with these interventions, both for the cat and the veterinary team.

30.1.3.3 For the Owner

Owners are often attuned to their cat's signs of fear, even if they struggle to recognize subtler indicators of discomfort, such as microstress signals (Mariti et al. 2016). The stress experienced by their pet can lead to fewer visits, particularly for preventive care. According to a study by Volk et al. (2011), 28% of owners feel stressed about bringing their cat to the veterinarian, and 38% believe their cat "hates" veterinary visits. This perception directly impacts their overall satisfaction and likelihood of returning to the same clinic (Dawson et al. 2016). Positive interactions between the veterinary team and their cat, along with the use of non-stressful restraint methods, are also significant factors for these owners when choosing a clinic (Beaver 2004).

When owners perceive that their pet's care is poorly managed, they may decide to switch clinics to find one where their cat receives better handling or delay the visit as long as possible, hoping the issue resolves itself. Volk et al. (2011) found that 22% of cat owners reported they would visit the veterinarian more often if their pet experienced less stress during visits. This highlights the issue: animal distress reduces the frequency of veterinary visits, especially for preventive care, which owners often consider less essential.

Additionally, owners now have access to a significant amount of information on stress management for cats, positive methods, and animal welfare. Increasingly, they are unwilling to tolerate coercive or forceful handling and actively seek respectful or cooperative care approaches for their cats (Zhang et al. 2022).

Moreover, owners often feel distress when observing their cat react poorly during visits. In such cases, they may attempt to shorten the consultation, aware that the situation is uncomfortable for both their cat and the veterinary team. This haste can hinder the owner's awareness of their cat's behavioral issues, as once the visit concludes, it becomes more challenging to propose solutions. Few owners will return spontaneously to seek help for this problem.

Summary of ▶ Sect. 30.1

Veterinary procedures often cause pain, fear, and frustration in cats, which are particularly sensitive to their environment.

Depending on their genetic vulnerabilities and history, a cat may develop healthcare phobia either following an objectively traumatic procedure or due to heightened sensitivity, even when handled by a compassionate veterinary team. In the latter case, an underlying behavioral disorder is often present.

Healthcare phobia frequently manifests as aggressive behavior toward caregivers. Managing phobic cats requires an approach that ensures the safety of the team while minimizing additional trauma. Behavioral intervention is essential, particularly for cats with comorbid conditions.

Failing to address healthcare phobia results in diminished quality of care, increased stress for the cat, veterinary staff, and owners, and ultimately compromises the trust-based relationship between the owner and the clinic.

30.2 Practical Solutions in Veterinary Clinics

A thorough understanding of feline behavior, emotions, and learning mechanisms, along with raising owner awareness about the benefits of positive methods and the dangers of coercive approaches, compels us to rethink how we welcome feline patients (Kogan et al. 2017; Lockhart et al. 2013). It is unrealistic to assume that the veterinary environment and procedures do not evoke emotions in cats (Caney et al. 2022). However, we now have many tools to improve their well-being and make the clinical environment less hostile.

Improving the clinical environment for cats is not just about kindness toward the animal; it also directly impacts the quality of care, visit frequency, and owner satisfaction. Owners are increasingly informed about methods that respect feline needs (Rodan et al. 2022). To achieve this, it is crucial to recognize that the best way to manage difficult feline patients is to avoid traumatizing them from the outset.

Intervention should not be limited to moments when a cat shows signs of fear or stress. Every feline patient should receive special attention to make their medical journey as pleasant as possible, not merely tolerable (Westlund 2015). By preserving their comfort and avoiding overly negative experiences, it becomes easier to perform occasional invasive or painful procedures without triggering phobic reactions (see ▶ Sect. 30.2.2.1). Each clinic visit should provide the cat with the opportunity to create a positive experience, thereby reducing the risk of care-related phobias (Herron and Shreyer 2014).

In ▶ Sect. 30.2.1, we will describe the tools and techniques available to the veterinary team to maximize feline well-being during their care journey. ▶ Section 30.2.2 will discuss specific protocols to prevent and treat healthcare phobia.

30.2.1 Available Tools

30.2.1.1 Environmental Tools

- **Facilities**

Minimizing Sensory Stimulation

The environment within veterinary facilities plays a crucial role in managing stress in cats. Sensory stimuli, which are omnipresent in these environments, are among the main sources of stress for feline patients. Implementing strategies to limit these stimuli is essential to make visits more tolerable for cats.

Regarding *visual stimuli*, cats are particularly sensitive to sudden movements and changes in their surroundings. Limiting their exposure to the comings and goings of people and other animals, particularly dogs, can help reduce stress levels. Visual barriers, such as curtains or panels between cages in the waiting area, can shield cats from excessive visual stimuli. Additionally, light intensity should be adjusted, as cats are more sensitive to light variations than dogs. It is advisable to dim the lighting in cat-specific areas to create a more calming environment (Rodan and Heath 2016).

Auditory stimuli are another significant stress factor for cats. Reducing ambient noise, such as conversations, ringing phones, and vocalizations from other animals, is essential to minimize their anxiety. Sudden or loud sounds can trigger fear responses. Research has shown that cat-specific music, such as classical music or sounds tailored to feline auditory preferences, has a more calming effect on cats than music designed for humans (Snowdon et al. 2015).

Olfactory stimuli are also highly significant. Strong smells, such as those from disinfectants or alcohol, can be perceived as unpleasant and associated with negative experiences. Additionally, cats can detect stress pheromones left by other cats, which may heighten their fear (Riemer et al. 2021). Proper cleaning and ventilation of consultation and waiting areas are therefore essential to avoid the accumulation of stressful odors. The use of calming pheromone diffusers, such as Feliway®, is also common in clinics and may help reduce stress in cats (Hargrave 2021), although the evidence supporting their effectiveness remains controversial (Frank et al. 2010).

Finally, *tactile stimuli* must be considered. Slippery or cold surfaces, such as stainless steel, can be uncomfortable for cats, so it is recommended to use mats or blankets in areas where they are handled.

When hospitalized, providing cats with hiding places or spots to retreat to, such as boxes or blankets, contributes to their well-being and reduces stress. Moreover, unnecessary handling, such as uninvited petting or contact by unfamiliar individuals, can become aversive for cats, especially when they are already anxious in a clinical environment (Ellis 2009; Herron and Tony Buffington 2012; Wojtaś et al. 2024).

Facility Layout and Design in a Veterinary Practice

Facilities can always be adapted to mitigate potential shortcomings. For teams working in pre-existing spaces, it may be sufficient to make simple, practical changes without undertaking extensive renovations, as improvements are always possible. That said, certain precautions should be considered when designing or arranging the various rooms in a veterinary clinic, especially when the team can contribute to its layout.

Unlike dogs, cats are transported in carriers, which exposes them to fewer stimuli outside the clinic.

The *waiting room* should be designed to provide ample space and minimize proximity to other animals, particularly dogs, which can cause stress in cats. Ideally, separate waiting areas should be allocated for cats, or transport carriers should be isolated in designated zones (Taylor et al. 2022). If this is not feasible, the room should be arranged to prevent interactions between animals, for example, by spacing chairs far apart and guiding owners to minimize patient interactions.

Regarding the placement and handling of the transport carrier in the waiting room, it is recommended to position the carrier at a height, ideally on a chair or table, so the cat feels more secure (Rodan and Heath 2016). Additionally, the carrier can be partially covered with a cloth to reduce visual stimuli, providing the cat with a quiet and safe space.

Ideally, lighting should be dimmed, and noise levels should be minimized as much as possible. Loud or sudden sounds, such as conversations or ringing phones, can increase feline stress. The use of classical music or calming sounds specifically designed for cats has been shown to have beneficial effects on their stress levels while waiting (Snowdon et al. 2015).

Consultation rooms should also be designed to minimize stress. Ample space should be provided to allow the cat to exit its carrier and explore the environment before the examination. Adding enrichments, such as toys or cat trees, can help the cat relax and better tolerate the exam. Hiding spots or elevated areas are also recommended to provide a sense of security, though care should be taken to ensure the cat can be easily retrieved if necessary.

If the examination table is adjustable, it should operate silently to avoid startling or scaring the cat. It should also be equipped with a non-slip and comfortable surface to prevent any discomfort. Some consultation rooms are specifically designed for feline patients, and in these cases, the presence of cat trees or elevated resting areas is often appreciated by cats (Taylor et al. 2022).

In *hospitalization settings*, cages should be positioned to provide cats with a calm and secure space, avoiding face-to-face configurations with other animals, which can increase their stress levels. Cages should be large enough to accommodate hiding spots or boxes where the cat can retreat, as this greatly contributes to their sense of calm (Ellis 2009).

Noise levels must also be minimized. A vocalizing animal not only disrupts the environment for both human and non-human individuals present, emotionally affecting others and diminishing the serenity of the space, but it is also an indication of distress. The impact of vocalizations should be mitigated, first by isolating noisy animals, and more importantly by using appropriate protocols (pheromones, anxiolytics, or sedatives if necessary) to reduce their occurrence. Staff should also avoid unnecessary conversations near the cages, particularly when patients are calm and resting.

Repeated care and monitoring of hospitalized patients inevitably generate high levels of sensory disturbances. While large veterinary hospitals often lead in terms of resources, patient care, and technical quality, they are also the most prone to noise and olfactory nuisances due to their size and the number of people working there.

Each facility, from small clinics to large veterinary hospitals, should assess the advantages and disadvantages inherent in their layout. By intelligently arranging waiting rooms, consultation rooms, and hospitalization areas, stress levels in feline patients can be significantly reduced, thereby improving the quality of care and patient cooperation.

Density of Individuals

Unlike dogs, cats are not inherently social animals and generally prefer to observe their environment before engaging in interactions, whether with humans or other animals. They respond better when given the opportunity to take their time exploring their surroundings and analyzing scents. In a veterinary setting, the density of individuals is therefore a particularly stressful factor for them.

It is advisable to allow the cat to exit its carrier at its own pace, giving it the chance to explore the environment and sniff different scents before being handled (Rodan et al. 2022). This helps the cat acclimate to the room. Unlike dogs, which may be stimulated by social interactions, the density of individuals, whether other cats, animals, or humans, should be minimized for cats to provide them with a sense of security and control over their environment (Taylor et al. 2022).

For cats, creating a calm environment and limiting interactions to those that are necessary is essential. Scheduling appointments during quieter times, such as early in the day, can help establish a more serene atmosphere. Additionally, veterinarians should pay close attention to a cat's body language and behavior to detect signs of stress or discomfort, allowing them to adapt their approach during the consultation (see ► Sect. 30.2.1.2). Because cats are highly sensitive to smells, it is also recommended to limit strong odors and the buildup of scents from other animals, as these can be disruptive.

Equipment

The equipment used in veterinary clinics should be thoughtfully planned to provide cats with a calm environment throughout their care journey. Various accessories and tools can significantly enhance feline well-being or ensure the safety of the veterinary staff. The goal during care is for the veterinary team to do everything possible to maintain the cat's "goodwill," which will inevitably be tested at some point by a more invasive or painful procedure.

Observing daily practices can reveal many areas for improvement to gradually reduce the overall aversive experience a cat may endure. The most effective improvements in care are not always the most obvious or expensive; enhancing well-being begins with implementing all the changes that are easy to make (Riemer et al. 2021).

Restraint

Cats feel more secure when they have the opportunity to hide or shelter themselves. Cat muzzles, which obscure the cat's vision and fasten with Velcro, can be helpful. For some cats, they prevent biting and often have a calming effect, likely because they also allow the practitioner to perform procedures more calmly, knowing they are safer.

Additionally, using towels for cat restraint is an effective method that also reduces stress. This technique, often referred to as "towel wrapping" or the "burrito method," secures both the cat and the veterinary staff while avoiding overly forceful immobilization (Rodan and Heath 2016).

It is recommended to learn proper towel-handling techniques to ensure effective restraint without causing frustration for the cat. This type of restraint is particularly useful for brief procedures such as physical examinations, injections, or taking a temperature (Williams 2016). At the same time, it is advisable to minimize additional handling during the examination and incorporate regular breaks to prevent emotional overload in the cat (Rodan et al. 2022).

There are also restraint bags, which offer an even higher level of safety than towels. It is very common for cats placed in these bags to become inhibited and, possibly, even calmed.

Finally, the option of performing gas anesthesia without removing the cat from its carrier, by placing the carrier into a large induction box, is an effective way to avoid forced handling. However, it does not appear that such anesthetic procedures are entirely stress-free for cats that are already defensive.

The overarching goal is always to minimize restraint to make the experience as positive as possible for the cat and to preserve cooperation for future visits. For this reason, restraint cages are traumatic for the animal and should be reserved for non-handled cats, such as stray cats brought in by animal welfare organizations for sterilization.

Food Rewards

Food rewards are a powerful tool to improve the veterinary care experience for cats. It is important to choose treats that are suitable for cats and ensure they are accepted by the animal. If a cat refuses a treat, it indicates a stress level too high for the cat to eat, even something it typically enjoys. As mentioned, for cats, a change in environment is already a significant stress factor, which should be counterbalanced from an early age with positive interactions.

Rewards should be provided throughout the consultation to help the cat remain calm, but they are especially crucial during the early stages of life, before the first handling sessions. At this age, kittens are less sensitive to environmental changes than they will be after puberty. Therefore, establishing reward habits well before puberty is essential for all cats.

The use of licking mats, smeared with wet food or pâté, is particularly effective for occupying cats during examinations and can be easily incorporated into all consultations (Martin and Lees 2023).

Pain-Specific Tools

Reducing pain during care is a critical aspect of feline veterinary practice. Using small-gauge needles for injections and blood draws is essential to minimize discomfort in cats. Finer needles, such as 25G, are often preferred to limit pain, and it is important to always change the needle after drawing from the vial to ensure a sharp, unblunted tip (Rodan and Heath 2016). For fluid therapy, the use of small-diameter catheters, such as 24G, is recommended.

The selection of adhesive bandages should prioritize those that can be easily removed, particularly when dressings are applied under anesthesia and are intended to be removed or changed while the animal is awake. Hypoallergenic dressings are less adhesive and thus less likely to cause discomfort. Spray products like *Eaze Off*® can facilitate painless removal of adhesive bandages.

Equipment to Reduce Restraint and Handling

Thermosensitive microchips allow for temperature measurement without physically touching the animal. These chips should become standard in veterinary practices, enabling their use on patients we have identified ourselves during their early years. For other cases, flexible thermometers are much better tolerated than rigid ones, and petroleum jelly should be readily available in every consultation room.

During surgery, sutures and bandages should be chosen based on the expected reaction of the animal during device removal. Long suture ends and loosely tied knots facilitate easier removal. Subcutaneous absorbable continuous sutures are preferable for difficult patients or those following an unfinished DS-CC protocol.

Elizabethan collars that are easy to assemble and remove should be prioritized. While velcro collars are unfortunately noisy during removal, they are very easy to put on. *Inflatable "moon" collars* can be used as an alternative to traditional cone collars for animals that cannot tolerate them, such as those that become overly agitated or inhibited to the point of anorexia.

The use of a "body suit" or "T-shirt" made from rolls of fabric typically used to protect the skin before bandaging or casting can protect wounds while being more comfortable for animals that do not tolerate cones or when cones risk injuring their owners.

In hospitalization settings, *extension lines* for IV tubing should be available. These allow intravenous care to be administered from outside the cage, minimizing disturbances to the animal (Taylor et al. 2022).

Calming Pheromones

Pheromones, such as *Feliway*®, are commonly used in veterinary clinics to reduce stress in cats. These pheromones mimic the natural substances cats produce to mark their territory, helping to calm them in stressful environments. While anecdotal reports and some studies have shown positive effects (Griffith et al. 2000; Shu and Gu 2022), a systematic review by D. Frank on the use of pheromones to treat undesirable behaviors in cats highlighted the lack of robust scientific evidence. According to this review, although some studies reported beneficial effects, they were often limited by small sample sizes or the absence of control groups, making the conclusions uncertain (Frank et al. 2010).

30.2.1.2 Behavioral Tools

■ Observation and Assessment of the Cat

A fundamental prerequisite for evaluating a cat's emotional state is a solid understanding of postures, behavioral sequences, and stress signals. Cats communicate their discomfort through body language and subtle signs, which must be recognized to minimize their stress during care (Ellis 2018). In a veterinary clinic, staff must be able to quickly identify these signs to prevent or mitigate a fear response (Pereira et al. 2014). Therefore, it is essential for practitioners and veterinary teams to be trained in interpreting feline body language, particularly calming signals such as nose licking or slow blinking (Horwitz and Rodan 2018; Rodan et al. 2022).

Equally important is educating cat owners, as many do not notice the subtle signs of stress in their cats (Mariti et al. 2017). However, owners are often a reliable source of information about changes in their cat's behavior, providing valuable insights into stress indicators. Communication between the veterinary team and the owners should be encouraged from the moment the cat arrives at the clinic to ensure optimal care (Mariti et al. 2016).

Productive and Inhibited Responses

30

In veterinary settings, stress responses in cats generally fall into two broad categories: productive responses and inhibited responses (Beaver 2004; Riemer et al. 2021). While inhibited responses are less dramatic, they must still be identified as they also have harmful consequences for the animal.

Productive responses include behaviors such as fleeing, motor agitation, loud vocalizations like meowing, growling, or hissing, and occasionally attempts to scratch or bite. These reactions are more visible, relatively easy to identify, and are often interpreted as clear signs of high stress in cats (Beaver 2004; Carlstead et al. 1993).

Inhibited responses, on the other hand, are more subtle and harder to detect. They may involve total or partial immobility, muscular tension, ears flattened backward, avoidance of eye contact, or refusal to leave the carrier. Additionally, an inhibited cat may refuse to eat, which is a strong indicator of stress. These inhibited responses are often misinterpreted by owners, who may perceive their cat as calm or cooperative when, in fact, it is frozen in fear (Mariti et al. 2016, 2017).

For these cats, the absence of productive behaviors is a strong indicator of stress. It is essential for both the veterinary team and the owners to recognize these signs and adjust their approach accordingly. An appropriate response to a cat exhibiting inhibited signs includes pausing all handling, temporarily stepping back, and then resuming care gradually, integrating gentle and incremental methods to prevent an escalation of stress into more violent productive responses (Rodan et al. 2022).

Physiological manifestations such as tachypnea, sweating of the paw pads, tachycardia, trembling, and piloerection are often present during inhibited responses and confirm the existence of distress in the cat. These physiological signs, though more subtle, are critical indicators of a cat's stress level during a consultation. When these overt expressions of fear occur during care, proposing a *desensitization protocol* or sedation is vital to limit the risk of worsening healthcare phobia in these patients. This approach helps preserve the cat's emotional state and ensures that future care is more manageable and safer.

Traffic Light Model

Observing behavior and its changes is a non-invasive method for assessing immediate well-being. The traffic light model is a tool for identifying a cat's emotional state based on its posture and habits, which can assist in understanding and implementing immediate procedures (◘ Fig. 30.4).

Monitoring the cat, evaluating its emotional state, and continuously applying this scale are essential throughout the cat's presence in the clinic, especially during handling (Ellis 2018).

A patient in the orange zone should be treated as potentially transitioning into a productive response (red zone) at any moment. Patients unable to leave the red zone should have their care deferred, with premedication if necessary, or undergo immediate sedation if the procedure is urgent.

▪ Communication Methods Within the Veterinary Structure

Type of Training Method

Any professional interacting with a domestic animal should exclusively base their actions on *positive reinforcement* and minimize the use of negative punishment as much as possible (Ziv 2017). The use of coercive methods increases stress and may trigger aggressive, anxious, or avoidant responses in cats. All forms of physical

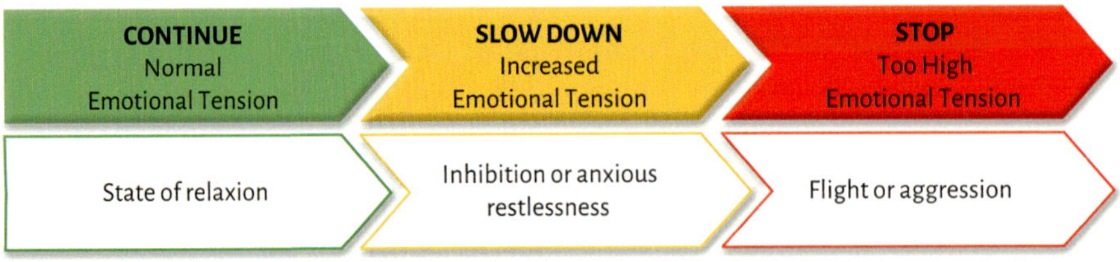

◘ **Fig. 30.4** Emotional state assessment model adapted from the WSAVA (World Small Animal Veterinary Association) (Ryan et al. 2019)

restraint that may cause pain or fear in the animal should be avoided whenever possible.

However, while no companion animal should ever be subjected to violence or forced into visibly unpleasant contact, they must be capable of tolerating friendly and non-violent handling, such as having a paw touched for examination, their back palpated, being brushed, or having their ears handled. For cats, this type of training is far less common than for dogs.

It is therefore essential for the veterinary team to educate owners about the potential for training their cats, provided that such training is carried out at home. For example, training a cat to become accustomed to a transport crate is a straightforward exercise for all cats, yet failing to do so can make crate confinement for a clinic visit a traumatic experience (Pratsch et al. 2018).

Similarly, gently examining the entire body of the animal, along with establishing a daily feeding ritual (such as providing wet food or treats), facilitates the administration of treatments to the cat throughout its life. The veterinary team is at the forefront of demonstrating how to implement this type of training using positive and cat-appropriate methods during consultations.

For example, by systematically utilizing lick mats for all cat-handling procedures from a young age, the veterinarian can show the owner that such gentle interactions enable treatments to be conducted under good conditions. Additionally, when the entire team consistently employs these methods, it helps build trust with the owner, who is often highly concerned about their cat's stress during visits (Karn-Buehler and Kuhne 2022).

Proxemics and Approaching the Animal

It is crucial never to force a cat into interaction. A gentle and respectful approach to the cat's personal space is key to reducing its stress level. Unlike dogs, which can tolerate friendly handling from the outset, cats should be approached slowly, ideally with time to explore their surroundings at their own pace, sniffing and becoming accustomed to the environment's scents before any attempt at physical contact (Rodan et al. 2022). Positive reinforcement, such as treats and toys, should be used to encourage desired behaviors and reduce the association between the clinic and unpleasant experiences (Ellis 2009).

The spatial approach to cats in veterinary environments differs significantly from that for dogs. Cats do not respond to human proximity in the same way and typically do not seek direct contact initially. Approaching a cat head-on or too quickly can provoke fear or even aggression. It is therefore preferable to give the cat time to exit its carrier voluntarily, such as while taking the owner's history, leaving the carrier open. When approaching the cat, it is essential to approach from the side, avoid sudden movements, and wait for signs of relaxation before initiating direct contact (Hammerle et al. 2015).

The best way to ensure a team is sensitive to postural communication is through practical training during care routines, with mutual adjustments until everyone can approach animals using reassuring postures. This includes avoiding a direct, rapid approach, pausing if the animal freezes, averting the gaze or blinking slowly to appear non-threatening, speaking calmly, and avoiding abrupt gestures or jerky contacts.

Restraint

The general rule of restraint is to avoid facing the cat directly and to perform all actions calmly. Immobilizing the cat for a few seconds is often necessary, and proper restraint is what allows this to happen.

Ideally, restraint would not be required at all, but in most cases, it is unavoidable. When restraint is necessary, it should always be carried out in a calm environment using gentle methods, such as wrapping the cat in a towel (the "burrito" technique) (◘ Fig. 30.5), which provides a sense of security and limits movement without causing additional stress (Williams 2016).

Unlike dogs, the presence of the owner is not always reassuring for the cat, as it does not derive the same sense of security from its relationship with its owner as it does from its territory. The staff should guide the restraint process with patience and gentleness, allowing the cat to gradually acclimate to its surroundings and giving it time to relax (Ellis 2009, 2018).

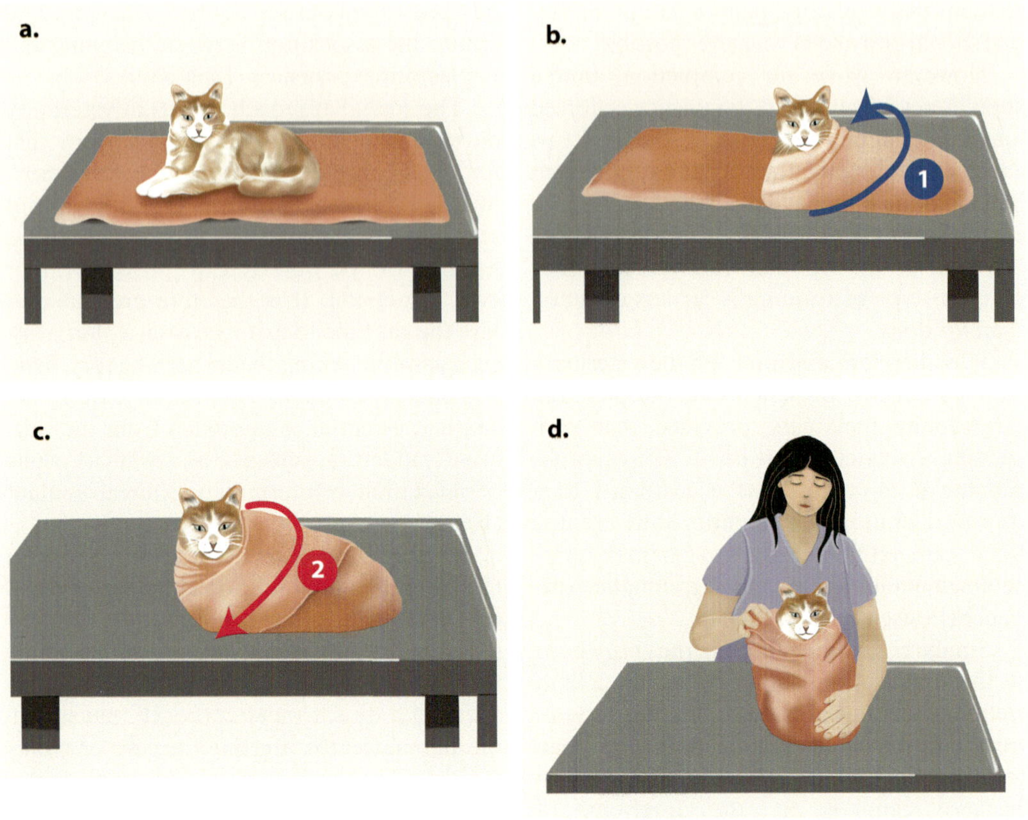

Fig. 30.5 Burrito technique for gentle cat restraint

When gentle restraint is insufficient for performing invasive or painful procedures, or for cats with healthcare phobias, the use of sedatives, administered intramuscularly, intravenously, or via gas induction, may be necessary. Sedation helps to reduce stress and prevents emotional trauma associated with the procedures.

Lastly, it is important to note that emotional contagion operates between cats and humans: a practitioner who interacts with a stressed cat while experiencing emotional tension themselves risks starting off poorly. Quick relaxation techniques, such as heart coherence or other stress-reduction methods, can help minimize tension for the team, the patient, and the owner. Without adhering to a rigid or time-intensive protocol, staff should make an effort to breathe deeply, lower the volume of their voice, and reduce abrupt movements, essentially managing their own stress to help mitigate that of the cat.

▪ Involvement of the Owner in the Care Process

Owners are increasingly seeking compassionate care, and the attitude of the staff toward their animal has become a growing priority (Couture et al. 2022).

However, unlike dogs, a cat's sense of reassurance during veterinary care does not necessarily come from its owner but rather from the stability of its environment. Removed from their familiar territory, cats may exhibit intense signs of stress that their owner's presence does not always alleviate (Behnke et al. 2021).

In some cases, cats may even cooperate better in the absence of their owner. Occasionally, the owner themselves may request to step out during a procedure. When this happens, it is crucial to prepare them for what they might hear from outside the room, such as whimpering or other sounds. Without this forewarning, their anxiety could become greater than if they had stayed to witness the procedure directly.

If a cat displays signs of freezing or inhibition when separated from its owner, it is essential for the veterinarian to proceed with care quickly, respecting the animal's boundaries. The state of inhibition should not be exploited to push the cat beyond its tolerance, as this could lead to panic or aggressive responses later on (Riemer et al. 2021).

The owner can still play an indirect but valuable role in the care process. They can provide important insights about their cat's habits, signs of stress at home, or preferences for treats or toys, which may be used to help soothe the animal during the examination.

Depending on the cat's reactions, it can also be beneficial to quickly reunite the owner with their pet after interventions or anesthesia to minimize the time the cat spends alone in the clinic. For cats hospitalized for longer periods, the owner's presence is often a key factor in encouraging the resumption of eating. Scheduling visits where the owner can spend time with their cat under calm and favorable conditions helps improve the cat's well-being throughout its care journey.

▪ Services Offered in the Clinic

Predictability is a key factor in reducing stress (Nibblett et al. 2015). Conversely, a lack of predictability is an additional aversive stressor for the animal (Bassett and Buchanan-Smith 2007). In simpler terms, a procedure that is predictable for the cat has less impact on its emotional state than one that is unfamiliar. This effect is evident when performing a lateral recumbency X-ray on a conscious patient for the first time: cats tend to panic during the initial attempt, requiring patience until they understand that the procedure is quick and painless. Under similar conditions and with the same gentle approach, they often react much better to a second attempt.

All principles of *medical training* and *consent to care* aim to increase predictability for the cat, helping to make veterinary procedures more manageable for the animal.

Medical Training and Consent to Care

Medical training in cats, though less common than in dogs, can significantly reduce stress during veterinary care. It is based on *habituation* and *conditioning*, teaching cats to progressively tolerate certain manipulations such as nail trimming, blood draws, or physical examinations. This type of conditioning is not feasible for cats already suffering from healthcare phobia, as these cases require a *desensitization* and *counter-conditioning* protocol, which demands a much greater number of sessions and cannot be implemented in emergency situations.

Learning to *tolerate care* is an important skill for cats and should be encouraged from an early age (see ▸ Sect. 30.2.2.1). Cats raised for feline exhibitions often exemplify this increased tolerance, as their training overlaps with many of the learning processes required for cooperative behavior. As a result, they exhibit less fear during medical examinations.

The primary emotions we aim to eliminate (or prevent) in cats during care are fear, pain, and the fear of pain. The cat's individual and immediate perception of the experience is central to this process. Veterinary teams can draw inspiration from cognitive or sensory distraction protocols studied in human pediatrics to apply similar methods to cats (Riemer et al. 2021).

Sensory distractions effectively reduce the perception of pain: the activation of sensory fibers through stimuli such as gentle scratching, tapping, or massage diminishes the activity of nociceptive fibers. A sensory distraction protocol might involve asking the owner to scratch the cat under the chin during a potentially painful procedure, such as placing a venous or urinary catheter or trimming nails.

Cognitive distractions, which engage and divert attention, also reduce the perception of pain. Food is an excellent distractor, and using a lick mat smeared with wet food, for instance, can enable a short procedure to be completed without the cat noticing. Offering treats during the procedure is even more effective if it is an activity the owner already performs at home, as it will carry positive associations for the cat and remain feasible even when the cat is out of its comfort zone, such as during a veterinary visit.

One of the simplest ways to implement cooperative care, where the cat signals discomfort by pausing its activity, is to use a lick mat smeared with wet food. If the cat stops licking, the practitioner or owner halts their action. Once the cat resumes licking, the care activity can continue. In this way, the lick mat serves as a marker of consent.

The implementation of medical training or cooperative care is highly recommended in veterinary practices. However, the team's schedule, often constrained by critical emergencies, the time required for longer consultations, and the limited number of available rooms, can limit the possibility of offering such services to all patients directly within the clinic.

Since cats are typically more comfortable in familiar environments, it is often more effective to delegate these training activities to owners so they can begin the process at home. It is crucial to discuss with owners the importance of habituating their cats to handling and care from an early age. Essential training includes getting the cat accustomed to being brushed, entering the carrier on command, and allowing the handling of their paws, eyes, and ears.

This approach works best when the entire team, including both veterinary assistants and veterinarians, embraces these practices and is trained to guide owners effectively. Demonstrating these techniques during consultations helps owners understand how to replicate them at home, fostering cooperation and reducing stress for both the cat and the care team.

Kitten School

Although still rare in France, implementing an early education program for kittens, similar to puppy school, can play a crucial role in their behavioral development. Such a program provides kittens with the opportunity to become accustomed to basic handling, which facilitates their cooperation during future veterinary examinations. Additionally, by holding these sessions within the veterinary clinic, kittens learn to associate visits with positive experiences, significantly reducing stress related to medical care.

Unlike dogs, cats do not need to learn commands, but it is vital to educate owners about feline behavior, their nonsocial nature, and the importance of their territory for overall well-being. Kitten school offers an ideal opportunity to teach owners how to accustom their cats to essential handling practices (grooming, brushing, nail trimming) while respecting the animal's natural limits.

It also serves as an excellent occasion to discuss normal feline development, such as the acquisition of bite and claw control, and to identify kittens exhibiting signs of excessive fear that may require early intervention.

Groomer

Grooming for cats, although rarely offered in veterinary practices, is often performed under sedation for certain cats, particularly long-haired ones with mats that are difficult to manage. This situation frequently highlights the owners' inability to perform regular brushing, which may result from the cat's aggression or intolerance to handling. It is therefore essential to discuss with owners the reasons behind these difficulties and, if necessary, consider a behavioral intervention.

Cats with a grooming phobia are particularly susceptible to generalizing this fear to other medical procedures, which can jeopardize their veterinary care. To prevent this, it is crucial to accustom cats, starting at a young age, to grooming-related handling, such as brushing, ear manipulation, or nail trimming.

The veterinarian should encourage owners to approach grooming-related handling in a gentle and gradual manner, using rewards such as treats or a lick mat to associate grooming with a positive experience. These grooming practices include progressive training in ear handling and nail trimming, which have a significant positive impact on the cat's cooperation during future veterinary visits.

This is particularly important if the cat has already had a negative experience, as desensitization and counterconditioning are key to addressing their reactions. It is also helpful to train owners to recognize the first signs of discomfort in their cat to avoid forcing interactions and risking further sensitization.

Ultimately, this gradual and positive approach to grooming not only enhances the cat's cooperation during future veterinary care but also minimizes the risk of developing a healthcare phobia.

30.2.1.3 Pharmacological Tools

As discussed in ► Sect. 30.1.2.1), healthcare phobia is very common in cats, partly because less effort is made to prevent its development compared to dogs. This phobia may appear in isolation in animals that were behaviorally balanced beforehand or develop in animals already affected by another behavioral condition, predisposing them to this additional phobia. In both cases, these animals are in a pathological psychological state, and medical intervention is essential to reduce their suffering and accelerate or enable a return to balance.

A study conducted by Barletta on dogs (2016) demonstrated that the use of chemical restraint is beneficial compared to physical restraint for several reasons. Firstly, the cooperation of dogs is significantly better with chemical restraint. This reduces the number of people needed during examinations or procedures, making it more cost-effective for the veterinary clinic. Additionally, the contact time is significantly shorter, making it less stressful for the animal. Although no similar studies have been conducted in cats, our experience suggests that the results would be comparable.

The occasional use of medications during re-education sessions is a consistent approach (◘ Table 30.2) for treating healthcare phobia in animals that were previously behaviorally balanced. For animals suffering from another behavioral condition in addition to healthcare phobia, medical treatment should be initiated after a behavioral diagnosis and target the primary condition; this is often supplemented by situational medication for re-education sessions. When addressing a healthcare phobia, practitioners must keep this distinction in mind to provide the best advice to their clients, as therapeutic strategies and prognoses differ accordingly.

In cases of previously balanced cats, most veterinary clinics are equipped to prescribe situational treatments (e.g., alpha-casozepine, alpha-lactalbumin, pregabalin, gabapentin, trazodone) to facilitate a medical training protocol conducted by a team member, potentially in collaboration with a qualified trainer. However, for cases of healthcare phobia associated with another behavioral disorder, situational medication alone is likely insufficient, and a comprehensive treatment approach is necessary to resolve the phobia. These pharmaceutical options assume that oral adminis-

◘ Table 30.2 Psychotropic medications for managing cats with healthcare phobia

Molecule	Dosages	Indications	Remarks
Alprazolam	0.125–0.25 mg per cat. Can increase up to 0.1 mg/kg	Useful for its amnesic effect after traumatic exposure. Limited benefit for habituation to care or addressing established phobia	Reserved for prevention in cats known to be sensitive or immediately after a trauma related to care in reactive cats
Pregabalin	5 mg/kg, i.e., 0.1 mL/kg of a 50 mg/mL solution, 1.5 to 2 h before exposure	Healthcare phobia, whether isolated or comorbid with another behavioral condition	Preferred molecule for use alongside a habituation or desensitization protocol for care
Gabapentin	30–50 mg/kg 2 h before exposure. May be reduced to 15 mg/kg if sedation is excessive	Healthcare phobia, whether isolated or comorbid with another behavioral condition	Preferred molecule for use alongside a habituation or desensitization protocol for care
Trazodone (not available in France)	5–10 mg/kg 2 h before exposure	Healthcare phobia, whether isolated or comorbid with another behavioral condition	Limited usefulness for habituation due to its strong sedative effect. Effective for performing care without sedation

tration of medication is feasible for the cat. This can be achieved if owners are educated during initial visits about the importance of establishing a feeding ritual with their cat (see box "feeding ritual for medication administration").

Finally, for non-compliant or even aggressive patients, any painful procedure that cannot be deferred should be performed under sedation, even if the cat is already undergoing a re-education program. Failing to do so would expose the cat to a traumatic experience, worsening the condition and nullifying the progress made through re-education. Moreover, the decision to use anesthesia should be made as early as possible in the treatment process to minimize the patient's stress level prior to induction, reduce the amount of anesthesia required, and thereby lower the associated anesthetic risk (Herron and Shreyer 2014).

Important Point of View

Feeding Ritual for Medication Administration

Administering medication to cats can be challenging, especially if the cat has not been accustomed to receiving treatments regularly. To prevent this, it is crucial to establish a pleasant ritual around food, progressively associating medication administration with a positive routine. Such a ritual helps avoid the development of lasting aversion to treatments while making the process easier for owners.

It is recommended to start by offering a small amount of wet food in addition to the usual kibble, without introducing any medication initially. This creates a positive experience for the cat, who associates the new food with an enjoyable moment. The chosen food should be palatable and practical, such as pâté, cheese, or specific cat treats, and enticing enough for the cat to consume it eagerly. The goal is to habituate the cat to receiving this treat daily without suspicion.

Once the cat is familiar with this routine, owners can begin gradually incorporating medications by hiding them in the food. It is often advisable to start with palatable treatments, such as deworming tablets, to make the initial experiences positive. This natural association between appetizing food and medication intake creates a seamless process for the cat.

It is vital to inform owners that this ritual should be established well before the onset of a condition requiring urgent treatment. A cat unaccustomed to taking medications may categorically refuse them if the first experience is unpleasant. By anticipating and incorporating this daily feeding routine, medication administration becomes significantly easier, providing lifelong benefits for the animal and greatly enhancing the quality of care.

It is possible to classify the molecules used in this context into several categories (Table 30.2):

- **Dietary supplements**: these can be used with young, reactive patients who are being habituated to medical care (e.g., alpha-lactalbumin, alpha-casozepine, hemp oil, etc.);
- **Situational medications**: these are used for visits and medical training protocols, as chemical restraints to perform procedures without resorting to sedation, or in hospitalization to calm the animal. These drugs act quickly but have a short duration: pregabalin (Lamminen et al. 2023), gabapentin (Kruszka et al. 2021; Pankratz et al. 2018), trazodone (Stevens et al. 2016).

The use of acepromazine or valium is contraindicated, primarily because their anxiolytic action is limited and, at low doses, they can have a disinhibiting effect, increasing the risk of aggression (Rodan et al. 2011).

All these molecules are compatible with long-term psychotropic treatments (Selective Serotonin Reuptake Inhibitors (SSRIs), Tricyclic Antidepressants (TCAs), Serotonin-Norepinephrine Reuptake Inhibitors (SNRIs), etc.), which are most often prescribed following a comprehensive psychiatric diagnosis. However, when a cat is on selegiline, the administration of other situational medications is contraindicated, except for gabapentin and pregabalin.

Medication prescription holds significant value, even for an issue as seemingly "contextual" as healthcare phobia. It increases the likelihood of successful management, improves the short- and medium-term well-being of the animals under our care, and, at the very least, prevents the onset of a phobia or limits its progression. No practitioner should avoid considering their use, even in a strictly general veterinary practice.

30.2.2 Medical Training in Veterinary Practices

30.2.2.1 HC Protocol: Habituation and Conditioning

The *HC protocol* represents a medical training approach aimed at habituation and conditioning. This protocol is particularly effective for cats that exhibit fear during their first exposure to a medical environment but have not yet developed sensitization or healthcare phobia. The objective of this protocol is to prevent the onset of healthcare phobias in cats that, while uncomfortable, remain in a normal adaptive or reactive state.

As extensively discussed, the best way to avoid dealing with cats suffering from healthcare phobias and their harmful consequences is to *prevent such phobias from developing*. For kittens, this involves implementing a medical training protocol that combines habituation to unfamiliar objects or stimuli with conditioning to facilitate care.

The inherently negative aspects of certain veterinary procedures, inevitably associated with pain or fear, have been previously addressed. Throughout their lives, all cats will face unsettling, unpleasant, or even genuinely painful experiences in the context of care. The goal of an HC protocol is to build a reservoir of positive emotions as a preventive measure, enabling the cat to tolerate unpleasant interventions when they become unavoidable during care.

The technique of *habituation*, involves exposing the cat to a potentially anxiety-inducing stimulus (such as the carrier or examination table) and allowing the cat to explore it at its own pace until there is no fearful reaction and, ideally, complete indifference to the stimulus (see box on learning the transport carrier). This process is particularly important for cats, given their sensitivity to changes in their environment and their tendency to associate such experiences with negative emotions if not gradually accustomed.

Conditioning relies on positive reinforcement. This means associating exposure to medical objects or procedures with pleasant elements such as food, play, or petting. This type of emotional conditioning fosters a positive association with the veterinary clinic and its staff (▶ Sect. 30.1.1.2). Initial visits, when made pleasant, serve as a protective factor against the development of healthcare phobia in cats, much like they do in dogs (Rodan 2010; Westlund 2015).

In practice, the habituation and conditioning components of the HC protocol are applied simultaneously, hence its name. This combined use enhances the effects of each part and significantly accelerates learning. The *habituation* aspect of the HC protocol involves exposing the cat, during its visit, to common elements of the clinic (microchip reader, stethoscope, examination table, etc.) while observing its reaction. If the cat is already indifferent, there is no further action required, and the practitioner can move on to the next step. However, if the cat shows even a slight withdrawal reaction or hesitant contact, it indicates that further effort is needed to achieve genuine indifference.

To achieve this, the practitioner can utilize various tools outlined in ▶ Sect. 30.2.1. They may recommend participation in a kitten school, if available in the clinic, to establish classical conditioning (associating pleasurable moments with the clinic environment and its staff). Alternatively, veterinary assistants can conduct individualized sessions aimed at teaching owners how to perform habituation exercises at home, including the use of treats or a licking mat.

Important Point of View

Training for the Transport Carrier

Habituation to the transport carrier is an often overlooked but essential aspect of minimizing stress in cats during veterinary visits. A recent study demonstrated the positive impact of a structured habituation program on cats' stress levels during transport and consultation (Pratsch et al. 2018). In the study, cats were gradually familiarized with the transport carrier over 6 weeks. Compared to a control group, habituated cats exhibited significantly lower stress levels, as measured by the *Cat Stress Score*. These cats were more cooperative during examinations, and the time required for handling was reduced. Taking the time to explain the benefits of this straightforward habituation process can motivate owners to implement it. Creating a technical guide or video that outlines the steps can further support owners in practicing the protocol.

As detailed in the previous chapter (see ▸ Chap. 29, ▸ Sect. 29.2.3.3), teaching owners the transport carrier habituation protocol involves a gradual process that uses positive reinforcement. This starts with exposing the cat to the open carrier and rewarding its acceptance with treats or toys. The procedure also includes steps to familiarize the cat with the carrier's movements and transportation, ensuring that these experiences are consistently associated with positive outcomes.

The *conditioning exercises* in the HC protocol involve teaching sequences that facilitate care and promote positive conditioned emotional responses, which in turn aid habituation. This approach helps the cat associate the veterinary clinic with positive experiences. Regular use of treats or lick mats can also alter the cat's olfactory perception, making the consultation room more appealing upon entry due to the smell of food (Rodan et al. 2022). The time dedicated to this preparation is an investment that allows the cat to adapt at its own pace, thus easing interaction with the veterinarian.

Unfortunately, this practice is often dismissed as unnecessary for friendly and easy-to-handle cats. However, the opposite is true: only by systematically using rewards with cooperative animals can we effectively prevent the development of healthcare phobia. For routine procedures like vaccinations or check-ups, offering a lick mat to all cats is beneficial. Not only does it make handling easier, but it also reduces the number of patients who might find the vaccine or examination unpleasant, thereby minimizing the risk of sensitization.

It's important to note that the establishment of a cat's territory during puberty often adds difficulty to veterinary visits. This underscores the importance of implementing these protocols early, starting with the first visits. Many kittens are easily compliant, but healthcare phobia can begin to manifest during the 1-year vaccination booster or shortly after sterilization. This shift occurs because puberty introduces additional factors to situations that were previously perceived as neutral or mildly unpleasant, rather than positive.

In the HC protocol, no medication is required since it is designed for animals in a normal adaptive state. On the contrary, this protocol is key to preventing the need for medicalization in the future due to unpleasant experiences. Notably, habituation occurs, by definition, very quickly, often within just a few exposures. A cat that does not reach a state of indifference after a few exposures should be considered as either already in a reactive state or misassessed as it likely has an underlying pathological condition. In other words, the HC protocol should yield rapid and effortless results within a few sessions. If this is not the case, the reasons should be investigated.

Many practitioners believe that implementing such measures is time-consuming and impractical (Arhant et al. 2019). However, the opposite is true in the medium term: time invested in young animals to create a positive association with visits translates to time saved in all future procedures. Cooperative patients are more likely to remain so, and their overall care is made significantly easier.

The HC protocol also serves to educate owners on the importance of observing and understanding their cat's reactions to new objects or situations. By watching how the veterinary team implements these methods, owners can learn to apply similar techniques at home, particularly for initial procedures like claw trimming or brushing. This approach helps to avoid negative experiences that could lead to healthcare phobia.

30.2.2.2 DS-CC Protocol: Desensitization and Counterconditioning

The DS-CC protocol corresponds to a medical training protocol combining desensitization and counter-conditioning. It is specifically designed for cats that are already sensitized and exhibit a healthcare phobia, whether isolated or as part of a broader behavioral disorder. This protocol aims to address the phobia by teaching the cat to tolerate handling while simultaneously reducing its fear response.

While the ideal scenario is to prevent such phobias through the habituation and conditioning protocol, it is inevitable that some cats will arrive at the veterinary clinic already exhibiting phobic behaviors. Some animals develop these phobias in adulthood, while others—particularly those with developmental disorders (e.g., deprivation syndrome, social phobia)—may develop a healthcare phobia despite all precautions taken (see Sect. 30.1.1.5). Their condition predisposes them to fear the veterinary care environment, a reaction that neither treats nor the team's kindness alone can overcome.

Desensitization involves procedures similar to habituation: it consists of gradually exposing the animal to the phobic stimuli step by step, waiting until it no longer exhibits a fear response (see ► Sect. 30.1.1.2). However, for sensitized cats targeted by desensitization, synaptic connections have already been formed in the basolateral amygdala. The goal is not simply to teach a neutral response to a given stimulus but to train the prefrontal cortex to regulate a fear response that can never be entirely erased.

This explains why desensitization takes significantly longer to implement compared to habituation. It also accounts for why, in cases of overly rapid or intense re-exposure to the stimulus, a sensitized cat's exaggerated fear response can resurface almost immediately. Following sensitization, the potential for a fear response never completely disappears; the amygdala and hippocampus can recall it very quickly. This defensive mechanism enables rapid reactions to perceived danger, especially when a negative experience occurred in the same context, even years earlier. Unfortunately, this self-protective mechanism underpins all phobic disorders and highlights why their recovery is so fragile.

The *counter-conditioning* component of the desensitization-counterconditioning (DS-CC) protocol is analogous to the conditioning (C) component of the habituation-conditioning (HC) protocol. Both rely on the same principle: the cat learns an exercise that allows it to respond differently when exposed to the phobic stimulus (e.g., resting its head on the owner's lap and staying still). The difference between C and CC also lies at the cerebral level.

In the first case (C), we are working with a "naive" brain where new synapses are created, a rapid process akin to teaching a puppy to sit. In the second case (CC), the sensitized brain has already formed numerous connections, enabling it to respond rigidly and quickly to a stimulus, for example, turning its head and threatening the owner at the sight of a brush. Learning a new response in such a brain requires significantly more time and steps: first, to avoid triggering the learned response (DS), and second, to teach a new response (CC).

For instance, the exercise might involve the cat licking a treat-filled mat (CC) while the owner approaches the brush very gradually and progressively (DS). To ensure this new response becomes the cat's primary reaction, it is essential to reinforce the associated brain connections until they are stronger than the previous ones. This involves extensive repetition without ever retriggering the initial reflex response being suppressed.

The DS-CC protocol is considered successful when, after the training, the owner can brush the cat while it licks the mat, even if the brushing inadvertently pulls on its fur, without the cat reverting to threatening, turning its head, or fleeing.

The DS-CC protocol is lengthy and requires gradual steps, reinforcing new neural connections so that the new response replaces the initial reflexive fear response. The goal is to achieve a stable reaction in which the cat tolerates or even accepts handling and medical care without reacting out of fear or aggression.

In practice, this protocol is far more complex than the HC protocol and often requires the assistance of an animal behavior professional. It is difficult to carry out entirely within a veterinary clinic due to the time and space required and because initially, the cat may not accept rewards in a perceived hostile environment, even with medication. Therefore, the commitment and willingness of the owners to implement this protocol are critical factors in its success. The desensitization process typically begins at home and is completed at the clinic when the cat is ready to tolerate a medical environment.

It is crucial to emphasize that cats with healthcare phobia are in a pathological state, and the use of psychotropic medications is essential to reduce their reactivity and facilitate rehabilitation (see ▶ Sect. 30.2.1.3). Psychotropics serve two purposes: they decrease the intensity of fear reactions, reduce the risk of accidental sensitization, and minimize aggression risk. They also shorten the time needed to progress through each step of an already lengthy protocol. Refusing to use psychotropic medications to accelerate progress during sessions would be as illogical as declining to prescribe anti-inflammatories for a sprain: in both cases, we promote recovery by ensuring greater comfort, which is central to our mission as veterinarians.

Furthermore, DS-CC protocols demand significant motivation and investment from owners. Failing to ease this challenging protocol with appropriate medication often leads to its abandonment or failure, as many owners may give up mid-way through the DS-CC process. Neglecting to provide pharmacological support alongside a DS-CC protocol represents a significant missed opportunity for the cat (Riemer et al. 2021; Rodan et al. 2022).

Integrating these protocols into the care pathway also strengthens the veterinarian-owner relationship (Karn-Buehler and Kuhne 2022; Mariti et al. 2016). Owners often take pride in seeing tangible improvements in their cat's behavior, which boosts their motivation to adhere to these procedures. They gain a better understanding of their cat's reactions and learn to manage its fear more positively.

Finally, while some owners may lack the time or patience to implement such protocols, it is the veterinarian's responsibility to explain the importance of these procedures and direct them to qualified professionals. Demonstrating that the clinic prioritizes the psychological well-being of cats is a valuable asset appreciated by owners, enhancing their trust in the clinic and the quality of care provided to their pets.

30.2.2.3 Comprehensive Management

A comprehensive behavioral approach is recommended for cats presenting with a comorbid behavioral disorder alongside healthcare phobia. This management strategy primarily targets the overarching behavioral condition, encompassing diagnosis, environmental and behavioral modifications, as well as medication therapy. Addressing healthcare phobia is then integrated into the broader therapeutic framework, becoming one of the specific goals, with the selected protocol supported by overall behavioral therapy. The implemented DS-CC protocol is further reinforced, in most cases, by additional situational medication to enhance its effectiveness.

Let us emphasize that cats exhibiting the most extreme reactions, including aggression during veterinary care, almost invariably suffer from a behavioral disorder extending beyond this specific situation. A singular focus on healthcare phobia is, therefore, destined to fail.

When practitioners face cats whose fear makes them dangerous, their primary responsibility is to ensure safety during the ongoing procedure by employing immediate chemical restraint (Kruszka et al. 2021; Lamminen et al. 2023; Stevens et al. 2016) or sedation. All actions and observations should be meticulously documented in the cat's medical file to protect team members who may handle the animal in the future. Additionally, a referral for a veterinary psychiatry consultation should be suggested to the owners, focusing on a broader behavioral management plan that includes guidelines specific to veterinary care.

In some instances, managing healthcare phobia itself may constitute the primary reason for seeking a veterinary psychiatry consultation. In many cases where aggression is a concern, it often manifests during veterinary care as well. Behavioral therapy is then structured to address this, incorporating the initial stages of the DS-CC protocol into the broader therapeutic plan. These initial steps are typically carried out at home to improve the cat's overall tolerance to care. These severe cases pose significant challenges to the entire team and demand a comprehensive and tailored approach that matches the severity of the cat's condition.

Whichever technique is chosen, HC, DS-CC, or a more comprehensive approach, the partnership with the owner is essential: the procedures are lengthy and require significant commitment, without which the results are often disappointing (Riemer et al. 2021). This partnership involves reaching a mutual agreement on the goals and the means to achieve them. These goals can vary greatly from case to case. For some cats, the ultimate objective might simply be getting them accustomed to the transport carrier and enabling the administration of medication before visiting the clinic for sedation. For others, it might be achieving full cooperation during all necessary examinations.

With experience, practitioners learn to assess the potential progress for each individual cat through careful observation of its behavior, combined with the rate of improvement observed during the process.

Summary of ▶ Sect. 30.2

Preventing healthcare phobia or managing a phobic cat requires the ability to observe and interpret feline postures and to adopt communication methods that reduce the patient's fear while facilitating care.

The organization of the clinic environment, the flow of individuals, and the equipment used must all be optimized to minimize pain and fear during handling. Medical training based on habituation and conditioning techniques is an excellent preventive tool that should be introduced early to all owners for every kitten.

For cats already experiencing phobia, medical support combined with a desensitization and counterconditioning protocol is recommended. A more comprehensive behavioral management approach is often necessary in such cases.

It is crucial to remember that treating a healthcare phobia is far more challenging than preventing its occurrence.

30.3 The Care Journey in Practice

This section outlines the journey of a cat through a veterinary clinic, applying the tools discussed in ▶ Sect. 30.2. The goal is to identify and minimize as many controllable stressors as possible. For every cat, whether cooperative or not, managing potential sources of stress is essential, even as a preventive measure. In cases of healthcare-phobic cats, recognizing and mitigating these stressors is crucial, not only to avoid worsening the phobia but also to ensure the safety of everyone involved, including the cat itself.

Each cat is unique, shaped by its behavioral traits and previous experiences with care. The veterinarian's role is to carefully observe the cat's behavior and reactions from the moment it arrives at the clinic, tailoring the approach and care to the specific needs of the animal throughout its journey.

30.3.1 Reception and Waiting Area

30.3.1.1 Entry into the Waiting Area and Emotional Assessment

Unlike dogs, cats typically enter the waiting area in their transport carriers, which complicates the evaluation of their emotional state. Observing their posture may be challenging, but certain signs, such as panting, high-pitched meowing, drooling, or complete immobility, can indicate a high level of stress. Owners are not always aware of these indicators, making it the responsibility of the reception staff to quickly assess the cat's condition upon arrival (Mariti et al. 2017).

If a cat displays clear signs of stress, it is advisable to direct them immediately to a quiet consultation room or, if that is not possible, to allow the owner to wait in a more isolated area, such as their vehicle. This helps avoid the sensory overload often experienced in the waiting area (Taylor et al. 2022).

30.3.1.2 Positioning of the Carrier in the Waiting Area

The placement of the transport carrier is crucial to minimizing the cat's stress. The carrier should not be placed directly on the floor, as this creates a position of vulnerability where the cat may see other animals and humans moving above them. Ideally, the carrier should be positioned at a higher level, such as on a chair, and covered with a lightweight cloth to reduce visual stimuli and provide a sense of security (Taylor et al. 2022). Owners should be encouraged to bring a cloth imbued with the scent of the home to help comfort their cat.

It is also important to prevent direct visual contact with other animals, particularly dogs. Many veterinary clinics provide separate waiting areas for cats and dogs or, at a minimum, zones divided by partitions to keep the two species apart (voir ◘ Fig. 30.6). While cats' acute sense of smell allows them to detect the presence of dogs, physical separation and ele-

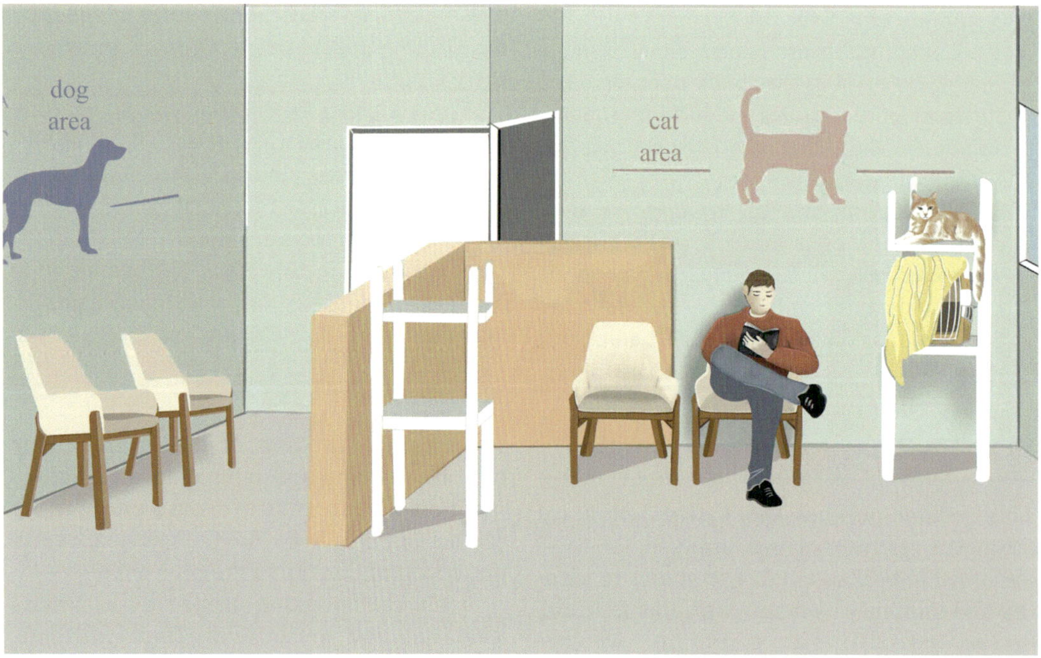

◘ **Fig. 30.6** Separate waiting areas for cats and dogs with provisions for placing cat carriers at an elevated height

vated, enclosed spaces help significantly reduce stress during the waiting period.

30.3.1.3 Interactions in the Waiting Areas

The approach to a cat in the waiting area must be gentle and indirect. Unlike dogs, attempting immediate contact through the bars of the carrier is rarely appropriate. Cats should remain in their secure space until they are in a calmer environment where they can be allowed to exit the carrier at their own pace. Avoiding sudden movements, loud noises, and minimizing activity around the carrier can significantly reduce stress. The lighting in the room should be subdued, and the use of calming music in the waiting area (Hampton et al. 2020) and pheromone diffusers may help establish a more soothing atmosphere for cats (Frank et al. 2010).

While direct interaction with the cat in the waiting area is limited, this is an ideal time for the veterinary team to assess whether the transport carrier is appropriate. If necessary, they can provide advice on selecting a suitable carrier (see *Transport carrier guide*). This moment can also be utilized to weigh the cat inside the carrier and note the carrier's weight (using tape to mark the empty carrier's weight during or after the consultation). This allows owners to bring their cat back for weight checks without needing to remove them from the carrier, which is particularly helpful for dietary monitoring and follow-up care.

Important Point of View

Transport Carrier Guide

The transport carrier is an essential tool for ensuring the comfort and safety of a cat during travel, particularly for veterinary visits. Choosing the right carrier and maintaining it properly is crucial. A good carrier should be sturdy and secure, with a reliable locking system to prevent accidental opening. Ideally, it should have a removable lid, which makes handling in the clinic easier without forcing the cat out of its refuge. This feature also helps minimize stress by allowing the cat to remain in a perceived safe space during examinations.

The size of the carrier is equally important. The cat should be able to stand, turn around, and lie down comfortably. It is best to select a carrier that is spacious enough to provide comfort but not so large that the cat feels exposed or vulnerable. Additionally, the carrier should be well-ventilated, with sufficient openings to ensure airflow while still serving as a sheltered space for the cat.

Regular cleaning of the carrier is often overlooked but is critical for the cat's well-being. After each use, the carrier should be thoroughly cleaned to remove any odors, especially stress pheromones the cat may have left behind. Using mild, nonaggressive cleaning products is recommended. After cleaning, the carrier should be aired out well to eliminate residual odors, ensuring that the cat does not develop an aversion to entering it for future trips.

30.3.1.4 Managing Difficult Patients

Some cats can exhibit extreme anxiety or aggression upon arriving at the clinic. To minimize stress, it is essential to reduce their waiting time in the reception area and avoid exposing them to stressful stimuli such as loud noises or interactions with other animals. It is advisable to schedule a swift consultation for these patients or immediately guide them to a quiet room upon their arrival.

Anxiolytics such as pregabalin, gabapentin, or trazodone, prescribed prior to the clinic visit, can also be considered for cats with high stress levels (Kruszka et al. 2021; Lamminen et al. 2023; Stevens et al. 2016). This approach helps manage their anxiety, ensuring a safer and more effective consultation for both the cat and the veterinary team.

30.3.2 The Consultation Room

During a consultation with a cat, the approach must differ from that with a dog, primarily due to cats' territorial tendencies and their greater reluctance to tolerate direct handling. In consultation rooms reserved for cats, the presence of furniture such as cat trees can help them acclimate after being taken out of their carrier. In all cases, the consultation room must be a calm and reassuring environment, as cats are particularly sensitive to changes in their surroundings and to handling. The room can include hiding spots, which appeal to cats who enjoy retreating to such spaces. However, it is important to ensure that retrieving them from these hiding spots can be done easily and without difficulty.

30.3.2.1 Exiting the Carrier and Establishing Contact

When entering the consultation room, it is crucial not to rush the cat out of its carrier. The cat should remain in this secure space until it feels sufficiently comfortable to exit, ideally on its own. Often, the carrier is placed on the examination table, at an elevated height, which is generally reassuring for cats not previously sensitized to veterinary care.

The consultation room can be a significant source of stress for the cat, and it is often necessary to give them time to acclimate to this new environment. Allowing the cat a period of exploration without direct interaction can help it relax. The use of food rewards or a licking mat placed near the carrier's exit can also facilitate its emergence. Staff should exercise discretion in their movements, avoiding sudden noises or rapid gestures.

If the cat does not come out spontaneously after a few moments, the veterinarian may opt for a more indirect approach by gently removing the carrier's lid. This allows easy access to the cat while still leaving it within its familiar refuge, thereby reducing anxiety. If the carrier's lid is not removable and the cat shows signs of anxiety, the carrier should be handled with care. Slightly tilting the carrier can sometimes encourage the cat to exit on its own. However, any coercion to remove the cat should be applied gently and progressively to minimize stress.

With highly stressed animals, maintaining contact is essential. Once the cat is out of its carrier, the veterinarian should guide it "into their hands" smoothly and avoid releasing it, as letting go may widen the aggression zone, making it impossible to reestablish contact.

For cats that display aggression from the outset, a quick assessment of the situation is necessary. If the veterinarian determines that removing the cat from the carrier poses a danger to staff, sedation is recommended to ensure everyone's safety and to avoid exacerbating the cat's stress. For more restless but manageable patients, using a towel and employing the "burrito" technique can be an effective solution to secure the cat while minimizing its stress. However, in severe cases, sedation remains essential to ensure the examination can be performed under optimal conditions.

30.3.2.2 Examination on the Table

- **Positioning and Handling**

When examining a cat on the table, it is crucial to consider its emotional state and how it perceives being handled. As with most species, a direct frontal approach should be avoided. The cat should be approached gently and gradually, ideally from the side or rear, to prevent any perception of threat. If the owner is present, they can be positioned next to the cat to provide reassurance, particularly for more stressed cats.

It is important not to restrain the cat too forcefully at the start of the examination. Whenever possible, the cat's movements should be followed, with light handling and minimal restraint. Using a nonslip surface, such as a mat or a soft blanket, helps the cat stabilize on the table and reduces its discomfort.

If needed, the "burrito" technique (see ► Sect. 30.3.2.2) can be used for slightly restless patients. This gentle restraint method allows examination of different parts of the body without exposing staff to potential scratches or bites, while also reassuring the cat.

Once again, it is essential to avoid releasing and then restraining the cat again. If it becomes necessary, the process must be carried out gently and delicately. The fact that the cat has accepted the first point of contact does not guarantee that it will permit a second one.

Procedures and Examination

The clinical examination should be conducted progressively, starting with the least stressful procedures, such as auscultation or abdominal palpation. More invasive procedures, such as ear or mouth examinations and rectal temperature measurement, should be reserved for the end of the consultation. If the cat has a microchip capable of measuring temperature, this method should be preferred, especially during routine check-ups. It is also recommended to use appropriate equipment, such as fine-gauge needles for injections, to minimize pain and discomfort during the examination.

Careful monitoring of stress signals throughout the examination is essential. If the cat exhibits signs of agitation, vocalizations, or rigid immobility, it is advisable to take regular breaks and offer food rewards to help the cat return to a calmer state more quickly. Naturally, the more accustomed the cat is to these manipulations and cognitive distractions, the easier it will be to get them to accept rewards during the consultation. Cats that were not introduced to licking mats or treats before adolescence rarely accept them after puberty.

For procedures requiring stricter restraint, such as blood sampling or catheter placement, the "burrito" technique can be extended.

For more stressed or agitated cats, the use of a clipnosis device may be considered (Moody et al. 2020; Nuti et al. 2016; Pozza et al. 2008). This method involves applying light pressure to the nape of the neck, which can help calm some cats. However, this technique must be used with caution and only for short periods, as prolonged application can cause stress. Care should also be taken not to lift the cat but to use the clip gently without exerting traction on the cat's skin, which could cause significant pain.

When to Stop?

It is essential to recognize when to halt a procedure before the cat's stress level reaches a tipping point. If the animal becomes overly agitated or aggressive, it is better to stop the examination and consider rescheduling or premedicating the cat before making another attempt.

Escalating restraint or making repeated attempts to immobilize the cat must be avoided at all costs, as this could reinforce healthcare phobia in the animal. Sedation should be considered as soon as the cat's emotional state no longer allows for safe handling.

Abandoning a procedure should not be viewed as a failure but rather as a thoughtful decision to protect the cat's well-being. Explaining the reasoning behind this choice to the owner and scheduling a follow-up appointment, potentially with light sedation, ensures that care can be provided under better conditions during the next visit.

After the Examination

Once the examination is complete, it is essential to help the cat return to a more relaxed emotional state. If the cat has been outside its carrier, most will instinctively seek to return to it after the exam, as the carrier is perceived as a secure space. It is important to allow the cat to re-enter the carrier on its own, without haste or force. The carrier can be repositioned on the examination table and left open to encourage the cat to re-enter at its own pace.

If the cat does not return to the carrier spontaneously, it can be gently guided by the veterinarian or the owner, avoiding any abrupt handling.

While the veterinarian writes prescriptions or discusses care with the owner, it is preferable to let the cat settle in its carrier. Allowing the cat to remain in this secure space promotes its return to a calmer state. If possible, offering a treat or some wet food can provide positive reinforcement for the conclusion of the consultation.

The goal is to ensure the cat leaves the clinic in a more relaxed emotional state, helping to create a more positive perception of future visits.

30.3.3 Hospitalization

Hospitalization is a particularly stressful experience for cats, as they are naturally territorial animals who are deprived of the ability to organize their space with the various activity zones that provide them reassurance

(Zeiler et al. 2014). It is therefore essential to minimize the patient's stress and take into account the specific behavioral traits of cats throughout their hospitalization.

30.3.3.1 Admission to Hospitalization

Hospitalization begins with the separation of the cat from its owner, which can cause significant anxiety, particularly in cats sensitive to environmental changes. Managing this transition in a gentle and gradual manner is essential. Ideally, the cat should remain in its carrier until it is placed in a calm and appropriately equipped hospital space.

If the clinic is designed accordingly, having a separate cattery away from the dog area is highly beneficial for feline patients, as it spares them exposure to the smells and vocalizations of dogs, which are a common source of stress for cats (Taylor et al. 2022). When the clinic is not designed this way, compensatory measures can still be implemented, such as providing hiding spaces for hospitalized cats by leaving their carrier inside the cage. If the cage is too small, a drape can be placed in front of it to allow the cat to hide.

The auditory, lighting, and olfactory environments should also be carefully managed to minimize overstimulation as much as possible during the cat's stay.

30.3.3.2 Handling Cats in the Cage

The confined space of a cage can become a significant source of stress for a hospitalized cat. Therefore, it is essential to avoid performing invasive procedures or care within the cage whenever possible. Such actions not only increase the cat's stress, leading it to associate negative experiences with its resting space, but also pose a danger to staff. The limited space and enclosed setting reduce the staff's options for withdrawal and heighten the risk of aggression.

As a result, handling the cat in the cage should be kept to an absolute minimum. If intervention within the cage is unavoidable, such as administering a quick injection, it is crucial to do so in a non-invasive manner. The use of infusion line extensions can be a valuable tool for these types of procedures.

For more complex procedures, such as wound cleaning or postoperative care, it is essential to move the cat out of the cage to a more suitable area. This space should be secure and equipped with the necessary tools for appropriate restraint if needed.

When removing the cat from its cage, techniques similar to those used during consultations should be applied. Allowing the cat time to sniff the staff's hand and avoiding a direct frontal approach often facilitates gentle handling without triggering aggressive reactions. Using calming signals, such as averting one's gaze or blinking slowly, can also help soothe the cat.

The goal is to ensure the cat does not perceive its cage as a place of threat or constraint, as this could complicate its hospitalization and increase its long-term stress.

30.3.3.3 Care, Examinations, and Procedures

The fact that the cat is not in its usual environment during procedures deprives it of its sense of security, thereby increasing its emotional discomfort. This discomfort often leads to reduced tolerance for care, complicating necessary handling. For this reason, all prior interactions aimed at building a bond with the veterinary team will have a positive effect on the patient's cooperation during procedures.

The anxiety caused by handling can generate a state of emotional tension, with reactions that may be delayed. For instance, irritation-induced aggression may be suppressed during the procedure itself but triggered upon release from restraint or return to the cage.

The methods of restraint used during hospitalization should be identical to those used during clinical examinations, favoring gentle and adaptive approaches. Positioning the handler at the cat's rear or side, as well as using towels for gentle and secure restraint (such as the "burrito" technique), are effective strategies for minimizing stress. Certain procedures, such as subcutaneous injections or superficial wound care, do not require complete immobilization. A slight degree of freedom can be tolerated as long as the restraint focuses on securing the cat's body without rigid immobilization.

The softest and most gentle restraint possible should always be prioritized. The use of treats as a distraction during care can be considered, although most cats are unlikely to eat during these moments, especially without prior training for handling. The additional time taken to minimize stress is often offset by improved future cooperation and enhanced animal well-being.

Important Point of View

Specific Considerations for Highly Aggressive Patients

Minimizing Interactions and Medication Management

For highly aggressive cats, minimizing direct interactions is essential to avoid escalating the animal's stress and to protect veterinary staff. Providing a calm and secure hospitalization environment with care limited to the strict minimum is often the best approach.

Pharmacological Approach

Extremely aggressive cats may require pharmacological management to reduce violent reactions and allow for necessary care during hospitalization. Short-term, fast-acting medications such as pregabalin, gabapentin, or trazodone (Kruszka et al. 2021; Lamminen et al. 2023; Stevens et al. 2016) can enable specific procedures that would otherwise be impossible.

For patients requiring long-term hospitalization, venlafaxine is an interesting option due to its ease of administration in food, which can sometimes make it the only viable option when tablets cannot be given (Masson et al. 2024b; Metz et al. 2021). Venlafaxine provides anxiolytic effects and helps control impulsivity over a longer period. This is particularly effective for young, abandoned kittens that were not socialized with humans, as it reduces aggression and enhances their overall well-being.

These medications allow cats to be handled with less resistance and reduce the need for more invasive physical methods. Medication administration should be tailored to the cat's behavior, respecting individual needs and responses to the treatment.

Use of Restraint Bag

Commercially available restraint bags designed for cats can be useful for performing simple procedures, such as measuring blood pressure or placing a catheter, in a calm manner. These bags, equipped with openings for the cat's head and limbs, often induce a state of relative immobilization in the animal, which appears to be better tolerated than certain forms of sedation. However, the cat must be sufficiently amenable to handling to be placed into the bag. The bag should be snug, or even tight, to help the cat relax once secured.

Use of Gas Anesthesia

Gas anesthesia can be relatively straightforward for these cats by keeping them in their carrier, which can be placed inside a slightly larger plastic container connected to the anesthesia machine. However, this approach does not address healthcare phobia, which must be managed separately. It does, nonetheless, provide a practical solution for treating cats that are highly aggressive from the outset.

Managing Care, Hospitalization Duration and Staff Safety

In many cases, managing highly aggressive cats can lead to shorter or less thorough care. When it becomes impossible to provide adequate care due to the cat's behavior, hospitalization must end. Although this decision is restrictive, it helps minimize the animal's stress and avoids dangerous handling for the veterinary team.

This limitation often serves as an opportunity to educate owners about the necessity of long-term behavioral management to prevent similar situations during future consultations or hospitalizations.

30.3.3.4 Discharge from Hospitalization

The scheduling constraints of owners often require setting a discharge time at the moment of admission for day hospitalization. However, when possible, it is preferable to release patients as soon as their post-anesthesia recovery is sufficient, minimizing the time spent in the cage, a significant source of stress.

For the most anxious cats, it is recommended to invite the owner to retrieve their cat directly from the cage, which helps reduce the patient's stress. The removal of catheters and application of dressings can be done at the end of anesthesia or in the presence of the owner after the cat is taken out of the cage.

30

Owners often respond positively to being allowed to pick up their cat in the hospitalization area. This allows them to see the conditions firsthand, clean and comfortable surroundings enhance their confidence in the veterinary team. Staff should ensure that other hospitalized animals are not disturbed during these visits.

30.3.3.5 Postoperative Instructions

Postoperative instructions should include practical tools to help owners implement them effectively. For an agitated or difficult-to-handle cat following complex surgery, or when strict immobilization is required, it is essential to prescribe appropriate medication to facilitate at-home management. Just as we provide an Elizabethan collar to prevent the cat from interfering with its sutures, it is our responsibility to prescribe psychotropics or analgesics to maintain calmness and reduce the risk of complications caused by excessive agitation.

In cases where a behavioral disorder is suspected, prescriptions should be tailored to prevent impulsive or anxious behaviors that could hinder recovery. Hospitalization provides an excellent opportunity to discuss with owners the potential for long-term behavioral management, explaining the benefits of comprehensive care or specialized consultations.

The medications prescribed in a postoperative context often allow owners to observe their pet as calmer and more relaxed, which may encourage them to consider behavioral interventions they might have previously dismissed (see box on postoperative medication protocols).

> **Important Point of View**
>
> **Postoperative Medication Protocols**
>
> - **Pregabalin**: 5 mg/kg every 12 h. Provides analgesic and mild sedative effects (can be combined with clomipramine or fluoxetine).
> - **Gabapentin**: 20–30 mg/kg every 8–12 h. Provides analgesic and mild sedative effects (can be combined with clomipramine or fluoxetine).
> - **Venlafaxine**: 1–2 mg/kg once daily, in granule form from the original brand capsule (EFFEXOR®), equivalent to 4–8 granules per kilogram. Used for its anxiolytic and anti-impulsive effects in postoperative care for anxious and impulsive patients, helping with procedures and ensuring rest compliance.
> - **Fluoxetine**: 2–4 mg/kg daily, to reduce impulsivity in postoperative patients who need to adhere to strict rest.
> - **Clomipramine**: 2 mg/kg once daily in the morning (or in the evening if agitation is more nocturnal), for a sedative effect via H1 histamine receptor blockade.
> - **Trazodone**: 7–10 mg/kg every 12 h. Provides sedative effects, allowing for the animal's immobilization. Can be combined with gabapentin.
>
> These protocols do not exclude the prescription of Non-Steroidal Anti-Inflammatory Drug (NSAIDs) or other pain relief treatments.

30.3.3.6 Long-Term Hospitalization

The absence of attachment references impacts both well-being and the healing process. While the team strives to build a bond with the cat and ensure its comfort, visits from the owners and time spent with them are significant factors that enhance well-being and promote recovery. Many cats exhibit notable behavioral improvements, such as eating better and relaxing more, in the presence of their owner. Therefore, it is recommended to encourage owners to visit their cat, scheduling these visits during calm periods of the day to minimize disturbances for other animals.

During these visits, the ideal setup is a quiet room separate from the noisy hospitalization environment, where the cat can interact with its owner in a soothing setting. Owners can take this time to offer food, as many hospitalized cats are reluctant to eat alone but willingly accept food from a familiar person. This approach also allows for reassessment of the cat's emotional and physical state in an environment that facilitates more natural behaviors than those observed in its cage.

Finally, for cases that do not require overnight monitoring or care, our experience suggests offering day hospitalization with a return home at night. The positive effects on the patient's mood, appetite, and recovery process are undeniable, provided proper arrangements are made (e.g., securing the catheter and providing clear instructions regarding permitted activities outside the hospital setting).

Summary of ▶ Sect. 30.3

From admission to hospitalization, including the consultation and waiting room, every step of the care journey can be designed to prevent the development of healthcare phobia and to optimally manage animals already experiencing such phobias. At each stage, understanding the cat's behavior, effective communication and handling by the veterinary team, appropriate restraint techniques, the use of tailored tools, and medication are all essential strategies for preserving the cat's well-being throughout the care process.

Summary of This Chapter

In veterinary practices, healthcare phobia is a major barrier to medical treatment. It jeopardizes the safety of all involved, including that of the cat. The best prevention of this phobia is to avoid "creating" a cat that develops a fear of care. Early habituation of kittens to care, from a young age, relies on habituation-conditioning (HC) protocols, which can be easily implemented in general veterinary practices, especially by educating owners on what can be done.

When healthcare phobia is already present, a desensitization-counterconditioning (DS-CC) protocol, supported by medication, must be initiated. However, this is far more complex and requires consistent involvement from the owner. For many phobic patients, a broader behavioral issue exists and must be addressed, or the DS-CC protocol will likely fail.

In a veterinary setting, numerous stressful stimuli arise throughout the care process. Real-time reading of the cat's emotions through direct observation is essential to assess the impact of our interventions. Without specific measures focused on well-being and fear reduction, sensitization is often inevitable. Many tools are available to combat this sensitization and can be implemented in any practice.

The last decade has seen a clear increase in pain management in veterinary settings. Managing psychological well-being during medical care will undoubtedly be a key issue for the next 10–20 years.

References

Arhant C, Hörschläger N, Troxler J (2019) Attitudes of veterinarians and veterinary students to recommendations on how to improve dog and cat welfare in veterinary practice. J Vet Behav 31. https://doi.org/10.1016/j.jveb.2019.01.004

Barletta M, Raffe M (2016) Behavioral response and cost comparison of manual versus pharmacologic restraint protocols in healthy dogs. Canad Vet J = La Revue Veterinaire Canadienne 57(3):258–264

Bassett L, Buchanan-Smith HM (2007) Effects of predictability on the welfare of captive animals. Appl

30

Anim Behav Sci 102(3):223–245. https://doi.org/10.1016/j.applanim.2006.05.029
Beaver BV (2004) Fractious cats and Feline aggression. J Feline Med Surg 6(1):13–18. https://doi.org/10.1016/j.jfms.2003.09.011
Behnke AC, Vitale KR, Udell MAR (2021) The effect of owner presence and scent on stress resilience in cats. Appl Anim Behav Sci 243:105444. https://doi.org/10.1016/j.applanim.2021.105444
Bracha HS, Vega EM, Vega CB (2006) Posttraumatic dental-care anxiety (PTDA): is "Dental Phobia" a misnomer? Hawaii Dent J 37(5):17–19
Brondani JT, Mama KR, Luna SPL, Wright BD, Niyom S, Ambrosio J, Vogel PR, Padovani CR (2013) Validation of the English version of the UNESP-Botucatu multidimensional composite pain scale for assessing postoperative pain in cats. BMC Vet Res 9(1):143. https://doi.org/10.1186/1746-6148-9-143
Caney SM, Robinson NJ, Gunn-Moore DA, Dean RS (2022) Happy cats: stress in cats and their carers associated with outpatient visits to the clinic. J Feline Med Surg 24(12):e551–e557. https://doi.org/10.1177/1098612X221121907
Carlson NR, Birkett MA (2017) Physiology of behavior, 12th edn, global edition. Pearson, Boston/Columbus/Indianapolis/New York City/San Francisco/Amsterdam/Cape Town/Dubai/London
Carlstead K, Brown JL, Strawn W (1993) Behavioral and physiological correlates of stress in laboratory cats. Appl Anim Behav Sci 38(2):143–158. https://doi.org/10.1016/0168-1591(93)90062-T
Couture M, Stellato AC, Moody CM, Niel L (2022) Owner perspectives of cat handling techniques used in the veterinary clinic. J Appl Anim Welf Sci 1–11. https://doi.org/10.1080/10888705.2022.2039144
Dawson LC, Dewey CE, Stone EA, Guerin MT, Niel L (2016) A survey of animal welfare experts and practicing veterinarians to identify and explore key factors thought to influence canine and feline welfare in relation to veterinary care. Anim Welf 25(1). https://doi.org/10.7120/09627286.25.1.125
Drobatz J, Smith KG (2003) Evaluation of risk factors for bite wounds inflicted on caregivers by dogs and cats in a veterinary teaching hospital. J Am Vet Med Assoc 223(3):312–316. https://doi.org/10.2460/javma.2003.223.312
Edwards PT, Smith BP, McArthur ML, Hazel SJ (2019) Fearful Fido: investigating dog experience in the veterinary context in an effort to reduce distress. Appl Anim Behav Sci 213. https://doi.org/10.1016/j.applanim.2019.02.009
Ellis SL (2009) Environmental enrichment: practical strategies for improving Feline welfare. J Feline Med Surg 11(11):901–912. https://doi.org/10.1016/j.jfms.2009.09.011
Ellis SL (2018) Recognising and assessing Feline emotions during the consultation: history, body language and behaviour. J Feline Med Surg 20(5):445–456. https://doi.org/10.1177/1098612X18771206
Frank D, Beauchamp G, Palestrini C (2010) Systematic review of the use of pheromones for treatment of undesirable behavior in cats and dogs. J Am Vet Med Assoc 236(12):1308–1316. https://doi.org/10.2460/javma.236.12.1308
Furgala NM, Moody CM, Flint HE, Gowland S, Niel L (2022) Veterinary background noise elicits fear responses in cats while freely moving in a confined space and during an examination. Behav Process 201:104712. https://doi.org/10.1016/j.beproc.2022.104712
Griffith CA, Steigerwald ES, Tony CA, Buffington. (2000) Effects of a synthetic facial pheromone on behavior of cats. J Am Vet Med Assoc 217(8):1154–1156
Hammerle M, Horst C, Levine E, Overall K, Radosta L, Rafter-Ritchie M, Yin S (2015) 2015 AAHA Canine and Feline behavior management guidelines*. J Am Anim Hosp Assoc 51(4):205–221. https://doi.org/10.5326/JAAHA-MS-6527
Hampton A, Ford A, Cox RE, Liu C-c, Koh R (2020) Effects of music on behavior and physiological stress response of domestic cats in a veterinary clinic. J Feline Med Surg 22(2):122–128. https://doi.org/10.1177/1098612X19828131
Hargrave C (2021) Pheromones and 25 years of pheromonotherapy: what are they and how do they work? Vet Nurse 12(3):116–122. https://doi.org/10.12968/vetn.2021.12.3.116
Herron ME, Shreyer T (2014) The pet-friendly veterinary practice. Vet Clin N Am Small Anim Pract 44(3):451–481. https://doi.org/10.1016/j.cvsm.2014.01.010
Herron ME, Tony Buffington CA (2012) Environmental enrichment for indoor cats: implementing enrichment. Compendium (Yardley, PA) 34(1):E3
Horwitz DF, Rodan I (2018) Behavioral awareness in the Feline consultation: understanding physical and emotional health. J Feline Med Surg 20(5):423–436. https://doi.org/10.1177/1098612X18771204
Johnson DC, Casey BJ (2015) Extinction during memory reconsolidation blocks recovery of fear in adolescents. Sci Rep 5(1):8863. https://doi.org/10.1038/srep08863
Karn-Buehler J, Kuhne F (2022) Perception of stress in cats by German cat owners and influencing factors regarding veterinary care. J Feline Med Surg 24(8):700–708. https://doi.org/10.1177/1098612X211041307
Kogan L, Kolus C, Schoenfeld-Tacher R (2017) Assessment of clicker training for shelter cats. Animals 7(10):73
Kruszka M, Graff E, Medam T, Masson S (2021) Clinical evaluation of the effects of a single oral dose of gabapentin on fear-based aggressive behaviors in cats during veterinary examinations. J Am Vet Med Assoc 1–7. https://doi.org/10.2460/javma.20.06.0307
Lamminen T, Korpivaara M, Aspegrén J, Palestrini C, Overall KL (2023) Pregabalin alleviates anxiety and fear in cats during transportation and veterinary

visits—a clinical field study. Animals 13(3):371. https://doi.org/10.3390/ani13030371

Lockhart J, Wilson K, Lanman C (2013) The effects of operant training on blood collection for domestic cats. Appl Anim Behav Sci 143(2–4):128–134. https://doi.org/10.1016/j.applanim.2012.10.011

Mariti C, Bowen JE, Campa S, Grebe G, Sighieri C, Gazzano A (2016) Guardians' perceptions of cats' welfare and behavior regarding visiting veterinary clinics. J Appl Anim Welf Sci 19(4). https://doi.org/10.1080/10888705.2016.1173548

Mariti C, Guerrini F, Vallini V, Bowen JE, Fatjó J, Diverio S, Sighieri C, Gazzano A (2017) The perception of cat stress by Italian owners. J Vet Behav 20:74–81. https://doi.org/10.1016/j.jveb.2017.04.002

Martin D, Lees RM (2023) Husbandry and veterinary care. In: Martin D, Shaw JK (eds) Canine and Feline behavior for veterinary technicians and nurses. Wiley, pp 245–274

Masson S, Bleuer-Elsner S, Muller G, Médam T, Chevallier J, Gaultier E (2024a) The causes and consequences of care in veterinary facilities. In: Veterinary psychiatry of the dog. Springer Nature Switzerland, Cham, pp 917–965

Masson, Sylvia, Delphine Metz, Stéphane Bleuer-Elsner, Françoise Schwobthaler. 2024b. « Retrospective study on the use of Venlafaxine in 176 cats diagnosed with behavioral disorders »

Metz D, Medam T, Masson S (2021) Double-blind, placebo-controlled trial of Venlafaxine to treat behavioural disorders in cats: a pilot study. J Feline Med Surg 1098612X2110367. https://doi.org/10.1177/1098612X211036792

Moir F, Van Den Brink A (2020) Current insights in veterinarians' psychological wellbeing. N Z Vet J 68(1):3–12. https://doi.org/10.1080/00480169.2019.1669504

Moody CM, Mason GJ, Dewey CE, Niel L (2020) Getting a grip: cats respond negatively to Scruffing and clips. Vet Rec 186(12):385–385. https://doi.org/10.1136/vr.105261

Nibblett BM, Ketzis JK, Grigg EK (2015) Comparison of stress exhibited by cats examined in a clinic versus a home setting. Appl Anim Behav Sci. https://doi.org/10.1016/j.applanim.2014.10.005

Nuti V, Cantile C, Gazzano A, Sighieri C, Mariti C (2016) Pinch-induced behavioural inhibition (Clipthesia) as a restraint method for cats during veterinary examinations: preliminary results on cat susceptibility and welfare. Anim Welf 25(1):115–123. https://doi.org/10.7120/09627286.25.1.115

Pankratz KE, Ferris KK, Griffith EH, Sherman BL (2018) Use of single-dose oral gabapentin to attenuate fear responses in cage-trap confined community cats: a double-blind, placebo-controlled field trial. J Feline Med Surg 20(6). https://doi.org/10.1177/1098612X17719399

Pereira DG, Gonçalo SF, Morais D, De Brito MTV, De Sousa L (2014) Comparison of interpretation of cat's behavioral needs between veterinarians, veterinary nurses, and cat owners. J Vet Behav 9(6):324–328. https://doi.org/10.1016/j.jveb.2014.08.006

Powell L, Watson B, Serpell J (2023) Understanding feline feelings: an investigation of cat owners' perceptions of problematic cat behaviors. Appl Anim Behav Sci 266:106025. https://doi.org/10.1016/j.applanim.2023.106025

Pozza ME, Stella JL, Chappuis-Gagnon A-C, Wagner SO, Tony Buffington CA (2008) Pinch-induced behavioral inhibition ('Clipnosis') in domestic cats. J Feline Med Surg 10(1):82–87. https://doi.org/10.1016/j.jfms.2007.10.008

Pratsch L, Mohr N, Palme R, Rost J, Troxler J, Arhant C (2018) Carrier training cats reduces stress on transport to a veterinary practice. Appl Anim Behav Sci 206:64–74. https://doi.org/10.1016/j.applanim.2018.05.025

Raja SN, Carr DB, Cohen M, Finnerup NB, Flor H, Gibson S, Keefe FJ, Mogil JS, Ringkamp M, Sluka KA, Song X-J, Stevens B, Sullivan MD, Tutelman PR, Ushida T, Vader K (2020) The revised International Association for the Study of Pain Definition of Pain: concepts, challenges, and compromises. Pain 161(9):1976–1982. https://doi.org/10.1097/j.pain.0000000000001939

Rand JS, Kinnaird E, Baglioni A, Blackshaw J, Priest J (2002) Acute stress hyperglycemia in cats is associated with struggling and increased concentrations of lactate and norepinephrine. J Vet Intern Med 16(2). https://doi.org/10.1111/j.1939-1676.2002.tb02343.x

Riemer S, Heritier C, Windschnurer I, Pratsch L, Arhant C, Affenzeller N (2021) A review on mitigating fear and aggression in dogs and cats in a veterinary setting. Animals 11(1):158. https://doi.org/10.3390/ani11010158

Rodan I (2010) Understanding feline behavior and application for appropriate handling and management. Top Companion Anim Med 25(4). https://doi.org/10.1053/j.tcam.2010.09.001

Rodan I (2011) Attract cat clients with friendly handling. Vet Med 106(10):486–487

Rodan I, Heath S (eds) (2016) Feline behavioral health and welfare. Elsevier, St. Louis

Rodan I, Sundahl E, Carney H, Gagnon AC, Heath S, Landsberg G, Seksel K, Yin S (2011) AAFP and ISFM Feline-Friendly Handling Guidelines. J Feline Med Surg 13(5). https://doi.org/10.1016/j.jfms.2011.03.012

Rodan I, Dowgray N, Carney HC, Carozza E, Ellis SLH, Heath S, Niel L, Denis KS, Taylor S (2022) 2022 AAFP/ISFM cat friendly veterinary interaction guidelines: approach and handling techniques. J Feline Med Surg 24(11):1093–1132. https://doi.org/10.1177/1098612X221128760

Ryan S, Bacon H, Endenburg N, Hazel S, Jouppi R, Lee N, Seksel K, Takashima G (2019) WSAVA Animal Welfare Guidelines. J Small Anim Pract 60(5):E1–E46. https://doi.org/10.1111/jsap.12998

Shu H, Gu X (2022) Effect of a synthetic feline facial pheromone product on stress during transport in domestic cats: a randomised controlled pilot study. J Feline Med Surg 24(8):691–699. https://doi.org/10.1177/1098612X211041305

Siragusi RH, Souza D, De Barros VTM, Fiorini EA, Bergamo TM, Siqueira RC, Manhoso FFR, Franco RP (2017) Measurement of serum lactate values in domestic cats (*Felis catus*) submitted to physical and chemical restraint. Braz J Vet Res Anim Sci 54(4):383–387. https://doi.org/10.11606/issn.1678-4456.bjvras.2017.132024

Snowdon CT, Teie D, Savage M (2015) Cats prefer species-appropriate music. Appl Anim Behav Sci 166:106–111. https://doi.org/10.1016/j.applanim.2015.02.012

Steagall PV, Robertson S, Simon B, Warne LN, Shilo-Benjamini Y, Taylor S (2022) 2022 ISFM consensus guidelines on the management of acute pain in cats. J Feline Med Surg 24(1):4–30. https://doi.org/10.1177/1098612X211066268

Stellato J, Dewey W, Niel. (2019) Effect of a standardized four-week desensitization and counter-conditioning training program on pre-existing veterinary fear in companion dogs. Animals. https://doi.org/10.3390/ani9100767

Stevens BJ, Frantz EM, Orlando JM, Griffith E, Harden LB, Gruen ME, Sherman BL (2016) Efficacy of a single dose of trazodone hydrochloride given to cats prior to veterinary visits to reduce signs of transport- and examination-related anxiety. J Am Vet Med Assoc 249(2). https://doi.org/10.2460/javma.249.2.202

Taylor S, Denis KS, Collins S, Dowgray N, Ellis SLH, Heath S, Rodan I, Ryan L (2022) 2022 ISFM/AAFP cat friendly veterinary environment guidelines. J Feline Med Surg 24(11):1133–1163. https://doi.org/10.1177/1098612X221128763

Vlaeyen, J.W.S., Crombez, G., Linton, S.J., 2016. The fear-avoidance model of pain. Pain 157, 1588–1589. https://doi.org/10.1097/j.pain.0000000000000574

Volk JO, Felsted KE, Thomas JG, Siren CW (2011) Executive summary of the Bayer veterinary care usage study. J Am Vet Med Assoc 238(10):1275–1282. https://doi.org/10.2460/javma.238.10.1275

Westlund K (2015) To feed or not to feed: counterconditioning in the veterinary clinic. J Vet Behav 10(5):433–437. https://doi.org/10.1016/j.jveb.2015.05.008

Willemse T, Vroom MW, Mol JA, Rijnberk A (1993) Changes in plasma cortisol, corticotropin, and α-melanocyte-stimulating hormone concentrations in cats before and after physical restraint and intradermal testing. Am J Vet Res 54(1):69–72. https://doi.org/10.2460/ajvr.1993.54.01.69

Williams L (2016) Cat handling and associated stress: a clinical nursing perspective. Vet Nurs J 31(3):88–93. https://doi.org/10.1080/17415349.2015.1128859

Wojtaś J, Czyżowski P, Kaszycka K, Kaliszyk K, Karpiński M (2024) The impact of environmental enrichment on the cortisol level of shelter cats. Animals 14(9):1392. https://doi.org/10.3390/ani14091392

Zeiler GE, Fosgate GT, Van Vollenhoven E, Rioja E (2014) Assessment of behavioural changes in domestic cats during short-term hospitalisation. J Feline Med Surg 16(6):499–503. https://doi.org/10.1177/1098612X13509081

Zhang L, Bian Z, Liu Q, Deng B (2022) Dealing with stress in cats: what is new about the Olfactory strategy? Front Vet Sci 9:928943. https://doi.org/10.3389/fvets.2022.928943

Ziv G (2017) The effects of using aversive training methods in dogs—a review. J Vet Behav Clin Appl Res 19:50–60. https://doi.org/10.1016/j.jveb.2017.02.004

Veterinary Psychiatry in the Daily Life of the General Practitioner

Contents

31.1 Preventive Care – 842
31.1.1 Vaccination Visits for Kittens – 842
31.1.2 Puberty Consultations – 849
31.1.3 Adult Cat Visits – 851
31.1.4 Senior Cat Visits – 853
31.1.5 Behavioral Balance Assessment Questionnaire – 853

31.2 General Medicine – 857
31.2.1 Painful Conditions – 857
31.2.2 Chronic Conditions – 859
31.2.3 End of Life, Palliative Care, and Euthanasia – 861

31.3 French Legislation (for informational purposes) – 865

References – 867

© The Author(s), under exclusive license to Springer Nature Switzerland AG 2026
S. Bleuer-Elsner et al., *Veterinary Psychiatry of the Cat*,
https://doi.org/10.1007/978-3-032-06260-4_31

This chapter highlights the importance of veterinary psychiatry in all consultations conducted by the general practitioner throughout the cat's life. It reviews preventive care consultations, followed by general medical consultations. The topic of euthanasia, an ultimate and highly emotional act, is also addressed.

As described in ▶ Chap. 14, there are close relationships between all fields of medicine and veterinary psychiatry. This chapter presents the key psychological aspects to assess in a feline patient during preventive care consultations or when managing chronic illnesses, ensuring that health evaluations are truly holistic and include the cat's emotional and behavioral aspects. The chapter is structured around various consultations conducted in general veterinary practice and also includes a special consultation: the euthanasia consultation.

31.1 Preventive Care

This section concerns cats that present for a routine check-up and are generally in good physical health. These preventive visits are crucial opportunities to educate and advise owners on their cat's overall health. They are also the ideal time to detect behavioral issues early or prevent them from developing.

Depending on the cat's age and, consequently, its level of brain development, the practitioner will aim to rule out or screen for the most common conditions. The following paragraphs outline specific areas of focus.

For young cats, the emphasis will be on ensuring there are no exaggerated fears, establishing proper control of biting and scratching, and getting the cat accustomed to the carrier and veterinary handling. For adult cats, the focus will shift towards establishing pheromone marking across different activity areas and managing relationships with other cats, humans, or other animals in their environment. For senior cats, it will be important to monitor for early signs of cognitive decline or behavioral changes related to chronic pain, such as from osteoarthritis or dental diseases.

Preventive consultations are typically short, often between 15 and 20 min, but it may be helpful to schedule more time for certain categories of cats, such as kittens or newly adopted adult cats. These visits should not solely address behavioral concerns but also cover other aspects of the cat's physical health. The goal is not to conduct a full behavioral consultation, but to quickly check for significant deviations from the expected norms for that age, just as one would assess joint conformation or testicular descent. The number of questions posed to the owner should be limited, focusing on the most concerning deviations from normal, but should apply to all patients to ensure broad screening. Of course, when the clinical picture is severe, a more specialized behavioral consultation should be offered.

31.1.1 Vaccination Visits for Kittens

31.1.1.1 From the Decision to Adopt to the First Veterinary Visit

Although it is sometimes the case that the practitioner is consulted before adopting a kitten, this is still an exception. However, such a preliminary consultation provides a valuable opportunity to guide future owners in choosing their kitten, taking into account their living environment, breed, sex, or the animal's behavioral predispositions. Emphasizing this preadoption consultation could prevent many future behavioral problems.

The kitten's birth environment plays a crucial role in its behavioral and emotional development. Breeders or kitten-raisers should provide a stimulating yet secure environment, allowing kittens to explore while remaining in contact with their mother and siblings until at least 10–12 weeks. Unlike recommendations for other species, such as dogs, it has been shown that early weaning before 12 weeks increases the risk of behavioral issues in cats (Ahola et al. 2017). In fact, it is between 8 and 12 weeks that self-control behaviors related to biting, scratching, and frustration management are learned through interactions with the mother.

This critical period before adoption also allows kittens to become familiar with various stimuli, including those from different species, and to socialize with humans, dogs, and other species they will encounter. If their environment does not expose them to these elements before 6 weeks, socialization with the species in question will be very difficult, if not impossible. This is why kittens found at this age in the street, born to feral mothers and not exposed to humans, have a hard time adapting to a normal family life. Their fear of humans is very high, and this is difficult to reverse due to the synaptic pruning that occurs very early, around 8 weeks in kittens (see ◘ Fig. 31.1).

These risks are amplified for kittens found abandoned or orphaned before 6 weeks of age, who are often bottle-fed by humans. While these compensatory care measures can save the kitten's life, they cannot replace the maternal influence and the socialization with siblings. Kittens raised without their mother and peers tend to develop issues with fear management, inadequate self-control, and anxious or aggressive behaviors in adulthood. The same issues arise for kittens that are overly stimulated (Campbell et al. 2024; Graham et al. 2024).

Another important factor not to be overlooked is the genetic role. Studies have shown that kittens born to wild fathers, even when raised in captivity in a human environment, have more difficulty adapting and are more likely to develop escape or aggression behaviors compared to those born to domesticated parents (McCune 1995). This difference highlights the influence of genetics on kitten socialization. However, in most cases, the father of the kittens is unknown, and for many adopted kittens, the mother is also not necessarily known.

Owners often tend to adopt a kitten without knowing the conditions of its early development, as they believe it is normal to adopt a free kitten and that a "found" or "free" cat will suffice. However, the lack of this knowledge often leads to disappointment when the cat does not meet expectations, particularly

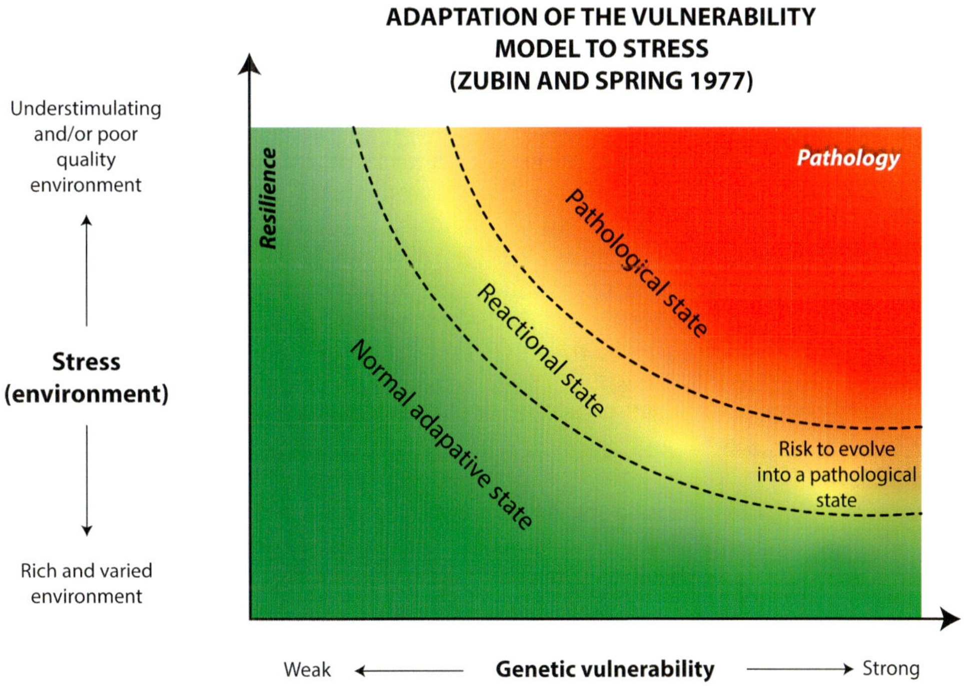

◘ **Fig. 31.1** The Epigenetic Model by Zubin and Spring (1977). (a) Green zone: normal adaptive state. (b) Orange and red zones: pathological state. (c) Yellow zone: environmental stress = allostatic load. This is the reactive state. Adaptation leads back to the green zone, while maladaptation leads to a shift into the orange or red zones those of children, who may become frustrated with a fearful or distant cat. A common misconception persists that a fearful kitten will eventually adapt with enough kindness and care. However, as shown by Zubin's stress vulnerability model (Zubin and Spring 1977), once fear responses are established and solidified, they tend to persist, even if the environment is favorable (◘ Fig. 31.1).

It is therefore essential to inform adopters about the specific behavioral development needs of kittens. Furthermore, it is important to emphasize the need to choose kittens raised in conditions conducive to their well-being, providing them with an environment that is rich in stimulation—without excess—and varied, resembling as much as possible the future home they will live in. A well-socialized kitten, from a breeding environment where interactions with humans and other cats are regular and positive, has a much higher chance of integrating harmoniously into a new home.

Finally, it is important to note that breeders of purebred cats have worked to create cats that are good companions and interact willingly with humans, as this is primarily what cat owners seek (Sinn 2016).

When the decision is made to adopt a cat from a breeder or simply from a family that has had a litter, adopters should be able to interact with the kittens to ensure they are neither huddling in a corner and rejecting contact nor, conversely, too excited and rough. Avoiding these extreme behaviors helps reduce the risk of choosing a kitten with significant behavioral imbalances. Being able to meet the kitten's parents to ensure they are also friendly with visitors is another important factor. Finally, the facilities should be clean, spacious, and the environment should be enriched (toys, activities).

Most often, once the kitten is adopted, the owners visit the veterinary clinic to establish contact and receive initial advice. This quick visit after acquisition should be encouraged, even if no vaccinations are immediately required, as it allows for early screening of any health or behavioral issues.

During this first visit, the veterinarian already has an important task: verifying identification, vaccinations, and antiparasitic treatments, as well as guiding the owners on these areas for the upcoming months. The veterinarian also provides nutritional advice.

For the behavioral aspect, a quick, systematic approach can be adopted, allowing for a rapid assessment of behavioral risk factors and the provision of priority advice. This involves quickly reviewing litter box training, the four behavioral axes with a focus on fear and self-control, and checking for any signs of aggressive behavior since adoption. The goal is not to be exhaustive but to be sufficiently attentive to identify elements that could significantly hinder the kitten's proper development. These checks should be repeated at each vaccination follow-up, up until puberty. They also signal to the owner that the veterinarian is the go-to person to consult in case of issues in this area. We will expand on these aspects in the following two sections.

31.1.1.2 Litter Box Training

Litter box training is generally already established by the time of adoption. However, owners often proudly announce that their kitten is "already" clean, especially if they are accustomed to dogs, for whom house training is a major concern during adoption.

Kittens that grow up observing their mother typically become clean around 4–6 weeks of age. However, in the absence of the mother or in situations where she does not directly demonstrate litter box usage, this learning can be delayed. For example, it has been observed that kittens raised without their mother or bottle-fed kittens, although they tend to use the litter box naturally because it is an appropriate place for elimination, still need to be praised and shown where the litter box is several times in order to fully adopt this habit. They are generally a few weeks behind in this acquisition.

During pediatric consultations, it is important to ensure that litter box training is well established. This allows for a quick discussion with the owners about key aspects of hygiene and the cat's behavior. It is also useful to address proper litterbox cleaning practices, explaining the importance of regular maintenance and quality litter. A dirty or unsuitable litter box can quickly become a source of inappropriate elimination, as cats, being sensitive to smells, may refuse to use it if it is not properly maintained.

Indeed, this is also an opportunity to explain the importance of scent in cats and its role in managing their territory. Although kittens do not yet fully structure their activity areas, this is also the time to explain to owners that, at puberty, cats begin marking their environment with facial pheromones, which helps them feel calm in their living space. These discussions allow owners to better understand their cat's behavior, particularly the significance of activity areas. They will better grasp why it is essential to keep elimination, feeding, and resting areas separate.

This understanding will help prevent inappropriate elimination behaviors. In fact, more than 50% of cases of urinary incontinence in cats are related to issues directly connected to the litter box (size, type, contents, cleaning, etc.) (Carney et al. 2014).

It is also a good opportunity to discuss methods of reprimanding a cat during consultations. If a kitten eliminates outside the litter box, it is important not to use punishment, as this would only add stress. Instead, it is recommended to gently interrupt the behavior without frightening the animal, and redirect the cat to the litter box. This approach, based on positive redirection, helps reinforce desired behaviors without creating additional anxiety. It is an excellent way to start a conversation with owners about the counterproductive nature of any positive punishment in cats and the lack of hierarchy within this species.

Even briefly discussing these concepts can help answer owners' questions and establish a constructive dialogue about their animal's behavior, positioning the veterinarian as the go-to expert for behavioral concerns.

31.1.1.3 Assessment of Behavioral Axes

It is important for the practitioner to assess the emotional state of the kitten presented to them in order to tailor advice to the owner. However, comments should be moderated for two reasons. First, regardless of the symptom observed on the day of the visit, change is always possible, and one should refrain from making definitive conclusions. Secondly, for elementary psychological reasons, it is not advisable to highlight a significant flaw outright. The owner is unaware of this flaw, and it is important to make them aware of the issue tactfully, then guide them toward a willingness to address it. A lack of finesse would immediately result in the owner seeking another practitioner and lead to failure.

The discovery of a significant heart murmur can illustrate this point. While this is a clearly measurable and objective symptom, the practitioner would instead express it as a suspicion to be confirmed, ensuring there is no concern going forward, in order to avoid causing panic or undue worry for the owner.

▪ Behavioral Component of the Overview

The behavioral axes are checked one by one, with particular attention given to those related to fear and self-control (▫ Fig. 31.2), as these are the ones that most often show deviations from the norm during the juvenile period. The normal benchmarks for these axes, based on developmental stages, are described in ► Chap. 4 of the first part.

To verify that the kitten appears balanced on the *fear axis*, it is useful to ask how the kitten reacts to unfamiliar humans, loud noises, and more generally in its everyday life context. Of course, the responses provided should be compared with the direct observation of the kitten during the consultation to assess its level of fear in response to the new stimuli it will encounter, such as the clinic staff, animals in the waiting room, new objects, etc.

At this age, a fearful reaction is normal when encountering a new stimulus, but exploration should resume quickly afterward. A normal kitten during the juvenile period is curious and able to adapt quickly to novelty. The practitioner should be concerned if there is an excessively intense fear reaction that does not decrease after several exposures, or if the fear leads to aggressive reactions, scratching, or an attempt to bite. Depending on the severity, the practitioner can offer advice to alleviate timidity or fear reactions. This advice

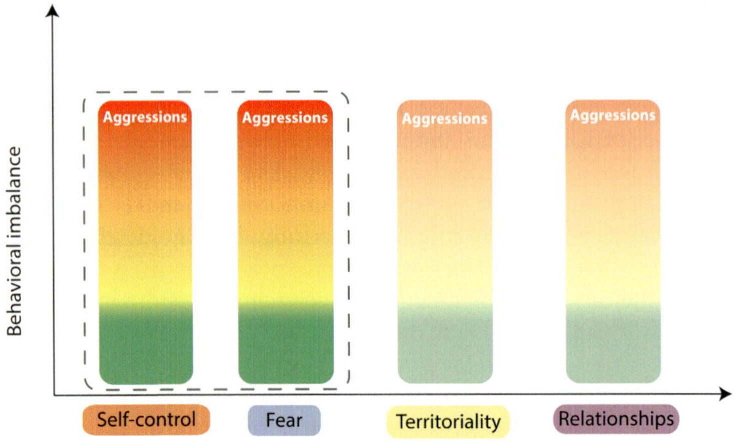

Fig. 31.2 Behavioral axes to focus on during the first vaccination visit

includes, for example, gradual exposure instructions, without force, using highly appealing treats to encourage exploration of the phobic stimulus.

The veterinarian should also educate owners that exposure to new stimuli should be controlled: the goal is to move the kitten out of its comfort zone and provoke a moderate reaction, but without placing it in an intensely fearful situation. To achieve this, owners must be able to read their kitten's behavior and adapt accordingly. If the fear axis is already severely affected and disabling, the practitioner should recommend a more specialized consultation.

At this age, adapting exposure to the kitten's profile is crucial: the advice provided must be tailored to each individual patient. This is not only part of our mission but also a key aspect of our added value.

To assess whether the kitten shows any vulnerability on the *self-controls axis*, the practitioner may inquire whether the kitten is able to rest spontaneously, whether it engages in oral exploration and chews non-food objects, and the intensity of its chewing when excited, especially during play. Various degrees of lack of self-control exist, and here too, the severity of the clinical picture allows the practitioner to decide whether to intervene preventively by providing advice on acquiring calmness (e.g., by isolating the kitten) (see ▶ Chap. 29, ▶ Sect. 29.3.2.3) or whether to recommend a specialized consultation in cases of injury to the owner's arms or if the kitten destroys or ingests non-food items, such as cables, toys, or pieces of string.

Regarding the *axes of territoriality* and *relationships*, whose deviations from the norm typically appear during puberty, screening is focused only on the most severe cases that show symptoms during the juvenile period. Severe abnormalities in these axes are ruled out if the kitten interacts amicably with familiar people, is easy to handle for routine care (such as brushing), and is clean.

A quick assessment of the kitten's behavior also encourages owners to ask more questions if any behavioral issues have been problematic since adoption.

Therapeutic Decision at the End of the Consultation

At the end of this quick and targeted behavioral assessment, the practitioner may encounter different scenarios. In case of simple doubt, the first step is to make a note in the animal's file to ensure that the issue is addressed at the next consultation.

When the practitioner detects *mild vulnerabilities*, they should prescribe measures to remedy or compensate for them. The use of pheromones, dietary supplements, or phytotherapy is justified if it helps facilitate learning. Additionally, the practitioner should address the associated behavioral therapy by prescribing exercises, which they should explain and summarize on a prescription or using a practical handout (**see the box below**). Just as they would for a medical suspicion, the practitioner may suggest a follow-up visit or plan to check the progress of the noted vulnerabilities at the next vaccination consultation.

Important Point of View

Importance of Practical Handouts

The use of practical handouts is quite common, especially in preventive care, to illustrate our points or simply as a written support for owners to have a summary of what was discussed, especially if they forget some details or want to review them thoroughly.

However, it is important to emphasize the need for a prior, clear explanation before handing the sheet to the client. Providing owners with pages of advice without explanation is ineffective—they are unlikely to read them. In contrast, it is much more effective to give an oral explanation, or even a practical demonstration on a specific topic, and then provide the corresponding handout at the end. In this case, the client has benefited from an explanation that engaged different sensory channels (auditory, tactile, visual), significantly increasing the likelihood that the information is understood and retained.

(Example of practical handouts below, see ◘ Fig. 31.3)

When faced with a kitten that appears unusually fearful, the practitioner might prescribe a dietary supplement and ask the owner to monitor the kitten's interactions with guests, advising the use of exceptional treats during visits. If there are no further issues, the concern is alleviated.

This attentive management of the kitten's vulnerabilities is optimal because it allows for the rapid correction of observed weaknesses while ensuring quick reactions in case of slow progress or worsening behavior.

In cases where there is a *clear anomaly in one of the behavioral axes*, an intervention commensurate with the suspected pathology should be proposed. This announcement should be made gently to avoid startling the new owners, who may be entranced by their kitten and could become defensive if the description provided does not align closely enough with their own perception of their pet. For this reason, it is very important to guide the owners to evaluate and articulate the problem themselves: "Are its fears impairing daily life?" "Do its biting behaviors interfere with your relationship?"

It should also be emphasized that with a very young animal, moderation is always necessary because, regardless of the symptom observed, change is still possible. While discussing a severe illness is highly anxiety-inducing and should be approached with great tact, failing to address it would be a professional error. Finding the right balance is crucial, and this is often acquired through experience and, unfortunately, some failures (**see the box below**).

When it becomes clear that the owners are not ready to receive an alarming diagnosis, even if it is justly made, the practitioner can set a few simple goals. These may be objectively insufficient to treat the case but are acceptable to the owners, to be completed before a follow-up consultation, which should be scheduled. Any failure to achieve these simple goals can then serve as a stronger lever for a more specialized and comprehensive intervention, which will be better accepted by owners who can objectively recognize their failure to meet the simple goals.

Choosing and cleaning your cat litter box properly

Why should you care about your cat's litter box properly?

Cats are very sensitive to the type, location, and cleanliness of their litter box, and it has been shown that these factors are responsible for more than 60% of cases of inappropriate elimination.

How to choose your cat's litter?

Size and shape of the litterbox

- ✓ **Width:** at least ***once*** the length of your cat.
- ✓ **Length**: at least ***one and a half*** times the length of your cat.
- ✓ The edges should be low so that your cat can easily access it.
- ✓ **Avoid lids.**
- ✓ If you choose a covered litter box, ***remove the entrance door.***

Number of litter boxes

Without access to the outside, the number of litter boxes should be **at least equal to the number of cats in the household *+1*.**

Litter contents

- ✓ Prefer a fine granulation
- ✓ **Avoid litters with deorizers.**

Litter box placement

The litter box should be placed:

- ➲ **Far from *resting areas.***
- ➲ **Far from *food bowl.***

How to clean your cat's litter box?

Cleaning the litter box

- ✓ Feces should be removed ***once a day.***
- ✓ The litter should be completely changed ***once a week.***

Cleaning the box

- ✓ The box should be cleaned with bleach.
- ✓ The box should be replaced ***once a year.***

Cleaning urine outside the litter box

- ✓ Prefer ***white vinegar*** or ***sparkling water.***
- ✓ Do not use bleach or quaternary ammonium.

If you encounter difficulties, contact your veterinarian

PSY4VET

Fig. 31.3 Litterbox handout

Important Point of View

Referring a Kitten for a Behavioral Consultation

When a kitten exhibits a significant imbalance on one or more behavioral axes, referring the owner for a behavioral consultation is essential, though this can be challenging to implement.

It is not enough to simply alert owners to the issue their kitten is displaying; it is our responsibility to guide them towards a behavioral consultation, either directly within the veterinary practice or by referring them to a colleague with this expertise.

Cost is often a major objection when it comes to managing a kitten's behavioral issues. On the one hand, the first year's expenses are already considerable; on the other hand, owners are influenced by the common misconception that "it will pass with age," which makes the expense seem unnecessary to them.

Certain clinical situations, such as scratches from playful behavior that mark the owner's arms on a young kitten of 3 or 4 months, are alarming to us, but often much less so for an owner who struggles to understand how these behaviors may escalate as the kitten grows. In other words, we often find it difficult to convey our prognostic vision to them.

There is also a significant risk of raising a concern that the owners are unaware of, which could make them defensive and prompt them to seek care elsewhere. The practitioner should ideally lead the owner to recognize the problem themselves, either by having them compare the kitten's behavior to that of their previous cats, by focusing the questions on the fluidity of the relationship, by identifying unusual behaviors for a kitten of that age, or by comparing the kitten's behavior to the owner's expectations…

For all these reasons, the specific behavioral consultation for a kitten should be priced at a level that is not prohibitive. These consultations are often shorter, as the history is quickly gathered. Furthermore, our experience shows that it is more beneficial to charge an accessible fee for these consultations and repeat them at 1–2-month intervals, rather than charging for a standard behavioral assessment of one to one and a half hours, which many owners would refuse, potentially leading to negative development for the kitten. Finally, it should be noted that behavioral medicine is covered by some health insurance plans, and of course, if owners have subscribed to one from the time of adoption, this issue is no longer relevant.

31.1.2 Puberty Consultations

Puberty consultations have been offered in an increasing number of veterinary practices in recent years. These consultations address the need for a growth assessment, but also provide an opportunity to discuss expected behavioral changes at puberty, talk about sterilization, and review dietary rations. The practice of puberty consultations is facilitated by new vaccination protocols, which recommend a vaccination visit between 6 and 12 months.

Puberty consultations are typically conducted between 6 and 10 months of age. At this stage, cats are in their adolescence. The kitten's age at the time of the puberty consultation requires the practitioner to adjust their behavioral evaluation based on how advanced the kitten is in the adolescent phase. For example, fights between conspecifics may arise when multiple cats coexist, as well as urinary marking in male cats who are not neutered, and the onset of pheromone marking.

First, it is important to remember that developmental disorders that were not identified or adequately addressed tend to worsen at the onset of puberty. On the fear axis, cats that were simply "shy and reserved," displaying inhibited fear responses, can shift towards more productive fear behaviors during adoles-

handling that owners can practice at home. The prescription of fast-acting medications like pregabalin or gabapentin for future visits is essential to slow down the sensitization process that is clearly occurring with each visit.

Several scenarios may explain the onset of a behavioral issue in adulthood.

First, the animal may have experienced a traumatic event, whether it occurred during the course of treatment or due to external circumstances, such as an accident or a move. In both cases, the concept of trauma is not objective; it should be assessed based on the cat's response, i.e., its vulnerabilities. In the presence of a clinical picture of phobia, anxiety, or depression secondary to an identifiable event in the history, a treatment plan should be proposed. Any painful treatments performed on an adult animal that appear to have triggered a change in its usual tolerance (for example, manifest fear following surgery or a painful procedure) require the practitioner to address the issue and propose appropriate management.

In other cases, the annual preventive visit reveals the worsening of a pre-existing condition. This is particularly true for issues related to the fear axis, where sensitization continues year after year until it reaches a pathological threshold that the practitioner must be able to identify for early management in veterinary psychiatry. A cat known to be shy or fearful that now displays intense fear reactions during the consultation should prompt a more thorough investigation. Carefully chosen open-ended questions aim to allow the owners to address the issue themselves: in the absence of a request from them, any abrupt and categorical diagnosis can generate strong resistance and a refusal of care, which would be counterproductive for the animal.

Among cats whose vulnerabilities shift them into a pathological state in adulthood, those exhibiting aggressive manifestations are the most concerning. These aggressive developments can appear in patients that had imbalances in the fear, self-control, or territorial axes. The aggressive evolution most often occurs around the time of puberty, but the onset of aggression during the annual visit may be deferred to the next year's visit, for example. Aggressive manifestations require a discussion with the owners, and the possibility of managing the issue depends again on the practitioner's ability to raise awareness and encourage a request, which may not be spontaneous from the owners, without alarming them. Just as it is impossible to ignore a diagnosis of morbid obesity, even if the owners do not request it, it is also impossible to accept a cat exhibiting aggressive reactions without attempting to address this issue, which concerns both the safety of people and the well-being of the cat or other cats in the household.

House soiling is also a common reason for behavioral consultations, and the issue of idiopathic cystitis is generally well known to veterinarians in internal medicine. However, they often feel less equipped when it comes to the behavioral management aspect. This is because they are aware that psychological causes are common in this condition, but in the absence of a precise diagnosis, treatment remains difficult, if not limited to the usual advice on litter box cleaning and trials of different sizes and substrates.

In practice, urinary incontinence needs to be distinguished from urinary marking. It is a common symptom of productive anxiety states, and therefore, it is a frequent symptom in many nosographic entities. However, the treatment will differ if the origin is territorial, such as the arrival of a new cat in the home, or if it stems from a deprivation syndrome that has been ongoing for several years but never addressed until now. In the absence of a clear response to the cause of the incontinence, it is crucial to take the time for a veterinary psychiatry consultation.

Questions such as, "We agree, it never urinates outside the litter box?" can sometimes be enough to initiate a more focused inquiry.

Thus, if behavioral questioning is objectively unnecessary for a cat that seems as well as it did the previous year and as well as its owners, the practitioner should, when that is not the case, conduct a quick screening by organizing open-ended questions around the four behavioral axes.

31.1.4 Senior Cat Visits

The increased lifespan of our patients, coupled with advancements in diagnostic techniques and available treatments, has made geriatric veterinary care an essential part of the general practitioner's role. In addition to the organic medical aspect, which represents a significant portion of the geriatric consultation, the veterinarian's task during these visits is to determine whether the cat is starting to enter the *spiral of aging*. In ▶ Chap. 13, we described four pathways into this spiral: emotions, mood, cognition, and self-control. It is crucial to intervene as early as possible, before aging becomes global, in order to slow down the progression of this process.

The key feature of the senior cat consultation is that the medical and behavioral aspects are closely intertwined, and the differential diagnosis must always include all physical conditions, particularly those related to aging, which may trigger, amplify, or sometimes result in the behavioral signs presented by the cat. For example, a cat exhibiting polyuria and polydipsia is more likely to be incontinent, and this will be even more pronounced if the condition is compounded by arthritis and cognitive decline.

Each component of the clinical picture must be addressed to achieve an overall improvement in the patient's health and well-being.

For owners, early cognitive disorders are often considered normal because they are attributed, quite justifiably, to aging. It is therefore the veterinarian's role to emphasize that these issues can be treated, and that their management will help slow the progression of cognitive symptoms, thereby providing a better quality of life for the aging animal.

Among the signs the veterinarian should look for when the owners do not report them spontaneously, these may include:

- Early loss of autonomy: the cat has difficulty moving, orienting itself, or seeking contact. These issues can be cognitive (loss of orientation), related to mood (depression), or stem from a variety of physical problems (arthritis, blindness, etc.).
- Loss of emotional regulation, manifested by new or worsened fears, or even aggression, may also suggest a decline in sensory organ function, an endocrine disorder, or a painful condition.
- Excessive vocalizations, which could indicate cognitive decline but may also be a sign of any physical illness that causes discomfort or sensory loss.
- Early incontinence, which owners may attribute to old age, while it could actually indicate a physical or psychological condition.

This screening work leads to the care of the aging cat. Therapeutic work with elderly cats is specific and delicate. It is a general practitioner's task, as it is essential for a trusting relationship to be established with the owners at the beginning of this therapy. Moreover, the organic and behavioral aspects of managing an elderly cat's care are inseparable, and the concept of a holistic approach is truly meaningful during geriatric consultations. Indeed, any chronic disease, of which there are many in older cats (diabetes, kidney failure, hyperthyroidism, arthritis, etc.), will make the cat more vulnerable to anxiety or depression. An organic starting point will always facilitate a shift towards a pathological psychological state, especially in individuals who were already fragile in their youth.

Thus, the practitioner and the owner must work together to support the aging cat, addressing each emerging condition until the point in aging and disease progression where end-of-life considerations begin to arise. This approach will be discussed further in ▶ Sect. 31.2.3.

31.1.5 Behavioral Balance Assessment Questionnaire

To assist the general practitioner in evaluating the emotional state of a cat presented for consultation, there are tools that help answer the owner's questions within a reasonable consultation time.

The questionnaire presented below (◘ Fig. 31.5) is available online at the following address: ► https://noledge.org/questionnaire/. Its goal is twofold: firstly, it can be given to owners to raise awareness about behavioral conditions, and secondly, it can be used by the practitioner to save time and get an idea of the cat's behavioral profile.

This questionnaire, developed by the authors of this work, evaluates each of the behavioral axes, fear, self-control, territoriality, relationships, and communication quality, in 4 to 5 questions. After completing the questionnaire, which takes the owner 5–7 min to fill out, the website displays the results in a graphical format: green indicates that the axis is balanced, yellow suggests that there may be one or more elements to monitor in this behavioral axis, and red indicates that the cat shows one or more signs highly suspicious of a behavioral issue in the considered axis.

Behavioral axis	N°	Questions and possible answers
		Questions and possible answers **Scoring and Interpretation** • **Green:** Normal findings; no immediate concern. • **Orange:** Attention required; preliminary advice should be sought. • **Red:** A pathological condition is suspected; a professional consultation is recommended. **Scoring Rules:** • If more than three yellow responses are recorded within a single functional axis, the axis should be classified as red. • A single red response within any axis is sufficient to classify the entire axis as red.
Self-controls	1	**How does your cat behave with its toys, your belongings, and furniture?** •It doesn't chew anything •It may chew objects or damage, even destroy, the objects it chews •It may ingest parts of objects or whole objects (fabric, elastic, etc.).
	2	**How does your cat behave with you during friendly interactions, especially petting?** •It is gentle, never bites, and never extends its claws. •It may extend its claws or nibble, but it doesn't hurt you. •It is less tolerant and may scratch or bite to the point of causing injury (bleeding).
	3	**How does your cat behave when you play with it?** •It is gentle, never extends its claws or bites, even when very excited. •It may extend its claws or bite when excitement increases, but not to the point of causing harm (or only rarely). •Excitement during play makes it rough quickly: it may scratch or bite enough to cause injury (bleeding).
	4	**How does your cat behave in its usual environment (moving person, object, presence of insects, etc.)?** •A noise or movement sparks interest but does not lead to direct interaction with the source. •It almost always reacts (jumping, hunting behavior) but does not scratch or bite people. •It almost always reacts, and may scratch or bite, injuring its target if it is a person (bleeding).
	5	**What is your cat's ability to stop when it is excited (playing, running, or jumping)?** •It can stop upon verbal command (no, stop, etc.), or spontaneously if ignored. •It is sometimes difficult to stop with a verbal command or by ignoring it, and it may need to be isolated for a few minutes to calm down. •It struggles to stop despite attempts to ignore it or verbal commands. Even when isolated, excitement quickly reappears.
	6	**Does your cat become aggressive (hissing, attempting to scratch or bite) when excited or frustrated?** •No, never. •Yes, it has happened once or twice but never repeated. •Yes, it has happened more than twice.
Fear	7	**How does your cat behave in the presence of noises (storms, fireworks, door slamming, household appliances)?** •It is indifferent or shows brief surprise. •It shows moderate fear (runs to another room, hides), but quickly overcomes its fear and returns within a few minutes. •It shows intense fear (runs, hides, becomes aggressive), and this may last even after the noise has stopped.

◘ **Fig. 31.5** Behavioral condition awareness for owners

	8	**How does your cat behave in the presence of unfamiliar humans?**
		• It is indifferent or tries to establish friendly contact. • It keeps its distance or hides. Eventually, it returns and visitors can usually pet it. • It freezes or hides for the entire visit (sometimes even afterward). If forced into contact, it tries to escape, freezes, or becomes aggressive (scratches or bites).
	9	**How does your cat behave in an unfamiliar environment?**
		• It takes 1 or 2 days to explore the new environment and is almost as comfortable as at home. • It attempts exploration and makes progress over time but is never as comfortable as at home. • It is very stressed (stays hidden, explores very little) unless it is at home. If removed from its refuge, it tries to escape, freezes, or becomes aggressive (scratches or bites).
	10	**How does your cat behave at the veterinarian's office?**
		• It is relaxed, exits its carrier spontaneously, explores easily, and accepts petting. • It is stressed, reluctant to leave its carrier. During the medical exam, it is very uncomfortable (frozen, tense). It has sometimes hissed or growled but never escalated. • It is highly stressed, refuses to leave its carrier (frozen or aggressive if the vet approaches). The medical exam is always difficult. It is nearly impossible to trim nails or take a blood sample without sedating it.
	11	**Can your cat show aggression (scratching or biting) when afraid?**
		• No, never. • Yes, it has happened once or twice but never repeated. • Yes, it has happened more than twice.
Relationships	12	**How does your cat communicate with the humans in your household?**
		• It has many rituals, initiates contact, seeks petting, and looks for proximity (e.g., shared time on the bed, sofa, etc.). • The desire for contact varies depending on the family members. It rarely approaches some but never aggresses them. • It does not spontaneously seek contact with anyone. It avoids proximity to one or more household members and has already attacked them (scratches or bites).
	13	**How does your cat react when you want to pet it?**
		• It accepts and enjoys contact. Refusals are rare, and in these cases, the cat calmly moves away. • It generally tolerates contact. Petting certain areas or a long time may cause it to flee or hiss/growl, but without scratching or biting. • It tolerates contact for a very limited time, more easily in certain areas. If you insist, it may scratch or bite.
	14	**How does your cat communicate with other cats in the house? (If there are multiple cats, describe the worst situation.)**
		• It frequently approaches others for mutual interactions (grooming, shared play, resting together) OR it does not live with other cats, so I don't know. • It rarely seeks contact with other cats in the house but is not aggressive. • It completely avoids contact with the other cats in the house or seeks contact aggressively (chasing or direct attacks).
	15	**How does your cat behave when another cat in the house seeks to approach it? (If there are multiple cats, describe the worst situation.)**
		• Contact is welcomed and appreciated. Refusals are rare, and in these cases, the cat calmly moves away. OR it doesn't live with other cats, so I don't know. • It may tolerate the presence of the other cat but tends to maintain some distance. If the other cat insists too much, it will flee, hiss, or growl, but never scratch or bite. • It tries to avoid contact and may become aggressive if the other cat insists (scratching, biting).
	16	**Can your cat be aggressive and cause injury during interactions with humans, dogs, or other cats in the house (scratching or biting)?**
		• No, never. • Yes, it has happened once or twice but never repeated. • Yes, it has happened more than twice.
Territoriality	17	**Is your cat likely to urinate outside its litter box or designated area?**
		• No, never. The few times this occurred, it was clearly related to a medical issue (e.g., urinary infection) or an inability to access the elimination site, or if it was very dirty. • There have been rare incidents of urinary incontinence (less than once a month).

Fig. 31.5 (continued)

		•There are frequent episodes of urinary incontinence (more than once a month).
	18	**Does your cat rub its cheeks on objects or furniture?**
		•Yes, it does this several times a day or week on various objects. •Yes, it does this, but the frequency or objects have changed recently. •No, I have never seen it do that.
	19	**Does your cat rub its cheeks on humans?**
		•Yes, it does this with both household members and some visitors. •Yes, but only with one or more household members. •No, I have never seen it do that.
	20	**Does your cat scratch objects other than its scratching post?**
		•No •Yes, on a few spots, often related to where it likes to rest. •Yes, on many objects, which it has damaged over time.
	21	**How does your cat behave in the presence of unknown cats?**
		•It is indifferent or tries to establish friendly contact OR it never encounters other cats. •It runs away or threatens, keeping its distance but does not attack. •It freezes, panics, or becomes aggressive (scratching, biting).
Communication and environment	22	**How does your cat generally behave when you raise your voice or make a noise to interrupt it?**
		•It stops its action and quickly moves away if I insist •It is very impressed and flees. After a few minutes, it accepts contact if I approach. •It flees to hide. After a few minutes, it either flees or becomes aggressive if I approach. It can also become aggressive (scratching or biting) as soon as I raise my voice.
	23	**How does your cat behave if you take it by the scruff of the neck or punish it with a spray or slap?**
		•I don't know: I never do this. •It is very impressed and flees. After a few minutes, it does not flee if I approach. •It flees to hide. After a few minutes, it either flees or becomes aggressive if I approach. It can also become aggressive (scratching or biting) as soon as I raise my voice.

Fig. 31.5 (continued)

Of course, as with all questionnaires, owners may be unsure and feel that no answer perfectly fits their pet; they should be instructed to choose the one that best represents their animal, even if it is not exact.

Thus, in the case of a yellow result, an assistant or the veterinarian can provide advice aimed at improving the issues raised in the suspected behavioral axis. In the case of a red result, the veterinarian may suggest a behavioral consultation or, at the very least, inform the owner that they believe it is necessary, as they strongly suspect a behavioral condition in the animal.

To interpret the table, you should look at the number of green, yellow, or red responses in each axis. If there is a red response or more than 3 yellow responses, the axis should be considered red. If there are no red responses and a maximum of 3 yellow responses, the axis should be considered yellow, and the practitioner should provide advice to the owner regarding that axis. Finally, if there are only green responses or at most one yellow response, the animal should be considered balanced (green) in that axis.

A global assessment can then be made to give an overview of the cat's behavioral balance. All axes in green indicate a balanced animal with a low likelihood of having a behavioral disorder. When one or more axes are yellow but none are red, an imbalance is suspected, and advice should be given to the owner for the concerned axis, along with an over-the-counter product to help restore balance. After a few weeks, it is important to check that the animal has returned to normal. Finally, when one or more axes are in red, a specialized behavioral consultation should be recommended as a pathological psychological condition is suspected.

This type of tool does not provide a diagnosis and does not replace a specialized consultation lasting one and a half hours. However, it offers a significant time-saving advantage for the practitioner, who can prescribe the questionnaire to owners in the waiting room. Moreover, in case of suspicion of a behavioral condition, the answers to the questionnaire can support or, conversely, dispel this impression.

Summary of ▶ Sect. 31.1

Preventive visits for young cats, from adoption to adulthood, are numerous and provide opportunities to assess their proper behavioral development. Using the four behavioral axes and a few questions, the kitten's balance can be evaluated. The communication and skills of the owners are also assessed. The fear and self-control axes are the most commonly dysfunctional during the juvenile period, while deviations from the norm are more likely to be observed in the territorial and relational axes during the puberty visit.

In adults, the behavioral state is usually stable. However, the evolution of pre-existing vulnerabilities or life events can disrupt this balance, and the annual vaccination reminder is a good opportunity to check it.

As the cat ages, brain imbalances may occur. It will therefore be necessary to look for signs of cognitive, emotional, or mood disorders.

Tools such as the behavioral assessment questionnaire are available to assist the practitioner in quickly evaluating the cat's behavioral balance, providing a useful overview of the young cat's behavioral development and facilitating the monitoring of any changes or imbalances throughout its life.

31.2 General Medicine

31.2.1 Painful Conditions

Throughout their lives, and outside of preventive visits, veterinarians see feline patients for diseases or painful conditions. While every veterinary consultation should, in a holistic approach, serve as an opportunity to detect behavioral signs that may require management, the most important challenge is navigating the paradox of the caregiving mission (see ▶ Chap. 30, ▶ Sect. 30.1.1) and successfully improving the patient's physical condition without negatively impacting their emotional state, both in the short and long term.

When a cat presents at a veterinary practice with an organic disease, it is experiencing pain or discomfort. In ▶ Chap. 30, we describe the increased risk of sensitization to care that is directly correlated with this pain. Thus, the more painful the cat is, the greater the risk of sensitization, and the practitioner must anticipate this to prevent the emergence of care-related phobia as much as possible. This can be achieved by using appropriate analgesics and, when necessary, resorting to sedation or anesthesia. Performing a very painful procedure with coercive restraint is not a viable choice for the patient's future medical care, even if it is completed quickly. The resulting memory of this experience would have a far greater future cost than a reasonably planned anesthesia, especially if it led to the development of a care phobia, requiring sedation even for simple, nonpainful procedures in the future.

It is also important to note that a cat presenting a given organic disease (e.g., otitis, pyoderma, gastritis…) is likely to experience it multiple times throughout its life. If pain or fear associated with care is not properly managed, the anticipatory process would hinder the ability to carry out similar future treatments.

Clinical Illustration

Tina is a 4-year-old cat brought in for a wound on her hindquarters. She had been limping on her left rear leg, which appeared swollen for 3 days, and since this morning, she had been licking the area frantically. When her owners tried to examine the area, she hissed and growled, not allowing them to touch her. At the veterinarian's office, Tina was very inhibited and had never liked visiting. She sometimes urinates in her carrier on the way to the clinic. She remained at the back of her cage and refused to come out on her own. When the vet opened the carrier by removing the top, Tina growled as soon as he attempted to touch her hindquarters. However, the vet noticed that her fur was wet and suspected an abscess but could not see the full extent of the opening, and Tina attempted to scratch as he tried to visualize the wound.

To ensure Tina's well-being, she was given analgesics, anti-inflammatory medication, and first-line antibiotics orally. A follow-up appointment was scheduled for the next morning, with fasting, and gabapentin prescribed. The cat would either be sedated for abscess cleaning or observed carefully if the inflammation had decreased enough to make the procedure manageable, based on her tolerance.

The medication and gabapentin helped reduce local inflammation, making Tina more manageable for examination. At the second appointment, the veterinarian could easily examine Tina, shave the wound without her reacting, and determine that the abscess could be treated without further surgery. A virtual follow-up appointment was scheduled 1 week later to avoid additional stress from transporting Tina. During the virtual check, the veterinarian confirmed that the wound had closed well and no further intervention was needed.

Had the examination not been possible during the second appointment, the veterinarian was prepared to sedate Tina, allowing for a thorough exam under optimal conditions. Thus, organizing a second appointment with the option of sedation, if necessary, before manipulating the wound, helped avoid causing Tina intense pain that could have hindered her cooperation in future treatments.

31

This observation applies to all feline patients, even those in perfect behavioral balance, but those who already have a vulnerability or a behavioral condition are particularly at risk.

First, cats with *a fear-related condition*, particularly social phobics, have a reduced tolerance for handling directly due to their behavioral condition. Any pain, even moderate, associated with handling can trigger a severe care phobia. Hospitalized phobic patients can pose a real danger to the veterinary staff, as they are more likely to trigger aggression due to irritation or uncontrolled emotional outbursts. When the team encounters this type of phobic behavior during the course of treatment, the difficulties faced in providing medical care should serve as an alert to the owner. It is essential to assess whether the behavioral symptoms occur only at the clinic or also in other contexts and to propose appropriate care.

Cats with a condition affecting the *self-controls axis* are, for some of them, hypersensitive, and this hypersensitivity extends to pain. They have little patience, tolerate restraint and immobility poorly, and have difficulty controlling their emotions. Pain, restraint, and frustration imposed by treatment, combined with their impulsivity, increase the risk of triggering aggression during the procedure.

Their imbalance also predisposes them to tear out sutures or collars, and they may struggle to tolerate a recovery period during which exercise must be limited. These difficulties hinder healing and prevent owners from following medical instructions. The postoperative care for these cats often requires addressing their behavioral problem medically to make the treatments possible (see ▶ Chap. 30, ▶ Sect. 30.3.3.5, box "Postoperative medication protocols"). It is not uncommon for this temporary prescription to open a dialogue about long-term management, especially once the owners have seen the positive effects of psychotropic treatment on their cat's quality of life and overall well-being.

In the case of *interspecific relationship disorders,* the cat may, with a lowered threshold for tolerance, engage in an aggressive sequence that prevents veterinary staff from intervening due to an anxious state combined with pain. When this occurs, and unlike social phobic animals who only react poorly to strangers, the owners cannot perform the necessary care at home either. All medical care for the case must take this fact into account: prescribing treatments that endanger the owners is not an option, and the medical prescription should be adapted to what is reasonably feasible with that cat. For example, a bi-daily ear cleaning is not a reasonable prescription if the cat becomes aggressive as soon as its head is handled. Furthermore, a *short-term psychotropic treatment* should be prescribed when necessary to help owners follow the instructions. As discussed with cats exhibiting imbalances in the self-control axis, recognizing the limits of the cat's cooperation or the beneficial effect of temporary psychotropic treatment can be entry points for suggesting a more comprehensive approach to ensure future treatments are carried out in better conditions.

Important Point of View

Medication Assistance During a Consultation

In cases where aggression seems to be a strategy that the cat is willing to adopt, and before it occurs (see ▶ Chap. 30, ▶ Sect. 30.2.1.2, "Observation and assessment of the cat"), the practitioner may use medications to carry out the necessary care. Pregabalin (5 mg/kg), gabapentin (20–50 mg/kg), or trazodone (3–10 mg/kg), administered 2 h before the clinical exam, can increase the threshold of tolerance and allow the procedure to be performed under better conditions. If sufficient tolerance and cooperation levels are not reached, or if the pain is too intense to protect the cat from the risk of sensitization and the staff from the risk of aggression, sedation should be scheduled.

31.2.2 Chronic Conditions

Chronic organic conditions, in addition to potentially being associated with varying levels of pain, require ongoing management, which leads to numerous follow-up visits. During these visits, exams and, in some cases, additional tests must be repeated. Managing patients with chronic conditions increases the risk of sensitization to care and can eventually lead to a phobia of veterinary treatments. Unlike acute painful conditions, it is often the frequency and recurrence of treatments that are problematic in these cases.

For cats, veterinary visits present a unique challenge: they are removed from their territory, which is a significant stress factor. This is different from dogs, which can calm down by being with one or more members of their social group. Additionally, as prey animals, cats are more vulnerable to fear than dogs. For these reasons, sensitization to care is almost inevitable unless careful attention is paid during each visit. Repeated treatments, therefore, pose an even more delicate challenge in cats than they do in dogs.

To prevent this unfavorable progression, it is important to identify manipulations that may lead to sensitization and to initiate a medical training protocol at the start of care. In cases where the cat has not yet been sensitized, this training serves a preventative purpose. The practice must be able to implement an HC (Habituation and Conditioning) protocol, which is easy to execute and accessible even without the presence of a veterinary behavior specialist (see ► Chap. 30, ► Sect. 30.2.2.1).

For cats already phobic of care who require regular treatments, a DS-CC (Desensitization and Counterconditioning) protocol becomes necessary (see ► Chap. 30, ► Sect. 30.2.2.2). This protocol should be supported by a sufficiently effective psychotropic treatment to enable at least the completion of the required treatments or be incorporated into a more comprehensive care plan if the animal has a concurrent overall condition.

The situation in which a cat phobic of care develops a condition that will require regular treatments is as common as it is complex in general practice. The behavioral profile must be considered as one of the factors in choosing the therapeutic strategy. For many of these patients, care phobia is an unfavorable factor in the prognosis, as it prevents certain necessary medical procedures. This must be identified as early as possible and applies equally to the portion of follow-up care done in the clinic and the ability of owners to perform treatments at home.

As with the previous situations described, it is not uncommon for significant difficulties in performing treatments, compounded by the frequency with which these treatments need to be done, to lead owners to agree to implement a behavioral management plan (see ► Chap. 28, ► Sect. 28.2.4.1, ► Fig. 28.3).

Clinical Illustration

Sacha is a 9-year-old European cat who lives alone with his owner in an apartment. He has always been difficult to treat, and his owner stopped taking him for vaccinations when he was 3 years old because the visits were so unpleasant. Sacha would growl at the veterinarian even before being taken out of his carrier, and the examination was nearly impossible. The veterinarian would ask an assistant to help with restraint using a towel, so the vet could administer the vaccine injection.

Since he was a kitten, Sacha has been very fearful of strangers, and only his owner can touch him. Visitors never see him because he runs into the bedroom and hides under the bed as soon as he hears the doorbell. Typically, he comes out about an hour after they leave. His owner has never seen the need for behavioral care, even though the vet mentioned it at the first appointment. She fears that medication might make him apathetic and does not want to "drug" her cat. For the same reason, she never wanted to give Sacha any medication before visits and preferred not to take him to the vet at all.

Today, Sacha has an appointment because he has been meowing all night, and his owner feels that he's getting worse. She thinks he has lost weight and has a poorly groomed coat because he has stopped grooming himself. The veterinarian needs to conduct further tests, starting with a blood sample. To do so, he must sedate Sacha. After the examination and tests, it turns out that Sacha has hyperthyroidism and early-stage kidney failure. The veterinarian explains to the owner the treatments she will need to administer and the regular blood tests that need to be done to monitor the disease's progression and treatment balance.

The owner tries, but soon Sacha hides under the bed whenever he sees her with the methimazole bottle. When she tries to get him out, he becomes aggressive. She has been bitten and scratched several times, even though her cat has never been aggressive with her before. She has tried various tricks with pâté and tuna, but Sacha refuses to eat. The owner is very upset because she loves Sacha dearly. She lives alone with him and cannot imagine losing him. Reluctantly, she agrees to schedule an appointment with a veterinary behaviorist.

Sacha's diagnosis is intermittent anxiety progressing toward generalization due to repeated but necessary veterinary visits, associated with a sensory deprivation syndrome. The behaviorist prescribes venlafaxine in granules, which will be easy to administer starting a week after the appointment, to allow the owner time to establish a positive feeding routine with Sacha. During this time, the hyperthyroidism treatment is also withheld. This approach works, and Sacha begins taking both the behavioral treatment and the methimazole in a teaspoon of salmon pâté every evening.

Once this step is completed, the behaviorist also prescribes gabapentin (10 mg/kg) for use during home training sessions. She explains to the owner what needs to be done to teach Sacha to position himself for blood draws, which will be performed by his owner in the jugular region. An agreement is made that if Sacha doesn't cooperate during the visit, he will need to be sedated for the blood draw. The owner will have to hold him during the visits, using the same positioning as at home. She also teaches her how to train Sacha to enter the carrier on command.

Finally, she asks the owner to have visitors come over while keeping the bedroom closed so that Sacha stays in the room during their visit. She instructs her to use treats to encourage Sacha to approach the visitors at his own pace.

After 4 months of treatment, the owner returns for a behavioral check-up, very proud of Sacha: the veterinarian had to sedate him during the first two visits, but last week they successfully performed the check-up while Sacha remained awake and was not aggressive at all. He even ate the treats she placed in his carrier during the drive back home. Additionally, about a month ago, he allowed the neighbor, who had come over for dinner, to pet him.

31.2.3 End of Life, Palliative Care, and Euthanasia

In the daily practice of a general veterinarian, the approach to end-of-life care is a common aspect of their work. It involves the announcement of an incurable or fatal disease, the organization of palliative care for animals whose conditions have surpassed the threshold that inevitably leads to death, as well as the euthanasia itself. While these consultations do not fall under the expertise of a veterinary psychiatrist, behavioral medicine can offer valuable insights, particularly in analyzing and managing emotions, as well as in the communication that the veterinarian needs to establish with owners who are often deeply affected.

Understanding the emotional dynamics during this stage, including the feelings of both the animal and its owner, is essential. The veterinarian must be sensitive to the emotional state of the owners, offering support and guidance while respecting their emotional journey. Communication strategies must be empathetic and clear to help owners navigate through these challenging decisions, making it an integral part of the veterinary care process during the end-of-life phase.

31.2.3.1 Anticipating End of Life and Organizing Palliative Care

Regardless of the nature of the relationship they have with their cat, all owners are affected when the prospect of end-of-life care arises. As this moment approaches, there is often an increase in emotional attachment, and it is common for the announcement of a serious illness leading to death to significantly *strengthen the bond* between the cat and its owner. It seems as though the cat's weakness and dependence create a special sense of duty in its owner.

The emotional dimension of the exchanges that take place with the veterinarian in consultations related to the end-of-life perspective generates a *particular sensitivity* in owners: tears or bursts of anger are easily triggered. The practitioner must show *empathy*, be understanding of certain excessive reactions, and downplay any accusations that may be directed at them. Additionally, *emotions induced by memories* may arise. The prospect of an animal's death inevitably brings up past losses for the owners, and these memories amplify their emotions. Finally, it is important to consider *loyalty mechanisms*. For example, when the cat that is passing is the pet of a deceased loved one, the situation experienced by the owner can become doubly painful.

This heightened sensitivity requires increased attention and technical skill because, just like the mechanism described for cats, the intensity of the negative emotions felt by owners will be linked to the context: the practitioner must be very careful about communication details and the execution of the cat's manipulations, as these will be definitively remembered if deemed clumsy. While an empathetic word can generate undeniable recognition from an owner, even a small procedural or communicative error can lead to strong and lasting rejection. In these moments, particularly in the context of euthanasia, judgments are harsh and extremely severe.

Veterinary teams are accustomed to reflecting on and discussing end-of-life care and euthanasia, but owners are often inexperienced in this area. Therefore, the practitioner must be *patient* enough to ensure mutual understanding of the inevitable aspect of euthanasia, acknowledging that the realization may come more slowly to the owners than to the practitioner. The clinical evidence of the animal's deteriorating state is not always sufficient for the owners. Additionally, owners, especially during such an emotional time, are influenced by their beliefs and feelings, which, though sometimes unfounded, must be taken into consideration. Our duty is to strike a balance between transmitting our knowledge, such as the objective clinical evaluation of an animal, and respecting the *beliefs of the owners* (e.g., the animal must see a specific person before dying, it must die in a certain place, if it still purrs sometimes, it doesn't want to die, etc.).

For some, care must be pursued as far as possible because the greatest guilt would be not showing enough commitment to their pet. For others, prolonging life is the most dreaded thing. Under the term "prolonging" may lie a refusal of any suffering or pain, and when owners open up about this, the practitioner may discover that this refusal often refers to the end-of-life experiences of a loved one, where intense pain became a source of suffering and helplessness for the family. Sometimes, refusal to prolong life can simply disguise a reluctance to invest emotionally, so it is important to clarify what the phrase means, as the objectives will be vastly different.

Thus, in order to establish what becomes a "care contract," whether written or tacit, surrounding the end of a cat's life, it is necessary to discuss in detail the owners' wishes and goals. This discussion is a fundamental part of the *therapeutic alliance.* The earlier this conversation takes place relative to the actual end of life, the more calmly it can be addressed.

As soon as the practitioner diagnoses a condition that will lead to death, they can address the topic of *end-of-life care*, which includes *palliative care*, i.e., compensatory care with no therapeutic effect, and the act of euthanasia itself. Sometimes, owners are surprised, as it seems premature to them, but this discussion often brings real relief once the key elements have been raised, particularly around the act of euthanasia itself. Indeed, this act carries many assumptions and concerns for owners who have never faced it before. Moreover, this open discussion is a good opportunity to establish a *palliative care plan* while the animal is still relatively healthy. It allows the owners to calmly reflect on what is desirable or not, to consider the level of care they are willing to provide, and what would feel like unnecessary prolongation of life. They can also discuss the costs of different options, any special requests, and highlight the emotional factors that are part of their family dynamics.

The primary goal of end-of-life care is the *well-being of the cat*. Collaboration between the owner and the practitioner allows for this objective to be achieved optimally, with the owner being the best at knowing their pet and the practitioner ensuring medical objectivity. The partnership that is established is reassuring for the owner, making it easier for them to express their doubts.

As in many therapies, we must teach owners to assess their cat. In end-of-life conditions, the animal's well-being and quality of life are the primary elements. In an elderly cat, often with multiple health issues, several sources of discomfort coexist: pain, fatigue, decreased sensory perception, etc. The owner must learn to detect these discomforts by identifying signs of suffering or pain. One idea to convey is, for example, to try to compensate for *discomfort* with *pleasure*. The goal is to ensure that every little suffering, every pain, and every frustration is compensated by an equal amount of pleasure, which the owner, as the expert on their cat, is best suited to choose.

In addition to achieving an acceptable level of well-being for the cat, this approach provides a *relational compensation* that helps the owner not feel helpless in the face of their pet's declining health.

Care can be maintained as long as the balance remains positive when weighing the well-being of both the cat and its owner against each other's discomfort. The day this balance becomes negative, it is legitimate and preferable to discontinue care. Sometimes, this negative shift happens very abruptly, and the owner's decision becomes inevitable from 1 day to the next. The practitioner must be prepared for this request and be ready to respond quickly, without delaying euthanasia.

31.2.3.2 Euthanasia Consultation

This final consultation is often dreaded, particularly because it is difficult to describe the steps involved beforehand in order to dedramatize the act itself. Most cat owners are very unfamiliar with the process of euthanasia, which makes it a highly anxiety-inducing experience. Being able to discuss it in advance, before being faced with it, is therefore particularly helpful and beneficial for everyone involved.

If this conversation took place when the practitioner and the client established the end-of-life care plan, the steps have been explained. The owner already knows who should be present for each part of the process, what their wishes are for the animal's body afterward, and any special requests they have expressed.

On the day of euthanasia, the owner is relieved from logistical decisions, and the veterinarian's role is to ensure that everything proceeds as previously agreed upon, allowing them to focus on carrying out the act itself.

Clinical Illustration

Ms. X arrives for the euthanasia of Marlow, who is suffering, and she can no longer bear to see him pleading with his eyes. She arrives at 10 a.m., in a panic, and wants to end it as quickly as possible. She is received by her veterinarian in an emergency. The veterinarian reminds her that she had requested her 10-year-old daughter, who is at school, to be present. It is then agreed to proceed with immediate hospitalization and the administration of morphine, which will allow the euthanasia of Bobby to be delayed until later in the afternoon, under very good conditions. This time allows Ms. to pick up her daughter from school and prepare her for this difficult moment. The owner is very grateful that this arrangement allowed her daughter to be present.

International recommendations suggest that the cat should be anesthetized prior to the lethal injection (Underwood and Anthony 2020). While each practitioner may have their own rules, therapeutic freedom does not exclude adhering to some essential principles for a proper procedure, presented here as theoretical and ideal guidelines.

The practitioner can verify what change has disrupted the balance of well-being and then *confirm the owner's choice*. This confirmation is the first key point to ensuring the euthanasia proceeds smoothly, as the main fear of the owner is making the wrong decision and prematurely ending the animal's life.

The practitioner should *describe the procedure* once again to demystify the act itself. They must be aware of and respect the choices the owners have made, as these provide a framework to guide the procedure, especially when emotions hinder clear thinking. However, the practitioner must also allow the owners the possibility to change their minds if they wish to. Many owners initially refused to stay with their animal at the time of the lethal injection but, after sedation, realize that they feel courageous enough to stay with their companion until their last breath. They must feel listened to and be able to express this new request.

The practitioner's actions must be professional and demonstrate great skill. They must then provide a clear indication or marker of the end of life. This is most often determined through auscultation, which allows the practitioner to declare that *"it is over."* The owner often needs to know the exact moment of death. They must also be informed about the potential for post-mortem movements.

Finally, depending on what was planned, the client will either remain alone with their cat or not, may collect some fur, or leave an object belonging to the cat for cremation... Respecting these personal wishes without judgment is a major source of appreciation for the veterinarian who has accompanied them through this moment.

The entire care team should be aware of what is happening in the room. The session should not be interrupted by a phone call or the unexpected entry of a nurse or colleague. The owners should be allowed to stay as long as they wish with their companion's remains.

However, even when everything is planned and scheduled in advance, some situations may disrupt the expected protocol. A common case is when euthanasia is required on a day when the cat's regular veterinarian is unavailable. In this case, the practitioner in charge of the procedure should refer to written guidelines in the animal's record (a transcription of the "care contract" established beforehand) and follow them as closely as possible while respecting the steps described in the previous paragraph. When euthanasia occurs earlier than expected, the owner may not have yet decided what will happen to the body. In such situations, it is important that the team is able to keep the deceased animal and allow the owner a few days to make their decision calmly.

For cats, more than for dogs, home euthanasia can be greatly appreciated by the owners, especially if they perceived their pet to be in significant distress during veterinary visits. The ability to "let them go" in a place where they felt calm is often highly valued by the owners, even if it is not objectively easy to organize from the veterinary practice's perspective.

Finally, euthanasia, even when it is fully justified and represents an opportunity to end the suffering of an animal sometimes in agony, is never an easy experience for the veterinary team (Quain 2021). Having the opportunity to discuss acceptable ethical boundaries within the team is crucial so that the veterinarian in charge of the decision can have the necessary calmness to accompany the cat and its owners, without carrying the burden of guilt due to any hesitation about the legitimacy of their action.

Summary of ▶ Sect. 31.2
Pain or discomfort that leads a cat to a consultation can have different consequences depending on any behavioral imbalances across various axes. Agitation, aggression, and difficulties during hospitalization are among the most common expressions. It is important to use medication when necessary to ensure that the animal and the healthcare team can proceed with care under the best possible conditions. Preventing care phobia can be particularly challenging, especially in cases of chronic conditions requiring multiple visits. In all cases, behavioral expressions of distress should trigger a discussion with the owners and a proposal for appropriate management, just as with any physical symptom.

Euthanasia is a unique consultation that requires flawless organization. Communication with the owners, listening, and empathy are essential, and the decision-making process leading to this final consultation helps reduce the suffering experienced by the owners.

31.3 French Legislation (for informational purposes)

Although this legislation is more frequently applied to dogs than to cats, in France, the Ministry of Agriculture, Fisheries, and Food requires a health follow-up for any cat that has bitten or scratched a person, regardless of whether it has been vaccinated against rabies. The primary legal texts governing this issue are as follows and concern Article 232-1 of the Rural Code, which states (articles consulted on ▶ https://www.legifrance.gouv.fr on 01.01.2025):

- Article 1: When an animal, whether domestic or wild, tamed or kept in captivity, vaccinated or not against rabies, is a biting or scratching animal as defined in Article R223-25***, point 5° of the Rural Code, and it can be seized without being euthanized, it must be placed under the supervision of a sanitary veterinarian at the owner's or keeper's expense.
- During the period of this supervision, the owner or keeper may not dispose of or euthanize the animal without the authorization of the director of veterinary services.
- If the owner or keeper is unknown or fails to comply with the demand to place their animal under the supervision of a sanitary veterinarian, the municipal authority will arrange for the animal to be supervised in a pound.

*** ***Article R223-25, point 5° of the Rural Code***

A biting or scratching animal is any animal susceptible to rabies that:

a) Anywhere, has bitten or scratched a person;

b) Or, in a department officially declared infected with rabies, has bitten or scratched either a domestic animal or a wild animal that has been tamed or kept in captivity;

c) Or, in a rabies-free department, has bitten or scratched either a domestic animal or an animal tamed or kept in captivity, and has come from a department officially declared infected with rabies, or from a country affected by enzootic rabies, within a period defined by a ministerial order from the Minister of Agriculture.

- Article 2: A biting or scratching animal is placed under the supervision of a sanitary veterinarian for a period of 15 days if it is a domestic animal.
  - During this period of supervision, the animal must be presented three times by its owner or keeper to the same sanitary veterinarian.
  - During this period of supervision, any rabies vaccination injection to the animal is prohibited.

  The first visit must take place within 24 h of the animal having bitten or scratched someone, and the second visit must occur no later than the seventh day after the bite or scratch.

In the absence of any symptoms suggesting rabies, the consulted sanitary veterinarian issues a provisional certificate after each of the first two visits, confirming that the animal shows no suspected signs of rabies at the time of the visit.

At the end of the third visit, on the 15th day for cats, the sanitary veterinarian prepares a final certificate confirming that the animal under observation did not show any symptoms that could suggest rabies during the observation period.

- Article 3: If the owner or keeper of an animal placed under the supervision of a sanitary veterinarian needs to travel before the end of the observation period, the director of the veterinary services may authorize the animal's regulatory visits to continue with a second sanitary veterinarian at the new place of residence. However, this transfer must be communicated in advance to: the person bitten or scratched, the director of the veterinary services of the receiving department, the first consulted sanitary veterinarian, and the authority with police powers who was informed of the incidents that led to the animal being placed under veterinary surveillance.
- Article 4: The failure to present the animal within the prescribed timeframes outlined in Article 2 of the present order must be immediately reported to the authority with police powers and to the director of veterinary services of the department by the sanitary veterinarian under whose supervision the animal has been placed.
- Article 5: The certificates in accordance with the models defined by the annex of this order are drawn up in *five copies* at the end of each visit for the animal. Three copies are given to the owner or keeper of the animal, who is responsible for sending one to:
  - The person who was bitten or scratched, or the owner of the animals that were bitten or scratched.
  - The authority with police powers who was informed of the events leading to the animal being placed under veterinary surveillance.

The fourth copy is sent by the consulted sanitary veterinarian, at the end of each visit, to the *director of veterinary services of the department* where the person or the domesticated or captive wild animal was bitten or scratched.

The fifth copy is kept by the consulted sanitary veterinarian *for* 1 *year.*

- Article 6: During the period of surveillance for the biting or scratching animal, the appearance of any sign of illness or the death of the animal, regardless of the cause, must immediately lead to the presentation of the animal or its carcass by its owner or keeper to the sanitary veterinarian under whose surveillance it is placed. Its disappearance must also be immediately reported to the veterinarian.

In the event of suspected rabies, the animal must be kept under observation, strictly isolated, and tied up, unless impossibility justifies its immediate euthanasia.

- Article 7: When, during the surveillance period, the biting or scratching animal dies or is euthanized, either with the authorization of the director of veterinary services or in cases of force majeure, the carcass, or at least the head, must be sent to the director of veterinary services for dispatch to the departmental veterinary laboratory for transmission to an accredited laboratory for rabies diagnosis.

Indeed, according to French law, a cat that has scratched or bitten should always undergo sanitary surveillance. However, this is very rarely the case in practice. Although France is considered rabies-free for domestic animals, there are still cases of rabies in certain wild animals, particularly bats. Therefore, the veterinarian must remain vigilant regarding this serious disease, include it in their differential diagnosis, and initiate sanitary visits at the slightest suspicion.

Summary of This Chapter

The approach to general medicine consultations outlined in this chapter is based on a quick yet systematic evaluation of the four behavioral axes presented in this book. This approach allows the practitioner to identify the most severe cases for referral, while also systematizing their practice to include behavioral evaluation of the cat in all consultations, whether in preventive or general medicine. The presence of aggressive sequences in the clinical picture observed by the practitioner highlights the severity of the concerned axis and provides a reference point for urgently referring the most severe cases.

The evaluation of communication quality and the environment complements the medical aspect of behavioral assessment. It remains crucial as it directly impacts the improvement or, conversely, the worsening of the behavioral sequences produced.

References

Ahola MK, Vapalahti K, Lohi H (2017) Early weaning increases aggression and stereotypic behaviour in cats. Sci Rep 7(1):10412. https://doi.org/10.1038/s41598-017-11173-5

Campbell GR, Arnott ER, Graham C, Niel L, Ward MP, Ma G (2024) Impact of early socialisation in foster care on kitten behaviour. Appl Anim Behav Sci 276:106306. https://doi.org/10.1016/j.applanim.2024.106306

Carney HC, Sadek TP, Curtis TM, Halls V, Heath S, Hutchison P, Mundschenk K, Westropp JL (2014) AAFP and ISFM guidelines for diagnosing and solving house-soiling behavior in cats. J Feline Med Surg 16(7):579–598. https://doi.org/10.1177/1098612X14539092

Graham C, Pearl DL, Niel L (2024) Too much too soon? Risk factors for fear behaviour in foster kittens prior to adoption. Appl Anim Behav Sci 270:106141. https://doi.org/10.1016/j.applanim.2023.106141

McCune S (1995) The impact of paternity and early socialisation on the development of cats' behaviour to people and novel objects. Appl Anim Behav Sci 45(1–2):109–124. https://doi.org/10.1016/0168-1591(95)00603-P

Quain A (2021) The gift: ethically indicated euthanasia in companion animal practice. Vet Sci 8(8):141. https://doi.org/10.3390/vetsci8080141

Sinn L (2016) Factors affecting the selection of cats by adopters. J Vet Behav 14:5–9. https://doi.org/10.1016/j.jveb.2016.06.001

Underwood W, Anthony R (2020) AVMA guidelines for the euthanasia of animals: 2020 edition. Retrieved on March 2013(30):2020–21

Zubin J, Spring B (1977) Vulnerability: a new view of schizophrenia. J Abnorm Psychol 86(2):103. https://doi.org/10.1037/0021-843X.86.2.103

Supplementary Information

Appendix 1 Diagnosis Criteria for Phobic and Anxious States – 871

Appendix 2 Diagnostic Criteria for Impulsive and Compulsive States – 873

Appendix 3 Diagnostic Criteria for Depressive, Manic, and Bipolar States – 875

Appendix 4 Diagnostic Criteria for Dissociative and Senile States – 877

Appendix 5 Disorders Associated with Dysregulation of the Fear Axis – 879

Appendix 6 Disorders Associated with Dysregulation of the Self-Controls Axis – 880

Appendix 7 Disorders Associated with Dysregulation of the Territoriality Axis – 882

Appendix 8 Disorders Associated with Dysregulation of the Relationship Axis – 884

Appendix 9 Essential Psychiatric Disorders – 885

Appendix 10 Aging Disorders – 886

Appendix 11 Elements of Differential Diagnosis Between Neurology and Psychiatry – 888

Appendix 12 Body Condition Score (BCS) Chart for Cats – 889

Appendix 13 Reported Signs in the Semiology of Feeding Behavior and Possible Interpretations – 890

Appendix 14 Reported Signs in the Semiology of Sleeping Behavior and Possible Interpretations – 891

© The Editor(s) (if applicable) and The Author(s), under exclusive license to Springer Nature Switzerland AG 2026

S. Bleuer-Elsner et al., *Veterinary Psychiatry of the Cat*,
https://doi.org/10.1007/978-3-032-06260-4

Appendix 15 Differential Diagnosis of Urinary and Fecal Inappropriate Elimination and Their Associated Characteristics – 892

Appendix 16 Reported Signs in the Semiology of Repetitive Behaviors and Possible Interpretations – 893

Appendix 17 Classification of Injurious Sequences in Cats Based on the Modified Moyer Classification – 894

Appendix 18 Semiology of Play and Mutual Grooming – 895

Appendix 19 Semiology of Exploratory Behavior and Marking – 896

Appendix 20 Summary Table of SSRIs – 897

Appendix 21 Summary Table of Tricyclic Antidepressants – 898

Appendix 22 Summary Table of MAOIsQ3 – 899

Appendix 23 Summary Table of Benzodiazepines – 899

Appendix 24 Summary Table of Sympatholytics – 900

Appendix 25 Summary Table for Antiepileptics – 900

Appendix 26 Summary Table of Tetracyclic NaSSAs – 902

Appendix 27 Summary Table of Neuroleptics – 903

Appendix 28 Summary Table of Other Psychotropic Medications – 905

Appendix 29 Summary Table of Complementary Hormonal Treatments – 906

Appendix 30 Psychotropic Medications for Managing Cats with Healthcare Phobia – 907

Appendix 31 Fear Behavioral Management General Principles – 907

Appendix 32 Self-Control Behavioral Management General Principles – 908

Appendix 33 Territoriality Behavioral Management General Principles – 909

Appendix 34 Relationships Behavioral Management General Principles – 909

Appendix 35 Behavioral Condition Awareness for Owners – 910

Index – 913

Appendix 1 Diagnosis Criteria for Phobic and Anxious States

Pathological state	Diagnostic criteria	State variations	Differential diagnosis
Phobic	*Phobic state* is a pathological mental state characterized by disproportionate fear responses during repeated exposure to a specific stimulus or group of stimuli. No adaptation occurs despite repeated exposures **Behavioral symptoms:** – Global motor inhibition – Avoidance and anticipation with hypervigilance – Hyperactivity related to flight – Irritation-based aggression in the context of fear or pain – Uncontrolled emotional aggression – Instrumental aggression – Appeasement signals – Fear postures **Organo-vegetative symptoms:** – Anal gland evacuation – Tachycardia – Tachypnea – Tremors – Paw pad sweating – Emotional urination and defecation	*Simple phobia* Limited number of phobogenic stimuli Minimal anticipation or generalization *Complex phobia* Expanded phobogenic stimuli Anticipation and generalization processes starting to worsen the phobic state	*Reactive state* Transient hypervigilance disappearing with habituation Quick return to equilibrium after exposure *Anxious state* Exaggerated fear symptoms without identifiable stimuli Frequent organic symptoms *Depressive state* Prolonged inhibition without stimuli Reduced motivation rather than fear *Impulsive state* High motivation Motor agitation linked to desire to react, not exaggerated fear

Pathological state	Diagnostic criteria	State variations	Differential diagnosis
Anxious	*Anxious state* is a pathological mental state where the cat exhibits inappropriate, recurrent, objectless fear reactions in response to changes in internal or external environments **Common symptoms** (in whole or in part): – Objectless fear reactions (4F) – Hypervigilance – Tachypnea – Cutaneous hyperesthesia – Digestive issues: dyspepsia, recurring gastritis, chronic intestinal disease – Territorial marking anomalies – Urine soiling, emotional cystitis – Alopecia	Paroxysmal crises Highly productive episodes (e.g., syncope, Rolling Skin Syndrome), not attributable to organic causes *Intermittent to Generalized Forms* Varying levels of anxiety from short episodes (intermittent) to pervasive behaviors affecting relationships and frequent startled awakenings (generalized) *Inhibited Type* Withdrawal, self-centered activities *Productive Type:* Agitation, irritability	*Reactive state* Transient hypervigilance resolved through habituation *Phobic state* Fear reactions tied to identifiable stimuli *Epilepsy* Organovegetative symptoms (e.g., polypnea, tremors) with or without convulsions; diagnosis relies on therapeutic outcomes *Depressive state* Context-independent anhedonia and indifference to the environment; grooming may cease entirely *Impulsive state* High motivation, motor agitation unrelated to hypervigilance *Compulsive state* Repetitive or stereotyped behavior triggered by identifiable supports (e.g., play, frustration, food)

Appendix 2 Diagnostic Criteria for Impulsive and Compulsive States

Pathological state	Diagnostic criteria	State variations	Differential diagnosis
Impulsive	*The impulsive state* is a pathological mental state characterized by an imbalance between activation and inhibition, or bottom-up versus top-down processes, leading to behavioral responses with a shortened appetitive phase and potentially increased intensity	*Contextual to invasive form* The level of intrusion in the impulsive state depends on the number of behaviors involved: – Modification of the behavioral sequence of one or more behaviors, with a shortened appetitive phase, high reaction intensity, and prolonged calming phase – These sequence modifications persist despite attempts at behavioral modification – When multiple stimuli are present, sequences may stop and a new appetitive phase may begin	Reactive state Environmental adjustments and meeting physiological needs lead to the rapid return to the normal adaptive state. Impulsive behaviors disappear completely Compulsive state The consummatory phase does not provide calming. Repetitive or stereotypical behaviors, often identifiable (e.g., play, frustration, eating) Phobic state Fear reactions (escape, aggression, inhibition, agitation) in the presence of a phobic stimulus. Absence of excitement or frustration. Motor agitation is linked to exaggerated fear response, not excess motivation Anxious state Fear reactions (escape, aggression, inhibition) even in the absence of a phobic stimulus. Absence of excitement or frustration. Motor agitation is related to hypervigilance, not excess motivation **Organic diseases** *Pain* may cause motor agitation but is rarely present from adoption. Once the organic issue is resolved, the impulsive behavior should disappear

Pathological state	Diagnostic criteria	State variations	Differential diagnosis
Compulsive	*The compulsive state* is a pathological mental state characterized by the production of invariable behavioral responses, with a delayed, shortened, or absent calming phase. Compulsion can be defined as the inability to stop a behavioral sequence, regardless of the action's consequences. It is as though the consummatory phase, which normally leads to calming through action execution, remains ineffective	*Contextual to invasive form* The level of intrusion in the compulsive state depends on the time spent on compulsive behavior: – Modification of the behavioral sequence of one or more behaviors, with an elongation of the consummatory phase, leading to the absence of a calming phase – These sequence modifications persist despite attempts at behavioral modification – When the compulsive behavior is performed, there is indifference to other environmental stimuli – Transformation of compulsive activity into stereotypy in the most intrusive cases	Reactive state Environmental adjustments and meeting physiological needs lead to the rapid return to the normal adaptive state. Compulsive behaviors disappear completely **Maladies organiques** **Organic diseases** *Neurological disorder* partial temporal epilepsy, syringomyelia *Pain* may lead to stereotypical behaviors *Maladie dermatologique* atopy, abscesses, wounds Once the organic issue is resolved, compulsive behavior should disappear Impulsive state The consummatory phase provides calming Anxious state Fear reactions (escape, aggression, inhibition), even in the absence of a phobic stimulus. Absence of excitement or frustration. Motor agitation is related to hypervigilance, not excess motivation Manic state Manic phases are not permanent, and there are periods of normality during which repetitive behavior is not produced Dissociative state The trigger for compulsive behavior is not identifiable (internal stimulus)

Appendix 3 Diagnostic Criteria for Depressive, Manic, and Bipolar States

Pathological state	Diagnostic criteria	State variations	Differential diagnosis
Depressive	*The depressive state* is a psychopathological condition where the animal stops responding to its environment and withdraws. Clinically, it is characterized by intense inhibition, social withdrawal, sleep and appetite disturbances, and anhedonia	Acute depressive state – Hyporexia or anorexia – Hypersomnia – Generalized decrease in motivation – Exclusion of organic conditions explaining these symptoms	Organic diseases Conditions presenting symptoms like hyperthermia, hypothermia, intense pain, or acute anemia must be ruled out Anxious state Hypervigilance in response to the environment, associated with productive behaviors or inhibition Contexts exist where pleasure and motivation remain intact Senile state Older animal with cognitive or emotional changes unrelated to the context
		Chronic depressive state – Appetite variations, with phases of hyporexia – Sleep disturbances, including quality and quantity changes – Generalized decrease in motivation associated with withdrawal – Irritability that may lead to irritation-based aggression – Exclusion of organic conditions explaining these symptoms	

Pathological state	Diagnostic criteria	State variations	Differential diagnosis
Manic	*The manic state* is a psychopathological condition where the individual experiences episodic crises of invariable obsessive behaviors, either triggered by the context or created by the animal itself Common symptoms: – Onset of obsessive episodes, referred to as manic crises, during puberty or early adulthood, alternating with periods of normalcy – During manic phases, the animal's perception of reality is altered, leading to disproportionate reactions to external stimuli – Unpredictable onset and end of episodes, independent of the context	*Contextual to invasive form* The level of pervasiveness depends on the frequency and intensity of symptoms: – Hypervigilance et agitation motrice – Gaze changes due to near-constant mydriasis – Hypervigilance and motor agitation – Obsessive motivation – Idiopathic aggression, often irritation-based and highly intense without control from the first episodes. This intensity is abnormally disproportionate to the triggering context – Hyperphagia – Hyposomnia	Phobic state Triggering elements are identifiable and generates fear Anxious state Hypervigilance is constant, making crises predictable Compulsive state The cat is continuously symptomatic; there are no normal phases when the object of compulsion is present Dissociative state Behaviors are incoherent and unrelated to the context Organic diseases *Hormonal diseases* causing increased hunger (e.g., hypercortisolism, diabetes) or heightened food motivation; diagnosed biologically *Neurological tumors.* Other symptoms are usually observable, diagnosed through imaging
Bipolar	Alternation between depressive and manic states		

Appendix 4 Diagnostic Criteria for Dissociative and Senile States

Pathological state	Diagnostic criteria	State variations	Differential diagnosis
Dissociative	The *dissociative state* is a pathological mental condition in which the animal exhibits behavioral responses to stimuli that are not objectively present in the environment Common symptoms - First symptoms appear in early adulthood - Evidence of repetitive activity (spinning, hallucinations, self-aggressions), unrelated to external stimuli, with no identifiable trigger - No organic cause can be identified through routine complementary tests (blood tests, MRI)	*Contextual to invasive form* The level of intrusiveness of the dissociative state depends on the time spent engaging in repetitive behavior. In parallel, a gradual loss of usual activities is observed. The cat becomes less responsive to external interventions	**Maladies organiques** *Pain* Managing pain allows for the immediate cessation of episodes Complementary blood tests may be abnormally indicative of conditions like insulinoma or porto-systemic shunt Imaging (MRI) helps exclude malformations, such as tumors Anxious state Repetitive behaviors occur as part of hypervigilance or behavioral inhibition. There is no alteration in the perception of reality. External interventions usually stop the repetitive behaviors, at least temporarily Manic state Obsessive behaviors are aggressive and occur in spontaneous or cat-provoked interactions. There is no alteration in the perception of reality Compulsive state One or more identifiable trigger stimuli are present. There is no alteration in the perception of reality Senile state Symptoms appear in senior cats. Stereotypies, while possible, are not central

Pathological state	Diagnostic criteria	State variations	Differential diagnosis
Senile	The *senile state* is a pathological condition characterized by various behavioral abnormalities resulting from involution and its effects on normal neural circuit functioning. This abnormal aging is marked by a self-perpetuating worsening, ultimately leading to a complete disorganization of brain functions Common symptoms: – The cat is over 8 years old – Management of somatic diseases does not resolve the symptoms	Loss of abilities occurs in one or more areas, with functional self-aggravation The observed symptoms are cognitive, with spatial or temporal disorientation and loss of learning. They often extend to emotional and relational disorders, even mood disturbances	Trouble organique et sensoriel Blindness, Deafness: To be excluded Chronic pain Intercurrent diseases. In these organic disorders, there is no loss of spatial and temporal orientation during normal aging. Management of these conditions leads to improvement Dissociative state The first symptoms generally appear around adolescence or early adulthood Anxious state No spatial or temporal disorientation No loss of learning Depressive state No spatial or temporal disorientation No loss of learning

Appendix 5 Disorders Associated with Dysregulation of the Fear Axis

Disorder's name	Subtype	State diagnosis	Diagnostic criteria	Differential diagnosis
Categorical phobias	Noise phobia	Phobic or anxious depressive	Pathological state linked to one or more noises (explosions, thunderstorms, household appliances, vehicles, etc.)	*Deprivation syndrome* Poor early developmental conditions *Phobie post-traumatique* Tangible and identifiable traumatic event *Self-controls axis disorders* Motor agitation not linked to fear but to excitement *Organic disease: pain, hyperthyroïdie, hypertension* These present with a truly sudden onset, and resolving the underlying condition eliminates the fear responses
	Healthcare phobia	Phobic	Pathological state linked to veterinary care or grooming	
	Transport phobia	Phobic	Pathological state linked to car travel	
	Interspecific social phobia	Phobic or Anxious Depressive	Pathological state linked to unfamiliar individuals (all or specific types)	
	Intraspecific social phobia	Phobic or Anxious Depressive	Pathological state linked to unfamiliar conspecifics (all or specific types)	
Deprivation syndrome		Phobic or anxious depressive	Fear reactions present from adoption Evidence of development in a hypostimulating environment. Intense fear reactions to any environmental changes. Learning capacities blocked by fear	*Self-controls axis disorders* Motor agitation not linked to fear but to excitement *Categorical phobias* Fear reactions limited to certain stimuli with good early developmental conditions *Post-traumatic phobias* tangible and identifiable traumatic event Organic disease: pain, joint or cardiac malformations, hyperthyroidism
Post-traumatic phobia		Phobic or anxious depressive	Pathological state appears following an objectively identifiable and tangible trauma	Organic disease: pain, hyperthyroidism *Categorical phobias* No objectively traumatic event *Deprivation syndrome* Poor early developmental conditions

Appendix 6 Disorders Associated with Dysregulation of the Self-Controls Axis

Disorder's name	Subtype	State diagnosis	Diagnostic criteria	Differential diagnosis
HSHA syndrome or feline ADHD	There are 3 forms: **Hypersensitive (HS) form** Lowered threshold for triggering motor and emotional responses. **Hyperreactive (HA) form** Increased intensity of motor and emotional responses **Mixed form** proportion of both HS and HA forms	Impulsive or compulsive	Diagnosis relies on the presence of several symptoms from the following list: – Delayed acquisition of control over biting and clawing – Oral exploration that may lead to ingestion of non-food objects – Delayed acquisition of locomotion control, manifested as visible clumsiness – Absence or deficit in food satiety – Rough play, with difficulties controlling excitement, often resulting in scratches or bites – Excessive demand for interactions – Hyposomnia not associated with an anxious state – Disordered exploration – Disordered pheromonal marking with difficulty structuring territories	*Insufficient covering of physiological needs* Symptoms disappear with the restoration of normal food intake, play, and social contact *Pain* Motor agitation arises from the search for an antalgic context (e.g., specific positions or locations) *Categorical phobias* Motor agitation linked to fear occurs only in specific phobogenic contexts *Addictions* Emergence during early adulthood A compulsive state is mandatory *Relational disorders* Impulsive reactions vary depending on the person The cat is capable of controlling its behavior in many situations

Disorder's name	Subtype	State diagnosis	Diagnostic criteria	Differential diagnosis
Addictions		Compulsive	Symptoms gradually appear and become noticeable in early adulthood. Repetition of the addictive behavior may occur spontaneously through intensive self-reinforcement or through intentional training by the owner The diagnosis is based on identifying some or all of the following symptoms: – Behavior is performed repetitively and automatically, without signs of pleasure – Inability to spontaneously stop the behavior when it is possible to perform it – Active search for the object of compulsion when it is absent, with evident withdrawal symptoms, such as meowing, stereotypical behaviors, and fixation if the object is visible – Intense searching prevents the animal from engaging in normal species-specific behaviors when the compulsive behavior is possible – Increased aggression due to irritation in frustration contexts, with possible redirected aggression	*Insufficient covering of physiological needs* Symptoms disappear with the restoration of normal food intake, play, and social contact *Pain* Motor agitation arises from the search for an antalgic context (e.g., specific positions or locations) *Categorical phobias* Motor agitation linked to fear occurs only in specific phobogenic contexts *HSHA syndrome* Symptoms appear at a young age and tend to improve with appropriate intervention and management

Appendix 7 Disorders Associated with Dysregulation of the Territoriality Axis

Disorder's name	State diagnosis	Diagnostic criteria	Differential diagnosis
Disorders resulting from territorial poverty	Anxious Depressive	Symptoms appear in early adulthood when the territory should be the cat's central point of comfort. The presence of an impoverished environment with a lack of stimulation, where the cat cannot structure its activity zones, is key. The diagnosis is based on several symptoms that express the cat's frustration in the face of an impoverished environment: - Aggression due to irritation in response to any moving object - Altered facial pheromonal marking, often increased, sometimes absent - Increased scratching around nonisolation zones - Early-stage urine marking, sometimes complicated by urine soiling - Compensatory binge eating - Abdominal licking - Idiopathic cystitis	*Organic disorders* Loss of sensory abilities Any painful condition *HSHA syndrome* Environmental enrichment does not resolve the condition, as the environment is never stimulating enough for cats suffering from HSHA syndrome *Territorial disorder secondary to another behavioral disorder* Environmental improvement does not significantly improve the clinical picture. Investigate another behavioral disorder that may explain the observed pathological state *Aging disorders* The animal is over 8 years old. The disorder appears at this age in an environment that has not become poorer and was previously well tolerated

Disorder's name	State diagnosis	Diagnostic criteria	Differential diagnosis
Disorders resulting from territorial disrupting	Anxious Depressive	Symptoms appear in early adulthood when the territory should be the central point of comfort for the cat. Identifying a living environment that has been modified, often enriched with new stimuli responsible for variations that disturb the cat, is essential. The diagnosis is based on the presence of several symptoms that express the cat's inability to make the territory comforting. These symptoms include: – Aggression due to irritation – Altered facial pheromonal marking, often increased, sometimes absent – Increased scratching around areas that are not designated isolation zones for the cat – Early-stage urine marking, sometimes complicated by urine soiling – Compensatory binge eating – Abdominal licking – Idiopathic cystitis	*Organic disorders* Loss of sensory abilities Any painful condition *Social phobia or deprivation syndrome* Evidence of fear reactions before the suspected environmental change that is thought to be the cause of the disorder. *Interspecific relational disorder* Poor relationships existed prior to the environmental change, and symptoms of aggression due to irritation towards certain individuals were observed beforehand *Intraspecific relational disorder* Poor relationships themselves constitute the environmental change, and symptoms of aggression due to irritation towards the other cat are observed beforehand *Aging disorders* The animal is over 8 years old. The condition appears at this age in an environment that has not become poorer and was previously well tolerated
Territoriality disorders secondary to another behavioral disorder	Anxious Depressive	Symptoms related to this condition appear in early adulthood, but symptoms of the primary condition are present earlier, with other possible pathological states emerging before The diagnosis relies on the inability to explain the entire clinical picture solely by the territorial disorder. While there is a territory that the cat cannot structure, other untreated symptoms are noted during behavioral observation, indicating the presence of the primary condition as the underlying cause or an aggravating factor of the clinical signs	*Organic disorders* Loss of sensory abilities Any painful condition *Disorders resulting from territorial disrupting* Before the environmental modification, the cat exhibits no behavioral symptoms and has had normal development. *Aging disorders* The animal is over 8 years old. The disorder appears at this age in an environment that has remained stable and was well tolerated prior to aging.

Appendix 8 Disorders Associated with Dysregulation of the Relationship Axis

Disorder's name	Subtype	State diagnosis	Diagnostic criteria	Differential diagnosis
Interspecific relational disorder	– Linked to a dysfunctional communication – Secondary to another behavioral disorder	– Phobic or anxious – Depressive	The diagnosis is based on the presence of one or more of the following symptoms toward one or more members of the group: – Identification of dysfunctional communication – Aggression due to irritation in a fearful context directed at one or more individuals in the household, becoming more pronounced and worsening over time – Anticipation of the actions of the individual, with a tendency to flee – Hypervigilance in the presence of the individual(s) involved	*Organic disorders* Loss of sensory abilities Any painful condition *Interspecific social phobia* The aggressions concern unknown individuals, whereas in relational affection, there is a genuine poor relationship between two particular cats. *Dysthymia* The aggressions become independent of context over time and are much more unpredictable and violent *HSHA syndrome* The triggering sequences occur during play or excitement. They do not lead to avoidance of a specific individual
Intraspecific relational disorder	– Linked to a dysfunctional communication – Secondary to another behavioral disorder	– Phobic or anxious – Depressive	The diagnosis is based on the presence of one or more of the following symptoms towards one or more members of the group: – Repeated conflicts between two or more cats in the household, characterized by aggression due to irritation in a fearful context and territorial aggression – Territorial aggression between two or more cats in the household – Anticipation of the actions of this individual, with a tendency to flee for one and aggression for the other – Hypervigilance in the presence of the conspecific(s) involved	*Organic disorders* Loss of sensory abilities Any painful condition *Intraspecific social phobia* The aggressions involve all unknown cats, whereas in relational affection, there is a genuine poor relationship between two particular cats.

Appendix 9 Essential Psychiatric Disorders

Disorder's name	State diagnosis	Diagnostic criteria	Differential diagnosis
Dissociative syndrome	Dissociative	First dissociative episodes appear in *adulthood* The diagnosis is based on identifying the following symptoms: – Onset of symptoms in early adulthood; – Evidence of repetitive activity (hallucinations, self-aggressions) unrelated to external stimuli, with no identifiable trigger – No organic cause identified through standard complementary tests (e.g., bloodwork, MRI) – During episodes, the animal appears disconnected from its environment – Increasingly difficult to interrupt episodes	*Somatic diseases* Pain The painful area can be identified, and pain management resolves the issue Partial temporal epilepsy Episodes are shorter and less invasive Dermatological hypersensitivity Pruritus treatment resolves the condition Brain tumor Imaging can exclude this, though early evolution may appear similar *Anxious states with self-centered behaviors* Behavior aligns with contextual triggers; anxiety management improves behavior Hyper-Sensitivity–Hyper-Activity syndrome (HSHA) *syndrome or feline ADHD* Episodes are short and occur only in contexts of excitement or frustration *Addictions* Triggers are real, and the focus of the addiction is not imaginary
Dysthymia	Manic or bipolar	First manic episodes appear in *early adulthood.* Diagnostic criteria include some or all of the following symptoms: – Altered gaze due to almost continuous mydriasis – Hypervigilance and motor agitation – Idiopathic aggression, often irritation-based or territorial, with disproportionate intensity to the triggering context – Appetite changes, often refusal to eat during episodes – Sexual dysthymia: abrupt behavioral change during estrus that does not resolve post-estrus Between episodes, the cat is normal (manic state) or depressed (bipolar state)	*Interspecific relational disorder* Reestablishing communication resolves or diminishes episodes. The same actions consistently produce the same effects, unlike dysthymia, where mood changes lead to varying reactions despite identical contexts *Dissociative syndrome* The cat appears disconnected from reality; episodes are not relational and are more invasive than those in dysthymia *Sexual dysthymia* Gestation Diagnosed via ultrasound. Maternal aggression appears post-parturition, not during gestation Pyometra Excluded via bloodwork and ultrasound. Aggression is rare and less dramatic despite pain. Clinical signs include apathy and PUPD (polyuria/polydipsia)

Appendix 10 Aging Disorders

Disorder's name	State diagnosis	Diagnostic criteria	Differential diagnosis
Cognitive dysregulation in elderly cats	Senile	Cat older than 8 years. Evidence of multiple cognitive errors, expressed through some or all of the following symptoms: – Spatial disorientation: searching for familiar environmental elements in the wrong place – Temporal disorientation: day/night inversion, requests at unusual times relative to expected cues – Loss of learned behaviors: loss of litter box habits	*Pain* Erratic behavioral responses related to pain *Organic disease* PUPD, hormonal disorders, decreased sensory acuity *Emotional dysregulation in elderly cats* Presence of worsening anxiety Loss of autonomy Absence of cognitive errors in the early stages *Mood dysregulation in elderly cats* Presence of depressive signs Absence of cognitive errors in the early stages *Self-controls dysregulation in elderly cats* Presence of signs of loss of inhibition Absence of cognitive errors in the early stages
Emotional dysregulation in elderly cats	Phobic or anxious	Cat older than 8 years. Evidence of a sudden worsening of emotional state and associated distress symptoms, including some or all of the following: – Daytime distress signs: vocalization, inappropriate elimination – Nocturnal distress signs: wandering and vocalization – Panic attacks – Restless sleep with frequent awakenings – Identification of pre-existing fear axis vulnerabilities before senescence – Worsening of a previously stable phobic state with generalization and progression into an anxious state	*Pain* Behavioral responses related to pain *Organic disease* PUPD, hormonal disorders, decreased sensory acuity *Cognitive dysregulation in elderly cats* Presence of cognitive errors in the early stages Absence of anxiety or worsening emotional state initially *Mood dysregulation in elderly cats* Presence of depressive signs Absence of anxious agitation in the early stages *Self-controls dysregulation in elderly cats* Presence of signs of loss of inhibition Absence of anxious agitation in the early stages

Disorder's name	State diagnosis	Diagnostic criteria	Differential diagnosis
Mood dysregulation in elderly cats	Depressive	Cat older than 8 years. Evidence of a general loss of motivation, expressed through some or all of the following symptoms: – Loss of interest in external events – Capricious appetite – Fluctuating sleep patterns – Persistent demotivation despite treatment of organic conditions	*Pain* Interaction avoidance due to pain *Organic disease* PUPD, hormonal disorders. deafness, and decreased sensory acuity *Cognitive dysregulation in elderly cats* Presence of cognitive errors in the early stages Absence of demotivation initially *Emotional dysregulation in elderly cats* Presence of worsening anxiety Loss of autonomy Absence of demotivation initially *Self-controls dysregulation in elderly cats* Presence of signs of loss of inhibition Absence of demotivation initially
Self-controls dysregulation in elderly cats	Impulsive or compulsive	Cat older than 8 years. Evidence of some or all of the following symptoms: – Reappearance of oral exploration – Recurrence of impulsive or compulsive behaviors previously present in youth – Loss of control in interactions, with a return to rough exchanges – Persistent motor agitation despite treatment of organic conditions	*Pain* Agitation caused by pain *Organic disease* PUPD, hormonal disorders *Cognitive dysregulation in elderly cats* Presence of cognitive errors in the early stages Absence of controls deficit *Emotional dysregulation in elderly cats* Presence of worsening anxiety Loss of autonomy *Mood dysregulation in elderly cats* Presence of depressive signs Absence of anxious agitation in the early stages Presence of general demotivation

Appendix 11 Elements of Differential Diagnosis Between Neurology and Psychiatry

Symptom	Neurological or neurologically expressed conditions	Psychiatric disorder	Clinical tools for differentiation
Aggressions	Brain tumor Hepatic encephalopathy Hydrocephalus Pain	Anxious states Derealization states Phobic states	Consistency of sequences and pathological pathway with the overall clinical picture *Additional tests:* MRI, biochemical panel
Generalized seizures	Epilepsy Hypoglycemia Insulinoma Toxins	Paroxysmal anxious state	Age, onset, progression *Additional tests:* biochemical panel
Partial seizures (tail-chasing or attacking the immediate environment)	Temporal lobe epilepsy Syringomyelia Hydrocephalus Brain tumor Cauda equina syndrome Phantom pain after amputation Hyperesthesia Toxins	Productive anxious state Impulsive state Compulsive state Dissociative state	Pathological pathway: onset, frequency, progression *Additional tests:* biochemical panel, MRI, specialized consultation
Spatial or temporal disorientation	Brain tumor Toxins	Cognitive dysregulation in elderly cats	*Additional tests:* MRI, biochemical panel
Hallucinations	Abnormal vision or blindness Toxins Brain tumor Encephalitis Anesthetic toxicity	Dissociative syndrome Dysthymia	Pathological pathway: onset, frequency, progression *Additional tests:* MRI, specialized consultation

Appendix 12 Body Condition Score (BCS) Chart for Cats

BCS 9-point scale		BCS 5-point scale		CHAT
1		1		**Extremely emaciated with very poor body condition** Ribs, spine, and hips very visible. No fat, severe muscle loss. Severely tucked abdomen
2	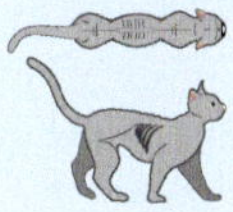			**Very thin** Ribs and spine easily palpable with very little fat. Prominent hips. Tucked abdomen
3		2		**Thin** Ribs easily palpable with a thin fat layer. Spine and hips visible. Slightly tucked abdomen
4				**Moderately thin** Ribs palpable with a very thin fat layer. Spine still noticeable. Less prominent hips. Slightly tucked abdomen
5		3		**Ideal** Ribs palpable under a thin layer of fat. Spine smooth and not protruding. Slightly tucked abdomen. Visible waist when viewed from above
6				**Slightly overweight** Ribs palpable under a thin fat layer. Smooth, non-prominent spine. Slightly tucked abdomen. Visible waist from above
7		4		**Overweight** Ribs difficult to feel under a fat layer. Spine and hips barely palpable. Rounded abdomen. Waist barely visible or absent
8				**Obesity** Ribs and spine difficult to feel under a thick fat layer. Well-rounded abdomen. Waist absent. Visible fat deposits
9		5		**Severe obesity** Ribs impossible to feel. Spine and hips covered by fat. Large abdomen with prominent fat deposits. Waist completely absent

Appendix 13 Reported Signs in the Semiology of Feeding Behavior and Possible Interpretations

Reported sign	Symptoms *In all cases, a detailed exploration of the frequency, intensity, and context in which the sign arises is necessary to qualify it as a symptom.*	Underlying dysfunction
Aggression during feeding related to an inadequate food distribution method and/or highly palatable food	Excessive motivation, which may escalate to irritation-induced aggression in a context of frustration, particularly with overly palatable food	Imbalance in the self-control axis
	Fractionated food distribution causing intense anticipation, which may trigger irritation-induced aggression in a context of frustration	Imbalance in the communication or self-control axis
Food fussiness	Decreased appetite in situations of intense fear	Imbalance in the fear axis
	Owner misinterprets an attention-seeking behavior as a food request	Imbalance in the communication axis
	Dysorexia with sometimes dramatic variations	Chronic depressive state
	Any organic condition causing reduced appetite	Organic disorder
	Medication-induced appetite suppression (e.g., fluoxetine, venlafaxine)	Iatrogenic
Anorexia	Climatic adaptation (high external temperature) Suspicion of an odor linked to the introduction of a medication	Reactive state
	Decreased appetite in situations of intense fear	Imbalance in the fear axis
	Following intense stress or sensory deprivation (likely more frequent)	Acute depressive state
	Any organic condition affecting appetite. Self-perpetuated hepatic lipidosis due to prolonged anorexia	Organic disorder
	Medication affecting appetite (Non-Steroidal Anti-Inflammatory Drugs (NSAIDs), certain antibiotics, fluoxetine)	Iatrogenic

Appendix 14 Reported Signs in the Semiology of Sleeping Behavior and Possible Interpretations

Reported sign	Symptoms *In all cases, a detailed exploration of the frequency, intensity, and context in which the sign arises is necessary to qualify it as a symptom*	Underlying dysfunction
Hypersomnia (more than 20 hours per day)	Behavioral inhibition due to overwhelming and persistent fear	Generalized inhibited anxious state
	Behavioral inhibition due to a general decrease in motivation	Chronic depressive state
	Chronic pain	Any organic disorder leading to chronic pain
Hyposomnia (less than 15 hours per day)	Lowered behavioral response threshold, hypersensitivity	Impulsive state
	Behavioral agitation due to generalized fears and increased vigilance	Intermittent or generalized productive anxious state
	Daytime or nighttime agitation related to pathological aging in a cat over 8 years old	Senile state with emotional and/or cognitive dysregulation in aging cats
Startle-induced awakenings	Hypervigilance	Intermittent or generalized productive anxious state
	Hypersensitivity	Impulsive or compulsive state
	Agitation linked to an underlying somatic condition, such as a painful disorder	Organic disease causing discomfort

Appendix 15 Differential Diagnosis of Urinary and Fecal Inappropriate Elimination and Their Associated Characteristics

Reported sign	Symptoms *In all cases, a detailed exploration of the frequency, intensity, and context in which the sign arises is necessary to qualify it as a symptom*	Underlying dysfunction
Urine elimination outside the litter box	Complete elimination sequence. Dysuria, stranguria, and hematuria may be observed, and the elimination sequence may become simplified due to pain or the presence of owners and associated punishment **Presence of crystals and/or bacteria**	Infectious cystitis and/or related to urinary crystals
	Complete elimination sequence. Dysuria, stranguria, and hematuria may be observed, and the elimination sequence may become simplified due to pain or the presence of owners and associated punishment **Sterile urine without crystals**	Idiopathic cystitis associated with an anxious state
	Complete elimination sequence. No signs of pain or inflammation at the onset	Litter aversion due to elements that have repelled or frightened the cat during elimination. Most often associated with dysfunction in the fear or relational axis, or age-related impairments
Urine marking	Urine spray on vertical surfaces without burying behavior	Sexual marking in intact male cats at puberty and in females in heat Persistent territorial competition Sudden environmental change associated with dysregulation of the self-control axis, territoriality axis, or cognitive dysregulation in aging cats
Feces outside the litter box	Normal stools and a complete elimination sequence with an attempt to bury. If pain is present, the cat often runs away immediately after defecation and may vocalize during elimination (this can also occur with urination)	Litter aversion due to elements that have repelled or frightened the cat during elimination. Most often associated with dysfunction in the fear or relational axis, or age-related impairments
	Soft or diarrheic stools, either continuous or intermittent	Intestinal disorder. Phobic or anxious state
	Normal stools not buried, usually found outdoors and at the edges of the territory	Fecal marking (rare)

Appendix 16 Reported Signs in the Semiology of Repetitive Behaviors and Possible Interpretations

Reported sign	Symptoms *In all cases, a detailed exploration of the frequency, intensity, and context in which the sign arises is necessary to qualify it as a symptom*	Underlying dysfunction
Ritualized activity	Repetitive behaviors diverted from their initial purpose, becoming a means of communication. The cat or its ritual partner can stop the behavior spontaneously	Positive reinforcement by the owner Contextual compulsive state
Substitutive activity	Redirection of behaviors toward repetitive activities providing a coping strategy or relief from internal or external stress that cannot be resolved: licking, sucking, grooming. These behaviors are contextual and intermittent	Pain Inhibited anxious state Chronic depressive state
Stereotypy	An invasive, repetitive behavioral sequence that cannot be stopped spontaneously and has lost its function	Invasive compulsive state Chronic depressive state Senile state Derealized state

Appendix 17 Classification of Injurious Sequences in Cats Based on the Modified Moyer Classification

Name of the sequence	Subtypes	Offensive/ Defensive	Target	Trigger	Intention	Level of control
Non-agressive injurious sequences						
Predation		Offensive	Prey	Prey	Kill to consume	
Vulnerant sequence during play		Offensive	Play partner	Excitement	Play	
Agressions						
Maternal aggression		Defensive	Outsider to the group	Approche des chatons	Defend her kittens + repel the threat	
Territorial aggression		Defensive	Outsider to the group	Intrusion du territoire	Defend against intrusions into the territory + repel the threat	
Irritation aggression	Frustration Excitement Fear Pain	Defensive	Indifferent but greater tolerance toward familiars	Approach of the individual	Repel the threat	
Uncontrolled emotional aggression		Defensive	More frequent with strangers but possible with anyone	Approach of the individual	Repel the threat	
Idiopathic aggression		Defensive or offensive	Indifferent	No specific rule	Unknown	

Appendix 18 Semiology of Play and Mutual Grooming

Reported sign	Symptoms *In all cases, a detailed exploration of the frequency, intensity, and context in which the sign arises is necessary to qualify it as a symptom*	Underlying dysfunction
Excessive and/or injurious play	The cat spends a large part of its time playing at the expense of sleep (less than 12 hours)	Dysregulation of the self-control axis (feline hypersensitivity-hyperactivity)
	The cat has a seemingly normal play duration, but play sequences with a partner frequently result in injurious interactions The movements of household members are the only stimulation triggers, leading to aggressive play	Dysregulation of the self-control axis (feline hypersensitivity-hyperactivity) Dysregulation of the territoriality axis due to environmental impoverishment
Absence of play	The cat plays very little or not at all	Dysregulation of the fear axis (sensory deprivation syndrome) Aging Relationship disorders Organic pathology
Aggression during contact	The cat tolerates contact but quickly turns to scratch or bite. Often described as "petting-induced aggression syndrome"	Dysregulation of the self-control axis (feline hypersensitivity-hyperactivity) Dysregulation of the inter- or intraspecific relational axis Dysregulation of the fear axis (social or contextual phobia due to past punishments). Pain

Appendix 19 Semiology of Exploratory Behavior and Marking

Reported sign	Symptoms *In all cases, a detailed exploration of the frequency, intensity, and context in which the sign arises is necessary to qualify it as a symptom*	Underlying dysfunction
Inhibited exploration	During consultation, the cat quickly seeks to flee or hide. In a confined situation, it may become aggressive. At home, it occupies only a limited area of the territory due to its inhibited behavior	Dysregulation of the fear axis (sensory deprivation syndrome, social phobia, healthcare phobia) Chronic inhibited anxious state Chronic depressive state Senile state
Disorganized and excessive exploration	The cat is highly agitated during consultation, reacting to all stimuli. It struggles with frustration, and restraint may be difficult, as its excitement rapidly escalates, potentially leading to aggression.	Dysregulation of the self-control axis (feline hypersensitivity-hyperactivity)
Facial marking	Excessive: – The cat faces significant territorial changes and attempts to restructure its environment – The cat fails to structure its territory through facial marking due to impulsivity Deficient: – The cat's behavior is inhibited, including territorial marking	Reactive state. Dysregulation of the self-control axis (feline hypersensitivity-hyperactivity) Chronic inhibited anxious state Chronic depressive state Senile state
Lack of allomarking	The cat does not recognize one or more individuals as part of its territory or as a source of reassurance The cat is unable to structure its territory globally	Dysregulation of the fear axis (sensory deprivation syndrome, social phobia). Dysregulation of the inter- or intraspecific relational axis Dysregulation of the self-control axis (feline hypersensitivity-hyperactivity)
Excessive scratching	The cat experiences significant territorial disruptions and scratches vertical surfaces unrelated to its isolation area, marking multiple locations in its territory The cat fails to structure its territory effectively	Dysregulation of the territoriality axis due to changes in the territory Intermittent productive anxiety state Dysregulation of the self-control axis (feline hypersensitivity-hyperactivity)
Urine marking	Except for sexual marking in intact male cats and females in heat, urine marking inside the territory is always pathological. The cat faces significant territorial upheavals and is unable to reorganize its territorial markers through facial marking or scratching	Dysregulation of the territoriality axis due to significant territorial changes

Appendix 20 Summary Table of SSRIs

Substance	Mechanism of action	Pathological states for which the molecule is indicated	Dosage	Cytochromes
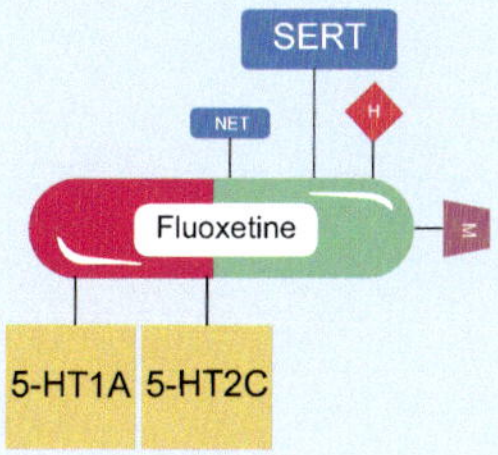	SERT inhibitor	Chronic depressive	0.25 to 1 mg/kg SID	CYP3A4
	5-HT1A desensitizer 5-HT2C antagonist	Phobic or anxious		CYP2C19 CYP2C9
	NET Inhibitor Histaminic antagonist	Impulsive or compulsive		CYP2D6
	Muscarinic antagonist			
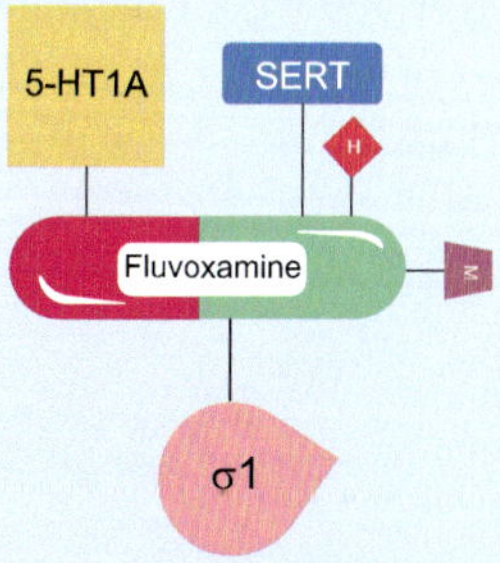	SERT inhibitor 5-HT1A desensitizer	Chronic depressive Phobic or anxious	0.25 to 0.5 mg/kg SID	CYP3A4 CYP1A1
	σ1 agonist Histaminic antagonist	Impulsive or compulsive		CYP2C9 CYP2C19
	Muscarinic antagonist			
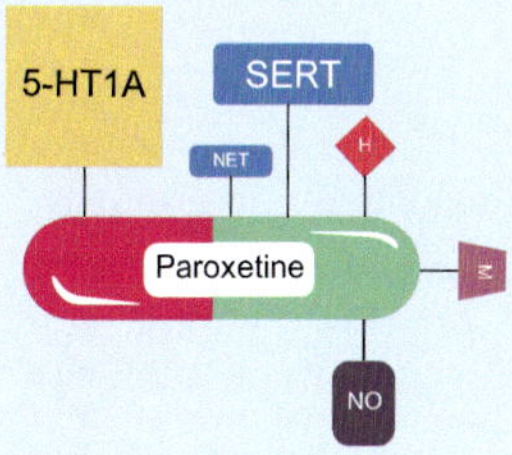	SERT inhibitor	Chronic depressive	0.25 to 1 mg/kg	CYP2C9
	5-HT1A desensitizer	Phobic or anxious	SID	CYP2D6
	NO inhibitor	Impulsive or compulsive		
	NET inhibitor			
	Histaminic antagonist			
	Muscarinic antagonist			

Substance	Mechanism of action	Pathological states for which the molecule is indicated	Dosage	Cytochromes
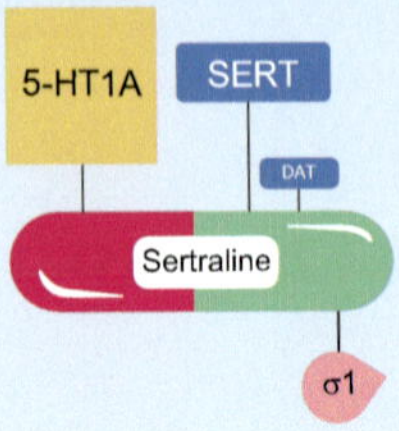	SERT inhibitor	Chronic depressive	0.5 to 4 mg/kg	CYP2C9
	5-HT1A desensitizer	Phobic or anxious	SID	CYP2D6
	DAT inhibitor	Impulsive or compulsive		
	σ1 agonist			

Appendix 21 Summary Table of Tricyclic Antidepressants

Substance	Mechanism of actions	Pathological states for which the molecule is indicated	Dosage	Cytochromes
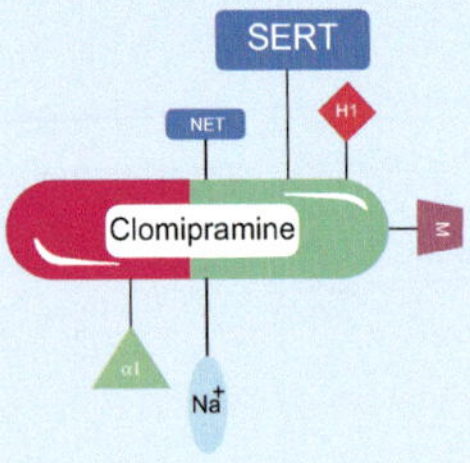	SERT inhibitor NET inhibitor H1 histamine receptor antago-nist M1 muscarinic receptor antago-nist α1-adrenergic receptor antago-nist Bloqueur des canaux Na^+	Chronic depressive Phobic or anxious Impulsive or compulsive	0.25 to 1 mg/kg SID. Prefer an SSRI.	CYP3A4 CYP2C19 CYP2C9 CYP1A2 CYP2D6
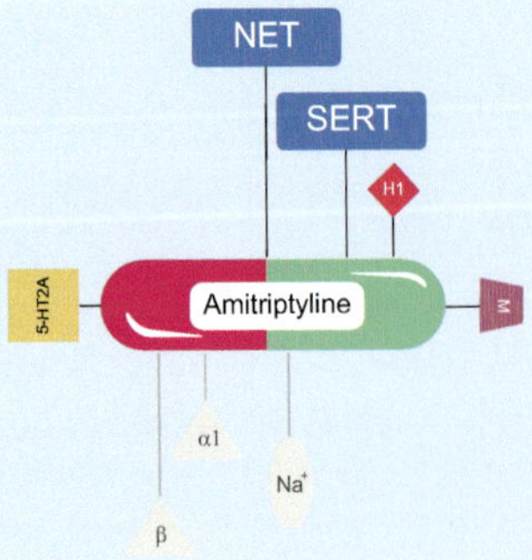	SERT inhibitor NET inhibitor H1 histamine receptor antago-nist M1 muscarinic receptor antago-nist α1-adrenergic receptor antago-nist Na^+ sodium channel blocker	Anxious Pain	0.5 to 1.5 mg/kg SID.	CYP2C19 CYP1A2 CYP3A4 CYP2D6

Appendix 22 Summary Table of Monoamine Oxidase Inhibitors (MAOIs)

Substance	Mechanism of action	Pathological states for which the molecule is indicated	Dosage	Cytochromes
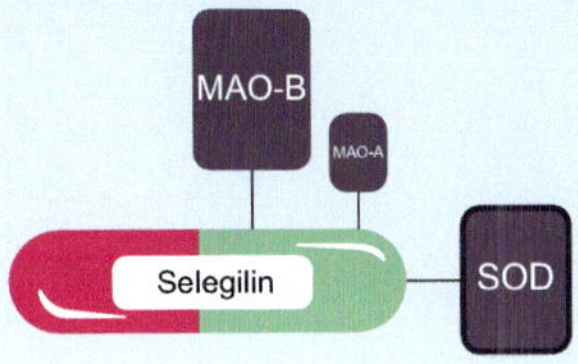	MAO-B inhibitor SOD activator (MAO-A inhibitor)	Senile	0.25–1 mg/kg SID	CYP2B6 CYP2A6 CYP3A4

Appendix 23 Summary Table of Benzodiazepines

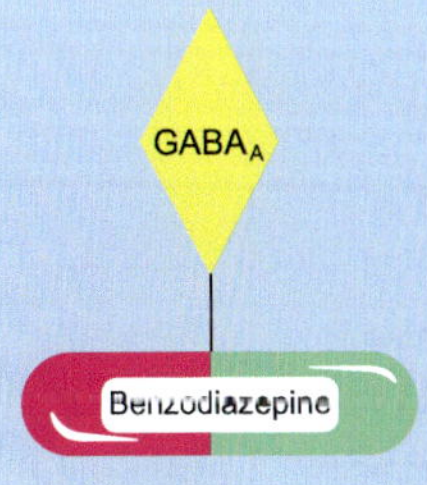	Mechanism of action	Pathological states for which the molecule is indicated	Dosage	Cytochromes
Alprazolam	Positive allosteric modulator of $GABA_A$	Phobic or anxious	0.125–1 mg per cat PRN or TID	CYP3A4 CYP3A5
Clonazepam	Positive allosteric modulator of $GABA_A$	Phobic (noise)	0.02–0.25 mg/kg PRN	CYP3A4
Clorazepate	Positive allosteric modulator of $GABA_A$	Phobic or anxious	0.2–0.5 mg/kg SID, BID, or TID	CYP3A4
Lorazepam	Positive allosteric modulator of $GABA_A$	Phobic or anxious	0.025–0.08 mg/kg PRN	CYP3A4
Oxazepam	Positive allosteric modulator of $GABA_A$	Phobic or anxious	0.2–0.5 mg/kg SID, BID, or TID	CYP3A4

Appendix 24 Summary Table of Sympatholytics

Substance	Mechanism of action	Pathological states for which the molecule is indicated	Dosage	Cytochromes
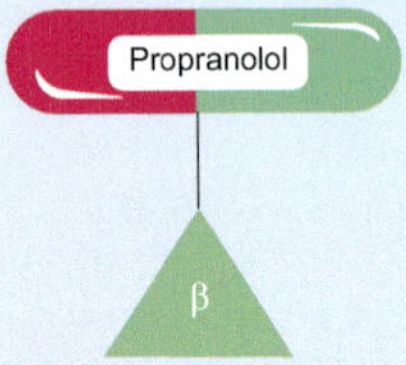	β-adrenergic antagonist	Phobic	0.2 to 0.8 mg/kg BID, TID	CPY2D15
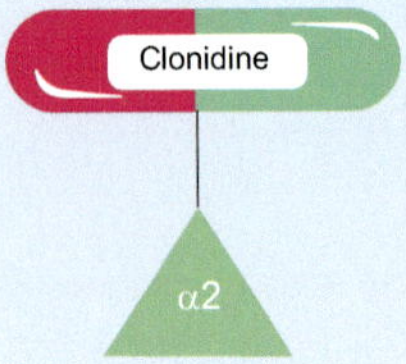	α_2-adrenergic agonist	Phobic	0.005 to 0.01 µg/kg SID, BID, or TID.	CYP2D6 CYP1A2
				CYP3A4
				CYP1A1 CYP3A5

Appendix 25 Summary Table for Antiepileptics

Substance	Mechanism of action	Pathological states for which the molecule is indicated	Dosages	Cyto-chromes
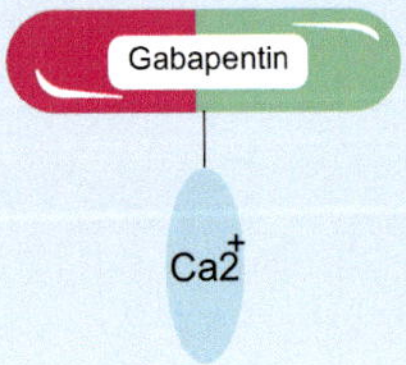	Voltage-dependent Ca^{2+} channel inhibitor	Phobic Anxious	10 to 20 mg/kg BID or TID 10 to 30 mg/kg PRN	No hepatic metabolism

Substance	Mechanism of action	Pathological states for which the molecule is indicated	Dosages	Cytochromes
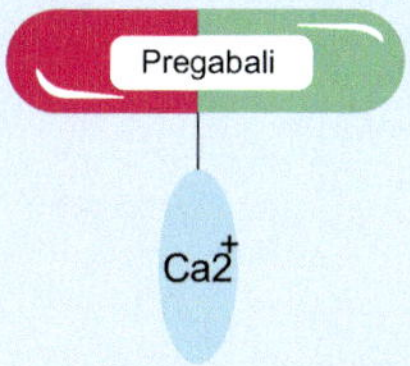	Voltage-dependent Ca^{2+} channel inhibitor	Phobic Anxious	5 to 10 mg/kg BID	No hepatic metabolism
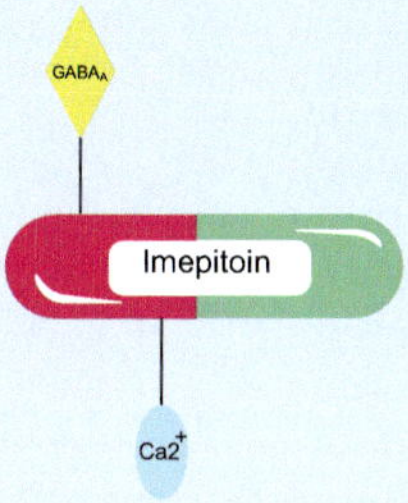	Voltage-dependent Ca^{2+} channel inhibitor Partial $GABA_A$ agonist	No current indications	30 mg/kg BID	–
	Voltage-dependent Ca^{2+} channel inhibitor	Manic	1 to 3 mg/kg BID	CYP2C19
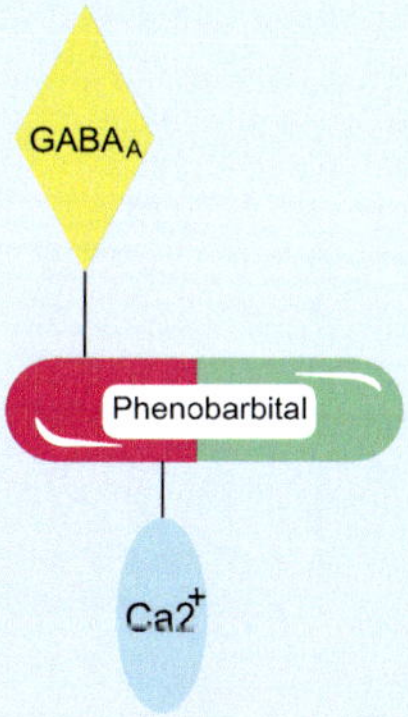	$GABA_A$ agonist	Compulsive Dissociated Neurological		
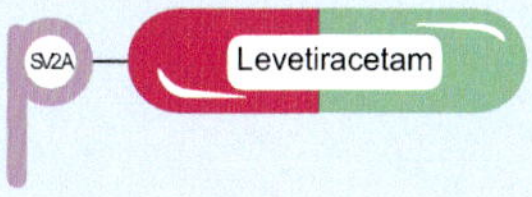	Inhibitor of synaptic vesicle release via SV2A	Neurological	20 to 25 mg/kg TID	No metabolism via cytochromes

Appendix 26 Summary Table of Tetracyclic NaSSAs

Substance	Mechanism of action	Pathological states for which the molecule is indicated	Dosage	Cytochromes
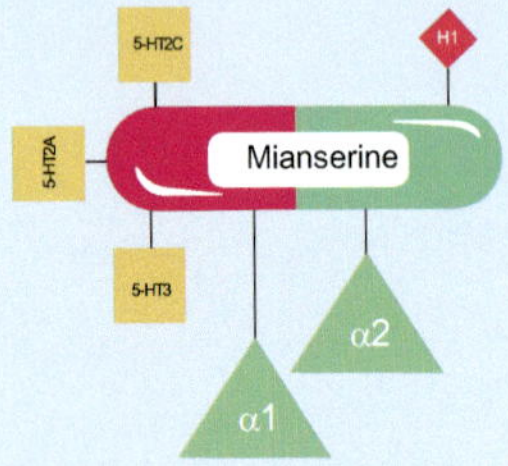	α1-adrenergic antagonist	Acute depressive Inhibited anxious	1 to 5 mg/kg *Semel in die* (Latin for once a day) (SID) to *Bis in die* (Latin for twice a day) (BID)	–
	α2-adrenergic antagonist			
	5-HT2A antagonist			
	5-HT2C antagonist			
	5-HT3 antagonist			
	Histaminergic H1 antagonist			
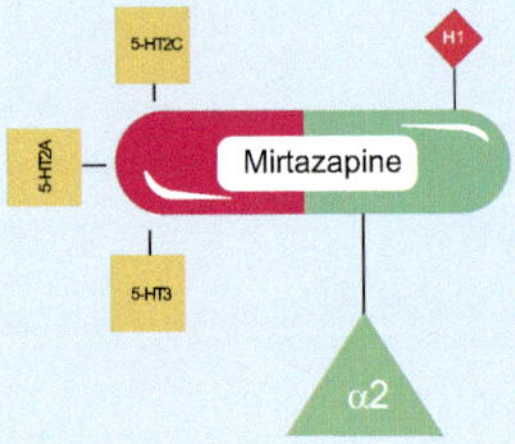	α2-adrenergic antagonist	Acute or chronic depressive inhibited anxious	1.88 mg/cat BID	CYP2D6 CYP1A2
	5-HT2A antagonist			CYP3A4
	5-HT2C antagonist			
	5-HT3 antagonist			
	Histaminergic H1 antagonist			

Appendix 27 Summary Table of Neuroleptics

Substance	Mechanism of action	Usage and dosage	Experience feedback	Cytochromes
Classical neuroleptics				
Acepromazine 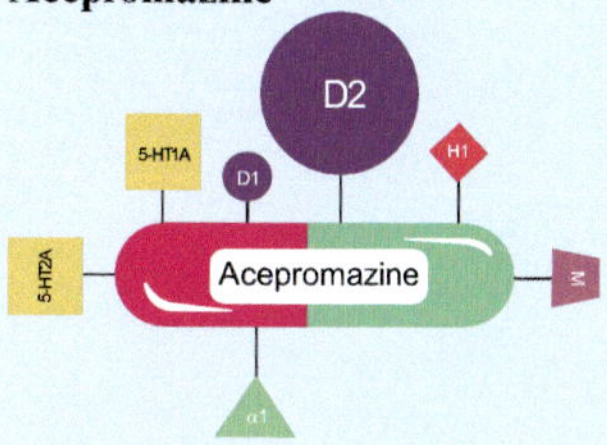 (Calmivet™, Vetranquil™, PromAce™)	**D2 antagonist** α1-adrenergic antagonist Histamine antagonist Muscarinic antagonist	Sedation, tranquilization Pruritus Nausea Urethral obstruction Dose: 0.5–1 mg/kg PO 1.1–2.2 mg/kg IV, Intramuscular injection into a muscle (IM), or (SC) Subcutaneous (injection under the skin) (SC) Pro re nata (Latin for as needed) (PRN) or Ter in die (Latin for three times a day) (TID)	Use discouraged	–
Azaperone 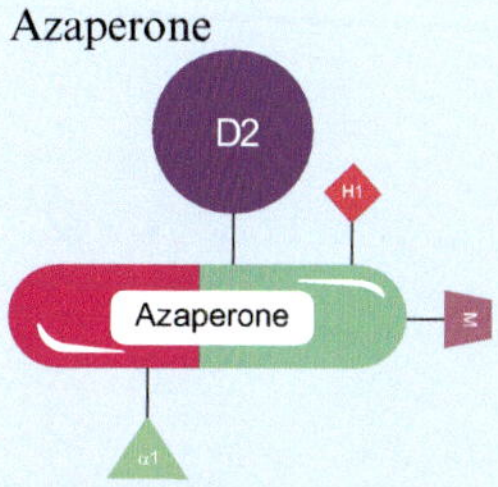 (*Stresnil*™)	**D2 antagonist** α1-adrenergic antagonist Histamine antagonist Muscarinic antagonist	Sedation, tranquilization	Use discouraged	–
Haloperidol 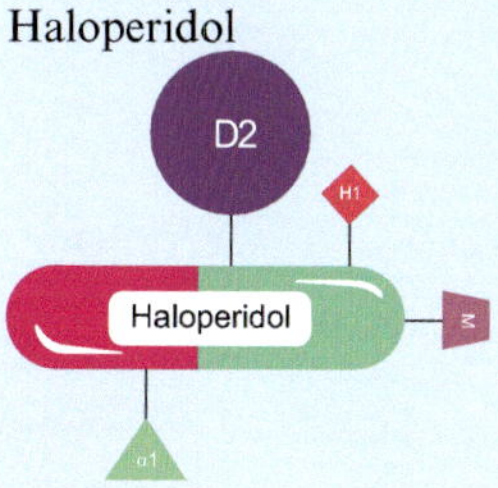 (*Haldol*™)	**D2 antagonist** α1-adrenergic antagonist Histamine antagonist Muscarinic antagonist	Impulsive, compulsive manic, bipolar dissociated	Not currently used	–

Substance	Mechanism of action	Usage and dosage	Experience feedback	Cytochromes
Pimamperone 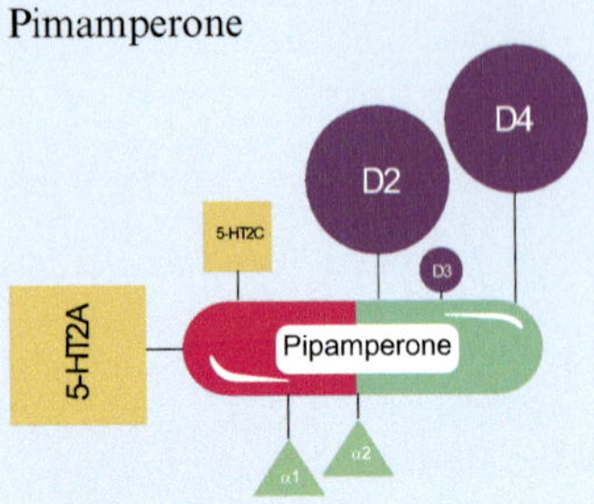 (*Dipiperon*™)	D2 antagonist 5-HT2A antagonist **D4 antagonist** D1, D3 antagonist α1- and α2-adrenergic antagonist Histamine and muscarinic antagonist	Impulsive, compulsive manic, bipolar dissociated	Not currently used	–
Sulpiride 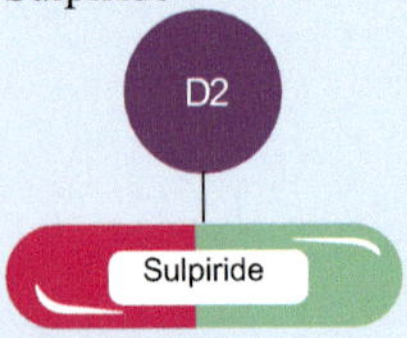 (*Dogmatil*™)	**D2 antagonist** (post and presynaptic)	Phobic, anxious	Not currently used	–
Atypical neuroleptics				
Clozapine 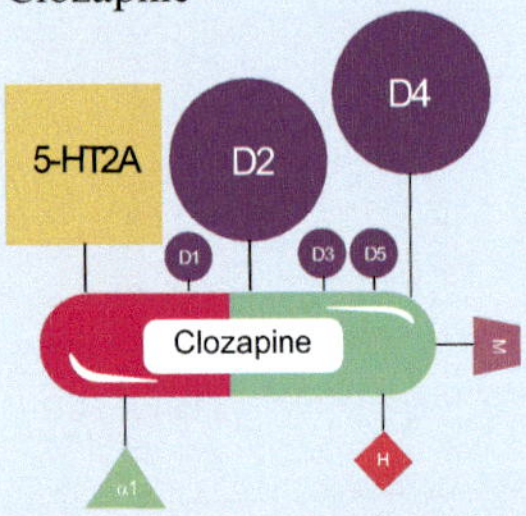 (Léponex[ND] ou Clozaril[ND])	D2 antagonist 5-HT2A antagonist D4 antagonist D1, D3, D5 antagonist α1-adrenergic antagonist Histamine antagonist Muscarinic antagonist	Impulsive, compulsive manic, bipolar dissociated	Not currently used	–
Risperidone 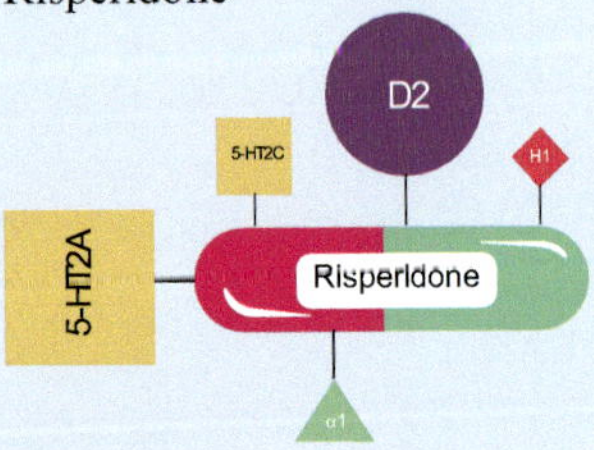 (*Risperdal*™)	D2 antagonist 5-HT2A antagonist 5-HT2C antagonist α1-adrenergic antagonist Histamine antagonist	Impulsive, compulsive manic, bipolar dissociated	Not currently used	CYP2D6
Aripiprazole 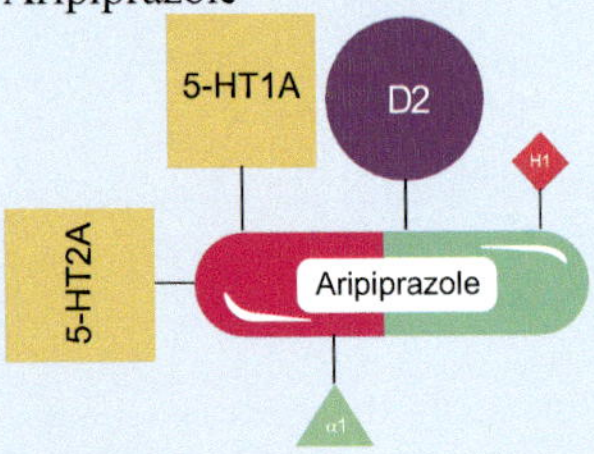 (Abilify™)	D2 antagonist 5-HT2A antagonist 5-HT1A antagonist α1-adrenergic antagonist Histamine antagonist	Impulsive, compulsive manic, bipolar dissociated	Not currently used	CYP2D6 CYP3A4

Appendix 28 Summary Table of Other Psychotropic Medications

Substance	Mechanism of action	Pathological states for which the molecule is indicated	Dosage	Cytochromes
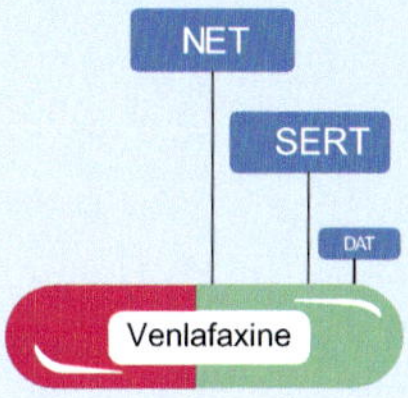	Inhibitor of SERT Inhibitor of NET Inhibitor of DAT	Anxious Impulsive or compulsive states	1 to 2 mg/kg BID	CYP2D6 CYP3A4 CYP2C19 CYP2C9
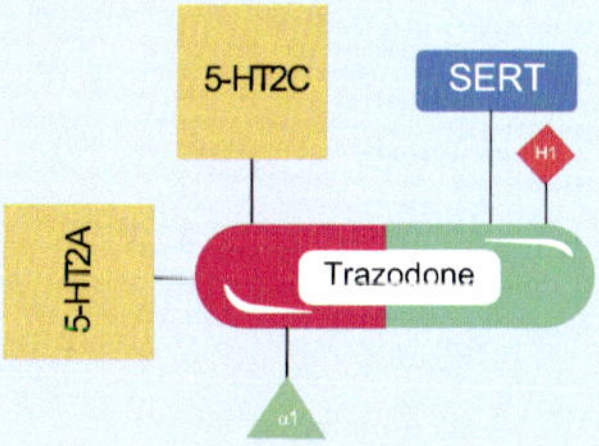	Inhibitor of SERT Antagonist of 5-HT2A Antagonist of 5-HT2C α1-adrenergic antagonist Histamine antagonist	Phobic Anxious Impulsive or compulsive states	7 to 10 mg/kg PRN (every 8 hours) In combination with fluoxetine, start at 5 mg/kg	CYP3A4 CYP2D6
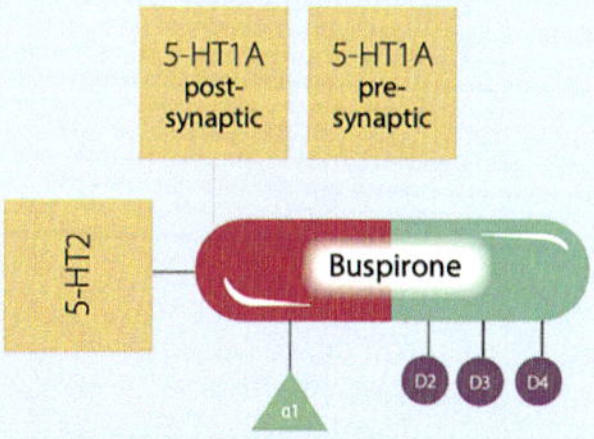	**Partial agonist of 5-HT1A (presynaptic)** **Partial agonist of 5-HT1A (postsynaptic)** Antagonist of 5-HT2 Partial antagonist of D2 D3, D4 Antagonist Partial α1-adrenergic agonist	Phobic or anxious	0.5 to 1 mg/kg BID or TID	CYP3A4

Substance	Mechanism of action	Pathological states for which the molecule is indicated	Dosage	Cytochromes
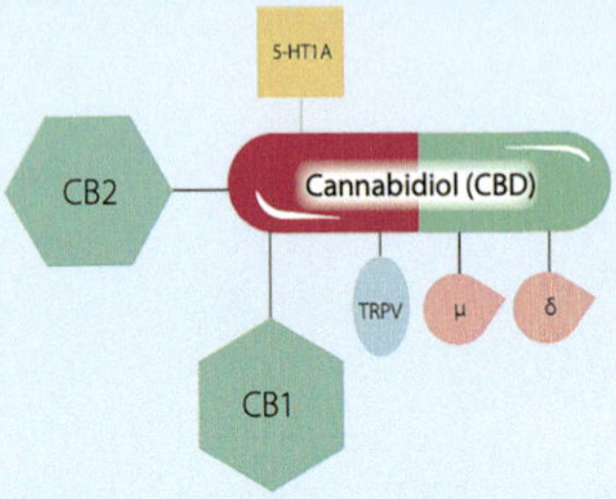	**Negative allosteric modulator of CB1** **Inverse agonist of CB2** Transient Receptor Potential Vanilloid (TRPV) desensitizer. Partial agonist of 5-HT1A Negative allosteric modulator of μ-opioids Negative allosteric modulator of δ-opioids	Phobic or anxious	1.5 to 2 mg/kg BID Increase as needed up to 10 mg/kg BID	CYP2C9 CYP2C19 CYP2D6 CYP3A4
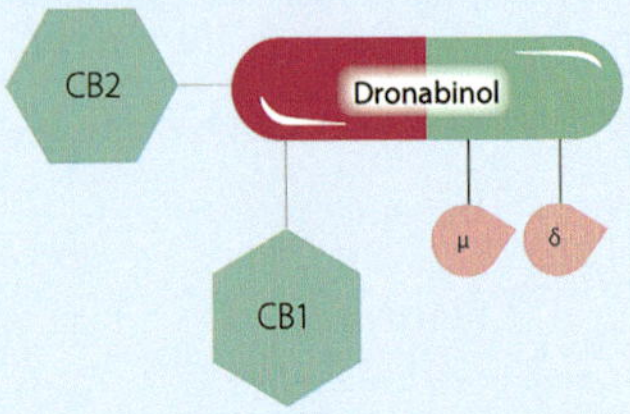	**Partial agonist of CB1** **Partial agonist of CB2** Negative allosteric modulator of μ-opioids Negative allosteric modulator of δ-opioids	Phobic or anxious Impulsive or compulsive	2.5 to 5 mg per individual (in humans)	

Appendix 29 Summary Table of Complementary Hormonal Treatments

Substance	Mechanism of action	Pathological states for which the molecule is indicated	Dosage	Cytochromes
Thiamazole	Prevents iodination of T3 and T4 precursors	Hyperthyroidism	Methimazole 2.5–10 mg/cat BID	CYP450
Megestrol acetate	Testosterone inhibitor GnRH inhibitor	Not indicated		CYP3A4
Deslorelin	GnRH agonist	Behavioral issues related to reproduction ("physiological tat")	Implant PRN	–

Appendix 30 Psychotropic Medications for Managing Cats with Healthcare Phobia

Molecule	Dosages	Indications	Remarks
Alprazolam	0.125 to 0.25 mg per cat. Can increase up to 0.1 mg/kg	Useful for its amnesic effect after traumatic exposure. Limited benefit for habituation to care or addressing established phobia	Reserved for prevention in cats known to be sensitive or immediately after a trauma related to care in reactive cats
Pregabalin	5 mg/kg, i.e., 0.1 ml/kg of a 50 mg/ml solution, 1.5 to 2 hours before exposure	Healthcare phobia, whether isolated or comorbid with another behavioral condition	Preferred molecule for use alongside a habituation or desensitization protocol for care
Gabapentin	30 to 50 mg/kg two hours before exposure. May be reduced to 15 mg/kg if sedation is excessive	Healthcare phobia, whether isolated or comorbid with another behavioral condition	Preferred molecule for use alongside a habituation or desensitization protocol for care
Trazodone (not available in France)	5 to 10 mg/kg per cat two hours before exposure	Healthcare phobia, whether isolated or comorbid with another behavioral condition	Limited usefulness for habituation due to its strong sedative effect. Effective for performing care without sedation

Appendix 31 Fear Behavioral Management General Principles

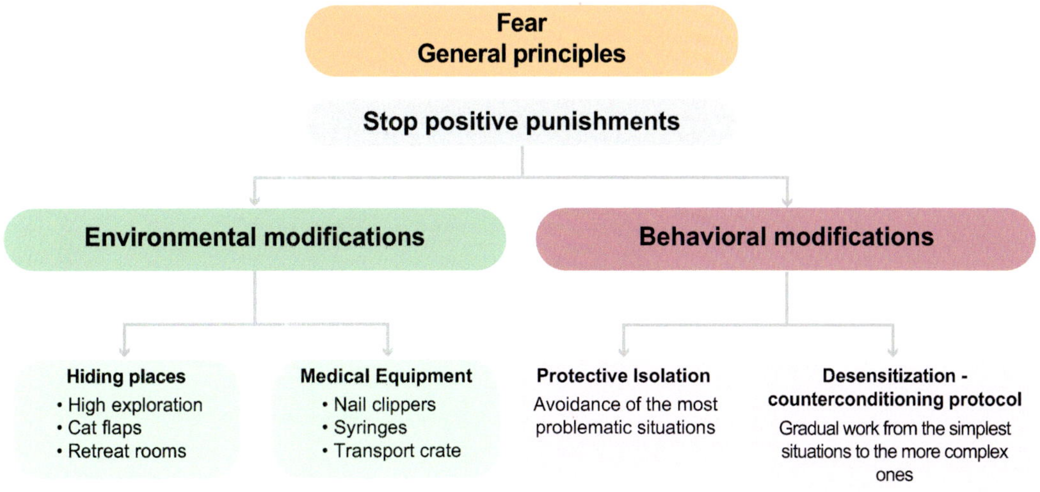

Appendix 32 Self-Control Behavioral Management General Principles

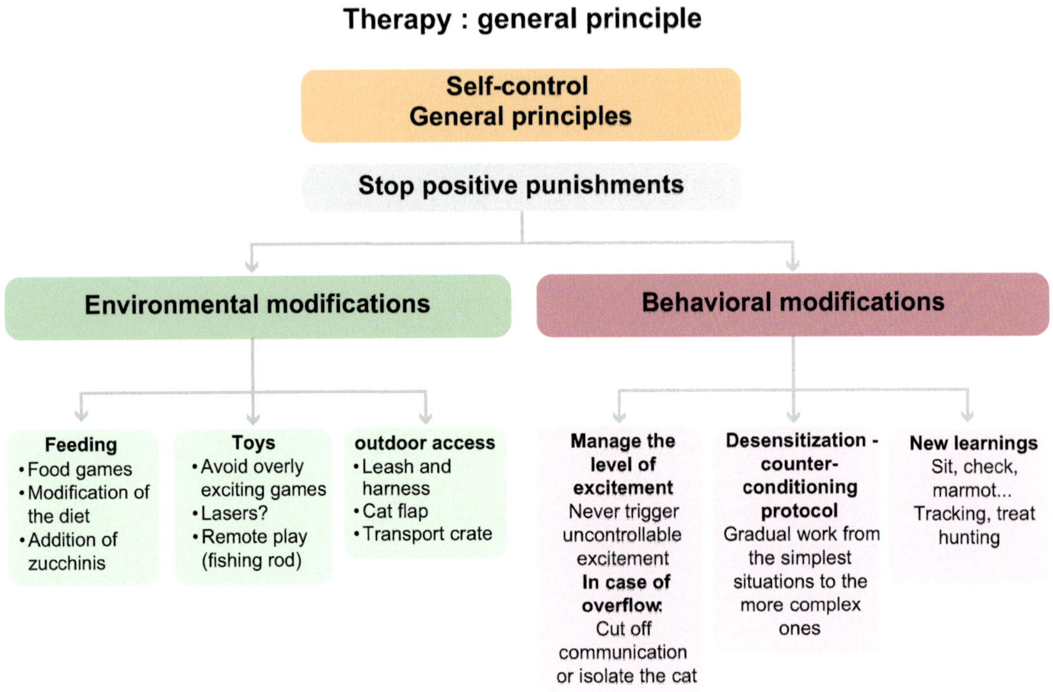

Appendix 33 Territoriality Behavioral Management General Principles

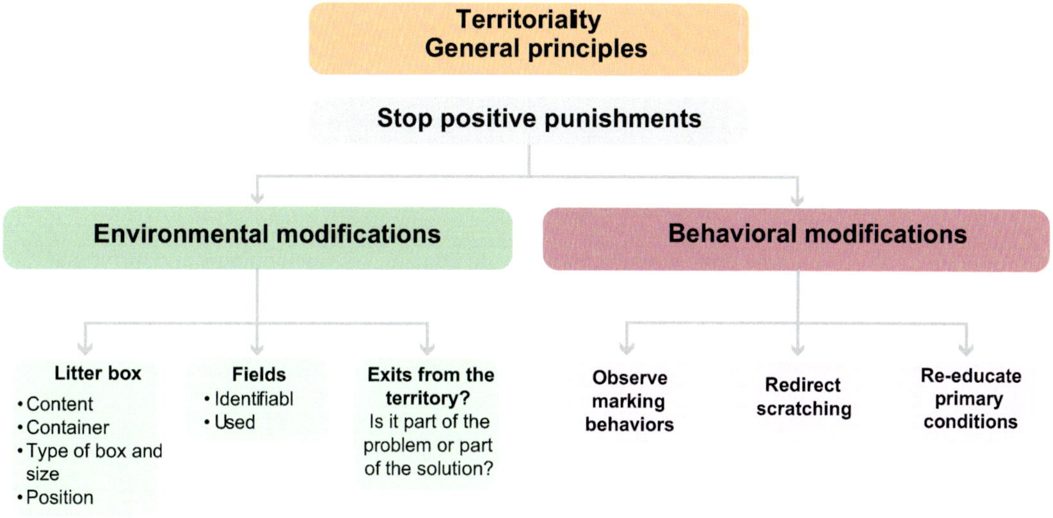

Appendix 34 Relationships Behavioral Management General Principles

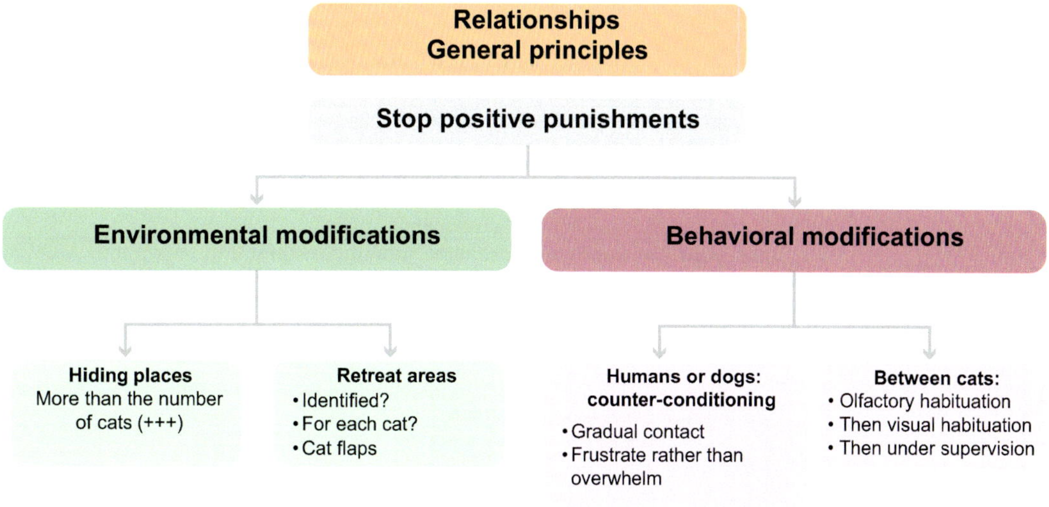

Appendix 35 Behavioral Condition Awareness for Owners

Behavioral axis	N°	Questions and possible answers
		Scoring and Interpretation • **Green**: Normal findings; no immediate concern. • **Orange**: Attention required; preliminary advice should be sought. • **Red**: A pathological condition is suspected; a professional consultation is recommended. **Scoring Rules:** • If more than three yellow responses are recorded within a single functional axis, the axis should be classified as red. • A single red response within any axis is sufficient to classify the entire axis as red.
Self-controls	1	**How does your cat behave with its toys, your belongings, and furniture?**
		• It doesn't chew anything • It may chew objects or damage, even destroy, the objects it chews • It may ingest parts of objects or whole objects (fabric, elastic, etc.).
	2	**How does your cat behave with you during friendly interactions, especially petting?**
		• It is gentle, never bites, and never extends its claws. • It may extend its claws or nibble, but it doesn't hurt you. • It is less tolerant and may scratch or bite to the point of causing injury (bleeding).
	3	**How does your cat behave when you play with it?**
		• It is gentle, never extends its claws or bites, even when very excited. • It may extend its claws or bite when excitement increases, but not to the point of causing harm (or only rarely). • Excitement during play makes it rough quickly: it may scratch or bite enough to cause injury (bleeding).
	4	**How does your cat behave in its usual environment (moving person, object, presence of insects, etc.)?**
		• A noise or movement sparks interest but does not lead to direct interaction with the source. • It almost always reacts (jumping, hunting behavior) but does not scratch or bite people. • It almost always reacts, and may scratch or bite, injuring its target if it is a person (bleeding).
	5	**What is your cat's ability to stop when it is excited (playing, running, or jumping)?**
		• It can stop upon verbal command (no, stop, etc.), or spontaneously if ignored. • It is sometimes difficult to stop with a verbal command or by ignoring it, and it may need to be isolated for a few minutes to calm down. • It struggles to stop despite attempts to ignore it or verbal commands. Even when isolated, excitement quickly reappears.
	6	**Does your cat become aggressive (hissing, attempting to scratch or bite) when excited or frustrated?**
		• No, never. • Yes, it has happened once or twice but never repeated. • Yes, it has happened more than twice.
	7	**How does your cat behave in the presence of noises (storms, fireworks, door slamming, household appliances)?**
		• It is indifferent or shows brief surprise. • It shows moderate fear (runs to another room, hides), but quickly overcomes its fear and returns within a few minutes.

Fear		• It shows intense fear (runs, hides, becomes aggressive), and this may last even after the noise has stopped.
	8	**How does your cat behave in the presence of unfamiliar humans?**
		• It is indifferent or tries to establish friendly contact. • It keeps its distance or hides. Eventually, it returns and visitors can usually pet it. • It freezes or hides for the entire visit (sometimes even afterward). If forced into contact, it tries to escape, freezes, or becomes aggressive (scratches or bites).
	9	**How does your cat behave in an unfamiliar environment?**
		• It takes 1 or 2 days to explore the new environment and is almost as comfortable as at home. • It attempts exploration and makes progress over time but is never as comfortable as at home. • It is very stressed (stays hidden, explores very little) unless it is at home. If removed from its refuge, it tries to escape, freezes, or becomes aggressive (scratches or bites).
	10	**How does your cat behave at the veterinarian's office?**
		• It is relaxed, exits its carrier spontaneously, explores easily, and accepts petting. • It is stressed, reluctant to leave its carrier. During the medical exam, it is very uncomfortable (frozen, tense). It has sometimes hissed or growled but never escalated. • It is highly stressed, refuses to leave its carrier (frozen or aggressive if the vet approaches). The medical exam is always difficult. It is nearly impossible to trim nails or take a blood sample without sedating it.
	11	**Can your cat show aggression (scratching or biting) when afraid?**
		• No, never. • Yes, it has happened once or twice but never repeated. • Yes, it has happened more than twice.
Relationships	12	**How does your cat communicate with the humans in your household?**
		• It has many rituals, initiates contact, seeks petting, and looks for proximity (e.g., shared time on the bed, sofa, etc.). • The desire for contact varies depending on the family members. It rarely approaches some but never aggresses them. • It does not spontaneously seek contact with anyone. It avoids proximity to one or more household members and has already attacked them (scratches or bites).
	13	**How does your cat react when you want to pet it?**
		• It accepts and enjoys contact. Refusals are rare, and in these cases, the cat calmly moves away. • It generally tolerates contact. Petting certain areas or a long time may cause it to flee or hiss/growl, but without scratching or biting. • It tolerates contact for a very limited time, more easily in certain areas. If you insist, it may scratch or bite.
	14	**How does your cat communicate with other cats in the house? (If there are multiple cats, describe the worst situation.)**
		• It frequently approaches others for mutual interactions (grooming, shared play, resting together) OR it does not live with other cats, so I don't know. • It rarely seeks contact with other cats in the house but is not aggressive. • It completely avoids contact with the other cats in the house or seeks contact

		aggressively (chasing or direct attacks).
	15	**How does your cat behave when another cat in the house seeks to approach it? (If there are multiple cats, describe the worst situation.)**
		• Contact is welcomed and appreciated. Refusals are rare, and in these cases, the cat calmly moves away. OR it doesn't live with other cats, so I don't know. • It may tolerate the presence of the other cat but tends to maintain some distance. If the other cat insists too much, it will flee, hiss, or growl, but never scratch or bite. • It tries to avoid contact and may become aggressive if the other cat insists (scratching, biting).
	16	**Can your cat be aggressive and cause injury during interactions with humans, dogs, or other cats in the house (scratching or biting)?**
		• No, never. • Yes, it has happened once or twice but never repeated. • Yes, it has happened more than twice.
Territoriality	17	**Is your cat likely to urinate outside its litter box or designated area?**
		• No, never. The few times this occurred, it was clearly related to a medical issue (e.g., urinary infection) or an inability to access the elimination site, or if it was very dirty. • There have been rare incidents of urinary incontinence (less than once a month). • There are frequent episodes of urinary incontinence (more than once a month).
	18	**Does your cat rub its cheeks on objects or furniture?**
		• Yes, it does this several times a day or week on various objects. • Yes, it does this, but the frequency or objects have changed recently. • No, I have never seen it do that.
	19	**Does your cat rub its cheeks on humans?**
		• Yes, it does this with both household members and some visitors. • Yes, but only with one or more household members. • No, I have never seen it do that.
	20	**Does your cat scratch objects other than its scratching post?**
		• No • Yes, on a few spots, often related to where it likes to rest. • Yes, on many objects, which it has damaged over time.
	21	**How does your cat behave in the presence of unknown cats?**
		• It is indifferent or tries to establish friendly contact OR it never encounters other cats. • It runs away or threatens, keeping its distance but does not attack. • It freezes, panics, or becomes aggressive (scratching, biting).
Communication and environment	22	**How does your cat generally behave when you raise your voice or make a noise to interrupt it?**
		• It stops its action and quickly moves away if I insist • It is very impressed and flees. After a few minutes, it accepts contact if I approach. • It flees to hide. After a few minutes, it either flees or becomes aggressive if I approach. It can also become aggressive (scratching or biting) as soon as I raise my voice.
	23	**How does your cat behave if you take it by the scruff of the neck or punish it with a spray or slap?**
		• I don't know: I never do this. • It is very impressed and flees. After a few minutes, it does not flee if I approach. • It flees to hide. After a few minutes, it either flees or becomes aggressive if I approach. It can also become aggressive (scratching or biting) as soon as I raise my voice.

Index

A

Acepromazine 903
Acetyl coenzyme a 73
Acetylcholine 73, 105
Acetylcholinesterase 73
Acromegaly 497
Action potential 65, 69
Activation 105
Activatory 96
Addiction 100, 262
Addictive behavior 100
Adolescence 144
Adoption 156
Adoption of another pet 372
Aggression 105
– idiopathic aggression 259, 426
– irritation aggression 335, 395
– irritation aggression in a context of fear 238, 247
– irritation aggression in a context of frustration 367
– non-aggressive injurious sequence 266, 335
– petting-induced aggression 173
– self-directed aggression 477
– territorial aggression 406
– uncontrolled emotional aggression 238
– vulnerant sequence during play 266
Aging 161, 437, 498
– abnormal aging 281
– normal aging 281
Agitation 252
Agression 103–104
Alertness 78
Allomarking 157
Allostatic load 229
Allostatic overload 229
Allosteric modulators
– positive allosteric modulators 604
Alopecia 252, 480
Alprazolam 607–608
Amino acid 71–74, 487
Amnesic effect 604, 605
Amygdala 60, 83–84, 92, 103, 236
Anal gland expression 238
Anhedonia 251
Anorexia 251, 489
Anticipation 85, 236, 292
Anticonvulsant 604
Antihistaminic 96
Anxiety 105
– generalized anxiety 241
– hypnagogic anxiety 459
– inhibited anxiety 241
– intermittent anxiety 241
– paroxysmal anxiety 241
– productive anxiety 241
Anxious dependence 312
Apartment 365
Apathy 252
Appetite 78
Arginine 487
Aripiprazole 904
Arousal 96
Atavistic fear 292
Atopic dermatitis 479
Attachment 61, 150, 174–177, 393
– disorganized attachment 174
– insecure-ambivalent attachment 174
– insecure-avoidant attachment 174
– maternal attachment 157
– secure attachment 174
– territorial attachment 157
Attention 61, 78, 96–97
Attention deficit 263
Atypical neurodermatitis 480
Auditory capacity 458
Autonomopathy 392
Avoidance 233
Azaperone 903

B

Barbiturate 72
Basal ganglia 59
Behavior
– addictive behaviors 78
– appeasment phase 263
– appetitive phase 263
– behavioral development 148–164
– calming phase 271
– consummatory phase 271
– feeding behavior 157
– stereotyped behavior 419
Benzodiazepine 72
Bile acid 491
Binge eating 366
Birth of a child 232
Biting 158, 171
Blood-brain barrier 61–63
Bottle-fed kitten 316
Bottom-up 330
Brain-Derived Neurotrophic Factor (BDNF) 91, 252
Brain maturation 338
Brain tumor 485
Breed 156, 172, 339
Breeding 270, 308, 317

Hyposomnia 266, 337
Hypothalamus 61, 63, 103
Hypothyroidism 494–495

I

Idiosyncratic hepatic insufficiency 606
Impulsive 263
Impulsivity 105, 171
Inconsistency in the environment 232
Inconsistent request 393
Ingestion of non-food object 265, 336, 354
Inhibition 88, 105, 170–172
Inhibitory 71, 75, 96
Injury 404
Instrumentalization 93
Insulin 494
Internal medicine 489–491
Intestinal obstruction 356
Involution 162, 280
Involutional depression 437
Irritability 247

K

Ketamine 71, 279, 418
Kibble 486
Kneading 149

L

Lack of food satiety 337, 355
Lactation 78
Laser pointer game 271
L-DOPA 74
Learned helplessness 300
Learning 74, 78, 91, 105
– associative learning 92
– perceptual learning 92
– relationnal learning 93
Licking 274, 366, 477
– excessive licking 337, 356
Lick mat 307
Limbic system 83
Litter box aversion 490
Long-term depression 96
Long-term potentiation 85, 95, 99, 146
Lorazepam 608
Loss
– cognitive loss 281
– loss of inhibition 464
– loss of orientation 282
– loss of spatial reference 447
– sensory loss 440
Lost opportunity 270
Lysine 487

M

Manic crisis 257
Manic episode 257
Marking 177–178
– excessive marking 365
– facial marking 157, 159
– urine marking 158, 373
Meal 486
Memory 60, 74, 78, 91, 105
– long-term memory 94
– sensory memory 92
– short-term memory 92
Memory loss 458
Mesencephalon 58
Methionine 487
Methylphenidate 76
Microbiota 491
Miliary dermatitis 480
Model of zubin and spring 146
Monoamine 74–82, 90
Monoamine oxidases (MAOs) 76, 78
Mood 78, 90–91, 105
Motion sickness 301
Motivation 60, 61, 170–172, 251
– behavioral motivation 170
Motor agitation 269, 335
Motor inhibition 238
Moving to a new home 177, 178, 234, 372
Mydriasis 258, 426
Myelin 65, 67

N

Neophobia 312
Neuroanatomy 58–106
Neurogenesis 144–146, 148
Neurological examination 482
Neurology 481–485
Neuromuscular junction 73
Neuron 64–82
– postsynaptic neuron 68
– presynaptic neuron 68
Neuronal loss 438, 443
Neurons 144
Neurophysiology 82–104, 144
Neurotransmitter 63–82
Neurotrophin 144
Neutering 158, 170, 292
Nipping 154
Nocturnal awakening 452
Node of ranvier 67
Norepinephrine 78, 105
Norepinephrine transporter (NET) 78
Nucleus accumbens 59, 99
Nursing period 150
Nutrition 486–489

O

Obesity 246, 366, 486
Obsession 89, 257, 274
Olfaction 177
Olfactory exploration 336
Onychophagia 246
Oral exploration 156, 336
Orexin 96
Organic condition 232
Oxazepam 608
Oxytocin 63, 176

P

Pain 78, 440, 476–478
– chronic pain 477
– pain assessment 479
– pain assessment grid 479
– pain grimace 479
Pallidum 59, 271
Pandora syndrome 490
Panic 238
Panic attack 604
Panic crisis 239, 241
Paradoxical reaction 606
Paroxysmal crisis 241
Pathway 160
– bottom-up pathway 101
– dopaminergic pathway 58, 99
– long pathway 89, 104
– mesocortical pathway 78
– mesolimbic pathway 76, 99
– nigrostriatal pathway 76
– noradrenergic pathway 58
– serotoninergic pathway 58
– short pathway 89
– top-down pathway 101
– tuberoinfundibular pathway 78
Pawing 149
Periaqueductal gray matter 103
Period
– adult period 161
– critical period 145, 148, 169
– juvenile period 155–158
– neonatal period 148–149
– prenatal period 148
– pubertal period 158–161
– senescent period 161–164
– socialization period 150–155
– transitional period 149–150
Pheromone 157
– appeasing pheromone 176
– F1 fraction 159
– F2 fraction 159
– F3 fraction 159, 177
– F4 fraction 159, 177
– F5 fraction 159
– pheromone diffuser 377, 379
Phobia
– categorical phobia 290–311
– dog phobia 298
– healthcare phobia 291, 300
– interspecific social phobia 291
– intraspecific social phobia 291
– noise phobia 291, 298
– post-traumatic phobia 253, 320–325
– social phobia 235, 301
– specific phobia 291
– storm phobia 235, 291, 298
– transport phobia 301
– travel phobia 291
Phylogenesis 63
Physical interaction 177
Pica 489
Pimamperone 903
Pituitary 61
– anterior pituitary 63
– pituitary portal system 63
– posterior pituitary 63
Plasticity 96, 146
Play 153, 155
Pleasure 78, 99, 355
Pollakiuria 489
Polyphagia 252
Polyuria-polydipsia 491
Pons 58
Population density 404
Postictal recovery phase 483
Post-traumatic stress disorder (PTSD) 320–325
Posture 240
Potassium 65
Potassium clorazepate 608
Predation 103
Predator 83
Predatory 154
Predictability 232
Prefrontal cortex 83
– ventromedial prefrontal cortex 88, 89, 101
Prey 83
Processes
– complex processes 167
– elementary processes 84–89, 92
Prodrome 483
Progesterone 496
Prolactin 78, 176
Prosencephalon 58
Protein requirement 487
Psychological state
– acute depressive state 251
– anxious state 83, 89, 241–251, 604
– bipolar state 91, 260
– chronic depressive state 251
– complex phobic state 236